Antenna Engineering Handbook

OTHER McGRAW-HILL HANDBOOKS OF INTEREST

Antenna Engineering Handbook

SECOND EDITION

Editors

Richard C. Johnson

Georgia Institute of Technology
Atlanta, Georgia

Henry Jasik

AIL Division of Eaton Corporation
Deer Park, Long Island, New York

McGraw-Hill Book Company
New York   St. Louis   San Francisco   Auckland   Bogotá   Hamburg
Johannesburg   London   Madrid   Mexico   Montreal   New Delhi   Panama
Paris   São Paulo   Singapore   Sydney   Tokyo   Toronto

The Library of Congress has cataloged this serial publication as follows:
Antenna engineering handbook. 1st– ed.
  New York, McGraw-Hill, 1961–
      v. illus., diagrs. 24 cm.
    Editor: 1961–   H. Jasik.

1. Antennas (Electronics)   I. Jasik, Henry, ed.
TK7872.A6A6   621.384135   59-14455

    567890   KGP/KGP   89876

ISBN 0-07-032291-0

The editors for this book were Harold B. Crawford and Beatrice E.
Eckes, the design supervisor was Mark E. Safran, the designer was Ink
Studios, and the production supervisor was Sally Fliess. It was set in
Times Roman by University Graphics, Inc.

Printed and bound by The Kingsport Press.

Contents

3 APPLICATIONS

4   TOPICS ASSOCIATED WITH ANTENNAS

Contributors

James S. Ajioka, Manager, Electromagnetics Laboratories, Hughes Aircraft Company, Fullerton, California. (Chap. 19)

William P. Allen, Jr., Staff Engineer, Lockheed-Georgia Company, Marietta, Georgia. (Chap. 37)

M. C. Bailey, Senior Research Engineer, NASA Langley Research Center, Hampton, Virginia. (Chap. 31)

Harold L. Bassett, Chief, Modeling and Simulation Division, Engineering Experiment Station, Georgia Institute of Technology, Atlanta, Georgia. (Chap. 44)

Judd Blass, Blass Consultants, Inc., New York City and Herzliah, Israel. (Chap. 8)

Donald G. Bodnar, Principal Research Engineer, Engineering Experiment Station, Georgia Institute of Technology, Atlanta, Georgia. (Chap. 46)

David F. Bowman, Antenna Consultant, Williamstown, Massachusetts. (Chap. 43)

Richard W. Burns, Manager, Microwave Techniques Department, Array Antenna Laboratory, Hughes Aircraft Company, Fullerton, California. (Chap. 20)

George G. Chadwick, Division Manager, Lockheed Missiles & Space Company, Inc., Sunnyvale, California. (Chap. 14)

Brian S. Collins, Technical Director, C & S Antennas, Ltd., Rochester, England. (Chap. 27)

James H. Cook, Jr., Principal Engineer, Scientific-Atlanta, Inc., Atlanta, Georgia. (Chap. 36)

William F. Croswell, Head of RF Design Section, Harris Corporation, Melbourne, Florida. (Chap. 31)

Lorne K. DeSize, Consultant, Littleton, Colorado. (Chap. 23)

Raymond H. DuHamel, Antenna Consultant, Los Altos Hills, California. (Chaps. 14 and 28)

Boynton G. Hagaman, Senior Engineer, Kershner & Wright Consulting Engineers, Alexandria, Virginia. (Chap. 24)

Maurice M. Hallum III, Chief of Systems Evaluation Branch, Systems Simulation and Development Directorate, U.S. Army Missile Command, Redstone Arsenal, Alabama. (Chap. 38)

Howard T. Head, Consulting Radio Engineer, A. D. Ring & Associates, Washington, D.C. (Chap. 25)

Gene K. Huddleston, Associate Professor, School of Electrical Engineering, Georgia Institute of Technology, Atlanta, Georgia. (Chap. 44)

Geoffrey Hyde, Senior Staff Scientist, COMSAT Laboratories, Clarksburg, Maryland. (Chaps. 17 and 45)

*Henry Jasik,** Vice President and Director of Antenna Systems Division, AIL Division, Eaton Corporation, Deer Park, Long Island, New York. (Chap. 2)

Richard C. Johnson, Principal Research Engineer, Engineering Experiment Station, Georgia Institute of Technology, Atlanta, Georgia. (Chap. 1)

Edward B. Joy, Professor, School of Electrical Engineering, Georgia Institute of Technology, Atlanta, Georgia. (Chap. 29)

Kenneth S. Kelleher, Consultant, Alexandria, Virginia. (Chaps. 17 and 18)

Hugh D. Kennedy, Director of Marketing, Technology for Communications International, Mountain View, California. (Chap. 39)

Howard E. King, Head, Antennas and Propagation Department, Electronics Research Laboratory, The Aerospace Corporation, Los Angeles, California. (Chap. 13)

Charles M. Knop, Director, Antenna Research, Andrew Corporation, Orland Park, Illinois. (Chap. 30)

Dennis J. Kozakoff, President, Millimeter Wave Technology, Inc., Atlanta, Georgia. (Chap. 38)

John D. Kraus, Director, Ohio State University Radio Observatory, Columbus, Ohio. (Chap. 41)

Edmund A. Laport,† Corporate Director of Communications Engineering, RCA Corporation, Princeton, New Jersey. (Chap. 11)

Allan W. Love, Principal Engineering Specialist, Space Operations/Integration & Satellite Systems Division, Rockwell International, Downey, California. (Chap. 15)

Roderic V. Lowman, Consultant, AIL Division, Eaton Corporation, Deer Park, New York. (Chap. 42)

John A. Lundin, Consulting Radio Engineer, A. D. Ring & Associates, Washington, D.C. (Chap. 25)

Mark T. Ma, Senior Research Engineer, Electromagnetic Fields Division, National Bureau of Standards, Boulder, Colorado. (Chap. 3)

Robert J. Mailloux, Chief, Antenna Section, Electromagnetic Sciences Division, Rome Air Development Center, Hanscom Air Force Base, Massachusetts. (Chap. 21)

Raj Mittra, Professor, Electrical Engineering Department, University of Illinois, Urbana, Illinois. (Chap. 10)

Robert E. Munson, Manager, Advanced Antenna Programs, Ball Aerospace Systems Division, Boulder, Colorado. (Chap. 7)

Josh T. Nessmith, Senior Scientific Adviser, Missile and Surface Radar, RCA Corporation, Moorestown, New Jersey. (Chap. 34)

Warren B. Offutt, Vice President—Technical Management, Eaton Corporation, Cleveland, Ohio. (Chap. 23)

Willard T. Patton, Manager, Advanced Antenna and Microwave Technology, Missile and Surface Radar, RCA Corporation, Moorestown, New Jersey. (Chap. 34)

---

*Deceased.

†Retired.

George D. M. Peeler, Consulting Engineer, Missile Systems Division, Raytheon Company, Bedford, Massachusetts. (Chap. 16)

Paul E. Rawlinson, Manager, Microwave & Antenna Department, Equipment Division, Raytheon Company, Wayland, Massachusetts. (Chap. 32)

Leon J. Ricardi, Head, MILSTAR Technical Advisory Office, Lincoln Laboratory, Massachusetts Institute of Technology, Los Angeles, California. (Chaps. 22 and 35)

Charles E. Ryan, Jr., Principal Research Engineer, Engineering Experiment Station, Georgia Institute of Technology, Atlanta, Georgia. (Chap. 37)

James M. Schuchardt, Manager, Radar Systems of Advanced Tactical Systems, The Bendix Corporation, Guidance Systems Division, Teterboro, New Jersey. (Chap. 38)

Glenn S. Smith, Associate Professor, School of Electrical Engineering, Georgia Institute of Technology, Atlanta, Georgia. (Chap. 5)

Vernon C. Sundberg, Section Head, GTE Systems, Mountain View, California. (Chap. 40)

Jean-Claude Sureau, Technical Director, Radant Systems, Inc., Stow, Massachusetts. (Chap. 33)

Chen To Tai, Professor of Electrical Engineering, University of Michigan, Ann Arbor, Michigan. (Chap. 4)

Raymond Tang, Manager, Array Antenna Laboratory, Communications and Radar Division, Hughes Aircraft Company, Fullerton, California. (Chap. 20)

Harold R. Ward, Consulting Scientist, Equipment Division, Raytheon Company, Wayland, Massachusetts. (Chap. 32)

William Wharton, Technical Consultant, Technology for Communications International, London, England. (Chaps. 26 and 39)

Harold A. Wheeler, Chief Scientist, Hazeltine Corporation, Greenlawn, New York. (Chap. 6)

Ronald Wilensky, Manager, High-Frequency Products, Technology for Communications International, Mountain View, California. (Chap. 26)

Jimmy L. Wong, Senior Scientist, Antennas and Propagation Department, Electronics Research Laboratory, The Aerospace Corporation, Los Angeles, California. (Chap. 13)

Daniel F. Yaw, Advisory Engineer, Electronic Warfare Systems Engineering, Westinghouse Defense and Electronic Systems, Baltimore, Maryland. (Chap. 40)

Hung Yuet Yee, Senior Member of Technical Staff, Texas Instruments, Inc., Dallas, Texas. (Chap. 9)

Francis J. Zucker, Physicist, Antenna Section, Electromagnetic Sciences Division, Rome Air Development Center, Hanscom Air Force Base, Massachusetts. (Chap. 12)

Preface

SECOND EDITION

In the decades since the first edition of the *Antenna Engineering Handbook* was published, a great many significant and far-reaching advances have been made in antenna technology. This second edition updates the excellent material originally presented by Henry Jasik and the contributors to the first edition.

The *Handbook* again is organized into four major parts:

1 Introduction and Fundamentals
2 Types and Design Methods
3 Applications
4 Topics Associated with Antennas

Part 1 presents basic concepts, defines parameters, and discusses fundamentals that are common to most antennas. Part 2 presents the primary types of antennas and the special techniques that currently are in use. Emphasis is on succinct descriptions of operating principles and design methods. Part 3 covers major applications of antennas. Emphasis is on how antennas are employed to meet the requirements of electronic systems. Design methods which are peculiar to the applications are presented. Part 4 deals with additional topics that are closely related to antennas.

Many new chapters have been added in this edition. In Part 2, they are "Small Antennas," "Microstrip Antennas," "Frequency-Scan Antennas," "Phased Arrays," "Conformal and Low-Profile Arrays," and "Adaptive Antennas." In Part 3, they are "Microwave-Relay Antennas," "Radiometer Antennas," "Tracking Antennas," "Satellite Antennas," "Earth Station Antennas," "Seeker Antennas," and "ECM and ESM Antennas." In Part 4, they are "Microwave Propagation" and "Materials and Design Data."

Thanks are due to the many publishers who have granted permission to use material from their publications. As was done in the first edition, we have tried to credit all sources of information by references; any omissions are due to oversight rather than intent.

The work of many outstanding engineers who prepared the individual chapters has made this *Handbook* possible. Their magnificent efforts enabled this project to be completed according to the initial schedule.

It is impossible to list all the other individuals who have made contributions to this book, so I thank them all as a group. However, I extend special thanks to Maurice W. Long and J. Searcy Hollis, who introduced me to the fascinating world of antennas.

RICHARD C. JOHNSON

Antenna Engineering Handbook

PART 1

Introduction and Fundamentals

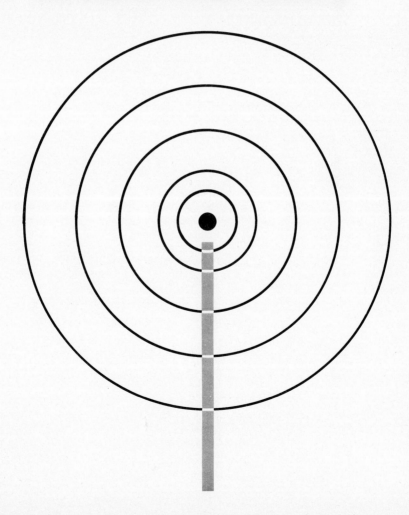

Chapter 1

Introduction to Antennas

Richard C. Johnson

Georgia Institute of Technology

1-1 FUNCTIONS AND TYPES

An *antenna* usually is defined as the structure associated with the region of transition between a guided wave and a "free-space" wave, or vice versa. The adjective free-space is in quotation marks because in practice there always is some interaction with the surroundings. On transmission, an antenna accepts energy from a transmission line and radiates it into space, and on reception, an antenna gathers energy from an incident wave and sends it down a transmission line.

When discussing an antenna, one usually describes its properties as a transmitting antenna. From the reciprocity theorem, however, we know that the directional pattern of a receiving antenna is identical with its directional pattern as a transmitting antenna, provided nonlinear or unilateral devices (such as some ferrite devices) are not employed. Thus, no distinction needs to be made between the transmitting and receiving functions of an antenna in the analysis of radiation characteristics. It should be pointed out, however, that the reciprocity theorem does not imply that antenna current distributions are the same on transmission as they are on reception.

A large variety of antennas have been developed to date; they range from simple structures such as monopoles and dipoles to complex structures such as phased arrays. The particular type of antenna selected for a certain application depends upon the system requirements (both electrical and mechanical) and, to a lesser extent, upon the experience of the antenna engineer.

1-2 BASIC CONCEPTS AND DEFINITIONS

Consider an antenna which is located at the origin of a spherical coordinate system as illustrated in Fig. 1-1. Suppose that we are making observations on a spherical shell having a very large radius r.

Assume that the antenna is transmitting, and let

- P_0 = power accepted by antenna, watts
- P_r = power radiated by antenna, watts
- η = radiation efficiency, unitless

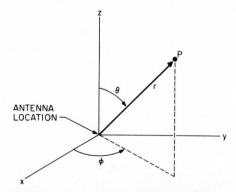

FIG. 1-1 An antenna in a spherical coordinate system.

The above quantities are related as follows:

$$\eta = \frac{P_r}{P_0} \tag{1-1}$$

Let

● $\Phi(\theta,\phi)$ = radiation intensity, watts/steradian

Note that since r was assumed to be very large, Φ is independent of r. This independence of r is a characteristic of the far-field region. The total power radiated from the antenna is

$$P_r = \int_0^{2\pi} \int_0^{\pi} \Phi(\theta,\phi) \sin \theta \; d\theta \; d\phi \tag{1-2}$$

and the average radiation intensity is

$$\Phi_{\text{avg}} = \frac{P_r}{4\pi} \tag{1-3}$$

Let

● $D(\theta,\phi)$ = directivity, unitless

Directivity is a measure of the ability of an antenna to concentrate radiated power in a particular direction, and it is related to the radiation intensity as follows:

$$D(\theta,\phi) = \frac{\Phi(\theta,\phi)}{\Phi_{\text{avg}}} = \frac{\Phi(\theta,\phi)}{P_r/4\pi} \tag{1-4}$$

The directivity of an antenna is the ratio of the achieved radiation intensity in a particular direction to that of an isotropic antenna. In practice, one usually is interested primarily in the peak directivity of the main lobe. Thus, if one says that an antenna has a directivity of 100, it is assumed that 100 is the peak directivity of the main lobe.

Let

● $G(\theta,\phi)$ = gain, unitless

The gain of an antenna is related to the directivity and power radiation intensity as follows:

$$G(\theta,\phi) = \eta D(\theta,\phi)$$
$$= \frac{\eta \phi(\theta,\phi)}{P_r/4\pi} \tag{1-5}$$

and from Eq. (1-1),

$$G(\theta,\phi) = \frac{\Phi(\theta,\phi)}{P_0/4\pi} \tag{1-6}$$

Thus, the gain is a measure of the ability to concentrate in a particular direction the power accepted by the antenna. Note that if one has a lossless antenna (i.e., $\eta = 1$), the directivity and the gain are identical.

Let

● $P(\theta,\phi)$ = power density, watts/square meter

The power density is related to the radiation intensity as follows:

$$P(\theta,\phi) = \frac{\Phi(\theta,\phi)\,\Delta\theta\,\Delta\phi}{(r\,\Delta\theta)(r\,\Delta\phi)}$$

or

$$P(\theta,\phi) = \frac{\Phi(\theta,\phi)}{r^2} \qquad (1\text{-}7)$$

Substituting Eq. (1-6) into Eq. (1-7) yields

$$P(\theta,\phi) = G(\theta,\phi)\frac{P_0}{4\pi r^2} \qquad (1\text{-}8)$$

The factor $P_0/4\pi r^2$ represents the power density that would result if the power accepted by the antenna were radiated by a lossless isotropic antenna.

Let

• $\qquad A_e(\theta,\phi)$ = effective area, square meters

It is easier to visualize the concept of effective area when one considers a receiving antenna; it is a measure of the effective absorption area presented by an antenna to an incident plane wave. The effective area is related[1] to gain and wavelength as follows:

$$A_e(\theta,\phi) = \frac{\lambda^2}{4\pi}\,G(\theta,\phi) \qquad (1\text{-}9)$$

Many high-gain antennas such as horns, reflectors, and lenses are said to be *aperture-type antennas*. The aperture usually is taken to be that portion of a plane surface near the antenna, perpendicular to the direction of maximum radiation, through which most of the radiation flows. Let

• $\qquad \eta_a$ = antenna efficiency of an aperture-type antenna, unitless

• $\qquad A$ = physical area of antenna's aperture, square meters

Then,

$$\eta_a = \frac{A_e}{A} \qquad (1\text{-}10)$$

The term η_a sometimes has been called *aperture efficiency*.

When dealing with aperture antennas, we see from Eqs. (1-9) and (1-10) that

$$G = \eta_a\frac{4\pi}{\lambda^2}\,A \qquad (1\text{-}11)$$

The term η_a actually is the product of several factors, such as

$$\eta_a = \eta\eta_i\eta_1\eta_2\eta_3 \cdots \qquad (1\text{-}12)$$

The term η is radiation efficiency as defined in Eq. (1-1). The term η_i is aperture illumination efficiency (or antenna illumination efficiency), which is a measure of how well the aperture is utilized for collimating the radiated energy; it is the ratio of the directivity that is obtained to the standard directivity. The standard directivity is obtained when the aperture is excited with a uniform, equiphase distribution. (Such a distribution yields the highest directivity of all equiphase excitations.) For planar apertures in which $A \gg \lambda^2$, the standard directivity is $4\pi A/\lambda^2$, with radiation confined to a half space.

The other factors, $\eta_1\eta_2\eta_3 \ldots$, include all other effects that reduce the gain of the antenna. Examples are spillover losses in reflector or lens antennas, phase-error losses due to surface errors on reflectors or random phase errors in phased-array elements, aperture blockage, depolarization losses, etc.

Polarization (see Refs. 2 and 3 for a more detailed discussion) is a property of a single-frequency electromagnetic wave; it describes the shape and orientation of the locus of the extremity of the field vectors as a function of time. In antenna engineering, we are interested primarily in the polarization properties of plane waves or of waves that can be considered to be planar over the local region of observation. For plane waves, we need only specify the polarization properties of the electric field vector since the magnetic field vector is simply related to the electric field vector.

The plane containing the electric and magnetic fields is called the *plane of polarization,* and it is orthogonal to the direction of propagation. In the general case, the tip of the electric field vector moves along an elliptical path in the plane of polarization. The polarization of the wave is specified by the shape and orientation of the ellipse and the direction in which the electric field vector traverses the ellipse.

The shape of the ellipse is specified by its *axial ratio*—the ratio of the major axis to the minor axis. The orientation is specified by the *tilt angle*—the angle between the major axis and a reference direction when viewed looking in the direction of propagation. The direction in which the electric field vector traverses the ellipse is the *sense of polarization*—right-handed or left-handed when viewed looking in the directions of propagation.

The polarization of an antenna in a specific direction is defined to be the polarization of the far-field wave radiated in that direction from the antenna. Usually, the polarization of an antenna remains relatively constant throughout the main lobe, but it varies considerably in the minor lobes.

It is convenient to define a spherical coordinate system associated with an antenna as illustrated in Fig. 1-2. The polarization ellipse for the direction (θ,ϕ) is shown inscribed on the spherical shell surrounding the antenna. It is common practice to choose $\mathbf{u}_\theta$ (the unit vector in the θ direction) as the reference direction. The tilt angle then is measured from $\mathbf{u}_\theta$ toward $\mathbf{u}_\phi$. The sense of polarization is clockwise if the electric field vector traverses the ellipse from $\mathbf{u}_\theta$ toward $\mathbf{u}_\phi$ as viewed in the direction of propagation and counterclockwise if the reverse is true.

In many practical situations, such as antenna measurements, it is convenient to establish a local coordinate system. Usually, the u_3 axis is the direction of propagation, the u_1 axis is horizontal, and the u_2 axis is orthogonal to the other two so that the unit vectors are related by $\mathbf{u}_1 \mathbf{X} \mathbf{u}_2 = \mathbf{u}_3$. The tilt angle is measured from $\mathbf{u}_1$.

When an antenna receives a wave from a particular direction, the response will be greatest if the polarization of the incident wave has the same axial ratio, the same sense of polarization, and the same spatial orientation as the polarization of the antenna in that direction. The situation is depicted in Fig. 1-3, where E_t represents a transmitted wave (antenna polarization) and E_m represents a matched incident wave. Note that the sense of polarization for E_t and that for E_m are the same when viewed in their local coordinate system. Also, note that the tilt angles are different because the directions of propagation are opposite. As depicted in Fig. 1-3, τ_t is the tilt angle

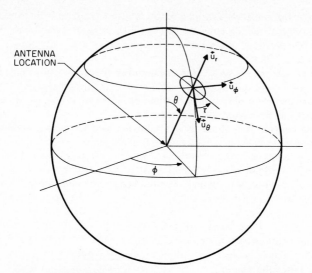

FIG. 1-2 Polarization ellipse in relation to antenna coordinate system. *(After Ref. 2.)*

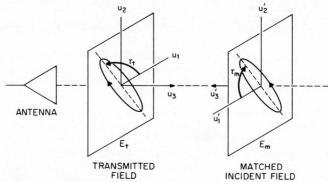

FIG. 1-3 Relation between polarization properties of an antenna when transmitting and receiving. *(After Ref. 2.)*

of the transmitted wave and τ_m is the tilt angle of the polarization-matched received wave; they are related by

$$\tau_m = 180° - \tau_t \qquad (1\text{-}13)$$

The polarization of the matched incident wave, as described above, is called the receiving polarization of the antenna.

When the polarization of the incident wave is different from the receiving polarization of the antenna, then a loss due to polarization mismatch occurs. Let

$\bullet$ η_p = polarization efficiency, unitless

The polarization efficiency is the ratio of the power actually received by the antenna to the power that would be received if the polarization of the incident wave were matched to the receiving polarization of the antenna.

The Poincaré sphere, as shown in Fig. 1-4, is a convenient representation of polarization states. Each possible polarization state is represented by a unique point on the unit sphere. Latitude represents axial ratio, with the poles being circular polarizations; the upper hemisphere is for left-handed sense, and the lower hemisphere is for right-handed sense. Longitude represents tilt angles from 0 to 180°. An interesting feature of the Poincaré sphere is that diametrically opposite points represent orthogonal polarizations.

The Poincaré sphere also is convenient for representing polarization efficiency. In Fig. 1-5, W represents the polarization of an incident wave, and A_r represents the receiving polarization of the antenna. If the angular distance between the points is 2ξ, then the polarization efficiency is

$$\eta_p = \cos^2 \xi \qquad (1\text{-}14)$$

1-3 FIELD REGIONS

The distribution of field strength about an antenna is, in general, a function of both the distance from the antenna and the angular coordinates. In the region close to the antenna, the field will include a reactive component. The strength of this reactive component, however, decays rapidly with distance from the antenna so that it soon

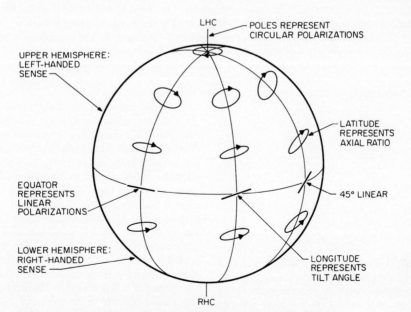

FIG. 1-4 Polarization states on the Poincaré sphere. *(After Ref. 2.)*

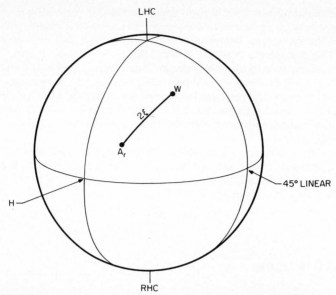

FIG. 1-5 Receiving polarization of an antenna A_r and polarization of an incident wave W.

becomes insignificant compared with the strength of the radiating component. That region in space in which the reactive component of the field predominates is called the *reactive near-field region,* and beyond this region the radiating field predominates.

That region in which the radiating field predominates is further subdivided into the *radiating near-field region* and the *radiating far-field region.* In the radiating near-field region, the angular distribution of radiated energy is dependent on the distance from the antenna, whereas in the radiating far-field region the angular distribution of radiated energy is essentially independent of distance from the antenna.

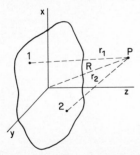

FIG. 1-6 Schematic representation of a planar-antenna aperture in the *xy* plane, an observation point *P*, and distances to the observation point from the origin and two elements of the antenna.

In the radiating near-field region, the relative phases and the relative amplitudes of contributions from various elements of the antenna are functions of the distance from the antenna. To visualize the situation refer to the schematic representation of Fig. 1-6. For simplicity, assume that the antenna is planar and is located in the *xy* plane; the distances to the observation point *P* from two arbitrary elements of the antenna are represented by $\bar{r}_1$ and r_2. Notice that as the observation point is moved farther from the origin, in a fixed angular direction, the relative distance to the arbitrary elements (r_2 minus r_1) changes; this causes the relative phases and amplitudes of contributions from elements 1 and 2 to change with distance from the antenna. By extending this argument to include all contributing elements of the antenna, one sees that the measured radiation pattern of the antenna will depend upon the radius to the observation point.

When the distance to the observation point gets very large, straight lines from any two contributing elements to the observation point (r_1 and r_2, for example, in Fig. 1-6) are essentially parallel and the relative distance to the elements (r_2 minus r_1) is essentially constant with changes in distance to the observation point. Thus, at large distances, the relative phases and amplitudes of contributions from the various elements change very slowly with distance, and the angular distribution of radiated energy measured at such large distances is essentially independent of the distances to the observation point. This condition is indicative of the radiating far-field region.

Thus, the space surrounding an antenna is composed of three regions: the reactive near-field region, the radiating near-field region, and the radiating far-field region.[4] These three regions are shown pictorially in Fig. 1-7. The boundaries between

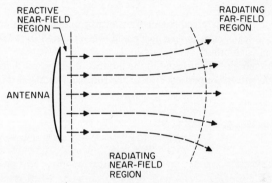

FIG. 1-7 Pictorial representation of the three regions surrounding an antenna.

the regions are not well defined, but for any antenna the reactive near-field region extends only a short distance.[5] The commonly accepted distance to the boundary between the reactive and radiating near-field regions is $\lambda/2\pi$. For electrically large antennas of the aperture type, such as that depicted in Fig. 1-7, the commonly used criterion to define the distance to the boundary between the radiating near-field and far-field regions is[6]

$$R = \frac{2D^2}{\lambda} \tag{1-15}$$

where D is the largest dimension of the aperture and λ is the wavelength.

Although the aforementioned criterion to define distance to the far-field region is generally accepted and is used quite widely, one must always remember that it is an arbitrary choice and that it is inadequate for some special situations. For example, if one must accurately measure patterns of antennas having very low sidelobes or if one must make accurate gain measurements of pyramidal horns which have large phase deviations across their apertures, the measurement distance may have to be much longer than $2D^2/\lambda$.

Arguments have been advanced for decreasing or for increasing the accepted distance to the boundary between the near-field and far-field regions; however, $2D^2/\lambda$ seems to be the most popular choice. The situation is analogous to trying to decide the ideal height for a stepladder. Most of us might agree that 2 m is an ideal height, but there will be special jobs which require a higher ladder and other jobs in which a shorter ladder is acceptable and more convenient.

It has been customary in the past to refer to field regions as Fresnel or Fraunhofer, after the approximations described. As pointed out by Hansen,[5] this practice should be discouraged. It is better to define field regions as reactive near field, radiating near field, and radiating far field as discussed earlier and illustrated in Fig. 1-7. Then, the terms Fresnel and Fraunhofer can be used more correctly to refer to analytical approximations.

1-4 POWER TRANSFER

Consider power that is transferred from a transmitting antenna to a receiving antenna; assume that the antennas are in free space and are separated by a large distance R (in the far field of each other).

The received power will be equal to the product of the power density of the incident wave and the effective aperture area of the receiving antenna; that is,

$$P_r = PA_e$$

Substituting from Eqs. (1-8) and (1-9),

$$P_r = \frac{G_t P_t}{4\pi R^2} \frac{\lambda^2 G_r}{4\pi}$$

or

$$P_r = \left(\frac{\lambda}{4\pi R}\right)^2 G_t G_r P_t \qquad (1\text{-}16)$$

The subscripts r and t refer to the receiving and transmitting antennas, respectively. Note in the above case that G_t is the gain of the transmitting antenna in the direction of the receiving antenna and that G_r is the gain of the receiving antenna in the direction of the transmitting antenna. A form of this equation was presented first by Friis,[7] and it usually is called the Friis transmission formula.

Similar arguments can be used to derive the radar equation. From Eq. (1-8), the power density at distance R from the antenna is

$$\frac{G_t P_t}{4\pi R^2}$$

The radar cross section σ of a target is a transfer function which relates incident power density and reflected power density. The term σ has units of area (e.g., m^2), and one might think of the process as having the target intercepting the incident power density over an area σ and then reradiating the power isotropically. The return power density at the radar is then

$$\frac{G_t P_t}{4\pi R^2} \frac{\sigma}{4\pi R^2}$$

and the received power is

$$\frac{G_t P_t}{4\pi R^2} \frac{\sigma}{4\pi R^2} A_e$$

where A_e is the effective area (or capture area) of the receiving antenna.

Using Eq. (1-9), the received power is

$$P_r = \frac{G_t P_t}{4\pi R^2} \frac{\sigma}{4\pi R^2} \frac{G_r \lambda^2}{4\pi} \tag{1-17}$$

If the same antenna is used for both transmission and reception of energy, then $G_t = G_r = G$, and the received power may be written as

$$P_r = \frac{G^2 \lambda^2 \sigma}{(4\pi)^3 R^4} P_t \tag{1-18}$$

which is a simple form of the radar equation.

1-5 RADIATION PATTERNS

When the power radiation intensity $\Phi(\theta,\phi)$ and the power density $P(\theta,\phi)$ are presented on relative scales, they are identical and often are referred to as the *antenna radiation pattern*. The main (or major) lobe of the radiation pattern is in the direction of maximum gain; all other lobes are called sidelobes (or minor lobes).

There are many types of antenna radiation patterns, but the most common are the following:

1 Omnidirectional (azimuthal-plane) beam
2 Pencil beam
3 Fan beam
4 Shaped beam

The omnidirectional beam is most popular in communication and broadcast applications. The azimuthal pattern is circular, but the elevation pattern will have some directivity to increase the gain in the horizontal directions.

The term *pencil beam* is applied to a highly directive antenna pattern consisting of a major lobe contained within a cone of small solid angle. Usually the beam is circularly symmetric about the direction of peak intensity; however, even if it is slightly fanned, it often is still called a pencil beam. A radar pencil beam, for example, is analogous to an optical searchlight beam.

A fan beam is narrow in one direction and wide in the other. A typical use of a fan beam would be in search or surveillance radar in which the wide dimension of the beam would be vertical and the beam would be scanned in azimuth. Another use of a fan beam would be in a height-finding radar in which the wide dimension of the beam would be horizontal and the beam would be scanned in elevation.

There are a number of applications that impose beam-shaping requirements upon the antenna. One such application is in an air search radar that is located on the ground or on a ship. The antenna for such an application is required to produce a narrow beam in azimuth and a shaped beam in elevation; azimuth coverage is obtained by scanning the beam.

The elevation shape of the beam must provide sufficient gain for detection of aircraft up to a certain altitude and angle of elevation and out to the maximum range of the system. To accomplish this without wasteful use of available power, the general shape of the coverage in the vertical plane should be as indicated in Fig. 1-8.

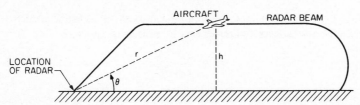

FIG. 1-8 Vertical beam shape (amplitude) of ground-based radar for detecting aircraft up to a certain altitude.

To maintain a fixed minimum of illumination on the aircraft at various points along the upper contour of the coverage diagram, it is necessary that the amplitude of the antenna pattern be proportional to the distance r from the antenna to the aircraft on the upper contour. In other words, the coverage contour of Fig. 1-8 can be taken to be the amplitude pattern of the antenna. Since $r = h \csc \theta$, the amplitude pattern must be proportional to $\csc \theta$, or the power pattern (or gain) must be proportional to $\csc^2 \theta$. Such proportionality must hold over the required coverage pattern; thus, such a pattern is said to have a $\csc^2$ shape.

In the case of a $\csc^2$ pattern, note that the gain is proportional to $\csc^2 \theta$ and the range is proportional to $\csc \theta$. Thus, we see from Eq. (1-18) that a target of constant cross section approaching at a constant altitude will produce a constant received-signal level.

Antenna radiation patterns are three-dimensional, but we have a need to describe them on two-dimensional paper. The most popular technique is to record signal levels along great-circle or conical cuts through the radiation pattern. In other words, one angular coordinate is held fixed, while the other is varied. A family of such two-dimensional patterns then can be used to describe the complete three-dimensional pattern.

Patterns usually are displayed as relative field, relative power, or logarithmic relative power versus angle on rectangular or polar diagrams. The rectangular diagram can easily be expanded along the angular axis by merely changing diagram speed relative to the angular rate of the antenna positioner; this is a big advantage when measuring patterns of narrow-beam antennas. Polar diagrams give one more realistic "pictures" of the radiation pattern, so they often are used for broad-beam antennas.

Another popular technique for displaying patterns is the *radiation distribution table*. In this type of display, signal levels are plotted in decibels at preselected intervals of the two angular coordinates, ϕ and θ. A contour appearance is obtained by printing only the even values of signal level and omitting the odd values.

With the advent of computers, a new type of "three-dimensional" pattern display has become popular. The two orthogonal far-field angles are represented by the base of the display, and relative gain is represented by height above the base. Such pattern displays can be generated from either calculated or measured gain data.

1-6   ESTIMATING CHARACTERISTICS OF HIGH-GAIN ANTENNAS

There are many occasions when it is desirable to make quick estimates of the beamwidths and gains of aperture antennas. A convenient rule of thumb for predicting 3-dB beamwidths is

$$\mathrm{BW_{3\text{-}dB}} = k\frac{\lambda}{D} \qquad (1\text{-}19)$$

where k is a beamwidth constant, λ is the wavelength, and D is the aperture dimension in the plane of the pattern.

Most antenna engineers seem to use a value of $k = 70°$. This is adequate for most rough estimates; however, more accurate estimates must take into account the fact that the value of k depends upon the aperture illumination function. Generally speaking (but not always), illumination functions that yield lower sidelobes result in a larger value of k.

Komen[8] reported on the variation of the beamwidth constant for reflector-type antennas. From computed patterns for various edge illuminations, he determined that

$$k = 1.05238I + 55.9486 \qquad (1\text{-}20)$$

where I is the absolute value of edge illumination (including space attenuation) in decibels and k is in degrees. (In practice, one normally would calculate k to only a few significant figures.) By applying Eqs. (1-19) and (1-20) to measured data from several antennas, Komen concluded that the relationship between beamwidth and edge illumination holds regardless of frequency, reflector size, reflector type, or feed type.

The beamwidth constant and approximate sidelobe level versus edge illumination for reflector-type antennas are illustrated in Fig. 1-9. The beamwidth constants for antennas having one of several special aperture distributions are illustrated in Figs. 46-7 and 46-9 in Chap. 46.

A convenient rule of thumb for predicting gain (of a relatively lossless antenna) is

$$G = \frac{K}{\theta_1\theta_2} \qquad (1\text{-}21)$$

where K is a unitless constant and θ_1 and θ_2 are the 3-dB beamwidths (in degrees) in the two orthogonal principal planes. The correct value of K for an actual antenna depends on the antenna efficiency. A popular value used by many antenna engineers is 30,000, but many other values are in use. For example, Stutzman and Thiele[10] suggest a value of 26,000, and Stegen[11] suggests a value of 35,000. One should subtract about 1½ dB from the estimated gain for antennas with cosecant-squared-shaped beams.

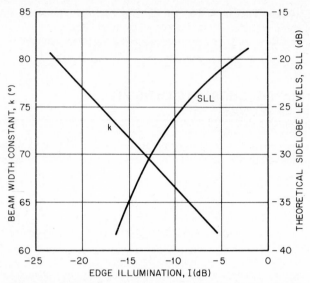

FIG. 1-9 Beamwidth constant and approximate first sidelobe level versus edge illumination (including space attenuation) for paraboloidal reflector-type antennas with plane-wave feeds. *(Data from Refs. 8 and 9.)*

REFERENCES

1 S. Silver, *Microwave Antenna Theory and Design,* McGraw-Hill Book Company, New York, 1949, sec. 2.14.
2 *IEEE Standard Test Procedures for Antennas,* IEEE Std. 149-1979, Institute of Electrical and Electronics Engineers, New York, 1979, sec. 11.
3 J. S. Hollis et al., *Techniques of Microwave Antenna Measurements,* John Wiley & Sons, Inc., New York, 1984.
4 R. C. Hansen et al., "IEEE Test Procedures for Antennas; Number 149 (Revision of 48 IRE 252) January 1965," *IEEE Trans. Antennas Propagat.,* vol. AP-13, May 1965, pp. 437–466.
5 R. C. Hansen, "Aperture Theory," in R. C. Hansen (ed.), *Microwave Scanning Antennas,* Academic Press, Inc., New York, 1964, pp. 31–32.
6 Silver, op. cit., sec. 6.9.
7 H. T. Friis, "A Note on a Simple Transmission Formula," *IRE Proc.,* May 1946, pp. 254–256.
8 M. J. Komen, "Use Simple Equations to Calculate Beam Width," *Microwaves,* December 1981, pp. 61–63.
9 E. M. T. Jones, "Paraboloid Reflector and Hyperboloid Lens Antennas," *IRE Trans. Antennas Propagat.,* vol. AP-2, July 1954, pp. 119–127.
10 W. L. Stutzman and G. A. Thiele, *Antenna Theory and Design,* John Wiley & Sons, Inc., New York, 1981, p. 397.
11 R. J. Stegen, "The Gain-Beamwidth Product of an Antenna," *IEEE Trans. Antennas Propagat.,* vol. AP-12, July 1964, pp. 505–506.

Chapter 2

Fundamentals of Antennas *

Henry Jasik

AIL Division of Eaton Corporation

*This chapter was written by the late Henry Jasik for the first edition; it has been shortened and edited slightly by Richard C. Johnson for inclusion in this second edition.

2-1 RADIATION FROM ELECTRIC CURRENT ELEMENTS

One of the types of radiators frequently used in antenna practice is some form of thin wire arranged in a linear configuration. If the current distribution on such a wire is known or can be assumed with a reasonable degree of accuracy, then the radiation pattern and the radiated power can be computed. This computation is based on the integration of the effects due to each differential element of the current along the wire. It is therefore of interest to set down the complete expressions for the fields at any distance due to a differential element of current oriented along the z axis as shown in Fig. 2-1. The rms electric and magnetic field components are given as follows:

$$E_r = 60\beta^2 I \, dz \left[\frac{1}{(\beta r)^2} - \frac{j}{(\beta r)^3} \right] \cos\theta \, e^{-j\beta r}$$

$$E_\theta = j30\beta^2 I \, dz \left[\frac{1}{\beta r} - \frac{j}{(\beta r)^2} - \frac{1}{(\beta r)^3} \right] \sin\theta \, e^{-j\beta r}$$

$$H_\phi = j \frac{\beta^2}{4\pi} I \, dz \left[\frac{1}{\beta r} - \frac{j}{(\beta r)^2} \right] \sin\theta \, e^{-j\beta r}$$

$$E_\phi = H_r = H_\theta = 0$$

where $I \, dz$ = moment of differential current element (I is given in rms amperes and dz is given in meters.)
r = distance, m, to observation point
$\beta = 2\pi/\lambda$
λ = wavelength, m
$j = |\sqrt{-1}$
E is given in volts per meter
H is given in amperes per meter

A time factor of $e^{j\omega t}$ has been omitted, since for all the cases in which we are interested it is assumed that we have a sinusoidally time-varying current of constant frequency.

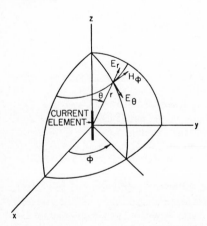

FIG. 2-1 Coordinate system for an electric dipole.

For most problems of interest it is only necessary to know the components in the far field, i.e., when r is very much greater than the wavelength. Under these conditions, the field components are simply given by

$$E_\theta = j\frac{30\beta I\, dz}{r}\sin\theta\, e^{-j\beta r}$$

$$= j\frac{60\pi I\, dz}{r\lambda}\sin\theta\, e^{-j\beta r}$$

$$H_\phi = j\frac{\beta I\, dz}{4\pi r}\sin\theta\, e^{-j\beta r}$$

$$= \frac{E_\theta}{120\pi}$$

These expressions apply only for a very short element of current having a constant value along its length. However, they may readily be used to determine the field from any wire having a known current distribution by integrating the field due to each of the differential current elements along the length of the antenna. Taking into account the variation of current and the phase differential due to the varying distance from the observation point to each current element, the general expression for the field of any current distribution becomes

$$E_\theta = j\frac{60\pi\sin\theta}{r\lambda}\int_{-\ell/2}^{\ell/2} I(z)\, dz\, e^{-j\beta r(z)}$$

where both $I(z)$ and $r(z)$ are now functions of z and the integration takes place along the length of the antenna from $-\ell/2$ to $+\ell/2$.

For very short antennas, the above expression can be simplified to

$$E_\theta = j\frac{60\pi\sin\theta}{r\lambda} I_0 L_e e^{-j\beta r}$$

where I_0 = current at center of antenna
L_e = effective length of antenna defined as

$$L_e = \frac{1}{I_0}\int_{-\ell/2}^{\ell/2} I(z)\, dz$$

The effective length is of interest in determining the open-circuit voltage at the terminals of a receiving antenna. It is also used on occasion to indicate the effectiveness of a transmitting antenna.

For a short top-loaded linear antenna which has uniform current distribution as shown in Fig. 2-2a, the effective length is simply equal to the physical length. For a short antenna which is much less than a half wave long, as shown in Fig. 2-2b, the current distribution is essentially triangular and its effective length is one-half of its physical length.

For antennas with an overall length greater than about a quarter wavelength, the variation of the phase term cannot be neglected and the integral must be evaluated by taking this term into account. The method will be very briefly illustrated for the case of a thin half-wave radiator which can be assumed to have a sinusoidal current

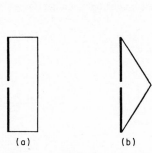

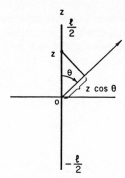

FIG. 2-2   Current distribution on a short linear antenna. (*a*) With top loading. (*b*) Without top loading.

FIG. 2-3   Coordinates for computing radiation from a half-wave dipole.

distribution so that $I(z)$ is given by $I_0 \cos \beta z$. The geometry for finding $r(z)$ is shown in Fig. 2-3, from which it is readily seen that

$$r(z) = r - z \cos \theta$$

The field for a half-wave dipole is then given by

$$E_\theta = j \frac{60\pi \sin \theta}{r\lambda} I_0 e^{-j\beta r} \int_{-\ell/2}^{\ell/2} \cos \beta z \; e^{j\beta z \cos \theta} \, dz$$

which reduces to

$$E_\theta = j \frac{60 I_0}{r} e^{-j\beta r} \frac{\cos\left(\frac{\pi}{2} \cos \theta\right)}{\sin \theta}$$

The relative-radiation pattern for a half-wave antenna is shown in solid lines in Fig. 2-4. For comparison purposes, the relative-radiation pattern of a very short dipole is shown in dotted lines. The patterns shown are those in a plane which contains the axis for the antenna. The pattern in the plane perpendicular to the antenna is perfectly circular because of symmetry.

There are a number of other properties for the half-wave dipole which are of considerable interest, such as radiation resistance, gain, and input impedance. These properties are discussed in Chap. 4.

The method of computing radiation patterns for thin linear radiators is basic regardless of the length or complexity of shape. As a matter of interest, the following formula gives the radiated field from a center-fed thin wire of arbitrary length ℓ with an assumed sinusoidal current distribution:

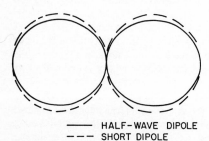

——— HALF-WAVE DIPOLE
- - - SHORT DIPOLE

FIG. 2-4   Radiation patterns.

$$E_\theta = j \frac{60 I_0 e^{-j\beta r}}{r} \frac{\cos\left(\frac{\beta \ell}{2} \cos \theta\right) - \cos \frac{\beta \ell}{2}}{\sin \theta}$$

It will be noted that the radiated field perpendicular to the antenna continues to increase as the length is increased until the overall length is about 1¼ wavelengths. Beyond this point the field starts falling off. When the overall length is two wavelengths, the field is zero normal to the axis of the antenna. For this length, the radiation pattern has broken up into two major lobes which are directed off the normal to the antenna. For a still longer length the antenna pattern will continue to break up into a large number of lobes whose positions depend on overall length of the antenna (Chaps. 4 and 11).

2-2 RADIATION FROM MAGNETIC CURRENT ELEMENTS

Another basic radiator which is frequently used in antenna practice is a magnetic current element. Although magnetic currents do not exist in nature, a number of configurations produce fields identical with those which would be produced by a fictitious magnetic current. For instance, a circular loop carrying electric current whose diameter is very small in terms of wavelengths will produce fields which are equivalent to those of a short magnetic dipole. The fields for any distance are given by the following expressions:

$$E_\phi = 30\beta^3 \, dm \left[\frac{1}{\beta r} - \frac{j}{(\beta r)^2} \right] \sin \theta \, e^{-j\beta r}$$

$$H_r = \frac{\beta^3}{2\pi} \, dm \left[\frac{j}{(\beta r)^2} + \frac{1}{(\beta r)^3} \right] \cos \theta \, e^{-j\beta r}$$

$$H_\theta = -\frac{\beta^3}{4\pi} \, dm \left[\frac{1}{\beta r} - \frac{j}{(\beta r)^2} - \frac{1}{(\beta r)^3} \right] \sin \theta \, e^{-j\beta r}$$

$$E_r = E_\theta = H_\phi = 0$$

where the coordinate system is as shown in Fig. 2-5, and dm is defined as the differential magnetic dipole moment. For a small-diameter loop, the magnetic moment of the loop is equal to the electric current I flowing through the loop times its area A.

For the far field, when r is very much greater than the wavelength, the field components reduce to

$$E_\phi = \frac{30\beta^2 \, dm}{r} \sin \theta \, e^{-j\beta r}$$

$$H_\theta = -\frac{\beta^2 \, dm}{4\pi r} \sin \theta \, e^{-j\beta r}$$

$$= -\frac{E_\phi}{120\pi}$$

It will be noted that the field expressions for the magnetic current element are almost exactly analogous to those for the electric current element except for the interchange of electric and magnetic quantities. The radiation of a short magnetic dipole or a small-diameter loop is also a doughnut pattern, as in the case of an electric dipole. For a small loop the radiation pattern in the plane of a loop is perfectly circular, while

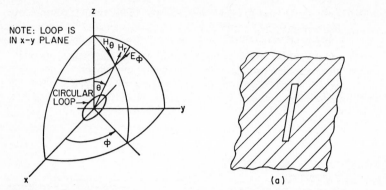

FIG. 2-5 Coordinate system for a magnetic dipole.

FIG. 2-6 (*a*) Thin slot in ground plane. (*b*) Complementary dipole.

the pattern in the plane through the axis of the loop is a figure of eight whose amplitude is proportional to $\sin \theta$. The expressions given are accurate for loop diameters which are considerably less than one-tenth wavelength. As a matter of fact, for very small loops, the radiation pattern does not depend on the exact shape of the loop, which may be square, rectangular, or some other shape, provided the overall circumference remains much less than a quarter wavelength.

For loops whose diameter is of the order of the wavelength, the radiation pattern can deviate considerably from the doughnut form, depending on the nature of the current distribution along the loop and diameter of the loop. These considerations are treated in Chap. 5.

Another antenna whose radiation characteristics are essentially similar to those of a magnetic dipole consists of a very thin slot in an infinitely large metallic ground plane, as shown in Fig. 2-6*a*. For this type of antenna, the electric field is applied across the narrow dimension of the slot. It is possible to show that the field radiated by this slot is exactly the same as would be radiated by a fictitious magnetic dipole with a magnetic current distribution *M* which is numerically equal to the distribution of electric voltage *V* across the slot. Thus the radiation pattern of a thin rectangular slot is identical with the radiation pattern of the complementary electric dipole which would just fill the slot, as shown in Fig. 2-6*b*. The only difference between the two types of radiators is the fact that the electric and magnetic quantities have been interchanged. This complementary relationship has been treated in considerable length in technical literature and will not be elaborated on in this section.

It is pertinent to point out a few precautions concerning the application of these complementary relationships. For one thing, the complementary relationship is based on the assumption that only electric field exists within the slot and that no magnetic field is present. This is true only for vanishingly thin slots. For slots whose width is appreciable in comparison with their length, the above assumption is no longer valid and the radiation pattern will be somewhat modified as compared with the case of the very thin slot. Another assumption on which the complementary relationship is based is that the size of the conducting ground plane is infinitely large. This, of course, is never true in practice, and for finite ground planes the size of the ground plane may exert a large influence on the radiation pattern, particularly at angles close to the plane of the sheet. Even for sheets which are fairly large in terms of wavelength, there are some minor modifications of radiation pattern.

It might also be mentioned that the complementary relationship holds only when the slot is cut in a large flat ground sheet. For slots cut in circular cylinders, the pattern can be considerably different from that predicted by the complementary relationship, particularly in the plane perpendicular to the axis of the cylinder. However, use of the complementary relationship is still very useful to engineers as an intuitive guide, and while in many cases the results will not be exact, they will certainly give a first approximation to the actual radiation pattern.

2-3   ANTENNAS ABOVE PERFECT GROUND

The characteristics of antennas operating near ground level will be modified by the effect of ground reflections. This is particularly true of antennas operating at frequencies below 30 MHz when the height of the antenna above ground may be less than one or two wavelengths. For airborne antennas or for large-aperture, narrow-beam antennas, in which the main beam is elevated upward at least several beamwidths, the ground may play a relatively small role. In all cases, the ground will play a part in the propagation between transmitter and receiver, and to compute its effect the characteristics of the ground and the geometry of the propagation path, as well as the pattern characteristics of the receiving and transmitting antennas, must be known.

For antennas at relatively small heights above ground, the ground is a basic part of the antenna system and will affect not only the radiation pattern of the antenna but also its impedance properties. To obtain a first-order estimate of the effect of ground, simple image theory can be applied to the case of a perfectly conducting ground surface. It is well known from electromagnetic theory that the tangential field on a perfect conductor must be zero and that the electric field must be normal to the conducting surface. To satisfy this requirement, a conducting charge above a conducting plane will induce a charge distribution on the plane exactly equivalent to that which would be produced by an equal charge of the opposite sign at the same distance below the plane, as shown in Fig. 2-7a. Since current is a movement of charge, it is readily possible to deduce the direction of the images for vertical, horizontal, and inclined wires above ground, as shown in Fig. 2-7b–d. The image for any other configuration can readily be determined by using the rule that vertical components are in the same direction while horizontal components are in opposite directions.

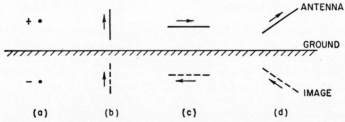

FIG. 2-7   Images above a perfectly conducting ground plane. (a) Point charge. (b) Vertical wire. (c) Horizontal wire. (d) Inclined wire.

The field in any direction above the ground plane can easily be determined by replacing the ground plane by the image and computing the resulting field due to the antenna plus its image. This is valid above the ground plane, since for a perfect conductor the field below the ground plane is zero.

Although the above method is rigorously true only for antennas above highly conducting ground, it does give good results in many cases of interest. The same technique can be used when the ground has arbitrary values of conductivity and dielectric constant by assuming an image current which is related to the antenna current by the ratio of the complex-reflection coefficient for the appropriate angle of incidence.

The effect of the ground on the impedance of an antenna can also be determined by image theory. For instance, the input impedance of an antenna above perfectly conducting ground is simply the input impedance of the antenna in free space plus the mutual impedance due to the image antenna. For arbitrary ground, the same sort of relation is still true except that the mutual impedance due to the image must be multiplied by the complex-ground-reflection coefficient for normal incidence.

A special case of considerable interest occurs when one end of the antenna above ground is exactly one-half of the input impedance of the antenna plus its image when driven in free space. For example, the input impedance of a quarter-wave dipole above ground is exactly one-half that of a half-wave dipole in free space.

2-4   RADIATION FROM APERTURES

The computation of radiation patterns for linear-wire antennas is relatively simple if the current distribution on the wire is known. The current distribution is not usually known exactly except for a few special cases. However, physical intuition or experimental measurement can often provide a reasonable approximation to the current distribution, and for many engineering purposes a sufficiently accurate result can be obtained. In theory, of course, an exact result can be derived from a boundary-value solution if the nature of the exciting sources is known. From a practical point of view, the amount of labor involved in obtaining numerical results is excessive even for those cases in which the geometry is relatively simple and a rigorous solution can be expressed in terms of a series of tabulated functions. It is therefore necessary in many situations to be able to compute the radiation pattern by alternative methods based on a reasonable assumption of the nature of the electromagnetic fields existing in the vicinity of an antenna structure.

The Equivalence Principle

One powerful technique for simplifying this type of computation makes use of an equivalence principle given by Schelkunoff.[1,2] Briefly stated, this principle supposes that a given distribution of electric and magnetic fields exists on a closed surface drawn about the antenna structure. These fields are then canceled by placing a suitable distribution of electric and magnetic current sheets on the closed surface so that the fields inside the closed surface are zero. The radiation is then computed from the electric and magnetic current sheets and, except for a difference in sign, is identical to the radiation which would have been produced by the original sources inside the closed surface. It will be noted that this principle is essentially a more rigorous formulation of Huygens' principle. If the fields on the closed surface are known exactly, then the resulting computation is also exact. The degree of approximation which can be obtained by this technique depends only on how accurately the fields across the closed surface may be estimated.

When the electric and magnetic field strengths are respectively given by E and H, then the equivalent densities are given by

Electric current density: $\mathbf{J} = \mathbf{n} \times \mathbf{H}$

Magnetic current density: $\mathbf{M} = -\mathbf{n} \times \mathbf{E}$

where both J and H are expressed in amperes per meter and M and E are expressed in volts per meter.

The vector cross product has been used to show that the electric current is perpendicular to direction of propagation and to the magnetic field vector. If the initial E and H fields are as shown in Fig. 2-8a, then the resulting electric and magnetic currents are directed as shown in Fig. 2-8b.

One way of visualizing the effect of a small portion of the wavefront is to consider the physical equivalent of the electric and magnetic current sheets. The electric current is equivalent to a short electric dipole, while the magnetic current is equivalent to a short magnetic dipole or a small electric current loop, as was discussed in Sec. 2-

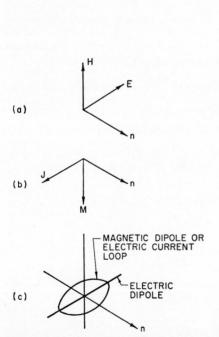

FIG. 2-8 Current sheet relations. (a) Initial E and H fields. (b) Resulting electric and magnetic currents. (c) Orientation of electric dipole and electric current loop.

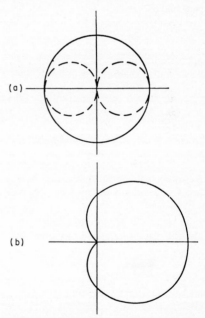

FIG. 2-9 Radiation patterns in plane of electric current loop. (a) ———— Pattern of electric current loop. –––– Pattern of electric dipole. (b) Cardioid pattern due to combination of electric current loop and electric dipole.

2. The electric dipole is oriented in the direction of the electric field E, while the electric current loop is located in the plane defined by the electric field E and the direction of propagation n, as shown in Fig. 2-8c.

If we look at the radiation pattern in the plane of the loop, the component due to the electric dipole will have a cos θ type of variation while the component due to the electric current loop will have a circular pattern as shown in Fig. 2-9a. For a portion of wavefront in free space in which the E and H fields are related as

$$E = 120\pi H$$

the relative amplitudes of the two components will be such that the radiation pattern of the combination is given by (1 + cos θ) or a cardioid pattern as shown in Fig. 2-9b.

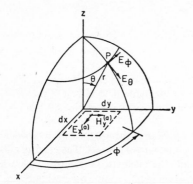

FIG. 2-10   Coordinates for current sheet computation.

Although the concept of the electric and magnetic current sheets is most useful for derivation purposes and for obtaining a physical picture of radiation from apertures, it is possible to compute the reradiated field from an aperture directly in terms of the tangential electric and magnetic field components in the aperture itself. With reference to Fig. 2-10, the aperture lies in the xy plane with field components $E_x^{(a)}$ and $H_y^{(a)}$. At a distant point P, the radiation components of the electric field are given by

$$E_\theta = -j[E_x^{(a)} + 120\pi H_y^{(a)} \cos \theta] \frac{\cos \phi}{2\lambda r} \, dx \, dy \, e^{-j\beta r}$$

$$E_\phi = +j[E_x^{(a)} \cos \theta + 120\pi H_y^{(a)}] \frac{\sin \phi}{2\lambda r} \, dx \, dy \, e^{-j\beta r}$$

The above is true for an aperture in which $E_x^{(a)}$ and $H_y^{(a)}$ take on arbitrary values. An important special case is that in which the field components are the same as exist in a free-space plane wave. For this case

$$E_x^{(a)} = 120\pi H_y^{(a)}$$

and the distant fields are now given by

$$E_\theta = -jE_x^{(a)}(1 + \cos \theta) \frac{\cos \phi}{2\lambda r} \, dx \, dy \, e^{-j\beta r}$$

$$E_\phi = jE_x^{(a)}(1 + \cos \theta) \frac{\sin \phi}{2\lambda r} \, dx \, dy \, e^{-j\beta r}$$

Applications of the Equivalence Principle

One typical application of these results is to the problem of determining the radiation patterns from an electromagnetic horn.[3] The tangential fields at the aperture of such a horn can be approximated by assuming that the fields at the aperture plane are the same as would be obtained if the guiding surfaces of the horn were extended to infinity. Although not exact, this assumption is quite good when both linear dimensions of the aperture are greater than one or two wavelengths. After this assumption of the electric and magnetic fields at the aperture, the far-field relationships are then formulated by taking the fields across the aperture and integrating over the aperture, taking into account the amplitude and phase variations of the field incident on the aperture and the phase differential due to the varying distance from the observation point to each area element of the aperture.

The above method gives results which are good to a high degree of approximation, particularly for large-aperture horns. The equivalence principle can also be used with small-aperture horns, but caution must be exercised with regard to the assumptions of the relative values of electric and magnetic field strengths. In small horns, it is not true that the aperture fields are those which would exist if the waveguide were extended to infinity. For instance, in a waveguide radiator whose E-plane width is small, the field at the aperture will be predominantly electric and the magnetic field may be quite small. For this condition, the equivalent current sheet will consist mainly of magnetic current, and for a small aperture this will have a substantial effect on the radiation pattern, particularly in the E plane. For those cases in which it is possible to measure the standing waves in the waveguide leading to the aperture, it is possible to estimate the relative values of electric and magnetic field strength at the aperture and to make a first-order correction to the radiation-pattern computation.

The equivalence principle can also be used to determine the radiation from the open end of a coaxial line. For this case, the field at the aperture is predominantly electric, with only a small component of magnetic field. The field variation in the aperture is assumed to be that of the dominant mode in the coaxial line. The problem is readily formulated in cylindrical coordinates.[4]

Another problem which can be solved by using the equivalent-current-sheet method is the case of reflection from a conducting sheet such as paraboloidal reflector or a plane reflector as employed in microwave relay applications. To formulate this problem, the tangential fields that would have existed if the reflector were not present are determined by using only the magnetic field term for the aperture field and integrating over the surface, taking into account variations of the aperture field amplitude and phase and the phase differential to the observation point.

2-5 IMPEDANCE PROPERTIES OF ANTENNAS

Self-Impedance

The input impedance of an antenna is a characteristic of considerable interest to engineers since they are concerned with the problem of supplying the antenna with the maximum amount of transmitter power available or in abstracting the maximum amount of received energy available from the antenna. Except for the simplest types of antenna configuration, the theoretical computation of the input impedance is an extremely arduous task, and for a large number of antenna types it is usually easiest to make a direct experimental measurement of the input impedance. However, for linear antennas which are relatively small in size, it is possible to make some reasonably good estimates as to the magnitude of the input resistance. It is also possible to assess, with a reasonable degree of accuracy, the mutual impedance between linear radiators for the purpose of estimating the input impedance of an individual element in an array of radiators.

In most practical cases, the input impedance of even a simple antenna is affected to a considerable degree by the terminal conditions at the point where the transmission line feeds the radiator. For most accurate results, it is therefore necessary to measure the various impedances involved and to use the calculated values primarily as a guide during the design procedure.

For very short wire dipoles, the radiation resistance is a quantity which is closely allied with the resistive component of the input impedance. The radiation resistance is normally defined as the ratio of the total power radiated by an antenna divided by the square of the effective antenna current referred to a specified point. For short antennas this is a useful quantity because it enables one to estimate the overall radiation efficiency of the antenna by separating the radiation component of the input resistance from the loss resistance due to the ground system or the loss resistance due to the impedance-matching elements.

To compute the radiation resistance, it is necessary to know the radiation field pattern of the antenna in terms of the current flowing at the point to which the radiation resistance is referred. The total radiated power is then computed by integrating the total power density passing through a sphere surrounding the antenna. This computation will be carried through very briefly for the case of a very short dipole having an effective length L_e and carrying a current I_0. The electric field intensity for this antenna is given in Sec. 2-1 as

$$|E_\theta| = \frac{60\pi I_0 L_e}{r\lambda} \sin \theta$$

The power density in the far field is given by the Poynting vector, which is equal to $E_\theta^2/120\pi$, where the electric field is given in rms volts per meter and the power density is expressed in watts per square meter. Integrating over a large sphere surrounding the antenna, we then obtain

$$P = \int_0^\pi \frac{E_\theta^2}{120\pi} 2\pi r^2 \sin \theta \, d\theta$$

$$= \frac{80\pi^2 I_0^2 L_e^2}{\lambda^2}$$

Dividing this result by I_0^2, we obtain

$$R_e = 80\pi^2 \frac{L_e^2}{\lambda^2}$$

where R_e is a radiation resistance in ohms. It will be noted that the important variable is the ratio of the effective length L_e to the wavelength and that the larger this ratio the greater the radiation resistance. Thus for a short antenna of physical length ℓ which is top-loaded so as to give a uniform current distribution, the effective length will be equal to the physical length and the radiation resistance will be $80\pi^2\ell^2/\lambda^2$. For an antenna with no top loading, the current distribution will be triangular (Sec. 2-1), and the effective length will be equal to one-half of the physical length, so that the radiation resistance will be $20\pi^2\ell^2/\lambda^2$. It can be seen that the radiation resistance for the uniform-current case is thus equal to 4 times that obtained in the case of triangular current distribution, despite the fact that the radiation pattern and directivity are the same for both antennas since their length is small with respect to the wavelength.

The problem of computing the input impedance of a dipole of finite diameter whose length is of the order of a half wavelength is one that has been considered at great length by a number of writers. The subject is treated in detail in Chap. 4, where a considerable amount of measured data is presented.

For the case of a very thin half-wave dipole, the input resistance may be computed by assuming a sinusoidal current distribution and integrating the total power radiated over the surface of a large sphere, in the same fashion as just described for the short current element. Using the far field for the half-wave dipole as given in Sec. 2-1 and performing the appropriate operations, the input resistance of the half-wave thin dipole is found to be 73.1 Ω. The reactive component of the input impedance cannot be determined by the far-field method since the reactance is governed primarily by the electromagnetic fields in the vicinity of the antenna itself. The input reactance is also a function of the relative diameter of the dipole and of the terminal conditions at the driving point.

Mutual Impedance

In an array of antennas, the driving-point impedance of an individual element may differ considerably from its self-impedance because of the effect of mutual coupling with other elements of the array.[5] In a multielement array, the relations between the currents and voltages are given by

$$V_1 = I_1 Z_{11} + I_2 Z_{12} + \cdots + I_n Z_{1n}$$
$$V_2 = I_1 Z_{12} + I_2 Z_{22} + \cdots + I_n Z_{2n}$$
$$\cdot \qquad \cdot \qquad \cdot \qquad \cdot$$
$$\cdot \qquad \cdot \qquad \cdot \qquad \cdot$$
$$V_n = I_1 Z_{1n} + I_2 Z_{2n} + \cdots + I_n Z_{nn}$$

where V_n = impressed voltage at nth element
  I_n = current flowing in nth element
  Z_{nn} = self-impedance of nth element
 $Z_{mn} = Z_{nm}$ = mutual impedance between mth and nth elements

The driving-point impedance for element 1, for instance, is found from the ratio of the impressed voltage to the current and is obtained from the above equations as follows:

$$Z_{1\text{input}} = \frac{V_1}{I_1} = Z_{11} + \frac{I_2}{I_1} Z_{12} + \cdots + \frac{I_n}{I_1} Z_{1n}$$

It is readily seen that the input impedance or driving-point impedance of a particular element is not only a function of its own self-impedance but also a function of the relative currents flowing in the other elements and of the mutual impedance between elements. In an array in which the current distribution in the elements is critical because of pattern requirements, it is necessary to determine the input impedance from the above relationship and to design the transmission-line coupling system to match the input impedance rather than the self-impedance. Some examples of this are given in Ref. 6.

An alternative method for accurately controlling the current distribution in certain types of arrays is to use a transmission-line distribution system which forces the required current to flow in an antenna element regardless of the effect of mutual impedance. For instance, the constant-current properties of a quarter-wave line are such that the current in a load at the end of a quarter-wave line is equal to the driving voltage divided by the characteristic impedance of the quarter-wave line regardless of the load impedance. This property is also true for a line whose length is an odd number of quarter wavelengths. Thus, for example, in order to feed an array of four dipole elements with exactly equal currents regardless of mutual coupling, the length of transmission line from the dipole to the junction would be an odd number of quarter wavelengths. By making use of the constant-voltage properties of a half-wavelength transmission line, it is possible to build up a distribution system to feed a large number of antenna elements by means of combinations of half-wave and quarter-wave lines. It is worth mentioning that although the uniform half-wave line behaves as a voltage transformer with a transformation ratio of 1, it is possible to obtain other transformation ratios by constructing the half-wave line of two quarter-wave sections of differing characteristic impedances.

In many situations it is not possible to sidestep the effects of mutual coupling and it is necessary to have a reasonably accurate estimate of the value of mutual impedance between antenna elements. It is possible to calculate the mutual impedance for very thin dipoles, and the results are given in Chap. 4 for several cases of interest. Although the finite diameter of a dipole does have some effect on the magnitude of the mutual impedance, the effect is a second-order one and for many computations may be neglected. This is not true for self-impedance, whose value is very definitely a function of the dipole diameter.

For antenna elements other than the simple dipole or slot radiator, little theoretical work is available on the magnitude of mutual-coupling effects and it is necessary to use experimental methods for determining the mutual impedance. Even in the case of dipole elements, it is frequently desirable to measure the mutual impedance, particularly for a dipole whose diameter is not small compared with its length.

Several experimental methods are available. When the antenna elements are identical and reasonably small physically, one simple method is to measure the input impedance when the element is isolated and then to repeat the measurement when a ground plane is placed near the element to simulate the effect of an image. The difference between the two impedance measurements is the mutual impedance for a distance corresponding to the distance between the driven element and its image. An alternative method when two elements are available is to measure the input impedance when one element is isolated, and then to repeat the measurement when the second element is in place and has a short circuit across its terminals.

2-6 DIRECTIVITY PATTERNS FROM CONTINUOUS LINE SOURCES

For antenna systems that have apertures which are very large in terms of wavelength, it is frequently desirable to use a continuous type of aperture distribution because of the relative simplicity, as compared with a discrete-element type of array which requires a large number of driven elements. For instance, a common form of large-aperture antenna is a paraboloidal reflector illuminated by a point-source feed. To replace an aperture of the order of 100 wavelengths in diameter by a discrete-element array would require more than 5000 individual radiating elements, each of which must be fed with current of the correct amplitude and phase.

For very large apertures it is apparent that a reflector type of antenna is considerably simpler than a discrete-element array. In addition, the reflector can be made to operate over a wide range of frequency simply by changing the feed, while a discrete-element array can operate efficiently only over a small bandwidth, on the order of 15 to 40 percent at most. Where very low sidelobe levels are required, discrete arrays have a substantial advantage over reflector-type structures since it is possible to obtain sidelobe levels of 40 dB down and greater by exercising considerable care in designing and constructing discrete-element arrays. By comparison, reflector-type antenna systems have sidelobe levels in the range of 18 to 25 dB down, and at the present state of the art it seems unlikely that sidelobe levels much lower than 35 dB down can be achieved with a reflector system. Use of a two-dimensional folded reflector system will offer some improvement because of the elimination of blocking by the feed and its supports, but even for this type of structure special techniques are necessary to achieve sidelobe levels lower than 30 dB down.

Although the continuous-aperture type of antenna has practical limitations with regard to very low sidelobe performance, it nevertheless is widely used because of its relative simplicity. For this reason, it is desirable to know what can be expected of various types of ideal distribution functions since these in effect place an upper limit on the potential performance of a continuous antenna. Also, for very long arrays of discrete elements, the continuous distribution may be used to obtain an excellent approximation to the element excitation coefficients.

Line-Source Distributions

For line sources, the current distribution is considered to be a function of only a single coordinate. The directivity pattern $E(u)$ resulting from a given distribution is simply related to the distribution by a finite Fourier transform,[7-9] as given below:

$$E(u) = \frac{\ell}{2} \int_{-1}^{+1} f(x) e^{jux} \, dx$$

where $f(x)$ = relative shape of field distribution over aperture as a function of x
$u = (\pi \ell / \lambda) \sin \phi$
ℓ = overall length of aperture
ϕ = angle measured from normal to aperture
x = normalized distance along aperture $-1 \leq x \leq 1$

The simplest type of aperture distribution is the uniform distribution where $f(x) = 1$ along the aperture and is zero outside of the aperture. The directivity pattern is given as

$$E(u) = \ell \, \frac{\sin u}{u} = \ell \, \frac{\sin \left[\left(\dfrac{\pi \ell}{\lambda} \right) \sin \phi \right]}{\left(\dfrac{\pi \ell}{\lambda} \right) \sin \phi}$$

This type of directivity pattern is of interest because of all the constant-phase distributions the uniform distribution gives the highest gain.[7] As in the case of the discrete-element uniform distribution, it also has high sidelobe levels, the intensity of the first sidelobe being 13.2 dB down from the maximum.

The intensity of the sidelobe levels can be reduced very considerably by tapering the aperture distribution in such a way that the amplitude drops off smoothly from the center of the aperture to the edges. There are an unlimited number of possible distributions. However, a few simple types of distributions are typical and illustrate how the beamwidth, sidelobe level, and relative gain vary as a function of the distribution. Table 2-1 gives the important characteristics of several distributions having a simple mathematical form.

TABLE 2-1   Line-Source Distributions

TYPE OF DISTRIBUTION $-1 \leq x \leq 1$	DIRECTIVITY PATTERN E(u)		HALF POWER BEAMWIDTH IN DEGREES	ANGULAR DISTANCE TO FIRST ZERO	INTENSITY OF 1st SIDELOBE db BELOW MAX.	GAIN FACTOR
$f(x)=1$	$l\,\dfrac{\sin u}{u}$		$50.8\dfrac{\lambda}{l}$	$57.3\dfrac{\lambda}{l}$	13.2	1.0
$f(x)=1-(1-\Delta)x^2$	$l(1+\mathcal{L})\dfrac{\sin u}{u}$ $\mathcal{L}=(1-\Delta)\dfrac{d^2}{du^2}$	$\Delta=$ 1.0	$50.8\dfrac{\lambda}{l}$	$57.3\dfrac{\lambda}{l}$	13.2	1.0
		.8	$52.7\dfrac{\lambda}{l}$	$60.7\dfrac{\lambda}{l}$	15.8	.994
		.5	$55.6\dfrac{\lambda}{l}$	$65.3\dfrac{\lambda}{l}$	17.1	.970
		0	$65.9\dfrac{\lambda}{l}$	$81.9\dfrac{\lambda}{l}$	20.6	.833
$\cos\dfrac{\pi x}{2}$	$\dfrac{\pi l}{2}\dfrac{\cos u}{(\frac{\pi}{2})^2-u^2}$		$68.8\dfrac{\lambda}{l}$	$85.9\dfrac{\lambda}{l}$	23	.810
$\cos^2\dfrac{\pi x}{2}$	$\dfrac{l}{2}\dfrac{\sin u}{u}\dfrac{\pi^2}{\pi^2-u^2}$		$83.2\dfrac{\lambda}{l}$	$114.6\dfrac{\lambda}{l}$	32	.667
$f(x)=1-|x|$	$\dfrac{l}{2}\left(\dfrac{\sin\frac{u}{2}}{\frac{u}{2}}\right)^2$		$73.4\dfrac{\lambda}{l}$	$114.6\dfrac{\lambda}{l}$	26.4	.75

Of considerable interest is the manner in which the sidelobes fall off as the angle from the main beam increases or as u increases. For the uniform distribution which has a discontinuity in both the function and its derivatives at the edge of the aperture, the sidelobes decrease as u^{-1}. For the gable distribution or for the cosine distribution, both of which are continuous at the edge of the aperture but which have a discontinuous first derivative, the far-out sidelobes fall off as u^{-2}. For the cosine-squared distribution which has a discontinuous second derivative, the far-out sidelobes fall off as u^{-3}.

Many distributions actually obtained in practice can be approximated by one of the simpler forms or by a combination of simple forms. For instance, suppose that it were desired to find the directivity pattern of a cosine-squared distribution on a pedestal, i.e., a combination of a uniform distribution and a cosine-squared distribution as given by

$$f(x) = C + \cos^2 \frac{\pi x}{2}$$

The resulting directivity pattern is then obtained directly by adding the two functions for the directivity pattern as follows:

$$E(u) = C\ell \frac{\sin u}{u} + \frac{\ell}{2} \frac{\sin u}{u} \frac{\pi^2}{\pi^2 - u^2}$$

It should be noted that the sidelobes and other characteristics of the pattern must be obtained from the new directivity pattern and cannot be interpolated from Table 2-1. It is of some interest to note that by choosing the proper relative intensities of a uniform distribution and a cosine-squared distribution, it is possible to obtain a theoretical sidelobe level which is very low. For instance, if $C = 0.071$, then the intensity of the largest sidelobe will be 43 dB below the maximum of the main beam with a half-power beamwidth given by $76.5\lambda/\ell$, a value which is somewhat lower than that for the cosine-squared distribution by itself.

In recent years, work has been done on line-source distributions which produce patterns approaching the Chebyshev type of pattern in which all the sidelobes have a constant level. Van der Maas[10] has shown that it is possible to find the element excitations for a discrete array with a large number of elements by determining the shape of the envelope function as the number of elements is increased indefinitely.

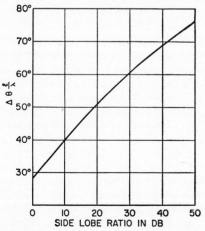

FIG. 2-11 Half-power beamwidth (multiplied by the aperture-to-wavelength ratio) versus sidelobe ratio for an ideal space factor.

This problem has also been investigated in detail by Taylor[11] for the case of a continuous line-source distribution. Of considerable interest is the relationship between the half-power beamwidth versus the sidelobe ratio for the ideal space factor (i.e., the equal-sidelobe type of space factor) since this relationship represents the limits of the minimum beamwidth that can be obtained for a given sidelobe level. Figure 2-11 is taken from Taylor's paper. For the details of computing the desired space factor, the reader is referred to the original paper.[11]

For many applications the equal-sidelobe-level pattern may be undesirable, and a more suitable pattern would be one in which the sidelobes decrease as the angle from the main beam, or the parameter u, increases. Work on this form of distribution has been carried out by

Taylor[12] in an unpublished memorandum. Basically, the approach is to represent the directivity pattern by a function of the form

$$E(u) = \frac{\sin \sqrt{u^2 - \pi^2 B^2}}{\sqrt{u^2 - \pi^2 B^2}}$$

where the ratio of the main beam to the first sidelobe is given by

$$R = 4.603 \frac{\sinh \pi B}{\pi B}$$

When B is equal to zero, the directivity pattern reduces to the familiar $\sin u/u$ for the case of a uniform distribution and the first sidelobe ratio is 4.603, or 13.2 dB.

As B is increased in value, the sidelobe ratio increases, and by the appropriate choice of B a theoretical sidelobe level as low as desired can be chosen. It will be noted that because of the expression for the directivity, at large values of u the sidelobes decrease as $1/u$. Hence we have a directivity pattern whose sidelobes decrease in a similar fashion to that of the uniform distribution but has the property that the maximum sidelobe level can be arbitrarily chosen.

Fortunately, the aperture distribution function for this type of pattern has a relatively simple form and is given by

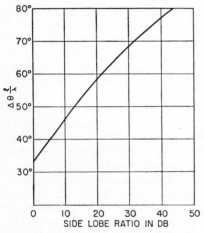

FIG. 2-12 Half-power beamwidth (multiplied by the aperture-to-wavelength ratio) versus sidelobe ratio for space factor defined by

$$E(u) = \frac{\sin \sqrt{u^2 - \pi^2 B^2}}{\sqrt{u^2 - \pi^2 B^2}}$$

$$f(x) = J_0(j\pi B \sqrt{1 - x^2}) \quad |x| \le 1$$
$$= 0 \quad |x| > 1$$

J_0 is a Bessel function of the first kind with an imaginary argument. Tables of this function are readily available.[13,14]

Taylor has tabulated the important characteristics of this type of distribution, and some of these data have been plotted in Fig. 2-12. By comparing Fig. 2-11 with Fig. 2-12 it can be seen that for a given maximum sidelobe level the half-power beamwidth of the distribution producing a decreasing sidelobe pattern is approximately 12 to 15 percent greater than the percent half-power beamwidth of the distribution which produces an equal-sidelobe pattern. This nominal loss in theoretical performance is a small price to pay when sidelobe levels must decrease as the angle from the main beam increases. When a decreasing sidelobe level which falls off inversely as the first power of the angle is desired, this type of distribution will result in a narrower beamwidth than that obtained by distributions which produce sidelobes falling off as a higher inverse power of the angle.

2-7 PATTERNS FROM AREA DISTRIBUTIONS

Rectangular Apertures

The directivity pattern of an area distribution is found in a similar manner to that used for line-source distributions except that the aperture field is integrated over two dimensions instead of one dimension. If the aperture distribution is given by $f(x,y)$, where x and y are the two coordinates, then the directivity pattern is given by

$$E(\theta,\phi) = \iint f(x,y)e^{j\beta \sin \theta (x \cos \phi + y \sin \phi)} \, dx \, dy$$

The difficulty of evaluating this expression will depend on the form of the distribution function. For many types of antennas, such as the rectangular horn, for example, the distribution function is separable; that is,

$$f(x,y) = f(x)f(y)$$

The directivity patterns in the principal planes are readily determined for the separable case since the pattern in the xz plane is identical with the pattern produced by a line-source distribution $f(x)$, while the pattern in the yz plane is identical with the pattern produced by a line-source distribution $f(y)$. If the distribution function is not separable, the integral must be evaluated either analytically or graphically.

Circular Apertures

An antenna that is frequently used in microwave applications is a paraboloid having circular symmetry. The radiation pattern may be computed by projecting the field distribution on the paraboloid to a plane at the opening of the paraboloid and computing the directivity pattern due to the plane aperture.

If the field in the aperture plane is a function of the normalized radius r and the aperture angular coordinate ϕ', then the directivity pattern is given by[7]

$$E(u,\phi') = a^2 \int_0^{2\pi} \int_0^1 f(r,\phi')e^{jur \cos (\phi-\phi')}r \, dr \, d\phi'$$

where a = radius at outside of aperture
ρ = radius at any point of aperture
$r = \rho/a$
$u = (2\pi a/\lambda) \sin \theta = (\pi D/\lambda) \sin \theta$
$D = 2a$ = aperture diameter
and $f(r,\phi')$ is the normalized aperture distribution function. The coordinates are as shown in Fig. 2-13.

The simplest forms of aperture distributions to evaluate are those in which the distribution is not dependent on the angular coordinate ϕ' but depends only on the radial coordinate r. The integral for the directivity pattern then becomes

$$E(u) = 2\pi a^2 \int_0^1 f(r)J_0(ur)r \, dr$$

When the distribution is constant, the integral becomes

$$E(u) = 2\pi a^2 \frac{J_1(u)}{u}$$

It is frequently desired to evaluate the directivity pattern for an illumination which tapers down toward the edge of the aperture. One function which is convenient for representing the aperture distribution is

$$f(r) = (1 - r^2)^p$$

This function behaves in a similar fashion to the nth-power distributions as discussed for the line-source case (Sec. 2-6). When the exponent increases, the distribution becomes more highly tapered and more concentrated in the center of the aperture. When the exponent decreases and approaches zero, the distribution approaches uniform illumination.

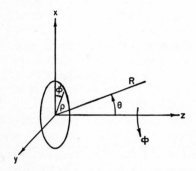

FIG. 2-13 Coordinates for a circular aperture.

TABLE 2-2 Circular-Aperture Distributions

TYPE OF DISTRIBUTION $0 \leq r \leq 1$	DIRECTIVITY PATTERN $E(u)$	HALF POWER BEAMWIDTH IN DEGREES	ANGULAR DISTANCE TO FIRST ZERO	INTENSITY OF 1st SIDELOBE db BELOW MAX.	GAIN FACTOR
$f(r)=(1-r^2)^0=1$	$\pi a^2 \dfrac{J_1(u)}{u}$	$58.9\dfrac{\lambda}{D}$	$69.8\dfrac{\lambda}{D}$	17.6	1.00
$f(r)=(1-r^2)$	$2\pi a^2 \dfrac{J_2(u)}{u^2}$	$72.7\dfrac{\lambda}{D}$	$93.6\dfrac{\lambda}{D}$	24.6	0.75
$f(r)=(1-r^2)^2$	$8\pi a^2 \dfrac{J_3(u)}{u^3}$	$84.3\dfrac{\lambda}{D}$	$116.2\dfrac{\lambda}{D}$	30.6	0.56

Evaluating the directivity pattern, we have

$$E(u) = 2\pi a^2 \int_0^1 (1 - r^2)^p J_0(ur)\, dr$$

$$= \pi a^2 \frac{2^p p! J_{p+1}(u)}{u^{p+1}} = \frac{a^2}{p+1} \Lambda_{p+1}(u)$$

Both the Bessel functions $J_{p+1}(u)$ and the lambda function $\Lambda_{p+1}(u)$ are available in tabular form.[13]

The principal characteristics of the directivity patterns are given in Table 2-2 for the cases $p = 0, 1, 2$. Comparison of the patterns of the uniformly illuminated circular aperture (i.e., when $p = 0$) with the results for the uniformly illuminated line source (Sec. 2-6) shows that the circular aperture has a lower sidelobe level and a broader beamwidth. This would be expected since projection of the circular-aperture illumination onto a line would produce an equivalent line source which is no longer uniform but has some degree of tapering.

Elliptical Apertures

In some applications an elliptically shaped reflector is used to permit control of the relative beamwidth in the two principal planes and to control the sidelobes by shaping the reflector outline. Computation of the directivity patterns for this aperture shape can be carried out from a knowledge of the Fourier components of the illumination function over the aperture.

2-8   EFFECTS OF PHASE ERRORS ON LINE SOURCES

The discussions on aperture distributions in Secs. 2-6 and 2-7 were concerned only with those aperture distributions in which the field was in phase across the entire array. In certain types of antenna systems, particularly those in which the beam is to be tilted, deviations from a uniform phase front do occur, so that it is desirable to evaluate the effects of phase errors on the directivity pattern.

For simplicity, the following discussions are limited to the case of a line-source distribution. Results will be first derived for the simple uniform amplitude distribution, and graphical results will be presented for a tapered amplitude distribution.

The most common phase-front errors are the linear, quadratic, and cubic phase errors. The linear phase error is expressed simply by

$$\Phi(x) = \beta_1 x$$

where β_1 = phase departure at edge of aperture
   x = aperture coordinate as defined in Sec. 2-6
The directivity pattern is given by

$$E(u) = \frac{\ell}{2} \int_{-1}^1 f(x) e^{jux} e^{-j\beta_1 x}\, dx$$

For a uniform illumination, the result is

$$E(u) = \ell \frac{\sin(u - \beta_1)}{u - \beta_1}$$

Thus the directivity pattern has the same form as for the in-phase case except that the pattern maximum is shifted in angle as defined by

$$u = \beta_1 \quad \text{or} \quad \sin \phi_0 = \frac{\beta_1 \lambda}{\pi \ell}$$

For other than a uniform illumination, the patterns will also be the same as for the in-phase case, except that $u - \beta_1$ is substituted for u in the expression for the directivity.

In the vicinity of the main beam, the directivity pattern is the same as that of an aperture tilted by ϕ_0 whose length is $\ell \cos \phi_0$. The half-power beamwidth is increased by the factor $1/\cos \phi_0$, while the gain is decreased by $\cos \phi_0$. For small angles of beam tilt, the pattern and gain are affected by only a minor amount.

The quadratic, or square-law, phase error is inherent in flared horn antennas. It also occurs in lens-type antennas and reflector antennas when the feed is defocused along the axis of symmetry. This type of phase error also appears when the directivity pattern of an antenna is measured at a finite distance.

The quadratic phase error is expressed by

$$\Phi(x) = \beta_2 x^2$$

and the directivity pattern is expressed by

$$E(u) = \frac{\ell}{2} \int_{-1}^{1} f(x) e^{jux} e^{-j\beta_1 x} \, dx$$

The integral is not readily evaluated except when $f(x)$ is constant or one of the cosine distributions. For these cases, the directivity can be expressed in terms of the Fresnel integrals. The directivity pattern for the uniformly illuminated case is given by[1,2]

$$E(u) = \frac{\ell}{2} \sqrt{\frac{\pi}{2\beta_2}} \{ C(m_2) - C(m_1) - j[S(m_2) - S(m_1)] \}$$

where

$$m_2 = \sqrt{\frac{2\beta_2}{\pi}} \left(1 - \frac{u}{2\beta_2} \right)$$

$$m_1 = \sqrt{\frac{2\beta_2}{\pi}} \left(-1 - \frac{u}{2\beta_2} \right)$$

and the Fresnel integrals are defined by

$$C(m) = \int_{0}^{m} \cos \left(\frac{\pi}{2} y^2 \right) dy$$

$$S(m) = \int_{0}^{m} \sin \left(\frac{\pi}{2} y^2 \right) dy$$

The expressions for the cosine and higher-order cosine distributions become increasingly complicated, and the problem of computation becomes exceedingly laborious. Some simplification in the computation problem can be obtained by the use of operational methods as developed by Spencer[8] and also by Milne.[15] Figure 2-14a–c is reproduced from Milne's paper.

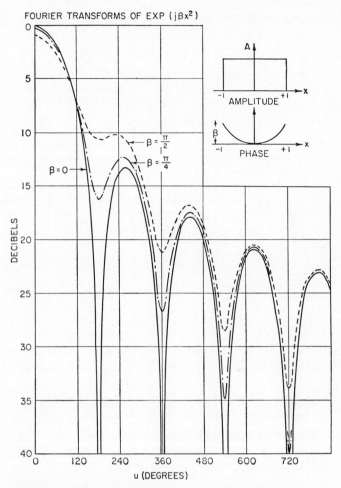

FOURIER TRANSFORMS OF EXP $(j\beta x^2)$

FIG. 2-14a Effects of square-law phase error in radiation patterns for uniform aperture illumination.

It will be noted that the principal effects of a quadratic phase error are a reduction in gain and an increase in sidelobe amplitude on either side of the main beam. For moderate amounts of phase error, the nulls between the sidelobes disappear and the sidelobes blend into the main beam, appearing as shoulders rather than as separate sidelobes.

The cubic phase error is expressed by

$$\Phi(x) = \beta_3 x^3$$

Computation of the directivity integrals for the cubic phase error becomes even more laborious than for the quadratic phase error, although the use of operational methods for computation simplifies the formal handling of the problem. Some typical results from Milne's paper are given in Fig. 2-15a–c.

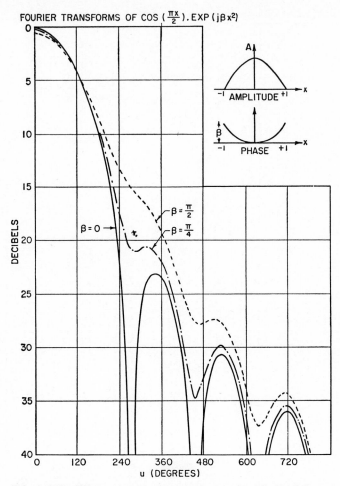

FIG. 2-14b Effects of square-law phase error in radiation patterns for cosine aperture illumination.

It will be observed that the cubic phase error produces a tilt of the beam in addition to a loss in gain. The sidelobes on one side of the beam increase in amplitude, while those on the other side diminish.

In general, when the feed is moved off axis to tilt the beam, both a linear phase term and a cubic phase term appear. The linear phase term causes a beam tilt which is a function of the geometry of the antenna system. For the case of a parabolic reflector, the cubic phase term causes a lesser amount of tilt in the opposite sense so that the resulting beam tilt is somewhat less than would be computed from geometrical considerations. For this case, the sidelobe increase appears on the side of the beam toward the axis.

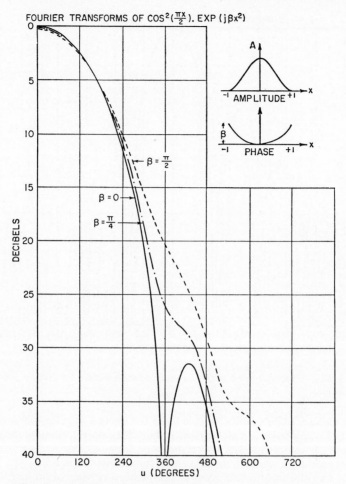

FIG. 2-14c Effects of square-law phase error in radiation patterns for cosine-squared aperture illumination.

2-9 EFFECTS OF RANDOM ERRORS ON GAIN AND SIDELOBES

The preceding section has discussed the effect of systematic phase errors on several different types of aperture distributions. In general, some of these errors are inherent in a given design and are a function primarily of the geometry of the antenna. In addition to systematic errors, a problem which commonly arises is that caused by random errors in both the phase and the amplitude across the aperture because of manufacturing tolerances.

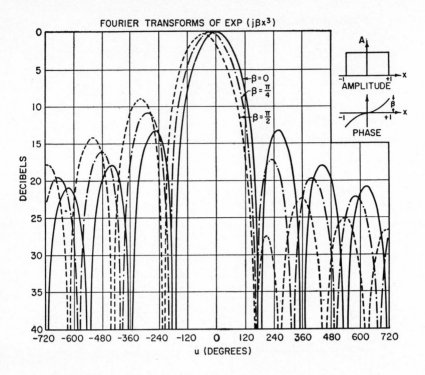

FIG. 2-15a Effects of cubic phase error on radiation pattern for uniform aperture illumination.

An estimate of the effect of random errors can be made in at least three different ways. The first method is to measure directly the radiation performance of the antenna in question. By comparison with the computed performance, it is possible to estimate the magnitude of the errors in the field distribution across the antenna.

The second method is to measure the amplitude and phase of the field distribution across the aperture. From these data, the effect of the errors on the antenna performance can be computed.

A third method is to compute the effect of manufacturing tolerances on the desired aperture distribution and from this, in turn, to compute the effect on the radiating properties of the antenna. This method is of considerable interest to the antenna designer, who must specify what sort of tolerances are necessary to achieve a desired result. As is true in all fabrication work, extremely tight tolerances result in excessive costs, while tolerances that are too loose can result in inadequate performance or failure of operation. It is therefore desirable to know how much accuracy is required for a given level of performance.

A theory of random errors is necessarily based on statistical considerations, so that the results predicted will not be on an absolute basis but rather in terms of a given probability level. The main use of this theory is to estimate what the effect will be on the average performance of a large number of antennas when a given tolerance level is specified.

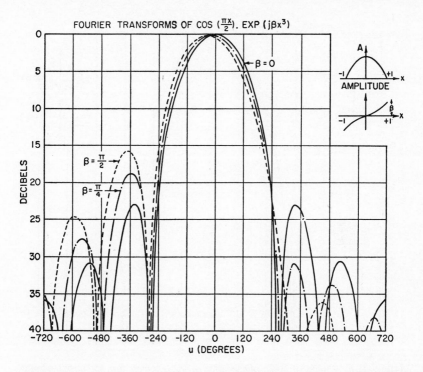

FIG. 2-15b Effects of cubic phase error on radiation pattern cosine aperture illumination.

Discrete-Element Arrays

The problem of the discrete-element array starts off with the premise that the ideal radiation pattern desired is specified by a uniformly spaced array of N elements, with each element carrying a specified current. If we denote the current in the nth element by I_n and the actual current by $I_n + \epsilon_n I_n$, then the radiation pattern due to the desired currents is given by

$$E(\phi) = \sum_{n=1}^{N} I_n \exp\left(j \frac{2\pi nd}{\lambda} \sin \phi\right)$$

where d = spacing between elements
ϕ = angle measured from normal to array
The radiation pattern due to the error terms is given by

$$R(\phi) = \sum_{n=1}^{N} \Delta_n I_n \exp\left(j\delta_n'\right)$$

where $\delta_n' = \delta_n + (2\pi nd/\lambda) \sin \phi$
Δ_n = ratio error of nth current
δ_n = phase error

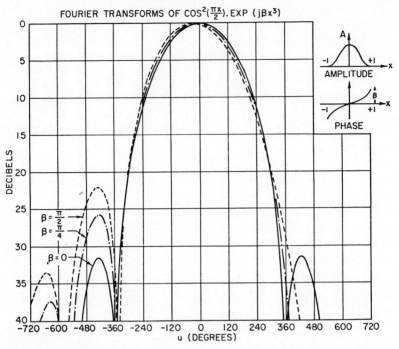

FIG. 2-15c Effects of cubic phase error on radiation pattern for cosine-squared aperture illumination.

It can readily be seen that the desired radiation pattern $E(\phi)$ is altered by the addition of the error-term radiation pattern $R(\phi)$. If measured data are available for the individual element currents, then it is possible to compute $R(\phi)$ exactly, even though this may be a rather laborious process.

As is often the case, experimental data are not available beforehand, and it is desirable to know what degree of precision is required in driving the array in order to achieve a given performance. For a large number of elements and for small errors, it is reasonable to assume that the individual errors are independent of one another and are distributed as a normal, or gaussian, distribution. The problem has been treated on a statistical basis by Ruze,[16] who has computed the sidelobe-level probabilities in terms of the rms error in the current values.

In statistical analysis, it is not possible to predict the exact pattern of a particular array with errors but rather the average pattern of a large number of similar arrays. If $\overline{P(\phi)}$ denotes the power pattern for an "average system" and $P_0(\phi)$ the power pattern for an antenna without errors, we have

$$\overline{P(\phi)} = P_0(\phi) + S(\phi)\overline{\epsilon^2}\frac{\Sigma I_n^2}{(\Sigma I_n)^2}$$

where $S(\phi)$ = a slowly varying function closely related to the power pattern of a single element

$$\overline{\epsilon^2} = \overline{\Delta^2} + \overline{\delta^2}$$

$\overline{\epsilon^2}$ = total mean square error
$\overline{\Delta^2}$ = mean square amplitude error
$\overline{\delta^2}$ = mean square phase error, rad^2

On the average, then, the effect of the random errors is to add to the pattern a constant power level which is proportional to the mean square error. For individual arrays and in particular directions, the sidelobe radiation will differ from this constant level in a fashion governed by the probability distribution for the particular array.

It is of interest to note that the spurious radiation is approximately proportional to $1/N$, so that for a given mean square error lower sidelobe levels are more readily obtained with larger antennas.

Computations have been made for a Dolph-Chebyshev type of array consisting of 25 elements with a design sidelobe level of 29 dB below the main beam. For an angular position where the no-error minor lobes have maxima, Fig. 2-16 shows the

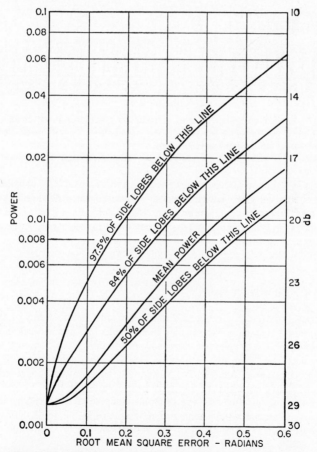

FIG. 2-16   Sidelobe distribution for 25-element broadside array, designed for 29-dB sidelobe suppression and computed at design-lobe maxima.

probability that the radiation will be below a specified number of decibels when a given mean error exists in the antenna currents.

As a check on the above theory, Ruze has computed the actual pattern of the 25-element antennas with a specific set of error currents. For a 0.40-rms error in each element with random phase, the radiation pattern was computed. Figure 2-17 shows the theoretical patterns for the cases with and without error. Analysis of the sidelobe magnitudes on this figure shows that their distribution compares very well with the theoretical distribution obtained on a statistical basis.

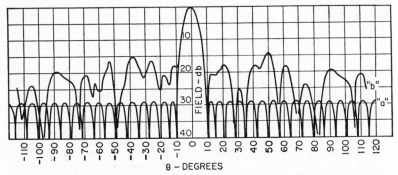

FIG. 2-17 Theoretical effect of error currents on radiation pattern of 25-element broadside array designed for 29-dB sidelobe suppression. (*a*) No error. (*b*) 0.40-rms error in each element at random phase.

The loss in gain can also be determined from this analysis. An excellent approximation for the actual gain G with errors as compared with the theoretical gain G_0 when no errors are present is given by

$$\frac{G}{G_0} \approx \frac{1}{1 + \left(\frac{3\pi}{4}\right)\left(\frac{d}{\lambda}\right)^2 \overline{\epsilon^2}}$$

where d = spacing between elements
λ = wavelength

The above analysis has concerned itself only with the accuracy of the current values required for the array. To translate these results into mechanical tolerances for a given antenna system depends on the nature of the individual elements, the type of feed system used, and a variety of other factors. One particular array which readily lends itself to this type of analysis is a waveguide-fed shunt-slot linear array. Some results for an X-band waveguide array have been obtained in Refs. 17 and 18. In the latter reference, computations for the probable sidelobe level have been made for a manufacturing tolerance with a standard deviation of 0.002 in (0.0508 mm) in a Dolph-Chebyshev array. The effect of the design sidelobe level and the number of elements is readily shown in Table 2-3.

It should be noted that the figures in Table 2-3 are based on a normal distribution in which the tolerances can take on all values but have a standard deviation of 0.002 in. In actual manufacturing practice, physical dimensions are not allowed to deviate

TABLE 2-3

Number of elements	Design sidelobe level, dB		
	20	30	40
12	18.6	25.9	29.1
24	19.0	26.7	31.8
48	19.1	27.3	33.3

NOTE: 84 percent of the sidelobe levels will be less than the tabulated value.

from the design value by more than an arbitrary amount so that the tolerance distribution is actually a truncated normal distribution. For this condition, the deterioration in sidelobe level is somewhat less than that given.

Two conclusions can be drawn regarding the effect of random errors in discrete-element arrays:

1 For a given sidelobe level, the effect of random errors becomes somewhat less critical as the number of elements in an array increases.

2 For a given number of array elements, the effect of random errors becomes more critical as the required sidelobe level is further suppressed.

As a practical matter, when designing low-sidelobe-level arrays, it is usually necessary to overdesign the array in order to be certain of achieving the desired sidelobe level in the presence of random errors. For instance, to attain a 26-dB sidelobe level, it may be desirable to design the aperture distribution for a 32-dB sidelobe level. Actually, the amount of overdesign required depends on a compromise between economical manufacturing tolerances and the loss in aperture efficiency due to the overdesign.

Continuous Apertures

The statistical analysis of continuous apertures, such as reflector-type antennas, is similar to that for discrete-element arrays, except for two important differences. For the discrete array, the error in one element has been assumed to be independent of the errors in adjacent elements. However, for a continuous-aperture array, a large error at one point implies that the error will be large in the immediate area around that point since the error could be due to warping of the reflector or a bump in the reflector. Also, the error will be purely a phase error since the amplitude distribution will be essentially unaffected by moderate changes in the reflector surface.

The fact that an error will extend over an area makes it convenient to use the concept of a correlation interval c. On the average, c is that distance in which the errors become essentially independent of one another. For instance, an error consisting of a bump extending over a large area implies a large value of c, while errors consisting of a number of bumps, each of which covers a small area, imply a small value of c.

In an analysis similar to that for the discrete array it has been shown[16] that the power pattern for an "average" system with small errors is given by

$$\overline{P}(\phi) = P_0(\phi) + S(\phi) \frac{4c^2\pi^2\overline{\delta^2}}{\lambda^2 G_0} \exp\left(- \frac{\pi^2 c^2 \sin^2 \phi}{\lambda^2} \right)$$

where the symbols have the same meaning as used earlier in this section, except for c, which is defined as the correlation interval. The effects of random errors on the sidelobes and gain of a circular paraboloid have been computed by Ruze. Some of his results for the case of a cosine-squared illumination are shown in Figs. 2-18 and 2-19.

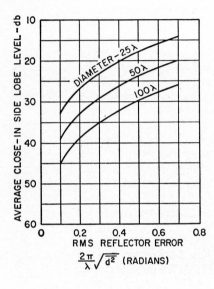

FIG. 2-18 Spurious radiation of paraboloid with cosine-squared illumination. Correlation interval $c = \lambda$.

The relationships for the loss in gain have also been worked out by Ruze. For small errors, simplified formulas have been obtained for the limiting cases of small and large correlation intervals, as follows:

$$\frac{G}{G_0} \approx 1 - \frac{3}{4}\overline{\delta^2}\frac{c^2\pi^2}{\lambda^2} \qquad \text{when } \frac{c}{\lambda} \ll 1$$

and

$$\frac{G}{G_0} \approx 1 - \overline{\delta^2} \qquad \text{when } \frac{c}{\lambda} \gg 1$$

Some results for the loss in gain are given in Fig. 2-20.

Periodic Errors in Aperture Illumination

In addition to random errors introduced in the manufacturing process, certain types of antennas may also introduce periodic errors due to the particular technique used in fabricating the antenna. For instance, in certain types of reflectors using the bulkhead-

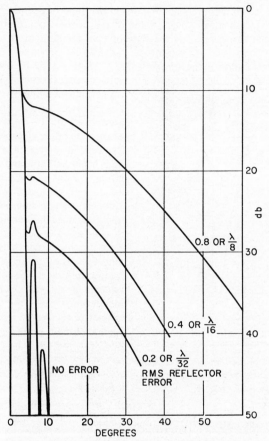

FIG. 2-19 Average system pattern for circular paraboloid with cosine-squared illumination. Correlation interval $c = \lambda$ calculated for reflector 24λ in diameter.

and-trusswork type of construction, mechanical stresses set up during fabrication are such that the surface has periodic errors at roughly equal intervals along the surface. This results in a periodic phase error along the aperture.

The effect of a sinusoidal phase error on the sidelobes has been treated in Ref. 19, which points out that a sinusoidal phase error will produce two equal sidelobes whose amplitude relative to the main-beam amplitude is equal to one-half of the peak phase error expressed in radians. The two sidelobes are symmetrically located on either side of the main beam at an angular distance of $m\lambda/\ell$ rad from the main beam, where m is the number of cycles of phase error along the aperture, ℓ is the total length of the aperture, and λ is the wavelength expressed in the same units as ℓ.

The effect of any type of phase error on the antenna gain has been treated by Spencer,[20] who shows that the fractional loss in gain is equal to the mean square phase error from the least-square plane-wave approximation to the phase front. Since a periodic phase error would not alter the direction of the unperturbed phase front, the loss in gain is equal to the mean square value of the periodic phase error. It is interesting to note that this represents the limiting case for random errors as shown in Fig. 2-20.

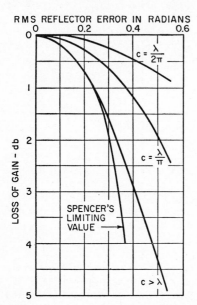

FIG. 2-20 Loss of gain for paraboloid as function of reflector error and correlation interval.

Another type of distribution error which can occur under certain conditions is a periodic amplitude error.[21] The sidelobe behavior of the amplitude-modulated distribution is very similar to that for the case of the sinusoidal phase error. Two equal sidelobes on either side of the main beam will appear for each sinusoidal component of the amplitude modulation. For a uniform-amplitude distribution with a sinusoidal ripple, the amplitude of each sidelobe due to the ripple will have a value relative to the main-beam amplitude which is equal to one-half of the ratio of the ripple amplitude to the amplitude of the constant term. As was true for the periodic phase error, the periodic amplitude error will produce two symmetrically located sidelobes on either side of the main beam at an angular distance of $m\lambda/\ell$ rad, where m is the number of cycles of amplitude variation across the aperture whose total length is ℓ.

2-10   METHODS OF SHAPING PATTERNS

A variety of radar system applications require that the radiation pattern of the antenna be shaped to meet certain operational requirements. One common requirement for the vertical-plane pattern is the cosecant-squared pattern, which produces a uniform ground return echo from an airborne radar antenna. There are, of course, other shapes of interest, but the basic principles of obtaining shaped beams are the same for a wide variety of shapes.

Methods Useful for Linear Arrays

One of the techniques useful with slot arrays or dipole arrays is to determine what aperture distribution will produce the desired radiation pattern and then design the array feed system so as to achieve the required aperture distribution.

There are several methods in use for determining the form of the aperture distribution. Historically, the first method is based on the Fourier approximation to a given function.[22] For an equally spaced array of radiating elements, the radiated field due to the array can be represented by a finite trigonometric series with a direct relationship between the current in each element and the coefficients of the trigonometric series. If the required radiation pattern is analyzed by standard Fourier methods, the coefficients of the Fourier series then determine the current amplitudes and phase for the array elements. The accuracy with which the finite series approximates the desired function depends on how many terms are used, that is, on how large an aperture is used. For a fixed number of terms, it is well known from the theory of Fourier series that the approximation is best in the sense that the mean square deviation is least. It is also a property of the Fourier approximation that at a point of discontinuity the function takes on a value that is the average of the value existing on either side of the discontinuity; i.e.,

$$\tfrac{1}{2}[E(\phi + 0) + E(\phi - 0)]$$

A second method of approximating the desired radiation pattern is one that enables the pattern to be specified exactly at a fixed number of points. This method, which was independently proposed by Woodward[23] and by Levinson, is particularly appealing in that it gives the designer some physical insight into the way in which the pattern is synthesized and in addition allows some control of the pattern at points of discontinuity.

Basically, the method consists of superposing a series of uniform-amplitude, linear-phase distributions across the aperture in such a way that the sum of the patterns produced by each of the distributions adds up to the desired pattern. The required aperture distribution is then found by adding up, in proper phase, the individual uniform distributions.

To simplify matters, the following discussion will deal with the case of a continuous aperture. The extension to the case of a uniformly spaced discrete-element array will be obvious from the discussion. As pointed out in Sec. 2-6, the radiation pattern for a uniform in-phase illumination is given by

$$E(\phi) = \frac{\sin u}{u}$$

where
$$u = \frac{\pi \ell}{\lambda} \sin \phi$$

The pattern will pass through zero whenever $\sin \phi = n\lambda/\ell$, $n = 1, 2, 3, 4$, etc., and the pattern has its maximum when $\phi = 0$. For a case in which $\ell = 10\lambda$, the pattern plotted against $\sin \phi$ will appear as in Fig. 2-21a.

It will be noted that when the pattern is plotted against $\sin \phi$, the zeros of the pattern are equally spaced except for the two zeros on either side of the main beam, which occupy two spaces. The width of one space is equal to λ/ℓ, the reciprocal of the aperture width in wavelengths.

A uniform illumination with a linear phase variation will have a field pattern given by

$$E(\phi) = \frac{\sin (u - \beta_n)}{u - \beta_n}$$

where β is one-half of the total phase variation across the aperture ℓ. The maximum of this pattern occurs when

$$\sin \phi_n = \frac{\beta_n \lambda}{\pi \ell}$$

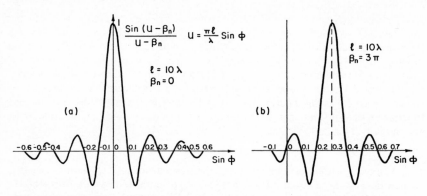

FIG. 2-21 Radiation pattern of uniformly illuminated aperture 10λ long. (*a*) $\beta_n = 0$. (*b*) $\beta_n = 3\pi$.

If the phase variation is chosen so that $\beta_n = n\pi$, then the pattern will be shifted by $n\lambda/\ell$, or n spaces when plotted against sin ϕ. A typical pattern for $\ell = 10\lambda$ and $\beta_n = 3\pi$ is shown in Fig. 2-21 *b*.

By combining a number of patterns shifted by integral numbers of spaces, it is possible to synthesize a pattern which can be uniquely specified at $2m + 1$ points for an aperture which is m wavelengths in extent. The total radiation pattern is given by

$$E(\phi) = \sum_n C_n \frac{\sin (u - \beta_n)}{u - \beta_n} = \sum_n C_n \frac{\sin (u - n\pi)}{u - n\pi}$$

and the corresponding normalized aperture distribution is given by

$$f(x) = \sum_n C_n e^{-j\beta_n x} = \sum_n C_n e^{-jn\pi x}$$

where $-1 \leq x \leq +1$.

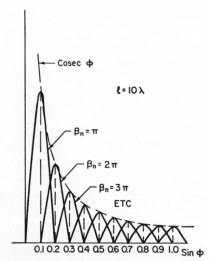

FIG. 2-22 $E(\phi)$ pattern for cosecant shaping.

As an illustration of this method, let us consider a problem in which it is desired to specify that the pattern have a cosecant shape over the range in sin ϕ from 0.1 to 1.0 and that it have no radiation for negative values of sin ϕ. For an aperture length of 10 wavelengths, sin ϕ is divided into spaces one-tenth wide from -1 to $+1$ and ordinates erected at each division with a height equal to $E(\phi)$. For the problem at hand, $E(\phi)$ is proportional to csc ϕ. A plot of the $E(\phi)$ diagram is shown in Fig. 2-22, where at each division an individual pattern is to be specified with a maximum value equal to each ordinate. The summation of the individual patterns is shown in Fig. 2-23, while the resulting total aperture distribution, consisting of the sum of each of the indi-

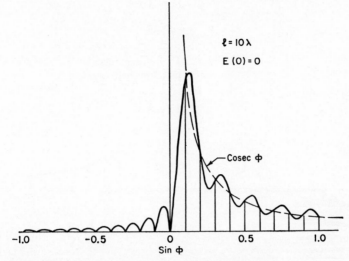

FIG. 2-23 Synthesis of cosecant pattern for $E(0) = 0$.

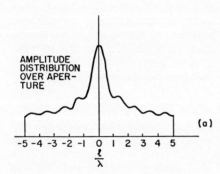

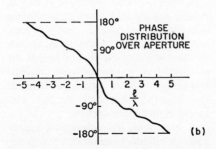

FIG. 2-24 Aperture distribution for Fig. 2-23. (*a*) Amplitude distribution. (*b*) Phase distribution.

vidual aperture distributions, is shown in Fig. 2-24.

For the case shown, $E(0)$ has been chosen to equal zero, and it will be noted that the resulting pattern has a fair amount of ripple in it. If the value of $E(0)$ is chosen to be 0.8, then the resulting pattern shown in Fig. 2-25 is considerably smoother although the aperture distribution is somewhat more peaked, as seen from Fig. 2-26.

This particular method of synthesis is quite useful because of the flexibility in arriving at the final radiation pattern and the designer's ability to adjust the theoretical pattern by graphical procedures. Although the example discussed has used only real values for C_n to simplify the calculations, there is nothing in the procedure which prevents the use of complex values for C_n. It is quite possible that a judicious choice of the phase angles for the C_n's might have given as effective a control over the ripple as did the changing the value of $E(0)$.

One note of caution should be added. Although beams can be added for values of $|\sin \phi| > 1$ in order to control

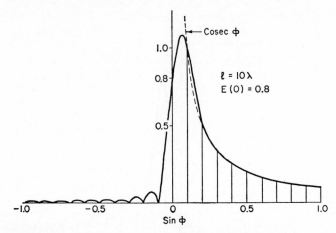

FIG. 2-25   Synthesis of cosecant pattern for $E(0) = 0.8$.

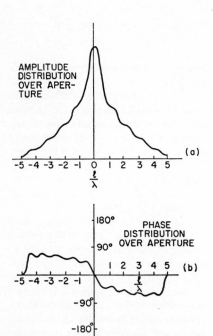

FIG. 2-26   Aperture distribution for Fig. 2-25. (*a*) Amplitude distribution. (*b*) Phase distribution.

the amplitude at points intermediate between the ordinates, the energy in these "imaginary beams" (when ϕ is complex) is primarily of a reactive nature and can result in large amounts of stored energy in the aperture, with consequent reduction of bandwidth and increase of losses. It is therefore desirable to minimize or eliminate those beams which radiate in directions specified by $|\sin \phi| > 1$.

Methods Useful for Reflector-Type Antennas

There are several methods useful for producing shaped patterns with reflector-type antennas. One technique, which is readily applied to a standard paraboloidal reflector, combines a number of narrow beams to form a wider beam, shaped in the vertical plane by the use of an extended feed system. In some respects, this technique has a certain similarity to the Woodward-Levinson method discussed earlier in the section.

In applying this technique, a rectangular or elliptical section of a full paraboloid is used as the reflector surface. The horizontal dimension of the reflector section is chosen to give the desired horizontal beamwidth, while the vertical dimension is chosen to control the rate of cutoff of the vertical beam. The vertical dimension also influences the feed-system problem, particularly if a small value is chosen, since the feed horn or dipole for each beam would have to have a sizable vertical dimension. The aspect ratio of the reflector commonly has a value in the range of 2:1 to 4:1.

A considerable amount of experimental effort is necessary in the design of an extended feed system. The basic principles can be seen from Fig. 2-27. For each feed in the vertical plane, there is a corresponding vertical beam whose position is related to the feed position. By exciting each feed with the appropriate amplitude and by choosing the feed separation and phasing, it is possible to obtain a fairly smooth vertical-plane pattern.

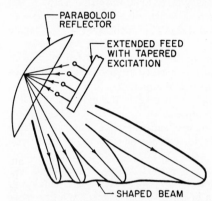

FIG. 2-27 Schematic of extended feed system for pattern shaping.

A number of examples of this technique are given in Ref. 7. Although most of the structures built to date have used reflector structures for collimating each beam, the same technique of combining beams is also applicable to lens-type antennas.

Another approach to the problem of generating shaped beams is the use of shaped-reflector systems. One method uses a line source to feed a shaped reflector, while a second method uses a point-source feed with a double-curvature reflector. The shaped-reflector approach is treated in Chap. 17, so no further discussion will be given here.

2-11 GAIN LIMITATIONS FOR AN APERTURE OF SPECIFIED SIZE

It has been shown that the gain of a uniformly illuminated aperture-type antenna without losses is expressed by

$$G = \frac{4\pi A}{\lambda^2}$$

where A is the area of the aperture. The value of gain obtained from this expression normally represents an upper limit which can be realized with practical structures.

Certain classes of aperture distributions offer the theoretical possibility of higher values of directivity than can be obtained with uniform distribution. In general, these theoretical distributions are characterized by reversals of phase over a distance short compared with the wavelength. One common feature of these superdirective distributions is the large amount of stored energy in the aperture region since very high values of field intensity are necessary to produce the same radiated field as would be produced by a uniformly illuminated aperture with much lower values of field intensity.

The large values of stored energy in the aperture region of a *superdirective antenna* cause a number of engineering problems which are severe enough to make this type of antenna completely impractical. The first problem is that of extremely high Q's, which limit the operating bandwidth to extremely small values. For instance, it has been stated by Taylor[24] that an antenna designed within a sphere of 50 wavelengths diameter will have a beamwidth of approximately 1°. If the same beamwidth is to be maintained while the diameter of the sphere is reduced to 45 wavelengths, the Q will rise to a value of 500. If the diameter is reduced to 40 wavelengths, the Q will rise to a value of 5×10^{10}. For further reductions in diameter, the value of Q rises to astronomical values. Since the bandwidth is of the order of the inverse of Q, it can be seen that the bandwidth diminishes rapidly.

As a result of the high stored energy, large values of circulating current flow in the antenna structure, and a point is very quickly reached at which the ohmic losses completely nullify any gain increase due to increased directivity.

Another concomitant of the superdirective antenna is the extreme precision required to achieve any substantial increase in directivity.

There is an abundant literature on the topic of superdirective antennas, and the antenna designer who may be tempted to build such antennas should consult these references.[25-27]

Lest the situation be considered completely hopeless, it should be mentioned that for certain end-fire arrays in which the values of gain are modest some increase in gain can be achieved. One particular design in which a four-element end-fire array has achieved a modest increase in gain and directivity at the expense of bandwidth has been given in the literature.[28] However, aside from the special case of end-fire arrays, the designer is to be discouraged from attempting to construct arrays which have gains higher than the value given for the uniform-illumination case. This caution is particularly true for broadside antennas in which the aperture is large in terms of wavelengths.

2-12   SCALE MODELS OF ANTENNAS

One of the most useful tools of antenna engineers is the ability to scale their designs. It is a direct consequence of the linearity of Maxwell's equations that an electromagnetic structure which has certain properties[29] at a given frequency f will have identically the same properties at another frequency nf, provided all linear dimensions are scaled by the ratio $1/n$. Thus an antenna design which works in one range of frequencies can be made to work at any other range of frequencies without additional redesign, provided an exact scaling of dimensions can be accomplished.

Quite aside from the ability to transfer design relationships is the ability to make radiation-pattern studies on scale models which are convenient in size. The aircraft antenna field is one in which full-size radiation studies are extremely awkward, time-consuming, and expensive. The possibility of studying aircraft antennas on a scale model[30] which may be as small as one-twentieth or one-fortieth of full size brings such studies within the realm of the laboratory rather than requiring an elaborate flight operation.

For most types of antennas, scaling is a relatively simple matter. Table 2-4 shows how the dimensions and electromagnetic properties vary as a function of the scale

TABLE 2-4

Quality	Full-scale system	Model system
Length	L_F	$L_M = L_F/n$
Frequency	f_F	$f_M = nf_F$
Dielectric constant	ϵ_F	$\epsilon_M = \epsilon_F$
Conductivity	σ_F	$\sigma_M = n\sigma_F$
Permeability	μ_F	$\mu_M = \mu_F$

factor. It will be noted that all the quantities except conductivity can be satisfactorily scaled. If the full-size antenna is constructed of copper or aluminum, then it is not possible to obtain materials which have conductivities that are an order of magnitude higher. Fortunately, conductivity losses affect the operation of most antennas to only a minor degree, so that the inability to scale the conductivity is not usually serious. This is not true for devices such as cavity resonators, for which the losses may be appreciable. For a few types of antennas, such as very long wire antennas, in which the conductivity losses of the antenna and of the ground may play a part in the radiating properties of the antenna, it may be necessary to proceed with considerable caution before making scale-model studies.

While it is true that an exact scale model will have exactly the same radiation patterns and input impedance as the full-scale antenna, it is not always possible to achieve perfect scaling. This is particularly true for such items as transmission lines, screw fastenings, etc. Slight discrepancies in scaling will usually affect the impedance properties much more than the radiation properties. It is therefore wise to consider scale-model impedance studies as primarily qualitative in nature, even though it is valid to consider scale-model radiation patterns as being highly accurate. Usually the general trend of impedance characteristics is determined from the scale model, but final impedance-matching work can be completed only on the full-size antenna.

REFERENCES

1   S. A. Schelkunoff, *Electromagnetic Waves,* D. Van Nostrand Company, Inc., New York, 1943.
2   S. A. Schelkunoff and H. T. Friis, *Antennas: Theory and Practice,* John Wiley & Sons, Inc., New York, 1952.
3   Schelkunoff, *Electromagnetic Waves,* p. 360.
4   S. A. Schelkunoff, "Some Equivalence Theorems of Electromagnetics and Their Application to Radiation Problems," *Bell Syst. Tech. J.,* vol. 15, 1936, pp. 92–112.
5   P. S. Carter, "Circuit Relations in Radiating Systems and Applications to Antenna Problems," *IRE Proc.,* vol. 20, June 1932, pp. 1004–1041.
6   G. H. Brown, "Directional Antennas," *IRE Proc.,* vol. 25, 1937, pp. 78–145.
7   S. Silver, *Microwave Antenna Theory and Design,* McGraw-Hill Book Company, New York, 1949.
8   R. C. Spencer and P. M. Austin, "Tables and Methods of Calculation for Line Sources," MIT Rad. Lab. Rep. 762-2, March 1946; see also Rep. 762-1.
9   J. F. Ramsay, "Fourier Transforms in Aerial Theory," *Marconi Rev.,* vol. 9, 1946, p. 139; vol. 10, 1947, pp. 17, 41, 81, 157.

10 G. J. van der Maas, "A Simplified Calculation for Dolph-Tchebyscheff Arrays," *J. App. Phys.*, vol. 25, January 1954, pp. 121–124.

11 T. T. Taylor, "Design of Line-Source Antennas for Narrow Beamwidth and Low Side-lobes," *IRE Trans. Antennas Propagat.*, vol. AP-3, January 1955, pp. 16–28; see also R. J. Spellmire, "Tables of Taylor Aperture Distributions," Hughes Aircraft Co. Tech. Mem. 581, Culver City, Calif., October 1958.

12 T. T. Taylor, "One Parameter Family of Line Sources Producing Modified Sin $\pi u/\pi u$ Patterns," Hughes Aircraft Co. Tech. Mem. 324, Culver City, Calif., September 1953.

13 E. Jahnke and F. Emde, *Tables of Functions,* Dover Publications, Inc., New York, 1943, p. 227.

14 *British Association Mathematical Tables,* vols. VI and X, Cambridge University Press, London, 1950 and 1952.

15 K. Milne, "The Effects of Phase Errors on Simple Aperture Illuminations," *Proc. Conf. Centimetric Aerials for Marine Navigational Radar,* June 15–16, 1950, H. M. Stationery Office, London, 1952.

16 J. Ruze, "Physical Limitations on Antennas," MIT Res. Lab. Electron. Tech. Rep. 248, Cambridge, Mass., October 1952; see also "The Effect of Aperture Errors on the Antenna Radiation Pattern," *Supplemento al Nuovo Cimento,* vol. 9, no. 3, 1952, pp. 364–380.

17 L. L. Bailin and M. J. Ehrlich, "Factors Affecting the Performance of Linear Arrays," *IRE Proc.,* vol. 41, February 1953, pp. 235–241.

18 H. F. O'Neill and L. L. Bailin, "Further Effects of Manufacturing Tolerances on the Performance of Linear Shunt Slot Arrays," *IRE Trans. Antennas Propagat.,* vol. AP-4, December 1952, pp. 93–102.

19 N. I. Korman, E. B. Herman, and J. R. Ford, "Analysis of Microwave Antenna Sidelobes," *RCA Rev.,* vol. 13, September 1952, pp. 323–334.

20 R. C. Spencer, "A Least Square Analysis of the Effect of Phase Errors on Antenna Gain," Air Force Cambridge Res. Cen. Rep. 5025, Bedford, Mass., January 1949.

21 J. Brown, "The Effect of a Periodic Variation in the Field Intensity across a Radiating Aperture," *IEE Proc. (London),* part III, vol. 97, November 1950, pp. 419–424.

22 I. Wolff, "Determination of the Radiating System Which Will Produce a Specified Directional Characteristic," *IRE Proc.,* vol. 25, May 1937, pp. 630–643.

23 P. M. Woodward, "A Method of Calculating the Field over a Plane Aperture Required to Produce a Given Polar Diagram," *IEE J. (London),* part IIIA, vol. 93, 1947, pp. 1554–1558.

24 T. T. Taylor, "A Discussion of the Maximum Directivity of an Antenna," *IRE Proc.,* vol. 36, September 1948, p. 1135.

25 L. J. Chu, "Physical Limitations of Omnidirectional Antennas," *J. App. Phys.,* vol. 19, December 1948, p. 1163.

26 P. M. Woodward and J. D. Lawson, "The Theoretical Precision with Which an Arbitrary Radiation Pattern May Be Obtained from a Source of Finite Size," *IEE J. (London),* part III, vol. 95, September 1948, pp. 363–370.

27 N. Yaru, "A Note on Super-Gain Arrays," *IRE Proc.,* vol. 39, September 1951, pp. 1081–1085.

28 A. Block, R. G. Medhurst, and S. D. Pool, "A New Approach to the Design of Superdirective Aerial Arrays," *IEE J. (London),* part III, vol. 100, September 1953, pp. 303–314.

29 G. Sinclair, "Theory of Models of Electromagnetic Systems," *IRE Proc.,* vol. 36, November 1948, pp. 1364–1370.

30 G. Sinclair, E. C. Jordan, and E. W. Vaughan, "Measurement of Aircraft Antenna Patterns Using Models," *IRE Proc.,* vol. 35, December 1947, pp. 1451–1462.

Chapter 3

Arrays of Discrete Elements *

Mark T. Ma

National Bureau of Standards

*Critical reviews by Dr. A. G. Repjar of the National Bureau of Standards and Dr. P. F. Wilson of the University of Colorado are acknowledged.

3-1 INTRODUCTION

A practical objective of directive communication is an improvement in received signal as measured relative to the prevailing noise. More precisely, directivity improves the signal-to-noise ratio. This improvement may be accomplished either at the transmitter side by using an antenna that projects the transmitted wave power in the form of a concentrated beam toward the distant receiver, or it may be accomplished at the receiver side by using a similar antenna to intercept a maximum wave power. Thus, any desired signal improvements may be taken either at the transmitter or at the receiver or be shared between the two.

One of the common methods of obtaining directive antenna characteristics is an arrangement of several individual antennas of the same kind so spaced and phased that their individual contributions add in one preferred direction while they cancel in others. Such an arrangement is known as an *array* of discrete antennas. The individual antennas, called *elements,* in an array may be arranged in various configurations such as straight lines, squares, rectangles, circles, ellipses, arcs, or other more innovative geometries. Years of experience have reduced the number of configurations that are considered practical to a relatively few.

Improvements in radiation characteristics by arraying may be described in a variety of ways. Sometimes the result takes the form of a plotted pattern which shows, at a glance, the relative signal level in various directions. In other cases the result is measured in terms of directivity or power gain. For purposes of this chapter, directivities or power gains are generally referenced to some accepted standard such as an isotropic source or an array element.

One of the practical arrays consists of a number of individual antennas set up along a straight line known as the *linear array.* For this particular configuration, the individual elements may be excited by currents in phase with equal or tapered amplitudes. Under this condition, radiation proceeds both forward and backward in directions perpendicular to the lines of the array. Such an array is said to be *broadside.* It may be made essentially unidirectional by adding a second identical array to the rear of the first and exciting the second array with appropriate phase difference.

The elements of a linear array may also be driven by currents with phase progressively varying along the array axis in such a way as to make the radiation substantially unidirectional. This array is referred to as an *end-fire array.* Both the broadside and the end-fire arrays of different elements are presented in Secs. 3-2 through 3-5. Necessary formulas without detailed derivations are given together with the first-hand design curves or tables with the number of elements, interelement spacings, amplitudes, phases, and other array characteristics as the parameters. A special array of arrays known as the *curtain antenna* is briefly discussed in Sec. 3-6.

When the general principle of arrays with isotropic elements is presented, no consideration of the mutual-coupling problem is needed. For the case involving simple practical antennas such as the short Yagi-Uda array (a small number of dipoles) for which methods of computing the mutual impedances between elements are available, the mutual impedances will be duly included.

3-2 UNIFORM LINEAR ARRAYS

When the identical elements of a linear array are equally spaced along the array axis and excited with uniformly progressive phases as shown in Fig. 3-1, the array pattern in free space may be considered, assuming that the principle of pattern multiplication applies,[1,2] as a product of the element pattern and an array factor,

$$E(\theta,\phi) = f(\theta,\phi)S(\theta) \tag{3-1}$$

where $E(\theta,\phi)$ is the far-field pattern expressed in terms of the ordinary spherical coordinates (θ,ϕ), $f(\theta,\phi)$ is the element pattern determined by the particular antenna element used in the array, and $S(\theta)$ is the array factor for a linear array placed along the z axis with θ measured from the array axis.

The array factor $S(\theta)$ may be explicitly written as

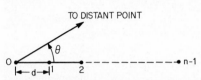

TO DISTANT POINT

FIG. 3-1 A linear array of n equally spaced elements.

$$S(\theta) = \sum_{i=0}^{n-1} I_i \exp(jiu) \tag{3-2}$$

where I_i represents the current amplitude excitation of the ith element and

$$u = kd(\cos\theta - \cos\theta_0) \tag{3-3}$$

with k being the wave number $(2\pi/\lambda, \lambda$ being wavelength), d the interelement spacing in wavelength, θ_0 the desired direction of the maximum radiation, and n the total number of elements in the array.

When $I_i = 1, i = 0, 1, \ldots, (n-1)$, a uniform linear array results, yielding

$$S_u(\theta) = \sum_{i=0}^{n-1} \exp(jiu) = \frac{1 - \exp(jnu)}{1 - \exp(ju)} \tag{3-4}$$

where the subscript u should not be confused with the symbol defined in Eq. (3-3).

Broadside Arrays of Isotropic Elements

If, in addition, the elements are isotropic and the desired maximum radiation is in the broadside direction, we have $f(\theta,\phi) = 1$, $\theta_0 = \pi/2$, $u = kd\cos\theta$, and $E(\theta,\phi) = S_u(\theta)$. The basic characteristics of this array may be summarized as follows.

Beam Maximum   It is clear that $S_u(\theta)_{\max} = n$ (the total number of elements) when $\theta = \pi/2$ or $u = 0$.

Nulls   If we express Eq. (3-4) in a different form such as

$$|S_u|^2 = S_u S_u^* = \frac{1 - \cos nu}{1 - \cos u} = \frac{\sin^2(nu/2)}{\sin^2(u/2)} \tag{3-5}$$

it is easy to see that the null positions $(S_u = 0)$ are given by

$$nu_m/2 = m\pi \tag{3-6a}$$

or

$$\cos\theta_m = m\lambda/nd, \quad m = \pm1, \pm2, \ldots, \pm M \leq nd/\lambda \tag{3-6b}$$

where $\theta_1, \theta_2, \ldots, \theta_M$ are the angular null positions on one side of the beam maximum in the forward direction and $\theta_{-1}, \theta_{-2}, \ldots, \theta_{-M}$ are null positions on the other side of the beam maximum, also in the forward direction. Of course, $\theta_{\pm m}$ are symmetrically located with respect to the beam maximum. It is clear from Eq. (3-6a) that $u = \pm\pi$ are always possible null positions when n is even and $d \geq \frac{1}{2}\lambda$. The exact number and locations of null positions as given by Eq. (3-6b) depend on n and d/λ (or approximately the overall array length). A designer may have choices of different arrays to produce a null in a particular direction in order to minimize possible interferences at this direction. For example, the arrays of four elements with $d = 3\lambda/8$ or $3\lambda/4$, of six elements with $d = \lambda/4, \lambda/2$ or $3\lambda/4$, etc., may be used to satisfy the purpose of having a null at $\theta = 48.19°$ (or $41.81°$ from the main beam). The final choice among these possibilities should be made from other design or cost considerations.

First-Null Beamwidth   One way of showing the directive pattern of a uniform broadside array is by the *first-null beamwidth,* which is defined as the angular space between the first nulls on each side of the main beam,

$$(\text{BW})_1 = \theta_{-1} - \theta_1 = 2\left(\frac{\pi}{2} - \theta_1\right) \qquad \text{rad} \qquad \textbf{(3-7a)}$$

or

$$(\text{BW})_1 = 2(90° - \theta_1) \qquad \textbf{(3-7b)}$$

where θ_1 in Eq. (3-7a) is in radians while that in Eq. (3-7b) is in degrees.

For very large arrays, we approximately have

$$(\text{BW})_1 \simeq \begin{cases} 2\lambda/nd & \text{in radians} \\ 360\lambda/\pi nd & \text{in degrees} \end{cases} \qquad \textbf{(3-8)}$$

Half-Power Beamwidth   Another more commonly used beamwidth is the angular space between the half-power points on each side of the main beam. The half-power point u_h may be determined by solving

$$|S_u(u_h)|^2 = \frac{\sin^2(nu_h/2)}{\sin^2(u_h/2)} = n^2/2 \qquad \textbf{(3-9)}$$

which, in turn, yields the corresponding θ_h in accordance with

$$\cos\theta_h = u_h/kd \qquad \textbf{(3-10)}$$

The half-power beamwidth then becomes

$$(\text{BW})_h = \begin{cases} 2\left(\frac{\pi}{2} - \theta_h\right) & \text{in radians} \\ 2(90° - \theta_h) & \text{in degrees} \end{cases} \qquad \textbf{(3-11)}$$

Sample results for $(\text{BW})_h$ are presented in Fig. 3-2.

Sidelobes   Precise locations of the sidelobes of the uniform linear broadside arrays may be found by setting the derivative of Eq. (3-5) to zero, yielding[3]

$$n\tan(u/2) = \tan(nu/2) \qquad \textbf{(3-12)}$$

It is clear that $u = \pm\pi$ are possible sidelobe positions when n is odd and $d \geq \lambda/2$. Other solutions of Eq. (3-12) for u in degrees are presented in Table 3-1. The corre-

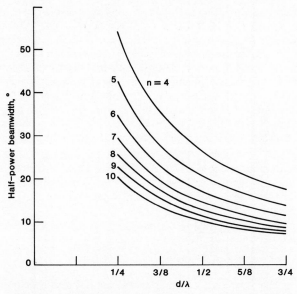

FIG. 3-2 Half-power beamwidths for uniform broadside arrays of isotropic elements.

sponding angular sidelobe positions in terms of θ can be determined by $u = kd \cos \theta$ when the element spacing is known.

The level of the ith sidelobe at u_i relative to the beam maximum may be computed as $|S_u(u_i)|/S_u(0) = |S_u(u_i)|/n$. The sidelobe levels in decibels are also included in Table 3-1.

Note from Table 3-1 that the sidelobe positions approach $u_i = \pm(2i + 1)\pi/n$ rad or $\pm(2i + 1)180/n°$, $i = 1, 2, 3, \ldots$ (or halfway between nulls) and that the sidelobe levels approach $|S_u(u_i)|/n = 2/(2i + 1)\pi$ for arrays with a very large number of elements. In particular, the first sidelobe level of large uniform arrays is $\frac{2}{3}\pi$ or -13.46 dB.

Directivity   The directive gain of an array in a given direction is defined as the ratio of the radiation intensity in that direction to the average intensity,

$$G(\theta,\phi) = \frac{|E(\theta,\phi)|^2}{W_0/4\pi} = 4\pi|E(\theta,\phi)|^2/W_0 \qquad \textbf{(3-13)}$$

where W_0 representing the total power radiated by the array is given by

$$W_0 = \int_0^{2\pi} \int_0^{\pi} |E(\theta,\phi)|^2 \sin \theta \; d\theta \; d\phi \qquad \textbf{(3-14)}$$

The directivity of the array is the maximum value of the directive gain,

$$D = G(\theta,\phi)_{\max} = 4\pi|E(\theta,\phi)|^2_{\max}/W_0 \qquad \textbf{(3-15)}$$

TABLE 3-1 Sidelobe Locations and Levels for Uniform Linear Broadside Arrays of Isotropic Elements

Sidelobe n	First Position, ° (u_1)	First Level, dB	Second Position, ° (u_2)	Second Level, dB	Third Position, ° (u_3)	Third Level, dB	Fourth Position, ° (u_4)	Fourth Level, dB
4	±131.81	−11.30						
5	±104.48	−12.04	±180.00	−13.98				
6	± 86.66	−12.43	±149.12	−15.25				
7	± 74.08	−12.65	±127.42	−15.98	±180.00	−16.90		
8	± 64.71	−12.80	±111.28	−16.43	±157.14	−17.89		
9	± 57.45	−12.90	± 98.79	−16.73	±139.48	−18.54	±180.00	−19.08
10	± 51.66	−12.97	± 88.84	−16.95	±125.41	−18.99	±161.82	−19.89
$\cdots$								
Very large	±540/n	−13.46	±900/n	−17.90	±1260/n	−20.82	±1620/n	−23.01

For the uniform linear broadside array of isotropic elements considered here, we have $f(\theta,\phi) = 1$, $|E(\theta,\phi)|^2 = |S_u|^2$, $u = kd \cos \theta$, and $|E(\theta,\phi)|^2_{max} = n^2$. The directivity becomes

$$D_b = \frac{(kd)n^2}{(nkd) + 2 \sum_{m=1}^{n-1} \frac{n-m}{m} \sin (mkd)} \tag{3-16}$$

Note that when $kd = p\pi$ or $d = p\lambda/2$, $p = 1, 2, \ldots$, $D_b = n$. It implies that under this condition the directivity is numerically equal to the total number of elements in the array. For other values of kd, the results are presented in Table 3-2. For a given

TABLE 3-2   Directivities for Uniform Linear Broadside Arrays of Isotropic Elements

n	d/λ $\frac{1}{8}$	$\frac{1}{4}$	$\frac{3}{8}$	$\frac{1}{2}$	$\frac{5}{8}$	$\frac{3}{4}$	$\frac{7}{8}$	1
4	1.28	2.16	3.11	4.00	4.84	5.58	5.29	4.00
5	1.45	2.70	3.83	5.00	6.12	6.97	7.68	5.00
6	1.67	3.17	4.63	6.00	7.30	8.54	9.52	6.00
7	1.93	3.64	5.34	7.00	8.61	10.10	11.22	7.00
8	2.20	4.16	6.10	8.00	9.84	11.55	12.82	8.00
9	2.48	4.68	6.86	9.00	11.07	12.99	14.42	9.00
10	2.74	5.17	7.59	10.00	12.41	14.43	16.10	10.00

n, the directivity is almost linearly proportional to d/λ in the range of $\frac{1}{4} \leq d/\lambda \leq \frac{3}{4}$ and then drops to n when $d = \lambda$ where the grating lobe at the same level as that of the main beam appears. For this reason, the element spacing for a fixed-beam broadside array is normally kept at less than a full wavelength. For arrays designed with beam-scanning capabilities, the element spacing is even smaller. However, the element spacing should not be chosen to be too small for the obvious reasons of obtaining a respectable directivity and minimizing the mutual coupling among the array elements. Thus, the actual value for the element spacing is usually a compromise of the above considerations.

Broadside Arrays with Practical Antennas as Elements

Broadside arrays with practical antennas may be achieved by placing the antenna elements in such a way that the maximum radiation of the element coincides with the desired broadside direction of the array. One example is to place the half-wave or short dipoles collinearly along the array axis as in the case for the well-known Marconi-Franklin antenna.[4] There are, of course, many other arrangements that satisfy the requirement. We consider only the specific example mentioned above.

Arrays of Half-Wave or Short Dipoles   The particular arrangement is shown in Fig. 3-3. Although the current distribution on two or more antennas in an array environment is not the same as when the elements are isolated because of the influence of

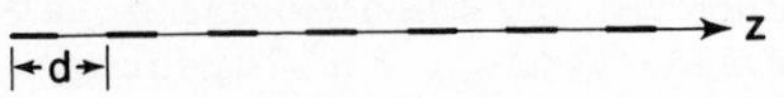

FIG. 3-3 An equally spaced linear array of collinear dipoles.

$|\!\leftarrow\!d\!\rightarrow\!|$

mutual couplings, the mutual impedance of collinear dipoles is, however, relatively small, and therefore the change in current distribution is not large. This is especially true when the element dipole length is near or less than $\lambda/2$.[4] Under the assumption of no mutual coupling, the element pattern may be written as[3,4]

$$f(\theta,\phi) = \frac{\cos(kh\cos\theta) - \cos kh}{\sin\theta} \tag{3-17}$$

where h is the half length of the individual dipole.

For half-wave dipoles, $h = \lambda/4$, $kh = \pi/2$, the element pattern reduces to

$$f(\theta,\phi) = \frac{\cos\left(\frac{\pi}{2}\cos\theta\right)}{\sin\theta} \tag{3-18}$$

The maximum radiation of Eq. (3-18) occurs at $\theta = \pi/2$, thus satisfying the requirement. The overall array pattern is then given by

$$E(\theta,\phi) = \frac{\cos\left(\frac{\pi}{2}\cos\theta\right)}{\sin\theta} |S_u| \tag{3-19}$$

where S_u is the array factor given in Eq. (3-4).

The null positions remain the same as those determined by Eq. (3-6) except for the introduction of additional nulls at $\theta = 0$ and π owing to the element pattern. Thus, the first-null beamwidth also remains unchanged. Locations and levels of the first few sidelobes do not change much either. The half-power beamwidth becomes naturally somewhat smaller than that with isotropic elements. The directivity is somewhat higher because the directivity for a single half-wave dipole is approximately 1.64 rather than 1.00 for the isotropic case.

For very short dipoles, $h \le \lambda/16$, $kh \le \pi/8$, the element pattern may be shown approximately to be proportional to

$$f(\theta,\phi) = \sin\theta \tag{3-20}$$

Thus, the overall array pattern is proportional to

$$E(\theta,\phi) = \sin\theta\,|S_u| \tag{3-21}$$

which, again, is not much different from S_u. The directivity in this case can be expressed as[3,5]

$$D_s = n^2/W_{0s} \tag{3-22}$$

where

$$W_{0s} = \frac{2n}{3} + \frac{4}{k^3d^3}\sum_{m=1}^{n-1}\frac{n-m}{m^3}\sin(mkd) \tag{3-23}$$

$$-\frac{4}{k^2d^2}\sum_{m=1}^{n-1}\frac{n-m}{m^2}\cos(mkd)$$

It is clear that the directivities for $d = \lambda/2$ and $d = \lambda$ are no longer equal to each other. Furthermore, the grating lobe at $\theta = 0$ for $d = \lambda(kd = 2\pi)$ owing to the array factor S_u is now eliminated by the null of the element pattern. The directivity for $d = \lambda$ is therefore expected to be higher than that for $d = \lambda/2$ in this case. Representative results for D_s with $n = 4$, 5, and 6 are presented in Table 3-3.

Note that the directivities for various cases given in Table 3-3 are numbers, which can be converted into decibels with respect to the isotropic element (dBi) by taking 10 log D_s. The directivity can also be expressed relative to that of a single element. Since the directivity of a single short dipole is 1.5, the directivity of arrays with short dipoles relative to a single short dipole in decibels is then given by $D_{sr} = 10 \log (D_s/1.5)$. Directivities for arrays with a larger number of short dipoles in this latter unit are given in Fig. 3-4.

TABLE 3-3 Directivities for Uniform Broadside Arrays of Collinear Short Dipoles

n	d/λ						
	¼	⅜	½	⅝	¾	⅞	1
4	2.45	3.39	4.29	5.21	6.05	6.84	6.95
5	2.94	4.05	5.30	6.45	7.55	8.59	8.86
6	3.44	4.87	6.29	7.81	9.15	10.37	10.77

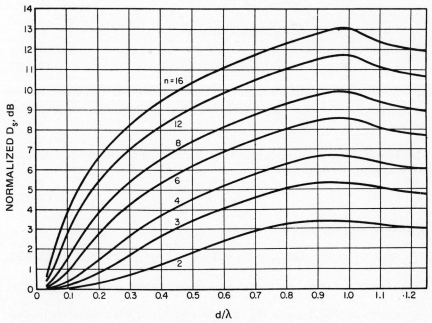

FIG. 3-4 Directivities for a broadside array of collinear short dipoles (relative to a single element).

Ordinary End-Fire Arrays of Isotropic Elements

When the condition $\theta_0 = 0$ is imposed upon Eq. (3-3), an ordinary end-fire array results. The visible range for $u = kd\,(\cos\theta - 1)$ corresponding to $0 \le \theta \le 180°$ is then $-2kd \le u \le 0$. The basic characteristics for the uniform ($I_i = 1$) ordinary end-fire array of isotropic elements [$f(\theta) = 1$] are again summarized first and then are compared with those for the broadside array presented in the subsection "Broadside Arrays of Isotropic Elements." Samples of end-fire arrays with properly phased practical elements or parasitic elements (short Yagi-Uda arrays of small number of dipoles) are discussed in the following subsection.

Beam Maximum   Since the array factor in terms of u remains unchanged, the beam maximum is still numerically equal to n, which occurs at $u = 0$ or $\theta = 0$.

Nulls   The expression [Eq. (3-6a)] for determining the null positions still applies provided that $m = -1, -2, \ldots$. The exact number of nulls depends on n and d/λ. In terms of θ, the null positions are given by

$$\cos\theta_m = 1 + \frac{m\lambda}{nd}, \, m = -1, -2, \ldots \qquad \textbf{(3-24)}$$

First-Null Beamwidth   The first-null beamwidth for the ordinary end-fire array is defined as

$$(\text{BW})_1 = 2\theta_{-1} = 2\cos^{-1}\left(1 - \frac{\lambda}{nd}\right) \qquad \textbf{(3-25)}$$

For very large arrays, θ_{-1} is small, yielding

$$(\text{BW})_1 \simeq 2\sqrt{\frac{2\lambda}{nd}} \quad \text{rad} \qquad \textbf{(3-26)}$$

By comparing Eqs. (3-26) and (3-8), we conclude that the first-null beamwidth for the ordinary end-fire array is always wider than that for the broadside array with the same n and d/λ.

Half-Power Beamwidth   Equation (3-9) is still valid for determining the half-power point u_h (taking the smallest negative solution), which in turn gives

$$\cos\theta_h = 1 + \frac{u_h}{kd} \qquad \textbf{(3-27)}$$

The half-power beamwidth thus becomes

$$(\text{BW})_h = 2\theta_h = 2\cos^{-1}\left(1 + \frac{u_h}{kd}\right) \qquad \textbf{(3-28)}$$

Sample results for $(\text{BW})_h$ for the uniform ordinary end-fire array are given in Fig. 3-5.

Comparing Fig. 3-5 with Fig. 3-2 also reveals that when the beam position changes from the broadside direction ($\theta_0 = 90°$) to the end-fire direction ($\theta_0 = 0$), the half-power beamwidth becomes broader for an array with the same n and d/λ.

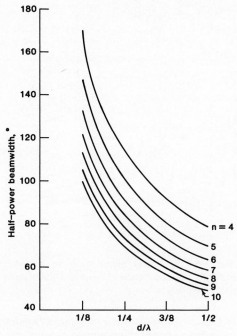

FIG. 3-5 Half-power beamwidths for uniform ordinary end-fire arrays of isotropic elements.

The degree of beam broadening depends on the array length and θ_0 (the position of the beam maximum). For beam-scanning applications in which θ_0 is being changed gradually, beam broadening should be an important factor to consider in the design. Changes in half-power beamwidths with respect to θ_0 are presented in Fig. 3-6 for $n = 10$ and various element spacings. The limit of scanning is also indicated in the figure.[6]

Sidelobes   Equation (3-12) and the first few sidelobes (positions and levels) given in Table 3-1 for the uniform broadside array are also good for the end-fire array, with the understanding that only the negative solutions for u are to be taken from Eq. (3-12) or Table 3-1. The corresponding sidelobe positions in terms of θ are then obtained from $u = kd (\cos \theta - 1)$. The element spacing d is normally kept less than $\lambda/2$ for an end-fire array because the grating lobe will appear with this particular value of d.

Directivity   The directivity for the uniform ordinary end-fire array may be expressed as[3]

$$D_e = \frac{(kd)n^2}{nkd + \displaystyle\sum_{m=1}^{n-1} \frac{n-m}{m} \sin (2mkd)} \qquad \textbf{(3-29)}$$

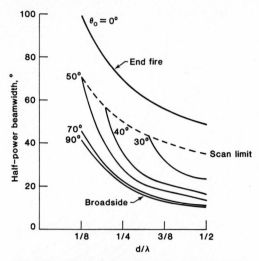

FIG. 3-6 Half-power beamwidths for a uniform array of 10 isotropic elements as a function of beam-maximum position.

Note that the second term in the denominator of Eq. (3-29) vanishes when $kd = p\pi/2$ or $d = p\lambda/4$, $p = 1, 2, \ldots$. Under this condition, $D_e = n$. For other values of kd, the results of D_e can be easily obtained from D_b given in Table 3-2. For the same n, D_e for a value of d is equal to D_b for $2d$. For example, for $n = 4$, $D_e(d = \lambda/8) = D_b$ $(d = \lambda/4)$, which is 2.16 in accordance with Table 3-2. This fact is obvious if Eqs. (3-16) and (3-29) are carefully examined.

Ordinary End-Fire Arrays of Practical Antennas as Elements

End-fire arrays of practical elements may be realized by proper arrangements of the antenna elements. One example which will be discussed here is to place short dipoles parallel to the x axis, forming an array along the z axis as shown in Fig. 3-7. Another method is to excite only one antenna element and leave the other elements unexcited (parasitic) with appropriate antenna lengths and spacings to produce approximate phase distributions among the elements required for end-fire radiation. This latter arrangement is known as the Yagi-Uda array.

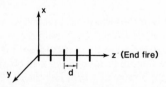

FIG. 3-7 An equally spaced linear array of parallel short dipoles.

Since the current amplitudes on the elements of a Yagi-Uda array are no longer the same ($I_i \neq 1$), the array cannot, strictly speaking, be considered as a uniform array. For this reason, characteristics of short Yagi-Uda arrays (with a small number of dipoles) are presented in a separate section. Long Yagis treated from the viewpoint of surface-wave antennas are given in Chap. 12.

For the array of parallel short dipoles as shown in Fig. 3-7, the general element pattern is given by $f(\theta,\phi) = (1 - \sin^2 \theta \cos^2 \phi)^{1/2}$. The element pattern in the xz plane ($\phi = 0$) reduces to $f(\theta) = \cos \theta$, which does have the maximum radiation at the end-fire direction ($\theta = 0$). When the elements are excited with equal amplitude and progressive phases, an end-fire array results. The overall array pattern then, neglecting the mutual-coupling effect, becomes

$$E(\theta) = \cos \theta \, |S_u(u)| \qquad \textbf{(3-30)}$$

where $u = kd \, (\cos \theta - 1)$.

Clearly, the element pattern produces a null at $\theta = \pi/2$ which may or may not be a null for S_u. Since $E_{max} = E(0) = n$, we obtain, using Eq. (3-15), the directivity as[3,5]

$$D_p = n^2/W_{0p} \qquad \textbf{(3-31)}$$

where $\qquad W_{0p} = \dfrac{2n}{3} + \dfrac{1}{kd} \displaystyle\sum_{m=1}^{n-1} \dfrac{n-m}{m} \left(1 - \dfrac{1}{m^2 k^2 d^2}\right) \sin(2mkd) \qquad \textbf{(3-32)}$

$$+ \dfrac{2}{k^2 d^2} \sum_{m=1}^{n-1} \dfrac{n-m}{m^2} \cos^2(mkd)$$

Representative results for D_p are presented in Table 3-4. Directivities relative to that of a single element, which are shown in Fig. 3-8, may be compared with those in Fig. 3-4.

TABLE 3-4   Directivities for
Uniform Ordinary End-Fire Arrays of
Parallel Short Dipoles.

n \ d/λ	⅛	¼	⅜	½
4	3.21	5.21	6.89	4.71
5	3.83	6.26	8.33	5.74
6	4.31	7.33	9.99	6.77

3-3   YAGI-UDA ARRAYS

The typical Yagi-Uda array consists of many parallel dipoles with different lengths and spacings as shown in Fig. 3-9. Only one of the dipoles is driven. All the other elements are parasitic and may function respectively as a reflector or as a director. Arrays of this kind were first described in Japanese by S. Uda[7] and subsequently in English by H. Yagi.[8] In general, the longest element, of the order $\lambda/2$ in length, is the reflector. The director elements are always shorter than the driven element in length. Although there may, in principle, be many reflectors, experience shows that little is gained by having more than one reflector. The reflector is usually spaced $\lambda/4$ to the rear of the driven element. Considerable gain can be realized by adding numerous directors. The simplest Yagi-Uda has only one director, making a three-element array.

Since the element patterns for each dipole are different, the principle of pattern multiplication certainly does not apply. The element pattern of the ith dipole in the xz plane is given by[3]

$$f_i(\theta) = \frac{\cos{(kh_i \sin{\theta})} - \cos{kh_i}}{\cos{\theta}} \qquad (3\text{-}33)$$

where h_i is the half length of the ith dipole.

The overall array pattern may be written as

$$E(\theta) = \sum_{i=1}^{n} I_i f_i(\theta) \exp{(jkd_{i-1}\cos{\theta})} \qquad (3\text{-}34)$$

where n is the total number of dipoles in the array, $d_0 = 0$, and I_i is the maximum current amplitude of the ith dipole, which may be determined by the available method.[3]

The power gain may be computed by $G(\theta,\phi) = 60|E(\theta)|^2/P_{in}$, where $P_{in} = \frac{1}{2}|I_{b2}|^2 R_{in}$ represents the input power. R_{in} is the input resistance, and I_{b2} is the base current of the driven element. As an example, the normalized pattern with $n = 3$, $h_1 = 0.26\lambda$, $h_2 = 0.25\lambda$, $h_3 = 0.23\lambda$, $d_1 = 0.25\lambda$, $d_2 = 0.45\lambda$ and dipole radius of 0.001 m is shown in Fig. 3-10 to demonstrate that the Yagi-Uda is indeed designed as an end-fire array (beam maximum at $\theta = 0$). For this example, the normalized base currents in amperes, with the assumption of a three-term current distribution,[9] are

$$I_{b1}/V_2 = 0.0038334 \exp{(j\,69.86°)}$$

$$I_{b2}/V_2 = 0.014546 \exp{(-j\,43.16°)}$$

$$I_{b3}/V_2 = 0.009361 \exp{(j\,166.97°)}$$

We see here that I_{b2} has an approximate phase lag of 113° relative to I_{b1}, although the phase difference should ideally be $-kd_1 = -90°$. Similarly, the phase difference between I_{b3} and I_{b1} is approximately $-263°$ rather than $-kd_2 = -162°$. Nevertheless, it is because of this kind of phase distribution approximately in the right direction over the Yagi-Uda elements that the end-fire pattern is pro-

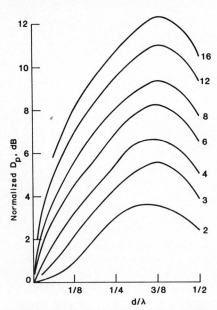

FIG. 3-8 Directivities for ordinary end-fire arrays of parallel short dipoles (relative to a single element).

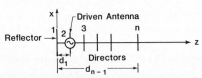

FIG. 3-9 A typical Yagi-Uda array.

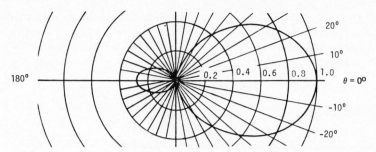

FIG. 3-10 Horizontal pattern for a three-element Yagi-Uda array with h_1 = 0.26λ, h_2 = 0.25λ, h_3 = 0.23λ, d_1 = 0.25λ, d_2 = 0.45λ, and dipole radius = 0.001 m.

duced. The input resistance for this example is approximately R_{in} = 50.1 Ω, and the power gain is about G = 5.01 or 7.0 dBi.

As another example, for n = 4, h_1 = 0.26λ, h_2 = 0.25λ, h_3 = h_4 = 0.23λ, d_1 = 0.25λ, d_2 = 0.52λ, d_3 = 0.88λ, and dipole radius = 0.001 m, we obtain R_{in} = 82.8 Ω, and G = 5.92 or 7.72 dBi.

An analytical method, as well as perturbational procedures, is available for maximizing the Yagi-Uda gain by adjusting the dipole lengths and spacings.[10,11] A gain in the order of 10 dBi can be easily designed with only a moderate number of dipoles in the array. Yagi-Uda arrays with a large number of dipoles are treated in Chap. 12 as surface-wave antennas.

3-4 IMPROVED END-FIRE ARRAYS

The ordinary end-fire array with u = kd (cos θ − 1) has been presented in Sec. 3-2. It has been known that an increase in directivity may be realized by increasing the progressive phase lag beyond that for an ordinary end fire by a phase quantity δ > 0.[12] More specifically, it was concluded that a maximum directivity which may be realized is given approximately by

$$D_m \cong \frac{7.28nd}{\lambda} \qquad (3-35)$$

when

$$\delta \cong \frac{2.94}{n-1} \quad \text{rad} \qquad (3-36)$$

which is known as the Hansen-Woodyard condition for maximizing the directivity of end-fire arrays and was reached on the basis of a formulation for line-aperture arrays with a continuous excitation. Its applicability to the equally spaced discrete array is therefore subject to the conditions that the number of elements in the array is very large and that the overall array size, $(n − 1)d$, is much greater than a wavelength. An exact formulation for determining the required optimum additional phase lag δ for the discrete array was made later to obtain the maximum directivity[13] for any n and d.

For the purpose of clarity, we have designated the one considered in the subsection "Ordinary End-Fire Arrays of Isotropic Elements" as the ordinary end-fire and call the case being presented here as the improved end-fire array.

The array factor in terms of u for the improved uniform ($I_i = 1$) end-fire array remains the same as that in Eq. (3-5). That is,

$$|S_u|^2 = \frac{\sin^2 (nu/2)}{\sin^2 (u/2)} \qquad (3\text{-}37)$$

provided that

$$u = kd (\cos \theta - 1) - \delta \qquad (3\text{-}38)$$

which is herein called the improved end-fire condition.

From Eq. (3-38), it can be seen that the location of the beam maximum has been shifted to $u = -\delta$, which corresponds to $\theta = 0$. The beam-maximum strength thus becomes

$$(S_u)_{\max} = \frac{\sin (n\delta/2)}{\sin (\delta/2)} \qquad (3\text{-}39)$$

which is smaller than n for the ordinary end-fire case. Naturally, for a given n, δ cannot be too large in order to have a respectable $(S_u)_{\max}$.

Null and sidelobe positions are similarly shifted. More specifically, the first-null position is determined by

$$\theta_1 = \cos^{-1} \left(1 - \frac{\lambda}{nd} + \frac{\delta}{kd} \right) \qquad (3\text{-}40)$$

The half-power point u_h can be obtained from

$$\frac{\sin^2 (nu_h/2)}{\sin^2 (u_h/2)} = \frac{\sin^2 (n\delta/2)}{2 \sin^2 (\delta/2)} \qquad (3\text{-}41)$$

which, in turn, yields

$$\theta_h = \cos^{-1} \left(1 + \frac{u_h + \delta}{kd} \right) \qquad (3\text{-}42)$$

Again the selection of the smallest negative solution from Eq. (3-41) is understood. Since $\delta > 0$, both the first-null and half-power beamwidths, $(BW)_1 = 2\theta_1$ and $(BW)_h = 2\theta_h$, obtained respectively from Eqs. (3-40) and (3-42), should be smaller than those for the ordinary end fire, where $\delta = 0$. Numerical results for $(BW)_h$ are shown in Fig. 3-11.

Although the sidelobe levels relative to the main beam are increased slightly, the overall directivity is always increased because of the improvement in beamwidth. For the improved uniform end-fire array of isotropic elements, the directivity may be computed as follows:[3]

$$D_i = \frac{kd \dfrac{\sin^2 (n\delta/2)}{\sin^2 (\delta/2)}}{nkd + 2 \displaystyle\sum_{m=0}^{n-1} \frac{n-m}{m} \sin (mkd) \cos (mkd + m\delta)} \qquad (3\text{-}43)$$

Sample results for D_i with a typical element spacing of $d = \lambda/4$ are presented in Table 3-5, where the optimum δ required to produce the maximum directivity can be easily

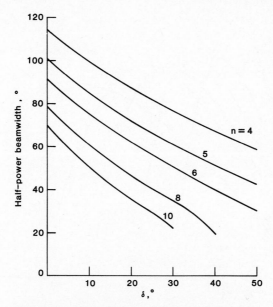

FIG. 3-11 Half-power beamwidths for improved end-fire arrays with $d = \lambda/4$.

TABLE 3-5 Directivities for Uniform Improved End-Fire Arrays of Isotropic Elements with $kd = \pi/2$ (δ in Degrees)

n \ δ	0	10	20	30	40	50	60	70
3	3	3.45	3.90	4.29	4.71	4.98	5.03	4.68
4	4	4.92	5.89	6.72	7.04	6.30		
5	5	6.37	7.72	8.68	8.37	5.57		
6	6	8.10	10.06	10.56	7.20			

observed. For larger numbers of elements, the directivities relative to those for the ordinary end-fire arrays are given in Fig. 3-12. The corresponding phases satisfied by the Hansen-Woodyard condition [Eq. (3-36)] are also indicated in the figure for comparison purposes.

3-5 LINEAR ARRAYS WITH TAPERED EXCITATIONS—EQUAL SIDELOBES

The first sidelobe level realized from linear arrays with a constant amplitude excitation approaches $2/(3\pi)$ or -13.46 dB when n is very large, as indicated in Table 3-1. If this sidelobe level is considered too high for specific applications, nonuniform amplitude or tapered excitations are normally used to reduce the sidelobe to an accept-

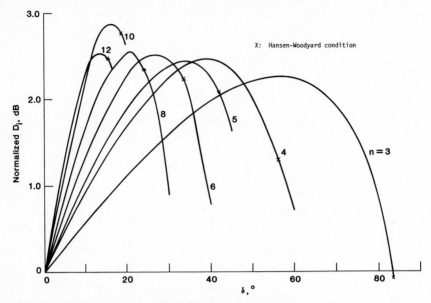

FIG. 3-12 Normalized directivities (relative to the ordinary end fire) for uniform improved end-fire arrays of isotropic elements with $d = \lambda/4$.

able level. One method for achieving this objective is the well-known Dolph-Chebyshev array,[14] in which all the sidelobes are set at the same level. This array is considered optimum in the sense that the first-null beamwidth is minimum for a specified sidelobe level or that the sidelobe level is minimum for a specified first-null beamwidth. However, the above statement is true for broadside arrays only with an element spacing no less than one-half wavelength or for ordinary end-fire arrays in which the element spacing is no less than one-quarter wavelength.[15] Many others have also devised methods for optimizing the sidelobe and first-null-beamwidth relationship for special cases.[16,17] All these methods are essentially based on manipulation of Chebyshev polynomials.

A New Approach for Treating Optimum Broadside and End-Fire Arrays

An alternative unified approach is to deal with the power pattern in the form of a polynomial of $(n-1)$th degree with real coefficients,[3,18] where n is the total number of elements in the array. This approach has advantages that it not only applies to the array yielding equal sidelobes with a specialized tapered amplitude excitation but also is valid for other linear arrays including the uniform array considered in Sec. 3-2. Basically, the power pattern for a linear array with n equally spaced isotropic elements may be expressed as

$$P(y) = \sum_{m=0}^{n-1} A_m y^m \qquad (3\text{-}44)$$

where
$$y = 2\cos u \qquad u = kd\,(\cos\theta - \cos\theta_0) \qquad (3\text{-}45)$$

and the coefficients A_m are real. When, in addition, the array has a symmetrical amplitude excitation, as is the case in practice, the power pattern takes the following product forms:

$$P_0(y) = \prod_{m=1}^{(n-1)/2} (y + b_m)^2 \qquad \text{for odd } n \qquad \textbf{(3-46)}$$

or

$$P_e(y) = (y + 2) \prod_{m=1}^{(n-2)/2} (y + b_m)^2 \qquad \text{for even } n \qquad \textbf{(3-47)}$$

where all b_m's are real.

Note that both $P_0(y)$ and $P_e(y)$ are nonnegative in the visible range $-2 \le y \le 2$, as representing the power pattern they should be. In addition, if the mth null is visible, we require $|b_m| < 2$. If all the nulls do not coincide, b_m's must also be distinct. Thus, there are at most $(n - 1)/2$ independent real nulls for odd n and $n/2$ independent real nulls for even n. Further, if $y = 2$ or $u = 0$ is the desired beam-maximum position, the combination of b_m's cannot be arbitrary.

The required amplitude excitations for the power patterns in Eqs. (3-46) and (3-47) may be shown as the coefficients of the following array polynomials:

$$S_0(z) = \prod_{m=1}^{(n-1)/2} (1 + b_m z + z^2) \qquad \text{for odd } n \qquad \textbf{(3-48)}$$

and

$$S_e(z) = (1 + z) \prod_{m=1}^{(n-2)/2} (1 + b_m z + z^2) \qquad \text{for even } n \qquad \textbf{(3-49)}$$

where

$$z = \exp(ju) \qquad \textbf{(3-50)}$$

For an example, when $n = 7$ (odd), we have

$$P_0(y) = (y + b_1)^2 (y + b_2)^2 (y + b_3)^2 \qquad \textbf{(3-51)}$$

and

$$\begin{aligned} S_0(z) &= (1 + b_1 z + z^2)(1 + b_2 z + z^2)(1 + b_3 z + z^2) \\ &= 1 + (b_1 + b_2 + b_3)z + (3 + b_1 b_2 + b_2 b_3 + b_1 b_3)z^2 \\ &\quad + (2b_1 + 2b_2 + 2b_3 + b_1 b_2 b_3)z^3 + (3 + b_1 b_2 + b_2 b_3 + b_1 b_3)z^4 \\ &\quad + (b_1 + b_2 + b_3)z^5 + z^6 \end{aligned} \qquad \textbf{(3-52)}$$

Thus, the required amplitude excitations for the linear array of seven elements are respectively

$$1, (b_1 + b_2 + b_3), (3 + b_1 b_2 + b_2 b_3 + b_1 b_3),$$
$$(2b_1 + 2b_2 + 2b_3 + b_1 b_2 b_3), \dots \qquad \textbf{(3-53)}$$

Note that the excitation coefficients are symmetrical.

As another example, when $n = 6$ (even), we have

$$P_e(y) = (y + 2)(y + b_1)^2 (y + b_2)^2 \qquad \textbf{(3-54)}$$

and

$$S_e(z) = (1 + z)(1 + b_1 z + z^2)(1 + b_2 z + z^2) \qquad \textbf{(3-55)}$$

which yields, after expansion, the required amplitude excitations as

$$1, (1 + b_1 + b_2), (2 + b_1 + b_2 + b_1 b_2), \ldots \qquad \textbf{(3-56)}$$

Again, the excitations are symmetrical.

   With Eqs. (3-46) and (3-47), a variety of arrays may be synthesized. For example, one may wish to produce three distinct nulls at desired directions, which essentially specifies b_1, b_2, and b_3 for an array of seven elements. From Eq. (3-52) or Eq. (3-53), the required amplitude excitations can be easily determined. Of course, the other array characteristics such as the sidelobe locations and levels and the directivity are also fixed accordingly. As a second example, if a uniform excitation is desired for an array of six elements, it equivalently requires, in accordance with Eq. (3-56), that $b_1 + b_2 = 0$ and $b_1 b_2 = -1$. The solutions are simply $b_1 = -1$ and $b_2 = 1$. This implies that the first two nulls are given by $y = 2 \cos u_1 = 1$ and $y = 2 \cos u_2 = -1$, which yields respectively $u_1 = \pm 60°$ and $u_2 = \pm 120°$. From Eq. (3-54), we see that a third null is given by $y = -2$ or $u_3 = \pm 180°$. The corresponding angular null positions and the other array characteristics can also be computed once the element spacing d and the beam-maximum position θ_0 are known or specified. The results of this example agree, of course, with those presented in Sec. 3-2.

   Equations (3-46) and (3-47) can also be used to synthesize arrays with equal sidelobes. Locations for the sidelobes may be obtained by setting $dP(y)/dy = 0$. Thus, $dP_0(y)/dy = 0$ for the odd-n case gives $(n - 3)/2$ values of y_i's, namely, $y_1, y_2, \ldots, y_{(n-3)/2}$. Note that $y_i \neq -b_m$, $m = 1, 2, \ldots, (n - 1)/2$, which have already been identified as the nulls. Note further that $y = -2$ ($u = \pm 180°$) is another sidelobe as evidenced by Eq. (3-12). To ensure that all the sidelobes are at the same level, we then set

$$P_0(y_1) = P_0(y_2)$$

$$P_0(y_2) = P_0(y_3) \qquad \textbf{(3-57)}$$

$$\ldots$$

$$P_0(y_{(n-5)/2}) = P_0(y_{(n-3)/2})$$

and

$$P_0(y_{(n-3)/2}) = P_0(-2)$$

or a total of $(n - 3)/2$ independent equations. If we require that $P_0(2)$ be the beam maximum (which is normally true) and have another condition such as $P_0(2)/P_0(-2) = K^2$ representing a desired sidelobe level relative to the main beam or, alternatively, a value for b_1 representing a desired first null, we will have enough equations to solve for the $(n - 1)/2$ unknowns, namely b_m, $m = 1, 2, \ldots, (n - 1)/2$.

   Similarly, for the even-n case $dP_e(y)/dy = 0$ gives $(n - 2)/2$ values of y_i's, namely, $y_1, y_2, \ldots, y_{(n-2)/2}$. Again $y_i \neq -b_m$. Note that $y = -2$ ($u = \pm 180°$) is a null in this case. We then set

$$P_e(y_1) = P_e(y_2)$$

$$P_e(y_2) = P_e(y_3)$$

and $$\ldots$$

$$P_e(y_{(n-4)/2}) = P_e(y_{(n-2)/2}) \qquad \textbf{(3-58)}$$

which consist of a total of $(n - 4)/2$ independent equations to ensure equality of all the sidelobes. If we also have another condition $P_e(2)/P_e(y_1) = K^2$ or a value for b_1 as in the odd-n case above, we then can solve for the $(n - 2)/2$ unknowns, namely, b_m, $m = 1, 2, \ldots, (n - 2)/2$.

Since the sidelobes are set to be equal, the null positions ($y = -b_m$) and the sidelobe positions ($y = y_i$) are all related. Specifically, we expressed all the parameters in the following condensed ratio form,

$$(2 - b_1):(2 - b_2): \ldots :(2 - b_{(n-1)/2}):(2 + y_1):(2 + y_2): \ldots :(2 + y_{(n-3)/2})$$

$$= \cos^2\left(\frac{\pi}{2(n-1)}\right):\cos^2\left(\frac{3\pi}{2(n-1)}\right): \ldots :\cos^2\left(\frac{(n-2)\pi}{2(n-1)}\right):\cos^2\left(\frac{\pi}{n-1}\right):$$

$$\cos^2\left(\frac{2\pi}{n-1}\right): \ldots :\cos^2\left(\frac{(n-3)\pi}{2(n-1)}\right) \qquad \text{for odd } n \qquad \textbf{(3-59)}$$

and

$$(2 - b_1):(2 - b_2): \ldots :(2 - b_{(n-2)/2}):(2 + y_1):(2 + y_2): \ldots :(2 + y_{(n-2)/2})$$

$$= \cos^2\left(\frac{\pi}{2(n-1)}\right):\cos^2\left(\frac{3\pi}{2(n-1)}\right): \ldots :\cos^2\left(\frac{(n-3)\pi}{2(n-1)}\right):\cos^2\left(\frac{\pi}{n-1}\right):$$

$$\cos^2\left(\frac{2\pi}{n-1}\right): \ldots :\cos^2\left(\frac{(n-2)\pi}{2(n-1)}\right) \qquad \text{for even } n \qquad \textbf{(3-60)}$$

where we have arranged the null and sidelobe positions in the following order:

$$2 > -b_1 > y_1 > -b_2 > y_2 > \ldots > y_{(n-3)/2}$$
$$> -b_{(n-1)/2} > -2 \qquad \text{for odd } n \qquad \textbf{(3-61)}$$

and

$$2 > -b_1 > y_1 > -b_2 > y_2 > \ldots > -b_{(n-2)/2}$$
$$> y_{(n-2)/2} > -2 \qquad \text{for even } n \qquad \textbf{(3-62)}$$

It is convenient to express, for a given n, b_m ($m \geq 2$) and y_i (all i) in terms of b_1 first. For the case in which K^2 is specified, this reduces the entire problem to only one equation, $P_0(2)/P_0(-2) = K^2$ or $P_e(2)/P_e(y_1) = K^2$, representing the desired sidelobe level, from which b_1 may be solved numerically. This should not be a difficult task since $y = -b_1$ represents the first null. For a very large n, b_1 is close to -2. On the other hand, for the case in which a desired first null (b_1) is specified, all the other b_m's and y_i's can be easily determined from Eq. (3-59) or Eq. (3-60), and the resultant sidelobe level can be obtained by computing $P_0(2)/P_0(-2)$ or $P_e(2)/P_e(y_1)$.

It is instructive to demonstrate in Fig. 3-13 how b_1 varies with K^2 and n. In general, for a given n, b_1 always increases with K^2. For one extreme situation when the sidelobe level is set to be minus infinity in decibels or equivalently when K^2 approaches infinity, b_1 and, in fact, all the other b_m's will approach 2 as their limits. Under this condition, the power pattern $P(y)$ will approach $(y + 2)^{n-1}$. The required amplitude excitations, according to Eq. (3-48) or Eq. (3-49), will be the expansion coefficients of $(1 + z)^{n-1}$, which is the so-called binomial array.[19] Thus, the arrays of equal sidelobes discussed herein include the binomial array as a special case in which all the sidelobes are equal to $-\infty$ decibels with respect to the beam maximum. Cer-

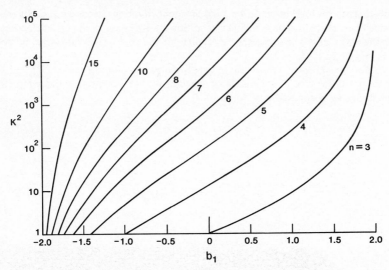

FIG. 3-13 First-null positions and sidelobe-level relationships for arrays with equal sidelobes.

tainly, the accomplishment of a very low sidelobe level is at the expense of wide beamwidth and highly tapered amplitude excitations. The other extreme situation occurs when the sidelobe level is set at a very high level or when K^2 approaches unity. Under this condition, the sidelobes are at the same level as that for the main beam. The value for b_1 may be determined from Fig. 3-13, and the other b_m's and y_i's can be obtained from Eq. (3-59) or Eq. (3-60). The required amplitude excitations computed from Eq. (3-48) or Eq. (3-49) will vanish except for the first and last elements, exhibiting a so-called edge distribution.[2] For the practical case, the sidelobes, beamwidths, and required amplitude excitations are somewhere between these two extremes.

The first-null position in θ can be determined easily from b_1 when the element spacing d and the beam-maximum direction θ_0 are known. That is,

$$kd(\cos\theta_1 - \cos\theta_0) = \cos^{-1}(-b_1/2) \qquad \textbf{(3-63)}$$

For the broadside array ($\theta_0 = 90°$) with $d \geq \lambda/2$ and the ordinary end-fire array ($\theta_0 = 0$) with $d \geq \lambda/4$, the first-null and sidelobe-level relationship shown in Fig. 3-13 is known as the optimum in the sense defined by Dolph.[14] Of course, the element spacing cannot be allowed to be too large in order to avoid the appearance of a grating lobe. For single-frequency operation, the maximum allowable element spacings for the optimum broadside array (i.e., equal sidelobes) are given in Table 3-6. The maximum allowable element spacings for the corresponding ordinary end-fire arrays of equal sidelobes are just one-half of those listed in Table 3-6.

The amplitude excitations required to produce the optimum array characteristics shown in Fig. 3-13 for $n = 3$ through 8, with the condition on the element spacing satisfied, are presented in Table 3-7. The results are valid for either broadside or ordinary end-fire arrays.

TABLE 3-6 Maximum Allowable Element Spacing in Wavelengths for an Optimum Broadside Array with Equal Sidelobes

n \ K^2	10 (-10-dB sidelobe level)	10^2 (-20-dB sidelobe level)	10^3 (-30-dB sidelobe level)	10^4 (-40-dB sidelobe level)
3	0.7438	0.6402	0.5796	0.5452
4	0.8179	0.7249	0.6566	0.6081
5	0.8601	0.7813	0.7169	0.6653
8	0.9182	0.8679	0.8216	0.7794
10	0.9365	0.8960	0.8584	0.8227

TABLE 3-7 Amplitude Excitations for Optimum Arrays with Equal Sidelobes

n \ K^2	1	10	10^2	10^3	10^4
3	1	1	1	1	1
	0	1.039	1.636	1.877	1.960
	1	1	1	1	1
4	1	1	1	1	1
	0	0.879	1.736	2.331	2.669
		symmetrical			
5	1	1	1	1	1
	0	0.724	1.609	2.413	3.013
	0	0.790	1.932	3.140	4.149
		symmetrical			
6	1	1	1	1	1
	0	0.608	1.437	2.312	3.087
	0	0.681	1.850	3.383	4.975
		symmetrical			
7	1	1	1	1	1
	0	0.519	1.277	2.151	3.008
	0	0.586	1.684	3.306	5.269
	0	0.610	1.839	3.784	6.275
		symmetrical			
8	1	1	1	1	1
	0	0.452	1.139	1.978	2.861
	0	0.510	1.509	3.097	5.200
	0	0.541	1.725	3.815	6.848
		symmetrical			

The directivity for the array with equal sidelobes may also be computed in accordance with Eq. (3-15). In terms of y, we have

$$D = \frac{2P(2)}{\displaystyle\int_0^\pi P(y) \sin \theta \, d\theta} = 2kd \, P(2)/W \qquad \textbf{(3-64)}$$

where

$$W = \int_{y_a}^{y_b} \frac{P(y)}{\sqrt{4 - y^2}} \, dy \qquad \textbf{(3-65)}$$

and

$$y_b = 2 \cos \left[kd(1 + \cos \theta_0) \right], \; y_a = 2 \cos \left[kd(1 - \cos \theta_0) \right] \qquad \textbf{(3-66)}$$

For the broadside array, $\theta_0 = 90°$, $y_a = y_b$, Eq. (3-65) should be replaced by

$$W = 2 \int_{y_c}^{2} \frac{P(y)}{\sqrt{4 - y^2}} \, dy \qquad \textbf{(3-67)}$$

where

$$y_c = 2 \cos kd \qquad \textbf{(3-68)}$$

The directivity for the optimum broadside array with various sidelobe levels has been extensively computed.[20] Sample results for $d = \lambda/2$ are presented in Fig. 3-14, from which we may conclude (1) that, for a specified sidelobe level, the directivity always increases with the total number of elements in the array and (2) that, for a given n, the directivity does not always increase when the sidelobe level is decreased. The results shown in Fig. 3-14 are, of course, also valid for the ordinary end-fire array with equal sidelobes and with $d = \lambda/4$.

Optimum Broadside Arrays When $d < \lambda/2$

It was noted previously that the first-null and sidelobe-level relationship shown in Fig. 3-13 is optimum in Dolph's sense for the broadside array of equal sidelobes only when

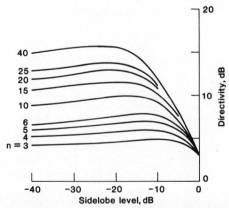

FIG. 3-14 Directivities for optimum broadside arrays of equal sidelobes when $d = \lambda/2$.

the element spacing is no less than one-half wavelength. When $d < \lambda/2$, the visible range of u will be $-\pi < -kd < u < kd < \pi$. Consequently, some of the nulls and sidelobes will disappear into the invisible region so that the final radiation pattern does not contain the maximum possible number of nulls and sidelobes. For this reason, the results shown in Fig. 3-13 and Table 3-7 no longer yield optimum characteristics (first-null-position and sidelobe-level relationships) for the broadside array. If, however, a transformation such as

$$y' = A_1 y + A_2 \tag{3-69}$$

with

$$A_1 = \tfrac{1}{2}(1 - \cos kd) \qquad \text{and} \qquad A_2 = 1 + \cos kd \tag{3-70}$$

is applied so that all the possible nulls and sidelobes originally existing under the condition $d = \lambda/2$ remain in the shrunk visible region of u, the array characteristics obtained after this transformation are still optimum in Dolph's sense.[3] The power pattern $P_0(y)$ for the odd-n case and the required amplitude excitations $S_0(z)$ will become respectively

$$Q_0(y') = \prod_{m=1}^{(n-1)/2} (y' + B_m)^2 \tag{3-71}$$

and

$$S_0'(z) = \prod_{m=1}^{(n-1)/2} (1 + B_m z + z^2) \tag{3-72}$$

where

$$B_m = A_1 b_m - A_2, \; m = 1, 2, \ldots, (n-1)/2 \tag{3-73}$$

and b_m's are still related in accordance with Eq. (3-59).

Note that B_m, since it involves A_1 and A_2, is now a function of d, although b_m is not.

The directivity becomes

$$D' = \frac{kd \, Q_0(2)}{\displaystyle\int_{y_c}^{2} \frac{Q_0(y') \, dy'}{\sqrt{4 - y'^2}}} \tag{3-74}$$

where

$$y_c' = 2 \cos kd \tag{3-75}$$

Numerical examples are now given to illustrate the point. If the excitation coefficients given in Table 3-7 are still maintained without using the transformation suggested in Eq. (3-69), the directivities for the broadside arrays with equal sidelobes for $n = 5$ and 7 are presented as solid curves in Fig. 3-15. Clearly, the directivity decreases rather fast when the element spacing becomes smaller than one-half wavelength. The corresponding directivities after the transformation [Eq. (3-69)] is used are presented as dashed curves. Although the directivity still decreases with d, it does not now decrease as fast. To achieve this improvement in directivity for small d, the required new excitation coefficients obtained in accordance with Eq. (3-72) are given

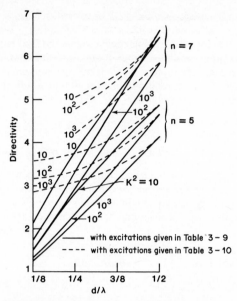

FIG. 3-15 Directivities for broadside arrays of equal sidelobes.

in Table 3-8 and may be compared with the corresponding cases in Table 3-7. The prices paid for the improvement in directivity are, therefore, the possible phase reversal for the even-numbered elements in the array, the requirement of higher accuracies in computing and maintaining the excitation coefficients, and much lower radiation intensities in the broadside direction. Thus, the overall radiation efficiency is very low for this kind of superdirective array.[3] Furthermore, the severe mutual couplings among the elements when they are so closely spaced will make the accurate maintenance of desired excitations more difficult. The practicality of superdirective arrays is therefore quite limited unless special care is exercised.[21] Superdirective arrays for the even-n case are not possible.[3]

The first-null beamwidths for $n = 5$ and 7 with and without the transformation [Eq. (3-69)] are given in Table 3-9 for reference purposes.

Optimum End-Fire Arrays When d Is Less Than the Maximum Allowable Spacing

The characteristics shown in Fig. 3-13 are also optimum for the ordinary end-fire array with equal sidelobes, provided that the element spacing is no less than $\lambda/4$ and no greater than the maximum allowable element spacing (d_{max}) which is required to avoid the grating lobe. The d_{max} values for the ordinary end-fire arrays with various n and K^2 are just one-half of those presented in Table 3-6. Under the above conditions, the sidelobe and first-null relations shown in Fig. 3-13 can be realized by the excita-

TABLE 3-8 Amplitude Excitations for Optimum Broadside Arrays with Equal Sidelobes when $d < \lambda/2$

n	K^2	10	10^2	10^3
	$d = \frac{3}{8}\lambda$	1	1	1
		0.0324	0.7874	1.4734
		1.0231	1.6343	2.3130
		symmetrical		
5	$d = \frac{1}{4}\lambda$	1	1	1
		-1.6378	-1.1956	-0.7937
		2.3350	2.1787	2.0788
		symmetrical		
	$d = \frac{1}{8}\lambda$	1	1	1
		-3.3081	-3.1786	-3.0609
		4.7072	4.5106	4.3355
		symmetrical		
	$d = \frac{3}{8}\lambda$	1	1	1
		-0.4354	0.2151	0.9561
		1.2395	1.6635	2.4046
		-0.6091	0.3336	1.6577
7		symmetrical		
	$d = \frac{1}{4}\lambda$	1	1	1
		-2.7403	-2.3593	-1.9252
		4.8773	4.3917	3.9266
		-5.6712	-4.8402	-3.9165
		symmetrical		

TABLE 3-9 First-Null Beamwidth in Degrees for Broadside Arrays with Equal Sidelobes

n	K^2	Without transformation [Eq. (3-69)]			With transformation [Eq. (3-69)]			
		10	10^2	10^3	10	10^2	10^3	
		59.1	82.3	107.4	53.6	73.3	92.4	$d = \frac{3}{8}\lambda$
5		95.4	161.5	.....	61.0	82.5	102.5	$d = \frac{1}{4}\lambda$
		....	.....	.....	65.2	87.3	107.4	$d = \frac{1}{8}\lambda$
		39.1	54.8	71.2	35.9	49.9	64.1	$d = \frac{3}{8}\lambda$
7		60.3	87.3	121.7	41.0	56.6	72.5	$d = \frac{1}{4}\lambda$
		....	....	.....	44.2	60.8	77.1	$d = \frac{1}{8}\lambda$

tions listed in Table 3-7. Naturally, for single-frequency operation, we should always use d_{max} to obtain the narrowest beamwidth for a given sidelobe level. When the actual element spacing is less than d_{max} for scanning or other considerations, parts of the maximum possible number of nulls and sidelobes will be shifted to the invisible region, yielding a wider beamwidth. Thus, the resultant array characteristics are no longer optimum in the Dolph sense. A method for optimizing the array characteristics even when $d < d_{max}$ is also available.[3,22] Basically, a nonordinary end-fire condition such as

$$u = kd \cos \theta + \alpha \qquad \alpha \neq -kd \qquad (3\text{-}76)$$

is used together with the transformation [Eq. (3-69)] so that the maximum possible number of nulls and sidelobes under the condition $d = d_{max}$ is still kept. The required phase α and the transformation coefficients A_1 and A_2 for Eq. (3-69) are given respectively by

$$\tan (\alpha/2) = \cot^2 (kd_{max}/2) \tan (kd/2) \qquad (3\text{-}77)$$

$$A_1 = -\sin^2 \left(\frac{\alpha + kd}{2} \right) \qquad (3\text{-}78)$$

and

$$A_2 = 2(1 + A_1) = 2 \cos^2 \left(\frac{\alpha + kd}{2} \right) \qquad (3\text{-}79)$$

Thus, given the total number of elements n and a desired sidelobe level represented by K^2, we determine first the maximum allowable element spacing d_{max}.[3] The actual element spacing d and d_{max} so determined are used in Eq. (3-77) to obtain α. These values of α and d are then used in Eqs. (3-78) and (3-79) to compute the required transformation coefficients A_1 and A_2. After the above procedures are completed and Eq. (3-69) is applied, the power pattern given in Eq. (3-46) for the odd-n case remains in the same form as Eq. (3-71). Therefore, the required excitations can still be computed by expanding Eq. (3-72). For the even-n case, the power pattern given in Eq. (3-47) will become

$$Q_e(y') = (2 - y') \prod_{m=1}^{(n-2)/2} (y' + B_m)^2 \qquad (3\text{-}80)$$

where B_m is also related to b_m by Eq. (3-73) but with $m = 1, 2, \ldots, (n - 2)/2$. The required excitations are then obtained by expanding

$$S_e'(z) = (1 - z) \prod_{m=1}^{(n-2)/2} (1 + B_m z + z^2) \qquad (3\text{-}81)$$

Once the transformation [Eq. (3-69)] has been obtained, the computation of directivity [Eq. (3-15)] for n both odd and even can be achieved. Since the beam maximum is at $\theta = 0$, or $u = kd + \alpha$, or $y' = 2 \cos (kd + \alpha)$, the directivity may be expressed as

$$D' = \frac{2kd Q(y_a')}{\displaystyle\int_{y_a}^{y_b} \frac{Q(y') \, dy'}{\sqrt{4 - y'^2}}} \qquad (3\text{-}82)$$

where $y'_a = 2 \cos (kd + \alpha)$ corresponds to $\theta = 0$ and $y'_b = 2 \cos (-kd + \alpha)$ corresponds to $\theta = \pi$ and where $Q(y')$ can be either $Q_0(y')$ or $Q_e(y')$.

Sample results for this newly formulated optimum end-fire array are presented in Figs. 3-16 through Fig. 3-18. First, the directivities computed for the end-fire array with equal sidelobes of -20 dB ($K^2 = 100$) in accordance with Eq. (3-82) are given in Fig. 3-16. The points marked with x correspond to the cases in which $d = d_{max}$.

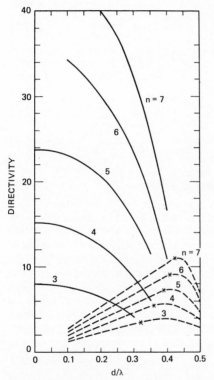

More specifically, $d_{max} = 0.3201\lambda$, 0.3625λ, 0.3906λ, 0.4099λ, and 0.4235λ respectively for $n = 3, 4, 5, 6$, and 7. Corresponding results for the ordinary end-fire array where $u = kd (\cos \theta - 1)$ are also included in the figure for comparison purposes.

Next, the first-null beamwidths for the optimum and ordinary end-fire arrays of equal sidelobes with $K^2 = 100$ are given in Fig. 3-17. Note that the beamwidth decreases with d for the optimum end-fire array (solid curves) and that it increases when d is decreased for the ordinary end fire (dashed curves).

The required phases α obtained from Eq. (3-77) are presented in Fig. 3-18. It can be easily verified that $\alpha \neq - kd$.

The excitations required to produce these characteristics are given in Table 3-10 and are quite different from the corresponding excitations in Table 3-7. Once again, the improvements in directivity and beamwidth for the small element spacings are made possible by the requirement of highly tapered amplitude excitations.

Thus far we have discussed various equally spaced linear arrays. Arrays with nonuniform spacings have also been designed for the purposes of reducing the average sidelobe level, eliminating the grating lobe, minimizing the total number of elements needed in the array, or broadening the frequency bandwidth. However, this subject is not pursued here.

FIG. 3-16 Directivities for end-fire arrays with equal sidelobes of -20 dB ($K^2 = 100$). *(After Ref. 3.)*

3-6 ARRAY OF ARRAYS

It often is desirable in practice to form an array out of a number of identical arrays, resulting in a two-dimensional array. If the mutual-coupling effect is again neglected and the principle of pattern multiplication applies when the same kind of antenna

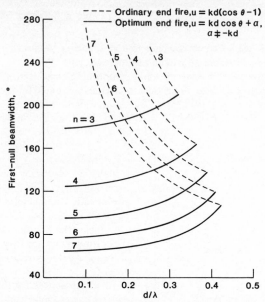

FIG. 3-17 First-null beamwidths for end-fire arrays with equal sidelobes of −20 dB ($K^2 = 100$).

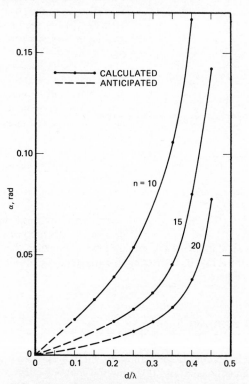

FIG. 3-18 Phases required for optimum end-fire arrays with equal sidelobes of −20 dB. *(After Ref. 3.)*

TABLE 3-10 Amplitude Excitations for the Optimum End-Fire
Arrays with Equal Sidelobes of -20 dB

n \ d/λ	0.4	0.35	0.3	0.25	0.2	0.15
3	.....	.....	1 1.6418	1 1.6923	1 1.7724	1 1.8586
			symmetrical			
4	.....	1 1.7430	1 1.8791	1 2.1106	1 2.3726	1 2.6212
			symmetrical			
5	.....	1 1.7279 2.1011	1 2.0686 2.6035	1 2.5033 3.2867	1 2.9579 4.0516	1 3.3756 4.8000
			symmetrical			
6	1 1.4496 1.8699	1 1.7621 2.3831	1 2.2938 3.3460	1 2.9195 4.6242	1 3.5576 6.0889	1 4.1379 7.5623
			symmetrical			
7	1 1.3587 1.8202 1.9972	1 1.8457 2.6974 3.0395	1 2.5513 4.1787 4.8749	1 3.3571 6.1746 7.4641	1 4.1707 8.5194 10.6410	1 4.9077 10.9290 14.0312
			symmetrical			

elements are used to form an array in the xy plane, the two-dimensional pattern may be simply written as

$$E(\theta,\phi) = f(\theta,\phi)S_x(\theta,\phi)S_y(\theta,\phi) \qquad (3\text{-}83)$$

where $f(\theta,\phi)$ is the element pattern, $S_x(\theta,\phi)$ is the array factor for the x-directed linear array, and $S_y(\theta,\phi)$ is the array factor for the y-directed linear array.

When, in addition, all the elements are equally excited and spaced to produce a beam maximum in the broadside direction ($\theta = 0$), Eq. (3-83) reduces to

$$E(\theta,\phi) = f(\theta,\phi) \frac{\sin(Mu_x/2)}{\sin(u_x/2)} \frac{\sin(Nu_y/2)}{\sin(u_y/2)} \qquad (3\text{-}84)$$

where $\qquad u_x = kd_x \sin\theta\cos\phi \qquad u_y = kd_y \sin\theta\sin\phi \qquad (3\text{-}85)$

d_x and d_y are respectively the element spacings for the x- and y-directed arrays, and M and N are respectively the number of elements in x and y directions. Thus, the total number of elements in the planar array is MN.

Arrays of this kind are sometimes referred to as curtains of antennas. The element pattern of various unidirectional types of practical radiating elements such as a

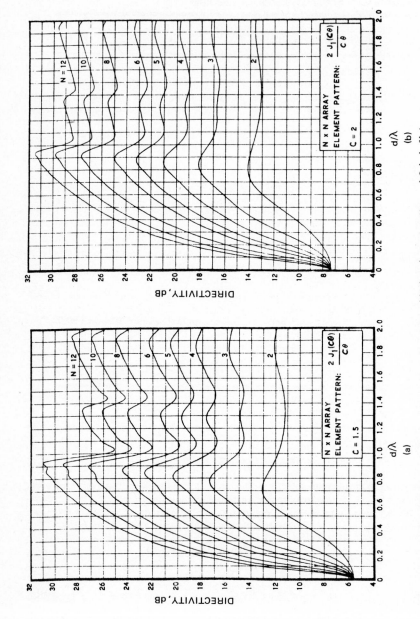

FIG. 3-19 Directivities for uniformly excited $N \times N$ arrays with the element pattern of $2J_1(c\theta)/(c\theta)$. (After Ref. 23.) (a) $c = 1.5$. (b) $c = 2$.

dipole in front of a reflector or a slot, helix, or log-periodic type may be adequately approximated by $2J_1(c\theta)/(c\theta)$, where c is a parameter and J_1 is the Bessel function of the first order. On the basis of this approximation, the directivity has been computed for a uniform square array ($M = N$ and $d_x = d_y = d$) as a function of the element spacing d.[23] Sample results are given in Fig. 3-19 to show a designer how to choose an array aperture to realize a desired level of directivity.

Two-dimensional phased arrays of closely spaced waveguides with thin walls in which the mutual coupling cannot be neglected are discussed in Chap. 20 and elsewhere.[24] Other two-dimensional arrays taking the form of a circle, concentric circles, or an ellipse are also possible. Details of these arrays may be found from the references.[3,25–27]

REFERENCES

1 E. C. Jordan, *Electromagnetic Waves and Radiating Systems*, Prentice-Hall, Inc., Englewood Cliffs, N.J., 1950.

2 J. D. Kraus, *Antennas*, McGraw-Hill Book Company, New York, 1950.

3 M. T. Ma, *Theory and Application of Antenna Arrays*, John Wiley & Sons, Inc., New York, 1974.

4 R. W. P. King, *The Theory of Linear Antennas*, Harvard University Press, Cambridge, Mass., 1956.

5 H. Bach, "Directivity of Basic Linear Arrays," *IEEE Trans. Antennas Propagat.*, vol. AP-18, January 1970, pp. 107–110.

6 R. C. Hansen, "Array Theory and Practice," *Microwave Scanning Antennas*, vol. II, Academic Press, Inc., New York, 1966.

7 S. Uda, "Wireless Beam of Short Electric Waves," *J. IEE (Japan)*, no. 452, March 1926, pp. 273–282; no. 472, November 1927, pp. 1209–1219.

8 H. Yagi, "Beam Transmission of Ultra Short Waves," *IRE Proc.*, vol. 16, June 1928, pp. 715–741.

9 R. W. P. King and T. T. Wu, "Currents, Charges, and Near Fields of Cylindrical Antennas," *Radio Sci.*, vol. 69D, March 1965, pp. 429–446.

10 C. A. Chen and D. K. Cheng, "Optimum Element Lengths for Yagi-Uda Arrays," *IEEE Trans. Antennas Propagat.*, vol. AP-23, January 1975, pp. 8–15.

11 D. K. Cheng and C. A. Chen, "Optimum Element Spacings for Yagi-Uda Arrays," *IEEE Trans. Antennas Propagat.*, vol. AP-21, September 1973, pp. 615–623.

12 W. W. Hansen and J. R. Woodyard, "A New Principle in Directional Antenna Design," *IRE Proc.*, vol. 26, March 1938, pp. 333–345.

13 T. M. Maher, "Optimum Progressive Phase Shifts for Discrete Endfire Arrays," Syracuse Univ. Res. Inst. Rep. EE 492-6002T8, Syracuse, N.Y., February 1960.

14 C. L. Dolph, "A Current Distribution for Broadside Arrays Which Optimizes the Relationship between Beam Width and Side-Lobe Level," *IRE Proc.*, vol. 34, June 1946, pp. 335–348.

15 H. J. Riblet, "Discussion of Dolph's Paper," *IRE Proc.*, vol. 35, May 1947, pp. 489–492.

16 R. H. DuHamel, "Optimum Patterns for Endfire Arrays," *IRE Proc.*, vol. 41, May 1953, pp. 652–659.

17 R. L. Pritchard, "Discussion of DuHamel's Paper," *IRE Trans. Antennas Propagat.*, vol. AP-3, January 1955, pp. 40–43.

18 M. T. Ma, "A New Mathematical Approach for Linear Array Analysis and Synthesis," Ph.D. dissertation, Syracuse University, Syracuse, N.Y., 1961.

19 J. S. Stone, U.S. Patents 1,643,323 and 1,715,433.

20 L. B. Brown and G. A. Sharp, "Tschebyscheff Antenna Distribution, Beamwidth, and Gain Tables," Nav. Ord. Lab. Rep. NOLC 383, Corona, Calif., February 1958.

21 A. Block, R. G. Medhurst, and S. D. Pool, "A New Approach to the Design of Super-Directive Aerial Arrays," *IEE Proc. (London),* vol. 100, part III, 1953, pp. 303–314.

22 M. T. Ma and D. C. Hyovalti, *A Table of Radiation Characteristics for Uniformly Spaced Optimum Endfire Arrays with Equal Sidelobes,* NBS Monograph 95, National Bureau of Standards, Boulder, Colo., December 1965.

23 J. L. Wong and H. E. King, "Directivity of a Uniformly Excited $N \times N$ Array of Directive Elements," *IEEE Trans. Antennas Propagat.,* vol. AP-23, May 1975, pp. 401–404.

24 N. Amitay, V. Galindo, and C. P. Wu, *Theory and Analysis of Phased Array Antennas,* John Wiley & Sons, Inc., New York, 1972.

25 J. D. Tillman, Jr., *The Theory and Design of Circular Antenna Arrays,* University of Tennessee Engineering Experiment Station, Knoxville, 1966.

26 C. O. Stearns and A. C. Stewart, "An Investigation of Concentric Ring Antennas with Low Sidelobes," *IEEE Trans. Antennas Propagat.,* vol. AP-13, November 1965, pp. 856–863.

27 D. K. Cheng and F. I. Tseng, "Maximisation of Directive Gain for Circular and Elliptical Arrays," *IEEE Proc. (London),* vol. 114, May 1967, pp. 589–594.

PART 2

Types and Design Methods

Chapter 4

Dipoles and Monopoles

Chen To Tai

The University of Michigan

4-1 INTRODUCTION

Since the publication of the first edition of this handbook several sections in the original chapter on linear antennas have become outdated. They have been deleted in this
revised edition. The availability of the computing program[1] in finding impedance and
other characteristics of antennas, particularly linear antennas, makes parametric tab-
ulation of limited usage. Only some essential formulas and designing data are therefore included in this chapter.

For the entire subject of linear antennas, the book by R. W. P. King[2] remains
authoritative. Another book[3] by the same author on the tables of antenna characteristics contains the most comprehensive data on the characteristics of cylindrical antennas. Calculations on circular-loop antennas and some simple arrays are also found
there.

A new section on the effective height of antennas is included in this chapter. The
usage of this parameter in describing the transmitting and receiving characteristics of
linear antennas and other simple structures is discussed in detail. In addition, material
on the general formulation of receiving antennas is included so that engineers can
apply the formulation for design purposes or for estimation of the coupling effect
between elements made of linear antennas and other antennas.

Antennas in lossy media are of great current interest. Unfortunately, the subject
cannot be covered in this chapter because of limited space. The book by King and
Smith[4] on antennas in matter can be consulted for this subject, particularly for linear
antennas embedded in a lossy medium.

4-2 CYLINDRICAL DIPOLES

Impedance as a Function of Length and Diameter

The impedance characteristics of cylindrical antennas have been investigated by many
writers. Theoretical work has mainly been confined to relatively thin antennas (length-
to-diameter ratio greater than 15), and the effect of the junction connecting the
antenna proper and the transmission line is usually not considered. Among various
theories, the induced-emf method[5] of computing the impedance of a cylindrical
antenna based upon a sinusoidal distribution is still found to be very useful. The formula derived from this method is extremely simple. It is, however, valid only when the
half length of a center-driven antenna is not much longer than a quarter wavelength.
In practice, this is the most useful range. To eliminate unnecessary computations, the
formula has been reduced to the following form:[6]

$$Z_i = R(k\ell) - j\left[120\left(\ln\frac{\ell}{a} - 1\right)\cot k\ell - X(k\ell)\right] \qquad \textbf{(4-1)}$$

where Z_i = input impedance, Ω, of a center-driven cylindrical antenna of total length
2ℓ and of radius a

$k\ell = 2\pi(\ell/\lambda)$ = electrical length, corresponding to ℓ, measured in radians

The functions $R(k\ell)$ and $X(k\ell)$ are tabulated in Table 4-1 and plotted in Fig.
4-1 for the range $k\ell \leq \pi/2$. When the length of the antenna is short compared with

TABLE 4-1 Functions $R(k\ell)$ and $X(k\ell)$ Contained in the Formula of the Input Impedance of a Center-Driven Cylindrical Antenna

$k\ell$	$R(k\ell)$	$X(k\ell)$	$k\ell$	$R(k\ell)$	$X(k\ell)$
0	0	0	0.9	18.16	15.01
0.1	0.1506	1.010	1.0	23.07	17.59
0.2	0.7980	2.302	1.1	28.83	20.54
0.3	1.821	3.818	1.2	35.60	23.93
0.4	3.264	5.584	1.3	43.55	27.88
0.5	5.171	7.141	1.4	52.92	32.20
0.6	7.563	8.829	1.5	64.01	38.00
0.7	10.48	10.68	$\pi/2$	73.12	42.46
0.8	13.99	12.73			

a wavelength but still large compared with its radius, the same formula reduces to

$$(Z_i)_{\text{short}} = 20(k\ell)^2 - j120(k\ell)^{-1}\left(\ln\frac{\ell}{a} - 1\right) \tag{4-2}$$

For antennas of half length greater than a quarter wavelength, a number of refined theories provide formulas for the computation of the impedance function. None of them, however, is simple enough to be included here. As far as numerical computation is concerned, Schelkunoff's method[7] is relatively simpler than Hallén's.[2] It should be emphasized that all these theories are formulated by using an idealized model in which the terminal condition is not considered.

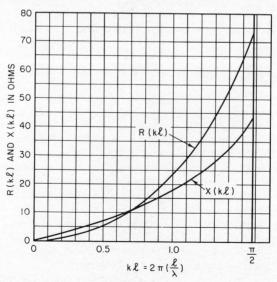

FIG. 4-1 The functions $R(k\ell)$ and $X(k\ell)$.

In practice, the antenna is always fed by a transmission line. The complete system may have the appearances shown in Fig. 4-2. The effective terminal impedance

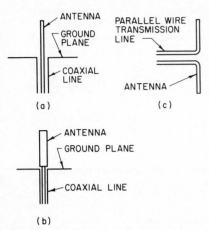

of the line (often referred to as the antenna impedance) then depends not only upon the length and the diameter of the antenna but also upon the terminal condition. In cases *a* and *b*, the impedance would also be a function of the size of the ground plane. For a given terminal condition the variation of the impedance of a cylindrical antenna as a function of the length and the diameter of the antenna is best shown in the experimental work of Brown and Woodward.[8] The data cover a wide range of values of the length-to-diameter ratio. Two useful sets of curves are reproduced in Figs. 4-3 and 4-4. The impedance refers to a cylindrical antenna driven by a coaxial line

FIG. 4-2   Driving an antenna by a pair of transmission lines.

through a large circular ground plane placed on the surface of the earth. The arrangement is similar to the one sketched in Fig. 4-2*a*. The length and diameter of the antenna are measured in degrees; i.e., a length of one wavelength is equivalent to 360°. If the effects due to the terminal condition and finite-size ground plane are neglected, the impedance would correspond to one-half of the impedance of a center-driven antenna (Fig. 4-2*c*). In using these data for design purposes, one must take into consideration the actual terminal condition as compared with the condition specified by these two authors. In particular, the maximum value of the resistance and the resonant length of the antenna may change considerably if the base capacitance is excessive.

Effect of Terminal Conditions

Many authors have attempted to determine the equivalent-circuit elements corresponding to different terminal conditions. Schelkunoff and Friis[9] have introduced the concepts of *base capacitance* and *near-base capacitance* to explain the shift of the impedance curve as the terminal condition is changed. Similar interpretations have been given by King[10] for a cylindrical antenna driven by a two-wire line or by a coaxial line and by Whinnery[11] for a biconical antenna driven by a coaxial line. The importance of the terminal condition in effecting the input impedance of the antenna is shown in Figs. 4-5 and 4-6. They are again reproduced from Brown and Woodward's paper. Because of the large variation of the effective terminal impedance of the line with changes in the geometry of the terminal junction, one must be cautious when using the theoretical results based upon isolated antennas. For junctions possessing simple geometry, the static method of Schelkunoff and Friis, King, and Whinnery can be applied to estimate the shunt capacitance of the junction. The latter then can be combined with the impedance of the antenna proper to evaluate the resultant impedance. For intricate junctions, accurate information can be obtained only by direct measurement.

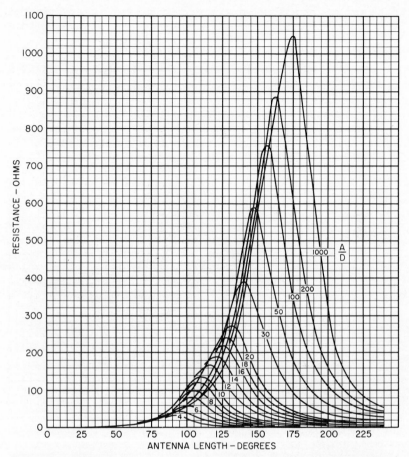

FIG. 4-3 Antenna resistance versus antenna length A when a constant ratio of length to diameter A/D is maintained. Here the length and diameter are held constant while the frequency is changed.

Equivalent Radius of Noncircular Cross Sections

As far as the impedance characteristics and radiation pattern are concerned, a thin cylindrical antenna with a noncircular cross section behaves like a circular cylindrical antenna with an equivalent radius. In stating this characteristic, the terminal effect is, of course, not considered. The equivalent radius of many simply shaped cross sections can be found by the method of conformal mapping.[12] For an elliptical cross section the following simple relation exists:

$$a_{eq} = \frac{1}{2}(a + b) \qquad \textbf{(4-3)}$$

where a = major axis of ellipse
b = minor axis of ellipse

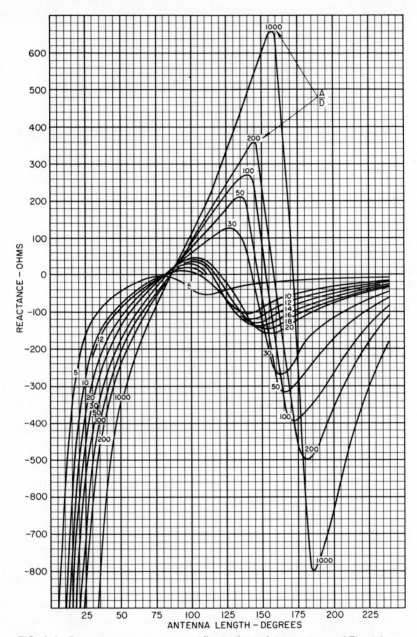

FIG. 4-4   Reactance curves corresponding to the resistance curves of Fig. 4-3.

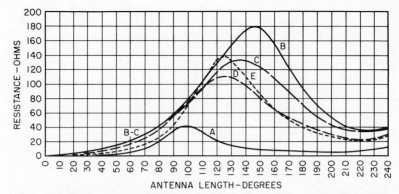

FIG. 4-5 Resistance as a function of antenna length *A*. The diameter *D* is 20.6°. Curve *A*: The arrangement shown in Fig. 4-2*b*. Curve *B*: The arrangement of Fig. 4-2*a* with the diameter of the outer conductor equal to 74°. The characteristic impedance of the transmission line is 77.0 Ω. Curve *C*: The outer-conductor diameter is 49.5°, and the transmission line has a characteristic impedance of 52.5 Ω. Curve *D*: The diameter of the outer conductor is 33°. The characteristic impedance is 28.3 Ω. Curve *E*: This curve was obtained by turning out the base reactance with an inductive reactance of 65.0 Ω.

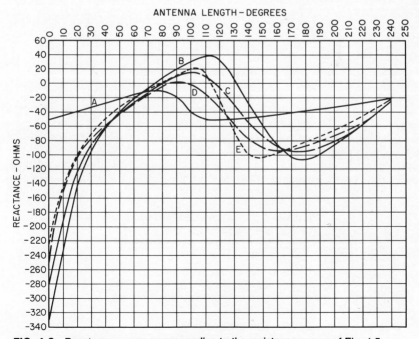

FIG. 4-6 Reactance curves corresponding to the resistance curves of Fig. 4-5.

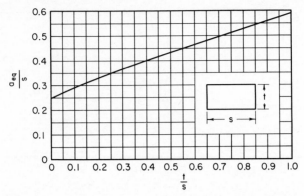

FIG. 4-7 Equivalent radius a_{eq} of a rectangle as a function of the ratio of thickness t to width s.

For a rectangular cross section the result is plotted in Fig. 4-7. In the case of a strip, Eq. (4-3) and Fig. 4-7 give the identical result. When the cross section has the form of a regular polygon, the result is tabulated in Table 4-2. The equivalent radius of two parallel cylinders of radius ρ_1 and ρ_2 separated by a distance d between the centers is given by[13]

$$\ln \rho_e = \frac{1}{(\rho_1 + \rho_2)^2} (\rho_1^2 \ln \rho_1 + \rho_2^2 \ln \rho_2 + 2\rho_1\rho_2 \ln d) \qquad \textbf{(4-4)}$$

Formulas for the equivalent radius of three cylinders and an angle strip are found in Ref. 13.

Patterns as a Function of Length and Diameter

In this subsection only the radiation pattern of center-driven cylindrical antennas is discussed. For base-driven antennas, the patterns depend very much upon the size of the ground plane. The subject will be discussed in Sec. 4-8.

The radiation pattern of a center-driven cylindrical antenna in general depends upon its length and thickness. The terminal condition which plays an important role in determining its impedance has a negligible effect on the pattern. For thin antennas, the calculated pattern obtained by assuming a sinusoidal current distribution is a good

TABLE 4-2   Equivalent Radius of a Regular Polygon

n	3	4	5	6
a_{eq}/a	0.4214	0.5903	0.7563	0.9200

n = number of sides.

a = radius of the outscribed circle.

approximation of the actual pattern. Thus, with an assumed current distribution of the form

$$I(z) = I_0 \sin k(\ell - |z|) \qquad +\ell \geq z \geq -\ell \qquad \textbf{(4-5)}$$

the radiation field, expressed in a spherical coordinate system, is given by

$$E_\theta = \frac{j\eta I_0 e^{jkR}}{2\pi R} \left[\frac{\cos (k\ell \cos \theta) - \cos k\ell}{\sin \theta} \right] \qquad \textbf{(4-6)}$$

where $\eta = (\mu/\epsilon)^{1/2} = 120\pi \quad \Omega$

θ = angle measured from axis of dipole, or z axis

The field pattern is obtained by evaluating the *magnitude* of the term contained in the brackets of Eq. (4-6). Some of the commonly referred-to patterns are sketched in Fig. 4-8. Comparing those patterns with the actual patterns of a thin cylindrical antenna

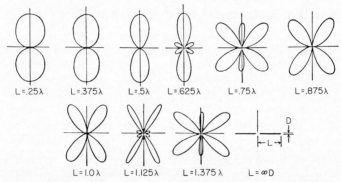

FIG. 4-8 Radiation patterns of center-driven dipoles if sinusoidal current distribution is assumed.

obtained by measurement, one finds that the theoretical patterns based upon a sinusoidal current distribution do not contain the following information:

1 The nulls between the lobes, except the *natural null* in the direction of the axis, are actually not vanishing.

2 The phase of the field varies continuously from lobe to lobe instead of having a sudden jump of 180° between the adjacent lobes.

3 The actual patterns vary slightly with respect to the diameter of the antenna instead of being independent of the thickness.

Depending upon the particular applications, some of the fine details may require special attention. In most cases, the idealized patterns based upon a sinusoidal current distribution give us sufficient information for design purposes.

When the half length ℓ of the antennas is less than about one-tenth wavelength, Eq. (4-6) is well approximated by

$$E_\theta = \frac{j\eta I_0 (k\ell)^2 e^{jkR}}{4\pi R} \sin \theta \qquad \textbf{(4-7)}$$

The figure-eight pattern resulting from the plot of the sine function is a characteristic not only of short cylindrical antennas but also of all small dipole-type antennas. Equations (4-6) and (4-7) are also commonly used to evaluate the directivity of linear antennas. The directivity is defined as

$$D = \frac{\text{maximum radiation intensity}}{\text{average radiation intensity}} \qquad (4\text{-}8)$$

For a short dipole, D is equal to 1.5. The directivity of a half-wave dipole ($\ell = \lambda/4$) is equal to 1.64. The half-wave dipole is often used as a reference antenna to describe the gain of more directive antennas, particularly arrays made of dipoles.

4-3 BICONICAL DIPOLES

Impedance as a Function of Length and Cone Angle

When the angles of a symmetrical biconical antenna (Fig. 4-9) are small, the input impedance of the antenna can be calculated by using Schelkunoff's formula.[7] Some

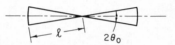

FIG. 4-9 A biconical dipole.

sample curves are shown in Fig. 4-10. While the biconical antenna is an excellent theoretical model for studying the essential property of a dipole-type antenna, small-angle biconical antennas are seldom used in practice. Wide-angle biconical antennas or their derived types such as discones, however, are frequently used as broadband antennas. The broadband impedance characteristics occur when the angle of the cones, θ_0 of Fig. 4-9, lies between 30 and 60°. The exact value of θ_0 is not critical. Usually it is chosen so that the characteristic impedance of the biconical dipole matches as closely as possible the characteristic impedance of the line which feeds the antenna. The characteristic impedance of a biconical dipole as a function of the angle is plotted in Fig. 4-11. For a conical monopole driven against an infinitely large ground plane, the characteristic impedance and the input impedance of the antenna are equal to half of the corresponding values of a dipole. Several formulas[15] are available for computing the input impedance of wide-angle biconical antennas. Actual computation has been confined to a very few specific values of θ_0.[16,17] More complete information is available from the experimental data obtained by Brown and Woodward.[18] Two curves are reproduced in Figs. 4-12 and 4-13. The case corresponding to $\alpha = 0°$ represents a cylindrical antenna having a diameter of 2.5 electrical degrees at a frequency of 500 MHz, since the feed point was kept fixed at that diameter.

Patterns of the Biconical Dipole

The radiation patterns of biconical dipoles have been investigated theoretically by Papas and King.[19] Figure 4-14 shows the patterns of a 60°-flare-angle ($\theta_0 = 30°$) conical dipole for various values of ka, where $k = 2\pi/\lambda$ and $a =$ half length of the

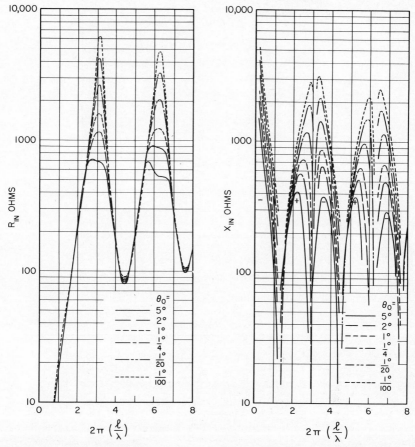

FIG. 4-10a Input impedance of small-angle biconical antennas (resistance).

FIG. 4-10b Input impedance of small-angle biconical antennas (reactance).

dipole, which is the same as the ℓ used in Fig. 4-9. Similar curves corresponding to different values of the flare angle have been obtained experimentally by Brown and Woodward.[18]

4-4 FOLDED DIPOLES

Equivalent Circuit of a Folded Dipole

A folded dipole is formed by joining two cylindrical dipoles at the ends and driving them by a pair of transmission lines at the center of one arm as shown in Fig. 4-15. The diameters of the two arms can be either identical or different. A simple analysis, based upon a quasi-static approach, of the operation of a folded dipole of arbitrary dimension has been given by Uda and Mushiake.[13] According to their method, the

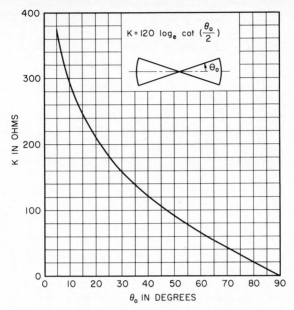

FIG. 4-11   Characteristic impedance of a biconical dipole.

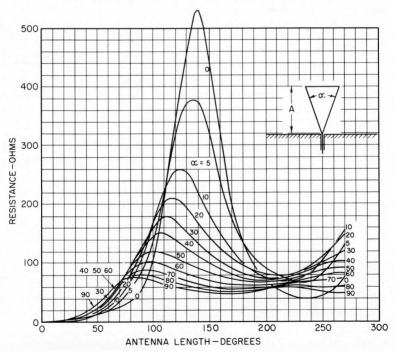

FIG. 4-12   Measured resistance curves of the conical unipole versus length in electrical degrees for various flare angles.

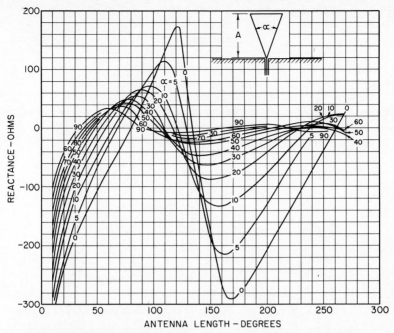

FIG. 4-13 Measured reactance curves of the conical unipole versus length in electrical degrees for various flare angles.

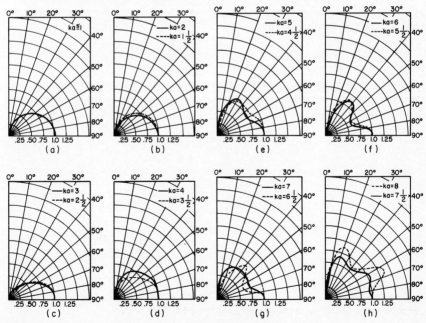

FIG. 4-14 Plots of the absolute values of the far-zone electric field as a function of the zenithal angle θ for various values of ka and with a flare angle equal to $60°$ $(\theta_0 = 30°)$.

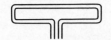

FIG. 4-15   Folded dipole.

excitation of a folded dipole can be considered as a superposition of two modes as shown in Fig. 4-16. The impedance of the symmetrical mode, characterized by two equal driving voltages, can be calculated by making use of the equivalent radius of

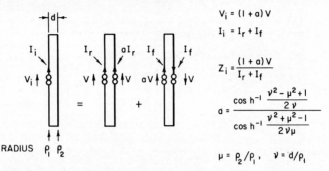

RADIUS ρ_1 ρ_2

$$V_i = (1 + a)V$$
$$I_i = I_r + I_f$$

$$Z_i = \frac{(1 + a)V}{I_r + I_f}$$

$$d = \frac{\cos h^{-1}\dfrac{\nu^2 - \mu^2 + 1}{2\nu}}{\cos h^{-1}\dfrac{\nu^2 + \mu^2 - 1}{2\nu\mu}}$$

$$\mu = \rho_2/\rho_1, \quad \nu = d/\rho_1$$

FIG. 4-16   Decomposition of the folded dipole into two fundamental modes.

two conductors as discussed in Sec. 4-2. The equivalence is shown in Fig. 4-17. The impedance function Z_r is therefore the same as the impedance of a cylindrical dipole with an equivalent radius ρ_e given by

$$\ln \rho_e = \ln \rho_1 + \frac{1}{(1 + \mu)^2} (\mu^2 \ln \mu + 2\mu \ln \nu) \qquad \textbf{(4-9)}$$

where the various parameters are explained in Fig. 4-16. The impedance of the asymmetrical mode, characterized by equal and opposite currents on the two arms, is the same as the shorted section of transmission line of length equal to ℓ; that is,

$$Z_f = \frac{(1 + a)V}{2I_f} = jZ_0 \tan k\ell \qquad \textbf{(4-10)}$$

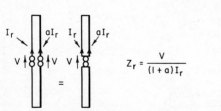

$$Z_r = \frac{V}{(1 + a)I_r}$$

FIG. 4-17   The equivalent representation of the symmetrical mode in computing Z_r.

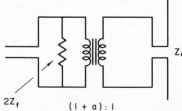

$2Z_f$ $(1 + a):1$ Z_r

FIG. 4-18   Equivalent circuit of a folded dipole.

where Z_0 is the characteristic impedance of the two-wire line. Expressed in terms of Z_r and Z_f, the input impedance of a folded dipole is given by

$$Z = \frac{V_i}{I_i} = \frac{(1 + a)V}{I_r + I_f} = \frac{2(1 + a)^2 Z_r Z_f}{(1 + a)^2 Z_r + 2Z_f} \qquad \textbf{(4-11)}$$

An equivalent circuit based upon Eq. (4-11) is shown in Fig. 4-18. For a folded dipole of length ℓ equal to $\lambda/4$, Z_f is very large compared with $(1 + a)^2 Z_r$; hence

$$Z_{\lambda/4} = (1 + a)^2 Z_r \qquad \textbf{(4-12)}$$

Impedance Transformation as a Function of the Ratio of Conductor Sizes

The step-up impedance ratio $(1 + a)^2$ as a function of μ and ν has been calculated by Mushiake.[20] The diagram is reproduced in Fig. 4-19 by using the formula for a given

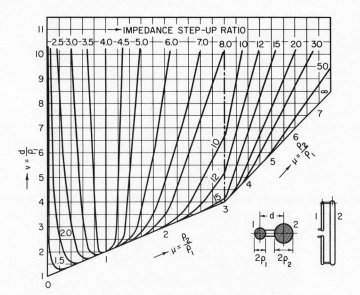

FIG. 4-19   Step-up transformation chart for a folded dipole.

in Fig. 4-16. When ρ_1 and ρ_2 are small compared with d, the value of a is given to a good approximation by

$$a = \frac{\ln (d/\rho_1)}{\ln (d/\rho_2)} \qquad \textbf{(4-13)}$$

This formula was first derived by Guertler.[21]

Another presentation[22] of the transformation ratio $(1 + a)^2$ in a logarithmic scale as a function of ρ_2/ρ_1 and d/ρ_1 is given in Figs. 4-20 and 4-21.

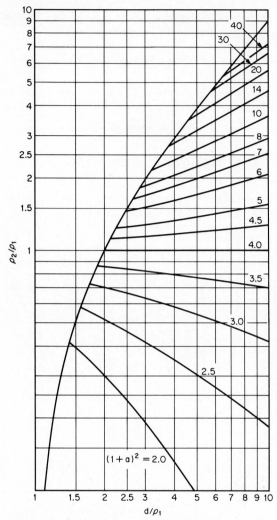

FIG. 4-20 ρ_2/ρ_1 versus impedance transformation ratio and d/ρ_1. (© *1982 IEEE.*)

4-5 SLEEVE DIPOLES

Equivalent Circuit of a Sleeve Dipole

The geometrical shape of a sleeve antenna, or a sleeve monopole, is sketched in Fig. 4-22*a*. If the image of the structure is included, then we have a sleeve dipole as shown in Fig. 4-22*b*. A sleeve dipole can therefore be considered as a doubly fed antenna in which the current is a relative maximum at the center of the dipole or at the base of the monopole. Antiresonances of the antenna impedance function take place when S

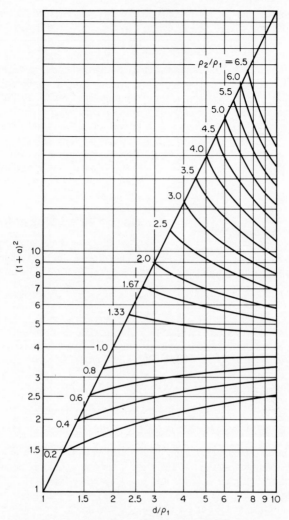

FIG. 4-21 Impedance transformation ratio versus ρ_2/ρ_1 and d/ρ_1. (© *1982 IEEE.*)

is approximately equal to an odd multiple of a quarter wavelength or L is a multiple of a half wave. A special case of interest is that in which $L + S$ is equal to a quarter wavelength. Then the current distribution along the structure is approximately cosinusoidal. At resonance the input resistance of the sleeve monopole or the sleeve dipole is approximately given by

$$R_S = R \left(\csc \frac{2\pi L}{\lambda} \right)^2 \qquad \textbf{(4-14)}$$

where R denotes either the input resistance of a resonant quarter-wave monopole or that of a half-wave dipole. The sleeve in this case plays the role of an impedance transformer.

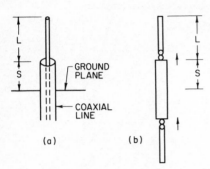

(a)   (b)

FIG. 4-22   The sleeve antenna.

Wong and King[23] have shown experimentally that open-sleeve dipoles properly designed exhibit a broadband voltage-standing-wave-ratio (VSWR) response and unidirectional radiation patterns over nearly an octave bandwidth when placed above a reflector. Figure 4-23 shows the VSWR response of open-sleeve dipoles for various dipole and sleeve diameters. Figure 4-24 shows the same with sleeve spacing as the parameter.

Open Folded Sleeve Monopole

The work of Uda and Mushiake[13] on folded dipoles can be extended to include a load at the unexcited arm as shown schematically in Fig. 4-25. The impedance of the loaded dipole is given by

$$Z_i = \frac{(Z_a + Z_L)Z_s + \left(\dfrac{a}{1 + a}\right)^2 Z_L Z_a}{Z_s + Z_L + \dfrac{1}{(1 + a)^2}\, Z_a} \qquad \textbf{(4-15)}$$

where $Z_a = jZ_c \tan \dfrac{2\pi \ell}{\lambda}$

$$Z_c = \frac{Z_0}{2\pi}\left[\cosh^{-1}\frac{1}{2\gamma}(1 + \gamma^2 - \mu^2) + \cosh^{-1}\frac{1}{2\gamma\mu}(-1 + \gamma^2 + \mu^2)\right]$$

Z_s = input impedance of the folded dipole when it is driven simultaneously by a common voltage at the base. It is approximately equal to that of a dipole with an equivalent radius ρ_e given by Eq. (4-4).

The parameter a in Eq. (4-15) and the parameters γ and μ in the expression for the characteristic impedance Z_c of a transmission line made of wires of unequal radius are the same as those defined for Fig. 4-16.

In the special case in which Z_L approaches infinity, Eq. (4-15) reduces to

$$Z_i = Z_S + \left(\frac{a}{1 + a}\right)^2 Z_a \qquad \textbf{(4-16)}$$

The structure then corresponds to an open folded dipole or monopole. Josephson[24] studied this structure both theoretically and experimentally. However, his analysis is correct only if the wires are of the same size. A sleeve version of an open folded monopole was also investigated by Josephson. The two structures are shown in Figs. 4-26 and 4-27.

The open folded monopole with a displaced feed point is equivalent to a folded

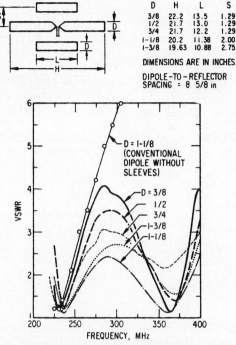

D	H	L	S
3/8	22.2	13.5	1.29
1/2	21.7	13.0	1.29
3/4	21.7	12.2	1.29
1-1/8	20.2	11.38	2.00
1-3/8	19.63	10.88	2.75

DIMENSIONS ARE IN INCHES

DIPOLE-TO-REFLECTOR
SPACING = 8 5/8 in

FIG. 4-23 VSWR response of open-sleeve dipoles for various dipole and sleeve diameters.

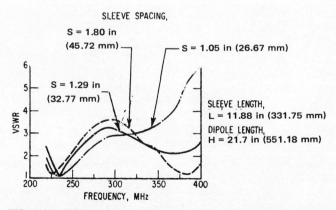

FIG. 4-24 VSWR for a ¾-in- (19-mm-) diameter open-sleeve dipole with sleeve spacing as the parameter.

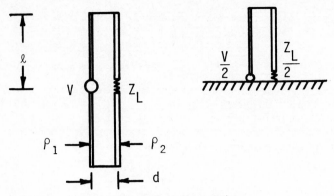

FIG. 4-25 Loaded folded dipole and monopole.

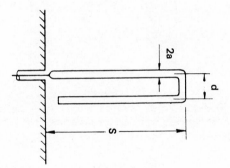

FIG. 4-26 Open folded monopole.

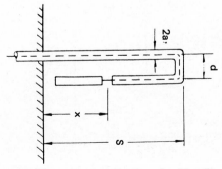

FIG. 4-27 Open folded monopole with a displaced feed point.

sleeve monopole. The input resistance at resonance of the open folded sleeve monopole is given approximately by

$$R_x = R_0 \left(\sin \frac{4\pi s}{\lambda} \right)^2 \Big/ \left(\sin \frac{2\pi x}{\lambda} \right)^2 \qquad \textbf{(4-17)}$$

where R_0 denotes the resonant input resistance of the open folded monopole and R_x is the input resistance of the folded sleeve monopole. The precise value of R_0 depends on the radius of the wires and their separation. A typical value is about 10 Ω. The open folded sleeve monopoles made of arms with different sizes are useful models in designing aircraft trail antennas or fin-type antennas on vehicles as illustrated by Josephson.

4-6   EFFECTIVE HEIGHT OF ANTENNAS

General Formula and Its Role in the Theory of Transmitting and Receiving Antennas

The radiation field of any antenna can always be written in the form

$$\mathbf{E} = \frac{-jKZ_0 I_i e^{-jkR}}{4\pi R} \mathbf{h} \qquad \textbf{(4-18)}$$

where $Z_0 = (\mu_0/\epsilon_0)^{1/2}$
$\qquad k = 2\pi/\lambda$
$\qquad \mathbf{h} =$ effective height of antenna
$\qquad I_i =$ input current to antenna

The effective height of an antenna was originally introduced by Sinclair.[25] It is related to the radiation vector defined by Schelkunoff to characterize the radiation field of an antenna, i.e.,

$$\mathbf{N}_t = I_i \mathbf{h} \qquad \textbf{(4-19)}$$

where $\mathbf{N}_t$ denotes the transversal part of Schelkunoff's radiation vector. The effective height is a very useful parameter in antenna engineering. For example, the open-circuit voltage of a receiving antenna can be expressed as

$$V_{\text{op}} = \mathbf{E}_i \cdot \mathbf{h} \qquad \textbf{(4-20)}$$

where $\mathbf{E}_i$ denotes the incident electric field. It is also an important parameter involved in the polarization-matching factor of a receiving antenna. In the theory of receiving antennas the receiving cross section or the effective aperture is defined by[26]

$$A = \frac{\lambda^2}{4\pi} Dpq \qquad \textbf{(4-21)}$$

where $\lambda =$ operating wavelength
$\qquad D =$ directivity of antenna
$\qquad p =$ polarization-matching factor
$\qquad = \dfrac{|\mathbf{h} \cdot \mathbf{E}_i|^2}{|\mathbf{h}|^2 \, |\mathbf{E}_i|^2}$
$\qquad q =$ impedance-matching factor

$$= 1 - \left| \frac{Z_L - Z_i^*}{Z_L + Z_i} \right|^2$$

Z_i = input impedance of antenna

Z_L = load impedance

The polarization-matching factor again involves the effective-height function. The effective-height function of a center-driven short dipole, if a linear current distribution along the dipole is assumed, is given by

$$\mathbf{h} = -\ell \sin \theta \hat{\theta} \tag{4-22}$$

where ℓ denotes the half length of the dipole which is assumed to be pointed in the vertical direction and θ denotes the polar angle measured between the axis of the dipole and the direction of observation or the direction of radiation. For a linearly polarized incident field making a skew angle α with the axis of the dipole, the polarization-matching factor is given by

$$p = \cos^2 \alpha \tag{4-23}$$

The effective-height functions for other simple antenna elements are listed below:

Antenna type	Effective height
Short dipole of length 2ℓ	$-\ell \sin \theta \hat{\theta}$
Half-wave dipole	$-\dfrac{\lambda}{\pi} \dfrac{\cos\left(\dfrac{\pi}{2} \cos \theta \right)}{\sin \theta} \hat{\theta}$
Small loop of radius a pointed in the z direction	$j \dfrac{2\pi^2 a^2}{\lambda} \sin \theta \hat{\phi}$
Half-wave folded dipole	$-\dfrac{2\lambda}{\pi} \dfrac{\cos\left(\dfrac{\pi}{2} \cos \theta \right)}{\sin \theta} \hat{\theta}$

4-7   COUPLED ANTENNAS

Circuit Relationships of Radiating Systems

When several antennas are coupled to each other, the input voltage and input currents to the antennas follow the same relationship as ordinary coupled circuits.[27] For a system of n antennas, the relationships are

$$V_i \sum_{j=1}^{n} Z_{ij} I_j \qquad i = 1, 2, \ldots, n \tag{4-24}$$

where Z_{ii} is called the self-impedance of antenna i and Z_{ij} or Z_{ji} is called the mutual impedance between antenna i and antenna j. In the case of linear radiators, Carter's method, or the induced-emf method based upon sinusoidal current distribution, is the simplest one to use in determining the various Z's. The method applies only to antennas shorter than a half wavelength. The self-impedance determined by this method is the same as that given by Eq. (4-1). The formulas for the mutual impedance of two parallel antennas of equal size are found in Carter's original paper or in Kraus's book.[28] Figure 4-28 shows the mutual impedance of two parallel half-wave antennas

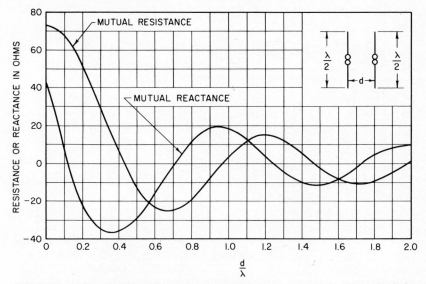

FIG. 4-28   Mutual impedance between two parallel half-wave antennas placed side by side.

placed side by side. Figure 4-29 shows the mutual impedance of two parallel collinear half-wave antennas. Mutual impedances of two parallel antennas of unequal sizes have been investigated by several authors.[29-31] The induced-emf method has also been applied to crossed or skewed antennas[32,33] to evaluate their mutual impedance. Refined calculations based upon Hallén's integral-equation technique are found in the works of Tai,[34] Bouwkamp,[35] and Uda and Mushiake.[13] The last two authors also evaluated the self-impedance and mutual impedance of parallel antennas of unequal sizes, which ultimately applies to the design of Yagi-Uda arrays.

For dipoles separated by a distance which is large compared with a wavelength, the mutual impedance between the two dipoles can be calculated by using the asymptotic formula

$$Z_{12} = \frac{jkZ_0 e^{-jkR}}{4\pi R} (\mathbf{h}_1 \cdot \mathbf{h}_2) \qquad (4\text{-}25)$$

where $k = 2\pi/\lambda$
$Z_0 = (\mu_0/\epsilon_0)^{1/2}$
R = distance between centers of dipoles
$\mathbf{h}_1, \mathbf{h}_2$ = effective height of dipoles
The formula is quite accurate for half-wave dipoles with a separation barely greater than one wavelength.

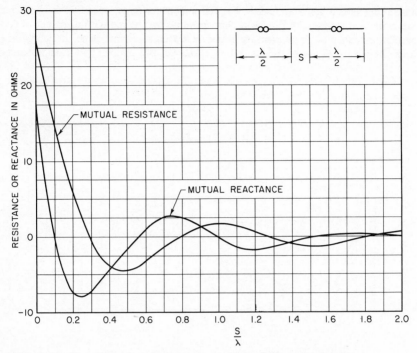

FIG. 4-29   Mutual impedance between two collinear half-wave antennas.

4-8   MONOPOLE ANTENNAS

Relationship to Balanced Antennas

When a monopole is mounted on an ideally infinite ground plane, its impedance and radiation characteristics can be deduced from that of a dipole of twice its length in free space. For a base-driven monopole, its input impedance is equal to one-half that of the center-driven dipole, and the radiation pattern above the infinite ground plane is identical with the upper half of the radiation pattern of the corresponding dipole. When the ground plane is of finite size, the image theorem does not apply.

Several methods have been devised to investigate the characteristics of a monopole mounted on a finite-size ground plane. The first method is due to Bolljahn,[36] who considers the problem from the point of view of symmetrical components. The decomposition is shown in Fig. 4-30a, in which the ground plane is assumed to be of the form

FIG. 4-30a   Monopole and finite-size ground plane and its decomposition into two modes of excitation.

of an infinitely thin conducting disk. For the symmetrical mode of excitation the presence of the disk has no effect upon the radiation of the two elements. The problem is therefore the same as if the two elements were placed in free space. The antisymmetrical pair of current elements excites equal currents on the top and the bottom sides of the disk. This mode is responsible for the variation of the input impedance of the antenna as a function of the disk diameter. It is also responsible for the asymmetry of the resultant radiation pattern with respect to the ground plane. Bolljahn's original work was developed by assuming a short monopole on a disk. The entire analysis is found in Schelkunoff's book *Advanced Antenna Theory*.[7] His study of the characteristics of large ground planes was later extended by Storer[37] to monopoles of arbitrary length.

Effect of Finite-Size Ground Plane on Impedance and Pattern

According to Storer, who used a variational method to formulate the problem, the change of the input impedance of a base-driven monopole erected upon a large circular ground plane can be written as

$$\Delta Z = Z - Z_0 = j \frac{60}{kd} e^{-jkd} \left| k \int_0^h \frac{I(z)}{I(0)} dz \right|^2 \qquad \textbf{(4-26)}$$

where Z_0 = impedance of monopole referred to an infinite ground plane, Ω
d = diameter of circular ground plane
$k = 2\pi/\lambda$
h = height of monopole
$I(z)$ = current-distribution function of monopole
$I(0)$ = base current or input current

The function $j(60/kd)e^{-jkd}$, which is independent of the current distribution, is plotted in Fig. 4-30b. The real and the imaginary parts of the function are respectively equal to $(R - R_0) \Big/ \left| k \int_0^h \frac{I(z)}{I(0)} dz \right|^2$ and $(X - X_0) \Big/ \left| k \int_0^h \frac{I(z)}{I(0)} dz \right|^2$. For a quarter-wave monopole, if we assume $I(z) = I(0) \cos kz$, then

$$\left| k \int_0^h \frac{I(z)}{I(0)} dz \right| = 1$$

Thus, with a ground plane of a diameter greater than 10 wavelengths, it is seen from Fig. 4-30b that the variation of the resistance of the reactance of a quarter-wave monopole is less than 1 Ω.

Radiation Pattern of a Monopole on a Circular Ground Plane*

While the effect of a ground plane upon the impedance of a monopole is not very great, the radiation pattern is affected considerably. The pattern of such a composite antenna

---

*The material contained in this subsection is condensed from a communication from Dr. Robert G. Kouyoumjian exclusively prepared for this handbook. The help of Dr. Kouyoumjian is gratefully acknowledged.

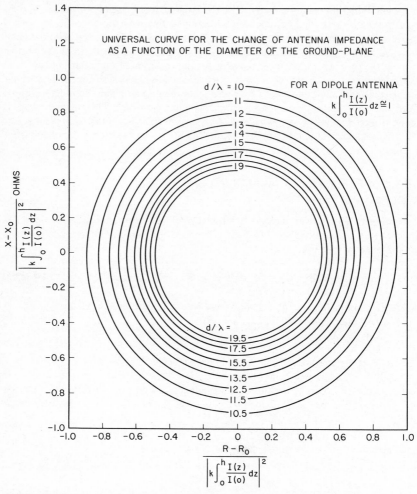

FIG. 4-30b Universal curve for the change of antenna impedance as a function of the diameter of the ground plane.

can be obtained quite accurately from the solution of the uniform geometrical theory of diffraction (GTD).[38,39] For a short monopole of length h positioned at the center of a circular disk of radius a, as shown in Fig. 4-31, the current on the monopole is assumed to be

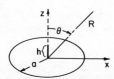

FIG. 4-31 A monopole on a circular disk.

$$I(z) = I_0 \frac{\sin k(h - z)}{\sin kh} \quad (4\text{-}27)$$

If the point of observation is not near the vertical axis (θ not close to 0 or π), the radiation field can then be expressed in the form

$$\mathbf{E} = (E_0 + E_{d1} + E_{d2})\hat{\theta} \tag{4-28}$$

where E_0 is the geometrical optics field given by

$$E_0 = \begin{cases} \dfrac{jZ_0I_0e^{-jkR}}{2\pi R \sin kh}\left[\dfrac{\cos{(kh\cos\theta)} - \cos kh}{\sin\theta}\right], 0 \le \theta < \dfrac{\pi}{2} \\ 0, \dfrac{\pi}{2} < \theta \le \pi \end{cases} \tag{4-29}$$

and E_{d1} and E_{d2} represent, respectively, the singly and doubly diffracted field. They are given by

$$E_{d1} = -\frac{E_0\left(a, \dfrac{\pi}{2}\right)a}{2\sqrt{2\pi ka}\sin\theta}\left\{ F\left[2ka\cos^2\left(\frac{\pi}{4} + \frac{\theta}{2}\right)\right] \sec\left(\frac{\pi}{4} + \frac{\theta}{2}\right) \right.$$

$$\left. e^{j[ka\sin\theta - (\pi/4)]} \mp \sec\left(\frac{\pi}{4} - \frac{\theta}{2}\right) e^{-j[ka\sin\theta - (\pi/4)]} \right\} \frac{e^{-jkR}}{R} \tag{4-30}$$

$$E_{d2} = -\frac{E_0\left(a, \dfrac{\pi}{2}\right)a}{2\sqrt{2\pi ka}\sin\theta}\frac{e^{-j(2ka - \pi/4)}}{2\sqrt{\pi ka}} \cdot \left\{ F\left[2ka\cos^2\left(\frac{\pi}{4} + \frac{\theta}{2}\right)\right] \sec\left(\frac{\pi}{4} + \frac{\theta}{2}\right) \right.$$

$$\left. e^{j[ka\sin\theta - (\pi/4)]} \mp \sec\left(\frac{\pi}{4} - \frac{\theta}{2}\right) e^{-j[ka\sin\theta - (\pi/4)]} \right\} \frac{e^{-jkR}}{R} \tag{4-31}$$

where

$$F(x) = 2j\sqrt{x}\,e^{jx}\int_{\sqrt{x}}^{\infty} e^{-jt^2}\,dt \tag{4-32}$$

$$E_0\left(a, \frac{\pi}{2}\right) = \frac{jZ_0I_0e^{-jka}}{2\pi a\sin kh}(1 - \cos kg) \tag{4-33}$$

In Eqs. (4-30) and (4-31), the upper sign associated with the secant functions is for $0 \le \theta < \pi/2$, and the lower sign is for $\pi/2 < \theta \le \pi$. Although the uniform GTD is a high-frequency method, which implies that $ka \gg 1$, in practice the result is valid even for ka as small as 2π. When the argument of the transition function $F(x)$ is greater than 10 or so, $F(x)$ is approximately equal to 1; then the sum of E_{d1} and E_{d2} yields

$$E_{d1} + E_{d2} = -E_0(a,\pi/2)\frac{a}{4}\sqrt{\frac{2}{\pi ka \sin\theta}}\left\{ \sec\left(\frac{\pi}{4} + \frac{\theta}{2}\right) e^{j[ka\sin\theta - (\pi/4)]} \mp \right.$$

$$\left. \sec\left(\frac{\pi}{4} - \frac{\theta}{2}\right) e^{-j[ka\sin\theta - (\pi/4)]}\left[1 + \frac{e^{-j[2ka - (\pi/4)]}}{2\sqrt{\pi ka}}\right] \right\} \frac{e^{-jkR}}{R} \tag{4-34}$$

In the axial region (θ close to 0 or π), the pattern can be found by means of equivalent edge current.[39,40] The total diffracted field in this case is given by

$$E_{d1} + E_{d2} = \mp jE_0\left(a, \frac{\pi}{2}\right)\frac{a}{2}\sec\frac{\pi}{4}J_1(ka\sin\theta)\left[1 + \frac{e^{-j(2ka - \pi/4)}}{2\sqrt{\pi ka}}\right] \tag{4-35}$$

This approximately equals E_{d1} and E_{d2} given by Eqs. (4-30) and (4-31) at $ka \sin \theta \simeq$ 5 for $\theta \ll \pi/2$. Thus, the pattern calculated from Eq. (4-35) joins smoothly with that calculated by Eqs. (4-31) and (4-32) when $ka > 5$.

Figures 4-32 and 4-33 show two typical patterns based on Eq. (4-28), the solid

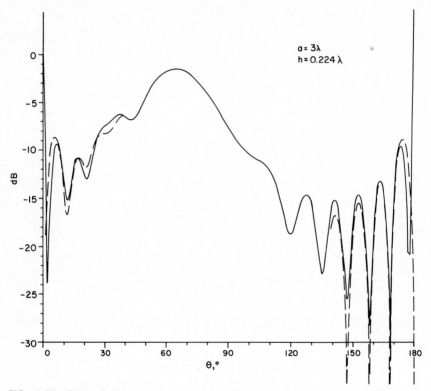

FIG. 4-32   The radiation pattern of a monopole above a circular disk having a radius of three wavelengths.

lines, and Eq. (4-35), the dashed lines. For practical purposes the term E_{d2} due to the doubly defracted rays is much smaller compared with E_{d1} except in the region where θ is close to $\pi/2$. By using a hybrid moment method jointly with GTD, it is possible to determine the input impedance to a monopole centered on a perfectly conducting circular disk.[41]

Monopole Mounted on the Edge of a Sheet

When a monopole is driven against an infinitely large conducting half sheet, the problem can be formulated conveniently with the aid of the dyadic Green's function pertaining to the half sheet.[42] By transforming the resultant series into definite integrals, Sawaya, Ishizone, and Mushiake[43] have been able to calculate the impedance of a monopole mounted on a half sheet in several orientations. Their results are shown in Figs. 4-34 through 4-36.

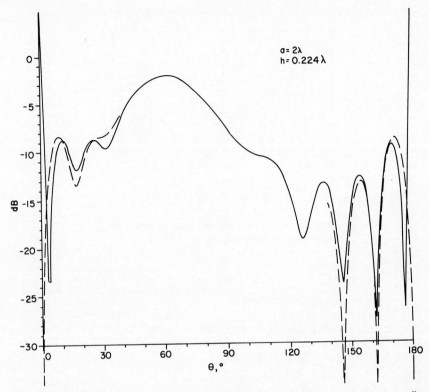

a = 2λ
h = 0.224λ

FIG. 4-33 The radiation pattern of a monopole above a circular disk having a radius of two wavelengths.

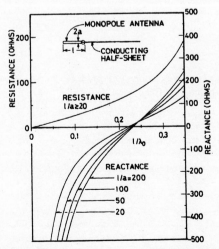

MONOPOLE ANTENNA

2a

CONDUCTING
HALF-SHEET

RESISTANCE
l/a≥20

REACTANCE
l/a=200
100
50
20

FIG. 4-34 Impedance of horizontal monopole antenna as function of ℓ/λ_0. (© *1981 IEEE.*)

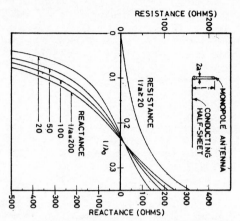

FIG. 4-35 Impedance of vertical monopole antenna as function of ℓ/λ_0. (© *1981 IEEE.*)

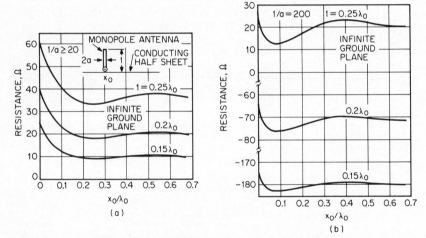

FIG. 4-36 Impedance of vertical monopole antenna on conducting half sheet as function of x_0/λ_0. (*a*) Resistance. (*b*) Reactance. (© *1981 IEEE.*)

These theoretical data can be used to estimate the impedance of monopoles mounted on large but finite conducting sheets. These authors also calculated the impedance of a notch antenna cut on a half sheet. The radiation patterns of short dipoles and small loops mounted on a half sheet are found in Ref. 42.

REFERENCES

1 G. J. Burke and A. J. Poggio, "Numerical Electromagnetic Code (NEC)," Tech. Doc. 116, Naval Ocean Systems Center, California, 1980.

2   R. W. P. King, *The Theory of Linear Antennas,* Harvard University Press, Cambridge, Mass., 1956.

3   R. W. P. King, *Tables of Antenna Characteristics,* Plenum Press, New York, 1971.

4   R. W. P. King and G. Smith, *Antennas in Matter,* The M.I.T. Press, Cambridge, Mass. 1981.

5   P. S. Carter, "Circuit Relations in Radiating Systems and Applications to Antenna Problems, *IRE Proc.,* vol. 20, 1932, pp. 1004–1041.

6   R. S. Elliott, *Antenna Theory and Design,* Prentice Hall, Inc., Englewood Cliffs, N.J., 1981, pp. 301–302.

7   S. A. Schelkunoff, *Advanced Antenna Theory,* John Wiley & Sons, Inc., New York, 1952.

8   George H. Brown and O. M. Woodward, Jr., "Experimentally Determined Impedance Characteristics of Cylindrical Antennas," *IRE Proc.,* vol. 33, 1945, pp. 257–262.

9   S. A. Schelkunoff and H. T. Friis, *Antennas: Theory and Practice,* John Wiley & Sons, Inc., New York, 1952, sec. 13.22, pp. 445–448.

10 R. W. P. King, "Antennas and Open-Wire Lines, Part I, Theory and Summary of Measurements," *J. Appl. Phys.,* vol. 20, 1949, pp. 832–850; "The End Correction for a Coaxial Line When Driving an Antenna over a Ground Screen," *IRE Trans. Antennas Propagat.,* vol. AP-3, no. 2, April 1955, p. 66.

11 John R. Whinnery, "The Effect of Input Configuration to Antenna Impedance," *J. Appl. Phys.,* vol. 21, 1950, pp. 945–956.

12 Y. T. Lo, "A Note on the Cylincrical Antenna of Noncircular Cross-section," *J. Appl. Phys.,* vol. 24, 1953, pp. 1338–1339.

13 S. Uda and Y. Mushiake, *Yagi-Uda Antenna,* Maruzen Co., Ltd., Tokyo, 1954, p. 19.

14 S. A. Schelkunoff, "Theory of Antennas of Arbitrary Size and Shape," *IRE Proc.,* vol. 29, September 1941, p. 493; see also Ref. 7, Chap. 2.

15 Ref. 7, Chap. 2.

16 C. H. Papas and R. King, "Input Impedance of Wide-Angle Conical Antennas Fed by a Coaxial Line," *IRE Proc.,* vol. 37, November 1949, p. 1269.

17 C. T. Tai, "Application of a Variational Principle to Biconical Antennas," *J. Appl. Phys.,* vol. 20, November 1949, p. 1076.

18 G. H. Brown and O. M. Woodward, Jr., "Experimentally Determined Radiation Characteristics of Conical and Triangular Antennas," *RCA Rev.,* vol. 13, no. 4, December 1952, p. 425.

19 C. H. Papas and R. King, "Radiation from Wide-Angle Conical Antenna Fed by a Coaxial Line," *IRE Proc.,* vol. 39, November 1949, p. 1269.

20 Y. Mushiake, "An Exact Step-Up Impedance Ratio Chart of a Folded Antenna," *IRE Trans. Antennas Propagat.,* vol. AP-3, no. 4, October 1954, p. 163.

21 R. Guertler, "Impedance Transformation in Folded Dipoles," *J. Brit. IRE,* vol. 9, September 1949, p. 344.

22 R. C. Hansen, "Folded and *T*-Match Dipole Transformation Ratio," *IEEE Trans. Antennas Propagat.,* vol. AP-30, no. 1, January 1982.

23 J. L. Wong and H. E. King, "An Experimental Study of a Balun-Fed Open Sleeve Dipole in Front of a Metallic Reflector," *IEEE Trans. Antennas Propagat.,* vol. AP-20, March 1972, p. 201. See also "Design Variations and Performance Characteristics of the Open-Sleeve Dipole," Aerosp. Rep. TR-0073 (3404)-2, Electronics Research Laboratory, The Aerospace Corporation, Los Angeles, Calif.

24 Bengt Josephson, "The Quarter-Wave Dipole," *IRE Wescon Conv. Rec.,* part I, San Francisco, August 1957, p. 77.

25 George Sinclair, "The Transmission and Reception of Elliptically Polarized Waves," *IRE Proc.,* vol. 38, 1950, p. 148.

26 C. T. Tai, "On the Definition of the Effective Aperture of Antennas," *IRE Trans. Antennas Propagat.,* vol. AP-9, 1961, p. 224.

27 P. S. Carter, "Circuit Relations in Radiating Systems and Applications to Antenna Problems," *IRE Proc.,* vol. 20, 1932, p. 1004.

28 J. D. Kraus, *Antennas,* McGraw-Hill Book Company, New York, 1950.

29   C. R. Cox, "Mutual Impedance between Vertical Antennas of Unequal Heights," *IRE Proc.,* vol. 35, November 1947, p. 1367.

30   G. Barzilai, "Mutual Impedance of Parallel Aerials," *Wireless Eng.,* vol. 25, November 1948, p. 347.

31   R. G. Medhurst, "Mutual Impedance of Parallel Aerials," *Wireless Eng.,* vol. 28, February 1951, p. 67.

32   L. Lewin, "Mutual Impedance of Wire Aerials," *Wireless Eng.,* vol. 28, December 1951, p. 352.

33   R. G. Medhurst, "Dipole Aerials in Close Proximity," *Wireless Eng.,* vol. 28, December 1951, p. 356.

34   C. T. Tai, "Coupled Antennas, *IRE Proc.,* vol. 36, April 1948, p. 487.

35   C. J. Bouwkamp, "On the Theory of Coupled Antennas," *Philips Res. Rep.,* vol. 3, June 1948, p. 213.

36   J. T. Bolljahn, "Antennas near Conducting Sheets of Finite Size," Univ. California Dept. Eng. Rep. 162, December 1949.

37   J. E. Storer, "The Impedance of an Antenna over a Large Circular Screen," *J. Appl. Phys.,* vol. 12, 1951, p. 1058.

38   R. G. Koujoumjian and P. H. Pathak, "A Uniform Geometrical Theory of Diffraction for an Edge in a Perfectly Conducting Surface," *IEEE Proc.,* vol. 62, 1974, pp. 1448–1461.

39   R. G. Kouyoumjian, "The Geometrical Theory of Diffraction and Its Application," in R. Mittra (ed.), *Topics in Applied Physics,* vol. 3: *Numerical and Asymptotic Techniques in Electromagnetics,* Springer-Verlag OHG, Berlin, 1975, p. 204.

40   E. F. Knott and T. B. A. Senior, "Comparison of Three High-Frequency Diffraction Techniques," *IEEE Proc.,* vol. 62, 1974, pp. 1468–1474.

41   G. A. Thiele and T. H. Newhouse, "A Hybrid Technique for Combining Moments with the Geometrical Theory of Diffraction," *IEEE Trans. Antennas Propagat.,* vol. AP-23, 1975, pp. 62–69.

42   C. T. Tai, *Dyadic Green's Functions in Electromagnetic Theory,* INTEXT, Scranton, Pa., 1971.

43   K. Sawaya, T. Ishizone, and Y. Mushiake, "A Simplified Expression for the Dyadic Green's Function for a Conducting Halfsheet," *IEEE Trans. Antennas Propagat.,* vol. AP-29, September 1981, p. 749.

Chapter 5

Loop Antennas

Glenn S. Smith
Georgia Institute of Technology

5-1 INTRODUCTION

The single-turn loop antenna is a metallic conductor bent into the shape of a closed curve, such as a circle or a square, with a gap in the conductor to form the terminals. A multiturn loop or coil is a series connection of overlaying turns. The loop is one of the primary antenna structures; its use as a receiving antenna dates back to the early experiments of Hertz on the propagation of electromagnetic waves.[1]

The discussion of loop antennas is conveniently divided according to electrical size. Electrically small loops, those whose total conductor length is small compared with the wavelength in free space, are the most frequently encountered in practice. For example, they are commonly used as receiving antennas with portable radios, as directional antennas for radio-wave navigation, and as probes with field-strength meters. Electrically larger loops, particularly those near resonant size (circumference of loop/wavelength $\approx$ 1), are used mainly as elements in directional arrays.

The following symbols are used throughout the chapter:

λ = wavelength in free space at the frequency $f = \omega/2\pi$, where the complex harmonic time-dependence exp $(j\omega t)$ is assumed

$\beta = 2\pi/\lambda$ = propagation constant in free space

$\zeta = \sqrt{\mu_0/\epsilon_0}$ = wave impedance of free space ($\approx$ 377 Ω)

b = mean radius of a circular loop or mean side length of a square loop

a = radius of loop conductor (All results presented are for thin-wire loops, $a/b \ll 1$.)

A = area of loop

N = number of turns

ℓ_c = length of solenoidal coil

5-2 ELECTRICALLY SMALL LOOPS

The axial current distribution in an electrically small loop is assumed to be uniform; that is, the current has the same value I_0 at any point along the conductor. For single-turn loops and multiturn loops that are single-layer solenoidal coils, measurements suggest that this is a good assumption provided the total length of the conductor ($N \times$ circumference) is small compared with the wavelength in free space, typically $\lesssim 0.1\lambda$, and the length-to-diameter ratio for the solenoidal coil is greater than about 3 ($\ell_c/2b \gtrsim 3.0$).[2] With a uniform current assumed, the electrically small loop antenna is simply analyzed as a radiating inductor.[3]

Transmitting Loop

The electromagnetic field of an electrically small loop antenna is the same as that of a magnetic dipole with moment $m = I_0 NA$:

$$E_\phi = \frac{\zeta\beta^2 m}{4\pi r}\left(1 - \frac{j}{\beta r}\right) e^{-j\beta r} \sin\theta \qquad (5-1)$$

$$B_\theta = \frac{-\mu_0 \beta^2 m}{4\pi r} \left(1 - \frac{j}{\beta r} - \frac{1}{\beta^2 r^2} \right) e^{-j\beta r} \sin\theta \qquad \textbf{(5-2)}$$

$$B_r = \frac{\mu_0 \beta^2 m}{2\pi r} \left(\frac{j}{\beta r} + \frac{1}{\beta^2 r^2} \right) e^{-j\beta r} \cos\theta \qquad \textbf{(5-3)}$$

where the plane of the loop is normal to the polar axis of the spherical coordinate system (r, θ, ϕ) centered at the loop, as shown in Fig. 5-1. In the far zone of the loop $(\lim \beta r \to \infty)$, only the leading terms in Eqs. (5-1) and (5-2) are significant, and the

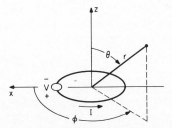

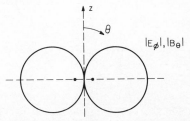

FIG. 5-1 Loop antenna and accompanying spherical coordinate system.

FIG. 5-2 Far-zone vertical-plane field pattern of an electrically small loop.

field pattern for both E_ϕ and B_θ in the vertical plane is the simple figure eight shown in Fig. 5-2.

The driving-point voltage and current are related through the input impedance of the loop, $V = ZI_0$. For electrically small loops, the impedance is the series combination of the reactance of the external inductance L^e with the radiation resistance R^r and the internal impedance of the conductor $Z^i = R^i + j\omega L^i$:

$$Z = R^r + Z^i + j\omega L^e = R^r + R^i + j\omega(L^e + L^i) \qquad \textbf{(5-4)}$$

In the equivalent circuit for the small loop, a lumped capacitance C is sometimes placed in parallel with Z to account for the distributed capacitance between the sides of a single turn and between the turns of a solenoid, as shown in Fig. 5-3. This capacitance is omitted here, since in practice a variable capacitance is usually placed in parallel with the loop to tune out its inductance; the capacitance of the loop simply decreases the value of the parallel capacitance needed. Note that a loop with a truly uniform current distribution would have no capacitance, since from the equation of continuity there would be no charge along the conductor of the loop.

The radiation resistance of the small loop is proportional to the square of the product of the area and the number of turns:

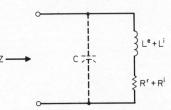

FIG. 5-3 Equivalent circuit for input impedance Z of an electrically small loop.

$$R^r = \frac{\zeta}{6\pi} \beta^4 (NA)^2 \qquad \textbf{(5-5)}$$

For single-turn loops and solenoidal coils whose turns are not too closely spaced, the internal impedance is approximately

$$Z^i = z^i \times \text{total length of conductor} \qquad \textbf{(5-6)}$$

where z^i is the internal impedance per unit length of a straight conductor with the same cross section as the loop conductor.[4] If the turns of the coil are closely spaced, the proximity effect must also be included in determining Z^i.[5]

The external inductance is determined from one of the many formulas available for the inductance of coils:[6]

For a single-turn circular loop

$$L^e = \mu_0 b[\ln (8b/a) - 2] \qquad \textbf{(5-7)}$$

and for a single-turn square loop

$$L^e = \frac{2\mu_0 b}{\pi} [\ln (b/a) - 0.774] \qquad \textbf{(5-8)}$$

The external inductance of a tightly wound single-layer solenoidal coil of length ℓ_c and a radius b is often approximated by Lorenz's formula for the inductance of a circumferentially directed current sheet.[6] Numerical results from this formula can be put in a form convenient for application:

$$L^e = K\mu_0 N^2 A/\ell_c \qquad \textbf{(5-9)}$$

where the factor K, known as Nagaoka's constant, is shown as a function of the ratio $\ell_c/2b$ (length of the coil to the diameter) in Fig. 5-4. Note that, for a long coil ($\ell_c/2b \gg 1$), $K \approx 1$. The use of Eq. (5-9) assumes that the turns of the coil are so closely spaced that the winding pitch and insulation on the conductors can be ignored; if highly accurate calculations of L^e are necessary, corrections for these factors are available in the literature.[6]

Receiving Loop

When the electrically small loop is used as a receiving antenna, the voltage developed at its open-circuited terminals V_{OC} is proportional to the component of the incident magnetic flux density normal to the plane of the loop B_z^i:

$$V_{OC} = j\omega NAB_z^i \qquad \textbf{(5-10)}$$

where the incident field is assumed to be uniform over the area of the loop. This simple relation between V_{OC} and B_z^i makes the small loop useful as a probe for measuring the magnetic flux density. If a relation between the incident electric and magnetic fields at the center of the loop is known, V_{OC} can be expressed in terms of the magnitude of the incident electric field E^i and an effective height h_e. This is the case for an incident plane wave with the wave vector $\mathbf{k}_i$ and the orientation shown in Fig. 5-5:

$$V_{OC} = j\omega NAB^i \cos \psi_i \sin \theta_i = h_e(\psi_i, \theta_i)E^i \qquad \textbf{(5-11)}$$

where

$$h_e(\psi_i, \theta_i) \equiv V_{OC}/E^i = j\beta NA \cos \psi_i \sin \theta_i \qquad \textbf{(5-12)}$$

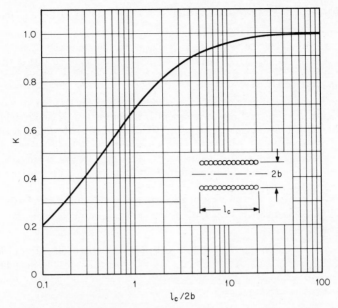

FIG. 5-4 Nagaoka's constant K for a solenoidal coil as a function of the coil length to the diameter, $\ell_c/2b$.

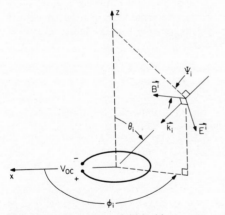

FIG. 5-5 Plane-wave field incident on receiving loop.

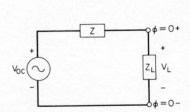

FIG. 5-6 Thévenin equivalent circuit for the receiving loop.

The voltage across an arbitrary load impedance Z_L connected to the terminals of the loop with input impedance Z is determined from the Thévenin equivalent circuit in Fig. 5-6:

$$V_L = V_{OC}Z_L/(Z + Z_L) \qquad \textbf{(5-13)}$$

Ferrite-Loaded Receiving Loop

The open-circuit voltage at the terminals of the electrically small receiving loop can be increased by filling the loop with a core of permeable material, usually a ferrite. The effect of the core is to increase the magnetic flux through the area of the loop, as illustrated in Fig. 5-7 for a solenoidal coil with a cylindrical core placed in a uniform axial magnetic field.

The ferrite material is characterized by a complex relative initial permeability $\mu_r = \mu/\mu_0 = \mu_r' - j\mu_r''$ and a relative permittivity $\varepsilon_r = \varepsilon/\varepsilon_0$.* The material is usually selected to have a loss tangent $p_m = \mu_r''/\mu_r'$ which is small at the frequency of operation, and consequently μ_r'' is ignored in the analysis except when the power dissipated in the core is being calculated. The dimensions of the core are also assumed to be small compared with the wavelength in the ferrite $\lambda_m \approx \lambda/\sqrt{\varepsilon_r \mu_r'}$ to prevent internal resonances within the core.[7]

The open-circuit voltage for a single-turn loop at the middle of a ferrite cylinder of length ℓ_r and radius b is increased by the factor μ_{rod} over the value for the same loop in free space:

$$V_{OC} = j\omega\mu_{\text{rod}}AB_z^i \qquad \textbf{(5-14)}$$

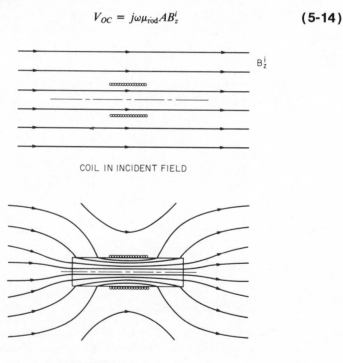

B_z^i

COIL IN INCIDENT FIELD

COIL WITH FERRITE CORE IN INCIDENT FIELD

FIG. 5-7   Effect of a cylindrical ferrite core on the magnetic flux through a solenoidal coil.

---

*The initial permeability is the derivative dB/dH in the limit as H is reduced to zero. Dielectric loss in the ferrite is ignored here, and the permittivity is assumed to be real.

Here the radius of the loop conductor a is ignored, and the mean radius of the loop and the core are assumed to be the same value b. The graph in Fig. 5-8 shows the apparent permeability μ_{rod} as a function of the length-to-diameter ratio for the rod $\ell_r/2b$ with the relative initial permeability of the ferrite μ'_r as a parameter.[8] Similar graphs for the apparent permeability of solid and hollow spheroidal cores are in the literature.[9]

For a single-layer solenoidal coil of length ℓ_c centered on the rod, an averaging factor F_V must be included in the open-circuit voltage to account for the decrease in the flux along the length of the coil from the maximum at the middle:

$$V_{OC} = j\omega\mu_{\text{rod}}F_VNAB_z^i \qquad \textbf{(5-15)}$$

The empirical factor F_V, determined from an average of experimental results, is shown in Fig. 5-9 as a function of the ratio ℓ_c/ℓ_r (length of the coil to length of the rod).[10,11] For a long rod of moderate permeability ($\ell_r/2b \gg 1$, $\mu_{\text{rod}} \approx \mu'_r$) covered by a coil of

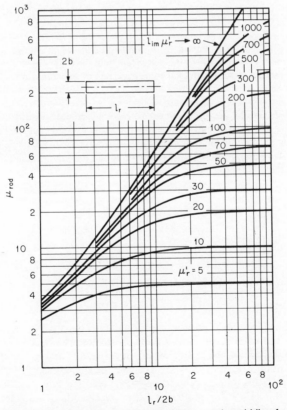

FIG. 5-8 The apparent permeability μ_{rod} at the middle of a cylindrical rod as a function of the length-to-diameter ratio $\ell_r/2b$ with the initial permeability μ'_r as a parameter.

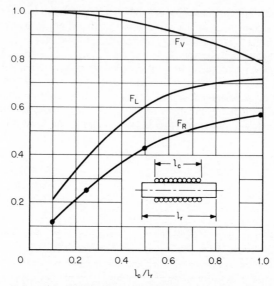

FIG. 5-9 The factors F_V, F_L, and F_R as functions of the ratio ℓ_c/ℓ_r (length of the coil to length of the rod). These factors were determined from averages of experimental data.

equal length ($\ell_c/\ell_r = 1$), the open-circuit voltage is increased by approximately the factor $0.8\ \mu'_r$ over the open-circuit voltage for the same coil without the core.

The equivalent circuit for the impedance of the ferrite-loaded solenoidal coil is that in Fig. 5-3 with an additional series resistor R^m included to account for the power dissipated in the core. The elements in the circuit are:

the radiation resistance

$$R^r = \frac{\zeta}{6\pi}\beta^4(\mu_{\text{rod}}F_V NA)^2 \qquad \textbf{(5-16)}$$

the resistance due to core loss

$$R^m = \omega(\mu_{\text{rod}}/\mu'_r)^2\mu''_r\mu_0 F_R N^2 A/\ell_c \qquad \textbf{(5-17)}$$

the external inductance of the loaded solenoidal coil

$$L^e = \mu_{\text{rod}}F_L\mu_0 N^2 A/\ell_c \qquad \textbf{(5-18)}$$

The internal impedance of the conductor Z^i is assumed to be the same as that for the unloaded loop. The empirical factors F_R and F_L in Eqs. (5-17) and (5-18), like F_V, were determined from an average of experimental results and are also shown as a function of the ratio ℓ_c/ℓ_r in Fig. 5-9.[10,11] It should be emphasized that the graphs for the three factors F_V, F_R, and F_L represent typical measured values and show only the dependence on the ratio ℓ_c/ℓ_r; some dependence on the other parameters describing the coil and the rod is to be expected.

Equations (5-15) through (5-18) provide a complete description of the electrically small ferrite-loaded receiving loop (single-layer solenoidal coil with a cylindrical core); other parameters of interest, such as the Q of the antenna, can be determined from these results. The permeability of a specific ferrite can be obtained from the manufacturer or from the extensive tables and charts in Ref. 11. The many parameters that are to be chosen for the ferrite-loaded loop, such as μ_r', ℓ_r, ℓ_c, N, etc., offer a great deal of flexibility in its design. There are several discussions in the literature that determine these parameters to optimize the performance for a particular application.[12]

The electromagnetic field of the ferrite-loaded transmitting loop is given by Eqs. (5-1) to (5-3) with the moment $m = \mu_{\text{rod}}F_V I_0 N A$. The ferrite-loaded loop, however, is seldom used as a transmitting antenna because of the problems associated with the nonlinearity and the dissipation in the ferrite at high magnetic field strengths.[13]

5-3   ELECTRICALLY LARGE LOOPS

As the electrical size of the loop antenna is increased, the current distribution in the loop departs from the simple uniform distribution of the electrically small loop. For single-turn loops, this departure has a significant effect on performance when the circumference is greater than about 0.1λ. For example, the radiation resistance of an electrically small circular loop with a uniform current, as predicted by Eq. (5-5), is about 86 percent of the actual resistance when $\beta b = 2\pi b/\lambda = 0.1$ and only about 26 percent of the actual resistance when $\beta b = 0.3$.

Of the possible shapes for an electrically large loop antenna, the single-turn thin-wire circular loop has received the most attention, both theoretical and experimental. The popularity of the circular loop is due in part to its straightforward analysis by expansion of the current in the loop as a Fourier series:

$$I(\phi) = I_0 + 2 \sum_{n=1}^{m} I_n \cos n\phi \qquad \textbf{(5-19)}$$

where the angle ϕ is defined in Fig. 5-1.[14] Measurements on electrically large loops with other shapes, such as the square loop, show that their electrical performance is qualitatively similar to that of the circular loop; therefore, only the circular loop will be discussed here.[15]

Circular-Loop Antenna

The theoretical model for the circular-loop antenna assumes a point-source generator of voltage V at the position $\phi = 0$, making the input impedance of the loop $Z = R + jX = V/I(\phi = 0)$. In practical applications, the full-loop antenna is usually driven from a balanced source, such as a parallel-wire transmission line, and the half-loop antenna, the analog of the electric monopole, is driven from a coaxial line, as in Fig. 5-10. The point-source generator of the theoretical model contains no details of the geometry of the feed point, and it is not strictly equivalent to either of these methods of excitation. However, theoretical current distributions, input impedances, and field patterns computed with the point-source generator and 20 terms in the Fourier series [Eq. (5-19)] are generally in good agreement with measured values.* Thus, the theory serves as a useful design tool.

In Figs. 5-11 and 5-12, the input impedance of a loop constructed from a perfect conductor is shown as a function of the electrical size of the loop $\beta b = 2\pi b/\lambda$ (cir-

*The theoretical results in Figs. 5-11, 5-12, 5-14 to 5-18, and 5-20 were computed by the author by using 20 terms in this series.

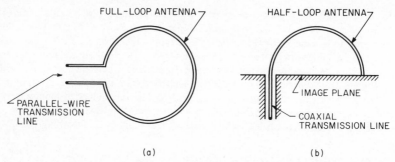

FIG. 5-10 Methods of driving the circular-loop antenna. (*a*) Full-loop antenna driven from parallel-wire transmission line. (*b*) Half-loop antenna driven from coaxial transmission line.

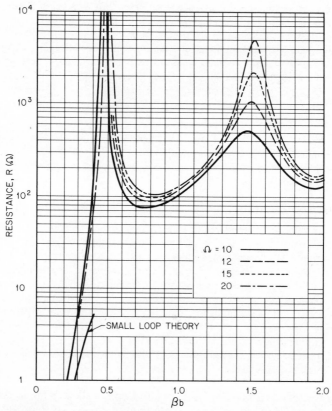

FIG. 5-11 Input resistance of circular-loop antenna versus electrical size (circumference / wavelength).

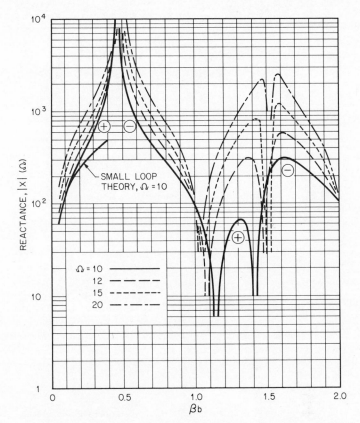

FIG. 5-12 Input reactance of circular-loop antenna versus electrical size (circumference/wavelength).

cumference/wavelength) for various values of the radius of the conductor, indicated by the *thickness parameter* $\Omega = 2 \ln (2\pi b/a)$. These impedances are for full-loop antennas; for half-loop antennas with the same radius and conductor size, impedances are approximately one-half of these values. The reactance X is seen to be zero at points near $\beta b = \frac{1}{2}, \frac{3}{2}, \frac{5}{2}, \ldots$ (antiresonant points) and $\beta b = 1, 2, 3, \ldots$ (resonant points). The resistance obtains relative maxima near the points of antiresonance and relative minima near the points of resonance. Impedances computed from Eqs. (5-5) and (5-7), which apply to electrically small loops, are also shown in Figs. 5-11 and 5-12; the inaccuracy of these formulas with increasing βb is evident.

When the electrical size of the loop is near that for resonance ($\beta b = 1, 2, 3, \ldots$), the dominant term in the Fourier series for the current [Eq. (5-19)] is the one with n = integer (βb). For example, near the first resonance $\beta b \approx 1$, the current in the loop is approximately $I(\phi) = 2I_1 \cos \phi$, and the loop is commonly referred to as a resonant loop. The resonant loop ($\beta b \approx 1$) is the most frequently used electrically large loop. It has a reasonable input resistance, $R \approx 100 \ \Omega$, for matching to a transmission line, particularly when compared with the resistance of the antiresonant loop ($\beta b \approx 0.5$), which may be larger than 10 kΩ.

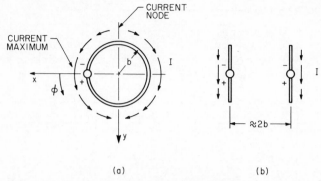

FIG. 5-13 Schematic of current distribution in resonant loop (*a*) and in the approximately equivalent pair of dipoles (*b*).

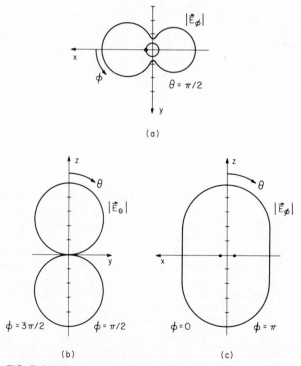

FIG. 5-14 Far-zone electric field for loop with $\beta b = 1.0$, $\Omega = 10$. (*a*) Horizontal-plane field pattern $|\mathbf{E}_\phi|$, $\theta = \pi/2$. (*b*) Vertical-plane field pattern $|\mathbf{E}_\theta|$, $\phi = \pi/2$, $3\pi/2$. (*c*) Vertical-plane field pattern $|\mathbf{E}_\phi|$, $\phi = 0$, π.

Resonant Circular Loop

The current in the resonant loop has maxima at the generator, $\phi = 0$, and at the diametrically opposite point, $\phi = \pi$, with nodes at $\phi = \pi/2$ and $3\pi/2$. On examination of Fig. 5-13, the current is seen to be roughly equivalent to that in a pair of parallel dipole antennas driven in phase and with a spacing approximately equal to the diameter of the loop.

The far-zone field patterns for the resonant loop shown in Fig. 5-14a–c are also similar to those for the pair of dipoles; they have little resemblance to the figure-eight pattern of the electrically small loop, Fig. 5-2. There are two components to the electric field, E_θ and E_ϕ; E_θ is zero in the horizontal plane $\theta = \pi/2$ and in the vertical plane $\phi = 0, \pi$, while E_ϕ is small in the vertical plane $\phi = \pi/2, 3\pi/2$. The amplitude patterns are symmetrical about the planes $\theta = \pi/2$ and $\phi = 0, \pi$ owing to the geometrical symmetry of the loop, and they are nearly symmetrical about the plane $\phi = \pi/2, 3\pi/2$ owing to the dominance of the term $2I_1 \cos \phi$ in the current distribution. At the maxima ($\theta = 0, \pi$) of the bidirectional pattern, the electric field is linearly polarized in the direction $\hat{\mathbf{y}}$.

To help us visualize the electric field, three-dimensional amplitude patterns for the electrically small loop and the resonant loop are presented in Fig. 5-15. Each drawing is a series of patterns on planes of constant angle ϕ; only the patterns in the upper hemisphere ($0 \leq \theta \leq \pi/2$) are shown, since those in the lower hemisphere are identical.

The directivity of the circular loop in the direction $\theta = 0$ or π is shown as a

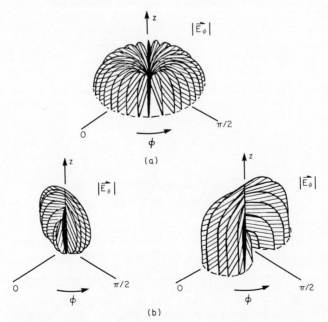

FIG. 5-15 Far-zone electric field patterns in upper hemisphere. (*a*) Electrically small loop, $\beta b \ll 1$. (*b*) Resonant loop, $\beta b = 1.0$.

function of the electrical size βb in Fig. 5-16; it is about 3.4 dB for $\beta b = 1.0$ and has a maximum of about 4.5 dB for $\beta b = 1.4$. The directivity is fairly independent of the parameter Ω for $\beta b \lesssim 1.4$.

The resonant loop antenna is attractive for practical applications because of its moderate input resistance and symmetrical field pattern with reasonable directivity.

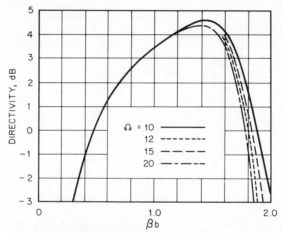

FIG. 5-16 Directivity of circular-loop antenna for $\theta = 0$, π versus electrical size (circumference/wavelength).

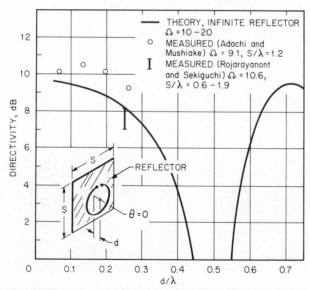

FIG. 5-17 Directivity of circular-loop antenna, $\beta b = 1.0$, for $\theta = 0$ versus distance from reflector d/λ. Theoretical curve is for infinite planar reflector; measured points are for square reflector.

The bidirectional nature of its pattern, however, is usually not desired, and a reflector or an array of loops is used to make the pattern unidirectional.

Circular Loop with Planar Reflector

The pattern of the resonant loop is made unidirectional and the directivity in the direction $\theta = 0$ is increased by placing the loop over a planar reflector. The theoretical results for an infinite perfectly conducting reflector (Fig. 5-17) show that the directivity is greater than 9 dB for spacings between the loop and the reflector in the range $0.05 \lesssim d/\lambda \lesssim 0.2$.[16] Over this same range of spacings, the input impedance $Z = R + jX$ (Fig. 5-18) has values which are easily matched; the resistance is reasonable ($R \lesssim 135\ \Omega$), and the reactance is small ($|X| \lesssim 20\ \Omega$).

The theoretical results for an infinite reflector are in good agreement with measured data for finite square reflectors of side length s. The directivities measured by Adachi and Mushiake[17] (Fig. 5-17) for a reflector with $s/\lambda = 1.2$ and $d/\lambda \leq 0.26$ are slightly higher than those for an infinite plane, while the input impedances measured by Rojarayanont and Sekiguchi[18] (Fig. 5-18) show variations with reflector size, $0.48 \leq s/\lambda \leq 0.95$, but general agreement with the results for an infinite plane.

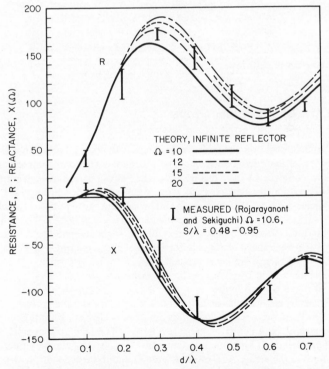

FIG. 5-18 Input impedance of circular-loop antenna, $\beta b = 1.0$ versus distance from reflector d/λ. Theoretical curves are for infinite planar reflector; measured points are for square reflector.

Electric field patterns measured by Rojarayanont and Sekiguchi[18] for resonant loops one-quarter wavelength, $d/\lambda = 0.25$, in front of square reflectors are shown in Fig. 5-19. The shaded area in each figure shows the variation in the pattern that is a result of changing the size of the square reflector from $s/\lambda = 0.64$ to $s/\lambda = 0.95$.

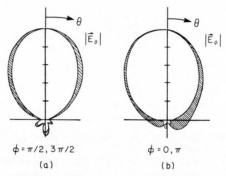

$\phi = \pi/2, 3\pi/2$ $\phi = 0, \pi$

(a)   (b)

FIG. 5-19 Measured far-zone electric field patterns for loop with $\beta b = 1.0$ over square reflector, $d/\lambda = 0.25$. Inner curve $s/\lambda = 0.95$; outer curve $s/\lambda = 0.64$. (a) Vertical-plane field pattern $|\mathbf{E}_\phi|$, $\phi = \pi/2, 3\pi/2$. (b) Vertical-plane field pattern $|\mathbf{E}_\phi|$, $\phi = 0, \pi$. (Measured data from Rojarayanont and Sekiguchi.)

Coaxial Arrays of Circular Loops

Loop antennas, like linear antennas, can be combined in an array to improve performance. The most common array of circular loops is the coaxial array in which all the loops are parallel and have their centers on a common axis; an example of a coaxial array is shown later in the inset of Fig. 5-21. The Fourier-series analysis for the single loop is easily extended to the coaxial array when all the driven loops are fed at a common angle, e.g., $\phi = 0$ in Fig. 5-1. The current distribution in each loop is expressed as a series of trigonometric terms like that in Eq. (5-19). The simplicity of the analysis results from the orthogonality of the trigonometric terms which makes the coupling between loops occur only for terms of the same order n. Thus, if all the driven loops in the array are near resonant size, $\beta b \approx 1$, the term $n = 1$ is the dominant one in the current distributions for all loops; i.e., the current is approximately proportional to $\cos \phi$ in all loops.

When all the elements in the loop array are driven, the same procedures that are used with arrays of linear elements can be applied to select the driving-point voltages to optimize certain parameters, such as directivity.[19] The feed arrangement needed to obtain the prescribed driving-point voltages, however, is very complex for more than a few elements in the array. As a result, a simpler and more economical arrangement, an array containing only one driven element and several parasitic loops, is often used (a *parasitic loop* is a continuous wire with no terminals).

When a single closely spaced parasite is used with a driven loop, the parasite

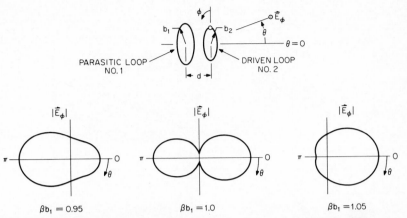

FIG. 5-20   Far-zone electric field patterns $|\vec{\mathbf{E}}_\phi|$ in plane $\phi = 0$, π for driven loop with single parasite, $\beta b_2 = 1.0$, $d/\lambda = 0.1$, $\Omega_1 = \Omega_2 = 20$.

may act as a director or as a reflector. This is illustrated in Fig. 5-20, in which electric field patterns are shown for a driven loop ($\beta b_2 = 1.0$) and a parasitic loop with the spacing $d/\lambda = 0.1$. For loops of the same electrical size ($\beta b_1 = \beta b_2 = 1.0$), the maxima in the pattern at $\theta = 0$, π are nearly equal. The parasitic loop that is slightly smaller than the driven loop ($\beta b_1 = 0.95$) acts as a director, producing a maximum in the pattern at $\theta = \pi$, while the parasitic loop that is slightly larger than the driven loop ($\beta b_1 = 1.05$) acts as a reflector, producing a maximum in the pattern at $\theta = 0$. This behavior is very similar to that observed for a resonant linear antenna with a closely spaced parasite.

The driven loop of electrical size $\beta b_2 = 1.2$ ($\Omega_2 = 11$, $a_1 = a_2$) with a single parasite was studied in detail by Ito et al.[20] In that study, the optimum director was determined to be a loop with $\beta b_1 \approx 0.95$ and spacing $d/\lambda \approx 0.10$; this produced a directivity of about 7 dB at $\theta = \pi$. The optimum reflector was a loop with $\beta b_1 \approx 1.08$ and a spacing $d/\lambda \approx 0.15$; this produced a directivity of about 8 dB at $\theta = 0$. Note that, for this case, the optimum director and the optimum reflector are both smaller than the driven loop.

A Yagi-Uda array of loops with a single reflector (element 1), an exciter (the driven element 2), and several directors of equal size βb and equal spacing d/λ is shown in the inset of Fig. 5-21.* As in its counterpart with linear elements, in the Yagi-Uda array of loops the reflector-exciter combination acts as a feed for a slow wave that propagates along the array of directors.[21] The lowest-order propagating wave (mode) exists for directors less than about resonant size ($\beta b \lesssim 1.0$) with spacings less than about a half wavelength ($d/\lambda \lesssim 0.5$).[22] An array supporting this mode has an end-fire pattern with a linearly polarized electric field at the maximum, $\theta = 0$.

The procedure for designing a Yagi-Uda array of loops is the same as for an

---

*In the literature of amateur radio the Yagi-Uda array of loops, usually square loops, is referred to as a *quad antenna*.

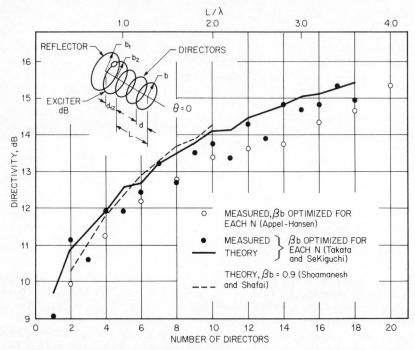

FIG. 5-21   Directivity of Yagi-Uda array of circular-loop antennas for $\theta = 0$ versus number of directors, director spacing $d/\lambda = 0.2$.

array with linear elements.[23] The isolated reflector-exciter combination is usually chosen to have maximum directivity in the direction $\theta = 0$. For example, the optimized two-element array described above might be used. The number, size, and spacing of the directors are then adjusted to obtain the desired performance, such as a specified end-fire directivity. The maximum end-fire directivity is determined by the electrical length of the array L/λ (L is the distance from the exciter to the last director). The larger the number of directors within the length L, the smaller the electrical size of the directors will be for maximum directivity, typically $0.8 \leq \beta b \leq 1.0$.

As an example, the directivity of a Yagi-Uda array of loops with the director spacing $d/\lambda = 0.2$ is shown as a function of the number of directors or the length of the array L/λ in Fig. 5-21. Two theoretical curves and two sets of measured data are shown. All the results agree to within about 1 dB, even though they are for different reflector-exciter combinations and slightly different director sizes.*

---

*The parameters used by the different investigators are: Appel-Hansen, $\beta b_1 = \beta b_2 = 1.10$, d_{12}/λ optimized for the isolated reflector-exciter, and βb optimized for each length L/λ; Takata and Sekiguchi, $\beta b_1 = 1.05$, $\beta b_2 = 1.20$, $d_{12}/\lambda = 0.15$, and βb optimized for each length L/λ; Shoamanesh and Shafai (1979), $\beta_1 b = 1.05$, $\beta b_2 = 1.10$, $d_{12}/\lambda = 0.1$, and $\beta b = 0.9$ for all lengths L/λ.

5-4 SHIELDED-LOOP ANTENNA

For certain applications, it is desirable to position the terminals of the loop antenna precisely so as to produce geometrical symmetry for the loop and its connections about a plane perpendicular to the loop. This can often be accomplished by using the so-called shielded loop; Fig. 5-22a is an example of a shielded receiving loop whose external surface is symmetrical about the yz plane.[24]

With reference to Fig. 5-22a, the thickness of the metal forming the shield is chosen to be several skin depths; this prevents any direct interaction between the currents on the internal and the external surfaces of the shield. The effective terminals of the loop antenna are at the ends of the small gap AB. The inner conductor and the

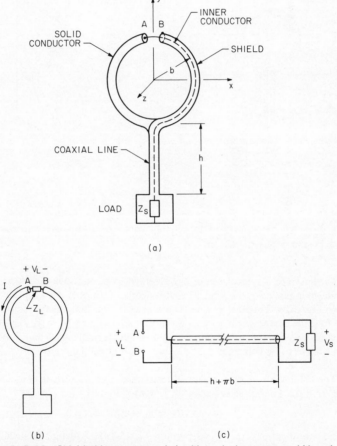

FIG. 5-22 Shielded-loop antenna (a) with equivalent antenna (b) and equivalent transmission line (c).

shield form a coaxial transmission line of length $h + \pi b$ connecting the gap with the load impedance Z_S. Thus, the effective load impedance Z_L at the gap is Z_S transformed by the length of transmission line $h + \pi b$.

The receiving antenna in Fig. 5-22a is easily analyzed by considering the loop, Fig. 5-22b, and the transmission line, Fig. 5-22c, separately. The incident field produces a current on the external surface of the shield; the current passes through the effective impedance Z_L, producing the voltage V_L, which for an electrically small loop can be determined from Eqs. (5-11) and (5-13). This voltage is transmitted over the coaxial line to become V_S at the load impedance Z_S.

Other examples of the shielded loop are shown in Fig. 5-23. A balanced version of the loop in Fig. 5-22a is in Fig. 5-23a, and a method for feeding a loop in front of a planar reflector is in Fig. 5-23b.

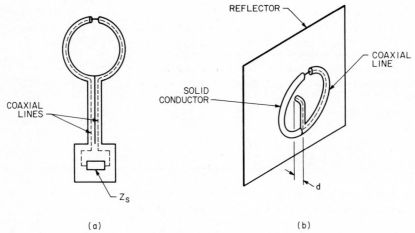

(a)   (b)

FIG. 5-23 (a) Balanced shielded-loop antenna and (b) method of feeding loop antenna in front of planar reflector.

To illustrate a typical use of the shielded loop, consider the electrically small receiving loop placed in an incident electromagnetic plane wave with the wave vector $\mathbf{k}_i$, as in Fig. 5-24. This is the same geometry as in Fig. 5-5, except that the terminals of the loop are at the angle $\phi = \phi_L$ instead of $\phi = 0$, and $\phi_i = \pi$, $\psi_i = 0$. The loop in this example might be an antenna in a direction finder with the direction of the incident wave to be determined by placing a null of the field pattern in the direction of $\mathbf{k}_i$.

The voltage at the open-circuited terminals of the electrically small loop, determined from the Fourier-series analysis, is approximately

$$V_{OC} = j\omega A B^i (\sin\theta_i - 2j\beta b \cos\phi_L) \qquad\qquad \textbf{(5-20)}$$

For many applications, the second term in Eq. (5-20) is negligible, since $\beta b \ll 1$ for an electrically small loop; in this event, Eq. (5-20) reduces to Eq. (5-11) with $N = 1$, $\psi_i = 0$. In other applications, however, this term may represent a significant contri-

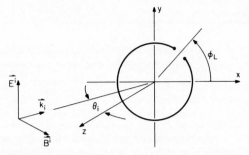

FIG. 5-24   Receiving loop in plane-wave incident field.

bution to the response. For example, the sensitivity of the antenna in the direction finder is decreased by this term because it fills in the nulls of the $\sin \theta_i$ field pattern (for $\beta b = 0.1$, $\phi_L = 0$, the minima in the pattern are only 14 dB below the maxima).

The second term in Eq. (5-20) can be made insignificant by reducing the electrical size of the loop βb; however, this will also decrease the sensitivity, since the area of the loop is decreased. An alternative is to make this term zero by placing the terminals of the loop precisely at $\phi_L = \pm \pi/2$ ($\cos \phi_L = 0$); this can be accomplished by using a shielded loop as in Fig. 5-22a or Fig. 5-23a.

5-5   ADDITIONAL TOPICS

The brevity of this review requires omission of many interesting topics concerning loop antennas. In recent years, there has been considerable study of loop antennas in close proximity to or embedded in material media such as the ocean, the earth, or a plasma. The electrical characteristics of loops in these instances can be quite different from those of loops in unbounded free space, as described in this review. The major applications of this work are in the areas of subsurface communication and detection (geophysical prospecting).

The loop antenna near a planar interface separating two semi-infinite material regions, such as the air and the earth, has been investigated extensively. When the loop is electrically small, it can be approximated by an elementary magnetic dipole, and the electromagnetic field away from the loop can be determined from the classical analysis of Sommerfeld.[25] If the field near the electrically small loop is required, the approximation by a magnetic dipole may no longer be adequate, and a loop with a finite radius and a uniform current must be considered.[26] For the electrically large loop near a planar interface, an analysis that allows a nonuniform current in the loop, such as the Fourier-series analysis for the circular loop,[27] must be used.

The performance of a loop embedded in a material can be altered significantly by placing the loop in a dielectric cavity, such as a sphere, to form an insulated loop. The electrical size and shape of the insulating cavity and the location of the loop in the cavity can be used to control the electromagnetic field and input impedance of the antenna.[28]

REFERENCES

1 H. Hertz, *Electric Waves,* Macmillan and Co., Ltd., London, 1893.

2 R. G. Medhurst, "H.F. Resistance and Self-Capacitance of Single-Layer Solenoids," *Wireless Eng.,* vol. 24, March 1947, p. 80. The measurements of Medhurst show that the self-resonance of a solenoid with $\ell_c/2b \geq 3$ occurs at a wavelength at which $N\beta b \geq 0.4$. The current distribution in the solenoid is assumed to be uniform well below self-resonance; i.e., $N\beta b \lesssim 0.1$.

3 The electrically small loop antenna in free space is discussed in many textbooks; see, for example, R. W. P. King, *Fundamental Electromagnetic Theory,* Dover Publications, Inc., New York, 1963, pp. 441–457; or S. A. Schelkunoff and H. T. Friis, *Antennas: Theory and Practice,* John Wiley & Sons, Inc., New York, 1952, pp. 319–324.

4 Formulas and graphs for the internal impedance per unit length of round conductors are in S. Ramo, J. R. Whinnery, and T. Van Duzer, *Fields and Waves in Communication Electronics,* John Wiley & Sons, Inc., New York, 1965, pp. 286–297.

5 G. S. Smith, "Radiation Efficiency of Electrically Small Multiturn Loop Antennas," *IEEE Trans. Antennas Propagat.,* vol. AP-20, September 1972, p. 656.

6 F. W. Grover, *Inductance Calculations: Working Formulas and Tables,* D. Van Nostrand Company, Inc., New York, 1946.

7 Internal resonance transverse to the axis of an infinitely long magnetic rod is discussed in L. Page, "The Magnetic Antenna," *Phys. Rev.,* vol. 69, June 1946, p. 645.

8 The graph in Fig. 5-8 was constructed by using the static demagnetizing factor for a cylindrical rod as presented in R. M. Bozorth and D. M. Chapin, "Demagnetizing Factors of Rods," *J. App. Phys.,* vol. 13, May 1942, p. 320; also R. M. Bozorth, *Ferromagnetism,* D. Van Nostrand Company, Inc., New York, 1951, pp. 845–849; and G. A. Burtsev, "Computing the Demagnetizion Coefficient of Cylindrical Rods," *Soviet J. Nondestructive Test. (Defektoskopiya),* vol. 5, September–October 1971, p. 499.

9 The receiving loop with a spheroidal core is discussed in R. E. Burgess, "Iron-Cored Loop Receiving Aerial," *Wireless Eng.,* vol. 23, June 1946, p. 172; J. R. Wait, "Receiving Properties of Wire Loop with a Spheroidal Core," *Can. J. Tech.,* vol. 31, January 1953, p. 9, and "The Receiving Loop with a Hollow Prolate Spheroidal Core," *Can. J. Tech.,* vol. 31, June 1953, p. 132; V. H. Rumsey and W. L. Weeks, "Electrically Small, Ferrite-Loaded Loop Antennas," *IRE Conv. Rec.,* part 1, 1956, p. 165; and E. J. Scott and R. H. DuHamel, "Effective Permeability of Spheroidal Shells," Tech. Rep. 9, Antenna Lab., University of Illinois, Urbana, 1956.

10 The customary procedure for analyzing the solenoidal coil with a cylindrical ferrite core is in H. van Suchtelen, "Ferroxcube Aerial Rods," *Electron. Appl. Bull.,* vol. 13, June 1952, p. 88. Extensions of this procedure and additional measured data are in J. S. Belrose, "Ferromagnetic Loop Aerials," *Wireless Eng.,* vol. 32, February 1955, p. 41; and J. Dupuis, "Cadres utilisant des ferrites," *L'onde électrique,* vol. 35, March–April 1955, p. 379.

11 E. C. Snelling, *Soft Ferrites: Properties and Applications,* CRC Press, Cleveland, 1969, pp. 182–192, 327–336.

12 There are many journal articles in addition to those in Refs. 9 through 11 that discuss the design and optimization of ferrite-loaded loop antennas for broadcast receivers; an incomplete list follows: H. Blok and J. J. Rietveld, "Inductive Aerials for Modern Broadcast Receivers," *Philips Tech. Rev.,* vol. 16, January 1955, p. 181; E. J. Maanders and H. van der Vleuten, "Ferrite Aerials for Transistor Receivers," *Philips Matronics Tech. Info. Bull.,* February 1961, p. 354; H. J. Laurent and C. A. B. Carvalho, "Ferrite Antennas for A.M. Broadcast Receivers," *IRE Trans. Broadcast Telev. Receivers,* vol. BTR-8, July 1962, p. 50; G. Schiefer, "A Small Ferroxcube Aerial for VHF Reception," *Philips Tech. Rev.,* vol. 24, 1962–1963, p. 332; I. D. Stuart, "Practical Considerations in the Design of Ferrite Cored Aerials for Broadcast Receivers," *IREE Proc. (Australia),* vol. 27, December

1966, p. 329; R. C. Pettengill, H. T. Garland, and J. P. Meindl, "Receiving Antenna Design for Miniature Receivers," *IEEE Trans. Antennas Propagat.*, vol. AP-25, July 1977, p. 528.

13   The ferrite-loaded transmitting loop is discussed in R. DeVore and P. Bohley, "The Electrically Small Magnetically Loaded Multiturn Loop Antenna," *IEEE Trans. Antennas Propagat.*, vol. AP-25, July 1977, p. 496.

14   The Fourier-series analysis for the circular loop has a long history dating back to the work of H. C. Pocklington in 1897 on the closed loop. Recent treatments and additional references are in R. W. P. King and G. S. Smith, *Antennas in Matter: Fundamentals, Theory, and Applications*, The M.I.T. Press, Cambridge, Mass., 1981, pp. 527–605; and R. W. P. King, "The Loop Antenna for Transmission and Reception," in R. E. Collin and F. J. Zucker (eds.), *Antenna Theory*, part I, McGraw-Hill Book Company, New York, 1969, pp. 458–482. Tables of input admittance are in R. W. P. King, *Tables of Antenna Characteristics*, Plenum Press, New York, 1971, pp. 151–160. The approach used by Japanese authors is described in N. Inagaki, T. Sekiguchi, and S. Ito, "A Theory of a Loop Antenna," *Electron. Commun. Japan*, vol. 53-B, March 1970, p. 62.

15   P. A. Kennedy, "Loop Antenna Measurements," *IRE Trans. Antennas Propagat.*, vol. AP-4, October 1956, p. 610.

16   The properties of a loop over an infinite image plane are obtained by using the theory of images and the analysis for an array of two loops; see, for example, K. Iizuka, R. W. P. King, and C. W. Harrison, Jr., "Self- and Mutual Admittances of Two Idential Circular Loop Antennas in a Conducting Medium and in Air," *IEEE Trans. Antennas Propagat.*, vol. AP-14, July 1966, p. 440.

17   S. Adachi and Y. Mushiake, "Directive Loop Antennas," *Res. Inst. Sci. Rep.*, ser. B, vol. 9, no. 2, Tōhoku University, Sendai, Japan, 1957, pp. 105–112.

18   B. Rojarayanont and T. Sekiguchi, "One-Element Loop Antenna with Finite Reflector," *Electron. Commun. Japan*, vol. 59-B, May 1976, p. 68.

19   The coaxial array of driven loops is discussed in M. Kosugi, N. Inagaki, and T. Sekiguchi, "Design of an Array of Circular-Loop Antennas with Optimum Directivity," *Electron. Commun. Japan*, vol. 54-B, May 1971, p. 67; and S. Ito, M. Kosugi, N. Inagaki, and T. Sekiguchi, "Theory of a Multi-Element Loop Antenna," *Electron. Commun. Japan*, vol. 54-B, June 1971, p. 95.

20   S. Ito, N. Inagaki, and T. Sekiguchi, "Investigation of the Array of Circular-Loop Antennas," *IEEE Trans. Antennas Propagat.*, vol. AP-19, July 1971, p. 469.

21   H. W. Ehrenspeck and H. Poehler, "A New Method for Obtaining Maximum Gain from Yagi Antennas," *IRE Trans. Antennas Propagat.*, vol. 7, October 1959, p. 379.

22   M. Yamazawa, N. Inagaki, and T. Sekiguchi, "Excitation of Surface Wave on Circular-Loop Array," *IEEE Trans. Antennas Propagat.*, vol. AP-19, May 1971, p. 433.

23   The design of Yagi-Uda arrays of loops is discussed in J. E. Lindsay, Jr., "A Parasitic End-Fire Array of Circular Loop Elements," *IEEE Trans. Antennas Propagat.*, vol. AP-15, September 1967, p. 697; J. Appel-Hansen, "The Loop Antenna with Director Arrays of Loops and Rods," *IEEE Trans. Antennas Propagat.*, vol. AP-20, July 1972, p. 516; L. C. Shen and G. W. Raffoul, "Optimum Design of Yagi Array of Loops," *IEEE Trans. Antennas Propagat.*, vol. AP-22, November 1974, p. 829; N. Takata and T. Sekiguchi, "Array Antennas Consisting of Linear and Loop Elements," *Electron. Commun. Japan*, vol. 59-B, May 1976, p. 61; A. Shoamanesh and L. Shafai, "Properties of Coaxial Yagi Loop Arrays," *IEEE Trans. Antennas Propagat.*, vol. AP-26, July 1978, p. 547; and A. Shoamanesh and L. Shafai, "Design Data for Coaxial Yagi Array of Circular Loops," *IEEE Trans. Antennas Propagat.*, vol. AP-27, September 1979, p. 711.

24   The shielded loop is discussed in L. L. Libby, "Special Aspects of Balanced Shielded Loops," *IRE Proc.*, vol. 34, September 1946, p. 641; and R. W. P. King, "The Loop Antenna for Transmission and Reception," in R. E. Collin and F. J. Zucker (eds.), *Antenna Theory*, part I, McGraw-Hill Book Company, New York, 1969, pp. 478–480. The shielded half loop as a current probe is treated in R. W. P. King and G. S. Smith, *Antennas in*

Matter: Fundamentals, Theory and Applications, The M.I.T. Press, Cambridge, Mass., 1981, pp. 770–787.

25 The analysis of elementary vertical and horizontal magnetic dipoles near a planar interface is discussed in A. Sommerfeld, *Partial Differential Equations in Physics,* Academic Press, Inc., New York, 1949, pp. 237–279; A. Baños, Jr., *Dipole Radiation in the Presence of a Conducting Half-Space,* Pergamon Press, New York, 1966; and J. R. Wait, *Electromagnetic Waves in Stratified Media,* Pergamon Press, New York, 1970.

26 J. Ryu, H. F. Morrison, and S. H. Ward, "Electromagnetic Fields about a Loop Source of Current," *Geophysics,* vol. 35, October 1970, p. 862; J. R. Wait and K. P. Spies, "Subsurface Electromagnetic Fields of a Circular Loop of Current Located above Ground," *IEEE Trans. Antennas Propagat.,* vol. 20, July 1972, p. 520; J. R. Wait and K. P. Spies, "Low-Frequency Impedances of a Circular Loop over a Conducting Ground," *Electron. Lett.,* vol. 9, July 26, 1973, p. 346.

27 L. N. An and G. S. Smith, "The Horizontal Circular Loop Antenna near a Planar Interface," *Radio Sci.,* vol. 17, May–June 1982, p. 483.

28 Bare and insulated electrically small loop antennas in dissipative media are discussed in J. R. Wait, "Electromagnetic Fields of Sources in Lossy Media," in R. E. Collin and F. J. Zucker (eds.), *Antenna Theory,* part II, McGraw-Hill Book Company, New York, 1969, pp. 438–514, and references therein. Bare and insulated loop antennas of general size are treated in R. W. P. King and G. S. Smith, *Antennas in Matter: Fundamentals, Theory and Applications,* The M.I.T. Press, Cambridge, Mass., 1981, pp. 527–605; and L. N. An and G. S. Smith, "The Eccentrically Insulated Circular Loop Antenna," *Radio Sci.,* vol. 15, November–December 1980, p. 1067, and vol. 17, May–June 1982, p. 737.

Chapter 6

Small Antennas

Harold A. Wheeler

Hazeltine Corporation

A *small antenna* is here defined as an antenna occupying a small fraction of one radiansphere in space. Typically its greatest dimension is less than one-quarter wavelength (including any image in a ground plane). Some of its properties and its available performance are limited by its size and by the laws of nature. An appreciation of these limitations has proved helpful in arriving at practical designs. This chapter is a revision of the author's 1975 paper summarizing the subject.[1]

The radiansphere is the spherical volume having a radius of $1/2\pi$ wavelength.[2] It is a logical reference here because, around a small antenna, it is the space occupied mainly by the stored energy of its electric or magnetic field. A small antenna is essentially an electric dipole C or a magnetic dipole L, or possibly a combination of both.[3,4]

Some limitations are peculiar to a passive network, in which the concepts of efficiency, impedance matching, and frequency bandwidth are essential and may be the controlling factors in performance evaluation. This discussion is directed mainly to these limitations in relation to small size. It centers in the term *radiation power factor* and its proportionality to volume.[3]

Figure 6-1 shows the principles of a small antenna exemplified by an electric dipole C. Its impedance over a bandwidth can be represented by a dummy antenna of constant parameters as follows:

- The principal reactance is $X_e = 1/\omega C$.
- The (much smaller) radiation resistance is $R_e \propto \omega^2$ (not indicated).
- The radiation resistance is simulated by an inductive reactance ($\omega L \ll X_e$) having in parallel a much greater constant resistance ($R \gg \omega L$). They are proportioned so that the radiation is represented by their effective series resistance, $R_e = (\omega L)^2/R \propto \omega^2$.
- The resulting (small) radiation power factor becomes:

$$p_e = R_e/X_e = \omega^3 CL^2/R \ll 1 \qquad \textbf{(6-1)}$$

These relations refer to the limiting case of a small antenna. In practice, they are taken to represent the behavior at the lowest frequency (ω_1) of an operating bandwidth. They may be relaxed at the highest frequency ($\omega_2 \gg \omega_1$) in the case of wideband operation with fixed tuning.

Figure 6-1 shows also the double tuning of the small antenna in a manner that is common and will receive further attention. It may be used to obtain a useful degree

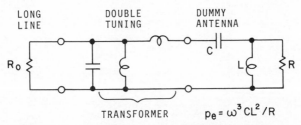

FIG. 6-1 A small antenna of C type with double tuning.

of matching of the antenna with a long line (simulated by R_0) connecting it with a transmitter and/or a receiver.

The design considerations will be presented, after which a variety of small antennas that are useful for some purposes will be described.

6-2   THE RADIATION POWER FACTOR

The term *radiation power factor* (PF) is a natural one introduced by the author in 1947.[3] It is descriptive of the radiation of real power from a small antenna taking a much larger value of reactive power. The term is applicable to either kind of reactor, and its (small) value is limited by some measure of the size in either kind.

Here PF is equal to $1/Q$, in the common parlance of networks. The term PF is preferred for describing radiation and losses because it is additive. Furthermore, the radiation PF is a positive description of what is desired (the opposite of Q).

The operating efficiency of a small antenna is limited by its radiation PF, which is proportional to its size. The nature of this limitation depends on the relative bandwidth of operation as compared with the radiation PF.

Radiation efficiency in the utilization of an antenna is relevant to transmission or reception and is defined as follows:

- In transmission, it is the fraction of available power from a generator that is radiated into space.
- In reception, it is the fraction of available power from space that is delivered to a load representing the receiver. It is a measure of the ability of a received signal to overcome the noise level in the circuits.

The generator or load is taken to have pure resistance (R_0), which may be the wave resistance of a transmission line connecting with the antenna. In either case, the greatest efficiency of power transfer requires the familiar impedance matching suited to any situation.

Figure 6-2 shows the circuit properties of the two types of small antenna, which are identified as the electric dipole C and the magnetic dipole L. The former behaves as a capacitor and the latter as an inductor. Either is to be resonated by a reactor of the opposite kind.

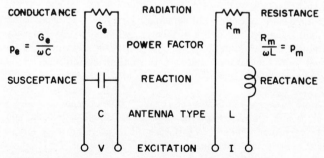

FIG. 6-2 The radiation power factor of a small antenna.

Representation of the radiation is shown in a manner consistent with circuit *duality*. (This is not the usual manner for *C*.) The radiation parameter is G_e or R_m, either one $\propto \omega^4$. Then the (small) radiation PF $\propto \omega^3$. As will be described, the power factor is proportional to the volume and may be evaluated in terms of *effective volume*.

Some amount of radiation PF (and hence size) may be required to achieve some measure of performance, such as radiation efficiency in transmission or its equivalent in reception. The following two situations are indicative of the two extremes:

- In narrowband operation, the relative bandwidth of operation is taken to be less than the radiation PF. Then efficiency is limited by dissipation or heat losses (loss PF) in the antenna and associated tuning reactors.

- In wideband operation, the relative bandwidth is taken to be much greater than the radiation PF. Then efficiency is limited by the ability of a passive network of fixed reactors to match the antenna to a fixed resistance.

Each of these situations will be developed by evaluating the efficiency of a model and by stating the radiation PF required for achieving this efficiency. Then a design procedure which can be adapted to an antenna of any type will be presented.

6-3 EFFICIENCY OVER LOSSES

With reference to Fig. 6-2, either type of antenna may be tuned to one frequency with an opposite reactor. Because the radiation PF is small, the efficiency may be reduced substantially by a small loss PF in the antenna and the tuning reactor.[3] From the transmitter viewpoint, it is assumed that the available power can be delivered to the tuned circuit including the antenna. Then the efficiency is simply stated:

$$\text{Radiation efficiency} = \frac{\text{radiation PF}}{\text{radiation PF} + \text{loss PF}} \qquad \textbf{(6-2)}$$

In general, greater size yields greater efficiency because it increases the radiation PF and decreases the loss PF. Conversely, a specified efficiency imposes a requirement of some size.

This situation is relevant to an antenna which has fixed or adjustable tuning to a narrowband signal. A transmitter of this type will realize this efficiency.

A simple case of this situation is the proximity fuse, in which the tuned antenna is integrated in an oscillator serving as both transmitter and receiver. The radiation efficiency becomes a measure of the reaction of a nearby object on the oscillator amplitude.

A simple case in a receiver is the tuned ferrite-core inductor used as a built-in antenna. Although the radiation efficiency is very small, it may be adequate for a purpose. In very-low-frequency (VLF) reception, the radio noise temperature is so great that the very small efficiency may be all that is useful.

6-4 EFFICIENCY OVER BANDWIDTH

A small antenna may be required to operate over a relative bandwidth (BW) much greater than its radiation PF. Matching with a line cannot be efficient over the entire BW. This is the subject of Ref. 5; it will be described here only briefly.

Over any BW (ω_1 to ω_2) the radiation PF at the low cutoff (p_1 at ω_1) is the most significant property of a small antenna for impedance matching. The matching efficiency e at any frequency is the fraction of available power that is delivered from generator to load through the matching network. For double tuning, as exemplified in Fig. 6-1, the matching efficiency may vary with frequency in the manner shown in Fig. 6-3.

The objective here is taken to be the highest-level floor of e over the BW. This may be termed the *maximin* of efficiency, as indicated. It is realized by minimizing

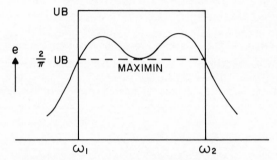

FIG. 6-3 The wideband matching efficiency.

the "useless" excess within the BW and outside. The maximin obtainable with double tuning is found to be

$$e = \frac{12p_1}{1 - (\omega_1/\omega_2)^3} \ll 1 \qquad p_1 = \frac{e}{12}\,[1 - (\omega_1/\omega_2)^3] \qquad \textbf{(6-3)}$$

A slightly higher level is obtainable by a higher order of tuning, but the theoretical upper bound (UB) is only $\pi/2$ times this level. For a required efficiency and BW, the second form gives the required radiation PF and hence the size.

6-5 THE EFFECTIVE VOLUME

For any shape of small antenna of either kind (C or L), the radiation PF at one frequency is proportional to the volume. Moreover, it is nearly equal for the two kinds if they occupy nearly equal volume. This statement needs explanation, because their configuration differs in accordance with the properties of different materials.

Figure 6-4 shows examples of the two kinds configured to occupy the space within cylinders of equal dimensions.[1,3] From familiar formulas for C, L, and radiation resistance, they have equal values of radiation PF except for two factors (k_a, k_b) which

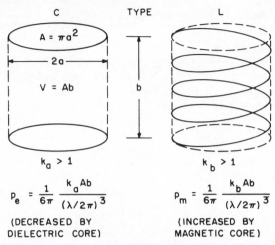

C   TYPE   L

$A = \pi a^2$

$2a$

$V = Ab$

b

$k_a > 1$

$$p_e = \frac{1}{6\pi} \frac{k_a Ab}{(\lambda/2\pi)^3}$$

(DECREASED BY
DIELECTRIC CORE)

$k_b > 1$

$$p_m = \frac{1}{6\pi} \frac{k_b Ab}{(\lambda/2\pi)^3}$$

(INCREASED BY
MAGNETIC CORE)

FIG. 6-4   The radiation power factor in terms of volume.

are somewhat greater than unity. Either of these factors multiplies the volume ($V = Ab$) to give the *effective volume* as here defined (V').

In Fig. 6-4, the effective volume is compared with the radian cube $(\lambda/2\pi)^3$.[3] It is more logical to compare it with the volume of the radiansphere:[1]

$$V_s = \frac{4\pi}{3}\left(\frac{\lambda}{2\pi}\right)^3 \tag{6-4}$$

Within this sphere, the stored energy or reactive power is predominant. Outside this sphere, the radiated power is predominant.

In terms of the effective volume ($V' = k_a Ab$ or $k_b Ab$ in Fig. 6-4), the radiation becomes

$$p = \tfrac{3}{8}V'/V_s \tag{6-5}$$

(The coefficient $\tfrac{3}{8}$ reflects some properties of the near field of either antenna.) The effective volume may be stated as a sphere of radius a':

$$V' = \frac{4\pi}{3}a'^3 \qquad p = \frac{2}{9}\left(\frac{2\pi a'}{\lambda}\right)^3 \qquad a' = \frac{\lambda}{2\pi}\left(\frac{9}{2}p\right)^{1/3} \tag{6-6}$$

It is noted that a certain shape of self-resonant coil radiates equally from both C and L, so the total radiation is double that of either one.[4] This is known as the helix radiator of circular polarization in the normal mode.

There is one theoretical case of a small antenna which has the greatest radiation PF obtainable within a spherical volume. Figure 6-5 shows such an antenna and its relation to the radiansphere (V_s).[1,2] It is a spherical coil with a perfect magnetic core. The effective volume of an empty spherical coil has a shape factor $\tfrac{3}{2}$. Filling with a perfect magnetic core ($k_m = \infty$) multiplies the effective volume by 3.

$$p_m = \frac{2}{9}\frac{(3)(3/2)\,V}{V_s} = \frac{V}{V_s} = \left(\frac{2\pi a}{\lambda}\right)^3 \tag{6-7}$$

This is indicated by the shaded sphere a.

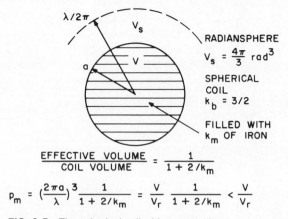

FIG. 6-5   The spherical coil with a magnetic core.

This idealized case depicts the physical meaning of the radiation PF that cannot be exceeded. Outside the sphere occupied by the antenna, there is stored energy or reactive power that conceptually fills the radiansphere, but there is none inside the antenna sphere.[6] The reactive power density, which is dominant in the radiation within the radiansphere, is related to the real power density, which is dominant in the radiation outside.

In a rigorous description of the electromagnetic field from a small dipole of either kind, the radiation of power in the far field is accompanied by stored energy which is located mostly in the near field (within the radiansphere).[2] The small spherical inductor in Fig. 6-5 is conceptually filled with perfect magnetic material, so there is no stored energy inside the sphere. This removes the *avoidable* stored energy, leaving only the *unavoidable* amount outside the inductor but mostly inside the radiansphere. This unavoidable stored energy is what imposes a fundamental limitation on the obtainable radiation PF.

For any actual antenna, the effective volume and its spherical radius are

$$V' = \frac{9}{2} p V_s \qquad a' = \frac{\lambda}{2\pi} \left(\frac{9}{2} p\right)^{1/3} \qquad \textbf{(6-8)}$$

This volume includes any image in an adjoining ground plane regarded as integral with the antenna. It is convenient to show any antenna configuration with its sphere of effective volume drawn to scale, as a rating of its radiation PF.

6-6   THE RADIATION SHIELD

It is difficult to measure the small radiation PF, and it is especially difficult to separate it from the loss PF of the antenna and its tuning reactor. The *radiation shield* was devised to separate the two.[6]

Figure 6-6 shows the concept of the radiation shield. Its purpose is to preserve the near field in the radiansphere while avoiding radiation farther out. Ideally, it is a

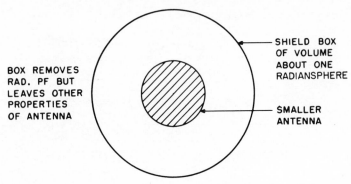

FIG. 6-6 The radiation shield for use in measuring the radiation power factor.

spherical box (of high conductivity) whose radius is $\lambda/2\pi$. Its dissipation is negligible as compared with free-space radiation.

The PF of the tuned antenna is measured with and without the shield, to give the ratio

$$\frac{\text{Loss PF only}}{\text{Loss PF + radiation PF}} \qquad\qquad \textbf{(6-9)}$$

From this ratio, the radiation PF can be evaluated as a fraction of the measured total. [See Eq. (6-2).]

The radiation shield is not critical as to size or shape. An open-ended circular or square cylinder is usually convenient. It should be large enough to avoid much disturbance of the field near the antenna and small enough to avoid cavity resonance, especially in any mode excited by the antenna. If open-ended, it should be long enough to attenuate radiation outside. A small shift of antenna resonance frequency is tolerable.

6-7 DESIGN PROCEDURE

In the design of a small antenna for operation over a frequency band (ω_1 to ω_2) a typical objective is one of these two:

- From a specified antenna, obtain the greatest efficiency by a practical circuit or less (as by economies) if sufficient for the purpose.
- To obtain specified efficiency, determine a practical circuit and antenna configuration that will require the smallest size.

For either objective, a design procedure will be outlined. Here we ignore the dissipation in the reactors of the antenna and matching network.

The properties of a specified antenna are evaluated by computation and/or measurement. The principal properties are the essential C or L and the radiation PF. The reactance is the dominant factor in designing a lossless matching network such as the double tuning in Fig. 6-1. The result to be expected is the double-peak graph in Fig. 6-3 and the matching efficiency from Eq. (6-3).

The other objective is essentially the reverse. The required efficiency is specified, and perhaps also a tolerance of the voltage standing-wave ratio (VSWR). The network configuration is specified as a practical constraint on complication. Figure 6-1 is taken as an example.

From the specified efficiency and BW, the required radiation PF can be computed by Eq. (6-3). The required effective volume is then given by Eq. (6-8), also its spherical radius a'. These are stated in terms of the radiansphere and radianlength at the lowest frequency (ω_1).

For any kind and shape of antenna, size is related to the effective volume or spherical radius. The latter will be shown to scale for some typical configurations.

6-8   TYPICAL SMALL ANTENNAS

A number of typical small antennas are here compared with the effective volume by diagrams showing to scale the sphere of this radius:

$$a' = \frac{\lambda}{2\pi}\left(\frac{9}{2}p\right)^{1/3} \tag{6-10}$$

It is drawn as a dashed circle.

Figure 6-7 shows some examples of an electric dipole with a linear axis of symmetry. A thin wire (a) and a thick conical conductor (b) differ greatly in occupied volume but much less in effective volume. The latter is influenced most by length and less by the smaller transverse dimensions.

Figure 6-7c shows a pair of separated disks like the basic electric dipole in Fig. 6-4. Their simple rating can be preserved by the use of a tuning inductor in the form of a coil distributed between the ends as shown.

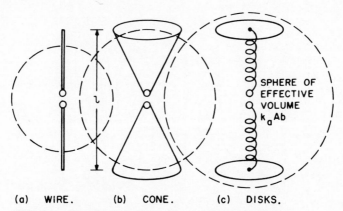

(a)   WIRE.   (b)   CONE.   (c)   DISKS.

FIG. 6-7   The effective volume of an axial electric dipole.

With reference to Fig. 6-4 for a pair of disks, the area factor and the effective volume are (within 5 percent)

$$k_a = 1 + \frac{4}{\pi} b/a \qquad V' = k_a \pi a^2 b = \pi a^2 b + 4ab^2 > 2\sqrt{4\pi a^3 b^3} \qquad \textbf{(6-11)}$$

Converting the last expression to the lower bound (LB) of effective radius (within 2 percent) yields

$$\text{LB } a' = 1.19 \sqrt{ab} \qquad \textbf{(6-12)}$$

This simple formula, being noncritical as to shape, is useful for estimating purposes. For the shape shown in Fig. 6-7c, the sphere diameter is slightly greater than the length.

The thin disks shown in Figs. 6-4 and 6-7c do not make the fullest use of the occupied cylindrical space. The fullest use would be made by adding to each disk a skirt going about halfway to the neutral plane of symmetry. The decrease of effective length would be more than compensated by the increase of C, to give greater effective volume.

Figure 6-8 shows some examples of a loop inductor on a square frame. A thin wire (a) and a wide strip (b) differ rather little in effective volume, because this is influenced most by the size of the square. A multiturn loop (c) has nearly the same effective volume as one turn occupying the same space. This is one of the principal conclusions presented in the writer's first paper.[3]

It happens that a cubic coil has a simple length factor ($k_b = \%$). The diameter of the sphere of effective volume is nearly equal to the diagonal of one face. Likewise, a spherical coil has an effective volume $\%$ times its enclosed volume.[6]

The effective volume of a square or circular coil may be stated by comparing the sphere radius with the coil radius (a = radius of circle or half side of square):

Circle: $\text{LB } a' = 1.12 \, a(b/a)^{1/6}$ $\textbf{(6-13)}$

Square: $\text{LB } a' = 1.25 \, a(b/a)^{1/6}$ $\textbf{(6-14)}$

As might be expected, the radiation PF is determined mainly by the radius.

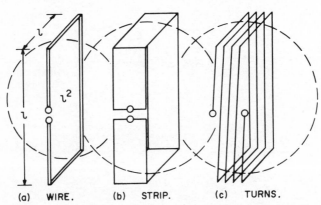

(a)   WIRE.   (b)   STRIP.   (c)   TURNS.

FIG. 6-8   The effective volume of a square loop.

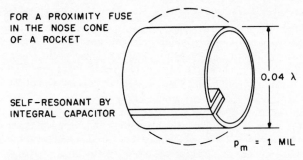

FOR A PROXIMITY FUSE
IN THE NOSE CONE
OF A ROCKET

0.04 λ

SELF-RESONANT BY
INTEGRAL CAPACITOR

p_m = 1 MIL

FIG. 6-9   A one-turn loop of wide strip.

Figure 6-9 shows a particularly effective small antenna with integral tuning. It is a one-turn loop of a wide strip which has remarkably small loss PF. In an extremely small size, its radiation efficiency was found to be about 50 percent.

One-half of any of these configurations can be imaged in a ground plane. The result is an equal effective volume, which may be represented as a hemisphere on the plane.

6-9   FLUSH ANTENNAS

A useful family of small antennas comprises those that are recessed in a shield surface such as a ground plane or the skin of an aircraft. Some may embody inherently flush designs, while others may be suited for operation adjacent to a shield surface, whether recessed or not. The antenna may be of the C or the L type, either radiating in a polarization compatible with the shield surface.

Figure 6-10 shows a flush disk capacitor (sometimes termed an *annular slot*). This capacitor in the flush mounting may be compared with a like capacitor just above the surface. The recessing somewhat reduces the radiation PF. The remaining effec-

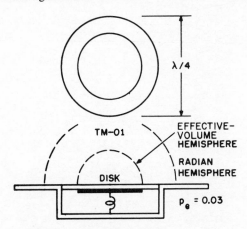

λ/4

TM-01

EFFECTIVE-
VOLUME
HEMISPHERE

RADIAN
HEMISPHERE

DISK

p_e = 0.03

FIG. 6-10   A flush disk capacitor.

tive volume is that of a hemisphere indicated by the dashed semicircle. Its size is comparable with that of the disk. The cylindrical walls may be regarded as a short length of waveguide beyond cutoff, operating in the lowest transverse magnetic (TM) mode (circular TM-01, as shown, or rectangular TM-11). The capacitor may be resonated by an integral inductor as shown. In any cavity, there is a size and shape of disk that can yield the greatest radiation PF. The primary factor is the size of the cavity.

The evaluation of a flush antenna includes the shield surface. It is necessary first to evaluate the radiation PF by some method of computation. Then the radiation PF can be stated in terms of a volume ratio. Here we consider the half space of radiation and show the hemisphere of $\frac{1}{2}V'$, which may then be compared with the half radiansphere, $\frac{1}{2}V_s$. The radii are retained (a' and $\lambda/2\pi$). An antenna located on the surface (not recessed) could be considered with its image to yield the complete sphere of V' to be compared with the radiansphere V_s. Then one-half of each may be shown above the shield plane, as for the flush antenna.

The disk capacitor radiates in the same mode as a small vertical electric dipole, by virtue of vertical electric flux from the disk. This is vertical polarization on the plane of the shield, with omnidirective radiation. The other examples of a flush antenna, to be shown here, radiate as a small horizontal magnetic dipole by virtue of magnetic flux leaving the cavity on one side and returning on the other side. This is vertical polarization but directive in a figure-eight pattern. Omnidirective radiation can be provided by quadrature excitation of two crossed modes in the same cavity. The radiation PF of either kind is reduced by recessing, but the magnetic dipole suffers less reduction.

Figure 6-11 shows an idealized cavity resonator which radiates as an inductor. The cavity is covered by a thin window of high-k dielectric, which serves two purposes. It completes the current loop indicated by the arrows I, and it also provides, in effect, series capacitance which resonates the current loop. The cylindrical walls and the aperture excitation may be regarded as the lowest (cutoff) transverse electric (TE) mode (circular TE-11 or rectangular TE-10 to TE-01, as shown). Each of these modes has two crossed orientations, of which one is indicated by the current loop. The con-

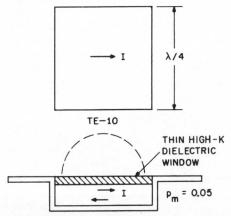

FIG. 6-11 A flush cavity inductor with a dielectric window.

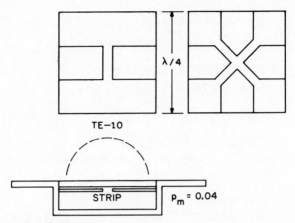

TE—10

STRIP $P_m = 0.04$

FIG. 6-12 A flush strip inductor.

tinuous dielectric sheet on a square or circular cavity resonates the two crossed modes. Because each resonance is in the lowest mode, it involves the smallest amount of stored energy relative to radiated power and therefore the greatest value of radiation PF.

Figure 6-12 shows some practical designs which yield nearly the same performance by the use of conductive strips on ordinary (low-k) dielectric windows (high-k dielectric is not required). Here the radiating inductor (strip) and the resonating series capacitor (gap) are apparent. The two alternatives, one mode and a pair of crossed modes, are shown. Practical designs about $\lambda/4$ square have been made with radiation PF about 0.04. This is about the largest size that follows the rules of a small antenna.

The required coupling with any of the resonant antennas in Figs. 6-10 to 6-12 may be provided by another (smaller) resonator located within the cavity. This enables the bandwidth of matching expected from double tuning. Each of these antennas is suited for self-resonance and requires some depth of cavity to hold down the extra amount of energy storage in this nonradiating space.

Figure 6-13 shows a flush inductor made of crossed coils on a thin magnetic disk. At medium or low frequencies (MF, LF, VLF) the available ferrite materials can provide a magnetic core that forms a return path nearly free of extra energy storage even in the thin disk. It also adds very little dissipation. The required depth of cavity is then only sufficient to take the disk thickness with some margin. The antenna is too small relative to the wavelength at the lower frequencies to enable high efficiency even at its frequency of resonance, so it is useful only for reception. A rotary coil or crossed coils can be used for a direction finder or omnidirectional reception. The principal application is on the skin of an aircraft.

Figure 6-14 shows the ferrite-rod inductor, the antenna most commonly used in a small broadcast receiver (MF, around 1 MHz). The ferrite rod greatly increases the effective volume of a thin coil, as indicated. The effective volume is then determined primarily by the length rather than the diameter of the coil. Like the ferrite disk, this antenna can be used close (parallel) to a shield surface or be recessed in the surface.

Here we may note that a long coil, with its small shape factor ($k_b \rightarrow 1$), can have its effective volume greatly increased by a ferrite core. On the other hand, a parallel-plate capacitor, with its small shape factor ($k_a \rightarrow 1$), can only have its effective volume decreased by a dielectric core. This is one respect in which the inductor offers greater

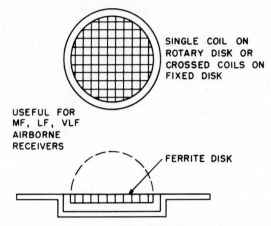

SINGLE COIL ON
ROTARY DISK OR
CROSSED COILS ON
FIXED DISK

USEFUL FOR
MF, LF, VLF
AIRBORNE
RECEIVERS

FERRITE DISK

FIG. 6-13   A flush inductor on a thin ferrite disk.

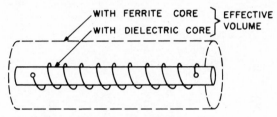

WITH FERRITE CORE ⎱ EFFECTIVE
WITH DIELECTRIC CORE ⎰ VOLUME

MAY BE LOCATED NEAR A SHIELD PLANE (OR FLUSH)
NEARLY DOUBLING EFFECTIVE VOLUME (BY IMAGE EFFECT)

FIG. 6-14   A long coil on a ferrite rod.

opportunity in design. In another respect, the number of turns can be used to set the impedance level, a freedom that may be desired but is unavailable in a simple capacitor.

If a long coil as a magnetic dipole were filled with perfect magnetic material, its effective volume would be comparable with that of an equally long conductor as an electric dipole.

6-10   ANTENNAS FOR VLF

The greater the wavelength, the more relevant may be the concept of a small antenna. Current activities go as low as 10 kHz with a wavelength of 30 km. Even the largest of the transmitter antennas is small in terms of this wavelength, or its radianlength of 5 km.

For reception, however, a much smaller antenna is adequate for one of two reasons:

● Above the surface (ground or sea) the radio noise level is so high that it still becomes the limiting factor in a small antenna with very low efficiency of radiation.

- Below the surface, in salt water like the ocean, the radianlength (equal to the skin depth) is only a few meters, so a small antenna may occupy a substantial fraction of the radiansphere.

For the former purpose, the magnetic antennas in Figs. 6-13 and 6-14 are suited for VLF. For the latter purpose, different factors become relevant.

The principles of the smallest effective underwater receiving antenna will be formulated with reference to Fig. 6-15. It is similar to a retractable antenna used on a submarine, which may be submerged to a small depth. This is an inductor in a hollow cavity (radome). It has greater radiation efficiency than a capacitor because it is subject to smaller near-field losses caused by the conductivity of the water. Also, it avoids the need for conductive contact with water.

Figure 6-15 shows an idealized small antenna in a submarine cavity.[6,7] It is a spherical coil with a magnetic core, as shown in Fig. 6-5. In the water, the radianlength is equal to the skin depth (δ). At 15 kHz, this is about 2 m. The size of the cavity is much less and that of the coil still less, so it is a small antenna in this environment. The radiation PF includes two quantities, the desired coupling to the medium and the undesired dissipation in the medium. The former is proportional to the coil volume and is increased by the magnetic core. The latter is decreased by increasing the cavity radius. The coil is in the vertical plane for vertical polarization. Crossed coils may be used for omnidirective reception and direction finding.

For a transmitter antenna, the capacitor is usually chosen for vertically polarized radiation in all directions on the horizon.

For efficient transmission at lower frequencies, one of the early simple types is that shown in Figure 6-16.[8] It is a flattop grid of wires forming a capacitor with ground as the lower conductor. The effective height h determines the radiation resistance. The capacitance enables the statement of an effective area ($k_a A$) as noted. The effective volume ($k_a A h$) in half space is compared with one-half radiansphere to determine the radiation PF. It is notable that the grid of many wires may provide an effective area greater than that of the grid despite the much smaller area of the conductor.

A large transmitting antenna for radiating high power presents different prob-

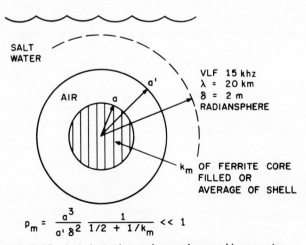

$$p_m = \frac{a^3}{a' \delta^2} \frac{1}{1/2 + 1/k_m} \ll 1$$

FIG. 6-15 An inductor in a radome submerged in seawater.

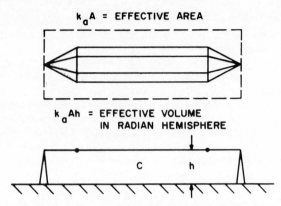

FIG. 6-16 A large flattop capacitor which is still small relative to the wavelength.

lems relating to power, current, and voltage. The compromise solutions of these problems require size as measured in these three ways:

- Effective volume for radiation PF to operate over a bandwidth and to compete with any loss PF
- Effective height squared for radiation of power proportional to current squared
- Effective area for radiation of power proportional to voltage squared

The first is the principal topic of this chapter. The second and third have been presented in Ref. 8 and are stated here as design requirements.

$$h = \frac{\lambda}{2\pi I} \sqrt{P/40} \qquad k_a A = \pi \left(\frac{\lambda}{2\pi}\right)^2 \left(\frac{3}{V}\right) \sqrt{40P} \qquad \textbf{(6-15)}$$

For example, the last relation states the effective area required if the voltage is limited by supporting insulators. If, in addition, the average voltage gradient on the wires is restricted by corona discharge to a value E_a (rms volts/meter), the required surface area of the wires (A_a) is proportional to the current (or $\sqrt{P}$):

$$A_a = \pi \left(\frac{\lambda}{2\pi}\right)^2 \frac{3}{hE_a} \sqrt{40P} \qquad \textbf{(6-16)}$$

Large VLF antennas are the principal topic of Chap. 24. Some are described in Refs. 1 and 9. The largest are the Navy stations at Cutler, Maine (NAA), and North West Cape, Australia (NWC). These qualify as small antennas by a large margin at their lowest frequencies of operation.

6-11 SYMBOLS

$p = 1/Q$ = radiation power factor (PF)

X = reactance

C = capacitance of a small electric dipole antenna

L = inductance of a small magnetic dipole antenna

R = series resistance

G = parallel conductance

$\omega = 2\pi f$ = radian frequency

$\lambda/2\pi$ = radianlength

λ = wavelength

$e = 1 - \rho^2$ = matching efficiency

A = area of cylinder

a = radius of cylinder or sphere

a' = effective radius of sphere of effective volume

b = axial length of cylinder

$V_s = (4\pi/3)(\lambda/2\pi)^3$ = volume of radiansphere

V' = effective volume of C or L

k_a = shape factor for effective area of a cylindrical capacitor

k_b = shape factor for effective axial length of a cylindrical or spherical inductor

k_m = magnetic constant (relative permeability)

REFERENCES

1 H. A. Wheeler, "Small Antennas," *IEEE Trans. Antennas Propagat.*, vol. AP-23, July 1975, pp. 462–469. Comprehensive review; background for this chapter.

2 H. A. Wheeler, "The Radiansphere around a Small Antenna," *IRE Proc.*, vol. 47, August 1959, pp. 1325–1331. Ideal sphere inductor; radiation shield.

3 H. A. Wheeler, "Fundamental Limitations of Small Antennas," *IRE Proc.*, vol. 35, December 1947, pp. 1479–1484. First paper on the radiation power factor of C and L radiators of equal volume.

4 H. A Wheeler, "A Helical Antenna for Circular Polarization," *IRE Proc.*, vol. 35, December 1947, pp. 1484–1488. Coil with equal E and M radiation PF.

5 H. A. Wheeler, "The Wideband Matching Area of a Small Antenna," *IEEE Trans. Antennas Propagat.*, to be published in 1983.

6 H. A. Wheeler, "The Spherical Coil as an Inductor, Shield or Antenna," *IRE Proc.*, vol. 46, September 1958, pp. 1595–1602; correction, vol. 48, March 1960, p. 328. Ideal sphere inductor; submarine coil.

7 H. A. Wheeler, "Fundamental Limitations of a Small VLF Antenna for Submarines," *IRE Trans. Antennas Propagat.*, vol. AP-6, January 1958, pp. 123–125. Inductor in a cavity; radiation power factor.

8 H. A. Wheeler, "Fundamental Relations in the Design of a VLF Transmitting Antenna," *IRE Trans. Antennas Propagat.*, vol. AP-6, January 1958, pp. 120–122. Effective area; radiation power factor.

9 A. D. Watt, *V.L.F. Radio Engineering*, Pergamon Press, New York, 1967. About one-fourth devoted to small antennas of various kinds; does not build on Wheeler, 1947.

BIBLIOGRAPHY

Chu, L. J.: "Physical Limitations of Omni-Directional Antennas," *J. App. Phys.,* vol. 19, December 1948, pp. 1163–1165.

Fano, R. M.: "Theoretical Limitations on the Broadband Matching of Arbitrary Impedances," *J. Franklin Inst.,* vol. 249, January–February 1950, pp. 57–83, 139–154. Tolerance and bandwidth; graphs, p. 144.

Hagaman, B. G.: "Low-Frequency Antennas," Chap. 24 of this handbook.

Schelkunoff, S. A., and H. T. Friis: "Small Antennas," *Antennas: Theory and Practice,* John Wiley & Sons, Inc., New York, 1952, pp. 302–329. Analytic treatment of various C and L configurations; does not build on Wheeler, 1947.

Walt, J. R.: "The Magnetic Dipole Antenna Immersed in a Conducting Medium, *IRE Proc.,* vol. 40, October 1952, pp. 1951–1952. In a spherical cavity.

Microstrip Antennas

Robert E. Munson

Ball Aerospace Systems Division

7-1 INTRODUCTION

Microstrip-antenna elements radiate efficiently as devices on microstrip printed-circuit boards. Microstrip-antenna arrays consist of microstrip-antenna elements, feed and phasing networks, and any other microstrip devices. Both microstrip-antenna elements and microstrip arrays are discussed in this chapter.

7-2 MICROSTRIP-ANTENNA-ELEMENT DESIGN PARAMETERS

This section can be used as a guide in selecting and designing microstrip elements which can be used alone as single radiators or as elements of microstrip phased arrays, which are discussed in Sec. 7-3.

The most commonly used microstrip element consists of a rectangular element that is photoetched from one side of a printed-circuit board (Fig. 7-1). The element is fed with a coaxial feed. The length L is the most critical dimension and is slightly less than a half wavelength in the dielectric substrate material.[1]

$$L \approx 0.49\lambda_d = 0.49 \frac{\lambda_0}{\sqrt{\epsilon_r}}$$

where L = length of element

ϵ_r = relative dielectric constant of printed-circuit substrate (The exact value is critical and is usually specified and measured by the manufacturer.)

λ_0 = free-space wavelength

The variation in dielectric constant and feed inductance makes it hard to predict exact dimensions, so usually a test element is built to determine the exact length.

The thickness t is usually much less than a wavelength (usually on the order of $0.01\lambda_0$). The selected value of t is based on the bandwidth over which the antenna must operate; it is discussed in greater detail later. The exact value of t is determined by commercially available board thicknesses, such as 0.005 in (0.127 mm), 0.010 in (2.54 mm), 1/64 in (0.397 mm), 1/32 in (1.588 mm), 3/64 in (1.191 mm), 1/8 in (3.175 mm), etc. Teflon-fiberglass boards are commercially available from the Minnesota Mining and Manufacturing Company, Rogers Corporation, Keene Corporation, and Oak Industries.

The width W must be less than a wavelength in the dielectric substrate material

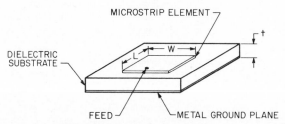

FIG. 7-1 Rectangular microstrip-antenna element.

so that higher-order modes will not be excited. An exception to this constraint, in which multiple feeds are used to eliminate higher-order modes, is discussed later.

Most microstrip elements are fed by a coaxial connector which is soldered to the back of the ground plane; the feed pin is soldered to the microstrip element as shown in Fig. 7-2. A direct-contact feed (rather than a probe, as in waveguide feeds) is always used. It is important that the feed pin be securely soldered to the microstrip element since most failures of microstrip antennas occur at this point.

A second microstrip-antenna element commonly used is the quarter-wave microstrip-antenna element (Fig. 7-3). It consists of a photoetched element in which the length L is about a quarter

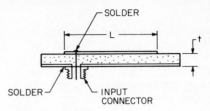

FIG. 7-2 Side view of microstrip element with coaxial-feed connection.

wavelength in the substrate material. Such an element is used for its broader E-plane beamwidth. A third microstrip-antenna element is a full-wavelength element, which is similar to that in Fig. 7-1 except that the feed is at the center and the length L is a full wavelength in the substrate material. Its radiation pattern is similar to that of a monopole. This element can also be round and of equal area to the equivalent square element.

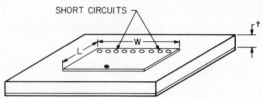

FIG. 7-3 Quarter-wave microstrip element.

Microstrip-Antenna Impedance

The surprising feature of a microstrip antenna is its efficient radiation despite its low profile. As the thickness of a microstrip antenna is reduced, its radiation resistance approaches a constant value. The source of radiation for the rectangular microstrip radiator shown in Fig. 7-1 is the electric field (Fig. 7-4) that is excited between the edges[1,2] of the microstrip element and the ground plane (excitation of a nearly infinitesimal slot with uniform E field). The fields are excited 180° out of phase between the opposite edges.

The input impedance of the microstrip element can be calculated from the equivalent circuit in Fig. 7-5; R_R (the radiation resistance of each slot) can be calculated for a uniformly excited slot as a function of width by

$$R_R = 120\lambda_0 / W$$

Since the microstrip element consists of two slots that combine in parallel, the input impedance is given as

$$R_{in} = 60\lambda_0 / W$$

The susceptance of the slots is shown[1] to combine at resonance to cause a shortening

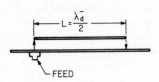

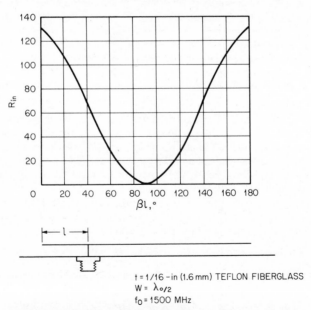

FIG. 7-4 Side view of rectangular micro-strip electric fields.

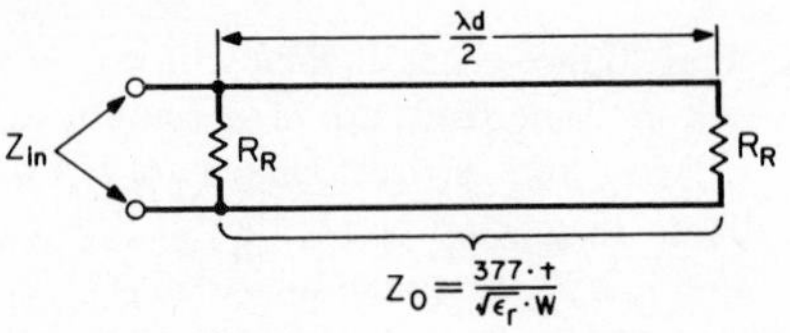

FIG. 7-5 Equivalent circuit for determining input impedance of rectangular microstrip element.

of the microstrip element to values typically less than a half wavelength, such as $L = 0.49\lambda_d$.

The input impedance can be matched to 50 Ω by using one of two techniques. For an element that is to be employed individually, a coaxial input connector is used. The 50-Ω-impedance point may be obtained by varying the distance from the edge of the element to the feed location ℓ, as shown in Fig. 7-6. The impedance as a function of the feed location for the rectangular microstrip element was calculated to be

$$R_{in} \simeq \frac{(120\lambda_0)^2 + \left(\dfrac{377t}{\sqrt{\epsilon_r}W}\right)^2 \left(\dfrac{\tan^2 \beta\ell + \tan^4 \beta\ell}{1 + \tan^2 \beta\ell}\right)}{240W\lambda_0(1 + \tan^2 \beta\ell)}$$

Note that the impedance of the element goes essentially to zero at the center of the element. Sometimes it is necessary to ground a microstrip element; the use of a rivet or a plated-through hole at the center of the element results in negligible effects on

$t = 1/16 - in (1.6 mm)$ TEFLON FIBERGLASS
$W = \lambda_0/2$
$f_0 = 1500$ MHz

FIG. 7-6 Input-impedance variation as a function of feed location for a rectangular microstrip element (side view).

patterns and a small change in resonant frequency. Likewise, for air-loaded microstrip elements a metal support is used at the center of the element. When an element is to be used in a microstrip array, it is fed at the edge, and a quarter-wave transformer is used to convert the input impedance to any desired level (Fig. 7-7).

The quarter-wave microstrip element has an impedance twice that of the rectangular microstrip element. The input impedance can be calculated from the equivalent

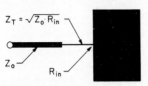

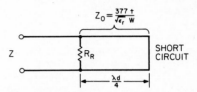

FIG. 7-7 Matching a rectangular microstrip element by using a monolithic quarter-wave transformer (top view).

FIG. 7-8 Equivalent circuit for a quarter-wave microstrip element.

circuit shown in Fig. 7-8. At resonance (when the distance from the short circuit to the radiating slot is a quarter wavelength) the short circuit transforms a quarter wavelength to become an open circuit. The open circuit combines in parallel with R_R, the radiation resistance of the single slot. The resulting input impedance is given by Z_{in} = R_R = $120\lambda_0/W$.

For a quarter-wavelength element that is a half wavelength wide the radiation resistance and input impedance are 240 Ω. This is very high, and a 5:1 voltage standing-wave ratio (VSWR) would occur if a 50-Ω coaxial connector were directly connected to the edge of the element. Therefore, the feed point is inset a distance ℓ from the edge or slot as shown in Fig. 7-9. The input impedance of a quarter-wave element may also be matched by using a quarter-wave transformer as discussed for the rectangular microstrip element. Sometimes the line widths of a quarter-wave transformer become too narrow, so an inset monolithic feed is used as shown in Fig. 7-10. The distance ℓ given in Fig. 7-9 is used as a starting point, but exact dimensions are usually determined experimentally. The input impedance of a quarter-wavelength element as a function of the feed location was calculated as

$$R_{in} \cong \frac{120\lambda_0 \tan^2 \beta(L - \ell)(1 + \tan^2 \beta\ell)}{W\,[\tan \beta\ell + \tan \beta(1-\ell)]^2}$$

Microstrip-Element Antenna Patterns

Microstrip antennas have radiation patterns that can be accurately calculated. The key to accurate calculation is the fact that the source of radiation is the electric field across a small gap formed by the edge of the microstrip element and the ground plane directly below. Since its dimension $t \ll \lambda_d/4$, the individual slots cannot exhibit any directionality. Each slot therefore radiates an omnidirectional pattern into the half space above the ground plane. Figure 7-11 shows a side view of a rectangular microstrip element and its associated source and radiating E fields.

The opposing slots are excited out of phase, but their radiation adds in-phase normal to the element. This occurs because the slots are inverted. The radiation of two slots excited in phase with equal amplitude is given by

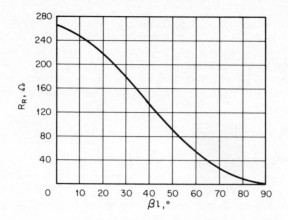

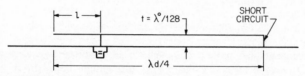

FIG. 7-9 Measured input impedance as a function of ℓ (ℓ is the distance to which the feed is inset into a $\lambda_d/4$ element).

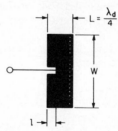

FIG. 7-10 Monolithic feed for a quarter-wave microstrip.

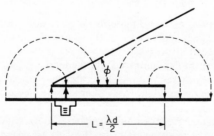

FIG. 7-11 Side view of a rectangular microstrip element and associated radiation.

$$E = K \cos \left(\frac{\pi L}{\lambda_0} \cdot \cos \phi \right)$$

Since the microstrip element is resonant at a half wavelength in the dielectric under the element,

$$E_\phi = K \cos \left(\frac{\pi}{2\sqrt{\epsilon_r}} \cdot \cos \phi \right)$$

This formula is valid for $0° < \phi < 180°$ (above the ground plane). The formula is not exact near the ground plane. Edge radiation from the end of the ground plane usually reduces theoretical radiation by 6 dB at $\phi = 0°$ and $180°$. Some radiation also will occur in the aft hemisphere $180° < \phi < 360°$. The exact amount of aft-hemisphere radiation diminishes rapidly as the ground plane becomes large in wavelengths. The theoretical E-plane patterns for two commonly used rectangular microstrip elements are shown in Fig. 7-12a and b.

The quarter-wavelength microstrip element radiates from a single slot; thus its E-plane radiation is uniform ($E_\phi = K$) for $0° < \phi < 180°$ for an infinite perfect conducting ground plane. The measured pattern for a quarter-wavelength element centered on a $6\lambda_0$ ground plane is shown in Fig. 7-13. Note that the radiation is reduced by about 6 dB at $\phi = 0°$ and $180°$. The backlobes and ripple in the radiation pattern would be reduced as the ground plane got larger.

The H-plane patterns of the rectangular and quarter-wavelength microstrip elements are given by the formula for a uniformly excited radiator:

$$E_H = K \tan \theta \sin \left(\frac{\pi W}{\lambda_0} \cdot \cos \theta \right)$$

where θ = angle above ground plane in the H plane.

The full-wavelength element has a radiation pattern similar to that of a monopole. The pattern of this element is given by

$$E_\phi = K \sin \left(\frac{\pi}{\sqrt{\epsilon_r}} \cdot \cos \phi \right)$$

The pattern of a center-fed, air-loaded, $\epsilon_r = 1.0$ element is shown in Fig. 7-14. The radiation pattern of a round full-wavelength microstrip element fed at the center is given by

$$E = K J_1 \left(\frac{2\pi}{\lambda_0} \cdot a \cdot \sin \theta \right)$$

where J_1 = Bessel function of first kind
a = radius of element

The round full-wavelength microstrip element has essentially the same pattern as the square full-wavelength microstrip element of equivalent area and identical dielectric-constant substrate.

Microstrip-Antenna Bandwidth

The bandwidth of microstrip antennas is proportional to the thickness of the substrate used. Since most substrates are very thin in terms of wavelengths ($t \ll \lambda_0/4$), the

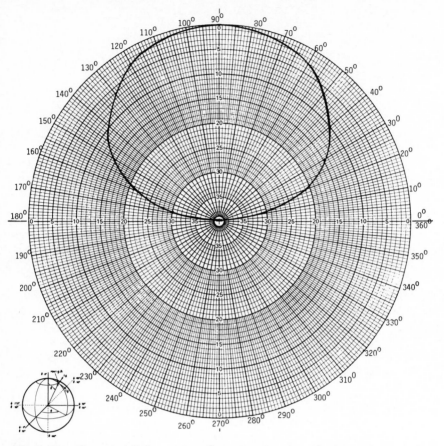

FIG. 7-12a   Theoretical *E*-plane pattern of a rectangular microstrip element for $\epsilon_r = 1.0$.

bandwidth is usually narrow. A useful formula for determining expected microstrip-element bandwidth is given by

$$\text{BW} = 4f^2 \left(\frac{t}{1/32} \right)$$

where BW = bandwidth, MHz, for a VSWR less than 2:1
     f = operating frequency, GHz
     t = thickness, in [Most board thicknesses are available in steps of ¹⁄₃₂ in
          (0.794 mm).]

     Since the feed networks used to feed most microstrip arrays are low-*Q*, the bandwidths of most arrays are given by the preceding formula. Exceptions are large series-fed microstrip arrays and microstrip phased arrays. With these two types of arrays, the pattern often degrades before the VSWR increases.

     Broadbanding techniques discussed in Chap. 43 have been applied to microstrip elements. In most cases, the bandwidth has been doubled, but it is very difficult to increase the bandwidth beyond this point. It is much easier to increase the thickness

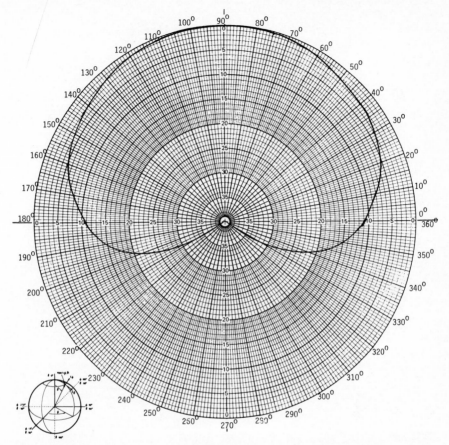

FIG. 7-12b   Theoretical *E*-plane pattern of a rectangular microstrip element for $\epsilon_r = 2.45$.

of the microstrip element. It seems certain that the bandwidth of a microstrip element is related directly to its volume.

Microstrip-Antenna Mutual Coupling

When microstrip elements are used in arrays, ultimate performance is attained for each design if mutual coupling is minimized. Three methods of coupling exist:

1   Coupling between microstrip elements (Figs. 7-15 and 7-16)

2   Coupling between microstrip transmission lines and microstrip elements (Figs. 7-17 and 7-18)

3   Coupling between microstrip transmission lines (Table 7-1)

The coupling between microstrip elements in Figs. 7-15 and 7-16 affects the pattern shape of the elements, radiated power and phase, and input VSWR. The coupling

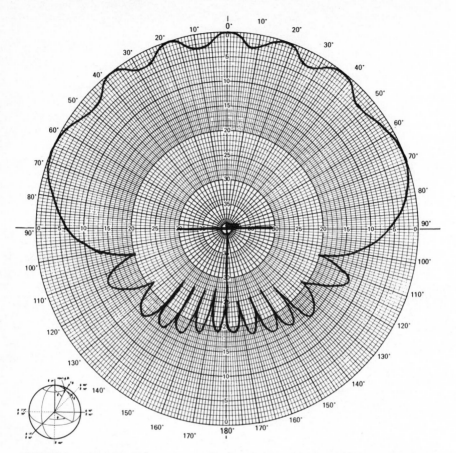

FIG. 7-13   Measured *E*-plane pattern of a quarter-wavelength element on a 6λ₀ ground plane.

between feed lines and microstrip elements affects the radiation patterns, radiation phase and amplitude, antenna impedance, and microstrip transmission-line match. The coupling between transmission lines affects the transmission-line match.

If the separations are maintained so that coupling is less than 20 dB (usually three substrate thicknesses *t*), the gain and VSWR of the antenna are not degraded. For a low-sidelobe antenna, isolations of 30 dB or more are required.

Microstrip-Antenna Efficiency

For a microstrip element, *efficiency* is defined as the power radiated divided by the power received by the input to the element. Factors that reduce efficiency are the dielectric loss, the conductor loss, the reflected power (VSWR), the cross-polarized loss, and the power dissipated in any loads involved in the elements. Most microstrip elements are between 80 and 99 percent efficient. For very thin elements the Q of the microstrip cavity becomes so high that the current losses become excessive, and the

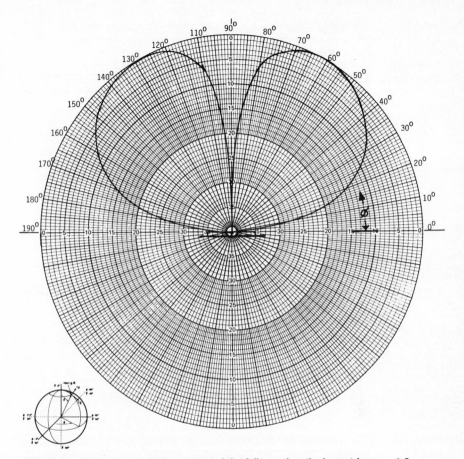

FIG. 7-14 Theoretical radiation pattern of the full-wavelength element for $\epsilon_r = 1.0$.

thickness t becomes so small that the conductance across the cavity yields excessive dielectric losses. This usually occurs when the thickness is reduced to about $\lambda_0/1000$. At this thickness, the resonance VSWR can be matched to 50 Ω, but the very narrow bandwidth leads to temperature instabilities; i.e., a slight change in temperature causes the VSWR to rise rapidly, and reflection losses reduce efficiency.

Dielectric losses are eliminated by using air as a substrate. Since most of these elements have a large separation between the element and ground plane t, the Q is also reduced and so are conductor and VSWR losses. Most of these elements have an efficiency of 95 to 99 percent. Air-loaded elements are usually built as separate entities and are not photoetched in large arrays. A single metal part, rivet, or bolt is used at the center of the element for support. To build a photoetched array by using a nearly air substrate, a honeycomb core is employed to separate the microstrip elements and feed network from the ground plane. A 35-ft by 8-ft (10.7-m by 2.4-m) array was built on a honeycomb core for the Seasat satellite program (1978) with an aperture efficiency of 60 percent. (This included all feed-network losses.)

Microstrip elements are not efficient if solid epoxy fiberglass is used as a sub-

TABLE 7-1 Isolation (dB) of Parallel Microstrip Transmission Lines

Board thickness, in	Board spacing (number of thicknesses)	Frequency			
		2.0 GHz	2.7 GHz	3.4 GHz	4.0 GHz
1/16	1	8.0	12.5	8.0	3.0
	2	11.0	13.0	12.5	5.0
	3	17.5	18.0	17.5	10.0
	4	25.0	25.0	26.0	17.5
1/32	1	10.0	12.0	10.0	2.5
	2	12.5	15.0	13.5	5.0
	3	17.5	17.5	15.0	9.5
	4	25.0	25.0	27.5	16.5
1/64	1	10.0	10.0	11.5	2.5
	2	10.0	12.5	12.0	5.0
	3	17.5	16.0	22.5	10.0
	4	25.0	24.0	22.5	17.5

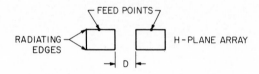

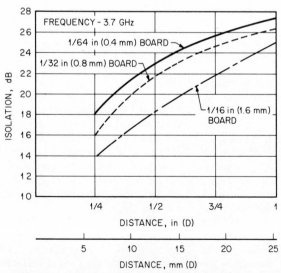

FIG. 7-15 Isolation of two half-wave microstrip patch radiators as a function of their *H*-plane separation.

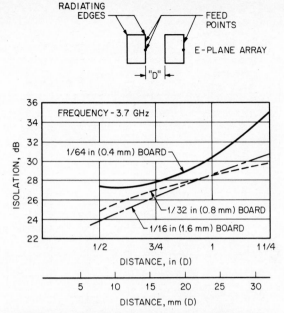

FIG. 7-16 Isolation of two half-wave microstrip patch radiators as a function of their *E*-plane separation.

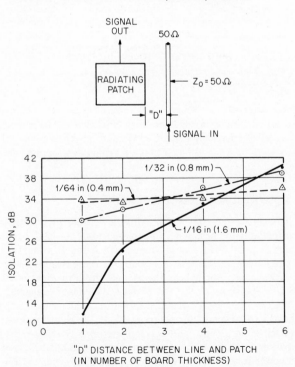

FIG. 7-17 Isolation between radiating patch and feed line along *H* plane.

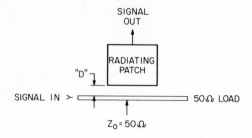

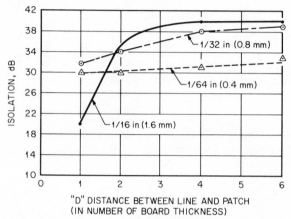

FIG. 7-18 Isolation between radiating patch and feed line along *E* plane.

strate. Repeated attempts to circumvent the use of expensive Teflon-fiberglass substrates have resulted in antenna elements with efficiencies of 10 percent.

The efficiency of microstrip phased arrays depends upon microstrip-element efficiency, feed-network losses, and phase and amplitude distribution. These efficiencies are discussed later.

Circularly Polarized Microstrip Elements

Microstrip elements are probably the simplest devices for producing circular polarization. Three methods for producing circular polarization have been developed. A microstrip element (Fig. 7-19) that is square with *L* and *W* equal to $\lambda_d/2$ will have two modes of radiation, vertical and horizontal. If these two modes are excited 90° out of phase with a monolithic hybrid, the square microstrip element will radiate vertical and horizontal polarization 90° out of phase. One input to the 90° hybrid power divider will result in right-hand circular polarization, while the other port of the 90° hybrid will have left-hand circular polarization. If the right-hand circular input is driven and the left-hand circular port is terminated in a 50-Ω load, the input VSWR will remain low for a bandwidth that extends beyond the bandwidth of the element. The reflected power is absorbed by the load. A spinning linear pattern of a circular polarized element is shown in Fig. 7-20. Note that the axial ratio remains low out to wide angles.

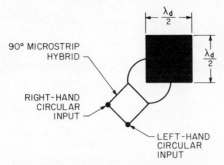

90° MICROSTRIP
HYBRID

RIGHT-HAND
CIRCULAR
INPUT

LEFT-HAND
CIRCULAR
INPUT

FIG. 7-19 Circularly polarized square microstrip element with 90° hybrid feed.

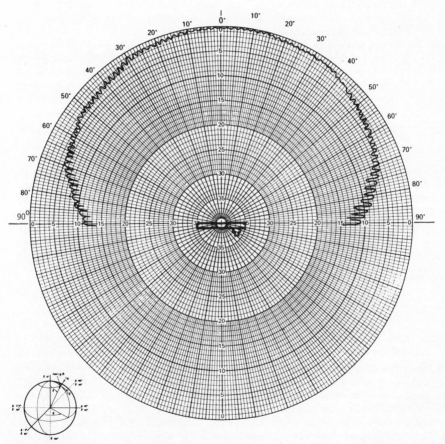

FIG. 7-20 Spinning dipole radiation pattern of a square microstrip element fed with a hybrid.

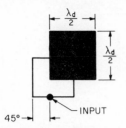

FIG. 7-21 Circularly polarized microstrip element with 45° offset feed.

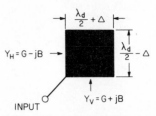

FIG. 7-22 Circularly polarized microstrip element with conjugate impedance matching and a single feed.

Another method for exciting circular polarization is shown in Fig. 7-21. This square element is fed in two orthogonal modes (vertical and horizontal) with a microstrip line. The phase of excitation of the microstrip-element modes is offset by 90° by offsetting the microstrip feed by 45°; i.e., a 90° lag is induced in the feed network. This design has a narrow circularly polarized bandwidth, and the VSWR bandwidth is about twice as wide as the element bandwidth.

A third method for exciting circular polarization is the microstrip element shown in Fig. 7-22. This element is adjusted slightly off resonance by $+\Delta$ in the vertical plane and $-\Delta$ in the horizontal plane. If the vertical dimension is 1 percent longer than $\lambda_d/2$, the horizontal dimension is adjusted to be 1 percent shorter than $\lambda_d/2$. This results in an increase in mode susceptance. The admittance in the vertical and horizontal planes respectively is given by

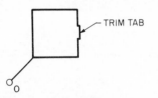

FIG. 7-23 Circularly polarized microstrip element with trim tab added.

$$Y_V = G + jB \text{ and } Y_H = G - jB$$

When $B = G$, the feed applies a uniform voltage to conjugate impedances. The two resulting modes of radiation are excited with equal power and are 90° out of phase. Exact dimensions for a design are determined empirically by trimming a square element or an element as shown in Fig. 7-23. The tab is added for ease of trimming to achieve circular polarization.

Dual-Frequency, Dual-Polarization Microstrip-Antenna Element

Dual-frequency microstrip elements may be stacked microstrip elements fed in series as shown in Fig. 7-24a. The most important feature to note in Fig. 7-24a, b, and c is that off resonance the microstrip element looks like a short circuit. This allows the resonant microstrip element to work independently of the nonresonant element since the nonresonant element's impedance adds in series with the resonant element as a short circuit (see Fig. 7-25). Microstrip elements cannot be stacked as dual-frequency elements if their resonant frequencies are separated by less than 10 percent (except for very-high-Q elements) or if their resonant frequencies are harmonically related.

Multiple elements can be stacked and fed in series to get a 3, 4, or N multiresonant microstrip array. It is important that each element be centered so that it is

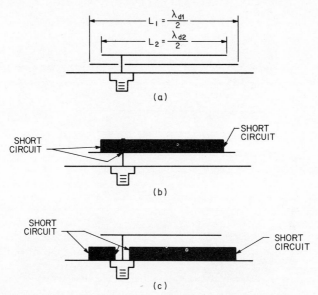

FIG. 7-24 Dual-frequency microstrip element (side view). (*a*) Fed in series. (*b*) Impedance model at f_1. (*c*) Impedance model at f_2.

fed at its 50-Ω-input point. Quarter-wavelength elements as well as full-wavelength elements can also be fed in series.

The exact dimension of a series-fed element and its feed-point location must be defined experimentally by an iterative process. This process usually consists of building an element to the dimensions defined by the design equations given for a rectangular microstrip element. The resonant frequency and impedance are measured; usually they differ slightly from prediction because of a variety of effects:

1   Mutual coupling

2   Feed-probe inductance

3   Slight effect owing to short-circuit approximation for impedance of nonresonant element

4   Dielectric-constant variation

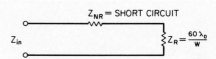

FIG. 7-25 Equivalent circuit of a stacked dual-frequency rectangular microstrip element. Z_{NR} = impedance of nonresonant rectangular microstrip element. Z_R = impedance of resonant rectangular microstrip element.

A linear adjustment of microstrip-element size and feed-point location is made to correct the resonant frequency and feed impedance. Multiple iterations may be required, but usually only one or two are necessary.

Dual-frequency circularly polarized microstrip elements can be built by feeding stacked circularly polarized microstrip elements along the diagonal of each element as shown in Fig. 7-26.

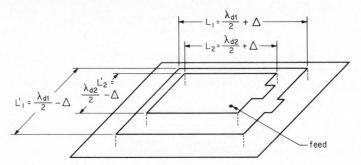

FIG. 7-26   Dual-frequency circularly polarized microstrip element.

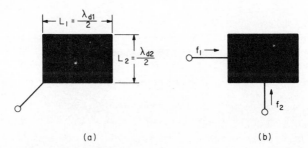

FIG. 7-27   Rectangular dual-polarization microstrip element.
(*a*) Horizontal at f_1, vertical at f_2 (same input). (*b*) Horizontal at f_1, vertical at f_2 (separate feeds).

Dual-polarized microstrip elements are rectangular microstrip elements in which the two dimensions are picked as resonant dimensions. As shown in Fig. 7-27*a*, the two dimensions are picked so that L_1 equals a half wavelength in the dielectric at the first resonant frequency and L_2 equals a half wavelength in the dielectric at the second resonant frequency. The element can also be fed with two independent feeds at the midpoint of each side as shown in Fig. 7-27*b*. In both cases, f_1 will radiate horizontal polarization and f_2 will radiate vertical polarization.

A square microstrip element can be excited in two orthogonal modes as shown in Fig. 7-28. It is important that the two feeds be located at the midpoint of the

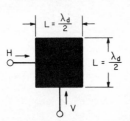

FIG. 7-28   Dual-polarized square micro-strip element.

edge of the element, for this is the low-impedance point for the orthogonal mode in each case. Coupling of 25 dB between ports is typical for square elements fed in this manner.

7-3 MICROSTRIP-ARRAY DESIGN PRINCIPLES

Section 7-2 dealt with microstrip elements, which can be used individually or in microstrip arrays. A *microstrip array* is the integration of microstrip elements with a microstrip feed network consisting of power dividers, transmission lines, phase lines, and active components, etc. The active components are devices such as phase shifters, amplifiers, oscillators, receivers, mixers, etc. It is important to note that microstrip feed lines are connected directly to a microstrip radiating element, affecting neither the radiation pattern nor the impedance of the radiator.[3,4,5] The feed line, being near the ground plane, is perpendicular to the electric field emanating from the microstrip radiator. Therefore, the feed line is a septum of equipotential, and radiating fields cannot excite currents in the feed lines.

The real advantages of microstrip antennas appear when all the elements of the array along with feed network are monolithically etched from one side of a printed-circuit board. At least four distinct advantages can be identified:

1 The process of photoetching hundreds or even thousands of microwave components in one process results in a low-cost antenna array.

2 The resulting printed-circuit board is very thin. Since the array is designed to operate from the ground plane on the back of the printed board, its performance is unaffected by mounting to a metallic surface such as an aircraft or a missile. The resulting design is doubly conformal. It is conformal to the underlying structure to which it can be bolted or laminated, and it is externally conformal aerodynamically because of minimum protrusion.

3 Microstrip arrays have high performance because an infinite variety and quantity of antenna elements, power dividers, matching sections, phasing sections, etc., can be added to the printed-circuit board without any cost impact (except the cost per square foot of the board). This gives the design engineer many components (for instance, 110-Ω quarter-wave transformers or unequal power dividers) that are not commercially available in separate packages.

4 The microstrip array is very reliable since the entire array is one continuous piece of copper. Other types of antennas most commonly have failed at interconnections within the antennas and at their input connectors.[6]

Omnidirectional Microstrip Arrays

Microstrip arrays were first used as omnidirectional arrays on missiles.[3] Two types of antennas are used:

● A continuous radiator for linear axial polarization (Fig. 7-29).

● An array of discrete radiators for omnidirectional circular polarization (Fig. 7-30). (Eight of the circularly polarized arrays in Fig. 7-30 fit end to end and are fed in phase to provide a large circularly polarized omnidirectional array.)

The antenna shown in Fig. 7-29 was designed to give omnidirectional linear (axial) polarization coverage. The printed antenna is shown wrapped around the 8-in (203-

FIG. 7-29   Continuous-radiator microstrip antenna for linear polarization.

mm) cylinder. Note that a ⅟₃₂-in- (0.794-mm-) deep lathe cut has been made in the 8-in cylinder so that the antenna presents a flush surface when wrapped around and attached to the cylinder.

The design procedure for the 8-in wraparound operating at 2270 MHz (Fig. 7-31) is calculated as follows:

1   The resonant length of the radiator is $L = \lambda_0/2\sqrt{\epsilon_r} = 5.2/(2\sqrt{2.45}) = 1.66$ in (42.16 mm).

2   The width of the radiator is $W = \pi D = \pi 8 = 25.1$ in (637.54 mm).

3   The number of feeds required N_F is at least 1 per wavelength in the dielectric, and a binary number 2, 4, 8, 16 ... is used to accommodate a corporate-feed network. The width of the element in terms of wavelengths in the dielectric is given by

$$W_{\lambda_d} = W/\lambda_0/\sqrt{\epsilon_r} = 7.5$$

Since the width in wavelengths does not exceed 8, the number of feeds required N_F is 8 (at least 1 feed per λ_d).

4   The input impedance R_{in} of the antenna at each feed point is the total impedance of the radiator times the number of feeds:

$$R_{in} = N_F \cdot 60 \cdot \lambda_0/W = 99 \ \Omega$$

5 The feed network is used to match the antenna while dividing the power. A quarter transformer is used next to the element to match the antenna impedance R_{in} to 100 Ω:

$$Z_0 \text{ quarter-wave transformer} = \sqrt{R_{in} \cdot 100} = 99.5 \ \Omega$$

At the first junction a the two 100-Ω impedances combine to give 50 Ω. A quarter-wavelength ($Z_0 = 70 \ \Omega$) transformer is used to transform the 50-Ω impedance to 100 Ω. This impedance is transferred down a 100-Ω line to junction b, where the process at junction a is repeated. At junction c the two 100-Ω impedances combine to provide a 50-Ω input for a coaxial input attached to the back of the printed-circuit board.

Microstrip-Antenna Pattern Coverage for Omniapplications The pattern coverage for the omniantenna shown in Figs. 7-29 and 7-31 depends on the diameter of the missile. The limiting factor in omnidirectional pattern coverage is a singular hole at the tip and tail of the missile (Fig. 7-32) which gets narrower as the diameter of the missile increases. For instance, a 15-in-diameter antenna produces a null along the missile axis of radius 1° at the −8-dB-gain level. The fractional area with gain below −8 dB is 0.0002. Conversely, the fraction of the area with gain above −8 dB is 0.9998, or 99.98 percent. The percent coverage increases without limit for larger diameters until a nearly perfect coverage is attained for a single linear polarization. The percent coverage is only a function of

FIG. 7-30 One-eighth of a circularly polarized omnidirectional array.

diameter and wavelength and is independent of antenna thickness. The theoretical and experimental pattern coverages, at S band, for microstrip antennas on a smooth cylinder are given in Fig. 7-33 for gain levels greater than −8 dB.

The roll-pattern variation for a circularly polarized wraparound microstrip antenna is a function of center-to-center element separation as shown in Fig. 7-34. For a uniform roll-plane pattern, the separation between elements should not exceed $0.7\lambda_0$. For separations approaching $0.35\lambda_0$ the mutual coupling between elements makes it difficult to obtain circular polarization with a tolerable axial ratio. The input impedance of an element is measured, and a corporate-feed network is designed, by using the procedure discussed earlier.

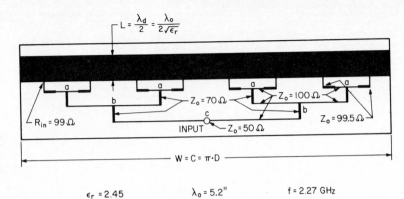

$$\epsilon_r = 2.45 \qquad \lambda_0 = 5.2'' \qquad f = 2.27 \text{ GHz}$$

FIG. 7-31 Printed-circuit-board design for an 8-in- (203-mm-) diameter cylinder.

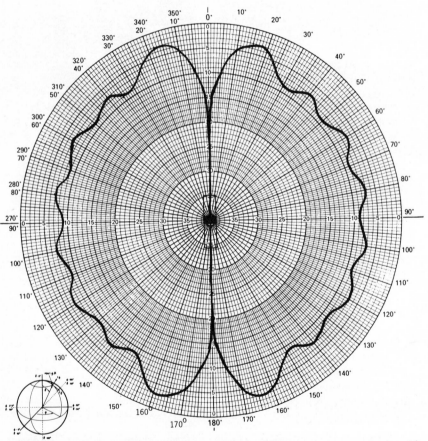

FIG. 7-32 Measured *E*-plane pattern of the 8-in (203-mm) wraparound microstrip antenna shown in Fig. 7-29. The antenna pattern is a figure of revolution about the missile axis.

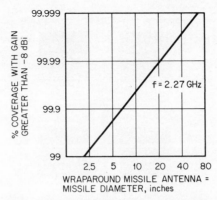

FIG. 7-33 Coverage versus diameter for an omnidirectional microstrip array.

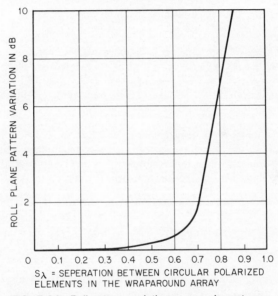

FIG. 7-34 Roll-pattern variation versus element spacing. (*Courtesy of Cliff Garvin, Ball Aerospace Systems Division.*)

Fixed-Beam Microstrip Arrays

High-gain directional fixed-beam antennas can be photoetched from one side of a printed-circuit board (Fig. 7-35). Most arrays of this type have a single beam perpendicular to the plane of the array. The microstrip elements are spaced slightly less than a free-space wavelength to prevent grating lobes. Spacings of less than a half free-space wavelength are generally not used because of feed-line crowding.

The example shown in Fig. 7-35 consists of four rectangular microstrip elements

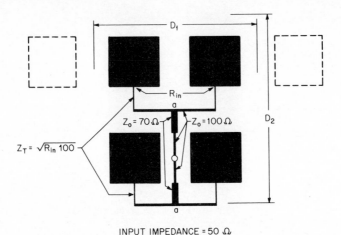

INPUT IMPEDANCE = 50 Ω

FIG. 7-35   Directional fixed-beam microstrip antenna.

in a two-by-two array. Each element is matched with a quarter-wavelength transformer to match its input impedance R_{in} to 100 Ω. Each pair of 100-Ω inputs is combined to give 50 Ω at point a. The 50-Ω impedance is transformed to 100 Ω by a 70-Ω quarter-wavelength transformer. The two 100-Ω-impedance transmission lines maintain the 100-Ω impedance until they are combined in parallel to provide 50 Ω at the input.

A fixed-beam microstrip array can be expanded to any 2^n-by-2^m array. The corporate-feed network is expanded to provide equal power and phase split; the symmetrical line length from the input to the microstrip elements forces all the elements to be fed in phase. Variations of frequency or dielectric constant do not cause a differential phase taper over the array when a corporate-feed network is used.

The maximum gain of a fixed-beam microstrip phased array is given by

$$G_{dB} = 10 \log \left(\frac{4\pi A}{\lambda_0^2} \right) - \alpha \cdot (D_1 + D_2)/2$$

where $A = D_1 \cdot D_2$

D_1 = effective width of the uniformly spaced array [It is defined as the sum of the distance between the centers of the edge elements plus one interelement spacing; $(n + 1) \times$ horizontal spacing.]

D_2 = height of array defined in the same manner as D_1 [$(m + 1) \times$ vertical spacing].

α = attenuation, dB per unit length, of a 50-Ω transmission line being used in the monolithic feed [A typical value of α is 0.4 dB/ft for a 50-Ω microstrip line on $\frac{1}{32}$-in (0.794-mm) Teflon fiberglass at 2.2 GHz.]

Unequal line lengths can be used to produce phase tapers which yield fixed beams that are scanned away from broadside. Phase tapers are also used to compensate for curvature of microstrip arrays that conform to the cylindrical surface of aircraft or missiles.

Microstrip Phased Arrays

A C-band microstrip phased array is shown in Fig. 7-36. This array is similar to the fixed-beam microstrip arrays discussed above except that a phase shifter has been added to the input of each microstrip element. The microstrip phase shifters, microstrip radio-frequency chokes, and direct-current bias lines are all photoetched on the same piece of copper as the radiators and the corporate-feed network.

Quarter-wavelength microstrip elements are used as radiators for two reasons. First, they require less area than the other types of microstrip radiating elements. Second, they have a uniform radiation pattern in the E plane and therefore do not contribute to loss of gain as a function of scan angle because of the element factor roll-off.

The array shown in Fig. 7-36 has a 3-bit phase shifter at the corporate-feed input to each element. Each 3-bit phase shifter consists of a $90°$ and an $180°$ switched-line phase shifter and a $45°$ loaded-line phase shifter. The phase shifters (Fig. 7-37) are pin-diode phase shifters. Forward bias short-circuits the upper two diodes and causes the signal to travel down the upper-path length $L + \phi$, and reverse bias short-circuits the lower two diodes and causes the signal to travel the shorter path L. The differential length is ϕ the desired phase shift. The $45°$ loaded-line phase shifter (Fig. 7-37b) is

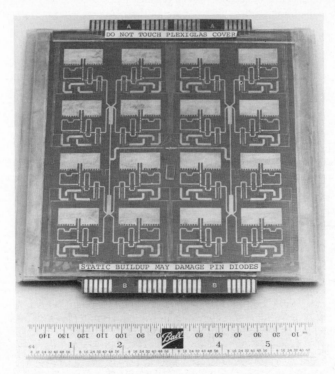

FIG. 7-36 Monolithic 5-GHz phased array. (*Courtesy of Frank Cipolla, Ball Aerospace Systems Division.*)

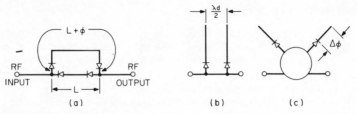

FIG. 7-37 Microstrip phased-array phase shifters. (a) Switched-line phase shifter. (b) Loaded-line phase shifter. (c) 90° hybrid phase shifter.

used for 45° or less induced phase shifts. These phase shifters are used because of their compact size and the fact that only two diodes are required. Each inductive stub produces half of the total loaded-line phase shift. The quarter-wavelength spacing between the stubs minimizes the reflection coefficient, and an impedance match is maintained. For more information on switched-line and loaded-line phase shifters, a more extensive reference on phase-shifter design should be consulted before attempting a design.[7,8] The 90° hybrid phase shifter (Fig. 7-37c) is not useful in monolithic microstrip phased arrays because of its large size.

The gain of a microstrip phased array of the type shown in Fig. 7-36 is given by

$$G = 10 \log \left(\frac{4\pi A}{\lambda_0^2}\right) - \frac{\alpha}{2}(D_1 + D_2) - \quad 1 \text{ dB} \quad - 10 \log (\cos \theta_E \cdot \cos^2 \phi_H)$$

$$\underbrace{\qquad\qquad}_{\text{aperture gain}} \underbrace{\qquad\qquad}_{\substack{\text{feed-line} \\ \text{losses}}} \bigg| \underbrace{\quad}_{\substack{\text{3-bit phase-} \\ \text{shifter loss}}} \underbrace{\qquad\qquad\qquad}_{\text{scanning loss}}$$

where D_1, D_2, α, and A = dimensions as defined in preceding subsection
θ_E = scan angle in the E plane
ϕ_H = scan angle in the H plane

Although the array shown in Fig. 7-36 had measured gain[9] that agreed with the preceding formula, larger arrays should consider additional mutual-coupling VSWR losses.

Series-Fed Microstrip Arrays

Two types of series-fed microstrip arrays are feasible: the resonant, or broadside, series-fed microstrip array and the nonresonant, or traveling-wave, series-fed microstrip phased array.

A resonant series-fed microstrip array is shown in Fig. 7-38a. The array is fed at its input with a 90° offset near the center so that radiation of both halves of the array will be in phase. The elements of the array are separated by a wavelength in the effective dielectric of the transmission line; the line consists of the radiating elements and the series-connected microstrip lines. For Teflon-fiberglass boards, the effective wavelength will be approximately $\lambda_0/\sqrt{\epsilon_r}$ in the elements and $1.2\lambda_0/\sqrt{\epsilon_r}$ in the microstrip transmission line. This will result in an effective separation S_λ of $\lambda_d \approx 1.1\lambda_0/\sqrt{\epsilon_r}$. Exact values are usually measured experimentally by constructing a test array and varying the frequency until the beam is centered at broadside. The conductance of each element is a function of its width. For a perfectly lossless feed, the power

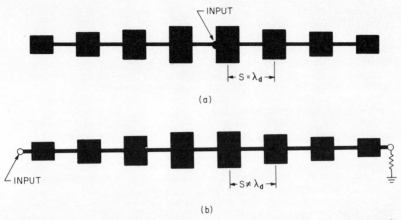

FIG. 7-38 Series-fed microstrip phased array. (*a*) Resonant series-fed microstrip array. (*b*) Traveling-wave series-fed microstrip array.

radiated by each element will be its conductance divided by the total sum of the conductance of all elements.

The sidelobe level and the beamwidth at resonance can be used to deduce the amplitude taper existing over the array. Near-field probe techniques can also be used to measure the amplitude taper. At resonance, the array has a beam at broadside with sidelobes of -20 to -30 dB. Off resonance, the main beam splits into two identical beams that scan away from broadside as a function of frequency.

A traveling-wave series-fed microstrip array is shown in Fig. 7-38*b*. The array is fed at one end, and varying widths of elements are used to control the radiation. The radiation from each element is proportional to the width of the element. The elements are separated by a dimension not equal to the effective dielectric constant of the traveling-wave array. For a separation S_λ less than an effective wavelength, the beam will be scanned backfire (toward the feed from broadside). For a separation greater than a wavelength, the array will produce an end-fire beam (scanning away from broadside as the frequency is increased). A grating lobe will appear for the end-fire beam.

Traveling-wave arrays have very good input impedance since the reflection coefficient due to each element cancels in the reverse direction. The dimensions of a series-fed array are predicted by the following authors: Derneryd,[10] Metzler,[11,15] Bahl,[12] Campi,[13] Huebner,[14] and Sanford.[16]

REFERENCES

1 R. E. Munson, "Conformal Microstrip Antennas and Microstrip Phased Arrays," *IEEE Trans. Antennas Propagat.*, vol. AP-22, no. 1, January 1974, pp. 74–78.

2 A. Derneryd, "A Theoretical Investigation of the Rectangular Microstrip Antenna Element," *IEEE Trans. Antennas Propagat.*, vol. AP-26, July 1978, pp. 532–535.

3 U.S. Patent 3,713,162.

4 U.S. Patent 3,810,183.

5 U.S. Patent 3,921,177.
6 Al Brejha, "Survey of Failure Mode in Antennas," unpublished.
7 Robert V. Garver, "Broadband Diode Phase Shifters," HDL-TR-1562, August 1971.
8 Joseph F. White, "Diode Phase Shifters for Array Antennas," *IEEE Trans. Microwave Theory Tech.,* vol. MTT-22, no. 6, June 1974, pp. 658–674.
9 Frank W. Cipolla, John D. Martinko, and Michael A. Weiss, "Microstrip Phased Array Antennas," RADC-TR-78-27, vol. I of final technical report, March 1978.
10 A. Derneryd, "Linear Polarized Microstrip Antennas," *IEEE Trans. Antennas Propagat.,* vol. AP-24, November 1976, pp. 846–850.
11 T. Metzler, "Microstrip Series Arrays," *Proc. Workshop Printed Circuit Antenna Tech.,* Oct. 17–19, 1979, New Mexico State University, Las Cruces, pp. 20-1–20-16.
12 I. J. Bahl and P. Bhartia, *Microstrip Antennas,* Artech House, Inc., Dedham, Mass., 1980, chaps. 5 and 7.
13 M. Campi, "Design of Microstrip Linear Array Antennas," *Proc. Antenna Appl. Symp.,* Robert Alerton Park, University of Illinois, Urbana, Sept. 23–25, 1981.
14 G. Oltman and D. Huebner, "Electromagnetically Coupled Microstrip Dipoles," *IEEE Trans. Antennas Propagat.,* vol. AP-29, no. 1, January 1981, pp. 151–157.
15 T. Metzler, "Microstrip Series Arrays," *Proc. Workshop Printed Circuit Antenna Tech.,* Oct. 17–19, 1979, New Mexico State University, Las Cruces, pp. 174–178.
16 U.S. Patent 4,180,817.

BIBLIOGRAPHY

Carver, K. R.: "A Modal Expansion Theory for the Microstrip Antenna," *Dig. Int. Symp. Antennas Propagat.,* Seattle, June 1979, pp. 101–104.
———— and J. W. Mink: "Microstrip Antenna Technology," *IEEE Trans. Antennas Propagat.,* vol. AP-29, no. 1, January 1981, pp. 25–38.
Derneryd, A. G.: "Linearly Polarized Microstrip Antennas," *IEEE Trans. Antennas Propagat.,* vol. AP-24, no. 6, November 1976, pp. 846–850.
————: "Analysis of the Microstrip Disk Antenna Element," *IEEE Trans. Antennas Propagat.,* vol. AP-27, no. 5, September 1979, pp. 660–664.
———— and A. G. Lind: "Extended Analysis of Rectangular Microstrip Resonator Antennas," *IEEE Trans. Antennas Propagat.,* vol. AP-27, no. 6, November 1979, pp. 846–849.
James, J. R., and G. J. Wilson: "Microstrip Antennas and Arrays, Part II: New Array-Design Techniques," *IEE J. MOA,* September 1977, pp. 175–181.
Kerr, J. L.: "Microstrip Antenna Developments," *Proc. Workshop Printed Circuit Antenna Tech.,* New Mexico State University, Las Cruces, Oct. 17–19, 1979, pp. 311–320.
Mailloux, R. J., J. McIlvenna, and N. Kernweis: "Microstrip Array Technology," *IEEE Trans. Antennas Propagat.,* vol. AP-29, no. 1, January 1981, pp. 25–38.
Proc. Workshop Printed Circuit Antenna Tech., 31 papers, New Mexico State University, Las Cruces, Oct. 17–19, 1979.
Sanford, G. G.: "Conformal Microstrip Phased Array for Aircraft Tests with ATS-6," *IEEE Trans. Antennas Propagat.,* vol. AP-26, no. 5, September 1978, pp. 642–646.
———— and L. Klein: "Increasing the Beamwidth of a Microstrip Radiating Element," *Dig. Int. Symp. Antennas Propagat.,* June 1979, pp. 126–129.

Chapter 8

Slot Antennas

Judd Blass

Blass Consultants, Inc.

8-1 INTRODUCTION

This chapter deals with the characteristics of small slot antennas. The metal surfaces in which the slots are cut are many wavelengths, but the slots themselves are less than one wavelength in extent. Such a slot may be excited by means of an energized cavity placed behind it, through a waveguide, or by a transmission line connected across the slot.

The simplest example of such an antenna consists of a rectangular slot cut in an extended thin flat sheet of metal with the slot free to radiate on both sides of this sheet, as shown in Fig. 8-1. The slot is excited by a voltage source such as a balanced parallel transmission line connected to the opposite edges of the slot or a coaxial transmission line.

The electric field distribution in the slot can be obtained from the relationship between the slot antennas and complementary wire antennas as established by Booker.[1] It has been shown that the electric field distribution (magnetic current) in the slot is identical with the electric current distribution on the complementary wire. In the rectangular slot illustrated the electric field is perpendicular to the long dimension, and its amplitude vanishes at the ends of the slot.

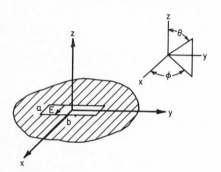

The electric field is everywhere normal to the surface of the slot antenna except in the region of the slot itself. The theoretical analysis of this configuration shows that the radiation of the currents in the sheet can be deduced directly from the distribution of the electric field in the slot. Consequently, the radiated field of an elementary magnetic moment within the slot boundaries should include the contributions of the electric current flowing on a metal surface.

FIG. 8-1 Rectangular slot.

A slot-antenna design will often require that the slot be cut in other than an extended flat sheet surface. Whatever the surface is, the electric field will be everywhere normal to it except in the region of the slot. The field due to the electric currents on this metal surface can be deduced from the exciting magnetic currents[2] in the slot, just as in the case of the flat metal sheet. This field can be combined with the exciting field so that the resultant is the total field due to a magnetic current on the given boundary surface. Thus the field of a thin rectangular slot cut in a circular cylinder differs from that of a slot cut in a flat metal sheet since the distribution of electric currents is different for the two cases. Sections 8-2 to 8-10 discuss the radiated fields of slot antennas cut in a variety of surfaces.

In general, the slot antenna is not free to radiate on both sides of the surface on which it is cut since one side is either completely enclosed, e.g., the slotted cylinder antenna, or it is desired that the radiation on one side be minimized. In these cases, the influence of the enclosed cavity region on the excitation and impedance of the slot antenna is significant to the antenna design. Aspects of this problem are discussed in Sec. 8-12 for the rectangular slot. The types of slot antenna considered in this chapter include cavity-backed rectangular, annular, edge-slot on circular cylinder, and notch antennas.

8-2 SMALL RECTANGULAR SLOT IN INFINITE GROUND PLANE

The theoretical properties of a radiating slot in a flat sheet can be obtained from Booker's extension[1] of Jacques Babinet's principle, which shows that the field of the slot can be deduced from those surrounding a dipole of the same dimensions by interchanging the electric and magnetic vectors. The radiation field of a small rectangular slot such as shown in Fig. 8-1 is given by

$$E_\theta = -j\overline{E}_x \frac{\cos\phi}{2r\lambda} \, ab \, e^{-jkr} \qquad (8\text{-}1a)$$

$$E_\phi = j\overline{E}_x \frac{\cos\theta \sin\phi}{2r\lambda} \, ab \, e^{-jkr} \qquad (8\text{-}1b)$$

where $\overline{E}_x$ is the average of the x component of the electric field in the slot, $k = 2\pi/\lambda$, r is the distance from the slot, θ and ϕ are as defined in Fig. 8-1, and it is assumed that the electric field is parallel to the x axis. In the case $E_x = E_x(0) \cos \pi y/b$, then $\overline{E}_x = 2/\pi E_x(0)$. The principal plane radiation patterns of this magnetic current dipole are presented in Fig. 8-2. It is seen that the radiation pattern in the xz plane is omnidirectional and the pattern in the yz plane varies as $\sin\theta$. It is important to observe that the phase of the radiated field reverses on the two sides of the ground plane although the patterns are otherwise identical.

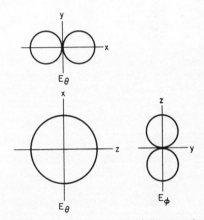

FIG. 8-2 Principal-plane field diagrams for a thin rectangular slot.

8-3 HALF-WAVE RADIATING SLOT IN INFINITE GROUND PLANE

Radiation Field

A thin rectangular slot ($a \ll b$) cut in a flat sheet of metal will be resonant when it is a half wavelength long. As in the case of the complementary wire antenna, the magnetic current distribution for the thin slot is approximately sinusoidal. The radiation pattern of the half-wave slot in the xz plane is the same as that given in Fig. 8-2. In the yz plane, however, the radiation pattern is given by Eq. (8-2):

$$E_\phi = E_\phi(0) \cos\left(\frac{\pi}{2}\sin\theta\right) \sec\theta \qquad (8\text{-}2)$$

where $E_\phi(0)$ is the field strength at the peak of the pattern.

Near Field

The near field of a slot antenna is of interest in determining the coupling of antennas to each other. Although the radiation pattern of a slot antenna is dependent upon the shape of the metal surface at large distances from the slot, the near fields attenuate rapidly for relatively short distances from the source. The near field of a half-wave slot antenna in a flat ground plane is thus approximately equal to the near field in any surface of large curvature. The fields due to a half-wave slot in an infinite conducting sheet are given by Eqs. (8-3a) to (8-3c):

$$H_z = -j \frac{E_0}{2\pi\eta} \left(\frac{e^{-jkr_1}}{r_1} + \frac{e^{-jkr_2}}{r_2} \right) \tag{8-3a}$$

$$H_\rho = j \frac{E_0}{2\pi\eta} \left(\frac{z - \lambda/4}{\rho} \frac{e^{-jkr_1}}{r_1} + \frac{z + \lambda/4}{\rho} \frac{e^{-jkr_2}}{r_2} \right) \tag{8-3b}$$

$$E_\phi = -j \frac{E_0}{2\pi} \frac{1}{\rho} (e^{-jkr_1} + e^{-jkr_2}) \tag{8-3c}$$

where ρ, ϕ, z are the coordinates of a cylindrical coordinate system in which the slot is coincident with the z axis, r_1 and r_2 are the distances from the ends of the slot to the point of observation, $\eta = 120\pi$ is the impedance of free space, E is expressed in volts per meter, and H is expressed in amperes per meter.

These theoretical expressions indicate that the electric field in the plane of the slot is zero along the axis of the slot inversely proportional to distance from the center along the x axis. Surfaces of equal phase are ellipsoids of revolution about the x axis.

The experimental work carried out by Putnam, Russell, and Walkinshaw[2] indicates that the distribution in a half-wavelength slot with a length-to-width ratio of 14:1 is closely sinusoidal but that within 20 percent of the resonant frequency there is a considerable departure from the sinusoidal distribution. Figure 8-3 illustrates the measured attenuation of the electric field along the sheet on the x axis. Figure 8-4a is a plot of the phase front of a half-wave slot at resonance, and Fig. 8-4b is a plot of the amplitude distribution at resonance. It should be noted that the phase velocity is greater than the free-space velocity near the center of the slot.

8-4 RADIATION CHARACTERISTICS OF A SLOT IN A FINITE FLAT SHEET

A typical slot antenna is an open-ended waveguide free to radiate from one side of a finite flat sheet. Theoretical radiation patterns are available for rectangular and circular waveguide apertures in a square ground plane.[3] These results have been derived by a hybrid solution which combines wedge diffraction with boundary-value analysis. The hybrid approach makes it possible to solve more complex problems than can be handled by the analytic boundary-value approach alone. Measured and theoretical E-plane patterns for the rectangular apertures in square ground planes with dimensions $L \times L$ are given in Fig. 8-5a and b. The patterns for a small circular aperture (not shown) are almost identical to those of the rectangular aperture.

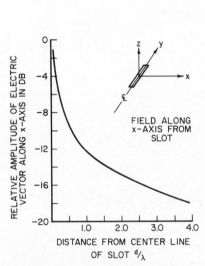

FIG. 8-3 Field attenuation near a half-wave slot.

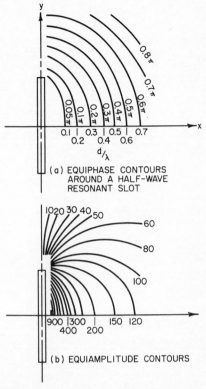

(a) EQUIPHASE CONTOURS AROUND A HALF-WAVE RESONANT SLOT

(b) EQUIAMPLITUDE CONTOURS

FIG. 8-4 Phase and amplitude contours near a half-wave slot.

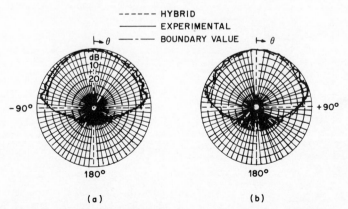

(a)　　　　　　　　(b)

FIG. 8-5 Rectangular waveguide ($a = 0.42\lambda$) in a flat ground plane ($L \times L$); xz plane, $\phi = 0°$. (a) $L/\lambda = 29.6$. (b) $L/\lambda = 11.8$.

The radiation in the shadow zone immediately behind the slot ($\theta \simeq 180°$) is appreciable even for large-diameter ground planes. In the illuminated region of the E-plane diagram ($0 < \theta < 90°$) it is seen that an interference pattern is generated by diffraction from the edges of the ground plane. The number of lobes in the interference pattern appearing in the forward region is equal to L/λ to the nearest integer, while those in the back region are equal to the nearest integer of $2L/\lambda$.

8-5   AXIAL SLOT IN A CIRCULAR CYLINDER

A cylinder of radius a is shown in Fig. 8-6, which is coaxial with a cylindrical coordinate system (ρ,ϕ,z). The radiation field of a slot cut along the z axis consists of E_ϕ and H_θ components referred to the indicated spherical coordinate system. If V_0 is the voltage across the center of a half-wave resonant slot,[4]

$$E_\phi = V_0 \frac{e^{-jkr} \cos\left(\dfrac{\pi}{2} \cos\theta\right)}{r \sin\theta} M(ka\sin\theta, \phi - \phi_0) \qquad \textbf{(8-4}a\textbf{)}$$

and

$$H_\theta = -\frac{E_\phi}{120} \qquad \textbf{(8-4}b\textbf{)}$$

The cylinder space factor $M(x,\phi)$ is given by

$$M(x,\phi) = \frac{1}{\pi^2} \sum_{m=0}^{\infty} \frac{\epsilon_m e^{jm\pi/2} \cos m\phi}{x H_m^{(2)\prime}(x)} \qquad \textbf{(8-5)}$$

where

$$\epsilon_0 = 1, \epsilon_m = 2 \qquad m \neq 0$$

For a small-diameter cylinder, that is, $x = ka\sin\theta \ll 1$,

$$M(ka\sin\theta,\phi) = \frac{1}{2\pi} e^{j\pi/2} \qquad \textbf{(8-6)}$$

The radiation pattern is omnidirectional, with phase 90° relative to V_0. The cylinder space factor $M(x,\phi)$ contains all the information pertinent to the radiation characteristic for a slot on a cylinder of any diameter. The relative amplitude of M is plotted as a radiation diagram in Fig. 8-7a to e as a function of ϕ with $ka\sin\theta$, that is, x, as a parameter; x takes on values between 0.1 and 21 in these graphs.

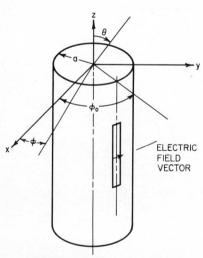

FIG. 8-6   Axial slot on a cylinder.

ELECTRIC FIELD VECTOR

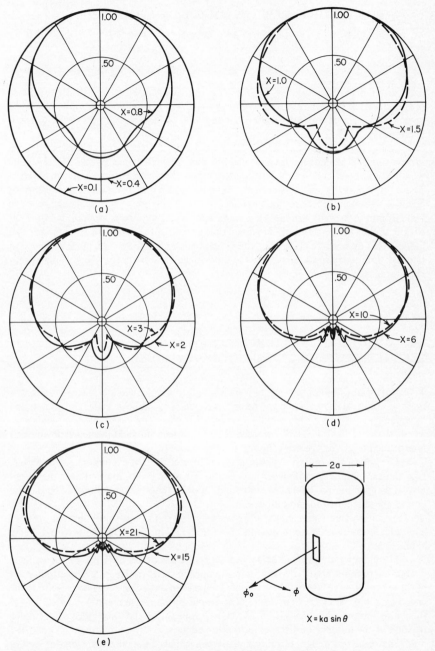

FIG. 8-7 Azimuth patterns of an axial slot on a circular cylinder: (*a*) *x* = 0.1, 0.4, 0.8. (*b*) *x* = 1.0, 1.5. (*c*) *x* = 2, 3. (*d*) *x* = 6, 10. (*e*) *x* = 15, 21. *(After Ref. 4.)*

The phase of M is α, which is given by

$$\alpha(x,\phi) = \alpha(x,0) - 57.3°\,x(1 - \cos \phi) + \Delta(x,\phi) \quad \text{for } \phi \leq 90° \qquad \textbf{(8-7}\textit{a}\textbf{)}$$

and

$$\alpha(x,\phi = \alpha(x,0) - x(57.3° + \phi - 90°) + \Delta(x,\phi) \quad 90° \leq \phi \leq 180° \qquad \textbf{(8-7}\textit{b}\textbf{)}$$

α is plotted directly in Fig. 8-8 for $x = 0.1, 0.4,$ and 0.8. Note that this is the phase of the radiation pattern relative to V_0 for $kr = 2N\pi$. Figure 8-9 is a family of graphs of Δ with x as a parameter between 1.0 and 15.0. The curves are displaced to avoid overlap.

Analysis of aperture radiation from an axially slotted circular conducting cylinder using the geometrical theory of diffraction (GTD) yields mathematical results which are in good agreement with the exact boundary-value solution for large ka.[5,6] The GTD solutions are economical (compared with the boundary-value solution) because they require less computer time.

Computed results for the TE_{10}-mode slot excitation are compared in Fig. 8-10 with boundary-value solutions. It is seen that agreement is better for smaller-size cylinders. This appears to contradict the GTD, which should become more accurate in the high-frequency limit. However, the discrepancy is inherent in the half-plane representation which has been used to model the slotted cylinder.

8-6 APERTURE RADIATION FROM AN AXIALLY SLOTTED ELLIPTICAL CONDUCTING CYLINDER

The equatorial radiation pattern of a parallel-plate TEM axial slot on an infinite cylinder of elliptical cross section has been computed by using both wedge diffraction and creeping-wave theory. The combined wedge-diffracted and creeping-wave fields represent the total field in the illuminated sector, whereas the field in the shadow region is produced solely by the creeping wave.

The equatorial radiation patterns ($\theta = 90°$) of a TEM-excited axial slot in various elliptic cylinders are shown in Fig. 8-11a, b, and c. The solid graphs are measured results using a short axial slot centered in a 30-wavelength-long cylinder. The dashed curves are calculated for an infinite slot.

An interesting comparison between measured data on a thin elliptical cylinder and calculations with a finite ground plane is shown in Fig. 8-12 (both with a TEM axial slot).

The agreement between theory and experiment is better for large cylinders since diffraction parameters are derived from asymptotic series for small λ. The attenuation of the creeping wave is greater for thinner-shape ellipses because these waves are influenced by rapid changes in curvature. The ripples in the region $-90° < \phi < 90°$ of the finite ground plane in Fig. 8-13 do not appear for the thin elliptical cylinder because of the absence of discrete diffraction edges.

The results of wedge-diffraction analysis applied to the elliptical cylinder for the TE_{10}-mode slot are shown in the radiation patterns in Fig. 8-13a and b for different ellipticity.

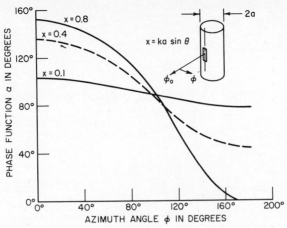

FIG. 8-8 Phase of cylinder space factor, $x = 0.1, 0.4, 0.8$. *(After Ref. 8.)*

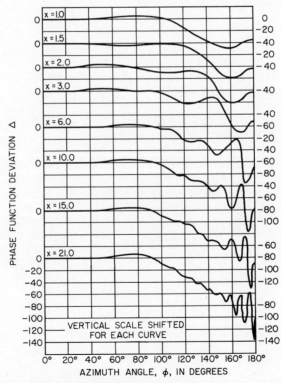

FIG. 8-9 Phase of cylinder space factor, $x = 1.0, 1.5, 2, 3, 6, 10, 15, 21$. *(After Ref. 4.)*

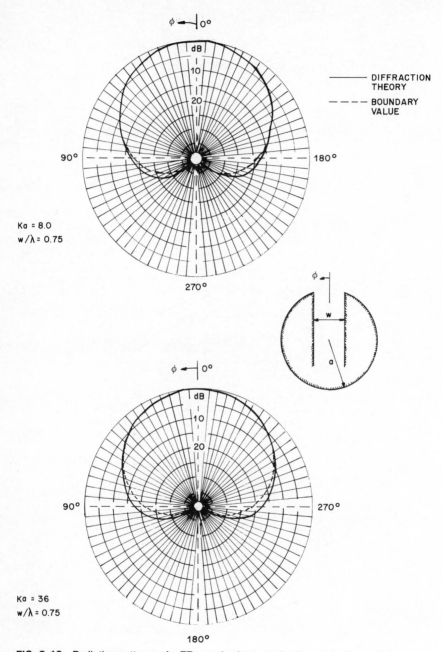

FIG. 8-10 Radiation patterns of a TE_{10}-mode slot on a circular conducting cylinder.

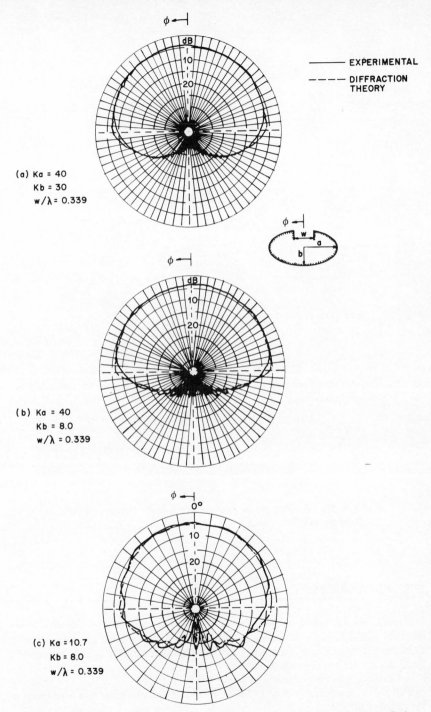

FIG. 8-11 Radiation patterns of an axial TEM slot on an elliptical conducting cylinder.

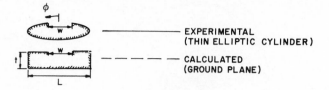

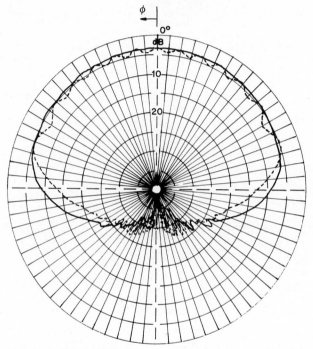

FIG. 8-12   Radiation patterns of a thin elliptical cylinder and finite ground plane (TEM mode).

8-7   ANNULAR SLOT

The radiation characteristics of an annular slot (cut in an infinite ground screen) are identical with those of a complementary wire loop with electric and magnetic fields interchanged. In the case of the small slot the radiation diagram is close to that of a small electric stub in the ground screen.

Consider a thin annular slot as shown in Fig. 8-14. The polar axis of a spherical coordinate system being normal to the plane of the slot, the magnetic component of the radiated field is

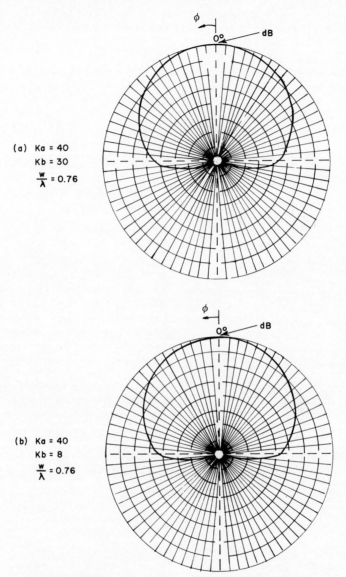

FIG. 8-13 Radiation patterns of a TE_{10}-mode slot on an elliptical cylinder by using diffraction theory.

(a) $Ka = 40$
   $Kb = 30$
   $\frac{w}{\lambda} = 0.76$

(b) $Ka = 40$
   $Kb = 8$
   $\frac{w}{\lambda} = 0.76$

$$H_\phi = \frac{aVe^{-jkr}}{120\pi\lambda r} \int_0^{2\pi} \cos(\phi - \phi')e^{jka \sin\theta \cos(\phi-\phi')} \, d\phi' \qquad \textbf{(8-8)}$$

where a = radius of slot
$\quad\quad V$ = voltage across slot
$\quad\quad k = 2\pi/\lambda$

For small values of a, that is, $a < \lambda/2\pi$,

$$H_\phi = j\frac{Ve^{-jkr}}{60r}\frac{A}{\lambda^2}\sin\theta \qquad \text{A/m} \qquad \textbf{(8-9)}$$

where $A = \pi a^2$ is the included area of the annular slot. Equation (8-9) is valid for small slots of arbitrary shape.

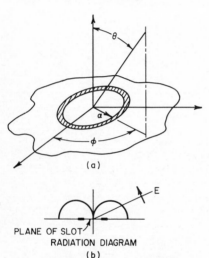

(a)

PLANE OF SLOT

RADIATION DIAGRAM

(b)

FIG. 8-14 Annular slot. (*a*) Coordinate system. (*b*) Vertical-plane pattern for a small-diameter slot.

The integral in Eq. (8-8) can be evaluated exactly as

$$H_\phi = j\frac{aVe^{-jkr}}{60r}J_1(ka \sin\theta) \qquad \textbf{(8-10)}$$

where J_1 is the Bessel function of the first kind and the first order. Figure 8-15*a* and *b* illustrates the vertical-plane patterns through the polar axis for values of $ka = 1, 2, 3.83,$ and 5. It will be noted that a null in the plane of the slot results for $ka = 3.83$, the zero of the Bessel function J_1.

The radiation characteristics on a large but finite ground screen are closely approximated by Eqs. (8-9) and (8-10). There are slight perturbations because of edge effects which result in energy radiated into the shadow region plus modulation of the main radiation pattern.

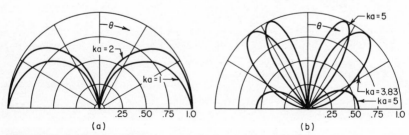

FIG. 8-15 Vertical-plane patterns of an annular slot on an infinite ground plane. (*a*) $ka = 1, 2.$ (*b*) $ka = 3.83, 5.$

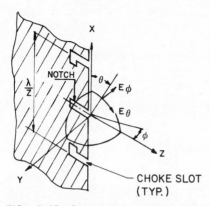

FIG. 8-16 Open-ended notch antenna (in conducting sheet).

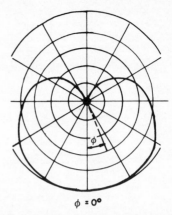

$\phi = 0°$

FIG. 8-17 Radiation field (E_θ) in yz plane for a notch antenna.

8-8 NOTCH ANTENNA

The notch, or open-ended, slot antenna (without choke slots) shown in Fig. 8-16 has been described.[7,8] The notch antenna is a broadband radiator which can be used to excite the empennage of aircraft, e.g., the leading edge of a wing or vertical stabilizer. The radiation patterns of the notch alone on a flat sheet will display interference diffraction lobes, which can be reduced by the use of choke slots. The choke slots are electrically $\lambda/4$ long and spaced $\lambda/4$ (approximately) from the excited notch. The radiation pattern in the xz plane is then similar to a dipole at an edge.

Figure 8-17 shows the pattern in the yz plane, which is a cardioid. Figure 8-18 illustrates the xz-plane pattern of the notch cut in a flat semicircular sheet of radius 1.75λ.

8-9 CONFORMAL DIELECTRIC-FILLED EDGE-SLOT ANTENNAS

A class of circumferential-slot antennas, designated as edge-slot antennas,[9] is suited to conformal mounting on conducting cylinders and cones. In its simplest form such an antenna is a disk of dielectric copper-plated on both sides. This disk is mounted between the mating two parts of the conducting body, which has been cut to accept the antenna. In this way the slot aperture coincides with the surface of the cylinder. The antenna is excited by a single coaxial stub at the center of the disk and is tuned for proper operating frequency by inductive posts that connect the two copper-plated sides of the disk (see Fig. 8-19). By varying the number and location of the inductive posts the operating frequency of a disk can be discretely tuned over a 6:1 range, but the instantaneous bandwidth is only 3 percent (typical). In practice, plated-through

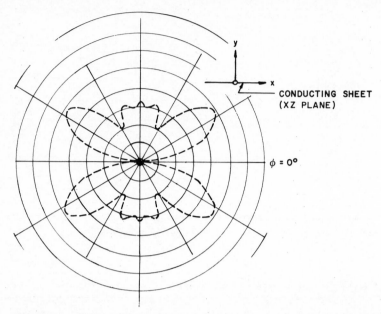

FIG. 8-18 Radiation field ($E\phi$) in *xy* plane for a notch antenna in a semicircular screen of 1.75λ radius.

holes are used as the inductive posts to provide mounting points and access holes for other equipment. The single coaxial feed excites azimuthally symmetric fields that are not significantly distorted by the symmetrically placed inductive posts, and the radiation patterns are omnidirectional. A typical directive gain is 2 dBi.

The resonant frequency and impedance of the edge-slot antenna are strongly affected by the number and location of the inductive tuning posts. No external network is needed to match the edge-slot antenna to a standard 50-Ω transmisssion line. A 76-mm-diameter antenna, for example, can be tuned from 660 to 3080 MHz by increasing the number and location of posts without changing its fundamental dimensions.

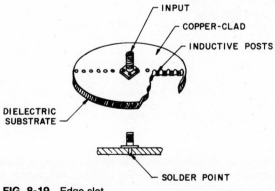

FIG. 8-19 Edge slot.

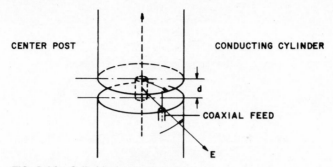

FIG. 8-20 Cylindrical gap antenna.

At least one pair of posts close to the edge is necessary for proper operation of the antenna at the lowest frequency. To lower the frequency further (for a fixed-dielectric-constant disk) the diameter of the antenna must be increased.

A cylindrical gap antenna[10] similar in appearance to the edge-slot antenna consists of two parallel matching faces of a cut cylinder separated by a small distance to form the gap. The two cylinders are also connected by a center post, and the edge gap is excited by a coaxial cable, as shown in Fig. 8-20. The antenna can be reinforced by using a cylindrical dielectric cover or by filling the gap with a dielectric. Center frequency is determined by the gap width d and the diameter of the center post. The equatorial-plane pattern is only omnidirectional within ± 2.5 dB, but for a cylinder of approximately $\lambda/2$ diameter a 10 percent bandwidth is possible with the addition of suitable matching elements at the junction of the feed and the use of a coaxial matching transformer.

8-10 RADIATION FROM CIRCUMFERENTIAL SLOT ON OR NEAR CONICAL CAP

The radiation patterns of a circumferential-slot antenna on a finite circular cylinder having conical or disk end caps can be computed by the techniques of wedge diffraction and edge currents.[11]

Figure 8-21a shows measured and calculated patterns for a circumferential-slot antenna mounted on the cylindrical section of a conically capped circular cylinder. The results of the calculations are in good agreement with measurement. Figure 8-21b shows measured and calculated patterns for the circumferential slot mounted on the conical section itself.

8-11 IMPEDANCE OF RECTANGULAR SLOT IN THIN FLAT METAL SHEET

The impedance Z_s of a slot antenna in a flat metal sheet, free to radiate on both sides, can be obtained directly from the admittance Y_D of the complementary wire antenna by using Babinet's principle:

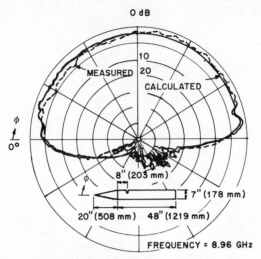

FIG. 8-21a Pattern of transverse $\lambda/2$ slot on a conically capped circular cylinder.

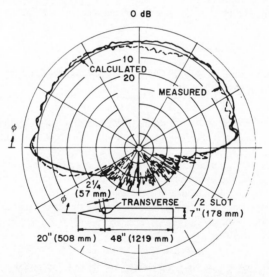

FIG. 8-21b Pattern of transverse $\lambda/2$ slot on a conical cap.

$$Z_s = \frac{1}{4} \zeta^2 \, Y_D \simeq 36{,}000 \, Y_D \qquad \textbf{(8-11)}$$

The solid-line graphs in Fig. 8-22*a* and *b* are the measured resistance and reactance respectively of a rectangular slot 25 cm long and 1 cm wide[12] cut in a large square metal plate (6.4 mm thick). Also shown are graphs of slot impedance based on Eq. (8.11). The complementary dipole admittance Y_D was derived by relating the

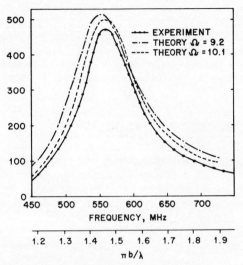

FIG. 8-22a Resistance of a slot antenna.

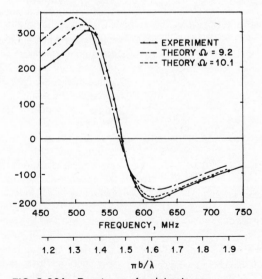

FIG. 8-22b Reactance of a slot antenna.

impedance of a thin strip of width d to an equivalent cylindrical antenna of radius a where $a = d/4$. It is seen that the theoretical curve for $\Omega = 9.2$ ($\Omega = 2 \ln b/a$) is in reasonable agreement with the measured results. An improved theoretical result is generated by compensating for the influence of the finite thickness of the plate on the effective slot width. A Schwartz-Christoffel transformation shows the effective width of the slot to be reduced by a factor of 1.55, i.e., from a thickness of 10 mm to 6.5 mm, i.e., $\Omega = 10.1$. The admittance graphs for $\Omega = 10.1$ are seen to be closer to the experimental results. The fit to the Babinet principle can be further improved by correcting for the small change in effective slot length also due to the finite thickness of the plate.

The resistance of the center-fed resonant slot antenna is a high impedance (400 Ω nominal). Since the characteristic impedance of a typical coaxial feeder is much less, e.g., 50 Ω, the frequency response of a coaxial-fed rectangular slot will also depend on the characteristics of the coaxial-to-slot-mode transducer.

8-12 CAVITY-BACKED RECTANGULAR SLOT

The electrical field on the coaxially fed, cavity-backed rectangular slot (Fig. 8-23 a) is neither sinusoidal nor complementary to a ribbon dipole antenna. This antenna is a cavity resonator energized by the coaxial transducer, which radiates from the slot aperture. The field distribution in the slot therefore is dependent on the excitation of higher cavity modes as well as the principal mode (TE_{10}). The equivalent circuit of a cavity-backed slot antenna is shown in Fig. 8-23 b; the shunt conductance is the radiation conductance of the slot. The conductance of the cavity-backed resonant half-wave slot is half of the open slot, free to radiate on both sides. That is, the shunt resistance is at least 800 rather than 400 Ω. The parallel susceptance shown in the equivalent circuit is the sum of the shunt susceptance of the slot radiator and the TE-mode susceptances of the cavity. The series-resonant circuit is the result of the energy stored in the TM modes in the cavity and feed structure.

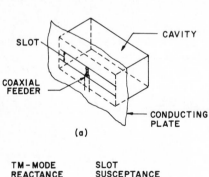

(a)

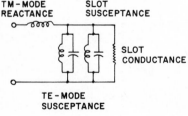

(b)

FIG. 8-23 Cavity-backed rectangular slot. (a) Pictorial representation. (b) Equivalent circuit.

To obtain the maximum radiation conductance a sinusoidal distribution of electric field (magnetic current) must be generated. This distribution will be achieved when the energy stored in the cavity in the vicinity of the slot is primarily in the TE_{10} mode, i.e., by making the cavity dimensions big enough so that the dominant mode is above cutoff. For small cavities, edge loading as in a highly capacitive slot will improve the field distribution.

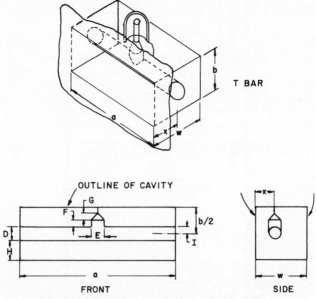

FIG. 8-24 Cavity-backed T-bar-fed slot antenna. Typical dimensions in inches (millimeters) for a frequency range of 0.5–1.2 GHz: *a*, 12.00 (304.8); *b*, 4.00 (101.6); *x*, 3.25 (82.55); *w*, 6.75 (171.45); *D*, 2.25 (57.15); *E*, 0.75 (19.05); *F*, 0.63 (16.002); *G*, 0.19 (5.826); *H*, 0.25 (6.35).

An important design parameter is the antenna Q, which is minimum when the stored energy is only in the dominant mode. The Q limits the inverse voltage-standing-wave-ratio (VSWR) bandwidth product; for a small cavity it is

$$Q > \frac{3}{4\pi^2}\left(\frac{1}{V}\right) \qquad\qquad \textbf{(8-12)}$$

where V is the volume of the cavity expressed in cubic wavelengths. This minimum Q is realized when the series reactance shown in Fig. 8-23*b* is eliminated through efficient feed and cavity design. For the simple capacitive slot-loaded cavity shown in Fig. 8-23*a*, higher TE and TM modes will be generated with attendant high Q.

A broadband cavity-backed antenna can be realized by using a T-bar feed[13] as shown in Fig. 8-24. A flat T bar instead of the illustrated circular cross section will generate the same impedance if its width is equal to the diameter D.

The nominal input impedance to the T bar is 125 Ω (approximate). To achieve the available bandwidth a broadband impedance transformer is needed between the 50-Ω coaxial transmission line and the T-bar junction.

The dimensions for a broadband *flat* T bar covering the frequency range 0.5–1.2 GHz are shown in the diagram of Fig. 8-24. The VSWR does not exceed 3:1 in the frequency band and is less than 2:1 over 90 percent of the band.

The resonant frequency can be lowered by the use of dielectric or ferrite loading in the cavity.[14] The reduction in cavity volume and aperture size results in increased Q, smaller bandwidth, and lower efficiency. The test results of loaded rectangular-

TABLE 8-1 Performance Comparison of Rectangular-Cavity Slot Antennas

Loading	Air	Ferrite powder	Ferrite solid	Dielectric
Size, in (mm)	$30 \times 7\frac{1}{2} \times 10$	$12 \times 3 \times 4$	$5 \times 2 \times 1\frac{1}{2}$	$12 \times 3 \times 5$
	$(762 \times 190.5 \times 254)$	$(304.8 \times 76.2 \times 101.6)$	$(127 \times 50.8 \times 38.1)$	$(304.8 \times 76.2 \times 127)$
Volume, in³ (cm³)	2250 (36,870)	144 (2360)	15 (246)	180 (2950)
Bandwidth, MHz, at VSWR = 3	. . . .	22	19	10
Bandwidth, MHz, at VSWR = 6	. . . .	50	34	18
Efficiency, percent	90	65	30	85
Directivity, dBi	5.8	5	5	5
Weight, lb (kg)	25.8 (11.7)	16.8 (7.6)	3.6 (1.6)	14.5 (6.6)
Frequency, MHz	300	320	352	316

cavity slot antennas are presented for powdered ferrite, solid ferrite, and solid dielectric material in Table 8-1. A coaxial probe transducer was used, but since its circuit Q is much less than that of the slot cavity itself, the probe does not significantly influence the total Q. The reduction in volume relative to air loading is evident. The characteristics of the materials at $20°C$ and 300 MHz are: ferrite powder, $\mu' = 8.3$, $\epsilon = 2.2$, magnetic $Q = 30$; solid ferrite, $\mu' = 6.63$, $\epsilon' = 12.6$, magnetic $Q = 30$; solid dielectric, $\epsilon' = 10$, electric $Q > 1000$.

8-13   CROSSED-SLOT RIDGED-CAVITY ANTENNA

A shallow ridged-cavity crossed-slot antenna[15] has been developed for wide-angle coverage in the ultrahigh-frequency range. The VSWR is dependent on the slot width, slot length, and cavity depth at the low end of the band. The ridge parameters tune the antenna in the midband and high-band frequencies. The VSWR is less than 2.7:1 from 240 to 279 MHz and under 2.1:1 from 290 to 400 MHz for cavity dimensions of 33 by 33 by 4 in.

An experimental square cavity-slot antenna (half scale) with crossed slots cut along the diagonal dimensions of the cavity is shown in Fig. 8-25. The cavity configuration, ridges, and crossed-slot arrangement are illustrated. The slots are excited by four symmetrically located feed probes near the center of the cavity. Each opposite pair is connected to a wideband $180°$ hybrid. For circular polarization, the input ports of the two $180°$ hybrids are connected to a wideband $90°$ hybrid. The 3-dB beamwidth varies from $120°$ at the low end of the band to about $40°$ at the high end.

8-14   ADMITTANCE OF ANNULAR SLOT

The optimum excitation of an annular slot, i.e., least stored energy and lowest Q, results when the magnetic current distribution is uniform around the slot. One method for obtaining this result is to feed the annular slot by a coaxial transmission-line structure which has the same inner and outer diameter as the annular structure. Figure 8-26a and b consists of graphs of the conductance and susceptance in the plane of the aperture relative to the characteristic admittance of the feed line as a function of the radian length ka of the inner radius a. It is seen that the slot is at all times nonresonant and has a capacitive susceptance.

8-15   NOTCH-ANTENNA IMPEDANCE[8]

The radiation resistance of a notch antenna at the top of the notch is

$$R = \frac{2\eta}{3[C(k\ell) - \cot k\ell \; S(k\ell)]^2} \tag{8-13}$$

where $\eta = 377 \; \Omega$
C and S = even and odd Fresnel integrals respectively

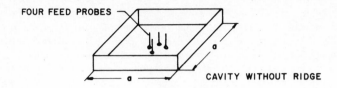

FOUR FEED PROBES

CAVITY WITHOUT RIDGE

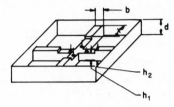

CAVITY WITH RIDGE

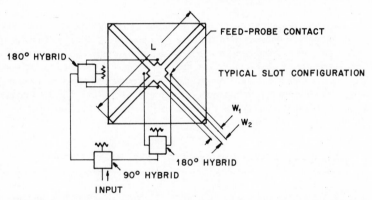

180° HYBRID

FEED-PROBE CONTACT

TYPICAL SLOT CONFIGURATION

180° HYBRID

90° HYBRID

INPUT

FIG. 8-25 Cavity-slot configuration with circular polarization of a ridged-cavity crossed-slot antenna. Typical dimensions in inches (millimeters) for a frequency range of 480–800 MHz: *a*, 16.00 (406.4); *d*, 2.00 (50.8); *b*, 1.25 (31.75); W_1, 2.50 (73.5); W_2, 0.65 (16.51); h_1, 1.50 (38.1); h_2, 1.75 (44.45); *L*, 22 (558.8).

The resistance appearing in shunt across the mouth of the notch is given by

$$X = jZ_\ell \tan k\ell \qquad \text{(8-14)}$$

where Z_ℓ, the characteristic impedance of the notch transmisssion line, is given by

$$Z_\ell = \frac{60\pi^2}{\log_e \dfrac{16\ell}{w} - 1} \qquad \text{(8-15)}$$

ℓ is the length of the notch, and *w* its width. The input impedance can be reduced by moving the feed point away from the top of the slot. The addition of the choke slots also influences the impedance of the antenna for $X = 0$. Figure 8-27 is a graph of the radiation resistance as a function of ℓ/λ.

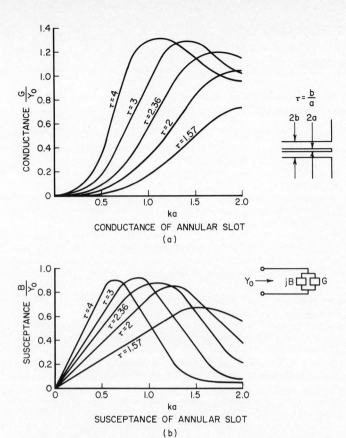

CONDUCTANCE OF ANNULAR SLOT
(a)

SUSCEPTANCE OF ANNULAR SLOT
(b)

FIG. 8-26 Admittance of coaxial-fed annular slot in infinite ground plane.

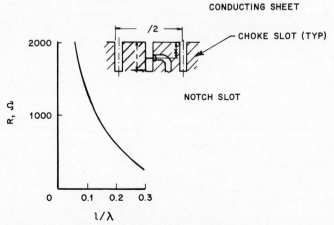

FIG. 8-27 Radiation resistance of notch antenna ($X = 0$); no choke slots.

REFERENCES

1 H. G. Booker, "Slot Aerials and Their Relation to Complementary Wire Aerials," *J. IEE (London),* part IIIA, vol. 93, 1946, pp. 620–626.

2 J. L. Putnam, B. Russell, and W. Walkinshaw, "Field Distributions near a Center Fed Half-Wave Radiating Slot," *J. IEE (London),* part III, vol. 95, July 1948, pp. 282–289.

3 C. A. Balanis, "Pattern Distortion Due to Edge Diffractions," *IEEE Trans. Antennas Propagat.,* vol. AP-18, no. 4, July 1970, pp. 561–563.

4 J. R. Wait, "Radiation Characteristics of Axial Slots on a Conducting Cylinder," *Wireless Eng.,* December 1955, pp. 316–323.

5 C. A. Balanis and L. Peters, Jr., "Aperture Radiation from an Axially Slotted Elliptical Conducting Cylinder Using Geometrical Theory of Diffraction," *IEEE Trans. Antennas Propagat.,* vol. AP-17, no. 4, July 1969, pp. 507–513.

6 C. A. Balanis and L. Peters, Jr., "Radiation from TE_{10} Mode Slots on Circular and Elliptical Cylinders," *IEEE Trans. Antennas Propagat.,* vol. AP-18, no. 3, May 1970, pp. 400–403.

7 R. H. J. Cary, "The Slot Aerial and Its Application to Aircraft," *IEE Proc. (London),* part III, vol. 99, July 1952, pp. 187–196.

8 W. A. Johnson, "The Notch Aerial and Some Applications to Aircraft Radio Installations," *IEE Proc. (London),* part B, vol. 102, March 1955, pp. 211–218.

9 D. H. Schaubert, H. S. Jones, Jr., and F. Reggia, "Conformal Dielectric-Filled Edge-Slot Antennas with Inductive-Post Tuning," *IEEE Trans. Antennas Propagat.,* vol. AP-27, no. 5, September 1979, pp. 713–715.

10 G. T. Swift, T. G. Campbell, and H. Hodara, "Radiation Characteristics of Cavity-Backed Cylindrical Gap Antenna," *IEEE Trans. Antennas Propagat.,* vol. AP-17, no. 4, July 1969, pp. 467–477.

11 C. E. Ryan, Jr., "Analysis of Antennas on Finite Circular Cylinders with Conical or Disk End Caps," *IEEE Trans. Antennas Propagat.,* vol. AP-20, no. 4, July 1972, pp. 474–476.

12 S. A. Long, "Experimental Study of the Impedance of Cavity-Backed Slot Antennas," *IEEE Trans. Antennas Propagat.,* vol. AP-23, no. 1, January 1975, pp. 1–7.

13 E. H. Newman and G. A. Thiele, "Some Important Parameters in the Design of T-Bar Fed Slot Antennas," *IEEE Trans. Antennas Propagat.,* vol. AP-23, no. 1, January 1975, pp. 97–100.

14 A. T. Adams, "Flush Mounted Rectangular Cavity Slot Antennas—Theory and Design," *IEEE Trans. Antennas Propagat.,* vol. AP-15, no. 3, May 1967, pp. 342–351.

15 H. E. King and J. L. Wong, "A Shallow Ridged-Cavity Crossed-Slot Antenna for the 240- to 400-MHz Frequency Range," vol. AP-23, no. 5, September 1975, pp. 687–689.

Chapter 9

Slot-Antenna Arrays

Hung Yuet Yee

Texas Instruments, Inc.

9-1 INTRODUCTION

Slot-antenna arrays have been used in many ground-based and airborne radar systems since the 1940s. Waveguide-fed slot-antenna arrays are used as resonant and traveling-wave antennas when precise amplitude and phase control are needed. A typical airborne radar resonant slotted array with four-lobe monopulse is shown below in Fig. 9-18. A sidelobe level of -28 dB within a 2 percent frequency bandwidth is the state of the art for this antenna. In recent years, demand for the operation of radar systems with lower sidelobe levels and greater efficiency has required more accurate estimation of waveguide slot characteristics, improved modeling of array effects, and greater manufacturing accuracies.

Discussions in this chapter will be limited to rectangular-waveguide-fed slot arrays. Transmission-line-fed slots using strip transmission line (triplate or microstrip) are reported in available literature.[1,2,3] The advantages of strip-transmission-line-fed slots are manufacturing simplicity, low cost, and low profile. However, the use of strip-line-fed slots for radar applications is limited by excessive transmission-line loss, power handling, and performance dependence on dielectric-material variations.

Both resonant and nonresonant (traveling-wave) waveguide-fed slot-antenna arrays can be configured into linear or planar arrays, but the majority of recent airborne radar antenna designs have used planar resonant slot arrays (flat plates). Design of linear slot arrays with uniform element spacing can be found in the literature.[4,5] Nonuniformly spaced array designs have received little attention in the literature and will be discussed in greater detail in this chapter. The linear-array design will be used as a basis for understanding the planar slot-array design in this chapter; however, emphasis will be placed on practical planar slot-array design techniques.

The following sections are organized to give the antenna designer the broad technical background necessary for waveguide-fed slot-array design. Various slot radiators with basic operating theories and mathematical representations are first introduced and then followed by analytical methods for slot characterization which can be used for future designs. Typical characteristics are presented for slots cut in a standard X-band RG52/U waveguide with a 0.0625-in (1.5875-mm) slot width and a frequency of 9.375 GHz. These data can be used to design a planar slotted array. Basic radiation characteristics and applications of resonant waveguide slotted arrays and traveling-wave slotted arrays are detailed with emphasis on practical design methods including mutual-coupling effects. The nonuniformly spaced traveling-wave slotted-waveguide array design is included. Finally, power-handling capacity, tolerance effects, and fabrication techniques are detailed for practical application.

9-2 SLOT RADIATORS

The slot is a commonly used radiator in antenna systems. An attractive feature is that slots can be integrated into the array feed system, such as a waveguide or stripline system, without requiring a special matching network. Low-profile high-gain antennas can be easily configured by using slot radiators, although their inherent narrow-frequency bandwidth can limit antenna performance in some applications.

A slot cut into the waveguide wall which interrupts the flow of currents will couple power from the waveguide modal field into free space. Depicted in Fig. 9-1 are the

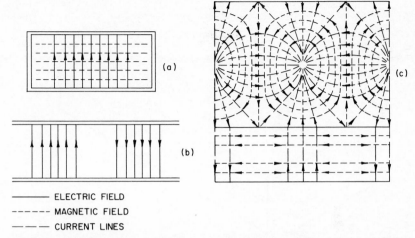

ELECTRIC FIELD
------ MAGNETIC FIELD
——— CURRENT LINES

FIG. 9-1 Surface-current distribution for rectangular waveguide propagating TE$_{10}$ mode. (*a*) Cross-sectional view. (*b*) Longitudinal view. (*c*) Surface view.

surface currents flowing along the rectangular-waveguide inner wall when only the TE$_{10}$ mode is excited in the waveguide. A singly moded waveguide is required for a slotted-waveguide array design to control the aperture illumination. Commonly used slot types are shown in Fig. 9-2. Slots cut in the rectangular waveguide as shown in Fig. 9-3 are nonradiating because they are parallel to the surface-current vector.

Slots are conveniently classified by shapes, locations in the rectangular wave-guide, and equivalent-network representations. Edge-wall slots (slots *b* and *c* in Fig.

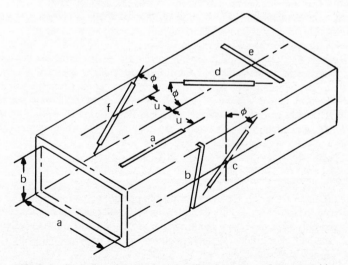

FIG. 9-2 Radiating slots cut in the walls of a rectangular waveguide.

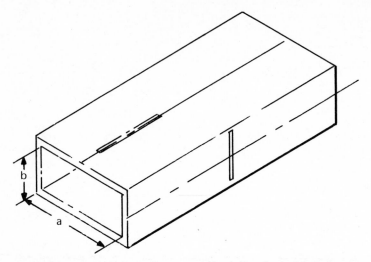

FIG. 9-3 Nonradiating slots cut in the walls of a rectangular waveguide.

9-2) are cut into the narrow wall and wrapped around the broad wall of the rectangular waveguide, where the resonant slot length is determined by the depth of the cutting. In designing a planar array using edge slots, the wrap around the broad wall contributes to both design and fabrication difficulties. In this circumstance, folded edge-wall slots such as C and I slots (see Fig. 9-4) can be used to replace the simple edge slot.

A longitudinal shunt slot (slot a in Fig. 9-2) and an inclined series slot (slot d in Fig. 9-2) are cut on the rectangular-waveguide broad walls; they are distinguished by their equivalent-impedance representations. A rectangular slot shape is convenient for analytical investigation, but in most applications the shape is chosen to simplify the fabrication process. Many arrays of broad-wall slots employ a rectangular shape with round ends formed by the milling process.

To design a slotted-waveguide antenna array, the basic characteristics for a given slot geometry, such as the radiated-field amplitude and phase plus mutual-coupling effects between slots and within the feed network, are needed. From the standpoint of array design and simple analysis, equivalent impedance is the most convenient slot characteristic. However, the scattering-matrix representation is required in the analysis and design of certain large slotted-waveguide arrays in which complicated feed networks and the slots are coupled.

Equivalent-Impedance Representation

The simple single-element equivalent circuit is a preferred slot representation conventionally used in slotted-waveguide design. This single-element-impedance representation is valid only for slot widths small compared with a wavelength and slot lengths less than one-half waveguide wavelength; otherwise, the more complicated equivalent T or pi network is required.[6]

Narrow slots such as slots a, b, and c in Fig. 9-2 interrupt only the transverse

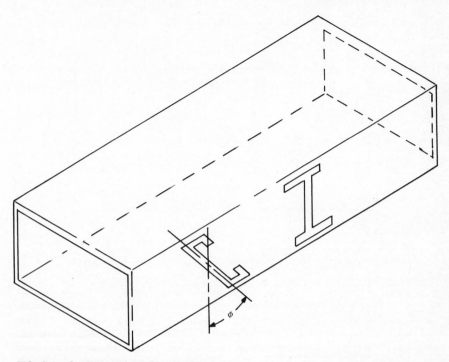

FIG. 9-4 Configurations of C and I slots.

currents and are represented by a simple two-terminal shunt admittance; therefore, they are identified as shunt slots. Similarly, slots d and e interrupt only the longitudinal currents and are represented by a simple series impedance; they are referred to as series slots. Slot f is coupled to both longitudinal and transverse currents, and a T- or pi-network representation is necessary even for a narrow slot width. The equivalent circuits for three frequently used slots are shown in Fig. 9-5.

Scattering-Matrix Representation

Waveguide-fed slots can also be represented by scattering matrices. A scattering-matrix representation is convenient for analyzing slot-antenna performance by taking into account the coupling of slots to other slots or to the feed network. An example is a resonant waveguide section with several series slots cut on the broad wall fed by an E-plane T junction as shown in Fig. 9-6. The characterization of the series feed with a series slot directly above the T junction requires a scattering-matrix representation. Let the waveguide-fed slot be represented by a stub guide branching out from the main guide. If the slot is facing a T junction on the opposite side of the guide, a four-port scattering matrix representing the T junction and the stub guide can be derived by solving the physical problem. Combining this scattering matrix with other slot scattering matrices, the power coupled out from each slot can be computed by taking into account the coupling between the slot and the feed junction.

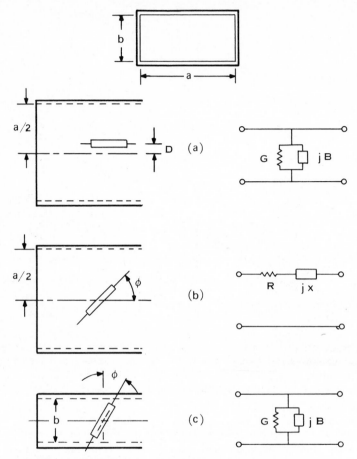

FIG. 9-5 Equivalent-network representations of slot radiators.

9-3 COMPUTATION METHODS FOR SLOT CHARACTERIZATION

Extensive analytical and experimental studies of the characteristics of waveguide-fed slots have been performed since the 1940s. Conventional impedance-measurement techniques and the moving lossy short techniques[7] have been successfully applied to measure slot characteristics. Since these techniques have been detailed in the literature, they will not be discussed here. The general approaches of four theoretical computation methods for slot characteristics are discussed in the following subsections.

Stevenson's Solution

Stevenson[8] pioneered the computation of slot characteristics by using the following assumptions: (1) a perfectly conducting thin wall, (2) a narrow slot, (3) a slot length

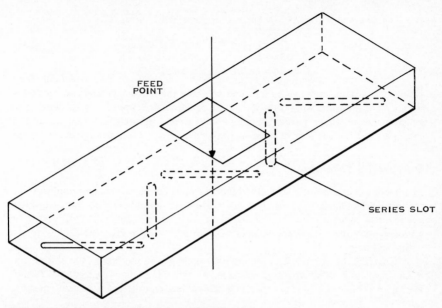

FIG. 9-6   Typical waveguide section with series slots.

nearly equal to half free-space wavelength, and (4) a perfectly conducting ground plane of infinite extent. Using transmission-line theory and the waveguide modal Green functions, Stevenson derived the values of the resonant resistance and conductance, normalized to the waveguide impedance, for various slots cut in a rectangular waveguide.

1   Conductance of the longitudinal shunt slot in the broad face (slot a in Fig. 9-5):

$$g = g_1 \sin^2 (D\pi/a) \qquad\qquad \textbf{(9-1)}$$

$$g_1 = (2.09 a\lambda_g/b\lambda) \cos^2 (\lambda\pi/2\lambda_g) \qquad\qquad \textbf{(9-2)}$$

where λ is the free-space wavelength, λ_g is the guide wavelength, D is the slot displacement from the guide centerline, and a and b are the waveguide width and height.

2   Resistance of the centered inclined series slot on the broad face (slot b of Fig. 9-5):

$$r = 0.131(\lambda^3/ab\lambda_g)[I(\phi) \sin \phi + (\lambda_g/2a)J(\phi) \cos \phi]^2 \qquad \textbf{(9-3)}$$

$$\left.\begin{matrix} I(\phi) \\ J(\phi) \end{matrix}\right\} = \frac{\cos (\pi\xi/2)}{1 - \xi^2} \mp \frac{\cos (\pi\zeta/2)}{1 - \zeta^2} \qquad\qquad \textbf{(9-4)}$$

$$\left.\begin{matrix} \xi \\ \zeta \end{matrix}\right\} = \frac{\lambda}{\lambda_g} \cos \phi \pm \frac{\lambda}{2a} \sin \phi \qquad\qquad \textbf{(9-5)}$$

where ϕ is the slot inclination angle with the guide centerline.

3 Conductance of edge slot on the narrow face (slot c of Fig. 9-5):

$$g = \frac{30\lambda^3\lambda_g}{73\pi a^3 b}\left[\frac{\sin\phi\,\cos\,(\pi\lambda\,\sin\,\phi/2\lambda_g)}{1-(\lambda\,\sin\,\phi/\lambda_g)^2}\right]^2 \tag{9-6}$$

These equations yield quite accurate results. However, they are valid only for a single-slot radiator and give no information about the reactive component. In particular, significant errors will result when these equations are applied to most two-dimensional arrays in which mutual-coupling effects cannot be neglected. Equations (9-1) through (9-6) are useful only when mutual-coupling effects are negligible.

Variational Technique

Application of the variational method to slot equivalent-circuit computation was first formulated by Oliner.[9] This method can account for mutual coupling and also reveal the reactive component of the slot impedance. Accurate impedance solutions can be obtained with the variational technique by using a slot-aperture field close to the actual slot field.

Consider the example of a longitudinal shunt slot in the broad face of a rectangular waveguide radiating into free space. By extending Oliner's variational formulation, an explicit impedance expression can be derived in terms of the slot-aperture admittance $Y_r = G_r + jB_r$.[10]

Let the length and the width of the rectangular slot be a' and b' respectively. The propagation constant and the characteristic admittance of the dominant mode in the waveguide are given by $\kappa = (k^2 - (\pi/a)^2)^{1/2}$ and $Y_0 = \kappa/\omega\mu$, where $k = 2\pi/\lambda$, ω is the angular frequency, and μ is the permeability. The rectangular slot on the thick waveguide wall is regarded as a stub guide with propagation constant κ' and characteristic admittance Y_0'. The input admittance at the stub-guide junction can be determined by

$$Y_{rj} = Y_0'\frac{Y_r + jY_0'\tan\kappa't}{Y_0' + jY_r\tan\kappa't} \tag{9-7}$$

where t is the thickness of the guide wall. From the standpoint of the feed waveguide, the equivalent circuit of a displaced longitudinal shunt slot can be represented by a shunt network as depicted in Fig. 9-7. Z_{rj} is the impedance due to the radiating aperture, while X_j is related to the stored power in the feed waveguide. Near resonance, these quantities can be approximated by the variational result[10]

$$\frac{R + jX}{Z_0} = \frac{Z_{rj} + jX_j}{Z_0} = \frac{a'b'}{2N_s^2}\frac{(Y_{rj} + jB_j)}{Y_0} \tag{9-8}$$

$$N_s^2 = (2/ab)\frac{2\pi^2 b'}{aa'\kappa}\left[\frac{\sin\,(\pi D/a)\,\cos\,(\kappa a'/2)}{(\pi/a')^2 - \kappa^2}\right]^2 \tag{9-9}$$

where D is the displacement of the slot from the center of the waveguide. The junction susceptance B_j is given by

$$B_j = \frac{4b'}{\omega\mu ab}\sum_{\substack{m=0\\n=0}}\left[\cos\left(\frac{m\pi}{2} + \frac{m\pi D}{a}\right)\text{sinc}\left(\frac{m\pi b'}{2a}\right)\right]^2 \text{Im}(W)/(\epsilon_m\epsilon_n) \tag{9-10}$$

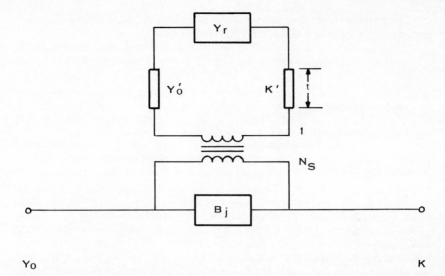

FIG. 9-7 Equivalent network of a longitudinal shunt slot.

where sinc $(x) = \sin x / x$, $W = j[Q_2(m,n) + Q_3(m,n)]$

$$Q_2(m,n) = \kappa'^2 / [(\pi/a)^2 - \kappa_{mn}^2]$$

$$Q_3(m,n) = \left| 2(\pi/a')^2 \frac{(m\pi/a)^2 + (n\pi/b)^2}{[(\pi/a')^2 - \kappa_{mn}^2]^2} \frac{1 + \exp(-ja'\kappa_{mn})}{ja'\kappa_{mn}} \right.$$

$$\kappa_{mn} = [k^2 - (m\pi/a)^2 - (n\pi/b)^2]^{1/2}$$

$$\epsilon_0 = 2$$

$$\epsilon_n = 1 \text{ if } n > 0$$

Im indicates the imaginary part. This solution can account for mutual-coupling effects if the slot-aperture admittance Y_{rj} is computed under the mutual-coupling environment.

Resonance can be defined as the condition in which slot admittance is a real quantity. To compute the slot resonant length, several slot admittances corresponding to several values of a' nearly equal to one-half wavelength are computed first. The slot resonant length can then be determined by interpolating the length which yields zero imaginary part for the admittance.

Note that the above solution is applicable to rectangular slots only. For round-ended slots, the resonant length can be approximated by

$$L = a' + Cb'(1 - \pi/4) \tag{9-11}$$

where a' is the resonant length of the equivalent rectangular slot and C is the correction factor.[10]

A similar variational formulation has been applied to compute the impedance of an inclined series slot in the rectangular-waveguide broad face. The slot inclination angle replaces the variable of slot displacement from the waveguide centerline in the derivation. However, the slow convergent series in the resultant reactance expression requires special treatment in the numerical computation.

Because of the complicated slot shape and the unknown field distribution in the slot aperture, variational solutions for edge slots and folded slots have not been derived.

Scattering Modal Analysis

Using an approach similar to Frank and Chu's formulation of the E-plane T solution,[11,12] Montgomery derived the slot-characterization scattering matrices and hence the shunt-slot admittance and series-slot impedance.[13] This scattering modal analysis technique requires a greater computation effort than Stevenson's and the variational solutions.

Basically, the slot is assumed to be a stub guide branched out from the main guide. The slot field is conventionally expressed by a series of stub-guide modal functions. The main-guide field is represented by a series of modal Fourier integrals in which the wave number corresponding to the guide axis is the integration variable. By matching the two tangential field expressions on both sides of the slot aperture, two integral equations are obtained. The expansion coefficients and the scattering parameters can be determined by solving these two integral equations by using the Ritz-Galerkin method. After the expansion coefficients have been computed, the junction scattering matrix of the slot in the waveguide can be calculated easily. By combining the junction scattering matrix with the slot-radiating-aperture reflection coefficient, the slot equivalent circuit can be determined.

This method has been applied to the computation of individual slot characteristics and to the analysis of coupling effects between two slots or between a slot and a waveguide junction. Numerical results of longitudinal-slot parameters computed by the variational method and the scattering modal analysis compare favorably.

Moment Method

The method of moments has been applied to characterize a narrow slot in the broad wall of a rectangular waveguide[14,15] and can be extended to obtain the solution of folded slots in the edge wall with additional effort. The moment method requires greater computational effort than any of the three other methods discussed here.

Vu Khac and Carson[14] formulate the slot characterization by using the field-equivalent principle to replace the slot aperture by short circuits and magnetic currents. The appropriate magnetic Green's functions in the feed guide, inside the slot cavity (taking into account the finite wall thickness), and in free space are employed to derive the basic integral equations. The slot field is expressed as a series of discrete impulse functions. By matching the tangential components at the slot aperture and using the Ritz-Galerkin method, a matrix equation is derived. Then the expansion coefficients can be determined by matrix inversion. The computational effort can be greatly reduced by using the sinusoidal basis functions as shown by Lyon and Sangster.[15] The computed results obtained by using the moment method closely match the variational solution and experimental results.

When the moment method is applied to characterize the inclined series slots, a slowly convergent series is encountered in the same manner as with the variational formulation and the scattering modal analysis. All the matrix elements used in the moment method involve one or two slowly convergent series compared with only one slowly convergent series in the variational technique; therefore, the moment method requires much more computation time.

9-4 DESIGN PARAMETERS

Many slotted-waveguide arrays have been fabricated since the 1940s. Early slot-antenna array designs depended primarily on measured slot characteristics. Accurately measured design data require a major effort to design and fabricate many precision test pieces. Analytical methods for characterizing broad-wall slots have since been developed into useful tools for practical slotted-waveguide array design. Only 2 to 3 h of computer time is required to obtain design parameters which would have required several months of experimentation. A planar slotted-waveguide array with more than 1000 longitudinal shunt slots has been fabricated successfully by using analytical data.

The measured and computed slot characteristics presented in this chapter, except that shown later in Fig. 9-13, are given for a standard X-band RG52/U waveguide with a 0.0625-in (1.5875-mm) slot width and a frequency of 9.375 GHz. This slot will subsequently be referred to as the *baseline slot*. A planar slotted-waveguide array can be designed at 9.375 GHz by using data presented in this section.

Longitudinal Shunt Slots

Watson[16] and Stegen[5] performed extensive experimental studies of isolated longitudinal shunt slots radiating into free space. From the measured data, Stegen found that the ratio of conductance to resonant conductance G/G_m and the ratio of susceptance to resonant conductance B/G_m versus the ratio of slot length to resonant length are independent of the slot displacement off the waveguide centerline. These universal curves of the baseline-slot admittance normalized by the resonant conductance versus slot length normalized by the resonant length are plotted in Fig. 9-8. The measured ratio of slot resonant length to wavelength and the measured ratio of slot resonant conductance to waveguide wave admittance versus the slot displacement off the waveguide centerline are shown in Figs. 9-9 and 9-10 respectively.

By using the variational result of Eqs. (9-7) through (9-10), the computed slot admittance normalized by the resonant conductance is also shown in Fig. 9-8. The aperture admittance of a slot radiating into the half space is approximated by the solution given by Kurss and presented in Ref. 9. Observe that the computed results agree very well with the measured data. However, the computed results show that the ratios G/G_m and B/G_m versus the normalized slot length L/L_0 are not exactly independent of the slot displacement off the guide centerline.

By using the same technique, the computed slot resonant length and resonant conductance are shown in Figs. 9-9 and 9-10 respectively for comparison with the measured results. Note that the round-ended-slot resonant length L_0 is approximated by Eq. (9-11). Choosing C equal to 1.7 and comparing the computed and measured results show that the computation method is sufficiently accurate for practical applications.

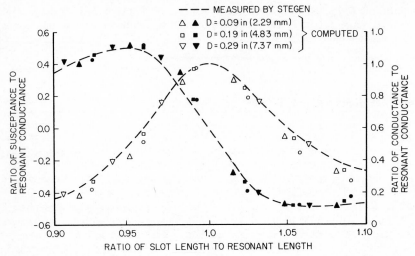

FIG. 9-8 Admittance of the baseline longitudinal shunt slot versus normalized slot length.

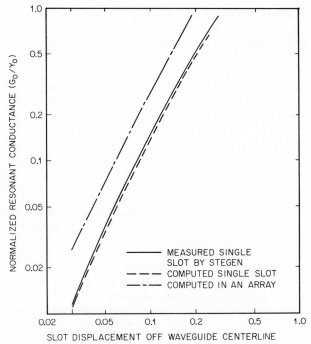

FIG. 9-9 Normalized resonant conductance of the baseline longitudinal shunt slot versus displacement off the waveguide centerline.

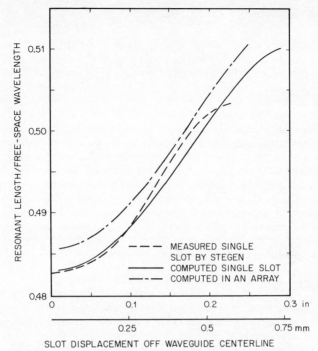

FIG. 9-10 Normalized resonant length of the baseline longitudinal shunt slot versus displacement off the waveguide centerline.

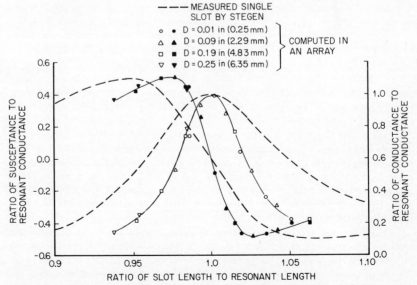

FIG. 9-11 Admittance of the baseline longitudinal shunt slot versus normalized resonant slot length. The solid curves are best fit to the computed points.

In airborne radar applications, large arrays of uniformly spaced slots are commonly used. The longitudinal-shunt-slot admittance of a slot in the central region of a uniform large array can be computed by the variational technique.[10] If the element spacing of a square grid is assumed to be equal to 0.71 wavelength, the slot-aperture admittance can be determined by an infinite array of open-ended rectangular waveguides.[17] The computed normalized admittance of the baseline slot in the array is compared with the measured results of the isolated slot in Fig. 9-11. The comparison shows that the mutual-coupling effects reduce the slot bandwidth performance.

By using the same dimensions, the resonant conductance and resonant length for a slot in a large array are shown in Figs. 9-9 and 9-10. Observe that the mutual-coupling effects also affect the slot resonant length as well as the slot resonant conductance. The small change in the slot resonant length in this case may not significantly affect antenna performance. However, an increase of the slot resonant conductance by a factor of 2 usually causes a 2:1 voltage standing-wave ratio (VSWR) at the center frequency. This has been observed in slot-antenna arrays designed by using measured single-longitudinal-slot conductances.

Edge-Wall Slots

The narrow-wall slots used for common linear slot-antenna arrays are cut into the narrow wall and wrapped around the broad wall of the rectangular waveguide (edge slots). Folded slots such as C and I slots are more convenient for planar arrays using edge-wall slots. Since analytical solutions for narrow-wall-slot characteristics, except for the moment-method solution of the C slot[33] and the partial solution for the I slot,[18] are not readily obtained, experimental results must be used to characterize these slots. Fortunately, the strong mutual-coupling effects can be measured by a waveguide section with a sufficient number of identical edge-wall slots. Watson[16] has defined the incremental conductance caused by the mutual-coupling effects as the increase in the conductance of a group of resonant half-wave-spaced identical edge-wall slots.

For both the edge slot and the C slot, the resonant conductance may be approximated by

$$G = G_0 \sin^2 \phi \qquad |\phi| \leq 15° \qquad \textbf{(9-12)}$$

where ϕ is the slot inclination angle. The slot resonant conductance is insensitive to the variation in the slot resonant length. Hence, if the incremental conductance is obtained by experiment for one value of angle, Eq. (9-12) can be used to obtain the resonant conductance for other slot inclination angles with resonable accuracy.

Plotted in Fig. 9-12 are the single-slot resonant conductance and the incremental conductance for edge slots cut on the standard RG52/U rectangular waveguide measured by Guptill and Watson.[16] The incremental conductance data are sufficiently accurate for slots in a linear array far removed from the ends. The conductance of the slots near the ends of the array lies between that of the single-slot data and that of the incremental conductance.

The edge-slot admittance variation as a function of the depth of cut into the rectangular-waveguide broad faces was investigated by Watson.[16] Measurements were performed on a slot cut in the standard RG48/U waveguide. Depicted in Fig. 9-13 are the measured results for a single 15° slot with slot width equal to 0.25 in (6.35 mm) at 2.8 GHz. These results can be translated to the frequency performance of the edge slot with the conclusion that the frequency bandwidth of edge slots is greater than that of longitudinal shunt slots.

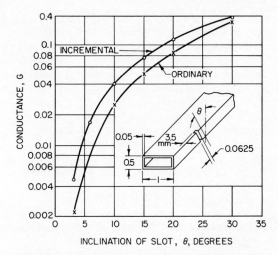

FIG. 9-12 Incremental and single-slot conductances (unmarked dimensions of waveguide section are in inches); frequency = 9.375 GHz, RG52/U waveguide.

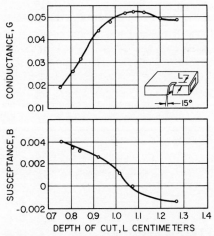

FIG. 9-13 Variation of slot admittance with depth of cut; frequency = 2.8 GHz, RG48/ U waveguide.

Series Slots

Centered inclined series slots in the rectangular-waveguide broad face radiating into free space can be used as radiating elements for slot-antenna arrays. The series-slot-radiation primary polarization is the same as that of the longitudinal shunt slot, but the series-slot cross-polarization radiation is much higher than that of the shunt slot. Hence the inclined series slot is not often used as a radiating element in practical designs. However, this slot is very useful for coupling radio-frequency power from one waveguide to another in a planar slot-array design such as that shown in Fig. 9-6.

The series-slot impedance can be computed by the variational formulation,[9] the scattering modal analysis technique,[13] or the moment method[14] as described above. By using either the variational formulation or the scattering modal analysis, the slot is first considered as a branched waveguide radiating into the main waveguide. The aperture impedance of the branched waveguide can be determined by assuming a cosine distribution across the slot aperture. The slot-aperture impedance facing the other waveguide can be computed in the same manner except that the slot inclination angle is the complementary angle. Properly combining these two solutions yields the series-slot impedance and hence the slot resonant resistance and resonant length. When round-ended series slots are used in the design, the slot resonant length can be determined by setting $C = 1$ in Eq. (9-11).

By using the scattering modal analysis solution, the computed resonant resistance and round-ended resonant length versus the inclination angle for the baseline slot are plotted in Figs. 9-14 and 9-15. The input and output waveguides are assumed to be perpendicular to each other for these computed data.

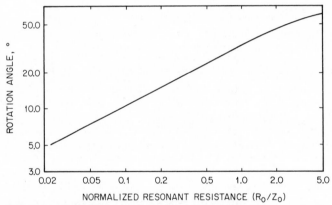

FIG. 9-14 Normalized resonant resistance versus rotation angle of a baseline inclined series slot.

9-5 BASIC SLOTTED-WAVEGUIDE ARRAY DESIGN METHODS

After obtaining the required slot characteristics, the next task in slotted-waveguide array design is to specify the slot locations and resonant conductances and/or resistances. Determination of the resonant conductances and/or resistances is an important part of antenna input matching and aperture illumination design.

Antenna requirements are usually specified in terms of desired gain, sidelobe level, beamwidth, bandwidth, polarization, input VSWR, cross-polarization level, power-handling capability, etc. All these requirements impact the slotted-waveguide

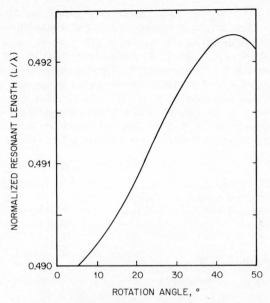

FIG. 9-15 Normalized resonant length versus rotation angle of a baseline inclined series slot.

array design. Aperture distribution is determined by aperture synthesis techniques,[19,20] taking the gain, sidelobe level, and beamwidth into consideration, and is assumed to be known for the following discussion.

The slot locations are governed by the relationship between the element spacing and the appearance of grating lobes. Adjacent slots are spaced apart by less than 1 free-space wavelength to avoid the appearance of grating lobes. For a resonant slotted-waveguide array, the desired aperture illumination is in phase across the aperture. To satisfy the grating-lobe restriction and the in-phase radiation requirement, the interelement spacing for slots on the same waveguide of a resonant array is therefore equal to one-half waveguide wavelength along the waveguide centerline. The interelement spacing of a nonresonant slotted-waveguide array is slightly larger or smaller than the spacing of resonant array elements, as will be detailed later.

Resonant conductances and/or resistances are determined by the given aperture illumination. To optimize bandwidth performance, the slots are designed to be resonant at midband for either resonant or nonresonant arrays. Basically, slot-radiated power is proportional to slot conductance or resistance. If P is the required fraction of slot-radiated power, normalized shunt-slot conductance and series-slot resistance are given by

$$g = 2P/V^2 \qquad r = 2P/I^2 \qquad \text{(9-13)}$$

where V and I are respectively the voltage and current across the slot. The transmission-line or scattering-matrix theories can be used to determine V and I and, therefore, g and r.

After slot locations and resonant conductances or resistances have been determined, the slot-array design can be completed by using the known slot characteristics or design curves to specify the slot geometry in the waveguide. If single-slot results, such as Eq. (9-1) with slot length equal to half wavelength, or the design curves depicted in Figs. 9-9, 9-10, and 9-13 for a single shunt slot radiating into free space are used to determine the slot geometry, the array aperture illumination will not be the desired distribution and the input VSWR will be higher than the requirement because of neglect of the mutual-coupling effects. There are four known approximate techniques to overcome this difficulty. Three of these methods take advantage of the known dipole array solution[21,22] and Babinet's principle, which relates slot-aperture admittance to the complementary dipole radiation impedance by

$$Y_r = 4b'\tilde{Z}/(\eta_0^2 a') \tag{9-14}$$

where $\tilde{Z}$ is the dipole impedance and η_0 is the free-space intrinsic impedance. Another method is the empirical technique originated by Watson.[16] The first two methods to be described apply only to longitudinal-shunt-slot arrays, while the last two methods are valid for all cases.

Active Slot Admittance

Elliott[23,24] derives two expressions for slot-array active admittances in terms of slot excitation voltages, slot parameters, and the complementary dipole array active impedances. By combining these two expressions with the corresponding dipole array solution, a small-array slot geometry can be specified by using the iterative procedure as described by Elliott.[23] The single-slot admittances of a single-slot radiator are used to model the slot admittance characteristics to reduce computational effort. This technique is quite accurate for small-array design as demonstrated by Elliott.[23] The required computation effort makes this technique unattractive for large-array applications.

Equivalent Self-Admittance

Slot equivalent self-admittances can be computed by Eqs. (9-7) through (9-10) by neglecting the edge effects, the small alternating displacements from the guide centerline, the slowly varying illumination, and the small slot-length variation. The aperture admittance is determined by either (1) the equivalent infinite waveguide array aperture admittance[17] or (2) the substitution of the complementary dipole array self-impedance[21] into Eq. (9-14). By using the computed admittance values corresponding to several slot lengths, the resonant conductance and resonant length versus slot displacement similar to that shown in Figs. 9-9 and 9-10 can be determined. By combining these design curves with the predetermined slot locations and resonant conductances, the slot-array antenna design is completed.

Note that these computed slot admittances take mutual-coupling effects into account but neglect edge effects, slot displacements from the guide centerline, non-uniform aperture illumination, and slot-length variation. For large arrays in general, displacements are small, illumination is slowly varying, and edge effects can be neglected. This technique is much more efficient than the active slot admittance method. For a small array, edge effects can be taken into account by several sets of resonant-conductance and resonant-length curves by using the corresponding small-dipole-array solutions. Dipole arrays corresponding to several uniformly alternating displacements can be used to compute the design curves for central elements and edge elements. For practical applications, neglecting alternating displacements in the equivalent dipole array computation results in negligible conductance error and in less than 1.2 percent resonant-length error.

A dielectric cover on the outer waveguide surface is frequently used to protect the antenna from the environment. In general, dielectric loading has little effect on the resonant conductance or resistance of a slot. However, the resonant length is greatly reduced by a dielectric cover and depends on the dielectric constant and thickness. Dielectric-loading effects can be taken into account by using the infinite waveguide array solution.[17]

Incremental Conductance

The two methods described above are applicable only to arrays of longitudinal shunt slots. Analytical solutions have not been well developed for edge-wall slots, and therefore the designs rely on experimentation.

Many waveguide test sections are required to characterize the slots, including mutual-coupling effects. Each of the test arrays has a sufficiently large number of identical conductance resonant slots so that the edge effects can be neglected. The moving lossy short technique has been successfully applied to the slot-characteristic measurements.[7,25] Design curves such as resonant conductance and resonant length versus inclination angle for edge slots can be determined as shown in Fig. 9-12. Observe that this technique is adequate for large linear-array designs but that inaccurate aperture illumination will be expected when it is applied to small-array designs.

Semiempirical Admittance

The three methods discussed above are not applicable to small arrays or unequally spaced slotted arrays of edge-wall shunt slots; hence the semiempirical admittance technique can be used. Basically, the dipole array solution is combined with the following equations to obtain the semiempirical slot characteristics. This technique is also applicable to design arrays of longitudinal shunt slots. By deduction from Eqs. (9-8) and (9-14), the slot resonant conductance can be expressed by

$$g = U'/\tilde{r} \qquad \textbf{(9-15)}$$

where $\tilde{r}$ is the complementary dipole active resistance normalized by the waveguide characteristic impedance and

$$U' = U(a,b,a',b') \begin{cases} \sin^2(\pi D/a') \\ \sin^2 \phi \end{cases} \qquad \textbf{(9-16)}$$

The upper and lower expressions apply to the longitudinal shunt slot and the edge-wall shunt slots respectively. In general, the change in slot resonant length due to mutual coupling is small, and U is a slowly varying function of a'. Thus, U can be determined approximately by experimentation. By combining the approximate value of U, the complementary dipole array solution, the resonant slot conductance, and the slot-location design procedure to be described later, the slot geometry can be determined in the following steps:

Step 1   Measure the single-slot characteristics.
Step 2   By substituting the complementary single-dipole resistance (zero-order

approximate value of r) and the measured single-slot resonant conductance into Eqs. (9-15) and (9-16), the values of U for a given slot length and width are determined.

Step 3   By combining the single-slot solution with the equations for specifying the slot locations and resonant conductances such as Eqs. (9-18), (9-19), and (9-32), the first-order design of the slot locations and displacements (or inclination angles) can be determined.

Step 4   The driving-point impedance of the complementary dipole array can be computed by the technique of Carter,[21] using the computed slot locations and displacements (or inclination angles) obtained from Step 3 or Step 5 for the lower-order approximation.

Step 5   By substituting the dipole array results from Step 4 into Eqs. (9-15) and (9-16) and combining with slot-location and conductance equations, the next-higher-order approximate solutions for slot locations and displacement (or inclination angles) can be obtained.

Step 6   Repeat Steps 4 and 5 until a satisfactory convergence solution is achieved.

Note that this design procedure yields no information on the resonant lengths of the slots in the array. Fortunately, the slot admittance phase variation introduced by mutual coupling is, in general, small compared with π. Suppose that the single-dipole driving-point impedance is represented by $Z_d' = R_d' + jX_d'$ and the final solution is $Z_d = R_d + jX_d$; then if the guide-wall thickness is negligibly small, the phase difference between the single-slot admittance and the admittance of the slot in the array ψ can be approximated by

$$\tan \psi = 2(X_d' - X_d)/(R_d' + R_d) \qquad \textbf{(9-17)}$$

The phase correction can be determined by experimentation or computation of the complementary dipole array by using the ratio of the incremental phase to the incremental slot length near the first resonance. This quantity is almost a constant and depends only on the slot and waveguide geometry.

The above iterative procedure has been found to converge rapidly when applied to a planar array of nonuniformly spaced C slots.

9-6   RESONANT SLOTTED-WAVEGUIDE ARRAY DESIGN

Both linear and planar resonant slotted-waveguide arrays have been designed for practical applications. The planar resonant array of longitudinal shunt slots is the most common antenna used in modern airborne radar. The resonant array can be used to produce a broadside pencil beam, with options to obtain two- or four-lobe monopulse. The basic characteristics common to all resonant slotted-waveguide arrays are summarized as follows:

1   All slots are resonant in the array; i.e., the susceptances or the reactances of the slot equivalent circuits vanish at the center frequency.

2   Waveguide standing-wave voltage maxima appear at the shunt slots and minima at the series slots.

3 All slots are spaced one-half waveguide wavelength away from adjacent slots in the same waveguide.

4 The main beam is normal to the array aperture (broadside).

In order to make two slots one-half waveguide wavelength apart radiate in phase, the adjacent longitudinal shunt slots are placed on opposite sides of the waveguide centerline; similarly, the adjacent edge-wall slots are inclined on opposite sides of the vertical centerline. When the rotational series-slot array is used to feed a group of linear arrays, the adjacent slots are rotated in the opposite direction across the longitudinal waveguide axis.

Since the planar array can be considered a composition of linear arrays, the design of a planar array is quite similar to the linear-array design. The mutual-coupling environment of these two types of antenna arrays differ; hence they are discussed separately.

Linear Array

To date the majority of linear slotted arrays have used either longitudinal shunt slots or edge-wall shunt slots as radiating elements. Longitudinal shunt slots are used to produce the radiation polarized perpendicular to the array axis, while edge-wall shunt slots are used to obtain the radiation polarized parallel to the array axis. The inclined series slot is not popular because of its high cross-polarization level compared with the longitudinal shunt slot.

There may be either one waveguide section or several waveguide sections in a linear-array design, depending on the required aperture length and frequency bandwidth. Conventionally, each waveguide section can be fed from one end or at the center as shown in Fig. 9-16. The number of slots in one waveguide section limits the frequency bandwidth in terms of input VSWR[26] and radiation pattern.

To ensure that the input is perfectly matched at the center frequency and the aperture excitation is the desired illumination, the following are required:

1 The sum of all the normalized slot resonant conductances is equal to 1 for end feed but equal to 2 for center feed.

2 The slot resonant conductance is proportional to the radiating power required for a given slot location.

Mathematically, for an array of N slots we have

$$\sum_{n=1}^{N} g_n = W \tag{9-18}$$

$$g_n = KA^2(n) \qquad n = 1, 2, \ldots N \tag{9-19}$$

where g_n is the normalized resonant conductance of the nth slot, $A(n)$ is the given aperture voltage distribution at the nth-slot location, K is the normalization constant, and W equals 1 for end-fed arrays or 2 for center-fed arrays. Equations (9-18) and (9-19) can be solved for g_n, and the design is completed by using either one of the design methods discussed previously to relate g_n with the physical slot geometry.

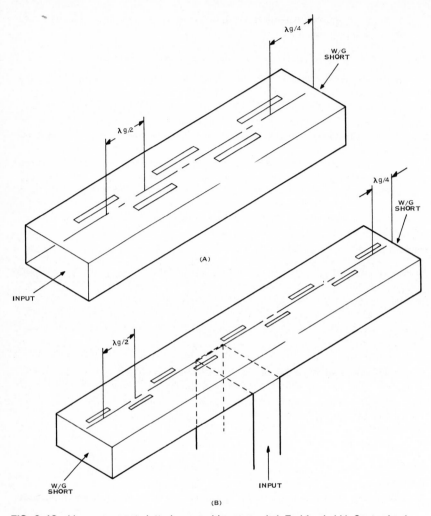

FIG. 9-16 Linear resonant slotted-waveguide arrays. (*a*) End feed. (*b*) Center feed.

Planar Array

Physically, a planar array is constructed by placing several linear arrays side by side as shown in Fig. 9-17. A large planar array may consist of several subarrays, and each subarray may consist of several linear arrays. The element spacing d along the direction perpendicular to the waveguide axes is limited by the physical geometry and the radiation performance. In practical applications the element spacing for a longitudinal-shunt-slot array is defined by

$$a + t \leq d \leq \lambda \qquad \textbf{(9-20)}$$

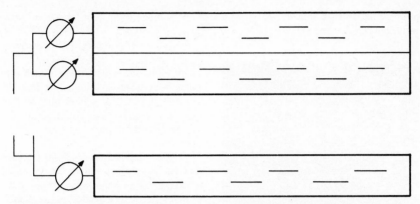

FIG. 9-17 Schematic diagram of an electronic steerable planar slot-antenna array.

where λ is the free-space wavelength and a and t are the waveguide width and wall thickness respectively. Choosing $d < \lambda$ avoids the appearance of grating lobes in visible space. For an edge-wall slot array, a in Eq. (9-20) is replaced by the waveguide height b.

The resonant planar slot-array design procedure is similar to that for a resonant linear slot array except that an additional feed system is required to feed the radiation shunt-slot waveguide. Two types of slot-array antennas commonly used in planar array design are the longitudinal-shunt-slot array and the edge-shunt-slot array.

Longitudinal-Shunt-Slot Array   This is the most popular slot-array antenna used in modern airborne radars. Two methods, (1) waveguide end feed and (2) series-slot central feed, have been used to feed the shunt-slot waveguides. In general, the series-slot center-feed system is a simpler and more compact design than the waveguide end feed for the same antenna size. Thus, the waveguide end feed is usually reserved for special-purpose applications such as one-dimensional electronic scan.

Waveguide end feed   Assume that the required aperture distribution $A(m,n)$ is given, where m and n indicate the mth row and nth column in the array respectively. The aperture distribution along the direction perpendicular to the waveguide axes is achieved by the waveguide manifold design. The longitudinal-shunt-slot planar-array design is similar to the linear array except that g_n and $A(n)$ in Eqs. (9-18) and (9-19) are replaced by $g_{m,n}$ and $A(m,n)$ and computation of mutual-coupling effects is performed for a planar array. By using any one of the four methods described above, the desired radiation performance can be realized if the planar-array mutual-coupling effects are taken into account correctly. However, for a large array the active slot admittance technique requires excessive computing efforts, and the incremental conductance technique demands intensive experimentation to complete the design. The most convenient technique for designing a large array of longitudinal shunt slots is the equivalent self-admittance method, which will be detailed in the series-slot center-feed discussion.

Series-slot center feed   Each shunt waveguide cavity or resonant linear array is fed by a rotational series slot. A slotted-waveguide array using the series-slot center feed is illustrated in Fig. 9-18. For a large planar array which consists of several subarrays, a waveguide manifold is required to feed the subarrays. A suitable comparator design such as that shown in Fig. 9-19 can accommodate the four-lobe monopulse radiation requirement. Details of the manifold design will not be discussed here.

FIG. 9-18   Typical flat-plate antenna.

Again, it is assumed that the aperture distribution $A(m,n)$ is given. The longitudinal-shunt-slot resonant-conductance design can be obtained by Eqs. (9-18) and (9-19). To achieve an accurate geometry design for a small array, the active slot admittance technique can be used.

The equivalent self-admittance method is more convenient for designing a large array of longitudinal shunt slots. For simplicity, by considering the slots as stub waveguides located on the shunt waveguide axes, the infinite-array solution of open-ended waveguides can be used to compute the slot-aperture admittance. This computation yields the slot-aperture admittance by taking into account approximate mutual-coupling effects (the alternating displacement of slots along the waveguide axes is neglected). By substituting the computed slot-aperture admittances for several slot lengths into Eqs. (9-7) through (9-10) and using the simple interpolation technique, the resonant conductance and resonant length versus displacement design curves such as those shown in Figs. 9-9 and 9-10 can be easily computed. Note that this solution is valid for slots located in the array central region, and the edge effects in a large array may be neglected. Thus, by using these design curves and Eqs. (9-19) and (9-20), an array of longitudinal slots can be designed.

The next step in the planar-array design using the series-slot center feed is to design the series slots which feed the shunt waveguides. Before doing this, the system for feeding the waveguides which excite the series slots must be defined. Again, two feed techniques, the waveguide end feed and the E-plane T center feed, have been used by antenna designers. The E-plane T center feed yields a compact feed system with a shorter series waveguide run, but coupling between the junction and the directly coupled series slot is a problem. The waveguide end feed does not exhibit the distortion caused by coupling between the junction and the adjacent series slot.

After the series-slot feed has been chosen, the design is quite similar to the shunt-slot design. By using the slot equivalent circuit and assuming that all the slots are at resonance, the series-slot resonant resistances in each series waveguide are determined by

$$\sum_{m=1}^{M} r_m = W \qquad\qquad\qquad \textbf{(9-21)}$$

$$r_m = K'B^2(m) \qquad m = 1, 2, \ldots M \qquad \textbf{(9-22)}$$

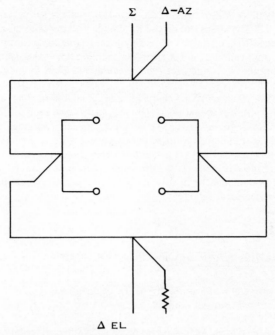

FIG. 9-19 Schematic diagram of a four-lobe monopulse comparator.

where r_m is the normalized resonant resistance of the mth slot, B is the required series-slot excitation function, K' is the excitation normalization constant, and $W = 1$ for end feed or $W = 2$ for center feed. $B(m)$ is given in terms of aperture illumination $A(m,n)$ by

$$B^2(m) = \sum_{n=1}^{N_m} A^2(m,n) \qquad \textbf{(9-23)}$$

where N_m is the number of shunt slots fed by the mth-series slot. Now, complete the series-slot design by solving Eqs. (9-21) through (9-23) for the resonant resistances and using the design curves of the slot resonant resistance and resonant length versus the inclination angle given in Figs. 9-14 and 9-15.

Edge-Wall Shunt-Slot Array The edge slot is cut into the narrow wall and wrapped around the broad walls of the rectangular waveguide. The slot length is determined by the depth cut into the broad wall. If a planar-array design using the edge slot is needed, the slot wrapped around into the broad wall can be a design and fabrication problem. Thus, planar slot-antenna arrays often use alternative approaches such as C or I slots to allow resonant slots on a planar surface.

There usually is no advantage for a planar slot-antenna array using edge-wall shunt slots. One of the few practical applications in a planar array is to achieve electronic scan in one direction when the required element spacing is too small for the longitudinal-shunt-slot design.

The edge-wall slotted waveguide fed from the center is not practical for the following reasons: (1) The shunt slot on the waveguide wall opposite the radiating shunt slot will introduce strong coupling effects and degrade antenna performance. (2) Analytical tools are not available for designing the coupling slot in this complicated geometry. Thus, the practical design uses the waveguide manifold to end-feed the slotted waveguides.

Since the analytical solution is not available for edge-wall shunt slots, the methods of active slot admittance and equivalent self-admittance used for the longitudinal slot are not applicable to design the slot geometry, and the incremental conductance method is used. Fortunately, the incremental conductance technique can be easily applied to this case since the strong mutual-coupling effects can be measured by a linear array of edge-wall slots. Also, the semiempirical admittance method can be applied in the same manner as in the case of linear-array antennas and is more efficient than the incremental conductance technique.

9-7 TRAVELING-WAVE SLOT-ARRAY DESIGN

The traveling-wave slot array differs from the resonant waveguide slot array in two ways: first, the resonant slots of the traveling-wave array are spaced by either more or less than one-half waveguide wavelength; and, second, the slotted waveguide is terminated by a matched load. As in the case of resonant waveguide slot arrays, resonant slots are used in traveling-wave arrays to maximize frequency-bandwidth performance.

Because slot spacing is close to one-half waveguide wavelength, adjacent longitudinal shunt slots are on opposite sides of the guide centerline, and adjacent edge-wall shunt slots are inclined to opposite sides of the vertical centerline. The alternating displacement or inclination with respect to the waveguide axis of successive slots may produce second-order beams.[27,28] If the main beam of the traveling-wave slotted array is pointing far away from broadside, the second-order beams may exceed the desired sidelobe level. The second-order beams can be effectively suppressed by choosing suitable element spacing.[27,28] The restrictions on element spacing given in these references can be relaxed when the element pattern is included in the computation.

The following are the basic characteristics common to all traveling-wave slot-waveguide-antenna arrays:

1 All slots in the array are resonant at the center frequency.

2 All slots in the same waveguide are spaced by either more or less than one-half waveguide wavelength from adjacent slots.

3 The beam pointing is off broadside and is frequency-dependent.

4 Matched loads are required to terminate the waveguides.

5 Array efficiency is less than unity.

The reflected wave from a termination with a VSWR greater than unity will produce a spurious beam scanned to the opposite side of the array normal from the main beam. Therefore, the reflected power from the termination must be kept sufficiently small so that the spurious beam is below the required sidelobe level.[29]

Two types of traveling-wave slot-antenna arrays have been successfully designed: (1) a uniformly spaced array is used to produce a low-sidelobe pencil beam, and (2) a nonuniformly spaced array is applied to shaped-beam designs. Longitudinal shunt slots, edge-wall shunt slots, and rotational series slots can be used in both traveling-wave types.

Again, the complex aperture illumination of a traveling-wave slot-antenna array is assumed to be predetermined by aperture synthesis techniques. Consider first the design of a traveling-wave array of shunt slots. With reference to Fig. 9-20, the ith-slot normalized conductance related to its radiated power as derived by Stegen[4] is

$$g_i = g_i^+ P_i / P_i^+ \tag{9-24}$$

where

$$P_i = P_{i+1}^-[e^{2\alpha\ell}i + 2|\Gamma_{i+1}|^2 \sin(2\alpha\ell_i)/(1-|\Gamma_{i+1}|^2] \tag{9-25}$$

The superscipts $^+$ and $^-$ indicate the quantity at the outgoing and incoming sides of the slot, $\gamma = \alpha + j\kappa$ is the complex propagation constant of the waveguide, ℓ_i is the interelement spacing, and Γ_i is the reflection coefficient at the ith element. By using the known aperture illumination, the design curves, Eqs. (9-24) and (9-25), and the transmission-line equation which relates the normalized admittances and element spacing,

$$y_i^+ = g_i^+ + jb_i^+ = \frac{y_{i+1}^- \cos h\gamma\ell_i + \sinh \gamma\ell_i}{\cos h\gamma\ell_i + y_{i+1}^- \sinh \gamma\ell_i} \tag{9-26}$$

where

$$y_i^- = y_i + y_i^+ \qquad |\Gamma_i| = |(1 - y_i)/(1 + y_i)| \tag{9-27}$$

the slot conductances of a traveling-wave-antenna array of shunt slots can be specified.

When the need for a traveling-wave array of series slots arises, Eqs. (9-24) through (9-27) are applicable, provided that the normalized admittance y is replaced by the normalized impedance $z = r + jx$. Other design procedures are similar to the ones used for the shunt-slot array design.

If the array aperture is sufficiently large so that each slot radiates only a small fraction of the incident power, a simplified method due to Dion[29] can be used to specify slot conductances or resistances. In this approximate method, both the waveguide

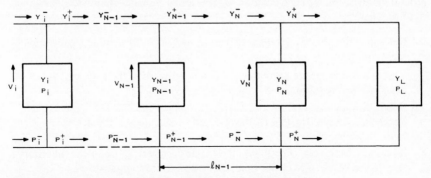

FIG. 9-20   Equivalent-network representation of a shunt-slot array.

losses and reflections from slots are neglected in the computation. Under this assumption

$$P_{i-1}^+ \approx P_i + P_i^+$$

This equation can be used to replace Eqs. (9-25) through (9-27) to design a traveling-wave array with a large number of slots.

Uniformly Spaced Traveling-Wave Slot-Array Design

A uniformly spaced traveling-wave slot array can be designed to produce a low-side-lobe pencil beam with larger bandwidth than that of the resonant waveguide slot array in terms of the array radiation pattern and input VSWR. If the reflections between elements are negligible, a constant phase difference between elements results in a progressive phase shift along the array aperture. The resulting phase front is inclined at an angle θ given by

$$\sin \theta = \lambda/\lambda_g - \lambda/2\ell \qquad \qquad \textbf{(9-28)}$$

where ℓ is the interelement spacing. In most traveling-wave slot-array designs, 2ℓ is chosen greater than λ_g and the beam moves in angle toward the load end as frequency is increased. For special situations such as limited aperture space, the array interelement spacing is chosen so that 2ℓ is less than λ_g and the beam moves in angle toward the feed end as the frequency is increased. Hence, the beam peak position of a traveling-wave array is frequency-dependent.

The nonresonant array achieves its large impedance bandwidth by virtue of the phase differences between the reflections from the various slots. The phase differences which arise from nonresonant spacing cause the resultant sum of all the reflected waves to be quite small. When the slot spacing is approaching one-half waveguide wavelength and the beam peak is very close to the broadside position, the input VSWR will increase rapidly as the slot admittances add in phase so that the input-impedance magnitude is much larger than unity. The matched load termination will absorb a fraction of the input power. The amount of absorption is a function of the frequency and slot conductances and is usually chosen to be about 5 to 10 percent of the input power at midband.

The design procedure for determining the slot conductances and locations of a general uniformly spaced traveling-wave array using resonant shunt slots is summarized as follows:

1 Determine the proper interelement spacing by Eq. (9-28) and the beam-scanning range and set $y_N^+ = y_L = 1$.

2 Choose the appropriate percentage of power delivered to the load; that is, set P_N^+ (usually equal to 0.05).

3 By substituting these values and the predetermined percentage of power radiated by the Nth slot into Eq. (9-24), the resonant conductance of the slot closest to the load can be determined.

4 By using Eqs. (9-25) and (9-26), the incident power P_{N-1}^+ and the admittance termination of the second slot from the load y_{N-1}^+ are computed.

5 Repeat Steps 3 and 4 by replacing the subscript N by i until the resonant conductance of the slot nearest the feed is computed.

This design procedure is applicable to a general uniformly spaced series-slot array provided that the conductance and admittance are replaced by the resistance and impedance.

For a long array with many slots and negligible waveguide losses, the approximate method of Dion is applicable and the design is much simpler. In this approximate method, the nth-slot resonant conductance is given by

$$g_i = P_i \bigg/ \left(1 - \sum_{n=1}^{i-1} P_n \right) \qquad \textbf{(9-29)}$$

where $\Sigma P_n = 1 - P_L$. Note that this approximate design is valid only if the largest slot resonant conductance g_{mx} satisfies the following relationship:

$$|g_{mx} \csc \kappa \ell| < 0.447 \qquad \textbf{(9-30)}$$

Through the proper choice of P_L the resultant slot conductances can usually be made to satisfy the criterion of Eq. (9-30), although undesirably high power may be dissipated at the termination.

In general, there are many slots in a traveling-wave array. Therefore, the active admittance technique is not very efficient for determining the slot-array configuration, and the equivalent self-admittance and the semiempirical techniques discussed previously are more attractive. The incremental conductance technique using the design curve shown in Fig. 9-12 is applicable with reasonable accuracy. By using a traveling-wave array of series slots to feed a resonant planar array of longitudinal shunt slots, the slot configuration can be completed by the design curves shown in Figs. 9-14 and 9-15. The longitudinal-shunt-geometry design is the same as in the resonant slot-array design, provided that the progressive phase shift is taken into account in computing the mutual-coupling effects.

Nonuniform Traveling-Wave Slot-Array Design

Antenna arrays that produce shaped beams are desired for some special applications. The aperture illumination of a shaped beam requires, in general, both amplitude and phase variation across the aperture. Again, this complex aperture distribution is assumed to be determined by some aperture synthesis technique.[20] A slot-array antenna which produces a shaped beam can be easily designed by using a properly fabricated manifold to feed a group of linear slot arrays. These linear slot arrays can be designed by the methods discussed in connection with the planar slot-array design provided that the phase variation is included in the computation. The manifold is designed to account for the required aperture amplitude and phase variations.

There are some special applications, such as in air traffic control, for which the shaped beam is required to scan over a required sector. Also, occasionally a linear array producing a shaped beam is desired. In these situations, it is desirable to know whether and how a linear slot array can produce a desired shaped beam.

As seen from the above discussion, the aperture illumination amplitude can be controlled by the slot conductances. However, phase variation could be introduced either by varying slot reactance or by slot spacing. Variation in slot admittance gives only limited phase control and results in a narrow frequency bandwidth. A better method for varying the phase is the use of resonant slots and nonuniform interelement spacing. The resonant slot conductance or resistance controls the aperture illumination amplitude, while the nonuniform element spacing provides the required phase distribution across the aperture.

Assume that a complex aperture illumination of a nonuniformly spaced slotted-waveguide array is given by

$$A(v) = A(v) \exp [j\psi(v)] \qquad (9\text{-}31)$$

where A and ψ are the complex aperture illumination amplitude and phase and v is the spatial coordinate along the waveguide axis. To simplify the presentation, waveguide losses and reflection from the slots are assumed to be negligible. Under these assumptions, the illumination phase distribution can be achieved by recognizing that

$$\psi(v_n) = \pi[2v_n/\lambda_g \pm (n - N)] \qquad (9\text{-}32)$$

where n signifies the nth slot and the choice of a $+$ or a $-$ sign depends on the phase taper reference. In general, a closed form for the slot locations cannot be derived from Eq. (9-32) for a general illumination phase function. Iterative techniques such as the design procedure listed below are required to determine the slot locations for a nonuniformly spaced slotted array.

Based on Eq. (9-32) and an iteration technique, a procedure for computing the slot locations of a nonuniformly spaced slotted array is practical:

1 Normalize the given illumination phase function so that the phase at the slot nearest the load is zero. Using this normalization and Eq. (9-32) yields the coordinate reference $v_N = 0$.

2 Set the nominal interelement spacing $\ell_i = \lambda/2$ for all i to compute the zero-order approximate values of slot locations.

3 By substituting the previously determined slot locations into the right-hand side of Eq. (9-32), the slot radiation phases are calculated.

4 Comparing these slot radiation phases with the desired illumination phase at these locations yields the incremental phase errors $\Delta\psi_n$.

5 The next-higher-order approximated slot locations can be obtained by

$$v_n^{(i)} = v_n^{(i+1)} + \Delta\psi_n\lambda_g/(2\pi) \qquad (9\text{-}33)$$

where the superscripts indicate the order of approximation.

6 Repeat Steps 3 through 5 until convergent slot locations are obtained.

The rate of convergence is quite rapid for the iterative procedure if the desired illumination phase function is a slowly varying function such as the phase function used in a cosecant-squared-beam synthesis. After the slot locations have been determined, the slot resonant conductances can be computed by Eq. (9-29) in the same manner as for the uniform traveling-wave array.

If the waveguide losses or the reflections from the slots are not negligible, Eq. (9-32) does not yield the correct radiation phase. In this circumstance, the transmission-line models of radiation phases and Eqs. (9-24) through (9-27) are required. The simple solutions obtained by neglecting the reflection from the slots and waveguide losses are employed as the first-order approximation in the iteration procedure. This iteration procedure is similar to the one based on Eq. (9-32), except that Eq. (9-32) is replaced by the transmission-line theory to compute the phases and Eq. (9-29) is replaced by Eqs. (9-24) through (9-27) to compute the slot resonant conductances. This general iteration procedure can be summarized as follows:

1 By substituting the previously determined slot locations and resonant conductances into transmission-line equations, the slot radiation phases can be computed.

2 By comparing these radiation phases with the desired illumination phases at these locations, the incremental phase errors $\Delta\psi_n$ are determined.

3 By substituting these incremental phase errors into Eq. (9-33), the approximate slot locations are calculated.

4 Compute the slot resonant conductances by Eqs. (9-24) through (9-27) by using the approximate slot locations, and repeat Steps 1 and 2 to obtain the new incremental phase errors.

5 By applying the linear interpolation or extrapolation to the two successive slot locations and incremental phase errors, the next-higher-order approximation of the slot locations is computed, and hence the slot resonant conductances.

6 Repeat Steps 1 through 5 until convergent results are obtained.

The slot geometry can be realized by the semiempirical method discussed above after the slot locations and resonant conductances have been determined. A compatible nonuniformly spaced dipole array solution, together with Eqs. (9-14) through (9-17), can be used to determine the slot configurations.

In general, if the nominal value of the unequal element spacing is larger than one-half free-space wavelength, higher sidelobes will appear in the region far off broadside. Therefore, it is recommended that nominal interelement spacing be less than one-half waveguide wavelength for the nonuniformly spaced slotted array.

9-8 POWER-HANDLING CAPABILITIES OF SLOTTED-WAVEGUIDE ARRAYS

Slotted-waveguide arrays are required to operate at high-power levels in many applications. In some cases the slot-antenna arrays can handle the high power without undue difficulty at sea-level atmospheric pressure, but they may fail to withstand the power at high altitudes where atmospheric pressure is much lower. When high-power capacity is required for an airborne antenna, the power-handling capability at the desired high altitude must be used in the designs.

The power-handling capacity of a slotted-waveguide array depends on both waveguide manifold design and slot design. Following the procedure established by Gould,[30] the waveguide manifold power-handling capability can be estimated with reasonable accuracy. As as example, consider a standard X-band RG52/U waveguide operating at a frequency of 9.375 GHz. On the basis of Gould's data, the continuous-wave power and the power of a 1-μs rectangular pulse versus the altitude are shown in Fig. 9-21. Observe that the maximum power level which can be handled by the waveguide decreases rapidly when the altitude increases and the waveguide pressure

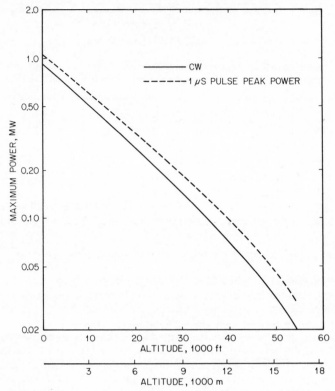

FIG. 9-21 Power-handling capacity of an RG52/U waveguide at 9.375 GHz.

follows the altitude pressure. Also note that these results apply to room temperature for a perfectly matched waveguide section. If this is not the case, the pressure at other temperature ρ is required for computing the maximum power level where

$$\rho = \rho_0(293/T) \tag{9-34}$$

T is the absolute temperature (degrees Kelvin), and ρ_0 is the pressure at room temperature.

The electric field strength E across the slot aperture for a given slot-radiated power P is needed to compute the slot power-handling capacity. By using Eq. (9-14), the electric field strength across the slot aperture is approximated by

$$E = (\eta_0/b')\sqrt{P/2R} \tag{9-35}$$

where R is the complementary dipole driving-point resistance. The baseline slot at 9.375 GHz is used to illustrate the slot power-handling capability. By combining Gould's results[30] with Eq. (9-35), the maximum continuous-wave and peak power versus the altitude for the baseline slot is computed and shown in Fig. 9-22. Equation (9-34) should be used for a temperature other than room temperature.

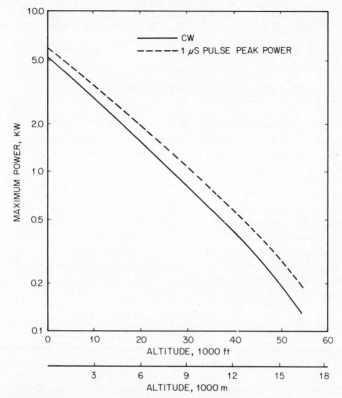

FIG. 9-22 Power-handling capacity of the baseline slot in the planar array.

Techniques which can be used to increase power-handling capability are as follows:

1 Avoid capacitive windows and sharp corners in the waveguide manifold.

2 Increase the slot width.

3 Cool and pressurize the antenna system.

The most effective technique for increasing power-handling capability is pressurization.[31] If a slot-antenna array is wholly encapsulated in a pressurized housing, the increase in power-handling capability is proportional to the achievable pressure. Occasionally, freon is used for pressurization, further increasing power-handling capability through the higher dielectric strength of this substance. This technique adds to the complexity of the mechanical design and fabrication of the antenna.

Alternatively, instead of using a fully pressurized housing, the antenna can be partially pressurized in only those areas where power levels are highest. This can be achieved by using pressure windows in the waveguides or dielectric covers on the slots. Dielectric covers, when used, must be taken into account in slot design. For a large array of longitudinal shunt slots, a dielectric cover can easily be included in the computation of the slot-aperture admittance by the infinite array of rectangular waveguides.

9-9 TOLERANCE AND FABRICATION TECHNIQUES

In practice, an antenna will include errors introduced by inaccuracies in the manufacturing process since high precision in manufacturing generally increases antenna costs. Thus, a systematic allocation of dimensional tolerances is usually a worthwhile part of a design.

The tolerance for waveguide width can be established easily by computing the incremental phase per unit change in the manifold and feed waveguide width. To determine the required dimensional tolerance for slots, the incremental conductance (or impedance) per unit change in the slot length from the resonant length is needed. These incremental values per unit change of slot parameters can be determined from given slot parameters discussed in preceding sections. Listed in Table 9-1 are the incremental phases of baseline shunt and series slots and the S-band edge slot due to the decrease of 1 percent slot length from resonant lengths. Note that the tolerance requirement for edge-slot arrays is less critical than for other slot arrays.

TABLE 9-1 Slot Incremental Phase Change with 1 Percent Change in Slot Length from Resonant Length

Longitudal shunt slot		Rotational series slot	Edge slot
Single	Coupled		Coupled
8.1°	19.3°	17.1°	2.7°

The difference in the incremental phase change indicates that the mutual-coupling effects reduce the longitudinal shunt-slot frequency-bandwidth performance and increase tolerance requirements. Sidelobe performance can be greatly degraded by a 1 percent variation in the slot length in a planar array of longitudinal shunt slots.

Fabrication of the commonly used edge-slot array is quite different from that of other slot arrays. Since the edge slot is cut into the narrow wall and wrapped around the broad faces of a rectangular waveguide, an edge-slot array can be easily fabricated by cutting (sawing) the slots at the proper angle and depth. The results are, in general, quite good. To generate a planar edge-slot array, slotted-waveguide sections are manifolded and assembled side by side to form a planar aperture.

To fabricate a planar array of longitudinal shunt slots or edge-wall folded shunt slots, a milling machine may be used to cut the slot on a thin, flat plate (face plate) such as shown in Fig. 9-18. The back plate is milled out of a thick metal plate. Two narrow walls of the shunt and series waveguides are fabricated on the opposite sides of the metal plate, and the series slots are cut on the common walls of the shunt and series guides. The shunt waveguides are formed by assembling the back plate to the face plate, and the series waveguides by securing the feed covers on the back plate.

Basically, there are three techniques for assembling the face plate to the back plate array of broad-wall slots:

1   Welding   Both laser welding and electron-beam welding can be used to join the face plate to the back plate. Electron-beam welding has been successfully applied to manufacture antennas made of magnesium. Because of the inherent properties of aluminum, applying the electron-beam welding technique to this metal in assembling the flat-plate antenna requires additional investigative effort. Laser welding of the face plate to the back plate is possible, but the process is better suited to welding very thin metal cross sections and has limited application to flat-plate antennas.

2   Brazing   Dip brazing has been widely used to secure two aluminum objects together. Many aluminum slotted-waveguide arrays are assembled by using dip brazing. The disadvantages of this technique are that the residue deposits left in the waveguides and on the slots and the overall material shrinkage must be accounted for. The fluxless-brazing technique for joining the aluminum face and back plates together in a retort with inert atmosphere has been investigated with successful results.[32]

3   Bonding   There are three commonly used bonding techniques: (*a*) thin-film bonding, (*b*) conductive-epoxy bonding, and (*c*) diffusion bonding. Thin-film bonding and conductive-epoxy bonding are quite popular in flat-plate-antenna applications.

Other miscellaneous techniques such as screw fasteners and twist tabs have been employed to join the face and back plates together. These two methods are either limited as to application or less accurate, adversely affecting antenna performance when compared with the above three assembling techniques.

The linear array is subjected to fewer tolerance and assembly problems than the planar array. To fabricate and assemble a planar array such as the flat-plate antenna, both before and after the process of joining the face and back plates together, attention must be paid to keep the antenna aperture flat and the plates aligned accurately. A 1 percent free-space wavelength warpage of an antenna can significantly degrade side-lobe performance, and an 0.7 percent misalignment of the free-space wavelength of the front and back plates will produce a grating-lobe level higher than -20 dB. Slot-length tolerance depends on the desired radiation performance.

Although a slot-antenna array can be designed and fabricated by using the design information given in this chapter, antenna performance may be less than expected. Some of the second-order effects, such as the slot alternating displacements, alternating inclined angles, edge effects, nonuniform aperture illumination, slot-length errors, and internal coupling through higher-order waveguide modes have not been addressed. The combined effects of neglecting second-order effects, manufacturing tolerances, and inaccurate manifold design usually significantly degrade practical slotted-array performance. For instance, a -35-dB-sidelobe-level array design may have only -28-dB sidelobes in practice. Thus, a thorough error analysis is recommended to define dimensional tolerances and to determine aperture illumination errors. Normally, overdesign is used to allow for design and manufacturing deficiencies, the degree of overdesign usually being determined by experience. Note that some of the degradation factors will be more noticeable for a -30-dB-sidelobe-level array design than for a -20-dB design.

REFERENCES

1 A. A. Oliner, "The Radiation Conductance of a Series Slot in Strip Transmission Line," *IRE Conv. Rec.*, vol. 2, part 8, 1954, pp. 89–90.

2 R. W. Breithaupt, "Conductance Data for Offset Series Slots in Stripline," *IEEE Trans. Microwave Theory Tech.*, vol. MTT-16, 1968, pp. 969–970.

3 Y. Yoshimura, "A Microstripline Slot Antenna," *IEEE Trans. Antennas Propagat.*, vol. AP-20, 1972, pp. 760–762.

4 R. J. Stegen, "Longitudinal Shunt Slot Characteristics," Hughes Aircraft Co. Tech. Mem. 261, Culver City, Calif., November 1951.

5 I. Kaminow and R. J. Stegen, "Waveguide Slot Array Design," Hughes Aircraft Co. Tech. Mem. 348, Culver City, Calif., 1954.

6 B. N. Das and G. S. Sanyal, "Investigations on Waveguide-Fed Slot Antenna: Equivalent Network Representation," *J. Inst. Telecom. Eng.*, vol. 14, no. 6, 1968, pp. 249–264.

7 H. M. Altschuler and A. A. Oliner, "Microwave Measurements with a Lossy Variable Short Circuit," Res. Rep. R-399-54, Polytechnic Institute of Brooklyn, Brooklyn, N.Y., 1954.

8 R. J. Stevenson, "Theory of Slots in Rectangular Waveguides," *J. App. Phys.*, vol. 19, 1948, pp. 24–38.

9 A. A. Oliner, "The Impedance Properties of Narrow Radiating Slots in the Broad Face of Rectangular Waveguide, Part I and II," *IRE Trans. Antennas Propagat.*, vol. AP-5, 1957, pp. 4–20.

10 H. Y. Yee, "Impedance of a Narrow Longitudinal Shunt Slot in a Slotted Waveguide Array," *IEEE Trans. Antennas Propagat.*, vol. AP-22, 1974, pp. 589–592.

11 L. Lewin, *Advanced Theory of Waveguides*, Iliffe & Sons, Ltd., London, 1951, pp. 106–114.

12 N. H. Frank and L. J. Chu, MIT Rad. Lab. Rep. 43-6 and 43-7, 1942.

13 J. P. Montgomery, unpublished work; private communication.

14 Vu Khac Thong, "Impedance Properties of a Longitudinal Slot Antenna in the Broad Face of a Rectangular Waveguide," *IEEE Trans. Antennas Propagat.*, vol. AP-21, 1973, pp. 106–114.

15 R. W. Lyon and A. J. Sangster, "Efficient Moment Method Analysis of Radiating Slots in a Thick Walled Rectangular Waveguide," *IEE Proc. (London)*, vol. 128, part H, 1981, pp. 197–205.

16 W. H. Watson, "Resonant Slots," *J. IEE (London)*, part IIIA, vol. 93, 1946, pp. 747–777.

17 S. W. Lee and W. R. Jones, "On the Suppression of Radiation Nulls and Broadband Impedance Matching of Rectangular Waveguide Phased Array," *IEEE Trans. Antennas Propagat.*, vol. AP-19, 1971, pp. 41–51.

18 R. J. Chignell and J. Roberts, "Compact Resonant Slot for Waveguide Arrays," *IEE Proc. (London)*, vol. 125, no. 11, November 1978, pp. 1213–1216.

19 T. T. Taylor, "Design of Line-Source Antennas for Narrow Beamwidth and Low Sidelobes," *IRE Trans. Antennas Propagat.*, vol. AP-3, 1955, pp. 16–28.

20 P. M. Woodward and J. D. Lawson, "The Theoretical Precision with Which an Arbitrary Radiation Pattern May Be Obtained from a Source of Finite Size," *IEE Proc. (London)*, vol. 95, part III, pp. 363–370.

21 P. S. Carter, "Circuit Relations in Radiating System and Applications to Antenna Problems," *IRE Proc.*, vol. 20, 1932, pp. 1004–1041.

22 V. W. H. Chang, "Infinite Phased Dipole Array," *IEEE Proc.*, vol. 56, 1968, pp. 1892–1900.

23 R. S. Elliott, *Antenna Theory and Design*, Prentice-Hall, Inc., Englewood Cliffs, N.J., 1981, chap. 8.

24 R. S. Elliott and L. A. Kurtz, "The Design of Small Slot Arrays," *IEEE Trans. Antennas Propagat.,* vol. AP-26, 1978, pp. 214–219.

25 M. G. Chernin, "Slot Admittance Data at *Ka* Band, *IRE Trans. Antennas Propagat.,* vol. AP-4, 1956, pp. 632–636.

26 T. Takeshima and Y. Isogai, "Frequency Bandwidth of Slotted Array Aerial System," *Electron. Eng.,* February 1969, pp. 201–204.

27 L. A. Kurtz and J. S. Yee, "Second Order Beams of Two-Dimensional Slot Arrays," *IRE Trans. Antennas Propagat.,* vol. AP-5, 1957, pp. 356–362.

28 R. E. Collin and F. J. Zucker (eds.), *Antenna Theory,* part I, McGraw-Hill Book Company, New York, 1969, chap. 14.

29 A. Dion, "Nonresonant Slotted Array," *IRE Trans. Antennas Propagat.,* vol. AP-6, 1958, pp. 360–365.

30 M. Gilden and L. Gould, *Handbook on High Power Capabilities of Waveguide Systems,* Microwave Associates, Inc., Burlington, Mass., June 1963.

31 J. Ciavolella, "Take the Hassle out of High Power Design," *Microwaves,* June 1972, pp. 60–62.

32 "Aluminum Brazed Antenna Manufacturing Methods," AFML Cont. F33615-75-5266, Antenna Dept., Hughes Aircraft Company, Culver City, Calif., 1975.

33 T. Shicopoulos, "C-Slot: A Practical Solution for Phased Arrays of Radiating Slots Located on the Narrow Side of Rectangular Waveguides," *IEE Proc. (London),* vol. 129, part H, April 1982, p. 49–55.

Chapter 10

Leaky-Wave Antennas *

Raj Mittra

University of Illinois

*Portions of this chapter were prepared originally by F. J. Zucker for Chap. 16 of the first edition.

10-1 DESIGN PRINCIPLES

Leaky waves arise when either a closed or an open waveguide structure supporting a guided wave is perturbed, either continuously or at periodic intervals. Typically, these antennas are designed so that the rate of leakage per unit wavelength is fairly low and the perturbation of the phase velocity is relatively little. This is done so that the antenna does not produce a large mismatch to the exciting source. The aperture of the antenna is frequently chosen to be many wavelengths long in order to assure that there is little or no energy left at the end of the antenna structure. The phase velocity along a leaky-wave antenna is a function of frequency, and consequently these antennas are frequency-scannable. This point is further discussed below.

Direction of Main Beam

A leaky antenna is typically characterized by a complex propagation constant $\alpha_z + j\beta_z$. If the length of the leaky-wave antenna is ℓ, the normalized far-field pattern $P(\theta)$ radiation intensity is given by

$$P(\theta) = \left| \frac{\sin\{(\pi\ell/\lambda)[\cos\theta - (\alpha_z + j\beta_z)/k]\}}{(\pi\ell/\lambda)[\cos\theta - (\alpha_z + j\beta_z)/k]} \right|^2 \qquad (10\text{-}1)$$

where $k = 2\pi/\lambda$, λ = free-space wave number, and the element pattern of the antenna has been suppressed in writing Eq. (10-1).

Under the approximation of small α_z, the direction of the main beam is given by

$$\theta = \cos^{-1}(\beta_z/k) = \cos^{-1}(\lambda/\lambda_z) \qquad (10\text{-}2)$$

$$\lambda_z = 2\pi/\beta_z$$

while the width of the beam is determined by the attenuation constant α_z. It now becomes clear from Eq. (10-2) that if β_z varies with frequency, then the main beam will scan as a function of frequency. Typically, λ/λ_z is a monotonically increasing function of frequency; hence the beam scans in the direction of decreasing θ, i.e., from broadside to end-fire as the frequency is increased. Frequently, the dispersion characteristics of leaky-wave antennas can be expressed as

$$\frac{\lambda}{\lambda_z} = \sqrt{1 - (\lambda/\lambda_c)^2} \qquad (10\text{-}3)$$

where λ_c is a function of geometry only.

Control of Aperture Distribution

The amplitude taper of a leaky-wave antenna is controlled by varying the leakage along it. The power distribution along the antenna can be expressed as

$$P(z) = P_{\text{in}} \exp\left[-2 \int_0^z \alpha_z(z)dz\right] \qquad (10\text{-}4)$$

Differentiating Eq. (10-4), we get

$$-\frac{dP(z)}{dz} = 2\alpha_z(z)P(z) \qquad (10\text{-}5)$$

The design of a leaky-wave antenna may be carried out with the aid of the following equations. Let us say that the desired aperture distribution to achieve a certain pattern is expressed in the form $A(z)\,e^{-j\beta_z z}$. Then $P(z)$ can be written in terms of $A(z)$ as

$$P(z) = P_{in} - \int_0^z |A(\xi)|^2\, d\xi \qquad (\textbf{10-6})$$

where P_{in} is the input power. If P_L is the residual power at the end of the antenna, then we have

$$P_{in} = \int_0^\ell |A(\xi)|^2\, d\xi + P_L \qquad (\textbf{10-7})$$

where ℓ is the length of the antenna. If R is the ratio of the residual power to the input power, i.e.,

$$R = P_L/P_{in}$$

then, after some rearrangement, we can write Eq. (10-6) as

$$P(z) = \frac{1}{1-R} \int_0^\ell |A(\xi)|^2\, d\xi - \int_Q^z |A(\xi)|^2\, d\xi \qquad (\textbf{10-8})$$

Differentiating Eq. (10-8) and making use of the relationship in Eq. (10-5), we obtain

$$\alpha_z(z) = \frac{(1/2)|A(z)|^2}{[1/(1-R)] \int_0^\ell |A(\xi)|^2\, d\xi - \int_0^z |A(\xi)|^2\, d\xi} \qquad (\textbf{10-9})$$

Thus, the desired taper along the antenna may be calculated from Eq. (10-9) once the amplitude taper and the residual power in the load have been specified.

Array Synthesis and Beam Shaping

The techniques of array synthesis outlined by Taylor[1] and Dunbar[2] can be employed to design leaky-wave antennas. Kelly[3] has shown that an amplitude distribution $A(z) = 1 - \pi z/\ell$ with linear phase progression (within ± 1.5 dB) results in a good approximation to the often-needed $\csc^2 \theta \cos^{1/2} \theta$ pattern. Some synthesis problems require θ, thus λ/λ_z, to be variable (function of z) along the aperture of the antenna.

Leaky-wave antennas can be designed for installation on a curved surface, a feature not possessed by surface-wave-excited arrays (since it is too complicated to control the pattern in the presence of both element radiation and curvature-induced leakage radiation); the change in β_z and α_z has been found to be negligible in the case of leaky-wave structures that are perturbations of shielded guides (e.g., a long slot), provided that the radius of curvature is larger than 20λ. It is clear from Fig. 10-1 that the relative phase velocity must vary along s to produce a beam that keeps pointing in the same direction in space (at right angles to the projected aperture z):

$$\frac{\lambda}{\lambda_s(s)} = \cos[\theta(s)]$$

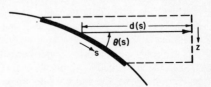

FIG. 10-1 Curved leaky-wave antenna.

where $\theta(s)$ is prescribed by the geometry. The required aperture distribution $A(s)$ in the curved aperture is obtained from the prescribed distribution $A(z)$ in the projected aperture by conservation of power. If the curved antenna is an area source (width transverse to the page), then

$$A^2(s) = A^2(z) \sin [\theta(s)]$$

where $A^2(s)$ and $A^2(z)$ have dimensions of power per unit area. If the curved antenna is a line source (not confined between parallel plates), energy spreads at all angles with the page and[4]

$$A^2(s) = A^2(z)d(s) \sin [\theta(s)]$$

where $A^2(z)$ has dimensions of power per unit area but $A^2(s)$ power per unit length (along s); $d(s)$ is prescribed by the geometry.

Beam shaping can also be accomplished if the antenna is placed on a surface bent in the shape of a polygon.[4]

Gain, Beamwidth, and Sidelobe Level

Assuming constant phase progression along the aperture of the antenna and constant (or moderately tapered) amplitude and using simple array theory, we obtain for the direction of the main beam

$$\cos \theta_m = \frac{\lambda}{\lambda_z} - \frac{\lambda}{s} m \qquad \qquad \textbf{(10-10}a\textbf{)}$$

and with phase reversal,

$$\cos \theta_n = \frac{\lambda}{\lambda_z} - \frac{\lambda}{s} (n + \tfrac{1}{2}) \qquad \qquad \textbf{(10-10}b\textbf{)}$$

where θ is the angle off end fire, s the spacing between elements, and m or $n = 0, 1, 2, \ldots$ the order of the beam. The spacing must be small enough so that no secondary principal lobes appear at real angles. The position of the principal lobes nearest the main lobe is found by replacing m by $m + 1$ and $m - 1$ in Eq. (10-10a [and likewise for n in Eq. 10-10b]). As s increases, principal lobes appear first in the end-fire or backfire ($\theta = 180°$) direction; they will not appear provided their angle is imaginary, or

$$\frac{s}{\lambda} \leq \frac{1}{1 + |\cos \theta_m|} \qquad \qquad \textbf{(10-11)}$$

(and likewise for n), where the equality sign holds only if the element pattern has nulls along the array axis. According to Eq. (10-11), the maximum permissible spacing increases from $\lambda/2$ at end fire and backfire to λ at broadside.

Because of mutual coupling between elements and to save on the total number in the array, we usually impose $s \geq \lambda/2$ as a lower bound.

Figure 10-2, which combines Eqs. (10-10) and (10-11), shows the dependence of main-beam angle on λ/λ_z and on the *permissible* spacings. [If $m = 0$ in Eq. (10-10a), s is arbitrary except for its upper bound, which is explicitly indicated in Fig. 10-2.] It is seen that provided λ/λ_z and s/λ values can be freely assigned, the $m = 0$ mode of operation allows the main beam to be positioned anywhere between end fire and broadside and all other modes anywhere between end fire and backfire. Very slow and

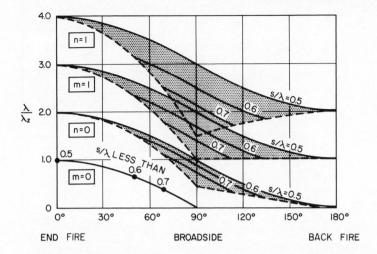

FIG. 10-2 Angle θ between main lobe and array axis of an array of discrete elements spaced s/λ apart and excited by a traveling wave with relative phase velocity λ/λ_z. Curves marked m are for identical consecutive elements; those marked n, for alternately phase-reversed elements [Eq. (10-10a and b)]. Range of curves satisfies Eq. (10-11) (no other principal lobes).

very fast traveling waves, however, are not easily realized, so that usually λ/λ_z lies in the vicinity of 1. In that case, the $m = 0$ mode is useful only for near-end-fire radiation (e.g., trough guide with periodic discontinuities that are not phase-reversed); the $n = 0$ mode, only for near-broadside radiation (e.g., phase-reversed dipoles); and the $m = 1$ mode, for radiation from near broadside all the way to backfire (e.g., proximity-coupled dipoles that are not phase-reversed). Higher modes of operation require λ/λ_z to be well above 1 (very slow waves), and no practical applications are in fact known for $m \geq 2$ and $n \geq 1$.

If $\lambda/\lambda_z = 1$ and $s = \lambda/2$ (no phase reversal), Fig. 10-2 shows that two main beams arise, one end fire ($m = 0$) and the other backfire ($m = 1$). To suppress one of these, the element pattern must have a null in the unwanted direction.

Arrays of Line Sources; Area Sources

To obtain directivity in both planes a flared horn can be attached to the leaky guide,[5] or several guides can be placed parallel to each other. The mutual interaction of parallel leaky-wave antennas depends on whether the elemental radiators are end-to-end or broadside. With transverse slots (Figs. 10-3, 10-6, etc.) the magnetic dipoles are end-to-end, and the coupling is slight; with TE-excited longitudinal slots (Figs. 10-4a and b, 10-5a) the magnetic dipoles are broadside, and although β_z is affected only negligibly, α_z is much perturbed.

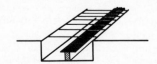

FIG. 10-3 Coaxially fed nonresonant slots. (*After Ref. 16.*)

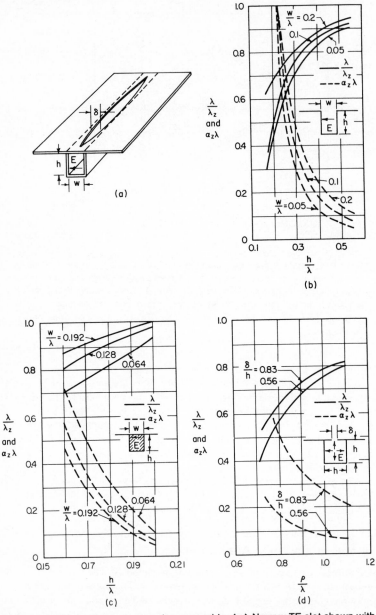

FIG. 10-4 Long slots in rectangular waveguide. (*a*) Narrow TE slot shown with lip of variable width. (*b*)–(*d*) Relative phase velocity λ/λ_z and leakage attenuation $\alpha_z\lambda$ on (*b*) TE channel, (*c*) dielectric-filled channel ($\epsilon = 2.56$) with hybrid (quasi-TE) excitation, (*d*) narrow TM slot. (*After Hines, Rumsey, and Walter*[19] *and Goldstone and Oliner.*[17])

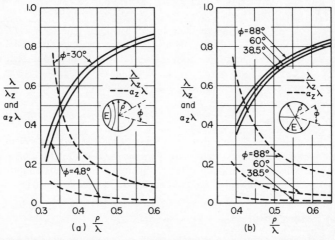

FIG. 10-5 Long slots in circular waveguide. (*a*) TE. (*b*) TM. (*After Harrington*[21] *and Goldstone and Oliner.*[22])

Detailed analysis in the case of two long slots excited in phase shows that for very small spacings between line sources α_z is almost twice its unperturbed value. As spacing increases, α_z decreases and then oscillates about its unperturbed value, the amplitude of oscillation remaining quite large even when the spacing is several wavelengths. If arrays of such slots are to be designed (although it would be far simpler to use an area source instead), the drastic change in the element pattern due to the presence of all other slots must be taken into account.[6]

Several area sources are described in Sec. 10-2. With two of these, one vertically and the other horizontally polarized (Figs. 10-7 and 10-9), extremely precise pattern control has been achieved, both in the elevation plane and in the plane perpendicular to it (in which the beam shape is controlled by the phase and amplitude distribution in the feed).

Feeds

If the transition from unperturbed to leaky guide is not too abrupt—and it need not be, since the α_z required near the feed is invariably small—there will be no detectable spurious feed radiation (a situation very unlike that with surface waves). There will also be no appreciable impedance mismatch, and the input impedance of a leaky-wave antenna is therefore largely that of the feed itself.

To decrease the power carried in the initial section of the leaky guide and thus lessen the danger of voltage breakdown in high-power application, a feeding guide can be coupled (through slots in a common wall) to the leaky guide along its entire length.[7] Distributed feeds of this type have also been explored, though not perfected, for pattern control and scanning.[8,9]

Suitable feeds for area arrays are either a (narrowband) broadside slot array or a (much wider-band) pillbox arrangement such as a hoghorn. Details on the latter are given by Honey.[10]

Termination

Resistance cards, e.g., three tapered 200-Ω cards arranged parallel to each other and to the waveguide sidewalls, make satisfactory loads in waveguide or parallel-plate systems. To test whether a design has been successful with respect to dissipating negligible power in the load, one can check the effect on the radiation pattern by short-circuiting the termination; the change should be insignificant.

Effect of Radome

Nishida[11] has shown that a thin dielectric cover placed over a long slot increases β_z very slightly and considerably reduces α_z. This means that the direction of the main beam shifts very little but the beam shape is impaired, and attenuation measurements should therefore be performed in the presence of the radome in an early design stage.

Circular Polarization

Two circularly polarized leaky-wave antennas have been proposed; they are described in Sec. 10-2 in connection with the structures illustrated in Figs. 10-6 and 10-8.

Scanning

Mechanical scanning, as with slot arrays, can be accomplished by moving a sidewall or by inserting a variable amount of dielectric loading in the guide.[12]

When frequency-scanned, the main beam moves as described in the discussion preceding Eq. (10-3). The scan angle is limited by mode cutoff at the low-frequency end and by the excitation of higher modes at the high-frequency end. If Eq. (10-3) applies—and it usually does—leaky-wave antennas exhibit the attractive property of constant beamwidth over the entire scanning range. The beamwidth depends on the antenna length ℓ_p/λ projected normally to the beam direction (half-power beamwidth $= 55\lambda/\ell_p$ ° for constant aperture illumination). If θ is the direction of the beam, then we have

$$\ell_p = \ell \sin \theta = \ell \sqrt{1 - \cos^2 \theta}$$

By using Eqs. (10-2) and (10-3),

$$\frac{\ell_p}{\lambda} = \frac{\ell}{\lambda} \sqrt{1 - \left(\frac{\lambda}{\lambda_z}\right)^2} = \frac{\ell}{\lambda_c} \qquad (10\text{-}12)$$

a ratio that depends only on the antenna geometry and is thus independent of frequency.[10] If, in addition, α_z and hence the aperture illumination do not change materially over the frequency range, the beamwidth and sidelobe level will remain constant. In the case of the inductive-grid antenna (Fig. 10-9) the beam scans from 30 to 70° over almost a frequency octave; the elevation beamwidth remains constant over the upper two-thirds of this band, increasing very slowly at lower frequencies because of a gradual change in α_z, and the sidelobe level remains below its preassigned value (23 dB) over the entire band.

Since the beamwidth in the plane perpendicular to the elevation pattern depends on the feed and thus necessarily broadens with decreasing frequency, the gain of the antenna must drop with frequency, but it will do so at a *slower* rate than the gain of a dish (about 4 dB over almost a frequency octave in the case of the inductive-grid antenna).

Mechanical scanning and frequency scanning have also been explored in connection with curved leaky antennas.[4]

10-2   SPECIFIC STRUCTURES

Leaky-wave structures are listed in accordance with the unperturbed waveguide from which they derive: first stripline, then ordinary waveguide (long slots in rectangular and circular, nonresonant slots in rectangular guide), then trough guide, and, last, parallel-plate guide (transverse and longitudinal strips, holes, and disks).

The first leaky-wave antenna was a long slot in waveguide, invented by Hansen[13] (a useful reference for early leaky-wave antenna work is Southworth[14]). Most of the structures listed here are, like the long slot, perturbations of shielded guides. The leaky trough guide (Fig. 10-8) can serve as a model for methods of introducing leakage in other open guides, for example, in H guide.[15]

Another open-waveguide leaky-wave antenna design utilizes a periodically perturbed planar dielectric waveguide and is discussed later in this section.

In choosing an antenna for design, one should first examine the data presented here to see whether the range of required α_z values is obtainable from the proposed structure, then whether these α_z values are compatible with the requirements imposed by the equation $\cos \theta = \lambda/\lambda_z$. In many instances, α_z and λ/λ_z are so coupled by the *same* geometric parameters as to preclude accurate beam shaping (except for special cases). True freedom in design is achieved only when the parameters that control α_z are entirely separate from those that control λ/λ_z (see the inductive-grid antenna, Fig. 10-9).

Some of the data shown are experimental, some theoretical; whenever both were available, those that seemed more reliable were plotted. With very accurate data, extremely precise pattern control can be achieved. The display is usually in terms of λ/λ_z and $\alpha_z\lambda$ as a function of the geometric parameters (z, as always, is in the direction of propagation along the guide), but because of the various ways in which parameters can be related to λ_z and α_z it is sometimes advantageous to arrange the data differently. The dimensionless constant $\alpha_z\lambda$ is in nepers per wavelength.

Leaky Coaxial Lines (Fig. 10-3)

Less dispersive than waveguide structures that satisfy Eq. (10-3), a leaky TEM-mode line is called for whenever one tries to hold the beam position reasonably constant over a frequency band. Leakage is controlled by varying the slot length and/or displacing the inner conductor from the center.[16] Good shaped patterns have been produced with this antenna, but no data are available on λ/λ_z and $\alpha_z\lambda$ as a function of the parameters.

Long Slots in Rectangular Waveguide[4,17,18,19]

By proper positioning of a longitudinal slot in the walls of rectangular waveguides, any one of the following field configurations can be excited:[19]

1   TE (no E_z in slot): slot must be narrow and placed where wall currents are perpendicular to it, e.g., Fig. 10-4a and 4b. The elemental radiation pattern is that of a z-directed magnetic dipole; hence end-fire radiation is zero.

2   TM (no H_z): slot must be placed where currents are parallel to it, e.g., Fig. 10-4d. Elemental pattern is that of an x-directed electric dipole.

3   Hybrid with negligible E_y: guide must be filled with dielectric ($\epsilon \geq 2.0$), e.g., Fig. 10-4c.

4   Hybrid with negligible H_y: guide must be *partially* filled with dielectric and excited in lowest-order hybrid mode (not illustrated).

Configurations 2 to 4 are sometimes termed *channel-guide antennas;* configuration 3 is known to support also waves with negligible leakage. Because of the presence of several field components in the channel-guide slots, the far-field patterns are linearly polarized only provided[19] $0.4 \leq w/\lambda \leq 0.9$, $1 \leq \epsilon \leq 2.5$, and $0.4 \leq \lambda/\lambda_z \leq 1.0$.

All the structures in Fig. 10-4 have been treated theoretically.[17,18,20] In the case of Fig. 10-4a, for example, λ/λ_z and $\alpha_z\lambda$ are explicitly related to the geometric parameters by[17]

$$\frac{\lambda}{\lambda_z} = \frac{\lambda}{\lambda_{z0}}\left(1 - \frac{\lambda_{z0}^2}{2\pi^2 hw}\frac{p}{1 + p^2}\right)$$

$$\alpha_z\lambda = \frac{\lambda\lambda_{z0}}{\pi hw}\frac{1}{1 + p^2}$$

where
$$p = \frac{2}{\pi}\left[\ln\left(\csc\frac{\pi\delta}{2w}\right) + \ln\left(1.526\frac{h}{\delta}\right)\right]$$

and λ_{z0} is the guide wavelength $(= \lambda/\sqrt{1 - (\lambda/2h)^2})$ in the unperturbed waveguide $(\delta = 0)$. These explicit expressions are based on a perturbation calculation that is valid only provided $\delta \ll w$. Zero ground-plane thickness is assumed. Though not strictly separable, λ/λ_z is primarily controlled by variations in h (through λ_{z0}) and $\alpha_z\lambda$ by variations in δ. Dielectric or other loading of the guide[4] will primarily affect λ/λ_z. Results for the other structures in Fig. 10-4 are displayed graphically.

Because they require a smaller variation in guide parameters to cover a prescribed range in $\alpha_z\lambda$, TM- and hybrid-excited slots lend themselves somewhat more easily to pattern control than TE-excited slots. Even so, good control has been achieved with the slot in Fig. 10-4a,[4,19] a typical result being that a gaussian amplitude distribution produces low sidelobes over a 2:1 frequency range in which the beam swings from 38° to 18° off end fire.

Long Slots in Circular Waveguide[19,21,22] (Fig. 10-5)

Perturbation formulas that relate the relative phase velocity and leakage explicitly to the geometric parameters are available.[22]

Closely Spaced Slots and Holes in Rectangular Waveguide

Like the long slot in Fig. 10-4a, the closely spaced slots in Fig. 10-6a are excited by the dominant TE waveguide mode. The element pattern in this instance is that of a

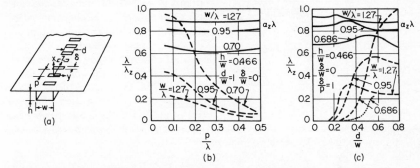

FIG. 10-6 Closely spaced slots in rectangular waveguide. *(After Hyneman.[23])* (a) Structure. (b) and (c) Relative phase velocity and leakage attenuation as a function of parameters (curves are based on theoretical values for infinitely many slots per wavelength; also valid experimentally for $\delta/w = 0.0033$, $\delta/\rho = 0.4$). Discrepancy in (c) between dotted curve *(after Kelly and Elliott[3])* and corresponding dashed line may be due to the fact that Kelly and Elliott's slots were milled to their required length while Hyneman's slots were milled across the entire width of the guide and then partially covered by metal foil.

transverse magnetic dipole (vertical polarization). Also as in the case of the long slot, λ/λ_z is primarily controlled by varying the width of the broad waveguide wall (h in Fig. 10-4a, w in Fig. 10-6a). $\alpha_z\lambda$ depends primarily on slot spacing and width (Fig. 10-6b and c)[23] and on the width of the narrow waveguide wall. Though not strictly separable, both λ/λ_z and $\alpha_z\lambda$ can be accurately controlled over a wide range of values by appropriate variations in the large number of available parameters.

Hyneman[23] points out that over a certain range of the parameters a surface wave exists together with the leaky wave; however, over much of this range the effects of the spurious mode—if it is excited at all—cannot be detected in the radiation pattern.[24]

A circularly polarized line source can be obtained by milling closely spaced transverse slots in all four sides of a square waveguide and exciting it with two orthogonal TE modes 90° out of phase.[25]

Theoretical results are available for closely spaced round holes in rectangular waveguide.[17]

Plane Array of Thick Transverse Slots[26]

The array of waveguides in Fig. 10-7 is TE-excited by means of a hoghorn or other arrangement. Each guide in turn excites an array of slots approximately one-quarter wavelength thick, which couple only a small amount of power from the guide and do not appreciably perturb the guide phase velocity. Since the interslot spacings s lie between 0.25λ and λ, this antenna is not really a leaky guide but a shielded-waveguide-excited array of discrete elements.

The main-beam direction is given by the $m = 0$ curve in Fig. 10-2. The pattern bandwidth is approximately 1.6:1, limited by mode cutoff at the low- and by Eq. (10-11) at the high-frequency end. In one particular application, the beam swings from 71° to 34° off end fire, and since Eq. (10-12) applies, the elevation-pattern (*E*-plane) beamwidth stays constant over the entire frequency range.

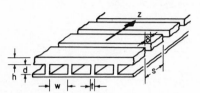

FIG. 10-7   Plane array of thick transverse slots. *(After Ref. 26.)*

At the frequency for which the equivalent electric thickness of the top wall is 0.25λ, the relative phase velocity is that of the waveguide ($\lambda/\lambda_z = \sqrt{1 - (\lambda/2w)^2}$). At lower frequencies, the slots load the guide capacitively, slightly decreasing λ/λ_z; at higher frequencies, λ/λ_z slightly increases. The electric thickness is 0.25λ provided

$$h = \frac{\lambda}{4} - \frac{\delta}{\pi}\left(1 + \ln\frac{2s}{\pi\delta} + \ln\frac{d}{2\delta}\right)$$

(see Fig. 10-7 for meaning of symbols).

The element normalized conductance is

$$g = 1.54\frac{w^2\delta^2}{\lambda_z sd(w + t)}$$

(*s* appears in this expression because the underlying analysis[26] takes into account the mutual impedance between slots.) *g* is controlled by varying δ, and to satisfy the first equation *h* must be variable along the array; flexing of the bottom surface in the *z* direction is avoided by holding $h + d$ constant.

Leaky Trough-Guide Antennas[27,28,29]

The design parameters of the asymmetric trough guide (Fig. 10-8*a*) are chosen as follows:

1   From Eqs. (10-2) and (10-3) determine λ/λ_z and $\alpha_z\lambda$, respectively.

2   Calculate

$$n = \frac{\alpha_z\lambda}{\lambda_z/\lambda - \lambda/\lambda_z}$$

3   Use the table to find values of *p* and *q* corresponding to *n*:

n	0.000	0.025	0.126	0.320	0.465	0.672	0.886	1.118
p	0.250	0.263	0.277	0.293	0.299	0.305	0.309	0.307
q	0.250	0.237	0.221	0.205	0.194	0.183	0.170	0.153

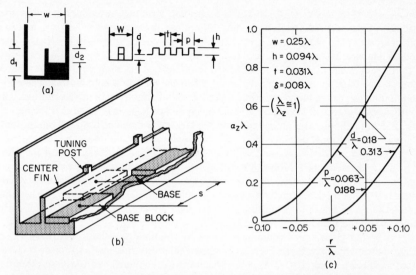

FIG. 10-8 Leaky trough-guide antennas. (*a*) Continuously asymmetric trough guide. *(After Ref. 28.)* (*b*) Periodically asymmetric trough guide. *(From Rotman and Oliner.[28])* (*c*) Leakage attenuation of trough guide with serrated center fin protruding a distance r/λ above the level of the sidewalls (negative values of r/λ correspond to center fin below sidewall level; δ = fin thickness). *(After Rotman and Karas.[29])*

4   Then

$$\frac{d_1}{\lambda} = \frac{p}{\sqrt{1 - (\lambda/\lambda_z)^2}} - \frac{\delta}{\lambda}$$

$$\frac{d_2}{\lambda} = \frac{q}{\sqrt{1 - (\lambda/\lambda_z)^2}} - \frac{\delta}{\lambda}$$

where $\delta \cong (w/\pi) \ln 2$ (assuming d_1 and $d_2 > w$; otherwise see Rotman and Oliner[28]). This procedure, which automatically "uncouples" the parameters controlling λ/λ_z and $\alpha_z\lambda$, is adapted from Rotman and Naumann[27]; formulas that give λ/λ_z and $\alpha_z\lambda$ explicitly (but coupled) in terms of the geometric parameters have been derived by Rotman and Oliner.[28] Our expressions assume that the frequency is not too near cutoff ($\lambda/\lambda_z > 0.5$) and that $\alpha_z\lambda$ is small; they also neglect the finite thickness of the center fin, which if taken into account is found slightly to increase λ/λ_z and decrease $\alpha_z\lambda$.[28]

By alternately reversing successive asymmetric sections (Fig. 10-8*b*)[27,28] and by serrating the center fin to obtain a wide range of λ/λ_z values, the main beam can be placed anywhere between end fire and just beyond broadside on the $n = 0$ curve in Fig. 10-2. The relative phase velocity of the periodically asymmetric trough guide is approximately the same as that of the corresponding continuously asymmetric one, but α_z is somewhat less because in the immediate vicinity of the interface between successive base blocks (Fig. 10-8b) the periodic structure resembles a symmetric trough guide and radiates less; α_z must therefore be measured.[27,28] (The radiation mechanism of the periodically asymmetric trough guide closely resembles that of the sandwich wire except that the radiation is distributed slightly more evenly along its length.) As in the case of the sandwich-wire antenna, small tuning posts (Fig. 10-8b) are needed to suppress the reflected wave when the radiation is broadside. The optimum dimensions are found experimentally by minimizing the input (VSWR) to a line with, say, 10 identical blocks, terminated in a matched load. It is found that the higher the block, the higher the tuning post and the larger the distance from the block interface at which it must be placed.[28]

The symmetric trough guide with serrated fin produces a leaky wave if the sidewalls are lowered appreciably.[29] From Fig. 10-8c, which shows $\alpha_z\lambda$ values measured in one particular instance, it appears that leakage begins as the sidewalls approach the level of the teeth and increases monotonically as they are lowered still further and also that $\alpha_z\lambda$ increases with the number of teeth per wavelength. λ/λ_z is only negligibly affected by the reduction in sidewall height.

Circular polarization could be obtained by combining the horizontally polarized asymmetric trough with the vertically polarized protruding-teeth structure.[29] Note that Eqs. (10-3) and (10-12) apply neither to the asymmetric nor to the serrated-fin leaky trough guide.

Inductive-Grid Antenna (Transverse Strips)[10] (also called Honey array, Fig. 10-9)

Unlike the plane array of thick transverse slots (see above), this antenna is horizontally polarized, and it is a true leaky-wave structure (many strips per wavelength). Because λ/λ_z and $\alpha_z\lambda$ can be independently controlled and are known very accurately, radiation patterns have been produced that differ by only 0.5 dB from the design value even in regions 40 dB below the peak. The full-pattern bandwidth, within which the beam scans from 70° to 30° off end fire, is 1.8:1. The elevation-pattern (H-plane) beamwidth remains constant over a 1.5:1 range, then increases at the low-frequency end of the band (by 35 percent at 70°).

λ/λ_z depends primarily on the grid-to-bottom-plate spacing d; $\alpha_z\lambda$, primarily on the grid parameters. The design procedure is to read the abscissa value in Fig. 10-9b corresponding to the desired λ/λ_z and $\alpha_z\lambda$ and then to use Fig. 10-9c for determining d. (The bottom plate is therefore necessarily flexed in the z direction.) Figure 10-9a shows flat strips that can be photoetched on Teflon-fiberglass laminate and supported by polyfoam. The curves in Fig. 10-9b and c, however, also apply to a grid of transverse round wires with diameter $t/2$, except that a correction term is needed if the wire-to-wire spacing p is an appreciable fraction of a wavelength.[10]

An omnidirectional radial-cylindrical variant of the inductive-grid antenna produced good patterns over a 1.5:1 frequency range.[30]

Longitudinal Strips[31] (Fig. 10-10)

Because it is difficult to achieve tight control over $\alpha_z\lambda$ when the leakage is small (as it must be in long arrays), this antenna is less satisfactory than the inductive grid. Formulas that give the explicit dependence of phase velocity and leakage on the parameters are available.[17]

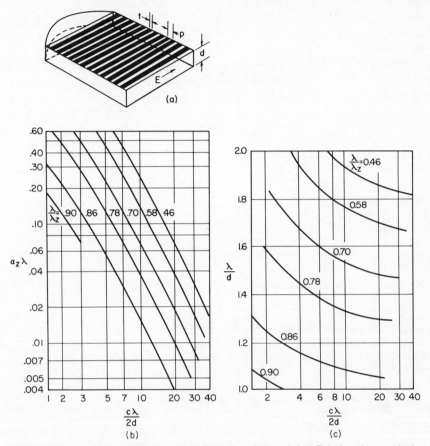

FIG. 10-9 Inductive-grid antenna. (*a*) Structure. (*b*) and (*c*) Relative phase velocity and leakage attenuation as a function of the parameters, with $c = 2\pi d/p$ ln [csc ($\pi t/2p$)]. (*After Honey.*[10])

Holey-Plate and Mushroom Antennas[32]

Since the center-to-center spacings between holes in Fig. 10-11*a* and *b* and between disks in Fig. 10-11*c* lie between 0.25λ and λ, these antennas are not really leaky waveguides but, like the plane array of thick transverse slots (see above), belong to the group of shielded-waveguide-excited discrete elements. We include them here for the sake of completeness and because they are, like the transverse and longitudinal strips, perturbations of parallel-plate structures.

If the holes in Fig. 10-11*a* and *b* are small ($\delta/\lambda < 0.3$), the normalized element conductance $g \cong 10(\delta/\lambda)$.[33] The plate spacing is close to a quarter wavelength: $d/\lambda \cong 0.25 + \sqrt{g}/4\pi$. The relative phase velocity is so close to 1 that end-fire radiation ($m = 0$ in Fig. 10-2) as well as radiation in the back quadrant ($m = 1$) is produced. Broussaud erects transverse vanes between rows of holes (Fig. 10-11*a*) and finds that this suppresses the end-fire radiation. The vanes presumably load the parallel-plate

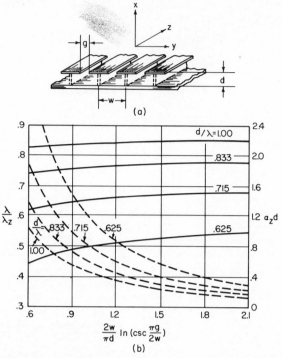

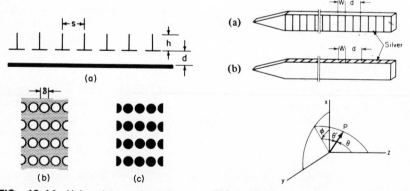

FIG. 10-10 Longitudinal strips. (*a*) Structure (*E* field in *y* direction). (*b*) Relative phase velocity (solid curves) and leakage attenuation (dashed curves) as a function of the parameters. (*After Honey.*[31])

FIG. 10-11 Holey-plate and mushroom antennas. (*a*) Top view. (*b*) Longitudinal cross section fo holey plate (with $h \cong \lambda/3$). (*c*) Top view of mushroom antenna. (*After Broussaud.*[32])

FIG. 10-12 Examples of leaky-wave antennas and reference system. (© *1981 IEEE.*)

structure inductively, thus increasing λ/λ_z slightly above 1 (Fig. 10-2); any other type of inductive loading would do as well.

The holey plate is excited by a TEM wave and produces an *E*-plane elevation pattern. Its Babinet equivalent, the array of disks (Fig. 10-11*c*), is TE-excited and produces an *H*-plane elevation pattern; because its *E* field is *y*-directed, the disks can be supported on metal stems (hence, *mushroom antenna*).

Dielectric-Rod Leaky-Wave Antennas

These antennas are periodically perturbed rectangular dielectric rods and have a potential for use[34] in millimeter-wave integrated systems which utilize planar dielectric-waveguide technology. The geometries of two such antennas are shown in Fig. 10-12. Silver strips are used as perturbations on dielectric rods with a relative dielectric constant of 2.33 at 81.5 GHz. The dielectric antenna is fed with an RG99/U waveguide. To avoid spurious radiation in the launching region, it is extremely important to design a proper transition between the waveguide and the antenna. The design of such a transition has been described by Trinh, Malherbe, and Mittra.[35]

The impedance match in the transition region is further improved by employing strips of varying widths in the launch region, with the strip widths slowly increasing from the feed to the antenna region. Likewise, radiation from the termination end of the antenna must be suppressed to reduce undesired radiation in the end-fire direction. To obtain a well-defined beam with low sidelobes, it is essential to couple the energy from the guided wave to the radiated wave in a gentle manner. This is accomplished by using narrow strips; the small coupling per unit length requires that a large number of strips and a relatively large aperture be used so that little or no energy is left at the end of the antenna. The larger aperture improves the gain and reduces the beamwidth.

The complex propagation constant along the perturbed dielectric rod can be theoretically calculated.[36] However, β_z, the real part of the wave number, remains rela-

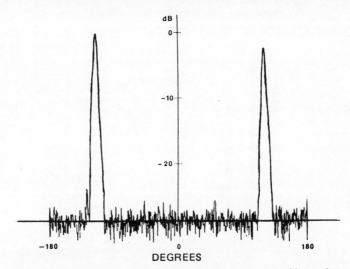

FIG. 10-13 Radiation pattern of leaky-wave antenna with varying strip widths (gain = 17.0 dB). (© *1981 IEEE.*)

tively unchanged when the perturbations are small. Hence, the scan angle of the antenna can be readily determined by using the propagation constant of the unperturbed dielectric waveguide. The field decay along the antenna has been measured to be approximately 4.5 dB/cm for strips 1.27 mm wide and spaced with a period of 2.54 mm on the broader side of the dielectric rod. This implies that with 30 strips there is only 0.05 of the energy left at the terminal end.

A design that employed varying strip width will now be described. Strips with widths equal to the width of the rod, 3.2 mm, and lengths of $(0.23 + 0.05 n)$ mm for

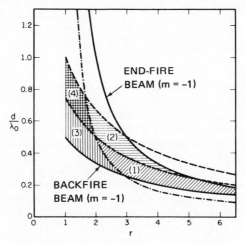

FIG. 10-14   Range of d/λ. (1) Main beam only. (2), (4) Main beam with partial backward grating lobe. (3), (4) Main beam with partial forward grating lobe. (© *1981 IEEE.*)

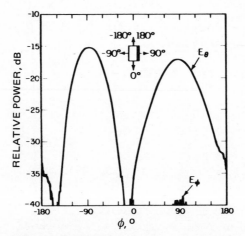

FIG. 10-15   Radiation pattern for different polarizations in main-beam cone (θ = constant). (© *1981 IEEE.*)

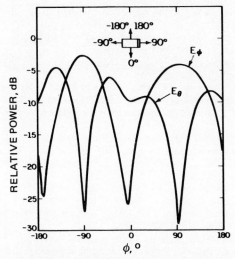

FIG. 10-16 Radiation pattern for different polarizations in main-beam cone ($\theta = 110°$). (© *1981 IEEE.*)

$1 \leq n \leq 21$ and of 1.28 mm for the remaining 40 strips were placed with a period of 2.09 mm. The pattern of this antenna is shown in Fig. 10-13. The range of d/λ that would avoid grating lobes can be determined from Fig. 10-14.

The pattern of the antenna is considerably different if the strips are placed on the narrow side of the rod and the polarization characteristics are complicated. This is seen with reference to Figs. 10-15 and 10-16, which show the azimuthal (ϕ) patterns for the two different antenna types, viz., one with the strips on the broad side (Fig. 10-12a) and the other as the narrow side of the dielectric rod (Fig. 10-12b). Note that the beam on the side of the strips is wider and has a lower gain than the beam on the opposite side of the strips.

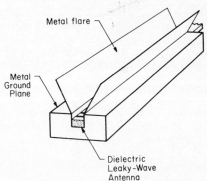

FIG. 10-17 Horn-dielectric-guide leaky-wave antenna.

Finally, the beamwidth of the azimuthal (ϕ) pattern can be improved[5] by embedding the leaky-wave antenna in a horn structure (see Fig. 10-17). Additionally, the radiation is now confined to one hemisphere, which is a desirable design feature.

REFERENCES

1 T. T. Taylor, "Design of Line Source Antennas for Narrow Beamwidths and Low Sidelobes," *IRE Trans. Antennas Propagat.*, vol. AP-3, 1955, p. 16.

2   A. Dunbar, "On the Theory of Antenna Beam Shaping," *J. App. Phys.,* vol. 23, 1952, p. 847.

3   K. C. Kelly and R. S. Elliott, "Serrated Waveguide—Part II: Experiment," *IRE Trans. Antennas Propagat.,* vol. AP-5 1957, p. 276.

4   C. H. Walter, "Curved Slot Antennas," Rep. 667-1, Cont. AF33(616)-3353, Ohio State University Research Foundation, Columbus, 1956.

5   T. N. Trinh, R. Mittra, and R. J. Paleta, Jr, "Horn Image-Guide Leaky-Wave Antenna," *IEEE Trans. Microwave Theory Tech.,* vol. MTT-29, 1981, p. 1310.

6   J. N. Hines, V. H. Rumsey, and T. E. Tice, "On the Design of Arrays," *IRE Proc.,* vol. 42, 1954, p. 1262.

7   J. A. Barkson, "Coupling of Rectangular Wave Guides Having a Common Broad Wall Which Contains Uniform Transverse Slits," *IRE Wescon Conv. Rec.,* part 1, 1957, p. 30.

8   W. L. Weeks, "Coupled Waveguide Excitation of Traveling Wave Antennas," *IRE Wescon Conv. Rec.,* part 1, 1957, p. 236.

9   R. H. MacPhie, "Use of Coupled Waveguides in a Traveling Wave Scanning Antenna," Antenna Lab. Rep. 36, University of Illinois, Urbana, 1959.

10   R. C. Honey, "A Flush-Mounted Leaky-Wave Antenna with Predictable Patterns," *IRE Trans. Antennas Propagat.,* vol. AP-7, 1959, p. 320.

11   S. Nishida, "Theory of Thin Dielectric Cover on Slitted Rectangular Waveguide Antenna," MRI Rep. R-754-59, PIB-682, Polytechnic Institute of Brooklyn, Brooklyn, N.Y., 1959.

12   S. Silver, *Microwave Antenna Theory and Design,* MIT Rad. Lab. ser., vol. 12, McGraw-Hill Book Company, New York, 1949.

13   W. W. Hansen, U.S. Patent 2,402,622, 1940.

14   G. C. Southworth, *Principles and Applications of Wave-Guide Transmission,* D. Van Nostrand Company, Inc., New York, 1950.

15   F. J. Tischer, "The H-Guide, a Waveguide for Microwaves," *IRE Conv. Rec.,* part 5, 1956, p. 44.

16   R. J. Stegen and R. H. Reed, "Arrays of Closely Spaced Non-Resonant Slots," *IRE Trans. Antennas Propagat.,* vol. AP-2 1954, p. 109.

17   L. O. Goldstone and A. A. Oliner, "Leaky Wave Antennas I: Rectangular Waveguides," *IRE Trans. Antennas Propagat.,* vol. AP-7, 1959, p. 307.

18   A. L. Cullen, "Channel Section Waveguide Radiator," *Phil. Mag.,* 7th ser., vol 40, 1959, p. 417.

19   J. N. Hines, V. H. Rumsey, and C. H. Walter, "Traveling-Wave Slot Antennas," *IRE Proc.,* vol. 41, 1953, p. 1624.

20   V. H. Rumsey, "Traveling-Wave Slot Antennas," *J. App. Phys.,* vol. 24, 1953, p. 1358.

21   R. F. Harrington, "Propagation along a Slotted Cylinder," *J. App. Phys.,* vol 24, 1953, p. 1366.

22   L. O. Goldstone and A. A. Oliner, "Leaky Wave Antennas II: Circular Waveguides," AFCRC-TN-58-141, Cont. AF19(604)-2031, Polytechnic Institute of Brooklyn, Brooklyn, N.Y., 1958.

23   R. F. Hyneman, "Closely-Spaced Transverse Slots in Waveguide," *IRE Trans. Antennas Propagat.,* vol. AP-7, 1959, p. 335.

24   A. F. Kay, unpublished work under AFCRC Cont. AF19(604)-3476, Technical Research Group, Inc., Somerville, Mass., 1959.

25   This suggestion is due to A. F. Kay, Technical Research Group, Inc., Somerville, Mass.

26   E. M. T. Jones and J. K. Shimizu, "A Wide-Band Transverse-Slot Flush-Mounted Array," AFCRC-TN-975, Cont. AF19(604)-3502, Stanford Research Institute, Menlo Park, Calif., 1959.

27   W. Rotman and S. J. Naumann, "The Design of Trough Waveguide Antenna Arrays," AFCRC-TR-58-154, Air Force Cambridge Research Center, Bedford, Mass., 1958.

28   W. Rotman and A. A. Oliner, "Asymmetrical Trough Waveguide Antennas," *IRE Trans. Antennas Propagat.,* vol. AP-7, 1959, p. 153.

29   W. Rotman and N. Karas, "Trough Waveguide Radiators with Periodic Posts," AFCRC-TR-58-356, Air Force Cambridge Research Center, Bedford, Mass., 1958.

30 R. V. Hill and G. Held, "A Radial Surface Wave Antenna," Tech. Rep. 27, University of Washington, Seattle, 1958.

31 R. C. Honey, "Horizontally-Polarized Long-Slot Array," Tech. Rep. 47, Cont. AF19(604)-266, Stanford Research Institute, Menlo Park, Calif., 1954.

32 G. Broussaud, "A New Antenna Type of Plane Structure" (in French), *Ann. Radioélectricité,* vol. 11, 1956, p. 70.

33 H. Ehrenspeck, W. Gerbes, and F. J. Zucker, "Trapped Wave Antennas," *IRE Conv. Rec.,* part 1, 1954, p. 25.

34 S. Kobayashi, R. Lampe, R. Mittra, and S. Ray, "Dielectric Rod Leaky-Wave Antennas for Millimeter-Wave Applications," *IEEE Trans. Antennas Propagat.,* vol. AP-29, no. 5, 1981, p. 822.

35 T. Trinh, J. Malherbe, and R. Mittra, "A Metal-to-Dielectric Waveguide Transition with Application to Millimeter-Wave Integrated Circuits," *Conf. Rec. Microwave Theory Tech. Soc. Int. Microwave Symp.,* 1980, pp. 205–207.

36 R. Mittra and R. Kastner: "A Spectral Domain Approach for Computing the Radiation Characterisitcs of a Leaky-Wave Antenna for Millimeter-Waves," *IEEE Trans. Antennas Propagat.,* vol. AP-29, no. 4, 1981, pp. 652–654.

Chapter 11

Long-Wire Antennas

Edmund A. Laport

RCA Corporation

11-1   INTRODUCTION

Long-wire antennas were in active development during the height of the high-frequency era, before 1940, and many different forms evolved. Since then development has been quiescent, the only surviving extant form being the horizontal rhombic antenna. Its continued popularity stems from its simple construction, low cost, and wideband capabilities. The rhombic antenna is uniquely adapted to the propagation of high frequencies via the ionosphere. Such later developments as have occurred have been in methods of engineering design either by graphical methods or by machine computation.

The configurations comprising many obsolete long-wire antennas, including the vertical Franklin arrays and the resonant horizontal types, are now history and can be found in the literature.

Some space in this chapter will be devoted to the Beverage antenna, which has desirable applications to the directive reception of low and very low frequencies.

11-2   SINGLE LONG WIRE AS A UNIT RADIATOR

Radiation Patterns

The following basic facts about the radiation patterns of long wires isolated in free space with idealized natural current distributions are useful for reference:

1   There is a lobe in the radiation pattern for each half wavelength of wire length. Each lobe is a cone of radiation centered on the wire.

2   With respect to the middle of the wire as center of the polar radiation pattern, half of the lobes are tilted forward and half of them backward. If the wire is an odd number of half wavelengths long, there is an odd number of lobes; so one of them will be normal to the wire.

3   The direction of the electric field reverses in each successive lobe. This is analogous to the reversals of phase of the currents in successive half-wavelength portions of the wire.

4   Between successive lobes are regions of little or no radiation, called nulls or zeros. Practically, these are minima because the field strengths at these angles never actually go to zero. These zeros are related analogously to the current zeros in the standing-wave current distribution or to the phase reversals in a traveling-wave distribution.

5   The angles of the *zeros* are symmetrically distributed about the plane normal to the middle of the wire.

6   For a given electrical length, the angles of the *zeros* are the same for both standing- and traveling-wave current distributions.

7   In the first quadrant the angles of the maxima for both current distributions are virtually coincident. The only exception applies to the direction of the main (first) lobe for wire lengths less than three wavelengths, as may be seen in Fig. 11-1.

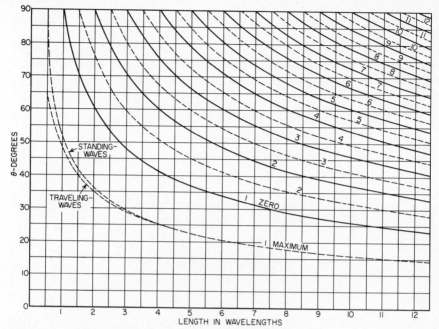

FIG. 11-1 Angles of maxima and zeros in the radiation patterns for isolated long wires with natural current distribution.

8 The largest lobe in the radiation pattern is the one forming the smallest angle (θ_1) to the wire. For the traveling-wave case, this lobe is in the direction of current flow only. For the standing-wave case the radiation pattern is symmetrical with respect to the middle of the wire and so has a complementary major lobe toward the other end of the wire.

9 With a standing-wave current distribution an integral number of half wavelengths long, the envelope of the polar plot of the field-strength distribution pattern is a straight line parallel to the wire or, more correctly, a concentric cylinder. For a nonintegral number of half wavelengths, the middle lobes are further decreased in amplitude. The maximum depression of inner lobes occurs when the current distribution consists of an odd integral number of quarter wavelengths (Fig. 11-2).

10 With traveling waves, the field-strength pattern has lobes of diminishing amplitude, the smallest being in the direction opposite that of current flow.

11 Typical deviations from idealized current distributions modify the relative lobe amplitudes slightly and fill in the zeros slightly but do not affect the *angles* of the maxima or the zeros in the radiation pattern.

12 The ratios of the amplitudes of successive lobes are higher for a traveling-wave system than for the equivalent standing-wave system (compare Figs. 11-3 and 11-4).

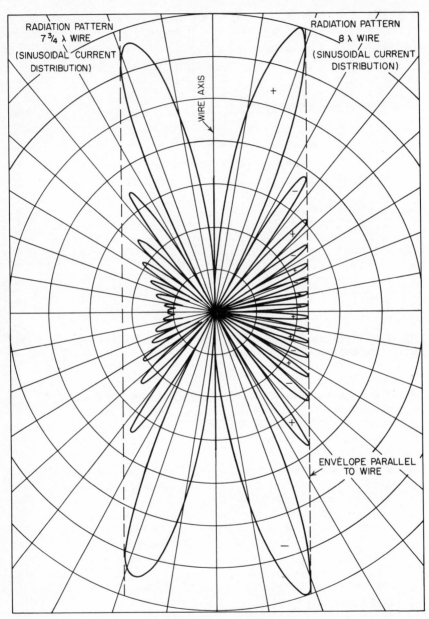

FIG. 11-2 Comparison between the polar radiation patterns of 7.75λ- and an 8λ-long wire with standing-wave current distribution, in terms of relative field strength.

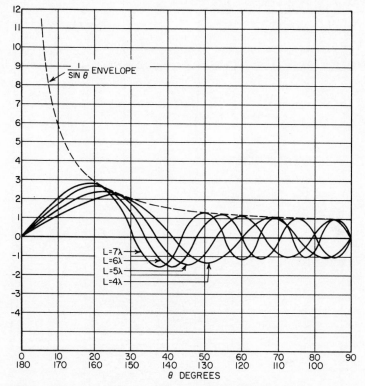

FIG. 11-3 Field-strength radiation patterns for standing-wave current distributions.

Equations for the Field-Strength Patterns

The far-field pattern for a long straight wire in free space with pure sinusoidal (standing-wave) current distribution is expressed by

$$E_\theta = \frac{60I}{r} \left[\frac{1}{\sin \theta} \cos \left(\frac{\pi L}{\lambda} \cos \theta \right) \right] \quad \text{V/m} \qquad \textbf{(11-1)}$$
$$\quad (A) \quad (B) \qquad (C)$$

(Cosine or sine in factor C is used when the number of half wavelengths is odd or even respectively.)

The coefficient (factor A) relates field strength to the antinode current I in amperes and the distance r in meters from the center of the wire. If I is expressed in rms values, the field strengths will also be expressed in rms values. The factor B by itself gives the envelope for the pattern lobes, θ being the angle to the wire axis. Factor C by itself oscillates from $+1$ to -1 and contains the information about the angles

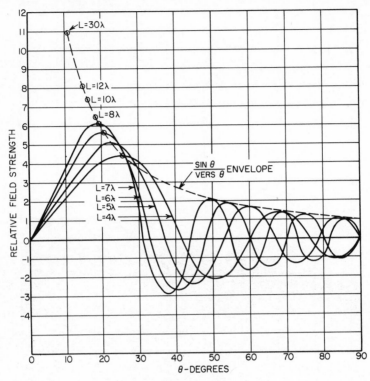

FIG. 11-4 Field-strength radiation patterns for traveling-wave current distributions.

of the zeros in the pattern. *B* and *C* together describe the *shape* of the pattern. The power flow at any angle θ is given by

$$W = \frac{E\theta^2}{377} = \text{power density, W/m}^2$$

For the case of a perfect unattenuated traveling-wave current distribution of any length,

$$E\theta = \frac{60I}{r} \frac{\sin\theta}{\text{vers }\theta} \sin\left(\frac{\pi L}{\lambda} \text{vers }\theta\right) \quad \text{V/m} \qquad \textbf{(11-2)}$$
$$\quad (A) \quad (B) \qquad\ (C)$$

As before, *A* is a scale factor. Factor *B* is the envelope factor, with vers $\theta \equiv 1 -$ cos θ. Factor *C* ranges in value from $+1$ to -1 and contains the information on the locations of the lobes and zeros. *B* and *C* together describe the complete *shape* of the pattern.

In these equations *L* is the wire length in meters and λ the wavelength in meters. Although the *C* factors in both equations are not identical, they do in fact have identical zeros, and the first quadrant maxima are virtually coincident, with the exception noted above.

The following equations are more convenient for the determinations of the angles of maxima and zeros in the far-field radiation patterns for long straight wires with traveling-wave current distributions.

For the angles of the series of maxima of diminishing amplitude starting with the largest at θ_{m1} nearest to the forward axis of the wire,

$$\theta_{mm} = \text{hav}^{-1} \frac{K}{2L/\lambda}$$

where K has successive values of 0.371, 1.466, 2.480, 3.486, and 4.495; then $m - \frac{1}{2}$ in order.

For the angle θ_{0m} for the mth zero with respect to the wire axis, from θ_{01} to 180°,

$$\theta_{0m} = \text{hav}^{-1} \frac{m}{2L/\lambda}$$

Haversines [$\equiv (1 - \cos \theta)/2$] are tabulated in *Handbook of Chemistry and Physics,* CRC Press, Boca Raton, Florida, and in many navigation handbooks.

Comparisons between standing-wave and traveling-wave field-strength patterns can be seen in Figs. 11-3 and 11-4, which plot the relative values of E_θ from Eqs. (11-1) and 11-2), using factors B and C only. They illustrate some basic differences between the two current distributions.

11-3   HORIZONTAL RHOMBIC ANTENNA

This antenna is constructed as an elevated diamond with sides from two to many wavelengths long. It is used for transmission and reception of high-frequency waves propagated via the ionosphere. When terminated with a resistance equal to its characteristic impedance at its forward apex, it functions as a traveling-wave antenna; the termination suppresses reflections of transmitted power and absorbs signals from the contrary direction.

Figure 11-5 illustrates optimum electrical parameters and the limits of frequency ratios within which performance is not unreasonably compromised. Note that with less than two wavelengths per leg the effectiveness of a rhombic antenna is generally regarded as unacceptable because its gain is too low, and in transmission a high percentage of the input power must be dissipated in a terminal resistor. With legs of seven and eight wavelengths, the terminal loss is down to 18 to 24 percent because of the radiation efficiency of such structures.

A resistance-terminated rhombic becomes essentially a transmission line, but this analogy is compromised by the expansion of the line toward the sides and its reconvergence to the forward apex. To improve the uniformity of distributed line constants and to reduce the rhombic's characteristic impedance to convenient values such as 600 Ω, three wires expand from and to each apex on each side. They are at their greatest spread at the sides. The side angles disturb the propagation of currents enough to cause small reflections, but these usually are negligible.

For receiving, many rhombics are made with a single wire for each leg. Such

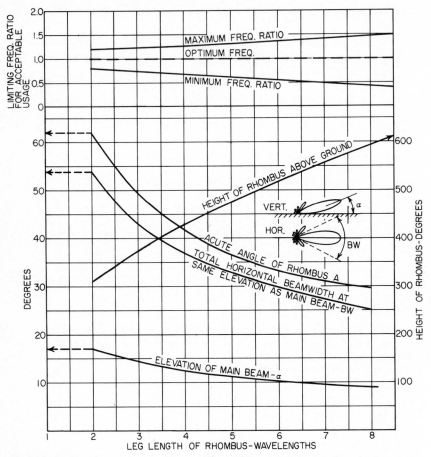

FIG. 11-5   Optimum design parameters for high-frequency horizontal rhombic antennas.

systems have characteristic impedances of the order of 800 Ω. If each apex is imped-ance-matched to a receiver, the same antenna can serve reciprocal directions.

Unless expense is a controlling factor, most professional operators prefer to use the three-wire rhombus for reception also, especially when the site is subject to severe precipitation static from sandstorms or hard-driven dry snow. In extremely noisy envi-ronments it is desirable to use insulated wires for all antenna and transmission lines so that charged particles cannot discharge directly to an exposed metallic surface. The lower impedance also contributes to static-noise reduction.

One can design a horizontal rhombic directly from data plotted in Fig. 11-5. If the design must be compromised for reasons of cost or for land restrictions, the use of stereographic charts enables the engineer to choose options wisely. Table 11-1 provides an idea of the relative amplitudes of the sidelobes for an idealized dissipationless rhombic element of various sizes.

Structures of poles, masts, wires, and standard electrical rigging hardware are

TABLE 11-1 Amplitudes of Secondary Lobes, in Decibels, Relative to a Fully Formed Main Beam, for Free-Space Rhombus

Order of maxima for first side	Order of maxima for second side								
	1	2	3	4	5	6	7	8	9
1	0								
2	−5.27	−10.6							
3	−7.7	−12.8	−15.6						
4	−9.0	−14.3	−16.5	−18.0					
5	−10.1	−15.4	−17.6	−19.1	−20.2				
6	−11.0	−16.3	−18.5	−20.0	−21.1	−21.9			
7	−11.7	−17.0	−19.2	−20.7	−21.8	−22.7	−23.4		
8	−12.3	−17.6	−19.9	−21.3	−22.3	−23.3	−24.0	−24.6	
9	−12.9	−18.1	−20.4	−21.9	−23.0	−23.8	−24.6	−25.2	−25.7

used in conventional ways. There is need for a direct ground wire between the center of the terminal resistance and a ground rod to act as a static drain on antenna and feeder.

It is usually not necessary to break up guy wires and stays with insulators. The external fields near the antenna are very weak except in the beam direction.

As a directive antenna, the rhombic is used for power gain. Best gain is realized when the side lengths, height factor, and angles are selected to produce in-phase fields in the desired direction at the design frequency. Gain is degraded at other frequencies. Figure 11-5 provides some guidance on reasonable tolerances. Beyond the indicated frequency ratios, the radiation patterns degrade rapidly.

The realized gain of a receiving rhombic would be identical to that of a transmitting antenna with the same parameters and frequency, provided the incoming equiphase wavefronts were ideally plane and were arriving along the main axis. High-frequency ionospherically propagated waves arrive with all degrees of turbulence and directions of arrival. Therefore, an areally extensive antenna such as a rhombic receives unequal illumination from the incoming waves in its various parts, and the induced currents are not coherent. Typically, then, the realized gain for receiving is highly variable, at moments approaching zero. Therefore, space-diversity reception is often essential for high-grade communication service.

A simple method for estimating the gain of a free-space rhombus is the following:

Take the total electrical length of one side (two legs) in degrees (360° per wavelength) and multiply that by the *average* current throughout its length, accounting for attenuation due to radiation, to obtain a value of degree-amperes that produces the maximum field strength on the main beam. The antenna input current will be $I = \sqrt{w/Z_0}$, where W is the input power and Z_0 is the input impedance, but this current, ideally, decreases exponentially to the far end of the antenna.

Since it is demonstrable (Ref. 1, page 4) that a doublet with a moment of one degree-ampere (1° · A) produces a maximum free-space field strength at 1 mi (1.6 km) of 325×10^{-6} V/m, the field strength produced by the rhombic (E_r) will be the product of its effective degree-amperes and 325×10^{-6} V/m on the nose of the beam.

This value, compared with that from a half-wave dipole with equal power input, gives the relative gain through the relation

$$G_{dB} = 20 \log \frac{E_r}{E_d}$$

E_d (dipole) with 1000-W input produces a field strength of 0.1376 V/m at 1 mi in free space. This is derived as follows: a half-wave dipole has a moment of $180° \times 0.636 = 114.48° \cdot A$ for 1 A of antenna current. With 1000 W radiated into a thin dipole having a radiation resistance of 73.1 Ω, the antenna current would be 3.699 A. The total moment with 1000 W radiated becomes $3.699 \times 114.48 = 423.467° \cdot A$. This value multiplied by 325×10^{-6} V/m at 1 mi gives the value 0.1376 V/m presented above.

11-4 STEREOGRAPHIC COMPUTING AIDS

Before the days of digital computers in every engineering office, the complicated multilobed radiation patterns for long-wire antennas were best computed graphically. The Foster[2] stereographic charts, giving a visual display of pattern formation and design-parameter adjustment, were developed for that purpose. Even now nothing could be simpler.

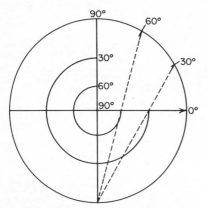

Stereographic projection of a sphere to a plane is conformal, meaning that all angular relationships are precisely maintained in the plane. Figure 11-6 illustrates the construction of a stereographic plane with latitude intervals of 30°.

Figure 11-7 illustrates the radiation pattern for a wire 4.5 wavelengths long in free space in stereographic coordinates. The wire is located on the diameter; for the case of a traveling wave with current flowing in the direction of the arrow, the resulting cones of radiation are shown by broken lines, their successive maxima

FIG. 11-6 Derivation of stereographic elevation angles from circular azimuth angles.

being marked M_1 (main lobe), M_2, etc. The intervening zeros are shown in solid lines. Along the diameter are ticks at 10° intervals in the third dimension, with 90° at the pole of the figure.

It is convenient to speak of the zeros, but physically they are in fact deep minima. The stereographic charts for any wire length can be plotted from the data of Fig. 11-1.

The free-space radiation pattern is seen when a transparent pair of stereographic charts is overlaid at the acute angle of the rhombus (or V). This is illustrated in Fig. 11-8, which is for 5.5-wavelength legs and an acute angle of 33°, producing a main beam of 13° from the plane of the rhombus.

Swinging the apex angle will show how the main-beam angle can be adjusted as desired at one frequency. If the apex angle is too wide, the main beam will shrivel and perhaps disappear.

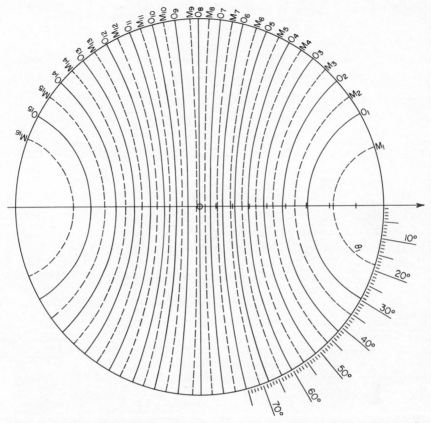

FIG. 11-7 Free-space radiation pattern for a straight wire with traveling-wave current distribution in stereographic coordinates.

Combining patterns stereographically is mathematically equivalent to multiplying the two patterns. A zero in one produces a zero in the combined pattern. This is also true if the height factor is included when locating the rhombus parallel to the ground to complete the horizontal rhombic antenna. The main beam is then horizontally polarized, whereas the various secondary lobes off the main axis are obliquely polarized.

Height factors for horizontal polarization can also be charted stereographically and used visually with the rhombus charts to show the full three-dimensional pattern.

11-5   OPTIMUM DESIGN AND BANDWIDTH FOR HORIZONTAL RHOMBIC ANTENNAS

This section will emphasize the electrical limitations of rhombic antennas. The frequency bandwidth for essentially uniform input impedance is vastly greater than the bandwidth for acceptable radiation patterns. Attempts to use the antenna over too broad a frequency band lead to excessive emissions in useless directions or employ-

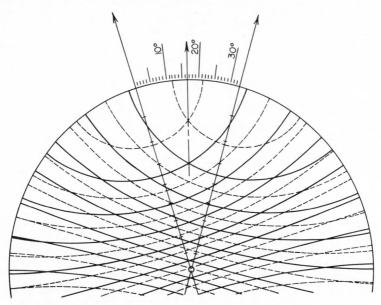

FIG. 11-8   Radiation patterns for a 5.5λ rhombus with an acute angle of 33°.

ment of excessive power to compensate for deficiencies in the antenna pattern. A sound operations rule is not to use a rhombic antenna over more than a 2:1 frequency range. New operators coming into old stations need to be made aware of such practices so as not to misuse the facilities.

11-6   LAND-SAVING ARRANGEMENTS FOR MULTIPLE RHOMBIC ANTENNAS

Large high-frequency radio stations employing many rhombic antennas can econo-mize on land area and feeder lengths in at least two ways: (1) a small rhombic can be located inside the area of a large one having about the same orientation, and (2) com-mon supports can be used for two or three antennas (at the corners).

These savings are possible because the mutual impedances between traveling-wave antennas are very much lower than for resonant systems.

11-7   SIDELOBE REDUCTION AND GAIN IMPROVEMENT WITH DOUBLE-RHOMBOID ANTENNAS

Particular high-frequency communication conditions that require the bandwidth of a rhombic antenna, but with substantial reduction in sidelobes and a gain improvement of 3 dB, may benefit by means of the double-rhomboid antenna.[3] This is a form of array employing either two coaxial rhombics of different sizes fed from a common

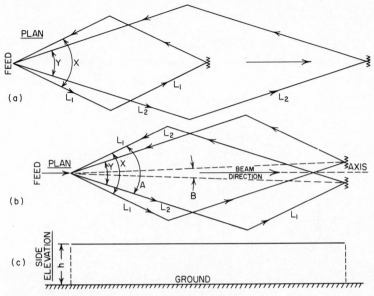

FIG. 11-9 Double-rhombic and double-rhomboid antennas having the same radiation patterns.

apex or the same leg lengths connected as two complementary rhomboids, which reduces the array length. The two geometries are illustrated in Fig. 11-9.

The radiation pattern is designed with stereographic charts, using two pairs for two different leg lengths. By choosing lengths that cause a better distribution of zeros in the sidelobe regions while aligning the main lobes, the growth of sidelobes is inhibited substantially. An example of a design that was scale-modeled and later constructed for transatlantic military traffic is shown in Fig. 11-10. Multipath transmission is reduced by transmitting with low beam angles that are essentially free of sidelobes. This assures fewer teleprinted-character errors and statistically better circuit reliability.

The input impedance of a double-rhomboid antenna is about half that of a single rhombus having the same wire structure.

11-8 IMPORTANCE OF PHYSICAL AND ELECTRICAL SYMMETRY

The proper operation of a rhombic antenna and its feeder depends on balanced potentials from all parts of the system to ground. Any unbalance produces a component of field operating against ground. Radiation from unbalanced components compromises the basic system performance. Care is needed to maintain symmetry through switching systems and in bends and corners in the feeders. Electrical balance is obtained only through structural symmetry, since there are no wideband methods of correcting unbalance otherwise.

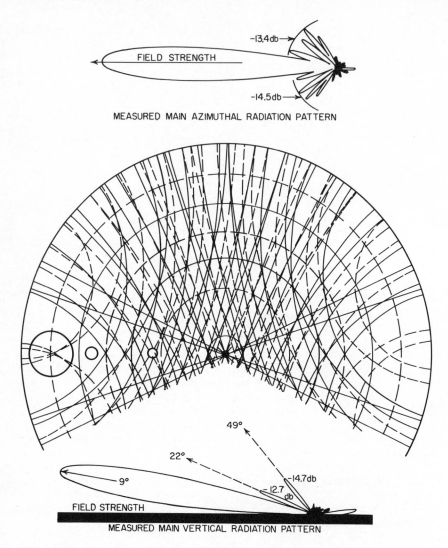

FIG. 11-10 Comparison of measured and stereographically computed radiation patterns for the array of Fig. 11-9b when $L_1 = 3.5\lambda$, $L_2 = 6.0\lambda$, $X = 50°$, and $Y = 35°$.

11-9 CROSSTALK BETWEEN FEEDERS

Long parallel feeder runs, nearby or on the same supports, can lead to crosstalk between feeders with consequent radiation of crosstalk power on the wrong frequency and antenna. Such systems lack the selectivity of resonant systems to filter out crosstalk. Therefore, precautions to control crosstalk must be included in system design, such as by transpositions or by rotating the plane of an adjacent open-wire feeder. For receiving, coaxial feeders matched to the antenna with balance-to-unbalance transformers are often employed.

11-10   CIRCUITAL PROPERTIES OF RHOMBIC ANTENNAS

Ideally, the antenna feeder matches the impedance of the antenna at all frequencies. The input impedance of the antenna varies somewhat with frequency owing to small internal reflections, but it remains close to resistive. The variations are minimized by using three spread side wires and by adjusting the far-end terminal resistance.

Balanced open-wire transmission lines can be configured for characteristic impedances from 200 to 650 Ω by appropriate combinations of sizes of wire, line cross sections, and numbers of wires (Ref. 1, pages 422–425). When a wideband transformer is needed to make a match between two different impedances, this can be done in a short length with an exponentially tapered line section. A linear taper may also make a satisfactory match if the tapered section is of sufficient length. Such transformations are usually performed in the vertical feed to the antenna input.

11-11   FEED-POINT IMPEDANCE

The input impedance for a one-wire rhombic is usually about 800 Ω. The conventional three-wire rhombus has an input impedance of the order of 600 Ω, and the two-wire intermediate design has an impedance of about 700 Ω. All these impedances are for terminated rhombics with optimum-value termination. (See Refs. 4 and 5 for good guiding values for well-constructed systems.)

11-12   TERMINATING RESISTANCE AND TERMINAL- POWER LOSS

Unradiated power arriving at the forward apex of a rhombic antenna will be reflected back to the input unless it is absorbed passively. If radiation loss is 8 dB on a long antenna, complete reflection of excess power will be attenuated by another 8 dB on the way back to the input. Whereas that loss may not change the input impedance intolerably, this backward power is radiated in a contrary direction, which could be disturbing to others.

If the terminal power is small, a fixed resistor of correct value may suffice, though it is preferable for the resistor to be composed of two units with the midpoint wires connected to a ground rod. Ceramic resistors of high impact strength are needed to withstand lightning impulses.

High-power transmissions often leave substantial power to be dissipated in the terminal resistance. For dissipating this power, sections of open iron-wire transmission lines of high dissipative capacity per unit length (obtained with high-μ iron wire) have been used. Length should be sufficient to have an attenuation of 6 or 13 or more. The far ends are then connected together to ground. Total residual reflection from the ground connection is attenuated sufficiently before it reaches the antenna.

Since most of the power to be dissipated is lost at the input end of the dissipation line, it is not unusual for that portion to be heated to a glow.

11-13   RHOMBIC ARRAYS

Arrays of rhombics may be useful in special circumstances. Because of the large spatial spread of one unit, the synthesis of an array must take account of centers of radiation to produce coherent array radiation patterns.

The largest rhombic-antenna array known, built during the days of the high-frequency transatlantic radiotelephone, was the multiple-unit steerable array described by Polkinghorn.[6]

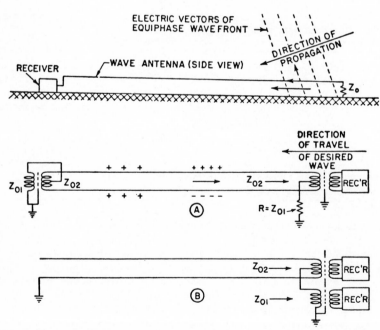

FIG. 11-11 Beverage antenna arrangements for one direction of reception, reversed-direction reception, and two-way reception.

11-14 BEVERAGE RECEIVING ANTENNA FOR LOW FREQUENCIES

This antenna, developed by Beverage, Rice, and Kellogg, is unique as a directive antenna for reception of low and very low frequencies. It was the first such antenna to use the traveling-wave principle.

It consists, in its simplest form, of a single-wire transmission line running in the direction of expected wave arrival and terminated in its characteristic impedance with a receiver at its near end. Reception depends on the tilt of the arriving vertically polarized wavefront caused by local ground losses. The electric vector of the equiphase front, tilted forward, produces a component of electric force parallel to the wire, inducing a current in the wire. This flows toward the receiver and is reinforced throughout the length of the antenna.

Wave tilt increases with frequency and with ground resistivity. The length of antenna depends on the available land, up to the limit at which induced currents tend to be out of phase with the advancing wavefront. Directivity increases with length up to the above limit. The wave transmitted along the antenna is a transverse magnetic wave.

Figure 11-11 illustrates the principles for antennas adapted for reception from either or both directions. The most extensive use of Beverage antennas was in an array of four units for overseas reception on the first transatlantic radiotelephone system operating at 50 and 60 kHz.[7]

REFERENCES

1 E. A. Laport, *Radio Antenna Engineering,* McGraw-Hill Book Company, New York, 1952.
2 Donald Foster, "Radiation from Rhombic Antennas," *IRE Proc.,* vol. 25, October 1937, p. 1327. A classic on rhombic-antenna theory and design.
3 E. A. Laport and A. C. Veldhuis, "Improved Antennas of the Rhombic Class," *RCA Rev.,* March 1960.
4 E. Bruce, A. C. Beck, and L. R. Lowry, "Horizontal Rhombic Antennas," *IRE Proc.,* vol. 23, January 1935, p. 24.
5 A. E. Harper, *Rhombic Antenna Design,* D. Van Nostrand Company, Inc., New York, 1941.
6 F. A. Polkinghorn, "Single-Sideband MUSA Receiving System for Commercial Operation on Transatlantic Radio Telephone Circuits," *Bell Syst. Tech. J.,* April 1940, pp. 306–335.
7 A. Baily, S. W. Dean, and W. T. Wintringham, "Receiving System for Long-Wave Transatlantic Telephony," *Bell Syst. Tech. J.,* April 1929, pp. 309–367.

BIBLIOGRAPHY

Booth, C. F., and B. N. MacLarty: "The New High-Frequency Transmitting Station at Rugby," *J. IEEE (London),* part B, vol. 103, October 1955. Rhombic antenna layouts, two rhombic antennas on one set of masts, and feeder crosstalk measurements.
Bruce, E.: "Developments in Short-Wave Directive Antennas," *IRE Proc.,* vol. 19, 1931, p. 1406; *Bell Syst. Tech. J.,* vol. 10, October 1951, p. 656. Evolution of the vertical-V antenna.

Carter, P. S., C. W. Hansell, and N. E. Lindenblad: "Development of Directive Transmitting Antennas by RCA Communications," *IRE Proc.*, vol. 19, October 1931, p. 1733. Evolution of standing-wave long-wire antenna development.

CCIR Antenna Diagrams, International Telecommunications Union, Geneva, 1954.

Christiansen, W. N.: "Directional Patterns for Rhombic Antennae," *AWA (Australia) Tech. Rev.*, vol. 7, 1946, p. 1. Study and pattern computations for horizontal rhombic antennas and four types of rhombic arrays; basic paper on rhombic arrays.

————, W. W. Jenvey, and R. D. Carmen: "Radio-Frequency Measurements on Rhombic Antennae," *AWA (Australia) Tech. Rev.*, vol. 7, no. 2, 1946, p. 131. Instrumentation and measuring techniques for determining current distributions on two-tier rhombic arrays.

Colebrook, F. M.: "Electric and Magnetic Fields for Linear Radiator Carrying a Progressive Wave," *J. IEE (London)*, vol. 89, February 1940, p. 169.

Dufour, Jean: "Diagrammes de réception d'antennes rhombiques dans un plan vertical: résultats expérimentaux," *Bull. Tech. PTT (Bern, Switzerland)*, vol. 31, March 1953, p. 65. Measurements of vertical-plane patterns with aircraft, including effects of rough terrain.

Harrison, C. W., Jr.: "Radiation from Vee Antennas," *IRE Proc.*, vol. 31, July 1943, p. 362.

————: "Radiation Field of Long Wires with Application to Vee Antennas," *J. App. Phys.*, vol. 14, October 1943, p. 537.

Iizuka, K.: "Traveling-Wave V and Related Antennas," *IEEE Trans. Antennas Propagat.*, vol. AP-15, March 1967.

Laport, E. A.: "Design Data for Horizontal Rhombic Antennas," *RCA Rev.*, vol. 13, March 1952, p. 71.

Wells, E. M.: "Radiation Resistance of Horizontal and Vertical Aerials Carrying a Progressive Wave," *Marconi Rev.*, no. 83, October–December 1946.

Chapter 12

Surface-Wave Antennas and Surface-Wave-Excited Arrays

Francis J. Zucker

Rome Air Development Center
Hanscom Air Force Base

12-1 INTRODUCTION

The two types of traveling-wave antennas discussed in this chapter are illustrated in Fig. 12-1. In Fig. 12-1a, a *surface wave* (also called a *trapped wave* because it carries its energy within a small distance from the interface) is launched by the feed F and travels along the dielectric rod to the termination T. Since a surface wave radiates only at discontinuities, the total pattern of this antenna (normally end-fire) is formed by the interference between the feed and terminal patterns.[1] The dielectric material could alternatively be an artificial one, e.g., a series of metal disks or rods (the Yagi-Uda antenna). On the other hand, discontinuities can be placed all along the surface-wave structure, producing radiation in the manner of a slotted waveguide except that the wave propagates on an open, not a shielded, guide. The array of dipoles proximity-coupled to a two-wire transmission line in Fig. 12-1b illustrates this second antenna type; here we are evidently using the term *surface wave* loosely to designate any wave mode that propagates along an open interface without radiating.

Two closely related antenna structures are discussed in other chapters: the helix (which involves a surface wave in one of its modes of operation; Chap. 13) and the frequency-independent antennas (Chap. 14). The short Yagi-Uda, in whose design surface waves play no role, is covered in Chap. 3. An additional discussion of surface-wave antennas can be found in Ref. 2.

Surface-wave antennas lend themselves to flush, or at least low-silhouette, installation. Their gain normally does not exceed 20 dB, and control over their pattern shape (including sidelobes) is *very limited*. Bandwidth is generally narrow. Frequency coverage extends from the popular Yagi-Udas at very high frequencies (VHF) to printed microwave (and perhaps millimeter-wave) antennas. By contrast, arrays excited by surface waves produce patterns that can in principle produce as high gain and be controlled as accurately as the patterns of nonresonant slot arrays.

12-2 PROPERTIES AND MEASUREMENT OF SURFACE WAVES

For design purposes, a surface wave is sufficiently characterized by one of its parameters, for example, the wavelength in the direction of propagation along the interface. The various parameters will now be defined and formulas given for their interrelationship; it will then be shown how they can be calculated if the surface impedance is known or how they can be determined by measurements.

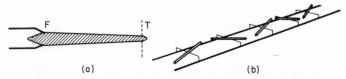

FIG. 12-1 Surface-wave-antenna types. (*a*) Dielectric rod. (*b*) Array of proximity-coupled dipoles (thin lines are dielectric supports).

Interrelationship of Parameters

The governing relation is the separability condition for the wave equation. In rectangular coordinates (Fig. 12-2a),

$$k_x^2 + k_y^2 + k_z^2 = k^2 \qquad\qquad \textbf{(12-1)}$$

where the k's, called wave numbers, are in general complex; for example, $k_x = \beta_x - j\alpha_x$, with β_x the phase constant in radians per unit length (inches, centimeters, or meters), and α_x the attenuation in nepers per unit length. The wave number k is that of the medium in which the wave travels, $k = \omega \sqrt{\epsilon_0 \mu_0}$, with ω the angular frequency, ϵ_0 the dielectric constant, and μ_0 the permeability. In air, k is pure real; it is related to the wavelength λ by $k = \beta = 2\pi/\lambda$ and to the phase velocity of light c by $k = \omega/c$. Similarly, the phase constant β_z along the surface is related to the surface wavelength λ_z by $\beta_z = 2\pi/\lambda_z$ and to the surface phase velocity v_z by $\beta = \omega/v_z$. The following ratios are therefore equivalent:

$$\frac{\beta_z}{k} = \frac{\lambda}{\lambda_z} = \frac{c}{v_z} \qquad\qquad \textbf{(12-2)}$$

When this ratio is greater than 1, we speak of a "slow" wave (i.e., slower than light, because $v_z < c$); when less than 1, of a "fast" wave ($v_z > c$). The surface wavelength λ_z of slow waves is shorter than λ; that of fast waves, longer than λ (as in shielded waveguide).

A *surface* wave is one that propagates parallel to the interface and decays vertically to it; that is, the phase constant β_x is zero, and $k_x = -j\alpha_x$ (pure attenuation), with α_x positive. Assume first that the wave extends indefinitely in the transverse direction so that $k_y = 0$ (Fig. 12-2a). Equation (12-1) requires that k_z be pure real;

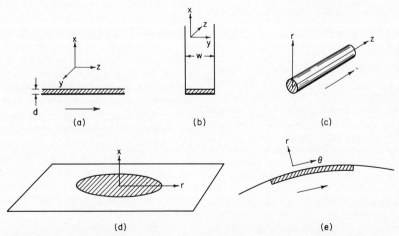

FIG. 12-2 Geometries for surface-wave propagation over dielectric sheets. (*a*) Infinite-plane sheet-on-metal. (*b*) Rectangular duct. (*c*) Axial-cylindrical rod. (*d*) Radial-cylindrical sheet-on-metal. (*e*) Azimuthal-cylindrical or spherical cap. Heavy lines indicate metal surfaces. Arrows point in the direction of propagation.

that is, there can be no attenuation in the direction of propagation, and we obtain the simple but basic relation

$$\beta_z^2 = k^2 + \alpha_x^2 \qquad (12\text{-}3a)$$

or equivalently, by using Eq. (12-2),

$$\frac{\lambda}{\lambda_z} = \frac{c}{v_z} = \sqrt{1 + \left(\frac{\alpha_x \lambda}{2\pi}\right)^2} \qquad (12\text{-}3b)$$

from which it follows that this is a slow wave. Equation (12-3b) is plotted as the dashed line in Fig. 12-3b. Slower-than-light surface waves are used chiefly in end-fire antenna design (Secs. 12-3 and 12-4). Figure 12-3a illustrates constant-phase and constant-amplitude fronts, which are at right angles to each other provided only that the medium above the interface is lossless. (Ohmic losses in the surface produce a slight forward tilt in the phase front and a small α_z.) The more closely the surface phase velocity approaches that of light, the smaller is α_x [Eq.(12-3b)] and the larger, therefore, the vertical extent of the surface wave in Fig. 12-3a.

All components of the total electromagnetic field of the surface wave, for example, E_z, are of the form

$$E_z(x,z,t) = E_z e^{-\alpha_x x} e^{-j\beta_{zz}} e^{j\omega t} \qquad (12\text{-}4)$$

The vertical decay of the E_z, E_x, and H_y components of a TM surface wave ($H_z = 0$) is shown in Fig. 12-4a, and the composite electric field lines over a full wavelength interval (longitudinal section) in Fig. 12-4b. It can be shown that E_z is in phase with H_y, E_x in phase quadrature. The first two components therefore carry all the power along the interface, while E_x and H_y form a vertically pulsating storage field. The dielectric slab-on-metal below the interface in Fig. 12-4 is used as an example. Since the form of Eq. (12-4), and thus of the surface-wave field lines, does not depend on the detailed structure of the medium below the interface, a multilayer dielectric or a corrugated sheet, etc., could have been shown equally well.

The field (or power) decay in decibels is plotted as a function of height x/λ above the surface in Fig. 12-3b [based on Eq. (12-3a)]. If we assume a relative surface wavelength λ/λ_z (or relative

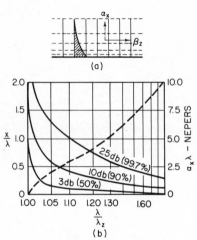

(a)

(b)

FIG. 12-3 Phase and amplitude contours of a plane surface wave. (a) Solid lines indicate constant-phase fronts; dashed lines, constant-amplitude fronts; shaded region shows amplitude decay. (b) Solid lines indicate height x/λ above surface of 3-dB, 10-dB, and 25-dB constant-amplitude or power contours as a function of relative surface wavelength λ/λ_z (use left-hand ordinate); numbers in parentheses refer to the percentage of total energy carried by the surface wave between the interface and each contour. The dashed line indicates the relation between relative surface wavelength and vertical attenuation $\alpha_x\lambda$ according to Eq. (12-3b) (use right-hand ordinate).

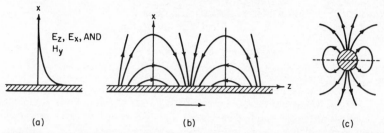

FIG. 12-4 Surface-wave field structure above interface. (*a*) TM-wave components on plane surface. (*b*) Composite electric field lines over a one-wavelength interval (arrow points in the direction of propagation). (*c*) Composite electric field lines of HE₁₁ wave on an axial-cylindrical surface (cross section). Dashed line shows the electric-image plane (Sec. 12-4).

phase velocity c/v_z) equal to 1.1, for example, the field at height $x = \lambda$ has decayed by 25 dB below its value at the interface.

Assume next that two parallel walls are erected normal to the surface, forming a duct in the z direction (Fig. 12-2*b*). [Examples other than the dielectric-loaded channel in this figure are the channel with a corrugated bottom and the trough guide (Sec. 12-6).] The transverse wave number is fixed by the duct width w as in conventional waveguide theory; from Eq. (12-1) we now have, instead of Eq. (12-3*a*),

$$\beta_z^2 = k^2 + \alpha_x^2 - \left(\frac{n\pi}{w}\right)^2 \qquad n = 0, 1, 2, \cdots \qquad \textbf{(12-5a)}$$

If $n = 0$ (no variation in y direction), Eqs. (12-3), and therefore also Fig. 12-3*b*, still apply. If $n = 1$ (half a sinusoid in y direction),

$$\frac{\lambda}{\lambda_z} = \frac{c}{v_z} = \sqrt{1 + \left(\frac{\alpha_x \lambda}{2\pi}\right)^2 - \left(\frac{\lambda}{2w}\right)^2} \qquad \textbf{(12-5b)}$$

Since the second term is usually smaller than the third, it follows that the wave is usually, though not necessarily, faster than light. Fast as well as slow surface waves can be used to excite arrays of discrete elements (Secs. 12-5 and 12-6). The field and power decay still follows Fig. 12-3*a* and *b*, provided that the abscissa in Fig. 12-3*b* is relabeled $\sqrt{(\lambda/\lambda_z)^2 + (\lambda/2w)^2}$ to account for the difference between Eqs. (12-3*b*) and (12-5*b*). The heights x/λ corresponding to the 25-dB contour in Fig. 12-3*b* are useful for estimating the extent to which the duct walls, which theoretically must be infinite in height, may be lowered without perturbing the surface wave.

Surface-wave geometries other than rectangular are shown in Fig. 12-2*c–e*. The separability condition for the wave equation in axial- and radial-cylindrical coordinates (Fig. 12-2*c* and *d*) reads

$$k_r^2 + k_z^2 = k^2 \qquad \textbf{(12-6)}$$

(note that k_θ does not appear), which is formally identical with Eq. (12-1) for the case $k_y = 0$. The interrelationships of the surface-wave parameters given in Eqs. (12-3) still hold if k_r is substituted for k_z. Above the surface, the field components decay like Hankel functions; in the axial-cylindrical case, for example,

$$E_z(r,z,t) = E_z \, H_o^{(2)} \, (\alpha_r r) e^{-j\beta_{zz}} e^{j\omega t} \tag{12-7}$$

The circumferential field dependence in the case of the HE_{11} mode—the one of principal interest in this geometry—is one full sinusoid. Composite E lines for this *hybrid* wave (so called because of the presence of both E_z and H_z) are shown in Fig. 12-4c. [The structure inside the circular interface could equally well be a metal rod with a dielectric mantle, an array of circular disks (cigar antenna), or even an array of dipoles (Yagi-Uda).] For large radii, the Hankel function is asymptotic to the exponential, and Eq. (12-7) differs negligibly from Eq. (12-4). As a consequence, the 25-dB contours in Fig. 12-3b still hold and the 10-dB contours nearly so; these curves are useful in estimating the degree of coupling between adacent end-fire line sources. In the radial-cylindrical case, the field components decay exponentially away from the surface so that Fig. 12-3b applies exactly. But the constant-amplitude fronts will now no longer be parallel to the surface because the field components decay radially like Hankel functions (asymptotically like $1/\sqrt{r}$), bringing the fronts closer to the surface with increasing r.

In the azimuthal-cylindrical or spherical case it can be shown that the wave numbers are always complex so that these surface waves are no longer completely trapped. Because the radius of curvature is usually large, it is convenient to treat azimuthal waves as a perturbation of Eqs. (12-3) for flat surfaces.

Calculation of Surface-Wave Parameters

Surface waves are modes on unshielded waveguides. As in the case of shielded waveguides, the boundary-value problem is often conveniently stated in terms of the *transverse resonance condition,* which requires that the equivalent network of a transverse section through the waveguide be resonant.[3] With Z_x the impedance looking straight up from the interface and Z_s the surface impedance looking straight down, the resonance condition reads

$$Z_x + Z_s = 0 \tag{12-8}$$

Z_x depends on the geometry and the mode; in rectangular coordinates,

$$Z_x = \frac{k_x}{\omega \epsilon_o} \qquad \text{for TM waves} \tag{12-9a}$$

$$Z_x = \frac{\omega \mu_o}{k_x} \qquad \text{for TE waves} \tag{12-9b}$$

In the case of a surface wave, $k_x = -j\alpha_x$; Z_x is therefore capacitive in Eq. (12-9a) and inductive in Eq. (12-9b). It follows from Eq. (12-8) that the TM surface wave requires an inductive surface, $Z_s = jX_s$, and the TE surface wave a capacitive surface, $Z_s = -jX_s$ (X_s positive). Equation (12-8) now reads

$$\frac{\alpha_x}{k} = \frac{X_s}{R_C} \qquad \text{for TM waves} \tag{12-10a}$$

$$\frac{\alpha_x}{k} = \frac{R_c}{X_s} \quad \text{for TE waves} \quad (\textbf{12-10}b)$$

where $R_c = \sqrt{\mu_0/\epsilon_0} = 377\ \Omega$. β_z or λ/λ_z follows immediately from Eqs. (12-3).

Knowledge of the surface impedance thus solves the boundary-value problem. It is instructive to calculate this impedance for two simple cases: surface waves on a dielectric slab-on metal and on a corrugated-metal sheet. Looking down from the air-dielectric interface in Fig. 12-2a, one sees a dielectric-loaded region short-circuited at the end. By using transmission-line theory, assuming that the wave is TM, and noting that a wave that is TM in the direction of propagation is also TM transverse to it,

$$X_s = \frac{k_x}{\omega\epsilon_1} \tan k_x d \quad (\textbf{12-11})$$

where k_x is the vertical wave number in the slab of dielectric constant ϵ_1. Since X_s must be positive real, $k_x = \beta_x$. [One could now write the field components below the interface in the form of Eq. (12-4), with a sinusoidal variation in x replacing the exponential decay; the field configuration below the interface, however, is rarely of interest to the engineer.] Substitution of Eq. (12-11) in Eq. (12-10a) gives one equation in two unknowns, α_x and β_x. α_x is a function of β_z through Eq. (12-3a); β_x can also be expressed as a function of β_z, since Eq. (12-1) reads

$$\beta_x^2 + \beta_z^2 = k_1^2$$

in the dielectric medium ($k_1 = \omega\sqrt{\mu_1\epsilon_1}$). The solution[4] shows that the lowest TM mode has no cutoff and that the phase velocity of the surface wave lies between c and $c\sqrt{\epsilon_1/\epsilon_0}$, approaching the free-space velocity in air for very thin sheets and the free-space velocity in the dielectric for thick sheets (numerical results are in Sec. 12-4). The analysis for a TE wave uses Eq. (12-10b) and results in a lowest mode that does have a cutoff. It is worth pointing out that below cutoff a surface wave does not become evanescent but ceases to exist altogether.

The interface of the corrugated-metal sheet (Fig. 12-5) is an imaginary plane through the top of the teeth. X_s is zero over the teeth and equals $R_c \tan kd$ over the grooves, which are short-circuited parallel-plane transmission lines (inductive in the range $0 < kd < \pi/2$). If there are many grooves per wavelength, the surface impedance over the teeth and grooves may be averaged; that is, $X_s \cong [g/(g + t)]R_c \tan kd$. Substitution in Eq. (12-10a) gives one equation in one unknown. The results of this approximation turn out to be good provided there are at least five grooves per wavelength. A more exact calculation takes into account the higher-order (below-cutoff) modes at the mouth of the grooves, which produce higher-order waves along the interface that modify the field lines sketched in Fig. 12-4a and b only in the close vicinity of the teeth.

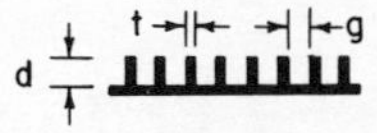

FIG. 12-5 Corrugated-metal surface. Arrow points in the direction of propagation.

Measurement of Surface-Wave Parameters

In the case of surface waves, a vertical metal plate large enough to reflect at least 90 percent of the incident energy (use Fig. 12-3b to determine the minimum size) produces deep nulls at half-wavelength intervals along the surface. Because of turbulence in the aperture field near the feed, measurements should be made at a minimum distance from the feed given by

$$d_{\min} \cong \frac{0.17\lambda}{(\lambda/\lambda_z) - 1}$$

If the frequency is so low that reflector size becomes a problem, the distance between nulls is found by feeding the probe signal into a phase-comparison circuit (using magic T and attenuator) that measures the traveling-wave phase with respect to a reference signal.

The probe used in these measurements should have a small cross section so as not to perturb the field. It needs chokes to suppress antenna currents on its outer conductor, and it must couple only into the E or the H field, with good discrimination against the unwanted one. A coaxially fed monopole or a waveguide horn pinched at the mouth (with dielectric loading to keep it above cutoff)[5] is very satisfactory; loops and horizontal dipoles, less so. The distance between probe and aperture should be as small as possible without perturbing the aperture field, and it should be held constant by providing a rigid probe carriage.

12-3   SURFACE-WAVE ANTENNAS: DESIGN PRINCIPLES (see Zucker in Bibliography)

Radiation of Surface-Wave Antennas

Before presenting design principles (for optimum gain, beamwidth, sidelobe level, and bandwidth), it is useful to examine how surface-wave antennas radiate. Two typical structures are shown in Fig. 12-6. The feed F (consisting of a monopole and reflector

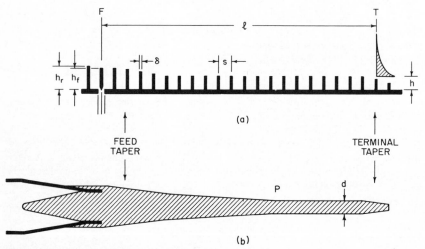

FIG. 12-6 Surface-wave antenna structures. (a) Yagi-Uda (row of monopoles) on ground plane, excited by dipole-reflector combination. (b) Dielectric rod, excited by circular or rectangular waveguide (broad wall in plane of paper).

in Fig. 12-6a and of a circular or rectangular waveguide in Fig. 12-6b) couples a portion of the input power into a surface wave, which travels along the antenna structure to the termination T, where it radiates into space. The ratio of power in the surface wave to total input power (efficiency of excitation) is usually between 65 and 75 percent. Power not coupled into the surface wave is directly radiated by the feed in a pattern resembling that radiated by the feed when no antenna structure is in front of it.

The tapered regions in Fig. 12-6 serve diverse purposes. The *feed taper* increases the efficiency of excitation and also affects the shape of the feed pattern. The *body taper* (extending to point P in Fig. 12-6b) suppresses sidelobes and increases bandwidth. Because a reflected surface wave spoils the pattern and bandwidth of the antenna, a *terminal taper* is employed to reduce the reflected surface wave to a negligible value.

The surface wave illuminates the terminal aperture (plane perpendicular to the antenna axis through T) in a circular region whose radius increases as the transverse attenuation of the surface wave diminishes or, equivalently, as the surface phase velocity approaches that of light [Eq. (12-3b)]. The larger the illuminated region, the higher the gain produced in the radiation pattern. Since the surface-wave phase front lies in the aperture plane, the pattern peaks in the end-fire direction. A simple but only approximate expression[2,6] for the terminal radiation pattern $T(\theta)$ is

$$T(\theta) \cong \frac{1}{(\lambda/\lambda_z) - \cos\theta} \quad \textbf{(12-12)}$$

where θ is the angle off end fire and λ_z is the surface wavelength at T (just to the left of the terminal taper, to be precise). This pattern has neither nulls nor sidelobes and, as expected, falls off more rapidly the closer λ/λ_z is to 1. More accurate expressions[2,7–9] and an experimental determination[10] indicate that the pattern is actually about 20 percent narrower than in Eq. (12-12). The dashed line in Fig. 12-7 shows this corrected terminal radiation pattern for $\lambda/\lambda_z = 1.08$, which is the optimum relative phase velocity for a maximum-gain antenna 4λ long [Eq. (12-14b)].

The radiation pattern of the feed shown in Fig. 12-6a is quite broad (approximately a cardioid in polar coordinates). In combination with the terminal radiation $T(\theta)$, it produces the total

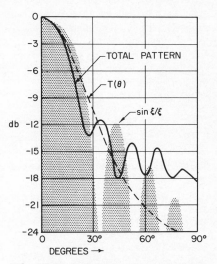

FIG. 12-7 Radiation pattern of 4λ-long surface-wave antenna adjusted for maximum gain in accordance with Eq. (12-14b) ($\lambda/\lambda_z = 1.08$). $T(\theta)$ is the *H*-plane pattern of the surface-wave distribution in the terminal plane. For comparison, $\sin\xi/\xi$ is also shown (shaded region).

surface-wave antenna pattern, shown by the solid line in Fig. 12-7 (experimental data for antenna length $\ell = 4\lambda$, $\lambda/\lambda_z = 1.08$). In the vicinity of end fire, the terminal pattern predominates; as θ increases, interference with the feed pattern first narrows the beam and then produces a sidelobe at 35°; beyond 45° the feed pattern predominates. Patterns of all surface-wave antennas adjusted for maximum gain look like those in Fig. 12-7, except that for antennas longer than 4λ the pattern spreads less in θ (first sidelobe closer to end fire than 35°) and for shorter structures it spreads more. The field beyond the first sidelobe falls off more rapidly the larger the end-fire gain of the feed radiation.

The curves in Fig. 12-7 are for the H plane, in which the element pattern of the Yagi-Uda monopoles (Fig. 12-6a) is omnidirectional. In the E plane, $T(\phi)$ (ϕ is the angle off end fire) is narrowed by multiplication with the E-plane element factor (a dipole pattern modified by the effects of mutual impedance), which for $\phi < 60°$ approximates $\cos \phi$ and then decays more gradually (no null at $\phi = 90°$). The total pattern in the E plane is therefore also narrower than in the H plane, and sidelobes are lower by 2 to 3 dB.

In the case of the HE_{11} mode on a dielectric rod (Fig. 12-6b), the E- and H-plane element factors, and consequently the beamwidths and sidelobe levels, are more nearly the same. The sidelobe level is now usually *higher* by 0.5 to 1.5 dB in the E plane than in the H plane, the precise amount depending on the relative shape of the E- and H-plane feed patterns.

Instead of being regarded as a combination of feed and terminal radiation, the surface-wave antenna pattern can be viewed as an integral over the field distribution along the antenna structure. What this distribution looks like can be seen in Fig. 12-8: the amplitude $A(z)$ along the row of monopoles (Fig. 12-6a) starts with a hump near the feed and flattens out as the surface wave peels itself out of the near field of feed and feed taper. The surface wave is well established at a distance ℓ_{min} from the feed where the radiated wave from the feed, propagating at the velocity of light, leads the surface wave by about 120°:

$$\ell_{min} k_z - \ell_{min} k = \frac{\pi}{3} \qquad (12\text{-}13)$$

The location of ℓ_{min} on an antenna designed for maximum gain is seen in Fig. 12-8 to be about halfway between feed and termination. Since the surface wave is fully developed from this point on, the remainder of the antenna length is used solely to bring feed and terminal radiation into the proper phase relation for maximum gain. (In the absence of a feed taper, ℓ_{min} occurs closer to F and the hump is higher.)

Were there no hump at all, the surface-wave antenna would radiate the familiar $\sin \xi / \xi$ pattern [$\xi = (\pi\ell/\lambda)(\lambda/\lambda_z - \cos \theta)$] produced by constant-amplitude illumination. This pattern is shown for comparison in Fig. 12-7 (shaded region); it is an unnatural pattern for parasitically excited antennas, which necessarily have feed turbulence.

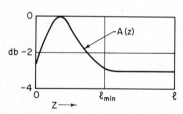

FIG. 12-8 Amplitude $A(z)$ of field along the surface-wave antenna structure.

Low-sidelobe and broadband designs require tapers that extend over a substantial part of the antenna (Fig. 12-6b). Although the field along one particular taper has been examined in detail,[11,12] the effect of tapers on the total radiation pattern is not well understood. When the taper is very gradual, the phase velocity at each point has the value one would expect from the local surface impedance; when it is sharp, a leaky wave which fills in the minima of the pattern, widens the beam, and reduces sidelobes is produced.

Design for Maximum Gain

The phase velocity along the antenna and the dimensions of the feed and terminal tapers in the maximum-gain design of Fig. 12-6a must now be specified.

If the amplitude distribution in Fig. 12-8 were flat, maximum gain would be obtained by meeting the Hansen-Woodyard condition (strictly valid only for antenna lengths $\ell \gg \lambda$), which requires the phase difference at T between the surface wave and the free-space wave from the feed to be approximately 180°:

$$\ell k_z - \ell k = \pi$$

or, equivalently,

$$\frac{\lambda}{\lambda_z} = 1 + \frac{\lambda}{2\ell} \qquad (12\text{-}14a)$$

which is plotted as the upper dashed line in Fig. 12-9.

Since the size and extent of the hump are a function of feed and feed-taper construction, the optimum terminal phase difference for a prescribed antenna length cannot easily be calculated. Experimental work on Yagi-Uda antennas (but without feed taper) by Ehrenspeck and Poehler[13] and Ehrenspeck[14] has shown that the optimum terminal phase difference lies near 60° for very short antennas, rises to about 120° for ℓ between 4λ and 8λ, then gradually approaches 180° for the longest antenna measured (20λ). Therefore

$$\frac{\lambda}{\lambda_z} = 1 + \frac{\lambda}{p\ell} \qquad (12\text{-}14b)$$

with p starting near 6 for $\ell = \lambda$ and diminishing to approximately 3 for ℓ between 3λ and 8λ and to 2 at 20λ. This relation is plotted as the solid curve in Fig. 12-9. In the presence of a feed taper, the optimum λ/λ_z values lie slightly *below* this curve.

If the efficiency of excitation is very high, interference between feed and terminal radiation is of minor importance, and the antenna need just be long enough so that the surface wave is fully established; that is, $\ell = \ell_{min}$ in Fig. 12-8. From Eq. (12-13),

$$\frac{\lambda}{\lambda_z} = 1 + \frac{\lambda}{6\ell} \qquad (12\text{-}14c)$$

which is plotted as the lower bound of the shaded region in Fig. 12-9.

Because feeds are more efficient when exciting slow surface waves than when the phase velocity is close to that of light, the solid line starts near the lower bound and ends at the upper. As long as λ/λ_z falls within the shaded region, its precise adjustment is not critical; with $\ell = 4\lambda$ and in the presence of a feed taper, the gain drops only 1 dB if λ/λ_z lies on the dashed lower bound instead of just below the solid upper bound.

Although this technique for maximizing the gain has been strictly verified only for Yagi-Uda antennas, data available in the literature on other structures suggest that optimum λ/λ_z values lie on or just below the solid curve in all instances. To pro-

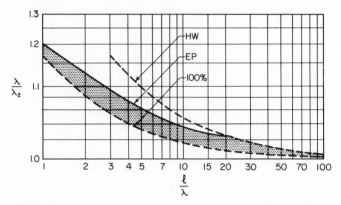

FIG. 12-9 Relative-phase velocity $c/v = \lambda/\lambda_z$ for maximum-gain surface-wave antennas as a function of relative antenna length ℓ/λ. HW = Hansen-Woodyard condition, Eq. (12-14b); EP = Ehrenspeck and Poehler experimental values, Eq. (12-14b); 100 percent = idealized perfect excitation, Eq. (12-14c).

duce these optimum values, the dependence of λ/λ_z on the structural parameters of the antenna must be calculated, measured, or sought in the literature.

The feed taper should begin at F with λ/λ_z between 1.2 and 1.3 and extend over approximately 20 percent of the full antenna length. Its exact shape is not important as long as a smooth transition is made to the main body of the antenna. The terminal taper should be approximately 0.5λ long, and to match the surface wave to space, λ_z at the end of the taper should be as close as possible to λ. In the case of a polyrod ($\epsilon = 2.56$), for example, Fig. 12-16 shows that λ_z is reasonably close to λ when

$$d/\lambda = 0.23$$

and the end of the rod is therefore blunt rather than pointed.

The *gain* above an isotropic radiator of a long ($\ell \gg \lambda$) uniformly illuminated end-fire antenna whose phase velocity satisfies Eq. (12-14a) was shown by Hansen and Woodyard to be approximately

$$G \cong \frac{7\ell}{\lambda}$$

If the design is based on the optimal phase velocities and taper dimensions just described, the gain for ℓ between 3λ and 8λ is

$$G \cong \frac{10\ell}{\lambda} \qquad (12\text{-}15)$$

which is 1.5 dB above the Hansen-Woodyard gain. For shorter lengths the gain may be 30 percent higher; for longer lengths, the proportionality factor slowly decreases because of ohmic loss and the difficulty of designing an efficient feed. Maximum gains reported in the literature are plotted in decibels as a function of ℓ/λ in Fig. 12-10 (solid-line margin of shaded region marked "gain"). These gains involve a slight degree of superdirectivity but not enough to produce excessively narrow bandwidths or high ohmic loss.

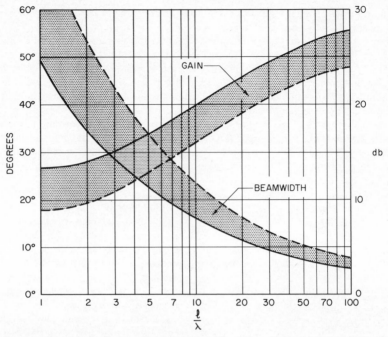

FIG. 12-10 Gain and beamwidth of a surface-wave antenna as a function of relative antenna length ℓ/λ. For gain (in decibels above an isotropic source), use the right-hand ordinate; for beamwidth, the left-hand ordinate. Solid lines are optimum values; dashed lines are for low-sidelobe and broadband design.

The half-power *beamwidth* BW of a maximum-gain design is approximately

$$BW = 55 \sqrt{\frac{\lambda}{\ell}} \quad \text{degrees} \qquad (12\text{-}16)$$

which lies just above the lower (solid-line) margin of the shaded region marked "beamwidth" in Fig. 12-10. Equation (12-16) gives an average figure, the beamwidth usually being slightly narrower in the E plane and slightly wider in the H plane.

The *sidelobe* level, for ℓ between 3λ and 8λ, is about 11 dB in the H plane and 10 or 14 dB in the E plane of a dielectric rod or Yagi-Uda, respectively (cf. above). For shorter antennas the sidelobes are somewhat higher; for longer antennas, lower.

The *bandwidth* within which the gain drops at most 3 dB is between ± 10 percent and ± 15 percent; below the design frequency, gain slowly decreases as beamwidth widens; above it, the pattern deteriorates rapidly as the main beam splits and sidelobes rise.

By comparing Eq. (12-15) with the gain of a paraboloidal dish (equal sidelobe level is assumed), the surface-wave-antenna length is found to be related to the diameter d of an equal-gain dish by

$$\frac{\ell}{\lambda} \cong \left(\frac{d}{\lambda}\right)^2 \qquad (12\text{-}17)$$

This trade-off between a broadside and an end-fire structure becomes very disadvantageous to the latter as ℓ/λ increases, and surface-wave antennas longer than 10λ are rarely used. This limits the gain of surface-wave line sources in practice to 20 dB. (Arrays of line sources and end-fire area sources are practical for much higher gains.)

The gain predicted by Eq. (12-15) is for feeds of the type shown in Fig. 12-6, with radiation patterns that are not very directive. Increasing the feed directivity increases the total antenna gain. For feeds whose radial extent is not much larger than the antenna cross section, it has been found experimentally[15] that a 1- or 2-dB increase in the end-fire gain of the feed produces an approximately equal increase in the total antenna gain. If the feed is much larger than the antenna cross section, the gain continues to rise, though at a slower rate. A large horn aperture, corner reflector, or dish can add as much as 3 or 4 dB. Structures of this sort are part surface-wave, part aperture antenna.

The *backfire* antenna, which also combines these two features, produces a gain about 4 dB above that of Eq. (12-15).[16] It consists of a surface-wave line source (for example, a Yagi-Uda, Fig. 12-11) terminated by a flat circular plate P, which reflects the surface wave launched by the feed back toward F, where it radiates into space. Its radiation mechanism is not completely understood. The optimum phase velocity is chosen with the aid of the solid curve in Fig. 12-9, but ℓ in this context must be reinterpreted as twice the physical antenna length ℓ' of the backfire antenna, since the surface wave traverses the antenna twice. By using Eq. (12-14b), the relation between surface phase velocity and ℓ' (for ℓ' between 1.5λ and 4λ) is $\lambda/\lambda_z = 1 + \lambda/6\ell'$. The feed taper should be much reduced in size or omitted. The plate diameter d is related to ℓ' by

FIG. 12-11 Backfire antenna. *(After Ref. 16.)*

$$\frac{d}{\lambda} \cong 1.5 \sqrt{\frac{\ell'}{\lambda}}$$

which implies that the plate is approximately as large as a paraboloidal dish whose gain equals that of the backfire antenna. For $\ell' = 2\lambda$, the gain is 19 dB, sidelobes are 12 dB down, and bandwidth is at least 20 percent. Easier and cheaper to build than a paraboloidal dish, the backfire antenna might be competitive for gains up to 25 dB (provided sidelobes do not have to be very low); it is superior to ordinary end-fire antennas as long as a low silhouette is not a requirement.

Design for Minimum Beamwidth

The approximate relation between E-plane beamwidth BW_E, H-plane beamwidth BW_H, and gain of a surface-wave antenna is

$$G \cong \frac{30,000}{BW_E BW_H}$$

The half-power beamwidth of a maximum-gain design [Eq. (12-16)] can be made 10 percent narrower by increasing the direct-feed radiation, which causes it to interfere destructively with $T(\theta)$ at a smaller angle θ than in Fig. 12-7. This is done by starting the feed taper with λ/λ_z less than 1.25. The narrowest beamwidths reported in the literature are plotted in Fig. 12-10 as the solid-line margin of the shaded region marked "beamwidth." (Usually E-plane patterns are slightly narrower; H-plane patterns, slightly wider.) Sidelobes are about 1 dB higher than in the maximum-gain case.

Design for Minimum Sidelobe Level

The fairly high sidelobes of the maximum-gain design can be reduced by increasing λ/λ_i to 1.35 at the start of the feed taper, letting $\lambda/\lambda_z \cong 1.2$ at the end of the feed taper, and continuing to taper over a large fraction of the total antenna length (to P in Fig. 16-6b). In the constant-cross-section region between P and T, λ/λ_z is chosen to lie on the lower margin of the shaded region in Fig. 12-9. If P is two-thirds of the distance from F to T, the sidelobes of a 6λ-long antenna are down at least 18 dB in the H plane and 17 or 21 dB in the E plane of a dielectric rod or Yagi-Uda, respectively, at the cost of a 1.5-dB drop in gain and a 10 percent rise in beamwidth.[17,18] As in the case of the optimum-gain design, the sidelobe level decreases monotonically with increasing antenna length.

The backlobe is suppressed by increasing the end-fire directivity of the feed pattern. A 25- to 30-dB level can be achieved without resorting to feeds that are much larger than the antenna cross section.

By tapering more sharply near F, then flattening out gradually to a value of λ/λ_z close to 1 at T, the H-plane sidelobe level can be reduced to 20 dB.[18] Gain and beamwidth then lie on the dashed margins of the shaded bands in Fig. 12-10. A further reduction to 30 dB in the H-plane sidelobe level is achieved by placing parasitic side rows on either side of the center array[19] (Fig. 12-12). The side-row length ℓ' is slightly longer than 0.5ℓ. λ/λ_z is the same as on the center array [use Eq. 12-14b)], and the spacing d is optimum when the side rows lie just beyond the 11-dB contour of the surface wave on the center array; or, using Fig. 12-3b and Eq. (12-14b),

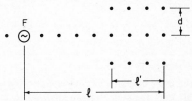

FIG. 12-12 Parasitic side rows for sidelobe suppression in one plane (top view; dots indicate dipoles or monopoles above a ground plane).

$$\frac{d}{\lambda} \cong 0.25 \sqrt{\frac{\ell}{\lambda}}$$

The gain is the same as for maximum-gain design [Eq. (12-15)]. To reduce sidelobes in all planes, parasitic side rows would have to be placed all around the center array.

Design for Broad Pattern Bandwidth[20]

The ± 15 percent bandwidth of most surface-wave antennas can be extended to as much as ± 33 percent (2:1) at the cost of a 2-dB gain decrease from optimum at midfrequency. At the feed, the midfrequency value of λ/λ_z should be 1.4. A uniform taper extends from the feed to the termination, as shown in Fig. 12-1a. At the low-frequency end of the band, gain and beamwidth lie close to the solid margins of the shaded strips in Fig. 12-10; at the high-frequency end, close to the dashed margins. Thus gain is constant within ± 1.5 dB and beamwidth within ± 15 percent. On a structure 13.5λ long at midfrequency, sidelobes were down at least 11 dB throughout the band.[20]

The bandwidth figures just quoted apply to the polyrod ($\epsilon = 2.56$) and to all surface-wave antennas whose dispersion curve (variation of λ/λ_z with frequency) resembles that of the polyrod. Dispersion increases with increasing dielectric constant; dielectric rods with $\epsilon < 2.56$ are therefore more broadband and rods with $\epsilon > 2.56$ less broadband than polyrods. The dispersion of artificial dielectrics varies; corrugated surfaces are comparable with polyrods, and Yagi-Udas are considerably more dispersive (precise bandwidth figures are not known).

The input impedance of most surface-wave antennas is slightly capacitive and changes slowly with frequency. A tapered dielectric section in waveguide (Figs. 12-1a and 12-6b) provides good matching over a 2:1 bandwidth if 1.5λ long at midfrequency and over a smaller bandwidth if shorter. To minimize the voltage standing-wave ratio (VSWR), an inductive iris is placed close to the insertion point of the rod. Equivalent techniques apply in the case of antennas other than dielectric rods and of input transmission lines other than waveguide.

Feeds

Design principles of feeds and feed tapers for optimum gain, sidelobes, and bandwidth have already been stated. Specific structures will now be described.

In Fig. 12-6a, the feeder and reflector monopoles are spaced 0.2λ apart, $h_r = 0.23\lambda$, and $\bar{h}_f = 0.21\lambda$.[13] These figures apply to $\delta = 0.048\lambda$; if the elements are thicker, they must be made slightly shorter. The backlobe is between 12 and 15 dB down. By replacing the reflector monopole with a semicircular plate (radius $= h_r$) or a small corner or paraboloidal reflector (height $= h_r$), one increases the gain of the feed by at least 1.5 dB and, therefore, the total antenna gain by approximately the same amount. The sidelobes are decreased by at least 2 dB; the backlobe is 25 to 30 dB down.

To couple from the dominant mode in rectangular waveguide, one replaces the feeder monopole by a metal half ring[21] (Fig. 12-13a); the excitation efficiency of this feed, with semicircular reflector plate, is around 80 percent. Another, slightly less efficient but simpler waveguide feed uses two or more slots to couple into the surface-wave structure. The waveguide should be dielectric-loaded so that its phase velocity approximates that in the surface-wave structure, and the slots are spaced about a quarter wavelength apart in the dielectric. Instead of cutting slots, the entire broad face of a dielectric-filled waveguide can be opened (Fig. 12-13b); by tapering the height of the waveguide over the appropriate length, feed and feed taper can be combined in a single design. In low-power applications, striplines can be slot- or dipole-coupled to the antenna.[22] In the absence of the ground plane one uses a center-fed dipole (coaxial input with built-in balun, Fig. 12-13c) or a folded dipole (usually preferred with two-wire input).

The coaxially fed metal cap (Fig. 12-13d), popular as a dielectric-rod feed or a cigar-antenna feed at ultrahigh frequencies (UHF) because of good efficiency combined with mechanical strength, forms the transition between feed types a and b of Fig. 12-6. Type b can handle much higher power than type a. In Fig. 12-6b the waveguide can be rectangular (with TE_{10} mode) or circular (TE_{11} mode). A variant of this

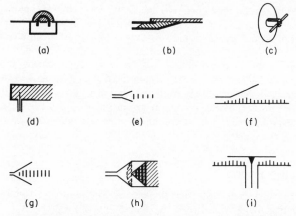

FIG. 12-13 Surface-wave-antenna feeds. (*a*) Semicircular
wire coupled to waveguide. *(After Ref. 21.)* (*b*) Dielectric-
filled waveguide with an open broad wall. (*c*) Dipole and
reflector feed. (*d*) Circular waveguide cap. (*e*) and (*f*) Horn
exciters. (*g*) Parallel-wire (distributed) feed. *(After Ref.
24.)* (*h*) Flared transition with lossy strips (with collimating
lens shown in horn aperture). *(After Ref. 28.)* (*i*) Coaxial
excitation of disk.

form, suitable for Yagi-Uda or cigar-antenna excitation, is shown in Fig. 12-13*a*. If
the horn aperture is considerably larger than the antenna cross section (Fig. 12-13*f*),
the surface-wave structure should start well inside, and it is in fact desirable to have
the horn extend just beyond the end of the feed taper, as indicated. (Theoretical work
shows that for their size horn apertures are not very efficient surface-wave exciters,[23]
but if the direct-feed radiation of the horn is well collimated—the design can be based
on Chap. 15—a large total antenna gain will be produced.)

Patterns that closely approximate the $\sin \xi / \xi$ shape have been obtained by flaring
a parallel wire into a $60°$ V (Fig. 12-13*g*)[24] or by running an unflared parallel wire
(loaded with transverse wire stubs to reduce phase velocity) alongside the antenna;[25]
both structures extend over a third of the antenna length. Feeds of this type are called
distributed. In the hope of obtaining good phase and amplitude control over the entire
antenna aperture, feeds have been studied that couple continuously from one end of
the antenna to the other, for example, a waveguide slot-coupled to a dielectric slab.[26,27]
These designs have not been very successful thus far, principally because energy flow
along two such closely coupled waveguides is difficult to control. If precise aperture
illumination is desired, one must abandon surface-wave antennas and turn to the end-
fire structures among the surface-wave-excited arrays of discrete elements (Secs. 12-
5 and 12-6).

Variants of Fig. 12-6*a* and *b* are also used to excite surface-wave area sources.
Rectangular dielectric or corrugated surfaces are fed by placing a slotted waveguide
alongside, or a hog horn, or a horn with correcting lens. The last is frequently plagued
by transverse standing waves, which produce sidelobes in azimuth (*yz* plane). To suppress these, a flared transition (1λ long at midfrequency) with transverse conducting

(Aquadag) strips can be used (Fig. 12-13h).[28] It keeps sidelobes below 20 dB over a frequency range of at least 2:1 at a cost of 1 or 2 dB because of the lossy strips.

Radial-cylindrical area sources (disks, Fig. 12-2d) can be center-fed with a simple monopole feeder; its efficiency of excitation turns out to be higher than with line sources. A higher-gain is shown in Fig. 12-13i, in which the coaxial input is excited in the TEM mode for vertical polarization on the disk and in the TE_{01} mode for horizontal polarization. A completely flush installation could be achieved by coupling through two or more annular slots. All the arrangements previously described for line sources can be carried over, and the optimum design principles for feed and feed taper still apply.

Arrays of Line Sources

Yagi-Udas, dielectric rods, and other surface-wave line sources can be arranged in a plane or volume array. The surface-wave illumination in the terminal plane of a plane array is scalloped (Fig. 12-14a), the overlapping dotted circles representing the 25-dB contours of the transverse-field decay. If identical end-fire elements and negligible interaction between them are assumed, the array gain G_a is the sum of the individual gains G. By letting n be the number of elements and using Eq. (12-15),

$$G_a = nG \cong 10n\frac{\ell}{\lambda} \qquad \textbf{(12-18)}$$

Interaction between adacent elements is found to be negligible if the crossover is at about the 12-dB contour; the minimum separation d is therefore [Fig. 12-3b and Eq. (12-14b)]

$$\frac{d}{\lambda} \cong 0.5 \ \sqrt{\frac{\ell}{\lambda}} \qquad \textbf{(12-19}a\textbf{)}$$

The exact value depends on the polarization (coupling is less in the E than in the H plane) and on the direct coupling between feeds. Too close a spacing produces a loss in gain. An upper limit on the spacing is obtained by requiring that the first principle sidelobe of the array pattern be strongly attenuated through multiplication with the element pattern. The angle with end fire of the array sidelobe must therefore be at

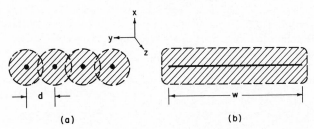

FIG. 12-14 Two-dimensional surface-wave antennas (transverse cross sections). (a) Array of line sources. (b) Area source.

least that of the first minimum in the element pattern, a criterion that corresponds to a crossover of about 22 dB, or a maximum separation

$$\frac{d}{\lambda} \cong \sqrt{\frac{\ell}{\lambda}}$$ (12-19b)

The principle of pattern multiplication, which has just been invoked, is useful only for a first design, since the element patterns are somewhat distorted by the presence of the other elements. The optimum spacing, which lies somewhere between the d/λ values given by Eqs. (12-19), must therefore be found empirically.

The beamwidth BW_a in the plane of the array (azimuth) is approximately

$$\text{BW}_a \cong \frac{65\lambda}{nd}$$ (12-20)

Sidelobes can be reduced by tapering the amplitude at the feeds. In the elevation plane, beamwidth and sidelobes are controlled by the element pattern. The bandwidth is less than for individual elements, probably about ± 5 percent.

Low-silhouette requirements often force the choice of a horizontal plane array of end-fire elements in place of a vertical cylindrical reflector (or a rectangular mattress array). The trade-off between them, if equal gain and sidelobes and equal width in the y direction (Fig. 12-14a) are assumed, is

$$\frac{\ell}{\lambda} \cong \left(\frac{h}{\lambda}\right)^2$$ (12-21)

[cf. Eq. (12-17)], where ℓ is the length of the end-fire elements and h the heights (in the x direction) of the equivalent reflector or broadside array. Because of space limitations, the line sources cannot always be equal in length. Equation (12-18) is then inapplicable, but the spacing between elements is still controlled by Eqs. (12-19).

A volume array of end-fire elements occupies almost as large an area in the transverse (xy) plane as the dish it replaces and has advantages in terms of transportability, ease of erection, and minimum weight needed for mechanical strength against certain kinds of stresses. Spillover is more easily controlled than with a conventional dish, but sidelobe control becomes more difficult and the bandwidth is much narrower. In the design of these arrays, the number of end-fire elements can be reduced by increasing their length; according to Eq. (12-18), $n = G\lambda/10\ell$. This relation clarifies the trade-off between an end-fire volume array and a broadside mattress array of dipoles: though one cannot increase the total gain by replacing the dipoles with higher-gain end-fire elements, one can reduce the number of elements (and thus of separate feeds) while keeping the gain constant.

Arrays of end-fire elements can be fed in cascade from a common transmission line (like slots in waveguide), but a corporate feed structure (branching transmission lines) produces better bandwidth.

Effect of Finite Ground Plane on the Radiation Pattern

Vertically polarized surface-wave line or area sources are often mounted on a ground plane whose finite length g (Fig. 12-15a) distorts the antenna pattern in two ways: by tilting the beam through an angle ψ away from end fire and by broadening it to a half-power beamwidth BW′. The tilt is maximum when $g = 0$ (no ground plane in front of antenna):[29,30]

$$\psi_{max} \cong 60 \frac{\lambda}{\ell} \quad \text{degrees}$$

(which is a full half-power beamwidth BW of the unperturbed pattern); the beam then broadens so that the 3-dB level lies in the end-fire direction, and BW′ is slightly over

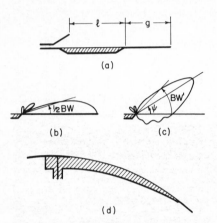

twice as wide as the unperturbed BW (Fig. 12-15*b* and *c*). The tilt can be reduced by slowing the surface wave down from its infinite-ground-plane optimum, but this increases sidelobes and decreases gain. The gain, on the other hand, is optimized by speeding the wave up at the expense of increased tilt.[30]

As g increases, the beam tilt decreases: when $g = \ell$, $\psi \cong 0.7\psi_{max}$; when $g = 3\ell$, $\psi \cong 0.5\psi_{max}$; when $g = 20\ell$, $\psi \cong 0.2\psi_{max}$—a very gradual approach to the infinite-ground-plane condition. The beamwidth approaches the unperturbed value more rapidly: when $g = \ell$, BW′ $\cong 1.25$BW.[29] Modifying the surface-wave velocity again has an opposite effect on beam tilt and gain.

If no beam tilting or loss in gain can be tolerated, flush mounting must be abandoned in favor of a full-size structure (pod-mounted to reduce drag, if necessary). A less drastic remedy is to

FIG. 12-15 Effect of finite ground plane on the radiation pattern of an end-fire antenna. (*a*) Finite ground plane. (*b*) Pattern on infinite ground plane. (*c*) Pattern in the absence of ground plane. (*d*) Antenna on a curved ground plane.

reduce the tilt angle, at a small cost in gain, by bending the antenna and ground plane into a cylindrical or spherical cap.[31–33] The curvature produces an attenuation in the surface wave, which can be enhanced by tapering the antenna along its entire length (Fig. 12-15*d*). If the total attenuation is such that about 50 percent of the power has leaked off before the surface wave reaches the termination, the tilt angle will be reduced to near zero no matter how short the ground plane.[30]

12-4   SURFACE-WAVE ANTENNAS: SPECIFIC STRUCTURES (See Zucker in Bibliography)

Numerical values of λ/λ_z and design details for specific structures are presented in this section. The λ/λ_z curves are displayed for several values of relative dielectric constant ϵ, ranging from 2.56 (polystyrene) to 165 (a calcium titanate ceramic). Hard, low-loss, high-temperature materials, for example, fused quartz ($\epsilon \cong 3.7$), pyroceram ($\epsilon \cong 5.6$), and alite (an aluminum oxide, very durable, $\epsilon \cong 8.25$), are of special interest.

The dielectric-loss tangent δ, which is temperature-dependent, produces a plane-wave attenuation

$$\alpha\lambda = \sqrt{\epsilon}\tan\delta$$

in an infinite medium, but since a large fraction of the surface wave propagates outside the medium, the surface-wave attenuation $\alpha_z\lambda$ is smaller. For $\lambda/\lambda_z = 1.2$, $\alpha_z \cong 0.7\alpha$; for $\lambda/\lambda_z = 1.03$, $\alpha_z \cong 0.1\alpha$.[34] As long as δ is smaller than 10^{-2}, the loss will therefore be less than 0.1 dB per wavelength. Data on long Yagi-Udas[35] indicate that, for λ/λ_z near 1, loss in artificial dielectrics is likewise negligible, provided all joints are firmly press-fit or soldered and the metal is of high conductivity. As the corrugations or dipole elements approach resonance, λ/λ_z increases and ohmic losses rise sharply. A typical figure for loss in a 6λ-long dielectric rod or Yagi-Uda is 0.5 dB. (Since the maximum-gain curve in Fig. 12-10 is based on experiment, the effect of ohmic loss is included.) Surface-wave losses can be measured by the resonator method.[36]

Because of low frequencies the metal elements of artificial dielectrics can be constructed of slender poles and chicken wire; they weigh far less than solid dielectrics. Large ground structures at UHF or below are therefore invariably artificial dielectrics. In the microwave range, the choice of medium depends on mechanical strength, temperature behavior, erosion resistance, and cost; ordinary dielectrics, some of which can be cheaply molded into cavities recessed in the skin, are usually preferred for flush airplane or missile installation.

Dielectric Rod[8,17,20,21,34,37–42]

Dielectric rod is termed *polyrod* and *ferrod* when made of polystyrene and ferrite material respectively. See Fig. 12-16*a* for phase velocity as a function of rod diameter. The higher the dielectric constant, the thinner the rod (for a given gain) and the lighter therefore the weight; also, however, the larger the dispersion and therefore the narrower the bandwidth.

Modes higher than the fundamental HE_{11} cannot propagate provided $d/\lambda < 0.626/\sqrt{\epsilon}$. The first two higher modes, TE_{01} and TM_{01}, produce a null in the end-fire direction.

Cross sections other than circular are often useful. To a good approximation,[34] the phase velocity depends only on the cross-section area A, and Fig. 12-16*a* therefore remains valid if d is replaced by $1.13 \sqrt{A}$. Sliced in half along the plane of electric symmetry (Fig. 12-4*c*)—which does not affect the phase velocity—a circular or rectangular rod can be placed on a ground plane (*dielectric-image line,* Fig. 12-17*a* and *b*); when recessed for flush mounting, this structure is called the *dielectric channel guide* (Fig. 12-17*c*). Another variant, the dielectric tube,[42] has also been studied but appears to have no advantages over ordinary rods.

Mallach[38] found that for optimum design the rod diameter at the feed end of a linearly tapered dielectric rod should be $d_{max}/\lambda \cong [\pi(\epsilon - 1)]^{-1/2}$, and at the termination $d_{min}/\lambda \cong [2.5\pi(\epsilon - 1)]^{-1/2}$. In the range $2.5 < \epsilon < 20$, this rule corresponds to letting λ/λ_z at the feed be roughly equal to 1.1 and at the termination to 1.0. The resulting patterns are often adequate but can be improved by using the methods described in Sec. 16-3.

A rough approximation to the rod pattern, due to Zinke,[38] consists in multiplying

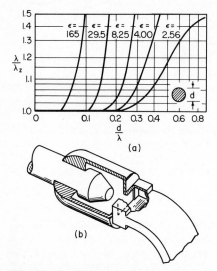

(a)

(b)

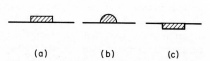

(a)   (b)   (c)

FIG. 12-16 Dielectric rod. (*a*) Relative phase velocity (see Refs. 37–39; in case of a noncircular cross section, replace d/λ by $1.13\sqrt{A}/\lambda$). (*b*) Conventional feed. (*Courtesy of G. C. Southworth and D. Van Nostrand Company, Inc.*[43])

FIG. 12-17 Dielectric image lines and channel guide. (*a*) and (*b*) Image lines (identical λ/λ_z if cross section, areas are the same). (*c*) Channel guide.

the $\sin\xi/\xi$ pattern shown in Fig. 12-7 by the factor $\cos[(\pi d/\lambda\sin\theta]$ (but the predicted sharp nulls do not in fact exist).

Figure 12-16*b* shows details of a conventional feed that has proved itself in practice. Other feeds are illustrated in Figs. 12-6 and 12-13. In the case of waveguide feeds, it has been found desirable,[17] both for snug fitting and for sidelobe suppression, to wrap dielectric tape around the feed point or, equivalently, to let the diameter of the waveguide feed be slightly smaller than that of the dielectric rod at the feed point, as shown in Fig. 12-6*b*. If $\epsilon > 8$, the rod can be fed over a narrow band by direct insertion in the narrow side of a waveguide, or in a cavity.[44]

Dielectric Channel Guide[28,45,46] (Fig. 12-17*c*)

Two modes are of interest: a vertically polarized one that is a deformation of the HE_{11} rod mode and a horizontally polarized one. The phase velocity of the vertically polarized mode is very slightly faster in Fig. 12-17*c* than in Fig. 12-17*a* or *b*. The mode may be slightly leaky (small attenuation in the axial direction), but this is not known with certainty. The structure can be viewed as a channel of the type shown in Fig. 12-2*b*, but with most of the sidewalls removed. The phase velocity is then determined as follows: we rewrite Eq. (12-5*b*) in the form

$$\frac{\lambda}{\lambda_z} = \sqrt{\left(\frac{\lambda}{\lambda_z'}\right)^2 - \left(\frac{\lambda}{2\omega}\right)^2}$$

where λ/λ_z' is the relative phase velocity on an infinitely wide slab-on-metal, as given in Fig. 12-19c. Removal of the sidewalls above the dielectric-air interface appears to have negligible effect on the phase velocity[28] and, at least as long as $\lambda/\lambda_s > 1$ (ω is not too narrow), introduces little if any leakage attenuation.

The horizontally polarized mode is the lowest-order TE mode in the channel of Fig. 12-2b (slightly perturbed by the removal of most of the sidewalls), which in turn is identical with the lowest-order TE mode on a dielectric slab-on-metal (solid curves, Fig. 12-19d).

The dielectric-filled channel also supports leaky waves, and care must be taken not to choose parameters that allow these to be excited.

Yagi-Udas[13,19,47−62]

These are sometimes referred to as *ladder* arrays. Figure 12-18 shows that the phase velocity is controlled by adjusting the spacing, height, and diameter of the monopole elements. The dispersion is more pronounced than on a dielectric rod, and the best-reported bandwidth (within which the gain drops no more than 3 dB) is only ± 10 percent; this can perhaps be improved by application of the broadband design method described in Sec. 16-3. The bandwidth of an array of Yagis is at best ± 5 percent.

The curves in Fig. 12-18 apply to the row of monopoles shown in Fig. 12-6a and equally to a row of dipoles (element height 2h). A center boom for mounting the dipole elements has negligible effect on the phase velocity provided the total dipole span 2h includes the boom diameter. Yagi-Udas can be photoetched on a copper-clad dielectric sheet;[50,51] since the dielectric slows down the wave, the element height needed to produce a given phase velocity is less than indicated in Fig. 12-18.

Dielectric Sheets and Panels[4,7,31−33,63−68]

Because of the imaging properties of vertically polarized waves, the lowest-order TM mode on a dielectric sheet-on-metal is identical with that on a panel of twice the sheet thickness (Fig. 12-19a–c). This wave has no cutoff frequency; its properties are discussed in Sec. 12-2, and Fig. 12-4 shows the field configuration.

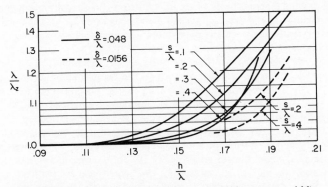

FIG. 12-18 Relative phase velocity on Yagi-Uda antenna. *(After Ehrenspeck and Poehler.*[13]*)* For the meaning of parameters, see Fig. 12-6a.

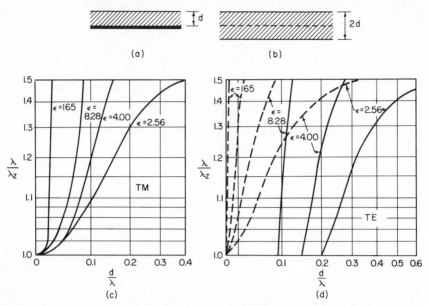

FIG. 12-19   Dielectric sheets and panels. (*a*) Sheet-on-metal. (*b*) Panel. (*c*) Lowest-order TM mode on sheet or panel. (*d*) Lowest-order TE mode on sheet (solid curves) and panel (dashed curves).

The lowest-order horizontally polarized wave on a dielectric sheet-on-metal produces a null in the end-fire direction and has a cutoff (solid curves, Fig. 12-19*d*). On a panel, on the other hand, the lowest-order horizontally polarized wave produces a maximum in the end-fire direction and has no cutoff (dashed curves, Fig. 12-19*d*). (The next higher TE mode on the panel is identical with the lowest TE mode on the sheet, but because of the end-fire null it is not a useful mode).

In designing tapered sheets on metal, a flush air-dielectric interface can be maintained by contouring the depth of ground plane. Good contact between dielectric and ground plane is needed, since air spaces affect the phase velocity.[7]

Circular dielectric disks[66] and spherical caps[32,33] are useful as omnidirectional beacon antennas. For design principles, see Sec. 12-3 (paragraphs on the effect of finite ground plane on the radiation pattern and on beam-shaping techniques). Figure 12-19*c* and *d* is still applicable, since λ/λ_z is only slightly affected by a decreasing radius of curvature. The leakage attenuation has been calculated by Elliott.[32] Patterns provide good null filling to 45° off end fire and have a bandwidth of ± 8 percent.[33]

Circular (or cross) polarization can be produced if the surface-wave structure supports a TM and a TE wave with identical phase velocities. One example is the two-layer dielectric sheet,[7,67] illustrated in Fig. 12-20*a* for the case in which the lower layer is air. Given λ/λ_z, Fig. 12-20*b* prescribes the layer thickness (with $\epsilon = \epsilon_l/\epsilon_0$ as a parameter). Another, structurally superior example is the simple-layer dielectric in which longitudinal metal vanes that short out the vertically polarized mode while leaving the horizontally polarized one unperturbed are embedded (Fig. 12-20*c*). The TM wave therefore "sees" a thinner sheet than the TE wave, and for a prescribed λ/λ_z, Fig. 12-20c and d gives the values (in first approximation) of the vane height and total layer thickness. For details, see Hansen.[68]

12-5 SURFACE-WAVE-EXCITED ARRAYS: DESIGN PRINCIPLES

The array of dipoles proximity-coupled to a two-wire line (Fig. 12-1b) has much in common with a nonresonant slotted waveguide. Both antennas are arrays of discrete elements excited by a traveling wave, one in shielded, the other on open waveguide. Both produce shaped patterns or narrow beams directed at any angle (other than broadside, where the cumulative impedance mismatch of the slots or dipoles is prohibitively high). If the elements are continuously distributed along the open waveguide, this antenna type has much in common with a leaky-wave antenna. Design techniques therefore closely follow those in Chaps. 9 and 10. Rarely competitive with waveguide slots in the microwave range (because less suitable for high-precision design), surface-wave-excited arrays of discrete elements are most useful at lower frequencies, where waveguides are too bulky; with distributed elements, these arrays may find application in the submillimeter range, where waveguides are too small. A thorough presentation of the theory of traveling (including surface) waves on continuous as well as periodic structures has been given by Hessel (see Bibliography).

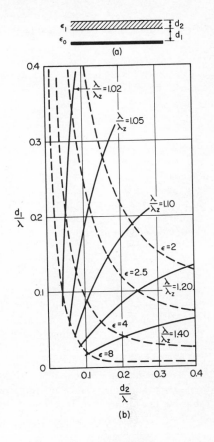

(a)

(b)

(c)

FIG. 12-20 Circular (or cross) polarization on dielectric sheets. (a) A simple double-layer structure. (b) Parameter combinations (with $\epsilon = \epsilon_1/\epsilon_0$) that allow (a) to support a circularly polarized surface wave with λ/λ_z between 1.02 and 1.40. (From Plummer and Hansen.[67]) (c) Longitudinal metal vanes embedded in dielectric. (After Ref. 68.)

Direction of Main Beam

As with slot arrays, successive radiating elements can be placed so that they are excited with the phase of the traveling wave or with an additional phase difference of 180° (*phase reversal*; Fig. 12-1b). Assuming constant phase progression along the array and constant (or moderately tapered) amplitude and using simple array theory

(Chap. 3), we obtain for the direction of the main beam, without phase reversal,

$$\cos \theta_m = \frac{\lambda}{\lambda_z} - \frac{\lambda}{s} m \qquad (12\text{-}22a)$$

and with phase reversal,

$$\cos \theta_n = \frac{\lambda}{\lambda_z} - \frac{\lambda}{s} (n + \tfrac{1}{2}) \qquad (12\text{-}22b)$$

where θ is the angle off end fire, s the spacing between elements, and m or $n = 0, 1,$ $2, \ldots$ the order of the beam. The spacing must be small enough so that no secondary principal lobes appear at real angles. The position of the principal lobes nearest the main lobe is found by replacing m by $m + 1$ and $m - 1$ in Eq. (12-22a) [and likewise for n in Eq. (12-22b)]. As s increases, principal lobes appear first in the end-fire or backfire ($\theta = 180°$) direction; they will not appear provided their angle is imaginary, or[69]

$$\frac{s}{\lambda} \leq \frac{1}{1 + |\cos \theta_m|} \qquad (12\text{-}23)$$

(and likewise for n), where the equality sign holds only if the element pattern has nulls along the array axis. According to Eq. (12-23), the maximum permissible spacing increases from $\lambda/2$ at end fire and backfire to λ at broadside. (See Fig. 10-2.)

Because of mutual coupling between elements and to hold down the total number in the array, we usually impose $s \geq \lambda/2$ as a lower bound.

As in the case of slot arrays, the surface-wave excited array can be made resonant by terminating it with a short. To hold the variation of input impedance with frequency to a minimum, such an array is usually center-fed. Figure 10-2 is still applicable, but two beams are now produced, one at θ and the other at $180° - \theta$; at broadside these merge into a single beam.

A detailed discussion of these array aspects can be found in the text by Elliott (Ref. 60, sec. II).

Control of Aperture Distribution

Although in general a four-terminal network is needed to represent a transmission-line discontinuity, a shunt conductance usually suffices, especially if the axial dimension of the element is much smaller than λ or if the element is resonant. The conductances must be so chosen as to produce the amplitude distribution prescribed over the array. The normalized element conductance $g_i = G_i/Y_c$ (Y_c is the characteristic admittance of the matched transmission line, Fig. 12-21) is much smaller than 1 if there are many elements in the array (array length $\ell \gg \lambda$). We can then approximate the normalized conductance per unit length, g_i/s, by the smoothed continuous function $g(z)$ [with $g(z_i) = g_i/s$]. The power loss due to radiation by element i is g_i times the power incident at i; for the continuous case, see Chap. 10.

Gain, Beamwidth, and Sidelobe Level of Linear Arrays

General array theory, as described in Chaps. 3 and 20, is fully applicable. If we assume dipole elements that are transverse to the array axis and constant amplitude

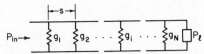

FIG. 12-21 Transmission line loaded with elements that can be represented as shunt conductances.

FIG. 12-22 Aperture length ℓ and its projection ℓ_p normal to the beam direction.

illumination, the gain of a line source of length ℓ (with $N > 20$) is

$$G \cong \frac{4\ell}{\lambda} \qquad (12\text{-}24)$$

above an isotropic source. Equation (12-24) is *independent* of angle of radiation. At end fire, the gain can be increased by 2 dB above that of Eq. (12-24) ($G \cong 7\ell/\lambda$) by satisfying the Hansen-Woodyard condition (Eq. 12-14a). If we assume dipole elements that are parallel to the array axis, the gain is $G \cong 2\ell/\lambda$ at broadside and decreases monotonically to zero at end fire.

The half-power beamwidth at broadside, with constant illumination assumed, is 51 λ/ℓ degrees; at angles other than broadside but not too near end fire, ℓ must be replaced by $\ell_p = \ell \sin \theta$, the geometric projection of ℓ in the plane perpendicular to θ (Fig. 12-22). Near end fire, where this relation no longer holds, a smooth transition is made to the end-fire half-power beamwidth, which is 105 $\sqrt{\lambda/\ell}$ degrees if $\lambda/\lambda_z = 1$ and 73 $\sqrt{\lambda/\ell}$ degrees if λ/λ_z is increased to its Hansen-Woodyard value. Details of the transition region are given by Bickmore.[70]

The sidelobe level is 13 dB with constant aperture illumination but can be lowered drastically by tapering. As with slot arrays, secondary lobes are produced by variations in the element pattern; these are significant only with phase-reversed elements, which are alternately oriented at opposite angles or displaced on opposite sides of the array axis.[71]

12-6 SURFACE-WAVE-EXCITED ARRAYS: SPECIFIC STRUCTURES

Numerical values of element conductances and design details for specific surface-wave-excited antennas are presented in this section. Direct-feed radiation, which is negligible in the case of parallel-wire and trough-guide-excited arrays, can never be completely suppressed in the case of dielectric-waveguide-excited arrays; in the context of the present section this is spurious radiation.

Two-Wire Line with Proximity-Coupled Dipoles[72–76]

(Figs. 12-1b and 12-23) Similar to the low-frequency "fishbone" antenna but with elements that are proximity-coupled rather than being tied to the line directly or through a capacitor, the array of dipoles excited by a two-wire line (called the Sletten array) is most useful in the VHF and UHF ranges and is well suited for high-power application. With the dipoles as shown in Figs. 12-1b and 12-23a, either m- or n-type

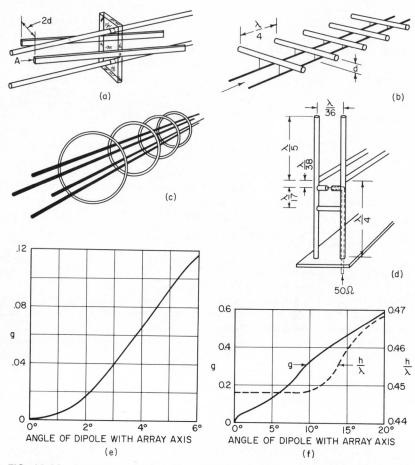

FIG. 12-23 Proximity-coupled dipoles. [*After C. J. Sletten et al., AFCRL Reps. TR-57-114 (1957) and TR-58-115 (1958).*] (*a*) Pairs of dipoles mounted on both sides of a two-wire line for good omnidirectional coverage in plane transverse to array. (*After Sletten and Forbes.*[72]) (*b*) End-fire array (spacing between dipoles and line shown variable). (*c*) Array of rings excited by orthogonal pair of two-wire lines for end-fire circular (or cross) polarization. [*(b) and (c) after Forbes.*[76]] (*d*) Balun construction for transition from coaxial to two-wire line. (*After Sletten et al.*[74]) (*e*) Normalized element conductance *g* as a function of the dipole angle of rotation for the antenna in Fig. 12-1*b*, with element spacing above wires $d = 0.03\lambda$, dipole length $= 0.446\lambda$, center-to-center wire separation $= \lambda/36$. (*After Sletten et al.*[74]) (*f*) Normalized element conductance *g* and dipole resonant length *h* as a function of the dipole angle of rotation for the antenna in Fig. 12-23*a*, with $A = 8.5 \times 10^{-5\lambda2}$, transverse element spacing $2d - 0.118\lambda$, center-to-center wire separation $= \lambda/36$. (*Based on data in Sletten and Forbes.*[72])

radiation patterns can be produced (Fig. 10-2), except near end fire, where the element pattern has a null. Arranged as in Fig. 12-23b, the dipoles cannot be phase-reversed, and the radiation is confined to near end fire (interdipole spacing = $\lambda/4$; see Sec. 12-5, under "Direction of Main Beam"). The two orthogonal pairs of wires in Fig. 12-23c can be fed 90° out of phase to produce circularly polarized end-fire radiation.[76]

Because of loading by the elements, λ/λ_z on the two-wire line is slightly larger than 1. The degree of coupling depends primarily on the distance d between dipole and line (Fig. 12-23a and b) and on the angle between dipole and array axis; secondarily, on dipole length and on spacing between wires. For equal coupling, d is an order of magnitude greater in Figs. 12-1b and 12-23a than in Fig. 12-23b (0.05λ versus 0.005λ for g = 0.25). Although there are exceptions,[75] the coupling generally decreases monotonically with increasing d. It rises to a maximum between 20 and 40° and then decreases again as the angle between dipole and array axis increases.

Figure 12-23a and f shows element conductance for the structures in Figs. 12-1b and 12-23a, respectively. Sufficiently large coupling is achieved with small dipole angles; the element patterns are therefore fairly uniform. The dipole resonant length is near 0.45λ. If $d \ll 0.05\lambda$, it becomes very difficult to find the dipole resonant length, and for transverse dipoles (Fig. 12-23b), resonance does not in fact exist.[76]

A vertically erected, center-fed, resonant VHF-UHF communications array with elements of the type shown in Fig. 12-23a gave good performance over a ± 10 percent band with a pattern omnidirectional in azimuth and shaped in elevation (VSWR = 1.2:1 at the design frequency).[72] In another interesting application of the two-wire line, a satellite-tracking antenna with elements of the type shown in Fig. 12-1b was fed from both ends; with the aid of a suitable comparison circuit, this allowed reception on both sum and difference patterns.[73,74] The type of balun used in this design is illustrated in Fig. 12-23d; note that balance does not depend on the height of the coaxial feed from the ground plane or on the tuning parameters.

Trough-Guide Antennas[77-79] (Fig. 12-24)

Characterized by mechanical simplicity and greater (by 50 percent) bandwidth for single-mode propagation than rectangular waveguide, the trough guide is well suited for exciting radiating discontinuities and has also found application as a leaky-wave antenna. It can be viewed as a strip transmission line operated in its first higher (a TE) mode; the plane bisecting the stripline is an electric null plane of this mode and can therefore be short-circuited (see inset in Fig. 12-24a). The mode obeys the familiar dispersion relation

$$\frac{\lambda}{\lambda_z} = \sqrt{1 - \left(\frac{\lambda}{\lambda_c}\right)^2}$$

with $\lambda_c = 2w$. It cannot propagate if $w/d > 9.6$, and the curves shown are in fact not quite reliable for $w/d > 4$.[77] Zero fin thickness is assumed, the effect of finite thickness being negligible for antenna design purposes. Figure 12-3b [with relabeled abscissa—see discussion in connection with Eq. (12-5b)] helps in estimating the extent to which the (theoretically infinite) sidewalls can be reduced in height without perturbing the mode. If the center fin is serrated (Fig. 12-24b),[78] one obtains a very large range of phase velocities, including slower-than-light velocities, and of dispersion

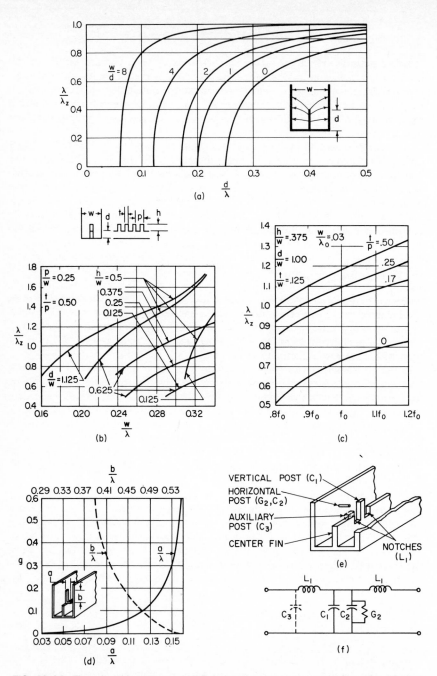

FIG. 12-24 Trough-guide antenna. (*a*) Relative phase velocity on trough guide. (*Based on data in Oliner[77] and Oliner and Rotman.[78] See text for restrictions on parameter ranges.*) (*b*) Relative phase velocity on trough guide with serrated center fin. (*After Oliner and Rotman.[78]*) (*c*) Dispersion around center frequency $f_0 = 2\pi/\lambda_0$ in trough guide with serrated center fin. (*After Oliner and Rotman.[78]*) (*d*) "L-rod" radiating element. (*After Rotman and Naumann.[79]*) (*e*) and (*f*) Notch element and its equivalent circuit. (*From Rotman and Naumann.[79]*)

characteristics (for which the above equation does not hold; see Fig. 12-24c). In this form, the trough guide is well suited for mechanical and frequency scanning. One method, due to W. Rotman, would consist in arranging three serrated center fins side by side (with negligible spacing in between); as the center fin is moved through the short distance of one serration period p (the outer fins remaining static), the phase velocity is swept through its entire range.

Successive discontinuities can be placed on the same side of the center fin (m modes, Fig. 10-2) or on alternately opposite sides (n modes). The "L rod" (Fig. 12-24d)[79] consists of a horizontal radiating and a vertical tuning element; it is a resonant radiator (therefore bandwidth-limited) and produces a good range of normalized conductance values. Except for a useful equivalent circuit, no data are available on the notch (with compensating posts) element (Fig. 12-24c),[79] which is also horizontally polarized. Because its equivalent circuit is a pi network (Fig. 12-24f), this element is nonresonant and could be used for broadside radiation.

Open Dielectric Waveguide Antennas[80,81]

Dielectric rods or strips periodically loaded with conducting rings or dielectric grooves are discussed at the end of Chap. 10 and in Refs. 80 and 81. With silicon ($\epsilon_r = 12$) or aluminum oxide ($\epsilon_r = 10$) as a substrate suitable at millimeter wavelengths, a 10 percent change in wavelength produces a scanning angle of about 20° in either the forward or backward direction.

An analysis by A. A. Oliner of continuously perturbed surface-wave structures ("H-guide," "groove-guide") suitable for extremely-high-frequency (EHF) applications soon will apear in the literature.

Acknowledgment

I am grateful to Dr. Richard C. Johnson for preparing a first abbreviated draft of this chapter (with augmented bibliography) using material from the first edition of this *Handbook* and for editing the final draft.

REFERENCES

1 F. J. Zucker and J. A. Strom, "Experimental Resolution of Surface-Wave Antenna Radiation into Feed and Terminal Patterns," *IEEE Trans. Antennas Propagat.,* vol. AP-18, May 1970, pp. 420–422.
2 F. J. Zucker, "Surface-Wave Antennas," in R. E. Collin and F. J. Zucker (eds.), *Antenna Theory,* part 2, McGraw-Hill Book Company, New York, 1968, chap. 21.
3 N. Marcuvitz (ed.), *Waveguide Handbook,* MIT Rad. Lab. ser., vol. 10, McGraw-Hill Book Company, New York, 1951.
4 S. S. Attwood, "Surface Wave Propagation over a Coated Plane Conductor," *J. App. Phys.,* vol. 22, 1951, p. 504.
5 J. H. Richmond and T. E. Tice, "Probes for Microwave Near-Field Measurements," *IRE Trans. Microwave Theory Tech.,* vol. MTT-3, 1955, p. 32.
6 H. Ehrenspeck, W. Gerbes, and F. J. Zucker, "Trapped Wave Antennas," *IRE Nat. Conv. Rec.,* part 1, 1954, p. 25.
7 J. B. Andersen, *Metallic and Dielectric Antennas,* Polyteknisk Forlag, Copenhagen, 1971.

8 J. Brown and J. O. Spector, "The Radiating Properties of End-Fire Aerials, *IEE Proc.,* part B, vol. 104, 1957, p. 25.

9 A. F. Kay, "Scattering of a Surface Wave by a Discontinuity in Reactance," *IRE Trans. Antennas Propagat.,* vol. AP-7, 1959, p. 22.

10 F. J. Zucker, and J. A. Strom, "Experimental Resolution of Surface-Wave Antenna Radiation into Feed and Terminal Patterns," *IEEE Trans. Antennas Propagat.,* vol. AP-18, May 1970, pp. 420–422.

11 L. Felsen, "Field Solutions for a Class of Corrugated Wedge and Cone Surfaces," MRI Electrophys. Group Mem. 32, Polytechnic Institute of Brooklyn, Brooklyn, N.Y., 1957.

12 V. I. Talanov, "On Surface Electromagnetic Waves in Systems with Nonuniform Impedance," *Izv. VUZ MVO (Radiofiz.),* vol. 2, 1959, p. 32. (Translated from the Russian by Morris P. Friedman, Inc., Air Force Cambridge Res. Cen. Rep. TN-59-768.)

13 H. W. Ehrenspeck and H. Poehler, "A New Method for Obtaining Maximum Gain from Yagi Antennas," *IRE Trans. Antennas Propagat.,* vol. AP-7, 1959, p. 379.

14 H. W. Ehrenspeck and W. Kearns, unpublished experimental data, Air Force Cambridge Research Center, Bedford, Mass., 1959.

15 J. -C. Simon and G. Weill, "A New Type of Endfire Antenna," *Annales de radioélectricité,* vol. 8, 1953, p. 183. (In French.)

16 H. W. Ehrenspeck, U.S. patent applied for, Air Force Cambridge Research Center, Bedford, Mass., 1959.

17 F. E. Boyd and D. H. Russell, "Dielectric-Guide Antennas for Aircraft," Nav. Res. Lab. Rep. 3814, 1951.

18 F. J. Zucker and W. Kearns, unpublished experimental data, Air Force Cambridge Research Center, Bedford, Mass., 1959.

19 H. W. Ehrenspeck and W. Kearns, "Two-Dimensional End-Fire Array with Increased Gain and Side Lobe Reduction," *IRE Wescon Conv. Rec.,* part 1, 1957, p. 217.

20 C. F. Parker and R. J. Anderson, "Constant Beamwidth Antennas," *IRE Nat. Conv. Rec.,* part 1, 1957, p. 87.

21 R. H. DuHamel and J. W. Duncan, "Launching Efficiency of Wires and Slots for a Dielectric Rod Waveguide," *IRE Trans. Microwave Theory Tech.,* vol. MTT-6, 1958, p. 277.

22 A. D. Frost, C. R. McGeogh, and C. R. Mingins, "The Excitation of Surface Waveguides and Radiating Slots by Strip-Circuit Transmission Lines," *IRE Trans. Microwave Theory Tech.,* vol. MTT-4, 1956, p. 218.

23 A. L. Cullen, "The Excitation of Plane Surface Waves," *IRE Proc.,* part IV, Monograph 93R, vol. 101, 1954, p. 225.

24 D. K. Reynolds, "Broadband Traveling Wave Antennas," *IRE Nat. Conv. Rec.,* part 1, 1957, p. 99.

25 D. K. Reynolds and R. A. Sigelmann, "Research on Traveling Wave Antennas," AFCRC-TR-59-160, final rep., Cont. AF19(604)-4052, Seattle University, Seattle, Wash., 1959.

26 W. L. Weeks, "Coupled Waveguide Excitation of Traveling Wave Antennas," *IRE Wescon Conv. Rec.,* part 1, 1957, p. 236.

27 R. R. Hodges, Jr., "Distributed Coupling to Surface Wave Antennas," Cont. AF33(616)-3220, University of Illinois, Urbana, 1957.

28 B. T. Stephenson and C. H. Walter, "Endfire Slot Antennas," *IRE Trans. Antennas Propagat.,* vol. AP-3, 1955, p. 81.

29 R. S. Elliott, "On the Theory of Corrugated Plane Surfaces," *IRE Trans. Antennas Propagat.,* vol. AP-2, 1954, p. 71.

30 R. A. Hurd, "End-Fire Arrays of Magnetic Line Sources Mounted on a Conducting Half-Plane," *Can. J. Phys.,* vol. 34, 1956, p. 370.

31 R. S. Elliott, "Azimuthal Surface Waves on Circular Cylinders," *J. Appl Phys.,* vol. 26, 1955, p. 368.

32 R. S. Elliott, "Spherical Surface-Wave Antennas," *IRE Trans. Antennas Propagat.,* vol. AP-4, 1956, p. 422.

33 R. E. Plummer, "Surface-Wave Beacon Antennas," *IRE Trans. Antennas Propagat.*, vol. AP-6, 1958, p. 105.

34 S. P. Schlesinger and D. D. King, "Dielectric Image Lines," *IRE Trans. Microwave Theory Tech.*, vol. MTT-6, 1958, p. 29.

35 A. F. Kay, unpublished work under AFCRC Cont. AF19(604)-3476, Technical Research Group, Inc., Somerville, Mass., 1959.

36 E. H. Scheibe, B. G. King, and D. L. Van Zeeland, "Loss Measurements of Surface Wave Transmission Lines," *J. App. Phys.*, vol. 25, 1954, p. 790.

37 G. E. Mueller and W. A. Tyrrell, "Polyrod Antennas," *Bell Syst. Tech. J.*, vol. 26, 1947, p. 837.

38 D. G. Kiely, *Dielectric Aerials*, Methuen & Co., Ltd., London, 1953.

39 L. W. Mickey and G. G. Chadwick, "Closely Spaced High Dielectric Constant Polyrod Arrays," *IRE Nat. Conv. Rec.*, part 1, 1958, p. 213.

40 R. W. Watson and C. W. Horton, "The Radiation Patterns of Dielectric Rods—Experiment and Theory," *J. App. Phys.*, vol. 19, 1948, p. 661; "On the Calculation of Radiation Patterns of Dielectric Rods," *J. App. Phys.*, vol. 19, 1948, p. 836.

41 R. Chatterjee and S. K. Chatterjee, "Some Investigations on Dielectric Aerials," part I, *J. Ind. Inst. Sci.*, vol. 38, 1956, p. 93; part II, *J. Ind. Inst. Sci.*, vol. 39, 1957, p. 134.

42 R. E. Beam, "Wave Propagation in Dielectric Tubes," final rep., Cont. DA36-039-sc-5397, Northwestern University, Evanston, Ill., 1952.

43 G. C. Southworth, *Principles and Applications of Wave-Guide Transmission*, D. Van Nostrand Company, Inc., Princeton, N.J., 1950.

44 F. Reggia, E. G. Spencer, R. D. Hatcher, and J. E. Tompkins, "Ferrod Radiator System," *IRE Proc.*, vol. 45, 1957, p. 344.

45 W. Rotman, "The Channel Guide Antenna," *Proc. Nat. Electron. Conf.*, vol. 5, 1949, p. 190.

46 D. K. Reynolds and W. S. Lucke, "Corrugated End-Fire Antennas," *Proc. Nat. Electron. Conf.*, vol. 6, 1950, p. 16.

47 J. O. Spector, "An Investigation of Periodic Rod Structures for Yagi Aerials," *IEE Proc.*, part B, vol. 105, 1958, p. 38.

48 A. D. Frost, "Surface Waves in Yagi Antennas and Dielectric Waveguides," final rep., Cont. AF19(604)-2154, AFCRC-TR-57-368, Tufts University, Medford, Mass., 1957.

49 D. L. Sengupta, "On the Phase Velocity of Wave Propagation along an Infinite Yagi Structure," *IRE Trans. Antennas Propagat.*, vol. AP-7, 1959, p. 234.

50 J. A. McDonough and R. G. Malech, "Recent Developments in the Study of Printed Antennas," *IRE Nat. Conv. Rec.*, 1957, p. 173.

51 R. G. Malech, "Lightweight High-Gain Antenna," *IRE Nat. Conv. Rec.*, part 1, 1958, p. 193.

52 H. E. Green, "Design Data for Short and Medium Length Yagi-Uda Arrays," *Elec. Eng. Trans. Inst. Eng. (Australia)*, March 1966, pp. 1–8.

53 R. J. Mailloux, "The Long Yagi-Uda Array," *IEEE Trans. Antennas Propagat.*, vol. AP-14, March 1966, pp. 128–137.

54 P. P. Viezbicke, "Yagi Antenna Design," NBS Tech. Note 688, National Bureau of Standards, U.S. Department of Commerce, December 1968.

55 G. A. Thiele, "Analysis of Yagi-Uda Type Antennas," *IEEE Trans. Antennas Propagat.*, vol. AP-17, January 1969, pp. 24–31.

56 D. K. Cheng and C. A. Chen, "Optimum Spacings for Yagi-Uda Arrays," *IEEE Trans. Antennas Propagat.*, vol. AP-21, September 1973, pp. 615–623.

57 C. A. Chen and D. K. Cheng, "Optimum Element Lengths for Yagi-Uda Arrays," *IEEE Trans. Antennas Propagat.*, vol. AP-23, January 1975, pp. 8–15.

58 N. K. Takla and L. C. Shen, "Bandwidth of a Yagi Array with Optimum Directivity," *IEEE Trans. Antennas Propagat.*, vol. AP-25, November 1977, pp. 913–914.

59 J. L. Lawson, series of papers in *Ham Radio:* "Yagi Antenna Design: Performance Cal-

culations," January 1980, pp. 22–27; "Yagi Antenna Design: Experiments Confirm Computer Analysis," February 1980, pp. 19–27; "Yagi Antenna Design: Performance of Multi-Element Simplistic Beams," May 1980, pp. 18–26; "Yagi Antenna Design: More Data on the Performance of Multi-Element Simplistic Beams," June 1980, pp. 33–40; "Yagi Antenna Design: Optimizing Performance," July 1980, pp. 18–31; "Yagi Antenna Design: Quads and Quagis," September 1980, pp. 37–45; "Yagi Antenna Design: Ground or Earth Effects," October 1980, pp. 29–37; "Yagi Antenna Design: Stacking," November 1980, pp. 22–34; "Yagi Antennas: Practical Designs," December 1980, pp. 30–41. See also corrections, September 1980, p. 67.

60   R. S. Elliott, *Antenna Theory and Design,* Prentice-Hall, Inc., Englewood Cliffs, N.J., 1981, secs. 8.7 and 8.8.

61   W. L. Stutzman and G. A. Thiele, *Antenna Theory and Design,* John Wiley & Sons, Inc., New York, 1981, sec. 5.4.

62   C. A. Balanis, *Antenna Theory, Analysis and Design,* Harper & Row Publishers, Incorporated, New York, 1982, sec. 9.3.3.

63   L. Hatkin, "Analysis of Propagating Modes in Dielectric Sheets, *IRE Proc.,* vol. 42, 1954, p. 1565.

64   F. E. Butterfield, "Dielectric Sheet Radiators," *IRE Trans. Antennas Propagat.,* vol. AP-3, 1954, p. 152.

65   R. L. Pease, "On the Propagation of Surface Waves over an Infinite Grounded Ferrite Slab," *IRE Trans. Antennas Propagat.,* vol. AP-6, 1958, p. 13.

66   E. M. T. Jones and R. A. Folsom, Jr., "A Note on the Circular Dielectric Disk Antenna," *IRE Proc.,* vol. 41, 1953, p. 798.

67   R. E. Plummer and R. C. Hansen, "Double-Slab Arbitrary-Polarization Surface-Wave Structure," *IEE Proc.,* part C, Monograph 238R, 1957.

68   R. C. Hansen, "Single Slab Arbitrary Polarization Surface Wave Structure," *IRE Trans. Microwave Theory Tech.,* vol. MTT-5, 1957, p. 115.

69   A. Dion, "Nonresonant Slotted Arrays," *IRE Trans. Antennas Propagat.,* vol. AP-6, 1958, p. 360.

70   R. W. Bickmore, "A Note on the Effective Aperture of Electrically Scanned Arrays," *IRE Trans. Antennas Propagat.,* vol. AP-6, 1958, p. 194.

71   H. Gruenberg, "Second-Order Beams of Slotted Waveguide Arrays," *Can. J. Phys.,* vol. 31, 1953, p. 55.

72   C. J. Sletten and G. R. Forbes, "A New Antenna Radiator for VHF-UHF Communications," AFCRC-TR-57-114, Air Force Cambridge Research Center, Bedford, Mass., 1957.

73   C. J. Sletten, F. S. Holt, P. Blacksmith, Jr., G. R. Forbes, L. F. Shodin, and H. J. Henkel, "A New Satellite Tracking Antenna," *IRE Wescon Conv. Rec.,* part 1, 1957, p. 244.

74   C. J. Sletten, F. S. Holt, P. Blacksmith, Jr., G. R. Forbes, L. F. Shodin, and H. J. Henkel, "A New Single Antenna Interferometer System Using Proximity-Coupled Radiators," AFCRC-TR-58-115, Air Force Cambridge Research Center, Bedford, Mass., 1958.

75   S. R. Seshadri and K. Iizuka, "A Dipole Antenna Coupled Electromagnetically to a Two-Wire Transmission Line," *IRE Trans. Antennas Propagat.,* vol. AP-7, 1959, p. 386.

76   G. R. Forbes, "An Endfire Array Continuously Proximity-Coupled," AFCRC-TR-59-368, Air Force Cambridge Research Center, Bedford, Mass., 1959.

77   A. A. Oliner, "Theoretical Developments in Strip Transmission Line," *Proc. Symp. Modern Advances Microwave Tech.,* Polytechnic Institute of Brooklyn, Brooklyn, N.Y., 1954, p. 379.

78   A. A. Oliner and W. Rotman, "Periodic Structures in Trough Waveguides," *IRE Trans. Microwave Theory Tech.,* vol. MTT-7, 1959, p. 134.

79   W. Rotman and S. J. Naumann, "The Design of Trough Waveguide Antenna Arrays," AFCRC-TR-58-154, Air Force Cambridge Research Center, Bedford, Mass., 1958.

80   S. T. Peng and A. A. Oliner, "Guidance and Leakage Properties of a Class of Open Dielec-

tric Waveguides, Part I: Mathematical Formulations," *IEEE Trans. Microwave Theory Tech.*, vol. MTT-29, September 1981, pp. 843–854.

81   A. A. Oliner, S. T. Peng, T. I. Hsu, and A. Sanchez, "Guidance and Leakage Properties of a Class of Open Dielectric Waveguides, Part II: New Physical Effects," *IEEE Trans. Microwave Theory Tech.*, vol. MTT-29, September 1981, pp. 855–869.

BIBLIOGRAPHY

Agrawal, N. K.: "A Radical Cylindrical Surface Wave Antenna," *J. Inst. Electron. Telecommun. Eng.* (India), vol. 23, July 1977, pp. 448–450.

Andersen, J. B.: "Low and Medium Gain Microwave Antennas," in A. W. Rudge, K. Milne, A. D. Olver, and P. Knight (eds.), *The Handbook of Antenna Design*, vol. 1, Peter Peregrinus Ltd., London, 1982, chap. 7.

Blakey, J. R.: "Calculation of Dielectric-Aerial-Radiation Patterns," *Electron. Lett.*, vol. 4, 1968, p. 46.

Bojsen, J. H., H. Schjaso-Jacobsen, E. Nilsson, and J. Bach Andersen: "Maximum Gain of Yagi-Uda Arrays," *Electron. Lett.*, vol. 7, Sept. 9, 1971, pp. 531–532.

Chen, K. M., and R. W. P. King: "Dipole Antennas Coupled Electromagnetically to a Two-Wire Transmission Line," *IEEE Trans. Antennas Propagat*, vol. AP-9, 1961, pp. 405–432.

Cho, S. H., and R. J. King: "Numerical Solution of Nonuniform Surface Wave Antennas," *IEEE Trans. Antennas Propagat.*, vol. AP-24, July 1976, pp. 483–490.

Hall, P. S., B. Chambers, and P. A. McInnes: "Calculation of the Radiation Pattern, Gain and Input Impedance of a Yagi Antenna, *Electron. Lett.*, vol. 11, June 1975, pp. 282–283.

Hessel, A.: "General Characteristics of Traveling-Wave Antennas," in R. E. Collin and F. J. Zucker (eds.), *Antenna Theory*, part 2, McGraw-Hill Book Company, New York, 1969, chap. 19.

Hirasawa, K.: "Optimum Gain of Reactively Loaded Yagi Antennas," *Trans. IECE of Japan*, vol. E63, no. 2 (Abstracts), February 1980, p. 139.

Hockham, G. A., and M. D. Byars: "A Broad Band, Compact Launcher for Surface Waves," *1969 European Microwave Conference*, IEEE, New York, 1970, pp. 294–297.

Inagaki, N., T. Sekiguchi, and M. Yamazawa: "Excitation of Surface Wave on Circular-Loop Array," *IEEE Trans. Antennas Propagat*, vol. AP-19, May 1971, pp. 433–435.

Kahn, W. K.: "Double-Ended Backward-Wave Yagi Hybrid Antenna," *IEEE Trans. Antennas Propagat.*, vol. AP-29, no. 3, 1981, pp. 530–532.

Kajfez, D.: "Nonlinear Optimization Reduces the Sidelobes of Yagi Antenna," *IEEE Trans. Antennas Propagat.*, vol. AP-21, September 1973, pp 714–715.

———: "Nonlinear Optimization Extends the Bandwidth of Yagi Antennas," *IEEE Trans. Antennas Propagat.*, vol. AP-23, March 1975, pp. 287–289.

Kamal, A. K., N. K. Agrawal, and S. C. Gupta: "Radiation Characteristics of Zig-Zag Antenna at 600 MHz," *J. Inst. Eng.* (India) *Electron. Telecommun. Eng. Div.*, vol. 59, August 1978., pp. 9–11.

———, S. C. Gupta, and R. A. Nair: "Dielectric Sphere-Loaded Radial Surface-Wave Antenna," *Int. J. Electron.* (Great Britain), vol. 44, May 1978, pp. 553–558.

King, R. W. P.: "Cylindrical Antennas and Arrays," in R. E. Collin and F. J. Zucker (eds.), Antenna Theory, part 2, McGraw-Hill Book Comapny, New York, 1969, chap. 11.

Kosta, S. P.: "A Theory of Helical Yagi Antenna," *Nachrichtentech. Z.* (Germany), vol. 25, July 1972, pp. 342–344.

Morozov, B. N., and V. V. Chebyshev: "Design of a Corrugated Surface-Wave Antenna with Low Sidelobe Level," *Telecommun. Radio Eng.*, part 1 (United States), vol. 25, April 1971, pp. 26–28.

Rojarayanot, B., and T. Sekiguchi: "Yagi-Uda Loop Antenna with Feeder System Composed of Two Elements Excited in Sequence," *Trans. IEEE of Japan*, vol. E61, April 1978, pp. 326–327.

Rotman, W., and A. A. Oliner: "Asymmetrical Trough Waveguide Antennas," *IRE Trans. Antennas Propagat.*, vol. AP-7, 1959, pp. 153–162.

Takata, N., and T. Sekiguchi: "Radiation of Surface Wave Antenna Consisting of Circular Loop Elements." *Trans. Inst. Electron. Commun. Eng. Japan, Sec. E*, vol. E60, November 1977, pp. 657–658.

Tilton, E. P.: *The Radio Amateur's VHF Manual*, American Radio Relay League, Inc., Newington, Conn., 1972, p. 155.

Tranquilla, J. M., and K. G. Balmain: "Resonance Phenomena on Yagi Arrays," *Can. Elec. Eng. J.* (Canada), vol. 6, April 1981, pp. 9–13.

Walter, C. H.: *Traveling Wave Antennas*, McGraw-Hill Book Company, New York, 1965.

Zucker, F. J.: "On the Design of Surface-Wave Antennas," 1984. (Additional material which is pertinent to the subject matter in Secs. 12-3 and 12-4 can be obtained by writing to the chapter author.)

Chapter 13

Helical Antennas

Howard E. King
Jimmy L. Wong

The Aerospace Corporation

A helical antenna consists of a single conductor or multiple conductors wound into a helical shape. Although a helix can radiate in many modes, the axial mode and the normal mode are the ones of general interest. The axial mode, the most commonly used mode, provides maximum radiation along the helix axis, which occurs when the helix circumference is of the order of one wavelength. The normal mode, which yields radiation broadside to the helix axis, occurs when the helix diameter is small with respect to a wavelength. Higher-order-radiation modes are also possible; e.g., when the helix dimensions exceed those required for the axial mode, a conical or multilobed pattern will result, as illustrated in Fig. 13-1.

The basic concepts of a helix antenna were established by Kraus[1,2] in 1947, and much of Kraus's early results was summarized by Harris.[3] Generally, helical antennas are wound with a single conductor. However, a helix can be designed with bifilar,[4] quadrifilar,[5,6] or multifilar[7] windings. Radiation characteristics of bifilar helices operating in the backfire mode have been described by Patton.[8] An advantage of a backfire helix is that it does not generally require a ground plane.

The helix-antenna parameters are defined as follows (see Fig. 13-2):

D = diameter of helix (center to center)

C = circumference of helix = πD

S = spacing between turns (center to center)

α = pitch angle = $\tan^{-1}(S/\pi D)$

N = number of turns

L = axial length of helix = NS

d = diameter of helix conductor

ℓ = length of one turn = $\sqrt{(\pi D)^2 + S^2}$

Although helical antennas are normally constructed with a circular cross section, elliptical helical antennas have also been investigated.[9]

This chapter presents detailed helical-antenna design information. Measured impedance, pattern, gain, and axial-ratio characteristics are shown for a variety of

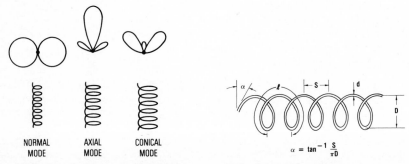

NORMAL MODE   AXIAL MODE   CONICAL MODE

$\alpha = \tan^{-1}\frac{S}{\pi D}$

FIG. 13-1 Three radiation modes of a helical antenna.

FIG. 13-2 Helix geometry.

helical-antenna configurations. Empirical relations which express the antenna radiation performance characteristics as a function of wavelength and the helix design parameters (diameter, pitch angle, and number of turns) are derived on the basis of measured data.

13-2   AXIAL-MODE HELICAL ANTENNAS

The helical beam antenna is a very simple structure possessing a number of interesting properties including wideband impedance characteristics and circularly polarized radiation. It requires a simple feed network, and its radiation characteristics are reasonably predictable. Either right-hand or left-hand circular polarization may be generated by a helical beam antenna. A helix wound like a right-hand screw radiates or receives right-hand circular polarization, while a helix wound like a left-hand screw radiates or receives left-hand circular polarization.

Helical antennas are generally constructed with a uniform diameter and operated in conjunction with a ground plane, cavity, or helical launcher.[10] However, as will be shown, nonuniform-diameter helical structures can be employed to widen the bandwidth of a uniform-diameter helical antenna and improve radiation performance characteristics.[11,12]

A typical uniform-helix configuration is shown in Fig. 13-3. The helix is backed by a circular cavity, rather than a conventional ground plane, to reduce the

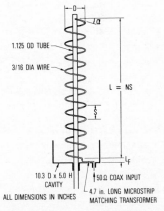

FIG. 13-3 Mechanical arrangement of a cavity-backed helix for 650 to 1100 MHz.

back radiation and enhance the forward gain. Although the helix of Fig. 13-3 was designed[11] specifically to operate in the ultrahigh-frequency (UHF) range from about 650 to 1100 MHz, the helix and the cavity dimensions can be scaled to other frequencies of operation as well. The helix is fed by a coaxial connection at the bottom of the cavity. The helical conductor can be made from round tubing or flat strip and supported with a lightweight foam dielectric cylinder or by radial dielectric rods. The center mechanical support can be constructed from either metal or dielectric material.

Impedance and VSWR

The impedance characteristics of a helical antenna have been established by Glasser and Kraus.[13] When the helix circumference is less than two-thirds wavelength, the terminal impedance is highly sensitive to frequency changes. However, when the circumference is of the order of one wavelength (axial mode), the terminal impedance is nearly a pure resistance and is given approximately (within $\pm$ 20 percent) by the empirical relation $R = 140\ C/\lambda\ \Omega$. The relatively constant terminal impedance may be explained by the rapidly attenuating character of the total outgoing wave near the input end and the total reflected wave near the open end.[14]

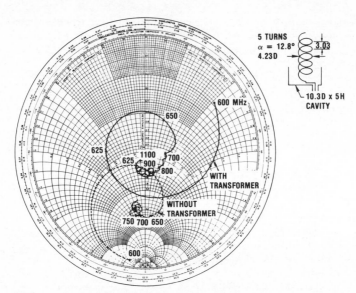

FIG. 13-4 Impedance characteristics of a five-turn helix with and without an impedance-matching transformer.

For the UHF helix configuration of Fig. 13-3, a 4.7-in- (119-mm-) long, linear-taper impedance transformer is used to match the 140-Ω helix to the 50-Ω coaxial line. The transformer is a microstripline constructed from Teflon-fiberglass printed-circuit board. The same transformer can be used for all axial-mode helices—uniform-, tapered-, or nonuniform-diameter. Another approach to attain a 50-Ω impedance for helical beam antennas has been described by Kraus.[15]

Figure 13-4 depicts the impedance characteristics of a five-turn helix with and without the impedance-matching transformer, and Fig. 13-5 depicts the voltage-

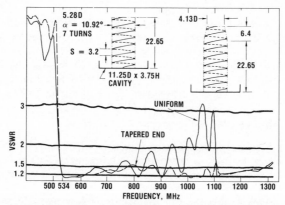

FIG. 13-5 VSWR of a seven-turn uniform helix and the same helix with two additional turns wound on a tapered diameter.

standing-wave-ratio (VSWR) response measured at the input of the matching trans-former for a seven-turn uniform helix and also for the same helix with two additional turns of tapered diameter, as illustrated by the inset. By adding the two-turn end taper, a significant reduction in VSWR over a wide frequency band can be achieved (dashed curve). The reduction in VSWR is due to the suppression of the reflected currents by the tapered end section.[16,17] Also, it is noted that the low-frequency char-acteristics are essentially unchanged with the cutoff at ~ 534 MHz, corresponding to $C/\lambda \sim 0.75$, where C is the circumference of the 5.28-in- (134.1-mm-) diameter helix. The low-frequency cutoff characteristics agree well with theoretical predictions.[1,2] The VSWR characteristics for longer helices are generally similar to those of Fig. 13-5.

Uniform-Diameter Helices

As a first approximation, the radiation pattern of an axial-mode helix may be obtained by assuming a single traveling wave of uniform amplitude along the conductor. By the principle of pattern multiplication, the far-field pattern is the product of the pattern of one turn and the pattern of an array of N isotropic elements with spacing S, where N equals the number of turns and S is the spacing between turns. If the pattern of a single turn is approximated by $\cos \theta$, where θ is the angle measured from the axis of the helix, the total radiation pattern becomes

$$E(\theta) = A \frac{\sin (N\psi/2)}{\sin (\psi/2)} \cos \theta \qquad (13\text{-}1)$$

where A = normalization factor

$$\psi = \frac{2\pi}{\lambda} S \cos \theta - \delta \qquad (13\text{-}2)$$

$\delta = \dfrac{2\pi}{\lambda} \dfrac{\ell}{(v/c)}$ = progressive phase between turns

ℓ = length of one turn

v = phase velocity along helical conductor

c = velocity of light in free space

Kraus[2] has shown that for an axial-mode helix the relative phase velocity, $p = v/c$, of the wave propagating along the helical conductor is in close agreement with that required to satisfy the Hansen-Woodyard condition for an end-fire array with increased directivity[18]; i.e.,

$$p = \frac{(\ell/\lambda)}{(S/\lambda) + (2N + 1)/2N} \qquad (13\text{-}3)$$

Thus, for the increased-directivity condition, the quantity ψ may be written as

$$\psi = \frac{2\pi}{\lambda} S (\cos \theta - 1) - \frac{\pi}{N} \qquad (13\text{-}4)$$

and

$$A = \sin (\pi/2N) \qquad (13\text{-}5)$$

The radiation is elliptically polarized. The ellipticity ratio or axial ratio (AR) of an N-turn helix operating in the axial mode for the increased-directivity condition is given by

$$AR = \frac{2N + 1}{2N} \qquad (13\text{-}6)$$

If N is large, the axial ratio approaches unity and the polarization is nearly circular. In a practical situation, the observed axial ratio may be expected to deviate somewhat from the theoretical value for various reasons such as methods of construction, tolerances, instrumentation, and range errors. For a uniform helix consisting of at least a few turns, an axial ratio of the order of 1 dB is not uncommon.

To provide parametric design equations for axial-mode helices, Kraus[2] has suggested the following relations for the half-power beamwidth (HPBW) and gain (G) as a function of C/λ and NS/λ for constant-pitch helices with $12° < \alpha < 15°$, $\frac{3}{4} < C/\lambda < \frac{4}{3}$, and $N > 3$:

$$HPBW = K_B/(C/\lambda) \sqrt{NS/\lambda} \qquad (13\text{-}7)$$

$$G = K_G(C/\lambda)^2(NS/\lambda) \qquad (13\text{-}8)$$

where K_B is the HPBW factor, K_G is the gain factor, $C = \pi D$ is the circumference, and NS is the axial length. On the basis of a number of pattern measurements on helices < 10 turns, Kraus quasi-empirically established $K_B = 52$. Also, he derived $K_G = 15$ for the directive gain (lossless antenna) based on the approximation $G = 41,250/(HPBW)^2$, where HPBW is in degrees. A gain-beamwidth product G $(HPBW)^2 < 41,250$ is generally expected for most practical antennas[19] because of minor-lobe radiation and beam-shape variations.

Additional data on the radiation characteristics of axial-mode helices are available in the literature.[11,20-22] The low-frequency and high-frequency limits were investigated by Maclean and Kouyoumjian[20] and by Maclean.[23] In a practical situation, the pattern and gain versus frequency characteristics are of interest to the antenna designer. Such data would allow the designer to optimize the helix parameters for operation over a specified bandwidth.

The discussions which follow summarize the results of an extensive study of the gain and pattern characteristics of uniform helical antennas, one to eight wavelengths long, in the UHF range from about 650 to 1100 MHz. With reference to Fig. 13-3, the helices were constructed by winding $\frac{3}{16}$-in diameter copper tubing about a Styrofoam cylindrical form that was concentric with the 1.125-in-diameter metallic support tube. The helix dimensions can be scaled to other frequencies of operation with essentially the same radio-frequency performance characteristics. All gain and pattern measurements were made with respect to the phase center of the helix, which was estimated to be one-fourth of the helix length from the feed point.[24]

Fixed-Length Helix   Figures 13-6 and 13-7 show the gain and HPBW versus frequency characteristics, respectively, of a 30-in- (762-mm-) long ($NS = 30$ in) and 4.3-in- (109.22-mm-) diameter helix for three helix pitch angles ($\alpha = 12.5$, 13.5, and $14.5°$). The helix with a smaller pitch angle (more turns per unit length) yields a higher peak gain and a lower cutoff frequency. With $N = 8.6$ to 10 turns, it appears that the gain-frequency slope is approximately proportional to f^3 and the HPBW-frequency slope is approximately proportional to $f^{-3/2}$, where f equals the frequency, which are in general agreement with Kraus for $C/\lambda < 1.1$ [see Eqs. (13-7) and (13-8)]. However, as will be shown later, experimental data indicate that the gain-frequency slope depends on the antenna length and is approximately proportional to $f^{\sqrt{N}}$.

Figures 13-8 and 13-9 show the gain and HPBW characteristics, respectively, of

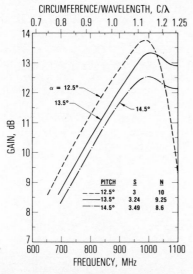

FIG. 13-6 Gain versus frequency of a fixed-length (NS = 30 in, or 762 mm) helix; α = 12.5°, 13.5°, and 14.5°, D = 4.3 in (109 mm).

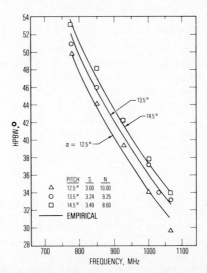

FIG. 13-7 Half-power beamwidth versus frequency of a fixed-length helix (NS = 30 in, or 762 mm); α = 12.5°, 13.5°, and 14.5°, D = 4.3 in (109 mm).

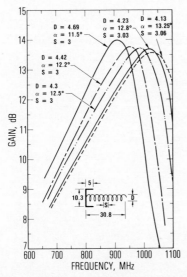

FIG. 13-8 Gain of a fixed-length (NS = 30 in, or 762 mm) helical antenna with different diameters.

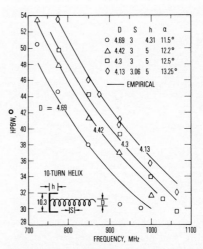

FIG. 13-9 Half-power beamwidths of a 10-turn fixed-length (NS = 30 in, or 762 mm) helix with different diameters.

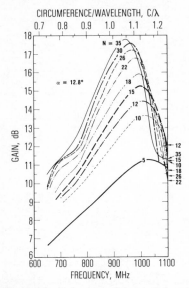

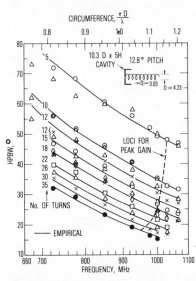

FIG. 13-10 Gain versus frequency for a 5- to 35-turn helix; $\alpha = 12.8°$, $D = 4.23$ in (107.4 mm).

FIG. 13-11 Half-power beamwidth versus frequency for a 5- to 35-turn helix; $\alpha = 12.8°$, $D = 4.23$ in (107.4 mm).

a 30-in-long ($N \approx 10$ turns) helix with variable diameter and pitch angle. The peak gain varies by less than 0.5 dB. In general, a slightly higher peak gain is observed for a larger-diameter helix with a smaller pitch angle, but the bandwidth is narrower than that of a smaller-diameter helix with a larger pitch angle.

The axial ratio, a measure of the purity of the circularly polarized wave, is generally less than 1.5 dB for $0.8 < C/\lambda < 1.2$. The axial ratio can be improved by tapering the last two turns of the helix, particularly at the high end of the band.[12,17,25]

Variable-Length Helix   Figures 13-10 and 13-11 show the gain and HPBW versus frequency, respectively, of helices consisting of 5 to 35 turns with constant pitch ($\alpha = 12.8°$, $S = 3.03$ in, or 76.96 mm) and constant diameter ($D = 4.23$ in, or 107.4 mm). These helices were designed to operate over the UHF test frequencies with the helix circumference varied from 0.75λ to 1.25λ. N was selected as 5, 10, 12, 15, 18, 22, 26, 30, and 35 turns. The gain is referred to a circularly polarized illuminating source. The gain curves reveal that the peak gain occurs at $C/\lambda = 1.55$ for $N = 5$ and at a lower value, $C/\lambda = 1.07$, for $N = 35$. The gain-frequency slope is not proportional to f^3 for all values of N [see (Eq. 13-8)]; e.g., for $N = 5$ the gain varies approximately as $f^{2.5}$, and for $N = 35$ the gain follows approximately an f^6 slope, where f is the frequency.

Typical measured radiation patterns for $N = 5$, 10, 18, and 35 are shown in Fig. 13-12. The axial ratio is ~ 1 dB over most of the measurement frequency range and is slightly higher at the band edges. The HPBWs are generally within $\pm 1°$ in the two orthogonal principal planes. At frequencies a few percent above the peak gain frequency, the patterns begin to deteriorate. The beamwidth broadens rapidly, and the first sidelobes merge with the main lobe as the operating frequency approaches the upper limit.

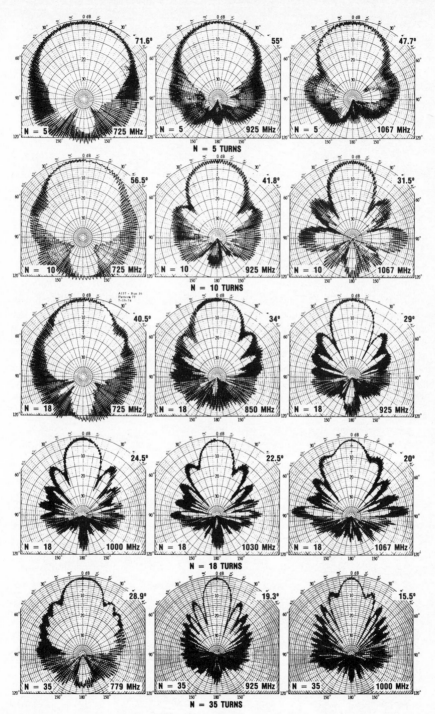

FIG. 13-12 Typical radiation patterns of a 5- to 35-turn helix; $\alpha = 12.8°$, $D = 4.23$ in (107.4 mm).

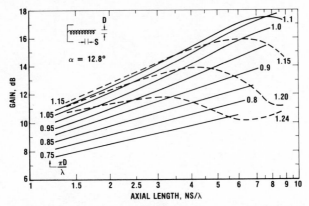

FIG. 13-13 Parametric helix-antenna gain curves as a function of axial length with circumference as a parameter.

Parametric helix characteristic curves—gain and halfpower beamwidth—for the 4.23-in- (107.4-mm-) diameter constant-pitch ($\alpha = 12.8°$) helix are shown in Figs. 13-13 and 13-14, respectively, as a function of axial length NS/λ with circumference $\pi D/\lambda$ as a parameter. Thus, for a specified length and diameter the helix gain and HPBW can be estimated.

Figure 13-15 depicts the gain-beamwidth product $K = G \, (\text{HPBW})^2$ based on the measured gain data of Fig. 13-10 and the HPBW data of Fig. 13-11. The quantity K is useful for estimating the gain when the HPBW is known, and vice versa. It should be noted, however, that the gain-beamwidth product is not constant but depends on N

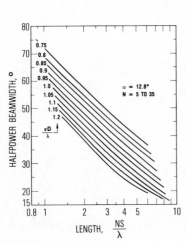

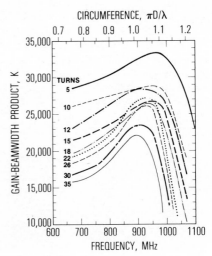

FIG. 13-14 Parametric helix-antenna half-power-beamwidth curves as a function of axial length with circumference as a parameter.

FIG. 13-15 Gain-beamwidth products of a 5- to 35-turn helix; $\alpha = 12.8°$, $D = 4.23$ in (107.4 mm).

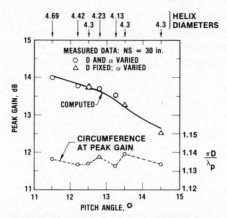

FIG. 13-16 Peak gain of a fixed-length helix (*NS* = 30 in, or 762 mm) as a function of a pitch angle.

and frequency. All curves have been smoothed to within ± 5 percent of the measured-data points.

Empirical Relations On the basis of the gain data of Figs. 13-6, 13-8, and 13-10, the peak gain may be empirically expressed as[11]

$$G_p = 8.3 \left(\frac{\pi D}{\lambda_p}\right)^{\sqrt{N+2}-1} \left(\frac{NS}{\lambda_p}\right)^{0.8} \left[\frac{\tan 12.5°}{\tan \alpha}\right]^{\sqrt{N}/2} \qquad (13\text{-}9)$$

where λ_p is the wavelength at peak gain. The computed values for the fixed-length helices (*NS* = constant) are within ± 0.1 dB of the measured data as depicted in Fig. 13-16. The data points indicated by the circle were obtained by varying the diameter and pitch angle while keeping the length constant with $N = 10$ turns, and those indicated by the triangle were obtained by varying the pitch angle while keeping the length and diameter constant ($N \approx 8.6$ to 10). The diameters of the various experimental helices are shown on the top of the figure. For these fixed-length helices, the peak gains occur at nearly the same circumference, $\pi D/\lambda \sim 1.135$.

Figure 13-17 is a plot of the peak gain versus N for a constant-pitch ($\alpha = 12.8°$)

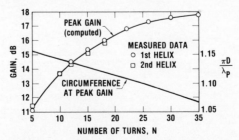

FIG. 13-17 Peak gain of a variable-length helix with $\alpha = 12.8°$.

helix. The corresponding values of C/λ_p are also shown in the figure, where λ_p is the wavelength at peak gain. The peak gain is not quite proportional to the number of turns; i.e., doubling the number of turns does not yield a 3-dB increase in the peak gain. The computed values for the peak gain, using Eq. (13-9) with $\alpha = 12.8°$ and $N = 5$ to 35 turns, are within $\pm$ 0.1 dB of the measured data.

A similar empirical expression for the HPBW is as follows:[26]

$$\text{HPBW} \approx \frac{K_B \left(\dfrac{2N}{N+5} \right)^{0.6}}{\left(\dfrac{\pi D}{\lambda} \right)^{\sqrt{N}/4} \left(\dfrac{NS}{\lambda} \right)^{0.7}} \left(\frac{\tan \alpha}{\tan 12.5°} \right)^{\sqrt{N}/4} \qquad (13\text{-}10)$$

where K_B is a constant in degrees. For the helices constructed by using the arrangement of Fig. 13-3, it was found that, with $K_B \simeq 61.5°$, Eq. (13-10) matches the measured data within $\pm$ a few percent over the useful operating frequency range of the helix as shown in Figs. 13-7, 13-9, and 13-11. For helices that employ a different construction technique (e.g., tape helices that are wound on a dielectric support instead of using a metallic central support rod), Eq. (13-10) can still be applied, but a slightly different value of K_B must be used. The measured HPBWs for the helices investigated are generally 10 to 20 percent wider than Kraus's formula [Eq. (13-7)]. It should be mentioned that Eqs. (13-9) and (13-10) are not unique; however, these relations are useful as a design tool.

The helix bandwidth may be defined as the operating frequency range over which the gain drops by an allowable amount. The -3-dB and -2-dB bandwidths as a function of N may be obtained by using the curves of Fig. 13-18. For example, if the -3-dB or -2-dB bandwidth is desired for a given N, one determines the values of $\pi D/\lambda_h$ and $\pi D/\lambda_\ell$ from Fig. 13-18, where λ_h and λ_ℓ are the free-space wavelengths corresponding to the upper frequency limit f_h and the lower frequency limit f_ℓ respectively. The choice of a -3-dB or -2-dB bandwidth depends upon the antenna designer's application. The frequency limits f_h and f_ℓ may be determined by using the mea-

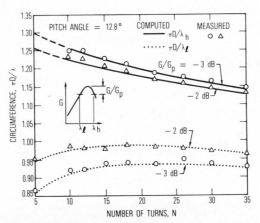

FIG. 13-18 Gain-bandwidth characteristics of a uniform helix with $\alpha = 12.8°$.

sured gain data of Fig. 13-10. Note that the gain varies approximately as $f^{\sqrt{N}}$ for $f < f_p/1.04$ and as $f^{-3\sqrt{N}}$ for $f > 1.03\, f_p$, where f_p is the frequency at peak gain. On the basis of these observations, the bandwidth frequency ratio may be empirically expressed as[11]

$$\frac{f_h}{f_\ell} \approx 1.07 \left(\frac{0.91}{G/G_p}\right)^{4/(3\sqrt{N})} \qquad \textbf{(13-11)}$$

where G_p is the peak gain from Eq. (13-9). The computed bandwidth characteristics for $G/G_p = -3$ dB and -2 dB agree reasonably well with the measured data as shown in Fig. 13-18. The bandwidth decreases as the axial length of the helix increases. This bandwidth behavior follows the same trends described by Maclean and Kouyoumjian,[20] although these authors employ a sidelobe criterion rather than a gain criterion. Beyond the $G/G_p = -3$ dB point the gain drops off sharply at the high-frequency end as the upper limit for the axial mode is approached.

Nonuniform-Diameter and Tapered Helices

The tapered-diameter[27,28] and nonuniform-diameter[12] helices represent additional types of axial-mode helical antennas. These helical configurations are capable of providing a wider bandwidth than a conventional uniform (constant-diameter) helix. A nonuniform helix consists of multiple uniform-diameter helical sections that are joined together by short tapered transitions. Some studies have been made to investigate the broadband-frequency response of nonuniform-diameter helical antennas.[29,30] The various types of helical-antenna configurations (uniform, tapered, and nonuniform-diameter) are shown in Fig. 13-19, and the radiation performance characteristics are described in Ref. 12.

The experimental helices were wound with thin copper strips 0.468 in (11.887 mm) wide. The plane of the strip (the wide dimensions of the strip) was wound orthogonally to the helix axis similarly to a Slinky toy. Helices wound with round conductors or with metallic tapes (wound so that the plane of the tape is parallel to the helix axis) yield similar results. A constant-pitch spacing of 3.2 in (81.28 mm) was selected, and the helix was backed by an 11.25-in- (285.75-mm-) diameter $\times$ 3.75-in- (95.25-mm-) high cavity. The feed arrangement and the metallic central support tube are similar to those of Fig. 13-3. The VSWR of a nonuniform helix is $\leq 1.5:1$ over the test frequency range of 650 to 1100 MHz (similar to the dashed curve of Fig. 13-5).

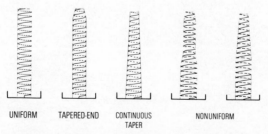

UNIFORM   TAPERED-END   CONTINUOUS   NONUNIFORM
TAPER

FIG. 13-19 Various axial-mode helical configurations.

Tapered-End Helix   For reference purposes, the gain and axial-ratio characteristics of an 18-turn uniformly wound helix, with D = 4.59 in (116.59 mm) and α = 12.5°, are shown in Fig. 13-20. Also shown in this figure are the gain and axial-ratio characteristics of a similar helix with the same overall length, but the diameter of the last two turns is tapered from 4.59 in to 2.98 in (75.69 mm). It can be seen that the end taper provides a marked improvement in axial-ratio characteristics,[12,16,17] although the peak gain is reduced slightly. Figure 13-21 compares the radiation patterns of a uniform and a tapered-end helix measured at the same frequency. The HPBWs are approximately the same, but the axial-ratio improvement is observed over the entire pattern (on axis and off axis). Similar results have been reported by Donn.[25]

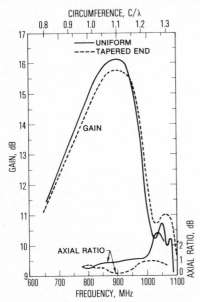

FIG. 13-20   Gain and axial-ratio characteristics of an 18-turn uniformly wound and tapered-end helix.

Continuously Tapered Helix   The gain and axial characteristics of a continuously tapered helix,[12] often referred to as a conical helix, are shown in Fig. 13-22. The helix consists of 17.64 turns with a constant-pitch spacing of 3.2 in (81.28 mm) and tapers from a 5.32-in (135.13-mm) diameter at the base to a 2.98-in (75.69-mm) diameter at the top (see inset). Typical measured patterns are shown in Fig. 13-23. The peak gain is slightly lower than that of the uniform helix, but the pattern characteristics and bandwidth performance are much better.

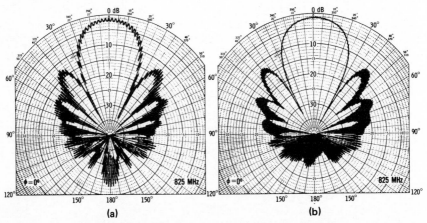

FIG. 13-21   Radiation patterns of an 18-turn (*a*) uniform and (*b*) tapered-end helix.

Nonuniform-Diameter Helix A nonuniform helix provides a unique approach for widening the bandwidth of a helical antenna with improved gain and pattern characteristics. It may be constructed with two or more uniform helix sections of different diameters or a combination of uniform and tapered sections. The dimensions of the different helical sections (diameter, number of turns, etc.) can be varied to synthesize an antenna with a specific gain-frequency response.[12,31]

Two nonuniform helical configurations and the corresponding gain and axial-ratio characteristics are illustrated in Fig. 13-24. Configuration *a* consists principally of two uniform-diameter sections [5.28 and 4.13 in (134.1 and 104.9 mm)] joined together by a short tapered transition. This helix may be described as 7-turn (5.28 *D*) + 2-turn taper (5.28 *D* to 4.13 *D*) + 6.64-turn (4.13 *D*) + 2-turn end taper (4.13 *D* to 2.98 *D*). A constant-pitch spacing of 3.2 in (81.28 mm) was maintained in all four helical sections. The performance of this nonuniform helix configuration was optimized over the low-frequency region. The gain is 14.7 ± 0.4 dB from 773 to 900 MHz and remains relatively flat (14.05 ± 0.25 dB) from 900 to 1067 MHz. Note that the gain is constant within ± 1 dB over a frequency ratio of 1.55:1 (1100/710 MHz) as compared with 1.26:1 for a uniform helix. Table 13-1 provides a comparison of the ± 1-dB bandwidth for the various axial-mode helical antennas. The axial ratio is < 1 dB. The beam-shape and sidelobe characteristics are improved over those

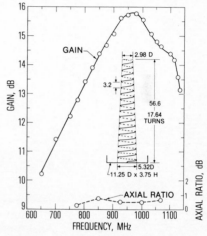

FIG. 13-22 Gain and axial-ratio characteristics of a 17.64-turn conical helix.

of a uniform helix as illustrated in the patterns of Fig. 13-25. Note that the high-frequency cutoff is limited not by the larger 5.28-in-diameter helical section (C/λ = 1.55 at 1100 MHz) but rather by the smaller 4.13-in-diameter helical section (C/λ = 1.21 at 1100 MHz). The HPBW is relatively constant, 33° ± 3° over the 773- to 1067- MHz test frequency range.[12]

Configuration *b* of Fig. 13-24 consists of a uniform section (5.28-in diameter) plus a tapered section from 5.28- to 2.98-in diameter. The ± 1.1-dB gain bandwidth is wider than that of configuration *a*, but the gain at the high-frequency end is lower.

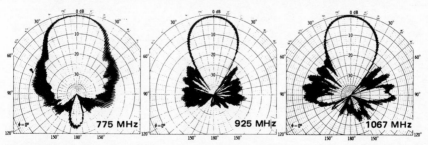

FIG. 13-23 Radiation patterns of a 17.64-turn conical helix.

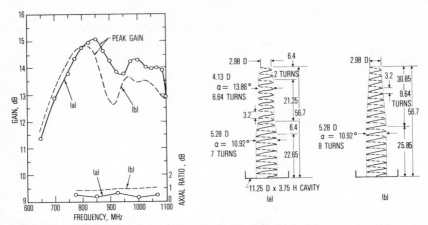

FIG. 13-24 Gain and axial-ratio characteristics of two 17.64-turn nonuniform-diameter helices.

TABLE 13-1

Type of helix	Frequency range with ±1-dB gain variation, MHz	Frequency ratio, f_{max}/f_{min}
Uniform	770–970	1.26:1
Tapered-end	770–980	1.27:1
Continuous-taper	820–1120	1.37:1
Nonuniform	710–1100	1.55:1

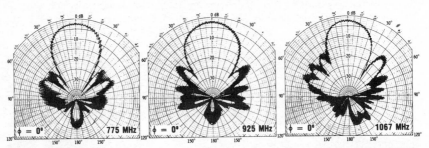

FIG. 13-25 Radiation patterns of a 17.64-turn nonuniform-diameter helix—configuration *a* of Fig. 13-24.

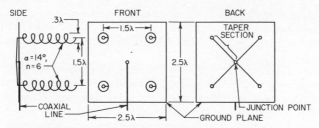

FIG. 13-26   2 x 2 array of six-turn helices.

13-3   ARRAY OF HELICAL ANTENNAS

A single axial-mode helical antenna produces moderate gain in the end-fire direction (see Fig. 13-17). When higher gain is desired, a linear or planar array of helices may be used to produce a fan-shaped beam or a pencil beam respectively. The array gain depends on the element spacing with respect to the operating wavelength and the element gain. To realize the advantage of the increased gain, the helical elements should be spaced so that the area occupied by each helix is approximately equal to its effective area. To the first-order estimate, this spacing is $\sim \sqrt{G_e/4\pi} \; \lambda$, where G_e is the power gain of the individual helices. As the spacing is increased to $> \sqrt{G_e/4\pi}\lambda$, the array gain approaches the asympototic value NG_e, where N is the number of elements.[32]

Some measured data on this type of array have been reported by Harris,[3,33] and the construction and the radiation characteristics of a typical 2 x 2 planar array of six-turn helices ($D = 0.3\lambda$ and $\alpha = 14°$) spaced 1.5λ are shown in Figs. 13-26 and 13-27 respectively. Also, shaped-beam and steerable arrays employing axial-mode helices have been designed for spacecraft applications.[22,34,35]

Generally, arrays employing high-gain elements are not suitable for wide-angle scanning because of grating lobes. However, when antenna gain and side-lobes (or grating lobes) are not important, such arrays can be phased to scan over a limited angular range.

For certain applications, such as FM and TV broadcasting, a moderately high-gain, azimuthally omnidirectional pattern is required. A vertical array of side-fire helices mounted concentrically on a conducting cylinder may be employed to produce a narrow-beam pattern in the elevation plane. The helix ele-

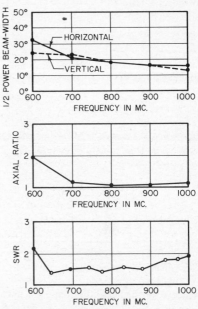

FIG. 13-27   Measured characteristics of a 2 x 2 array of six-turn helices.

ments are generally excited in the second- or higher-order modes which produce maximum radiation broadside (normal) to the helix axis and a null along the helix axis. The helix diameter and pitch are chosen so that the length of one turn is equal to an integral number of wavelengths.

If horizontal polarization is desired, the radiating element may be constructed with two oppositely wound helices (one right-handed and the other left-handed), each consisting of a few turns, placed end to end and fed in the center. Design data on this type of vertical array are available in the literature.[3,36,37]

Vertical helical-array antennas designed to produce omnidirectional, circularly polarized radiation in the horizontal plane for VHF-UHF TV broadcasting (Chap. 28) have been described by DuHamel.[38] In this case, the radiating elements are end-fed, four-turn bifilar helices operating in various modes which are selected so that the helix circumference is about $\lambda/2$ greater than the circumference of the central support cylinder. Depending on the operating frequency, horizontal circularity of ± 0.5 dB to ± 1 dB and an axial ratio of ~ 1.5 dB have been reported.

13-4 NORMAL-MODE HELICAL ANTENNAS

When the dimensions of the helix are small compared with wavelength, the maximum radiation is in a direction normal to the helix axis. Kraus[2] has shown that the radiation from a short-axial-length helix can be calculated by assuming that the helix is composed of small loops of diameter D and short dipoles of length S. The far field of a short dipole has only an E_θ component,

$$E_\theta = \frac{jBe^{-jkr}}{r}\left(\frac{S}{\lambda}\right)\sin\theta \qquad (13\text{-}12)$$

and the far field of a small loop has only an E_ϕ component,

$$E_\phi = \frac{Be^{-jkr}}{2r}\left(\frac{\pi D}{\lambda}\right)^2\sin\theta \qquad (13\text{-}13)$$

where r is the distance and B is a constant. The E_θ and E_ϕ components are 90° out of phase, and the far field of a small helix is, in general, elliptically polarized. The axial ratio is given by

$$\text{AR} = \frac{|E_\theta|}{|E_\phi|} = \left(\frac{2S}{\lambda}\right) \Big/ \left(\frac{\pi D}{\lambda}\right)^2 \qquad (13\text{-}14)$$

When $\pi D = \sqrt{2S\lambda}$, the axial ratio becomes unity and the radiation is circularly polarized.[39] On the other hand, the polarization of the radiated field will be predominantly horizontal ($\text{AR} \to 0$) when $\pi D \gg \sqrt{2S\lambda}$ or predominantly vertical ($\text{AR} \to \infty$) when $\pi D \ll \sqrt{2S\lambda}$.

For a normal-mode helix whose dimensions are small compared with wavelength, the current distribution along the helix is approximately sinusoidal.[2] The terminal impedance is very sensitive to changes in frequency, and the bandwidth is narrow. Nevertheless, a normal helix has been used as an effective means for reducing the length of thin-wire-type (whip) antennas for personal radio and mobile communications systems in the HF and VHF bands.[40-43] Also, balanced-fed dipole antennas can be constructed by using short-axial-length normal-mode helices when a reduced dipole length is desired.[44]

When a short normal-mode helix is used in conjunction with a ground plane, the polarization is predominantly vertical and the radiation pattern is similar to that of a monopole. The radiation resistance of a short resonant helix above a perfect ground is approximately given[40] by $(25.3 \ h/\lambda)^2$, where h is the axial length or height above the ground plane.

Typical feeding arrangements for a helical monopole are shown in Fig. 13-28. In the series-fed arrangement, the helix is connected directly to the coaxial input, and an impedance transformer or matching network may be required. In the shunt-fed arrangement, the helix provides a self-matching network by tapping a small portion of the helix. A feed

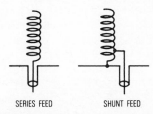

SERIES FEED   SHUNT FEED

FIG. 13-28 Typical feed arrangements for a helical monopole.

arrangement which employs a bifilar helix to increase the input impedance is described in Ref. 41, and another design which utilizes a short helical monopole and top-loading wire to produce a self-resonant antenna over the 2–30-MHz frequency range without the need of a matching device is described in Ref. 43.

13-5 OTHER HELICAL ANTENNAS

Fractional-Turn Resonant Quadrifilar Helices A class of resonant quadrifilar helices, also referred to as volutes, capable of radiating circular polarization with a cardioid-shaped pattern has been described by Kilgus.[45-47] The antenna consists of two orthogonal fractional-turn (one-fourth to one turn) bifilar helices excited in phase quadrature. Each bifilar helix is balun-fed at the top, and the helical arms are wires or metallic strips of resonant length ($\ell = m\lambda/4, \ m = 1, 2, 3, \ldots$) wound on a small diameter with a large pitch angle. The ends of the helices are open-circuited at the base when m = odd and short-circuited when m = even. When properly excited, the volute produces a very broad beamwidth with relatively low backlobes and good axial-ratio characteristics over a wide angular range. Since the volute is a resonant structure, the impedance bandwidth is narrow. Typically, the VSWR is < 2:1 over a 3 to 5 percent bandwidth. Generally a wider bandwidth can be achieved with larger-diameter wires for a given helix design. This type of antenna has been designed for various applications requiring a wide beamwidth over a relatively narrow frequency range.[48,49]

Because of its low-backlobe characteristics, a volute can be operated as an isolated antenna or with a ground plane. A half-turn $\lambda/2$-volute antenna and a typical measured pattern obtained with a rotating linearly polarized source at 488 MHz are shown in Fig. 13-29. The pattern was measured with a 90° hybrid connected to the coaxial input ports of the two bifilar helices. The half-turn, half-wavelength volute is of particular interest because the input impedance of each bifilar can be matched to a 50-Ω coaxial input simply by minor adjustment of the helical-arm lengths without

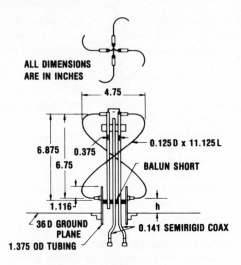

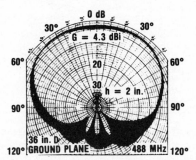

FIG. 13-29 Half-turn, half-wavelength volute antenna and measured pattern at 488 MHz.

the need of a transformer. The measured VSWR is < 2:1 over ~ 5 percent bandwidth centered at 488 MHz.*

Short Axial-Mode Helices For certain applications requiring a wideband and broadbeam antenna with a relatively small physical size, such as reflector feeds, a short axial-mode helix (two to four turns) housed in a conical or circular cavity similar to that of Fig. 13-3 can be employed. As discussed in the subsection "Nonuniform-Diameter and Tapered Helices," the end of the helix may be slightly tapered to improve the VSWR and axial-ratio performance. A similar cavity-backed short helix designed for flush-mounting applications is described by Bystrom and Berntsen.[50] A dual-helix feed system which utilizes two concentric helices to provide operation over two narrow-frequency bands is described by Holland.[51]

---

*J. L. Wong and H. E. King, Electronics Research Laboratory, The Aerospace Corporation, unpublished results.

A helical feed design for a paraboloidal reflector which utilizes a four-turn back-fire bifilar helix is described in Ref. 52. This feed was designed for a spacecraft antenna operating in the 240- to 270-MHz band.† An attractive feature of the back-fire bifilar-helix feed is that it does not require a cavity or ground plane, thus minimizing blockage effects and improving reflector-antenna efficiency.

Helicone Antennas   An axial-mode helix can be operated in conjunction with a conical horn to provide a broadband antenna with low sidelobes.[53] The axial length of the helix and the horn are approximately the same. Sidelobe levels of the order of −25 dB have been reported. Design and performance data are available in the literature.[54]

REFERENCES

1 J. D. Kraus, "Helical Beam Antennas," *Electronics,* vol. 20, April 1947, pp. 109–111.
2 J. D. Kraus, "The Helical Antenna," *Antennas,* McGraw-Hill Book Company, New York, 1950, chap. 7.
3 E. F. Harris, "Helical Antennas," in H. Jasik (ed), *Antenna Engineering Handbook,* 1st ed., McGraw-Hill Book Company, New York, 1961, chap. 7.
4 A. G. Holtum, Jr., "Improving the Helical Beam Antenna," *Electronics,* Apr. 29, 1960.
5 A. T. Adams, R. K. Greenough, R. F. Wallenberg, A. Mendelovicz, and C. Lumjiak, "The Quadrifilar Helix Antenna," *IEEE Trans. Antennas Propagat.,* vol. AP-22, March 1974, pp. 173–178.
6 C. C. Kilgus, "Shaped-Conical Radiation Pattern Performance of the Backfire Quadrifilar Helix," *IEEE Trans. Antennas Propagat.,* vol. AP-23, May 1975, pp. 392–397.
7 C. W. Gerst and R. A. Worden, "Helix Antennas Take Turn for Better," *Electronics,* Aug. 22, 1966.
8 W. T. Patton, "The Backfire Bifilar Helical Antenna," Ph.D. dissertation, University of Illinois, Urbana, October 1963.
9 J. Y. Wong and S. C. Loh, "Radiation Field of an Elliptical Helical Antenna," *IRE Trans. Antennas Propagat.,* vol. AP-7, January 1959, pp. 46–52.
10 B. A. Munk and L. Peters, "A Helical Launcher for the Helical Antenna," *IEEE Trans. Antennas Propagat.,* vol. AP-16, May 1968, pp. 362–363.
11 H. E. King and J. L. Wong, "Characteristics of 5- to 35-Turn Uniform Helical Antennas," Aerospace Corp. Tech. Rep. TR-0078(3724-01)-2, DDC AD A046487, June 1977; or "Characteristics of 1 to 8 Wavelength Uniform Helical Antennas," *IEEE Trans. Antennas Propagat.,* vol. AP-28, March 1980, pp. 291–296.
12 J. L. Wong and H. E. King, "Broadband Quasi-Taper Helical Antennas," Aerospace Corp. Tech. Rep. TR-0077(2724-01)-2, DDC AD A046067, Sept. 30, 1977; and *IEEE Trans. Antennas Propagat.,* vol. AP-27, January 1979, pp. 72–78.
13 O. J. Glasser and J. D. Kraus, "Measured Impedances of Helical Antennas," *J. App. Phys.,* vol. 19, February 1948, pp. 193–197.
14 J. A. Marsh, "Measured Current Distributions on Helical Antennas," *IRE Proc.,* vol. 39, June 1951, pp. 668–675.
15 J. D. Kraus, "A 50-Ohm Input Impedance for Helical Beam Antennas," *IEEE Trans. Antennas Propagat.,* vol. AP-25, November 1977, pp. 913–914.
16 H. Nakano, J. Yamauchi, and H. Mimaki, "Tapered Balanced Helices Radiating in the Axial Mode," *Dig. Int. Symp. Antennas Propagat.,* 1980, pp. 700–703.

---

†By TRW Systems Group, Redondo Beach, Calif.

17 D. J. Angelakos and D. Kajfez, "Modifications on the Axial-Mode Helical Antenna," *IEEE Proc.,* vol. 55, April 1967, pp. 558–559.

18 W. W. Hansen and J. R. Woodyard, "A New Principle in Directional Antenna Design," *IRE Proc.,* vol. 26, March 1938, pp. 333–345.

19 R. J. Stegen, "The Gain-Beamwidth Product of an Antenna," *IEEE Trans. Antennas Propagat.,* vol. AP-12, July 1964, pp. 505–506.

20 T. S. M. Maclean and R. G. Kouyoumjian, "The Bandwidth of Helical Antennas," *IRE Trans. Antennas Propagat.,* vol. AP-7, special supplement, December 1959, pp. S379–S386.

21 G. C. Jones, "An Experimental Design Study of Some S- and X-Band Helical Aerial Systems," *IEE Proc.,* vol. 103, part B, November 1956, pp. 764–771.

22 K. G. Schroeder and K. H. Herring, "High Efficiency Spacecraft Phased Arrays," AIAA Pap. 70-425, *AIAA 3d Comm. Satellite Syst. Conf.,* Apr. 6–8, 1970.

23 T. S. M. Maclean, "An Engineering Study of the Helical Aerial," *IEE Proc.,* vol. 110, January 1963, pp. 112–116.

24 S. Sander and D. K. Chang, "Phase Center of Helical Beam Antennas," *IRE Conv. Rec.,* part I, 1958, pp. 152–157.

25 C. Donn, "A New Helical Antenna Design for Better On-and Off-Boresight Axial Ratio Performance," *IEEE Trans. Antennas Propagat.,* vol. AP-28, March 1980, pp. 264–267.

26 J. L. Wong and H. E. King, "Empirical Helix Antenna Design," *Dig. Int. Symp. Antennas Propagat.,* 1982, pp. 366–369.

27 J. S. Chatterjee, "Radiation Field of a Conical Helix," *J. App. Phys.,* vol. 24, May 1953, pp. 550–559.

28 H. S. Barsky, "Broadband Conical Helix Antennas," *IRE Conv. Rec.,* part I, 1959, pp. 138–146.

29 K. F. Lee, P. F. Wong, and K. F. Larm, "Theory of the Frequency Responses of Uniform and Quasi-Taper Helical Antennas," *IEEE Trans. Antennas Propagat.,* vol. AP-30, September 1982, pp. 1017–1021.

30 J. M. Tranquilla and G. B. Graham, "Swept-Frequency Radiation Pattern Anomalies on Helical Antennas," *Can. Elect. Eng. J.,* 1982.

31 J. L. Wong and H. E. King, "Broadband Helical Antennas," U.S. Patent 4,169,267, Sept. 25, 1979.

32 H. E. King and J. L. Wong, "Directivity of a Uniformly Excited $N \times N$ Array of Directive Elements," *IEEE Trans. Antennas Propagat.,* vol. AP-23, May 1975, pp. 401–403.

33 E. F. Harris, "Helical Beam Antenna Performance," *Commun. Eng.,* July–August 1953, pp. 19–20, 44–45.

34 C. T. Brumbaugh, A. W. Love, G. M. Randall, D. K. Waineo, and S. H. Wong, "Shaped Beam Antenna for the Global Positioning Satellite System," *Dig. Int. Symp. Antennas Propagat.,* 1976, pp. 117–120.

35 C. Donn, W. A. Imbriale, and G. Wong, " A S-Band Phased Array Design for Satellite Applications," *Dig. Int. Symp. Antennas Propagat.,* 1977, pp. 60–63.

36 L. O. Krause, "Sidefire Helix UHF TV Transmitting Antenna," *Electronics,* August 1951, p. 107.

37 H. G. Smith, "High-Gain Side Firing Helical Antennas," *AIEE Trans.,* part I, *Commun. & Electronics,* vol. 73, May 1954, pp. 135–138.

38 R. H. DuHamel, "Circularly Polarized Helix and Spiral Antennas," U.S. Patent 3,906,506, Sept. 16, 1975.

39 H. A. Wheeler, "A Helical Antenna for Circular Polarization," *IRE Proc.,* vol. 35, December 1947, pp. 1484–1488.

40 A. G. Kandoian and W. Sichak, "Wide Frequency Range Tuned Helical Antennas and Circuits," *IRE Conv. Rec.,* part 2, 1953, pp. 42–47.

41 L. H. Hansen, "A New Helical Ground Plane Antenna for 30 to 50 MHz," *IRE Trans. Veh. Commun.,* vol. VC-10, August 1961, pp. 36–39.

42 D. A. Tong, "The Normal-Mode Helical Aerial," *Radio Commun.,* July 1974, pp. 432–437.

43 M. Eovine, "Helical-Monopole HF Antenna," Dept. of Army Proj. 3F37-01-001-04, Chu Associates, Oct. 15, 1962.

44 Y. Hiroi and K. Fujimoto, "Practical Usefulness of Normal Mode Helical Antenna," *Dig. Int. Symp. Antennas Propagat.,* 1976, pp. 238–241.

45 C. C. Kilgus, "Multielement, Fractional Turn Helices," *IEEE Trans. Antennas Propagat.,* vol. AP-16, July 1968, pp. 499–501.

46 C. C. Kilgus, "Resonant Quadrifilar Helix," *IEEE Trans. Antennas Propagat.,* vol. AP-17, May 1969, pp. 349–351.

47 C. C. Kilgus, "Resonant Quadrifilar Helix Design," *Microwave J.,* vol. 13, December 1970, pp. 49–54.

48 C. C. Kilgus, "Spacecraft and Ground Station Applications of the Resonant Quadrifilar Helix," *Dig. Int. Symp. Antennas Propagat.,* 1974, pp. 75–77.

49 R. W. Bricker and H. H. Rickert, " A S-Band Resonant Quadrifilar Antenna for Satellite Communication," *Dig. Int. Symp. Antennas Propagat.,* 1974, pp. 78–82.

50 A. Bystrom, Jr., and D. G. Berntsen, "An Experimental Investigation of Cavity-Mounted Helical Antennas," *IRE Trans. Antennas Propagat.,* vol. AP-4, January 1956, pp. 53–58.

51 J. Holland, "Multiple Feed Antenna Covers L, S, and C Band Segments," *Microwave J.,* vol. 24, October 1981, pp. 82–85.

52 H. E. King and J. L. Wong, "240–400 MHz Antenna System for the FleetSatCom Satellites," *Dig. Int. Symp. Antennas Propagat.,* 1977, pp. 349–352.

53 K. R. Carver, "The Helicone—A Circularly Polarized Antenna with Low Sidelobe Level," *IEEE Proc.,* Vol. 55, no. 4, April 1967, p. 559.

54 K. R. Carver and B. M. Potts, "Some Characteristics of the Helicone Antenna," *Dig. Int. Symp. Antennas Propagat.,* 1970, pp. 142–150.

Chapter 14

Frequency-Independent Antennas

Raymond H. DuHamel

Antenna Consultant

George G. Chadwick

Lockheed Missiles & Space Company, Inc.

The types of antennas described in this chapter are capable of bandwidths of operation that were believed impossible several decades ago. They represent the development of relatively simple yet powerful ideas.

The first idea[1] evolves from the observation that the impedance and pattern properties of an antenna are determined by its shape and dimensions expressed in wavelengths. If by an arbitrary scaling the antenna is transformed into a structure equal to the original one, its properties will be independent of the frequency of operation. The antenna then satisfies the *angle condition,* which means that its form can be specified entirely by angles only and not by any particular dimension. There are two classes of antennas satisfying this condition: *conical antennas,* made up of infinite cones of arbitrary cross section having a common apex; and *equiangular antennas,* with surfaces generated by equiangular spirals having a common axis and the same defining parameter. Since equiangular spirals are also called logarithmic spirals, the name *log-spiral antenna* is quite often used.

The second idea[2] is that if a structure becomes equal to itself by a particular scaling τ of its dimensions, it will have the same properties at frequencies f and τf. The consequence is that the antenna characteristics are a periodic function, with the period log τ, of the logarithm of frequency. Antennas obtained from this principle are called *log-periodic.* By making τ close enough to 1, the variation of the properties over the frequency band $(f, \tau f)$, and therefore everywhere, may be small for some types of structures. In practice, even with τ not very close to 1, good frequency-independent behaviors are observed.

Log-spiral antennas have been referred to as continuously scaled structures wherein the scaling is equivalent to a rotation of the structure. A scaling of $\tau = \exp(2\pi a)$, where a is defined later, leaves the structure unchanged (i.e., there is no rotation of the pattern) so that in this sense it is a log-periodic structure.

To conform to the conditions stated above, equiangular as well as log-periodic structures should extend from the center of expansion 0, which is also the feed point, to infinity. A practical antenna is obtained by taking a section of the ideal infinite structure contained between two spheres of center 0 and of radii r_1 and r_2 respectively.

Inside the smaller sphere is the feed region, where the infinite structure, which ideally should converge to the point 0, is replaced by some coupling mechanism to the feed transmission line or waveguide. The length r_1 determines the highest frequency of operation f_1 of the antenna by the condition that r_1 must be small enough, compared with the wavelength λ_1, to make the exact shape of the coupling mechanism have little influence on the impedance or the pattern of the antenna. Ultimately the dimension r_1 is fixed by the size of the waveguide or transmission line connected to the structure.

The outside dimension of the antenna, specified by the radius r_2, determines the lowest frequency of operation by the following considerations. It is observed, for some structures, that the currents (or the field components near the material boundaries of the structure) decrease rapidly with distance from the center. Thus it becomes possible to cut off the structure at the distance where these currents are so small with respect to the feed-point current that they can be suppressed without changing appreciably

---

*Portions of the material in this chapter were written by Professor G. A. Deschamps for the first edition of the *Antenna Engineering Handbook.*

the radiation pattern or the impedance. For a given frequency the distance at which this happens is proportional to the corresponding wavelength. Thus, the lower cutoff frequency may be decreased at will by increasing the size of the antenna.

The rapidity of decrease of the current with distance depends on the particular structure considered. For conical antennas excited at the center point, the field is transverse to the radial direction and decreases only as the inverse of the distance r to the center. When a finite portion of such an antenna, however large, is taken, the end effect will produce variations of the radiation pattern. Equiangular and some log-periodic antennas, on the other hand, exhibit a decrease of current with distance faster than $1/r$. The end effect becomes negligible at a distance in wavelengths that depends on the exact shape of the structure. Thus some antennas will be more compact than others for a given lowest frequency f_2 and a given relative importance of end effect.

A property common to both types of antennas and related to the absence of end effect is that the radiation field must be substantially zero in the direction of the conductors forming the structure, as seen from the center point.

Some relations between the two types of antennas should be noted. Equiangular antennas are special cases of log-periodic antennas with a period of log τ. This property is not destroyed if the cones of revolution on which the spirals lie are distorted or given different axes. For example, one cone may be flattened, or several equiangular spiral arms with the same apex but different axes may form an array having the log-periodic property.

A class of log-periodic elements is obtained by taking a section of solid equiangular structure by a plane passing through its axis. Besides reproducing themselves by expansion in the ratio τ, these structures will also reproduce by a reflection through the axis followed by an expansion in the ratio $\tau^{1/2}$. Antennas made of such elements usually exhibit smoother impedance characteristics than those that are simply log-periodic.

The Archimedean spiral antenna is another type of antenna that has extremely wideband characteristics which are essentially equivalent to those of the log-spiral antenna. The Archimedean spiral, with equal spacing between adjacent arms, can be considered as a spiral with the pitch angle varying with radius.

Although none of the finite-size antennas considered is truly frequency-independent over the range 0 to infinity, the name is justified because possibilities for increasing the bandwidth are almost unlimited. Without changing the design but simply by adding some turns to the spirals or some sections to the log-periodic antenna, any desired bandwidth can be achieved.

Strictly speaking, none of these antennas is frequency-independent over finite bandwidths because of the dispersive characteristic of the radiated fields. The phase of the radiated field of the circularly polarized spiral antennas is delayed an extra 360° per period of frequency because of the pattern rotation. The same phenomenon is observed for linearly and circularly polarized log-periodic antennas. The early history of the development of frequency-independent antennas is given in Ref. 3.

14-2 SPIRAL ANTENNAS

Both log-spiral and cavity-backed Archimedean spirals are discussed in the following subsections.

Modal Characteristics of Multiarm Spiral Antennas

Two-arm cavity-backed planar spirals and conical spirals are used to produce a unidirectional single-lobe pattern. Three or more arms may be used for direction-finding applications[4] and, in addition, to obtain larger bandwidths for dual-polarization spirals.

In general, consider an N-arm spiral with rotational symmetry such that a rotation of $2\pi/N$ rad about the axis of the conical spiral leaves the structure unchanged. Figure 14-1 shows the coordinate system for a left-hand four-arm spiral. Because of the rotational symmetry, it is convenient and very useful to decompose an arbitrary excitation into normal modes. There are $N - 1$ independent normal modes for an N-arm structure. For any mode M, the arms are excited with equal magnitudes and with phases varying as $-2\pi n M/N$, where n is the number of the arm. The variation of the radiation fields with the azimuth angle ϕ may be expressed as the summation of terms of the form $\exp[-j(M + q)\phi]$, where q is an integer. For a good spiral design the term for $q = 0$ dominates so that the azimuth field varies essentially as $\exp(-j M\phi)$. This occurs when there are only outward-traveling waves on the arms and large attenuation through the first active region.

The elevation patterns are also functions of the mode number. All modal patterns except $M = 1$ have nulls on the z axis. Figure 14-2 shows the computed elevation patterns[5] for the fundamental mode excitations of a planar-sheath spiral antenna. The minimum beamwidth that can be achieved for mode 1 is about 70°.

For a good spiral design, most of the radiation takes place in the first active region, which occurs in a small radial region on the cone where the circumference of the cone is approximately M wavelengths. This region is called an $M\lambda$ ring and occurs when the path-length differences to adjacent arms (same ϕ direction) are such that the currents are in phase. For the sum pattern mode ($M = 1$), radiation takes place in the λ ring. The pattern is approximately that of a traveling-wave current in a ring of one-wavelength circumference. For the first difference pattern ($M = 2$), the radiation occurs in the 2λ ring.

If the attenuation through the first active region is not large, as in the case of loosely wound spirals, then the waves will travel with little attenuation to the second active region, which is the $(M + N)\lambda$ ring (provided the spiral is large enough), where an azimuth pattern of the type $\exp[-j(M + N)\phi]$ is radiated. The total pattern is the superposition of the M and $(M + N)$ modes. The presence of the higher-order mode may be easily detected by measuring "conical cut" patterns about the boresight axis. The bandwidth over which the $(M + N)$

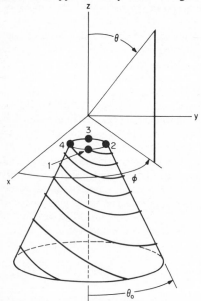

FIG. 14-1 A four-arm conical spiral and its associated coordinates.

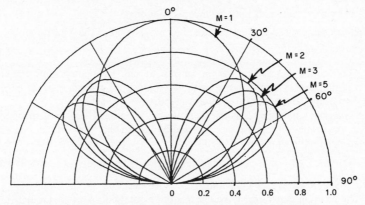

FIG. 14-2 Mode patterns for a planar-sheath log-spiral antenna.

λ ring does not exist increases with N, but this requires a more complicated feed network. If the spiral is too small to have an $(M + N)$ ring and there are no resistive terminations on the arms, then the waves are reflected and will radiate in the $(N - M)$ λ ring (if it exists). The azimuth variation will be exp $[j (N - M) \phi]$, and the sense of circular polarization will be reversed since the waves are traveling in the opposite direction in the ring. If the $(N - M)$ λ ring does not exist, the waves will flow back into the feed circuit.

The sense of circular polarization for spiral antennas may be determined as follows. Wrap the fingers of the right hand around the cone with the fingers pointing in the direction in which the currents are flowing. The spiral radiates right-hand circular polarization (RHCP) in the direction to which the thumb points and left-hand circular polarization in the opposite direction. The spiral of Fig. 14-1 is RHCP in the $\phi = 0$ direction, which is also the direction of the beam for mode 1.

Broadband direction-finding (DF) systems may be achieved by making use of two or more of the normal modes. These normal modes may also be used for conical arrays of log-periodic antennas.

Log-Spiral Antennas

General Characteristics   The structure of an antenna satisfying the angle condition must be such that if it is expanded in an arbitrary ratio τ about the feed point 0, the configuration obtained becomes congruent to the original one. Either it will coincide with the original structure, or it will coincide with a structure deduced from the original one by a rotation about some axis D passing through the point 0 (see Fig. 14-3).

The first condition is satisfied by conical structures having a common apex. Antennas obtained by taking finite sections of such structures (the discone antenna, for example) indeed have broadband properties. However, they are not frequency-independent in the sense discussed in a preceding subsection since any finite portion, however large, will show end effects. The bandwidth therefore cannot be increased at will by increasing the size.

The other type of structure (equiangular) has been characterized by Rumsey.[1] He has shown that the following conditions must be satisfied:

1   The axis of rotation D must be independent of τ.

2   The angle of rotation about D must be proportional to the logarithm of τ.

$$\tau = e^{a\phi} \qquad\qquad (14\text{-}1)$$

These equiangular structures, in contrast to the conical ones, show proper attenuation of the current with distance (with varying degree, according to the parameters of the structure) and lead to frequency-independent antennas.

Geometry   By using D as the z axis in a conventional system of spherical coordinates, the property may also be expressed by stating that the surfaces bounding the antenna must have equations of the form

$$F\,(\phi,\ re^{-a\phi}) = 0 \qquad\qquad (14\text{-}2)$$

where F is an arbitrary function and a is a constant.

The rotation expansion T^ϕ resulting from a rotation about D through the angle ϕ, followed (or preceded) by an expansion about 0, in the ratio $\tau = e^{a\phi}$ will carry the structure onto itself. (The exponential notation T^ϕ is justified by the obvious property $T^\phi\,T^\psi = T^{\phi+\psi}$.) Starting from an arbitrary point M, the point $T^\phi\,(M)$, as ϕ is varied, will describe an equiangular spiral, either in the plane $z = 0$ or on a cone of revolution having D as an axis. All surfaces of an antenna satisfying Rumsey's condition will be generated by such spirals having a common axis and the same parameter a.

Figure 14-3 shows an equiangular spiral arm. It is drawn on a cone of revolution about D,

$$\theta = \theta_0 \qquad\qquad (14\text{-}3)$$

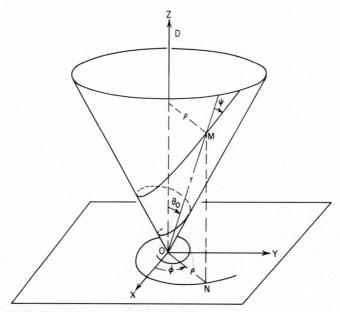

FIG. 14-3   Geometry of the log-spiral.

and its equation is

$$r = r_0 e^{a\phi} \qquad (14\text{-}4)$$

$$\rho = \rho_0 e^{a\phi} \qquad (14\text{-}5)$$

The spiral is called equiangular because it makes a constant-pitch angle ψ with the radius vector. The pitch angle is related to the cone angle and the expansion coefficient by

$$\tan \psi = \frac{\sin \theta_0}{a} \qquad (14\text{-}6)$$

This angle is the complement of the angle commonly defined as the pitch angle for helical antennas. The arc length s from the origin 0 is finite and proportional to the distance r.

$$s = \frac{r}{\cos \psi} \qquad (14\text{-}7)$$

The projection of the spiral on the xy plane is also equiangular with the same parameter a.

Pattern rotation   When the infinite structure is fed at the center by a constant-current source of unit intensity, the electric field at frequency f, at some point M, is $E(M, f)$. If the frequency is changed to $f = f/k$, which amounts to a scaling k of the dimensions, the field becomes

$$E\left(M, \frac{f}{k}\right) = \frac{1}{k^2} T^\phi E(T^{-\phi} M, f) \qquad (14\text{-}8)$$

where the rotation expansion T^ϕ corresponds to $\phi = (1/a) \ln k$. When the frequency is changed from f to f/k, the pattern is rotated about D through the angle $(1/a) \ln k$. For finite structures this has been verified experimentally over the bandwidth of operation of the antenna.

Practical structures   Simple practical structures are made of conducting wires or of metallic sheets. Although ideally the wires or the sheets should be tapered according to distance to satisfy the equiangular condition, in practice a wire of constant radius or a sheet of constant thickness may be used for moderately wide frequency bands (5:1 or 10:1).

The structures which are most often used are variants of the dipole (in which the two halves have been twisted into a pair of equiangular spirals extending from the feed point). Multiarm structures are also widely used.

Planar Structures   For mode 1 the planar structures usually are made of two symmetrical equiangular arms[6] as shown in Figs. 14-4 and 14-5.

Parameters defining the infinite structure are:

1   The rate of spiral $1/a$, which defines the expansion ratio corresponding to the rotation ϕ. It also determines the angle $\psi = \tan^{-1}(1/a)$ at which radial lines cut the spirals.

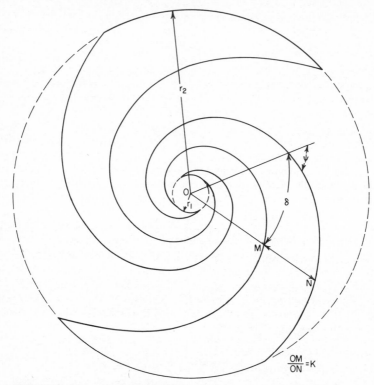

FIG. 14-4 Two-arm planar log-spiral antenna: outline and parameters.

FIG. 14-5 Two-arm log-spiral antenna cut in a metal plate.

FIG. 14-6 Feed-region detail for the antenna of Fig. 14-5.

14-8

2   The arm width specified either by an angle δ or by the ratio $K = e^{-a\delta}$. When δ > 90°, the structure may be considered as a slot of angular width 180° − δ rather than as a dipole.

The finite antenna is described by the cutoff radii r_1 and r_2 and the exact manner in which it is terminated outside this section. If the structure is large enough to allow complete radiation in a given frequency band, the exact shape of the termination does not matter.

The structures shown have bidirectional patterns with rotationally symmetric beams pointing normal to the plane of the structures. For most applications, which require a unidirectional pattern, a cavity is placed on one side of the spiral. The frequency-independent-pattern bandwidth in this case is limited to the order of an octave for lossless cavities. Decade bandwidths may be achieved by placing absorber in the cavities. This topic is expanded in the subsection "Cavity-Backed Archimedean Spiral Antennas."

The antenna may be fed by a balanced line brought into the center point along the axis. An alternative approach, which does not disturb the field as much, consists of using a coaxial cable embedded into one arm of the spiral. Because of the rapid attenuation of the fields with distance along the spiral, no appreciable current flows in the free parts of the cable. For symmetry a dummy cable is sometimes soldered to the other arm of the spiral in a similar configuration. This arrangement has been called an *infinite balun*. Figure 14-6 shows the feed region of the antenna shown in Fig. 14-5.

The decrease of currents on the antenna is less rapid for narrow-arm or narrow-slot structures and increases with the rate of spiral $1/a$. The phase velocity for the currents along the antenna is close to free-space velocity at the center and increases with distance.

Impedance   The impedance for the infinite equiangular structure is real and independent of frequency. This is also true of the impedance matrix if the structure has more than two terminals.

Practical antennas differ from this ideal because they are finite and because the sheets or the wires are not tapered. In spite of this difference, standing-wave ratios (SWR) with respect to the characteristic impedance of the spiral less than 1.5:1 may be obtained over the band of operation.

Thin two-armed symmetrical planar antennas without a cavity and with $\delta =$ 90°, which is self-complementary, should have an impedance of 60π Ω. Values observed experimentally (about 120 Ω) are lower because of the presence of the cable feeding the structure. Narrow-arm structures have higher impedances.

Radiation pattern   The measured patterns of a two-arm log-spiral are close to those calculated for an infinite-arm sheath spiral for mode 1 as shown in Fig. 14-2. The patterns are rotationally symmetric if a is less than about 0.2 (or ψ > 79°). For larger a the pattern shows asymmetry, and its rotation with frequency may be observed. The theoretical beamwidth varies from 88 to 76° as a varies from 1 to 0.1 and remains at 76° for smaller values of a.

Axial ratios of less than 2 dB may be obtained over the θ range of 0 to 70° for tightly wound spirals.

The currents on the antenna have odd symmetry with respect to the axis $D;$ the field everywhere has the same symmetry and therefore does not contain any multipole components with $e^{jm\phi}$ variation, where m is even.

Design considerations The problem is to design an antenna of the type described with frequency-independent behavior over a band f_2 to f_1. The high-frequency cutoff f_1 being given, the radius of the feed region must be small compared with the wavelength ($r_1 \approx \lambda/14$, for example). If the feed line is a coaxial cable embedded in one of the arms, the arm width at the radius r_1 must be at least equal to the coaxial-cable diameter d.

The low-frequency cutoff f_2 is linked to the attenuation along the spiral, which in turn depends on a and δ. The diameter of the antenna in wavelengths, $2r_2/\lambda_2$, at the lower edge of the band becomes a function of a and δ once the performance of the antenna at frequency f_2 has been specified. It is somewhat greater than λ_2/π. Detailed design information is given in Ref. 7, page 18-8, and in the subsection "Cavity-Backed Archimedean Spiral Antennas."

The ratio $(r_2/r_1)/(f_1/f_2)$ = dimension ratio/frequency ratio may be taken as a figure of merit of a particular structure, indicating how compact the antenna can be made for a given bandwidth. The ratio is normally close to 1 and larger.

Conical Structures Multiple-arm conical spiral antennas are employed to obtain unidirectional patterns without the use of a cavity or a reflector. Excellent computer programs[8,9,10] that calculate the patterns, current distribution, and impedance have been developed. When fed from the apex, the phasing of the currents in the arms along a radial line on the cone is such that radiation first takes place in a backfire mode, which is essential for frequency-independent operation. Most research and development efforts have been expended on conical log-spiral antennas, but Archimedes-type conical spiral antennas with uniform spacing between arms can also be used. Half-cone angles of 25° or less are usually required to achieve a good unidirectional pattern.

Two-arm conical spirals fed in a balanced manner (mode 1) are commonly used to provide a single-lobe unidirectional circularly polarized pattern. Four-arm conical spirals fed in mode 2 ($+, -, +, -$ excitation) are used to provide a circularly polarized, rotationally symmetric difference pattern. When the axis of the cone is oriented vertically, the antenna produces wideband omnidirectional circularly polarized coverage. Conical spiral structures with four or more arms may be excited by networks of hybrid circuits (similar to Butler networks) to provide sum and difference patterns or tilted beams for DF and homing applications.

Two-arm conical spirals Extensive experimental[11] and theoretical[9] results have been reported. Figure 14-7 shows a two-arm spiral consisting of

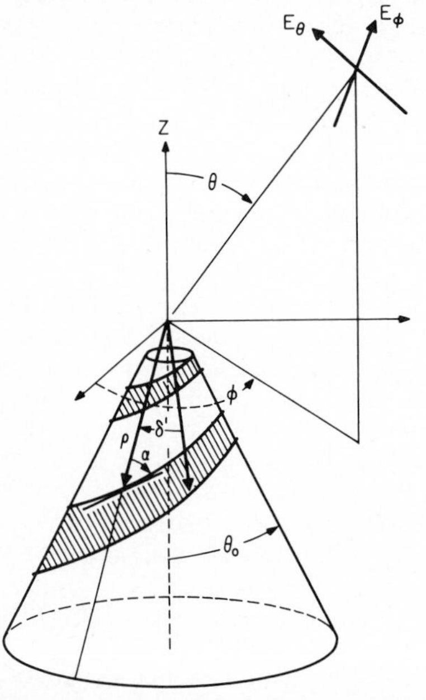

FIG. 14-7 Parameters of the conical log-spiral antenna. (© *1965, IEEE.*)

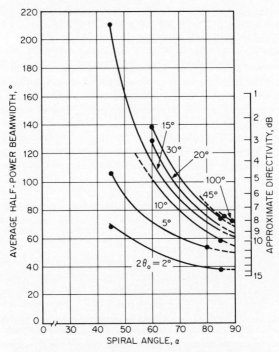

FIG. 14-8 Average half-power beamwidth and approximate directivity of conical log-spiral antennas ($\delta = 90°$). (© *1965, IEEE.*)

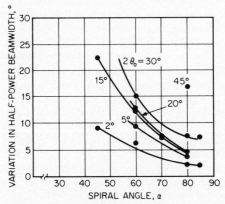

FIG. 14-9 Approximate variation in half-power beamwidth for wideband operation. (© *1965, IEEE.*)

angular strips lying on the surface of a cone and defines the various parameters which specify the spiral structure. Tapered rods can be used for the arms, and even constant-diameter rods may be used for surprisingly wide bandwidths.

Figure 14-8 shows average half-power beamwidth (HPBW) and directivity as a function of the spiral pitch angle for various cone angles as measured by Dyson.[11] The variation in HPBW to be expected over wide bandwidths is shown in Fig. 14-9. Notice that the variation decreases as the pitch angle increases. The variation is due to radiation from regions other than the one-wavelength ring, such as the three-wavelength ring, which produces a nonrotationally symmetric pattern. Pitch angles greater than 75° are required for small variations. However, this necessitates a tightly wound spiral with longer arms than for smaller pitch angles. The front-to-back ratio of the unidirectional pattern increases with pitch angle and decreases with increasing cone angles. For $\alpha = 80°$, the front-to-back ratio increases from 6 to 25 dB as $2\theta_o$ decreases from 45 to 15°. The axial ratio is on the order of 2 to 4 dB for θ varying from 0 to 90° for a pitch angle of 45° with a $2\theta_o = 20°$ cone. For larger pitch angles the axial ratio approaches 0 for $\theta = 0$ but increases rapidly for $\theta > 60°$ (such as 4 dB or more). This is caused by the predominance of horizontal-polarization radiation from the more horizontal arms of tightly wound spirals.

The radiation mechanism of the spiral may be explained as follows. A generator excites out-of-phase traveling-wave currents at the vertex of the two-arm spiral. These currents travel in a nonradiating or transmission-line mode until they reach the *active region*. There is little radiation since the currents in adjacent arms are approximately out of phase. In the active region, currents in adjacent arms are nearly in phase. Calculations by Mei[10] have shown that attenuation in the active region is in the range of 7 to 10 dB per wavelength along the spiral arms. Dyson[11] has shown that if the cone is truncated at a point where the total attenuation in the active region is 15 dB, then radiation and impedance characteristics are essentially the same as those of the infinite structure. This point occurs when the circumference of the cone is somewhat less than 1.5λ, depending upon α and θ_0. The transmission-line region may be truncated at a point where the circumference of the cone is less than 0.4λ. Detailed information for designing spirals for specified bandwidths is given by Dyson.[11]

Figure 14-10 shows the characteristic impedance of the conical spiral as a function of the angular arm width δ for several cone angles and one with an infinite-balun feed. In the latter case the addition of the coaxial line to the arm reduces the characteristic impedance. A voltage standing-wave ratio (VSWR) less than 1.5:1 with respect to the Z_0 of the spiral may be obtained over the bandwidth.

Four-arm conical spirals   A four-arm spiral fed in mode 2 $(+, -, +, -$ excitation) produces a difference pattern with rotational symmetry,[12] provided there is sufficient attenuation through the active region, which implies pitch angles greater than 60°. Figure 14-11 shows the computed direction and beamwidth[10] of the beams versus secant α. To obtain good coverage in the horizontal plane (ψ close to 90°), α must be in the range of 45 to 50° for $2\theta_0 \approx 20°$.

A photograph of an omnidirectional conical spiral is shown in Fig. 14-12. The antenna is fed by a four-conductor system running up the axis of the cone, which is excited by the feed network shown in the right side of the photograph. For pitch angles of less than 60°, there is considerable end effect so that the azimuth and elevation patterns are scalloped and the axial ratio is increased. However, the patterns are satisfactory for surveillance applications. A VSWR < 2:1 may be achieved over wide bandwidths.

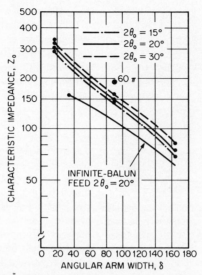

FIG. 14-10 Approximate characteristic impedance of the conical log-spiral antenna as a function of the angular arm width. (© *1965, IEEE.*)

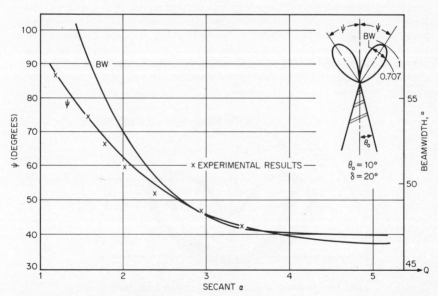

FIG. 14-11 Half-power beamwidth and beam orientation versus secant α for mode 2, $\theta_0 = 10°$, $\delta = 20°$. (© *1971, IEEE.*)

FIG. 14-12 Four-arm conical log-spirals excited in mode 2 for circularly polarized omnidirectional pattern, 2 to 12 GHz. *(Courtesy of GTE Sylvania Systems Group.)*

Measured mode 1 (sum) and mode 2 (difference) patterns,[4] with rotational symmetry, are shown in Fig. 14-13 for a four-arm conical spiral with $2\theta_0 = 20°$ and $\alpha = 80°$. A broadband hybrid-feed network may be used to provide two isolated inputs for modes 1 and 2. Since the phase of the azimuth patterns varies as ϕ and 2ϕ for modes 1 and 2 respectively, the azimuth direction of arrival may be determined by measuring the relative phase of the two modes. Ideally, the elevation direction of arrival may be determined by amplitude comparison of the two modes. Problems are involved in the azimuth direction finding. First, since the patterns rotate with frequency, a computation must be made to take this into account. Second and more serious, the phase centers for modes 1 and 2 are located approximately at the one- and two-wavelength rings of the conical spiral. Thus, the relative phases of modes 1 and 2 are a function of the elevation angle. These problems and others are discussed for multiple arm structures in Ref. 4.

Cavity-Backed Archimedean Spiral Antennas

General Characteristics   The cavity-backed spiral can properly contend as one of the most proliferate radiators since its conception in early 1950. Its unidirectional, broad-bandwidth characteristics are attractive features, particularly when combined with its mechanical simplicity and compact size. The general features of this radiator are shown in Fig. 14-14. The surface etching may be equiangular or archimedean. The cavity may be a hollow metal-based cylinder for near-octave bandwidths or absorber-loaded (at the expense of insertion loss) for a greater-than-decade band-

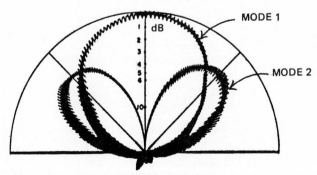

FIG. 14-13 Electric field radiation patterns for mode 1 and 2 excitations of a conical four-arm log-spiral antenna; $2\theta_0 = 20°$, $\alpha = 80°$, $\delta = 45°$. (© *1971, IEEE.*)

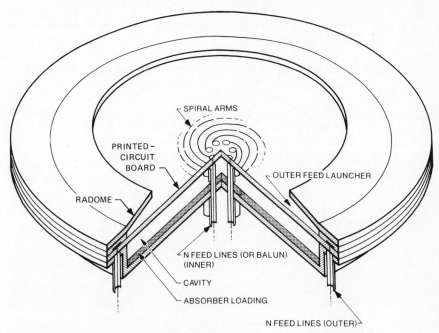

FIG. 14-14   Cavity-backed archimedean spiral.

width. The balun may be infinite, printed-circuit, lumped-constant, waveguide-fed, or matrix-excited. Dual circular polarization may be achieved by feeding both the central and the peripheral spiral terminals, and the number of arms may be varied from 2 to N, thereby allowing $N - 1$ instantaneous patterns (of one polarization).

The log-spiral radiator was described earlier. It differs from an archimedean spiral in that the separation between windings depends on the radius ρ to the winding traces. The archimedean spiral trace is defined by

$$\rho = a\phi + b \qquad (14\text{-}9)$$

where the coordinates are identified in Fig. 14-15. The two curves defining a spiral arm have the same value of a but different values of b (i.e., b_1 and b_2). The N conductors are identical except that their starting points in θ are displaced by $2\pi/N$. The radiator line width W for the self-complementary case is given by

$$W = b_2 - b_1 = \frac{a\pi}{N} \qquad (14\text{-}10)$$

Two-arm archimedean-spiral characteristics   Spirals radiate a unidirectional sum mode when the phase progression through all feed terminals is 2π or, more directly, the interarm phase progression is $2\pi/N$ (or π rad in the case of a two-arm spiral). The sum mode will radiate most efficiently at a diameter of λ/π independent of the number of arms. This condition occurs as a result of the in-phase condition of adjacent arms which occurs at this circumference. The derivations by Mei[10] et al. may be used to estimate approximately the archimedean spiral's performance if

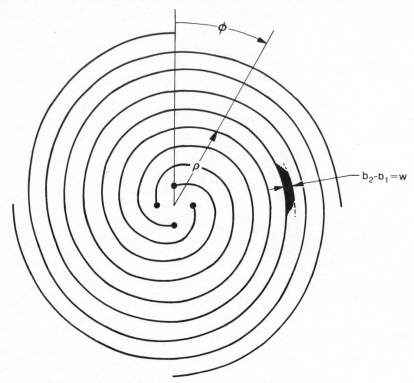

FIG. 14-15 Geometry of the archimedean spiral.

1 The same value of a is used.

2 The method of images is applied to allow for the cavity base.

Unfortunately, the cavity walls, mounting flange, and balun assembly all influence the pattern. It is also true that the image radiator must retransmit through the original spiral, further modifying the predicted results.

An empirical estimation of the two-arm archimedean spiral is provided by Wolfe[13] et al. Figure 14-16 illustrates the impact of cavity diameter on gain and axial ratio. The gain curve has a knee slightly above λ/π but does not asymptote until a cavity diameter of 0.5λ is reached, implying an active radiating band of 0.18λ width. The gain is relative to that of a linear isotropic source and is normalized to the average axial ratio. These data, as well as those of Fig. 14-17, represent a compilation from a large number of spirals operating in the 0.2- to 4.0-GHz region. Figure 14-16b illustrates the cavity diameter's effect on axial ratio for the case in which the end terminations are loaded. Deterioration of the axial ratio with diminishing diameter is not surprising, since energy reflected from the end points radiates the opposite sense of circular polarization compared with the desired polarization excited from the center of the antenna.

Figure 14-17a shows that the effect of cavity depth is approximately as predicted

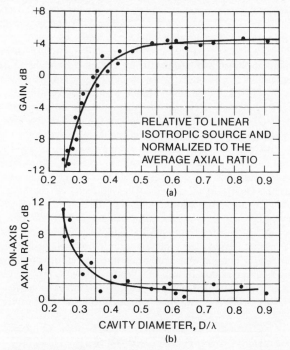

FIG. 14-16 Effect of cavity diameter on gain and axial ratio.

by image theory. This curve shows a maximum gain at a cavity depth of 0.25λ, dropping rapidly for depths greater than 0.4λ. Increasing the number of windings increases the radiation efficiency slightly, as shown in Fig. 14-17b. These probe data show the radial field for both an 8-turn and a 16-turn cavity-backed spiral on a 4.6-in- (116.8-mm-) diameter surface. The peak field magnitude occurs at an equivalent circumference of 0.85λ in both cases, but the 16-turn spiral clearly has more in-band energy

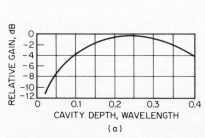

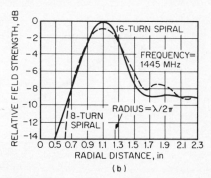

FIG. 14-17 Effect (a) of cavity depth on gain and (b) of two windings on field distribution.

near the 1λ circumference. Two-arm spirals typically can operate over an octave with SWR levels of less than 2.0, axial ratios of less than 2 dB, and beamwidths and pattern shapes typified by those shown in Fig. 14-18.

Broadband characteristics with absorber loading Two-arm spiral cavity radiators are limited to an octave bandwidth by the metallic cavity base. However, if absorbing material is interposed between the cavity base and the radiating surface, as suggested by Fig. 14-14, greater-than-decade bandwidth is achievable.

Figure 14-19 depicts a typical measured pattern obtained with an absorber-loaded 2- to 18-GHz spiral. The modulation on the pattern envelope represents the axial ratio at any given point. The quality of the patterns is excellent, and axial ratios of less than 2 dB over the full decade are achievable. While absorber removes the cavity-imposed bandwidth limitation, it results in an insertion loss compared with the unloaded-cavity case. This loss depends on the percentage of the cavity occupied by the absorber, the cavity depth, and the characteristics of the absorber itself. The absorber layer generally fills the lower volume of the cavity to within 80 or 90 percent of its depth. Absorber losses in these instances range from approximately 1 dB for 0.15λ depths to near 3 dB for frequencies at which the cavity depth exceeds 0.5λ.

Balun exciters Baluns are typically required to achieve the plus-minus phase excitation needed for unidirectional radiation in the axial direction. An octave-bandwidth balun was described by Bawer and Wolfe;[14] they utilized a hybrid microstrip-stripline printed-circuit implementation. This balun, based on earlier work by Roberts,[15] can provide an SWR of less than 1.7 over a full octave for cavity-backed spirals

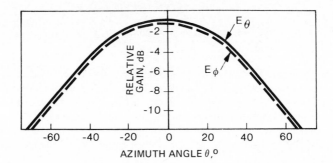

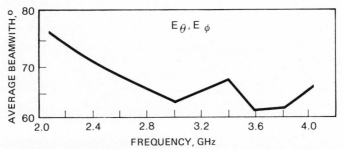

FIG. 14-18 Typical pattern shapes and beamwidths of two-arm archimedean cavity-backed antennas.

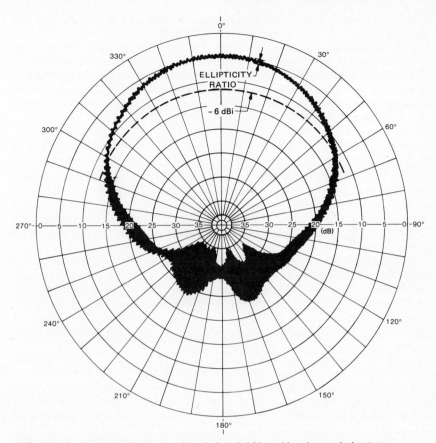

FIG. 14-19 Typical patterns of a decade-bandwidth archimedean spiral antenna.

in the 0.1- to 10-GHz region, yet it is formed of a single photoetched part. Lumped-circuit configurations are more size-practical below the lower limit.

Discontinuity effects rapidly become more troublesome above 10 GHz. Furthermore, the radiating ring above 10 GHz is physically small because the balun must confine itself to a diameter of less than 0.14λ to avoid introducing pattern perturbations. Saul[16] has developed a microstrip infinite balun which allows SWR levels of less than 2:1 to 40 GHz. One of the spiral arms is fed from its outer periphery by a coaxial-cable microstrip transition. The microstrip conductor uses the spiral radiating arm, printed on the opposite side of the substrate, as its ground plane. At the center point, the microstrip conductor simply transforms into the second arm to form the infinite balun.

Very-wide-band baluns are required for absorber-loaded cavity-backed spiral antennas. Many such radiators have used coaxial implementations based on the designs proposed by Marchand.[17] This balun provides excellent results for 10:1 bandwidths provided the upper limit is less than 20 GHz. SWR levels of less than 3:1 have been achieved by Greiser[18] et al., who also describe an alternative printed-circuit version which operates from 0.4 to 18 GHz with an SWR of less than 2:1 and with excellent phase ($\pm 4°$) and amplitude (± 0.5) dB balance.

Polarization diversity and axial ratio   Spirals radiate one sense of polarization when excited from their central feed points and the opposite sense when excited from their peripheral terminations. Energy which radiates when fed from the central terminals must be totally absorbed by terminations at the spiral ends to avoid deterioration of the axial ratio. The normalized voltage axial ratio A is approximately defined by

$$A = \frac{1 + (1 + \eta)^{1/2}\rho t}{1 - (1 - \eta)^{1/2}\rho t} \qquad (14\text{-}11)$$

where η is the radiation efficiency and ρ_t is the reflection coefficient at the termination. Equation (14-11) may be used to estimate the radiation efficiency of the spiral antenna by measuring the axial ratio for both an open-circuited ($\rho_t = 1$) and a terminated case and then solving for η. The sensitivity of a spiral axial ratio to end conditions can be estimated from knowledge of the spiral current distributions by using the method of Mei[10] et al.

Terminating loads may be achieved by microstrip transitions to matched coaxial loads. These terminations are usually located on a peripheral flange with the spiral etching on the top surface forming the microstrip conductor, the spiral substrate forming the separating dielectric, and the flange forming the ground plane. This method is adequate to at least 2 GHz but is awkward at higher frequencies. In these latter instances, any of a variety of resistive paints or cards applied to a termination length of at least one-half wavelength will provide an acceptable load. An alternative method, described by Yamauchi[19] et al., uses a zigzag or sawtoothed section of approximately π-rad length for the last section of the spiral radiating arms.

Dual-polarized radiation may be obtained by simultaneously feeding the spiral at both its central and its peripheral terminals. However, the spiral size for a two-arm structure should not exceed a diameter of $2\lambda/\pi$ to ensure against the generation of undesired modes when feeding from the outer rim. Given the earlier-cited radiating width (approximately 0.18λ), the operating bandwidth is restricted to $(2\lambda/\pi)/(\lambda/\pi + 0.18\lambda)$, or 1.28:1. More bandwidth is achievable by increasing the number of spiral arms, a subject to be discussed in later paragraphs.

Figure 14-20 illustrates a practical method for peripherally feeding the outer arms of a spiral. The spiral winding size gradually makes a transition from its radiating width to a stripline width which matches a desired characteristic impedance. This transition should be at least one-half wavelength long from the point where the spiral arm is just in the radiating region to the point where it is just inside the stripline region. The flared cavity-wall transition of Fig. 14-20 is a further aid to minimize mismatch. Since both inner and outer terminals of the spiral arms can be excited simultaneously, it is possible to provide polarization diversity by combining both outputs into a single port by using a variable coupler and by further providing a variable phase shifter in one arm.

Multiple-arm cavity-backed spirals   An N-arm spiral can excite N-1 pattern modes. The total phase progression across all spiral arms is $2\pi M (M = 1, 2, \cdots N - 1)$. Thus, the interarm phase progressions have the value $2\pi M/N$. When $M = 1$, the result is always a unidirectional pattern independent of N. All other pattern modes are symmetric about the boresight axis and have a null on axis. Aside from the uniqueness of the $M = 1$ or sum-channel mode, all modes are principally differentiated by

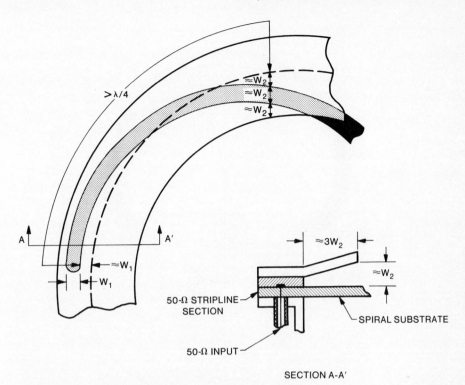

FIG. 14-20 Outside-feed detail for a dual-polarized spiral.

their phase progression about the normal axis (θ variant). This phase, as a function of mode number, was defined early in this section.

The secondary patterns may be computed by using the methods of Mei,[10] provided the corruptive effects of the ground plane, cavity edges, and balun structure are included. Figure 14-2 contained a computed set of patterns through the normal axis for the fundamental modes.

The first two modes are commonly used in a cavity-backed spiral to feed a paraboloid. The resultant sum and difference patterns form a two-channel monopulse pair. These outputs may be processed in the manner suggested by Shelton and Chadwick[20] to formulate a tracking system. Figure 14-21 illustrates typical measured patterns for a spiral-cavity four-arm center-fed antenna which operates about a center frequency of 15 GHz. The pattern modulation is a measure of the ellipticity ratio at any given pattern angle. When used in a paraboloidal reflector, the element patterns provide excellent secondary pattern characteristics for both sum and difference modes. Yaminy[21] et al. have achieved greater-than-20-dB sidelobes for both sum and delta patterns over an octave. These investigators utilized a six-arm inner- and outer-fed cavity-backed spiral to excite a 6-ft (1.8-m) paraboloid. This performance cannot be matched by a simple four-horn monopulse array in either its sidelobe-level or its bandwidth characteristics.

For self-complementary multiple-arm planar spiral (and log-periodic) structures the characteristic impedance of each arm with respect to ground is given by

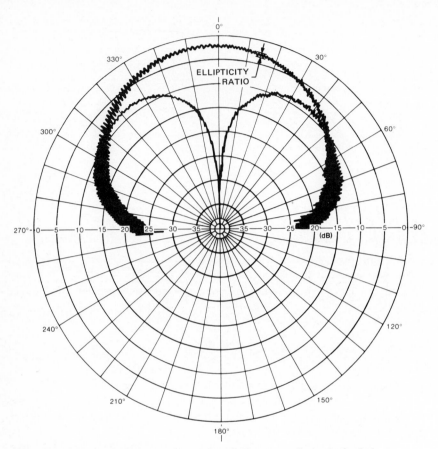

FIG. 14-21   Sum and difference patterns from a four-arm cavity-backed spiral.

$$Z_M = \frac{30\pi}{\sin\left(\pi\,\dfrac{M}{N}\right)}$$

where M is the mode number and N is the number of arms. This applies to thin-sheet structures for free-space conditions. Dielectric substrates will lower the impedances. Refer to Ref. 7 for a more detailed discussion, which includes simultaneous excitations of several modes.

The characteristic impedances are different for each mode, and compromises are necessary in selecting a feeder impedance. For example, a reasonable compromise for a four-arm spiral operating in the $M = 1$ and $M = 2$ modes, for which Z_1 and Z_2 are 133.3 and 94.2 Ω respectively, would be 112 Ω at the feed point. This value would limit the inherent mismatch to less than 1.19 for both modes. Broadband transformers are used to match to the characteristic impedance of the feed network.

When multimode spirals are center-fed, the excitation of higher-order modes on a large spiral surface can be avoided by proper structure design to ensure full radiation in the desired mode. When these spirals are fed from their periphery, however, as in the case of dual-mode spirals, the higher-order radiating bands are encountered first and thus impose a bandwidth limit. While the first-order radiating rings of a spiral are $M\lambda$, the next-order radiating-band circumference for the same excitation is $(M + N)\lambda$, with the boundary between the undesired and the desired modes occurring at a mean circumference of $(M + 0.5 N)\lambda$. The circumference required for the highest-order mode desired (M_h) at the lowest frequency f_ℓ is $(M_h + \pi\Delta)\lambda_\ell$, where Δ is the radiating width of the spiral ring. At the high end of the band the maximum allowable size occurs for the lowest-order mode (M_ℓ), or $(M_\ell + 0.5 N)\lambda_h$; thus, the bandwidth $(\lambda_\ell/\lambda_h = B)$ is related to the number of arms by the expression

$$N \geqq 2B(M_h + \pi\Delta) - 2 \qquad (14\text{-}12)$$

where the lowest mode M_ℓ is assumed as 1. For $\Delta = 0.18$ and for a dual-mode spiral $(M_h = 2)$, then

$$N \geqq 5.13B - 2 \qquad (14\text{-}13)$$

Equation (14-13) states that approximately eight arms are required to generate an octave-band, dual-mode, dual-polarized spiral which is totally free of higher modes.

An interesting variant of a dual-polarized spiral is described by Kim and Dyson,[22] who excited a four-arm spiral with both $M = 1$ and $M = 3$ modes. The spiral is designed to radiate only the $M = 1$ mode when center-fed. Thus, the $M = 3$ mode reflects from the periphery and radiates at the one-wavelength circumference, but with the opposite sense of polarity.

An N-arm spiral requires an N-arm matrix. Fortunately, Shelton,[23]

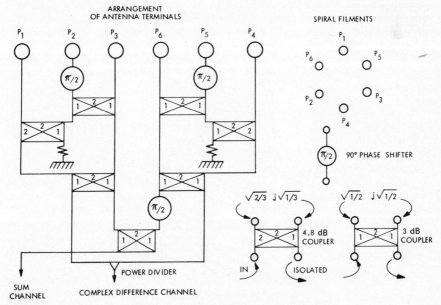

FIG. 14-22 A six-port symmetric network for exciting a six-arm spiral.

McFarland,[24] et al. have developed such devices. The most common of these are for $N = 2$ (or a balun), 4, and 6, although other values of N are realizable. An interesting variant of these networks consists of symmetric networks such as those developed by Chadwick and Shelton.[25] These networks provide perfectly stable difference-channel nulls independent of theoretical coupling or phase-shifter errors. A six-port example of these devices is pictured in Fig. 14-22. This network requires only mechanical symmetry in paths leading to opposite arms to achieve theoretically perfect null stability.

14-3 LOG-PERIODIC ANTENNAS

The first successful log-periodic (LP) antenna was introduced by DuHamel[2] and was a self-complementary planar-sheet structure consisting of two angular strips supporting curved teeth. It produced a bidirectional pattern. Isbell[26] then demonstrated that a unidirectional pattern could be obtained by using a nonplanar arrangement of the two halves of the antenna. Then it was found that the self-complementary condition was not required, and a variety of sheet and wire LP designs were introduced by DuHamel and Ore.[27] Isbell[28] then introduced the LP dipole array, the most commonly used LP antenna. Since that time, many types of LP antennas have been studied with varying degrees of intensity. Space does not permit a discussion of all the types or a presentation of detailed design data.

General Characteristics   The geometry of LP antenna structures is chosen so that the electrical properties must repeat periodically with the logarithm of the frequency. Although this appears to be a backward approach to the broadband-antenna problem, nevertheless frequency independence can be obtained when the variation of the properties over one period, and therefore all periods, is small. The LP design principles are illustrated in Fig. 14-23, which shows a metal-sheet LP structure with trapezoidal

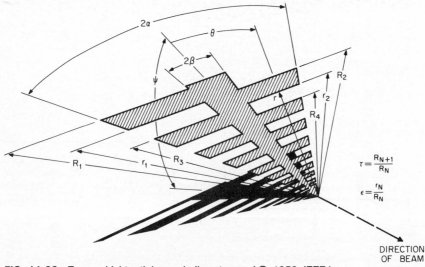

FIG. 14-23   Trapezoidal-tooth log-periodic antenna. (© *1958, IEEE.*)

teeth. The *H*-plane array of the two LP elements are fed against each other by a generator placed between their vertices. The four sets of teeth are defined by similar curves, the equations for which may be written in polar coordinates as $\theta = g(r)$, where $g(r)$ is some function of r. If θ is plotted versus $\ln r$ in rectangular coordinates, that is, $\theta = f(\ln r)$, then LP principles demand that f be a periodic function. (A more general but rather academic definition of LP structures for nonplanar and three-dimensional structures is given by DuHamel.[29]) Let $\tau = R_{n+1}/R_n$, where R_n is the distance from the vertex to the outer edge of a tooth. Let $r_n = \varepsilon \, R_n$. Then ε is a measure of the tooth width (the figure is drawn for $\varepsilon = \tau$, but usually $\varepsilon > \tau$). It may be seen that the logarithmic principle implies two conditions. The first is that all similar sets of dimensions, such as R_1, R_2, R_3, etc., must form a geometric sequence with the same geometric ratio τ. The second is that angles are used to a considerable extent in defining the antenna. For example, the extremities of the teeth and the triangular supporting section of the teeth are defined by angles.

Now if the antenna structure of Fig. 14-23 were infinitely large and infinitely precise near the feed point, it is easily reasoned that the structure must look exactly the same to the generator every time that the frequency is changed by the factor τ. The current distribution on the structure at τf_0 is identical with that at τ_0 except that everything is moved out one step along with a 180° rotation. Thus, since the wavelength has changed by a proportionate amount, the input impedance must be the same at τf_0 and f_0. Thus, the input impedance varies periodically with the logarithm of frequency with a period of $\ln (1/\tau)$. Because of the special left-right asymmetry of the structure, the period of the radiation pattern is $2 \ln (1/\tau)$ rather than $\ln (1/\tau)$.

If the variation of the impedance and pattern is small over a period and, therefore, over all periods because of the repetitive characteristics, it is seen that the result is essentially a frequency-independent antenna. Fortunately, some finite log-periodic structures provide frequency-independent operation above a certain low-frequency cutoff, which occurs when the longest tooth is approximately one-quarter wavelength long. Frequency-independent operation above the cutoff frequency is possible because some log-periodic antennas display little end effect. It is found that the currents on the structure die off quite rapidly past the region where a quarter-wave tooth exists. This means that a smaller and smaller portion of the antenna is used as the frequency is increased, which is another way of saying that the effective electrical aperture (the aperture measured in wavelengths) is essentially independent of frequency. Since it is not possible to extend the antenna to the origin because of the presence of the feed transmission line, a high-frequency cutoff occurs when the shortest tooth is less than one-quarter wavelength long.

The structure of Fig. 14-23 is horizontally polarized and has a bidirectional beam for $\psi = 180°$, with the beams pointing into and out of the paper. If the input impedance is plotted on a Smith chart over a frequency range of several periods, it will be found that the locus forms an approximate circle with the center lying on the zero-reactance line. The characteristic impedance of the log-periodic structure is defined as the geometric mean of the maximum and minimum real values on the locus. The VSWR referred to this characteristic impedance is then simply equal to the ratio of the maximum impedance to the characteristic impedance.

The operation of this antenna is explained briefly as follows. The two angular strips of width 2β form a uniform biconical TEM transmission line. The generator excites a wave on this line, which in turn excites currents on the *monopoles* (trapezoidal-shaped monopoles). Most of the radiation takes place in the active region where the monopoles are about $\lambda/4$ long. Attachment of monopoles to opposite sides of a strip introduces a 180° phase shift of the currents in the monopole. The additional phase delay in the transmission line between monopoles is sufficient to produce back-fire radiation, which is essential for frequency-independent performance.[30] Each of the two elements produces a unidirectional pattern pointing in the direction of the element. The region before the active region is called the transmission-line region, where there is little radiation. The monopoles present shunt capacitive loading of the line. Near the feed point, the loading per unit length is constant, and the characteristic impedance of the antenna is determined by the amount of the loading. It is always lower than the characteristic impedance of the two angular strips, which are commonly called the feeder.

Log-Periodic Dipole Antennas   If β approaches zero and ε approaches unity for the trapezoidal-tooth structure, then the antenna becomes two monopole arrays supported by thin angular strips. The strips form an angular transmission line whose characteristic impedance is independent of r. If ψ approaches zero, the antenna becomes a dipole array excited by a balanced transmission line. Rather than use tapered transmission-line strips, it is preferable to use constant-width strips or rods[28] with constant spacing, illustrated in Fig. 14-24. It is the most popular and most thoroughly studied LP antenna. A schematic circuit for the LP dipole array is illustrated in Fig. 14-25, wherein transpositions of the feeder between dipoles are shown. The parameter σ is defined as

$$\sigma = \frac{(1 - \tau)}{4} \cot \alpha \qquad (14\text{-}14)$$

and is the spacing in wavelengths between a $\lambda/2$ dipole and the adjacent shorter dipole. Once two of the three parameters α, τ, and σ are specified, the third may be determined from this formula. The ratio of the half-dipole length to the radius of the dipole is defined as h/a, and the transmission-line (feeder) characteristic impedance is Z_{0f}. The antenna may be considered as a cascade of cells consisting of sections of transmission line shunt-loaded at the center by the dipole impedance (taking into account the coupling between dipoles). For small α and large τ, there is essentially a traveling wave propagating away from the generator on the left. The transmission-line region of the antenna is the first portion where the dipole lengths are considerably less than a half wavelength. Because of the phase reversal between dipoles and the small spacing, the dipoles are phased to radiate in the invisible region, which means that

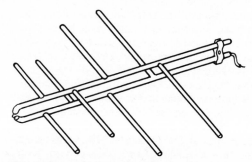

FIG. 14-24   Log-periodic dipole antenna.

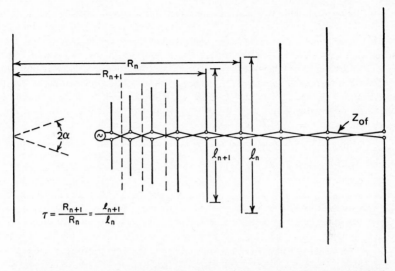

FIG. 14-25 Schematic diagram of a log-periodic antenna. The three dashed lines indicate typical locations of electric ground planes between each pair of dipoles.

there is little radiation. The first active region occurs when the dipoles are about one-half wavelength. Radiation occurs in the backfire direction[30] (to the left) because of the phase reversal between dipoles. If the phase reversals are not included, radiation will occur in the end-fire direction (to the right), which illuminates the larger portions of the array and produces scalloped patterns and erratic impedance behavior. This is called *end effect* and has been observed many times.

The attenuation through the active region due to radiation and reflection loss is a complicated function of τ, σ, Z_{0f}, and h/a. The shunt loading (or attenuation) of the line by a resonant dipole increases with the feeder impedance, and the bandwidth over which the dipole loads the line is inversely proportional to the Q of the dipole, which increases with a/h. The attenuation increases as τ is increased, provided σ does not become very small. An attenuation of about 20 dB through the active region is desired to keep the end effects small. The energy that passes through the active region will radiate to some extent in the region where the dipole lengths are between $\lambda/2$ and $3\lambda/2$, but predominantly in the second active region, where the dipoles are $3\lambda/2$ long. The radiation direction will depend on τ and σ and usually produces frequency-dependent patterns. If the frequency is such that the longest dipole is less than $3\lambda/2$ long, then the energy will be reflected (unless there is a resistive termination on the antenna) and will radiate in the active region in the backfire direction, which produces a back-lobe in the antenna pattern.

The directivity of the antenna increases with the length of the active region, which increases as τ as increased with σ held constant (α is decreased), since there are more dipoles near a length of $\lambda/2$. The length of the active region increases as Z_{0f} is decreased, since the attenuation per cell is decreased. The E- and H-plane beamwidths vary with α approximately as shown below in Fig. 14-29 for a single-wire trapezoidal-tooth LP element.

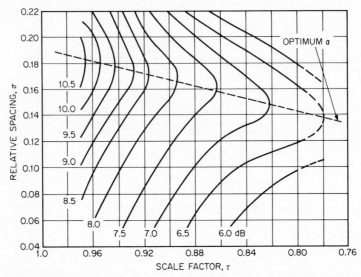

FIG. 14-26 Constant-directivity contours in decibels versus τ and σ; Z_{0f} = 100 Ω, h/a = 125. Optimum σ indicates maximum directivity for a given value of τ. (*After Ref. 31.*)

Carrel[31] performed the first and an extensive theoretical study of the LP dipole array wherein he assumed sinusoidal currents on the dipoles. Later, Cheong and King[32] and De Vito and Stracca[33,34] performed theoretical studies by assuming a three-term current distribution, which they claimed gave much better accuracy and gains of 1 or 2 dB less than that reported by Carrel. Still later, Butson and Thomson[35] pointed out that Carrel had made an error in the calculation of the E-plane pattern. When it was corrected, they showed that the gain calculations for the two methods were nearly identical. Figure 14-26 shows directivity as a function of τ and σ for Z_{0f} = 100 Ω and h/a = 125, as reported by Carrel. His gain numbers have been reduced by 1.5 dB, which is an average correction for the error. Since the E-plane beamwidth is determined mostly by the dipole pattern and is approximately 60°, the H-plane beamwidth may be determined from the beamwidth formula for directivity, i.e., D = $41,253/(\mathrm{BW}_E\mathrm{BW}_H)$. De Vito and Stracca give extensive results for impedance and gain for a wide range of Z_{0f}, τ (with σ optimum), and h/a. Increased gains may be achieved by using an E- or an H-plane array of LP monopole elements.

The characteristic impedance of the antenna may be computed from the image impedance of a cell for small dipole lengths. It is given by

$$Z_0 = Z_{0f} \left[1 + \frac{Z_{0f}}{2Z_{0a}X} \right]^{-1/2} \qquad (\textbf{14-15})$$

where X is the ratio of the cell length to the monopole length and is

$$X = \frac{2\,\tau\,(1-\tau)}{(1+\tau)\tan\alpha} = \frac{8\,\tau\,\sigma}{(1+\tau)} \qquad (\textbf{14-16})$$

and Z_{oa} is the characteristic impedance of the monopole. For small spacings, adjacent dipoles are 180° out of phase. Then, the characteristic impedance of a monopole is approximately the same as that of a rod of radius a placed between electric ground planes as illustrated by the dashed lines in Fig. 14-25.

The characteristic impedance of the monopole is then given by

$$Z_{0a} = 60 \ln(2hX/\pi a) \qquad (14\text{-}17)$$

Equation (14-15) may be solved for Z_{0f} in terms of the other parameters in order to achieve a desired Z_0.

Asymmetrical Log-Periodic Structures   Some of the useful LP elements, which are asymmetricl about their centerline, are shown in Fig. 14-27. They have the property that a sealing by the factor τ and a rotation of 180° about the centerline leads to an identical structure. Hence, the impedance is unchanged, but the pattern is rotated 180°. It follows that the phase of the radiated field changes by 180° as the frequency is changed by τ. Each of the elements has a unidirectional beam in the direction toward which the element points. One of the elements may be fed against ground, or two or more elements may be fed against each other to form various types of frequency-independent arrays (such as circular or E-plane and/or H-plane arrays). These asymmetrical structures tend to perform naturally and are much less critical to the design parameters than the symmetrical structures discussed in the next subsection.

The first four elements of Fig. 14-27 consist basically of a trnasmission-line shunt loaded by various types of monopoles which are resonant when their length is about $\lambda/4$. The remaining elements are various types of zigzag elements and are similar to traveling-wave antennas. The active region occurs when the width of the structure is about $\lambda/2$. In order to obtain sufficient attenuation through the active region (and little end effect), α should be less than about 20° and τ greater than about 0.85. The restrictions on the shunt-loaded elements are much less severe. Two of the linear monopole elements of Fig. 14-27a placed in an H-plane array with ψ near zero form the LP dipole array which was discussed previously. The monopoles may be bent forward[36,37] to form a more compact structure but at the expense of increased E- and H-plane beamwidths. However, the LP V-dipole antennas have been used very successfully as television receiving antennas. The antenna is truncated at the front so that the active region occurs in the $\lambda/2$ and $3\lambda/2$ dipoles for the low- and high-VHF bands,

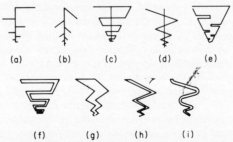

(a)   (b)   (c)   (d)   (e)

(f)   (g)   (h)   (i)

FIG. 14-27 Log-periodic elements which are invariant to a scaling by τ and a rotation of 180° about the centerline.

respectively, which provides for higher gain in the high band along with a more compact antenna.

The elements of Fig. 14-27c and d are wire outlines of trapezoidal and triangular sheet monopoles.[27] Because of the wider bandwidth of these thick monopoles compared with thin-wire monopoles, the pattern characteristics for a given τ_0 are about the same for the thin-wire monopoles with $\tau = \sqrt{\tau_0}$. Thus, the amount of wire is about the same, but the construction of the thin-wire structure is simpler.

The zigzag elements of Fig. 14-27g and h have been studied by Kuo and Mayes[38] for $\alpha < 20°$. E- and H-plane beamwidths in the range of 40 to 55° were obtained for an H-plane two-element array with $\psi = 2\alpha$. Limited studies have been reported on the sinusoidal type of LP zigzag. However, studies on periodic zigzags show that better performance is obtained with curved bends, since this practice reduces reflections from the bends. The reflections add up in the active region since the bends are spaced about $\lambda/2$.

For microwave applications the LP elements are commonly printed on the opposite sides of a dielectric sheet. For the shunt-loaded elements it is preferable to use the trapezoidal-tooth sheet structure of Fig. 14-23, which allows a smaller τ than the wire-monopole element. A constant-width feeder is used since the spacing between the feeder lines is constant. The tooth-width parameter ε should be somewhat greater than τ to prevent overlap of teeth on opposite sides of the sheet. If the teeth overlap, they tend to act like shunt transmission lines rather than radiating elements.

Symmetrical Log-Periodic Structures   Many applications require a vertically polarized frequency-independent antenna over a ground plane with a height of about $\lambda/4$ at the low-frequency cutoff. One-half of a symmetrical structure can be fed against a ground plane. The problem is how to obtain the extra 180° phase shift between adjacent radiating elements, which comes naturally with the asymmetrical elements. Figure 14-28 shows several techniques for accomplishing this, but all are quite sensitive designs.

Fig. 14-28a shows an LP monopole array fed by a transmission line over ground. The blocks represent 1:1 transformers with a phase reversal. This is not a sensitive design, but the cost and losses of the transformers may be prohibitive. If the transformers are not included, backfire radiation does not occur in the active region and broadband performance is not obtained. The monopole and zigzag elements may be bent along their centerlines and fed against the ground plane as illustrated in Fig. 14-28b and c. Half of the monopole elements act like open-circuited transmission lines whose lengths must be adjusted within a few percent to obtain frequency-independent performance. Berry and Ore[39] give extensive design data. The shunt transmission lines may be replaced by series LC circuits. The shunt loading produces the extra phase delay required for backfire radiation. The bent zigzag studied by Greiser[40] achieves the extra delay by the portions of the line lying over the ground plane. This antenna design is less sensitive than the bent-monopole antenna. Another scheme studied by Greiser[41] is the trapezoidal-type zigzag with parasite monopoles placed as shown in Fig. 14-28d. Apparently, the monopoles load the line enough to produce the correct phasing. Barbano[42] introduced parasitic monopoles between the elements of a monopole array as illustrated in Fig. 14-28e to achieve backfire radiation. The height of the parasitic element is approximately the geometric mean of the adjacent driven-element heights and must be adjusted to within 1 percent of the required height. Another technique devised by Wickersham,[43] using capacitive coupling to trapezoidal-sheet monopoles, is shown in Fig. 14-28f.

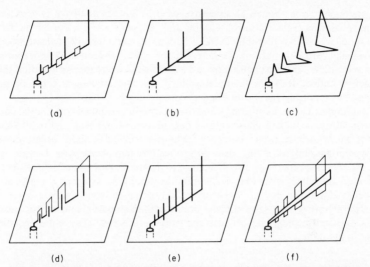

FIG. 14-28 Log-periodic elements which are symmetrical when combined with a ground-plane image.

Ingerson and Mayes[44] used delay lines between elements of a monopole array to achieve backfire radiation. Since the lines were longer than $\lambda/2$ in the active region, they had to devise a method of eliminating the stop band in the region of the antenna where the lines were $\lambda/2$ long. They achieved this by adding reflections in the lines by means of changing the impedance of a portion of the line. This was called a *modulated-impedance feeder*. They also used this technique to feed a log-periodic array of cavity-backed slots.

For these structures, τ is usually defined as the ratio of heights of adjacent vertical radiators. Hence the phase of the radiated field changes by 360° as the frequency changes by τ.

Special Considerations   Initial investigation of LP antennas over extreme bandwidths is greatly simplified because it is necessary only to establish that there is little end effect and then to determine performance over several periods of frequency. If there is little end effect, the impedance and pattern will be log-periodic. It is advisable to use swept frequency in order to detect anomalies as discussed below. It is also advisable that the structure have two active regions ($\lambda/2$ and $3\lambda/2$ widths) so that end effects will show up in the radiation pattern.

For wire LP antennas, wire-antenna computer programs making use of moment methods may be employed to calculate performance (patterns, polarization, gain, impedance, and current distribution) with good accuracy and great economy compared with experimental techniques. For more complex structures such as sheet and slot elements experimental techniques are used.

Theoretical and experimental studies of the periodic counterparts of LP structures which involve the determination of the Brillouin ($k - \beta$) diagram have led to considerable insight regarding operation of the antennas,[45,46] especially the determination of nonsuccessful structures. However, the technique gives little insight into the variation of performance of an LP antenna with the α angle.

Balmain[47,48] performed an extensive swept-frequency study of compressed LP antennas showing the variation of patterns and gain with α and τ. For a feeder impedance of 100 Ω, which is normally used for 50-Ω antennas, anomalies consisting of very-narrow-band drops in gain were detected for τ less than a certain value which increased with α. This was caused by end effect due to decreased attenuation through the active region. Balmain later determined that the anomalies could be greatly reduced by increasing the feeder impedance.

Asymmetries[49] in LP elements due to construction errors and/or mutual coupling to adjacent elements can cause serious degradation of patterns and gain over very narrow bands. Again, swept-frequency techniques should be used to determine that this type of anomaly does not occur. A serious problem occurs with E-plane arrays. Because of mutual coupling, the impedance of the monopoles on one side of the feeder is affected differently from that of those on the other side. For an LP dipole element it is believed that this sets up $++$ currents on the feeder which radiate broadside to the feeder. However, the same dropout occurs with monopole elements. The best solution is to increase the angle between elements in the array. The same phenomenon occurs in pyramidal arrays.

14-4 ARRAYS OF LOG-PERIODIC ELEMENTS

The pattern of an array of LP elements may be considered as the simple superposition of the patterns of the individual elements if it is assumed that the presence of other elements does not affect the pattern of an element.[50] To obtain frequency-independent operation with an array, it is necessary that the locations of the elements with respect to each other be defined by angles rather than by distances. This implies that all the elements have their vertices or feed points at a common point.

LP arrays are unique in two respects. First, although the element patterns are identical in shape (if it is assumed that design parameters are identical), they point in different directions. Second, since the radiation from an element may be considered to emanate from the phase center, the array sources lie on the surface of a sphere for a two-dimensional array and on a circle for a one-dimensional array. Thus the array theory is more complex and much less amenable to synthesis procedures than that for linear arrays of isotropic elements.

Element Characteristics   Figure 14-29 shows the variation of the E- and H-plane beamwidths of a single-wire trapezoidal-tooth element as a function of 2α. For a given α angle, there is a minimum value of the design ratio τ which can be used. For values of τ smaller than this minimum, the pattern breaks up considerably, and for larger values the beamwidths will decrease. The approximate minimum value of τ which can be used is plotted in Fig. 14-29 (solid curve) as a function of the parameter 2α. The E- and H-plane beamwidths for this minimum value of τ are also shown as solid lines. The dashed curves for the E- and H-plane beamwidths correspond to the dashed curve for τ, which is somewhat larger than the minimum value. It is noticed that the E-plane and especially the H-plane beamwidths decrease as α is decreased and τ is increased. For a given lower-frequency limit, which implies that the last transverse-element length remains fixed, decreasing α and increasing τ mean that the length of the structure and the number of transverse elements, respectively, are increased. The pattern behavior for the other types of elements is similar to that shown in Fig. 14-29.

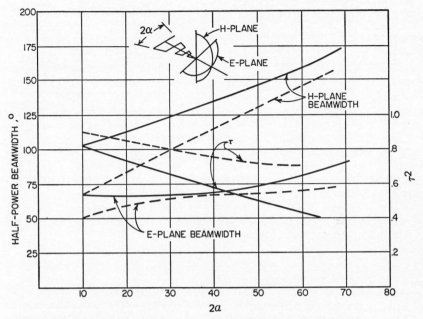

FIG. 14-29 Pattern characteristics of a wire trapezoidal-tooth element for ————, approximate minimum value of τ; ————, a larger value of τ.

The phase center of an LP element does not lie at the vertex; rather, it lies some distance d behind the vertex. Figure 14-30 shows the distance of the phase center from the vertices as a function of 2α for wire trapezoidal-tooth structures. For a fixed structure the distance as measured in wavelengths is essentially independent of frequency. For values of 2α less than $60°$ and for values of τ above the minimum value of τ given in Fig. 14-29, the position of the phase center is essentially independent of τ. Measurements[52] indicate that d depends upon τ in a complex manner for 2α greater than $60°$. The phase center lies on the centerline of the half structure at a point near where a half-wave transverse element exists.

Two-Element Arrays   Both the E-plane and the H-plane arrays may be considered as an array of two elements with the element patterns pointing in different directions. The pattern can easily be calculated once the element pattern and distance from the vertex to the phase center are known. A schematic representation of a two-element array is given in Fig. 14-31. The radial lines separated by the angle ψ represent the two elements, and d is the distance to the phase center. The direction to a distant field point is given by ϕ. Typical element patterns are shown in the schematic. The pattern of the array is given by

$$
E = \cos^n\left(\frac{\phi}{2} + \frac{\psi}{2}\right)\exp\left(j\beta d\,\sin\frac{\psi}{2}\,\sin\phi\right) +
$$

$$
\cos^n\left(\frac{\phi}{2} - \frac{\psi}{2}\right)\exp\left(-j\beta d\,\sin\frac{\psi}{2}\,\sin\phi\right) \quad \textbf{(14-18)}
$$

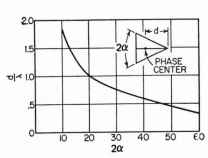

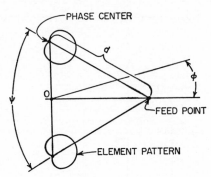

FIG. 14-30 Distance from vertex to phase center as a function of 2α.

FIG. 14-31 Schematic representation of a two-element array.

where $\cos^n (\phi/2)$ is an assumed functional form for the element pattern. The exponent n is related to the half-power beamwidth BW by

$$n = -0.35/\ln\left(\cos\frac{\mathrm{BW}}{2}\right)$$

For an *H*-plane array (like Fig. 14-23) the *H*-plane beamwidth is used, and for an *E*-plane array the *E*-plane beamwidth is used to determine n. The beamwidths may be obtained from Fig. 14-29. Although this procedure neglects the effect of the presence of one half structure on the pattern of the other, it will give fairly accurate results, especially for values of α smaller than 60°.

Figure 14-32 shows the variation of the *H*-plane beamwidth, gain, and front-to-back ratio with the angle ψ for a wire trapezoidal-tooth *H*-plane array with $2\alpha = 60°$ and $\tau = 0.77$. The *E*-plane beamwidth is nearly independent of the angle ψ, and its value is approximately 63°. Of course, for $\psi = 180°$ a bidirectional beam is produced.

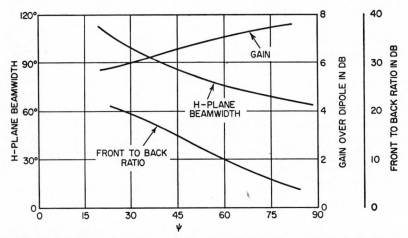

FIG. 14-32 Effect of angle ψ on pattern characteristics for an antenna with $2\alpha = 60°$ and $\tau = 0.77$.

Notice that if both high gain and high front-to-back ratio are desired, a compromise value of ψ must be chosen. As is to be expected, the H-plane beamwidth decreases rapidly with increasing ψ since this increases the H-plane aperture of the antenna.

The characteristic impedance increases as the angle between the two elements increases; it will be in the range of about 100 to 300 Ω and will also depend upon the type of element. Thus, it may be necessary to use broadband transformers to match to the transmission-line feed.

Multielement Arrays A schematic representation of an array of N elements as viewed from the top is illustrated in Fig. 14-33. The radial lines defined by δ_n represent the elements of the array. The α and τ parameters for the N elements are made identical so as to assure identical element patterns. The xy-plane radiation pattern of the array is given by

$$E(\phi) = \sum_{n=1}^{N} A_n f(\phi - \delta_n)\exp\left[-j\left(\beta d \cos(\phi - \delta_n) - \gamma_n\right)\right] \quad \textbf{(14-19)}$$

where $f(\phi)$ is the element pattern and $\beta d \cos(\phi - \delta_n)$ represents the phase advance of the phase center relative to the origin. The function f may take the same form as that used previously, that is, $f(\phi) = \cos^n(\phi/2)$. The value of the feed-point current for the nth element is given by A_n. In nearly all practical cases, $A_n = 1$ for all n. This is accomplished in practice by connecting half of the elements together and feeding them against the other half. The parameter γ_n is the relative phase of the field radiated from the nth element. It may be controlled by expanding or contracting the element according to the phase-rotation principle to be described later.

The assumptions made in Eq. (14-18) are that the element patterns and input impedances are identical. Although mutual effects can make these assumptions invalid, good correlation between theory and experiment has been obtained. Cut-and-try synthesis procedures may be used with Eq. (14-19).

A basic characteristic of logarithmically periodic antennas is the phase-rotation phenomenon. It has been verified experimentally that if the phase of the electric field received at a distant dipole (Fig. 14-34) is measured relative to the phase of the cur-

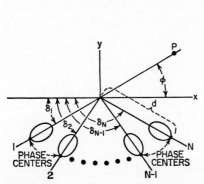

FIG. 14-33 Geometry for an array of end-fire elements.

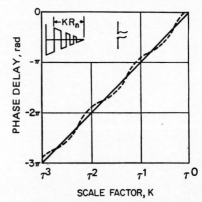

FIG. 14-34 Illustration of the phase-rotation phenomenon for log-periodic antennas.

rent at the feed point of the structure, the phase of the received signal will be delayed by 180° as the structure is expanded through a period. In Fig. 14-34, the distance to an arbitrary transverse element is given by KR_n. The expansion of the structure through a period is accomplished by letting K increase from 1 to $1/\tau$. During this expansion all lengths involved in the structure are multiplied by K. In Fig. 14-34 the phase delay in radians is plotted versus K on a logarithmic scale. The ideal phase variation is given by the solid straight line. Measurements have indicated that the actual phase variation is somewhat like the dashed line. The approximate measurements made to date indicate that the deviation of the dashed line from the straight line is not more than 15°. The relation between the phase γ_n and K_n is given by

$$K_n = \tau^{\gamma_n/\pi} \qquad (14\text{-}20)$$

Fortunately, the phase center and the element patterns are independent of the expansion or contraction of a logarithmically periodic element.

This phase rotation appears not only in the radiation field but also in the reflected wave on the feeder for log-periodic antennas and log-periodic transmission-line circuits.[51] It produces dispersion of transmitted or received signals, and the dispersion increases as τ approaches unity.

The information given in Figs. 14-29, 14-30, and 14-34 is sufficient for predicting the pattern of an array of similar end-fire elements. The method can be extended to cover a combination of E- and H-plane arrays. The phase-rotation phenomenon is extremely important since it allows a frequency-independent method of phasing the elements of the array.

Experimental and theoretical patterns for a six-element phased H-plane array of wire trapezoidal-tooth elements are given in Fig. 14-35. The values of the design

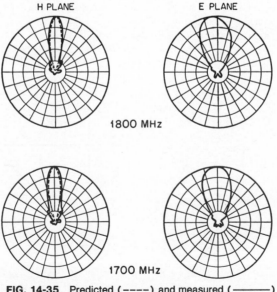

FIG. 14-35 Predicted (----) and measured (——) patterns of a six-element phased array.

parameters for this array were $\alpha = 19°$, $\tau = 0.94$, $N = 6$, $\delta_n - \delta_{n-1} \approx 17°$ for all n, and $d/\lambda = 1.95$. The elements were phased to produce a beam-cophasal condition. The gain of the array was 14 $d\beta$ over a dipole.

Although gains of up to 18 $d\beta$ are feasible by using a combined E- and H-plane array, the size and complexity of the antenna become great since it is necessary to use very small α angles and τ values near unity.

An approximate design procedure for arrays is given in Ref. 7.

14-5 CIRCULARLY POLARIZED STRUCTURES

Circular polarization may be obtained by using crossed LP dipole arrays. Figure 14-36 is a front view of a practical configuration. Feeders 1 and 2 are used to excite the vertical and horizontal dipoles respectively. The solid lines represent the dipoles in one cell and the dashed lines the dipoles in the next cell. It is desirable to make the spacing between the feeder lines as small as practicable in order to simulate planar arrays. If the two crossed arrays are identical, it is necessary to excite the two feeders with a broadband 90° phase difference circuit or a broadband quadrature hybrid. In the latter case, both left and right circularization are obtained simultaneously. Alternatively, one dipole array may be scaled by $\tau^{1/2}$ with respect to the other, producing a 90° phase shift of the radiated field. The two feeders may be fed in or out of phase for one sense of circular polarization or by a broadband 0–180° hybrid for both senses of circular polarization.

Another approach is to use two crossed dipole arrays, one scaled by $\tau^{1/2}$, and excite them by a single feeder as illustrated in Fig. 14-37. It is quite surprising that this gives the 90° phasing between the two arrays. Of course, only one sense of circular polarization may be obtained.

Since the H-plane beamwidth is usually about 50 percent greater than the E-plane beamwidth, the axial ratio is low only for directions close to the axis of the array. To obtain low axial ratios over wide angles, a traveling-wave ring-type radiator is desired rather than crossed dipoles. The simplest solution is a spiral antenna. Another approach is to use four LP elements placed on the sides of a pyramid. Because of the asymmetrical-coupling problems between adjacent elements, the shunt-loaded LP elements of Fig. 14-27 do not work unless the spacing between the tips of adjacent elements is large. In this case the ψ angle is large, and in turn the back radiation is large. The traveling-wave structures of Fig. 14-27 may be used since they are much less susceptible to the asymmetrical coupling. However, these structures are limited to small α angles, which leads to a long antenna.

Some LP monopulse antennas have been used as feeds for reflectors. They consist of a circular array of six or eight

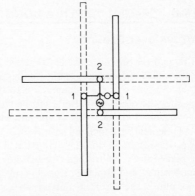

FIG. 14-36 Feeder configuration for crossed LP dipole antennas.

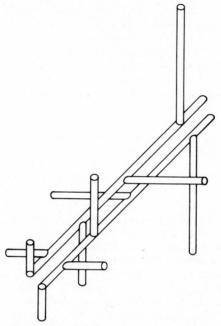

FIG. 14-37 View of circularly polarized crossed LP dipole arrays with a single feeder.

elements of the type shown in Fig. 14-28 placed on the surface of a metal cone and excited in modes 1 and 2 (Σ and Δ patterns) by a broadband hybrid-feed network. Again, the axial ratio is poor off boresight because of the large difference in beamwidths of the E_θ and E_ϕ polarizations.

14-6   SPECIAL APPLICATIONS

Arrays over Ground

Although the antennas discussed above have radiation patterns which are essentially independent of frequency, many applications demand that the antennas be placed near ground. If one of these antennas is placed with its feed point above ground, it is apparent that the resultant pattern will be frequency-dependent since its electrical height above ground changes with frequency. However, the above array theory suggests that if LP elements are inclined with respect to ground and their feed points are at ground level, the resultant radiation pattern will be frequency-independent. The elements can be placed so that with their images they form either H-plane or E-plane arrays or a combined E- and H-plane array. For an equivalent H-plane array there will, of course, be a null in the resultant pattern on the horizon. The array theory may be used to calculate the resultant pattern by adjusting the phase of the image elements. For an H-plane array the phase of the image element will be 180° different from that of the element above ground, whereas for an *E*-plane array the phase of the image element would be the same as that of the element.

A very important application for this type of antenna is in point-to-point communication circuits.[52] Figure 14-38 shows a two-element array above ground oriented so that these elements and their image elements form an *H*-plane array. Theoretically, the feed point of the antenna should be at ground level, but in practice the feed point is placed a small height above ground to protect personnel from high radio-frequency voltages. Except for very short distances, high-frequency point-to-point communication is accomplished by the reflection of the radio waves from the ionosphere. The vertical angle of arrival or departure (from the ground) depends upon the distance between the points and the height of the reflecting layer. Although its value ranges from 70° down to a few degrees for various circuits, its value for a particular circuit is relatively constant since the height of the reflecting layer does not change by a great amount. However, because of changing ionospheric conditions during the sunspot cycle and from night to day, it is necessary to change the operating frequency over bandwidths of 4 or 6:1. Thus it is most desirable to have an antenna for which the vertical angle of the main lobe is independent of frequency. Some antennas, such as the dipole, rhombic, billboard, and discone, do not satisfy this requirement.

The direction of the main lobe for the structure of Fig. 14-38 may be controlled by the α angle of the individual element and the angles of the elements with respect to ground. The size of the structure is determined by the lower frequency limit. Figure 14-39 shows the vertical-plane pattern for a structure with $\alpha = 14°$ and $\tau = 0.75$. The two half structures are oriented at angles of 32 and 48° with respect to ground. The dimensions of the antenna at the lowest frequency are given in wavelengths in the figure. The lower half structure is scaled so that its radiation leads that of the upper by 130°. This particular phasing increases the gain by 5 dB over that obtained with no difference in phasing. Although the gain for this type of antenna is only moderate (13.4 dB over a dipole), maximum use of this gain is made over the complete fre-

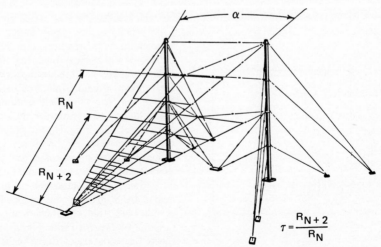

FIG. 14-38 Log-periodic high-frequency antenna with a frequency-independent elevation pattern.

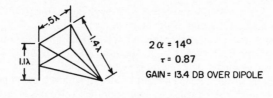

$$2\alpha = 14°$$
$$\tau = 0.87$$
GAIN = 13.4 DB OVER DIPOLE

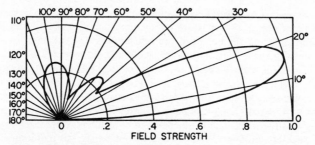

FIELD STRENGTH

FIG. 14-39   Elevation pattern and design parameters for the typical antenna of Fig. 14-38.

quency range. The gain for this structure can be increased by 3 dB by adding two additional elements to the side to form a combined *E*- and *H*-plane array.

If vertical polarization is required rather than horizontal polarization (which is obtained for the structure above), then the two-element array can be rotated 90° about its centerline so that the plane of each of the half structures would be normal to the ground plane.

Feeds for Reflectors and Lens[53]

Many applications call for high-gain antennas which will work over extremely wide frequency ranges. Although lens- or reflector-type antennas are ideally suited, their bandwidths have been limited in the past by that of the primary feed. Ideally, the radiation pattern and input impedance of the primary radiator should be independent of frequency; thus unidirectional LP or equiangular antennas are well suited for this application. The required design information for LP feeds is the primary pattern, phase-center and input-impedance.

Figure 14-40 shows the angle subtended by a circular-parabolic reflector as observed from the focal point as a function of the f/d ratio. Figure 14-41 shows the variation of the *E*-plane and *H*-plane 10-dβ beamwidths and the side-lobe level as a function of ψ for several

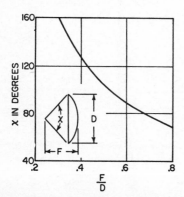

FIG. 14-40   Angle subtended by a parabolic reflector as a function of f/d.

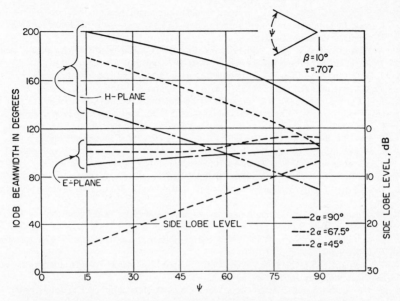

FIG. 14-41 Pattern characteristics of a sheet trapezoidal-tooth structure.

different values of α for two-element sheet trapezoidal-tooth structures. The largest sidelobe nearly always occurs in the opposite direction of the beam. If a 10-dβ taper is desired, it is seen that a two-element structure can be designed so as to give fairly good illumination tapers for f/d ratios from about 0.35 to 0.6.

REFERENCES

1   V. H. Rumsey, "Frequency Independent Antennas," *IRE Nat. Conv. Rec.,* part I, 1957, pp. 114–118.
2   R. H. DuHamel and D. E. Isbell, "Broadband Logarithmically Periodic Antenna Structures," *IRE Nat. Conv. Rec.,* part I, 1957, pp. 119–128.
3   E. C. Jordan, G. A. Deschamps, J. D. Dyson, and P. E. Mayes, "Developments in Broadband Antennas," *IEEE Spectrum,* vol. 1, April 1964, pp. 58–71.
4   G. A. Deschamps and J. D. Dyson, "The Logarithmic Spiral in a Single-Aperture Multimode Antenna System," *IEEE Trans. Antennas Propagat.,* vol. AP-19, January 1971, pp. 90–96. An extensive bibliography is included.
5   B. R. S. Cheo, V. H. Rumsey, and W. J. Welch, "A Solution to the Frequency-Independent Antenna Problems," *IRE Trans. Antennas Propagat.,* vol. AP-9, November 1961, pp. 527–534.
6   J. D. Dyson, "The Equiangular Spiral, *IRE Trans. Antennas Propagat.,* vol. AP-7, no. 2, April 1959.
7   H. Jasik (ed.), *Antenna Engineering Handbook,* 1st ed., McGraw-Hill Book Company, New York, 1961, chap. 18.
8   Y. S. Yeh and K. K. Mei, "Theory of Conical Equiangular-Spiral Antennas, Part I— Numerical Technique," *IEEE Trans. Antennas Propagat.,* vol. AP-15, September 1967, pp. 634–639.

9 Y. S. Yeh and K. K. Mei, "Theory of Conical Equiangular-Spiral Antennas, PART II—Current Distributions and Input Impedances," *IEEE Trans. Antennas Propagat.*, vol. AP-16, January 1968, pp. 14–21.

10 A. E. Atia and K. K. Mei, "Analysis of Multi-Arm Conical Log-Spiral Antennas," *IEEE Trans. Antennas Propagat.*, vol. AP-19, May 1971, pp. 320–331.

11 J. D. Dyson, "The Characteristics and Design of the Conical Log-Spiral Antenna," *IEEE Trans. Antennas Propagat.*, vol. AP-13, July 1965, pp. 488–499.

12 J. D. Dyson and P. E. Mayes, "New Circularly-Polarized Frequency-Independent Antennas with Conical Beam or Omnidirectional Patterns," *IRE Trans. Antennas Propagat.*, vol. AP-9, July 1961, pp. 334–342.

13 R. Bawer and J. J. Wolfe, "The Spiral Antenna," *IRE Int. Conv. Rec.,* 1960.

14 R. Bawer and J. J. Wolfe, "A Printed Circuit Balun for Use with Spiral Antennas," *IRE Trans. Microwave Theory Tech.,* vol. MTT-8, no. 3, May 1960, p. 319.

15 W. K. Roberts, "A New Wide-Band Balun," *IRE Proc.,* vol. 45, December 1957, p. 1628.

16 D. L. Saul, "Cavity-Backed Spiral Has Microstrip Feed," *Microwaves,* October 1981, pp. 88–89.

17 Marchand, N., "Transmission Line Conversion Transformers," *Electronics,* vol. 17, December 1944, pp. 142–145.

18 J. W. Greiser and M. L. Wahl, "New Spiral-Helix Antenna Developed," *Electron. Warfare,* May–June 1975, pp. 67–70.

19 J. Yamauchi and H. Nakano, "Axial Ratio of Spiral Antennas," IEEE Antennas Propagat. Soc. meeting, May 1981.

20 G. G. Chadwick and J. P. Shelton, "Two Channel Monopulse Techniques—Theory and Practice," paper presented at Int. Conv. Milit. Electron, September 1965.

21 R. R. Yaminy, "A Two Channel Monopulse Telemetry and Tracking Antenna Feed," *Int. Telemetering Conf. Proc.,* vol. VI, 1968.

22 O. K. Kim and J. D. Dyson, "A Log Spiral Antenna with Selectable Polarization," *IEEE Trans. Antennas Propagat.,* vol AP-19, September 1971, pp. 675–677.

23 J. P. Shelton, "Multiple-Feed Systems for Objectives," *IEEE Trans. Antennas Propagat.,* vol. AP-13, November 1965.

24 J. L. McFarland, "The Triangular Spaced $3N^2$ Multiple Beam Array Family," Antennas Propagat. Soc., Santa Clara, Calif., November 1978.

25 G. G. Chadwick and J. P. Shelton, "Sum Difference Feed Network," U.S. Patent 3,346,861, October 1967.

26 D. E. Isbell, "Nonplanar Logarithmically Periodic Antenna Structures," Univ. Illinois Antenna Lab. TR 30, Cont. AF 33(616)-3220, Feb. 20, 1958.

27 R. H. DuHamel and F. R. Ore, "Logarithmically Periodic Antenna Designs," *IRE Nat. Conv. Rec.,* part 1, 1958, pp. 139–151.

28 D. E. Isbell, "Log Periodic Dipole Arrays," *IRE Trans. Antennas Propagat.,* vol. AP-8, no. 3, May 1960, pp. 260–267.

29 R. H. DuHamel, "Log Periodic Antennas and Circuits," in E. C. Jordan (ed.), *Electromagnetic Theory and Antennas,* Pergamon Press, New York, 1963, pp. 1031–1050.

30 P. E. Mayes, G. A. Deschamps, and W. T. Patton, "Backward-Wave Radiation from Periodic Structures and Application to the Design of Frequency-Independent Antennas," *IRE Proc.* (correspondence), vol. 49, May 1961, pp. 962–963.

31 R. L. Carrel, "The Design of Log-Periodic Dipole Antennas," *IRE Int. Conv. Rec.,* vol. 1, 1961, pp. 61–75.

32 W. M. Cheong and R. W. P. King, "Log-Periodic Dipole Antenna," *Radio Sci.,* vol. 2, November 1967, pp. 1315–1325.

33 G. De Vito and G. B. Stracca, "Comments on the Design of Log-Periodic Dipole Antennas," *IEEE Trans. Antennas Propagat.,* vol. AP-21, May 1973, pp. 303–308.

34 G. De Vito and G. B. Stracca, "Further Comments on the Design of Log-Periodic Dipole Antennas," *IEEE Trans. Antennas Propagat.,* vol. AP-22, September 1974, pp. 714–718.

35 P. C. Butson and G. T. Thomson, "A Note on the Calculation of the Gain of Log-Periodic Dipole Antennas," *IEEE Trans. Antennas Propagat.,* vol. AP-24, January 1976, pp. 105–106.
36 P. E. Mayes and R. L. Carrel, "Logarithmically Periodic Resonant V-Arrays," Univ. Illinois Antenna Lab. Rep. 47, Urbana, 1960.
37 K. K. Chan and P. Sylvester, "Analysis of the Log-Periodic V-Dipole Antenna," *IEEE Trans. Antennas Propagat.,* vol. AP-23, May 1975, pp. 397–401.
38 S. C. Kuo and P. E. Mayes, "An Experimental Study and the Design of the Log-Periodic Zig-Zag Antenna," TR AFAL-TR-65-328, University of Illinois, Dept. of Electrical Engineering, Urbana, February 1966.
39 D. G. Berry and F. R. Ore, "Log-Periodic Monopole Array," *IRE Int. Conv. Rec.,* part 1, 1961, pp. 76–85.
40 J. W. Greiser and P. E. Mayes, "The Bent Backfire Zig-Zag—A Vertically Polarized Frequency Independent Antenna," *IEEE Trans. Antennas Propagat.,* vol. AP-11, May 1964, pp. 281–299.
41 J. W. Greiser, "A New Class of Log Periodic Antennas," *IEEE Proc.,* vol. 52, May 1964, p. 617.
42 N. Barbano, "Log-Periodic Yagi Uda Array," *IEEE Trans. Antennas Propagat.,* vol. AP-14, March 1966, pp. 235–238.
43 A. F. Wickersham, Jr., "Recent Developments in Very Broadband End-Fire Arrays," *IRE Proc.* (correspondence), vol. 48, April 1960, pp. 794–795; A. F. Wickersham, Jr., R. E. Franks, and R. L. Bell, "Further Developments in Tapered Ladder Antennas," *IRE Proc.,* (correspondence), vol. 40, January 1961, p. 378.
44 P. G. Ingerson and P. E. Mayes, "Log-Periodic Antennas with Modulated Impedance Feeders," *IEEE Trans. Antennas Propagat.,* vol. AP-16, November 1968, pp. 633–642.
45 R. Mittra and K. E. Jones, "Theoretical Brillouin $(k-\beta)$ Diagrams for Monopole and Dipole Arrays and Their Application to Log-Periodic Antennas," *IEEE Trans. Antennas Propagat.,* vol. AP-12, 1964, pp. 533–540.
46 E. Hudock and P. E. Mayes, "Near Field Investigation of Uniform Periodic Monopole Array," *IEEE Trans. Antennas Propagat.,* vol. AP-13, November 1965, pp. 840–855.
47 C. C. Bantin and K. G. Balmain, "Study of Compressed Log-Periodic Dipole Antennas," *IEEE Trans. Antennas Propagat.,* vol. AP-18, March 1970, pp. 195–203.
48 K. G. Balmain, C. C. Bantin, C. R. Oakes, and L. David, "Optimization of Log-Periodic Dipole Antennas," *IEEE Trans. Antennas Propagat.,* vol. AP-19, March 1971, pp. 286–288.
49 K. G. Balmain and J. N. Nkeng, "Asymmetry Phenomenon of Log-Periodic Dipole Antennas," *IEEE Trans. Antennas Propagat.,* vol. AP-24, July 1976.
50 R. H. DuHamel and D. G. Berry, "Logarithmically Periodic Antenna Arrays," *Wescon Conv. Rec.,* part 1, 1958, pp. 161–174.
51 R. H. DuHamel and M. E. Armstrong, "Log-Periodic Transmission-Line Circuits—Part I: One-Port Circuits," *IEEE Trans. Microwave Theory Tech.,* vol. MTT-14, June 1966, pp. 264–274.
52 R. H. DuHamel and D. G. Berry, "A New Concept in High Frequency Antenna Design," *IRE Nat. Conv. Rec.,* part 1, 1959, p. 42.
53 R. H. DuHamel and F. R. Ore, "Log Periodic Feeds for Lens and Reflectors," *IRE Nat. Conv. Rec.,* part 1, 1959, p. 128.

BIBLIOGRAPHY

Collin, R. E., and F. J. Zucker, *Antenna Theory,* part 2, McGraw-Hill Book Company, New York, 1969, chap. 22.

Jordan, E. C., and K. G. Balmain, *Electromagnetic and Radiating Systems,* Prentice-Hall, Inc., Englewood Cliffs, N.J., 1968, chap. 15.

———, G. A. Deschamps, J. D. Dyson, and P. E. Mayes, "Developments in Broadband Antennas," *IEEE Spectrum,* vol. 1, April 1964, pp. 58–71.

Rumsey, V. H.: *Frequency Independent Antennas,* Academic Press, Inc. New York, 1966.

Chapter 15

Horn Antennas

Allan W. Love

Rockwell International

Types and Uses

The horn antenna in general is a device which effects a transition between waves propagating in a transmission line, usually a waveguide, and waves propagating in an unbounded medium such as free space. They are constructed in a variety of shapes for the purpose of controlling one or more of the fundamental properties: gain, radiation pattern, and impedance. The types most often used with rectangular and round guides are shown in Figs. 15-1 and 15-2.

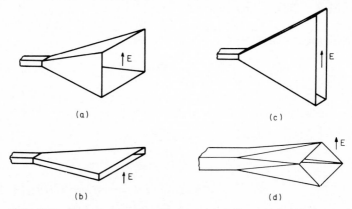

FIG. 15-1 Rectangular-waveguide horns. (*a*) Pyramidal. (*b*) Sectoral *H*-plane. (*c*) Sectoral *E*-plane. (*d*) Diagonal.

The pyramidal horn of Fig. 15-1*a* is most commonly used as a primary gain standard since its gain may be calculated to within 0.1 dB from its known dimensions. Independent control of the beamwidths in the two principal planes is possible by varying the rectangular-aperture dimensions.

The sectoral horns of Fig. 15-1*b* and *c* are special cases of the pyramidal horn, being flared in only one plane. They radiate fan-shaped beams, broad in the plane orthogonal to the flare. The diagonal horn of Fig. 15-1*d* is another special case in which the horn aperture is square but the electric field is essentially parallel to a diagonal. This horn radiates a pattern with a high degree of rotational symmetry but does suffer from a relatively high level of cross-polarization in the intercardinal planes.

The conical horn of Fig. 15-2*a* is, by virtue of its axial symmetry, capable of handling any polarization of the exciting dominant TE_{11} mode. It is particularly well suited for circular polarization. Despite its axial symmetry, however, the beam widths in the principal planes are generally unequal. Like the pyramidal horn, the conical horn may be used as a primary standard since its gain is accurately calculable.

The dual-mode and corrugated (hybrid-mode) conical horns of Fig. 15-2*b* and *c* overcome the lack of axial symmetry in the radiation pattern and at the same time achieve suppression of cross-polar radiation in the intercardinal planes to a very low level. In the dual-mode horn, some of the exciting TE_{11} mode is converted to TM_{11} at

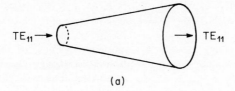

$TE_{11} \rightarrow$ $\rightarrow TE_{11}$

(a)

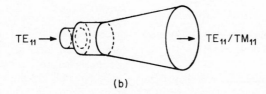

$TE_{11} \rightarrow$ $\rightarrow TE_{11}/TM_{11}$

(b)

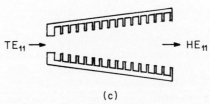

$TE_{11} \rightarrow$ $\rightarrow HE_{11}$

(c)

FIG. 15-2 Round-waveguide horns. (*a*) Conical. (*b*) Dual-mode conical. (*c*) Corrugated conical.

the step discontinuity in diameter. In the horn of Fig. 15-2*c* the corrugations force propagation in a hybrid mixture of TE_{11} and TM_{11} waves called the HE_{11} mode. In both cases the aperture field of the horn is modified in such a way as to produce axisymmetry in the radiation patterns.

A collection of important historical papers dealing with horn antennas will be found in Ref. 1.

Effect of Horn Flare

The radiation characteristics of horns with aperture dimensions greater than about one wavelength may be calculated with reasonable accuracy by using Huygens' principle along with the known aperture field in the horn mouth. For single dominant-mode horns, the aperture field is usually taken to be identical with the transverse electric field of that mode as it would exist in a cylindrical waveguide of the same cross section and size as the horn mouth. In a waveguide which flares to a horn, it is clear that the boundary conditions require the existence of higher-order nonevanescent modes in addition to the fundamental mode. It is customary to assume that the flare angle is small enough to permit the higher-mode field components to be neglected in comparison with those of the dominant mode.

Even when the flare angle is not particularly small, the above treatment gives surprisingly accurate results if the effect of the curvature of the dominant-mode wavefront in the aperture is taken into account. In Fig. 15-3 the field lines are shown as though the dominant mode were expanding as a cylindrical wave (in a sectoral horn) or a spherical wave (in a conical horn) with a radius of curvature at the aperture equal to ℓ, the slant length of the horn. The path difference between this curved wavefront and an ideal plane wave in the aperture constitutes a phase error which requires that the planar fields of the dominant mode be modified by the inclusion of a phase factor of approximately quadratic form, exp $(jk\Delta x^2)$, where x is the normalized aperture coordinate. From the geometry it is easy to see that the total excursion of the path error is Δ, where

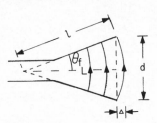

FIG. 15-3 Phase error due to wavefront curvature in the horn aperture.

$$\Delta = \ell - L = \frac{1}{2} d \tan \frac{\theta_f}{2} = 2\ell \sin^2 \frac{\theta_f}{2} \qquad (15\text{-}1)$$

L and ℓ are the axial and slant lengths respectively, and θ_f is the semiflare angle.

The quadratic phase form implicitly assumes that the small-angle approximations may be applied to Eq. (15-1), giving the equivalent simplified expressions

$$\Delta \doteq d^2/(8L) \qquad \text{or} \qquad d^2/(8\ell) \qquad (15\text{-}2)$$

The ambiguity implied in Eq. (15-2) arises only in cases in which the quadratic phase approximation is indiscriminately applied when θ_f is not small. This can be a source of confusion and error. In this section the approximation $\Delta = d^2/(8\ell)$ will be used, ℓ being the slant length, unless otherwise noted.

The consequences of quadratic phase error are loss in gain accompanied by broadening of the radiation pattern, an increase in sidelobe levels, and obliteration of the nulls. In the rest of this section all pattern calculations are based on the above approximate model with quadratic-error correction applied as necessary.

Obliquity Factors

Radiation patterns of dominant (i.e., TE-mode) horns differ in the principal planes for two reasons. The basic transverse-field distributions and the associated obliquity factors differ in the two orthogonal directions. The obliquity factor is a pattern function multiplier which takes account of the effects of mode propagation constant and reflection at the horn aperture.

In the E plane the obliquity factor, denoted by F_e, is given by,[2]

$$F_e = \frac{1}{2}\left[1 + \frac{\lambda}{\lambda_g}\cos\theta + \Gamma\left(1 - \frac{\lambda}{\lambda_g}\cos\theta\right)\right] \qquad (15\text{-}3)$$

while in the H plane

$$F_h = \frac{1}{2}\left[\cos\theta + \frac{\lambda}{\lambda_g} + \Gamma\left(\cos\theta - \frac{\lambda}{\lambda_g}\right)\right] \qquad (15\text{-}4)$$

where λ_g is the wavelength appropriate to the guide dimension at the horn mouth and Γ is the reflection coefficient presented by the aperture discontinuity. θ is the usual far-field polar angle.

In small-aperture horns F_e and F_h may be significantly different. Moreover, because Γ is generally an unknown complex number, they are not calculable, a fact which partially accounts for a notable lack of success in predicting the patterns of small horns.

For apertures larger than about 2λ the aperture discontinuity is small, so that $\Gamma \to 0$, while $\lambda/\lambda_g \to 1$. Under these circumstances F_e and F_h are nearly identical:

$$F_e \doteq F_h \doteq \tfrac{1}{2}(1 + \cos \theta) = \cos^2 \theta/2 \qquad \textbf{(15-5)}$$

When the aperture is very large, the main beam is narrow and radiation is largely confined to small values of θ. For this reason $\cos \theta$ is often taken to be unity in Eq. (15-5), and the obliquity factors, both then unity, are suppressed altogether.

15-2   DOMINANT-MODE RECTANGULAR HORNS*

Sectoral Horns

The radiation patterns of sectoral horns may be calculated in the plane of the flare on the assumption that the horn is excited by the dominant TE_{10} mode in rectangular waveguide.[2] The results of such calculations are shown in Figs. 15-4 and 15-5 for E- and H-plane flares respectively. In each case there is a family of curves corresponding to various values of the cylindrical-wave error Δ due to the flare. The ordinate is the normalized far-zone electric field strength, and the abscissa is the reduced variable $b/\lambda \sin \theta$ or $a/\lambda \sin \theta$, where b and a are the dimensions of the horn mouth (b in the E plane, a in the H plane).

The effects of phase error due to the flare are seen in the various curves corresponding to the parameters s and t, which are normalized path errors in wavelengths,

$$s = \frac{\Delta_e}{\lambda} = \frac{b^2}{8\lambda \ell_e} \qquad t = \frac{\Delta_h}{\lambda} = \frac{a^2}{8\lambda \ell_h} \qquad \textbf{(15-6)}$$

For the same values of s and t the phase error due to the flare has a much more pronounced effect in the E plane than in the H plane. The reason is simply that the field is constant across the aperture of a sectoral horn flared in the E plane but varies sinusoidally for one flared in the H plane. The patterns also exhibit much higher minor-lobe levels for the former case.

The curves in both Fig. 15-4 and Fig. 15-5 are reasonably accurate when either a or b is at least several wavelengths. For smaller apertures it will be necessary to multiply by the appropriate obliquity factor, F_e or F_h. The approximate form given by Eq. (15-5) is reasonably accurate for aperture dimensions greater than about 1.5λ.

Expressions for the field in the aperture of a sectoral horn, upon which radiation-pattern calculations depend, are given in Sec. 10.9 of Silver.[2] In Sec. 10.10 it is remarked that there is disagreement between theory and experiment for E-plane sectoral horns. In part, this is due to an error in Eqs. (45a) and (45b) on pages 357 and

---

*For 10-dB beamwidths of small horns, see also Fig. 46-13.

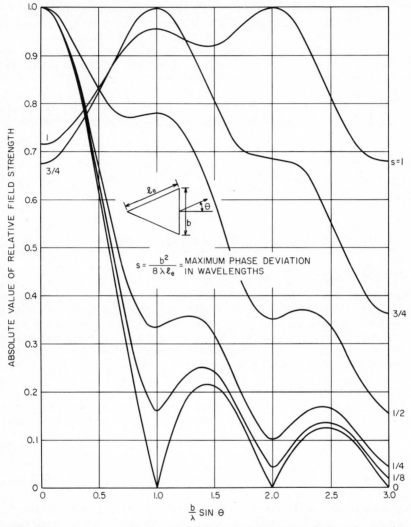

FIG. 15-4 Universal radiation patterns of horns flared in the *E* plane (sectoral or pyramidal).

358 of Ref. 2. This error was pointed out by Narasimhan and Rao[3] and arose inadvertently through the use of a left-handed coordinate system in Fig. 10.9a on page 351. To correct the error the factors $E_\theta - Z_0 H_x \cos(\Theta - \theta)$ and $E_\theta \cos\Theta - Z_0 H_x \cos\theta$, which appear in the integrands of Eqs. (45a) and (45b), should be changed to read $E_\theta + Z_0 H_x \cos(\Theta - \theta)$ and $E_\theta \cos\Theta + Z_0 H_x \cos\theta$.

The aperture field method is incapable of predicting the radiation pattern in the rear sector of a horn antenna and is not accurate in the wide-angle-sidelobe region because it neglects diffraction at the aperture edges. The prediction of wide-angle and

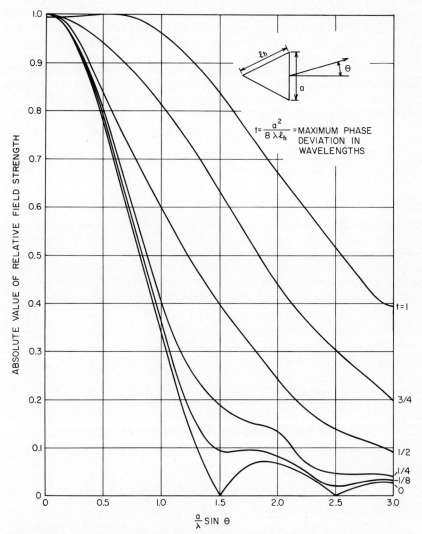

$$t = \frac{a^2}{8 \lambda \ell_h} = \text{MAXIMUM PHASE DEVIATION IN WAVELENGTHS}$$

FIG. 15-5 Universal radiation patterns of horns flared in the *H* plane (sectoral or pyramidal).

rearward-radiated fields is now possible by using the techniques of the geometrical theory of diffraction (GTD). It is impractical to attempt such a treatment here; the interested reader should consult Refs. 4, 5, and 6.

Pyramidal Horns

The pyramidal horn of Fig. 15-1*a* flares in both the *E* and the *H* planes. Its radiation patterns in these planes are the same as those of the corresponding flared sectoral horns. Thus the E-plane patterns of a pyramidal horn are as shown in Fig. 15-4, while its H-plane patterns are those of Fig. 15-5. For aperture dimensions greater than about 1λ the patterns in the two planes may be controlled independently by adjustment of the mouth dimensions.

The design of a horn for specified beamwidths in the principal planes must be done iteratively because of the need to fit the throat of the horn to a given waveguide. The patterns of Figs. 15-4 and 15-5 are first used to select possible values of a, b, ℓ_e, and ℓ_h. Only values which satisfy the condition

$$(a - a_0)^2[(\ell_h/a)^2 - \tfrac{1}{4}] = (b - b_0)^2[(\ell_e/b)^2 - \tfrac{1}{4}] \qquad (15\text{-}7)$$

may be retained if the horn is to be physically realizable. Here, a_0 and b_0 are the H- and E-plane waveguide dimensions respectively.

Gain

The gain of sectoral horns may be determined[7] from the normalized gain curves of Figs. 15-6 and 15-7, provided the narrow dimension is at least one wavelength. It is seen that the gain initially increases as the aperture dimension increases, then reaches a maximum before falling off. The decrease in gain is due to increasing cylindrical-wave phase error as the aperture increases while slant length remains constant.

The points of maximum gain in each case occur approximately where

$$s = b^2/(8\lambda\ell_e) = \tfrac{1}{4} \qquad (15\text{-}8)$$

$$t = a^2/(8\lambda\ell_h) = \tfrac{3}{8} \qquad (15\text{-}9)$$

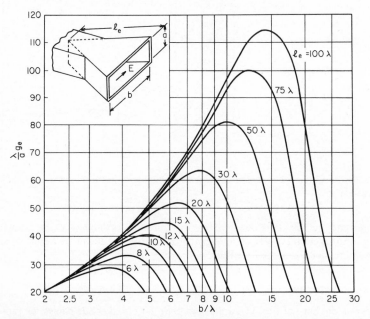

FIG. 15-6   Gain of E-plane sectoral horn ($a \leq \lambda$).

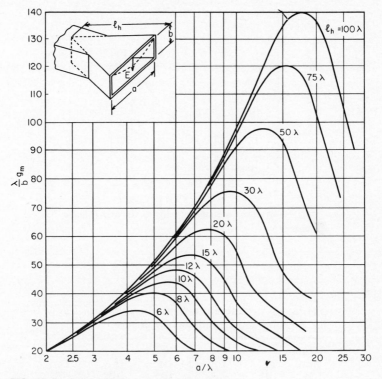

FIG. 15-7   Gain of *H*-plane sectoral horn ($b \le \lambda$).

for *E*- and *H*-plane sectoral horns respectively, corresponding to phase errors of 90° and 135°.

The gain of both sectoral and pyramidal horns[7] may be determined from

$$G = G_0 R_e R_h \qquad (15\text{-}10)$$

where the area gain G_0 is given by

$$G_0 = 32ab/(\pi\lambda^2) \qquad (15\text{-}11)$$

relative to a fictitious isotropic radiator. The factors R_e and R_h, both equal to or less than unity, account for the reduction in gain due to phase error caused by the flare; thus R_e is a function of *s*, while R_h is a function of *t*. Both R_e and R_h involve Fresnel integrals.

By combining Eq. (15-10) with Eq. (15-11) and converting to decibel form, the gain is

$$\text{Gain (dB)} = 10.08 + 10 \log_{10} ab/\lambda^2 - L_e - L_h \qquad (15\text{-}12)$$

where

$$L_e = 10 \log_{10} R_e \qquad L_h = 10 \log_{10} R_h \qquad (15\text{-}13)$$

The factors L_e and L_h, shown graphically in Fig. 15-8 and tabulated in Table 15-1,[8] are functions of the parameters *s* and *t* given in Eq. (15-6). For a sectoral horn either

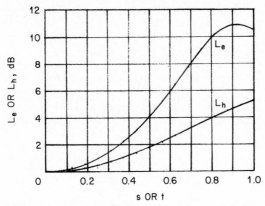

FIG. 15-8 Gain-reduction factor for *E*- and *H*-plane flares.

Table 15-1 Gain-Reduction Factors, L_e or L_h, for Sectoral and Pyramidal Horns

16*s* or 16*t*	L_e, dB	L_h, dB	16*s* or 16*t*	L_e, dB	L_h, dB
1.0	0.060	0.029	14.0	10.783	4.486
1.5	0.134	0.064	15.0	10.849	4.892
2.0	0.239	0.114	16.0	10.502	5.252
2.5	0.374	0.179	18.0	9.474	5.819
3.0	0.541	0.257	20.0	8.847	6.210
3.5	0.738	0.349	22.0	8.901	6.504
4.0	0.967	0.454	24.0	9.637	6.785
4.5	1.229	0.573	26.0	10.938	7.102
5.0	1.525	0.705	28.0	12.430	7.460
5.5	1.854	0.850	30.0	13.312	7.831
6.0	2.218	1.007	32.0	13.052	8.175
6.5	2.618	1.176	34.0	12.251	8.460
7.0	3.054	1.357	36.0	11.666	8.684
7.5	3.527	1.547	38.0	11.607	8.869
8.0	4.037	1.748	40.0	12.121	9.047
9.0	5.166	2.175	42.0	13.104	9.243
10.0	6.427	2.630	44.0	14.221	9.462
11.0	7.769	3.101	46.0	14.851	9.692
12.0	9.081	3.577	48.0	14.619	9.911
13.0	10.163	4.043	50.0	13.937	10.101

Courtesy IEE; see Ref. 8.

ℓ_e or ℓ_h will be infinite, and the corresponding loss factor $L_{h,e}$ will be zero. An alternative way of calculating the gain of a pyramidal horn is to take the product of the gains of the corresponding sectoral horns, obtained from Figs. 15-6 and 15-7, and then to multiply by the factor $\pi/32$.

In the above discussion of gain, far-field conditions are implicitly understood. It has been shown, however,[8] that Eqs. (15-10) and (15-12) may be used at finite range (in the Fresnel zone) by a simple redefinition of terms. This is possible because the effect of finite range is simply to modify the already existing quadratic phase error in the aperture due to the flare. G_0, R_e, and R_h are determined as before, but the parameters s and t are now taken to be

$$s = b^2/(8\lambda\ell_e') \qquad t = a^2/(8\lambda\ell_h') \qquad \textbf{(15-14)}$$

in which $\qquad 1/\ell_e' = 1/\ell_e + 1/r \qquad 1/\ell_h' = 1/\ell_h + 1/r \qquad \textbf{(15-15)}$

r being the distance from the aperture to the field point.

When the a dimension in a sectoral or pyramidal horn is small ($a \lesssim \lambda$), the close proximity of the magnetic walls causes errors in the gain equations, (15-10) and (15-11). A more accurate equation has been derived:[9]

$$G = R_e R_h \frac{16ab}{\lambda^2(1 + \lambda_g/\lambda)} \exp\left[\frac{\pi a}{\lambda}\left(1 - \frac{\lambda}{\lambda_g}\right)\right] \qquad \textbf{(15-16)}$$

in which R_e and R_h are as before except that in calculating R_e the parameter s must be slightly modified. Thus

$$s = b^2 \bigg/ \left(8\lambda_g \ell_e \cos^2\frac{\theta_f}{2}\right) \qquad \textbf{(15-17)}$$

where θ_f is the semiflare angle (Fig. 15-3). There is no such corresponding modification to t, and λ_g is the guide wavelength appropriate to the dimension a, i.e.,

$$\lambda/\lambda_g = [1 - (\lambda/2a)^2]^{1/2} \qquad \textbf{(15-18)}$$

Equation (15-10), which may be called Schelkunoff's gain formula, includes the geometrical-optics field in the aperture of the horn and the effects of singly diffracted fields at the aperture edges. It does not, however, include effects due to multiple-edge diffractions or diffracted fields which are reflected from interior horn walls. These effects manifest themselves in the form of small oscillations about the monotonic curve predicted by Schelkunoff's equations, (15-10) and (15-11). The oscillations are of the order of ±0.5 dB for pyramidal horns with a gain of about 12 dB. They decrease to about ±0.2 dB for horns with a gain of about 18 dB, and they appear to be less than ±0.1 dB when the gain exceeds 23 dB.[10,11]

Optimum-Gain Horns

An optimum-gain horn is a pyramidal horn whose dimensions are such that it produces the maximum far-field gain for given slant lengths in the E and H planes. From what has been said above, this condition will occur when the parameters s and t are respectively about ¼ and ⅜. Thus an optimum-gain horn will have aperture dimensions given approximately by

$$a = \sqrt{3\lambda\ell_e} \qquad b = \sqrt{2\lambda\ell_h} \qquad \textbf{(15-19)}$$

The effective area of an optimum-gain horn is very close to 50 percent of its actual aperture area; its gain is

$$\text{Gain (dB)} = 8.1 + 10 \log_{10} ab/\lambda^2 \qquad \textbf{(15-20)}$$

Such a horn will have full half-power beamwidths (HPBWs) and first-sidelobe levels given approximately by

$$2 \sin^{-1} (0.45\lambda/b) \qquad -9 \text{ dB} \qquad E \text{ plane}$$
$$2 \sin^{-1} (0.7\lambda/a) \qquad -16 \text{ dB} \qquad H \text{ plane}$$

The H-plane sidelobe is not well defined, amounting to no more than a shoulder on the main beam.

In designing a standard horn, one generally knows the desired gain, wavelength, and dimensions of the feeding waveguide. If the horn is also to be optimum, these quantities, together with the restrictions discussed above, completely determine the final dimensions of the horn. Let a_0 and b_0 be the H- and E-plane waveguide dimensions respectively. Then the following equation must be satisfied if the horn is to be physically realizable (that is, fit the feeding waveguide), be optimum, and have numerical gain G:

$$\left(\sqrt{\frac{2\ell_e}{\lambda}} - \frac{b_0}{\lambda}\right)^2 \left(\frac{2\ell_e}{\lambda} - 1\right) = \left(\sqrt{\frac{3K\lambda}{\ell_e}} - \frac{a_0}{\lambda}\right)^2 \left(\frac{4K\lambda}{3\ell_e} - 1\right) \qquad \textbf{(15-21)}$$

where $K = (G/15.75)^2$.

This equation may be easily solved by trial and error for ℓ_e. A rough approximation which can serve as a first trial is $\ell_e = \lambda/\sqrt{K}$. After ℓ_e has been determined, ℓ_h is given by

$$\ell_h = K\lambda^2/\ell_e \qquad \textbf{(15-22)}$$

Impedance

The impedance characteristics of sectoral horns depend on the mismatch between the horn mouth and free space, the length of radial guide between the horn aperture and throat, and the mismatch at the junction between the uniform guide and the throat. The input voltage standing-wave ratio (VSWR) of an E-plane horn in general varies periodically between 1.05 and 1.5 as the horn slant length is varied, with minima occurring every half-guide wavelength.[2,12] The guide wavelength in the E-plane sectoral horn is equal to that in the rectangular waveguide.

The input impedance of an H-plane sectoral horn differs somewhat from that of the E-plane horn, in that the mismatch at the junction of the horn and rectangular waveguide is much smaller than that at the horn aperture. The result is that the input VSWR to an H-plane sectoral horn is essentially constant in magnitude and varies only in phase as the length of the horn is changed.[2]

A broadband match is best obtained in the E-plane sectoral horn by treating the two discontinuities separately. The mismatch at the junction of the waveguide and horn may be eliminated with a reactive window (usually inductive) at the junction. The horn aperture is frequently matched to free space by using a plastic radome of proper thickness and dielectric constant. If no radome is used, a match may be obtained by the use of small reactive discontinuities at the aperture.

The problem of obtaining a broadband match with the H-plane horn is somewhat easier because of the small discontinuity at the horn throat. Generally, only the aperture needs to be matched, and the techniques mentioned for the E-plane horn may be employed.

The impedance characteristics of pyramidal horns are somewhat more complicated than those of sectoral horns. However, horns having a gain of 20 dB or more and moderate flare angles, such as optimum horns, are usually well matched by a rectangular waveguide supporting only the dominant mode. For example, an optimum horn with 28-dB gain at 3000 MHz fed by 1½- by 3-in (38- by 76-mm) waveguide has an input VSWR of about 1.03. Where the overall match must be improved, techniques similar to those described for sectoral horns may be used.

15-3 DOMINANT-MODE CONICAL HORNS

Radiation Patterns

The conical horn excited by a circular waveguide in the TE_{11} mode is the counterpart of a pyramidal horn with TE_{10} excitation in rectangular guide, and in many respects its behavior is quite similar. The modes in an infinite conical horn are exactly expressible in terms of a combination of spherical Hankel and Legendre functions. Rigorous application of vector diffraction theory is possible but leads to such complexity that approximate solutions must be sought. Considerable simplification can be achieved by use of asymptotic expressions for the fields, as shown in Ref. 13. One of the earliest applications of GTD to the conical horn is given in Ref. 14, in which contributions from as many as seven rays, including multiple diffractions, are summed. Better results are claimed in Ref. 15, in which the slope diffraction technique is combined with GTD. Results of such computations are in reasonably good agreement with observed patterns, but the number of test cases is very few and accuracy limits cannot, in general, be given. As a result, no reliable universal pattern curves, similar to those of Figs. 15-4 and 15-5, are available for the conical horn even today.

Experimental data on conical-horn patterns are also limited, and it still seems necessary to rely on the observations made by Bell Laboratories investigators[16] as far back as 40 years ago. The now familiar conical-horn patterns given by King[17] in 1950 are reproduced in Fig. 15-9 for aperture diameters in the range $1.4 < d/\lambda < 4.3$. The spherical-wave phase-error parameter $s = d^2/(8\lambda\ell)$ ranges from 0.03 to 0.6. It should be noted that in King's original paper[17] the parameter was given as $d^2/(8\lambda L)$, where L is the axial horn length. To conform with the first edition of this handbook s has been recomputed in terms of slant length ℓ rather than axial length L. The lengths ℓ and L can differ significantly because the semiflare angle is not small except for case a. This can be a source of confusion, as pointed out in Sec. 10-1. From Fig. 15-3 it is seen that the semiflare angle is given by

$$\tan \theta_f = d/(2L) \quad \text{or} \quad \sin \theta_f = d/(2\ell) \qquad \textbf{(15-23)}$$

The horns in Fig. 15-9a and b have aperture dimensions less than optimum. The horn in c is close to optimum size, while g shows the patterns of an optimum horn in greater detail. *Optimum* is again used to refer to a horn whose aperture dimension is such as to yield maximum gain for a given length; in this case the diameter is denoted

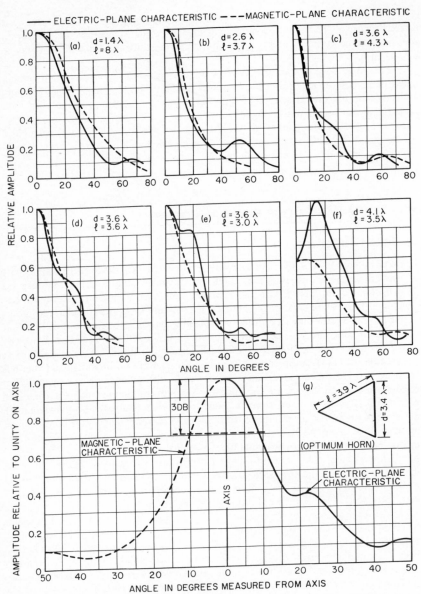

FIG. 15-9 Experimentally observed patterns of conical horns of various dimensions.

by d_m. Although only six different patterns are shown in Fig. 15-9, they can be used for approximate estimation of pattern shapes for a variety of horns, provided the proportions of the horn in question are such as to match (or nearly match) the phase-error parameter s of one of those in Fig. 15-9. As an example, consider a horn with aperture diameter $d = 6\lambda$, which does not appear in the figure. If the length of this horn is $\ell = 10\lambda$, then its parameter s will match that of horn d; i.e., $s = 0.45$. As a consequence, the patterns of this horn will have nearly the same shapes as those in Fig. 15-9d, with only a scale change in the angle θ. The new angle θ_1 is related to the θ in d by the relation $d_1/\lambda \sin \theta_1 = d/\lambda \sin \theta$, whence $\sin \theta_1 = 0.6 \sin \theta$.

Gain

The curves in Fig. 15-10 show how the gain (in decibels above isotropic) of a conical horn varies with its dimensions. Note that the parameter L/λ in this figure is the axial

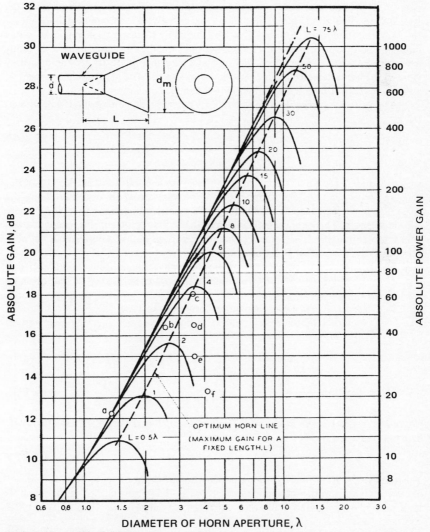

FIG. 15-10 Calculated gain of a conical horn as a function of aperture diameter with axial length as parameter. (© 1950 IEEE; see Ref. 17.)

length. Conversion to ℓ/λ has not been made in this case, since the resulting fractional values would hinder interpolation. The curves are taken from King's paper,[17] and he attributed them to the unpublished work of his Bell Laboratories coworkers M. C. Gray and S. A. Schelkunoff. They are believed to have been derived theoretically; since the nature of any approximations used is not known, it is difficult to assess the limits of accuracy associated with these curves. Numerical gain calculations based on a GTD analysis in Ref. 14 are said to be in agreement with gain values in Fig. 15-10 to within ± 0.1 dB, but the ranges of d/λ and L/λ over which this is true are not stated. A closed-form expression involving Lommel functions is derived in Ref. 13, from which a family of curves with parameter values the same as those in Fig. 15-10 is given. Comparison of the two sets of curves shows differences ranging from $+0.4$ to -0.8 dB. The same authors in another paper[18] give a very simple expression for the gain as a function of L/λ and d/λ. Unfortunately the expression applies only to a circular aperture with uniform illumination and quadratic phase; it most certainly does not apply to a conical horn. The points a to f in Fig. 15-10 are the measured gains of the horns depicted in Fig. 15-9, and they appear to be in reasonable agreement with values obtained by interpolation between curves.

The gain of a conical horn in terms of its aperture diameter may clearly be written as

$$G = (\pi d/\lambda)^2 R \qquad (15\text{-}24)$$

or

$$\text{Gain (dB)} = 20 \log_{10}(\pi d/\lambda) - L \qquad (15\text{-}25)$$

where R and L are factors analogous to those for sectoral horns in Sec. 15-2 that account for loss in gain due to spherical-wave phase error. Figure 15-11 shows the value of L in decibels as a function of the phase-error parameter $s = d^2/(8\lambda\ell)$, where ℓ is slant length. This curve, which appeared in the first edition of this handbook, was based on a gain expression involving Lommel functions that was derived but not published by W. C. Jakes* of Bell Laboratories.

The accuracy with which gain predictions can be made from Eq. (15-24) and Fig. 15-11 is not known with certainty, but it appears to be as good as that given by Fig. 15-10.

The case of $s = 0$ corresponds to a horn with no flare, i.e., to an open-ended circular-waveguide radiator. If only the TE_{11} mode is present, then the gain is given by Ref. 2 (page 341),

$$G = (\pi d/\lambda)^2 \frac{\left| \sqrt{\dfrac{\lambda}{\lambda_g}} + \sqrt{\dfrac{\lambda_g}{\lambda}} + \Gamma\left(\sqrt{\dfrac{\lambda_g}{\lambda}} - \sqrt{\dfrac{\lambda}{\lambda_g}} \right) \right|^2}{2(p_{11}'^2 - 1)(1 - |\Gamma|^2)} \qquad (15\text{-}26)$$

where $p_{11}' = 1.841$ is the first root of $J_1'(p) = 0$ and Γ is the complex reflection coefficient at the aperture. For large apertures

$$G \rightarrow 0.837(\pi d/\lambda)^2 \qquad (15\text{-}27)$$

This shows that $R = 0.837$ for long horns with small flare angles; also, $L = 0.773$ dB, corresponding to the intercept at $s = 0$ in Fig. 15-11. In contrast to the sectoral-horn case, the parameter R for the conical horn is always less than unity. This is

---

*Private communication: W. C. Jakes wrote the chapter "Horn Antennas" in the first edition of this handbook.

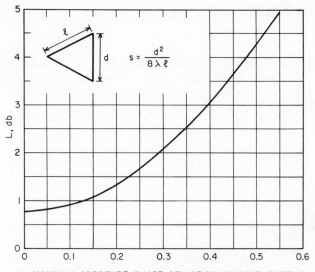

s, MAXIMUM APERTURE PHASE DEVIATION IN WAVELENGTHS

FIG. 15-11   Gain-correction factor for conical horns.

simply a consequence of a slight change in definition [compare Eq. (15-24) with Eq. (15-10)] such that for the conical horn the effect of amplitude taper as well as phase taper is included in R.

Optimum-Gain Horns

In Fig. 15-10 the dashed line connecting the maxima of the gain curves defines the gain and dimensions, d/λ, L/λ, of optimum conical horns. An empirical equation for this line over the range of variables shown in the figure is

$$\text{Gain (dB)} = 7.0 + 20.6 \log_{10} d/\lambda \qquad \textbf{(15-28)}$$

When the quantity $d^2/(8\lambda L)$, L being the axial length, is evaluated at the maxima in Fig. 15-10, it is found to vary from 0.56 for $L = 0.5\lambda$ to 0.31 for $L = 75\lambda$. The slant length ℓ can easily be determined from the known values of d and L, for example, by using Eqs. (15-23) or, more simply, $\ell = (L^2 + d^2/4)^{1/2}$. If this is done and the parameter $s = d^2/(8\lambda\ell)$ is computed, it is found to vary only from about 0.30 to 0.375. Thus the dimensions of optimum conical horns can be expressed by

$$d = \sqrt{n\lambda\ell} \qquad \textbf{(15-29)}$$

where n lies between 2.4 and 3.0.

The effective area of an optimum horn is between 52 and 56 percent of its actual aperture area.

Impedance

The remarks regarding the impedance of pyramidal horns in Sec. 15-2 also apply to conical horns if "dominant-mode round waveguide" is read for "rectangular guide."

15-4   MULTIMODE HORNS

General

The dominant single-mode horn, whether conical (TE_{11}) or pyramidal (TE_{10}), radiates a pattern in which the E plane differs significantly from the H plane so that axial-beam symmetry does not exist in general. The reason for this is that the electric field in the horn aperture is heavily tapered in the H plane (since it must vanish on the conducting walls) but tapered very little (in the conical case) or not at all (in the pyramidal case) in the E plane. The patterns in the two planes can be made identical or nearly so by introducing appropriate additional modes into the horn structure, which then propagate to the horn aperture along with the dominant mode. The aperture field is then a superposition of the fields of the individual modes. In most practical applications of this technique only one additional mode is required; such horns are called dual-mode.

Diagonal Horn

The simplest example of a dual-mode horn is the diagonal horn,[19] in which two spatially orthogonal dominant modes, the TE_{10} and the TE_{01}, are excited with equal amplitude and phase in a square waveguide which flares to a pyramidal horn. In this case the electric field in the aperture is aligned with one of the diagonals. A simple transformation from rectangular guide to diagonal horn is shown in Fig. 15-12. Another method of excitation is the use of coaxial-fed orthogonal probe couplers in square waveguide along with a suitable power splitter.

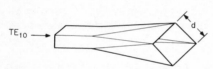

FIG. 15-12 Transformation from rectangular waveguide to diagonal horn.

For apertures in which $d \gtrsim 2\lambda$ the E-, H-, and intercardinal-plane patterns will be essentially identical to well below the -10-dB level, as is evident in the universal patterns in Fig. 15-13. True axial symmetry is not achieved, however, because the intercardinal-plane patterns differ markedly in sidelobe structure from those in the principal planes. Moreover, a significant amount of cross-polarization is manifested in the intercardinal planes, as shown by the dashed curves in Fig. 15-13.

By writing $u = (\pi d/\lambda) \sin \theta$, the principal-plane patterns of the diagonal horn are given by

$$E_{co} = \left\{ \begin{matrix} F_e \\ F_h \end{matrix} \right\} \frac{\sin u/\sqrt{2}}{u/\sqrt{2}} \cdot \frac{\cos u/\sqrt{2}}{1 - 2u^2/\pi^2} \qquad E_{cross} = 0 \qquad \textbf{(15-30)}$$

while in the intercardinal planes

$$\left. \begin{matrix} E_{co} \\ E_{cross} \end{matrix} \right\} = \frac{1}{2} \left(F_e \cdot \frac{\sin u}{u} \pm F_h \cdot \frac{\cos u}{1 - 4u^2/\pi^2} \right) \qquad \textbf{(15-31)}$$

F_e and F_h are the obliquity factors discussed in Sec. 15-1. When $d \gtrsim 2\lambda$, it was observed that $F_e \rightarrow F_h \rightarrow 1$; this is the case for the patterns shown in Fig. 15-13.

Prime-focus feed horns have small apertures ($d \doteq \lambda$) for commonly used f/D ratios. The diagonal horn is at a disadvantage here because the disparity between F_e

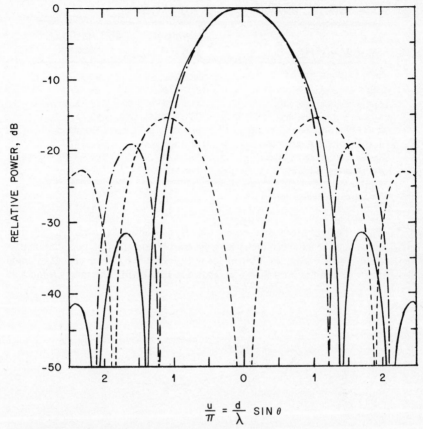

$$\frac{u}{\pi} = \frac{d}{\lambda} \text{ SIN } \theta$$

FIG. 15-13 Principal- and intercardinal-plane patterns of diagonal horn ($d \gg \lambda$): ——
——, E and H planes; —·—·—·—, intercardinal, copolar; --------, intercardinal, cross-
polar. (*From* Microwave Journal; *see Ref. 19.*)

and F_h destroys the circular symmetry of the beam. It is an observed fact, however, that in the neighborhood of $d = 1.23\lambda$ a diagonal horn does possess axial-beam symmetry with full widths of 40 and 73° at the -3- and -10-dB levels. Apparently, in this case the values of Γ and λ/λ_g (see Sec. 15-1) happen to be such that F_e and F_h are equal.

Important characteristics of large-aperture diagonal horns are given in Table 15-2.

Dual-Mode Conical Horn

In the case of the conical horn the addition of the higher-order TM_{11} mode[20] to the dominant TE_{11} mode can create a radiation pattern with good axial-beam symmetry and with a low level of cross-polarization as well. When properly phased at the horn aperture, the TM_{11} mode will cancel the ϕ component of magnetic field due to the

Table 15-2 Characteristics of Diagonal Horns

Parameter	Principal planes	45° and 135° planes
3-dB beamwidth, °	58.5 λ/d	58.0 λ/d
10-dB beamwidth, °	101 λ/d	98 λ/d
Angular position, first null, °	81 λ/d	70 λ/d
Angular position, first lobe, °	96 λ/d	92 λ/d
Level of first sidelobe	31½ dB	19 dB
Angular position, second null, °	122 λ/d	122 λ/d
Angular position, second lobe, °	139 λ/d	147 λ/d
Level of second sidelobe	41½ dB	24 dB

From *Microwave Journal;* see Ref. 19.

TE_{11} mode at the aperture boundary, causing H_ϕ as well as E_ϕ to vanish. This is a condition that leads to axial symmetry and low cross-polarization in the radiation pattern. In the original version, the TM_{11} mode was generated by a step change in guide radius from a to b in the horn throat, as shown in Fig. 15-14. The radius a must be

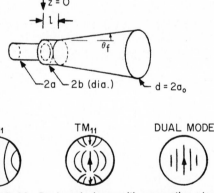

FIG. 15-14 Dual-mode horn with generating step and its approximate aperture field distribution.

large enough to support the TE_{11} mode but small enough to ensure that the TM_{11} is cut off ($1.84 < ka < 3.83$). The amount of TM_{11} so generated increases with the ratio b/a, but b should not be so large as to permit the TE_{12} mode to propagate ($3.83 < kb < 5.33$). Owing to the axial symmetry, TE_{mn} or TM_{mn} modes with $m > 1$ will not be excited. Figure 15-14 also indicates qualitatively the distribution of the electric field in the aperture for the two modes both individually and in combination. In the central region of the aperture the electric field of the TM_{11} mode reinforces that of the TE_{11}. Near the aperture boundary the two fields oppose one another. Thus the resulting electric field may be heavily tapered in the E as well as in the H plane.

A set of patterns in the E, H, and intercardinal planes is shown in Fig. 15-15 for

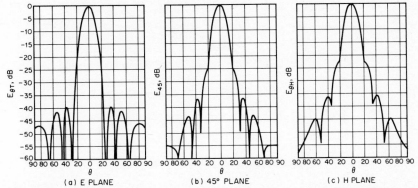

FIG. 15-15   Radiation patterns of dual-mode conical horn, $d = 4.67\lambda$. (*From* Microwave Journal; *see Ref. 20.*)

a horn with aperture diameter $2a_0 = 4.67\lambda$ and a small semiflare angle of $6.25°$. In this horn $a = 1.02\lambda$, $b = 1.30\lambda$, and $\ell = 0.20\lambda$. The short constant-diameter section of length ℓ is used to ensure that the two modes will be in the same phase at the center of the aperture at the design frequency. This length depends upon the flare angle and the distance from the step to the aperture as well as on the launch phase of the TM_{11} mode relative to the TE_{11}.

The TE_{11} mode generates θ and ϕ components of E in the far field, while the TM_{11} mode generates only the θ component. In terms of the usual variable, $u = ka_0 \sin \theta$, and with the assumption that the aperture is reasonably large ($a_0 \gtrsim \lambda$) so that the reflection coefficient Γ is approximately zero, the radiated far-field components are

$$\left. \begin{aligned} E_\theta &= \left[1 + \frac{\lambda}{\lambda_g} \cos \theta - \alpha \frac{\lambda/\lambda'_g + \cos \theta}{1 - (3.83/u)^2} \right] \frac{J_1(u)}{u} \cdot \cos \phi \\ E_\phi &= \left(\frac{\lambda}{\lambda_g} + \cos \theta \right) \frac{J'_1(u)}{1 - (u/1.84)^2} \cdot \sin \phi \end{aligned} \right\} \quad \textbf{(15-32)}$$

where λ/λ_g and λ/λ'_g refer to the TE_{11} and TM_{11} modes, respectively, and α, the mode-content factor, is a measure of the field strengths of the two modes at the aperture center. For $\alpha = 0.65$ the E- and H-plane HPBWs are equalized, and the phase centers in the two planes are coincident. Other values of α, in the neighborhood of 0.6, lead to suppression of the first sidelobe in the E plane or to suppression of the cross-polar radiation in the intercardinal planes. The presence of the TM_{11} mode has no effect in the H plane, where this mode does not radiate. The beamwidth of a dual-mode conical horn will vary somewhat with the mode-content factor; typically HPBW $= 72 \lambda/d$ °.

Mode conversion at a discontinuous step in round waveguide has been investigated theoretically and experimentally.[21,22] Figure 15-16 shows the mode conversion coefficient and excitation phase of the TM_{11} mode relative to the TE_{11} mode on the guide centerline in the plane of the step at $z = 0$. If ϕ_{ex} represents the launch phase, ϕ_{ph} the differential phase in the constant-diameter phasing section, and ϕ_{fl} the differential phase of the modes in the flared section of Fig. 15-14, then the condition for reinforcement of the electric fields at the aperture center is

$$\phi_{ph} + \phi_{fl} - \phi_{ex} = 0 \text{ or } 2m\pi \quad \textbf{(15-33)}$$

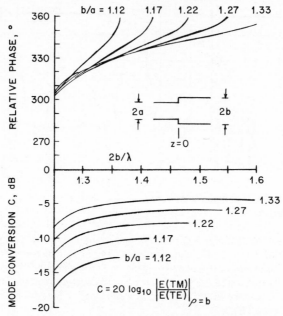

FIG. 15-16 Mode conversion coefficient C and launch phase of TM_{11} mode relative to TE_{11} at step discontinuity (TM_{11} lags TE_{11}). (© *1970 IEEE; see Ref. 21.*)

If λ_g and λ_g' are the guide wavelengths of the TE_{11} and the TM_{11} modes respectively, then

$$\phi_{ph} = 2\pi\ell(1/\lambda_g - 1/\lambda_g') = 2\pi\ell/\lambda(\lambda/\lambda_g - \lambda/\lambda_g') \qquad (15\text{-}34)$$

where λ_g and λ_g' are constant and depend on the diameter $2b$ in the phasing section. In the flared horn the differential phase shift is closely given by

$$\phi_{fl} = 2\pi \int \left[\frac{1}{\lambda_g(z)} - \frac{1}{\lambda_g'(z)} \right] dz \qquad (15\text{-}35)$$

in which the guide wavelengths are now functions of the axial coordinate z and where integration is taken from $z = \ell$ to the z value at the aperture plane. Note that ϕ_{ph} and ϕ_{fl} are positive angles because $\lambda_g' > \lambda_g$. These equations have been computed in Ref. 20 and are shown in Fig. 15-17. In the case of ϕ_{fl} the z coordinate has been eliminated by introducing the semiflare angle θ_f.

As an example of the use of these design curves, consider the horn whose patterns are given in Fig. 15-15. With $a_0/\lambda = 2.33$, $b/\lambda = 0.65$, and $\theta_f = 6.25°$, Fig. 15-17a gives $\phi_{fl} = 620°$, while b with $\ell/\lambda = 0.20$ gives $\phi_{ph} = 40°$. The expected excitation phase is obtained from Fig. 15-16; $\phi_{ex} = -40°$. When these values are substituted into Eq. (15-33), the equation is satisfied with a small error of 20°. From the same figure the TM_{11}-mode conversion coefficient appears to be about -7.5 dB; evidently this corresponds closely to the value $\alpha = 0.6$ for the mode-content factor defined in Ref. 20.

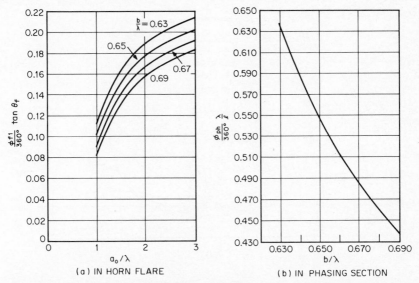

FIG. 15-17 Differential phase shifts in (*a*) the flared horn and (*b*) the phasing section. (*From* Microwave Journal; *see Ref. 20.*)

The phasing condition of Eq. (15-33) can be satisfied only at a single frequency because of the different dispersion characteristics of the two modes. Thus the dual-mode horn is frequency-sensitive (the horn of Fig. 15-15 has a useful bandwidth of only 3 to 4 percent), and the sensitivity increases with horn length. For horns that are not too long some improvement can be obtained by loading the step discontinuity with a dielectric ring.[21]

Use of a step discontinuity near the horn throat is not the only way to convert TE_{11} to TM_{11} energy in a horn. A technique described by Satoh[23] employs a thin dielectric band placed internally in contact with the horn wall. When the length of the band, its thickness, and its position are correctly determined, the resulting bandwidth of this dual-mode horn is said to be 25 percent (defined so that all sidelobes are below -20 dB and E- and H-plane beamwidths differ by no more than 5 percent).

Conversion from the TE_{11} mode to the TM_{11} and other higher modes (TE_{12}, TM_{12}, TE_{13}, ...) at a relatively large-diameter junction between round and conical waveguides has been investigated analytically by Tomiyasu.[24] In this case a discontinuous step (as in Fig. 15-14) is not needed. Using this technique, Turrin[25] has designed multimode horns of the type shown in Fig. 15-18; typical E-plane patterns are shown in Fig. 15-19 for two different aperture sizes: (*a*) 1.31λ and (*b*) 1.86λ. In this type of horn the input guide should be sized to propagate the TE_{11} mode but not the TM_{11} mode ($1.84 < ka < 3.83$). Conversion takes place at the plane AA' in Fig. 15-18, where the diam-

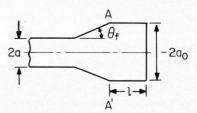

FIG. 15-18 Stepless dual-mode horn. (© *1967 IEEE; see Ref. 25.*)

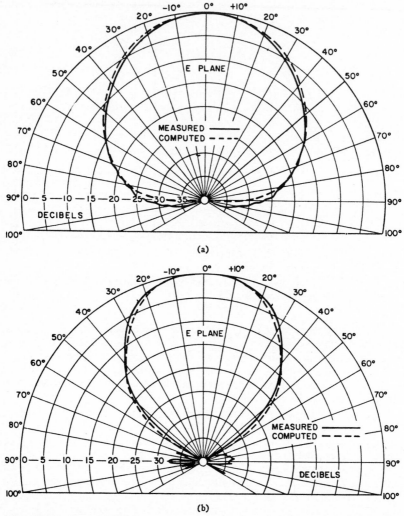

FIG. 15-19 *E*-plane patterns of stepless dual-mode horn. (*a*) $2a = 0.71\lambda$, $2a_0 = 1.31\lambda$, $\ell = 1.37\lambda$, $\theta_f = 30°$. (*b*) $2a = 0.71\lambda$, $2a_0 = 1.86\lambda$, $\ell = 3.75\lambda$, $\theta_f = 28°$. (© *1967 IEEE; see Ref. 25.*)

eter $2a_0$ is large enough to support this mode; i.e., $ka_0 > 3.83$. The next-higher modes with the proper symmetry are the TE_{12}, which requires $ka_0 > 5.33$, and the TM_{12}, requiring $ka_0 > 7.02$. Since these modes are not wanted, the aperture diameter should be restricted so that $3.83 < ka_0 < 5.33$; i.e., $1.22 < 2a_0/\lambda < 1.70$. This condition is satisfied by the horn in Fig. 15-19*a* but not *b*. In the latter case the aperture must contain TE_{12} as well as TE_{11} and TM_{11} modes; evidently the amount of TE_{12} is small and does not significantly affect the radiation pattern.

At the plane AA' in Fig. 15-18 the higher-order TM_{11} mode is 90° out of phase with the TE_{11} mode; hence the phasing section ℓ should have a differential length of 270° for the two modes to ensure correct phasing at the aperture; thus

$$\ell \left(1/\lambda_g - 1/\lambda_g'\right) = 0.75 \qquad (15\text{-}36)$$

where $\quad \lambda/\lambda_g = \left[1 - \left(\dfrac{1.84}{ka_0}\right)^2\right]^{1/2} \quad$ and $\quad \lambda/\lambda_g' = \left[1 - \left(\dfrac{3.83}{ka_0}\right)^2\right]^{1/2}$

To ensure that the correct amount of TM_{11} relative to TE_{11} power is generated at the conical junction, the flare angle θ_f should be chosen to satisfy

$$\theta_f = 44.6 \,\lambda/2a_0 \qquad \text{degrees} \qquad (15\text{-}37)$$

This simple equation comes from equating two different expressions for the mode power ratio, one given in Ref. 24 and the other in Ref. 25. If Eqs. (15-36) and (15-37) are applied to the horns in Fig. 15-19, they predict $\ell = 1.42\lambda$, $\theta_f = 34°$ for horn a and $\ell = 3.86\lambda$, $\theta_f = 24°$ for horn b, in reasonably good agreement with the actual values. Small horns constructed in accordance with these principles appear to yield a useful bandwidth of about 12 percent.

There is conversion of TE_{11} energy to TM_{11} at the aperture of the horn in Fig. 15-14 (see subsection "Multimode Pyramidal Horns") as well as at the step. The TM_{11} mode generated at the aperture is in quadrature with the TE_{11} component, and its amplitude is not negligible unless the flare angle is very small. The existence of two physically separate sources of TM_{11}-mode generation complicates the design of these horns, and recourse to empirical adjustment of step size and phasing length is usually necessary.

Multimode Pyramidal Horns

Beam shaping in a pyramidal horn with square aperture can be accomplished in much the same way as with the conical horn. In this case the dominant mode is the TE_{10}, while the necessary higher-order mode is a hybrid mixture of TE_{12} and TM_{12} modes, as shown by Jensen.[26] These two modes, which have the same propagation constant, thus can exist as a hybrid pair, and their relative amplitudes can be adjusted to give a purely linearly polarized aperture field. When added in the proper phase to the TE_{10} field in the aperture, the resulting distribution is tapered in the E as well as in the H plane, as in Fig. 15-20. If one assumes that dual-polarization capability is required, the input waveguide in the figure will be square with a side of dimension d_1 such that the TE_{12}/TM_{12} pair cannot propagate; i.e., $2d_1 < \sqrt{5}\lambda$. At the mode-generating step both dimensions of the guide increase to d_2, where $2d_2 > \sqrt{5}\lambda$. The ratio d_2/d_1 is in the range 1.2 to 1.3, which means that d_1 is large enough to support 11, 02, and 20

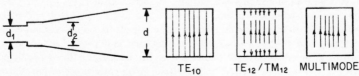

TE$_{10}$ TE$_{12}$/TM$_{12}$ MULTIMODE

FIG. 15-20 Dual-mode pyramidal horn.

modes as well as the dominant TE_{10} (or TE_{01}). For this reason a tapered transition is needed to transform from purely dominant-mode guide (for which $\lambda < 2d_0 < \sqrt{2}\lambda$) to the input guide of side d_1. As for the conical case, a constant-dimension phasing section of length ℓ is used to ensure correct phasing at the aperture. The frequency sensitivity is much like that of the dual-mode conical horn, and a pyramidal horn with a 3λ aperture has a useful pattern bandwidth of 3 to 4 percent.

If the aperture is large enough to make the obliquity factors essentially equal to unity ($d \gtrsim 2\lambda$) and the semiflare angle small enough to render spherical-wave error negligible, the patterns of a multimode horn are given by

$$
\left.
\begin{array}{ll}
E \text{ plane} & E = \sin u/(u) \left(1 + \beta\sqrt{2}\, \dfrac{u^2}{\pi^2 - u^2} \right) \\[2em]
H \text{ plane} & E = \cos u/\left[1 - \left(\dfrac{2u}{\pi} \right)^2 \right]
\end{array}
\right\}
\qquad \textbf{(15-38)}
$$

where $u = (\pi d/\lambda) \sin \theta$ and β^2 represents the fractional TE_{12}/TM_{12}-mode power content relative to unity for the TE_{10}. The interesting range for β is between 0.45 and 0.55, corresponding to a power level in the TE_{12}/TM_{12} mode between 7 and 5 dB below that in the dominant mode. The patterns shown in Fig. 15-21 were calculated from Eqs. (15-38) for $\beta = 0.464$. Note that the main-beam shapes are nearly identical in the two planes down to the -20-dB level, while the first E-plane sidelobe level is very low, approximately -45 dB. The second sidelobe in this plane has a level of -31 dB. The H-plane first sidelobe has a level of -23 dB. The full HPBW in all planes is $69\ \lambda/d\ °$.

Flare-angle changes may also be used in a pyramidal horn to generate the hybrid TE_{12}/TM_{12} pair, as shown by Cohn[27] and illustrated in Fig. 15-22. It is assumed that only the dominant mode propagates at the throat. Conversion to TE_{12}/TM_{12} occurs at the point where the flare angle changes from θ_f to zero. Relative to unity for the TE_{10} mode, the TE_{12}/TM_{12} amplitude is

$$
A = j2d\theta_f/(3\lambda_g) \qquad \textbf{(15-39)}
$$

providing θ_f is small, $\lesssim 23°$, and where λ_g is the TE_{10}-guide wavelength corresponding to the waveguide dimension d. It is seen that the hybrid-mode phase leads that of the dominant mode by $90°$. Therefore, the constant-diameter section must create a $270°$ differential shift to obtain correct phasing in the aperture. The length ℓ can be calculated from Eq. (15-36), but in this case

$$
\lambda/\lambda_g = [1 - (\lambda/2d)^2]^{1/2} \quad \text{and} \quad \lambda/\lambda_g' = [1 - 5(\lambda/2d)^2]^{1/2}
$$

If no phasing section is used (i.e., $\ell = 0$), mode conversion still occurs and Eq. (15-39) still holds. This means that the TE_{12} and TM_{12} energy is improperly phased relative to the TE_{11} and performance is that of a simple flared horn with spherical-wave error in the aperture. This complication must be reckoned with in any design.

If it is desired to design a horn to give the patterns of Fig. 15-21, it is required that $|A| = \beta\sqrt{2}$ with $\beta = 0.464$. Thus the condition is $2d\theta_f/(3\lambda_g) = 0.656$. For a small-aperture horn (e.g., $d = 2\lambda$) this condition leads to a relatively large value for θ_f, which may violate the small-angle condition used in deriving Eq. (15-39). In this case two or more separate flare changes may be used, since the A coefficients are additive, but frequency sensitivity will increase. Good pattern performance over a 30 percent band has been reported for the single-flare-change case.

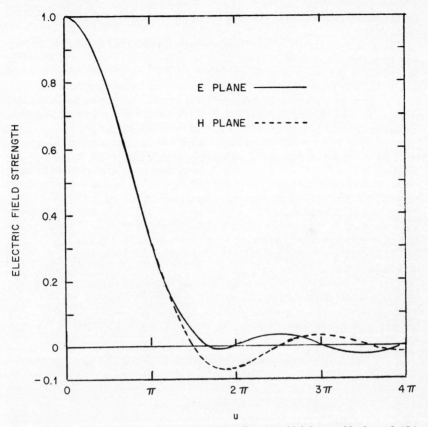

FIG. 15-21 Field-strength patterns of multimode pyramidal horn with $\beta = 0.464$ (calculated).

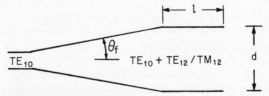

FIG. 15-22 Section through square-aperture pyramidal horn with flare-angle change.

Gain and Impedance

The gain of any of the multimode horns described here can be calculated from

$$G = \eta 4\pi A/\lambda^2 \qquad (A = \text{aperture area})$$

$$G = \eta k^2 a^2 \qquad \text{(conical horn, aperture radius } a)$$

$$G = \eta 4\pi d^2/\lambda^2 \qquad \text{(pyramidal horn, square aperture, side } d)$$

but of course it is necessary to calculate or at least estimate the value of the aperture efficiency η.

In the diagonal horn the power is split equally between the TE_{10} and TE_{01} modes, which are everywhere in phase. Each of these modes has $\eta = 8/\pi^2 = 0.811$; hence this is also the gain factor for a diagonal horn.

In horns employing higher modes the efficiency will be less than that of the dominant mode in all cases described above. This is true because the higher modes that are used for beam shaping do not radiate in the direction $\theta = 0$ and hence contribute nothing to the axial gain given above. Thus, for the pyramidal horn of Fig. 15-21 the aperture efficiency for the dominant TE_{10} mode is 0.81, but this mode carries only 1/ $(1 + \beta^2)$ of the power, namely, 82.3 percent for $\beta = 0.464$. Hence the net aperture efficiency is $\eta = 0.823 \times 0.811 = 0.667$. For a dual-mode conical horn the basic efficiency of the dominant TE_{11} mode is 0.837 (Sec. 15-3). Hence the net aperture efficiency for this horn can be expected to be slightly higher, about 69 percent, than for the pyramidal horn. These figures are upper limits and will be lower if the spherical-wave phase error due to horn flare is nonnegligible. In this case the gain may be estimated approximately by reference to the curve for $\kappa a = 2.405$ of Fig. 15-33 in Sec. 15-5, dealing with corrugated horns. This curve is roughly applicable to a dual-mode conical horn.

Horns, both conical and pyramidal, that use flare-angle change to generate the higher modes are usually quite well matched, with VSWR typically less than 1.03. However, those using a step discontinuity at the throat may have a VSWR of 1.2 to 1.4. This may be matched out by an iris or a dielectric plug inserted in the input dominant-mode waveguide. The matching device should have axial symmetry if dual polarizations are needed.

15-5   CORRUGATED HORNS

Hybrid Modes

The desirable properties of axial-beam symmetry, low sidelobes, and cross-polarization levels can be obtained over a wide frequency range (1.6:1 or more) in a conical horn or open-ended round waveguide that supports the hybrid HE_{11} mode. This mode is again a mixture of TE_{11} and TM_{11} modes of the form $TM_{11} + \gamma TE_{11}$, where γ is a parameter called the mode-content factor. However, the inner surface of the horn or guide may no longer be a smooth metallic conductor. Instead, the surface must be anisotropic in such a way that it has different reactances, X_ϕ in the azimuthal direction and X_z in the axial direction, with $X_\phi \to 0$ while $X_z \to \infty$. If these conditions are realized, the TE and TM components become locked together as a single (hybrid) mode and they propagate with a unique common velocity.

Such an anisotropic impedance can be achieved by cutting circumferential slots into the surface, as suggested by Simmons and Kay[28] and described by Minnett and Thomas.[29] Thus, in Fig. 15-23, which represents a longitudinal section through a cylindrical corrugated waveguide, E_ϕ will be zero on the surface of the teeth at $\rho = a$. Each slot acts as a radial transmission line that is short-circuited at $\rho = b$. If the line

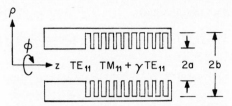

FIG. 15-23 Cylindrical corrugated waveguide with TE$_{11}$-mode excitation.

length, $b - a$, is one-quarter guide wavelength, then an open circuit appears across each gap at $\rho = a$. Providing there are several slots per wavelength, then the conditions $X_\phi \to 0$, $X_z \to \infty$ will be realized in an average sense. Looked at in another way, the open-circuit gaps ensure that no axial current can flow at $\rho = a$. Consequently, H_ϕ, the magnetic field which induces axial current flow, must also vanish. On average, the same boundary conditions now prevail for the azimuthal components of both E and H fields at the guide wall. This was conjectured by Simmons and Kay[28] and later shown by Rumsey[30] to be a condition on the aperture field of a horn that leads to symmetry and freedom from cross-polarization in its radiation pattern.

The properties of corrugated waveguides supporting such hybrid modes are thoroughly treated in the literature.[31,32,33,34] Nevertheless, it is not easy to abstract the fundamentals in a useful and comprehensive way. This is partly due to differences in approach and in notation on the parts of the authors and partly to the inherent complexity of the analysis. As an example of the latter, when the radiation pattern of a hybrid-mode horn is calculated by the Kirchhoff-Huygens method, the condition for symmetry and zero cross-polarization is found to be $\gamma = 1$, but when calculated by the Fourier transform technique the condition obtained is $\gamma = \beta/k$ (where $\beta/k = \lambda/\lambda_g$). Experimentally, it is found that the aperture diameter must be fairly large to obtain the desired performance. This means that β/k must be near unity, and so, of course, the condition becomes $\gamma \to 1$, but subject to the restriction to reasonably large aperture size. The discussion that follows, rather more tutorial than is customary in a handbook, is presented as a brief eclectic summary of the more important features necessary to an understanding of the radiation characteristics of corrugated horns. Only those modes that vary azimuthally as $\cos \phi$, $\sin \phi$ are treated. These correspond to fields that vary radially as $J_m(\kappa\rho)$, $J_m'(\kappa\rho)$, where m is restricted to the value unity.

With reference to Fig. 15-23, the dominant TE$_{11}$ mode is introduced in the smooth-wall section of round waveguide at the left. Some reflection of this mode occurs at the abrupt junction between the smooth wall and the first slot. In the corrugated section the new boundary conditions ensure that the propagating mode is the hybrid combination symbolically represented by TM$_{11}$ + γTE$_{11}$. This mode has the cylindrical components

$$\left.\begin{array}{ll} E_\rho = e_\rho(\rho) \cos \phi & Z_0 H_\rho = h_\rho(\rho) \sin \phi \\[2mm] E_\phi = e_\phi(\rho) \sin \phi & Z_0 H_\phi = h_\phi(\rho) \cos \phi \\[2mm] E_z = e_z(\rho) \cos \phi & Z_0 H_z = h_z(\rho) \sin \phi \end{array}\right\} \quad \textbf{(15-40)}$$

where $\exp(-j\beta z)$ is understood, and

$$
\begin{aligned}
e_\rho(\rho) &= -j\left[\frac{\beta}{k}J_1'(\kappa\rho) + \gamma\frac{J_1(\kappa\rho)}{\kappa\rho}\right] \\[2mm]
e_\phi(\rho) &= j\left[\frac{\beta}{k}\frac{J_1(\kappa\rho)}{\kappa\rho} + \gamma J_1'(\kappa\rho)\right] \\[2mm]
e_z(\rho) &= \frac{\kappa}{k}J_1(\kappa\rho) \\[2mm]
h_\rho(\rho) &= -j\left[\frac{J_1(\kappa\rho)}{\kappa\rho} + \gamma\frac{\beta}{k}J_1'(\kappa\rho)\right] \\[2mm]
h_\phi(\rho) &= -j\left[J_1'(\kappa\rho) + \gamma\frac{\beta}{k}\frac{J_1(\kappa\rho)}{\kappa\rho}\right] \\[2mm]
h_z(\rho) &= \gamma\frac{\kappa}{k}J_1(\kappa\rho)
\end{aligned}
\qquad\text{(15-41)}
$$

β and κ are the propagation constants in the z and ρ directions, respectively, and are related to the free-space wave number by

$$
\beta^2 + \kappa^2 = k^2 \qquad (k = 2\pi/\lambda) \qquad\text{(15-42)}
$$

By the use of Bessel-function recurrence formulas and a transformation to rectangular coordinates, the x and y components of field are found to be

$$
\left.
\begin{aligned}
E_x &\sim (1 + \gamma k/\beta) \cdot J_0(\kappa\rho) - (1 - \gamma k/\beta) \cdot J_2(\kappa\rho) \cdot \cos 2\phi \\
E_y &\sim (1 - \gamma k/\beta) \cdot J_2(\kappa\rho) \cdot \sin 2\phi
\end{aligned}
\right\}
\quad\text{(15-43)}
$$

These will be the transverse components of field in the aperture of the open waveguide in Fig. 15-23 if it is assumed that the aperture is reasonably large so that reflection is negligible. E_x and E_y are respectively the copolar and cross-polar components, and it is clear that the cross-polarization will be zero if $\gamma = \beta/k$. Furthermore, this condition ensures that E_x tapers radially with ρ but does not vary azimuthally with ϕ. However, if $\gamma = -\beta/k$, the aperture field is a mixture of E_x and E_y with strong azimuthal variations.

To satisfy the condition $X_\phi = 0$, E_ϕ must vanish at $\rho = a$, and Eqs. (15-41) then give

$$
\gamma = -\frac{\beta}{k} \cdot \frac{J_1(\kappa a)}{\kappa a J_1'(\kappa a)} = \frac{\beta}{k} \cdot \frac{J_2(\kappa a) + J_0(\kappa a)}{J_2(\kappa a) - J_0(\kappa a)} \qquad\text{(15-44)}
$$

Thus it is observed that the condition $\gamma = \beta/k$ occurs whenever $J_0(\kappa a)$ vanishes, which happens for $\kappa a = 2.405, 5.520, 8.654 \ldots$. In this case the modes are designated HE_{11}, HE_{12}, $HE_{13} \ldots$, respectively, and all these modes create purely copolar aperture fields. The condition $\gamma = -\beta/k$ is seen to occur whenever $J_2(\kappa a)$ vanishes; i.e., for $\kappa a = 0, 5.136, 8.417, \ldots$, and these modes are designated $EH_{11}, EH_{12}, EH_{13} \ldots$.

The longitudinal reactance X_z (which is infinite only at the slot resonant frequency) is given in the general case by $X_z/Z_0 = je_z/h_\phi$ evaluated at $\rho = a$. Hence, from Eqs. (15-41)

$$
X_z/Z_0 = \frac{-\kappa^2 a J_1(\kappa a)}{k\kappa a J_1'(\kappa a) + \gamma\beta J_1(\kappa a)} \qquad\text{(15-45)}
$$

Combining Eqs. (15-44) and (15-45) gives

$$\frac{1}{\gamma} - \gamma = \frac{\kappa^2 a}{\beta} \cdot Z_0/X_z \qquad \textbf{(15-46)}$$

and in the special case of resonant slots ($X_z = \infty$) it follows that $\gamma = \pm 1$. This is called the *balanced* hybrid condition, and the designation HE is used for $\gamma = +1$ and EH for $\gamma = -1$. Because Eq. (15-46) is a quadratic in γ, it has two roots in the general case and the designations HE and EH are still applicable even under unbalanced conditions.

The characteristic equation for HE or EH hybrid modes is obtained from Eq. (15-45) by substituting for γ from Eq. (15-44). Solutions are shown graphically in Fig. 15-24 in the form given by Chu and Legg,[35] in which the quantity κa is plotted as a function of guide circumference ka, with normalized reactance X_z/Z_0 as a parameter. The dispersion characteristics of the corrugated-waveguide modes then follow when β/k (i.e., λ/λ_g) is determined from these solutions for κa, along with the use of Eq. (15-42). β/k is shown in Fig. 15-25 as a function of ka for a number of HE and EH

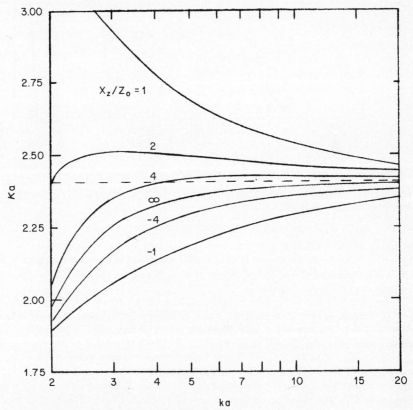

FIG. 15-24 Solutions to characteristic equation; κa versus ka with X_z/Z_0 as a parameter. (© *1982 IEEE; see Ref. 35.*)

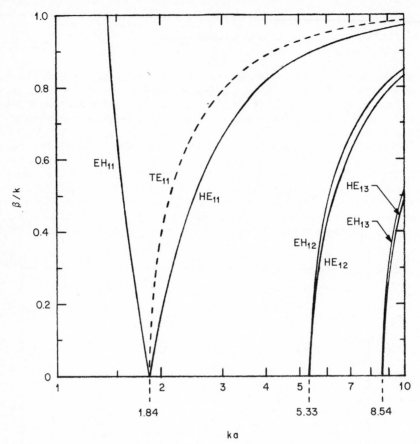

FIG. 15-25 Dispersion characteristics of balanced hybrid modes (TE$_{11}$ mode also shown for comparison).

modes under the condition $\gamma = \pm 1$. The dashed line, shown for comparison, is the dispersion curve for the dominant TE$_{11}$ mode, for which κa has the constant eigenvalue 1.841. The behavior of the hybrid modes is complicated by the fact that κa is not constant even in the balanced case.

Figure 15-24 shows that the desired condition on the aperture field, namely, $\kappa a = 2.405$, is approached asymptotically when ka is large. In that case β/k approaches unity, and from Eq. (15-44) γ must approach $+1$. The restriction to large values of ka was pointed out by Knop[33] in the form $ka \gtrsim 2\pi$. On the other hand, Fig. 15-24 also shows that when ka is large, κa will be close to the desired value of 2.405 even when X_z is far from resonance. This means that $J_0(\kappa a)$ will be small and that $\gamma k/\beta$ will not depart greatly from unity despite the fact that X_z changes drastically with frequency. This was noted by Dragone,[36] who also verified experimentally that operation is possible over a frequency range of more than 1.5:1. The value of ka in this case was 12 at the lowest frequency in the band, corresponding to an aperture diameter of about 4λ. Slot depth at the aperture varied from about 0.21λ to 0.31λ, and in the throat from 0.34λ to 0.51λ. Operation was limited at the high end of the band by the appearance of the unwanted EH_{11} mode.

Slot resonance ($X_z = \infty$) occurs when slot depth is one-quarter wavelength in the radial line of radii a, b. This depth approaches one-quarter free-space wavelength when ka is large, as shown in Fig. 15-26. The tooth width t in Fig. 15-27 should be small compared with slot width w, say, $t < 0.2w$, in order to reduce frequency sensitivity. There should be several slots per wavelength, preferably four or more, although in wideband horns it is permissible to have only two per wavelength at the high-frequency end of the band.

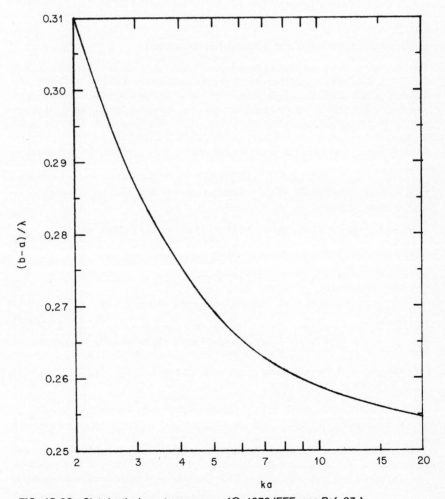

FIG. 15-26 Slot depth, *b-a*, at resonance. *(© 1978 IEEE; see Ref. 37.)*

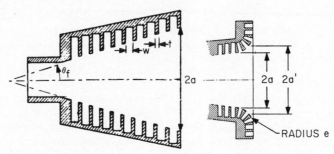

FIG. 15-27 Small-flare-angle corrugated horn at left; corrugations extended into flange at right. (© *1978 IEEE; see Ref. 37.*)

Conical Corrugated Horn (Small-Flare-Angle)

Figure 15-27 shows a corrugated horn in which slot depth varies in accordance with Fig. 15-26 in order to satisfy the resonance condition at some midband frequency. The assumption of a small flare angle permits the use of the approximation that the aperture field is just that of a corrugated waveguide of diameter $2a$, but with the superposition of a spherical-wave error of the kind discussed in Sec. 15-1. Thus the phase of the field in the aperture will vary quadratically with radius, and the expressions given by Eqs. (15-43) must be multiplied by $\exp(-jk\Delta r^2)$, where $r = \rho/a$ and $\Delta = a^2/(2\ell)$.

The far field of this aperture distribution can be calculated by the Fourier transform method, as in Ref. 33. If the obliquity factors of Sec. 15-1 are suppressed, the field at a distant point P having coordinates (R, θ, ϕ)[*] is

$$\left. \begin{aligned} E_x(P) &\sim (1 + \gamma k/\beta) \cdot g(u,\Delta) + (1 - \gamma k/\beta) \cdot h(u,\Delta) \cdot \cos 2\phi \\ E_y(P) &\sim (1 - \gamma k/\beta) \cdot h(u,\Delta) \cdot \sin 2\phi \end{aligned} \right\} \quad \textbf{(15-47)}$$

where $u = ka \cdot \sin\theta$ $r = \rho/a$ and

$$\left. \begin{aligned} g(u,\Delta) &= \int_0^1 J_0(\kappa ar) J_0(ur) \exp(-jk\Delta r^2) \cdot r\,dr \\ h(u,\Delta) &= \int_0^1 J_2(\kappa ar) J_2(ur) \exp(-jk\Delta r^2) \cdot r\,dr \end{aligned} \right\} \quad \textbf{(15-48)}$$

When $\Delta = 0$, corresponding to the open waveguide of Fig. 15-23, Eqs. (15-48) are directly integrable and yield

$$\left. \begin{aligned} g(u,0) &= [(\kappa a)^2 - u^2]^{-1} \cdot [\kappa a J_1(\kappa a) J_0(u) - u J_0(\kappa a) J_1(u)] \\ h(u,0) &= [(\kappa a)^2 - u^2]^{-1} \cdot [u J_1(u) J_2(\kappa a) - \kappa a J_1(\kappa a) J_2(u)] \end{aligned} \right\} \quad \textbf{(15-49)}$$

These equations show that $h(u,0)$ vanishes, while $g(u,0)$ has its maximum in the axial direction $u = 0$. The function $h(u,0)$, which determines the shape of the cross-polar pattern, has its first maximum at $u = 3.65$. Thus the cross-polar component E_y is zero

---

[*]This far-field spherical-coordinate system is distinct from the aperture coordinate frame (ρ, ϕ, z).

on axis and in the principal planes and has peaks in the intercardinal planes ($\phi = \pm 45°$) situated approximately where $\theta = \sin^{-1}(0.58\lambda/a)$.

Under balanced hybrid conditions and large ka, i.e., with $\kappa a \doteq 2.405$, Eqs. (15-47) and (15-48) simplify to

$$E_x(P) \sim \frac{J_0(u)}{1 - (u/2.405)^2} \qquad E_y(P) = 0 \qquad \textbf{(15-50)}$$

confirming that there is zero cross-polarization and that the copolar pattern is axially symmetric for the HE_{11} mode. This is also true for the higher-order HE_{12}, HE_{13}, ... modes.

The balanced EH modes correspond to $J_2(\kappa a) = 0$, $\gamma = -\beta/k$, and the far field is then

$$\left.\begin{array}{l} E_x(P) \sim (1 - u^2/\kappa^2 a^2)^{-1} \cdot J_2(u) \cdot \cos 2\phi \\[2mm] E_y(P) \sim (1 - u^2/\kappa^2 a^2)^{-1} \cdot J_2(u) \cdot \sin 2\phi \end{array}\right\} \qquad \textbf{(15-51)}$$

These modes therefore radiate with a null in the axial direction ($\theta = 0$) and a state of polarization that may be linear, elliptical, or circular, depending on ϕ. They are almost invariably undesirable in antenna engineering applications.

A set of universal patterns for the small-flare horn of Fig. 15-27 is shown in Fig. 15-28 under balanced hybrid conditions and large ka. The patterns were computer-

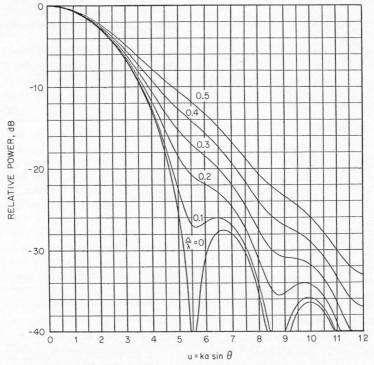

FIG. 15-28 Universal patterns for small-flare-angle corrugated horns under near-balanced conditions.

generated from the first of Eqs. (15-48) and are reasonably accurate for a half-flare angle less than about 20°. According to Thomas,[37] as ka decreases below about 12, the sidelobe level for a given Δ gradually decreases but the shape of the main lobe is virtually unaffected until ka reaches about 6. This is confirmed in Ref. 33. For $ka <$ 6 the flange at the aperture begins to affect the pattern. The patterns become invalid for $ka < 4$.

When the flange affects the pattern shape (i.e., $ka \gtrsim 6$), a remedy recommended in Ref. 37 is to extend the corrugations into the plane of the flange as indicated in Fig. 15-27. This will preserve pattern symmetry, but it increases the effective aperture size slightly. The new aperture radius is approximately $a' = a + e/2$, where e is the radius of the curved portion of the flange.

Wide-Flare-Angle Scalar Horn

A wide-flare horn is shown in Fig. 15-29, and it is apparent that the path error in the aperture can become quite large. In this case Eqs. (15-2) are inaccurate, and the expressions in Eqs. (15-48) and the patterns of Fig. 15-28 are no longer applicable. Such a horn nevertheless may still exhibit axial symmetry and low cross-polarization in its radiation pattern, but pattern calculation becomes considerably more complex, involving spherical rather than cylindrical hybrid modes.[32,38] Note that the slots are customarily cut perpendicularly to the horn wall, as in Fig. 15-29, rather than to the horn axis, as in Fig. 15-27. The term *scalar horn* was coined[28] to emphasize the fact that the far-zone field can be represented by a single scalar quantity under ideal conditions when there is no cross-polarization. Although this is equally true for the small-flare horn, the term scalar horn is usually reserved for the wide-flare case.

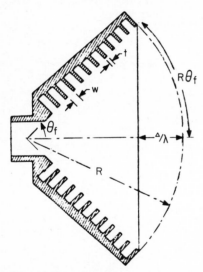

FIG. 15-29 Wide-flare scalar horn. *(© 1978 IEEE; see Ref. 37.)*

A set of universal patterns for scalar horns under near-balanced hybrid conditions is shown in Fig. 15-30, taken from Thomas.[37] It is important to note that the abscissa is now θ/θ_f, where θ is the far-field polar angle and θ_f is the semiflare angle of the horn. Since θ/θ_f is independent of frequency, it is clear that the scalar-horn patterns are, in large measure, frequency-insensitive. This is in direct contrast to the small-flare-horn case, in which beamwidth is proportional to aperture size ka. The transition is determined by the value of Δ as defined in Fig. 15-29 by

$$\Delta = R \sin \theta_f \cdot \tan \theta_f/2 \qquad \textbf{(15-52)}$$

For $\Delta < 0.4\lambda$ the beamwidth is controlled mainly by ka and is therefore frequency-dependent. As Δ increases beyond 0.5λ, this dependence becomes less and less, and when Δ exceeds about 0.75λ, the nearly frequency-independent conditions shown in

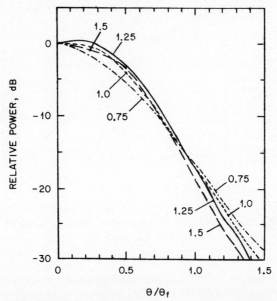

FIG. 15-30 Universal patterns for wide-flare scalar horns ($\theta_f < 70°$) with Δ/λ as a parameter. (© *1978 IEEE; see Ref. 37.*)

Fig. 15-30 result. These patterns are applicable for values of θ_f up to about 70°. As Δ increases beyond 0.75λ, the beam acquires a flattened top and the skirts become steeper. At $\Delta = 1.25\lambda$ there is actually a slight dip in the axial direction. Figure 15-31 shows the effect of Δ on the beamwidth of scalar horns at various power levels. A good rule of thumb is that the beam angle at the -12-dB level is very close to 0.8 θ_f, while at the -15-dB level it is very close to 0.9 θ_f for any Δ between 0.75λ and 1.5λ.

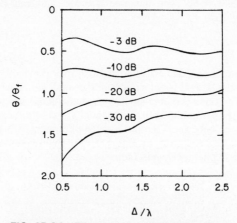

FIG. 15-31 Effect of Δ/λ on the beamwidth of scalar horns at various power levels. (© *1978 IEEE; see Ref. 37.*)

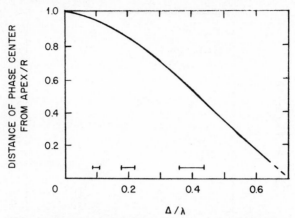

FIG. 15-32   Distance of phase center, normalized relative to slant length R from apex. (© *1978 IEEE; see Ref. 37.*)

The spherical-wave error Δ has a marked effect on the location of the phase center of the horn, as can be seen in Fig. 15-32. When Δ is very small, the phase center is on the horn axis very close to the plane of the aperture. As Δ increases, the phase center moves along the axis toward the throat and eventually becomes fixed at the horn apex for scalar horns with $\Delta \gtrsim 0.7\lambda$. The bars show the effect of a frequency change of ± 10 percent. The beam efficiency of scalar horns is high. When Δ exceeds 0.75λ, the beam efficiency at the -10-dB level is greater than 87 percent and at the -20-dB level it is greater then 98 percent. This is a valuable asset when such a horn is used as a feed in a reflector antenna since spillover loss will be small.

To design a scalar horn with a given beamwidth at some relative power level, say, -6-dB, one determines θ/θ_f from Fig. 15-30. It is apparent that $\theta/\theta_f = 0.6$ in this case and that the parameter Δ/λ may have any value between 1.0 and 1.5. If the desired beamwidth is $2\theta = 60°$, one finds that $\theta_f = 50°$, and from Eq. (15-52) R may lie between 2.8λ and 4.2λ.

Gain

The conical corrugated horn has certain advantages over the dominant-mode pyramidal horn as a gain standard. It is not subject to the small gain uncertainties[11] caused by E-plane edge diffraction, its ohmic loss is very small, and it has considerably lower sidelobe levels than the pyramidal horn. The gain of a small-flare corrugated horn has been calculated[35] as a function of the phase-error parameter Δ/λ by making use of the aperture field expressions given by Eqs. (15-43). Relative to a uniform circular aperture of the same diameter the gain factor is

$$\eta = \frac{\left| 2(1 + \gamma k/\beta) \int_0^1 J_0(\kappa ar) e^{-jk\Delta r^2} r\,dr \right|^2}{(1 + \gamma k/\beta)^2 [J_0^2(\kappa a) + J_1^2(\kappa a)] + (1 - \gamma k/\beta)^2 [J_2^2(\kappa a) - J_1(\kappa a)J_3(\kappa a)]}$$

(15-53)

This expression has been computed[35] as a function of Δ/λ for three values of κa, namely, 2.3, 2.405, and 2.5. The results are given in Table 15-3 and are shown graphically in Fig. 15-33. It is seen that the gain is not highly sensitive to deviations from the zero cross-polar condition, $\kappa a = 2.405$, and that this is particularly true in the vicinity of $\Delta = 0.5\lambda$. It is clear that Eq. (15-53) is based on the quadratic phase approximation and hence is valid only for small-flare angles ($\theta_f \gtrsim 20°$).

The actual gain of the corrugated horn is, of course, $G = \eta(2\pi a/\lambda)^2$. Using the small-flare-angle approximation, $\Delta = a^2/(2\ell)$, allows the gain to be written as $G = 8\pi^2\eta\ell\Delta/\lambda^2$, which shows that for a fixed ℓ/λ the gain is maximized when the product $\eta\Delta/\lambda$ is maximized. This product is plotted in Fig. 15-34 by using the data in Table 15-3. The optimum-gain condition is thus seen to be $\Delta = 0.5\lambda$, and in this region it is clear that the gain is insensitive both to frequency and to small deviations from the condition $\kappa a = 2.405$.

Experimental measurements reported in Ref. 35 on a conical corrugated horn with an aperture diameter $2a = 127$ mm and $\theta_f = 13.5°$ are in excellent agreement with values calculated from Eq. (15-53). At 19.04 and 28.56 GHz the calculated values were 24.4 and 25.1 dB respectively. The corresponding measured values were 24.40 ± 0.08 dB and 24.98 ± 0.08 dB.

Table 15-3   Calculated Gain Factor of Small-Flare Corrugated Horn, dB

Spherical-wave phase error Δ/λ	Eigenvalue parameter κa		
	2.3	**2.405**	**2.5**
0	−1.27	−1.59	−1.99
0.05	−1.30	−1.61	−2.01
0.10	−1.37	−1.68	−2.06
0.15	−1.50	−1.79	−2.15
0.20	−1.68	−1.95	−2.28
0.25	−1.92	−2.15	−2.45
0.30	−2.20	−2.40	−2.65
0.35	−2.54	−2.69	−2.90
0.40	−2.94	−3.04	−3.18
0.45	−3.39	−3.42	−3.50
0.50	−3.89	−3.86	−3.86
0.55	−4.45	−4.34	−4.26
0.60	−5.07	−4.86	−4.70
0.65	−5.74	−5.43	−5.18
0.70	−6.46	−6.05	−5.70
0.75	−7.22	−6.70	−6.27
0.80	−8.02	−7.39	−6.87
0.85	−8.84	−8.11	−7.50
0.90	−9.66	−8.84	−8.16
0.95	−10.46	−9.57	−8.85
1.00	−11.19	−10.28	−9.55

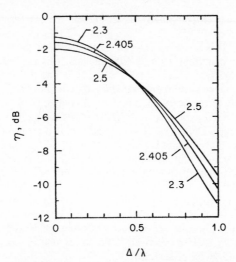

FIG. 15-33 Gain factor η versus spherical-wave error Δ/λ for conical corrugated horn with κa as a parameter. (© *1982 IEEE; see Ref. 35.*)

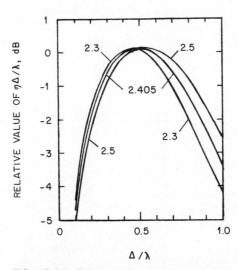

FIG. 15-34 Relative value of $\eta\Delta/\lambda$ versus Δ/λ (i.e., relative gain for fixed ℓ/λ) for a conical corrugated horn with κa as a parameter. (© *1982 IEEE; see Ref. 35.*)

Impedance

A good match between the TE_{11} and the HE_{11} modes at the point at which the corrugations begin can be achieved by making the first slot one-half wavelength deep, then gradually reducing the slot depth until the resonant depth determined from Fig. 15-26 is reached at the band-center frequency. If the depth is tapered over a distance of a wavelength or more, the match will remain good over a wide (1.5:1) frequency range. By making the first few slots close to one-half wavelength deep, the reactance X_z will be close to zero, thus minimizing the discontinuity between the smooth conducting boundary and the corrugated wall.

To prevent generation of the unwanted EH_{11} mode at the junction of the smooth and slotted guides the reactance X_z should not be positive ($X_z \not> 0$). Hence it is advisable to make the first slot one-half wavelength deep at the highest frequency in the band.

By following the above practices it should be possible to obtain a VSWR less than about 1.25 over a 1.5:1 frequency range in a scalar horn and to do considerably better in a small-flare horn over a narrower frequency range.

Two-Hybrid-Mode Horns

Additional beam shaping is possible in a corrugated horn by generating higher-order hybrid modes along with the HE_{11}. The desired higher-order modes are the HE_{12}, $HE_{13}, \ldots$, all of which have axially symmetric patterns that are free of cross-polarization under near-balanced conditions, i.e., large aperture and $\gamma \to 1$. However, the higher-order modes have different dispersion characteristics, as seen in Fig. 15-25, and this causes such horns to be quite frequency-sensitive. For this reason and also because of difficulties in generating the higher modes in the correct amplitude ratios, only the two-hybrid-mode horn has seen any practical use.

The two-mode horn utilizes only the HE_{11} and the HE_{12} modes, both of which can be generated by a step discontinuity in the throat of a corrugated horn as shown in Fig. 15-35. The smooth-wall input guide supports only the TE_{11} mode, so that $1.84 < ka < 3.83$. In the corrugated region beyond the step the radius is a_0, and it must be large enough to support the HE_{12} as well as the HE_{11} mode but not so large as to support the HE_{13} mode. Thus, Fig. 15-25 shows that $5.33 < ka_0 < 8.54$. Thomas[39] states that a signifi-

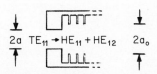

FIG. 15-35 Two-hybrid-mode generation at a step discontinuity.

cant improvement in the match near cutoff can be obtained by a gradual tapered increase in the diameter of the input guide near the step. This does result, however, in a slight reduction in the relative power of the HE_{12} mode for a given diameter a_0 in the corrugated region. The HE_{12}-mode power relative to that of the HE_{11} mode increases as the ratio a_0/a increases; unfortunately a small amount of the EH_{12} mode is also produced. The presence of this mode is undesirable since it causes a significant increase in cross-polar radiation.

The radius a_0 in the corrugated section should be chosen so that ka_0 is close to 8.54, the cutoff for the HE_{13} mode, at the uppermost frequency in the band of interest.

If the mismatch at the upper frequency is then too large, the radius a_0 must be decreased somewhat. A compromise also must be made in regard to the TE_{11} input-guide diameter. Large values of a (ka near 3.83) lead to a low VSWR and a low level of the unwanted EH_{12} mode but may not result in sufficient power in the HE_{12} mode. Values between 2.5 and 3.0 for the ratio a_0/a have provided satisfactory results.

Because the two modes have different propagation constants (Fig. 15-25), the two-hybrid-mode horn must usually incorporate a constant-diameter ($2a_0$) phasing section before it flares to its final aperture diameter. This is to ensure that the modes are in the correct phase at the aperture and is exactly analogous to the use of a phasing section in the dual-mode conical horn of Fig. 15-14.

The pattern of a two-hybrid-mode horn having $ka_0 = 8.17$ at the aperture (there is no flare) is shown in Fig. 15-36.[40] In this case the length of the corrugated section (of diameter $2a_0$) is such as to antiphase the main lobes of the HE_{11} and HE_{12} radiation patterns; this creates the dip in the resultant pattern in the direction $\theta = 0$. The phase relationship will change with frequency, and so, of course, will the pattern shape. This horn was designed as a highly efficient feed for a radiotelescope reflector antenna having an opening angle $\theta = 63°$ at the rim. Similar feeds described in Ref. 39 have bandwidths of a few percent. The relative power levels in the two modes are about equal. The pattern in Fig. 15-36 obviously has desirable characteristics in earth-coverage use from a satellite in a high-altitude circular orbit.

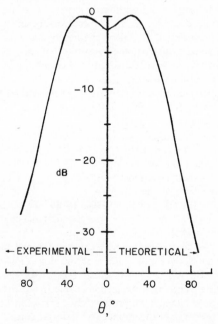

FIG. 15-36 Experimental and theoretical patterns of a two-hybrid-mode horn with ka_0 = 8.17. *(Courtesy IEE; see Ref. 40.)*

Rectangular Corrugated Horn

Sectoral and pyramidal horns excited by the TE_{10} mode in a rectangular waveguide may also benefit from the use of corrugations in the electric (i.e., broad) walls of the horn.[41,42] The effect of the corrugations is to make the magnetic field small or zero at the electric walls, analogously to the vanishing electric field at the magnetic walls. As a result the E-plane beamwidth is broadened, and the sidelobes are greatly suppressed.

In a square-aperture horn with corrugations in the electric walls the E- and H-plane patterns will be very nearly identical, at least to the -20-dB level, and this can be achieved over nearly a 2:1 frequency range.[42] Such a horn, however, is restricted to a single polarization, namely, that in which E is orthogonal to the corrugated walls.

In principle, a dual or circularly polarized horn can be constructed by corrugating all four walls of a square-aperture pyramidal horn. In practice, these horns have generally been found to suffer from impedance irregularities and deviations from axial-beam symmetry when wideband (1.5:1) operation is desired. They apparently have not found the favor accorded conical corrugated horns and will not be further discussed. However, dielectric-lined rectangular and square waveguides have been shown[43] to support hybrid modes of the type HE_{11}. This offers promise that some of the drawbacks noted above for the square-aperture hybrid-mode horn may be overcome.

REFERENCES

1 A. W. Love (ed.), *Electromagnetic Horn Antennas,* IEEE Press, New York, 1976.
2 S. Silver, *Microwave Antenna Theory and Design,* MIT Rad. Lab. ser., vol. 12, McGraw-Hill Book Company, New York, 1949.
3 M. S. Narasimhan and V. V. Rao, "A Correction to the Available Radiation Formula for E-Plane Sectoral Horns," *IEEE Trans. Antennas Propagat.,* vol. AP-21, November 1973, pp. 878–879.
4 P. M. Russo, R. C. Rudduck, and L. Peters, "A Method for Computing E-Plane Patterns of Horn Antennas," *IEEE Trans. Antennas Propagat.,* vol. AP-13, March 1965, pp. 219–224.
5 J. S. Yu, R. C. Rudduck, and L. Peters, "Comprehensive Analysis for E-Plane of Horn Antennas by Edge Diffraction Theory," *IEEE Trans. Antennas Propagat.,* vol. AP-14, March 1966, pp. 138–149.
6 C. A. Mentzer, L. Peters, and R. C. Rudduck, "Slope Diffraction and Its Application to Horns," *IEEE Trans. Antennas Propagat.,* vol. AP-23, March 1975, pp. 153–159.
7 S. Schelkunoff, "On Diffraction and Radiation of Electromagnetic Waves," *Phys. Rev.,* vol. 56, Aug. 15, 1939, pp. 308–316.
8 E. V. Jull, "Finite Range Gain of Sectoral and Pyramidal Horns," *Electron. Lett.,* vol. 6, Oct. 15, 1970, pp. 680–681.
9 E. V. Jull and L. E. Allan, "Gain of an E-Plane Sectoral Horn—A Failure of the Kirchhoff Method and a New Proposal," *IEEE Trans. Antennas Propagat.,* vol. AP-22, March 1974, pp. 221–226.
10 E. V. Jull, "On the Behavior of Electromagnetic Horns," *IEEE Proc.,* vol. 56, January 1968, pp. 106–108.
11 E. V. Jull, "Errors in the Predicted Gain of Pyramidal Horns," *IEEE Trans. Antennas Propagat.,* vol. AP-21, January 1973, pp. 25–31.

12   L. Lewin, *Advanced Theory of Waveguides,* Iliffe & Sons, Ltd., London 1951.
13   M. S. Narasimhan and B. V. Rao, "Modes in a Conical Horn: New Approach," *IEE Proc.,* vol. 118, February 1971, pp. 287–292.
14   M. A. K. Hamid, "Diffraction by a Conical Horn," *IEEE Trans. Antennas Propagat.,* vol. AP-16, September 1968, pp. 520–528.
15   M. S. Narasimhan and M. S. Sheshadri, "GTD Analysis of the Radiation Patterns of Conical Horns," *IEEE Trans. Antennas Propagat.,* vol. AP-26, November 1978, pp. 774–778.
16   G. C. Southworth and A. P. King, "Metal Horns as Directive Receivers of Ultra-Short Waves," *IRE Proc.,* vol. 27, February 1939, pp. 95–102.
17   A. P. King, "The Radiation Characteristics of Conical Horn Antennas," *IRE Proc.,* vol. 38, March 1950, pp. 249–251.
18   M. S. Narasimhan and B. V. Rao, "Radiation from Conical Horns with Large Flare Angles," *IEEE Trans. Antennas Propagat.,* vol. AP-19, September 1971, pp. 678–681.
19   A. W. Love, "The Diagonal Horn Antenna," *Microwave J,* vol. V, March 1962, pp. 117–122.
20   P. D. Potter, "A New Horn Antenna with Suppressed Sidelobes and Equal Beamwidths," *Microwave J.,* vol. VI, June 1963, pp. 71–78.
21   K. K. Agarwal and E. R. Nagelberg, "Phase Characteristics of a Circularly Symmetric Dual-Mode Transducer," *IEEE Trans. Microwave Theory Tech.,* vol. MTT-18, December 1970, pp. 69–71.
22   W. J. English, "The Circular Waveguide Step-Discontinuity Mode Transducer," *IEEE Trans. Microwave Theory Tech.,* vol. MTT-21, October 1973, pp. 633–636.
23   T. Satoh, "Dielectric Loaded Horn Antennas," *IEEE Trans. Antennas Propagat.,* vol. AP-20, March 1972, pp. 199–201.
24   K. Tomiyasu, "Conversion of TE_{11} Mode by a Large Diameter Conical Junction," *IEEE Trans. Microwave Theory Tech.,* vol. MTT-17, May 1969, pp. 277–279.
25   R. H. Turrin, "Dual Mode Small Aperture Antennas," *IEEE Trans. Antennas Propagat.,* vol. AP-15, March 1967, pp. 307–308.
26   P. A. Jensen, "A Low-Noise Multimode Cassegrain Monopulse Feed with Polarization Diversity," *IEEE NEREM Rec.,* November 1963, pp. 94–95.
27   S. B. Cohn, "Flare Angle Changes in a Horn as a Means of Pattern Control," *Microwave J.,* vol. 13, October 1970, pp. 41–46.
28   A. J. Simmons and A. F. Kay, "The Scalar Feed—A High Performance Feed for Large Paraboloid Reflectors," *IEE Conf. Publ. 21,* 1966, pp. 213–217.
29   H. C. Minnett and B. M. Thomas, "A Method of Synthesizing Radiation Patterns with Axial Symmetry," *IEEE Trans. Antennas Propagat.,* vol. AP-14, September 1966, pp. 654–656.
30   V. H. Rumsey, "Horn Antennas with Uniform Power Patterns around Their Axes," *IEEE Trans. Antennas Propagat.,* vol. AP-14, September 1966, pp. 656–658.
31   B. M. Thomas, "Theoretical Performance of Prime Focus Paraboloids Using Cylindrical Hybrid Modes," *IEE Proc.,* vol. 118, November 1971, pp. 1539–1549.
32   P. J. Clarricoats and P. K. Saha, "Propagation and Radiation Behavior of Corrugated Feeds," *IEE Proc.,* vol. 118, parts 1 and 2, September 1971, pp. 1167–1186.
33   C. M. Knop and H. J. Wiesenfarth, "On the Radiation from an Open-Ended Corrugated Pipe Carrying the HE_{11} Mode," *IEEE Trans. Antennas Propagat.,* vol. AP-20, September 1972, pp. 644–648.
34   C. Dragone, "Reflection, Transmission and Mode Conversion in a Corrugated Feed," *Bell Syst. Tech., J.,* vol. 56, July–August 1977, pp. 835–867.
35   T. S. Chu and W. E. Legg, "Gain of Corrugated Conical Horns," *IEEE Trans. Antennas Propagat.,* vol. AP-30, July 1982, pp. 698–703.
36   C. Dragone, "Characteristics of a Broadband Microwave Corrugated Feed: A Comparison between Theory and Experiment," *Bell Syst. Tech. J.,* vol. 56, July–August 1977, pp. 869–889.

37 B. M Thomas, "Design of Corrugated Conical Horns," *IEEE Trans. Antennas Propagat.,* vol. AP-26, March 1978, pp. 367–372.

38 J. K. M. Jansen, M. E. J. Jeuken, and C. W. Lambrechtse, "The Scalar Feed," *Arch. Elek. Übertragung,* vol. 26, January 1972, pp. 22–30.

39 B. M. Thomas, "Prime Focus One and Two Hybrid Mode Feeds," *Electron. Lett.,* vol. 6, July 23, 1970, pp. 460–461.

40 T. B. Vu and Q. H. Vu, "Optimum Feed for Large Radiotelescopes: Experimental Results," *Electron. Lett.,* vol. 6, Mar. 19, 1970, pp. 159–160.

41 R. E. Lawrie and L. Peters, "Modifications of Horn Antennas for Low Sidelobe Levels," *IEEE Trans. Antennas Propagat.,* vol. AP-14, September 1966, pp. 605–610.

42 W. F. Bahret and L. Peters, "Small Aperture Small Flare Angle Corrugated Horns," *IEEE Trans. Antennas Propagat.,* vol. AP-16, July 1968, pp. 494–495.

43 C. Dragone, "Attenuation and Radiation Characteristics of the HE_{11} Mode," *IEEE Trans. Microwave Theory Tech.,* vol. MTT-28, July 1980, pp. 704–710.

Chapter 16

Lens Antennas

George D. M. Peeler

Raytheon Company

Lenses and reflectors (Chap. 17) are used as collimating elements in microwave antennas. Reflectors have a single degree of freedom (the reflector surface), have no internal loss, have low reflection loss, have no chromatic aberration, are relatively easy to support, and can be perforated to reduce weight and wind loads. Lenses have up to four degrees of freedom [inner surface, outer surface, index of refraction n, and (for constrained lenses) inner- versus outer-surface radiator positions], have no aperture blockage by the feed, have internal and surface-reflection losses, must be edge-supported, and are relatively heavy and bulky. In general, if a reflector can provide the required performance for a given application, it should be used. However, since lenses are more versatile, especially in wide-scan-angle performance, they are used in many applications that do not require the greater flexibility of a phased array.

The methods for determining radiation-pattern characteristics from an aperture (amplitude and phase) distribution, including deviations from desired distributions, are derived and discussed in Chap. 2. Lens applications are discussed in several other chapters and are only suggested in this chapter. References 1 to 6 contain summaries of most lens types and list a number of references. The material in Ref. 1 is the basis of this chapter, and many sections are only updated; however, the section on artificial dielectric materials is not repeated here.

Some microwave lenses are adapted directly from optics, but microwave antennas generally use only a single element since multielement designs common in optics are generally too bulky and heavy for microwave applications. However, several techniques are available at microwave frequencies that permit specialized designs which are difficult, if not impossible, to use at optical frequencies; these techniques include nonspherical lens surfaces, artificial dielectric materials, constrained and geodesic media, and variable n with position in the media.

Lenses are designed to collimate one wavefront into another by using ray tracing on the basis of the law of the optical path that all rays between wavefronts (or phase fronts) have equal optical path lengths and of the application of the Fresnel equations (Snell's law plus polarization effects) at the lens surfaces. The lenses discussed in this chapter collimate a spherical or cylindrical wavefront produced by a point or line source feed into a planar or linear wavefront; i.e., they are focused at infinity. In practice, however, complex feeds can be used since performance does not deteriorate rapidly with small off-axis feed displacement. Other lens designs have limited application at microwave frequencies.

Figure 16-1 shows two ways of achieving lens designs; in methods a and b rotational lenses are used in conjunction with a point-source feed (the term *rotational* means that the surfaces of the lens are obtained by revolution of a curved line about the lens axis), while in methods c and d cylindrical lenses (i.e., lenses whose surfaces are generated by moving a straight line perpendicularly to itself) are used in conjunction with a line-source feed.

At microwave frequencies, natural homogeneous dielectric media always have $n > 1$ (i.e., a phase velocity less than that of light in free space), which leads to the convex lens shapes of Fig. 16-1a and c. However, artificial or fabricated media may be constructed with a range of n from $n \gg 1$ to $n \ll 1$. Examples of lenses constructed with $n < 1$ are shown in Fig. 16-1b and d, where it is seen that a concave shape is required to focus a beam. In general, the types having $n < 1$ are highly dispersive

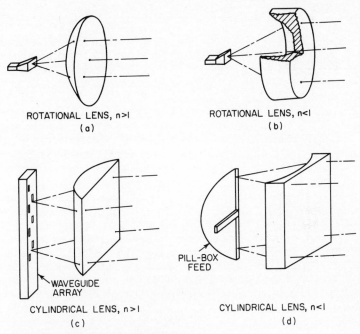

ROTATIONAL LENS, n>l
(a)

ROTATIONAL LENS, n<l
(b)

WAVEGUIDE
ARRAY

CYLINDRICAL LENS, n>l
(c)

PILL-BOX
FEED

CYLINDRICAL LENS, n<l
(d)

FIG. 16-1 Basic lens configurations.

(i.e., n varies rapidly with frequency), while those having $n > 1$ are nondispersive. Thus lenses having $n < 1$ are usually limited to small frequency bandwidths, while lenses having $n > 1$ may be designed to operate over an octave or more.

16-2 LENS-SURFACE FORMULAS FOR $n > 1$

Common shapes for $n > 1$ media lenses are shown in Fig. 16-2; all are single-degree-of-freedom, single-focal-point lenses because either they refract at only one surface or one surface is fixed during the design. The formulas given in the figure apply to both rotational and cylindrical surfaces.

Single Refracting Surface, $n > 1$

The lenses in Fig. 16-2a–d refract at only one surface. In lenses a to c, refraction occurs at the inner surface (adjacent to the focal point) and in lens d at the outer surface. In each case, the nonrefracting lens surface is parallel to a wavefront. The lens-surface formulas in Fig. 16-2 are derived by equating the optical-path length of a general ray to that of the central ray from an on-axis point at a distance f from the inner surface to a planar wavefront perpendicular to the axis; for example, for the lens in Fig. 16-2a, the relation is $r = f + n(r \cos \theta - f)$, which can be easily manipulated

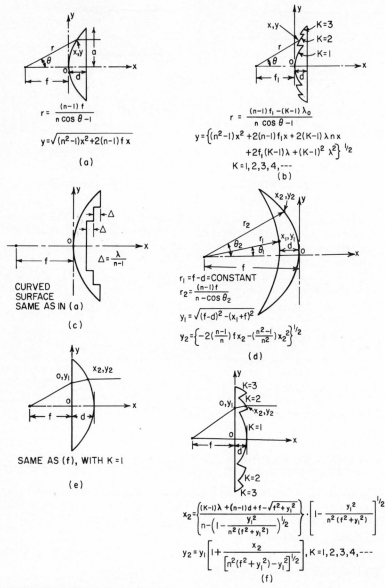

FIG. 16-2   Lens-shape designs for $n > 1$ media.

into the form given in Fig. 16-2a, a hyperbola. The feed is located at the focal point farthest removed from the inner surface.

A disadvantage of this type of lens is that reflections from the nonrefracting surface will converge at the focal point, since the surface is coincident with a wavefront, and cause a feed-line mismatch approximately equal to the surface-reflection coefficient. It is desirable that this surface reflection be either prevented from entering the

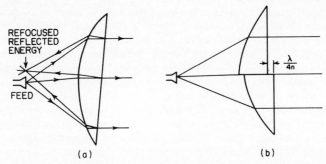

FIG. 16-3 Reduction of feed VSWR by (*a*) lens tilt and (*b*) quarter-wavelength displacement of half of the lens.

feed line or eliminated. A high feed-line voltage standing-wave ratio (VSWR) may be avoided by using a ferrite isolator, by tilting the lens slightly (as in Fig. 16-3*a*) so that the reflected energy will be refocused off the feed, or by displacing half of the lens a quarter wavelength (in the lens medium) along the axis with respect to the other half (as in Fig. 16-3*b*) so that reflections from the two lens halves are 180° out of phase and create a null at the feed center. These methods have been found to be effective in reducing the feed-line VSWR; however, they do not eliminate the other effects of surface reflection, such as loss of gain and increase in sidelobe level. Whenever these factors are important, the surface-matching techniques to be described later are recommended. Avoiding feed-line mismatch by the technique of Fig. 16-3*a* depends on low wavefront distortion for small off-axis feed displacements. Analysis[7] of aberrations due to feed displacement from the focal point shows that coma is the predominant term that limits performance, that coma is reduced as focal length and *n* are increased, and that the coma for a large-*n* hyperbolic lens is twice that for an equivalent-focal-length paraboloidal reflector. Although this is not usually necessary for small lenses, if the technique of Fig. 16-3*b* is used for large-diameter lenses, the two lens halves should be designed with different focal lengths.

These lenses may be constructed with natural dielectric materials and artificial delay media. Kock[8] developed a number of artificial delay media and constructed lenses of the type shown in Fig. 16-2*a* and *b* with these media.

Two Refracting Surfaces, *n* > 1

Microwave lenses that refract at both surfaces have not been used as frequently as the single-refracting-surface types because their performance is not sufficiently better to justify the more difficult design problems.

Lens designs that meet the Abbe sine condition (to eliminate coma for small off-axis feed displacements) have been developed.[9] The planoconvex single-focal-point lenses in Fig. 16-2*e* and *f* almost satisfy the Abbe sine condition for *n* = 1.6 and an *f*/*D* = 1. An experimental 50-wavelength-aperture lens had good performance over a total scan angle of 20°.

A lens with two refracting surfaces has two degrees of freedom that should permit designs with two focal points. Several investigators[10-12] have developed bifocal design techniques that are iterative and do not lead to closed-form solutions. These lens designs are two-dimensional (cylindrical) and symmetrical about the lens axis.

One technique[10] assumes prisms at the lens edges and represents surfaces with even-order polynomials to permit iterative ray tracing. Another technique[11] involves a lattice method wherein points on the two lens surfaces are determined by alternating between the two focal points for ray tracing. Brown[12] uses prisms at the lens edges and the lattice method for determining the surfaces. For $n = 1.6$ and focal points at $\pm 20°$, the resulting lens is almost planoconvex (the almost linear surface is slightly concave), very nearly that obtained in Ref. 10. Zoning is achieved by designing separate lenses for each of the zones. Three-dimensional lenses were obtained by rotating the two-dimensional designs about the lens axis. In these lenses, astigmatism is large and limits the beam-scan angles to about 20°. Brown[12] shows experimental data for 50-wavelength-diameter lenses with $n = 1.6$ and focal points at $\pm 20°$ for f/D's from 1 to 2.

Zoning Formulas, $n > 1$

In the case of physically large lenses, the use of continuous surfaces (as in Fig. 16-2a, d, and e) results in a massive structure that is difficult to produce without imperfections, and the long path lengths in the medium make its index of refraction highly critical. To alleviate these disadvantages, use is commonly made of discontinuous, or zoned, surfaces as in Fig. 16-2b, c, and f. Ray paths through any two adjacent zones are designed to differ from each other by exactly 360° (or a whole multiple thereof) at the design frequency so that a plane wavefront results on the outer side of the lens.

Figure 16-2b and c gives the surface-shape formulas for zoned versions of the lens of Fig. 16-2a, while Fig. 16-2f shows the zoned counterpart of Fig. 16-2e; the use of integers $K = 1, 2, \ldots$ gives a family of lenses that has a focal point at $x = -f$ and produces plane wavefronts that differ from each other in phase by integral multiples of 360°. In the use of these formulas, a thickness d on the axis is chosen, and the central zone of the surface is computed for $K = 1$. At the radius for which the lens thickness is equal to the minimum allowed by mechanical considerations, the second zone is computed for $K = 2$. In a similar manner, additional zones are computed with $K = 3$, etc., until the desired lens diameter is attained. If the minimum allowable thickness is d_{min}, the maximum thickness is approximately $d_{min} + \lambda/(n - 1)$, where λ is the design wavelength.

Effects of Zoning on Aperture Illumination, $n > 1$

The aperture illumination of a lens is determined by a number of factors, one of which is the lens-material loss that is proportional to lens thickness. In an unzoned lens with $n > 1$, this loss is larger for on-axis rays where the lens has maximum thickness than for edge rays. This leads to a more nearly uniform amplitude distribution than would be estimated without considering loss. Since a zoned lens is thinner and therefore has less loss, the aperture illumination is less strongly affected.

Certain defects in the aperture illumination usually occur with zoned lenses and should be considered carefully in the lens design. For example, Fig. 16-4a shows how shadowed bands (without energy) occur in the aperture field of a refracting outer surface. Inspection of the figure reveals that there is no way in which the zonal boundary can be shaped to eliminate the shadow. These nonilluminated bands produce aperture illumination discontinuities that increase sidelobe levels and decrease gain.

A second type of defect occurs when a first refracting surface is zoned, as in Fig.

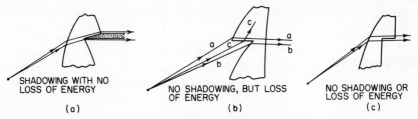

SHADOWING WITH NO
LOSS OF ENERGY

(a)

NO SHADOWING, BUT LOSS
OF ENERGY

(b)

NO SHADOWING OR
LOSS OF ENERGY

(c)

FIG. 16-4 Effects of zoning on aperture illumination, $n > 1$.

16-4b. In this case, the rays between rays a and b are not properly refracted by the lens and are scattered in undesired directions. Although there is no shadowing in this case, the aperture illumination is perturbed at the zone boundaries and deteriorates the expected radiation pattern.

However, zoning without shadowing or energy loss is possible if it is done on a nonrefracting surface of the lens. Figure 16-4c shows an example for the case of a plane outer surface. The surface of the step between zones must be perpendicular to the zoned surface. This type of zoning should provide the most satisfactory aperture illumination, but disturbances still exist along the zone boundaries because of phase differences between the rays just inside and outside the dielectric surface. In the special instance of a cylindrical lens in which the **E** vector is everywhere perpendicular to the step surfaces, this disturbance may be avoided by covering the zone boundaries with thin conducting sheets.

Bandwidth of Zoned Lenses, $n > 1$

The surface-shape design of an unzoned constant-n lens is independent of frequency. However, as may be seen from the formulas in Fig. 16-2, zoning introduces a frequency dependence so that a zoned lens has a limited bandwidth of satisfactory operation. Rays through adjacent zones will differ by exactly 360° only at the design frequency. If the lens has a total of N wavelength steps between the central and edge zones, the effective path-length difference between a ray through the edge zone and a ray through the central zone is $N \lambda_0$ at the design wavelength λ_0 and $N \lambda$ at any other wavelength λ, where N is 1 less than the K of the outermost zone. The phase difference, in wavelengths, of the edge zone compared with the central zone is $-N \, d\lambda$, where $d\lambda = \lambda - \lambda_0$. Since the total bandwidth is approximately $100 \, (2d\lambda/\lambda)$ percent, if a phase difference of $\pm\lambda/8$ is allowed, the bandwidth of a lens having N wavelength steps is $25/N$ percent. Thus a zoned lens has an approximate bandwidth of 5 percent for $N = 5$ if only zoning effects are present. However, since there are other sources of phase error and it is necessary to allocate the total allowable phase error among all error sources, the bandwidth may be smaller than indicated above.

16-3 FACTORS AFFECTING GAIN AND SIDELOBE LEVELS OF LENS ANTENNAS

The gain of a lens antenna is reduced by a number of factors that include the amount of feed energy not incident upon the lens (spillover loss), the effects of lens shape on aperture illumination, dissipation loss in the lens medium, and reflection from the surfaces.

Spillover Loss

Spillover may be avoided if the feed horn is extended to the lens edges; however, this is usually not done, since lens-surface reflections are reflected by the horn walls and radiated from the lens in undesired directions, thereby increasing sidelobe levels. The horn walls can be covered with absorbing material to reduce the energy in the reflected waves, but with this technique the antenna gain is reduced since feed energy is also absorbed.

In applications in which low sidelobe levels are important, the use of a multielement (array) feed to tailor the aperture illumination and reduce spillover is recommended. Also, surface matching is recommended to permit the use of an extended horn, increase gain, and reduce sidelobe levels.

Effect of Lens Shape on Aperture Illumination

The aperture amplitude distribution is a function of the feed radiation pattern, the shape of the lens, lens-surface reflections, and lens losses. References 1 and 2 derive the relations for conversion from the power per solid angle radiated by the feed to the power per unit area in the aperture for several lens types and plot the conversion factors for several sets of lens parameters. In general, for $n > 1$ lenses, the conversion factor is < 1; for $n < 1$ lenses, the factor is > 1; therefore, if a given feed is used for the two lens types, the $n < 1$ lens will have a higher edge illumination than the $n > 1$ lens.

Dissipation Loss in Lens Media

The dissipative attenuation constant for a dielectric medium is approximately 27.3 (tan δ) n (dB/wavelength), where n is the index of refraction and tan δ is the loss tangent of the medium. Since the maximum thickness of a zoned $n > 1$ lens is approximately $\lambda/(n - 1)$, the upper limit for lens attenuation is approximatley 27.3 (tan $\delta)n/(n - 1)$ (dB). Therefore, for most practical materials the maximum dissipation loss is several tenths of a decibel. Reference 13 contains extensive dielectric-constant and loss-tangent data for many dielectric materials over a large range of frequencies and temperatures.

Reflections from Lens Surfaces

The air-to-dielectric interface at each lens surface produces reflected and transmitted waves. The amplitude and phase of these waves are obtained by the application of Fresnel's equations and depend on n, the angle of incidence, and the polarization relative to the incidence plane (a plane through the incident ray and the surface normal). Since there are reflections at both lens surfaces, the effects of internal multiple reflections are determined by the ray path lengths between the surfaces. The thickness of a typical lens varies appreciably over its aperture so that, with the transmitted waves adding in various phases, there are a number of almost sinusoidal variations in amplitude and phase over the aperture. For most lenses, incidence angles are less than 45°

so that reflection losses can be averaged over all incident polarizations and angles to obtain an approximate lens reflection loss of $8.69(n - 1)^2/(n + 1)^2$ dB. For an $n = 1.6$ lens, this reflection loss is less than 0.5 dB. However, sidelobe levels are seriously increased and for most lenses are limited to values about -20 dB.

These effects may be reduced by matching the lens surfaces.[14-16] For near-normal incidence angles, this can be accomplished by adding to each surface a quarter-wavelength-thick coating (in the material) with an index of refraction of $(n)^{1/2}$, where n is the lens-material index of refraction. If it is necessary to match for nonnormal incidence angles, the coating thickness and index of refraction may be modified as discussed in Ref. 1.

For many lens materials, a coating material with the correct index may not be available; in that case, surface matching may be performed by machining the lens surfaces to the configurations shown in Fig. 16-5 and discussed in detail in Refs. 1 and 14-16. The improvement

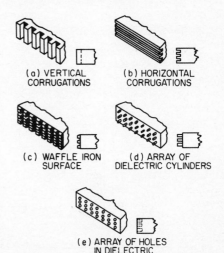

(a) VERTICAL CORRUGATIONS

(b) HORIZONTAL CORRUGATIONS

(c) WAFFLE IRON SURFACE

(d) ARRAY OF DIELECTRIC CYLINDERS

(e) ARRAY OF HOLES IN DIELECTRIC

FIG. 16-5 Simulated quarter-wavelength matching transformers for lens surface.

that can be obtained in lens performance by lens-surface matching[16] is shown in Fig. 16-6; however, these lenses were matched with artificial dielectrics not discussed in this chapter.

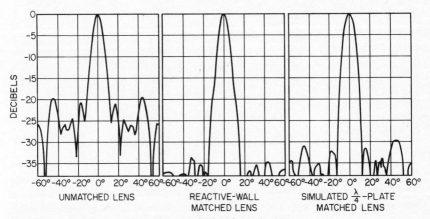

UNMATCHED LENS

REACTIVE-WALL MATCHED LENS

SIMULATED $\frac{\lambda}{4}$-PLATE MATCHED LENS

FIG. 16-6 Radiation patterns of unmatched and matched lenses.

16-4 LENS-SURFACE FORMULAS FOR $n < 1$

The lens formulas given in Fig. 16-2 apply for $n < 1$ as well as for $n > 1$; however, for convenience, the lens-surface formulas for $n < 1$ media planoconcave lenses are

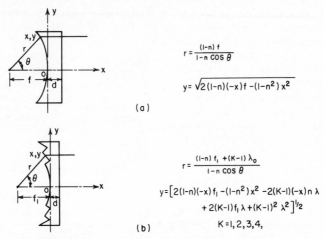

$$r = \frac{(1-n)\,f}{1-n\cos\theta}$$

$$y = \sqrt{2(1-n)(-x)f - (1-n^2)\,x^2}$$

(a)

$$r = \frac{(1-n)\,f_1 + (K-1)\,\lambda_0}{1-n\cos\theta}$$

$$y = \left[2(1-n)(-x)f_1 - (1-n^2)\,x^2 - 2(K-1)(-x)n\,\lambda \right.$$
$$\left. + 2(K-1)f_1\,\lambda + (K-1)^2\,\lambda^2\right]^{1/2}$$

$$K = 1, 2, 3, 4,$$

(b)

FIG. 16-7a and b   Lens-shape designs for $n < 1$ media.

shown in Fig. 16-7 and apply to both rotational and cylindrical surfaces. These are single-refracting-surface, single-focal-point lenses since the planar outer surface is fixed during the design to lie on a wavefront. The inner surface is an ellipse with foci on the axis; the feed is located at the focus farthest removed from the inner surface. Reflections from the outer surface converge at the focal point as for $n > 1$ lenses and can be eliminated by techniques used for those lenses.

For the stepped inner surface in Fig. 16-7b, shadow bands at the steps cannot be avoided, as may be seen in Fig. 16-8. Although all the rays emanating from the feed

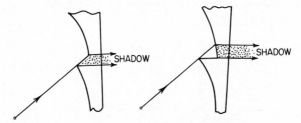

FIG. 16-8   Shadows introduced by zoned lens in Fig. 16-7b.

are collimated by the lens, the steps cause discontinuities in the aperture illumination. As in the case of $n > 1$ lenses, zoning can be accomplished without shadowing or energy loss only if the steps are formed in the equiphase outer surface, but the resulting meniscus-type lens shape is more difficult to manufacture and to support than the flat shape of Fig. 16-7b.

The maximum thickness of the stepped lens for $n < 1$ is equal to $d_{\min} + \lambda/(1 - n)$, where $d_{\min}$ is the minimum thickness permitted in the mechanical design. The bandwidth limitation due to zoning alone is the same as for $n > 1$ lenses. However, the bandwidth of the actual lens is considerably less than this when the frequency sensitivity of n is taken into account, since all known media having $n < 1$ exhibit a large rate of change of n with frequency. Thus, in the common case of a metal-plate zoned lens having the shape shown in Fig. 16-7b, if $\pm\lambda/8$ phase error is allowed, the bandwidth is approximately $25n_0/(1 + (N + 1)/n_0)$ percent, where n_0 is the n at the design frequency and N is the number of one-wavelength steps between the central and edge zones. In the case of frequency-sensitive media, zoning increases the bandwidth of a lens over that of an equivalent unzoned lens by a factor of 2 or 3 because the ray path lengths in the dispersive medium are greatly reduced by zoning. The bandwidth of a given lens antenna may be extended considerably if the antenna application permits moving the feed position along the lens axis as a function of frequency, since defocus is the principal effect of a frequency change.

The lenses in Fig. 16-7 may be constructed with artificial dielectric, metal-plate, or waveguide media. An excellent summary of artificial dielectric media is contained in Ref. 1. Metal-plate and waveguide media are discussed in the next section. Kock[17] used a number of techniques to construct metal-plate lenses with both circular and rectangular apertures. Figure 16-9 is a photograph from Ref. 8 of a stepped 96-wave-

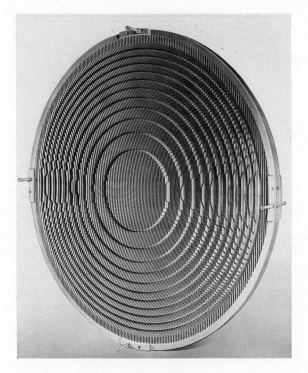

FIG. 16-9 An $F/D = 0.95$ zoned lens.

length-aperture, $F/D = 0.96$, metal-plate lens that has a useful bandwidth of about 5 percent. Most early metal-plate lenses used the designs and techniques of Kock. Additional $n < 1$ lenses are discussed in Sec. 16-6, "Waveguide Lenses."

16-5 METAL-PLATE AND WAVEGUIDE MEDIA

Rays incident on natural dielectric and many artificial dielectric media obey Snell's law at the surface. A metal-plate medium (composed of parallel, equally spaced plane metal sheets) constrains rays in the medium to paths parallel to the plates; therefore, Snell's law is not obeyed for incident rays in a plane normal to the plates (see Fig. 16-10), while it is obeyed for rays incident in a plane parallel to the plates (see Fig. 16-

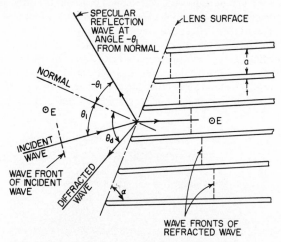

FIG. 16-10 Constrained refraction in the *H* plane.

11*a*). A waveguide medium constrains rays in the medium to paths parallel to the waveguide axis (see Fig. 16-11*b*); therefore, Snell's law is not obeyed for any incidence angle.

Both metal-plate and waveguide media operate in the fundamental TE mode of propagation, and since the phase velocity exceeds the velocity of light in free space, n < 1. If the cutoff wavelength is λ_c, then $n = [1 - (\lambda/\lambda_c)^2]^{1/2}$ for an operating wavelength λ. For both parallel plates spaced a apart and a waveguide of width a, $\lambda_c = 2a$. For a regular hexagon of width a between parallel sides, $\lambda_c = 1.792a$. For a circular waveguide of diameter a, $\lambda_c = 1.705a$.

When parallel plates are used, an E-field component perpendicular to the plates must be avoided, since this excites a TEM mode that propagates between the plates with the velocity of light ($n = 1$) and the energy would not be focused. For square, circular, and hexagonal cross-section waveguide media, n is independent of incident polarization, and these structures can be used for any polarization. A rectangular cross-section waveguide medium has different values of n for the two principal linear polarizations.

Limitations on Metal-Plate Spacing and Rectangular-Waveguide Width

A basic limitation on the choice of plate spacing a results from the fact that the medium must operate between the cutoff frequencies of the fundamental and the next

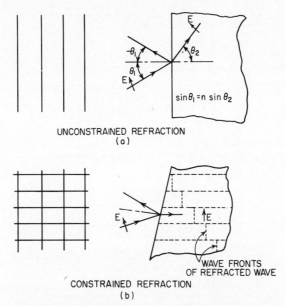

UNCONSTRAINED REFRACTION
(a)

$$\sin \theta_1 = n \sin \theta_2$$

CONSTRAINED REFRACTION
(b)

WAVE FRONTS
OF REFRACTED WAVE

FIG. 16-11 Unconstrained (a) and constrained (b) refraction in the E plane.

higher TE modes. For air-filled structures, this requires that $0.5 < a/\lambda < 1.0$; this range of a/λ corresponds to a range in n of $0 < n < 0.866$. However, a further restriction on the plate spacing must be imposed if the existence of diffracted waves due to the grating effect is to be avoided. It will be shown later how such diffracted waves cause considerable loss of transmitted power, distortion of the aperture illumination function, and consequent loss of gain and increased sidelobe levels. The presence of one diffracted wave is shown in Fig. 16-10. This occurs in the direction in which the diffracted waves from the individual plate edges combine in phase. It can only occur when the plate spacing exceeds a particular value and will be avoided if the following condition is met:

$$[\lambda \sin \alpha / (a + t)] - 1 > |\sin \theta_1| \qquad \textbf{(16-1)}$$

where t is the wall thickness and the other symbols are defined in Fig. 16-10. This is also the condition for avoiding a diffracted wave when the incident wave is in the medium and the transmitted ray emerges at angle θ_1. For $\alpha = 90°$, this is the familiar expression for avoiding grating lobes from a phased array: $s = a + t < \lambda/(1 + \sin \theta_1)$. The limiting values of Eq. 16-1 are shown graphically in Fig. 16-12, where the angle θ_1 of the ray in free space is plotted versus $(a + t)/\lambda$ for different values of α from 30 to 90°. For each particular value of α, the region to the lower left of the curve gives the ranges of θ_1 and $(a + t)/\lambda$ in which diffracted waves are avoided.

Reflection and Transmission at an Interface

For the case of the E field parallel to the plane of incidence and parallel to the plates (Fig. 16-11a), the power-reflection coefficient is

$$r^2 = \sin^2 (\theta_1 - \theta_2) / \sin^2 (\theta_1 + \theta_2) \qquad \textbf{(16-2)}$$

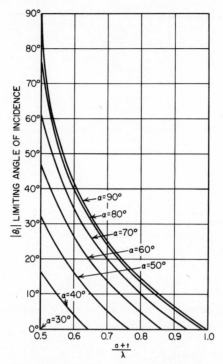

FIG. 16-12 Limiting angle of incidence for pure specular reflection versus plate spacing.

where θ_1 is the ray angle in free space and θ_2 the ray angle in the parallel-plate medium, both angles being measured from the normal to the surface. The plates are assumed to be very thin; however, thick plates are treated in Ref. 18 for $\theta_1 = 0$ and $\alpha = 90°$. Equation (16-2) also holds true for a wave incident from the metal-plate region. The reflected ray is then within the metal-plate region at an angle $-\theta_2$. It is of interest to note that Eq. (16-2) also applies for a solid dielectric with E perpendicular to the plane of incidence. Equation (16-2) and Snell's law of refraction hold for all angles of incidence [if $0.5 < (a + t)/\lambda < 1.0$], and diffracted waves do not occur. However, this is not true of parallel polarization incident on a constrained medium (Fig. 16-11b), and a solution for that case is not available.

Considerable published information is available for the case of E perpendicular to the plane of incidence and parallel to the edges of a set of very thin plates (Fig. 16-10). The basic analysis is due to Carlson and Heins,[19] while further theoretical work was done by Lengyel,[20] Berz,[21] and Whitehead.[22] In the range of θ_1 free of diffracted waves, as given by Eq. (16-1) and Fig. 16-12, the power-reflection coefficient for a wave incident on an array of plates is

$$|r|^2 = \left[\frac{\cos(\theta_1 + \psi) - n}{\cos(\theta_1 + \psi) + n} \right] \left[\frac{\cos(\theta_1 - \psi) - n}{\cos(\theta_1 - \psi) + n} \right] \qquad \textbf{(16-3)}$$

where $\psi = 90° - \alpha$ is the angle between the normal to the boundary and the plates. The specularly reflected wave propagates at angle $-\theta_1$. Because of reciprocity, this formula also applies to waves in the plate region incident upon the free-space boundary, where the waves between the plates are phased to produce a transmitted wave propagating at angle θ_1 from the normal. For normal incidence and $\alpha = 90°$, Eq. (16-3) reduces to $r^2 = (1 - n)^2/(1 + n)^2$, which also applies to a solid-dielectric boundary. In the range of θ_1 where diffracted waves may exist, Eq. (16-3) is no longer valid. Formulas for this range have been derived by Lengyel,[20] Berz, [21] and Whitehead[22] and are much more complicated than Eq. (16-3).

Figure 16-13 shows the percentage of incident power transmitted into a metal-plate medium for the conditions of Fig. 16-10 and $\alpha = 90°$. To the left of the break

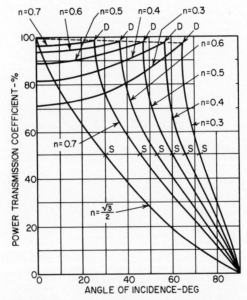

FIG. 16-13 Percentage of power transmitted into metal-plate medium, $\psi = 0$.

in each curve (points D), the nontransmitted power is reflected in the specular reflected wave, while to the right of the break the nontransmitted power is divided between the specular reflected wave and a diffracted wave. The power loss in the latter region is large and should be avoided in lens design. At points S, the angle θ_d of the diffracted wave is equal to θ_1 of the incident wave (Fig. 16-10). When $\alpha \neq 90°$, the reflection and transmission quantities are not symmetrical about $\theta_1 = 0$.

The discontinuity effect at the boundary of the metal-plate medium also results in a phase change. Formulas giving this change have been derived by Lengyel,[20] Berz,[21] and Whitehead.[22] Graphical plots of this phase change versus the angle of incidence are given by Berz and Whitehead and show a moderate variation with the angle of incidence. Calculations by Whitehead for a typical planoellipsoidal lens show that substantial curvature of the wavefront occurs at the aperture when the feed element is at the focal point but that the wavefront may be made almost plane by moving the element slightly away from the focal point.

The above-described data for perpendicular polarization incident on a set of parallel plates apply equally well to an array of thin-walled square or rectangular tubes (Fig. 16-11*b*). However, for parallel polarization the constraint offered by the array of tubes has a large effect on its reflection characteristics. Wells[23] shows that, at the angle of incidence for which the diffracted wave first appears, a sharp peak of specular reflection also occurs.

Methods of Reducing Reflection Losses

Within the angle-of-incidence range given by Eq. (16-1) and Fig. 16-12, the reflection from the surface of a metal-plate or waveguide lens may be canceled through the introduction of an additional reflecting discontinuity. Possible structures are shown in Fig. 16-14, where in *a* a dielectric sheet is placed outside the lens and in *b* obstacles

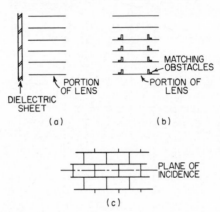

FIG. 16-14 Techniques for reducing surface reflections from metal-plate lenses.

are placed inside the lens. In both cases the magnitude and phase of the additional reflection are adjusted to cancel the reflection from the plates. The magnitude depends upon the dielectric constant and the thickness of the dielectric sheet or the shape and size of the obstacles, while the proper phase relationship is obtained through choice of the spacing of the dielectric sheet or obstacles from the plate edges. The correct design for a given lens may be obtained experimentally; however, a theoretical design may also be computed through the use of graphical and tabulated data given by Lengyel[20] for the magnitude and phase of the reflection coefficient of the boundary.

In a practical lens design, it is difficult to avoid the ranges of θ_1 for which diffracted waves occur. Many early designs of metal-plate lenses did not do this, and as a result considerable loss of gain resulted. One method of avoiding this condition, as may be seen from Fig. 16-12, is to reduce the plate spacing; however, this reduces n and increases the reflection loss at angles near normal incidence and reduces the bandwidth. A further possibility is to fill the regions between the plates with dielectric material or to introduce ridges in the waveguide parallel to the axis (Fig. 16-14*b*) in a manner analogous to ridge waveguide.[24,25] In this way, the plate spacing may be reduced to a point at which diffraction is no longer a problem while n remains at a reasonable value.

Another method of avoiding diffracted waves in one plane is to use a staggered arrangement of waveguide channels,[23] as in Fig. 16-14c, now a proven technique in space-fed phased arrays. The spacing between scattering edges is effectively cut in half, thus doubling the frequency at which a diffracted wave first appears. This staggered arrangement also eliminates the sharp peak of specular reflection that occurs with the nonstaggered array of tubes for parallel polarization.

16-6 WAVEGUIDE LENSES

Three-dimensional unzoned waveguide lenses of the type shown in Fig. 16-7a have been designed. The inner surface is an ellipsoid of revolution about the lens axis, and the outer surface is a plane perpendicular to the axis. One such lens[26] has a 36-in (914-mm) diameter and a 31.2-in (792-mm) nominal focal length and is constructed of 0.68-in- (17.3-mm-) inside-diameter square waveguide to provide a nominal $n = 0.5$ at 10 GHz. The lens was fed with a circular-aperture-waveguide horn that produced circular polarization. The reference has a number of plots of experimental data: gain, beamwidth, n, and focal length versus frequency; gain, beamwidth, and first sidelobe level versus feed-horn diameter; radiation patterns as a function of beam-scan angle; and gain, beamwidth, and first sidelobe level versus beam-scan angle. The sidelobe level increases from -22.5 dB on axis to -11 dB at $13°$ beam scan and remains at about that level for beam scan to $22°$. The gain decreases by 3 dB (from the on-axis value of 36 dB) at a beam-scan angle of $18°$, or 7.65 beamwidths.

Another design[27] has a 24-in (610-mm) diameter and a 36-in (914-mm) focal length constructed of $\frac{5}{32}$-in (4.0-mm) cell honeycomb to provide an $n = 0.642$ at 55 GHz. Because the honeycomb cells deviated from perfect hexagons, the performance varied with the linear-polarization orientation. The aperture efficiency was only 18.3 percent because random variations in the cell size produced large aperture phase variations.

To obtain wide-scan-angle performance, Ruze[28] investigated bifocal cylindrical constrained lenses in which the two focal points are equally displaced from the lens axis at P_1 and P_2 as shown in Fig. 16-15. The inner surface is found by equating optical-path lengths for the central ray and a general ray from each of the focal points. It is an ellipse, with foci at the two focal points, independent of n or lens thickness, given by

$$(x + af)^2/(af)^2 + y^2/f^2 = 1 \qquad (16\text{-}4)$$

where $a = \cos\phi$. Imposing the two focal points uses two of the three degrees of freedom available. Ruze considered several alternatives for choosing the third condition: constant n, constant thickness, linear outer surface, third correction point, and no second-order error; for the last four, n varies with y. The deviations from a linear phase front for a general feed point were expanded into a power series and examined for each case. For the on-axis feed point there are no odd-order terms, and the second-order deviation (defocus) can be corrected by moving the feed point along the axis;

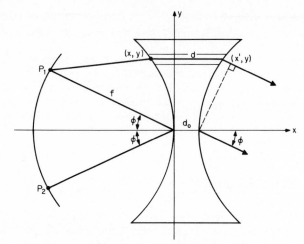

FIG. 16-15 Bifocal metal waveguide lens.

however, this correction is not needed for all cases. The feed arc is chosen as a circle through the corrected on-axis feed position and the two focal points. Several of these designs permit a beam scan of 110° for $\phi = 50°$.

The constant-n lens has been the most popular. The thickness is given by

$$d = d_0 + (m\lambda + ax)/(n - a) \qquad (16\text{-}5)$$

where m is the number of wavelength steps in the outer surface. Reference 29 analyzed this type of lens with results near those of Ruze. The constant-n lens was expanded to three dimensions by Ref. 30; the inner surface is an ellipsoid of revolution about the line through the focal points, and the outer surface is an ellipsoid with unequal principal radii. The performance is well behaved for feed points between the two focal points, but performance deteriorates rapidly for feed points beyond the focal points. Performance also degrades rapidly as the feed point is moved away from the xy plane.

A constant-n lens was used in an application[31] for which the scanning feed system was most conveniently designed for a planar feed aperture. The phase errors were minimized for the on-axis feed point by a method of stepping the outer surface. A constant-n lens was also used for a satellite antenna,[32] but with the two correction points collapsed to a single on-axis point so that the inner surface was a segment of a sphere and the outer surface a segment of a spheroid. The effects of zoning the outer surface were computed. The radiation-pattern effects of zoning were also computed in Ref. 33.

A linear-outer-surface two-dimensional lens between parallel plates was constructed[34] for an X-band multibeam application by milling waveguides of unequal width (to obtain variable n with y) in metal blocks. The lens had good performance, and the tolerances obtained appeared sufficient to permit this technique to be used for frequencies up to 40 GHz.

Waveguide lenses have been designed to provide a larger bandwidth by designing for equal time delay[35,36] for all rays through the lens. The phase delay in each waveguide is adjusted with a phase shifter to provide a planar wavefront. These lenses have a spherical inner surface so that they satisfy the Abbe sine condition for off-axis performance. Reference 36 has both single- and dual-frequency designs which obtain bandwidths that vary from 40 to 20 percent for lens diameters from 20 to 100 wavelengths.

16-7   BOOTLACE-TYPE LENSES

Bootlace lenses can have four degrees of freedom that permit designs with four focal points to obtain wide-angle performance for either multiple simultaneous beams or a scanning beam; however, most designs fix the outer surface and have three focal points. These constrained lenses have radiators on the inner and outer surfaces connected by TEM-mode transmission lines (usually coaxial or stripline), which give the appearance of untightened bootlaces, hence the name. Typically, all rays in these lenses have equal time delays from a focal point to the corresponding linear wavefront, and therefore they are inherently broadband, limited only by component bandwidths and the performance of a fixed aperture size.

Gent[37] identified most of the useful properties of this type of lens, including (1) four degrees of freedom, (2) the fact that one surface may be displaced and/or rotated relative to the other (i.e., the surfaces may have independent coordinate systems), (3) the ability to add phase shifters, attenuators, and/or amplifiers in the lens transmission lines to provide outer-surface aperture illumination control and beam scan, (4) the fact that the line lengths can be stepped by multiple wavelengths, analogously to the physical stepping shown in Fig. 16-2 without creating shadowing but with reduced bandwidth, (5) the ability of radiators on the two lens surfaces to have different polarizations, and (6) the fact that the spacing between aperture radiators must be less than $\lambda/(1 + \sin \theta_{max})$ to avoid grating lobes, where θ_{max} is the maximum beam-scan angle from the aperture normal.

The Rotman-Turner[38] lens is a two-dimensional bootlace lens with a linear outer surface perpendicular to the lens axis and three focal points, one on axis and two equally displaced from the lens axis, shown in Fig. 16-16 as points G $(-g, 0)$, F_1 $(-\cos \phi, \sin \phi)$, and F_2 $(-\cos \phi, -\sin \phi)$, respectively, where central rays from F_1 and F_2 subtend an angle ϕ with the axis. Desired wavefronts are displaced from the vertical by angles 0, ψ, and $-\psi$ respectively, where $\psi = \phi$ in this design. All dimensions are normalized to the F_1-to-origin distance. Pairs of inner- and outer-surface radiators are connected by coaxial lines with the general pair located at $P(x,y)$ and $Q(t,u)$ respectively. Let $w = s - s_0$ be the normalized electrical path-length difference between the general and the central line lengths. Equating the optical-path lengths for a general ray and the central ray for each focal point produces

$$y = u (1 - w) \tag{16-6}$$

$$x^2 + y^2 + 2dx = w^2 + e^2u^2 - 2w \tag{16-7}$$

$$x^2 + y^2 + 2gx = w^2 - 2gw \tag{16-8}$$

where $d = \cos \phi$

$e = \sin \phi$

x and y can be eliminated to obtain

$$aw^2 + bw + c = 0 \tag{16-9}$$

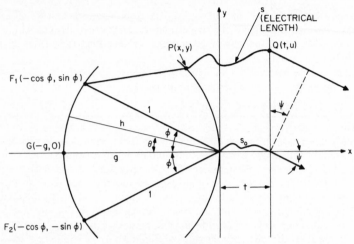

FIG. 16-16 Rotman-Turner lens nomenclature.

where $a = 1 - u^2 - (g-1)^2/(g-d)^2$
$b = 2g(g-1)/(g-d) - e^2u^2(g-1)/(g-d)^2 + 2u^2 - 2g$
$c = ge^2u^2/(g-d) - e^4u^4/(4(g-d)^2) - u^2$
w as a function of u is determined from Eq. (16-9) for fixed parameters ϕ and g. Substitution into Eqs. (16-6), (16-7), and (16-8) determines x and y. The technique of Ruze[28] (for the linear-outer-surface lens) for minimizing aberrations from a linear wavefront for a general feed point θ between $\pm\phi$ is used to determine $g = 1 + \phi^2/2$, for ϕ in radians. The feed path is chosen as a circle that intersects the three focal points.

Reference 38 contains computations of w, x, and y as a function of u for $\phi = 30°$ and computations of deviations from a linear wavefront for $0 \le \theta \le 40°$ and several values of g. Also shown are lens shapes for four values of g. For $\phi = 30°$ and $g = 1.137$ (the optimum value of g for $\phi = 30°$), the normalized path-length error is <0.00013 for $-0.55 \le u \le 0.55$ (i.e., aperture $= 2u_{max} = 1.10$ and $g/2u_{max} = 1.035$). If a phase error of $\pm\lambda/8$ is allowed, an aperture of 1000 wavelengths can be used to obtain a beamwidth $< 0.075°$ that can be scanned more than 60°, or more than 800 beamwidths.

An experimental lens at 3.0 GHz with $\phi = 30°$, $g = 1.137$, $u_{max} = 0.6$, and $D/\lambda = 18$ was constructed[38] by using RG-9/U coaxial cables for the lens, a TEM-mode parallel-plate structure between the feed position or positions and the inner lens surface, a flared parallel-plate aperture that contained the linear outer surface, and quarter-wavelength probes into the parallel-plate regions.

The Archer lens[39] is a Rotman-Turner lens with additional features: (1) The parallel-plate region between the feeds and the inner surface is filled with a dielectric material of relative dielectric constant ϵ_r to reduce all linear dimensions by a factor of $n = (\epsilon_r)^{1/2}$, thereby permitting its use in a smaller volume. (2) Stripline or microstrip lens and feed-port connecting lines are printed on an extension of the dielectric material, thereby removing connectors and probes and the associated mismatches and permitting construction to tight tolerances and wide-bandwidth operation. (3) The beam angle ψ can differ from the feed angle θ so that $\sin\psi/\sin\theta = K$ (where $K = 1$ for the Rotman-Turner lens);[40] $K > 1$ permits obtaining large beam-scan angles with practical feed positions, and $K < 1$ permits reduction of the lens size for limited scan-angle applications. These features increase the lens versatility but do not increase the design degrees of freedom.

There is a nomenclature difference between the Archer and Rotman-Turner lenses. In the Archer lens, the printed-circuit lines that form the lens appear to be just connecting lines, while the dielectric region between the feed ports and the inner surface appears to be a lens and is usually referred to as such.

Archer lenses[39] have been designed and constructed with materials with ϵ_r's from 2.5 (Teflon fiberglass) to 233 (cadmium titanate). Figure 16-17 shows a 20-radiator,

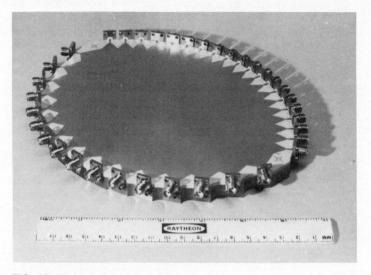

FIG. 16-17 Archer lens, $\epsilon_r = 38$.

16-beam lens constructed in microstrip on a barium tetratitanate ceramic substrate with $\epsilon_r = 38$. Most designs have a beam coverage of 120°, range from a 4-radiator, 4-beam lens to a 140-radiator, 153-beam lens, and have a typical operating frequency band of almost two octaves. Designs usually have adjacent beams overlap at the 3-dB points (with feeds at equal intervals in sine space) at the maximum operating frequency where the gain and beamwidth are consistent with that obtainable from the full aperture. At lower frequencies, the beams broaden and overlap at a higher level, and gain decreases more rapidly than the beamwidth broadening would indicate owing to beam-coupling loss. Low sidelobe patterns[41] are obtained by feeding several adjacent beam ports with weighted amplitudes.

Alternative designs[39] are obtained by fixing the outer surface as a circle, with an arc up to 60°, that is more adaptable to conformal applications. A circular aperture eases the design problem of eliminating aperture resonances that are common in linear arrays, increases the beam coverage with reduced gain roll-off with off-axis beam scan traded against on-axis gain, and reduces the effects of outer-surface mismatch since reflections do not have a constant phase difference from radiator to radiator as with a linear array.

The linear-aperture bootlace lenses discussed above produce fan beams. Multiple pencil beams in a plane can be produced by illuminating a parabolic-cylinder reflector[38] or cylindrical lens (see Fig. 16-1) with the linear aperture. A two-dimensional cluster of pencil beams may be formed[39] by stacking vertical linear-aperture lenses, so that the outer surface of each lens provides a column of radiators, and by feeding the beam ports of these lenses with the output ports of a stack of horizontal lenses. The beam ports can be used separately, adjacent beam ports may be combined to provide monopulse-type outputs, and multiple beam ports may be combined after weighting to provide low sidelobe beams. Three-dimensional bootlace-type lenses are discussed in Refs. 37, 39, and 42–45.

Gent[37] found a special case of the two-dimensional bootlace lens that has feed points and the inner surface on a circle of radius R and the outer surface on a circle of radius $2R$ with equal line lengths between the inner- and outer-surface radiators and recognized that it is equivalent to the R-$2R$ geodesic scanner.[46,47] It is proved in Ref. 48 that the R-$2R$ bootlace lens is unique. Bootlace R-$2R$ lenses have been constructed[39] by using printed-circuit techniques on dielectric materials. The R-$2R$-type lens has been extended into an R-KR lens,[39] shown in Fig. 16-18, that has equal

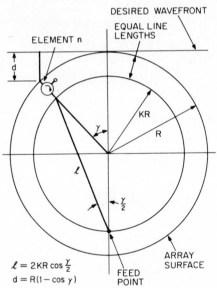

FIG. 16-18 *R-KR*-array geometric relationships.

radial line lengths between radiators on a circle of radius KR and radiators on a circle of radius R. With $K = 1.92$, the optimum value, it is the printed-circuit version of a constant-index lens. It provides a nearly linear wavefront for an arc of approximately 120° on the circle with radius R. By adding a circulator in each of the lines to permit feed inputs, beams covering 360° can be obtained simultaneously.

16-8 PHASED-ARRAY LENSES

Phased arrays are discussed in Chap. 20; in this section space-fed phased-array lenses are discussed for comparison with other lenses.

Space-fed arrays[49-55] have constrained lenses that usually are constant-thickness[49,52] with planar surfaces. Elements (or modules), usually perpendicular to the surfaces, contain inner- and outer-surface radiators, phase shifters, and in some cases other components such as phase-shifter drivers and polarization switches. The inner-surface radiator has the polarization of the feed or feeds and needs to be matched only for the incidence angles from the feed or feeds. The outer-surface radiator can have any polarization or polarizations that may be switchable, must be matched for the range of beam-scan angles used, and must withstand the environment. Most phase shifters are digital, modulo 360°, in which the number of bits is determined by the required beam-pointing accuracy and sidelobe levels. Either diode- or ferrite-type phase shifters are used, the type for a given application being determined by the operating frequency, radio-frequency power levels, required phase-state-switching speed, and switching and/or phase-state-holding power.

Beam steering and collimation (the conversion of the feed wavefront to a plane wave) are obtained by the phase shifters; the computed phase shifts required for the two functions are superimposed. The phase shift required for collimation is determined from ray tracing as for other lenses. Phase-shift commands may be computed and commanded on an element-by-element basis or by rows and columns.[50] This collimation method reduces the number of phase-shifter bits required to obtain a given beam-pointing error, since the otherwise-present regularity of phase errors over the lens outer surface is destroyed.

In many pulsed-radar applications, it is possible to change the state of the phase shifters between transmit and receive; in these cases, separate transmit and receive feeds may be used. If sufficient isolation between feeds is obtained, the need for a transmit-receive switch is eliminated. In addition, separate feeds permit the lens illumination to be optimized for both transmit and receive; e.g., to obtain maximum gain on transmit and low sidelobes on receive.

For applications that require large beam-scan angles and large instantaneous-frequency bandwidth, lenses permit the total angular coverage to be divided into sectors, each with a separate feed, to obtain more nearly equal time delay for all beam positions. The feeds may be used to obtain simultaneous beams,[51] or one feed may be selected by a switching network.

Some applications of space-fed-array lenses are discussed in Refs. 52 to 55.

16-9 DOME ANTENNAS[56-63]

The dome antenna was developed to obtain an antenna to scan a beam over a full hemisphere or more for hemispherical-coverage applications as an alternative to the three or four planar arrays usually considered.

In its simplest form, the dome antenna (Fig. 16-19) has a constrained constant-thickness lens (the dome) fed by a planar array, both symmetric about the vertical axis. The technique is to illuminate an aperture (a portion of the dome) that has an appreciable area perpendicular to any of the desired beam directions to obtain antenna gain in those directions. The feed array can produce convergent or divergent rays to illuminate the required lens area. The lens has a phase-delay gradient in the elevation plane to refract rays in the θ' direction when fed at an angle θ. $K = \theta'/\theta$ is defined as the scan amplification factor.

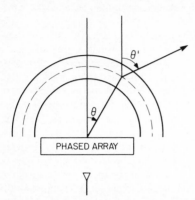

The dome is constructed of modules normal to the surfaces which have radiators on each surface connected by transmission lines that contain the required phase delay. The phase delay could be provided by switchable phase shifters,[56] but in practice this method is too expensive, so fixed delays are used. There are a number of methods for implementing the delays; in one module configuration,[57] the lengths of two dielectric-filled circular waveguides of different diameters in series are varied to obtain the phase delay in 20° increments. The modules may have phase delays modulo 360° (with a bandwidth penalty) and have different polarizations on the two lens surfaces. The spacing between modules must be small enough to suppress grating lobes.

FIG. 16-19   Dome-antenna nomenclature.

Phase delays are generally equal in all modules that have the same cone angle θ. There are several design techniques for determining dome-module phase delays. Since the dome is in the near field of the feed array, ray-tracing methods may be used. For a hemispherical dome of radius R and a constant $K \neq 1$, one technique[56] equates the path lengths for rays from the center of the feed array at angles θ and $\theta + d\theta$ and, after integrating, obtains the required phase delay, which is $(2\pi/\lambda)R[1 - \cos(\theta' - \theta)]/(K - 1)$. For these fixed phase delays, rays from the desired planar wavefront are traced through the lens to the feed array to determine the required feed-array wavefront, which in general is not planar. The gain (relative to the feed-array broadside gain without a dome) of an antenna designed in this manner is shown in Fig. 16-20 for several values of K. It is evident that beam angles greater than 90° can be obtained for $K > 1$ and that maximum gain occurs for beam angles $\theta' > 90°$ for $K > 2$.

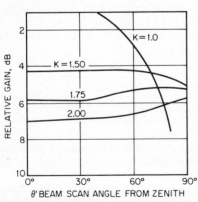

It is desirable to be able to design for prescribed antenna gain versus θ' profiles as, for example,[56] that shown in Fig. 16-21. For some applications, it is also helpful if the dome has an elongated shape along the axis; i.e., the dome is a cylinder generated by a noncircular arc. For these cases, the lens-module phase delays can be designed[56] by allowing a

FIG. 16-20   Dome-antenna relative gain versus θ' as a function of K.

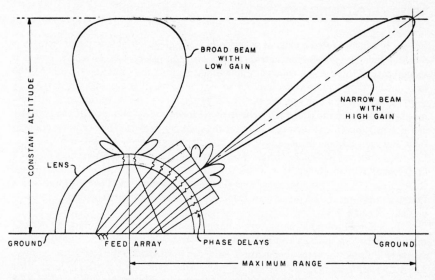

FIG. 16-21 Typical gain versus θ' profile for a dome antenna.

variable K with θ and applying the techniques used for designing a shaped-beam reflector (see Ref. 2, pages 497–500).

These techniques have been implemented[57] at C band with a 4-ft- (1.2-m-) diameter hemispherical dome composed of 3636 modules (dielectric-loaded circular waveguides with a phase-delay increment of 20°) and an F/D = 0.75, space-fed planar phased array with 805 elements. Dome loss was less than 1 dB over a 10 percent frequency band. Two gain-versus-scan-angle-profile designs were measured. One version with $K < 2$ with maximum gain at approximately 60° had peak sidelobes below -18 dB and average sidelobes below -25 dB. It had a 13 percent bandwidth for a zenith beam and a 6.3 percent bandwidth for a 60° beam position. The gain profile matched predictions within 0.3 dB for scan angles up to 60°. Another version with K = 2.6 with maximum gain at 110° demonstrated good performance in a limited amount of data. Another implementation,[58] also at C band, had a hemisphere on a cylinder dome 76 in (1.93 m) in diameter and 51 in (1.30 m) high that had 8511 modules, each a thin-walled metal tube that contained a serrated cruciform to provide the phase delay in increments of 20° between modules. The feed array was 47 in (1.19 m) in diameter and used 1120 elements. The gain profile had maximum gain at θ' = 90°.

An alternative to a constrained lens is the use of a graded-thickness (in elevation) homogeneous dielectric material with quarter-wavelength matching layers on each surface.[56,63] Several techniques for designing to desired gain versus scan-angle profiles and for determining limits on attainable profiles have been analyzed.[59–61] A two-dimensional antenna with K = 1.5 that used a concave-outer-surface Archer lens for the feed has been analyzed[61] and measured;[62] it improves the wide-scan-angle side-lobes, operates over a frequency band of more than an octave, and can provide multiple simultaneous beams.

16-10 LUNEBURG LENSES

Luneburg[64] investigated a class of spherically symmetric, variable-index-of-refraction n lenses that image two concentric spheres on each other. The *Luneburg lens* is a special case of this class in which one sphere has an infinite radius and the other is the lens surface. For a unit-radius lens, the n as a function of the radius r is $n = (2 - r^2)^{1/2}$. When fed at a point on the surface, it produces a plane wavefront, and by moving the feed on the surface, beam scan over all space can be obtained (if the presence of the lens support and feed-scanning mechanism is ignored). The scanning applications and many design principles of Luneburg lenses are discussed in Chap. 18.

A number of authors[65-73] have extended Luneburg's work. References 65 to 68 derive small-feed-circle lenses in which the feed radius is smaller than the lens radius. References 68 to 72 consider variations that include conical and beam-shaping wavefronts in addition to the plane wavefront. Reference 73 considers a design to use a defocused feed.

Luneburg-lens radiation-pattern computations have been considered by most of the Luneburg-lens authors and in particular by Refs. 74 and 75. The Luneburg-lens aperture illumination is the feed pattern times sec ϕ, where ϕ is the feed-pattern angle. The increase in energy density near the aperture edges produces (relative to other lenses) high gain and small beamwidths and limits the sidelobe levels to -17 to -18 dB.

Early Luneburg lenses were two-dimensional (or cylindrical) slices through the center of the sphere because techniques were not available for constructing a sphere. One model[76] was a 36-in- (914-mm-) diameter lens constructed of almost parallel plates filled with polystyrene that operated in the TE_{10} mode at X band. The plate spacing was $a = \lambda/[2(\epsilon_r - 2 - r^2)^{1/2}]$, where ϵ_r is the dielectric constant of the filler material and r is the normalized radius. In addition to the aperture-edge energy increase, two-dimensional Luneburg lenses produce a saddle-shaped wavefront because of the circular aperture. This can be corrected by adding to the aperture parallel plates with a linear aperture, but this limits the beam-scan angle. Calculated and experimental radiation patterns show about -17-dB sidelobes in the E plane and a flat-top pattern in the H plane. Jones[77] constructed an air-filled, almost parallel-plate TE_{10}-mode lens. Walter[78] constructed surface-wave lenses with tapered-thickness dielectric on a ground plane, metal posts on a ground plane, and a "holey-plate" structure.

The symmetry of the Luneburg lens can be exploited by adding planar reflectors through the lens center to obtain a virtual-source antenna.[79] This permits eliminating large segments of the lens to reduce weight and allows the lens to be supported more easily, but it introduces aperture blocking by the feed.

Spherical Luneburg lenses have been constructed by using several techniques. Crushed Styrofoam[80] was used for a 24-in- (610-mm-) diameter lens. Expandable polystyrene beads,[81-83] expanded and solidified in heated hemispherical molds, were used for a 10-step, equal-ϵ_r increment, 18-in- (457-mm-) diameter lens. Reference 84 provides performance of a 10-step Styrofoam 18-in- (457-mm-) diameter lens. Foamed glass[85] was used to fabricate a 10-step 12-in- (305-mm-) diameter lens to obtain high-power operation. Calculations[86] of the heat distribution in a Luneburg lens indicate that the feed region reaches the highest temperature under high-power operation. A ray-tracing method for checking the design of a stepped-n Luneburg lens is provided in Ref. 87.

An important use of the Luneburg lens is as a reflector,[83,88] obtained by placing a reflecting cap on the lens surface. An incoming plane wave is focused at a point on the cap and reflected; after transit through the lens the second time, it is a plane wave propagating in the opposite direction. The angular extent of the cap is chosen as a compromise between obtaining wide-angle performance and minimizing partial blockage of incoming rays by the cap.

For many applications that do not require the performance of a Luneburg lens, Refs. 89 to 93 investigate the performance of constant-n spherical lenses that are more easily constructed than Luneburg lenses.

Another class of two-dimensional Luneburg lenses is the geodesic lens, composed of a pair of nonplanar metal plates spaced a constant distance apart with a spacing small enough so that only a TEM mode can propagate. Myers[47] proved that rays in this medium follow geodesic paths on the mean surface. Rinehart[94,95] derived the surface shape that is required for a two-dimensional Luneburg analog whose edge has a tangent normal to the aperture plane. References 96 and 97 derived surfaces with the edge tangent parallel to the aperture plane. Kunz[98] generalized the analysis and described several designs. Rudduck et al.[99-101] describe small-feed-circle designs and also designs that radiate at a displaced angle from the aperture plane. Johnson[102] provides an excellent summary of geodesic lenses and distinguishes between tin-hat, helmet, and clamshell types. Reference 103 describes a feed-scanning implementation for limited beam-scan angles with a geodesic Luneburg lens. Johnson[104] shows radiation patterns of a 24-in- (610-mm-) diameter, 4.3-mm-λ lens with extension plates to a linear aperture that provides a total beam scan of 60°. The beamwidth is 0.5°, and sidelobes are below -25 dB for most beam positions. Johnson[105] demonstrates that any two-dimensional lens can have a geodesic analog.

REFERENCES

1 H. Jasik (ed.), *Antenna Engineering Handbook,* 1st ed., McGraw-Hill Book Company, New York, 1961. Chapter 14, "Lens-Type Radiators," by Seymour B. Cohn, is the basis for this chapter.

2 S. Silver, *Microwave Antenna Theory and Design,* MIT Rad. Lab. ser., vol. 12, McGraw-Hill Book Company, New York, 1949. Chapter 11, by J. R. Risser, contains considerable basic information on dielectric, metal-plate, and waveguide lenses.

3 J. Brown, *Microwave Lenses,* Methuen & Co., Ltd., London, 1953.

4 R. C. Hansen, *Microwave Scanning Antennas,* vol. 1, Academic Press, Inc., New York, 1964. Chapter 3, "Optical Scanners," by R. C. Johnson, includes material on Luneburg and bifocal dielectric lenses.

5 M. I. Skolnik, *Introduction to Radar Systems,* McGraw-Hill Book Company, New York, 1962. Chapter 7, "Antennas," has a section on lenses and many references.

6 M. I. Skolnik, *Radar Handbook,* McGraw-Hill Book Comapny, New York, 1970. Chapter 10, "Reflectors and Lenses," by D. C. Sengupta and R. E. Hiatt, discusses a number of lens types.

7 R. W. Kreutel, "The Hyperboloidal Lens with Laterally Displaced Dipole Feed," *IEEE Trans. Antennas Propagat.,* vol. AP-28, July 1980, pp. 443–450. Computes lens wavefront aberrations for off-axis feed positions.

8 W. E. Kock, "Metallic Delay Lenses," *Bell Syst. Tech. J.,* vol. 27, January 1948, pp. 58–82. This is the basic reference on artificial delay media.

9 F. G. Friedlander, "A Dielectric-Lens Aerial for Wide-Angle Beam Scanning," *J. IEE*

(London), part IIIA, vol. 93, 1946, p. 658. Discusses the design of dielectric lenses to satisfy the Abbe sine condition.

10   R. L. Sternberg, "Successive Approximation and Expansion Methods in the Numerical Design of Microwave Dielectric Lenses," *J. Math. & Phys.*, vol. 34, January 1956, pp. 209–235.

11   F. S. Holt and A. Mayer, "A Design Procedure for Dielectric Microwave Lenses of Large Aperture Ratio and Large Scanning Angle," *IRE Trans. Antennas Propagat.*, vol. AP-5, January 1957, pp. 25–30.

12   R. M. Brown, "Dielectric Bifocal Lenses," *IRE Conv. Rec.*, vol 4, part 1, 1956, pp. 180–187.

13   A. R. von Hippel, *Dielectric Materials and Applications*, John Wiley & Sons, Inc., New York, 1954.

14   E. M. T. Jones and S. B. Cohn, "Surface Matching of Dielectric Lenses," *J App. Phys.*, vol. 26, April 1955, pp. 452–457.

15   T. Morita and S. B. Cohn, "Microwave Lens Matching by Simulated Quarter-Wave Transformers," *IRE Trans. Antennas Propagat.*, vol. AP-4, January 1956, pp. 33–39.

16   E. M. T. Jones, T. Morita, and S. B. Cohn, "Measured Performance of Matched Dielectric Lenses," *IRE Trans Antennas Propagat.*, vol. AP-4, January 1956, pp. 31–33.

17   W. E. Kock, "Metal Lens Antennas," *IRE Proc.*, vol. 34, November 1946, pp. 828–836. This is the basic metal-plate-lens reference.

18   R. I. Primich, "A Semi-Infinite Array of Parallel Metallic Plates of Finite Thickness for Microwave Systems," *IRE Trans. Microwave Theory Tech.*, vol. MTT-4, July 1956, pp. 156–166. It is shown, for normal incidence and given n, that finite plate thickness can reduce reflections.

19   J. F. Carlson and A. E. Heins, "The Reflection of an Electromagnetic Plane Wave by an Infinite Set of Plates," *Q. App. Math.*, vol. 4, 1947, pp. 313–329. Presents a mathematical solution for polarization perpendicular to the plane of incidence.

20   B. A. Lengyel, "Reflection and Transmission at the Surface of Metal-Plate Media," *J. App. Phys.*, vol. 22, March 1951, pp. 265–276. Extends Carlson and Heins's theory to take account of diffracted beams; presents theoretical formulas and experimental verification.

21   F. Berz, "Reflection and Refraction of Microwaves at a Set of Parallel Metallic Plates," *IEE Proc. (London)*, part III, vol. 98, January 1951, pp. 47–55. Presents analysis similar to that of Lengyel but with the restriction that the boundary surface is perpendicular to the plates.

22   E. A. N. Whitehead, "The Theory of Parallel-Plate Media for Microwave Lenses," *IEE Proc. (London)*, part III, vol. 98, March 1951, pp. 133–140. Presents analysis similar to that of Lengyel; contains many theoretical graphs for different angles of plate stagger and for typical lens shapes.

23   E. M. Wells, "Some Experiments on the Reflecting Properties of Metal-Tube Lens Medium," *Marconi Rev.*, vol. 17, 1954, pp. 74–85. Gives experimental data for both polarizations for arrays of square tubes.

24   E. K. Proctor, "Methods of Reducing Chromatic Aberration in Metal-Plate Microwave Lenses," *IRE Trans. Antennas Propagat.*, vol. AP-6, July 1958, pp. 231–239. Slot-loaded ridge waveguides make possible greater bandwidth, improved scanning, and elimination of stepping.

25   R. L. Smedes, "High-Efficiency Metallized-Fiberglas Microwave Lens," IRE Nat. Conv., New York, March 1956. A waveguide-type lens that utilizes dielectric and ridge loading, surface matching, and a variable refractive index is described.

26   H. E. King, J. L. Wong, R. B. Dybdal, and M. E. Schwartz, "Experimental Evaluation of a Circularly Polarized Metallic Lens Antenna," *IEEE Trans. Antennas Propagat.*, vol. AP-18, May 1970, pp. 412–414.

27   A. R. Dion, "An Investigation of a 110-Wavelength EHF Waveguide Lens," *IEEE Trans. Antennas Propagat.*, vol. AP-20, July 1972, pp. 493–496.

28   J. Ruze, "Wide-Angle Metal-Plate Optics, *IRE Proc.*, vol. 38, January 1950, pp. 53–59.

Gives design equations and experimental data for bifocal waveguide lenses designed for wide-scan-angle performance.

29 E. K. Proctor and M. H. Rees, "Scanning Lens Design for Minimum Mean-Square Phase Error," *IRE Trans. Antennas Propagat.,* vol. AP-5, October 1957, pp. 348–355. A calculus-of-variations analysis leads to design parameters close to those of Ruze.

30 E. Fine and G. Reynolds, "A Point Source Bi-Normal Lens," Air Force Cambridge Res. Cen. Rep. E-5095, Bedford, Mass., May 1953.

31 G. D. M. Peeler and W. F. Gabriel, "Volumetric Scanning GCA Antenna," *IRE Conv. Rec.,* vol. 3, part 1, 1955, pp. 20–27.

32 A. R. Dion and L. J. Ricardi," A Variable-Coverage Satellite Antenna System," *IEEE Proc.,* vol. 59, February 1971, pp. 252–262.

33 H. S. Lu, "On Computation of the Radiation Pattern of a Zoned Waveguide Lens," *IEEE Trans. Antennas Propagat.,* vol. AP-22, May 1974, pp. 483–484.

34 C. F. Winter, "A TE-Mode Parallel Plate Lens," USAF Antenna Symp., University of Illinois, Monticello, October 1972.

35 J. S. Ajioka and V. W. Ramsey, "An Equal Group Delay Waveguide Lens," *IEEE Trans. Antennas Propagat.,* vol. AP-26, July 1978, pp. 519–527.

36 A. R. Dion, "A Broadband Compound Waveguide Lens," *IEEE Trans. Antennas Propagat.,* vol. AP-26, September 1978, pp. 751–755.

37 H. Gent, "The Bootlace Aerial," *Roy. Radar Establishment J.,* Malvern, England, no. 40, October 1957, pp. 47–58.

38 W. Rotman and R. F. Turner, "Wide-Angle Microwave Lens for Line Source Applications," *IEEE Trans. Antennas Propagat.,* vol. AP-11, November 1963, pp. 623–632. Derives and discusses the Rotman-Turner lens.

39 D. Archer, "Lens-Fed Multiple-Beam Arrays," *Electron. Prog. (Raytheon Co.),* winter, 1974, pp. 24–32; and *Microwave J.,* October 1975, pp. 37–42. Describes the Archer, *R-2R,* and *R-KR* lenses and methods of stacking lenses to obtain beam clusters.

40 D. H. Archer, R. J. Prickett, and C. P. Hartwig, "Multi-Beam Array Antenna," U.S. Patent 3,761,936, Sept. 25, 1973. This is the patent for the Archer lens.

41 D. T. Thomas, "Multiple Beam Synthesis of Low Sidelobe Patterns in Lens Fed Arrays," *IEEE Trans. Antennas Propagat.,* vol. AP-26, November 1978, pp. 883–886. Describes techniques for feeding several beam ports in an Archer lens to obtain low sidelobe radiation patterns.

42 J. B. L. Rao, "Multifocal Three Dimensional Bootlace Lenses," *IEEE PGAP Nat. Conv. Rec.,* 1979, pp. 332–335. Provides design equations for two-, three-, and four-focal-point bootlace lenses.

43 J. P. Shelton, "Focusing Characteristics of Symmetrically Configured Bootlace Lenses," *IEEE Trans. Antennas Propagat.,* vol. AP-26, July 1978, pp. 513–518. Derives design equations for a new type of antenna that has two (symmetrical) bootlace lenses, one on the feed-port side and one for the aperture.

44 J. P. Shelton, "Three-Dimensional Bootlace Lenses," *IEEE PGAP Nat. Conv. Rec.,* 1980, pp. 568–571. Provides design equations for several three-dimensional bootlace lenses.

45 W. Rotman and P. Franchi, "Cylindrical Microwave Lens Antenna for Wideband Scanning Applications," *IEEE PGAP Nat. Conv. Rec.,* 1980, pp. 564–567. Provides designs for a bootlace-type lens with phase shifters and a limited number of feeds to provide coverage over wide-scan angles without requiring a feed for each beam position.

46 H. B. Devore and H. Iams, "Microwave Optics between Parallel Conducting Sheets," *RCA Rev.,* vol. 9, December 1948, pp. 721–732. Discusses several microwave optics devices including the *R-2R* geodesic scanner.

47 S. B. Myers, "Parallel Plate Optics for Rapid Scanning," *J. App. Phys.,* vol. 18, no. 2, February 1947, p. 211. Discusses the general figure-of-revolution scanner formed as the mean surface of a parallel-plate region. This article discusses the scanner of Devore and Iams and considers other scanners in which the mean surface is a cone.

48 M. L. Kales and R. M. Brown, "Design Considerations for Two-Dimensional Symmetric

Bootlace Lenses," *IEEE Trans. Antennas Propagat.*, vol. AP-13, July 1965, pp. 521–528. Proves that the *R-2R* lens of Gent is unique.

49 P. J. Kahrilas, "HAPDAR—An Operational Phased Array Radar," *IEEE Proc.*, vol. 56, November 1968, pp. 1967–1975.

50 B. R. Hatcher, "Collimation of Row-and-Column Steered Phased Arrays," *IEEE Proc.*, vol. 56, November 1968, pp. 1787–1790.

51 R. Tang. E. E. Barber, and N. S. Wong, "A Wide Instantaneous Bandwidth Space Fed Antenna," *Eascon Rec.*, 1975, pp. 104A–104F. The Butler matrix provides multiple feeds for equal time delays.

52 E. J. Daly and F. Steudel, "Modern Electronically-Scanned Array Antennas," *Electron. Prog. (Raytheon Co.)*, winter, 1974, pp. 11–17.

53 D. K. Barton, "Radar Technology for the 1980's," *Microwave J.*, November 1978, pp. 81–86.

54 T. E. Walsh, "Military Radar Systems: History, Current Position, and Future Forecast," *Microwave J.*, November 1978, pp. 87–95.

55 D. K. Barton, "Historical Perspective on Radar," *Microwave J.*, August 1980, pp. 21–38.

56 J. J. Stangel and P. A. Valentino, "Phased Array Fed Lens Antenna," U.S. Patent 3,755,815, Aug. 28, 1973. This is the basic patent on the dome antenna.

57 L. Schwartzman and J. Stangel, "The Dome Antenna," *Microwave J.*, October 1975, pp. 31–34. Describes and gives experimental data on a particular dome-antenna design.

58 L. Schwartzman and P. M. Liebman, "A Report on the Sperry Dome Radar," *Microwave J.*, March 1979, pp. 65–69. Presents experimental data on one design of the dome antenna integrated into a radar.

59 L. Susman and H. Mieras, "Results of an Exact Dome Antenna Synthesis Procedure," *IEEE PGAP Nat. Symp. Rec.*, 1979, pp. 38–41. Presents a synthesis of gain versus scan angle for dome antennas.

60 H. Steyskal, A. Hessel, and J. Shmoys, "On the Gain-versus-Scan Tradeoffs and the Phase Gradient Synthesis for a Cylindrical Dome Antenna," *IEEE Trans. Antennas Propagat.*, vol. AP-27, November 1979, pp. 825–831.

61 D. T. Thomas, "Design Studies of Wide Angle Array Fed Lens," *IEEE PGAP Nat. Symp. Rec.*, 1979, pp. 340–343.

62 D. T. Thomas and S. D. Bixler, "Hardware Demonstration of a 2-Dimensional Wide Angle Array Fed Lens," *IEEE PGAP Nat. Symp. Rec.*, 1979, pp. 344–347.

63 P. A. Valentino, C. Rothenberg, and J. J. Stangel, "Design and Fabrication of Homogeneous Dielectric Lenses for Dome Antennas," *IEEE PGAP Nat. Symp. Rec.*, 1980, pp. 580–583. Discusses a nonconstrained dome lens of graded-thickness dielectric materials.

64 R. K. Luneburg, *Mathematical Theory of Optics*, University of California Press, Berkeley, 1964; and mimeographed lecture notes, Brown University Press, Providence, R.I., 1944. This is the basic reference on the Luneburg lens.

65 J. Brown, "Microwave Wide Angle Scanner," *Wireless Eng.*, vol. 30, no. 10, October 1953, pp. 250–255. Discusses the small-feed-circle Luneburg lens.

66 A. S. Gutman, "Modified Luneburg Lens," *J. Appl. Phys.*, vol. 25, 1954. Discusses the small-feed-circle Luneburg lens.

67 A. F. Kay, "The Impossibility of Certain Desirable Luneburg Lens Modifications," *IRE Trans. Antennas Propagat.*, vol. AP-4, January 1956, pp. 87–88.

68 J. E. Eaton, "An Extension of the Luneburg-Type Lenses," Nav. Res. Lab. Rep. 4110, 1953. Discusses the small-feed-circle Luneburg lens and lenses that produce wavefronts other than the plane wave.

69 S. P. Morgan, "General Solution of the Luneburg Lens Problem," *J. Appl. Phys.*, vol. 29, September 1958, pp. 1358–1368.

70 J. R. Huynen, "Theory and Design of a Class of Luneburg Lenses," *IRE Wescon Conv. Rec.*, vol. 2, part 1, 1958, pp. 219–230.

71 A. F. Kay, "Spherically Symmetric Lens," *IRE Trans. Antennas Propagat.*, vol. AP-7, January 1959, pp. 32–38.

72 S. P. Morgan, "Generalizations of Spherically Symmetric Lenses," *IRE Trans. Antennas Propagat.,* vol. AP-7, October 1959, pp. 342–345.

73 D. H. Cheng, "Modified Luneburg Lens for Defocused Source," *IRE Trans. Antennas Propagat.,* vol. AP-8, January 1960, pp. 110–111.

74 H. Jasik, "The Electromagnetic Theory of the Luneburg Lens," Air Force Cambridge Res. Cen. Rep. TR-54-121, Bedford, Mass., November 1954.

75 E. H. Braun, "Radiation Characteristics of the Spherical Luneburg Lens," *IRE Trans. Antennas Propagat.,* vol AP-4, April 1956, pp. 132–138.

76 G. D. M. Peeler and D. H. Archer, "A Two-Dimensional Microwave Luneburg Lens," *IRE Trans. Antennas Propagat.,* vol, AP-1, July 1953, pp. 12–23. Discusses a two-dimensional Luneburg lens constructed of almost parallel plates filled with dielectric material.

77 S. S. D. Jones, "A Wide Angle Microwave Radiator," *IEE Proc. (London),* part III, vol. 97, 1950, p. 225.

78 C. H. Walter, "Surface-Wave Luneburg Lens Antennas," *IRE Trans. Antennas Propagat.,* vol. AP-8, September 1960, pp. 508–515.

79 G. D. M. Peeler, K. S. Kelleher, and H. P. Coleman, "Virtual Source Luneburg Lenses," *IRE Trans. Antennas Propagat.,* vol. AP-2, July 1954, pp. 94–99. Discusses the Luneburg lens with metallic planes through the center.

80 G. P. Robinson, "Three-Dimensional Microwave Lens," *Tele-Tech & Electronic Ind.,* November 1954, p. 73.

81 G. D. M. Peeler and H. P. Coleman, "Microwave Stepped-Index Luneburg Lenses," *IRE Trans. Antennas Propagat.,* vol. AP-6, April 1958, pp. 202–207.

82 M. C. Volk and G. D. M. Peeler, " A Three-Dimensional Microwave Luneburg Lens," URSI meeting, Washington, 1955. Discusses a lens constructed of hemispherical shells of increasing radius and decreasing dielectric constant.

83 E. F. Buckley, "Stepped-Index Luneburg Lenses," *Electron. Design,* vol. 8, Apr. 13, 1960, pp. 86–89.

84 R. E. Webster, "Radiation Patterns of a Spherical Luneburg Lens with Simple Feeds," *IRE Trans. Antennas Propagat.,* vol. AP-6, July 1958, pp. 301–302.

85 L. C. Gunderson and J. F. Kauffman, "A High Temperature Luneburg Lens," *IEEE Proc.,* May 1968, pp. 883–884.

86 D. S. Lerner, "Calculation of Radiation Heating in a Microwave Luneburg Lens," *IEEE Trans. Antennas Propagat.,* vol. AP-12, January 1964, pp. 16–22.

87 H. F. Mathis, "Checking Design of Stepped Luneburg Lens," *IRE Trans. Antennas Propagat.,* vol. AP-8, May 1960, pp. 342–343.

88 J. I. Bohnert and H. P. Coleman, "Applications of the Luneburg Lens," Nav. Res. Lab. Rep. 4888, Mar. 7, 1957.

89 G. Bekefi and G. W. Farnell, "A Homogeneous Dielectric Sphere as a Microwave Lens," *Can. J. Phys.,* vol. 34, August 1956, pp. 790–803.

90 S. Cornbleet, "A Simple Spherical Lens with External Foci," *Microwave J.,* May 1965, p. 65.

91 T. L. ap Rhys, "The Design of Radially Symmetric Lenses," *IEEE Trans. Antennas Propagat.,* vol. AP-18, July 1970, pp. 497–506.

92 L. C. Gunderson, "An Electromagnetic Analysis of a Cylindrical Homogeneous Lens," *IEEE Trans. Antennas Propagat.,* vol. AP-20, July 1972, pp. 476–479.

93 W. R. Free, F. L. Cain, C. E. Ryan, Jr., C. P. Burns, and E. M. Turner, "High-Power Constant-Index Lens Antennas," *IEEE Trans. Antennas Propagat.,* vol. AP-22, July 1974, pp. 582–584.

94 R. F. Rinehart, "A Solution of the Rapid Scanning Problem for Radar Antennae," *J. App. Phys.,* vol. 19, September 1948, pp. 860–862. Derives a mean surface in a parallel-plate region that is an analog of a two-dimensional Luneburg lens.

95 R. F. Rinehart, "A Family of Designs for Rapid Scanning Radar Antennas, *IRE Proc.,* vol. 40, no. 6, June 1952, pp. 686–687. Generalizes the Luneburg analog for a small feed circle.

96 F. G. R. Warren and S. E. A. Pinnell, "Tin Hat Scanning Antennas," Tech. Rep. 6, RCA

Victor Co., Ltd., Montreal, 1951. Presents an analog of the Luneburg lens that has tangents at its periphery in the lens-aperture plane.

97 F. G. R. Warren and S. E. A. Pinnell, "The Mathematics of the Tin Hat Scanning Antenna," Tech. Rep. 7, RCA Victor Co., Ltd., Montreal, 1951.

98 K. S. Kunz, "Propagation of Microwaves between a Parallel Pair of Doubly Curved Conducting Surfaces," *J. Appl. Phys.,* vol. 25, May 1954, pp. 642–653. Discusses geodesic path-length lenses equivalent to the Luneburg lens.

99 R. C. Rudduck and C. H. Walter, "A General Analysis of Geodesic Luneburg Lenses," *IRE Trans. Antennas Propagat.,* vol. AP-10, July 1962, pp. 444–450.

100 R. C. Rudduck, C. E. Ryan, Jr., and C. H. Walter, "Beam Elevation Positioning in Geodesic Lenses," *IEEE Trans. Antennas Propagat.,* vol. AP-12, November 1964, pp. 678–684.

101 G. A. Thiele and R. C. Rudduck, "Geodesic Lens Antennas for Low-Angle Radiation," *IEEE Trans. Antennas Propagat.,* vol. AP-13, July 1965, pp. 514–521.

102 R. C. Johnson, "The Geodesic Luneburg Lens," *Microwave J.,* August 1962, pp. 76–85.

103 J. S. Hollis and M. W. Long, "A Luneburg Lens Scanning System," *IRE Trans. Antennas Propagat.,* vol. AP-5, January 1957, pp. 21–25.

104 R. C. Johnson, "Radiation Patterns from a Geodesic Luneburg Lens," *Microwave J.,* July 1963, pp. 68–70.

105 R. C. Johnson and R. M. Goodman, Jr., "Geodesic Lenses for Radar Antennas," *Eascon Rec.,* 1968, pp. 64–69.

Chapter 17

Reflector Antennas

Kenneth S. Kelleher

Consultant

Geoffrey Hyde

COMSAT Laboratories

17-1 CONVENTIONAL REFLECTOR ANTENNAS

The reflector antennas treated in the first half of this chapter are those which have been used for decades. The very-high-gain reflectors used in earth stations, which have a more sophisticated design, are treated in the last half of the chapter. The conventional reflector designs are (1) planar, (2) parabolic-arc, (3) double-curvature, (4) paraboloid, and (5) shielded reflectors. In addition, passive reflectors used in communications and radar are described. Sec. 17-2 then considers the sophisticated large reflectors.

Planar Reflectors

Corner Reflector[1]   The corner-reflector antenna is made up of two plane reflector panels and a dipole element (Fig. 17-1). This antenna is useful in obtaining gains of the order of 12 dBi. Higher gains can be obtained by using large reflectors and larger spacing of the dipole to the panel intersection. The gain also depends upon the corner angle. Figure 17-2 shows the gain as a function of dipole spacing for four values of corner angle. Figures 17-3 through 17-6 show the radiation patterns for E and H planes. One of the angles is 180°, and the data therefore include the planar reflector.

The size of reflector panels is a compromise between gain and antenna wind load. The wind load is reduced by replacing the panels by parallel rod reflectors with spacing of $\lambda/10$. The rod length should be 0.6λ. The size of the reflector panels must be increased as the dipole spacing is increased. A spacing of $\lambda/3$ requires a panel length of 0.6λ, a spacing of $\lambda/2$ requires λ, etc. The impedance of the corner reflector depends upon the spacing and upon the dipole impedance. Spacings of less than $\lambda/3$ for the 90° corner are not recommended. Larger spacings have less effect on impedance.

An increase in the corner-reflector gain[2] is obtained by adding a third panel which serves as a ground plane for a 0.75λ monopole feed. The long monopole gives a peak signal above the ground plane (about 45°). The panel lengths are 2λ, and the monopole spacing is about one wavelength. As in the conventional corner reflector, the gain depends upon corner angle and spacing: 90° and 0.9λ give 17 dBi, 60° and 1.2λ give 19 dBi, and 45° and 1.6λ give 21 dBi.

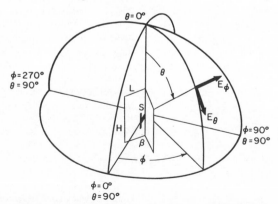

FIG. 17-1 Corner-reflector antenna and coordinate system.

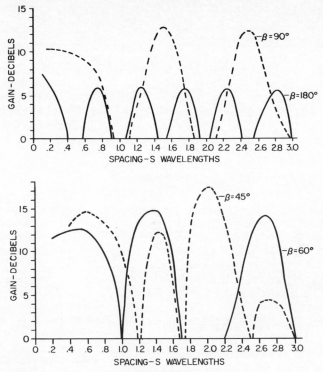

FIG. 17-2 Gain of corner reflectors.

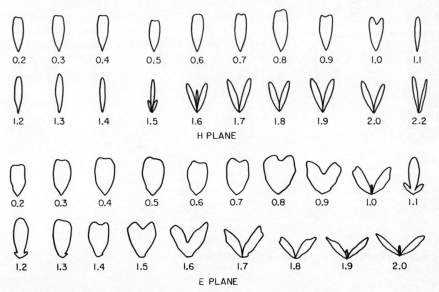

FIG. 17-3 Patterns of 60° corner, S_λ variable.

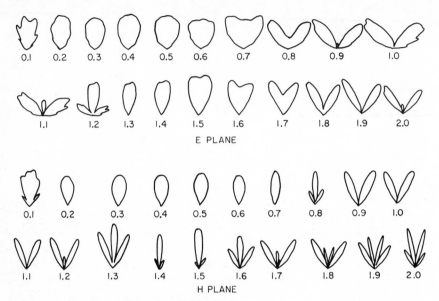

FIG. 17-4 Patterns of 90° corner, S_λ variable.

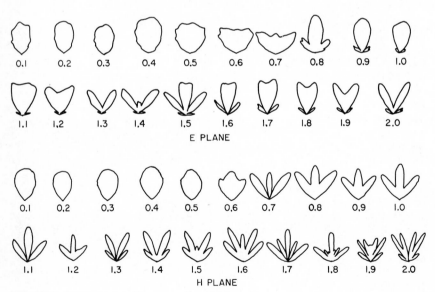

FIG. 17-5 Patterns of 120° corner, S_λ variable.

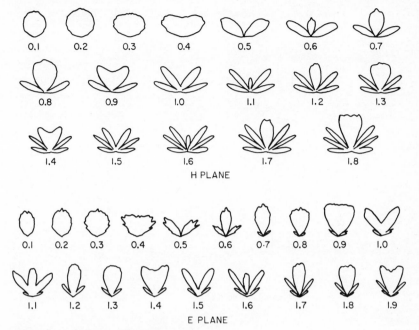

FIG. 17-6 Patterns of flat-sheet, S_λ variable.

An increase in corner-reflector bandwidth[3] can be achieved with a broadband dipole feed. An open-sleeve dipole with the center section of coaxial line in the sleeve replaced by slab line gave almost an octave bandwidth with a voltage standing-wave ratio (VSWR) of 2.5:1. The gain was 11.5 ± 1 dBi. The match was achieved by a cut-and-try procedure.

Planar Reflectors   The antennas in this subsection give good gain for limited size. Their characteristics are shown in Fig. 17-7. The size range is from $\lambda/2$ to 5λ. They appear to have a large reactive field, so good conductors and dielectrics are needed in the design.

0.5λ by 0.5λ The skeleton-slot antenna[4] has provided a gain of 9 dBi over the 225- to 400-MHz band. At the lowest frequency, the reflector has a side dimension of one-half wavelength. The driven element is a rectangular loop which forms the edge of a slot. The slot ends are firmly anchored to the ground plane to give a more rugged structure than that of a plane reflector with a dipole as the driven element.

λ by λ The plantenna[5] has a very close spacing of the dipole and reflector and employs bent edges of the reflector to give directivity of 11 dBi. This antenna operates over a 1.5:1 band and usually employs a printed-circuit dipole.

2λ by 2λ The short backfire antenna[6,7] is the best candidate for high gain over a restricted band. It uses a plane disk reflector of 2λ diameter and a rim at the edge of 0.25λ. When the dipole driver is in the plane of the rim and the dipole reflector is spaced 0.6λ from the 2λ reflector, the gain is 15 dBi. For large reflectors the edges are removed, and the gain[8] is reduced to 12 dBi. When the dipole reflector was removed and the edge retained, the gain was reduced to 11.5 dBi. The dipole had a length of

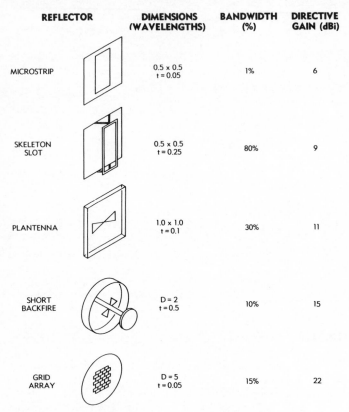

REFLECTOR	DIMENSIONS (WAVELENGTHS)	BANDWIDTH (%)	DIRECTIVE GAIN (dBi)
MICROSTRIP	0.5 x 0.5 t = 0.05	1%	6
SKELETON SLOT	0.5 x 0.5 t = 0.25	80%	9
PLANTENNA	1.0 x 1.0 t = 0.1	30%	11
SHORT BACKFIRE	D = 2 t = 0.5	10%	15
GRID ARRAY	D = 5 t = 0.05	15%	22

FIG. 17-7 Plane-reflector antennas. Dimension *t* is the depth of the antenna.

0.4λ, and the dipole reflector was a disk with a diameter of 0.55λ. In another program,[9] the reflector and the rim were replaced by loops of metal. Six loops were required to replace the reflector, and another six loops to replace the rim. The rim length for best performance was increased to 0.6λ.

5λ *by* 5λ   This reflector, with a driven element formed as a wire grid,[10] will give gains in excess of 20 dBi. The grid[11] is made up of rectangular building blocks which are λ/2 by λ. The element is very close to the ground plane so that only the λ/2 elements radiate; the remainder of the grid serves as the transmission line. Experimental data showed that the λ sides gave a cross-polarized pattern 20 dB below the copolarized (λ/2) patterns. The wire cross sections can be varied to control the currents on the λ/2 elements. One design showed a gain of 21 dBi and sidelobes of −22 dB.

Reflectors with Parabolic Arc

The parabolic arc can be used as a generating curve for a number of useful reflector surfaces. When moved so that the focal point travels along a straight line, the curve generates a parabolic cylinder. With a limited motion along this line, the reflector for a pillbox is generated. When the focal point is moved along a circle enclosing the arc, an hourglass surface is generated, and for a circle which does not enclose the arc a parabolic torus is generated. Figure 17-8 shows the basic parameters of these surfaces.

Parabolic Cylinder   The elements of the cylinder are all perpendicular to the plane of the parabolic arc. Because this is a singly curved surface, problems of construction and maintaining tolerances are somewhat easier than those encountered in the case of the paraboloid. The cylinder can be constructed of tubes or slats which are straight-line elements of the surface, or it can use identical parabolic arcs formed of tubes or slats positioned so that the planes of the parabolic arcs are perpendicular to the elements of the cylinder.

The widest use of the cylinder is with a line-source feed, which may be formed in a number of ways. The line-source aperture is made to coincide with the focal line of the cylinder (the line made by joining the focal points of all arcs). The aperture of the line source can contain energy in phase along the length or have a linear-phase variation along the length. The first produces a cylindrical wavefront, while the linear-phase variation produces a conical wavefront. The cylindrical wave is focused into a plane wave in the direction of the axis of a parabolic arc. The conical wavefront produces a plane wave tilted from the parabolic axis by an angle equal to the half angle of the cone.

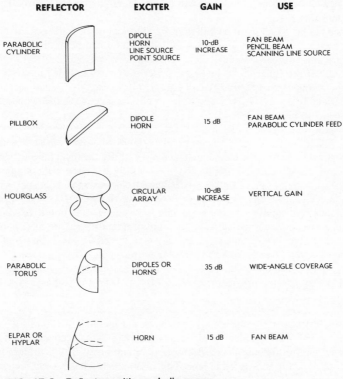

REFLECTOR		EXCITER	GAIN	USE
PARABOLIC CYLINDER		DIPOLE HORN LINE SOURCE POINT SOURCE	10-dB INCREASE	FAN BEAM PENCIL BEAM SCANNING LINE SOURCE
PILLBOX		DIPOLE HORN	15 dB	FAN BEAM PARABOLIC CYLINDER FEED
HOURGLASS		CIRCULAR ARRAY	10-dB INCREASE	VERTICAL GAIN
PARABOLIC TORUS		DIPOLES OR HORNS	35 dB	WIDE-ANGLE COVERAGE
ELPAR OR HYPLAR		HORN	15 dB	FAN BEAM

FIG. 17-8   Reflectors with parabolic curves.

The illumination problems of a line-source feed with the cylinder are related to those of the paraboloid. The decrease in energy in the cylindrical or conical wave is proportional to distance rather than to the square of the distance. The space-attenuation factor in decibels is one-half of that given below in Fig. 17-19. The sidelobes present a problem in the parabolic cylinder since the relative gain between the line source and the final antenna system is not sufficient to suppress the sidelobes of the source. Care must be given to the design of the aperture of the line source, particularly in control of the E-plane sidelobes.

The feed-blocking problem for a line source and parabolic cylinder is serious since the line source presents a large blockage. For this reason, it is usually desirable to use an offset reflector in which the line-source feed structure is removed from the path of the reflected radiation.

The parabolic cylinder can be fed by a point-source feed horn. The horn illumination function is defined by energy which falls off with the square of distance. The reflected wavefront will be a cylinder with elements perpendicular to the elements of the parabolic cylinder. In one plane the point source is focused, and in the other plane the cylinder reproduces the pattern of the feed horn. The result can be a narrow beam in azimuth and a broad-coverage beam in elevation. The best work on such an antenna[12] used a cylinder with its straight-line elements tilted from vertical. Close-in sidelobes were controlled, and the largest sidelobes (of the order of -28 dB) were removed from the main-beam region. This reflector operated at 9.3 GHz and measured 5 ft by 1 ft (1.5 m by 0.3 m), with a focal length of 1.5 ft (4.5 m). The tilt of the reflector was 12°.

Since the horn and cylinder will give a cylindrical wave and since the parabolic cylinder focuses the cylindrical wave, a second reflector can be used. Use of a point-source feed and two parabolic cylinders gives a final wavefront identical to that of a paraboloid.[13] If the equation of the first cylinder is $y^2 = 4fx$, with f the focal length, the second cylindrical surface is given by $(x + f)^2 = 4f(z + f)$, and the final rays are all parallel to the z axis. The plane curves of intersection for the cylinders before development are given in Ref. 13.

The parabolic cylinder can also be used as the reflector in the pillbox, or cheese, antenna. The pillbox is formed by two parallel planes which cut through a parabolic cylinder perpendicular to the cylinder elements. Typically the focal line of the cylinder is positioned in the center of the aperture formed by the open ends of the parallel plates. When a feed is placed at the focal line, it blocks a significant portion of the open region. The blockage causes large sidelobes in the pattern of the pillbox. The feed backlobes are significant, and a portion of the feed's radiation is reflected back into itself to produce a standing wave. An improvement in performance is obtained when a half pillbox is used in the same fashion as the offset parabolic cylinder. The arc of the parabola extends from the vertex to the 90° point. The feed horn is pointed at the 45° point, and although the illumination is asymmetrical, good sidelobes are obtained.[14] The maximum sidelobe of -26 dB was from the spillover past the 90° point of the parabolic arc.

Hourglass Reflector[15]   The hourglass reflector is generated by rotating a parabolic arc (symmetrical or offset) about a vertical axis which is on the convex side of the arc. The feed system for the hourglass reflector is a circular array. The hourglass serves to give increased gain in the vertical plane. The circular array has the characteristics of the Wullenweber system,[16] which provides scanning beams, fixed beams, and beams from a hybrid matrix system. The sidelobes in the elevation plane are controlled by controlling the elevation pattern of the array elements. In the horizontal plane the sidelobes are controlled by tapering the energy in the array elements.

Parabolic Torus[17]   This reflector is formed by rotating the parabolic arc about a vertical axis positioned on the concave side of the arc. It can be formed by a symmetrical arc or by the more widely used offset arc. This reflector is fed by multiple feed horns, each of which uses portions of the reflector surface used by adjacent horns. With multiple usage of the surface, a relatively compact structure can be obtained. The torus is discussed in greater detail in Sec. 17-2. The final surfaces with a parabolic arc are the elpar and hypar surfaces with an ellipse and a hyperboloid in the second principal plane. These reflectors are discussed in the next subsection.

Reflectors Other Than Plane or Parabolic

General Surface   There are a number of useful surfaces in addition to the plane and the surfaces with a parabolic arc. The coordinates of these surfaces are determined by the required phase and amplitude at points on the reflected wavefront. The phase is associated with the length of the incident and reflected ray paths,[18] and the amplitudes depend upon the density of the rays.[19] The most widespread use of general surfaces has been in shaped beams and Cassegrain antennas, but other useful surfaces are included here.

Sphere   The first surface considered is the sphere (Fig. 17-9), which was shown in Ref. 18 to have the phase error $2(1 - a^2)^{1/2} + 2(da^2 - 1)$, where d is the distance from the sphere center to the feed and a is the distance from the centerline to the

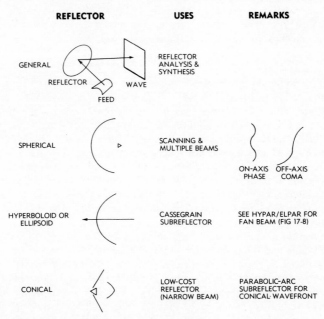

FIG. 17-9   Reflectors other than parabolic.

spherical surface. The sphere has unit radius. A practical sphere is limited in extent. If an attempt is made to scan it to one side, the phase errors do not remain the same but increase as the beam is moved off axis. That is, the value of r in the expression for phase error must increase on one side and decrease on the other. Consider a plane through the sphere containing the axis. The phase-error expression can be plotted in Fig. 17-10 for a number of values of the feed position d. The reflector edges move across the phase-error curves as the feed horn is moved off axis. The error will decrease slightly on one side and increase very rapidly on the other side. A careful analysis of the error will show that it has a strong cubic term (coma), just as the paraboloid has a strong cubic term. The sphere and the paraboloid have similar off-axis characteristics.

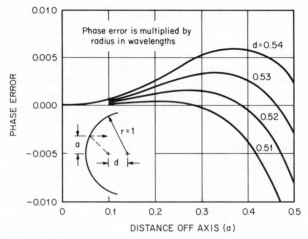

FIG. 17-10   Phase error in a spherical reflector.

If the spherical reflector can have an efficiency of the order of 10 percent, it will give patterns which do not change with the beam moved off axis.[20] In this instance, the sphere is fed from a large feed horn which confines its radiation to a small portion of the surface. The reflector edges which were the problem in the normal sphere are then eliminated. In effect, this solution uses a very-long-focal-length system in which the paraboloid and the sphere become very much alike.

The design of the sphere is based upon the phase-error formula. The error curve is used out to the point at which the phase error is zero. An argument can be made that some phase error beyond this point can be accepted. A tapered illumination from the feed will minimize these errors. In practical design, however, the phase error increases so fast that only a marginal improvement is achieved by considering the feed illumination. Some fine tuning can be done with illumination, but the basic design depends upon the simple use of the two points on the phase-error curves, namely, the maximum value and the edge (zero) value which defines the maximum value of the aperture radius r. A simple ratio, namely, maximum phase error divided by r, is used. Let the phase error be $\lambda/16$ and the radius be 10λ (aperture diameter of 20λ). The ratio is 160:1. The design consists of selecting a specific curve using the parameter d which gives a ratio of 160:1. Note that the y coordinate in the figure is greatly expanded compared with the r coordinate.

Point-Source to Point-Source Reflectors

For some applications, it is desirable to produce a virtual source from a given real source. For example, it may be desirable to position a real source at the vertex of a paraboloidal reflector and utilize some surface between the vertex and the focal point which images the real source into a virtual source located at the focal point. The resulting antenna system, a counterpart of the optical Cassegrain system, will then have a focused beam, since the paraboloid is energized by a source which is apparently at the focal point.

It is first obvious that the reflector surface desired must be symmetrical about the axis joining the two sources. If a cross section of the reflector is known, the complete surface can be formed by rotating this reflector arc about the axis. Figure 17-11 indicates the geometry of the two-dimensional problem. The reflected rays should

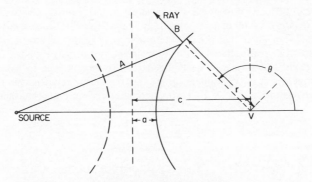

FIG. 17-11   Geometry of hyperbola reflection.

appear to originate at the point v, or, in other words, rays from the real source shall be reflected to form a spherical wavefront with its center at v. Since in the figure we require that $A + B$ be a constant and $r + B$ should be a constant, we find that the surface is determined by the fact that $A - r$ is a constant. From analytic geometry, it is known that the locus of points satisfying this condition is a hyperbola. This curve can then be written as

$$r = \frac{c^2 - a^2}{a \pm c \cos \theta}$$

The quantities here are indicated in the figure. The plus-or-minus sign corresponds to two possible reflector surfaces, one concave and one convex.

Another possible reflector for translating a point source into a point source is an ellipsoid. The ellipse cross section is given by

$$r = \frac{a^2 - c^2}{a + c \cos \theta}$$

where the quantities are the same as in Fig. 17-11.

For application to the paraboloidal reflector mentioned in the first paragraph of this subsection, three surfaces, namely, the ellipsoid and the two hyperboloids, are possible. For practical application, the ellipsoid, which must be mounted at a greater distance from the reflector, is not satisfactory. Of the two hyperboloid surfaces, the one which can be mounted farther from the reflector illuminates typical reflectors better than the nearer, which requires a paraboloid with large f/D ratio. When the real source is moved from one focal point of the hyperboloid system, the image source will move away from the other focal point but will, at the same time, become distorted. Any great motion of the real source produces a poor image.

Point-Source to Line-Source Reflectors

It has been mentioned that the parabolic-cylinder reflection will convert a point source into a line source. However, this property is not unique for this reflector. Other surfaces[21] capable of accomplishing the same result have been designated as hypar and elpar surfaces. These have principal plane sections which are hyperbolas, parabolas, and ellipses. It can be seen that in the plane of the parabola rays from the feed source can be focused to be parallel, whereas in the plane of a hyperbola (or an ellipse) the rays can be reflected so that they appear to come from a virtual-source point. This does not prove that the reflected wave is a cylindrical wavefront but indicates only that the cross sections of this wave in the two principal planes correspond to those of a cylindrical wave. A general reflector can be obtained by assuming that the point source lies at the coordinate value $(d, 0, 0)$ and that the line source coincides with the z axis as in Fig. 17-12. By taking a general ray not lying in either principal

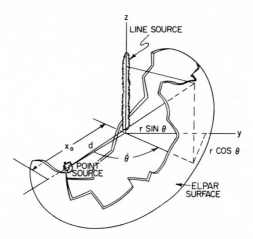

FIG. 17-12 Geometry of elpar surface.

plane and requiring that it have a constant path length between the point source and the line source, it can be shown that the reflector surface is given by the following expression:

$$\sqrt{(r \sin \theta)^2 + (r \cos \theta - d)^2 + z^2} + r = 2x_0 - d$$

or

$$\sqrt{(x - d)^2 + y^2 + z^2} + \sqrt{x^2 + y^2} = 2x_0 - d$$

All quantities in this expression are shown in Fig. 17-12. If one sets θ equal to a constant, it can be shown that all such sections of the reflector are parabolas, whereas setting z equal to a constant yields ellipses in those planes.

A reflector directly related to that indicated above involves cross sections which are hyperbolas rather than ellipses. The mathematics is similar. Using the quantities of Fig. 17-12, we have the expression

$$\sqrt{(r \sin \theta)^2 + (r \cos \theta - d)^2 + z^2} - r = \pm(2x_0 - d)$$

The plus-or-minus signs indicate two possible surfaces, one concave and one convex. Here it can be shown that the planes θ = constant intersect at the surface in parabolas and the planes z = constant intersect it in hyperbolas.

Conical Reflector[22]   The conical reflector is excellent in applications in which the complexity of a doubly curved large aperture is to be avoided. A simple feed with spherical wavefront is reflected from a unique surface into a conical wave which is reflected from the conical surface as a plane wave. Figure 17-13 shows a cross section of the three elements which are the basis of this system. The elements are the small horn, the small parabolic-arc reflector, and the conical reflector, which in the figure becomes a straight line. The complete system is obtained by rotating the cross sections about the axis. It can be seen that the tilted straight line when rotated about the axis becomes a cone with the center area open. The parabolic arc when rotated produces an unusual surface with a pointed vertex like a cone.

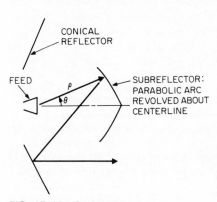

FIG. 17-13 Conical reflector.

The feed structure (small aperture and parabolic arc) blocks the center of the cone to give large sidelobes. An offset version of this system will eliminate the blocking, but the additional mechanical complexity reduces the advantage of using the simple cone.

The design of the conical reflector is based upon two angles, (1) the cone angle and (2) the angle which defines the subreflector. The cone angle is found to be 26°; the angle defining the subreflector is about 50°. The equation of the generating curve of the subreflector is $\rho = 2f/[1 + \cos(52° - \theta)]$. The zero coordinate of this curve coincides with the phase center of the feed. Reference 22 gives detailed design data which show that the optimum configuration has a usable annular ring of aperture with width 1.25 times the radius of the parabolic-arc reflector. Experimental measurements showed a gain in excess of 40 dBi with sidelobes of -12 dB.

The conical reflector can also be used as a magnifier system in which a smaller aperture radiates a plane wave toward a conical subreflector of equal aperture. The reflected conical wavefront then is used to feed the same conical reflector of Fig. 17-13. The concept is essentially the same as that of the figure except that the initial wavefront is a plane rather than the spherical wave from the horn antenna. The magnification gives an increased directivity of the order of 2.5:1.

Shaped-Beam, Singly Curved Reflectors

This problem is essentially two-dimensional, so that ray paths can be restricted to a plane and the cylindrical-reflector surface can be obtained from its plane-curve generator. The basic concepts for the design of a singly curved reflector are simple; the desired shaped-beam pattern is obtained point by point from the primary-feed pattern. The problem presented is that of forming the reflector to make any particular segment of primary-pattern energy appear at the desired point in space. Two basic conditions must be satisfied. The first is that of the energy correspondence between primary $I(\phi)$ and secondary pattern $P(\theta)$, which can be expressed as

$$\frac{\displaystyle\int_{\theta_1}^{\theta} P(\theta)\, d\theta}{\displaystyle\int_{\theta_1}^{\theta_2} P(\theta)\, d\theta} = \frac{\displaystyle\int_{\phi_1}^{\phi} I(\phi)\, d\phi}{\displaystyle\int_{\phi_1}^{\phi_2} I(\phi)\, d\phi}$$

where ϕ and θ are the primary- and secondary-pattern angles and the subscript values correspond to the reflector limits. The second condition relates the angle of incidence to the radius vector defining the surface:

$$\frac{\rho'}{\rho} = \tan i = \tan \frac{\phi - \theta}{2}$$

The relationship between θ and ϕ obtained from the first expression can be substituted in the second expression so that a differential equation is obtained between the radius vector ρ and the feed angle ϕ; this relationship defines the desired reflector curve.

The most difficult part of the shaped-beam problem is that of computing the desired curve from the above expressions. Quite often the first expression must be solved graphically so that the differential equation also requires a graphic solution. If the functions $P(\theta)$ and $I(\phi)$ are integrable, then the first expression can be solved for $\theta(\varphi)$ and the second expression becomes

$$\ln \frac{\rho}{\rho_1} = \int_{\varphi_1}^{\varphi} \tan \frac{\varphi - \theta(\varphi)}{2}\, d\varphi$$

and the integral is evaluated by numerical methods. If the pattern functions are not integrable directly, numerical methods are used. First, a plot is made of the pattern functions. Next, the four integrals in the first expression are evaluated by obtaining the corresponding areas under the curves. For the integrals with variable upper limits, many different values of the area must be found, corresponding to different values of the variable. With this information, it is possible to obtain the two curves of Fig. 17-14. These curves are used to obtain the desired value of θ for any given value of ϕ. With this known, $\tan \{[\varphi - \theta(\varphi)]/2\}$ is plotted and the integral evaluated so that $\rho(\varphi)$, the desired curve, can be obtained.

Doubly Curved Reflectors[23]

The problem of beam shaping becomes somewhat more complicated when the reflector is to shape the beam in one principal plane and focus it in the other plane. Since the problem is a three-dimensional one, the possibility also exists that the feed might not produce a simple spherical wavefront but might also yield a cylindrical surface, a foot-

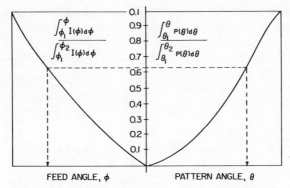

FIG. 17-14 Relationship between feed angle and pattern angle in a shaped reflector.

ball-type surface, or any one of a number of other surfaces. Fortunately, all feed systems of interest have a circular wavefront in the plane of beam shaping, so that techniques similar to that employed in the preceding subsection can be utilized to determine the surface cross section in that plane. The existing techniques consist of forming the reflector surface from the plane-curve cross section, which serves as a spine, and a series of other plane curves which are attached as ribs to the spine. Some question has arisen regarding this general technique, but it has been found to produce satisfactory experimental results for cases in which beam shaping is desired over a limited angle. This problem may be studied more carefully for application to beam shaping over wider angles.

The plane curve, identified as a spine above, has been called the central-section curve. This curve is found in a manner quite similar to that described in the preceding subsection. The design equations for the feed with spherical wavefront are obtained, using the same quantities as before:

$$\frac{\int_{\theta_1}^{\theta} P(\theta)\, d\theta}{\int_{\theta_1}^{\theta_2} P(\theta)\, d\theta} = \frac{\int_{\phi_1}^{\phi} \frac{I(\phi)}{\rho}\, d\phi}{\int_{\phi_1}^{\phi_2} \frac{I(\phi)}{\rho}\, d\theta} \qquad \frac{\rho'}{\rho} = \tan\frac{\phi - \theta}{2}$$

The major difference between these expressions and the previous ones occurs in the presence of ρ under the integral sign. The integration can be carried out as before if first ρ is assumed to be a constant over the shaped portion of the central-section curve; then follow a parabolic arc [$\rho = \sec^2(\phi/2)$] over the region which produces the main beam. With this assumption, the procedure is identical with that of the last subsection for obtaining the function $\rho(\varphi)$. This function represents a closer approximation to the correct value than the original assumption. When it is used as the value under the integral sign and the entire procedure is repeated, a new value of $\rho(\varphi)$ that is as close to the true value as necessary in any practical case is obtained.

A discussion of the rib curves will indicate the limitations in the present techniques. To obtain the ribs, it is customary to consider an incoming bundle of rays from the secondary-pattern angle θ. For beam shaping in the vertical plane, this bundle of rays is confined to a plane making an angle θ with the horizontal plane. It is required that the reflector surface direct these rays to the feed position. Figure 17-15 shows two parallel rays in this plane which intersect the reflector in the curve AA'. The rays

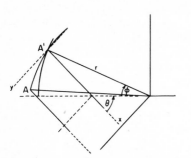

between the points A and A' are reflected to lie on a conical surface with apex at the feed.

If we considered the transmit case and a plane sheet of rays emanating from the feed point, they would intersect the reflector in a curve AA' and the reflected rays would lie on a cylindrical surface tilted at an angle of θ with the horizontal plane. It should be apparent that although they have the same end points, the plane curves AA' are not the same for the transmit and receive cases. This can be understood since it is known that if focusing is desired, the reflector surface

FIG. 17-15 Geometry of a three-dimensional shaped reflector.

in the neighborhood of the points of ray incidence must be a paraboloid. The plane curve cut by this paraboloid will be different if we consider the intersection caused by the transmit plane passing through a focal point and that caused by the receive plane inclined at an angle θ to the horizontal. It is difficult to make a choice between the two curves, although the great majority of existing efforts have chosen the receiving-plane case. A compromise solution might be the plane bisecting the angle formed by the transmitting and receiving planes. The parabolic curve for the receive case mentioned above is

$$y^2 = 4z\rho \cos^2 \frac{\theta + \varphi}{2}$$

Here z is the coordinate measured from the point A' away from the reflector, and y is the orthogonal coordinate. It should be noted that each of the rib curves is a parabola lying in a different plane; when the reflector surface is formed, the planes of various parabola ribs must be inclined at the corresponding angles.

All the previous expressions yield only normalized coordinates for the surface; these must be converted to usable dimensions before the reflector can be constructed. It is desirable to make the central-section curve as large as possible in order to minimize the diffraction effects on the shaped-beam pattern. In the other plane, the reflector size is chosen to produce the desired beamwidth. The feed angles associated with the reflector are chosen in a manner similar to those of the simple paraboloid. It is desirable that the aperture illumination be 10 dB down at the reflector edges.

The most advanced technique[24] for defining the reflector for beam shaping permits beam control in both principal planes. A two-variable generalized far-field pattern is used to define the reflector surface. The generalization is achieved by replacing the ray-to-ray correspondence by a curve of incident ray directions related to a curve of reflected ray directions.

The caution in the use of any ray relationship is that it ignores diffraction effects. Therefore the reflector defined by geometrical optics should be checked by defining the phase and amplitude across the reflector aperture and integrating for the far-field patterns.

Paraboloidal-Type Reflectors

The most useful of the high-gain reflector antennas utilizes the paraboloidal surface, which is formed by rotating the arc of a parabola about the line joining the vertex and the focal point. The geometrical relations for such a reflector can be obtained from Fig. 17-16, which represents a typical cross section through the reflector in a plane containing the axis. The major problems associated with this reflector antenna are

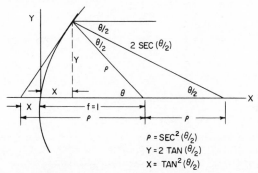

FIG. 17-16 Geometric relationships in parabola with unit focal length.

related to obtaining an efficient combination of reflector and feed. If it is assumed, for convenience, that the typical feed pattern can be described as a cosine function of the angle measured away from the peak of the beam, some general conclusions can be found. The first is that the reflector cannot intercept all energy from the feed, as might be desirable for maximum gain. A second consideration is that in attempting to intercept all this energy, the reflector size becomes large and its outer portion, which represents a considerable amount of the reflector area, receives a low-intensity field from the feed. This tends to minimize the abrupt field discontinuity at the reflector edges and so minimizes sidelobes. Since reflector size is limited from the mechanical-design standpoint, it is often desirable to minimize reflector regions of low intensity. A compromise must therefore be reached between making the reflector sufficiently large to capture the feed energy and making it too large to be useful from a mechanical standpoint.

In a typical design the reflector size is chosen to be as large as practical, and then the feed is designed for efficient illumination. Normally, the design is based on obtaining either the maximum gain from the antenna or a reduction in sidelobes at the expense of a slight decrease in gain. For maximum gain, it has been found that the energy reflected should be distributed so that the field at the reflector edges is approximately 10 dB below that of the center; for good sidelobe performance, the field at the edge should be about 20 dB down.

It has been found useful in feed design to employ a set of curves which will indicate the beamwidth of the feed necessary to produce the desired edge illumination on a particular reflector. One of the curves utilized is a normalized feed pattern (Fig. 17-17). This pattern was obtained through a comparison of many experimentally mea-

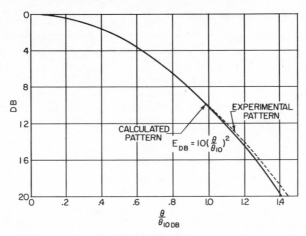

FIG. 17-17 Universal feed-horn pattern.

sured data points. It can be closely approximated over all but the lower-intensity regions of the beam by a simple quadratic function. The beamwidth at any point is proportional to the square root of the decibel reading at that point. For example, the beamwidth is doubled when the decibel reading is increased by a factor of 4. This curve is necessary because of the fact that beamwidth information is normally available only at a particular decibel value, such as 3 or 10 dB, whereas in actual application we are interested in the beamwidth at the points where rays are directed to the edges of the reflector.

To relate the reflector geometry to the pattern beamwidth, a second curve is necessary. Normally, the information available on the paraboloidal-reflector surface is its focal length and aperture dimension, so that we can immediately use the ratio of these quantities, designated by f/D. From the information of Fig. 17-16 it is possible to produce a curve relating the f/D ratio to the feed angle θ. This relationship is given graphically in Fig. 17-18.

Once the feed angle corresponding to the reflector edges has been determined, it would appear possible to utilize Fig. 17-17 directly in order to obtain a value of edge illumination. However, one additional factor must be considered, the divergence factor $(\cos^4 \theta/2)$. This function can be converted to decibels and plotted in Fig. 17-19. It is evident from the figure that the space-attenuation factor is negligible for small values of feed angle and becomes relatively large for feed angles of the order of 90°.

To understand the use of Figs. 17-17 to 17-19, a typical problem can be solved. If it is required that the edge illumination should be 20 dB in a reflector whose f/D ratio is 0.5, we first enter the curve of Fig. 17-18 and find that the total feed angle is 106°. From Fig. 17-19, the space attenuation at this angle (53° from the axis) is 1.9 dB. The desired edge illumination can then be found by noting that the feed horn should have a signal 18.1 dB down at an angle of 53° from the peak of the beam. Since the feed-horn design curves are given in terms of the 10-dB beamwidth, it is necessary to determine what 10-dB width is associated with 18.1 dB at 53°. Since, as previously stated, the beamwidth ratio is proportional to the square root of the decibel ratio, the desired width is given by $(10/18.1)^{1/2} \times 53° = 39.5°$, so that the total 10-dB width is equal to 79°.

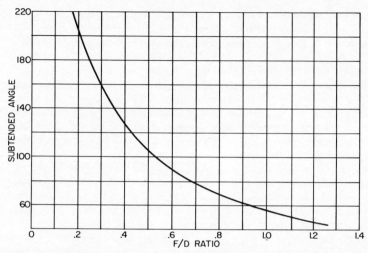

FIG. 17-18 Ratio of focal length to aperture diameter versus subtended angle at focal point.

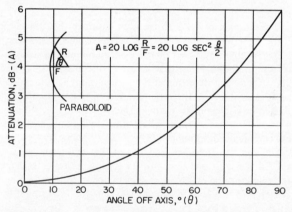

FIG. 17-19 Space attenuation versus feed angle.

Some slight improvement in accuracy might be obtained if the quadratic nature of the beamwidth were ignored and the standard pattern of Fig. 17-17 used. From this figure, it is found that the ratio in beamwidth between 10 and 18.1 dB is equal to 0.73, which corresponds to a feed whose 10-dB width is 77.5°. Because of the negligible difference between these two answers, it can be seen that for most practical applications the quadratic approximation to the standard pattern can be employed.

Interaction between Reflector and Feed From the considerations of the previous section, it is evident that the design of the reflector cannot be entirely divorced from that of the feed. An even more significant relationship arises when one considers interaction between these two elements. Because of this interaction, the feed disturbs the radiation pattern of the reflector and the reflector affects the match of the feed. A third problem of considerable significance in smaller reflectors is the contribution of the backward radiation from the feed to the pattern formed by the reflector. This effect tends to raise the sidelobe level of the radiation pattern as the reflector aperture is decreased below an aperture of 10 wavelengths.

The effect of the horn on the radiation pattern of a large reflector can be analyzed in a very simple manner. The horn, with its associated waveguide, represents an obstacle in the radiating aperture. The final radiation pattern can be shown to consist of the pattern from the unobstructed aperture minus the pattern of the obstruction. This very simple analysis has been shown to be satisfactory and to compare very closely with measured experimental results. Since, in general, the feed design is determined from mechanical considerations and from the reasoning of the preceding subsection, a minimization in the effect of feed blocking is obtained only as the aperture size in wavelengths increases. This is due to the fact that the feed geometry depends primarily on the angular sector of the dish and so is unchanged as the reflector size in wavelengths is increased. It should be noted that the feed obstruction, that is, the feed horn plus waveguide, is normally limited to a single plane. The effect of this obstruction on sidelobe performance is then most pronounced in the plane perpendicular to the feed obstruction where the obstruction pattern is very broad. In the other plane, the obstruction pattern has a much narrower beamwidth, which approaches that of the reflector and so is relatively weak in the regions where it might contribute an increase in sidelobe level.

The above discussion of feed obstruction assumes that the dimension of this obstruction in the direction of propagation is relatively small; however, if this dimension becomes appreciable, the effect of the obstruction on the feed pattern becomes more complex, and in almost every case a further increase in sidelobe level is found. Experimentally, it has been determined that the minimum value of sidelobe level is obtained if the feed structure in the direction of propagation is streamlined. This has been accomplished by enclosing all parts of the feed structure, with the exception of the feed horn itself, in a thin cylinder whose cross section approximates an ellipse of large ratio between major and minor axes.

The effect of the reflector on the match of the feed horn can be analyzed in a simple manner if it is understood that most of the energy reflected back into the feed comes from the region in the neighborhood of the reflector vertex. Since the incident wave was divergent, a focal length large in wavelengths would tend to minimize the amount of signal reflected back to the feed. For a given reflector, an increase in the gain of the feed, which corresponds to an increase in peak signal compared with signals from other directions, will result in greater energy incident upon the reflector vertex and therefore a greater reflected wave and a greater mismatch. From these considerations it can be shown that the reflection coefficient introduced by the presence of the paraboloid is given by $\Gamma = g\lambda/4\pi f$, where g is the gain of the feed in the direction of the vertex.

It is of interest to determine the proper value of focal length for minimum mismatch if the reflector aperture and illumination are held fixed. Since for a circular aperture the feed gain is inversely proportional to the beamwidth squared, it can be replaced in the formula by the square of an angle, which except for space attenuation is directly related to the angle θ, subtended by the aperture at the focal point. Since the aperture diameter is given by $D = 2f\tan(\theta/2)$ (Fig. 17-16), we can write $\Gamma \propto [\tan(\theta/2)]/\theta^2$, which indicates a smaller value of reflection coefficient for large θ, that is, for short focal lengths. The minimum reflection is limited by the decreased value of reflector gain realized when a short-focal-length reflector is used.

Paraboloid Edge Configuration   The paraboloid shape can be circular, elliptical, or diamond-shaped or of any other configuration. The edge configuration can be used to control sidelobes in the same manner that the feed taper controls sidelobes. The sidelobes in any plane through the aperture are determined by considering the illumination function along the line where the plane intersects the aperture. Figure 17-20 shows how the illumination function is determined when the line of interest is the x axis. The value at a point x depends upon the energy along the line perpendicular to x. For a given feed illumination, the energy depends upon the length of this line.

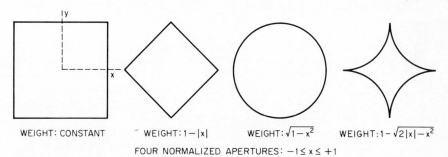

FOUR NORMALIZED APERTURES: $-1 \leq x \leq +1$

FIG. 17-20   Aperture illumination weighting.

That is, the longer the strips of energy, the greater the total energy at the point x. The diamond-shaped aperture has a taper in the line length as x increases, so that the illumination of a given feed horn is tapered by the use of a diamond configuration. The rectangular configuration gives no additional taper to the feed-horn illumination. The circle gives some taper and is the same for all planes through the aperture. It was found experimentally by the author in the late 1940s that a diamond with concave edges gave the best performance for horizontal-plane patterns. It should be noted that the sidelobes in a 45° plane will have much higher levels. The same analysis of the length of line perpendicular to the aperture line will give the illumination taper for any plane cutting the aperture.

The reason for caution in the use of the edge configuration to control sidelobes is that the spillover from the horn may be increased, which reduces the gain and gives sidelobes far removed from the main beam.

The illumination taper can also be controlled by using nonreflecting material at the edges of the aperture. The surface-impedance variation near the edges of the aperture will have a significant effect[25] on the far-out sidelobes. The surface-impedance approach can also be applied to the edges of the subreflectors used in Cassegrain systems.

Another illumination characteristic of interest is related to varying the illumination function in one of the principal planes while holding the taper fixed in the other plane. As expected, the increased illumination taper decreases the near-in sidelobes. However, this taper in one plane will increase sidelobe levels in the other plane.[26] The reason for this can be traced back to the explanation of Fig. 17-20. The illumination along the aperture line is weighted by the field in a strip perpendicular to the line. The taper in one plane (for example, the horizontal plane) tends to decrease the field at the center of the other (vertical) principal plane. A decrease in illumination in the center of the vertical line through the aperture means less illumination taper in that plane and so higher sidelobes (Fig. 17-21).

Shielded-Aperture Reflectors

In some applications it is desirable to have very low sidelobes from a pencil-beam reflector. In this instance, considerable improvement can be obtained by the use of metal shielding around the reflector aperture (Fig. 17-22). The typical sidelobes are at about 0 dBi, which for most reflectors represents a value of the order of -30 to

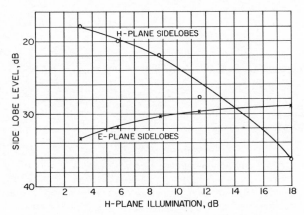

FIG. 17-21   Sidelobe level as a function of illumination.

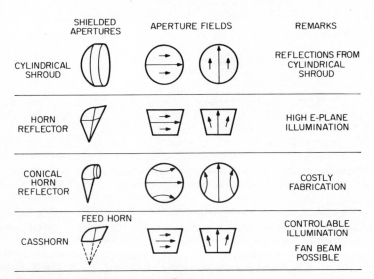

FIG. 17-22   Shielded-aperture reflectors.

−40 dB below the peak gain. With a shielding technique, the far-out sidelobes can be reduced to −80 dB.

The simplest approach to shielding the reflector is a cylindrical "shroud," or tunnel, of metal around the edge of a circular reflector (Fig. 17-22). If the aperture is elliptical in cross section, an elliptical cylinder can be used. The tunnel forms a waveguide of large cross section with the paraboloid at one end. The radiated energy has low levels of current on the outside edge of the tunnel and so has little radiation at wide angles from the beam. There is energy in the waveguide which leaves the feed horn to reflect from the inside of the cylinder. This energy is then reflected by the paraboloid into angles away from the axis.

An improved shielding system which eliminates the reflection of the feed-horn energy is shown in Fig. 17-23 as the *horn reflector*. The two versions employ a pyramidal horn and a circular-waveguide horn. These antennas are discussed in Sec. 30-5, "Conical and Pyramidal Cornucopias," in Chap. 30. The variation among cornucopia horns is associated with the variation in aperture illumination. In many applications, the low sidelobe capability is associated primarily with the horizontal plane where interfering sources are found. The circular aperture will give reduced lobes, and the diagonal aperture (see discussion of aperture illumination in the preceding subsection) will give the best performance. The references for experimental cornucopia design are given in Sec. 30-5.

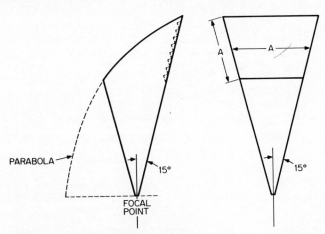

FIG. 17-23   Geometry of horn reflector.

The final shielded horn to be considered is the casshorn,[27] which uses a secondary reflector to replace the horn structure in the cornucopia. The secondary reflector is designed in the same manner as the Cassegrain system. In the lowest-cost version the reflector is plane.[28] The aperture of the casshorn can be formed in a variety of ways according to the illumination concepts discussed earlier. An elliptical aperture can be designed as easily as a trapezoid or circular aperture. The feed for the casshorn is mounted at one edge of the aperture (Fig. 17-22). The problems in the design are associated with making sure that the feed pattern is not distorted by the nearby reflector surface.

A comparison of the casshorn and the cornucopia indicates that the free-space feed of the casshorn can be used to provide more illumination taper for the aperture (particularly in the E plane) than the cornucopia. However, the casshorn has not proved to be superior to the cornucopia in sidelobe performance. Greater application of shielded antennas can be expected as the number of microwave installations increases.

Passive Reflectors

There are reflector systems which do not contain a feed source. These reflectors, which are called passive, usually receive and reflect a plane wave. There are three general types of passive reflectors: (1) those used in microwave links, (2) those used as radar targets, and (3) those used in a beam waveguide.

Microwave Relay   The passive reflectors[29,30,31] used in microwave links can be classified as repeaters or periscopes. The first reflector system controls a plane wave in the horizontal plane, and the second controls the plane wave in the vertical plane. The design information for repeater reflectors given in Fig. 17-24 shows the geometry of a

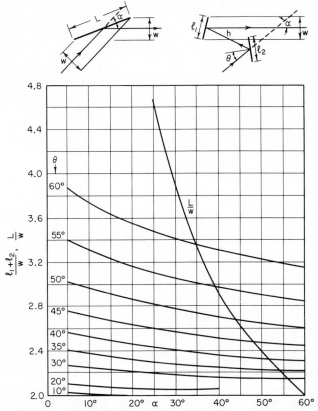

FIG. 17-24   Variation of (normalized) mirror length with α and θ for single- and double-mirror passive repeaters.

single-reflector and a dual-reflector repeater system. The reflector surfaces should be within $\lambda/16$ of a plane. The extent of the plane surface should be minimized to reduce costs. The figure shows dual and single reflectors with the dimensions required for various angles of the ray paths. The path loss for the two systems can be calculated by assuming that the projected area seen by the transmitter and receiver is the same (and assuming that in the case of the two-reflector system all the transmitter energy captured by one reflector is reflected from the second reflector). If the ranges from the transmitter and receiver are given by d_t and d_r, then the path loss is

$$10 \log \frac{\lambda^4 d_1^2 d_2^2}{A_T A_R A}$$

where the areas are designated for transmitter, receiver, and projected area of the repeater plane surfaces.

The periscope reflector system uses a large reflector at the top of a tower and a smaller reflector on the ground. This antenna can employ a plane reflector or a curved reflector at the top of the tower. The gain obtainable from the two type surfaces is shown in Figs. 17-25 and 17-26.

Radar Targets[32]   Radar targets are passive reflectors which have a reflected-signal distribution similar to the patterns of an antenna. A number of classical targets are given in Fig. 17-27. The basic problem in the design of radar targets is that of maximizing the target return. Mechanical tolerances of the reflector surface limit the performance. For targets of the order of 20λ in extent, a $\lambda/3$ error in the edges will give a loss of 3 dB in the returned signal compared with a perfect target. Figure 17-28

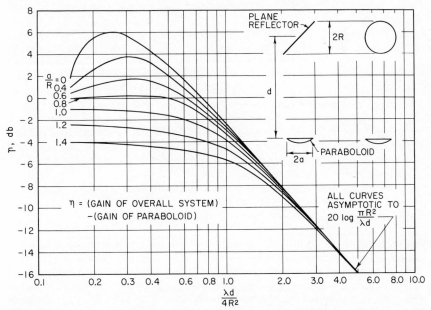

FIG. 17-25 Relative gain of a periscope antenna system employing a plane elevated reflector.

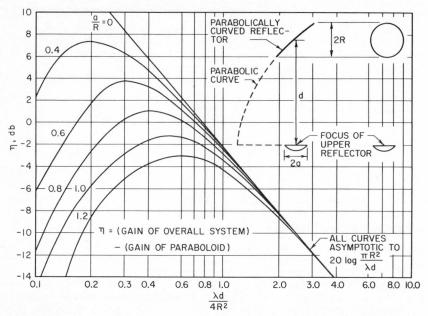

FIG. 17-26 Relative gain of a periscope antenna system employing a curved elevated reflector.

shows the loss for three different corner reflectors for angular errors of the order of 2°.

The simple corner reflector is usable over an angular sector of about 30°. For a wider-coverage angle, an array of corner reflectors can be employed, or the Luneburg lens can be used with a spherical cap reflector[33] (see Fig. 17-29). The figure shows that the maximum-coverage angle is of the order of 120°. Another lens[34] which gives a theoretical return from all directions has been developed. This lens was fabricated by Emerson & Cuming, Canton, Massachusetts. The index variation is given by $n^2 = (2 - r)/r$. The dielectric constant required for small r becomes very large (at quarter radius the value is 7.5). It is possible to replace the center of the lens by a metallic sphere in order to avoid the large dielectric constant. The center portion represents only a small portion of the aperture, so that an operable system could be achieved without the expense of specialized dielectrics.

For coverage in all directions of space, a combination of 20 corner reflectors (duodecahedron) has been used successfully tested.[35] The difficulty in obtaining reproducible lenses has made this reflector approach more satisfactory.

Beam Waveguide   The final use of passive reflectors is in earth stations, where typical waveguide losses cannot be tolerated. The beam-waveguide concept uses a series of large-aperture reflectors to transmit energy with little loss. Initial work on such waveguides[36] used lens-type apertures. The best existing beam waveguide uses a series of paraboloids, hyperboloids, or ellipsoids. The energy is carried along the guide from one focal point to another, and at each focal point it is received from one reflector aperture and transmitted to the next reflector.

TYPE	DIMENSIONS	MAXIMUM		ANGULAR RESPONSE	
		A_T	σ	θ	ϕ
SPHERE	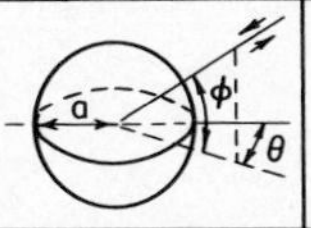	$\dfrac{a\lambda}{2}$	πa^2	360°	360°
CYLINDER	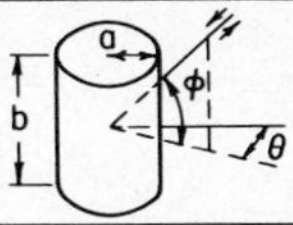	$b\sqrt{\dfrac{a\lambda}{2}}$	$\dfrac{2\pi ab^2}{\lambda}$	360°	SHARP
FLAT PLATE	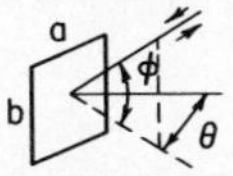	ab	$\dfrac{4\pi a^2 b^2}{\lambda^2}$	SHARP	SHARP
DIHEDRAL CORNER	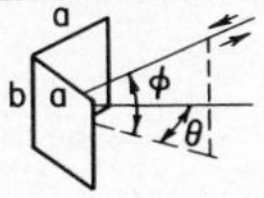	$\sqrt{2}ab$	$\dfrac{8\pi a^2 b^2}{\lambda^2}$	± 30° (TO −10 DB ECHO LEVELS)	SHARP
TRIANGULAR TRIHEDRAL	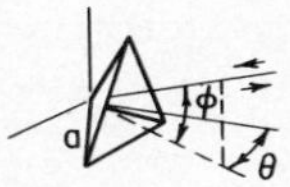	$\dfrac{a^2}{\sqrt{3}}$	$\dfrac{4\pi a^4}{3\lambda^2}$	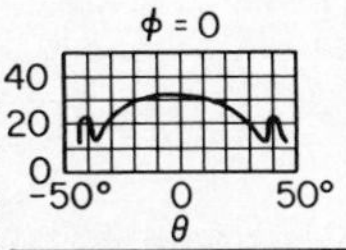 (plot: RELATIVE ECHO IN DB vs θ, $\phi=0$, −50° to 50°)	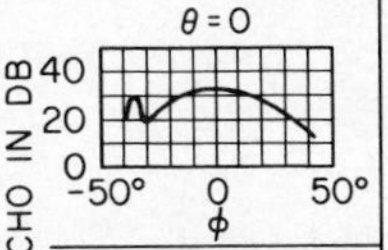 (plot: RELATIVE ECHO IN DB vs ϕ, $\theta=0$, −50° to 50°)
SQUARE TRIHEDRAL		$\sqrt{3}a^2$	$\dfrac{12\pi a^4}{\lambda^2}$	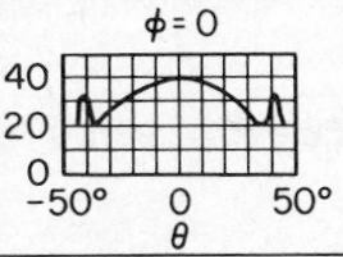 (plot: RELATIVE ECHO IN DB vs θ, $\phi=0$, −50° to 50°)	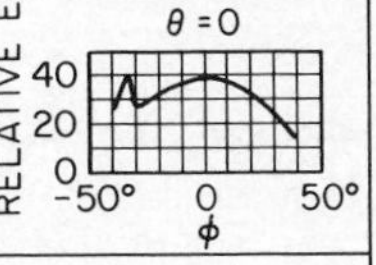 (plot: RELATIVE ECHO IN DB vs ϕ, $\theta=0$, −50° to 50°)

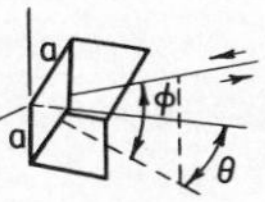

NOTES

A_T = EQUIVALENT FLAT-PLATE AREA OF TARGET

σ = SCATTERING CROSS SECTION OF TARGET

$\sigma = 4\pi\dfrac{A_T^2}{\lambda^2}$

$P_R = P_T\dfrac{A_R^2 A_T^2}{\lambda^4 d^4} = P_T\dfrac{A_R^2\,\sigma}{4\pi\,\lambda^4 d^4}$ (FREE-SPACE TRANSMISSION)

P_T = POWER EMITTED BY RADAR

P_R = ECHO POWER COLLECTED BY RADAR

d = DISTANCE FROM RADAR TO TARGET

A_R = EFFECTIVE AREA OF RADAR ANTENNA = $\dfrac{G\lambda^2}{4\pi}$, WHERE G = GAIN

λ = WAVELENGTH IN SAME UNITS AS d, $\sqrt{A_R}$ AND $\sqrt{A_T}$

ALL DIMENSIONS ARE ASSUMED LARGE IN WAVELENGTHS

θ AND ϕ ARE ANGLES BETWEEN THE DIRECTION TO THE RADAR ANTENNA AND THE MAXIMUM RESPONSE AXIS OF THE TARGET

FIG. 17-27 Characteristics of various types of radar targets. The columns labeled A_T and σ give the maximum values of these quantities for optimum orientation of the target. G is the gain of an antenna over an isotropic radiator.

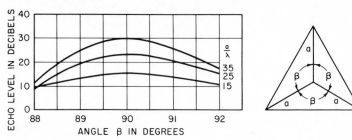

FIG. 17-28   Effect of error in all three corner angles upon a trihedral echo response. Incident radiation is parallel to the symmetric axis.

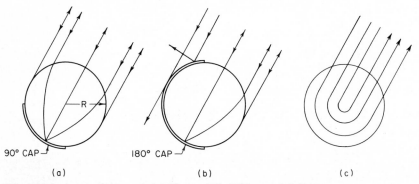

FIG. 17-29   Lens reflectors. (*a*) 90° cap. (*b*) 180° cap. (*c*) No cap.

Figure 17-30 shows the geometry of the most effective beam waveguide. The reflectors used are either paraboloid-paraboloid or ellipsoid-hyperboloid.[37,38] With the configuration chosen, it is possible to eliminate rotary joints in the transmission line. There is very little loss in beam waveguides, and they can be shielded from external signals by the use of large metallic cylinders which encase the beams as they travel between reflectors.

Polarization-Sensitive Reflectors[39]

One or more reflectors in an antenna system can be made up of linear parallel reflectors which will reflect one polarization and transmit the orthogonal polarization. Such reflectors are called *transreflectors*. When they are aligned with the linear reflecting elements at 45° to the incoming polarization, half of the energy is passed and half is reflected. A second solid reflector will redirect the transmitted energy back through the linear elements with a phase which depends upon the spacing between the linear elements and the solid reflector. A spacing of one-eighth wavelength creates a 90° phase shift and circular polarization. A spacing of one-quarter wavelength creates a half-wave phase shift and a reflected polarization at 90° from the incident polarization. This system is called a *twist reflector*.

A transreflector can be used as a focusing device which has 360° coverage in the azimuth plane. The linear elements are aligned at 45°, so that half of the incident linear or circular polarization is passed and the remainder is scattered. The reflector

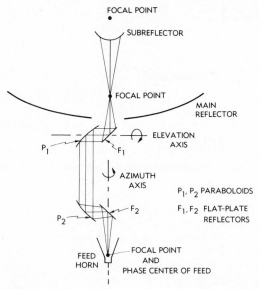

FIG. 17-30 Geometry of a beam-waveguide feed system.

appears as a dome which in the azimuth plane shows the near-side linear elements to be orthogonal to those on the far side. Therefore, the energy which passes through the near side of the dome is reflected at the far side. If the surface contour matches a sphere (helisphere) or parabolic torus, the energy which reflects from the far side is focused.

The transreflector in a planar configuration can be used to provide two-feed systems in a single reflector. That is, a horizontally polarized signal can pass through vertically polarized reflecting elements. The transreflector can be placed between a paraboloidal reflector and the horizontally polarized feed at the focus. A second, vertically polarized feed can be placed at the image of the focal point in the plane reflector. This has been termed *focus splitting*.

A unique use of the transreflector and twist reflector lies in providing a thin focusing system with a very low f/D ratio.* The paraboloid with focus in face has an f/D ratio of 0.25. A transreflector mounted across the face of the curved reflector will return the energy to the curved paraboloidal surface, where it is rereflected. If this surface is a twist reflector, the reflected radiation is shifted through 90° in polarization so that it now passes through the transreflector. The parallel beam of rays from the system was created by reflection at two curved surfaces (two uses of the twist-reflector surface), and therefore the surface curvature required is reduced. The f/D ratio is reduced to 0.125. A further reduction in the f/D ratio is obtained if a Cassegrain subreflector is added and the feed is placed at the vertex. The f/D ratio then is 0.08.

The transreflector or the twist reflector is made up of about 10 linear elements per wavelength. Accurate planar surfaces can be easily obtained by printed-circuit-

---

*See J. P. Shelton, U.S. Patent 4,228,437, Oct. 14, 1980.

card techniques. Curved surfaces are best formed with the same technology, and for this reason polarization-sensitive reflectors are most often used for physically small surfaces.

Frequency-Sensitive Reflectors[40]

It is possible to design a reflector which will pass signals of one frequency band and reflect signals of other frequencies. These reflectors are called *dichroic* or frequency-sensitive surfaces (FSS). Such reflectors are made up of a large array of closely spaced resonant elements. Slots or dipoles can be used. The dipoles reflect the chosen frequency band, and the slots transmit this band. When the dipoles become small compared with a wavelength of interest, there is limited reflection and most of the energy is transmitted. When the slots become small compared with the wavelength of interest, the energy is almost totally reflected. By a proper choice of dipoles or slots, the passband can be at a higher or lower frequency than the second band of frequencies.

The most important constraint in the design is the spacing between the two bands of frequencies. If there is an octave band or less, the design becomes difficult and greater losses are introduced. If two octaves separate the bands, the design restrictions are not severe. For sharper response between frequency bands, it is possible to use two or more surfaces, in a manner analogous to transmission-line filter design. The remaining serious design problem is associated with polarization. The surface should preserve the incoming polarization for any angle of incidence. Most surfaces have a different response to polarization in the plane of incidence than to orthogonal polarization.

Polarization limitations have ruled out simple patch and crossed dipole elements in the frequency-sensitive array. The two most widely used elements are the *tripole* and the *Jerusalem cross* (Fig. 17-31). These elements can be more tightly packed in an array. The support of these elements presents a practical design problem since the substrate dielectric will distort the design. Thin Mylar or Kevlar is used for the array, and foam dielectric is used for support of the reflector array.

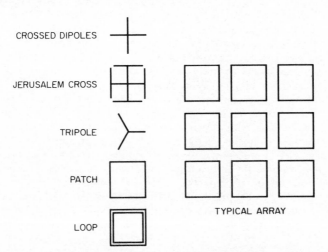

FIG. 17-31   Elements of frequency-sensitive surfaces.

17-2 LARGE-APERTURE REFLECTOR ANTENNAS

Introduction

Reflector antennas with electrically large apertures, greater than 60 wavelengths, became relatively common in the 1970s. The design of such antennas is dominated by considerations of geometrical optics,[41,42] which are often refined by considerations of physical optics.[43,44] There remain parameters of concern: gain, beamwidth, sidelobe level, polarization and cross-polarization, antenna noise temperature, and, if more than one beam position is needed, the variation of these parameters with scan angle. The feed design or, in the case of more complex antenna optics, the design of the feed system and subreflector intimately affect the parameters through illumination control, feed polarization properties, and spillover past the subreflector or the main reflector. Finally, the very structural design affects electrical performance through the scattering of energy from surface-panel irregularities and interpanel gaps, from subreflector and/or feed-support spars, and from the correlation of this scattering arising from the structural design and alignment procedures.

In general, electrically large reflectors are fed by simpler antennas such as horns or arrays of horns or by folded-optics systems involving subreflectors and feeds. In many instances, the antennas are designed to produce two coaxial beams with orthogonal polarizations over near-octave bandwidths. The feed systems may then terminate in complex microwave circuits involving polarizers and orthomode transducers to separate or combine orthogonally polarized beams as well as in microwave multiplexers to separate or combine different frequencies.

Electrically large reflector antennas have their principal applications in radar (Chaps. 32 and 34), where their high gain and narrow beamwidths enhance radar range and angular resolution; in radio astronomy (Chap. 41) and other deep-space applications, where the same parameters enhance sensitivity and resolution of stellar radio sources; and in microwave communications (Chaps. 30 and 36), where high gain, low noise, and sidelobe control enhance effective radiated power relative to isotropic (EIRP), receive sensitivity, and isolation from interference. For reflector antennas of diameter greater than 60 wavelengths, gains range from 40 dBi to about 70 dBi, half-power beamwidths (HPBWs) from about 1.5° to about 0.05°, and average wide-angle sidelobe levels down to −20 dBi (almost 90 dB below the peak of the main beam); dual-polarized radiation-pattern orthogonality provides isolations greater than 33 dB over the HPBWs. Of course, these properties depend upon the feed system as well as the reflectors.

Pencil-Beam Reflector Antennas

Front-Fed Paraboloidal-Reflector Systems The paraboloidal-reflector antenna and its design are described in Sec. 17-1. Since the design is geometry-controlled and since the larger electrical apertures considered here make geometrical-optics considerations even more valid, the choices of f/D ratio, edge taper, etc., remain the same. However, the simple front-fed reflector has disadvantages due to the blocking of the aperture by the feed system and by the feed supports.[45,46] Figure 17-32 shows two views of a front-fed paraboloidal-reflector antenna with feed supports extending from the reflector edge. It is seen that the feed supports intercept radiation reflected from the paraboloid. (Supports which are not on the edge intercept energy from the feed

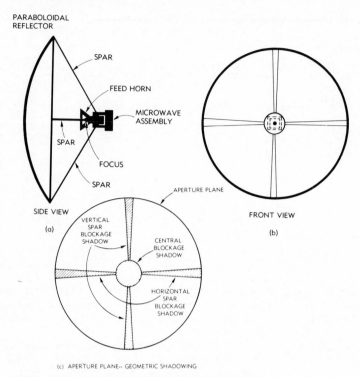

FIG. 17-32   Front-fed paraboloidal-reflector antenna.

before it reaches the reflector.) In a geometrical-optics sense, the feed supports cast a shadow upon the aperture plane, shown in the third view. This shadow, called *aperture blockage,* reduces gain and affects near-in sidelobes, raising some and lowering others. In addition, the blocked radiation is scattered by the feed supports and can significantly increase wide-angle sidelobes. In general, the front-fed design with a simple feed horn has less blockage than the folded-optics designs. Additionally, the feed subtends a wider angle at the main reflector than a feed illuminating a subreflector does, and therefore the feed is a simpler, lower-gain design. However, if the feed system is complex and the microwave package abutting it is large and heavy [containing polarizers, orthomode transducers (OMTs), low-noise amplifiers (LNAs), etc.], the blocking advantage is negated somewhat. Further, the focal point is an awkward location for such electronic gear, both structurally, because the weight is not desirable at the end of a large-moment arm, and from the point of view of equipment servicing. Finally, in electrically large reflectors, in which reflector size must be minimized, the folded-optics system can be made more efficient. The front-fed reflector has an efficiency of 55 to 60 percent. Folded-optics designs can employ shaped subreflectors to increase efficiency.

Folded Optics   Pencil-beam reflectors may be illuminated by feed systems in which the path of energy from the feed (which terminates the guided-wave–waveguide assembly) is reflected by one or more subreflectors. Figure 17-33 portrays the geometry of the Cassegrain reflector system, whence it is seen that the incoming-ray trajectories are folded back in the direction of the vertex of the paraboloid P by reflection from the hyperboloidal subreflector H. The focusing properties of the system derive from the colocation of one focus of the hyperboloid H with that of the paraboloid at F_1. An incoming plane wave from the direction of the paraboloid axis is reflected from the paraboloid and from the hyperboloid and focuses at F_2, the other focus of the hyperboloid H. A feed with its phase center at F_2 will then receive the incident energy.

Figure 17-34 portrays the geometry of the gregorian reflector system. Here the subreflector is an ellipsoid E with its near focus colocated at F_1 with the focus of the paraboloid, the ellipsoid lying on the side away from the paraboloid vertex. The phase center of the feed should then lie at F_2, the other focus of the ellipsoid. Another con-

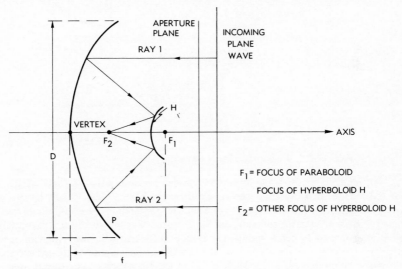

FIG. 17-33 Cassegrain geometry.

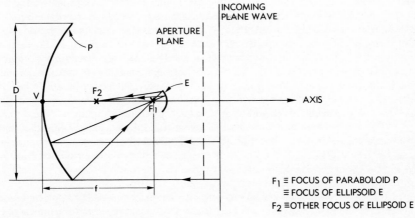

FIG. 17-34 Gregorian geometry.

figuration is shown in Fig. 17-35. Here the subreflector is paraboloidal, with its focus colocated at F with the focus of the paraboloidal main reflector. The result of the reflection of the rays representing the incoming plane wave is a collimated cylinder of rays parallel to the axis. This creates the possibility of another, smaller aperture plane parallel to XX'. A feed system associated with this aperture would have a plane wavefront, or the parallel rays could be associated with an appropriate beam-waveguide system.

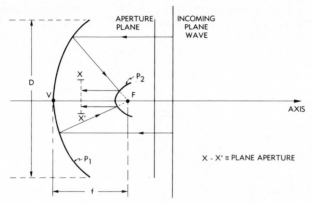

FIG. 17-35   Planar-feed field geometry.

An immediate advantage of all three folded systems is that the feed system is now located nearer the vertex of the main reflector and, in fact, can usually be reached from access provisions at the vertex. More subtly, introduction of the additional reflection at the subreflector provides an additional degree of design freedom. Aperture distribution can be controlled to increase efficiency and decrease blockage by shaping the subreflector and the main reflector so that the surfaces deviate from the hyperboloid, ellipsoid, and paraboloid.

The Cassegrain system is more compact than the gregorian system for the same focal length. There have been claims that gregorian systems provide improved wide-angle sidelobes in apertures of the order of 200 wavelengths. However, because of blockage of the feed system, subreflector, and supports, the sidelobes of these folded antennas are not especially low. By careful design and construction, the sidelobe envelope can be made to conform to international and Federal Communications Commission (FCC) requirements. For example, INTELSAT Standard A antennas, which have gain in excess of 57 dBi (generally 60 dBi), require sidelobe envelopes to be less than $32 - 25 \log \theta$, from 1° to 48°, and less than -10 dBi beyond 48°. (θ is the pattern angle measured in a band of $\pm 1°$ about the geostationary arc.)

One of the design principles used in minimizing subreflector and feed blockage is to make them equal, as shown in Fig. 17-36. The subreflector blocks incoming rays, and the feed blocks rays that would hit the main reflector and be reflected to the subreflector. For the case shown, the shadows of the feed and subreflector intersect on P, the paraboloid main reflector, presenting the ideal case.

It has been found that keeping the subreflector diameter small (10 percent of the diameter of the main reflector) will keep the central-blockage sidelobes from exceeding the specifications given above. If this is not done, the sidelobes are high in the 1°

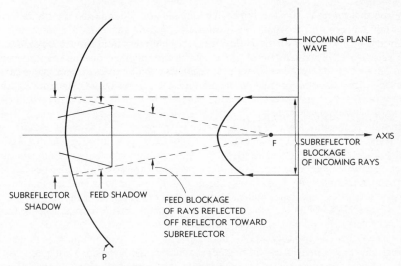

FIG. 17-36   Blockage by feed and subreflector.

to 10° region. The subreflector support spars tend to create sidelobe problems in a region of 5° to 20° in the plane perpendicular to the spars. It has been found useful to use a spar quadruped with all spars at 45° to the plane of concern. However, the best way to reduce sidelobes is to eliminate the blockage by the use of offset reflectors.

Sidelobes in Electrically Large Antennas   Sidelobes in a folded-optics system arise from a number of sources. These include (1) radiation-pattern sidelobes from the aperture illumination, (2) spillover and edge diffraction from the subreflector illuminated by the feed system, (3) spillover and edge diffraction from the main reflector illuminated from the subreflector, (4) central blockage (subreflector and feed system), (5) spar (subreflector support) blockage and scattering, (6) deviation of reflector surfaces from the desired contours, and (7) effects of gaps between panels.

Departure of the aperture illumination in amplitude and phase from that desired can arise from feed-aperture phase errors[45] and defocusing.[46] These will cause the radiation (diffraction) pattern to depart from the design values but are correctible by careful feed design and positioning. Spillover and edge-diffraction effects can be avoided by illumination of the subreflector with a more deeply tapered feed pattern to about 20 dB down at the edge. For the shaped subreflector, it is possible to obtain sharp illumination taper at the main-reflector edge.[47,48] Spillover and edge diffraction of feed radiation will give increased sidelobes in the 10° to 20° region of the pencil-beam axis.

Minimizing the central blocking and optimizing the illumination are not compatible, since the first requires smaller feed systems and the second requires larger feed systems. Equalizing the subreflector and horn blockage as discussed in the preceding subsection is the best approach. Then if the subreflector diameter can be kept to less than 10 percent of the main-reflector diameter, shaping is possible and the blockage lobes will meet the sidelobe criterion of the preceding subsection. Central blockage casts a wide beam inversely proportional to the diameter of the central blockage and thus typically 10 times the beamwidth and 20 dB down from the main beam. It is out of phase with the main beam and thus subtracts a fraction of 1 dB from the main beam, adds 2 to 3 dB to the first sidelobe, subtracts from the second sidelobe, etc. Spar blockage and scattering[45,49] are not circularly symmetrical and depend upon spar shape and spar-reflector-subreflector geometry. If the spars are not mounted at the dish edges, shadows are cast on the main reflector to reduce efficiency and blockage of the reflected energy is still present. Spar scattering effects are strongest in the forward hemisphere at an angle of $180° - \xi$, where ξ is the angle between the spar axis and the aperture plane.[45] The sidelobes are strongest in the forward direction in the plane containing the spar and the beam axis. This typically occurs in the region at 5° to 15° off the main-beam axis. Spar design can minimize but not eliminate this effect. Streamlining the spar cross section is useful for the polarization perpendicular to the spar. Orienting the spars in a quadruped at 45° to the horizontal is effective for earth stations in keeping spar-induced sidelobes out of the region of the geostationary arc. Figure 17-37 shows the effects of central blockage, spar blockage, and forward (subreflector) spillover on a radiation pattern.

The remaining two sources of sidelobe energy derive from the manner in which electrically large antennas are fabricated, assembled, and installed. Most electrically large reflectors are constructed of segments (panels) assembled onto a backup struc-

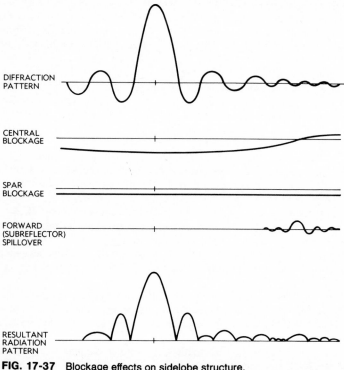

FIG. 17-37   Blockage effects on sidelobe structure.

ture or with the backup structure as an integral part of the panels. The subreflectors are generally single units.

The rms errors associated with fabricating a single panel or a subreflector can be controlled with care to 0.1 mm (4 mils) for a reflector or panel up to 3 m across; however, fabrication techniques commonly used permit rms errors of 0.25 to 1.27 mm (10 to 15 mils) for panels of this size. Assembly and installation errors are each of the same order, provided careful checkout procedures are used. A rule of thumb for communication and radar reflectors in the 8- to 30-m class is that the ratio of rms error to reflector diameter is 2 to 5 times 10^{-5}. Some electrically large antennas have been built to better tolerances for radio astronomy, but they usually require special protection from the environment. The pattern effects of the errors depend upon their statistical nature.[45,50,51,52,53]

It has been shown[54] that when error correlation intervals are small, the results obtained by Ruze[50,51] for axial gain apply (see Fig. 17-38 and Chap. 2). However, because panel-fabrication error contributions tend to be of the same order or smaller than assembly and installation error contributions, the large correlation errors of the assembly can dominate. Therefore, one cannot reasonably assume small correlation intervals in most large antennas. Bao[54] has shown that the effect of these large correlation intervals is to reduce the gain loss and that the sidelobes are significantly lower than predicted under Ruze's assumptions regarding correlation. In a similar fashion, the distortions due to wind and gravity have long correlation intervals and hence less effect than predicted by the small-correlation-interval theory. The large

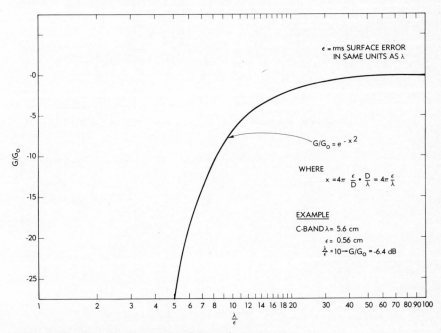

FIG. 17-38 Normalized loss due to finite surface error.

correlation intervals increase the sidelobes in the region near the main beam, while the small correlation intervals (random errors) contribute significantly to wide-angle sidelobes.

The best work on earth station sidelobes was given by Korvin and Kreutel,[46] who presented results of both theoretical and measured studies for apertures ranging from 100 to 600 wavelengths. Figure 17-39 presents horn of these results.

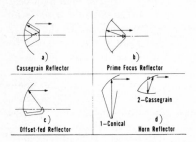

BASIC CONFIGURATIONS OF ANTENNAS

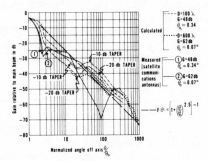

CASSEGRAIN-FED REFLECTOR RADIATION DIAGRAMS

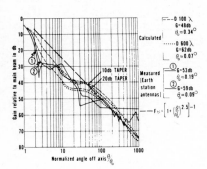

PRIME-FOCUS-FED PARABOLIC REFLECTOR

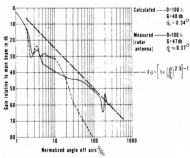

OFFSET-FED PARABOLIC REFLECTOR

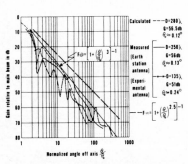

HORN REFLECTOR (CONICAL) ANTENNA

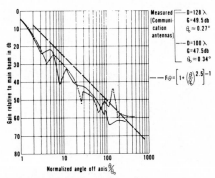

HORN REFLECTOR (CASSEGRAIN) ANTENNA

FIG. 17-39 Practical sidelobe-level envelope.

High-Efficiency Dual-Reflector Systems

Electrically large pencil-beam antennas are costly. Cost is almost a linear function of aperture area (reflector surface). The higher the efficiency, the smaller the aperture required to collect incoming waves or to provide gain for a transmitted wave. Antenna efficiency is the product of a number of factors, each of which represents the degree with which the energy is protected against diversion from the desired direction or polarization by a particular mechanism. The factors include effects of (1) aperture illumination (η_i), (2) spillover around the main reflector and subreflector (η_{s1}, η_{s2}), (3) blockage (η_b), (4) reflector-surface deviations (η_r), (5) polarization purity (η_p), and (6) losses in waveguide and feed (η_0). These efficiency factors are defined in Table 17-1.

TABLE 17-1

$\eta_i = \dfrac{1}{A} |\int F dA|^2 / \int |F|^2 dA$ (F is illumination over aperture A.)

η_{s1} = power on main reflector/(power from subreflector − scattered power)

η_{s2} = power on subreflector/power from feed system

η_b = power blocked/total power from main reflector

$\eta_r = \exp[-(4\pi\epsilon/\lambda)^2]$ (ϵ = rms error. This is worst-case approximation.)

η_0 = power radiated by feed/power input to feed system

η_p = aperture power of desired polarization/total aperture power

---

Spillover, blockage, and surface tolerance effects have been discussed earlier in terms of sidelobe control. Improvement in efficiency and reduction of sidelobe effects go hand in hand for these factors. Polarization purity is controlled by feed-system polarization properties, by curvature of the reflector, by symmetry, by scattering properties of the blockage elements, and, for circular polarization, by internal reflections in the waveguide and feed system.

Control of aperture illumination[55,56] is a major factor in improving efficiency. Techniques for controlling illumination can also minimize the effects of central blockage, as will be seen. If maximum gain is desired, the object is to shape the subreflector and the main reflector so that the aperture illumination is as nearly uniform as possible, with minimum spillover and with minimum energy blocked.

Three basic conditions are used to shape the two reflectors: (1) Path length from phase center of feed to aperture plane is constant for all paths. For the jth ray, as shown in Fig. 17-40, $(\overline{F_2 S_j} + \overline{S_j M_j} + \overline{M_j A_j})$ = path length from feed point to aperture plane = $r_j + z_j + (r_j \cos\theta_j - l + z_j)/\cos\theta_j$ = constant. (2) Aperture illumination is constant. It can be shown[56,57] that this requirement leads to

$$x_j^2 / x_{\max}^2 = \left(\int_0^{\theta_j} E_f^2 \sin\theta \, d\theta \right) \Big/ \left(\int_0^{\theta_{\max}} E_f^2 \sin\theta \, d\theta \right)$$

where $E_f = E_f(\theta)$ is the circularly symmetric pattern of the feed horn. (3) Snell's law must apply. This leads to the expressions

$$\frac{1}{r}\frac{dr}{d\theta} = \tan\left(\frac{\theta + \psi}{2}\right) \qquad\qquad -\frac{dz}{dx} = \tan\frac{\psi}{2}$$

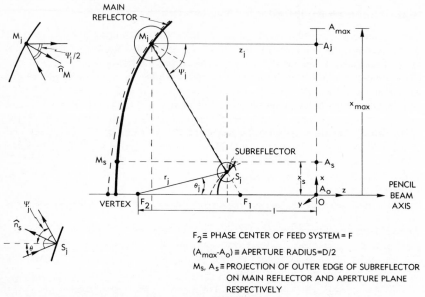

$F_2 \equiv$ PHASE CENTER OF FEED SYSTEM = F

$(A_{max} - A_o) \equiv$ APERTURE RADIUS = D/2

$M_s, A_s \equiv$ PROJECTION OF OUTER EDGE OF SUBREFLECTOR
ON MAIN REFLECTOR AND APERTURE PLANE
RESPECTIVELY

FIG. 17-40   Shaping of main reflector and subreflectors.

These three conditions (four equations) can be solved on a computer to yield the two surfaces desired. But the general nature of the solution can be deduced from the physical problem. It is desirable to make the aperture illumination more uniform when it is too high in the center. If the slope of the subreflector near the center were altered to reflect central rays from the feed away from the center of the main reflector, this would have the desired effect. The dotted curve in front of the subreflector represents such a distortion. The main reflector must now be reshaped to maintain the focused beam. This results in the dotted curve behind the main reflector.

If this process is exaggerated near the middle of the subreflector, a hole would result in the illumination of the main reflector. The energy here is normally blocked by the subreflector so that there is no basic change in the radiation pattern near the main beam. There will be a reduction in the energy scattered by the subreflector and, therefore, lower scattered sidelobes at wider angles.

At the edge of the subreflector the surface can be shaped to direct energy inward and so give a shorter and deeper illumination taper at the edge of the main reflector to reduce spillover. The shaping of the subreflector edges can compensate for deeper tapering of the energy incident upon this surface. Therefore a system with less spillover past the subreflector and with underillumination of the main reflector can use the shaping of the subreflector edges to compensate for the underillumination. Thus, it is seen that reflector and subreflector shaping not only can improve the aperture illumination but can also reduce spillover past both reflectors.

It has been found that the simple geometrical-optics analysis of shaping is not sufficiently accurate.[58] Improved performance can be obtained by considering diffraction effects at the subreflector.[48] The system improvement has been limited in bandwidth, and a third approach[48] modifies geometrical-optics design to give good performance over a wider bandwidth.

Wide-Angle, Multiple-Beam Reflector Antennas

Many applications in radar, astronomy, and communications require the use of more than one beam from a single reflector. The problem first arose in radar systems, and the solutions considered were based upon counterparts from optics.[59,60,61,62,63,64,65,66,67] Microwave-optics problems were similar for systems with multiple beams or with single beams and multiple-beam positions. The solutions employed simple front-fed paraboloids, folded-reflector systems, and spherical and toroidal reflectors.

Wide-Angle Characteristics of Large Paraboloidal Reflectors   When the feed

of a paraboloidal reflector is displaced from the focal point and off the axis of the system, the main beam of the radiation pattern moves in the opposite direction. Crude ray tracing, as shown in Fig. 17-41, demonstrates that rays from F_1, the transversely displaced feed above the axis, reflect in a downward direction. More precise ray tracing to the aperture plane A would show that for small feed displacements the ray paths are almost constant to a new aperture A_1 tilted with respect to A. If the feed at F_1 is displaced through an angle θ_f, then the main beam is tilted at an angle θ_b in the opposite direction. For practical values of antenna parameters, θ_b is less than θ_f. The ratio of these angles is called the beam-deviation factor (BDF). For a shallow reflector and for small displacements, Lo[68] has shown that

$$\frac{\sin \theta_b}{\tan \theta_f} \approx \left[\frac{1 + k(D/4f)^2}{1 + (D/4f)^2} \right] \equiv (\text{BDF})_0$$

and

$$\text{BDF} \approx (\text{BDF})_0 \left\{ 1 + \frac{1}{3}\left(\frac{d}{f}\right)^2 \left[\frac{1}{2}(\text{BDF})_0^2 + 1 \right] \right\}$$

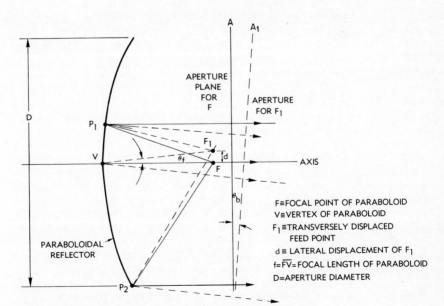

FIG. 17-41 Paraboloid with transverse displacement of feed.

Here k can vary from 0.3 to 0.7 as a function of the feed pattern and edge taper. Increased taper emphasizes the flatter central portion of the reflector to imply a longer effective f/D ratio. The f/D ratio is the most important factor for BDF, with the beam-deviation factor increasing rapidly to 0.9 as the f/D ratio goes from 0.25 to 0.5.

As the feed horn is moved off axis, the beam is scanned with a decrease in gain of the beam and with an increase in sidelobes. These changes are depicted in Fig. 17-42. There is a reduction in gain, ΔG. The first sidelobe nearest the system axis (coma lobe) is increased, and the opposite sidelobe is decreased. Figure 18-16 in Chap. 18 shows this process for a paraboloidal reflector of f/D ratio of 0.5- and 20-dB edge illumination.[61]

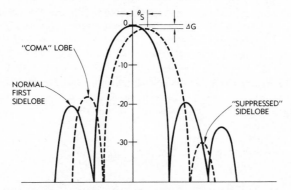

FIG. 17-42   Radiation patterns for focused and laterally defocused feeds.

It has been shown[69] that simple lateral displacement creates not only coma and astigmatism[70,71] but also curvature of the field and higher-order aberrations. These manifest themselves by filling in the nulls in the pattern and by a broadening of the main beam.

Ruze[69] provides computed graphs of HPBW, gain loss, beam-deviation factor, and level of coma lobes for a range of f/D ratios and aperture illuminations. Loss of gain ΔG can be predicted by the equation

$$26.75[\Delta G(\text{dB})]^2 - 4.75[\Delta G(\text{dB})]^4 \approx \theta_s \left(\frac{D}{f}\right)^2 \bigg/ \left[1 + 0.02\left(\frac{D}{f}\right)^2\right]$$

Physical-optics considerations are used by Rusch and Ludwig[72] to determine maximum scan-gain contours. These tend to be flat and close to the transverse plane through the focus. They initially curve toward the vertex and for wider-scan angle bend away from the vertex.

An equally important factor is the demonstration[72] that maximum gain depends upon the feed axis remaining parallel to the reflector axis rather than pointing the feed to the vertex. Gain loss and gain optimization are not a closed subject.[73] Data taken from six authors in the time period 1947–1973 are compared and shown to have significant differences which are attributed to differences in illumination.

Wide-Angle Scanning of Electrically Large Folded-Feed Reflectors

Cassegrain and gregorian antennas derive from the front-fed paraboloid through considerations of geometrical optics. Therefore, scanning of the reflectors can be related to scanning of front-fed reflectors. The concept of the equivalent paraboloid[65,66] provides the desired link. If, in a Cassegrain antenna system (Fig. 17-43) with F the focal point for the actual feed, the subreflector subtends an angle $2\phi'_f$ and the paraboloid subtends an angle $2\phi_r$, then one can define a magnification m in terms of focal lengths:

$$m = \frac{f'}{f} = \frac{\tan(\phi_r/2)}{\tan(\phi'_f/2)} = \frac{e+1}{e-1}$$

where e is the eccentricity of the hyperboloid subreflector and f' is the focal length of the equivalent paraboloid. It subtends the angle $2\phi'_f$ at F' and intercepts the same marginal rays as the main reflector (has the same axis and diameter D). Since $e > 1$, $f' > f$ and $F'/D > f/D$. As noted earlier, the beam deviation factor, coma aberration, and gain loss are strong functions of f/D, the larger f/D being preferred. Therefore, the longer effective f/D ratio of the Cassegrain reflector offers potential for enhanced scanning performance. For the gregorian system, $f'/f = -(e+1)(e-1)$, so that one may choose $f' > f$. This system also therefore offers potential for scanning.

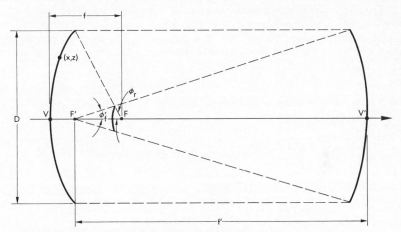

FIG. 17-43 Geometry of the equivalent paraboloid.

The design of scanning systems (Chap. 18) sometimes employs the Abbe sine condition to minimize aberrations for small scan angles; this is for the ray incident at the general point (x, z) in Fig. 17-43, $x/\sin\phi' = $ constant. Neither the front-fed paraboloid, nor the Cassegrain system, nor the gregorian system satisfies the Abbe sine condition. However, the latter two systems show better results for this criterion than the simple paraboloid. The best dual-reflector system for scanning is the Schwarzschild.[67] Studies[66,74,75] indicate that for electrically large antennas the Schwarzschild system is promising. Other dual-reflector systems with promise include the two-point correction[73,76,77] and the spheroboloidal-gregorian dual-reflector system.[73,77]

Spherical and Toroidal Wide-Angle, Multiple-Beam Reflectors   The spherical reflector offers symmetry which can be used for scanning systems. The design problem lies in designing a feed system which will effectively illuminate the aperture with the proper phase characteristics. Ashmead and Pippard[60] and Li[20] have provided useful information on point-source feeds. The data depend upon the sphere radius R, the aperture in wavelengths D/λ, and the allowable maximum phase error Δ/λ.

$$(D/R)^4_{max} = k(\Delta/\lambda)/(R/\lambda)$$

Ashmead and Pippard found a value of $k = 250$, and Li found a value of 235.2. These data show that point-source feeds require a small D/R value which is not practical. However, when efficiency is not the most significant parameter, as in receive-only earth stations, the spherical reflector has been used.

Improvement of the spherical-reflector scanning system requires correction of the phase error of Fig. 17-10. Analysis of the focal-region field (receiving antenna) by geometrical[78,79] and physical optics[80] shows a calculable phase progression along the line between the focal point and the reflector. Each small region along this axis is illuminated by concentric zones on the spherical reflector (Fig. 17-44). Line-source feeds with an element for each zone[81,82] have been effectively used to correct the spherical aberration represented by the phase of each zone in the 1000-ft (305-m) reflector at Arecibo, Puerto Rico. These feeds are bandwidth-limited and mechanically limited.

The alternative to correction along a line between the feed and the reflector is to correct in the transverse focal region.[83,84,85,86] Efforts at transverse correction[87,88,89] resulted in one promising design[87] shown in Fig. 17-45. The shape of the gregorian subreflector, $G(u, v)$, using the notation of Holt and Bouche, is given by the coordinates u and v:

$$u = c - [(c^2 - 2c + 2p)(2c^2 - 1)]/[4(pc^2 - c) + 2]$$
$$v = \pm s(2cu - 1)/(2c^2 - 1)$$

where $c = \cos\theta$, $s = \sin\theta$.

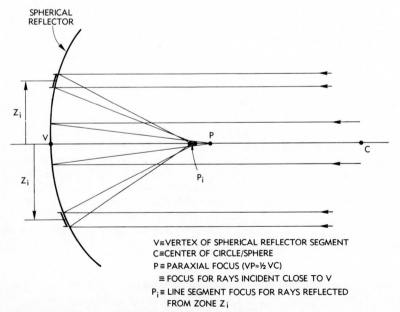

V ≡ VERTEX OF SPHERICAL REFLECTOR SEGMENT
C ≡ CENTER OF CIRCLE/SPHERE
P ≡ PARAXIAL FOCUS (VP=½ VC)
   ≡ FOCUS FOR RAYS INCIDENT CLOSE TO V
P_i ≡ LINE SEGMENT FOCUS FOR RAYS REFLECTED
      FROM ZONE Z_i

FIG. 17-44   Axial line focusing for a spherical reflector.

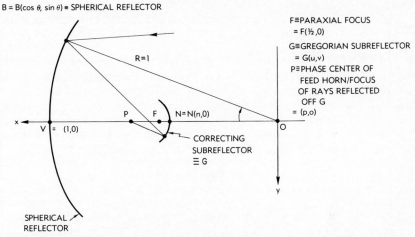

B = B(cos θ, sin θ) ≡ SPHERICAL REFLECTOR

R=1

F≡PARAXIAL FOCUS
= F(½,0)

G≡GREGORIAN SUBREFLECTOR
= G(u,v)

P≡PHASE CENTER OF
FEED HORN/FOCUS
OF RAYS REFLECTED
OFF G
= (p,o)

P   F   N=N(n,0)

x

V = (1,0)

CORRECTING
SUBREFLECTOR
≡ G

O

y

SPHERICAL
REFLECTOR

FIG. 17-45   Geometry of a gregorian correcting subreflector for a spherical reflector.

The toroidal-reflector family also provides scan by symmetry, but the symmetry is limited to a single plane. The generalized geometry for toroidal reflectors is shown in Fig. 17-46. M is a conic section with axis along the z coordinate, and z' is the axis of rotation lying in the xz plane at an angle of α to the z axis. When M is rotated about z', the toroidal surface is swept out. If M is a parabola of focal length f ($f \le \overline{OV}/2$) and R is the radius of the circular arc ($R = \overline{OV}$) and α is 90°, the torus

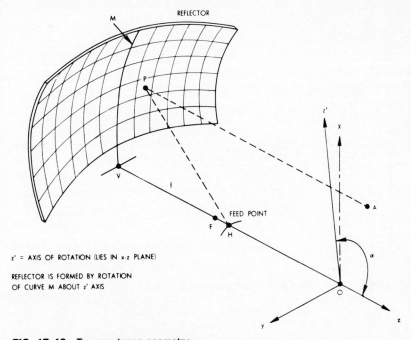

M   REFLECTOR

P

z'

x

V

f

FEED POINT

F   H

A

α

z' = AXIS OF ROTATION (LIES IN x-z PLANE)

REFLECTOR IS FORMED BY ROTATION
OF CURVE M ABOUT z' AXIS

O

Y   z

FIG. 17-46   Torus-antenna geometry.

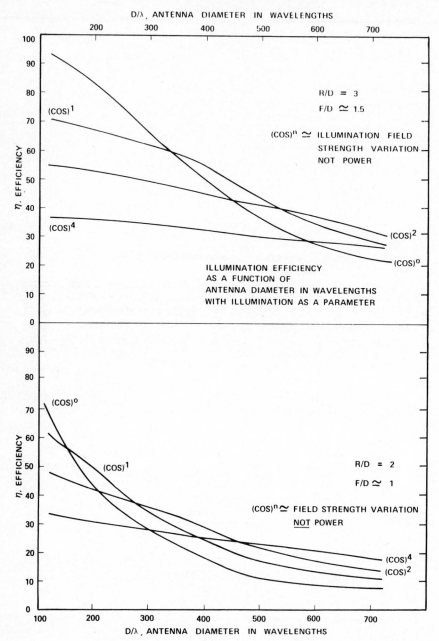

FIG. 17-47 Conical-torus illumination efficiency.

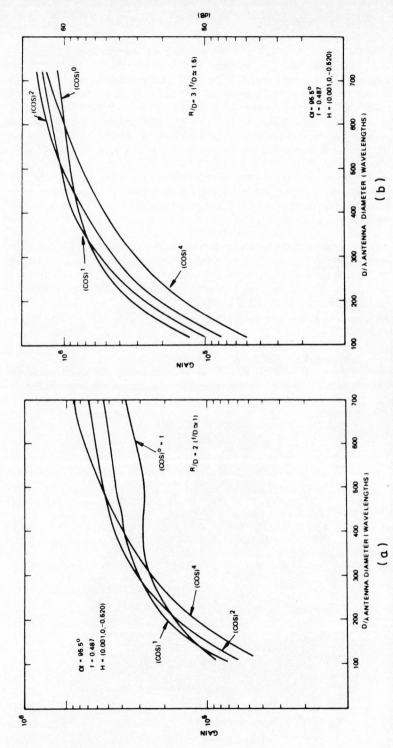

FIG. 17-48 Gain for a torus antenna. (a) $R/D = 2$, $f/D = 1$. (b) $R/D = 3$, $f/D = 1.5$.

17-47

reflector of Kelleher and Hibbs[90] is obtained. Replacing the parabola with an appropriate ellipse gives the Peeler-Archer torus.[91] For $\alpha \neq 90°$, the axis of the conic section moves on the surface of a cone.[92] For $\alpha = 90°$, the cone degenerates into a plane. This case will be termed the *rectangular torus*. Such a reflector has been used for very large radar reflectors with an organ-pipe-scanner feed system (Chap. 18).

For earth stations in temperate latitudes which are scanning the geostationary arc, $\alpha = 95.5°$ can be shown to give an optimum torus design. Studies of the aperture fields, far-field patterns, and efficiency led to the choice of a parabolic section with a focal length to radius f/R of 0.487 and feed location H, given by x,y,z coordinates as 0.001 R, 0.0, -0.5 R. Some of these considerations[93] are shown in Figs. 17-47 and 17-48. On the basis of these design parameters, a scale model and a full-scale antenna were fabricated and shown to give excellent performance[94] as part of the unattended earth terminal (UET). Test results verified the design process using the finite-element approach embodied in the GAP (General Antenna Program) computer program.[95] Both the scale model and the final torus antenna used a corrugated scalar horn[96,97,98] as a feed because this horn has a well-defined, frequency-invariant phase center near the throat. The horn has very symmetric patterns with low sidelobes that appear to match well the focal-region fields of large reflector antennas.[99] The implementation of the corrugated horn used in the UET is shown in Fig. 17-49. The 30- by 55-ft (9.1-m by 16.8-m) UET torus antenna operating in the 6- and 4-GHz fixed-satellite bands is shown in Fig. 17-50.

Finally, it should be noted that the aberration present in the conical-torus antenna system can be removed by an aberration-correcting subreflector,[100,101] designed by using the GAP[95] to define the surface. The geometry for the torus with subreflector is shown in Fig. 17-51, the scale-model subreflector is shown in Fig. 17-

FIG. 17-49   Corrugated feed horn.

FIG. 17-50 Multibeam torus antenna (MBTA).

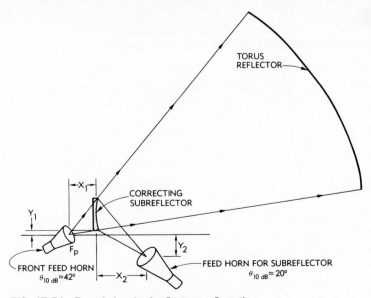

FIG. 17-51 Front-fed and subreflector configurations.

52, and the performance is given in Fig. 17-53. The subreflector has a hyperbola arc in the vertical plane of symmetry (as in the Cassegrain) and a somewhat elliptical arc in the other plane (gregorian). It is called *cassegorian*. Gain performance follows frequency squared beyond 60 dBi, whereas comparable uncorrected torus performance shows aberration losses above 53 dBi. Measured efficiency for an assumed circularly

FIG. 17-52   Aberration-correcting subreflector for a torus reflector.

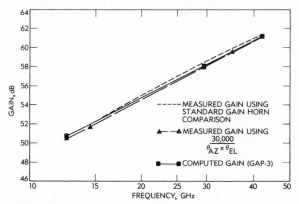

FIG. 17-53   Comparison of measured and calculated gain versus frequency.

illuminated aperture exceeded 70 percent at 12.4 GHz and approached 74 percent at 42 GHz.

REFERENCES

1   E. F. Harris, "An Experimental Investigation of the Corner-Reflector Antenna," *IRE Proc.*, vol. 41, May 1953, p. 645.
2   N. Inagaki, "Three Dimensional Corner Reflector Antennas," *IEEE Trans. Antennas Propagat.*, vol. AP-22, July 1974, p. 580.
3   J. L. Wong and H. E. King, "A Wideband Corner Reflector Antenna for 240-400 mHz," *IEEE Antennas Propagat. Symp. Dig.*, Quebec, 1980, p. 320.
4   C & S Antennas, Rochester, England: A Complete Antenna Service.
5   K. S. Kelleher, U.S. Patent 3,239,838.
6   H. W. Ehrenspeck, U.S. Patent 3,122,745.
7   H. W. Ehrenspeck, "The Short Backfire Antenna," *IEEE Proc.*, August 1975, p. 1138.
8   G. K. Hartmann and W. Engelhardt, "Dispersion Measurements of One-Element Short Backfire (SBF) Antennas," *IEEE Trans. Antennas Propagat.*, vol. AP-23, March 1975, p. 289.
9   A. Shoamanesh and L. Shafai, "Backfire Antennas Constructed with Coaxial Circular Loops," *IEEE Trans. Antennas Propagat.*, vol. AP-28, September 1980, p. 633.
10 J. D. Kraus, "A Backward Angle Endfire Antenna," *IEEE Trans. Antennas Propagat.*, vol. AP-12, January 1964, p. 48.
11 R. Conti et al., "The Wire Grid Microstrip Antenna," *IEEE Trans. Antennas Propagat.*, vol. AP-29, January 1981, p. 157.
12 D. G. Kiely, "Parabolic Cylinder Aerials," *Wireless Eng.*, March 1957, p. 73.
13 R. C. Spencer et al., "Double Parabolic Cylinder Pencil Beam Antenna," *IRE Conv. Rec.*, part 1, *Antennas and Propagation,* 1954.
14 D. G. Kiely et al., "Cheese Aerials for Marine Navigational Radar," *IEE Proc. (London),* part III, vol. 98, no. 51, January 1951.
15 N. N. Fullilove et al., "The Hourglass Scanner," *IRE Conv. Rec.*, part I, 1959, p. 190.
16 R. C. Benoit and W. M. Furlow, "Wullenweber-Type UHF Radio DF," *IRE Conv. Rec.*, 1955.
17 K. S. Kelleher, "A New Wide Angle Microwave Reflector," *Tele-Tech & Electronic Ind.*, June 1953.
18 K. S. Kelleher, "Relations Concerning Wavefronts and Reflectors," *J. App. Phys.*, vol. 21, no. 6, June 1950, p. 573.
19 H. J. Riblet and C. B. Barker, "A General Divergence Formula," *J. App. Phys.*, vol. 19, no. 1, January 1948, p. 63.
20 T. Li, "A Study of Spherical Reflectors as Wide-Angle Scanning Antennas," *IRE Trans. Antennas Propagat.*, vol. AP-7, 1959, p. 223.
21 G. Stavis and A. Dorne, "Horns and Reflectors," in *Very High Frequency Techniques,* vol. 1, McGraw-Hill Book Company, New York, 1947, chap. 6.
22 A. C. Ludwig, "Conical Reflector Antennas," *IEEE Trans. Antennas Propagat.*, vol. AP-20, March 1972, p. 146.
23 A. S. Dunbar, "Calculations of Doubly-Curved Reflectors for Shaped Beams," *IRE Proc.*, vol. 3b, October 1948, p. 1289.
24 F. Briskell and B. S. Wescott, "Reflector Design as an Initial-Value Problem," *IEEE Trans. Antennas Propagat.*, vol. AP-24, July 1976, p. 531.
25 O. M. Bucci et al., "Control of Reflector Antenna Performance by Rim Loading," *IEEE Trans. Antennas Propagat.*, vol. AP-29, September 1981, p. 733.

26 R. J. Adams et al., "Preliminary Design of SPS-2 Radar," *Nav. Res. Lab. Rep.* 3412, 1949.

27 S. R. Jones and K. S. Kelleher, "A New Low-Noise High Gain Antenna," *IEEE Conv. Rec.*, part I, 1963.

28 K. S. Kelleher, U.S. Patent 3,283,331.

29 W. C. Jakes, Jr., "A Theoretical Study of an Antenna-Reflector Problem," *IRE Proc.*, vol. 41, February 1953, p. 272.

30 R. E. Greenquist and A. J. Orlando, "An Analysis of Passive Reflector Antenna Systems," *IRE Proc.*, vol. 42, July 1954, p. 1173.

31 E. Bedrosian, "The Curved Passive Reflector," *IRE Trans. Antennas Propagat.*, vol. AP-3, October 1955, p. 168.

32 S. D. Robertson, "Targets for Microwave Radar Navigation," *Bell Syst. Tech. J.*, vol. 26, October 1947, p. 852.

33 K. S. Kelleher, U.S. Patent 2,866,971.

34 E. M. Lipsey, "The Theory of an Omni-Directional Radar Reflector," report written at the Air Force Institute of Technology in 1959.

35 R. Berg, U.S. Army MERADCOM, Fort Belvoir, Md., private communication. Note C. H. Lockwood, U.S. Patent 3,039,093.

36 G. Goubau and F. Schwering, "On the General Propagation of Electromagnetic Wave Beams," *IRE Trans. Antennas Propagat.*, vol. AP-9, May 1961, p. 248.

37 M. Mizusawa and T. Kitsuregawa, "A Beamwaveguide Feed Having a Symmetric Beam for Cassegrain Antennas," *IEEE Trans. Antennas Propagat.*, vol. AP-21, November 1973, p. 884.

38 B. Claydon, "Beam Waveguide Feed for a Satellite Earth Station Antenna," *Marconi Rev.*, vol. XXXIX, no. 201, second quarter, 1976, pp. 81–116.

39 L. K. DeSize and J. F. Ramsay, "Reflecting Systems," in R. C. Hansen (ed.), *Microwave Scanning Antennas*, vol. 1, Academic Press, Inc., New York, 1964, chap. 2.

40 V. D. Agrawal and W. A. Imbriale, "Design of a Dichroic Cassegrain Subreflector," *IEEE Trans. Antennas Propagat.*, vol. AP-27, no. 4, July 1979.

41 F. S. Holt, "Application of Geometric Optics to Design and Analysis of Microwave Antennas," Air Force Cambridge Res. Lab. Tech. Rep. AFCRL-67-0501, Bedford, Mass., September 1967.

42 M. Kline and I. W. Kay, *Electromagnetic Theory and Geometric Optics,* Interscience Pure and Applied Mathematics ser., vol. XII, Interscience Publishers, a division of John Wiley & Sons, Inc., New York, 1965.

43 W. V. T. Rusch, "A Comparison of Geometrical and Integral Fields from High Frequency Reflectors," *IEEE Proc.*, vol. 62, no. 11, November 1974, pp. 1603–1604.

44 W. V. T. Rusch and P. D. Potter, *Analysis of Reflector Antennas,* Academic Press, Inc. New York, 1970.

45 A. D. Monk, "The Prediction and Control of the Wide-Angle Sidelobes of Satellite Earth Station Antennas," *Marconi Rev.*, vol. XLIII, no. 217, second quarter, 1980, pp. 74–95.

46 W. Korvin and R. W. Kreutel, "Earth Station Radiation Diagram with Respect to Interference Isolation Capability: A Comparative Evaluation," *AIAA Progress in Astronautics and Aeronautics: Communications Satellites for the 70's: Technology,* vol. 25, The M.I.T. Press, Cambridge, Mass., 1971, pp. 535–548.

47 W. F. Williams, "High Efficiency Antenna Reflector," *Microwave J.*, vol. 8, no. 7, July 1965, pp. 79–82.

48 P. J. Wood, "Reflector Profiles for the Pencil-Beam Cassegrain Antenna," *Marconi Rev.*, second quarter, 1972, vol. XXXV, no. 185, pp. 121–138.

49 G. Hyde and R. W. Kreutel, "Earth Station Antenna Sidelobe Analysis," *IEEE ICC Conf. Rec.*, 1969, pp. 32-7–32-14.

50 J. Ruze, "Physical Limitations on Antennas," Ph.D. thesis, MIT Res. Lab. Electron. Tech. Rep. 248, October 1952.

51 J. Ruze, "The Effect of Aperture Errors on the Antenna Radiation Pattern," *Supplemento al Nuovo Cimento*, vol. 9, no. 3, 1952, pp. 364–380.

52 J. Robieux, "Influence de la précision de fabrication d'une antenne sur ses performances," *Annales de radioélectricité,* vol. XI, no. 43, January 1956, pp. 29–56.

53 H. Zucker, "Gain of Antennas with Random Surface Deviations," *Bell Syst. Tech. J.,* vol. 47, 1968, pp. 1637–1651.

54 Vu The Bao, "Influence of Correlation Interval and Illumination Taper in Antenna Tolerance Theory, *IEE Proc.,* vol. 116, no. 2, February 1969, pp. 195–202.

55 V. Galindo, "Design of Dual-Reflector Antennas with Arbitrary Phase and Amplitude Distributions," *IEEE Trans. Antennas Propagat.,* vol. AP-12, no. 4, July 1964, pp. 175–180.

56 W. P. Williams, "High Efficiency Antenna Reflector," *Microwave J.,* vol. 8, no. 7, July 1965, pp. 79–82.

57 K. Miya (ed.), *Satellite Communications Technology,* KDD Engineering and Consulting, Inc., Tokyo, 1980, chap. 5, sec. 5.4, pp. 177–180.

58 W. V. T. Rusch, "Phase Error and Associated Cross-Polarization Effects in Cassegrainian-Fed Microwave Antennas," *IEEE Trans. Antennas Propagat.,* vol. AP-14, no. 3, May 1966, pp. 266–275.

59 A. S. Dunbar, "Optics of Microwave Directive Systems for Wide-Angle Scanning," Nav. Res. Lab. Rep. R-3312, Sept. 7, 1948.

60 J. Ashmead and A. B. Pippard, "The Use of Spherical Reflectors as Microwave Scanning Aerials," *J. IEE,* vol. 93, part IIIA, 1946, pp. 627–632.

61 K. S. Kelleher and H. P. Coleman, "Off-Axis Characteristics of the Paraboloidal Reflector," Nav. Res. Lab. Rep. 4088, Dec. 31, 1952.

62 R. C. Gunter, F. S. Holt, and C. F. Winter, "The Mangin Mirror," Air Force Cambridge Res. Cen. Tech. Rep. TR-54-161, Bedford, Mass., April 1955.

63 A. E. Marston and R. M. Brown, Jr., "The Design of Mirror-Lenses for Scanning," Nav. Res. Lab. Rep. 5173, Aug. 22, 1958.

64 J. H. Provencher, "Experimental Study of a Diffraction Reflector," Air Force Cambridge Res. Cen. Tech. Rep. TR-59-126, Bedford, Mass., April 1959. Zoned reflector.

65 P. W. Hannan, "Microwave Antennas Derived from the Cassegrain Telescope," *IEEE Trans. Antennas Propagat.,* vol. AP-9, no. 2, March 1961, pp. 140–153.

66 L. K. DeSize, D. J. Owen, and G. E. Skahill, "Final Report Investigation of Multibeam Antennas and Wide-Angle Optics," Airborne Instruments Lab. Rep. 7358-1, January 1960; also RADC-TR-60-93. Multibeam Cassegrain system.

67 K. Schwarzschild, "Investigations in Geometric Optics," *Gesellschaft der Wissenschaften zu Göttingen Mathematisch-Physikalische Klasse,* new ser., vol. IV, 1905; English translation by W. A. Miller, Eng. Memo. 62-51, RCA Laboratories, Rocky Point, N.Y., Feb. 9, 1954.

68 Y. T. Lo, "On the Beam Deviation Factor of a Parabolic Reflector, *IRE Trans. Antennas Propagat.,* vol. AP-8, no. 3, May 1960, pp. 347–349.

69 J. Ruze, "Lateral Feed Displacement in a Paraboloic," *IEEE Trans. Antennas Propagat.,* vol. AP-13, no. 5, September 1965, pp. 660–665.

70 M. Born and E. Wolf, *Principles of Optics,* Pergamon Press, London, 1959.

71 F. A. Jenkins and H. E. White, *Fundamentals of Optics,* 3d ed., McGraw-Hill Book Company, New York, 1957.

72 W. V. T. Rusch and A. C. Ludwig, "Determination of the Maximum Scan-Gain Contours of a Beam-Scanning Paraboloid and Their Relation to the Petzval Surface," *IEEE Trans. Antennas Propagat.,* vol. AP-21, no. 3, March 1973, pp. 141–147.

73 P. J. B. Clarricoats, "Some Recent Advances in Microwave Reflector Antennas, *IEE Proc.,* vol. 26, no. 1, January 1979, pp. 9–25.

74 B. Claydon, "The Schwarzschild Reflector Antenna with Multiple or Scanned Beams," *Marconi Rev.,* vol. XXXVIII, no. 196, first quarter, 1975, pp. 14–43.

75 W. D. White and L. K. DeSize, "Scanning Characteristics of Two Reflector Antenna Systems," *IRE Conv. Rec.,* part I, 1962, pp. 44–70.

76 B. Claydon, "The Scanning-Corrected Cassegrain Reflector Antenna," *Marconi Rev.,* vol. XXXVIII, no. 197, second quarter, 1975, pp. 77–94.

77 G. Tong, P. J. B. Clarricoats, and G. I. James, "Evaluation of Beam Scanning Dual-Reflector Antennas," *IEE Proc.,* vol. 124, no. 12, December 1977, pp. 1111–1113.

78 R. C. Spencer, "Theoretical Analysis of the Effect of Spherical Aberration on Gain," Air Force Cambridge Res. Cen. Rep. E-5082, Bedford, Mass., December 1951.

79 R. C. Spencer, C. J. Sletten, and J. E. Walsh, "Correction of Spherical Aberration by a Phased Line Source, *Nat. Electron. Conf. Proc.,* vol. 5, 1950, pp. 320–333.

80 A. C. Schell, "The Diffraction Theory of Large-Aperture Spherical Reflector Antennas," *IEEE Trans. Antennas Propagat.,* vol. AP-11, July 1963, pp. 428–432.

81 A. F. Kay, "A Line Source Feed for a Spherical Reflector," TRG Rep. 131–Air Force Cambridge Res. Lab. Rep. AFCRL 529, Bedford, Mass., May 1961.

82 A. W. Love, "Spherical Reflecting Antennas with Corrected Line Sources," *IRE Trans. Antennas Propagat.,* vol. AP-10, no. 4, 1962, pp. 529–537.

83 Roy C. Spencer and G. Hyde, "Studies of the Focal Region of a Spherical Reflector: Geometric Optics," *IEEE Trans. Antennas Propagat.,* vol. AP-16, no. 3, May 1968, pp. 317–324.

84 G. Hyde and R. C. Spencer, "Studies of the Focal Region of a Spherical Reflector: Polarization Effects," *IEEE Trans. Antennas Propagat.,* vol. AP-16, no. 4, July 1968, pp. 399–404.

85 G. Hyde, "Studies of the Focal Region of a Spherical Reflector: Stationary Phase Evaluation," *IEEE Trans. Antennas Propagat.,* vol. AP-16, no. 6, November 1968, pp. 646–656.

86 L. J. Ricardi, "Synthesis of the Fields of a Transverse Feed for a Spherical Reflector," *IEEE Trans. Antennas Propagat.,* vol. AP-19, no. 5, May 1971, pp. 310–320.

87 F. S. Holt and E. L. Bouche, "A Gregorian Corrector for Spherical Reflectors," *IEEE Trans. Antennas Propagat.,* vol. AP-12, no. 1, January 1964, pp. 44–47.

88 C. J. E. Phillips and P. J. B. Clarricoats, "Optimum Design of a Gregorian-Corrected Spherical-Reflector Antenna," *IEE Proc.,* vol. 117, no. 4, April 1970, pp. 718–734.

89 P. H. Masterman, "A Spherical Reflector Antenna for Communications Satellite Earth Stations," SRDE Rep. 72031, 1972.

90 K. S. Kelleher and H. H. Hibbs, "A New Microwave Reflector," Nav. Res. Lab. Rep. 4141, May 11, 1953.

91 G. D. M. Peeler and D. H. Archer, "A Toroidal Wave Reflector," *IRE Conv. Rec.,* 1954, pp. 242–247.

92 G. Hyde, "A Novel Multiple-Beam Earth Terminal Antenna for Satellite Communications," *IEEE ICC Conf. Rec.,* 1970, pp. 38-24–38-33.

93 G. Hyde, W. Korvin, and R. Price, "A Fixed Multiple Beam Toroidal Reflector Antenna Operating at Frequencies above 12 GHz," *IEEE ICC Conf. Rec.,* 1971, pp. 27-11–27-17.

94 G. Hyde, R. W. Kreutel, and L. V. Smith, "The Unattended Earth Terminal Multiple-Beam Torus Antenna," *COMSAT Tech. Rev.,* vol. 4, no. 2, fall 1974, pp. 231–262.

95 W. L. Cook III, "A General-Purpose Conversational Software System for the Analysis of Distorted Multisurface Antennas," Ph.D. thesis, George Washington University, Washington, Feb. 19, 1973.

96 A. F. Kay, "The Scalar Feed," Air Force Cambridge Res. Lab. Rep. AFCRL-64-347, Bedford, Mass., Mar. 30, 1964.

97 P. J. B. Clarricoats and P. K. Saha, "Propagation and Radiation Behaviour of Corrugated Waveguides—Part I—Corrugated Waveguide Feed, Part II—Corrugated-Conical-Horn Feed," *IEE Proc.,* vol. 118, no. 9, September 1971, pp. 1167–1186.

98 R. Price, "High Performance Corrugated Feed-Horn for the Unattended Earth Terminal, *COMSAT Tech. Rev.,* vol. 4, no. 2, fall, 1974, pp. 283–301.

99 H. C. Minett and B. M. Thomas, "Fields in the Image Space of Symmetrical Focussing Reflectors," *IEE Proc.,* vol. 115, no. 10, October 1968, pp. 1419–1430.

100 G. Hyde, "Multiple Beam Ground Station Antennas," *Abstracts of the URSI XVIIIth General Assembly,* Commission VI, Lima, Peru, 1975.

101 G. Hyde, "Aberration Correcting Subreflectors for Toroidal Reflector Antennas," U.S. Patent 3,922,682, Nov. 25, 1975.

Chapter 18

Electromechanical Scanning Antennas

Kenneth S. Kelleher

Consultant

The material covered in this chapter is limited to lens or reflector systems capable of producing a scanning beam without a motion of the lens or reflector or a motion of the entire antenna assembly. It therefore excludes the simple scanning obtained from mount rotation or the conical scanning from feed rotation. These scanner types have been included in earlier literature.[1]

General Types of Scanners

The simplest type of scanner consists of a focusing objective and a point-source feed. The focusing objective can be the conventional paraboloid or the lens with hyperboloid-plane surfaces. The scanning action is then as depicted in Fig. 18-1. When the point source is at the focal point of either objective, a plane wave which propagates along the system axis is produced. When the point source is moved off axis, the wave propagates in a direction at some angle to the axis. The motion of the source moves the antenna beam.

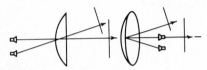

FIG. 18-1 Scanning with focusing objectives.

Another entirely different method of scanning keeps both the source and the objective fixed and adjusts the phase in the path between the source and the objective. This type of phase adjustment was first done in the Foster scanner (Sec. 18-6). At the present time, it is best done by phased-array techniques (Chap. 20). A variation of the phasing between the objective and the feed source is obtained when a large number of fixed feeds are used and phase correction is achieved in the feed complex. This approach can be accomplished with the Butler matrix or with bit phase shifters. Both techniques are *phased arrays*.

The material in this chapter is limited to wide-angle focusing objectives and the feed-motion systems used with such objectives. The best-known focusing device is the Luneburg lens, which is treated first. Among other devices are other lenses, pillbox scanners, and reflector scanners. The chapter concludes with a discussion of feed mechanisms for rapid motion of the source.

18-2 LUNEBURG*-TYPE LENSES

Spherical Luneburg Lens

The best-known symmetrical scanning system is a lens described by Luneburg[2] and investigated further by others.[3-7] In its complete form, the lens is a sphere with the property that energy from a feed source at any point on the spherical surface is propagated through the sphere and focused into parallel rays emerging from the other side of the sphere. Perfect focusing is obtained for all feed positions on the surface.

This lens is formed as a nonhomogeneous medium in which the index of refrac-

---

*For years this name has been misspelled because of an error in publishing Luneburg's notes of Ref. 2.

tion n varies with lens radius r according to the expression $n^2 = 2 - r^2$ for a unit-radius sphere. A central cross section of the sphere is shown in Fig. 18-2, together with typical ray paths through the lens.

The ray paths are sections of ellipses which are given in polar (r,θ) coordinates by the expression $r^2 = \sin^2 \alpha / [1 - \cos \alpha \cos (2\theta - \alpha)]$, where α is the feed angle defining a particular ray. Because the lens is a symmetrical structure, certain relationships between angles in the system are evident from Fig. 18-2. The most important is the equiangular relationship between the following: the angle formed between the ray and the radius vector at the point where the ray leaves the lens; the polar angle defined by the radius vec-

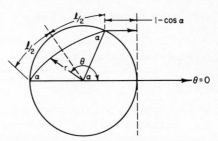

FIG. 18-2 Geometry of Luneburg-lens cross section.

tor to the point at which the ray leaves the lens; and the feed angle, i.e., the angle measured at the source point between the central ray and the general ray. Another point of interest is the fact that the radius vector normal to the ray path bisects the ray path within the lens. Further geometrical information obtained from Fig. 18-2 shows that the path length of a ray within the lens can be obtained as a function of the feed angle. For maximum feed angle of 90°, the ray travels along the lens periphery for a distance of $\pi/2$. A general path length within the lens can be determined from the fact that the optical path length equals $\pi/2 + \cos \alpha$. From the path-length variation it is possible to give an expression for the variation in phase across the semicircular output arc of the lens cross section as $1 - \cos \alpha$, where α is the polar angle.

Another significant property of the Luneburg lens is the fact that the rays emerging from the feedhorn do not appear in a uniform manner across the aperture, but instead tend to spread out in the center and approach a theoretically infinite concentration at the edges. Because of this fact, the analytical aperture illumination is obtained from the original feed pattern multiplied by the factor $\sec \alpha$.

Single-Refractive-Index Lens

An approximation to the Luneburg lens can be obtained by the use of a constant-refractive-index material for the sphere. This design sets the path length of the central ray equal to the path length of the ray at 30° from the axis. (Other angles can be chosen by the designer.) For a sphere of unit radius, the equation is $2n = 2n \cos \theta + 1 - \cos 2\theta$. This expression is solved for refractive index:

$$n = \frac{1}{2}\left[\frac{1 - \cos 2\theta}{1 - \cos \theta}\right] = 1.87$$

With this value of refractive index, the phase error normalized to the sphere radius is

$$\Delta(\theta) = 1 - \cos 2\theta - 3.74(1 - \cos \theta).$$

The phase error gives a reduction in gain and an increase in sidelobes. For a radius of 10 wavelengths, the maximum phase error becomes one-tenth wavelength, an acceptable value. Higher values of radius give greater error. Besides the phase error, the lens has a surface reflection not found in the Luneburg.

Double-Refractive-Index Lens

A variation of the spherical lens uses two indices of refraction to give more flexibility. A two-shell lens[8] was used to collect rays from across the entire aperture, whereas the lens described above used about 90 percent of the aperture dimension. This two-shell design requires indices of 3.6 in the outer shell and 2.75 in the inner shell. It has a much greater reflection coefficient at the lens surface. The design of the constant-index-lens shells is straightforward, and the designer can employ whatever compromise is needed between the use of the entire aperture and the increase in refractive index required.

Two-Dimensional Luneburg Lens

Many interesting variations of the Luneburg lens have been analyzed. The simplest to consider is one in which only a plane section of the lens is utilized.[9] The ray paths through this section are, of course, identical with those of Fig. 18-2. However, the emerging wavefront is not a plane but a saddle-shaped surface. This surface is the envelope of the Huygens wavelets, with centers on the semicircular aperture. In rectangular coordinates, this surface is given by the parametric expression

$$x = \beta$$

$$y = \frac{\sqrt{1 - \alpha^2}}{\alpha} (1 - \beta - \alpha)$$

$$z = \left[(1 - \alpha^2)^2 - (\beta - \alpha)^2 - \frac{1 - \alpha^2}{\alpha^2} (1 - \beta - 2\alpha)^2 \right]^{1/2}$$

Because of this distorted wavefront, certain limitations exist in the radiation pattern. This pattern has been carefully analyzed in Ref. 9, in which it is shown that a sidelobe level of 17 to 18 dB exists for all normal feed-horn illuminations. This problem can be circumvented by introducing a linear aperture, as shown in Fig. 18-3. A cylindrical wavefront is produced by this system, so the pattern expected is similar to that

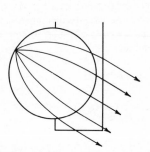

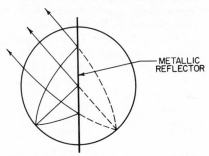

FIG. 18-3 Luneburg lens with a linear aperture.

FIG. 18-4 Ray paths in a virtual-source Luneburg lens.

obtained from an ordinary line source. It should be pointed out that the introduction of the linear aperture destroys the symmetry of the lens and so limits the system to narrower angles of scan. Experimental models using the linear aperture have been shown to have sidelobes of 25 dB.

Virtual-Source Luneburg Lens

Another variation of the Luneburg lens involves the addition of plane metallic reflectors passing through the center of the lens.[10] The addition of such reflectors produces virtual sources whose positions depend on the orientation of the real feed source and the metallic reflector. Figure 18-4 shows a lens cross section with a single reflector in place. From a consideration of the ray paths, it is evident that a perfect virtual image of the real source is formed. It should be noted from the figure that not all the energy from the real source which passes through the lens will strike the reflector. Therefore, this antenna will produce two focused beams, one from the real source and one from the virtual source.

Small-Feed-Circle Luneburg Lens

Another variation of the spherical Luneburg lens has produced a system with a smaller radius of the feed circle and a limited angle of scan. Three general expressions are available.

The first expression was obtained by Eaton in Ref. 3. He considered a sphere of unit radius, with refractive index equal to unity on the surface and with a feed position at any distance, less than or equal to unity, from the center of the sphere. If the radius of the feed circle is denoted by a, then the variation in index is given by the following expression:

$$n^2 = \frac{2a - r^2}{a^2} \qquad \text{for } 0 \le r \le a$$

$$n^2 = \frac{2 - r}{r} \qquad \text{for } a \le r \le 1$$

From the expression it can be seen that if the feed-circle radius is one-half of the radius of the sphere, the index at the center is 2, the index at the feed circle is 1.7, and at the edge of the lens the index is unity. The variation in index can be shown to be continuous with discontinuous slope at the point a.

The second expression for a small-feed-circle Luneburg lens was derived by J. Brown.[4] He indicated that the problem could be attacked by assuming a certain variation in index between the feed circle and the outer surface and then computing the variation within the feed circle which would yield the desired focusing properties. He considered the problem of bringing all rays incident upon the lens surface into the feed point and displayed two solutions to this problem. The first solution involved choosing a refractive index which was constant in the outer region. In terms of the variables used above, the refractive index was chosen as $1/a$. With this constant value of refractive index, the index variation in the inner region was found by a numerical-integration process. For a feed-circle radius of one-half of the lens radius, the results showed that the index had a maximum value of 2.34 at the lens center and decreased monotonically to the value of 2 at the feed circle. The index in the outer region had the constant value of 2.

A third small-feed-circle Luneburg lens was obtained by Gutman.[7] He selected an index variation given by the expression $n^2 = (1 + a^2 - r^2)/a^2$ and then showed that with the feed at a distance a from the lens center the outgoing rays would be parallel. It is obvious from this expression that the index and its slope are continuous functions of the radius. For a feed-circle radius equal to one-half of the lens radius, the index varies from 2.24 at the lens center to 2 at the feed-circle radius and to unity at the surface.

Parallel-Plate Luneburg Lens

The Luneburg can be used in parallel-plate form with either the TEM[11,16] or the TE[9,16] mode. The formulas are:

TE_{10}:

$$\text{plate spacing } s = \frac{\lambda}{2\sqrt{n^2 - 2 + r^2}}$$

TEM: $$\tanh\left[\frac{2\pi}{\lambda}(s - t)\sqrt{1 - r^2}\right]$$

$$-\frac{1}{n^2}\sqrt{\frac{n^2 - 2 + r^2}{1 - r^2}}\tan\left(\frac{2\pi t}{\lambda}\sqrt{n^2 - 2 + r^2}\right)$$

where t is dielective thickness and λ is free-space wavelength.

Surface-Wave Luneburg Lens

The Luneburg lens can be configured by using a TM or TE mode in a surface wave (see Chap. 12 for a discussion of surface-wave media). A number of surface-wave lenses[12] have been described. Their performance has been limited only by the difficulty of controlling the pattern in the plane perpendicular to the lens surface. The lens configuration is similar to that of the TEM lens with the top cover removed. Reference 12 gives design and experimental data.

Geodesic Analog of Luneburg Lens

Another variation of the Luneburg lens is obtained by using the relationship between geodesics on a surface and ray propagation in a plane variable-index medium. The relationship can be understood by considering Fig. 18-5, which shows geodesics on a surface and the corresponding rays in a plane. A given element of geodesic-arc length on the surface corresponds to an element of optical-path length in the plane. Since the required index variation in the plane is known, it is possible, by relating arc-length elements, to determine the required surface containing the geodesics.[13] It can be shown that this surface is a surface of revolution whose generating curve is given in rectangular coordinates by the expression

$$Z = \sqrt{P^2 + \frac{R^2}{4}} + P - \frac{1}{3}\ln\left(\frac{\sqrt{3}}{2}\sqrt{1 + P} + \frac{\sqrt{1 + 3P}}{2}\right)$$

where Z = coordinate along axis of revolution
$P^2 = 1 - R^2$

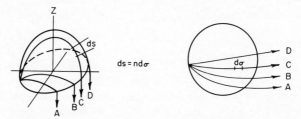

FIG. 18-5 Rays on a surface and in the corresponding variable-index region.

$$R^2 = X^2 + Y^2$$

A major problem arises in constructing the geodesic analog of the Luneburg lens. From a consideration of Fig. 18-5, it can be seen that the rays leave the dome-shaped surface along a cylindrical surface and are not focused in a plane, as required in the original lens. To correct for this fact, that the analog surface has a vertical tangent at its periphery, a toroidal bend is introduced. This bend produces some defocusing of the family of rays.

To avoid the defocusing introduced by the toroidal bend, Warren[14] developed an analog of the Luneburg lens whose surface had horizontal tangents at the periphery. The surface is obtained by a method of approximation, which amounts to a construction as a series of conical sections, the slope of each cone being chosen to focus the ray at a particular feed angle. Warren developed another analog in which a toroidal bend was introduced in the feed region so that the feed horn could be mounted and rotated under the dome of the lens.

Further work on the geodesic Luneburg lens is given in Refs. 15 and 16. The latter reference presents an excellent discussion of these antennas.

18-3 LENS SCANNERS

Dielectric Lens

The simplest type of scanning lens is that which uses a constant index of refraction, a hyperboloid as the initial surface, and a plane as the secondary surface.[17] The design of this lens is based upon placing the source at the focal point of the hyperboloid. Rays from the feed are bent at the initial surface to be parallel to the system axis. The lens cross section, shown in Fig. 18-6, is a hyperbola whose eccentricity is equal to the refractive index of the dielectric material. In the notation of the figure, the focal length of the system is $f = c + a$, and the index is $n = c/a$. For a given focal length f and index n, one can solve for c and a so that the hyperbola in rectangular coordinates is

$$x^2 - y^2 (n^2 - 1) = \frac{f^2}{(n + 1)^2}$$

Since the lens is designed only for on-axis focusing, exceptional scanning qualities are not expected.

Another dielectric lens which has been used for scanning involves a symmetrical element with identical initial and secondary surfaces.[18] A feature of this lens is that

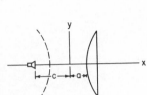

FIG. 18-6   Geometry of a dielectric lens.   **FIG. 18-7**   Symmetrical dielectric lens.

two scanning feeds can be used simultaneously, one on either side of the lens. The two beams can cover sectors 180° apart in space.

The design of this lens is based on the assumption that the lens can be closely approximated by a prism between the points where a ray enters and leaves the surface. The coordinates of the lens surface can be found from the following expression, all the parameters of which are given in Fig. 18-7:

$$x = \frac{[f(1 - \sec \varphi) + 2a(n - 1)](2 \cos^2 \alpha \sin \omega - \sin \beta)}{(n - \cos \alpha) \sin \beta + (n \sec \alpha - \sec \varphi)(2 \cos^2 \alpha \sin \omega - \sin \beta)}$$

$$y = (f - x) \tan \varphi$$

The three angles β, φ, and α are known in terms of the ω parameter:

$$\beta = 2\omega - \cos^{-1}\left(\frac{\cos \omega}{n}\right)$$

$$\varphi = \omega - \cos^{-1}(n \cos r_1)$$

$$\alpha = \cos^{-1}\left(\frac{\cos \omega}{n}\right) - \omega$$

The surface is found by selecting values of the parameter ω, computing the three angles, and substituting their values into the expression for x and y.

Wide-Angle Dielectric Lenses

The Abbe sine condition has been used in designing the dielectric lens for scanning. This condition applies to a symmetrical optical system with its source on axis. It concerns a matching of the ray from the source (spherical wave) with its corresponding ray (in a plane wave) emerging from the system. The reflections and refractions within the system are ignored, and the condition asks that the intersection of corresponding rays lie on the spherical wavefront. In a cross section containing the axis, an initial ray at an angle α meets a final ray at a distance d from the axis on a circle defined by $d = r \sin \alpha$, where r is the radius of the circle.

The greatest amount of work using this condition was done by Friedlander.[18] He found that he could closely approximate the circle of intersection between initial rays and final rays by using a relatively simple lens. Instead of using two curved surfaces, his lens had a plane face for the initial surface and a curved face for the final surface.

With a refractive index of 1.6 and $f/D = 1$, he found a deviation from the sine condition of the order of 0.33 percent of the focal length. An experimental lens of 50λ aperture gave satisfactory patterns over a 20° total scan angle.

A lens system which gives promise of wide-angle focusing is one that uses two curved surfaces to obtain perfect focusing (at two focal points) of waves from two conjugate directions.[19,20,21] The two most useful design methods are the *algebraic curve* and the *lattice*. The first assumes that the two-dimensional-lens faces are algebraic curves and then determines the coefficients of the algebraic expressions by selected ray tracing. The second is a step-by-step procedure, adapted to computer use, in which points on the two lens surfaces are found by considering alternately the two focal points. Some effort has been made toward designing a three-dimensional lens by using a two-dimensional design to form the lens faces as surfaces of revolution, but the resulting lens has serious astigmatism.

Metal-Plate Lenses

The simple metal-plate lens described in Chap. 16 has been used successfully as a scanning antenna. This lens design provides perfect focusing for a single position of the feed; wide-angle performance is obtained by choosing a long focal length. Some problems which arise at off-axis feed positions are related to the basic nature of the metal-plate lens. Since the lens is a periodic structure, diffraction lobes[22] which increase with an increase in scan angle appear; these lobes are present in addition to the normal coma lobes. To prevent them from becoming objectionable, the angle which the incident rays make with the lens surface must be limited; this limitation in turn limits the feed displacement and the angle of scan.

Metal-plate lenses can be designed for improved performance by using the Abbe sine condition. One such lens design uses square-waveguide elements which constrain the incident energy to travel through the lens in a direction parallel to the lens axis. Such a design simplifies considerably the analysis of the system, since the refraction follows a much simpler expression than that given by Snell's law. One lens design offered[23] not only satisfied the Abbe sine condition but also, by proper choice of the waveguide medium, yielded a system which would operate over a broader band than that possible with the normal metal-plate lens.

One of the earlier metal-plate scanning lenses was constructed between parallel plates with the general form shown in Fig. 18-8.[24] In this lens, the initial surface was curved, and the index of refraction was varied in the dimension. The design of the system then employed three basic variables: (1) the lens surface, (2) phase delay between input and output of a given lens channel, and (3) constraint of energy entering at a given point to leave at a desired point.

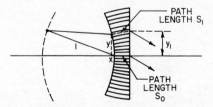

FIG. 18-8 Lens using path-length correction.

The lens surface is given by the circle $(x - \frac{1}{2})^2 + y^2 = (\frac{1}{2})^2$, the coordinates of which are shown in Fig. 18-8. The phase delay for a particular channel is found in terms of the delay in the central channel and the coordinate z. The required expression is $s_1 - s_0 = 1 - \sqrt{1 - x}$. The final variable, the distance from the axis to a general emerging ray, is given by the expression $y_1 = \sqrt{x}$.

Experimental results on this lens showed that it maintained a constant beam-width of 2.2° over a total scan angle of 40°; no information is available on the sidelobe level. In taking these data, the feed horn was confined to a circle of unit radius.

Metal-Plate Lenses with Two-Point Correction

A decided improvement in scanning has been achieved when the Abbe sine condition of optics is ignored and, instead, the lens is designed to provide perfect focusing at two points.[25] Figure 18-9 indicates the geom-etry of such a design in a two-dimen-sional lens. This design employs a con-strained lens of constant refractive index. The lens surfaces are chosen so that plane waves arriving from the directions $+\alpha$ and $-\alpha$ are exactly focused at the points 0 and 0' respectively. By using the design geometry of the figure, the first lens sur-face is an ellipse $(x/a + 1)^2 + y^2 = 1$, and the second lens surface is obtained by determining the lens thickness at any point. An expression involving the lens thickness as a function of the lens constants and the coordinate x is given as $ax = (n - a)(d - d_0)$, where n is the refractive index and the other parameters are given in the figure.

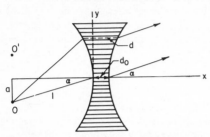

FIG. 18-9   Two-point-correction lens.

A second two-dimensional lens utilized a variable index but a plane outer sur-face. In this case, the inner surface was identical with that of the lens in the preceding paragraph, and the variation in the refractive index was given by the expression $(n - n_0)d_0 = nx - m\lambda$. The term $m\lambda$ has been introduced into this expression to obtain wavelength steps in the lens. This is necessary to minimize the range over which the refractive index must vary. Since waveguide techniques are utilized, the refractive index is limited to values between 0.5 and 0.9.

A three-dimensional lens which produces perfect focusing at two points can be obtained. The inner lens surface is an ellipsoid, given by the expression

$$\left(\frac{x}{a} + 1\right)^2 + y^2 + \left(\frac{z}{a}\right)^2 = 1$$

Here the symbols are those of Fig. 18-9, and the coordinate z is measured normal to the xy plane. Once the inner surface is found, the outer surface is obtained from an expression involving the lens thickness:

$$ax = (n - a)(d - d_0)$$

An important improvement in the performance of a constrained lens with two-point correction has been obtained by Peeler.[26] His design utilized the basic lens together with steps which were arranged to minimize the phase errors in the on-axis position. He thus had corrections at three points and an overall reduction in phase errors at all scan positions. Computations on the final design showed phase errors of less than one-tenth wavelength over the aperture throughout the scanning of a ½° beam through 20°.

18-4 PARALLEL-PLATE SCANNERS

Surface-of-Revolution Scanners

The scanner considered here is a parallel-plate region whose mean surface is a surface of revolution. This surface is intended to satisfy the following conditions.[13]

1 It contains a circle (the feed circle).

2 It contains a straight-line segment (the aperture which can become the focal line of a parabolic cylinder).

3 All geodesics joining a fixed point of the feed circle to points of the aperture meet the aperture at a constant angle (optical requirement).

4 All geodesics normal to the feed circle pass through a fixed point (the center of illumination) of the aperture (illumination requirement).

At the present time, no such surface has been found. If requirement 4 is eliminated, a solution which is known as the R-and-$2R$ scanner[24] is possible. The action of this scanner can be understood by considering the mean surface laid out in Fig. 18-10. Here the source is placed on the feed circle of radius R. Energy passes across this mean surface to the opposite side of the circle, where it encounters the output region with boundaries which include a circle of radius $2R$ and the linear aperture. From simple geometry, the path length of a ray in the initial region is given by $2R \cos \alpha$, where α is the feed angle. The ray path in the output region is given by $2R(1 - \cos \alpha)$, so it is evident that all path lengths from the source to the aperture are equal. For another position on the feed circle, because of the symmetry in the final structure, all rays meet the aperture at a constant angle so that perfect focusing is available. For feed positions other than that of Fig. 18-10, the central ray does not pass through the center of the aperture, so that the utility of the scanner is limited. As the feed horn scans, the maximum intensity appears at various positions in the aperture, and during the greater part of the scan the energy lies at either end of the aperture.

A lens of the type described above was constructed by introducing an imaging reflector in the initial parallel-plate region and rolling the plates into a cone with the feed arc formed into a complete circle on the base of the cone. The on-center radiation pattern had a beamwidth of 0.6° and a sidelobe level of 16 dB. This two-dimensional lens has a three-dimensional counterpart which has scan limited by astigmatism. Rays in the plane of scan have perfect focusing, but rays in the other principal plane are not perfectly focused.

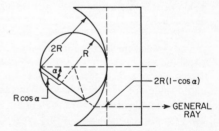

FIG. 18-10 Geometry of the general ray in the R-and-$2R$ scanner.

The majority of the other surface-of-revolution scanners are an outgrowth of the work of Myers.[27] The surface considered has been either a cone or a cylinder. Since both of these are developable surfaces, it is possible to analyze the ray paths with simple geometry. The general problem is indicated in Fig. 18-11, which shows a devel-

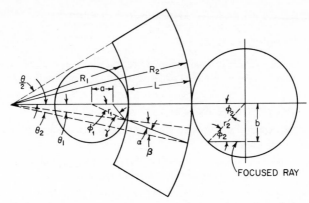

FIG. 18-11 Geometry of the conical surface-of-revolution scanner.

oped cone having a sectoral angle designated as θ. From a consideration of arc lengths on the circular top and bottom planes, one has that

$$2\theta R_2 = 2\pi r_2 \qquad 2\theta R_1 = 2\pi r_1$$

and therefore,

$$\frac{r_2}{R_2} = \frac{r_1}{R_1}$$

Further investigation of the arc lengths along the discontinuities in the conical surface shows that $R_1\theta_1 = r_1\varphi_1$ and $R_2\theta_2 = r_2\varphi_2$. It can therefore be seen that $\theta_2/\varphi_2 = \theta_1/\varphi_1$. Because geodesics on the surface form equal angles with the normal to a surface discontinuity, one can write that $\alpha = \gamma$ and $\beta = \varphi_2$.

For a cone in which r_1 and r_2 are nearly equal, $\alpha \approx \beta$, and some consideration can be given to the maximum distance b at which a ray can be normal to the aperture. It can be shown that $b = r_2 \sin \alpha$. Since r_2 is a constant, the maximum value of b occurs for large $\sin \alpha$ or large values of the feed-circle radius a. If a is set at its maximum value of r_1, a simple lens design can be evolved: for a length L of the cone, if a ray crossing the aperture at b is in phase with the central ray, then the length L must be $r_2 \cos \varphi_2$. This design is exact only when the disk radii r_1 and r_2 are equal, i.e., when the cone becomes a cylinder.

Double-Layer Pillbox

Rotman[28] has investigated several experimental models of the cylindrical scanners indicated in Fig. 18-12. The half-cylinder unit shown on the left possesses the general characteristics required in the surface-of-revolution scanner. The model on the right represents an improvement in symmetry obtained by eliminating the requirement for a linear aperture and permitting the energy to radiate from the circular aperture.

Rotman divided the double-layer pillbox antennas by size in wavelengths: (1) small, $<10\lambda$, with no correction needed; (2) medium, to 20λ, using parallel-plate contours; and (3) large, to 40λ, using plate contours and multiple feeds. Rotman's early

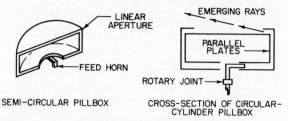

FIG. 18-12 Cylindrical surface-of-revolution scanner.

work was followed[29] by correcting in parallel-plate structures through using the bootlace principle of Sec. 16-10 in Chap. 16.

Concentric-Lens Scanner

The concentric-lens scanner is related to the scanners described immediately above. It is made up of a cylinder or half cylinder whose height is very small compared with its diameter. This scanner contains a lens whose surfaces are concentric with the cylinder surface; this lens can be mounted in either the upper or the lower pillbox region.

Rotman studied a concentric-lens system in which the lens was mounted in the lower (feed-horn) region, as shown in Fig. 18-13. Kales and Chait at the Naval Research Laboratory studied a scanner in which the lens was mounted in the upper (output) region. They found that if the radius of the cylinder was R and the lens radii were R_1 and R_2, then a useful lens was obtained by using the relation $1/R = 1/R_1 - 1/R_2$.

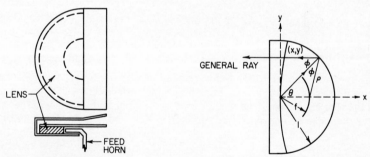

FIG. 18-13 Pillbox-concentric-lens system.

FIG. 18-14 Geometry of the Schmidt lens.

Schmidt Lens

This system consists of a spherical reflector, together with a nearly plane correcting lens, which passes through the center of the sphere. A derivation of the design of this system is contained in Ref. 30. The result is that the lens cross section in Fig. 18-14 can be defined in parametric form by the following expression:

$$x = \frac{2 \cos \theta - f \cos 2\theta - c \cos \psi}{1 - n \cos \psi}$$

$$y = 2 \sin \theta - f \sin 2\theta - \frac{c \sin \phi - n \sin \psi (2 \cos \theta - f \cos 2\theta)}{1 - n \cos \psi}$$

where $\psi = \theta - \phi$
$\quad c = 2 - f$
$\sin \phi = (f \sin \theta)/\sqrt{1 + f^2 - 2f \cos \theta}$

Several experimental models of the Schmidt system have been built in parallel-plate media. Some of these have involved a single layer of parallel plates with a circular reflector, while others have employed a double layer and a toroidal-bend reflector. Excellent scanning performance has been obtained on all models which have utilized a dielectric lens. Metal-plate lenses have been less satisfactory.

Schwarzschild System

This system employs multiple mirrors rather than a lens and a mirror. It is a two-mirror device based on a general optical method developed by Schwarzschild. Such a system, in optical terminology, has no spherical aberration or coma. The only experimental system using this principle was built in the form of a triple-layer parallel-plate region.[1] The feed horn moved on a 90° arc of a circle in the first layer. Energy from the feed was reflected at the first toroidal-bend reflector, passed through the middle layer, was reflected from the second bend, and left the parallel plates through a linear aperture in the third layer.

The surfaces of the first and second reflectors were calculated, respectively, from the equations

$$x_1 = 0.5y^2 - 0.03125y^4 - 0.22y^6$$

$$x_2 = 0.75y^2 - 5.96025y^4 + 40.53y^6$$

where the coordinate x is parallel to the direction of the optical axis.

Zeiss-Cardioid System

This system is similar to the Schwarzschild in that it employs two reflectors. The major difference is that it uses simple reflecting surfaces and yields perfect focusing for the feed on the axis. The mirrors used are a circle and a cardioid. The design of this system can be understood by considering Fig. 18-15. If the position of the feed horn is taken as the origin, then the reflecting surfaces have the following equations:

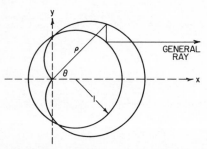

FIG. 18-15 Geometry of the Zeiss-cardioid reflector system.

Reflector 1: $\quad\quad \rho = 1 + \cos \theta$
Reflector 2: $(x - \frac{1}{2})^2 + y^2 = 1$

An X-band model of this reflector system with an aperture of 36 in (914.4 mm) had a half-power beamwidth (HPBW) of 3.3°, using a wavelength of 32 mm. This antenna, investigated by W. Rotman at the Air Force Cambridge Research Center, proved capable of scanning through a total angle greater than 15 beamwidths with a loss in gain of less than 1 dB.

18-5 SCANNING WITH REFLECTORS

The simplest type of focusing objective is the reflector. This is a lightweight component, relatively easy to construct, and so has found wide application in the scanning field. The majority of the reflectors have been paraboloids; others of interest are a section of spherical surface, a torus, and a parabolic torus. These will be considered in turn in the discussion which follows.

Paraboloidal Reflector

The paraboloidal reflector has been used as a scanner in many forms, of which the simplest is the symmetrical one with the vertex of the paraboloid at the center of the reflecting surface (circular aperture). Other reflectors were formed of offset sections of the paraboloid surface which did not include the vertex point or which included this point at some off-center position. The aperture of these offset reflector types does not fit any simple plane curve, although they could normally be approximated by an ellipse. Since there are so many different offset reflectors, it is impossible to describe the scanning performance of each.

The parameters available in the circular-aperture reflector are the aperture width, the aperture illumination, and the focal length. An analysis of the pattern of the circular aperture with feed displaced from the focal point showed minimum variation of pattern shape with moderate changes of the aperture width in wavelength.[31] For the aperture widths of interest, it is necessary only to consider the illumination and the focal length; Ref. 31 shows data for two illuminations and four focal lengths.

Most of the data of Ref. 31 were taken with the feed horn moved off axis in the H plane; a check of the characteristics of the feed horn moved off in the E plane showed agreement within experimental error. The implication is that the variation in scanning characteristics is a scalar problem independent of polarization. A check on the variation in the cross-polarized component over the angle of scan showed that it was relatively constant over the angle of scan and dropped in intensity at the edges of the scan sector.

The data for all four reflectors showed a decrease in gain and an increase in sidelobe level as the beam was scanned away from the axis. The sidelobe on the axial side of the pattern increased rapidly, while that on the other side of the pattern was merged into a main beam. Figure 18-16 shows a typical pattern variation over the angle of scan. Some difference was noted in the pattern characteristics with a change in aperture illumination. With 10-dB illumination, the sidelobe level increased more rapidly than with 20-dB illumination. Scanning characteristics improved with an increase in focal length; best results were obtained with the longest focal length, which corresponded to an f/D ratio of 0.75.

A study was made of the orientation of the feed during scanning. The two possibilities considered involved pointing the feed toward the vertex of the reflector and pointing it so that its axis was parallel to the axis of the reflector. Only minor differences in the patterns were found with changes in feed orientation, so that it was impos-

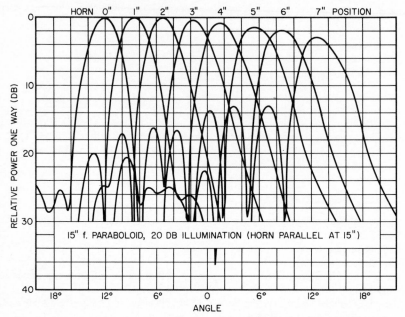

FIG. 18-16 Patterns from the paraboloid with feed displaced from the axis.

sible to say that one or the other orientation was superior. It was found that best scanning results were obtained when the feed was moved along a line perpendicular to the reflector axis.

Most deductions concerning scanning performance are based on radiation patterns taken in the plane of scan. Patterns taken in the other principal plane showed that the beam varied only slightly with feed positions and that for the larger scan angles the first sidelobe was merged into the main beam.

One very interesting effect arises from the study of the scanning performance of the paraboloid. It is found that if the feed horn is moved through a certain angle away from the axis, then the radiated beam is displaced through a smaller angle from the axis. The ratio between the angle which a radiated beam makes with the axis and the angle which the feed displacement from the axis subtends at the vertex is called the *beam factor*. It was found that the beam factor was dependent upon the focal length, the aperture distribution, and the position off axis. Of these three parameters, the focal length was the most significant. For moderate displacements from the axis, the beam factor showed only slight variation with feed position. Figure 18-17 shows a plot of the beam factor obtained from the experimental data.

Reflector Tilt

The paraboloidal reflector can achieve twice the angle of scan for a given beam degradation if the reflector is moved and the feed held fixed. In effect, one half of the scan comes from moving the feed off axis, and the other half comes from the motion of the focusing objective. The motion of the reflector can be very effective in millimeter antennas, in which reflectors are small.

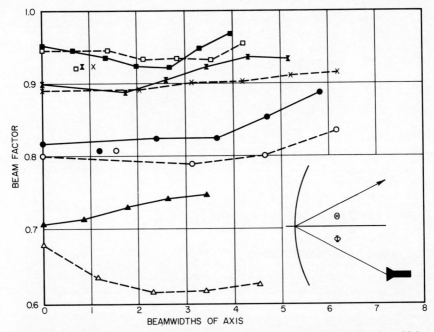

FIG. 18-17 Beam factor as a function of focal length and edge illumination in a 30-in (762-mm) paraboloid: ———— 20-dB illumination; --------- 10-dB illumination. Δ = 7.5-in (190-mm) focal length, 0 = 10.6-in (269-mm) focal length, X = 15-in (381-mm) focal length, □ = 22.5-in (571.5-mm) focal length.

Another type of reflector scanning uses a plane reflector[32] as indicated in Fig. 18-18. Here the paraboloidal reflector is linearly polarized so that it will reflect horizontal polarization and transmit vertical polarization. The plane reflector is a twist reflector which twists the incident polarization.

Spherical Reflector

The spherical reflector is not a replacement for the paraboloid when wide-angle scanning is needed. For long focal lengths, the two reflectors have similar performance. For short focal lengths and a restricted use of the spherical aperture, the spherical reflector gives excellent performance. However, this performance comes with the drawback of very poor aperture efficiency caused by the use of only one-tenth of the available aperture.[33]

A direct comparison of the spherical reflector and the paraboloid reflector was made in Ref. 31 for values of focal length ranging from 0.25 to 0.75. The reflectors were illuminated with edge values of 10 dB and 20 dB. The sphere showed a performance similar to the paraboloid except for higher sidelobes. A simple analysis of the wavefronts from the two surfaces shows that both have third-order phase errors, and so both have the asymmetry in sidelobes sometimes called coma lobes.

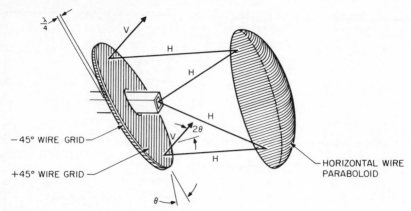

FIG. 18-18   Scanning with a plane reflector.

Corrected Spherical Reflector

There is a wide variety of approaches to correcting the spherical reflector so as to increase aperture efficiency. A correcting lens or reflector can be added to the system. The most noteworthy is the Mangin mirror technique. Here the lens is in contact with the reflector. The lens surface nearest the feed is a sphere whose radius can be varied to optimize performance. Computer programs permit a wide variety of designs. Correction of the aberration by a reflector[34] has been defined, but most of the effort has been devoted to corrected feed systems. The feed configuration of end-fire line sources[35,36,37] has been studied for many years, and the Arecibo antenna in Puerto Rico uses such a source.[38] The alternative configuration of feeds consists of broadside sources.[39,40,41] With phased-array techniques, the broadside feed array can be made effective for multiple scanning beams.

Torus Reflector[42]

Another reflector which has been suggested for scanning applications is the torus, obtained by rotating an arc of a circle completely about an axis which lies in the plane of the circle. For it to be useful, the reflecting surface must be made up of rods at 45° to the axis of revolution. A feed horn which yields polarization parallel to the reflecting elements is placed within the torus surface. The energy focused by the reflector appears at the opposite surface of the torus with a polarization normal to the reflecting elements so that it passes through the torus surface and forms a radiated pattern in space. Because of the symmetry of this reflector, the feed horn can be moved on a circle of constant radius to yield a scanning of the focused beam.

    The parameters available in the design of this reflector are the radius of the feed circle, the distance from the axis of revolution to the vertex of the circular arc, and the axial extent of the surface. With one choice of parameters, the torus degenerates into a section of a spherical surface. Other variations upon the basic design would include the substitution of any plane curve for the circular arc. In this way, beam shaping in the plane normal to the plane of scan can be achieved.

The basic limitation on the torus scanner is similar to that of the spherical reflector: perfect focusing cannot be achieved. However, if the radiation characteristics obtained are satisfactory, then this scanner offers a lightweight, compact structure capable of sweeping a beam through 360° in azimuth.

Parabolic-Torus Reflector[43]

One variation of the torus scanner, which has been investigated in some detail, is that obtained by replacing the circular arc with a parabolic arc and by limiting the rotation of the parabolic arc about the axis of revolution to an angle of less than 180° so that the reflector does not require the polarization properties of the previous torus. The parameters available in the design of this system are the focal length of the parabola, the distance from the axis of revolution to the vertex of the parabola (torus radius), the angle of revolution, and the axial length.

A study was made of the wavefront produced by the parabolic torus when illuminated by a point-source feed. A satisfactory reflector was the half torus, which used a parabolic arc, one of whose end points was the parabola vertex. With a ratio of focal length to radius of about 0.45, a minimum deviation from a plane wave was obtained. The patterns measured with this reflector are shown in Fig. 18-19. It is evident that reasonable sidelobe levels can be obtained in the plane of scan and that good sidelobe levels can be obtained in the other plane. However, one important characteristic of this type of reflector is the higher sidelobes which appear in the planes at 45° to the principal planes. Data obtained on this reflector show lobes as high as 13 dB in these planes. For applications in which these lobes are not objectionable, this reflector is superior to the paraboloid and sphere.

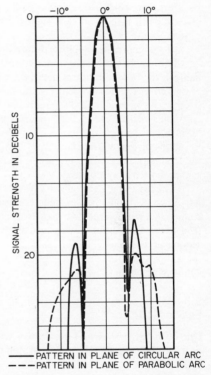

——— PATTERN IN PLANE OF CIRCULAR ARC
— — — PATTERN IN PLANE OF PARABOLIC ARC

FIG. 18-19 Patterns from parabolic torus.

Another parabolic-torus reflector was the full torus, which had a circular aperture similar to the paraboloids previously described. This reflector showed radiation characteristics unlike that of the half torus; the sidelobes were merged into the main beam to produce a tent-shaped structure. The maximum sidelobes were in the principal planes. An optimization of the torus reflector produced an elliptical torus, described in Ref. 44.

18-6   FEED-MOTION SYSTEMS

The scanning systems described here are used together with one of the focusing objectives of the earlier sections. These scanners satisfy the general requirements of moving a point source rapidly over a curve or a surface in space. The great majority of the feed-motion systems produce a beam moving through a plane in space so that the feed path is a plane curve. If a scan over a volumetric region is desired, it can be obtained by a combination of several scans, each confined to a different plane.

There are two general methods of scanning over a plane, namely, *oscillating* scan and *sawtooth* scan. In the first, the beam sweeps across a plane and then returns by the same route; this can be accomplished simply by mechanical oscillation of the feed horn. In the sawtooth scan the beam sweeps across the plane of scan and then steps back instantaneously to its initial point. This normally requires a more complex feed system, but it is more efficient since the beam equally covers all positions in the plane of scan.

If we ignore the simple feed-horn oscillation, all existing feed-motion systems are based on mechanical rotary motion. These systems can be placed in three categories, namely, the virtual-source scanner, the organ-pipe scanner, and the moving of a linear or curved waveguide element. In practical application, the virtual-source scanner utilizes a source produced by a reflection or a refraction at some optical component between a real source and the focusing objective. In most of these scanners, the feed horn, a real source, is rotated continuously in front of an imaging reflector, which forms the virtual source. However, it is possible to obtain the same effect by holding the feed horn fixed and rotating a refractive prism. In either case, the apparent feed-source motion is approximately sawtooth in nature.

The organ-pipe scanner is a system in which rapid, rotary motion of a small feed horn is converted to motion along the desired feed path through the simple process of connecting, with waveguide channels, points on the feed circle to points on the desired feed path. A variation in this system occurs when the feed horn is held fixed and a section of the waveguide channel is rotated to accomplish the same purpose. The remaining type of feed-motion system employs a moving portion of a waveguide wall, cylinder, or channel.

Lewis Scanner[45]

One of the earliest virtual-source scanners was the Lewis scanner, which employed a parallel-plate region. The action of this scanner is evident from Fig. 18-20, in which rays from the real source strike a reflector strip at 45° to the antenna aperture and so produce a virtual source which feeds the lens. The straight-line feed path is formed into a circle by rolling the parallel-plate region into a cylinder. A rotation of the feed about the base of the cylinder produces a virtual source which appears to be moving along a straight line. Some dead time occurs when the feed horn is opposite the position on the cylinder where the two ends of the

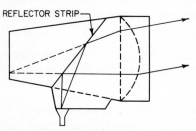

REFLECTOR STRIP

FIG. 18-20   Lewis scanner.

feed path are joined. The amount of dead time depends on the length of the feed path and the size of the feed horn; normally, it represents about 20 percent of the scanning time.

Robinson Scanner[45]

The Robinson scanner is a virtual-source system made up in parallel plates in a manner similar to that of the Lewis scanner. The major differences are the absence of the focusing lens and the use of a more complex imaging reflector. This scanner uses a developable surface which, in planar form, is an isosceles trapezoid. When the surface is developed, the smaller base is bent to lie above the larger base, and then the two ends of the base are brought together to form a circle (Fig. 18-21). Since the surface was developed from a plane, ray paths can be found from a consideration of the rays in a plane. Rays emanating from a horn moving on the feed circle leave the aperture in the same manner as those rays which might come from a feed moving along the smaller base of the trapezoid. The rotary horn motion then produces an apparent linear motion of a virtual source.

The dead time found in this feed-motion system occurs when the feed is opposite the end points of the smaller trapezoid base; the system's value depends on the feed-horn size and the circumference of the feed circle and so is similar to that of the Lewis scanner.

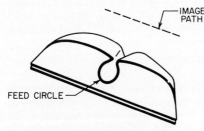

FIG. 18-21   Robinson scanner.

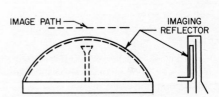

FIG. 18-22   Virtual-source scanner.

Virtual-Source Scanner

In this system, the feed horn rotates in the lower level of a two-layer parallel-plate region.[46] Energy from the feed is reflected into the upper layer by a parabolic cylinder placed at the junction between the upper and the lower layers (Fig. 18-22). The reflected rays appear to come from a virtual source which moves along an approximately linear path while the feed horn moves through an arc of 120°.

The center of the feed-horn circle corresponds to the focal point of the parabolic cylinder; its radius is about 80 percent of the focal length. Since this feed is used for only 120° of rotation, a complete system would require three feed horns and a suitable switching mechanism. The dead time for this system will then depend on the amount of the rotation cycle required to switch from one feed to the next.

Tilting-Plane Scanner

A variation of the virtual-source scanner is obtained when the feed horn is held fixed as it directs the energy into a plane reflector. The reflected rays appear to come from a virtual source behind the reflector. The position of this virtual source depends on the orientation of the plane. Consider an initial position with the plane vertical and the source in a horizontal plane a distance d from the reflector. If the plane is tilted about a horizontal line at a distance from the horizontal plane (Fig. 18-23), then the equation of the virtual-source path is given by $\rho = 2(d \cos \alpha + b \sin \alpha)$.

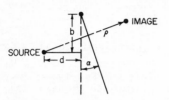

FIG. 18-23   Tilting-plane scanner.

Prism Scanner

In its simplest form, the prism scanner would employ an element placed between the feed and the objective. Figure 18-24 shows two possible prisms: (1) the circular element, in which scanning is obtained by simple rotation; and (2) the triangular element, in which scanning is achieved by an oscillation of the triangle.

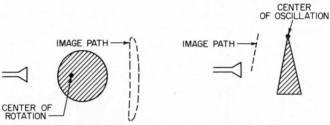

FIG. 18-24   Prism scanners.

The design problems associated with this type of scanner are concentrated in obtaining a good virtual source and minimizing the curvature of the path followed by the virtual source during the scan cycle. From the nature of the problem, most of the design work is experimental.

Foster Scanner

A scanner related to the prism scanners was invented by Foster.[47] It can be understood by considering Fig. 18-25, which shows a cross section and the complete antenna. At any cross section, the phase is varied by rotating the inner cone. The phase variation is a minimum when the input and aperture are aligned and is a maximum when the energy travels counterclockwise for almost 180°, then through the center region, and

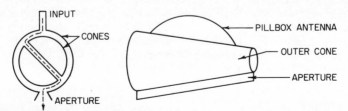

FIG. 18-25   Foster scanner.

finally clockwise for almost 180°. The maximum phase occurs immediately before the input lines up with the aperture (minimum position). This means that the system rapidly switches from maximum condition to minimum condition. Now when the cross section is moved along the cone, the amount of phase shift varies with the cone diameter. Therefore, rays at the smaller end of the cone are shifted in phase less than those at the larger end of the cone. The differential phase shift causes a tilt in the beam from the aperture as the inner cone is rotated. The development work after Foster was directed at implementing a simple configuration. In one instance the line-source input was placed on the rotating cone so that energy traveled only in the clockwise region of Fig. 18-25. The best Foster scanner[48] used a choke barrier within the conical structure and eliminated an earlier problem associated with directing the energy in the clockwise direction between the cones.

NRL Organ-Pipe Scanner[49]

The action of this scanner can best be understood by considering Fig. 18-26. It can be seen that energy is introduced into certain waveguide channels at the feed circle. It

FIG. 18-26   NRL organ-pipe scanner.

propagates through these waveguides, which are all of equal length to the aperture of the system. A rotation of the feed horn produces a change in the channels which are fed and so produces a motion of the radiated energy across the scanner aperture.

The problems associated with this scanner are centered in the transition region between the feed horn and the waveguide channels. By proper choice of dimensions here, the transition loss can be held to less than 0.25 dB at 9.3 GHz. The voltage standing-wave ratio (VSWR) is minimized by maintaining the intersection between channels to a value less than 1/64 in (0.397 mm). It was found that the VSWR varied with feed rotation, depending on the number of channels which were fed at any one time; best results were obtained by using a horn which energized 2½ waveguide channels.

A variation in this basic structure would permit the scanner aperture which did not lie along a straight line but fitted any curve. Another variation would involve a change in the spacing between the channel outputs; such a change would produce a nonuniform scanning with a constant rotation of the feed horn. The waveguide can be replaced by a coaxial line to give simpler packaging but increased transmission-line losses.

Sector Organ Pipe

If a circular feed path is desired, the radius of the feed circle can be minimized by introducing a sector of an organ pipe. In this system, channels on the feed circle follow radial lines to the desired scan circle. For constant feed rotation, several horns and a waveguide switch are required.

RCA Organ Pipe[50]

A variation of the organ-pipe principle is obtained when the feed horn is fixed and a section of the waveguide channels is rotated. One of the possible channel-rotation sys-

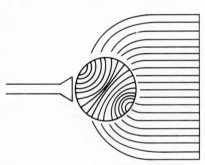

tems is shown in Fig. 18-27. From a consideration of the figure, it is seen that with rotation of the cylindrical-channel region the radiated energy is moved along the scanner aperture.

Tape Scanner[51]

The simplest type of moving-waveguide-wall scanner involves the physical motion of slots in one face of a waveguide. One system considered is shown in Fig. 18-28.

FIG. 18-27 RCA organ-pipe scanner.

Here energy in a linear length of waveguide is radiated through two slots which move along the waveguide lengths. To couple out all energy within the guide it is necessary to move an effective waveguide short circuit in synchronism with the slots.

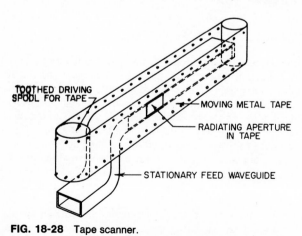

FIG. 18-28 Tape scanner.

This is done by gearing to the tape drive a rotary phase shifter which is terminated in a short circuit.

The major problem associated with this system is that of obtaining a tape which will fit tightly against the fixed waveguide section and yet will be flexible enough to pass around the rotor. One additional problem relates to the means of providing radio-frequency chokes around the periphery of the open-sided waveguide. Successful choke designs were obtained and good results were found by using metal plate in place of the tape.

Helical-Slot Scanner

Another type of moving-slot scanner utilizes a similar waveguide system with one waveguide wall removed. The slot radiator is formed between the waveguide opening and a helical opening on a circular cylinder (Fig. 18-29). As the cylinder is rotated, the effective radiating aperture moves along a straight line corresponding to the position of the opening in the waveguide.

FIG. 18-29   Helical-slot scanner.

The major difficulty encountered in the design of the scanner was traced to the problem of the radio-frequency chokes. Any chokes parallel to the waveguide opening were inefficient at the position of the radiating slot because of the lack of necessary cover plate. Various remedies were attempted to improve this situation, including the use of both rectangular and circular waveguide. It was found impossible to obtain satisfactory radiation patterns because of the choke problem. It appeared that the rectangular waveguide was superior to the circular one, since the field within this guide held the same polarization whereas in the circular waveguide polarization varied with the motion of the slot.

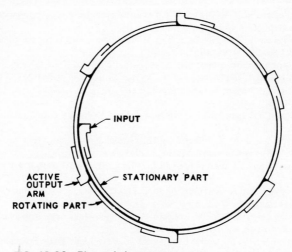

FIG. 18-30   Ring-switch scanner.

Ring Switch

The ring switch is made up of a circular section of waveguide with a fixed input and a number of moving outputs. Figure 18-30 shows a typical configuration. The waveguide between the input and the output is divided so that one portion is moving and the other portion is fixed. Choke arrangements sometimes are required to create radiofrequency continuity between the moving and the fixed portion. If the guide can be split on the center of the two broad faces, chokes may not be necessary. But such a guide might be difficult to match at the input and output ports. Several of these split-guide switches are described in Refs. 52 through 56. A ring switch designed for use with a multiple-reflector antenna (Ref. 57) is the best design available. This scanning switch uses a coupling cavity between the two waveguides, each of which is split along the centerline of the broad face; chokes normally are not necessary.

REFERENCES

1 W. M. Cady, M. B. Karelitz, and L. A. Turner, *Radar Scanner and Radomes,* McGraw-Hill Book Company, New York, 1948.
2 R. K. Luneburg, *Mathematical Theory of Optics,* Brown University Press, Providence, R.I., 1944.
3 J. E. Eaton, "An Extension of the Luneburg-Type Lenses," Nav. Res. Lab. Rep. 4110, 1953.
4 J. Brown, "Microwave Wide Angle Scanner," *Wireless Eng.,* vol. 30, 1953, p. 250.
5 S. P. Morgan, "Generalizations of Spherically Symmetric Lenses," *IRE Trans. Antennas Propagat.,* vol. AP-7, 1959, p. 342.
6 A. F. Kay, "Spherically Symmetric Lenses," *IRE Trans. Antennas Propagat.,* vol. AP-7, 1959, p. 32.
7 A. S. Gutman, "Modified Luneburg Lens," *J. App. Phys.,* vol. 25, 1954, p. 855.
8 G. Toraldo di Francia, "Spherical Lenses for Infrared and Microwaves," *J. App. Phys.,* vol. 32, 1961, p. 2051.
9 G. D. M. Peeler and D. H. Archer, "A Two-Dimensional Microwave Luneburg Lens," Nav. Res. Lab. Rep. 4115, 1953.
10 G. D. M. Peeler, K. S. Kelleher, and H. P. Coleman, "Virtual Source Luneburg Lenses," Nav. Res. Lab. Rep. 4194, 1953.
11 L. F. Culbreth, M. W. Long, and A. H. Schaufelberger, "Two-Dimensional Microwave Luneburg Lenses," ASTIA No. 72211, Georgia Institute of Technology, Atlanta, 1955.
12 C. H. Walters, "Surface-Wave Luneburg Lens Antennas," *IRE Trans. Antennas Propagat.,* vol. AP-8, 1960, p. 504.
13 R. F. Rinehart, "A Solution of the Rapid Scanning Problem for Radar Antennas," *J. App. Phys.,* vol. 19, 1948, p. 860.
14 F. G. Warren and J. E. A. Pinnel, "Tin Hat Scanning Antennas," Rep. 6 and 7, Cont. 2-1-44-4-3, RCA Victor Company, Montreal, 1951.
15 K. S. Kung, "Generalization of the Rinehart-Luneburg Lens," *J. App. Phys.,* vol. 25, 1954, p. 642.
16 R. C. Johnson, "Optical Scanners," in R. C. Hansen (ed.), *Scanning Antennas,* vol. 1, Academic Press, Inc., New York, 1964, chap. 3.
17 E. M. T. Jones, "Paraboloidal Reflector and Hyperboloid Lens Antennas," *IRE Trans. Antennas Propagat.,* vol. AP-2, 1954, p. 119.
18 F. C. Friedlander, "A Dielectric-Lens Aerial for Wide-Angle Beam Scanning," *J. IEE,* part IIIa, vol. 93, no. 4, 1946, p. 658.

19 F. S. Holt and A. Mayer, "A Design Procedure for Dielectric Microwave Lenses of Large Aperture Ratio and Large Scanning Angle," *IRE Trans. Antennas Propagat.,* vol. AP-5, 1957, p. 25.

20 R. L. Sternberg, "Successive Approximation and Expansion Methods in the Numerical Design of Microwave Dielectric Lenses," *J. Math. Phys.,* vol. 34, 1956, p. 209.

21 R. M. Brown, "Dielectric Lenses for Scanning," *Georgia Tech.–SCEL Symp. Scanning Antennas,* ASTIA No. AD132769, Georgia Institute of Technology, Atlanta, 1956.

22 B. A. Lengyel, "Physical Optics of Metal Plate Media, Part 1," Nav. Res. Lab. Rep. 3534, 1949.

23 N. I. Korman and J. R. Ford, "An Achromatic Microwave Antenna," *IRE Proc.,* vol. 38, no. 12, December 1950, p. 1445.

24 H. B. Devore and H. Iams, "Microwave Optics between Parallel Conducting Sheets," *RCA Rev.,* vol. 9, December 1948.

25 J. Ruze, "Wide Angle Metal Plate Optics," *IRE Proc.,* vol. 38, January 1950, p. 53.

26 G. D. M. Peeler and W. F. Gabriel, "Volumetric Scanning Antenna," *IRE Conv. Rec.: Antennas,* 1955.

27 S. B. Myers, "Parallel Plate Optics for Rapid Scanning," *J. App. Phys.,* vol. 18, 1947, p. 221.

28 W. Rotman, "A Study of Microwave Double Layer Pillbox, Part 3, Line Source Radiators," Air Force Cambridge Res. Cen. Rep. AFCRC-TR-54-102, July 1954.

29 W. Rotman and R. F. Turner, "Wide Angle Microwave Lens for Line Source Applications," *IEEE Trans. Antennas Propagat.,* vol. AP-11, 1963, p. 263.

30 H. N. Chait, "Wide Angle Scan Radar Antenna," *Electronics,* vol. 26, January 1953, p. 28; also C. N. Chait, "A Microwave Schmidt System," NRL Rep. 3989, May 14, 1952.

31 K. S. Kelleher and H. P. Coleman, "Off-Axis Characteristics of the Paraboloid Reflector," Nav. Res. Lab. Rep. 4088, 1952.

32 E. O. Houseman, Jr., "Millimeter Wave Polarization Twist Reflector," *IEEE AP-5 Symp. Dig.,* May 1979, p. 51.

33 T. Li, "A Study of Spherical Reflectors as Wide-Angle Scanning Antennas," *IRE Trans. Antennas Propagat.,* vol. AP-7, 1959, p. 223.

34 A. Ishimaru, "Double-Spherical Cassegrain Reflector Antenna," *IEEE Trans. Antennas Propagat.,* vol. AP-21, November 1973, p. 744.

35 R. C. Spencer, "Correction of Spherical Aberration by a Phased Line Source," *Nat. Electron. Conf. Proc.,* vol. 5, 1949, p. 320.

36 A. W. Love, "Spherical Reflecting Antennas with Corrected Line Source," *IRE Trans. Antennas Propagat.,* vol. AP-10, 1962, p. 529.

37 E. E. Altshuler, "Primary Pattern Measurements of a Line Source Feed for a Spherical Reflector," *IRE Trans. Antennas Propagat.,* vol. AP-10, 1962, p. 214.

38 L. M. LaLonde and D. E. Harris, "Aberration Correcting Line Source Feed for Arecibo Spherical Reflector," *IEEE Trans. Antennas Propagat.,* vol. AP-18, January 1970, p. 41.

39 C. J. Sletten and W. G. Mavroides, "A Method of Sidelobe Reduction," *Proc. Sidelobe Conf.,* Nav. Res. Lab. Rep. 4043, 1952, pp. 1–12.

40 W. Rotman, "Wide Angle Scanning with Microwave Double Layer Pillboxes," *IRE Trans. Antennas Propagat.,* vol. AP-6, 1958, p. 96.

41 N. Amitay, "Phase Aberration Correction Using Planar Array Feeds," *IEEE Trans. Antennas Propagat.,* vol. AP-20, January 1972, p. 49.

42 J. M. Flaherty and E. Kadak, "Early Warning Radar Antennas," *IRE Conv. Rec.,* part 1, 1958, p. 158.

43 K. S. Kelleher, "A New Microwave Reflector," *IRE Conv. Rec.,* part 2, 1953, p. 56.

44 G. D. M. Peeler and D. H. Archer, "A Toroidal Microwave Reflector," *IRE Conv. Rec.,* part 1, 1956, p. 242.

45 C. V. Robinson, *Radar System Engineering,* vol. 1, McGraw-Hill Book Company, New York, 1947, secs. 9-13–9-25.

46 K. S. Kelleher, "A Virtual Source Scanner," Nav. Res. Lab. Rep. 3957, 1952.

47 J. S. Foster, "A Microwave Antenna with Rapid Sawtooth Scan," *Can. J. Phys.*, vol. 36, 1958, p. 1652.

48 R. C. Honey and E. M. T. Jones, "A Mechanically Simple Foster Scanner," *IRE Trans. Antennas Propagat.*, vol. AP-4, 1956, p. 40.

49 K. S. Kelleher and H. H. Hills, "An Organ Pipe Scanner," Nav. Res. Lab. Rep. 4141, 1953.

50 H. B. Devore and H. Iams, "Microwave Optics between Parallel Conducting Sheets," *RCA Rev.*, vol. 9, December 1948.

51 W. F. Gabriel, "Some Waveguide Slot Scanners," *Third Symp. Scanning Antennas*, November 1950. Naval Research Laboratory report.

52 R. E. Honey et al., "Two Beam 16,000 Mc Modified Luneburg Lens Scanning System," ASTIA No. AD-17816, Georgia Institute of Technology, Atlanta,

53 J. S. Hollis and M. W. Long, "A Luneburg Lens Scanning System," *IRE Trans. Antennas Propagat.*, vol. AP-6, 1958, p. 21.

54 L. D. Breetz, U.S. Patent 2,595,186, Apr. 29, 1952.

55 G. D. M. Peeler and W. F. Gabriel, "A Volumetric Scanning GCA Antenna," *IRE Conv. Rec.*, part 1, 1955, p. 20.

56 K. Tomiyasu, "A New Annular Waveguide Rotary Joint," *IRE Proc.*, vol. 44, 1956, p. 548.

57 R. C. Johnson et al., "A Waveguide Switch Employing the Offset Ring-Switch Junction," *IRE Trans. Microwave Theory Tech.*, vol. MTT-8, 1960, p. 532.

Chapter 19

Frequency-Scan Antennas

James S. Ajioka

Hughes Aircraft Company

19-1 INTRODUCTION

A *frequency-scan antenna* may be defined as any antenna in which the direction of the radiated beam is controlled by changing the operating frequency. The antenna itself is completely passive and is by far the least costly antenna for inertialess beam scan. Simplicity, low cost, and high reliability are the virtues of frequency scan. For these reasons, frequency-scan radars far outnumber all other radars that have inertialess beam scannning. The U.S. Navy S-band AN/SPS-26 frequency-scan radar shown in Fig. 19-1 dates back to the mid-1950s. The AN/SPS-32 ultrahigh-frequency (UHF) frequency-scan radar and the AN/SPS-33 S-band 3-D track radar, which uses phase scan in azimuth and frequency scan in elevation, date back to 1960. In addition to these radars, the U.S. Navy AN/SPS-52 (Fig. 19-2) and AN/SPS-48 (Fig. 19-3) radars, both at S band, the Marine Corps AN/TPS-32 S-band radar, and the U.S. Army X-band mortar-locating radar, the AN/TPQ-36 radar, all utilize frequency scan.

Since 1950 much work has been done in the development of frequency-scan antennas. Most of this work has been government-sponsored, and information about it is contained in government and company reports and technical memoranda and thus is not readily available to the general public. At the present time probably the most extensive discussion on frequency-scan antennas was presented by N. A. Begovich[1]; an extensive bibliography is included. Begovich et al.[2] and Strumwasser et al.[3] are examples of patents resulting from work done in the 1950s. Other examples of early work are Dion,[4] Spradley,[5] Bystrom et al.,[6] Shelton,[7] and Ishimaru and Tuan.[8]

It is virtually impossible to come even close to an all-inclusive list of significant contributors to or references on frequency-scan antennas. For reasons of convenience and ease of access, a larger than is probably justified portion of the contents of this chapter is taken from the work done at Hughes Aircraft Company. However, it is felt that the basic concepts, designs, and practical implementation are representative of those of the industry as a whole.

19-2 THEORY OF OPERATION

The frequency-scan principle is very simple, in that it is based on the fact that the phase delay through a length of transmission line changes with frequency. For TEM lines the phase shift is directly proportional to frequency and line length. For dispersive transmission lines, it is proportional to the wave number $k_g = 2\pi/\lambda_g$ and line length. As stated before, most frequency-scan antennas fall in the category of traveling-wave antennas (see Chap. 9). For radiating elements fed serially along a transmission line, there will be a phase delay between elements. This interelement phase delay is linearly progressive and is a function of frequency. Hence, the beam will scan with frequency. A general schematic diagram of a series-fed frequency-scan antenna is shown in Fig. 9-4.

To form a beam in the direction θ_m from broadside, the following relationship must be satisfied:

$$kd \sin \theta_m = k_g s - 2m\pi \qquad (19\text{-}1)$$

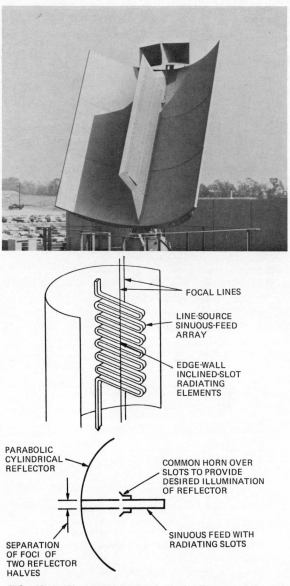

FIG. 19-1 Photograph and schematic diagram of the AN/SPS-26 frequency-scan antenna.

FIG. 19-2   AN/SPS-52 radar antenna.

where θ_m = angle from broadside of the beam maximum
$k = 2\pi/\lambda$, the free-space wave number
$k_g = 2\pi/\lambda_g$, the wave number for the feed line
d = distance between radiating elements
s = length of feed line between elements
m = integer (or odd half integer if the phase of the element couplers is alternately $0°$ and $180°$)

$$\sin \theta_m = \frac{1}{kd}(k_g s - 2m\pi) \tag{19-2}$$

Let λ_{gm} be defined as the feed-line wavelength at the frequency where s is m guide wavelengths long (which also defines the corresponding wave number k_{gm} as $k_{gm}s = 2m\pi$). At these frequencies, the right side of Eq. (19-2) becomes zero and the beam is at broadside ($\theta_m = 0$). With this definition, Eq. (19-2) becomes

$$\sin \theta_m = \frac{s\lambda}{d}\left(\frac{1}{\lambda_g} - \frac{1}{\lambda_{gm}}\right) \tag{19-3}$$

which is the usual form of the *frequency-scan equation*. The interelement phase delay $k_g s$ increases monotonically with frequency and cycles through multiples of 2π rad. This is the beam-controlling phase shift that causes a beam to scan.

FIG. 19-3 AN/SPS-48 radar antenna.

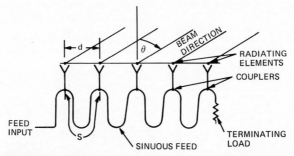

FIG. 19-4 General schematic diagram of a series-fed frequency-scan antenna.

Scan Bands

An mth beam can be identified with its broadside frequency corresponding to the broadside condition $k_{gm}s = 2m\pi$. By changing the frequency about its broadside frequency by an amount corresponding to one 2π interelement phase cycle, the mth beam will scan all space, which, depending on the element spacing, may include some invisible space as well as visible space. This frequency band is called the mth *scan band*. As the frequency is increased, a beam will scan from end fire toward the input end of the array, through broadside, to end fire toward the terminating load end of the array; this will be followed by another beam and another beam, etc.

Grating Lobes

As stated before, if the frequency band is great enough, there will be many beams at different frequencies, all pointing in the same direction in visible space. Similarly, if the interelement spacing d is great enough, there will be many beams in different directions in visible space, all at the same frequency. If one of these beams is the *desired beam,* all others are called *extraneous beams,* or *grating lobes.*

  Grating lobes are caused by the fact that for widely spaced elements, as the spatial angle is changed, there will be several directions in which the radiation from all the elements will add in phase. In other words, as the spatial angle is changed monotonically, the spatial phase delay from element to element will cycle through several multiples of 2π with corresponding grating lobes, just as when frequency changes, $k_g s$ cycles through multiples of 2π with corresponding scan bands. To differentiate between the origin of multiples of 2π the subscript n will be used for grating lobes instead of m, which was used for scan bands. From Eq. (19-1), let θ_n be the direction of the nth lobe and θ_{n+1} the direction of the adjacent grating lobes which are at the same frequency (same k, same k_g). Then

$$\sin \theta_n - \sin \theta_{n+1} = \frac{\lambda}{d} \qquad (19\text{-}4)$$

The beam spacing expressed in $\sin \theta$ is λ/d. This is shown in the diagram of Fig. 19-5.

  For only one beam to be in visible space at the same frequency requires $|\sin \theta_n| \le 1$ to ensure that θ_n is in visible space and $|\sin \theta_{n+1}| > 1$ to ensure that θ_{n+1} is not in visible space. The limit in element spacing for only one beam at a time to be in

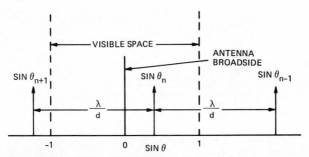

FIG. 19-5 Diagram of beam separation in $\sin \theta$.

visible space is that θ_n and θ_{n+1} are at least π rad apart ($\theta_n = \pi/2$ and $\theta_{n+1} = -\pi/2$). This leads from Eq. (19-4) to the well-known result $d \leq \lambda/2$ for grating-lobe suppression over all visible space.

In practice, it is usually not necessary to scan the beam a full $180°$ in any plane. Owing to aperture-impedance mismatch to free space for wide-angle scan and to a further decrease in gain due to projected aperture loss, planar antennas are seldom designed to scan more than a total of $120°$ in one plane. Many applications require only a limited amount of scan—up to $10°$ or so. In these limited-scan cases, it is desirable for economic reasons to space the elements as far apart as possible without excessive performance degradation.

From the diagram of Fig. 19-5 it can be seen that as beam n (the desired beam in visible space) is scanned to the right, the grating lobe θ_{n+1} will approach visible space from the left. When the grating lobe θ_{n+1} just reaches visible space ($\theta_{n+1} = -\pi/2$), Eq. (19-4) becomes $\sin \theta_n + 1 = \lambda/d$, and if θ_{n+1} is to be kept out of visible space, $\lambda/d > 1 + \sin \theta$. The diagram of Fig. 19-5 is symmetrical in that the beam spacing is λ/d to the right or to the left, so that if the beam θ_n is scanned to the left, beam θ_{n-1} will approach visible space from the right and reach visible space for the same angle off broadside ($-\theta_n$) of the desired beam as for scanning to the right (θ_n). This leads to the well-known condition that

$$\frac{d}{\lambda} < \frac{1}{1 + |\sin \theta_n|} \qquad (19\text{-}5)$$

for no grating lobes in visible space when the desired beam is scanned over a maximum scan range of $\pm \theta_n$ from broadside.

Since the beams are of finite width, even if the above criterion were satisfied, a portion of a grating lobe could appear in visible space although the beam peak is not in visible space. For low-sidelobe designs, this would not be desirable. Also, the aperture-impedance match to free space generally deteriorates rapidly when a grating lobe approaches free space (see Chap. 20). Therefore, it has become a rule of thumb to design the antenna so that the grating lobe is out of visible space by at least a null beamwidth (θ_{BW}) when the main beam is scanned to the maximum value. Equation (19-5) could be modified to

$$\frac{d}{\lambda} < \frac{\lambda}{1 + \sin \theta_{\text{BW}} + |\sin \theta_n|} \qquad (19\text{-}6)$$

19-3 SCAN BANDS AND MULTIPLE BEAMS

The greatest disadvantage of simple frequency scan is that the entire available bandwidth is used to steer the beam and that each direction in space is associated with a definite frequency. This allows a jammer to concentrate its energy over a narrow frequency band. The simple frequency scan also limits the instantaneous bandwidth of the system. The jammer problem could, in principle, be somewhat alleviated by employing more than one scan band, so that each direction in space would have as many frequencies as scan bands. This can be done by extending the tunable bandwidth of the system to include other scan bands or by lengthening the interelement feed-line lengths sufficiently so that there are multiple scan bands in the same nominal frequency band. These methods are not practical because extending the tunable bandwidth enough to cover several scan bands requires broadband components such as very-wide-bandwidth transmitters, receivers, and antenna components such as couplers and slot radiators.

By increasing the feed-line length s, the instantaneous bandwidth is severely restricted because the beam-scan rate with frequency is increased to a point at which not all the antenna beams corresponding to the different frequency components in a short pulse will point in the same direction. That is, the antenna beam pattern is "smeared out" in angular space, thereby broadening the effective beamwidth of the antenna and reducing its gain and angular resolution. The antenna is a narrow bandpass filter in that the frequency spectrum of the pulse radiated toward a target is modified from the spectrum entering the antenna. The spectrum is narrowed by the weighting applied to the various frequency components by their corresponding frequency-scanned beams. The frequency component corresponding to the beam scanned in the direction of the target is most heavily weighted, and each of the other frequency components is reduced (in the direction of the target) by the amount by which its *antenna pattern* is scanned off from the direction of the target.

On receive, the spectrum of the reflected pulse from the target is further modified in the antenna by the same transformation. There is a loss in gain, and consequently in radar range, because of the smearing out of the antenna pattern or, alternatively, because of the loss in power contained in the spectral components filtered out by the antenna. The power in the filtered-out frequency components is either reflected back into space or absorbed in the various terminating loads in the antenna. The modifying of the frequency spectrum results in the pulse being smeared out in time; thus range accuracy and range resolution are reduced. These and transient effects are discussed by Bailin,[9] Enenstein,[10] and Begovich.[1]

A more practical method for achieving greater frequency agility, due to E. C. DuFort and R. F. Hyneman,* combines the features of scan bands with the features of grating lobes so that any of several grating lobes can be selected as the desired beam while suppressing all others. This method uses a multiple-beam-forming network in conjunction with multiple interleaved arrays. This scheme has been practically implemented for two beams in the Hughes AN/SPS-33 shipborne radar antenna. The number of beams is equal to the number of interleaved arrays. Figure 19-6a shows an example of four interleaved arrays. The theory of this technique is described in the following paragraphs.

Suppose that there are q interleaved arrays, so that the spacing between radiating elements in the composite array is d but the spacing between elements fed by any one of the interleaved feed lines is qd. In the example of Figure 19-6a, $q = 4$. The spacing d is chosen so that the composite array does not have extraneous beams. However, each of the individual arrays with elements spaced qd will have q beams or grating lobes in the visible region of space corresponding to q values of n in Eq. (19-4). Now the entire composite array can be considered as a q-element array with the component arrays as the elements. In the example, $q = 4$. It is a four-element array of component arrays. The elements of the component arrays are marked by 1s for the first array, 2s for the second array, etc. The antenna pattern of the whole composite

---

*Private communication (1960) with E. C. DuFort. Both DuFort and Hyneman are with the Hughes Aircraft Company, Ground Systems Group, Fullerton, Calif.

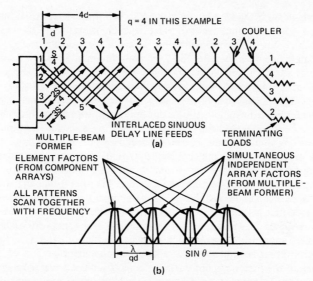

FIG. 19-6 Multiple beams and interleaved scan bands.

array is a product of the *element factor,* which is the pattern of a single-component array (each with q beams) and an array factor which is a pattern of q point sources spaced d apart.

Refer to Fig. 19-6b. This *array factor* (the broad pattern) picks out the desired beam while suppressing all the others. The maximum of the array factor coincides with that of the desired beam and has nulls at the peaks of all the other beams. This can be seen by applying Eq. (19-1) first to the element factor and then to the array factor. For the element factor which has element spacing qd, Eq. (19-1) becomes

$$kqd \sin \theta_m = k_g s - m2\pi$$

or

$$\tag{19-7}$$

$$kd \sin \theta_m = k_g \frac{s}{q} - \frac{m}{q} 2\pi$$

and for the array factor which has element spacing d, Eq. (19-1) becomes

$$kd \sin \theta_m = k_g s' - m'2\pi \tag{19-8}$$

For the element factor and the array factor to frequency-scan together, it is required that s', the incremental line length between linear arrays, be equal to s/q, and this also makes $m' = m/q$. Now the array factor whose element spacing is d has only one beam in visible space, while the element factor has q beams or grating lobes corresponding to q values of n in Eq. (19-4). By superposing the proper linear progressive phase to the array factor, the array factor could be superposed exactly on any one of the grating lobes corresponding to a particular value of n in Eq. (19-4). This would pick out that beam and suppress all others. If an additional linear phase of $[2\pi(q - 1)]/q$ over the array aperture were provided, the array factor would then pick out the beam corresponding to $n + 1$, etc., up to $n = q$.

An orthogonal beam-forming network such as a Butler matrix will supply all the proper phase gradients simultaneously. That is, each beam port of the multiple beam former will pick out its corresponding beam, and the frequency range to scan that beam over all space may be called the *corresponding scan band*. To do this requires that the array factor beams be separated by the same amount as the grating lobes (which is λ/qd in sin θ space) and also that there be nulls at the peaks of all the other grating lobes. This condition is automatically satisfied by an orthogonal beam-forming network (see Chap. 20). An optical-type multiple-beam former (multiple-beam reflector or lens) would not satisfy the beam-spacing criterion for all frequencies because the beam spacing would be independent of frequency.

In the preceding paragraph the frequency range to scan a beam over all space was referred to as a scan band corresponding to that beam. It should be noted that these scan bands overlap or are interleaved, so that the total frequency range to make full use of all the scan bands is only $(1 + 1/q)$ times as great as for the simple single-scan-band case first discussed. This requirement contrasts with the q times as much bandwidth required for multiple scan bands using frequency extension. Also, the instantaneous bandwidth is decreased at most by a factor of $q/(q + 1)$ as compared with $1/q$ for multiple scan bands using a longer interelement feedline of length qs.

Because of the factor s/q and m/q in Eq. (19-7), the term *fractional scan bands* has been suggested. However, the term *overlapped* or *interleaved scan bands* is probably more descriptive because it is related to the interleaved or overlapped array structure and, also, if a scan-band diagram of sin θ versus frequency were plotted, the scan bands would be seen to overlap.

19-4   MONOPULSE IN THE FREQUENCY-SCAN PLANE

For greater angular accuracy in tracking radars, monopulse antennas have long been employed. However, they have not been used extensively in frequency-scan antennas, probably because subpulse beam switching is easily implemented and is adequate for most applications. In subpulse beam switching, a single pulse is divided into two parts different in frequency by a small amount so that the two beams corresponding to the two frequencies overlap. Amplitude comparison between the two beams is used for improved angular accuracy. Although in principle monopulse can be implemented in the frequency-scan plane, there is an inherent physical asymmetry in end-fed traveling-wave arrays, which include the simplest series-fed frequency-scan antennas. The practical implications of the asymmetry of series-fed arrays will be discussed later. Some representative monopulse schemes are described in the following paragraphs.

Parallel-Fed Monopulse Array

Figure 19-7 shows a parallel-fed frequency-scan array fed by a conventional monopulse feed. While schematically this is quite simple and straightforward, it has the disadvantages of complexity and cost. Also, the feed-line attenuation varies from element to element owing to the different lengths of delay line. This circumstance must, of course, be taken into account in the design of the feed network to achieve the desired aperture distributions for the sum and difference modes. However, because of its inherently greater symmetry, this configuration may be the preferred design for applications in which an extremely high degree of monopulse tracking accuracy is required.

Tandem-Fed Monopulse Array

Figure 19-8 depicts a tandem-sinuous-feed technique for monopulse in the frequency-scan plane. This technique can generate two beams as in a Blass network with two frequency-scan sinuous delay lines in tandem as shown. The beam squint between the monopulse beams can be achieved by employing a progressive phase shift in the lines

between the tandem feeds or by making the interelement delay line length S slightly different for the two sinuous feeds. Alternatively, one feed may be designed for the monopulse sum and the other for the monopulse difference. As a whole, the two feeds do not couple owing to the orthogonality of even and odd distributions corresponding to the sum and difference modes. The end feeding destroys the physical symmetry that is inherent in center feeding. Symmetry is desirable for the monopulse difference operation to achieve a purely odd difference distribution with the resulting deep, distinct null in the difference pattern.

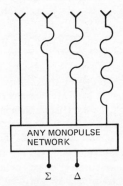

FIG. 19-7 Parallel-fed frequency-scan monopulse antenna.

Dummy-Snake Monopulse Technique

To illustrate some of the severe problems due to the symmetry of serially fed antennas for monopulse operation, a *dummy-snake monopulse technique* will be described.

With reference to Fig. 19-9, the array halves will be arbitrarily designated as first half and second half. The array halves must, of course, have the same feed design

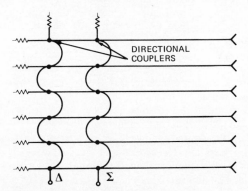

FIG. 19-8 Tandem-fed frequency-scan monopulse antenna.

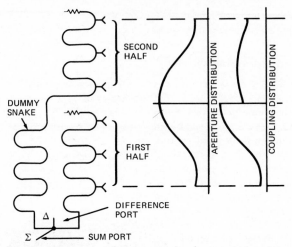

FIG. 19-9 Dummy-snake frequency-scan monopulse antenna.

so that they frequency-scan synchronously. In addition, for monopulse application they must accurately phase-track as the frequency is changed. To accomplish this, a dummy snake must be added to the second half as shown in the figure. The dummy snake must have the same phase delay as the antenna feed.

There are certain practical difficulties in using this method. First, the dummy snake must have the identical phase versus frequency characteristics of the feed in the first half. A sinuous feed loaded by radiating elements does not have exactly the same phase versus frequency characteristics as an unloaded line, and care must be exercised to design the dummy snake properly so that it will accurately phase-track the feed of the first array. A more difficult problem lies in the design of the aperture distributions of the two halves. It is apparent that the coupling distributions of the two halves must be radically different, especially for tapered-aperture distributions.

For the first half, the coupling of the first element must be very low because here the *aperture* distribution is lowest and the power in the feed line is highest, and the coupling must be nearly unity at the end of the first half because here the aperture distribution is highest and the power in the feed line is lowest. For the second half, the first element requires the highest power at the point where the power in the feed line is also highest. For a tapered-aperture distribution the conductance distribution in the second half is more nearly uniform because of the natural taper due to the attenuation caused by radiation and ohmic loss.

The general nature of the aperture and coupling distributions is shown in Fig. 19-9. Because of the widely differing conductance distributions of the two halves, it is difficult to maintain the desired symmetry of the aperture distribution of the halves for good monopulse operation. The efficiency of the two halves may also be difficult to match. Because of the differing coupling distributions, the power wasted into the terminating loads would be difficult to keep equal over a broad frequency range. It would require a very high range of coupling values, i.e., a very low value at the input end of the first half and a very high value at the output of the first half. Another disadvantage is that independent sum and difference aperture distributions are not possible. That is, if the sum distribution is optimized for low sidelobes, the difference patterns will have a step discontinuity at the center, resulting in a high sidelobe difference pattern.

19-5 PRACTICAL FREQUENCY-SCAN ANTENNAS

Frequency-scan antennas, like other antenna systems, can take many forms. In the frequency-scan plane, the feeding structure uses sinuous delay lines, but for pencil beams, beam collimation in the orthogonal plane can use linear arrays, reflectors, lenses, multiple-beam structures, etc. Figure 19-1 shows the AN/SPS-26 radar antenna, which is a waveguide sinuous-feed slot array feeding a parabolic cylindrical reflector. Figure 19-2 shows the AN/SPS-52 radar antenna, which is a waveguide sinuous feed feeding an array of traveling-wave slot arrays (a schematic diagram is given later in Fig. 19-12). Antennas that frequency-scan in one plane with multiple simultaneous beams in the other plane can be implemented by using stacks of planar multiple-beam antennas with multiple sinuous feeds. The multiple-beam structure can be a stack of multiple-beam lenses such as parallel-plate dielectric or geodesic lenses, constrained lenses, or any multiple-beam-forming network.

The AN/SPS-33 radar antenna is a good example of a practical phase-frequency-scan antenna with dual scan bands. A schematic diagram of the antenna is shown in Fig. 19-10.

The feed in the phase-scan plane is a folded pillbox[11] with monopulse capability. A folded pillbox is a two-layer parallel-plate power divider operating in the TEM

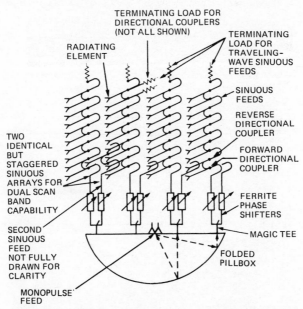

FIG. 19-10 Schematic diagram of a phase-frequency-scan antenna with dual scan bands.

mode, with a monopulse feed in the feed layer of the pillbox and the exit or aperture in the other layer of the pillbox. The two layers are separated by a common plate septum and joined by a 180° *E*-plane bend that has a parabolic contour in the plane of the plates. This feed has a very low loss (typically, a few tenths of a decibel for a 1° aperture) and can handle 2 MW peak and 4 kW average without pressurization at S band.

With simple open-ended waveguide or cavity-type feeds the aperture illumination for sidelobe levels below −30 dB is easily achieved over more than a 10 percent bandwidth.[12] The output of the pillbox has a stepped transformer in parallel plate to effect an impedance match from the parallel-plate region of the pillbox to an array of waveguides in the *H* plane. Each of these output waveguides is power-split 1 to 2 by an *H*-plane folded magic (hydrid) T. Each of the outputs from a magic T feeds a high-power ferrite phase shifter, which, in turn, feeds two interleaved sinuous feeds that feed the open-ended-waveguide radiating elements via directional couplers. The two interleaved sinuous feeds provide the two-scan-band capability.

The desired scan band can be chosen by the relative phasing between adjacent ferrite phase shifters. If adjacent phase shifters are in phase (except for the linear progressive phase for phase-scanning the beam), one scan band is picked out, and if the adjacent phase shifters are alternately phased 0° and 180°, the other scan band is picked out. It should be noted that the interleaved sinuous feeds are not physically interleaved but are displaced in the plane orthogonal to the frequency-scan plane. For a large array of sinuous feeds, this lateral displacement is inconsequential. In fact, it is an advantage because the resultant array element lattice becomes triangular, which reduces the number of radiating elements required for the same volume of beam coverage compared with a rectangular lattice.[13,14]

Figure 19-11 is an isometric drawing of a section of the sinuous feed. The sinuous

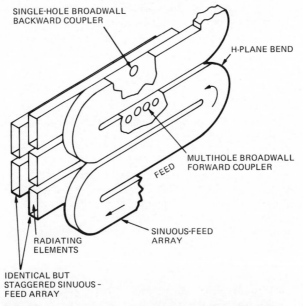

FIG. 19-11 Staggered sinuous feeds for dual scan bands.

feed is a waveguide of reduced height feeding open-ended-waveguide radiating elements with directional couplers in the broad walls of the waveguides. Both forward and reverse couplers are required.The forward couplers are multihole couplers, and the backward couplers are single Bethe hole couplers with a capacitive button on the wall opposite the hole to improve directivity. The coupling coefficients are controlled by the hole sizes. In each straight section of waveguide between the coupling holes and the radiating apertures, there is an array of capacitive buttons which correct a small amount of phase that is a function of the coupling coefficient. These designs gave essentially constant coupling in amplitude and phase over greater than a 10 percent frequency range at S band.

The monopulse feed to the folded pillbox is a multimode single-aperture feed which is fed by the two waveguide outputs of a folded *H*-plane magic T. The common multimode aperture is dielectrically loaded with a solid dielectric plug. The dielectric plug fulfills the dual purpose of providing one of the parameters for controlling the relative mode amplitudes and phases of the multimode aperture and of serving as a pressure seal. The pillbox itself is not pressurized. The pillbox illumination from the feed is such as to result in better than −32-dB sum-pattern sidelobes in the phase-scan plane of the antenna and −20-dB sidelobes for the difference patterns.

19-6   DESIGN CONSIDERATIONS

To illustrate some of the problems associated with frequency-scan-antenna design, an antenna of the design generally depicted in Fig. 19-12 will be discussed. This is a two-dimensional planar pencil-beam antenna that is designed to frequency-scan primarily in one plane, say, the elevation plane, with mechanical rotation for azimuth coverage. Such a design basically comprises a sinuous delay line with numerous 180° bends, couplers from the sinuous feed to branch lines, and linear arrays that collimate the beam in the plane orthogonally to the sinuous feed. Each of these parts and their effect on antenna performance are discussed in turn.

Sinuous Feed

The sinuous delay line must produce a large amount of interelement phase delay with a minimum of ohmic loss and be able to handle high power. For frequencies at and above S band, a rectangular waveguide operating in the dominant TE_{10} mode is the usual choice. At lower frequencies, in which the waveguide is too large and cumbersome, a large coaxial line operating in the TEM mode is used for high-power application. For low-power application, stripline or microstrip might be appropriate. The frequency-scan antenna for the AN/SPS-32 long-range search radar, operating in the 200-MHz band, uses 3⅛-in dielectric bead-supported coaxial line for the delay line to achieve minimum loss and high-power capability. For purposes of discussion, the waveguide sinuous feed will be used.

Bends

For compactness and a low voltage standing-wave ratio (VSWR), waveguide sinuous feeds wrapped in the *E* plane with 180° *E*-plane bends are most often used, and 180° *E*-plane circular bends with small inductive irises or posts having a VSWR less than

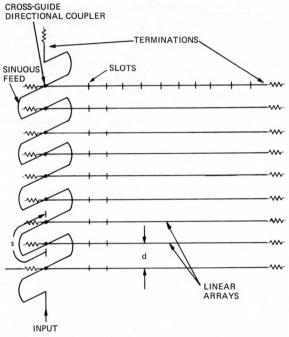

FIG. 19-12   Frequency-scanning sinuous feed feeding an array of traveling-wave linear arrays.

1.03 over a 10 percent frequency band are commercially available in most standard waveguide bands. As will be discussed later, at the frequency in which the beam is at broadside the bend VSWR should be less than 1.02:1 to produce an input VSWR on the order of 1.5:1 in a typical feed design. For the lower frequencies in which coaxial line is used, the 180° bends are made up of two well-matched 90° bends that are spaced $\lambda/4$ apart at the broadside frequency so that their reflections cancel.

Couplers

There are a number of choices for couplers from the sinuous delay line to the branch lines. The couplers can be nondirective three-port reactive couplers such as pure series or pure shunt couplers, nondirective matched couplers, or directional four-port couplers. Nondirective couplers are simple and inexpensive to fabricate. Two typical designs of nondirective waveguide shunt couplers are shown in Figs. 19-13 and 19-14. Figure 19-13 shows resonant inclined slots in the narrow dimension of the sinuous-feed waveguide which couple to the branch waveguides. The degree of coupling is controlled by the angle of slot inclination. Figure 19-14 shows nonresonant longitudinal slots in the narrow wall of the sinuous-feed waveguide which couple to the branch waveguides. An inductive iris or post is used to resonate the coupler. The degree of coupling is controlled by the length of the nonresonant slot. Figure 19-15 shows offset shunt slots in the bends of the sinuous feed.

The slot serves the dual purpose of coupler and radiating element. The degree of

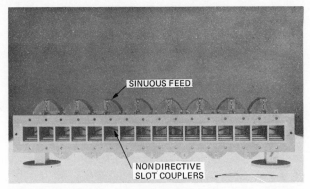

FIG. 19-13 Waveguide sinuous feed with reactive nondirective inclined-slot couplers to branch guides.

coupling is determined by the lateral offset of the slot from the centerline of the bend. Simple series or shunt couplers are inexpensive to fabricate, but they have very poor performance at or very near broadside frequencies. At broadside frequencies, the couplers are an integral number of half-guide wavelengths apart so that for shunt couplers the admittances are all in parallel and for series couplers the impedances are all in series. As a result, there is a high input VSWR, and the antenna power aperture distribution is equal to the admittance or impedance distribution, which for efficient arrays is quite asymmetrical. The smaller the power dissipated in the terminating load of the feed, the greater the asymmetry.[4] This asymmetrical distribution has a large odd component which does not contribute to antenna gain but causes increased sidelobes and null filling.

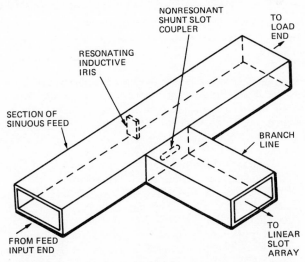

FIG. 19-14 Section of waveguide sinuous feed with nondirective, nonresonant shunt-slot couplers to branch guides.

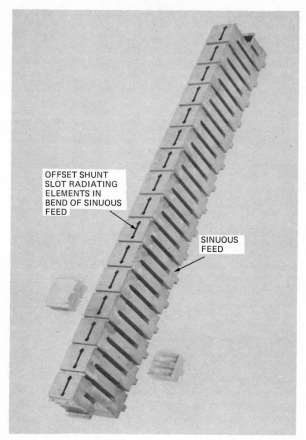

OFFSET SHUNT
SLOT RADIATING
ELEMENTS IN
BEND OF SINUOUS
FEED

SINUOUS
FEED

FIG. 19-15   Waveguide sinuous feed with offset shunt-slot radiators.

The curve of input VSWR versus frequency has a sharp peak at resonance and is generally low elsewhere, with only small minor lobes. It has an obvious analogy to an antenna pattern in which the independent variable is spatial angle instead of frequency. For tapered antenna distributions, the main beam broadens and the sidelobes decrease with the degree of tapering. Likewise, for tapered coupling distributions, the VSWR resonance curve broadens and the VSWR sidelobes decrease with the degree of tapering. Also, the larger the aperture distribution, the narrower the antenna beam, and the longer the feed line, the narrower the resonance curve.

Because of these resonance effects, the sinuous-feed bends and couplers must have extremely low reflections; otherwise operation at or near resonance is precluded. However, in many radar applications, beam scan through broadside is required, and the VSWR must be kept to a tolerable low level, on the order of 1.5:1 or less.

To allow operation at broadside frequencies, Kurtz and Gustafson[15] investigated the possibility of impedance-matching nondirective couplers as seen from the feed end of the coupler. This technique resulted in considerable success. The coupler used was a shunt coupler of the design shown in Fig. 19-14. The admittance in the main line at a resonant coupler is $1 + g$, where g is the coupler conductance which is small compared to 1. It is clearly seen on a Smith chart that going toward the generator slightly under $\frac{3}{8}\lambda_g$, the unit conductance circle is intercepted with a small capacitive susceptance approximately equal in magnitude to the coupler conductance g. This small capacitive susceptance is canceled by a thin inductive post near the wall opposite the coupler. The $\frac{3}{8}\lambda_g$ spacing is far enough from the coupling aperture that the coupler conductance is not affected. By this matching technique an intolerably high VSWR at the resonant frequenty was reduced to less than 1.4:1.

Without matched couplers, an antenna pattern taken slightly off the resonant frequency when reflections are still quite high produces a *reflected lobe* that interferes with the main antenna beam to cause severe pattern degradation. The reflected lobe occurs at an angle symmetrically opposite from broadside with respect to the main beam. Since the frequency is close to the resonant frequency, the reflection beam and the main beam overlap so as to interfere with each other, causing an asymmetrical pattern with a split-beam appearance. After the couplers have been matched, this effect virtually disappears.

For extremely low (-30-dB or lower) sidelobe designs, directional couplers with the isolated ports of the couplers terminated in well-matched loads are preferred because spurious reflections in the sinuous feed, regardless of their origin, will be absorbed in the terminating loads of the couplers and will not radiate into space to degrade the antenna pattern. A directional coupler is a four-port junction that is inherently matched in all ports. Figure 19-16 shows a typical directional coupler in a section of a sinuous feed. The first-order reflected power in the sinuous feed will split; some of it will go into the terminating load of the directional coupler, and the rest will continue back down the feed line with part of that going into the terminating load of the preceding directional coupler, etc., so that in general very little reflected power gets back to the feed input to affect the input VSWR appreciably. Because of the directivity of the directional coupler none of this first-order reflection is radiated. Also, broadband directional-coupler designs with a wide range of coupling values are readily available. As an example, the design of a waveguide sinuous feed using cross-slot–cross-guide couplers is discussed. An isometric view of a waveguide sinuous feed using

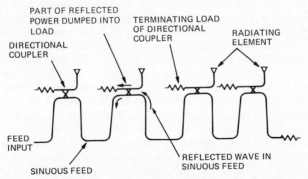

FIG. 19-16 Schematic diagram of a sinuous feed using directional couplers.

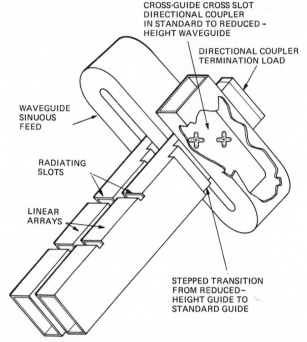

FIG. 19-17   Waveguide sinuous feed with cross-slot–cross-guide directional couplers feeding waveguide slot arrays.

cross-slot–cross-guide directional couplers is shown in Fig. 19-17. Figure 19-18 gives the coupling coefficient versus cross-slot parameters.

Couplers, in general, introduce phase shift in the main line past the coupler and in the coupled branch lines. Figure 19-19 gives phase shift past the coupler and phase shift in the branch lines as a function of coupling coefficient for a cross-slot–cross-guide directional coupler. The coupling is from standard WR-384 S-band waveguide to a reduced-height guide of the same width. This design was shown in Fig. 19-17. In this coupler, the phase shifts in the main and the coupled arms are phase delays that vary monotonically with the degree of coupling. Both phase shifts can be compensated for by shortening the interelement line length S in the sinuous feed accordingly. The phase correction in the sinuous feed has an additional benefit in that it breaks up the exact periodicity in the locations of the $180°$ bends in the sinuous feed and thereby improves the VSWR of the sinuous feed at and near the broadside frequency. To see the effect of bend mismatch on the input VSWR, calculations were made for a 60-element sinuous feed using cross-slot–cross-guide directional couplers designed for a -30-dB-sidelobe $(\bar{n} = 4)$ Taylor distribution for various VSWR values of an individual bend. Coupler phase compensation was included in the calculations. The results showed that to achieve an input VSWR less than 1.5:1 at the resonant frequency required a bend VSWR of less than 1.02:1. This is easily achieved with standard waveguide bend designs. For frequencies off resonance the bend VSWR can be as high as 1.3:1 and still maintain a feed input VSWR of less than 1.5:1.

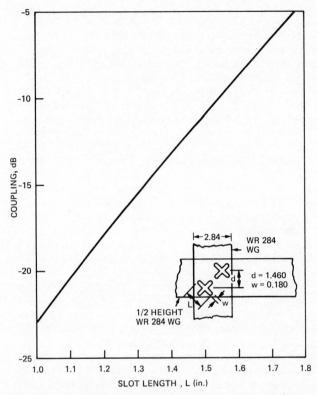

FIG. 19-18 Coupling coefficient versus cross-slot parameters.

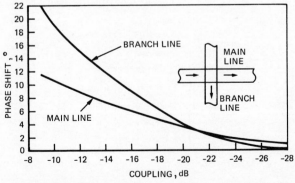

FIG. 19-19 Phase shifts of cross-slot–cross-guide directional couplers.

Linear Array

To collimate the beam in the plane orthogonal to the plane of the sinuous feed, waveguide linear slot arrays were used. The rod-excited, polarization-pure slot elements used noninclined transverse slots in the narrow wall of the waveguide.[16] Ordinarily these slots do not radiate because they are parallel to the radio-frequency current in the wall of the waveguide. The slots are excited by thin rods or wires inside the waveguide as shown in Fig. 19-20. Metallic irises have also been used to excite waveguide slots that otherwise do not radiate.[17,18] Iris techniques for narrow-wall slots were investigated but were not used in this application because of the high cost of fabrication and because of the high power arcing at the irises. The rod-excited design was tested to 40-kW peak and 1-kW average power in WR-284 S-band waveguide without pressurization.

For highly efficient very-low-sidelobe designs, this technique has a decided advantage over the more conventional inclined slot because it introduces no cross-polarization. For inclined slots there is a cross-polarized component proportional to the tangent of the inclination angle. This results in cross-polarization lobes in the antenna pattern. Since most two-dimensional slot arrays are designed so that the cross-polarization of adjacent elements and the cross-polarization of adjacent arrays cancel on the principal planes, the cross-polarization lobes occur in the off-principal planes. These relatively high cross-polarization lobes defeat the objective of very-low-sidelobe designs. Also, the more efficient the array, the larger the cross-polarization. The reason for this is that for lower power dissipation in the terminating load (and thus higher efficiency), higher conductance values are required,[4,19] which means greater slot inclination and hence greater cross-polarization.

To suppress this cross-polarization, closely spaced metallic strips or baffles that are perpendicular to the electric field of the principal polarization are used. The spacing between the baffles is much less than a half wavelength, so the baffles form waveguide regions that are below cutoff to the cross-polarization. The AN/SPS-52 radar antenna shown in Fig. 19-2 uses the cross-polarization-suppressing baffles shown in Fig. 19-21. These baffles add considerable weight and cost to the antenna. For these reasons, the polarization-pure transverse rod-excited slots in the narrow wall of the waveguide are used to design a −40-dB-sidelobe array.

Refer to Fig. 19-20. The degree of coupling of the rod-excited transverse slot is controlled by the distances of the ends of the rods from the inside corner of the wave-

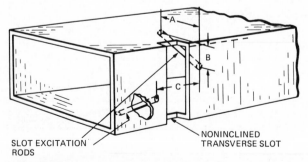

FIG. 19-20 Rod-excited transverse slot in narrow wall of waveguide.

FIG. 19-21 Inclined-slot arrays with cross-polarization-suppressing baffles.

guide adjacent to the slot as shown in the figure. The rods couple the slot to the electromagnetic wave in the waveguide. The rods are tilted so that they have a component of electric field from the TE_{10} mode in the waveguide that is parallel to the rods but in opposite directions for the upper and lower rods. Thus the rods act like a two-wire transmission line feeding the slot.

The coupling to the slot increases with an increase in the dimension A or B, or both. The slot coupling is also affected by the spacing C between the two rods. The coupling decreases with an increase in C. This dimension C is held constant at a convenient value of about $\lambda_g/8$ with the slot centered so that reflections from the wires tend to cancel. For convenience and ease of data taking, the dimension A is held constant, leaving B the only variable. With A held constant at 1.45 in (36.83 mm), a wide range of coupling can be achieved. For arrays with a large number of elements, the maximum coupling value can be kept under 0.1 with 5 percent of the power dissipated into the terminating load.

Figure 19-22 gives the slot-coupling coefficient versus the B dimension of Fig. 19-20. There are two curves, the dotted curve showing initial values and the solid curve the final values after an iterative design procedure to be explained later. The slot structure including the rods is made resonant at the design frequency by adjusting the overall length of the slots—in this case, the amount of slot cut back into the broad wall. This resonant (zero-phase-shift) condition depends on the slot coupling. Figure 19-23 gives slot coupling and slot cutback versus slot position in the array.

The design of a two-dimensional antenna aperture composed of an array of linear traveling-wave arrays is, in general, quite complex. The aperture must be impedance-matched to free space over a wide frequency band and over a wide range of scan angles. Aperture matching of phased arrays has received a great deal of attention both analytically and experimentally since the early 1960s. (See Chap. 20 of this handbook for detailed discussions and references.) Two-dimensional traveling-wave slot arrays are particularly difficult to analyze for the following reasons: (1) The feeding network,

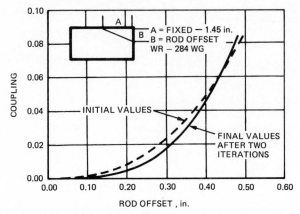

FIG. 19-22 Slot coupling versus rod offset for rod-excited transverse slot in narrow wall.

the element couplers, and the radiating elements cannot be treated separately because they often share a common physical and functional structure. For waveguide slot arrays, the waveguide with its slots is the feed network, and the slots serve the dual function of coupler and radiating element. (2) Owing to the serial feeding, there is a cumulative effect, in that errors in preceding elements affect the performance of succeeding elements. This has a great effect on aperture distribution and results in the loss of power into the terminating loads.

As with all phased-array apertures, for good impedance match to space mutual coupling between elements must be taken into account. There is mutual coupling between the elements exterior to the slots and also coupling interior to the waveguide. There is also strong mutual coupling between linear arrays that is a function of the scan angle, which, of course, for frequency scan is also a function of frequency. This

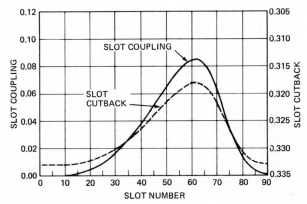

FIG. 19-23 Slot cutback and slot coupling versus slot number for −50-dB Taylor $\bar{N} = 8$ aperture distribution.

mutual coupling affects the coupling coefficients of each slot so that the aperture distribution and power lost in the terminating loads can be drastically different from the intended design if the mutual-coupling effects are not properly taken into account. This has been dramatically verified by taking a linear slot array that has very good performance by itself and arraying two such arrays, three such arrays, etc. It has been found that the measured aperture distributions, radiation patterns, and power lost in the termination, in general, vary widely from each other. It has been found that after five or more linear arrays have been arrayed, the aperture distribution and percent power into the loads become nearly constant. This has led to the conclusion that at least five linear arrays are required to approximate the environment of a large array.

Because of these problems and the difficulty of a priori prediction of performance for extremely-low-sidelobe designs, a semiempirical iterative design procedure was developed at Hughes Aircraft Company, Ground Systems Group, Fullerton, California. An outline of this procedure is given by an example of a -40-dB-sidelobe-level design that meets all MIL specification requirements and is fabricated by using production personnel and production methods. The rod-excited, polarization-pure slot array was used.

The design goal:

41-dB gain

-40-dB sidelobes

$2° \times 1°$ beamwidth

15 percent bandwidth at S band

Design parameters:

Taylor -50-dB $\bar{n} = 8$ aperture distribution

90 slots in linear arrays

98 linear arrays

15- by 20-ft overall aperture

Design Procedure

1   Design a linear array by standard techniques such as that due to V. T. Norwood.[19] In this example, slot data have been scaled from a previous design at a slightly different frequency band using standard WR284 S-band waveguide. The new design used thin-wall WR284 waveguide. Even approximately good slot data can be used as a starting point.

2   Fabricate a minimum of five linear arrays to make a small array, using any convenient feed which allows the required radio-frequency measurements.

3   Using a near-field probe facility,[20] measure the amplitude and phase along the slots of each linear array over the desired frequency range and over the desired scan angles. Measure the input power and the power into the terminating load. If the basic design concept is reasonably sound, data measured over frequency ranges and scan angles should not be widely different. Array spacing and radome

thickness and spacing from the slot aperture can be used to minimize the sensitivity to frequency and scan angle. From the measured input power, aperture distribution, and power into the load, the *actual* slot coupling and phase data of each slot in the very nearly actual environment, including all mutual-coupling effects, can be determined. Various weighted averages over frequency and scan angle can be constructed, depending on the system-performance weighting. For example, the best performance at the scan angle corresponding to the horizon scan may be more heavily weighted.

4   These new coupling data take into account all mutual-coupling effects in the actual array environment and encompass the entire range of coupling values required. A new and more accurate curve of coupling coefficient and phase versus the *B* dimension can be constructed for various frequencies and scan angles. By using these new data, a new set of at least five linear arrays is constructed and tested with the near-field setup, and another set of slot data is collected.

5   The process is repeated until the desired result is obtained. It has been found that two iterations are sufficient to converge to the optimum design. In this example, -40-dB-maximum sidelobes with 5 percent or less power dissipated in the terminating loads were actually achieved over a 12 percent frequency range in several production antennas. Figure 19-24 presents typical patterns over the frequency range. All the design goals were met with two iterations.

This near-field technique can also be used in the design of modifications to the antenna that might greatly alter the slot coupling. Modification to the aperture for fragmentation protection is an example.

19-7   MILLIMETER-WAVE APPLICATION

Electronic beam scanning using phase shifters for millimeter waves has not been practical because of the extreme difficulty in fabricating the phase shifters. In principle, ferrite phase shifters are feasible for millimeter waves, but the small size and the extremely high tolerance requirements make them impractical. However, because of their simplicity, frequency-scan antennas are practical for millimeter-wave application.

Frequency-scanning waveguide slot arrays at 60 GHz have been developed at Hughes Aircraft Company. These arrays use slot design data scaled from the X-band frequency range. Figure 19-25 shows an edge-wall inclined-slot array with cross-polarization-suppressing cavities to give longitudinal polarization. Figure 19-26 shows a broad-wall offset slot array for transverse polarization. Borowick, Bayka, Stern, and Babbit of the U.S. Army Electronics Research and Development Command, Fort Monmouth, New Jersey,[21] have developed a frequency-scan linear array in dielectric waveguide operating in the HE_{11} mode at 35 GHz. Slots cut in the dielectric waveguide perturb the guided wave to cause radiation, and the slots become the radiating elements. Since this design does not allow phase reversal, the slots are spaced one guide wavelength apart at the frequency corresponding to the broadside scan angle. However, because of the high-dielectric-constant material used in the dielectric waveguide, the guide wavelength is less than a free-space half wavelength, and there are

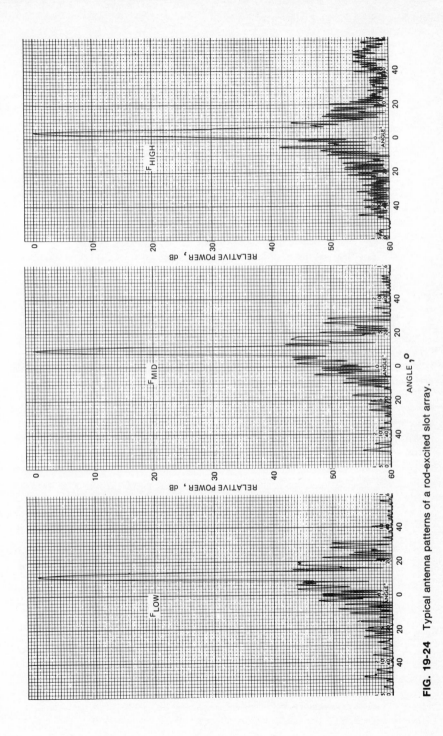

FIG. 19-24 Typical antenna patterns of a rod-excited slot array.

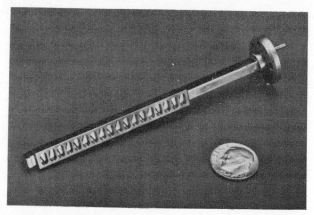

FIG. 19-25 60-GHz frequency-scan narrow-wall inclined-slot array for longitudinal polarization with cross-polarization-suppressing cavities.

no grating lobes even with the λ_g spacing. Figure 19-27 shows the geometry of the dielectric-waveguide slot array. Dielectric waveguides using metallic strips instead of slots have also been used as frequency-scanning arrays.[22,23]

Both the metallic-waveguide slot arrays and the dielectric-waveguide slot and metal-strip arrays are very simple in structure, are easily fabricated, and perform very well. For greater scan versus frequency, the metal-waveguide slot array can be put in the form of a sinuous feed. This may be somewhat more difficult with the dielectric array because of the problem at the bends. However, this may not be necessary for many applications because the scan versus frequency is enhanced by the square root of the dielectric constant.

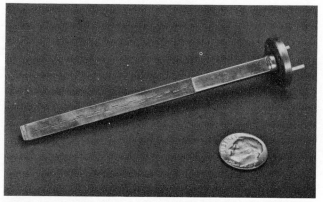

FIG. 19-26 60-GHz frequency-scan broad-wall offset slot array for transverse polarization.

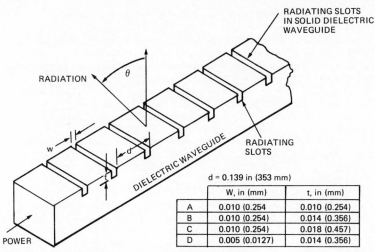

RADIATING SLOTS
IN SOLID DIELECTRIC
WAVEGUIDE

RADIATION

θ

w

DIELECTRIC WAVEGUIDE

RADIATING
SLOTS

t

POWER

d = 0.139 in (353 mm)

	W, in (mm)	t, in (mm)
A	0.010 (0.254	0.010 (0.254)
B	0.010 (0.254)	0.014 (0.356)
C	0.010 (0.254)	0.018 (0.457)
D	0.005 (0.0127)	0.014 (0.356)

FIG. 19-27 35-GHz dielectric-waveguide slot array.

REFERENCES

1 N. A. Begovich, in R. C. Hansen (ed.), *Microwave Scanning Antennas,* vol. III, Academic Press, Inc., New York, 1966, chap. 2.

2 N. Begovich et al., "Radar Scanning System," U.S. Patent 3,017,630, January 1962.

3 E. Strumwasser et al., "Frequency-Sensitive Rapid-Scanning Antenna," U.S. Patent 3,039,097, June 1962.

4 A. Dion, "Nonresonant Slotted Arrays," *IRE Trans. Antennas Propagat.,* vol. AP-6, 1958, pp. 360–365.

5 J. L. Spradley, "A Volumetric Electrically Scanned Two-Dimensional Microwave Antenna Array," *IRE Conv. Rec.,* part I, 1958, p. 204.

6 A. Bystrom, R. V. Hill, and R. E. Metter, "Ground Mapping Antennas," *Electronics,* vol. 33, May 1960.

7 P. Shelton, "Applications of Frequency Scanning to Circular Arrays," *IRE Wescon Conv. Rec.,* part I, 1960, p. 83.

8 A. Ishimaru and H. Tuan, "Theory of Frequency Scanning," *IRE Trans. Antennas Propagat.,* vol. AP-10, 1962, p. 144.

9 L. L. Bailin, "Fundamental Limitations of Long Arrays," Tech. Memo. 330, Hughes Aircraft Company, Culver City, Calif., October 1953.

10 N. H. Enenstein, "Transient Build-Up of the Antenna Pattern in End Fed Linear Arrays," *IRE Conv. Rec.,* part II, 1953, p. 49.

11 L. J. Chu and M. A. Taggart, "Pillbox Antenna," U.S. Patent 2,638,546, May 12, 1953.

12 J. S. Ajioka, "Development of an Integral MK X IFF Antenna for Low Frequency Radar," U.S. Navy Electron. Lab. Rep., January 1955.

13 E. C. DuFort and M. B. Rapport, "Spacing Elements for Planar Arrays," internal memorandum, Hughes Aircraft Company, Fullerton, Calif., March 1958.

14 E. C. Sharp, "Triangular Arrangement of Elements Reduces the Number Needed," *IRE Trans. Antennas Propagat.,* vol. AP-9, 1961, pp. 126–129.

15 L. A. Gustafson, "S-band Two-Dimensional Slot Array," Tech. Memo. 462, Hughes Aircraft Company, Culver City, Calif., March 1957.

16 J. S. Ajioka and D. M. Joe, "Rod Excited Transverse Slot on Narrow Wall of Waveguide," Patent Disclosure 79144, Hughes Aircraft Company, Fullerton, Calif.

17 R. Tang, "A Slot with Varying Coupling and Its Application to a Linear Array," *IRE Trans. Antennas Propagat.*, vol. AP-8, 1960, pp. 97–101.

18 D. Dudley, "An Iris-Excited Slot Radiator in the Narrow Wall of Rectangular Waveguide," *IRE Trans. Antennas Propagat.*, vol. AP-9, July 1961, pp. 361–364.

19 V. T. Norwood, "Note on a Method for Calculating Coupling Coefficients of Elements in Antenna Arrays," *IRE Trans. Antennas Propagat.*, vol. AP-3, May 1955.

20 A. E. Holley, "Near Field Antenna Measurement System," final report, Cont. DAAB07-77-C-0587, Hughes Aircraft Co., January 1982.

21 J. Borowick, R. A. Stern, R. W. Babbit, and W. Bayka, "Inertialess Scan Antenna Techniques for Millimeter Waves," DARPA Millimeter Wave Conf., U.S. Army Missile Command, Redstone Arsenal, Ala., Oct. 20–22, 1981.

22 K. L. Klohn, R. E. Horn, H. Jacobs, and E. Friebergs, "Silicon Waveguide Frequency Scanning Linear Array," *IEEE Trans. Microwave Theory Tech.*, vol. MTT-26, no. 10, October 1978.

23 S. Kobayashi, R. Lampe, R. Mittra, and S. Ray, "Dielectric Rod Leaky-Wave Antennas for Millimeter-Wave Applications," *IEEE Trans. Antennas Propagat.*, vol. AP-26, no. 5, September 1981.

Chapter 20

Phased Arrays

Raymond Tang
Richard W. Burns

Hughes Aircraft Company

In this chapter, the theoretical and practical considerations involved in the design of phased-array antennas are given. The intent is to provide the antenna designer with an overview of the basic design concepts of phased arrays and the means to determine the proper antenna parameters for a given radar application.

Basic Concept of Phased Arrays

Basically, the phased-array antenna is composed of a group of individual radiators which are distributed and oriented in a linear or two-dimensional spatial configuration. The amplitude and phase excitations of each radiator can be individually controlled to form a radiated beam of any desired shape (directive pencil-beam or fan-beam shape) in space. The position of the beam in space is controlled electronically by adjusting the phase of the excitation signals at the individual radiators. Hence, beam scanning is accomplished with the antenna aperture remaining fixed in space without the involvement of mechanical motion in the scanning process; in other words, this is inertialess scanning.

Advantages of Phased Arrays

The capability of rapid and accurate beam scanning in microseconds permits the radar to perform multiple functions either interlaced in time or simultaneously. An electronically steered array radar is able to track a large number of targets and illuminate some of these targets with radio-frequency (RF) energy and guide missiles toward them. It can perform complete hemispherical search with automatic target selection and hand over to tracking. It may even act as a communication system, directing high-gain beams toward distant receivers and transmitters. Complete flexibility is possible. Search and track rates may be adjusted to best meet the particular situation within the limitations set by the total use of time. The antenna beamwidth may be changed electronically by means of phase spoiling to search certain areas more rapidly but with less gain. Frequency agility can be achieved through changing the frequency of transmission at will from pulse to pulse or, with coding, within a pulse. Very high powers may be generated from a multiplicity of generators distributed across the aperture. Electronically controlled phase-array antennas can give radars the flexibility needed to perform all the various functions in a way best suited to the specific task at hand. The functions can be programmed rapidly and accurately with digital beam-steering computers.

Overview of the Chapter

The first part of this chapter describes the theory of phased-array antennas, with emphasis on the basic relationships between the array performance, the excitation coefficients (amplitude and phase) of each radiator, and the physical parameters of the antenna (see Sec. 20-2). These relationships are the design formulas which an antenna designer can use to determine the required array parameters that satisfy a given set of radar system requirements. After determining the array parameters such as the number of required radiators, the spacing between the radiators, and the size of the radiating aperture, the next step in the design would be the selection of the type of radiator, phase shifter, and beam-forming feed network to implement physically the required excitations at the radiating aperture. A detailed discussion of the various types of radiators, phase shifters, and feed networks that are commonly used is given in Secs. 20-3, 20-4, and 20-5 respectively. Corresponding to the various types of phase shifters and feed networks, the frequency-bandwidth characteristics of the phased-array antenna are described in Sec. 20-6.

20-2 THEORY OF PHASED ARRAYS

We shall begin the discussion of array theory with the simplest configuration possible: an array of isotropic radiators (radiating elements) equally spaced along a line (linear in a geometric sense). It will be seen that the more complex array theory can be framed as an extension of this simple case. For example, two-dimensional arrays are often constructed of interconnected linear arrays. From the linear-array discussion, we will then proceed to the theory of two-dimensional arrays. The theory of arrays of isotropic radiators will then be extended and modified to apply to arrays of real radiators.

To simplify and condense the material to be presented, the following assumptions will be adhered to in this discussion:

1 For practical radar purposes, targets of interest will lie in the far-field region of the antenna, commonly defined by ranges R, satisfying

$$R \geq \frac{2L^2}{\lambda}$$

where L is the largest dimension of the antenna and λ is the operating wavelength. In this region, the antenna pattern is insensitive to range except for a scale factor of $1/R^2$ in power. This scale factor is usually ignored.

2 The reciprocity theorem[1] will be used extensively to justify analysis of array far fields from either the transmitting or the receiving viewpoint as convenience dictates and usually without explicit mention. By virtue of this theorem, the pattern will be the same in either case if no nonreciprocal devices are used. When such devices are used, one can usually perform pattern analyses which neglect their presence.

3 For simplicity, interest will be restricted to the responses of arrays to continuous-wave (CW) signals.

Radiation Pattern of Linear Arrays

The theory of linear arrays is described in detail by Silver,[1] Schelkunoff,[2] and Allen.[3] Let us consider the elementary array of Fig. 20-1, consisting of N isotropic radiators, equally spaced at a distance d apart. On receive, if a plane wave is incident upon the array from a direction making an angle θ with the array normal, the current in the nth element will be of the form

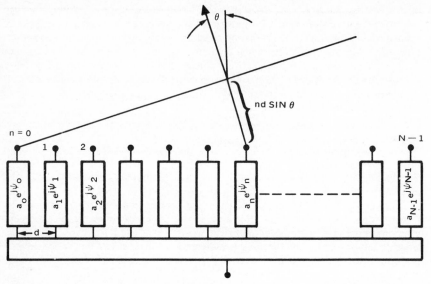

FIG. 20-1   Basic linear-array configuration.

$$i'_n = Ae^{jnkd \sin \theta} \qquad \textbf{(20-1)}$$

where A is a complex constant related to the instantaneous amplitude and phase of the plane wave and k is the wave number:

$$k = \frac{2\pi f}{c} = \frac{2\pi}{\lambda}$$

where f = operating frequency
c = velocity of light

Equation (20-1) shows that the current in the nth element leads that in the $(n + 1)th$ element by a phase shift given by $\Delta\psi = kd \sin \theta$. This phase shift corresponds to the difference in time of arrival τ of the plane wavefront of $\tau = d/c \sin \theta$. If we place a control element behind each radiator, as indicated in Fig. 20-2, with a transfer coefficient for the nth element given by

$$\frac{i''_n}{i'_n} = a_n e^{j\psi_n}$$

where a_n and ψ_n are the *real* current gain and phase shift of the control element respectively, the summing network produces an output

$$E_a(\theta) = \sum_{n=0}^{N-1} a_n e^{j(\psi_n + nkd \sin \theta)} \qquad \textbf{(20-2)}$$

where we now neglect the constant A of Eq. (20-1). This relationship gives the response of the array of Fig. 20-2 to a signal arriving from a direction θ in terms of the set of a_n's and ψ_n's. The set of coefficient a_n is usually called the *array-amplitude taper,* while the ψ_n's are called the *phase taper.* The expression of Eq. (20-2) is called the *array factor.* To combine the received signals from all the radiators in phase to produce a maximum response in the scan direction of θ_0, the ψ_n's must have the form

$$\psi_n = -nkd \sin \theta_0 \tag{20-3}$$

This expression shows that the required phase taper across the array aperture is a linear taper (constant phase differential between adjacent radiators). On transmit, when the phases of the control elements such as phase shifters are set to the phase taper of Eq. (20-3), the signals radiated from all the radiators will add up in phase to produce a main beam in the direction of θ_0. Hence, the array factor of Eq. (20-2) is the same for both transmit and receive. Substituting Eq. (20-3) into Eq. (20-2), we have for the array factor

$$E_a(\theta) = \sum_{n=0}^{N-1} a_n e^{jnkd(\sin \theta - \sin \theta_0)} \tag{20-4}$$

For the special case of a uniformly illuminated array, $a_n = 1$ for all n, the array factor for an N-element array becomes

$$E_a(\theta) = \frac{\sin \left[N\pi \dfrac{d}{\lambda} (\sin \theta - \sin \theta_0) \right]}{N \sin \left[\pi \dfrac{d}{\lambda} (\sin \theta - \sin \theta_0) \right]} \tag{20-5}$$

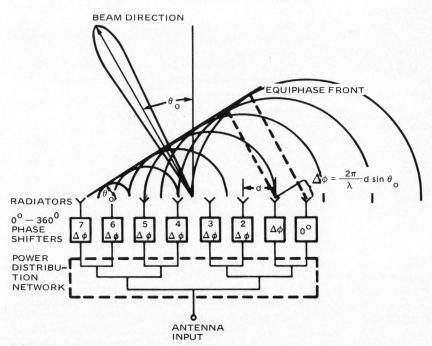

FIG. 20-2 Beam-steering concept using phase shifters at each radiating element.

Element Spacing to Avoid Grating Lobes

The array factor of Eq. (20-4) can also be expressed in terms of the variable $\nu = \sin \theta$ as follows:

$$E_a(\nu) = \sum_{n=0}^{N-1} a_n e^{jnkd(\nu-\nu_0)} \qquad \textbf{(20-6)}$$

where the beam-point direction ν_0 is related to the differential phase $\Delta\psi$ by $\Delta\psi = -kd\nu_0$. It is obvious that $E_a(\nu)$ and $E_a(\theta)$ are related by a one-to-one mapping in the region $|\nu| \leq 1$, which is often referred to as the *visible space* corresponding to the real angles of θ. It is also apparent that $E_a(\nu)$ is a periodic function of ν of period

$$\frac{2\pi}{kd} = \frac{1}{d/\lambda} = \frac{\lambda}{d}$$

and that Eq. (20-6) is in the form of a Fourier-series representation, which is readily analyzable and easy to visualize. The maxima of $E_a(\nu)$ occur whenever the argument of Eq. (20-6) is a multiple of 2π; i.e., $kd(\nu - \nu_0) = 2i\pi$, where $i = 0, \pm 1, \pm 2, \dots$, or

$$\nu_i - \nu_0 = \frac{i}{d/\lambda}$$

When $\nu_i = \nu_0$ or $i = 0$, this maximum is generally referred to as the *principal lobe* or main beam, and the other maxima are known as the *grating lobes* from the corresponding phenomena with optical gratings. In the design of phased arrays, it is imperative that the grating lobes be eliminated within the visible space since these lobes reduce the power in the main beam and thus reduce the antenna gain. This means that the element spacing d must be chosen to avoid the grating lobes over the range of ν from -1 to $+1$. When the main beam is scanned to ν_0, the closest grating lobe to the visible space is located at $\nu_i = \nu_0 - 1d/\lambda$ (see Fig. 20-3). This grating lobe will just appear in visible space (at end-fire direction of the array) when $\nu_0 - 1d/\lambda = -1$, or

$$\frac{d}{\lambda} = \frac{1}{1 + \sin |\theta_0|}$$

Thus, the element-spacing criterion stated in terms of the desired maximum scan angle θ_{0max} is

$$\frac{d}{\lambda} < \frac{1}{1 + \sin |\theta_{0max}|} \qquad \textbf{(20-7)}$$

If the equality sign is used in Eq. (20-7), we might actually incur rather than avoid the grating lobe. For narrow beams, however, the error committed by applying Eq. (20-7) is frequently small enough (particularly when the actual patterns of the real elements instead of the isotropic radiators are included) to justify using it rather than the more cumbersome exact formula which accounts for the finite width of the grating lobe. In actual practice, the value of d/λ is reduced by a few percent to assure that the grating lobe is positioned beyond end fire. Figure 20-4 presents a plot of Eq. (20-7).

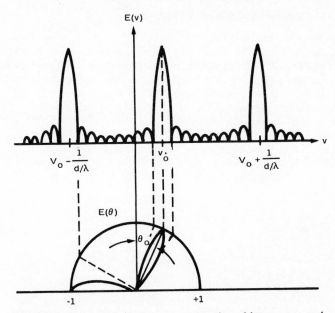

FIG. 20-3 Grating-lobe location as a function of beam scan and element spacing.

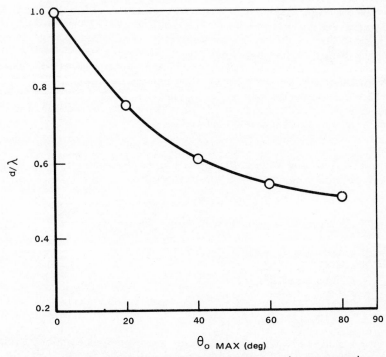

FIG. 20-4 Maximum allowable element spacing versus maximum scan angle.

Amplitude Tapers for Sidelobe Control

The far-field-pattern properties of most frequent concern to the array designer are the array sidelobe level, array gain, and beamwidth. All these properties depend upon the amplitude taper applied to the excitation coefficients a_n's of Eq. (20-4). In general, a stronger amplitude taper across the array aperture has the effect of reducing the sidelobes at the expense of increased beamwidth and reduced aperture efficiency (antenna gain). There are four principal classes of amplitude tapers which will be outlined and compared: (1) powers of cosine on a pedestal, (2) Taylor distributions,[4] (3) the modified $\sin \pi u/\pi u$ of Taylor distribution,[5] and (4) Dolph-Chebyshev distribution.[6]

The *powers-of-cosine distributions* are the class of tapers that resemble closely those provided by an optical feed system such as the primary illumination pattern of a feed horn illuminating a parabolic-reflector antenna. The continuous taper is of the form

$$a(x) = h + (1 - h) \cos^m \frac{\pi x}{L} \qquad \textbf{(20-8)}$$

where x = distance measured from center of array aperture
L = total length of aperture
h = normalized pedestal height

The most frequently used distributions of this class are the cosine ($m = 1$) and the cosine-squared ($m = 2$). The lowest first sidelobes achievable with the cosine are -23 dB when $h = 0$. This yields an illumination efficiency of 0.81 and a half-power beamwidth (HPBW) coefficient of $69°$.

The cosine-squared taper reduces the peak sidelobe level of -32 dB when $h = 0$. In general, the gain decreases and the beamwidth increases with increasing n. After the highest lobe, the sidelobes decay monotonically. The first sidelobe level, aperture efficiency, and HPBW for varous cosine tapers with $h = 0$ are shown in Table 20-1.

The Taylor distribution provides a certain number of equal sidelobes symmetrically located on both sides of the main beam with the amplitudes of the remaining sidelobes decreasing monotonically. For a given design sidelobe level, the Taylor distribution provides a narrower beamwidth than that of the cosine distributions. The Taylor amplitude distribution is of the form

$$A(x,A,\bar{n}) = \frac{1}{2\pi} \left\{ F(0,A,\bar{n}) + 2 \sum_{n=1}^{\bar{n}-1} F(n,A,\bar{n}) \cos \frac{n\pi x}{L} \right\} \qquad \textbf{(20-9)}$$

$$F(n,A,n) = \frac{[(n-1)!]^2 \prod_{m=1}^{\bar{n}-1} \left(1 - \frac{n^2}{\sigma^2[A^2 + (m - \frac{1}{2})^2]}\right)}{(\bar{n}-1+n)!\,(\bar{n}-1-n)!}$$

where x = distance from center of aperture
L = total length of aperture
$A = 1/\pi$ arc cosh R
R = design sidelobe voltage ratio
$\sigma = \dfrac{\bar{n}}{\sqrt{A^2 + (\bar{n} - \frac{1}{2})^2}}$
$\bar{n}$ = number of equiamplitude sidelobes adjacent to main beam on one side

TABLE 20-1 Pattern Characteristics for (cos)^m Distributions

m	First sidelobe level, dB	Aperture efficiency	Half-power beamwidth, °	Main-beam null width, °
0	13.2	1	50.4 λ/L	57.3 λ/L
1	23.0	0.810	68.76λ/L	85.95λ/L
2	32.0	0.667	83.08λ/L	114.6 λ/L
3	40.0	0.575	95.12λ/L	143.2 λ/L
4	48	0.515	110.59λ/L	171.9 λ/L

Corresponding to this amplitude distribution, the beamwidth of a one-wavelength source is given by

$$\beta = \sigma\beta_0 \tag{20-10}$$

where

$$\beta_0 = \frac{2}{\pi}\sqrt{(\text{arc cosh } R)^2 - \left(\text{arc cosh } \frac{R}{\sqrt{2}}\right)^2}$$

For an array with aperture length L, the beamwidth is given by

$$\text{Beamwidth} = \sigma\beta_0\frac{\lambda}{L} \quad \text{rad} \tag{20-11}$$

Beamwidth as a function of the design sidelobe level for various values of $\bar{n}$ is given in Table 20-2.

The modified sin $\pi u/\pi u$ taper of Taylor distribution produces monotonically decreasing sidelobes. The radiation pattern has a main lobe of adjustable amplitude and a sidelobe structure similar to that of the radiation from a uniformly illuminated source. The amplitude distribution is of the form

$$a(x) = \frac{1}{2\pi} J_0\left[j\pi\beta\sqrt{1 - \left(\frac{2x}{L}\right)^2}\right] \tag{20-12}$$

where x = distance measured from center of aperture
L = total length of aperture
J_0 = zero-order Bessel function of first kind
β = parameter fixing ratio R of main-beam amplitude to amplitude of first sidelobe by $R = 4.60333 \sinh \pi\beta/\pi\beta$

The array pattern corresponding to the above amplitude distribution is given by

$$E_a(u) = \frac{\sin \pi\sqrt{u^2 - \beta^2}}{\sqrt{u^2 - \beta^2}} \tag{20-13}$$

where $u = (L/\lambda)\sin\theta$

The beamwidth and aperture efficiency as a function of the design sidelobe ratio are shown in Table 20-3. Typical amplitude distributions corresponding to radiation patterns of 10-, 20-, and 30-dB sidelobe levels are shown in Fig. 20-5.

The taper of Dolph-Chebyshev distribution[6,7,8] is optimum in the sense that it

TABLE 20-2 Design Sidelobe Level and Beamwidth for Taylor Distributions

Design sidelobe level, dB	R (sidelobe voltage ratio)	$\dfrac{180\beta_0}{\pi}$ (degrees)	A^2	Values of the parameter								
				$\bar{n}=2$	$\bar{n}=3$	$\bar{n}=4$	$\bar{n}=5$	$\bar{n}=6$	$\bar{n}=7$	$\bar{n}=8$	$\bar{n}=9$	$\bar{n}=10$
15	5.62341	45.93	0.58950	1.18689	1.14712	1.11631	1.09528	1.08043	1.06969	1.06112	1.05453	1.04921
16	6.30957	47.01	0.64798	1.17486	1.14225	1.11378	1.09375	1.07491	1.06876	1.06058	1.05411	1.04887
17	7.07946	48.07	1.6267	1.13723	1.11115	1.11115	1.09216	1.07835	1.06800	1.06001	1.06367	1.04852
18	7.94328	49.12	0.77266	1.15036	1.13206	1.10843	1.09050	1.07724	1.06721	1.05942	1.05328	1.04815
19	8.91251	50.15	0.83891	1.13796	1.12676	1.10563	1.08879	1.07609	1.06639	1.05880	1.05273	1.04777
20	10.00000	51.17	0.90777	1.12549	1.12133	1.10273	1.08701	1.07490	1.06554	1.05816	1.05223	1.04738
21	11.2202	52.17	0.97927		1.11577	1.09974	1.08518	1.07367	1.06465	1.05750	1.06172	1.04697
22	12.5893	53.16	1.05341		1.11009	1.09668	1.08329	1.07240	1.06374	1.05682	1.05119	1.04654
23	14.1254	54.13	1.13020		1.10430	1.09352	1.08135	1.07108	1.06280	1.05611	1.05064	1.04610
24	15.8489	55.09	1.20965		1.09840	1.00029	1.07934	1.06973	1.06183	1.05538	1.05007	1.04565
25	17.7828	56.04	1.29177		1.09241	1.08598	1.07728	1.06834	1.06083	1.05463	1.04948	1.04518
26	19.9526	56.97	1.37654		1.08632	1.08360	1.07517	1.06690	1.05980	1.05385	1.04888	1.04669
27	22.3872	57.88	1.46395		1.08015	1.08014	1.07300	1.06543	1.05874	1.05305	1.04826	1.04420
28	25.1189	58.78	1.55406			1.07661	1.07078	1.06392	1.05765	1.05223	1.04762	1.04368
29	28.1838	59.67	1.64683			1.07300	1.06851	1.06237	1.05653	1.05139	1.04696	1.04316
30	31.6228	60.55	1.74229			1.06934	1.06619	1.06079	1.05538	1.05052	1.04628	1.04262
31	35.4813	61.42	1.84044			1.06561	1.06382	1.05916	1.05421	1.04963	1.04559	1.04206
32	39.8107	62.28	1.94126			1.06182	1.06140	1.05751	1.05300	1.04872	1.04488	1.04149
33	44.6684	63.12	2.04472				1.05893	1.05581	1.05177	1.04779	1.04415	1.04091
34	50.1187	63.96	2.15092				1.05642	1.05408	1.05051	0.04684	1.04341	1.04031
35	56.2341	64.78	2.25976				1.05386	1.05231	1.04923	1.04587	1.04264	1.03970
36	63.0957	65.60	2.37129				1.05126	1.05051	1.04792	1.04487	1.04186	1.03907
37	70.7946	66.40	2.48551					1.04868	1.04658	1.04385	1.04107	1.03843
38	79.4328	67.19	2.60241					1.04681	1.04521	1.04282	1.04025	1.03777
39	89.1251	67.98	2.72201					1.04491	1.04382	1.04176	1.03942	0.03711
40	100.0000	68.76	2.84428					1.04298	1.04241	1.04068	1.03808	1.03643

TABLE 20-3 Design Sidelobe Level, Beamwidth, and Aperture Efficiency for Modified sin $\pi u / \pi u$ Distributions

Sidelobe ratio, dB	R (sidelobe voltage ratio)	B	B^2	$180\beta/\pi$, beamwidth of a one-wavelength source, °	β/β_0, ratio of beamwidth to ideal beamwidth	η, aperture efficiency
0	1.00000	$-i\,0.812825$	-0.660684	33.307	1.1626	0.3452
5	1.77828	$-i\,0.686524$	-0.471315	40.000	1.1597	0.7110
10	3.16288	$-i\,0.459645$	-0.211274	46.620	1.1559	0.9635
15	5.62341	0.355769	0.126572	52.884	1.1514	0.9931
20	10.00000	0.738600	0.545530	58.659	1.1464	0.9330
25	17.7828	1.02292	1.046365	63.938	1.1410	0.8626
30	31.6228	1.27616	1.628584	68.775	1.1358	0.8014
35	56.2341	1.51363	2.291076	73.232	1.1305	0.7509
40	100.000	1.74148	3.032753	77.378	1.1254	0.7090

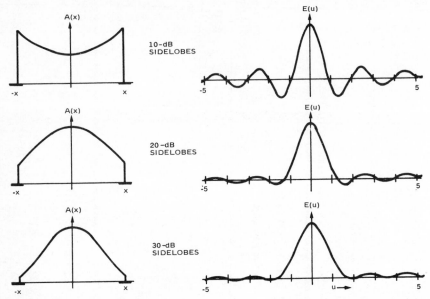

FIG. 20-5 Amplitude distributions and array patterns for three representative sidelobe levels.

yields the narrowest beamwidth for a given sidelobe ratio, and vice versa. It is not optimum in the terms of aperture efficiency for a given sidelobe level. A special property of this taper is that it yields an array function with equal-amplitude sidelobes, as shown in Fig. 20-6. This fact and its effect on the amplitude taper lead to some secondary considerations regarding the use of this taper. Since the sidelobes do not decay, the percentage of power in the main beam varies with the number of elements N in the array for a given sidelobe level. Hansen[9] has shown that for this reason, in order to maintain high efficiency, large Dolph-tapered arrays *must* use low-sidelobe tapers.

An amplitude taper which produces the array pattern of Fig. 20-6a is shown in Fig. 20-6b. The peaked-edge excitation currents are typical of a large array and high-sidelobe design (a 30-dB-sidelobe taper would have edge excitation currents approximately the same as neighboring elements). Such peak distributions are prone to error effects that arise in edge elements of arrays using real radiators. Hence, this distribution is not commonly employed in practice.

Array-Pattern Directivity of Linear Arrays

Array directivity is defined as the ratio of power density per unit solid angle at the peak of the main beam to the average power radiated per unit solid angle over all space. From this definition, it follows that the directivity of the main beam at angle θ_0 is given by

$$D(\theta_0) = \frac{|E(\theta_0)|^2}{\dfrac{1}{4\pi} \displaystyle\int_{\text{all space}} |E(\theta)|^2 \, d\Omega} \qquad (20\text{-}14)$$

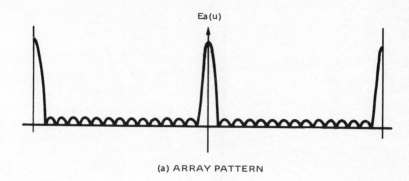

(a) ARRAY PATTERN

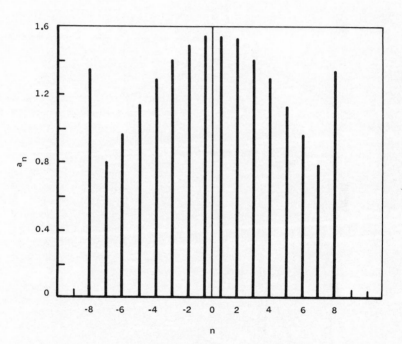

(b) ILLUMINATION COEFFICIENTS

FIG. 20-6 Illumination coefficients and array pattern for 20-dB Dolph taper applied to a 16-element array.

where $d\Omega \doteq \cos \theta \ d\theta \ d\phi$. For a linear array of isotropic radiators arranged as shown in Fig. 20-7, $E(\theta)$ is independent of ϕ. Therefore, $\int |E|^2 \ d\Omega = 2\pi \int |E|^2 \cos \theta \ d\theta$. Using $\nu = \sin \theta$ and $d\nu = \cos \theta \ d\theta$, we have

$$D(\nu_0) = \frac{2|E(\nu_0)|^2}{\displaystyle\int_{-1}^{+1} |E(\nu)|^2 \ d\nu} \tag{20-15}$$

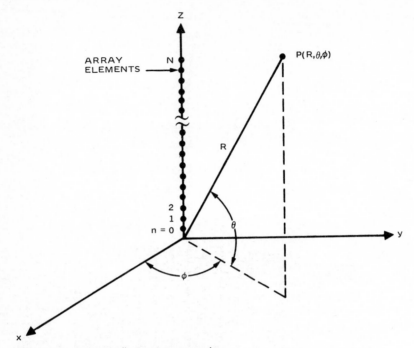

FIG. 20-7 *N*-element linear-array geometry.

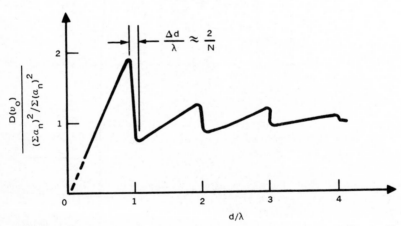

FIG. 20-8 Variation of directivity $D(v_0)$ of a linear array of isotropic radiators with element spacing d/λ. (*After Ref. 3.*)

Using Eq. (20-6) and integrating yield

$$D(\nu_0) = \frac{\left| \sum_{n=0}^{N-1} a_n \right|^2}{\sum_{n=0}^{N-1} \sum_{m=0}^{N-1} a_m a_n \dfrac{\sin 2\pi(d/\lambda)(n-m)}{2\pi(d/\lambda)(n-m)}} \qquad (20\text{-}16)$$

The denominator of Eq. (20-16) prevents a simple analysis of the directivity. Only when $2d/\lambda$ is an integral number is the result simplified:

$$D(\nu_0) = \frac{\left(\sum_{n=0}^{N-1} a_n \right)^2}{\sum_{n=0}^{N-1} (a_n)^2}$$

However, a detailed analysis of Eq. (20-16) shows that the directivity varies with element spacing approximately as indicated in Fig. 20-8 for beam shapes that concentrate most of the power in the main lobe.[10] For spacings less than λ, for which only one principal lobe exists in visible space when the main beam is at broadside, we can write the simple result

$$D(\nu_0 = 0) \approx \frac{2d}{\lambda} \frac{\left(\sum_{n=0}^{N-1} a_n \right)^2}{\sum_{n=0}^{N-1} (a_n)^2} \qquad (20\text{-}17)$$

Letting

$$\frac{\left(\sum_{n=0}^{N-1} a_n \right)^2}{\sum_{n=0}^{N-1} (a_n)^2} = \eta N \qquad (20\text{-}18)$$

where η is defined as the aperture efficiency, we can write

$$D(\nu_0 = 0) \approx \frac{2\eta N d}{\lambda} \qquad (20\text{-}19)$$

For a given N and d/λ, η alone determines the directivity. The value of η is determined by the amplitude taper applied across the aperture. The maximum value of η is unity, corresponding to a uniformly illuminated array. The values of η for various amplitude tapers of the modified sin u/u distributions are shown in Table 20-3.

Radiation Pattern of Planar Arrays

To provide beam scanning in two angular dimensions, a planar array of radiating elements must be used. In a spherical coordinate system, the beam position is defined by the two coordinates θ and ϕ as shown in Fig. 20-9. Also shown in Fig. 20-9 is the layout of the element lattice in the planar array. For a rectangular lattice, the mnth

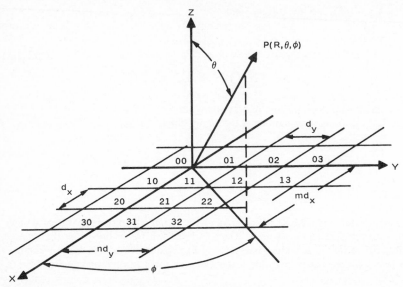

FIG. 20-9 Element geometry of a two-dimensional array.

element is located at $x_m = md_x$ and $y_n = nd_y$. For a triangular lattice, the element grid can be thought of as a rectangular grid in which every other element has been omitted. In this case, only every other value of mn contains an element. The element locations can be defined by requiring that the $(m + n)$ be even.

The array factor of a two-dimensional array may be calculated by summing the vector contribution of each element in the array at each point in space. The array factor can be written in terms of the directional cosines, $\cos \alpha_x$ and $\cos \alpha_y$, of the spherical coordinate system as follows:

$$E_a(\cos \alpha_x, \cos \alpha_y) = \sum_m \sum_n i_{mn} e^{jk(md_x \cos \alpha_x + nd_y \cos \alpha_y)} \qquad (20\text{-}20)$$

where

$$\cos \alpha_x = \sin \theta \cos \phi \qquad (20\text{-}21)$$
$$\cos \alpha_y = \sin \theta \sin \phi$$

For a uniformly illuminated ($i_{mn} = 1$) rectangular array, we have

$$E_a(\cos \alpha_x, \cos \alpha_y) = \sum_{m=-(M-1)/2}^{(M-1)/2} e^{jkmd_x \cos \alpha_x} \sum_{n=-(N-1)/2}^{(N-1)/2} e^{jknd_y \cos \alpha_y} \qquad (20\text{-}22)$$

Each sum can be evaluated, producing a result analogous to Eq. (20-5) for a uniformly illuminated linear array:

$$E_a(\cos \alpha_x, \cos \alpha_y) = \left[\frac{\sin \left(\pi M \dfrac{d_x}{\lambda} \cos \alpha_x \right)}{M \sin \left(\pi \dfrac{d_x}{\lambda} \cos \alpha_x \right)} \right] \left[\frac{\sin \left(\pi N \dfrac{d_y}{\lambda} \cos \alpha_y \right)}{N \sin \left(\pi \dfrac{d_y}{\lambda} \cos \alpha_y \right)} \right]$$

$$(20\text{-}23)$$

Beam scanning with planar arrays is accomplished by linear phasing along both array coordinates. To scan the beam to the angular position corresponding to the directional cosines $\cos \alpha_{x0}$ and $\cos \alpha_{y0}$, a linear phase taper is introduced at each element so that the excitation at the mnth element is given by

$$i_{mn} = a_{mn}e^{j(kmd_x \cos \alpha_{x0} + knd_y \cos \alpha_{y0})} \qquad \textbf{(20-24)}$$

where $kd_x \cos \alpha_{x0}$ = element-to-element phase shift in the x direction
$kd_y \cos \alpha_{y0}$ = element-to-element phase shift in the y direction
This form of steering phase indicates that the phase of the mnth element is the sum of a row phase $mkd_x \cos \alpha_{x0}$ and a column phase $nkd_y \cos \alpha_{y0}$.

The array factor of a rectangular planar array of M by N elements is then given by

$$E_a(\cos \alpha_x, \cos \alpha_y) = \sum_m \sum_n a_{mn}e^{jk[md_c(\cos \alpha_x - \cos \alpha_{x0}) + nd_y(\cos \alpha_y - \cos \alpha_{y0})]} \qquad \textbf{(20-25)}$$

Element Spacing and Lattice of Planar Arrays

As in the case of the linear array, the array factor of the planar array [Eq. (25)] has an infinite number of grating lobes in the directional cosine space. For example, the maxima of E_a occur whenever the argument of Eq. (20-25) is a multiple of 2π. Since there is a one-to-one correspondence between the directional cosine space ($\cos \alpha_x$ and $\cos \alpha_y$ space) and the visible space (θ and ϕ space within the boundary defined by $\cos^2 \alpha_x + \cos^2 \alpha_y = 1$), the number of grating lobes that can be projected from the directional cosine space into the visible space depends upon the parameters d_x/λ and d_y/λ. To avoid the formation of grating lobes in the visible space, the element spacings d_x/λ and d_y/λ must be chosen so that there is only one maximum from Eq. (20-25), namely, the main beam, in the visible space (real space). In the planar array, the element lattice and spacing can be chosen to shape the grating-lobe contour (location pattern of grating lobes) to fit the required scanning volume so that the total required number of elements in the planar array is minimized. To accomplish this optimization, it is more convenient to plot the position of the grating lobes when the main beam is phased for broadside and observe the motion of these lobes as the beam is scanned. Figure 20-10 shows the grating-lobe locations for both rectangular and triangular spacings. For a rectangular lattice, the grating lobes are located at

$$\cos \alpha_x - \cos \alpha_{x0} = \pm \frac{\lambda}{d_x} \cdot p$$

$$p,q = 0,1,2,\dots \qquad \textbf{(20-26)}$$

$$\cos \alpha_y - \cos \alpha_{y0} = \pm \frac{\lambda}{d_y} \cdot q$$

The grating-lobe pattern of Fig. 20-10 must be mapped onto the surface of the unit sphere, as shown in Fig. 20-11, to give a true spatial distribution; therefore, only the portion of the pattern of Fig. 20-10 inside a unit circle centered at $\cos \alpha_x = \cos \alpha_y = 0$ lies in visible space. The lobe at $p = q = 0$ is the main beam. For a conical scan volume, the triangular grid is more efficient for the suppression of grating lobes than a rectangular grid,[11] so that for a given aperture size fewer elements are required. If the triangular lattice contains elements at md_x and nd_y, where $m + n$ is even, then the grating lobes are located at

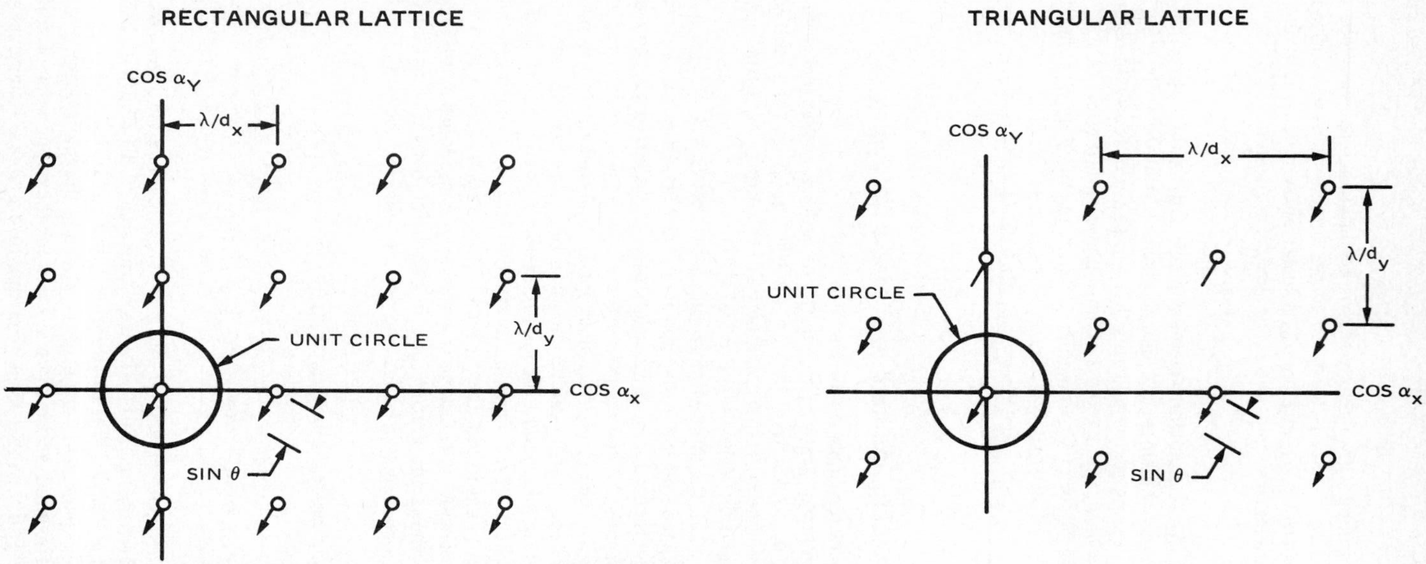

FIG. 20-10 Grating-lobe locations for rectangular and triangular element lattices.

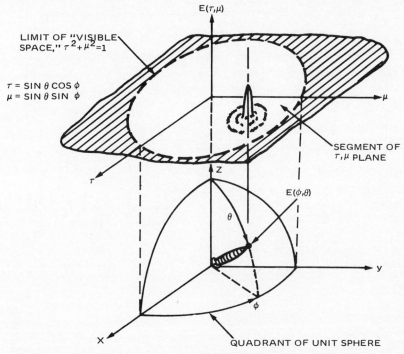

FIG. 20-11 Construction of array factor $E(\phi,\theta)$ by projection of array function $E(\tau, \mu)$ onto unit sphere.

$$\cos \alpha_x - \cos \alpha_{x0} = \pm \frac{\lambda}{2d_x} \cdot p$$

$$p, q = 0, 1, 2, \ldots \qquad \textbf{(20-27)}$$

$$\cos \alpha_y - \cos \alpha_{y0} = \pm \frac{\lambda}{2d_y} \cdot q$$

where $p + q$ is even.

As the array is scanned away from broadside, each grating lobe (in directional cosine space) will move a distance equal to the sine of the angle of scan and in a direction determined by the plane of scan. To ensure that no grating lobes enter visible space (real space), the element spacing must be chosen so that for the maximum scan angle θ_m the movement of a grating lobe by $\sin \theta_m$ does not bring the grating lobe into visible space. If a scan angle of 60° from broadside is required for every plane of scan, no grating lobe may exist within a circle of radius $1 + \sin \theta_m = 1.866$. The square lattice that meets this requirement has

$$\frac{\lambda}{d_x} = \frac{\lambda}{d_y} = 1.866 \qquad \text{or} \qquad d_x = d_y = 0.536\lambda$$

Here, the area per element is

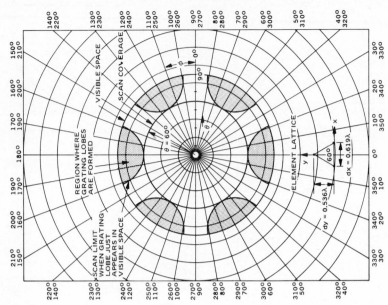

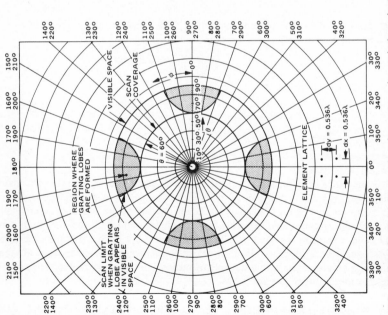

FIG. 20-12 Grating-lobe contours for rectangular and triangular lattices.

$$d_x \cdot d_y = (0.536\lambda)^2 = 0.287\lambda^2$$

For an equilateral-triangular lattice, the requirement is satisfied by

$$\frac{\lambda}{d_y} = \frac{\lambda}{\sqrt{3}d_x} = 1.866 \qquad \text{or} \qquad d_y = 0.536\lambda, \ d_x = 0.309\lambda$$

If we recall that elements are located only at every other value of mn, the area per element is

$$2d_xd_y = 2(0.536\lambda)(0.309\lambda) = 0.332\lambda^2$$

For the same amount of grating-lobe suppression, the saving in the number of elements for the triangular versus the rectangular lattice is approximately 14 percent. The grating-lobe-contour plots illustrating the above example are shown in Fig. 20-12.

Amplitude Taper for Planar Arrays

If markedly different patterns are desired in the two orthogonal planes ($\cos \alpha_x$ and $\cos \alpha_y$ planes), it is often advantageous to use rectangularly shaped arrays with sides parallel to the specified planes and separable illumination; that is, $a_{mn} = a_m a_n$, so that

$$
\begin{aligned}
E_a(&\cos \alpha_x, \cos \alpha_y) \\
&= \sum_m \sum_n a_{mn} e^{jk[md_x(\cos \alpha_x - \cos \alpha_{x0}) + nd_y(\cos \alpha_y - \cos \alpha_{y0})]} \\
&= \sum_m a_m e^{jkmd_x(\cos \alpha_x - \cos \alpha_{x0})} \sum_n a_n e^{jknd_y(\cos \alpha_y - \cos \alpha_{y0})} \\
&= E_{a1}(\cos \alpha_x) \cdot E_{a2}(\cos \alpha_y)
\end{aligned}
\qquad \textbf{(20-28)}
$$

where E_{a1} and E_{a2} are synthesized by linear-array techniques. In this case, the amplitude distributions described in the subsection "Amplitude Tapers for Sidelobe Control" can be applied along each orthogonal axis of the planar array to achieve the desired sidelobe control. One special feature of this separable illumination is that the sidelobe levels along the $\cos \alpha_x$ plane are those of $E_{a1}(\cos \alpha_x)$ alone, since E_{a2} is a maximum. Away from both the $\cos \alpha_x$ and the $\cos \alpha_y$ axes (diagonal planes), the sidelobes are lower as a result of the product of E_{a1} and E_{a2}, both less than their maxima. For example, the array factor for a uniformly illuminated rectangular aperture was given by Eq. (20-23). The array factor yields first sidelobes of about -13.2 dB below the main beam along the two orthogonal axes. Off the axes, the sidelobes are lower. In particular, along the 45° diagonals, the first sidelobes are twice as low in decibels, -26.4 dB.

Consequently, for a given peak sidelobe level, the taper efficiency ($\eta = \eta_1\eta_2$) is lower for a separable illumination than that which can be realized by nonseparable means, although the average sidelobe level should be comparable to tapers of like values of η.

If a circularly symmetric pattern is desired, an array shape approximating a circle is preferred. Here, the array symmetry refers to only the symmetry of the beam shape and near-in-sidelobe structure. The element grid will still govern the grating-lobe location. In most cases, the amplitude distributions for circular arrays are nonseparable. The two common distributions are $(1 - r^2)^P$ distributions[1] and circular

Taylor distributions.[12,13] The sidelobe level, beamwidth, and aperture efficiency for the above two types of distributions are given in Tables 20-4 and 20-5 respectively.

For the special case of a uniformly illuminated circular array, corresponding to $p = 0$ in the $(1 - r^2)^P$ distribution, the array pattern is given by

$$E_a(\theta) \approx \frac{J_1(ka \sin \theta)}{ka \sin \theta}$$

where $J_1(x)$ is the first-order Bessel function. The function has an HPBW coefficient of 58.5°, defined by

$$\text{Beamwidth} = 58.5 \frac{\lambda}{2a}$$

where a is the radius of the circular aperture. The first sidelobes are approximately -17.5 dB.

Array Directivity and Gain of Planar Arrays

The directivity of a planar array is defined as in Eq. (20-14) with appropriate notation changes to include both spatial angles:

$$D(\theta_0, \phi_0) = \frac{|E_a(\theta_0, \phi_0)|^2}{\frac{1}{4\pi} \int_{\text{all space}} |E_a(\theta, \phi)|^2 \, d\Omega} \qquad (20\text{-}29)$$

where $d\Omega = \sin \theta d\theta d\phi$.

For a large array, the array factor of a discrete array closely approximates that of a continuous array near the main lobe. Thus, if the array factor is such that almost all the power is in the main beam and the first few sidelobes and there is no grating lobe in the visible space, the directivity expression for a continuous aperture [Ref. 1, Eq. (19), page 177] may be used:

TABLE 20-4   Design Sidelobe, Beamwidth, and Aperture Efficiency of $(1 - r^2)^P$ Distributions for a Circular Aperture

P	First-sidelobe dB below peak intensity	Half-power width	Position of first zero	Aperture efficiency
0	17.6	$1.02 \dfrac{\lambda}{D}$	$\sin^{-1} \dfrac{1.22\lambda}{D}$	1.00
1	24.6	$1.27 \dfrac{\lambda}{D}$	$\sin^{-1} \dfrac{1.63\lambda}{D}$	0.75
2	30.6	$1.47 \dfrac{\lambda}{D}$	$\sin^{-1} \dfrac{2.03\lambda}{D}$	0.56
3	35.8	$1.65 \dfrac{\lambda}{D}$	$\sin^{-1} \dfrac{2.42\lambda}{D}$	0.44
4	40.6	$1.81 \dfrac{\lambda}{D}$	$\sin^{-1} \dfrac{2.79\lambda}{D}$	0.36

TABLE 20-5 Design Sidelobe and Beamwidth of Circular Taylor Distributions

Design sidelobe level, dB	η (sidelobe voltage ratio)	Ideal beamwidth β_0, °	A^2	Values of the parameter σ							
				$\bar{n}=3$	$\bar{n}=4$	$\bar{n}=5$	$\bar{n}=6$	$\bar{n}=7$	$\bar{n}=8$	$\bar{n}=9$	$\bar{n}=10$
15	5.62341	45.93	0.58950	1.2382	1.1836	1.1485	1.1244	1.1069	1.0937	1.0833	1.0751
16	6.30957	47.01	0.64798	1.2330	1.1809	1.1469	1.1233	1.1061	1.0931	1.0829	1.0747
17	7.07946	48.07	0.70901	1.2276	1.1781	1.1452	1.1222	1.1053	1.0925	1.0825	1.0743
18	7.94328	49.12	0.77266	1.2220	1.1752	1.1434	1.1210	1.1045	1.0919	1.0820	1.0740
19	8.91251	50.15	0.83891	1.2163	1.1723	1.1416	1.1198	1.1037	1.0913	1.0815	1.0736
20	10.00000	51.17	0.90777	1.2104	1.1692	1.1398	1.1186	1.1028	1.0906	1.0810	1.0732
21	11.2202	52.17	0.97927	1.2044	1.1660	1.1379	1.1173	1.1019	1.0899	1.0805	1.0728
22	12.5893	53.16	1.05341	1.1983	1.1628	1.1359	1.1160	1.1009	1.0892	1.0799	1.0723
23	14.1254	54.13	1.13020	1.1920	1.1594	1.1338	1.1146	1.1000	1.0885	1.0793	1.0719
24	15.8489	55.09	1.20965	1.1857	1.1560	1.1317	1.1132	1.0990	1.0878	1.0788	1.0714
25	17.7828	56.04	1.29177	1.1792	1.1525	1.1296	1.1118	1.0979	1.0870	1.0782	1.0709
26	19.9526	56.97	1.37654	1.1726	1.1489	1.1274	1.1103	1.0969	1.0862	1.0775	1.0704
27	22.3872	57.88	1.46395	1.1660	1.1452	1.1251	1.1087	1.0958	1.0854	1.0769	1.0699
28	25.1189	58.78	1.55406	1.1592	1.1415	1.1228	1.1072	1.0946	1.0845	1.0762	1.0694
29	28.1838	59.67	1.64683	1.1524	1.1377	1.1204	1.1056	1.0935	1.0836	1.0756	1.0688
30	31.6228	60.55	1.74229	1.1455	1.1338	1.1180	1.1039	1.0923	1.0827	1.0749	1.0683
31	35.8107	61.42	1.84044	1.1385	1.1298	1.1155	1.1022	1.0911	1.0818	1.0742	1.0677
32	39.8107	62.28	1.94126	1.1315	1.1258	1.1129	1.1005	1.0898	1.0809	1.0734	1.0671
33	44.6684	63.12	2.04473	1.1244	1.1217	1.1103	1.0987	1.0885	1.0799	1.0727	1.0665
34	50.1187	63.96	2.15092	......	1.1176	1.1077	1.0969	1.0872	1.0790	1.0719	1.0659
35	56.2341	64.78	2.25976	......	1.1134	1.1050	1.0951	1.0859	1.0779	1.0711	1.0653
36	63.0957	65.60	2.37129	......	1.1091	1.1023	1.0932	1.0846	1.0769	1.0703	1.0647
37	70.7946	66.40	2.48551	......	1.1048	1.0995	1.0913	1.0832	1.0759	1.0695	1.0640
38	79.4328	67.19	2.60241	......	1.1005	1.0967	1.0894	1.0818	1.0748	1.0687	1.0633
39	89.1251	67.98	2.72201	......	1.0961	1.0938	1.0874	1.0803	1.0737	1.0678	1.0627
40	100.0000	68.76	2.84428	......	1.0916	1.0910	1.0854	1.0789	1.0726	1.0670	1.0620

$$D(\theta_0,\phi_0) = \frac{4\pi}{\lambda^2} \cos\theta_0 \frac{\left|\int\int i(x,y)\,dxdy\right|^2}{\int\int |i(x,y)|^2\,dxdy} \quad\quad (20\text{-}30)$$

where the integrals are understood to extend over the entire aperture of the antenna. For a uniformly illuminated array, $i(x,y) = 1$, the directivity is given by

$$D(\theta_0,\phi_0) = \frac{4\pi A}{\lambda^2} \cos\theta_0 \quad\quad (20\text{-}31)$$

For all tapered illuminations, the directivity would be less than that of a uniformly illuminated array. Making use of the Schwartz inequality,[12]

$$\left|\int fg\,dxdy\right|^2 \le \int f^2\,dxdy \cdot \int g^2\,dxdy$$

where f and g are any two functions; by taking $f = i(x,y)$ and $g = 1$, we find

$$\left|\int_A i(x,y)\,dxdy\right|^2 \le A \int |i(x,y)|^2\,dxdy$$

Therefore,

$$\frac{\left|\int\int i(x,y)\,dxdy\right|^2}{\int\int |i(x,y)|^2\,dxdy} = \eta A \quad\quad 0 \le \eta \le 1 \quad\quad (20\text{-}32)$$

where A is the physical area of the array and η is, as in the case of linear arrays, the taper efficiency or illumination efficiency; η is unity for uniform illumination and is tabulated for other tapered illuminations as shown in Table 20-5.

Substituting Eq. (20-32) into Eq. (20-30), we have

$$D(\theta_0,\phi_0) = \frac{4\pi A}{\lambda^2} \eta \cos\theta_0 \quad\quad (20\text{-}33)$$

For a large array of discrete elements, the directivity can be expressed in terms of the actual taper coefficients by noting that

$$\int\int i(x,y)\,dxdy \approx \sum_m \sum_n i_{mn}\,dxdy$$

for $i_{mn} = i(md_x,nd_y)$. By a similar equivalence in the denominator of Eq. (32), it follows that

$$\frac{\left|\int\int i(x,y)\,d_xd_y\right|^2}{\int\int |i(x,y)|^2\,d_xd_y} \approx \frac{\left|\sum\sum i_{mn}\,d_xd_y\right|^2}{\sum\sum |i_{mn}|^2\,d_xd_y} \approx \eta N\,d_xd_y = \eta A$$

where N is the total number of elements in the array. Thus, for a large array

$$D(\theta_0,\phi_0) = 4\pi\eta N \frac{d_xd_y}{\lambda^2} \cos\theta_0 \qu\quad\quad (20\text{-}34)$$

The directivity expressions as shown in Eqs. (20-33) and (20-34) are for an array aperture which is impedance-matched for all beam-scan angles; i.e., the aperture is transparent for a plane-wave incidence at any scan angle θ_0 without reflections. However, in practice, the aperture is mismatched at some scan angles. Therefore, the array gain of the antenna is given by

$$G(\theta_0,\phi_0) = \frac{4\pi A}{\lambda^2}\, \eta \cos \theta_0 [1 - |\Gamma(\theta_0,\phi_0)|^2] \qquad (20\text{-}35)$$

where $|\Gamma(\theta_0,\phi_0)|$ is the amplitude of the array aperture reflection coefficient at the scan angle θ_0,ϕ_0.

In general, the antenna beam-forming network behind the array aperture has ohmic losses such as the losses in the phase shifters, power combiners, etc. Hence, the net antenna gain is given by

$$G(\theta_0,\phi_0) = \frac{4\pi A}{\lambda^2}\, \eta \cos \theta_0 (1 - |\Gamma(\theta_0,\phi_0)|^2) - \text{all other ohmic losses} \qquad (20\text{-}36)$$

20-3 RADIATOR DESIGN FOR PHASED ARRAYS

The discussion on array theory in Sec. 20-2 is based on radiators with isotropic patterns. In practice, however, the radiation patterns of real radiators are nonisotropic, and the impedance of the radiators varies as a function of scan caused by the mutual coupling between the radiators. In fact, the pattern of an element in the environment of an array is markedly different from the pattern of an isolated element in amplitude, phase, and, perhaps, polarization as well.

One of the functions performed by the antenna is to provide a good match between the radar transmitter and free space. This means that the radiation impedance (driving-point impedance) looking into the radiators in the array environment must be matched to their generator impedances. If the antenna aperture is not matched to free space, power will be reflected back toward the generator, resulting in a loss in radiated power. In addition, a mismatch produces standing waves on the feed line to the antenna. The voltage at the peaks of these standing waves is $(1 + |\Gamma|)$ times greater than the voltage of a matched line, where Γ is the voltage reflection coefficient of the radiation impedance. This corresponds to an increased power level that is $(1 + |\Gamma|)^2$ times as great as the actual incident power. Therefore, while the antenna is radiating less power, individual components must be designed to handle more peak power. In a scanning array, the impedance of a radiating element varies as the array is scanned, and the matching problem is considerably more complicated. In some instances, spurious lobes may appear in the array pattern as a consequence of the mismatch. Furthermore, there are conditions in which an antenna that is well matched at broadside may have some angle of scan at which most of the power is reflected.

To illustrate this mutual-coupling effect on the radiation impedance, the couplings from several elements to a typical central element, element 00, is shown in Fig. 20-13. $C_{mn,pq}$ is the mutual-coupling coefficient relating the voltage (amplitude and phase) induced in the mnth element to the voltage excitation at the pqth element. The coupled signals add vectorially to produce a wave traveling toward the generator of

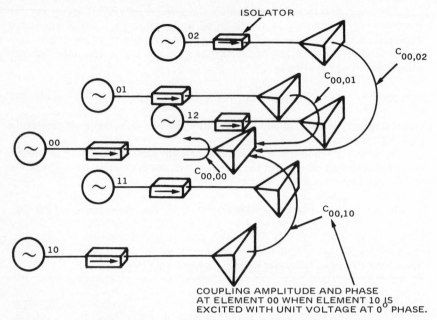

COUPLING AMPLITUDE AND PHASE
AT ELEMENT 00 WHEN ELEMENT 10 IS
EXCITED WITH UNIT VOLTAGE AT 0° PHASE.

FIG. 20-13 Coupled signals to a central element from neighboring elements.

element 00, which appears to be a reflection from the radiator of element 00. As the phases of the neighboring elements are varied to scan the beam, the vector sum of the coupled signals changes and causes an apparent change in the impedance of element 00. For some scan angles the coupled voltages will tend to add in phase, causing a large reflection and possibly the loss of the main beam. Large reflections often occur at scan angles just prior to the emergence of a grating lobe into real space, but in some instances such reflections may occur at smaller scan angles.

The mutual-coupling effects described above have been studied in detail for dipole and open-ended waveguide radiators.[14,15,16,17,18,19]

Active-Element Pattern

For an array with independently driven elements as shown in Fig. 20-13, the far-field array pattern at any angle (θ,ϕ) is obtained by vector addition of the contributions from all the elements at that angle with each single element driven one at a time while others are terminated in their generator impedances. The far-field array pattern is given by

$$E(\theta,\phi) = \sum_m \sum_n i_{mn} E_{e,mn}(\theta,\phi) \; e^{jk[md_x(\sin\theta\cos\phi-\sin\theta_0\cos\phi_0)+nd_y(\sin\theta\sin\phi-\sin\theta_0\sin\phi_0)]}$$

(20-37)

where $E_{e,mn}(\theta,\phi)$ = active-element pattern obtained when *only* the *mn*th element is driven, with all other elements terminated in their usual generator impedance

i_{mn} = current in amperes that *would* be flowing in the *mn*th feed line *if* all other elements were passively terminated in their usual generator impedance (i.e., the current in the *mn*th feed line due solely to the *mn*th-element generator, not including that coupled into the *mn*th element from any other generators)

The utility of these definitions arises from the fact that both computationally and experimentally one driven element in a passively terminated array is easier to cope with than all elements simultaneously driven. Furthermore, the active-element pattern has the significance of describing the array gain as a function of beam-scan angle.

Since $E_{e,mn}(\theta,\phi)$ is actually the pattern of the entire array when the excitation of the array is that due to the mutual couplings from the *mn*th element to all the neighboring elements, it is a function not only of the particular radiator configuration but also of the relative location of the driven radiator with respect to the other elements of the array. In general, the expressions of the active-element pattern, $E_{e,mn}(\theta,\phi)$, for any radiator type are quite complicated. Nevertheless, all the active-element patterns as defined in Eq. (20-37) could be individually measured or perhaps numerically calculated. Fortunately, this individual-element-pattern measurement is not usually necessary if the array is regularly spaced. For an infinite array of regularly spaced elements, every element in the array sees exactly the same environment and all the elements have the same active-element pattern. The radiation impedance of an element measured in waveguide simulators[20] corresponds to the element impedance in an infinite array. This infinite-array model can predict with good accuracy the array impedance and impedance variations with scan for a finite array. Even arrays of modest proportions (less than 1000 elements) have been in reasonable agreement with the results predicted for an infinite array.[21] The reason for this is that, in practice, the mutual coupling between the radiators decays as $1/x^2$, where x is the distance between the radiators. Hence, the effect of mutual coupling is quite localized in cases of practical interest. All the elements with the exception of the edge elements in the array behave as though they are in an infinite-array environment. For larger arrays, the effect of edge elements on array performance is negligible. Therefore, the array factor of Eq. (20-37) can be written as

$$E(\theta,\phi) = E_e(\theta,\phi) \sum_m \sum_n i_{mn}\, e^{jk[md_x(\sin\theta\cos\phi-\sin\theta_0\cos\phi_0)+nd_y(\sin\theta\sin\phi-\sin\theta_0\sin\phi_0)]}$$

$$= E_e(\theta,\phi) \cdot E_a(\theta,\phi) \qquad\qquad (20\text{-}38)$$

Equation (20-38) states that the radiation pattern of an array is the product of the element pattern and the array factor. The *array factor* is determined by the geometric placement of the elements and their relative phasing on the assumption that the elements are isotropic and that there is no mutual coupling. Its peak value is independent of the scan angle. The *active-element pattern* is the actual radiation pattern of an element in the array taken in the presence of all other elements and taking into account all mutual-coupling effects and mismatches. It follows from this definition that as the array is being scanned, the peak of the radiation pattern $E(\theta,\phi)_{max}$ will trace out the element pattern. The active-element pattern may also be obtained experimentally by exciting one central element in a small test array and terminating all the other elements in matched loads. The number of terminated elements used in practice is the same as the number used for providing the central element with a large-array environment. For example, in using dipole radiators above a ground plane, an element in the center of a 7-by-7 array may be taken as typical of an element in a large array.

For an array of open-ended waveguides, a 9-by-9 array should suffice. If there is any position of scanning where a main beam does not form or where a large loss in gain occurs, it will show up as a null in the active-element pattern.

As shown in Eq. (20-35), the gain of a *perfectly matched* array will vary as the projected aperture area, as given by

$$G(\theta_0, \phi_0) = \frac{4\pi A}{\lambda^2} \eta \cos \theta_0 \qquad (20\text{-}39)$$

where the term $|\Gamma(\theta_0, \phi_0)|$ is equal to zero. If we assume that each of the N elements in the array shares the gain equally and $\eta = 1$, the gain of a single element is

$$G_e(\theta, \phi) = \frac{4\pi A}{N\lambda^2} \cos \theta_0 \qquad (20\text{-}40)$$

Therefore, the ideal active-element pattern for a perfectly matched array is a cosine pattern.

Design Criteria for Radiators

The most commonly used radiators for phased arrays are dipoles, slots, open-ended waveguides (or small horns), spirals, and microstrip disk or patch elements. The selection of a radiator for a particular application must be based upon the following considerations:

1   The required area of the element is small enough to fit within the allowable element spacing and lattice without the formation of grating lobes. In general, this limits the element to an area of a little more than $\lambda^2/4$.

2   The active-element pattern of the element provides the appropriate aperture matching over the required scan coverage.

3   The polarization and power-handling capability (both peak and average power) meet the radar system requirements.

4   The physical construction of the radiator must be able to withstand environmental requirements such as thermal, shock, and vibration requirements.

5   The radiator must also be inexpensive, reliable, and repeatable from unit to unit since many hundreds or thousands of radiators are required in a large phased-array antenna.

Since the impedance and the pattern of a radiator in an array are determined predominantly by the array geometry and environment, the radiator may be chosen to best suit the feed system and the physical requirements of the antenna. For example, if the radiator is fed from a stripline or microstrip phase shifter, a stripline or microstrip dipole would be a logical choice for the radiator. An example illustrating the combination of a microstrip dipole and phase shifter is shown in Fig. 20-14. If the radiator is fed from a waveguide phase shifter, an open-ended waveguide would be most convenient. An example of this type is shown in Fig. 20-15. At the lower frequencies, where coaxial components are prevalent, dipoles have been favored for the radiating element. At the higher frequencies, open-ended waveguides and slots are frequently used.

For circular polarization, an open-ended circular waveguide or spiral may be

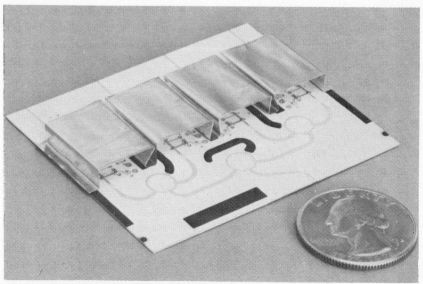

FIG. 20-14 Integrated subarray module containing four dipole radiators, four micromin diode phase shifters, and a power divider.

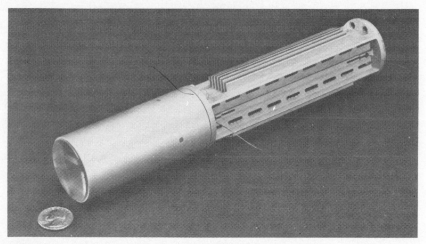

FIG. 20-15   Waveguide radiator and phase-shifter combination.

used. If polarization diversity is required, either crossed dipoles or circular waveguides may be used. With suitable feed networks, both are capable of providing vertical and horizontal polarization independently and may be combined to provide any desired polarization. An example of a dual-polarized circular-waveguide radiator is shown in Fig. 20-16.

Development Procedure for Radiators

After selecting the proper type of radiator on the basis of the criteria as stated above, the next step is to optimize the performance by following the procedure outlined below:

1   Select an element configuration and spacing that do not support a surface wave which generates grating lobes.

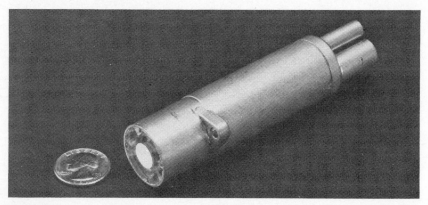

FIG. 20-16   Dual-polarized circular-waveguide radiator.

2 Formulate an analytical model of the radiating element in the array aperture, and optimize the performance by varying the design parameters of the radiator such as the spacing between the radiators, the aperture size, or the distance of the radiator above ground. The radiation impedance can be calculated by using modal expansion methods.[22,23,24] Furthermore, the aperture matching of the radiator can sometimes be improved by using inductive iris in the opening (aperture) of the waveguide radiator[22] or by using dielectric sheets in front of the radiating aperture.[23]

3 Using the analytical results from Step 2, fabricate waveguide simulators[20] to measure the radiation impedance at discrete scan angles. The design parameters of the element may have to be modified in order to optimize element performance. This measurement also verifies the calculated results of Step 2.

4 Fabricate a small test array and measure the active-element pattern of the central element over the required frequency band. The active-element pattern describes the variation in array gain as a function of beam-scan angle. Large aperture mismatches at any scan angle or scan plane will show up as dips in the active-element pattern.

5 The active-element-pattern measurement of Step 4 in conjunction with the impedance measurement of Step 3 establishes the performance characteristics of the final radiator design.

20-4 SELECTION OF PHASE SHIFTERS FOR PHASED ARRAYS

Selection Criteria

Over the past two decades considerable effort has gone into the development of phase shifters for phased-array applications. Only a cursory overview of the results of this work is given in this section. A more expansive description of phase-shifter theory and practice is covered in the literature.[25,26,27,28,29,30]

There are currently two types of electronic phase shifters suitable for practical phased arrays: the ferrite phase shifter and the semiconductor-diode phase shifter. Selection of the type of phase shifter to use in a particular application depends strongly on the operating frequency and the RF power per phase shifter. Above S band, waveguide ferrite phase shifters have less loss than diode phase shifters. In situations in which RF powers are low and size and weight constraints dictate a micromin design, diode phase shifters are preferred. Other factors which have a bearing on the selection, if a phased array with a phase shifter per radiating element is assumed, are listed below.

1 **Insertion Loss** Insertion loss should be as low as possible. It results in a reduction of generated power on transmit and of lower signal-to-noise ratio on receive. It also produces phase-shifter-heating problems.

2 **Switching Time** The time to switch should be as short as possible. A long switching time increases minimum radar range when nonreciprocal phase shifters are used and when a burst of pulses is transmitted in different directions. Times on the order of microseconds are adequate for most applications.

3   **Drive Power** Drive power should be as small as possible. A large amount of drive power generates heat and also may require power supplies that are too large for a mobile system. Large drive power may also require expensive driver-circuit components. Diode phase shifters require holding power as well as switching power. Ferrite phase shifters can be latched; that is, they do not consume drive power except when switching.

4   **Phase Error** Phase error should be as small as possible. It should not reduce antenna gain substantially or raise sidelobes in the radiation pattern. One cause of phase error is the size of the least significant bit of a digital phase shifter. Other phase errors are due to manufacturing tolerances in the phase shifter and driver.

5   **Transmitted Power: Peak, Average** The required power per phase shifter depends upon the maximum radar range and data rate of the system design. Typical values are from 1 to 100 kW peak and 1 to 500 W average. Power levels above 10 to 15 kW peak generally require a ferrite device.

6   **Physical Size** The phase shifter should fit within a $\lambda/2$-by-$\lambda/2$ cross section so that it can be packaged behind each element; otherwise an expensive fan-out feed is required.

7   **Weight** Phase-shifter weight should be minimized in mobile, and especially in airborne or spacecraft, installations.

8   **Cost and Manufacturing Ease** For arrays with thousands of elements the unit cost must be low. Manufacturing tolerances must be as large as possible, consistent with allowable system phase and amplitude errors.

Figure 20-17 compares ferrite and diode phase shifters as a function of operating frequency and insertion loss. In general, ferrite devices have higher power capacity with less loss above S band. Diode phase shifters predominate at the lower microwave frequencies.

Ferrite Phase Shifter

Construction of ferrite phase shifters falls into two general categories: those phase shifters enclosed by a waveguide structure and those built by using a microstrip configuration. While the construction of the microstrip ferrite phase shifter is extremely simple, the performance of the waveguide ferrite phase shifter is far superior, and this phase shifter is generally used for most applications.

Ferrite phase shifters use ferrimagnetic materials that include families of both ferrite and garnets which are basically ceramic materials with magnetic properties.

Ferrite phase shifters can be either nonreciprocal or reciprocal. The toroidal phase shifter is the best example of the nonreciprocal type, and the Reggia-Spencer and Faraday rotator phase shifters are the best representatives of reciprocal ferrite

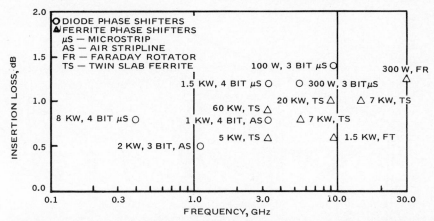

FIG. 20-17 Insertion loss of Hughes Aircraft Company diode and ferrite phase shifters at various frequency bands.

phase shifters. When nonreciprocal phase shifters are employed in a system that uses a single antenna for both transmitting and receiving, the device must be switched between these two modes of operation. This is accomplished by rapidly switching the phase shifter to the receiving state with a driving pulse w... se polarity is opposite that used in the transmitting state.

Toroidal Phase Shifter This nonreciprocal phase shifter uses a toroid geometry which consists of a ferrimagnetic toroid located within a section of waveguide as shown in Fig. 20-18. This phase shifter is sometimes referred to as a *twin-slab device* since phase shift is provided by the vertical toroid branches (parallel to the E field), while the horizontal branches are used to complete the magnetic circuit. The toroid is threaded with a drive wire, which in turn is connected to a driving amplifier that can supply either a positive or a negative current pulse. When the wire is pulsed with a current of sufficient amplitude, a magnetic field that drives the toroid material into saturation is produced. When the pulse is released, the material becomes latched at one of the two remanent-induction points, $+\beta_r$ or $-\beta_r$, depending on the polarity of the pulse.

The two remanent states of magnetization due to the positive and negative pulses correspond to a different electrical length, the difference being the differential phase shift. The electrical length for a forward-traveling wave latched to one state is the same as that for a backward-traveling wave latched to the opposite state. A complete digital phase shifter (Fig. 20-19) contains several lengths of ferrites adjusted to give differential phase shifts of 180°, 90°, 45°, etc., depending on the number of bits required.

This basic construction can be extended to an analog latching phase shifter by using the same toroid cross section as for the digital latching device and replacing the various toroid bits with a single toroid capable of producing at least 360° of phase

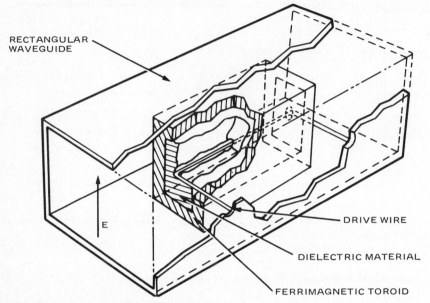

RECTANGULAR
WAVEGUIDE

E

DRIVE WIRE

DIELECTRIC MATERIAL

FERRIMAGNETIC TOROID

FIG. 20-18 Ferrite toroidal phase shifter.

BIT SIZES

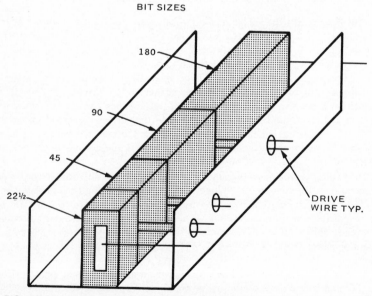

180

90

45

22½

DRIVE
WIRE TYP.

FIG. 20-19 Digital ferrite phase shifter using toroids.

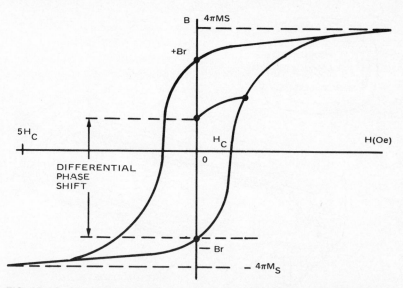

FIG. 20-20 Toroid *B-H* magnetic hysteresis loop illustrating the characteristics of drive.

shift. Phase shift is obtained by first pulsing the drive wire with a *clear* pulse (to set the induction at $-\beta_r$ and eliminate hysteresis errors), followed by a *set* pulse of opposite polarity. The set pulse raises the level of magnetization, and, when released, the magnetization falls back to the $H = 0$ line, as shown in Fig. 20-20. Since the amount of partial remanent magnetization depends on the amplitude of the set pulse and since phase shift depends on magnetization, an analog phase-shifting device is achieved.

Because the device always operates at partial remanent-magnetization levels, it can be readily adapted to a flux-drive technique. Flux driving tends to minimize the temperature sensitivity of a phase shifter that would normally be dependent on the temperature sensitivity of the material. It is particularly useful for applications requiring high average power or wide ambient-temperature variations.

Reggia-Spencer Phase Shifter   The Reggia-Spencer reciprocal phase shifter, shown in Fig. 20-21, consists of a bar of ferrimagnetic material located axially within a section of waveguide. A solenoid is wound around the waveguide, and when energized with a current, it produces a longitudinal magnetic field. The magnetic field causes a variation in the permeability of the material, which in turn produces a variation in the propagation constant of the RF energy. This results in an RF phase shifter whose phase shift can be controlled by a driving current.

When compared with latching ferrite phase shifters, the Reggia-Spencer device has disadvantages in terms of switching time and switching power. Switching time is increased by the large inductance of the solenoid and by the shorted-turn effect of the waveguide within the solenoid. Switching time is further increased since a saturation

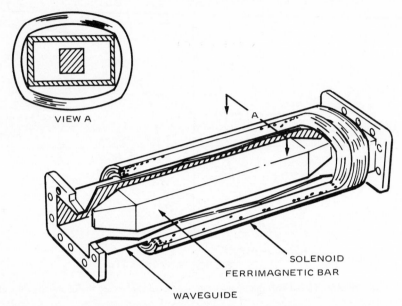

VIEW A

A

SOLENOID
FERRIMAGNETIC BAR

WAVEGUIDE

FIG. 20-21 Typical Reggia-Spencer phase-shifter configuration.

pulse must be applied to the phase-shifter solenoid before the next value of phase-shift-setting current is applied to eliminate hysteresis errors.

Because the device is nonlatching, holding currents must be employed, resulting in a relatively high drive-power requirement. As an example, at a switching rate of 500 Hz, the input drive power for an S-band Reggia-Spencer phase shifter is 21 W, whereas the input drive power for an S-band toroidal digital latching phase shifter is only 2.2 W.

Faraday Rotator Phase Shifter   This reciprocal phase shifter uses a Faraday rotator section in conjunction with quarter-wave plates at each end. The device employs a small, square waveguide completely filled with ferrite. The metal walls are plated directly on the ferrite bar. An axial coil is wound around the waveguide. The magnetic circuit is completed externally to the thin waveguide wall with ferrimagnetic yokes (Fig. 20-22). The total magnetic circuit has only the small gap of the plated waveguide wall; thus the device is latching, since high remanent magnetization is produced when the magnetizing pulse is reduced to zero.

From a microwave RF viewpoint the device consists of a phase-shift section with a nonreciprocal quarter-wave plate at each end. The nonreciprocal quarter-wave plate converts energy in a rectangular waveguide to either right- or left-hand circularly polarized energy in a quadratically symmetric waveguide, depending on the direction of propagation. The insertion phase of the energy is changed by a variable axial magnetic field supplied by the coil around the waveguide.

Semiconductor-Diode Phase Shifters

Digital diode phase shifters use a diode junction of the pin type as the control element in a microwave circuit. The pin diode can have high breakdown voltage and can be

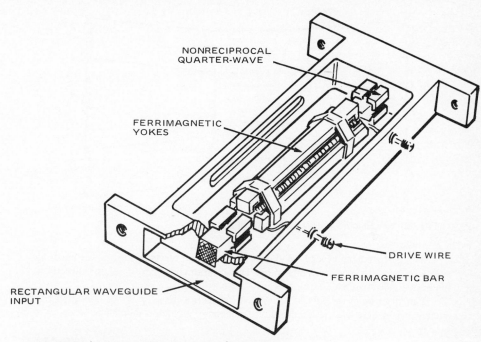

NONRECIPROCAL
QUARTER-WAVE

FERRIMAGNETIC
YOKES

DRIVE WIRE

FERRIMAGNETIC BAR

RECTANGULAR WAVEGUIDE
INPUT

FIG. 20-22 Reciprocal Faraday rotator phase shifter.

designed to have relatively constant parameters in either of two states, forward or reverse bias. It is therefore used in high-power digital phase shifters and is switched between forward and reverse bias. Switching times for diode phase shifters vary from a few microseconds with high-voltage diodes to tens of nanoseconds with low-voltage diodes.

Physically, the pin diode consists of a region of high-resistivity (intrinsic) silicon semiconductor material situated between thin, heavily doped, low-resistivity p^+ and n^+ regions, as shown in Fig. 20-23. The intrinsic region, which may have a volume resistivity of 200 to 3000 Ω-cm, behaves as a slightly lossy dielectric at microwave frequencies, and the heavily doped regions behave as good conductors. In either the unbiased or the reverse-biased state the diode equivalent circuit is a fixed capacitance in series with a small resistance, which varies somewhat with bias. The resistance increases with decreasing reverse bias and rises to a maximum value at a forward-bias voltage equal approximately to the barrier potential of the semiconductor, which is about 0.5 V for silicon (Fig. 20-24). The peak-resistance value is approximately equal to one-half of the reactance of the reverse-biased diode. At larger forward biases a plasma floods the intrinsic region, producing low resistance.

With the addition of external reactive tuning elements, as shown in Fig. 20-25, switching action between forward and reverse bias results; that is, the impedance presented between the network terminals varies between large and small values. The tuning elements can be arranged to give either an open or a closed switch in either bias state. A capacitor in series with the diode gives a *forward-mode* switch or an RF short circuit in forward bias. A *reverse-mode* switch, producing an RF short circuit in reverse bias, is obtained by adding an inductance in series with the diode.

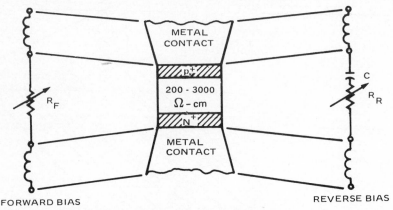

FORWARD BIAS   REVERSE BIAS

FIG. 20-23   Electrical equivalent circuit of a pin diode.

The performance of a phase shifter is specified in terms of loss, power-handling ability, bandwidth, and switching time. These parameters can be related directly to the basic diode switch.

For all circuits the loss of a diode switch varies directly as the operating frequency and inversely as the diode cutoff frequency f_c, which is defined as

$$f_c = \frac{1}{2\pi C \sqrt{R_R R_F}} \qquad (20\text{-}41)$$

where C is the microwave capacitance in reverse bias and R_R and R_F are the reverse and forward series resistances at the chosen bias levels. The variation of loss within a given band is not severe, but this relation indicates that the diode loss increases for the higher-frequency bands. Typical cutoff frequencies for high-voltage pin diodes range from 250 to 700 GHz.

The diode cutoff frequency is also a function of peak-power level because of high-level limiting in reverse bias, causing an increase in R_R. This occurs when the forward excursion of an applied RF voltage swings into the region of high resistance near zero bias. This effect is illustrated in Fig. 20-26. Under small RF signals the average resistance per cycle is small; thus no limiting occurs. With large signals the average RF series resistance increases because of the positive cycle of the RF voltage swinging into the high-series-resistance region near zero bias, and the switch loss will rise. The pin diode does not switch into the conducting or forward-biased state at microwave frequencies because the time required to create a plasma in the intrinsic region is of the order of the diode switching time, which is approximately 1 μs.

The maximum peak power that a diode can handle, without consideration being given to junction heating, is limited by the diode breakdown voltage. A generally safe procedure is to limit the sum of the RF voltage amplitude and bias level to a value less than the direct-current breakdown voltage. For some pin diodes, the sum of the two voltages can exceed the direct-current voltage rating without producing breakdown. This occurs when the direct-current breakdown is determined by a diode surface effect, which is slow to produce breakdown as compared with the length of an RF cycle. The operating RF voltage rating of a particular pin diode is a function of

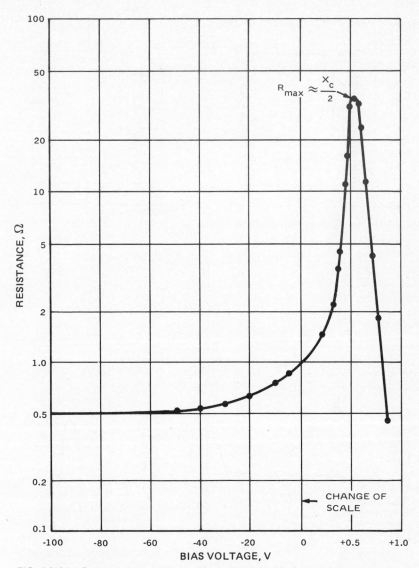

FIG. 20-24 Series resistance versus bias voltage for Microwave Associates silicon pin diode Type MA 4573 at 3.2 GHz; capacitance at -100 V bias $= 0.75$ pF.

diode construction, temperature of the device, pulse length, and operating frequency and generally must be determined experimentally.

The average-power rating of pin diodes is primarily determined by the temperature rise of the junction, which in turn depends on the diode heat sink and power dissipated.

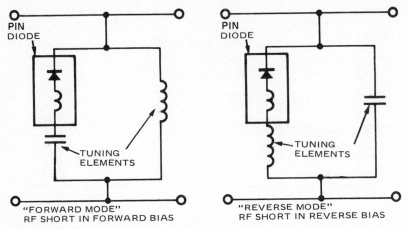

FIG. 20-25 Basic diode switching circuits.

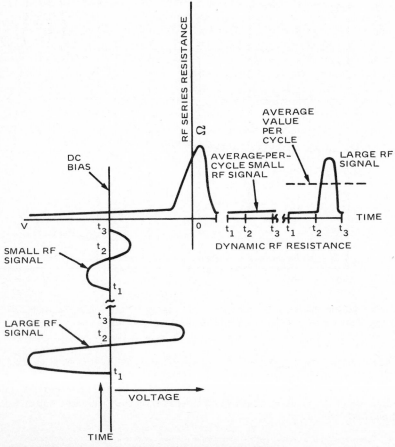

FIG. 20-26 High-level limiting in reverse-biased pin diodes.

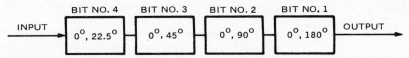

FIG. 20-27 Schematic of transmission 4-bit phase shifter.

Diode Phase-Shifter Circuits A schematic diagram of a general 4-bit phase shifter which gives 16 steps in increments of 22.5° is shown in Fig. 20-27. Any of the phase bits in the figure may take on three basic circuit forms (shown below):

1 Switched-line phase bit (Fig. 20-28)

2 Hybrid-coupled phase bit (Fig. 20-29)

3 Loaded-line phase bit (Fig. 20-31)

In the loaded-line-circuit form a phase bit may consist of a number of identical sections. The choice of circuit form of the phase bit depends on factors such as the number of diodes required, power level, insertion loss, and fabrication ease, which relates to cost. A comparison of these three circuit forms with regard to their loss and peak-power capacity is given in Table 20-6. For comparison purposes a minimum-loss condition has been assumed for all circuit forms. This implies equal loss in each bit state.

Table 20-7 compares the diode loss and number of diodes required for 4-bit-transmission phase shifters. The loss of the entire phase-shifter package will be greater than these values since the circuit loss must be added.

Switched-Line Phase Bit The switched-line phase shifter is shown in Fig. 20-28. This circuit consists basically of two single-pole, double-throw switches and two line lengths for each bit. The minimum number of diodes per bit is four. The line lengths are arbitrary; hence the circuit can be used for either a 0° to 360° phase shifter or a time delayer. The isolation per switch in the off branch must be greater than about 20 dB to avoid phase errors and sharp loss peaks,[31] which are due to a resonance effect occurring between the off diodes for critical line lengths of multiples of approximately $\lambda/2$.

TABLE 20-6 Comparison of Diode Phase-Bit Types for Peak Power and Diode Loss

	Loaded line		Hybrid-coupled	Switched line
	Stub-mounted	Mounted in main line		
Peak-power capacity*	$\dfrac{V_d^2}{8Z_0 \sin^2\dfrac{\phi}{2}}$	$\dfrac{V_d^2}{8Z_0}$	$\dfrac{V_d^2}{4Z_0 \sin^2\dfrac{\phi}{2}}$	$\dfrac{V_d^2}{2Z_0}$
Loss, dB	$17.4\dfrac{f}{f_c}\tan\dfrac{\phi}{2}$ (per section)	$17.4\dfrac{f}{f_c}\tan\dfrac{\phi}{2}$ (per section)	$17.4\dfrac{f}{f_c}\sin\dfrac{\phi}{2}$ (per bit)	$17.4\dfrac{f}{f_c}$ (per bit)

*V_d = peak radio-frequency voltage rating of diode; ϕ = phase shift.

TABLE 20-7   Theoretical Diode Loss and Number of Diodes Required for 4-Bit Transmission Phase Shifters of Different Types

	Loaded line*	Hybrid-coupled	Switched line
Total diode insertion loss	$52\dfrac{f}{f_c}$	$40\dfrac{f}{f_c}$	$69\dfrac{f}{f_c}$
Number of diodes	30	8	16

*Fifteen sections of 22.5° phase shift each are assumed.

Hybrid-Coupled Bit   The hybrid-coupled-bit phase shifter (Fig. 20-29) has a 3-dB hybrid junction with balanced phase bits connected to the coupled arms. Alternatively, a circulator with a single phase bit could be used if a nonreciprocal phase shifter is desired. The breakdown voltage required of the diodes depends on the bit size in which the diode is used, if equal power incident on all cascaded bits is assumed. The requirement is highest for the 180° bit but is reduced by a factor of $\sqrt{\sin(\phi/2)}$ for smaller bits. Insertion loss is also a function of bit size. If the loss of the 180° bit is L_0, the loss of the smaller bits is $L_0 \sin(\phi/2)$. The hybrid-coupled-bit phase shifter has the least loss of the three types and uses the smallest number of diodes.

An analysis for the hybrid-coupled 180° phase bit shows that the peak-power capacity and optimum impedance level for equal loss in both switch states are given by

$$P = \frac{1}{4}\frac{V_d^2}{Z_0} = \frac{1}{4}\frac{V_d^2}{X_c}\sqrt{\frac{R_R}{R_F}} \qquad (20\text{-}42)$$

$$Z_0 = X_c\sqrt{\frac{R_F}{R_R}} \qquad (20\text{-}43)$$

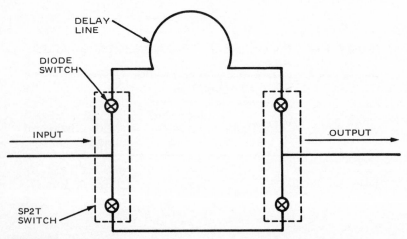

FIG. 20-28   Switched-line phase bit.

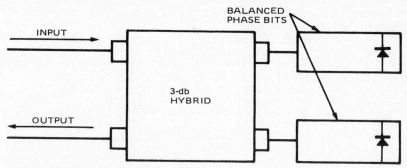

FIG. 20-29 Hybrid-coupled phase bit.

This shows that to obtain high-power capacity, high capacitance, and high breakdown voltage pin diodes operating at low impedance levels should be used.

The shorter bits can be obtained from the 180° bit by using the transformed-switch method.[32] This approach maintains equal loss in the two states of the bit and also gives less loss and greater peak power as the phase-step size is decreased. The transformed-switch method can also be designed to give either a constant or a linear phase-frequency characteristic. The transformed-switch technique in concept uses an ideal transformer placed an eighth wavelength in front of the 180° bit (Fig. 20-30). The impedance-transformation ratio of the transformer is varied to produce the various phase-bit sizes.

Loaded-Line Phase Bit The loaded-line circuit uses switched loading susceptances spaced a quarter wavelength along a transmission line (Fig. 20-31). Adjacent loading susceptances are equal and are switched into either a capacitive or an inductive state. Impedance-matched transmission for both states is maintained by correctly choosing the impedance level of the transmission-line section between diodes.

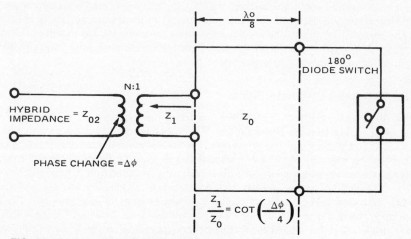

FIG. 20-30 Transformed-switch circuit.

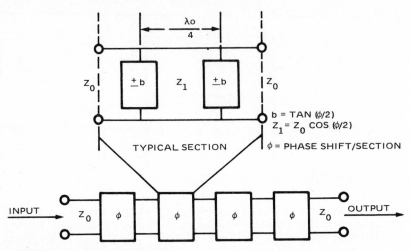

FIG. 20-31 Loaded-line phase bit.

The diodes can be either directly mounted or stub-mounted across the main line.[33] In the directly mounted configuration no peak-power advantage is achieved by using many small phase steps. Power in this circuit is limited by the breakdown voltage of the diode and the level to which the characteristic impedance of the line can be reduced, as indicated in Table 20-6. In the stub-mounted configuration the diode switches are mounted in shunt stubs which are in shunt with the main line. The peak-power capacity for the configuration, as shown in Table 20-6, is a function of the voltage rating of the diode and phase-step size. For equal insertion loss in each state of the phase step, the peak-power capacity for the loaded-line phase shifter is one-half of that of the hybrid-coupled-bit design. The insertion loss of n small steps cascaded to give 180° phase shift is $\pi/2$ times as great as the loss of the 180° hybrid-coupled-bit circuit. To achieve high-power capacity the loaded-line phase-shifter circuit uses many diodes and small phase increments. In applications in which maximum peak-power capacity is not a primary goal, the number of diodes can be greatly reduced. The phase-shifter design then reduces to a filter problem, and the number of diodes required becomes a function of the maximum voltage standing-wave ratio (VSWR) permitted in the phase-shifter passband.

Performance Characteristics

Ferrite Phase Shifter Table 20-8 compares the performance of two types of toroidal analog latching phase shifters. The low-power unit uses air cooling and is rated at 10 kW peak and 100 W average. The high-power unit uses liquid cooling and has a power capacity of 100 kW peak and 400 W average. Insertion losses for these types of units are generally under 1 dB at operating frequencies at X band and below.

Figure 20-32 shows an early high-power S-band Reggia-Spencer phase shifter. This unit used cold-plate liquid cooling and was rated at 50-kW peak and 150-W average power. Figure 20-33 shows a direct liquid-cooling technique[34] using a Teflon encapsulating jacket around the Reggia-Spencer phase-shifter ferrite bar. Table 20-9 lists the characteristics of these devices at S and C bands.

TABLE 20-8 Performance Characteristics of a Low- and a
High-Power Toroidal Analog Latching Phase Shifter*

	Low power	High power
Frequency	5.4–5.9 GHz	5.4–5.9 GHz
Maximum phase shift	360°	360°
Phase-shift increments	5⅝°	22.5°
Peak power	10 kW	100 kW
Average power	100 W	400 W
Type of cooling	Natural convection	Liquid-cooled cold plate
Insertion loss	0.7 dB	0.9 dB
VSWR	1.20 max	1.25 max
Switching time	10 μs	10 μs
Switching energy	800 μJ	950 μJ

*Courtesy of Hughes Aircraft Company.

Table 20-10 shows the typical performance of Faraday rotator phase shifters. One is an S-band liquid-cooled 110-kW-peak device,[35] and the other is an air-cooled X-band unit. The insertion losses for the two units are 1.0 and 0.9 dB respectively.

Semiconductor-Diode Phase Shifters   Figure 20-34 shows a typical air-strip-line-diode-phase-shifter module. The unit consists of a 4-bit phase shifter, a power divider, and an additional subarray steering bit, all integrated in a common package. The 4-bit phase shifter uses hybrid-coupled circuits for the three larger bits and a loaded-line circuit for the small bit. The unit is rated at 1 kW peak, 40 W average.

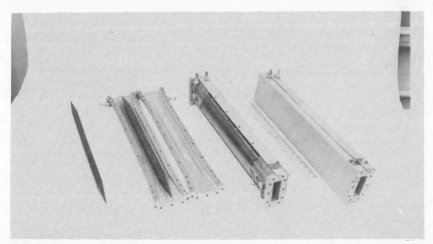

FIG. 20-32   An S-Band Reggia-Spencer phase shifter, the first large-scale application of ferrite phase shifters used in the AN/SPS-33 phase-frequency electronic-scanning radar system. The unit had an insertion loss of 1.6 dB and was rated at 50-kW peak and 150-W average power.

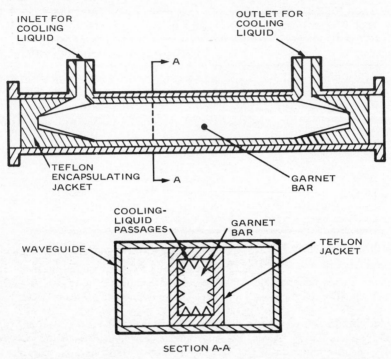

FIG. 20-33 Direct dielectric liquid-cooling technique used for high-average-power handling. *(After Ref. 34.)*

TABLE 20-9 Typical Parameters of S- and C-Band High-Power Reggia-Spencer Phase Shifters*

	S Band	C Band
Frequency	2.95–3.25 GHz	5.275–5.725 GHz
Maximum phase shift	360°	360°
Phase-shift increments	Analog	Analog
Peak power	60 kW	110 kW
Average power	600 W	600 W
Type of cooling	Direct dielectric liquid	Direct dielectric liquid
Insertion loss	1.0 dB	0.9 dB
VSWR	1.20 max	1.25 max
Switching time	700 μs	125 μs
Switching power at 300-Hz switching rate	19 W	16 W

*Courtesy of Hughes Aircraft Company.

TABLE 20-10   Typical Performance of Faraday
Rotator Phase Shifters

	S band, high power*	X band, low power*
Phase shift	360°	360°
Bandwidth	3 percent	10 percent
Peak power	110 kW	1.5 kW
Average power	1500 W	15 W
Type of cooling	Liquid	Air
Insertion loss	1.0 dB	0.9 dB

*After Ref. 35.

Figure 20-35 shows a low-power 5-bit integrated-circuit phase shifter using a microstrip construction. Switched lines are used for the 3 large bits, and series-coupled loaded-line circuits are used for the 2 smaller bits. The main advantage of this type of unit is size: 1 by 2 in (25.4 by 50.8 mm). The average insertion loss is 1.8 dB.

Figure 20-36 shows a 4-bit C-band loaded-line phase shifter. A total of 30 diodes mounted in shunt stubs are used. The unit has a peak power capacity of 15 kW and

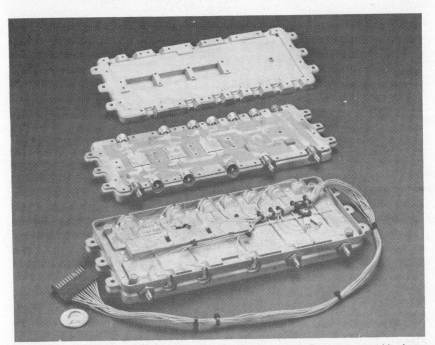

FIG. 20-34   Air-stripline-diode-phase-shifter module. The unit incorporates a 4-bit phase shifter, a power divider, and an additional loaded-line bit in one package. A total of 10 pin diodes are used.

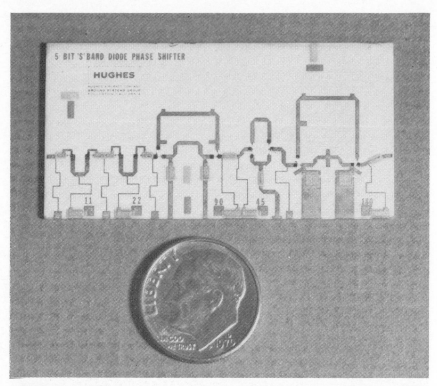

FIG. 20-35 Five-bit micromin phase shifter fabricated with thick film printing on a 1- by 2-in (25.4- by 50.8-mm) alumina substrate. The 3 long bits are frequency-compensated switched-line circuits, and the 2 short bits use series-coupled loaded-line circuits. A total of 16 pin diodes are used in the device.

200 W average. This phase shifter operates over a 15 percent bandwidh with an average insertion loss of 1.3 dB.

Figure 20-37 shows an ultrahigh-frequency phase shifter using a microstrip construction on a low-dielectric-constant substrate. The unit uses hybrid-coupled circuits for the 90° and 180° bits. Loaded-line circuits are employed for the 22° and 45° bits. The unit operates at a peak power of 8 kW, 240 W average, with an insertion loss of 0.7 dB.

20-5 FEED-NETWORK DESIGN FOR PHASED ARRAYS

This section presents a survey of the various methods of feeding the array elements to form the required beam patterns in space. For simplicity, linear-array configurations will be used as illustrations. Extrapolation to a planar array can be accomplished either with a single technique or by using a combination of two different techniques, for example, one technique for rows and the other for columns. In general, most of the beam-forming feeds for phased arrays can be categorized into two basic groups,

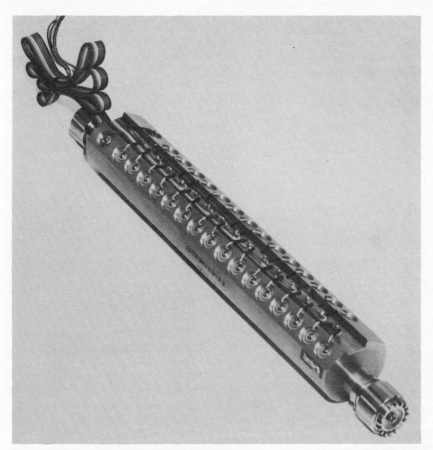

FIG. 20-36 Four-bit C-band coaxial loaded-line phase shifter. The unit uses 30 diodes mounted in shunt stubs along the main transmission line. The power capacity is 15 kW peak, 200 W average. *(Courtesy of J. F. White, M/A-COM, Inc.)*

(1) optical space feeds and (2) constrained feeds. A description of the various types of feeds in each of these groups is given below.

Optical Space Feeds

Illumination of the array elements in an optically space-fed antenna is accomplished by optically distributing the source signal (transmitter) through space, illuminating an array of pickup horns (elements) which are connected to the radiating elements. The advantage of an optical beam-forming feed over the constrained-feed system is simplicity. Disadvantages are a lack of amplitude-tapering control and an excessive volume of physical space required to accommodate the feed system. Two basic types of optical space feeds, the transmission type and the reflection type, are shown in Fig. 20-38.

In the transmission type, the array elements are the radiating elements of a feed-

FIG. 20-37 UHF diode phase shifter with cover removed. A microstrip circuit on a ⅛-in-(3.18-mm-) thick low-dielectric-constant board is used. A total of eight diodes are used in the circuit.

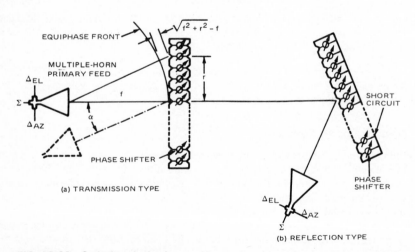

FIG. 20-38 Optical space feeds.

through lens as shown in Fig. 20-38a. These radiating elements are coupled to the pickup elements of the feed-through lens by phase shifters. Both surfaces of the lens require matching to optimize performance. The pickup surface is illuminated by a primary feed placed at a focal distance f behind the lens. The radiation pattern of the primary feed must be optimized to provide an efficient aperture illumination with little spillover loss. If desired, separate transmit and receive feeds may be used, with the feeds separated by an angle α as shown in Fig. 20-38a. The antenna is then rephased between transmitting and receiving so that in both cases the beam points in the same direction. The phasing of the antenna must include a correction for the spherical phase front of the feed. The required phase correction is given by

$$\phi_c = \frac{2\pi}{\lambda}(\sqrt{f^2 + r^2} - f) = \frac{\pi}{\lambda}\frac{r^2}{f}\left[1 - \frac{1}{4}\left(\frac{r}{f}\right)^2 + \cdots\right]$$

With a sufficiently large focal length, the spherical phase front may be approximated by that of two crossed cylinders, permitting the correction to be applied simply with row and column steering commands. The correction of the spherical phase front can be accomplished by the phase shifters. Space problems may be encountered in assembling an actual system, especially at higher frequencies, since all control circuits must be brought out at the side of the aperture. Multiple beams may be generated by adding additional primary feed horns. All the multiple beams will be scanned simultaneously by equal amounts in sin θ space. Monopulse sum-and-difference patterns may be generated by taking the sum and difference of two adjacent multiple beams with a magic T or by using multimode feed horns.

In the reflection type, the phased-array aperture is used as a reflector as shown in Fig. 20-38b. The same radiating element collects and reradiates the signal after it has been reflected from the short circuits terminating the phase shifters. The phase of the reflected signal is determined by the phase-shifter setting. The phase shifter must be reciprocal so that there is a net controllable phase shift after the signal passes through the device in both directions. This requirement rules out nonreciprocal phase shifters. Ample space for phase-shifter control circuits exists behind the reflector. To avoid aperture blocking, the primary feed may be offset as shown. As in the case of a transmission lens, transmit and receive feeds may be separated and the phases separately computed for the two functions. Multiple beams are again possible with additional feeds.

To achieve more precise control in amplitude tapering, a multiple-beam feed array, instead of a single feed horn, must be used to illuminate the feed-through lens. One approach is shown in Fig. 20-39, in which the output of a multiple-beam Butler matrix is used to feed the radiating elements of a feed array.[36,37,38] Each input terminal of the multiple-beam Butler matrix controls effectively the amplitude illumination of a portion of the feed-through lens (radiating aperture). Hence, the amplitude tapering across the array aperture can be controlled by adjusting the signal levels at the inputs of the multiple-beam Butler matrix. The feed system also provides wide instantaneous bandwidth (~ 10 percent) by using time-delay phase shifters at the inputs to the multiple-beam matrix.

Constrained Feeds

The various types of constrained feeds can be classified into two groups: the series feeds and the parallel feeds. In each group, the feeds can be designed to provide either

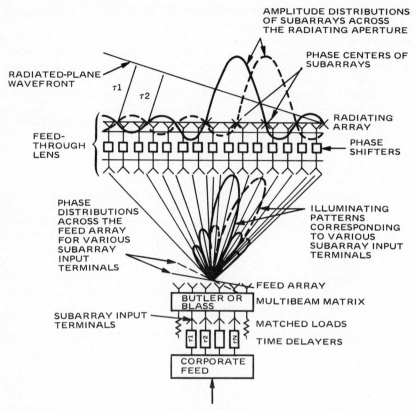

FIG. 20-39 Wide instantaneous-bandwidth feed system using completely overlapped subarrays.

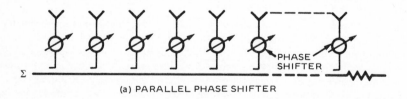

(a) PARALLEL PHASE SHIFTER

(b) SERIES PHASE SHIFTERS

FIG. 20-40 End-fed series feeds.

a single pencil beam or monopulse sum-and-difference beams. A description of the various types of feeds in each group follows.

Series Feeds   Figure 20-40 illustrates several types of series feeds. In all cases the path length to each radiating element must be computed as a function of frequency and taken into account when setting the phase shifter. Figure 20-40*a* and *b* illustrates two examples of end-fed serial feed systems in which the elements are arranged serially along the main line. They consist of a main transmission line from which energy is tapped (in the case of transmission) through loosely coupled junctions to feed the radiating elements. To steer the beam, phase shifters are added in either the branch lines feeding the radiating elements, as shown in Fig. 20-40*a*, or in the main line, as shown in Fig. 20-40*b*. The amplitude taper is established by properly designing the couplers at the junctions. For example, if all the junction couplers are identical, the amplitude-taper envelope will be approximately exponential.

Mechanical simplicity is, perhaps, the greatest advantage of these two configurations over others to be discussed. They are easy to assemble and to construct. The configurations are easily adapted to construction in waveguides, using cross-guide directional couplers as junctions. They are potentially capable of handling full waveguide power at the input (within the limitations of the phase shifters). However, they suffer losses associated with the corresponding length of waveguide plus that of the

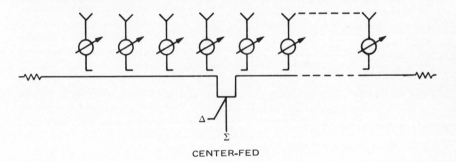

CENTER-FED

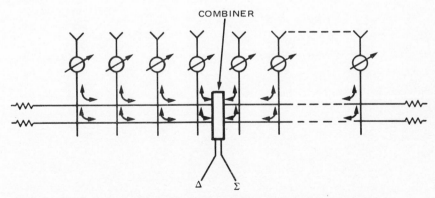

FIG. 20-41   Center-fed series feeds with monopulse sum-and-difference beams.

phase shifters. Perhaps the most severe limitation of this feed system is its dependence of pointing angle on frequency.

The configuration of Fig. 20-40a places lower power-handling demands on the phase shifters and also results in a lower system loss for a given phase-shifter loss. On the other hand, the configuration of Fig. 20-40b has an advantage in that all phase shifters will have identical phase shifts for a given pointing angle—a property which results in simplified array control. However, the total system loss of Fig. 20-40b is higher than that of Fig. 20-40a.

Monopulse sum-and-difference beams can be formed by feeding the array in the middle instead of from the end (see Fig. 20-41a). This approach, however, would not be able to provide good sum-and-difference patterns simultaneously. For example, if the couplers in the feed are designed to provide low sidelobes for the sum pattern, the resultant pattern for the difference beam would have high sidelobes. At the cost of some additional complexity this sidelobe problem can be solved by the method shown in Fig. 20-41. Two separate center-fed feed lines are used and combined in a network to give sum-and-difference-pattern outputs.[39] Independent control of the two amplitude distributions is possible. For efficient operation the two feed lines require distributions that are orthogonal to each other; that is, the peak of the pattern of one feed line coincides with the null from the other. The aperture distributions are respectively even and odd.

The bandwidth of a series feed can be increased by making the path lengths from the input to each output of the branch lines all equal as shown in Fig. 20-42. However, if the bandwidth is already limited by the phase shifters and the couplers, very little benefit can be derived from this approach at the cost of a considerable increase in size and weight. The network of Fig. 20-42 simplifies the beam-steering computation since the correction for path-length differences is no longer necessary.

Parallel Feeds   The frequency dependence associated with series feeds can be reduced by the use of parallel feeds at a cost of a slightly more complex mechanical structure. The frequency dependence (change in beam-pointing angle with frequency)

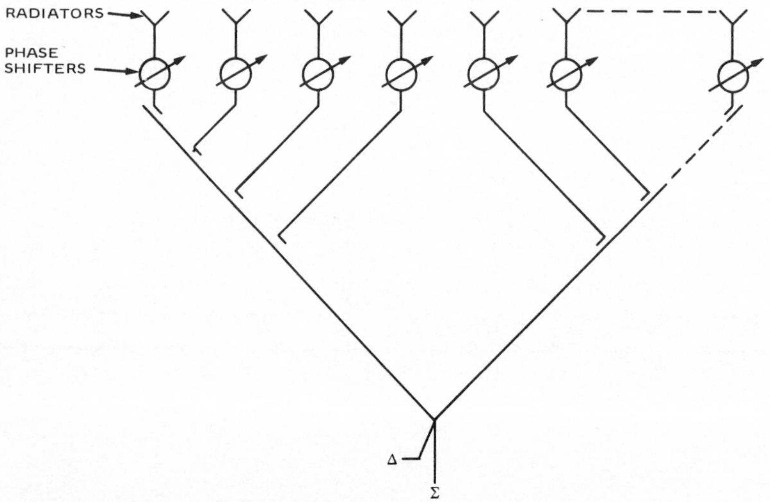

FIG. 20-42   Equal-path-length feed.

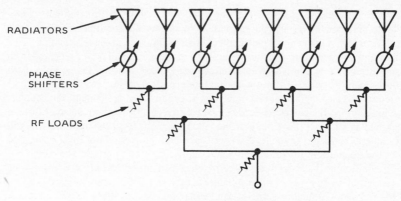

RADIATORS

PHASE
SHIFTERS

RF LOADS

(a) CORPORATE FEED

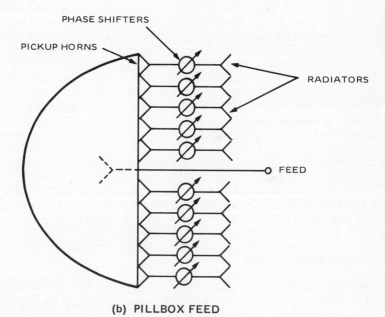

PHASE SHIFTERS

PICKUP HORNS

RADIATORS

FEED

(b) PILLBOX FEED

FIG. 20-43 Constrained parallel feeds. *(After Ref. 40.)*

of parallel-feed configurations depends mainly on the frequency characteristics of the phase shifters and the couplers at the junctions of the feed, if the line lengths from the transmitter to the radiators are all made equal. Thus, if variable time-delay phase shifters were used in place of the constant-phase type of phase shifters at each radiating element, the beam-pointing angle would be essentially independent of frequency. The frequency dependence of a parallel feed using constant-phase-type phase shifters is discussed in Sec. 20-6.

The most common technique for realizing parallel feeds is the *matched-corporate-feed structure* of Fig. 20-43 *a*. In this configuration, the corporate feed is assem-

bled from matched hybrids (four-port junction devices with the isolated port terminated into matched loads). The out-of-phase components of mismatch reflections from
the aperture and of other unbalanced reflections are absorbed in the terminations. The
in-phase and balanced components are returned to the input, and power reflected from
the aperture is not reradiated. When nonreciprocal phase shifters (the two-way path
length through the phase shifter is a constant independent of the phase-shifter setting)
are used in place of reciprocal phase shifters at each radiating element, the couplers
at the junctions of the corporate feed can be reactive T's (three-port devices) instead
of matched hybrids. In this case, however, the reflections from the aperture will add
up in phase at the input to the feed. Therefore, a high-power isolator must be used to
protect the transmitter and/or a high-power limiter be used to protect the receiver.

   In addition to the transmission-line parallel feed, an optical *pillbox feed* can be
used to illuminate an array of pickup elements to feed a linear array as shown in Fig.
20-43 *b*. The array of pickup elements can be matched to the pillbox so that no reflections occur; hence, all the energy is collected by the pickups, and the loss incurred is
only the normal RF loss. By the use of a folded pillbox,[40] the feed can be made not to
interfere with the pickup structure. If the back surface of the pillbox is made spherical
instead of parabolic, the feed can provide many multiple simultaneous beams with the
use of multiple feed horns. The cost of substituting the pillbox for a corporate feed is
an exchange of a certain amount of freedom in the choice of amplitude tapers and
packaging configurations for a simpler mechanical structure.

   A method of achieving sum-and-difference monopulse beams for a parallel feed
is shown in Fig. 20-44. The signals from a pair of radiating elements, which are
located symmetrically opposite from the centerline of the array, are combined in a
magic T (matched hybrid) to form their sum-and-difference signals. The sum signals
from all the pairs across the array aperture are then combined in a power-combiner
network to form a sum beam. The desired amplitude distribution for the sum beam
can be achieved by proper weighting of the signals in the power-combiner network.
The difference signals from all the pairs are combined in a separate power-combiner
network to form a difference beam. Amplitude weighting in the two combiner networks can be made differently to obtain low sidelobes for both the sum and the difference beams. For example, the low-sidelobe amplitude distribution for a sum beam
is in the form of a Taylor distribution, whereas a Bayliss distribution is used for low-
sidelobe difference beams.[41] By using this method, independent control of both the sum
and the difference beams is achieved.

Multiple-Beam Feeds   Another class of constrained feeds for phased arrays consists of the multiple-beam-forming feed networks. These feeds can generate multiple
simultaneous beams covering a large sector of space. For the same beam-pointing
direction, each beam has essentially the gain of a single-beam array of the same size
and illumination except for a possible increase in RF circuit losses over a single-beam
array. A separate beam terminal is provided for each beam of the multiple-beam-
forming feed network.

   Various types of multiple-beam-forming networks can also be classified into two
groups: (1) the serial type and (2) the parallel type. One example of a serial multiple-
beam-forming feed is shown in Fig. 20-45. This beam-forming network, known as the
Blass matrix,[42] is based on the serial feed of Fig. 20-40 *b*. To generate multiple beams,
several feed lines are coupled to the branch lines by using directional couplers. Each
feed line generates one beam in space. If a transmitter is connected to any of the beam

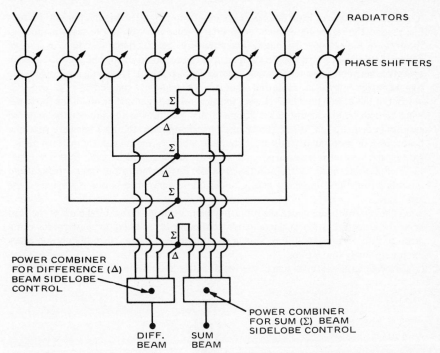

FIG. 20-44 Constrained monopulse feed with independently controllable sum-and-difference-beam sidelobes.

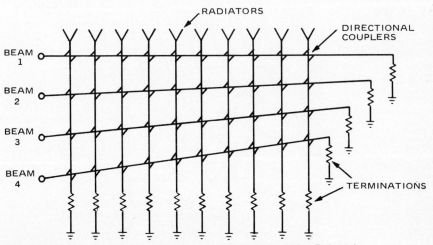

FIG. 20-45 Series-fed simultaneous beam-forming matrix. *(After Ref. 42.)*

terminals (feed lines), each coupler leaks off a fraction of the input signal in the feed line toward the radiating element connected to that junction in the same manner as the array of Fig. 20-40 *b*. In fact, if it were not for the additional feed lines in front of a given feed line (toward the array´aperture), there would ideally be no difference in operation between this configuration and that of Fig. 20-40. However, a portion of the signal coupled toward the radiating element is subsequently coupled into the feed lines in front of the driven feed line. Hence, this cross-coupling effect between the feed lines must be taken into account in the design of this feed network to provide the required amplitude distribution. When the beams corresponding to the feed lines are spaced a beamwidth or more apart, this cross-coupling effect is minimized, the pattern generated by each effect is minimized, and the pattern generated by each feed line is not significantly degraded by the additional feeds. Furthermore, the signals leaked into the other feed line sums very nearly to zero, and very little power is wasted in the terminations.

The phase shift between the radiating elements is governed by the *tilt* of the feed lines. This configuration is particularly adaptable to rectangular waveguides using cross-guide directional couplers. Such a waveguide feed is capable of high-power operation with fairly low RF loss.

A wide-bandwidth version of the Blass matrix is shown in Fig. 20-46. The branch

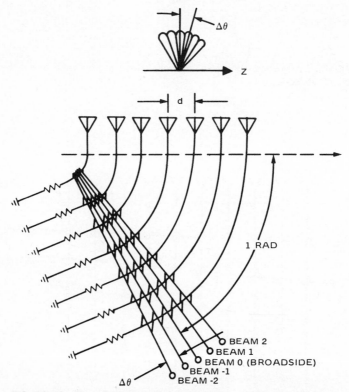

FIG. 20-46   Time-delayed multiple-beam matrix. *(After Ref. 42.)*

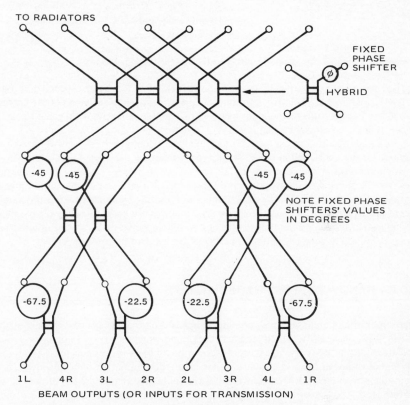

FIG. 20-47 Multiple-beam Butler-matrix feed. *(After Ref. 43.)*

lines to the radiating elements are circumferences of circles of radii nd. The feed line for the broadside beam is placed at an angle of 1 rad from a line parallel to the antenna face. Additional feed lines are spaced angularly by multiples of $\Delta\theta$, so that the signal travels a different distance $nd\Delta\theta$ to reach the nth radiating element. If the transmission lines have a constant velocity of propagation v, the beam-pointing angle of the mth beam ψ_m is given by the solution of $\sin\psi_m = c/v(m\Delta\theta)$. When $v = c$, this feed network becomes a true time-delay feed and the bandwidth is limited only by the directional couplers.

An example of a parallel-fed multiple-beam-forming feed is shown in Fig. 20-47. This feed configuration is commonly known as the Butler-matrix feed.[43,44] The feed network uses directional couplers that have a 90° phase-shift property (e.g., branch-line couplers) and fixed phase shifters as indicated in Fig. 20-47.

A signal injected at any of the beam input terminals excites all the radiating elements equally in amplitude with phase differentials of odd multiples of $180°/N$, where N is the total number of radiating elements or beam terminals. The array patterns of the beams generated are of the form $\sin Nx/\sin x$ since the amplitude taper is uniform. Specifically, they are

$$E_m(\nu) = \frac{\sin \dfrac{N}{2} \left[kd - (2m + 1)\dfrac{\pi}{N} \right]}{\sin \dfrac{1}{2} \left[kd - (2m + 1)\dfrac{\pi}{N} \right]} \qquad (20\text{-}44)$$

with N main beams completely covering one period of ν. The beams specified by Eq. (20-44) cross over at a relative amplitude of $2/\pi$ (≈ 4 dB), and the peaks of the beams are located at the nulls of the other beams. Since these beams are orthogonal to each other, there is no cross-coupling loss between beams.

The construction of the parallel multiple-beam configuration is not as straight-forward as the construction of the serial type because of the large number of trans-mission-line crossovers required. If the formation of a large number of beams is desired, the parallel configuration is preferred since it uses fewer couplers for a given number of elements. If N is the total number of radiating elements and B is the number of beams required, the serial configuration requires NB couplers, while the parallel configuration requires essentially $N/2 \log_2 N$ couplers regardless of the number of beams required as long as the total number of beams does not exceed N.

20-6   BANDWIDTH OF PHASED ARRAYS

The bandwidth of a phased-array antenna depends upon the types of components, such as radiators, phase shifters, and feed networks, comprising the array. In practice, most radiators for phased arrays are matched over a broad band of frequencies. Therefore, the radiator design is not a primary factor in the determination of bandwidth. The limitations on bandwidth are determined by the frequency characteristics of the phase shifters and feed networks.[45,46] In general, the effects due to the phase shifters and feed networks are additive, so that if the phase shifter causes the beam to scan by an amount equal to $\Delta\theta_p$ and the feed causes the beam to scan by $\Delta\theta_f$, then the total beam scan is given by $\Delta\theta_p + \Delta\theta_f$. To establish the bandwidth capability of a phased array, these two effects must be examined. A detailed evaluation of the bandwidth limita-tions of phased arrays is discussed by Frank.[45]

Phase-Shifter Effects

To evaluate the effects caused by the phase shifter alone, a corporate feed (equal-line-length parallel feed) is used to illuminate all the radiating elements (see Fig. 20-48). This corporate feed exhibits no feed effects since a signal at the input illuminates all the radiating elements with the same phase regardless of frequency. The bandwidth in this case is completely determined by the type of phase shifter used in the array. In this discussion, we will consider the effects of two basic types of phase shifters: (1) time-delay phase shifters and (2) constant-phase-type phase shifters. A detailed description of these two types of phase shifters is given in Sec. 20-4.

When time-delay phase shifters are used at each radiating element, the signals received by the elements from an incident wavefront at an angle θ_0 are appropriately time-delayed so that they all arrive at the output terminal of the corporate feed at the same time. For example, the amount of time delay at the first element of the array shown in Fig. 20-48 is equal to the additional time required for the wavefront to travel

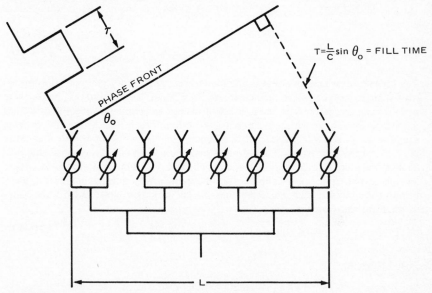

FIG. 20-48 Equal-line-length corporate feed. (*After Ref. 45.*)

to the last element after arriving at the first element. This time delay T, known as aperture-fill time, is given by

$$T = \frac{L}{c} \sin \theta_0$$

where L = total length of array aperture
   c = velocity of light
   θ_0 = angle of incidence of wavefront from array normal

In this case, the phase distribution across the array aperture produced by the time-delay feed matches that of the incident wavefront independently of frequency. Consequently, the beam position remains stationary with frequency change, and the array has infinite bandwidth. When the angle of incidence of the incoming wavefront changes, the amounts of time delays at each element must be changed accordingly to maintain the bandwidth. This time-delayed feed network is commonly known as a time-delayed beam-steering feed. The above discussion is valid for either a CW or a pulsed incidence signal. For the pulsed incidence signal, the time-delay feed preserves the shape of the pulse without any distortion. Various methods of time-delay beam steering are described in Ref. 46.

When constant-phase-type phase shifters (phase shifters whose phase shift is independent of frequency) are used at each element, the output phase distribution of the feed matches that of the incident phase front *only* at *one* frequency f_0 and for a particular incidence angle θ_0. At a different frequency f_1, the output phase distribution of the feed network remains fixed; hence the array is phased to receive at a difference incidence angle θ_1. The amount of beam squint with frequency is given by the following relationship:

$$f_1 \sin \theta_1 = f_0 \sin \theta_0 \qquad\qquad \textbf{(20-45)}$$

For a small change in frequency, Eq. (20-45) shows that the change in scan angle is given by

$$\Delta\theta_0 = -\left(\frac{\Delta f}{f}\right)\tan\theta_0 \qquad (20\text{-}46)$$

The above expression shows that the amount of beam squint depends upon the original scan angle as well as on the percent frequency change. At broadside ($\theta_0 = 0$), there is no scanning regardless of the amount of change in frequency, and the array has infinite bandwidth. When the beam scans away from broadside, the amount of beam squint with frequency increases with scan angle. Therefore, the bandwidth of an array must be specified in terms of the desired maximum scan angle. For most practical applications, the desired maximum scan angle is $\pm 60°$ from array broadside. If we assume that the maximum allowable beam squint is ± 1 quarter beamwidth away from the desired direction, corresponding to a one-way gain loss of 0.7 dB, the bandwidth of the array is given by

$$\text{Bandwidth (percent)} = \text{beamwidth (degrees)} \qquad (20\text{-}47)$$

Feed Effects

When a feed other than an equal-length parallel feed is used, phase errors due to the feed alone are produced across the array aperture. An example is the end-fed series feed shown in Fig. 20-49. The total phase shift across the length of the feed is $\Phi = (2\pi/\lambda)L$ rad with free-space propagation assumed. When the frequency is changed, the change in phase across the array aperture will be

$$\Delta\Phi = \frac{2\pi L}{c}\Delta f$$

This linear change in phase across the aperture scans the beam just as phase shifters would. To observe just how far the beam is scanned, let us examine the way in which an aperture is scanned with phase. For a given scan angle θ_0, the required phase across the array is

$$\psi = \frac{2\pi L}{\lambda}\sin\theta_0$$

The required change in ψ for a change in scan angle is

$$\frac{d\psi}{d\theta} = \frac{2\pi L}{\lambda}\cos\theta$$

or

$$\Delta\psi = \frac{2\pi L}{\lambda}\cos\theta_0\Delta\theta_0 = \frac{2\pi Lf}{c}\cos\theta_0\Delta\theta_0$$

when the change in phase across the array is induced by the feed, $\Delta\psi = \Delta\Phi$, or

$$\frac{2\pi L}{c}\Delta f = \frac{2\pi Lf}{c}\cos\theta_0\Delta\theta_0$$

Hence the amount of beam scan for a change in frequency is given by

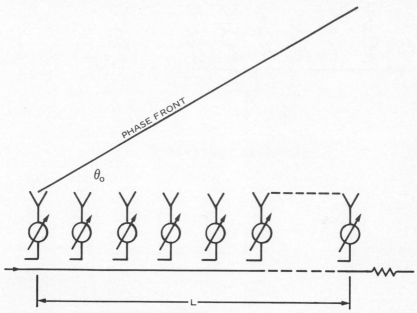

FIG. 20-49   End-fed series feed. *(After Ref. 45.)*

$$\Delta\theta_0 = \left(\frac{\Delta f}{f}\right)\frac{1}{\cos\theta_0} \qquad\qquad \textbf{(20-48)}$$

For beams scanned in the direction of the feed load, the phase change between the radiating elements due to the feed line is partially offset by the required phase change to maintain the beam position at a given scan angle θ_0. In fact, for an end-fire beam toward the load, the required phase change with frequency to maintain the beam at end fire is exactly the same as that due to the feed line. Hence, the bandwidth of a series feed at the end-fire-beam position toward the load is infinite. However, this end-fire-beam position is not practical to implement because of the drastic reduction in the gain of the radiating elements. Furthermore, in most applications the array is required to scan in both directions from broadside. For beams scanned in the direction of the feed input, the phase changes in space and the feed line become additive instead of canceling each other. As a result, the beam scans rapidly with frequency. For a 60° scan angle, the scanning caused by the feed alone is slightly greater than that caused by the aperture. Therefore, the bandwidth of the series feed is essentially half of that of the equal-line-length parallel feed.

In waveguide, the signal propagates more slowly; this is equivalent to having a longer feed line. This causes the feed to scan more rapidly, and the series-feed performance degrades accordingly. The slowing in waveguide is proportional to λ/λ_g, and the array scans by

$$\Delta\theta \cong \frac{\lambda_g}{\lambda}\left(\frac{\Delta f}{f}\right)\frac{1}{\cos\theta_0} \qquad \text{rad} \qquad\qquad \textbf{(20-49)}$$

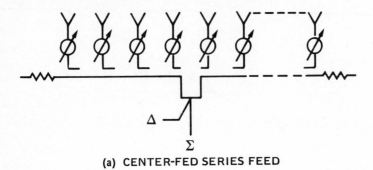

(a) CENTER-FED SERIES FEED

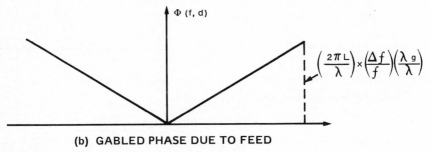

(b) GABLED PHASE DUE TO FEED

FIG. 20-50   Center-fed series array. *(After Ref. 45.)*

For $\pm 60°$ scanning from broadside, the bandwidth of a series feed under the same conditions as that of a parallel feed is given by

$$\text{Bandwidth (percent)} = \frac{1}{\left(1 + \dfrac{\lambda_g}{\lambda}\right)} \quad \text{beamwidth (°)} \quad \textbf{(20-50)}$$

A center-fed series array[47] as shown in Fig. 20-50a may be thought of as two end-fed series arrays. At band center the phase shifters are set so that all the elements radiate in phase. As the frequency is changed, each half scans in the opposite direction because of the gabled phase slope as shown in Fig. 20-50b. For the broadside-beam case, this results in a broadening of the beam with no change in direction. If the two beams were to move far enough apart (Fig. 20-51), the beams would split. However, by restricting the bandwidth to reasonable values, the result is a loss in gain caused by the beam broadening.

When the array is scanned away from broadside, aperture scanning due to the phase shifter (constant phase with frequency) is also introduced, and it is superimposed on the feed scanning. As shown in Fig. 20-51, for one-half of the array (solid curve) aperture scanning and feed scanning tend to cancel each other, while for the other half (dashed curve) they tend to reinforce each other. The performance of the center-fed array is worse than that of the parallel-fed array at broadside but quite

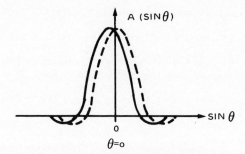

(a) FEED EFFECTS AT BROADSIDE

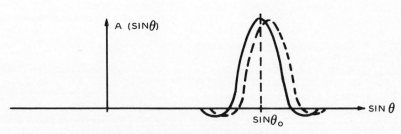

(b) APERTURE AND FEED EFFECTS

FIG. 20-51 Aperture scanning superimposed on feed scanning. *(After Ref. 45.)*

comparable at large scan angles. At a scan angle of 60°, the center-fed array will have approximately ¼ dB more loss than the parallel-fed array. For simplicity, the same bandwidth formula [Eq. (20-47)] can be used for both feeds.

The discussion above has assumed that the center feed used nondispersive transmission with free-space propagation. If waveguide is used, the performance of the center-fed array degrades and the bandwidth is given by

$$\text{Bandwidth (percent)} = \left(\frac{\lambda}{\lambda_g}\right)\text{beamwidth (degrees)} \qquad \textbf{(20-51)}$$

REFERENCES

1 S. Silver, *Microwave Antenna Theory and Design,* MIT Rad. Lab. ser., McGraw-Hill Book Company, New York, 1949.

2 S. A. Schelkunoff, "A Mathematical Theory of Linear Arrays," *Bell Syst. Tech. J.,* vol. 22, 1943.

3 J. L. Allen, "The Theory of Array Antennas," Tech. Rep. 323, Lincoln Laboratory, Massachusetts Institute of Technology, July 1963.

4 T. T. Taylor, "Design of Line-Source Antennas for Narrow Beamwidth and Low Sidelobes," *IRE Trans. Antennas Propagat.,* vol. AP-3, 1955, pp. 16–28.

5 T. T. Taylor, "One Parameter Family of Line Source Producing Modified sin $\pi\mu/\pi\mu$ Patterns," Tech. Memo. 324, Hughes Aircraft Company, Microwave Lab., Culver City, Calif., Sept. 4, 1953.

6   C. L. Dolph, "A Current Distribution for Broadside Arrays Which Optimizes the Relationship between Beamwidth and Sidelobe Level," *IRE Proc.*, vol. 34, 1946, pp. 335–348.

7   L. L. Bailin et al., "Empirical Approximations to the Current Values for Large Dolph Tchebycheff Arrays," Tech. Memo. 328, Hughes Aircraft Company, Culver City, Calif., Oct. 6, 1953.

8   T. T. Taylor, "Dolph Arrays of Many Elements," Tech. Memo. 320, Hughes Aircraft Company, Culver City, Calif., Aug. 18, 1953.

9   R. C. Hansen, "Gain Limitations of Large Antennas," *IRE Trans. Antennas Propagat.*, vol. AP-8, no. 5, 1960, pp. 490–495.

10   H. E. King, "Directivity of a Broadside Array of Isotropic Radiators," *IRE Trans. Antennas Propagat.*, vol. AP-7, no. 2, 1959, pp. 197–201.

11   E. D. Sharp, "Triangular Arrangement of Planar Array Elements That Reduces Number Needed," *IRE Trans. Antennas Propagat.*, vol. AP-9, no. 2, March 1961, pp. 126–129.

12   T. T. Taylor, "Design of Circular Apertures for Narrow Beamwidth and Low Sidelobes," *IRE Trans. Antennas Propagat.*, vol. AP-8, no. 1, 1960, pp. 17–23.

13   R. C. Hansen, "Tables of Taylor Distributions for Circular Aperture Antennas," *IRE Trans. Antennas Propagat.*, vol. AP-8, no. 1, 1960, pp. 23–27.

14   P. W. Hannan, "The Element-Gain Paradox for a Phased Array Antenna," *IEEE Trans. Antennas Propagat.*, vol. AP-12, July 1964, pp. 423–424.

15   L. Stark, "Radiation Impedance of a Dipole in an Infinite Planar Phased Array," *Radio Sci.*, vol. 1, no. 3, March 1966, pp. 361–377.

16   C. C. Chen, "Broadband Impedance Matching of Rectangular Waveguide Phased Arrays," *IEEE Trans. Antennas Propagat.*, vol. AP-21, May 1973, pp. 298–302.

17   A. A. Oliner and R. G. Malech, "Mutual Coupling in Infinite Scanning Arrays," in R. C. Hansen (ed.), *Microwave Scanning Antennas*, Academic Press, Inc., New York, 1966, chap. 3, pp. 195–335.

18   R. Tang and N. S. Wong, "Multi-Mode Phased Array Element for Wide Scan Angle Impedance Matching," *IEEE Proc.*, special issue on electronic scanning, November 1968, pp. 1951–1959.

19   S. W. Lee and W. R. Jones, "On the Suppression of Radiation Nulls and Broadband Impedance Matching of Rectangular Waveguide Phased Arrays," *IEEE Trans. Antennas Propagat.*, vol. AP-19, January 1971, pp. 41–51.

20   P. W. Hannan and M. A. Balfour, "Simulation of a Phased Array Antenna in a Waveguide," *IEEE Trans. Antennas Propagat.*, vol. AP-13, no. 3, May 1965, pp. 342–353.

21   J. Frank, "Phased Array Antenna Development," APL/JHU TP-882, March 1967.

22   B. L. Diamond, "A Generalized Approach to the Analysis of Infinite Planar Array Antenna," *IEEE Proc.*, vol. 56, November 1968, pp. 1837–1851.

23   N. Amitay, V. Galindo, and C. P. Wu, *Theory and Analysis of Phased Array Antennas*, Interscience Publishers, a division of John Wiley & Sons, Inc., New York, 1972.

24   G. V. Borgiotti, "Modal Analysis of Periodic Planar Phased Arrays of Apertures," *IEEE Proc.*, vol. 56, November 1968, pp. 1881–1892.

25   J. F. White, *Semiconductor Control*, Artech House, Inc., Dedham, Mass., 1977, chap. IX.

26   L. R. Whicker and D. M. Bolle, "Annotated Literature Survey of Microwave Ferrite Control Components and Materials for 1968–1974," *IEEE Trans. Microwave Theory Tech.*, vol. MTT-23, no. 11, November 1975, pp. 908–918.

27   L. Stark, "Microwave Theory of Phased Array Antenna—A Review," *IEEE Proc.*, vol. 62, no. 12, December 1974.

28   L. R. Whicker (guest ed.), *IEEE Trans. Microwave Theory Tech.*, vol. MTT-22, no. 6, special issue on microwave control devices for array antenna systems, June 1974.

29   L. Stark, R. W. Burns, and W. P. Clark, "Phase Shifters for Arrays," in M. Skolnik (ed.), *Radar Handbook*, McGraw-Hill Book Company, New York, 1970, chap. 12.

30   W. J. Ince and D. H. Temme, "Phasers and Time Delay Elements," *Advances in Microwave*, vol. 4, Academic Press, Inc., New York, 1969.

31 E. J. Wilkinson, L. I. Parad, and W. R. Connerney, "An X-Band Electronically Steerable Phased Array," *Microwave J.,* vol. VII, February 1964, pp. 43–48.

32 R. W. Burns and L. Stark, "PIN Diodes Advance High-Power Phase Shifting," *Microwaves,* vol. 4, November 1965, pp. 38–48.

33 J. F. White, "High Power, PIN Diode Controlled, Microwave Transmission Phase Shifters," *IEEE Trans. Microwave Theory Tech.,* vol. MTT-13, March 1965, pp. 233–242.

34 W. P. Clark, "A High Power Phase Shifter for Phased Array Systems," *IEEE Trans. Microwave Theory Tech.,* vol. MTT-13, November 1965, pp. 785–788.

35 C. R. Boyd, Jr., "A Dual Mode Latching Reciprocal Ferrite Phase Shifter," *IEEE G-MTT Int. Microwave Symp. Dig.,* 1979, pp. 337–340.

36 R. Tang, "Survey of Time-Delay Beam Steering Techniques," *Proc. 1970 Phased Array Antenna Symp.,* Artech House, Inc., Dedham, Mass., 1972, pp. 254–260.

37 F. McNee, N. S. Wong, and R. Tang, "An Offset Lens-Fed Parabolic Reflector for Limited Scan Applications," *IEEE AP-S Int. Symp. Rec.,* June 1975, pp. 121–123.

38 R. J. Mailloux, "Subarraying Feeds for Low Sidelobe Scanned Arrays," *IEEE AP-S Int. Symp. Dig.,* June 1979, pp. 30–33.

39 A. R. Lopez, "Monopulse Networks for Series Feeding an Array Antenna," *IEEE Antennas Propagat. Int. Symp. Dig.,* 1967.

40 W. Rotman, "Wide Angle Scanning with Microwave Double-Layer Pillboxes," *IRE Trans. Antennas Propagat.,* vol. AP-6, no. 1, 1958, pp. 96–106.

41 E. T. Bayliss, "Design of Monopulse Antenna Difference Patterns with Low Sidelobes," *Bell Syst. Tech. J.,* May–June 1968, pp. 623–650.

42 J. Blass,"The Multi-Directional Antenna: A New Approach to Stacked Beams," *IRE Conv. Proc.,* vol. 8, part 1, 1960, pp. 48–51.

43 J. Butler and R. Lowe, "Beamforming Matrix Simplifies Design of Electronically Scanned Antennas," *Electron. Design,* vol. 9, no. 7, April 1961, pp. 170–173.

44 J. P. Shelton and K. S. Kelleher, "Multiple Beams from Linear Arrays," *IRE Trans. Antennas Propagat.,* vol. AP-9, 1961, pp. 154–161.

45 J. Frank, "Bandwidth Criteria for Phased Array Antennas," *Proc. Phased Array Antenna Symp.,* Polytechnic Institute of Brooklyn, Brooklyn, N.Y., June 1970.

46 R. Tang, "Survey of Time-Delay Beam Steering Techniques," *Proc. Phased Array Antenna Symp.,* Polytechnic Institute of Brooklyn, Brooklyn, N.Y., June 1970.

47 R. R. Kinsey and A. L. Horvath, "Transient Response of Center-Fed Series Feed Array," *Proc. Phased Array Antenna Symp.,* Polytechnic Institute of Brooklyn, Brooklyn, N.Y., June 1970.

Chapter 21

Conformal and Low-Profile Arrays

Robert J. Mailloux

Rome Air Development Center
Hanscom Air Force Base

21-1 INTRODUCTION

A number of applications require antenna arrays conforming to a nonplanar surface. Primary among these are requirements for scanning antennas on aircraft or missiles, where aerodynamic drag is reduced for a conformal flush-mounted geometry, and for arrays that conform to surfaces that provide some coverage advantage, for example, to a hemispherical surface for hemispherical-pattern coverage or to a cylinder for 360° coverage in the azimuth plane.

Conformal arrays can be grouped into two broad categories that have very different design goals and present significantly different technological challenges. Arrays in the first category are small with respect to the radius of curvature of the conformed body, as depicted in Fig. 21-1 *a*. These proportions are representative of flush-mounted and low-profile aircraft arrays at superhigh-frequency (SHF) frequencies and above. Such arrays are relatively flat and behave nearly like planar arrays. The technological problems in this case are to design an array to scan over very wide angles (for example, zenith to horizon) or to make the array so thin that it can conform to the outer surface of the aircraft without requiring large holes in the fuselage.

The second category of conformal arrays, depicted in Fig. 21-1 *b*, is the type more commonly regarded as conformal and is comparable or large with respect to the radius of curvature of the mounting body. Such arrays include cylindrical arrays developed to provide radiation from spinning missiles and for 360° coverage of ground-based radars. Conical and spherical arrays have also been developed for ground, airborne, and missile applications.

Among the special features that make it difficult to design arrays that are large or highly curved in this sense are usually listed[1,2] the following: (1) Array elements point in different directions, and so it is often necessary to switch off those elements that radiate primarily away from the desired direction of radiation. (2) Conformal-array synthesis is very difficult because one cannot factor an element pattern out of the total radiation pattern. (3) Mutual coupling can be severe and difficult to analyze because of the extreme asymmetry of structures like cones and because of multiple coupling paths between elements (for example, the clockwise and counterclockwise

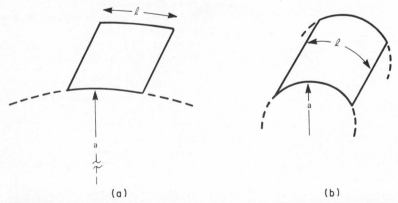

(a)   (b)

FIG. 21-1 Conformal-array apertures. (*a*) Aperture dimensions much less than local radius of curvature. (*b*) Aperture dimensions comparable with local radius of curvature.

paths between elements on a cylinder). (4) Cross-polarization effects arise because of the different pointing directions for elements on curved surfaces, which cause the polarization vector projections to be nonaligned. (5) Even for a cylindrical array there is a need to use different collimating phase shifts in the azimuth plane for an array scanned in elevation because steering in azimuth and elevation planes is not separate. Another phenomenon related to mutual coupling is the evidence of ripples on the element patterns of cylindrical arrays.

21-2 FUNDAMENTAL PRINCIPLES

Properties of Isolated Elements on Finite Planes and Cylinders

As indicated in Chap. 8, radiating elements behave differently according to the structure on which they are mounted. Examples of elementary slot antennas radiating from circular cylinders show significant pattern changes at angles near or below the horizon for a slot located at the top of the cylinder.

Figure 21-2 shows the radiated power pattern in the upper hemisphere ($\theta \leq 90°$) for an infinitesimal slot in a cylinder of radius a. Using the results of Pathak and

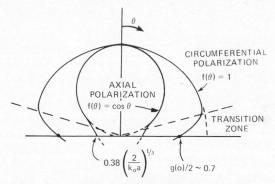

FIG. 21-2 Approximate pattern of slot on cylinder of radius a.

Kouyoumjian,[3] one can express the radiation pattern of elementary slots on cylinders in terms of three distinct regions, depending upon whether the point of observation is in the upper hemisphere (called the *illuminated region*), in the lower hemisphere (called the *shadow region*), or in the vicinity of the horizon (called the *transition region*). The angular extent of the transition region is of the order of $(k_0 a)^{-1/3}$ on each side of the shadow boundary. Figure 21-2 illustrates that within the illuminated region the circumferentially polarized radiation is nearly constant for an infinitesimal element but that the axially polarized radiation has a cos θ element pattern. Moreover, within the transition region these patterns are modified so that the circumferentially polarized component drops to a value of about 0.7 in field strength, or -3.2 dB as compared with its value of zenith, while for axial polarization the horizon radiation is not zero but approximately

$$0.4\left(\frac{2}{k_0 a}\right)^{1/3}$$

Thus, for a thin slot with circumferential polarization the horizon gain is independent of the size of the cylinder, but for an axially polarized slot the horizon gain varies directly as a function of $(k_0a)^{-1/3}$. For cylinders of radius approaching 50λ, the horizon power is thus approximately 23 dB below zenith gain.

Pattern-Coverage Limitations for Conformal Arrays

In addition to the diffraction effects observed in the preceding subsection, purely geometrical effects play a major part in determining the pattern characteristics of conformal arrays. Figure 21-3 shows a conical array of slots from several angles and illustrates the polarization distortion that occurs for an array near the point of a cone. This

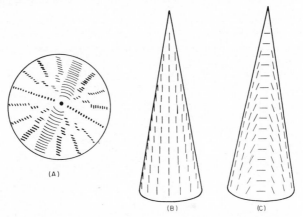

FIG. 21-3   Conical array. (*a*) View from axial direction. (*b*) Side view. (*c*) Side view rotated 90° from *b*. (*After Kummer.*[1])

figure, due to Kummer,[1] shows three views of a conical array with radiators located along the generatrices and oriented to give linear polarization in the axial direction. Clearly this array would radiate different polarization for each chosen angle of radiation.

Figure 21-3 also shows that for nearly any given radiation angle, especially for angles away from the tip of the cone, only a relatively small number of radiators have element patterns that point in the required direction. It is therefore usually necessary to devise a sophisticated network to switch an illuminated sector to the proper location on the conformal array for efficient radiation. Many such networks have been devised for cylindrical arrays, but the problem is more severe for cones, spheres, and more general surfaces.

Figure 21-4*a* shows the radiation pattern of several cylindrical arrays compared as a function of the angle subtended by the array. As pointed out by Hessel,[4] for an array of $k_0a = 86$ with 0.65λ separation between elements, doubling the number of excited elements in the array (from 60° arc to 120° arc) increases the array gain by only about 1 dB because the edge elements do not radiate efficiently toward the beam peak. The large sidelobes near $\phi = 100°$ are associated with grating lobes of the elements grouped at the array ends.

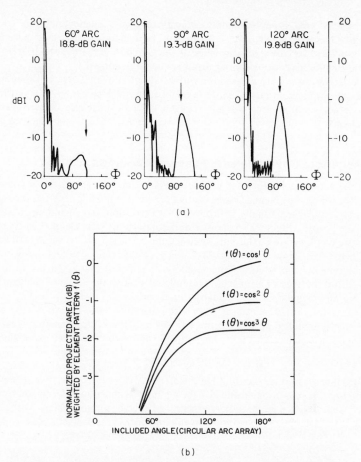

FIG. 21-4 Radiation characteristics of arc arrays. (*a*) Maximum-gain *E*-plane patterns for arc array of $k_{0a} = 86$, $b/\lambda = 0.65$. *(After Hessel.[4])* (*b*) Normalized projected area weighted by element patterns.

These results are not equally limiting, for closer element spacing can relieve the grating-lobe problem and can broaden the element patterns, as will be described in Sec. 21-3, but Fig. 21-4*b* shows that the gain restrictions are fundamental limitations. Consider the projected area of a cylindrical array weighted by the element-pattern radiation in the main-beam direction. In this simple case the optimum $\cos \theta$ and more realistic $\cos^2 \theta$ and $\cos^3 \theta$ element power patterns are assumed, and it is also assumed that the element spacing approaches zero (the continuous-aperture case). The peak forward radiation, normalized to the projected area, is the integral of the $\cos^n \theta$ functions to the maximum θ value and can be considered roughly proportional to gain. Figure 21-4*b* shows that there is little advantage to exciting much more than a 90° sector of the cylinder.

A second fundamental result obtainable from the same very simplified analysis is that, for a four-faced square array with a perimeter equal to the circular-array circumference, the average gain over a 90°-scan sector, normalized again to the cylinder projected area, is approximately -1.5 dB for a $\cos \theta$ element pattern, -1.9 dB for a $\cos^2 \theta$ element pattern, and -2.3 dB for $\cos^3 \theta$ element patterns. Figure 21-4b shows that these are essentially the same as the gain of a $90°$-arc circular array with the same element pattern. A similar result is quoted by Hessel.[4]

The above results are derived for uniformly illuminated arrays. When arrays have severe illumination tapers, as, for example, low-sidelobe arrays, it is possible to occupy larger circular arcs without reduced efficiency because of the low edge illumination.

Hemispherical Coverage with Fuselage-Mounted Arrays

One of the major applications for vehicle arrays is for aircraft-to-satellite communications, for which the antenna system is often required to provide coverage throughout the entire hemisphere, usually with circular polarization and including substantial gain at the horizon. Figure 21-5 shows several options for fully electronically steered arrays, two of which are conformal, while the third is a low-profile four-faced array.

The use of a single array (Fig. 21-5a) is very appealing provided that gain-contour requirements can be met. Unfortunately, this means that the array must scan from zenith to horizon in all planes. The following illustrates some of the gain limitations for an array scanned from zenith to the horizon.

In an infinite two-dimensional array with no grating lobes, the pattern directivity for a perfectly matched aperture varies like $\cos \theta$. If the array mismatch is not corrected as a function of scan angle, the gain will vary according to the expression

$$(1 - |\Gamma|^2) \cos \theta$$

For finite arrays the beamwidth in the plane of scan varies much like $\sec \theta$ except near end fire. The elevation beamwidth for an array scanned to $90°$ (horizon) can be obtained from well-known directivity values[5] and is given approximately by the equation below for column and two-dimensional arrays:

$$\delta = \sqrt{\frac{4\pi}{C}\left(\frac{\lambda}{L_1}\right)}$$

Here L_1 is the length of the array in the end-fire direction, and C is the constant which varies between 3.5 and 7, depending upon phase velocity and the array distribution. If one assumes an azimuth beamwidth of approximately λ/L_2, when L_2 is the array dimension perpendicular to the plane of scan, the directive gain at end fire is approximately

$$D = \sqrt{4\pi C}\left(\frac{L_2}{\lambda}\right)\sqrt{\frac{L_1}{\lambda}}$$

Since the directivity is multiplied by 4 for the array over a perfect ground plane (and the beamwidth is halved), the ratio of directive gain at end fire to directive gain at broadside (with perfectly conducting ground) is

$$\frac{D}{D_0} = \sqrt{\frac{C}{\pi}}\sqrt{\frac{\lambda}{L_1}}$$

This equation shows that for any given broadside directive gain D_0 it is advantageous to reduce L_1 (the array projection in the end-fire direction) as much as possible, for this minimizes scan loss. It also indicates that larger arrays undergo increased scan loss. For example, a square array with 20-dB gain at broadside should exhibit a directivity degradation of 1 dB at end fire, but for a 30-dB square array a 4.5-dB falloff is expected.

Unfortunately this decrease in directivity is only one of the factors tending to make scanning to end fire inefficient. Other major factors are diffraction and scattering due to the vehicle on which the array is mounted (see Sec. 21-1), element-pattern narrowing, and array mismatch due to mutual coupling.

Other options for hemispherical coverage shown in Fig. 21-5 include the use of multiple conformal arrays (Fig. 21-5*b*), one on each side of the vehicle to provide good coverage except in conical regions near the vehicle nose and tail, or a four-faced array (Fig. 21-5*c*), perhaps with adequate coverage throughout the hemisphere with a relatively streamlined geometry. Multiple planar arrays have been used for ground-based-radar coverage, and Knittel[6] gives optimum tilt angles and array orientations for that case.

Figure 21-6 shows contours of constant projected area for any of the configurations of Fig. 21-5, using the coordinate relationships below, for a planar array tilted an angle δ from the vertical:

$$\Theta = \cos^{-1}\left[v\cos\delta + c\sin\delta\right]$$

$$\Phi = \tan^{-1}\left[\frac{u}{-v\sin\delta + c\cos\delta}\right]$$

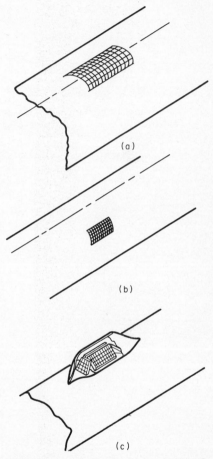

FIG. 21-5 Conformal and low-profile arrays for hemispherical coverage. (*a*) Large array on top of fuselage. (*b*) Array conformal to side of fuselage. (*c*) Four-faced "tent array."

where $u = \sin\theta\cos\phi$, $v = \sin\theta\sin\phi$, and $c = \cos\theta$ is the array projection factor. In general, the array projected area does not equate directly to relative gain because of array mismatch, element-pattern variations, and polarization effects, but it does provide a useful upper bound. This optimum-gain contour, normalized to peak, is given by

$$\frac{P}{P_0} = 10\log c$$

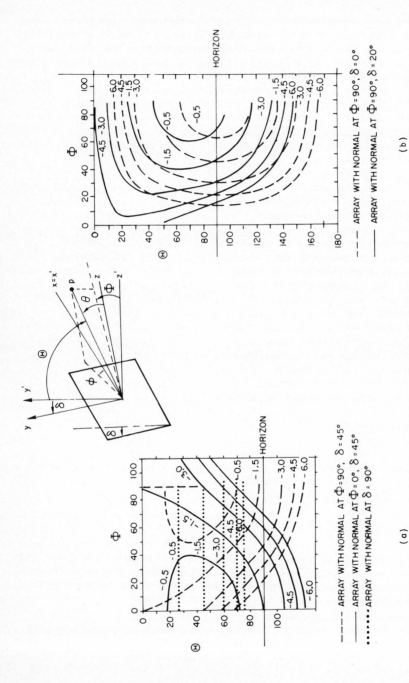

FIG. 21-6 Coverage diagrams for a variety of tilted planar-array faces. (a) Arrays tilted to $\delta = 45°$ with array normals at $\Phi = 0°$ and 90° and array with normal at zenith ($\delta = 90°$). (b) Arrays with normal at $\Phi = 90°$ and at $\delta = 0°$, 20°.

Figure 21-6a shows three contour plots of this ratio, two for arrays mounted at $\delta = 45°$ and one with $\delta = 90°$. The dotted line is the gain contour for an array with its boresight at zenith ($\delta = 90°$) and corresponds approximately to Fig. 21-5a. The solid curve corresponds to an array located as in Fig. 21-5b or a side face of the four-faced array of Fig. 21-5c mounted with $\delta = 45°$, $\Phi = 0°$. The dashed curve is for an array with its boresight in the $\Phi = 90°$ plane and corresponds to the end face of Fig. 21-5c. Combining the solid and dashed curves, as is done for the four-faced array, indicates that an array with these four faces would cover the upper hemisphere with only a 3-dB projected area falloff near $(\Theta, \Phi) = (90°, 45°)$, while the array with $\delta = 90°$ would have less gain for all $\Theta > 60°$ and thus has its lowest gain over the large area near the horizon.

A more streamlined four-faced array is obtained by reducing the front and rear faces to about one-half of the size of the side faces. In this case it is convenient to mount these arrays with their boresight direction closer to the horizon because the larger side faces provide sufficient gain at the higher elevation angles in the $\Phi = 90°$ or $180°$ planes. Figure 21-6 shows two gain contours for full-size arrays with $\delta = 20°$ (solid) and $\delta = 0°$ (dashed) for a front-face $\Phi = 90°$. By evaluating the coverage contours in conjunction with those of the solid curve from Fig. 21-6a (side face), one can compare the available coverage for a four-faced array with a variety of array sizes and elevation angles. If the front and rear faces are selected to be smaller than the side faces, the corresponding number of decibels (for example 3 dB for one-half of the area) must be subtracted from the values on the contour lines.

As an example, one can show that, with half-area front and rear faces and using $\delta = 20°$ for these end faces, the array system can scan throughout the hemisphere with a minimum projected area of approximately -4.5 dB with respect to the maximum.

Figure 21-6 can also be used with $\cos^2 \theta$ element patterns by doubling all the decibel ratios given in the figures or with any other assumed element patterns that are independent of Φ by making the appropriate scale change.

21-3 CYLINDRICAL ARRAYS

Synthesis of Omnidirectional Coverage

Several applications, including missile communication, require quality omnidirectional pattern coverage in the equatorial plane from an antenna system mounted on a cylindrical metal body of large diameter. Studies[7,8,9] have shown that an array of slots equally spaced around the vehicle circumference can produce patterns with very low ripple. Croswell and Knop[8] have obtained extensive numerical data by using realistic element patterns for slots on a perfectly conducting plane. Figure 21-7a gives data from an array computation[8] showing a ripple level for a cylinder with circumference C between 37 and 50 wavelengths for various numbers of slots S. The calculations show, for example, that for an array on a 39.5λ cylinder the choice of anywhere between 48 and 54 elements gives a ripple level below ± 0.05 dB, or $D = 0.1$ in the figure. The selection of 54 elements leads to broader-band performance, however, as indicated by the low ripple level over a range of C near the 39.5λ design point. In this case, the interelement spacing is approximately 0.73λ. Spacings that approach λ lead to high ripple levels that are, in general, unacceptable for omnidirectional-coverage

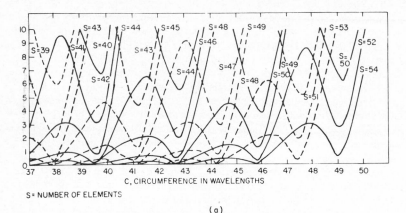

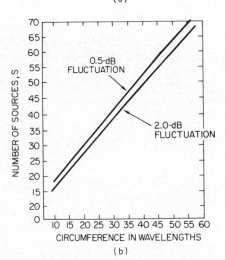

FIG. 21-7 (*a*) Pattern-fluctuation dependence on number of elements *S* and cylinder circumference *C*. (*After Croswell and Knop.*[8]) (*b*) Number of sources as a function of circumference and wavelengths for 0.5- and 2.0-dB fluctuation. (*After Croswell and Cockrell.*[9])

applications. Figure 21-7*b* gives the number of required sources as a function of the circumference in wavelengths for 0.5- and 2.0-dB fluctuation.[9]

Several approaches for implementation of omnidirectional missile antennas include the use of wraparound arrays as developed from the broadband Collings radiator,[10] employing plated-through holes, or from the basic microstrip elements[11] described in greater detail in Chap. 7. In either case, considerations like the above dictate the requisite number of feed points.

Element Patterns for Arrays on Cylinders

When the array size approaches the vehicle radius of curvature, the behavior of the array element pattern becomes very different from either the isolated element pattern or the element pattern in a planar array. This property is well documented in the case of circular and cylindrical arrays, which have been the subject of more detailed studies than other conformal arrays. It has long been known that by using the symmetry properties of the array one can avoid the solution of N simultaneous integral equations for an N-element circular array and, instead, superimpose the solution of N independent integral equations, one for each of the solutions with periodicity $2\pi/N$. This method has been used by Tillman[12] and Mack[13] for dipole and monopole arrays and by Borgiotti[14] for waveguide arrays with active elements covering a sector on an infinite cylinder. The results of these and other theoretical studies have revealed the existence of potentially severe ripples in cylindrical-array element patterns, as shown in Fig. 21-8a. Sureau and Hessel[15] have used asymptotic methods to show that the radiated pattern can be considered the superposition of a space-wave contribution and several creeping waves which propagate around the cylinder and radiate away from the cyl-

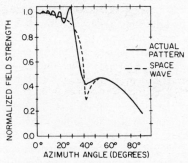

$k_0 a = 185$
$b = 0.6 \lambda$ = ELEMENT SPACING (CIRCUMFERENTIAL PLANE)
$d = 0.8 \lambda$ = ELEMENT SPACING (AXIAL PLANE)
APERTURE SIZE $0.32 \lambda \times 0.75 \lambda$, CIRCUMFERENTIAL POLARIZATION

(a)

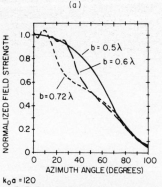

$k_0 a = 120$
b = ELEMENT SPACING (CIRCUMFERENTIAL PLANE)
$d = 0.72 \lambda$ ELEMENT SPACING (AXIAL PLANE)

(b)

FIG. 21-8 Circumferential element patterns. (a) Waveguide array on cylinder. (After Sureau and Hessel.[15]) (b) Dipole array in cylinder. (After Herper et al.[16])

inder in the forward direction. The space-wave and actual patterns are shown in the figure for radiation at two different angles θ measured from the cylinder axis. The ripple amplitude decreases with cylinder radius and increases with frequency. Recent work by Herper et al.,[16] indicated in Fig. 21-8b, has shown that smoother element patterns can be obtained by limiting element spacing to $\lambda/2$. The improvement is attributed to the fact that the interference of the fast creeping-wave grating lobe[16] with the direct ray decreases as the azimuth spacing is reduced and disappears at 0.5λ. This work strongly suggests that low-sidelobe circular arrays are feasible with proper concern for element design.

Technology of Arrays Conformal to Cylinders

Early experimental and theoretical studies concerned arrays built with dipole or waveguide[2] elements, with the array extending completely around the cylinder. Switching was accomplished by a variety of networks that were contained inside the cylinder, some of which are described in a following subsection. Figure 21-9 shows a circular array of waveguide horns developed by the Naval Electronics Laboratory Center[17] and used in a study of fundamental radiating properties as well as a comparison of several kinds of switching networks. Other studies dealing with full cylindrical or ring arrays are described in a number of survey and basic reference articles.[18–20]

More recently there has been substantial interest in developing microstrip conformal arrays so thin that they can be attached to the vehicle surface. Described in greater detail in Chap. 7, these array systems vary from the one- or two-dimensional receive-only arrays that have array, phase shifters, and power-divider network on one side of a thin circuit board to the use of multiple-layer construction with each of these functional pieces on a separate board. Figure 21-10 shows an eight-element circularly

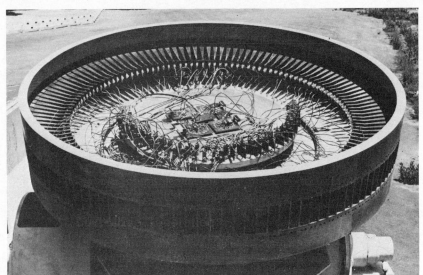

FIG. 21-9 Ring array. *(After Boyns et al.[17])*

FIG. 21-10   Conformal microstrip array. *(After Sanford.[21])*

polarized receive array developed for aircraft tests with the ATS-6 satellite. This L-band array was 0.36 cm thick and was scanned in one plane by means of 3-bit switched-line diode phase shifters. The array axial ratio was less than 2 dB over the scan sector.[21]

Low-Profile Arrays on Cylinders

The great difficulty in obtaining hemispherical coverage with a single fixed conformal array has stimulated interest in low-profile alternatives for SHF satellite-communication-aircraft antennas.

The low-profile array subsystem in Fig. 21-11 developed by the Hazeltine Corporation provides hemispherical coverage for an airborne satellite-communication terminal, the SESAST (small SHF-EHF satellite-communication terminal), by a combination of mechanical rotation in azimuth and electronic scanning in elevation. The array provides full duplex operation over the SHF terminal bands of 7.25 to 7.75 GHz (receive) and 7.90 to 8.40 GHz (transmit). It consists of 21 rows which are individually phased to steer the beam in elevation. Each row comprises 32 slots, providing a narrow azimuth beam. The design incorporates a dielectric sheet above the aperture for wide-angle impedance matching and delay progression along the aperture to steer the quiescent beam to 60° above the horizon. This feature minimizes scan losses at the scan extremes (0° and 90°). The array has independent ferrite phase shifters for receive and transmit and a low-loss radial power divider to provide the 21-way power split.

Arrays mounted above a cylindrical surface are subject to pattern deterioration owing to multipath scattering from the surface. Detailed pattern calculations for a

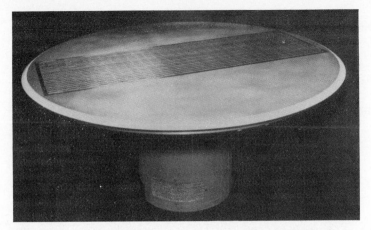

FIG. 21-11 Low-profile array. *(Courtesy of Hazeltine Corporation.)*

four-faced array (Fig. 21-5*c*) have been carried out by using geometrical optics[22] following the work of Kouyoumjian and Pathak.[23] Figure 21-12 shows the computed gain for both polarizations by using a small array mounted at 45° from the top of a large cylinder. The indicated pattern ripples are the result of a specular multipath from the cylinder surface, and the figure shows that in this particular instance these ripples do not materially alter the elevation-plane coverage of the array.

One last category of low-profile array that deserves some brief mention consists

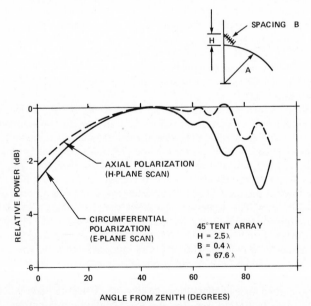

FIG. 21-12 Multipath effects on a tilted four-faced array.

of the use of an array plus some heavy dielectric or corrugated surface to enhance low-angle radiation. Such an antenna provides coverage over most of the hemisphere, operating as a conventional array for angles near the zenith and coupling into a surface-wave mode in the dielectric or corrugated surface to achieve strong radiation near end fire. Although successful in laboratory studies,[24-28] these techniques appear to offer solutions for gain up to about 20 dB and with relatively narrow bandwidth (tunable bandwidth up to about 6 percent[28]).

Feed Systems and Commutators for Cylindrical Arrays

When the array extends completely around the cylindrical surface, it is necessary to devise a feed network to rotate the illuminated spot continuously through multiples of 360°. As noted by Provencher,[2] Hill,[29] and Holley et al.,[30] a number of mechanical and electrical networks have been devised to accomplish that task. Mechanical commutators such as the waveguide commutator of Fig. 21-13a are appropriate for systems with continuous scanning over multiple 360° increments. Such commutators are often seen as viable low-cost solutions for a number of applications, and they can provide scanned patterns relatively free of modulation effects if care is taken to assure that the rotating (rotor) feed elements have spacings different from those of the stationary (stator) pickup elements. This design procedure can assure that a minimum of elements is misaligned at any one time and so reduces radiation modulation. Power distribution within the rotor is achieved by using coaxial or waveguide corporate networks or radial power dividers.

A number of investigators have considered switching networks for commuting an amplitude and phase distribution around a circular array. Giannini[31] describes a technique that uses a bank of switches to bring a given illumination taper to one sector of the array (usually a 90° to 120° arc) and a set of switches to provide fine beam steering between those characteristic positions determined by the sector-switching network. The circuit for a 32-element array is shown in Fig 21-13b; it requires 8 phase shifters and 12 transfer switches (double-pole, double-throw) and achieves sector selection by using 8 single-pole, four-throw switches. This network excites an 8-element quadrant of the array that can be moved in increments of 1 element to provide coarse beam steering. Fine steering is provided by the phase controls.

Several lens-fed circular arrays have been implemented by using R-$2R$ Luneburg and geodesic lenses.[30] The R-$2R$ lens (Fig. 21-13c) described by Boyns et al.[17] forms as many beams as there are elements in the array but, like all lens systems, does not admit to fine steering (selecting angles with a separation less than the angular separation between radiating elements on the cylinder) unless additional phase controls are added to each element.

As pointed out by Holley et al.,[30] lens systems can be adapted for fine steering by using an amplitude illumination with a movable phase center. The net result of a progressive phase tilt at the input of the Butler matrix is to synthesize intermediate beams from a composite of the available lens beams and so to provide high-quality fine steering of the lens-radiated pattern.

There have been a number of developments in the area of multimode electronic commutators for circular arrays. These systems derive from techniques similar to that first used by Honey and Jones[32] for a direction-finding-antenna application in which several modes of a biconical antenna were combined to produce a directional pattern with full 360° azimuthal rotation. Recent efforts by Bogner[33] and Irzinski[34] specifi-

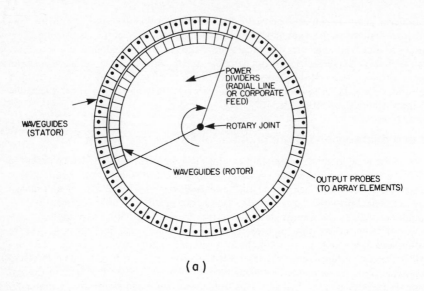

(a)

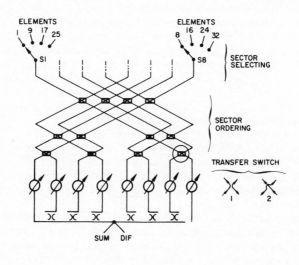

(b)

FIG. 21-13 Circular-array feed systems. (a) Commutator (waveguide). (b) Switch network. (After Giannini.[31]) (c) R-2R lens. (After Boyns et al.[17]) (d) Matrix scanning system. (After Sheleg.[36])

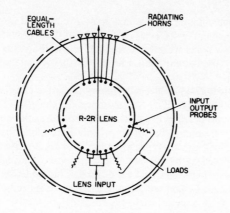

EQUAL-
LENGTH
CABLES

RADIATING
HORNS

R-2R LENS

INPUT
OUTPUT
PROBES

LOADS

LENS INPUT

(c)

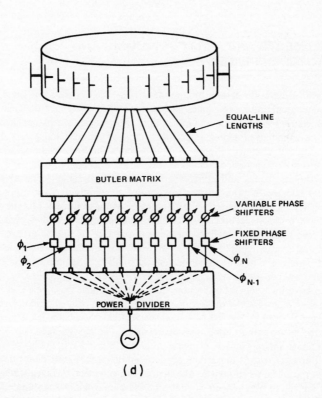

EQUAL-LINE
LENGTHS

BUTLER MATRIX

VARIABLE PHASE
SHIFTERS

FIXED PHASE
SHIFTERS

ϕ_1
ϕ_2

ϕ_N

ϕ_{N-1}

POWER DIVIDER

(d)

21-17

cally address the use of such commutators, which must be combined with phase shifters and switches at each element of the array. The phase shifters provide collimating and fine steering, and the switches are used to truncate the illumination so that only a finite sector of the array is used at any time, a procedure that is required for sidelobe control.

Several more sophisticated types of electronic switches for circular arrays are based on a concept, originally proposed by Shelton[35] and developed by Sheleg,[36] that uses a matrix-fed circular array with fixed phase shifters to excite current modes around the array and variable phase shifters to provide continuous scanning of the radiated beam over 360°. The geometry is shown in Fig. 21-13d. A more recent extension of this technique proposed by Skahil and White[37] excites only that part of the circular array that contributes to formation of the desired radiation pattern. The array is divided into a given number of equal sectors, and each sector is excited by a Butler matrix and phase shifters. With either of these circuits, sidelobe levels can be lowered by weighing the input excitations to the Butler matrix. The technique of Skahil and White was demonstrated by using an 8×8 Butler matrix, eight phase shifters, and eight single-pole, four-throw switches to feed four 8-element sectors of a 32-element array. The design sidelobes were -24 dB, and measured data showed sidelobes below -22 dB.

21-4   ELEMENTS AND ARRAYS ON CONIC AND GENERALIZED SURFACES

The dramatic illustration chosen by Kummer[1] (Fig. 21-3) points out the difficulty of synthesis on conical and more generalized surfaces that arise because such structures project differently into different angular regions. For example, a planar array projects as a plane independent of aspect angle, but a cone or a sphere looks very different as the aspect angle is changed. Figure 21-3 also points out the severe polarization distortion resulting from scanned conical arrays.

Nevertheless, cones and spheres are important surfaces for conformal-array development because of their obvious application to missile and certain hemispheric scanning structures. The experimental studies of Munger et al.[38] provide some data on the characteristics of several conical arrays, and theoretical studies by Balzano and Dowling[39] give rigorous expressions for element patterns in conical arrays. Golden et al.[40] evaluated mutual-coupling coefficients for waveguide slot elements on sharp cones, and Thiele and Donn[41] have performed analytical studies of relatively small cones excited by loop elements using a wire-grid model.

There have been several important studies of antennas on spheres. However, the dome antenna[42] is by far the most important application of a spherical array to date. This structure, shown in Fig. 21-14a, uses the vertical projection of a spherical dome to achieve increased gain at low angles of elevation. The passive spherical lens is made up of fixed phase shifters and a conventional planar phased array (Fig. 21-14b) which steers an illuminated spot to various portions of the lens. Fixed phase delays in the lens are computed so as to convert the array scan direction θ_s into a factor K times that angle and so to obtain scan in excess of the planar-array scan angle. The array phase-shifter settings are determined to form the nonlinear phase progression required to scan the searchlight-type beam to various spots on the lens. Although the radiated

(a)

DOME-ANTENNA CONCEPT

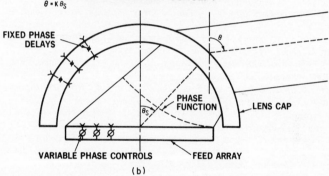

$\theta = \kappa\theta_S$

FIXED PHASE
DELAYS

θ

PHASE
FUNCTION

θ_S

LENS CAP

VARIABLE PHASE CONTROLS

FEED ARRAY

(b)

FIG. 21-14 Hemispherical dome array. (*a*) Dome array. (*b*) Dome-antenna concept. *(Courtesy of Sperry Corporation.)*

beamwidth varies with scan angle, the dome can achieve scanning over sectors larger than a hemisphere and, in fact, has achieved scan to $\pm 120°$ from zenith. Steyskal[43] gives equations for the gain limits of a given circular cylindrical dome based upon allowable scan angles for the feed array and shows that the ratio of the average gain to the broadside feed-array gain is bounded by a parameter that depends upon the feed-scan limit.

REFERENCES

1 W. H. Kummer, "Preface to Special Issue on Conformal Antennas," *IEEE Trans. Antennas Propagat.,* vol. AP-22, no. 1, January 1974, pp. 1–3.

2   J. H. Provencher, "A Survey of Circular Symmetric Arrays," in A. Oliner and G. Knittel (eds.), *Phased Array Antennas,* Artech House, Inc., Dedham, Mass., 1972, pp. 315–322.

3   P. Pathak and R. Kouyoumjian, "An Analysis of the Radiation from Apertures in Curved Surfaces by the Geometrical Theory of Diffraction," *IEEE Proc.,* vol. 62, no. 11, November 1974, pp. 1438–1447.

4   A. Hessel, "Mutual Coupling Effects in Circular Arrays on Cylindrical Surfaces—Aperture Design Implications and Analysis," in A. Oliner and G. Knittel (eds.), *Phased Array Antennas,* Artech House, Inc., Dedham, Mass., 1972, pp. 273–291.

5   C. H. Walter, *Traveling Wave Antennas,* McGraw-Hill Book Company, New York, 1965, pp. 121–122.

6   G. H. Knittel, "Choosing the Number of Faces of a Phased-Array Antenna for Hemisphere Scan Coverage," *IEEE Trans. Antennas Propagat.,* vol. AP-13, no. 6, November 1965, pp. 878–882.

7   T. S. Chu, "On the Use of Uniform Circular Arrays to Obtain Omnidirectional Patterns," *IRE Trans. Antennas Propagat.,* vol. AP-7, October 1959, pp. 436–438.

8   W. F. Croswell and C. M. Knop, "On the Use of an Array of Circumferential Slots on a Large Cylinder as an Omnidirectional Antenna," *IEEE Trans. Antennas Propagat.,* vol. AP-14, no. 3, May 1966, pp. 394–396.

9   W. F. Croswell and C. R. Cockrell, "An Omnidirectional Microwave Antenna for Use on Spacecraft," *IEEE Trans. Antennas Propagat.,* vol. AP-17, no. 4, July 1969, pp. 459–466.

10   E. V. Byron, "A New Flush-Mounted Antenna Element for Phased Array Applications," in A. Oliner and G. Knittel (eds.), *Phased Array Antennas,* Artech House, Inc., Dedham, Mass., 1972.

11   R. E. Munson, "Conformal Microstrip Antennas and Microstrip Phased Arrays," *IEEE Trans. Antennas Propagat.,* vol. AP-22, January 1974, pp. 74–78.

12   J. T. Tillman, Jr., "The Theory and Design of Circular Array Antennas," Eng. Experiment Sta. Rep., University of Tennessee, Knoxville, 1968.

13   R. W. P. King, R. B. Mack, and S. S. Sandler, *Arrays of Cylindrical Dipoles,* Cambridge University Press, London, 1968, chap 4.

14   G. V. Borgiotti, "Modal Analysis of Periodic Planar Phased Arrays of Apertures," *IEEE Proc.,* vol. 56, November 1968, pp. 1881–1892.

15   J. C. Sureau and A. Hessel, "Realized Gain Function for a Cylindrical Array of Open-Ended Waveguides," in A. Oliner and G. Knittel (eds.), *Phased Array Antennas,* Artech House, Inc., Dedham, Mass., 1972, pp. 315–322.

16   J. C. Herper, C. Mandarino, A. Hessel, and B. Tomasic, "Performance of a Dipole Element in a Cylindrical Array—A Modal Approach," *IEEE AP-S Inst. Symp.,* June 1980, pp. 162–165.

17   J. E. Boyns, C. W. Gorham, A. D. Munger, J. H. Provencher, J. Reindel, and B. I. Small, "Step-Scanned Circular Array Antenna," *IEEE Trans. Antennas Propagat.,* vol. AP-18, no. 5, September 1970, pp. 590–595.

18   J. C. Sureau, "Conformal Arrays Come of Age," *Microwave J.,* October 1973, pp. 23–26.

19   D. E. N. Davies, "Circular Arrays, Their Properties and Potential Applications," Second Int. Conf. Antennas Propagat., Part 1, "Antennas," Apr. 13–16, 1981.

20   R. G. Fenby and D. E. N. Davies, *IEE Proc.,* vol. 115, no. 1, 1968.

21   G. G. Sanford, "Conformal Microstrip Phased Array for Aircraft Tests with ATS-6," *IEEE Trans. Antennas Propagat.,* vol. AP-26, no. 5, September 1978, pp. 642–646.

22   R. J. Mailloux and W. G. Mavroides, "Hemispherical Coverage of Four-Faced Aircraft Arrays," RADC-TR-79-176, May 1979.

23   R. G. Kouyoumjian and P. H. Pathak, "A Uniform Geometrical Theory of Diffraction for an Edge in a Perfectly Conducting Surface," *IEEE Proc.,* vol. 62, November 1974, pp. 1448–1461.

24   R. J. Mailloux, "Phased Array Aircraft Antennas for Satellite Communication," *Microwave J.,* vol. 20, no. 10, October 1977, pp. 38–42.

25   G. V. Borgiotti and Q. Balzano, "Mutual Coupling Analysis of a Conformal Array of Ele-

ments on a Cylindrical Surface, *IEEE Trans. Antennas Propagat.,* vol. AP-18, January 1970, pp. 55–63.

26 G. V. Borgiotti and Q. Balzano, "Analysis of Element Pattern Design of Periodic Array of Circular Apertures on Conducting Cylinders," *IEEE Trans. Antennas Propagat.,* vol. AP-20, September 1972, pp. 547–553.

27 Q. Balzano, L. R. Lewis, and K. Swiak, "Analysis of Dielectric Slab-Covered Waveguide Arrays on Large Cylinders," AFCRL-TR-73-0587, Cont. F19628-72-C-0202, Scientific Rep. 1, August 1973.

28 W. G. Mavroides and R. J. Mailloux, "Experimental Evaluation of an Array Technique for Zenith to Horizon Coverage," *IEEE Trans. Antennas Propagat.,* vol. AP-26, no. 3, May 1978, pp. 403–406.

29 R. T. Hill, "Phased Array Feed Systems: A Survey," in A. Oliner and G. Knittel (eds.), *Phased Array Antennas,* Artech House, Inc., Dedham, Mass., 1972, pp. 197–211.

30 A. E. Holley, E. C. DuFort, and R. A. Dell-Imagine, "An Electronically Scanned Beacon Antenna," *IEEE Trans. Antennas Propagat.,* vol. AP-22, no. 1, January 1974, pp. 3–12.

31 R. S. Giannini, "An Electronically Scanned Cylindrical Array Based on a Switching and Phasing Technique," *IEEE Antennas Propagat. Int. Symp. Dig.,* December 1969, pp. 199–207.

32 R. C. Honey and E. M. T. Jones, "A Versatile Multiport Biconical Antenna," *IRE Proc.,* vol. 45, October 1957, pp. 1374–1383.

33 B. F. Bogner, "Circularly Symmetric R. F. Commutator for Cylindrical Phased Arrays," *IEEE Trans. Antennas Propagat.,* vol. AP-22, no. 1, January 1974, pp. 78–81.

34 E. P. Irzinski, "A Coaxial Waveguide Commutator Feed for a Scanning Circular Phased Array," *IEEE Trans. Microwave Theory Tech.,* vol. MTT-29, no. 3, March 1981, pp. 266–270.

35 P. Shelton, "Application of Hybrid Matrices to Various Multimode and Multibeam Antenna Systems," IEEE Washington Chap. PGAP Meeting, March 1965.

36 B. Sheleg, "A Matrix-Fed Circular Array for Continuous Scanning," *IEEE Proc.,* vol. 56, no. 11, November 1968, pp. 2016–2027.

37 G. Skahil and W. D. White, "A New Technique for Feeding a Cylindrical Array," *IEEE Trans. Antennas Propagat.,* vol. AP-23, March 1975, pp. 253–256.

38 A. D. Munger, G. Vaughn, J. H. Provencher, and B. R. Gladman, "Conical Array Studies," *IEEE Trans. Antennas Propagat.,* vol. AP-22, no. 1, January 1974, pp. 35–42.

39 Q. Balzano and T. B. Dowling, "Mutual Coupling Analysis of Arrays of Aperture on Cones," *IEEE Trans. Antennas Propagat.,* vol. AP-22, January 1974, pp. 92–97.

40 K. E. Golden et al., "Approximation Techniques for the Mutual Admittance of Slot Antennas in Metallic Cones," *IEEE Trans. Antennas Propagat.,* vol. AP-22, January 1974, pp. 44–48.

41 G. A. Thiele and C. Donn, "Design of a Small Conformal Array," *IEEE Trans. Antennas Propagat.,* vol. AP-22, no. 1, January 1974, pp. 64–70.

42 L. Schwartzman and J. Stangel, "The Dome Antenna," *Microwave J.,* October 1975, pp. 31–34.

43 H. Steyskal, A. Hessel, and J. Shmoys, "On the Gain-versus-Scan Trade-Offs and the Phase Gradient Synthesis for a Cylindrical Dome Antenna," *IEEE Trans. Antennas Propagat.,* vol. AP-27, November 1979, pp. 825–831.

Chapter 22

Adaptive Antennas

Leon J. Ricardi

Lincoln Laboratory
Massachusetts Institute of Technology

22-1 INTRODUCTION

Adaptive antennas were first used in an intrinsic form as radar antennas with sidelobe-canceling characteristics. The sidelobe-canceler antenna consists of a conventional radar antenna whose output is coupled with that of much-lower-gain auxiliary antennas. The gain of the auxiliary antennas is slightly greater than the gain of the maximum sidelobe of the radar antenna. Adding the weighted signals received by the auxiliary antenna to those received by the radar antenna permitted suppression of interfering sources located in directions other than the main beam of the radar antenna. This early use of adaptive antennas evolved to adapted array and multiple-beam antennas.

This chapter presents a description of the major components of an adaptive antenna and develops general relationships and rules governing its performance. A fundamental definition of *resolution* is developed, enabling one to determine the minimum tolerable angular separation between desired and interfacing signals. This resolution leads naturally to the definition of *degrees of freedom* and to a conceptual method of determining the potential and actual degrees of freedom available to a given adaptive antenna. It is shown that often an adaptive antenna with N degrees of freedom can suppress more than $N - 1$ sources.

The algorithm governing the adaptation process can be based on one of three classical methods. Two methods, named for the inventors, use a feedback weight-control system. The first method requires prior knowledge of the actual or potential location of the desired sources and is commonly referred to as the power-inversion or Applebaum-Howells algorithm. The second weight-determining method using feedback control is called the Widrow algorithm and requires a known characteristic of the desired signal in order to maximize the signal-to-noise ratio of signals received from a single desired source. The third, or sample-matrix-inversion (SMI), algorithm is a completely computational process that uses the signals received at the N ports of an adaptive antenna and the desired-source angular locations to suppress interfering signals in the computed output. This SMI algorithm does not use feedback to correct for implementation inaccuracies.

Finally, some general remarks about transient effects and the comparative performance of planar-array and multiple-beam antennas are presented.

22-2 FUNDAMENTAL CHARACTERISTICS

Adaptive antennas are used primarily for receiving signals from desired sources and suppressing incident signals from undesired or interference sources. The basic configuration shown in Fig. 22-1 consists of an antenna with N ($N > 1$) ports, N complex weights, a signal-summing network, and a weight-determining algorithm. The N ports of the antenna can be the output terminals (ports) of the elements of an array antenna, or the ports of a multiple-beam antenna (MBA), or a mixture of these. A complex weight, in general, attenuates the amplitude and changes the phase of the signals passing through it and is usually assumed to have a frequency-independent transfer function. The summing network is often a corporate-tree arrangement of four-port T's. Each T sums the in-phase components of the signals at the two input ports in one

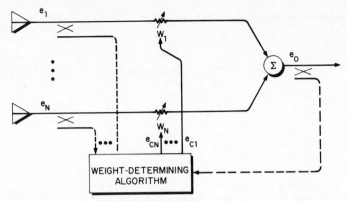

FIG. 22-1   Fundamental adaptive nulling circuit.

output port and has a matched load at the other output port to dissipate those components of the input signal that are 180° out of phase. The weight-determining algorithm (or simply the algorithm) makes use of a priori or measured information, or both, and specifies the complex weight W_n applied to the signal received at the nth port as

$$W_N = g e_{cn} \qquad (22\text{-}1)$$

As shown in Fig. 22-1, the algorithm uses information derived solely from the output signal e_0 (as in a weight-dither algorithm), or solely from the input signals e_n (as in the matrix-inversion algorithm), or from both e_0 and e_n as in the Applebaum-Howells and Widrow algorithms. An intrinsic algorithm need not use either e_0 or the e_n; hence these inputs to the weight-determining algorithms are indicated by dashed lines in Fig. 22-1. Note that these inputs are indicated as being "loosely" coupled, as opposed to being connected directly, to the e_n and e_0 signal ports.

Clearly the antenna (array, MBA, or hybrid) is that essential component of an adaptive-antenna system which is uniquely related to the disciplines of antenna design. The weight and summing circuits can operate at the antenna operating frequency f_0 or at an intermediate frequency (IF) or be part of a digital signal processor. Although the design and development of the antenna, the weights, and the summing network are important, the weight-determining algorithm almost always determines the uniqueness of an adaptive antenna. Consequently, this chapter will only briefly describe the antenna, the weights, and the summing network. Considerably more attention will be given to the fundamental limitations (i.e., degrees of freedom and resolution) of all classes of adaptive antennas and to a conceptual description of the classical weight-determining algorithms.

22-3   ANTENNA

Although adaptive-antenna systems use antennas of various types and configurations, it is possible to separate these into three basic classes: phased arrays,[1] multiple-beam antennas,[2] and a mixture of these two. Each configuration has several ports where received signals e_n appear in response to sources located in the antenna's field of view (FOV). Phased arrays characteristically have many identical elements, each of which has a port where the output signal can be represented by

$$e_n = \sum_{m=1}^{M} I_m F_n(\theta_m, \phi_m) e^{jH_n(\theta_m, \phi_m)} \qquad (22\text{-}2)$$

where $I_m = \dfrac{\lambda^2 P_m G_m}{(4\pi R_m)^2}$

P_m = power radiated by mth source
G_m = gain of antenna used by mth source
R_m = distance between mth source and adaptive antenna
λ = operating wavelength

The amplitude and phase that relate I_m to a signal at the antenna port are given by F_n and H_n respectively. The angular location of the mth source is given by the coordinate pair θ_m, ϕ_m measured in a suitable spherical coordinate system.

It is important to note that in most if not all phased-array adaptive antennas the F_n are identical, whereas H_n is generally different for all elements of the array. For signals at the output ports of an MBA, the H_n are nearly equal and the F_n differ. It will be shown later that this fundamental difference between a phased array and an MBA results in the inherently larger bandwidth of the latter.

Phased arrays are either filled or thinned. Filled arrays do not have significant grating lobes or high sidelobes within the FOV of an array element of the array, whereas thinned arrays characteristically have either grating lobes or high sidelobes (~ 10 dB) in the FOV of the array element. Grating-lobe characteristics are extremely important because even when interference and desired sources are separated by a large angle (i.e., compared with the array half-power beamwidth, or HPBW), the desired source is suppressed by approximately the same amount as the interference source when an interference source is located on a grating lobe.

Although there is not a suitable formal definition of thinned or filled arrays, it is convenient to consider the thinned array as having an interelement spacing S greater than S_T, where

$$S_T = \frac{\lambda}{\sin(\theta_e/2)} \qquad (22\text{-}3)$$

and θ_e is the HPBW of an element of the array. Unfortunately, Eq. (22-3) cannot be used in all cases, but it does serve as a conceptual definition.

Hybrid-antenna systems usually consist of an array antenna with each element of the array configured as an MBA. The directive gain of the MBA permits thinning of the array without forming harmful grating lobes while retaining the capability of scanning the instantaneous FOV over a larger FOV. Chapters 3, 16, 17, 20, 21, and 35 discuss some applicable antennas in greater detail.

22-4   WEIGHTS

Just as the excitation of phased arrays (see Chap. 3) determines the antenna radiation pattern, weighting of signals received at the ports of an adaptive antenna determines

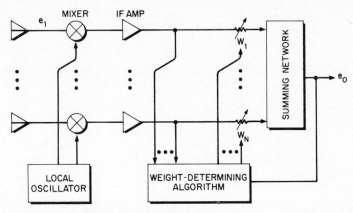

FIG. 22-2 IF adaptive nulling circuit.

the antenna's directional response to incident signals. Since the discussion of adaptive antennas presented here assumes a receiving as opposed to a transmitting antenna system, it is necessary to introduce and use the concept of weighting and combining receiving signals instead of "exciting" the array or MBA to produce a radiation pattern. However, it is often much clearer to represent the antenna's angular discrimination of received signals in the form of "radiation" patterns. The reader is cautioned to ignore the literal meaning of radiation patterns and think of the F_n and related functions as receiving-antenna patterns.

Weights attenuate and alter the phase of received signals. They are designed to be either frequency-independent or adaptively varied as a function of frequency. Some adaptive antennas operate entirely at the received frequency and use radio-frequency (RF) weights as indicated in Fig. 22-1. Still others have a mixer-amplifier at each antenna port, and the weights operate at a lower IF, as indicated in Fig. 22-2. The latter customarily uses lossy weights consisting of attenuators and hybrid power dividers as indicated in Fig. 22-3. The use of lossy weights in this way does not degrade system performance since the mixer–IF amplifier establishes the local signal-to-noise (S/N) ratio such that subsequent losses in the weights and combining, or summing, network attenuate both signal and noise without altering the S/N. For RF weighting, the weight circuit is usually a corporate configuration of "lossless" variable power dividers and a single phase shifter at each antenna port, such as shown in Fig. 22-4. Four-port variable power-dividing junctions (Fig. 22-4) are customarily used in weight and summing networks to prevent the resonant effect encountered with the use of three-port junctions. This effect is produced by small but significant reflected waves between input and output ports. Dissipative loads at the fourth port of a four-port junction (Fig. 22-4b) absorb reflected waves that do not couple to the input port; reflected waves at the input do, however, degrade the input match of the circuit. Although this effect is not uniquely associated with adaptive antennas, it is an important and significant consideration in the design of all adaptive-antenna systems.

All antenna systems have a frequency-dependent response to incident signals. Whenever frequency-independent weights are used, suppression of undesired signals will vary with frequency. This inherent performance characteristic cannot be succinctly and accurately described because it is scenario-dependent. However, it can be

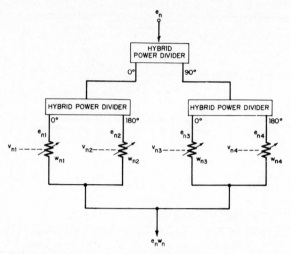

FIG. 22-3 Typical IF weight circuit.

shown[3] that, for frequency-independent weights, the cancellation C (i.e., suppression of an interfering signal) of an adaptive-array antenna is limited as

$$C \leq 20 \log \left[K \left(\frac{2\pi}{c} \right) DW \sin \theta_m \right] \qquad \mathrm{dB} \qquad\qquad (22\text{-}4)$$

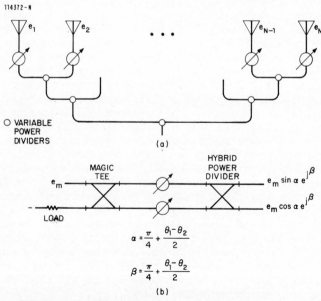

FIG. 22-4 (*a*) RF weight circuit. (*b*) Variable power-divider circuit.

where $2\theta_m$ = maximum angle subtended by FOV
   D = maximum dimension of antenna aperture
   W = nulling bandwidth
   c = velocity of light in free space
The constant K is a function of the array configuration and the particular scenario of interfering and desired sources; it ranges between ~ 5 and ~ 15. It is important to note that C varies as the square of W and the square of the maximum-scan angle θ_m and that Eq. (22-4) takes only antenna-related dispersive effects into account; it assumes frequency-independent weights and identical antenna-array elements. When the antenna array uses time-delay circuits, which in effect reduces θ_m, C increases accordingly.

One form of frequency-dependent weight consists of a delay line with two or more taps. The received e_n is delivered to the output of each tap, weighted, and summed with the output of all N taps (Fig. 22-5). The weights are frequency-independent, and the delay between successive taps[3] depends on W, D, θ_m, λ, and the distribution of interfering sources. Observing that the maximum delay $\tau_{\max}$ occurring between elements located at opposite edges of an array aperture is given by

$$\tau_{\max} = (D/c) \sin \theta_m \qquad \textbf{(22-5)}$$

The number of taps N_τ should exceed

$$N_\tau > \tau_{\max} W = (WD/c) \sin \theta_m \qquad \textbf{(22-6)}$$

Although the use of frequency-dependent weights will increase with the development of the associated technology, adaptive antennas using frequency-independent weights are currently most popular. Hence the discussions in this chapter are limited accordingly; the weight shown in Fig. 22-5 is introduced primarily because of its potential importance to broadband operation of large phased arrays.

22-5 RESOLUTION

Since the primary purpose of an adaptive antenna is to suppress interfering sources while maintaining adequate response to desired sources, the system designer needs to know the minimum tolerable angular separation between desired and interfering

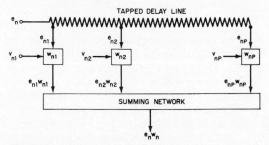

FIG. 22-5 Frequency-dependent adaptive weights.

sources. In other words, the inherent ability of the antenna to resolve desired and undesired sources is fundamental to the system design. Clearly, the antenna's *resolving power* increases with its aperture size D. Quantitative resolving characteristics of adaptive antennas will be discussed in this section.

The signals e_n received at the ports of the antennas (Fig. 22-1) are "multiplied" by a weight W and summed to produce the output signal e_0. That is,

$$e_0 = \sum_{n=1}^{N} W_n e_n \qquad (22\text{-}7)$$

Adaptive antennas select the W_n so as to place *nulls* in the radiation pattern in the direction of interfering sources. By using matrix notation and representing the weights as the vector

$$\mathbf{W} = [W_1, W_2, \ldots W_N] \qquad (22\text{-}8)$$

it can be shown[4] that the optimum set of weights $\mathbf{W}_0$ is given by

$$\mathbf{W}_0 = M^{-1}\mathbf{V} \qquad (22\text{-}9)$$

where $\mathbf{V}$ is called the steering vector and represents the quiescent weights which establish the antenna receiving pattern in the absence of interfering sources.

In Eq. (22-9), M is the covariance matrix of the e_n received signals. That is,

$$M = \begin{bmatrix} \overline{e_1 e_1^*} & \cdots & \overline{e_1 e_N^*} \\ & & \\ & & \\ & & \\ \overline{e_N e_1^*} & \cdots & \overline{e_N e_N^*} \end{bmatrix} \qquad (22\text{-}10)$$

where the overbar indicates averaging over the correlation interval and the asterisk indicates the complex conjugate. Clearly all adaptive antennas strive to determine or estimate M or M^{-1} which enables $\mathbf{W}_0$ to be determined from Eq. (22-9) or an equivalent algorithm. This optimum produces that spatial discrimination, available with the associated antennas, which maximizes the ratio of the desired signal power to the interfering plus internally generated noise power. Now the covariance matrix M and its inverse M^{-1} are key to the determination of the fundamental properties of an adaptive antenna. For example, M^{-1} can be written in the form

$$M^{-1} = \sum_{n=1}^{N} \frac{\mathbf{u}_n \mathbf{u}_n^{\dagger}}{\lambda_n} \qquad (22\text{-}11)$$

where $\dagger$ indicates the complex-conjugate transpose, $\mathbf{u}_n$ is the nth eigenvector of M with eigenvalue λ_n, and the N eigenvalues of M represent the antenna output power $e_0 e_0^*$ obtained when $\mathbf{W} = \mathbf{u}_n$. Hence, if a single source is in the antenna's FOV, one eigenvalue (i.e., λ_1) will be much larger than all other eigenvalues. If two sources are in the FOV, at least one and maybe two eigenvalues will be larger than the others. Since the $\mathbf{u}_n$ ($n = 1, 2, \ldots, N$) are an orthogonal set forming the basis of M^{-1}, two equal eigenvalues will result when two sources of equal intensity are located at spatially orthogonal positions in the FOV. In other words, if source S_1 is located in a null of the radiation pattern resulting when $\mathbf{W} = \mathbf{u}_2$, two equal-value eigenvalues will result.

The resolution of an adaptive antenna can be determined by first placing two sources, S_1 and S_2, with intensity P at the same point in the antenna's FOV. The resulting covariance matrix will have only one large eigenvalue, λ_1, indicating that a single degree of freedom of the array is used, and setting $\mathbf{W} = \mathbf{u}_1$ will maximize the power delivered to the output port of the adaptive antenna. Conversely, setting $\mathbf{W}$ so that $\mathbf{W} \cdot \mathbf{u}_1 = 0$ will reduce the output power to zero.

Now as the angular separation between two sources increases from zero, λ_1 decreases, and a second significant eigenvalue, λ_2, increases. When S_1 and S_2 are separated by an angle α_0, $\lambda_1 = \lambda_2$ and two of the N degrees of freedom are required to establish a radiation pattern which maximizes the output power. A detailed analysis[5] of this process indicates that when S_1 and S_2 are at the same location $\lambda_1 = 2P$ and when they are separated at an angle α_0, $\lambda_1 = \lambda_2 = P$. The angular resolution α_0 is approximately equal to the HPBW measured in the plane R containing the two sources and the adaptive antenna. The HPBW is calculated from the known antenna aperture dimension in the plane R and an assumed uniform excitation of all elements of a phased array or excitation of a single port of an MBA.

The results presented in the preceding paragraph can also be derived mathematically by substituting Eq. (22-11) into Eq. (22-9) to obtain

$$\mathbf{W}_0 = \sum_{n=1}^{N} \frac{\mathbf{u}_n \mathbf{u}_n^{\dagger}}{\lambda_n} \cdot \mathbf{V} \qquad (22\text{-}12)$$

where the $\mathbf{u}_n$ span the weight space. The radiation pattern produced by the N eigenvectors forms a complete orthogonal set whose appropriately weighted combination will form any radiation pattern that the adaptive antenna is capable of forming. (The concept of a radiation pattern is introduced to clarify the pattern synthesis implied. When receiving signals, the receiving pattern is identical to the radiation pattern for $W_n = I_n^*$, where I_n is the excitation coefficient of the nth port of the adaptive antenna.) Note that when λ_1 is the only large eigenvalue (i.e., S_1 and S_2 are at the same location), $\mathbf{W}_0$ does not include $\mathbf{u}_1$ and the output of the adaptive antenna will not contain energy received from S_1 and S_2; only weak signals or noise determines the value of the other eigenvalues, $\mathbf{W}_0$ and the antenna receiving pattern. It follows that since $\mathbf{W}_1 = \mathbf{u}_1$ maximizes the signal from S_1 and S_2, the resulting receiving pattern, obtained by using $\mathbf{W}_1$ for the weights, must be identical to the radiation pattern that maximizes the adaptive antenna's directivity in the direction of the location of S_1 and S_2.

As described in the foregoing, separating the sources gives rise to two large eigenvalues equal to P when the angular separation equals α_0. From Eq. (22-12), when $\lambda_1 = \lambda_2 = P$ and all other $\lambda_n \ll P$,

$$\mathbf{W}_0 \cdot \mathbf{u}_1 = \mathbf{W}_0 \cdot \mathbf{u}_2 = 0 \qquad (22\text{-}13)$$

minimizes the output power and suppresses the interference from S_1 and S_2. However, if S_2 is a source of desired signals and its radiated power $\ll P$, the angular separation between S_1 and S_2 must be equal to or greater than α_0 if the receiving pattern is to have a null in the direction of the interfacing source and maximize the signal received from the desired source.

Since any adaptive antenna must have internal noise sources, all λ_n of M must be greater than zero; consequently, $\mathbf{W}_0$ given by Eq. (22-12) is always finite. It is, however, customary to normalize $\mathbf{W}_0$ so that their magnitude varies between 0 and 1 and their phase varies between $0°$ and $360°$

In summary, the resolution α_0 of any adaptive antenna in a given direction θ_0,ϕ_0 is approximately equal to the minimum HPBW of the radiation pattern which can be generated by the antenna and has maximum directivity in the direction θ_0,ϕ_0. The HPBW is determined in the plane containing the resolution angle α_0. It follows that larger-aperture antennas have higher resolution. Further, when an interfering source and a desired source have angular separation α_0 or larger, an adaptive antenna with resolution α_0 can suppress the interfering source without altering its receiving pattern or directivity (i.e., with respect to the interference-free values) in the direction of the desired source.

22-6 DEGREES OF FREEDOM

Clearly any adaptive antenna with N ports and N weights has N degrees of freedom (DOF). However, the number of DOF required to satisfy a given scenario is not easily determined. Often is is understood mistakenly that an N-port antenna can suppress only $N - 1$ interfering sources. Nothing could be further from the truth. It is in fact difficult, and sometimes impossible, to find that set of N interference sources that can disable an adaptive antenna with N DOF.

In particular, consider an adaptive antenna with but two degrees of freedom and the ability to suppress an interfering source P_J, 30 dB with respect to the desired signal. Assume that the interference-free signal-to-noise power ratio (S/N), at the output of the adaptive antenna, is 13 dB in the absence of adaptive nulling and that it decreases to -10 dB in the presence of signals from the interference source. Spatial discrimination of the adaptive-antenna pattern will increase the S/N to slightly less than 13 dB. The addition of four additional interfering sources, each with effective radiated power ERP $= P_J$ at the same location, will reduce S/N to 10 dB.

The set of N eigenvectors into which the covariance matrix can be decomposed is unique only for a given set of signals e_n. Determining the strength and location of N sources which will capture the N eigenvalues of the covariance matrix M (or the N DOF of an N-port adaptive antenna) is in general difficult and often impossible. However, an intelligent method for locating the interfering sources may result in disabling an N-port adaptive antenna with only slightly more than N interference sources. Consequently, it is helpful to derive some rule of thumb which enables the antenna designer to determine how many DOF are potentially available for suppressing sources in a specified FOV. The following guidelines are based on the foregoing and on a conceptual, as opposed to a rigorous, mathematical derivation.

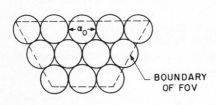

BOUNDARY OF FOV

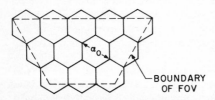

BOUNDARY OF FOV

FIG. 22-6 Resolution cells and DOF.

Although an N-port adaptive antenna has N DOF, the antenna aperture may not be large enough to use all N degrees of freedom in the specified FOV. By using the foregoing definition of α_0, it is conceptually acceptable to assume that an N-port adaptive antenna with a circular aperture has N circular resolution cells, each of which subtends an angle α_0 measured at the antenna aperture. The composite of all N ($N = 12$) resolution cells "covers" the antenna FOV as indicated in Fig. 22-6a, where α_0 is the angular diameter of each cell. As for the small but finite area at the intersection of three adjacent cells, one might consider hexagonal cells with diagonals equal to α_0; however, this carries the concept far beyond its intended accuracy.

If the FOV is larger than that indicated in Fig. 22-6, increasing N accordingly would still guarantee one DOF for each resolution cell within the FOV. Conversely, if the FOV is smaller than the composite area formed by the N resolution cells, either the antenna aperture size must be increased or some of the antenna ports can be eliminated without affecting the adaptive-antenna performance over the FOV.

For those adaptive antennas designed to serve a circular FOV with a circular aperture, the foregoing concept can be placed in mathematical form. That is, the number N_c of equal-diameter nonoverlapping circles that can be inscribed in a larger circle is maximum when the centers of the "smaller" circles are on a hexagonal grid and is given by

$$N_c = 1 + \sum_{m=1}^{M-1} 6m \qquad \textbf{(22-14)}$$

Setting the diameter of the smaller circles equal to α_0 gives

$$M = \text{odd integer of } [\gamma_0/\alpha_0] \qquad \textbf{(22-15)}$$

where γ_0 is the angular diameter of the larger circle defining the FOV. By recalling that α_0 is the HPBW when the antenna weights are chosen to produce maximum directivity, α_0 can be approximated as

$$\alpha_0 \approx \frac{60\lambda}{D} \qquad \textbf{(22-16)}$$

where D is the antenna aperture diameter. Substituting Eq. (22-16) into Eq. (22-15) gives

$$M = \text{odd integer of } [(D/\lambda)(\gamma_0/60)] \qquad \textbf{(22-17)}$$

Then $D/\lambda = 60M/\gamma_0$. For example, if $M = 5$, $N_c = 19$, the antenna D/λ must be greater than $300/\gamma_0$ if 19 degrees of freedom are required. If the antenna has more than 19 ports, only 19 will be useful in forming nulls within the FOV. If there are fewer than 19 ports, say, 10, and the antenna aperture is not thinned, the adaptive antenna still has 19 degrees of freedom, 9 of which are not available to the adaptive algorithm; however, they can be used to shape the quiescent (interference-free) receiving pattern. In particular, an adaptive antenna with a 19-beam MBA, 9 of whose ports are short-circuited, will have 19 DOF within the FOV defined by γ_0 if its aperture diameter $D = 300(\lambda/\gamma_0)$. Notice, however, that signals originating from 9 "short-circuited" resolution cells couple to the adaptive antenna only through the sidelobes of the receiving pattern associated with the 10 ports that are not short-circuited.

22-7 WEIGHT-DETERMINING ALGORITHMS

To achieve the desired antenna performance, an algorithm for choosing $\mathbf{W}_0$ must be formulated. The algorithms must take the form of hardware or software, and they range from the very simple and straightforward to the complicated and sophisticated. The algorithms all have the same common goal: they must derive the desired $\mathbf{W}_0$. It is helpful to separate these algorithms into two general classes, those with and those without feedback. Specifically, the latter, or feed-forward, systems determine the desired $\mathbf{W}_0$ from a given set of input data and install the weights with some small but finite error. They are not further corrected by examining the output signals, as is done in an algorithm which uses feedback. Truly adaptive algorithms adjust the weights by observing the output or some other metric related to system performance.

The algorithm for deriving weights is undoubtedly the central issue in determining the value and performance of any adaptive-antenna system. Consider first the simplest algorithm and then the more sophisticated methods for deriving $\mathbf{W}_0$. Most adaptive algorithms operate when the interfering-signal power is much larger than the desired-signal power, when both are measured in the nulling band W. This assumes that the interfering signal occupies a bandwidth W which is much larger than the instantaneous bandwidth of the desired signal.

All algorithms depend on two basic and distinct components, namely, the inverse of the covariance matrix of the received signals e_n and the steering vector $\mathbf{V}$. First, recall that the voltages e_n at each antenna terminal are produced by thermal noise, interfering sources, and desired sources in the FOV; that is, $e_n = e_{\text{noise}} + e_{in} + e_{dn}$. The goal of any algorithm is to choose weights so that the total power received from the interference sources is minimized and the output voltage e_0 is dominated by the desired signals. Since e_{noise}, e_{in}, and e_{dn} are not correlated with one another, the covariance matrix M, formed by determining the time average value of $e_n e_q^*$, equals the sum of the covariance matrices of the thermal noise, interfering-signal sources, and desired sources; that is, $M = M_N + M_I + M_D$. In most adaptive-antenna applications, the interfering signals dominate M [i.e., $M_I \gg (M_N + M_D)$], and a good estimate of M_I can be obtained from the voltages e_n. In the absence of interfering signals (i.e., $M_I = 0$), the weights are chosen to maximize the signals received from the desired sources. Choosing the desired weights $\mathbf{W}_0$ in accordance with Eq. (22-9) will maximize the desired signal-to-total-noise-power ratio [i.e., $P_D/(P_I + P_N)$] at the output of the adaptive antenna. This maximization will occur for all user signals if $\mathbf{W}_0$ is chosen appropriately.

Consider a common scenario to illustrate this basic algorithm. Assume that all signal-source locations and the relative intensity of each desired-signal-source ERP are known. Let us further assume that the antenna response to a signal source located anywhere in the FOV can be calculated. Following the

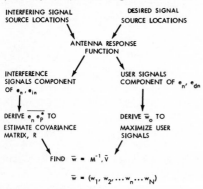

FIG. 22-7 Basic algorithm.

flow chart given in Fig. 22-7, the desired-signal and interference-signal components of e_n can be calculated from the foregoing information. The interference covariance matrix M_I and its inverse can be calculated from e_{in}. The covariance matrix M_D can be used to calculate $\mathbf{V}$, which will maximize the power received from each desired-signal source. Choosing the weights $\mathbf{W}_0$ equal to the product $M^{-1}\mathbf{V}$ will minimize the difference between the actual and the desired antenna directive gain in the direction of the desired-signal sources while suppressing the interfering signals. It is conceptually helpful[6] to assume that, in the absence of interfering signals, the desired signals are maximized when $\mathbf{W}_0 = \mathbf{V}$. Introduction of interfering signals causes the adaptive antenna to form a radiation pattern that is a least-mean-square fit to the quiescent receiving pattern, with nulls in the direction of the interfering sources.

All adaptive nulling algorithms, in their steady state, attempt to choose $\mathbf{W}_0 = M^{-1}\mathbf{V}$. Consequently, specific performances among various algorithms are manifested in their transient behavior, their choice of $\mathbf{V}$, their hardware and/or software implementation, and the degree to which they are sensitive to errors.

Power-Inversion Algorithm

Assume that the locations of the interference signals are unknown and that their ERP is much greater than the ERP of the desired signals; that is $M_I \gg (M_D + M_N)$. Further, assume that $\mathbf{V}$ is selected to provide the desired antenna directive gain in the known direction of each desired-signal source. Since e_{in} is unknown, M_I cannot be calculated. However, direct measurement of the e_n or $\overline{e_n e_p^*}$ would enable M or M^{-1} to be estimated; this estimate becomes more accurate as the interfering signals become much stronger than the desired signals. By choosing the nulling bandwidth appropriately, one can usually guarantee that troublesome interference will result in $M_I \gg M_d$ and that the algorithm will be very effective in choosing $\mathbf{W}_0$. Conversely, an inappropriate choice of nulling bandwidth, effective noise level, etc., will result in an undesirable reduction in the desired signals if it (they) and the interference signal (signals) have approximately the same amplitude (i.e., $|e_{in}| \sim |e_{sn}|$).

This algorithm is often referred to as the *power inversion* or Applebaum-Howells algorithm.[7] It is one of the best-known analog algorithms. A schematic representation of the fundamental circuit is shown in Fig. 22-8. The antenna-element, or beam-port, output signals e_n' are indicated for an N element (beam) array (multiple-beam) antenna. A mixer followed by a preamplifier (and perhaps by appropriate filtering) establishes e_n over the nulling band W. For the purpose of the present discussion, assume that e_n is a frequency-translated, band-limited representation of e_n' and consider the "loop" that is connected to antenna terminal 1. The signals e_n are weighted by W_n and summed to give an output signal e_0; that is,

$$e_0 = \sum_{n=1}^{N} W_n e_n \qquad \textbf{(22-18)}$$

Thus far everything is exactly as described in the foregoing. However, in this circuit the complex weights W_n are proportional to the complex control voltages e_{cn}. As in any analog adaptive algorithm, the derivation of e_{cn} is of interest. Note that the signal e_1 is mixed with e_0. The output of the mixer is filtered and amplified, giving a complex voltage proportional to correlation of e_1 with e_0. The correlator's output is subtracted from a beam-steering voltage V_1 to give e_{c1}. For the moment, assume that

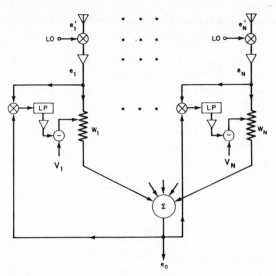

FIG. 22-8   Applebaum-Howells circuit.

$W_1 = V_1$. If e_0 is correlated with e_1, the low-pass filter integrates the output of the correlator to produce $e_{c1} \neq 0$, which changes the weight W_1 so as to reduce the correlation between e_0 and e_1, decreasing the output of the correlator. Similar responses in the other loops reduce e_0 to a minimum. Combined noise in the circuit and in e_1 prevents e_0 from vanishing. Hence we see that any signal (in e_1) above the effective noise level will be sensed by the loop and weighted to reduce e_0; the combined effect of all N loops is to reduce e_0 below the effective noise level of the correlators. Furthermore, in the absence of signals (interfering or desired) greater than the front-end noise level, the weights will be determined by the noise. However, if the V_n are not zero, they will determine the weights, and the antenna receiving pattern will assume its quiescent shape.

The quiescent weights are determined a priori from known or expected locations, etc. It can be shown that when $M_I \gg M_N > M_D$, $M \sim M_I$, and the steady-state weights assume the optimum values given by Eq. (22-9). The transient performance and convergence to stable steady-state weights that equal $\mathbf{W}_0$ are of paramount concern and will be discussed later.

In summary, the Applebaum-Howells[7] circuit establishes $\mathbf{W}_0$ when the desired-signal locations and antenna receiving pattern are known and used to determine $\mathbf{V}$. It is also necessary to choose the nulling bandwidth W so that the level of the desired-signal received power S is approximately equal to or less than the front-end noise power in the band W (i.e., $M_S \leq M_N$). This circuit senses the locations of all interfering sources and reduces their received signals below the front-end noise level. If the interfering and desired sources are located near one another, the desired signal may also be reduced, but the ratio e_{in}/e_{dn} will not be increased; it will probably be decreased.

Maximization of Signal-to-Noise Ratio

When there is a single desired-signal source present in an environment of one or many interfering sources, the definition of S/N is clear. This is not true when more than one desired source is present; however, the Applebaum-Howells circuit, discussed in the preceding subsections, allows for optimum adaptation when several desired sources and several interfering sources exist simultaneously. When there is only a single user and its location is known, the Applebaum-Howells circuit will maximize S/N if the quiescent weights $\mathbf{V}$ are chosen to form a single receiving pattern with maximum directivity in the direction of this desired source. By comparison, the circuit shown in Fig. 22-9 will form and steer a high-directivity beam in an unknown desired-signal-source direction while simultaneously placing a null on all interfering sources. This is accomplished without knowledge of the location of either the desired or the interfering sources. However, it is necessary to derive a reference signal that is a suitable replica of the expected desired signal. This can lead to the dilemma that if the desired signal must be known prior to adaptation, there is no need to send it to the receiver. However, the desired signal might have a deterministic component (i.e., a pseudo-noise sequence or a frequency-hopped "carrier") and a random component (i.e., that modulation on the carrier that contains unknown information). The former would be used to permit the adaptation circuit to recognize a desired signal and maximize the associated S/N.

The circuit shown in Fig. 22-9 is identical to the Applebaum-Howells circuit (Fig. 22-8) discussed in the preceding subsection except for the output portion and the absence of steering weights. A reference signal e_{REF} is subtracted from e_0' to generate an error signal with which the e_n are correlated and the W_n determined. Some sort of

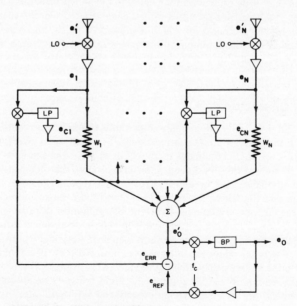

FIG. 22-9 Widrow algorithm.

spread-spectrum modulation is assumed so that a reference signal can be obtained by despreading e_0' and bandpass-filtering it in order to increase the S/I of e_{REF} compared with the S/I of e_0'. Amplifying the demodulated e_0' (i.e., to obtain e_0) and modulating it with the known spread-spectrum modulation sequence results in a reference signal e_{REF}.

To understand the operation of this circuit, assume that the interfering component of e_0' is much larger than the desired-signal component. Consequently, the error signal is dominated by the interference signal, and the correlators drive the weights to reduce e_0 by placing nulls in the direction of the interfering source. Reduction of the interfering source will not, in general, reduce the desired signal. Because the reference signal is principally an amplified replica of the desired signal, the correlators drive the weights to *increase* the desired signal in order to reduce the error signal e_{ERR}. This adjustment of the weights is identical to steering a receiving pattern with maximum directivity toward the desired source while simultaneously placing nulls in the direction of the interfering source.

This circuit (Fig. 22-9) is commonly referred to as the Widrow circuit after its inventor.[8] It is often recommended for use in time-division–multiple-access (TDMA) communication systems employing pseudo-noise modulation as spread-spectrum protection against interfering sources. The Widrow circuit is limited to multiple-interference-source scenarios with a single desired source because the beam will be acquired only by a desired signal that is in synchronism with the spread-spectrum modulation.

It is interesting to note that if there is one desired source in the FOV, the Applebaum-Howells and Widrow circuits produce the same steady weights if the **V** are chosen to steer the beam in the direction of the desired source. That is, the Widrow circuit contains a closed-loop determination of the **V**, whereas the Applebaum-Howells circuit requires that **V** be an externally generated open-loop constraint. Consequently, the former is applicable for scenarios with a single desired source, whereas the latter is applicable for any scenario for which a prescribed quiescent receiving pattern is essential, such as satellite-communication-system-spacecraft antennas that must have an earth-coverage or multiple-area-coverage receiving pattern to serve multiple users simultaneously.

Transient Characteristics

All analog and essentially all computational nulling algorithms require a finite time to respond to the onset of an interfering signal. These transient effects and a computational algorithm which eliminates them will be discussed in this subsection.

It can be shown[4] that for a single interfering source the time constant of the nth cancellation loop (Fig. 22-8 or Fig. 22-9) is inversely proportional to the product of the loop gain and the power P_n received at the nth port. Hence, increasing the loop gain decreases the time to adapt; however, the loop gain must be low enough to prevent unstable operation. Therefore, setting the loop gain in accordance with the strongest expected interference-signal source prevents loop instability and determines the shortest possible time required for the antenna system to complete its adaptation function. If two interfering sources, one strong and one weak, are present, the stronger source is at first reduced rapidly to a level somewhat higher than its steady-state level. Then at a much slower rate its signal strength is reduced until both interfering signals are reduced to their steady-state level at the *same* time. Consequently, it is important to set the loop gain so that the weakest expected interfering-signal strength will result in a tolerable loop time constant. If the spread in strength of all troublesome interference sources is sufficiently large, it may not be possible to choose a suitable loop gain. This situation is often referred to as an excessive spread in the eigenvalues of the covariance matrix M.

Sample Matrix Inversion

Although the circuits shown in Figs. 22-8 and 22-9 indicate the use of analog devices, it is not uncommon to use digital devices between the output of the correlator and the control terminal of the weights. Within the scope of this report, these changes do not modify the preceding discussions. However, consider a processor which converts the antenna terminal voltages e_n to a digital representation and completes the adaptation process entirely within a computational processor. For example, the processor indicated in Fig. 22-10 (LPF = low-pass filter) first demodulates the spread-spectrum modulation of the received signals and establishes an IF bandwidth that retains the necessary power differences between interfering and desired signals. The signals are amplified, to establish the system thermal noise level, and divided into two identical channels except for a relative phase of 90°. The signals are then converted to baseband and divided into a narrow *signal* band and a wider *nulling* band. All signals are then converted to a digital representation; the wideband signals are used to compute the covariance matrix M, its inverse M^{-1}, and the optimum weights. The narrowband

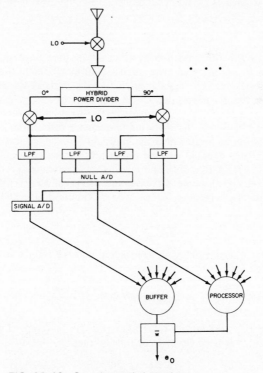

FIG. 22-10 Sample matrix inversion.

signals are stored in a buffer while the weights are being computed. As the narrow-band signals leave the buffer, they are then weighted, summed, and filtered to yield a digital representation of the desired signals. Notice that this processor eliminates the transient performance of the weights and has the linearity characteristics of a digital computer. However, the signals must undergo a delay while the weights are computed, and there is not a feedback loop as with the Applebaum-Howells and Widrow algorithms. As with all feed-forward systems, implementation must have accuracy commensurate with the desired signal suppression. In other words, the covariance matrix M must be estimated from the measured "samples" of the received signals. The performance of this algorithm is determined by the accuracy with which M is estimated, the rate at which the computations can be made, and the speed and accuracy of the analog-to-digital converters.

22-8   MULTIPLE-BEAM ANTENNA VERSUS PHASED ARRAY

Historically, adaptive antennas have used either an array antenna or an MBA. It is often important to be aware of the fundamental differences of these devices and how they affect the adaptive-antenna-system performance. In this section, the inherent bandwidth is discussed as an aid in choosing one antenna instead of the other.

Bandwidth

Classically and historically, phased-array antennas are focused to receive signals from a given direction by adjusting an array of phase shifters. Specifically, the differential time delay τ_n associated with signals arriving at the ports of the array elements is corrected by inserting a variable delay τ_n in the range $0 < \tau_n < \tau - p\lambda/c$, where p is an integer. Consequently, the array is perfectly "focused" at the design frequency f_0 and becomes defocused as the operating frequency f varies from f_0.

For a signal source located near the boresight direction of the array, p is small, and the array antenna does not become appreciably defocused even for a large bandwidth $\Delta f = |f_0 - f|$. A rule of thumb relating Δf to antenna aperture D and θ, the angle between the signal-source direction and a normal to the plane of a planar-antenna array, can be derived as follows. The maximum differential path delay S_{max} is given by

$$S_{max} = (D/\lambda) \sin \theta \qquad \textbf{(22-19)}$$

The phase shifter inserts a delay S_1 less than λ, and the error ε in differential path delay is given by

$$\varepsilon = \text{integer}[(D/\lambda) \sin \theta] \cdot \frac{\Delta f}{2f_0} \qquad \textbf{(22-20)}$$

For $(D/\lambda) \sin \theta \gg 1$, the integer-value operation can be removed without seriously affecting the results. Solving Eq. (22-20) for the fractional bandwidth yields

$$\frac{\Delta f}{f_0} = \frac{2\varepsilon\lambda}{D \sin \theta} \qquad \textbf{(22-21)}$$

If one assumes that ε must be less than 0.1 (i.e., a path-length error = $\lambda/10$), θ = 10°, and D = 120λ, Eq. (22-21) indicates a 1 percent frequency bandwidth. Calculated results[3] indicate that for ε = 0.1 the interference signal will be suppressed $\sim$ 20 dB; doubling or halving Δf changes the signal suppression to 14 dB or 26 dB respectively. That is, signal suppression varies approximately as Δf^2.

Most MBAs use a lens or a paraboloid which focuses received signals by introducing differential path delays (as opposed to phase shifting) and translating a feed located in the focal region of the lens or paraboloid. Consequently, the system remains focused over a very wide frequency band. However, the sidelobes and receiving pattern shape change with frequency and tend to alter the amplitude instead of the phase of the received signals. Studies have indicated that this effect of varying frequency does not degrade the associated adaptive-antenna performance as much as that of an "equivalent" planar array because each beam of the MBA performs like a phased array with its receiving beam in the boresight direction. Since the sidelobes of an MBA do not dominate the determination of the weight applied to a beam port, an MBA with an aperture D = 120λ can suppress interfering signals more than 20 dB for $\Delta f/f \sim 5$ percent.

In some applications, an array antenna is preferred to an MBA. By using the foregoing, the operating bandwidth can be estimated. If the expected $\Delta f/f_0$ meets or exceeds system requirements, the phased array may be the best choice. If the estimated $\Delta f/f_0$ is less than required, an MBA may be the best choice.

Channel Matching

Adaptive antennas suppress interfering signals in accordance with the qualitative relationship[3] given in Sec. 22-3 and the similarity among the transfer functions L_n of each of the N channels. In large-aperture (D) antennas operating over a large bandwidth W, such that $DW/C > 1$, the natural dispersion of array antennas may limit the cancellation ratio C (see Sec. 22-3). C can also be limited by channel mismatch

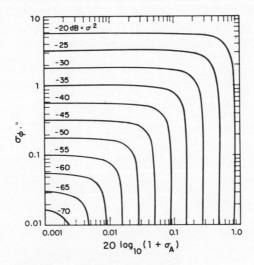

FIG. 22-11 Estimated cancellation.

$\Delta L_{mn} = L_n - L_m \neq 0$ when frequency-independent weights are used. The use of frequency-dependent weights (Fig. 22-5) improves C in the presence of both antenna dispersion and channel mismatch. Rigorous simulation is usually required to determine the expected value of C. However, as a guide to the tolerable rms variation of ΔL_{mn}, it is useful to refer to Fig. 22-11, where the required rms amplitude and phase variation of L_n of a two-channel system are given as a function of the cancellation ratio C. The reader is cautioned to apply these results only to obtain a qualitative estimate of the expected performance of adaptive antennas with more than a few elements or beam ports.

REFERENCES

1   *IEEE Trans. Antennas Propagat.*, vol. AP-12, special issue on active and adaptive antennas, March 1964.
2   J. T. Mayhan, "Adaptive Nulling with Multiple-Beam Antennas," *IEEE Trans. Antennas Propagat.*, vol. AP-26, no. 2, March 1978, p. 267.
3   J. T. Mayhan, A. J. Simmons, and W. C. Cummings, "Wideband Adaptive Antenna Nulling Using Tapped Delay Lines," Tech. Note 1979-45, Lincoln Laboratory, Massachusetts Institute of Technology, Lexington, Mass., June 26, 1979.
4   W. F. Gabriel, "Adaptive Arrays—An Introduction," *IEEE Proc.*, vol. 64, February 1976, p. 239.
5   A. J. Fenn, "Interference Sources and DOF in Adaptive Nulling Antennas," Symp. on Antennas, Allerton Park, Ill., Sept. 23–25, 1981.
6   J. T. Mayhan and L. J. Ricardi, "Physical Limitations on Interference Reduction by Antenna Pattern Shaping," *IEEE Trans. Antennas Propagat.*, vol. AP-23, no. 5, September 1975, p. 639–646.
7   S. P. Applebaum, "Adaptive Arrays," Rep. SPL TR 66-1, Syracuse University Research Corporation, Syracuse, N.Y., August 1966.
8   B. Widrow et al., "Adaptive Antenna Systems," *IEEE Proc.*, vol. 55, 1967, pp. 2143–2159.

Chapter 23

Methods of Polarization Synthesis

Warren B. Offutt

Eaton Corporation

Lorne K. DeSize

Consultant

The plane of polarization, or simply the polarization, of a radio wave is defined by the direction in which the electric vector is aligned during the passage of at least one full cycle. In the general case, both the magnitude and the pointing of the electric vector will vary during each cycle, and the electric vector will map out an ellipse in the plane normal to the direction of propagation at the point of observation. In this general case (shown in Fig. 23-1 a), the polarization of the wave is said to be elliptical. The minor-to-major-axis ratio of the ellipse is called the *ellipticity* and will be expressed in this chapter in decibels. (Although the axis ratio is less than unity, when expressing ellipticity in decibels, the minus sign is frequently omitted for convenience. The term *axial ratio* is also in common use. It is the reciprocal of ellipticity.) The direction in which the major axis lies is called the *polarization orientation* and in this chapter will be measured from the vertical (Fig. 23-2).

The two special cases of ellipticity of particular interest are (1) an ellipticity of ∞ dB (minor-to-major-axis ratio zero), which is *linear polarization,* and (2) an ellipticity of 0 dB (minor-to-major-axis ratio unity), which is *circular polarization.* A linearly polarized wave is therefore defined as a transverse electromagnetic wave whose electric field vector (at a point in a homogeneous isotropic medium) at all times lies along a fixed line. A circularly polarized wave is similarly defined as a transverse electromagnetic wave for which the electric and/or magnetic field vector at a point describes a circle. In attempting to produce a linearly polarized wave, elliptical polarization is thought of as imperfect linear polarization, while in attempting to produce a circularly polarized wave, elliptical polarization is thought of as imperfect circular polarization.

Confusion occasionally results in the use of mental pictures similar to Figs. 23-1 and 23-2 when one overlooks the fact that although the electric vector

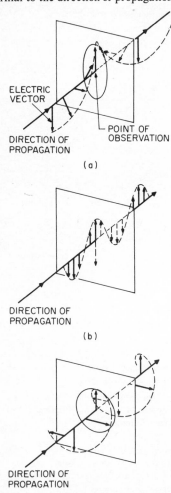

ELECTRIC
VECTOR

DIRECTION OF
PROPAGATION

POINT OF
OBSERVATION

(a)

DIRECTION OF
PROPAGATION

(b)

DIRECTION OF
PROPAGATION

(c)

FIG. 23-1 Diagrammatic illustration of waves of various polarization. (*a*) Elliptical polarization. (*b*) Linear polarization. (*c*) Circular polarization (right-hand).

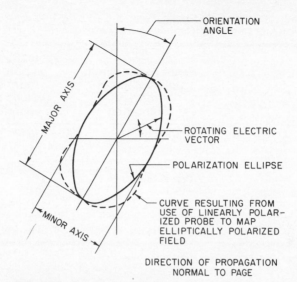

ORIENTATION
ANGLE

MAJOR AXIS

ROTATING ELECTRIC
VECTOR

POLARIZATION ELLIPSE

CURVE RESULTING FROM
USE OF LINEARLY POLAR-
IZED PROBE TO MAP
ELLIPTICALLY POLARIZED
FIELD

MINOR AXIS

DIRECTION OF PROPAGATION
NORMAL TO PAGE

FIG. 23-2   Polarization ellipse.

makes one complete revolution (Fig. 23-2) per cycle, it does not rotate at a uniform rate except in the special case of a circularly polarized wave. In this special case, rotation occurs at the rate of ω rad/s.

Figure 23-1c shows a circularly polarized wave having a right-hand sense. It is also possible, of course, to have left-hand circularly polarized waves. The definition of right-hand circular polarization as standardized[1] by the Institute of Electrical and Electronics Engineers and as used in this chapter is as follows: for an observer looking in the direction of propagation, the rotation of the electric field vector in a stationary transverse plane is clockwise for right-hand polarization. Similarly, the rotation is counterclockwise for left-hand polarization.

One simple way of determining experimentally the sense of rotation of a circularly polarized wave is to make use of two helical beam antennas of opposite sense. A right-hand helical antenna transmits or receives right-hand polarization, while a left-hand helical antenna transmits or receives left-hand polarization. If a circularly polarized wave is received first on a right-hand helical antenna and then on a left-hand helical antenna, the antenna which receives the greater amount of signal will have a sense which corresponds to the sense of the received wave. In the case of an elliptically polarized wave, the sense will be taken to be the same as that of the predominant circular component.

Not all workers in the field use the same definition of sense. The work of different authors should be compared with this precaution in mind.

Typical Applications

Although there is reasonably general agreement on the above definitions, the use of the words *circular polarization* does not always have quite the same meaning to different workers in the field. The difference lies in the permissible departure (for the application at hand) from precise polarization circularity before the circularly polar-

TABLE 23-1 Two Sample Applications of Circularly Polarized Antennas and Some Typical Characteristics

	Response to linear polarization of arbitrary orientation	Precipitation-clutter suppression in radar service (values based on search radar)
Ideal ellipticity*	0 dB	0 dB
Satisfactory ellipticity*	−3 dB	−0.5 dB
Unsatisfactory ellipticity*	−10 dB	−1.5 dB
Azimuth beamwidth	Omnidirectional	1–5°
Elevation beamwidth	20°	20°
Bandwidth	Greater than 50 percent	5–15 percent
Standing-wave ratio	5	1.15
Power rating	Milliwatt level	>100 kW, <10 MW

*Over all or almost all of the radiation pattern.

ized antenna becomes unsatisfactory. The application of a circularly polarized antenna often falls in one of two categories. The first category is the use of a circularly polarized antenna to provide response to a linearly polarized wave of arbitrary orientation. The second category is the use of a circularly polarized antenna for suppression of precipitation clutter in radar service.[2] Table 23-1 shows typical characteristics for each of the applications.

Orthogonal circularly polarized antenna pairs are also in use for frequency-reuse applications with communications satellites.[3] Channel isolation in excess of 20 dB is readily attainable. Reference 4 describes a practical polarization-ellipticity field measurement technique. Further discussion is in Sec. 23-2.

Polarization Synthesis

Any wave of arbitrary polarization can be synthesized from two waves orthogonally polarized to each other. For example, a circularly polarized wave will be produced by the coexistence of a vertically and a horizontally polarized wave, each having the same amplitude and with a 90° phase difference between them. If they have other than the same amplitude and/or other than a 90° relationship, the resulting wave will be elliptically polarized. If, for example, the amplitude of the vertically polarized wave is zero, the resulting wave is linearly polarized and has a horizontal orientation. Further, if the two waves have equal amplitude but 0° phase difference, the resulting wave is linearly polarized with 45° orientation. There are two possible combinations of vertically and horizontally polarized waves that can produce a wave of some specific ellipticity. One of the combinations will produce a predominantly left-hand wave, and the other combination will produce a predominantly right-hand wave. Figures 23-3 and 23-4 show the two cases. Separate illustrations are provided to emphasize the need for care in specifying the coordinate system in this work and the need for consistency in adhering to the selected coordinate system while performing and interpreting any theoretical or experimental work.

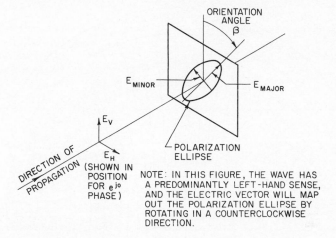

$$E_V = \frac{1}{2}\left[E_{MAJOR}\left(1 + e^{j2\beta}\right) + E_{MINOR}\left(1 - e^{j2\beta}\right)\right]$$

$$E_H = \frac{e^{j\frac{\pi}{2}}}{2}\left[E_{MAJOR}\left(1 - e^{j2\beta}\right) + E_{MINOR}\left(1 + e^{j2\beta}\right)\right]$$

FIG. 23-3 Linear-component synthesis of an elliptically polarized wave having a predominantly left-hand sense.

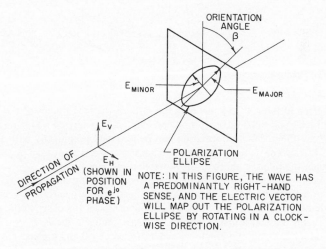

$$E_V = \frac{1}{2}\left[E_{MAJOR}\left(1 + e^{j2\beta}\right) - E_{MINOR}\left(1 - e^{j2\beta}\right)\right]$$

$$E_H = \frac{e^{j\frac{\pi}{2}}}{2}\left[E_{MAJOR}\left(1 - e^{j2\beta}\right) - E_{MINOR}\left(1 + e^{j2\beta}\right)\right]$$

FIG. 23-4 Linear-component synthesis of an elliptically polarized wave having a predominantly right-hand sense.

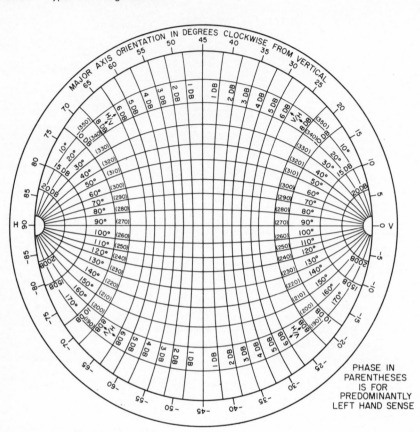

FIG. 23-5 Polarization chart.

It is often of interest to know only the ratio of the magnitudes of the vertically and horizontally polarized waves forming an elliptically polarized wave and the phase angle between them. Such information can be obtained graphically with the aid of Fig. 23-5. In this chart, the phase angle shown is the relative phase of the vertical element when the relative phase of the horizontal element is zero.

Example   Given an ellipticity of 3.0 dB and an orientation of $\beta = 37°$, find the ratio of the magnitudes of the vertically and horizontally polarized waves and the phase angle between them. Draw a line from the center of the chart to the periphery at the 37° point. Lay off a distance from the chart center to the $V/H = 3.0$-dB line measured on the horizontal diameter. Read $V/H = 0.8$ dB and a phase angle of 71° for predominantly right-hand sense or 289° for predominantly left-hand sense.

It is also possible to synthesize any elliptically polarized wave from two circularly polarized waves having opposite senses. For example, a linearly polarized wave will be produced by the coexistence of a right-hand and left-hand circularly polarized wave of the same amplitude. The orientation of the resulting linearly polarized wave will be determined by the phase difference between the two circularly polarized waves. Figure 23-6 provides the formulas for calculating the values.

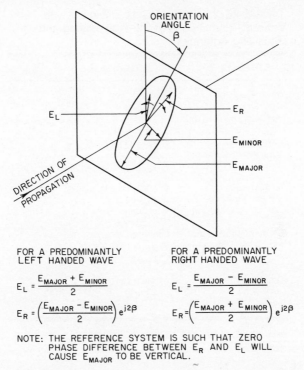

FOR A PREDOMINANTLY
LEFT HANDED WAVE

$$E_L = \frac{E_{MAJOR} + E_{MINOR}}{2}$$

$$E_R = \left(\frac{E_{MAJOR} - E_{MINOR}}{2}\right) e^{j2\beta}$$

FOR A PREDOMINANTLY
RIGHT HANDED WAVE

$$E_L = \frac{E_{MAJOR} - E_{MINOR}}{2}$$

$$E_R = \left(\frac{E_{MAJOR} + E_{MINOR}}{2}\right) e^{j2\beta}$$

NOTE: THE REFERENCE SYSTEM IS SUCH THAT ZERO
PHASE DIFFERENCE BETWEEN E_R AND E_L WILL
CAUSE E_{MAJOR} TO BE VERTICAL.

FIG. 23-6 Circular-component synthesis of an elliptically polarized wave.

Transmission between Two Elliptically Polarized Antennas

In order that a receiving antenna may extract the maximum amount of energy from a passing radio wave, it must have a polarization identical with that of the passing wave. For example, a vertically polarized receiving antenna should normally be used to receive a signal from a vertically polarized transmitting antenna. (The discussion assumes no polarization distortion in the transmission path.) Similarly, a right-hand circularly polarized antenna should be used for the reception of waves from a right-hand circularly polarized transmitting antenna. In general, maximum transmission will result between two elliptically polarized antennas when:

1 Their axis ratios E_{minor}/E_{major} are the same.

2 Their predominant senses are the same.

3 Their ellipse orientations are translated by a minus sign; that is, $\beta_{trans} = -\beta_{rec}$.

Note that Condition 3 will result in parallel-ellipse major axes in space and implies that only in three special cases will identical antennas at each end of the circuit with the same orientation yield maximum transmission. These three cases are:

TABLE 23-2 Transmission Efficiency with Various Polarizations at the Transmitting and Receiving Antennas

Antenna 1	Antenna 2	P/P_{max}
Vertical, e.g.: $E_R = 1$ $E_L = 1e^{j0}$ $\beta = 0$	Vertical, e.g.: $E_R = 1$ $E_L = 1e^{j0}$ $\beta = 0$	1
Vertical, e.g.: $E_R = 1$ $E_L = 1e^{j0}$ $\beta = 0$	Horizontal, e.g.: $E_R = 1$ $E_L = 1e^{j\pi}$ $\beta = 90°$	0
Vertical, e.g.: $E_R = 1$ $E_L = 1e^{j0}$ $\beta = 0$	Circular (right), e.g.: $E_R = 1$ $E_L = 0$	½
Circular (right), e.g.: $E_R = 1$ $E_L = 0$	Circular (left), e.g.: $E_R = 0$ $E_L = 1$	0
Circular (left), e.g.: $E_R = 0$ $E_L = 1$	Circular (left), e.g.: $E_R = 0$ $E_L = 1$	1

1 $\beta = 0°$ or $180°$.

2 $\beta = \pm 90°$.

3 $E_{minor}/E_{major} = 1$, so that β has no significance.

Identical antennas can always be made to yield maximum transmission, however, if they are rotated so as to cause the major axes of their waves to be parallel in space.

When two arbitrarily polarized antennas are used, the normalized output power from the receiving-antenna terminals will be

$$\frac{P}{P_{max}} = \frac{|(E_{R1}E_{R2} + E_{L1}E_{L2})|^2}{(|\mathbf{E}_{R1}|^2 + |\mathbf{E}_{L1}|^2)(|\mathbf{E}_{R2}|^2 + |\mathbf{E}_{L2}|^2)} \qquad \textbf{(23-1)}$$

All four variables in Eq. (23-1) are vector quantities. Similar expressions in terms of linear components, or axial ratios, may be found in References 5 and 6.

The values of E_R and E_L may be obtained with the aid of Fig. 23-6. Table 23-2 shows some typical combinations of interest.

Figure 23-7 shows a graphical means developed by Ludwig to display the maximum and minimum values of loss due to mismatched polarization between two antennas of arbitrary polarization. As an example of the use of the graph, if a predominantly right-hand polarized (RCP) transmitting antenna has an axial ratio of 8 dB and a predominantly left-hand polarized (LCP) receiving antenna has an axial ratio of 4 dB, draw two straight lines which intersect at the point P shown on the graph. This point is near the minimum-loss curve of 4 dB and the maximum-loss curve of 16 dB, and therefore the loss will vary (approximately) between 4 and 16 dB as one or

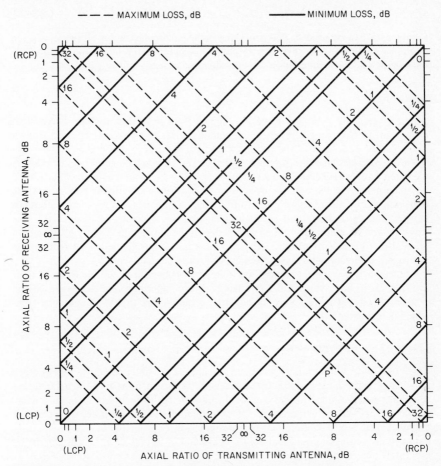

FIG. 23-7   Polarization loss between two elliptically polarized antennas.

the other antenna is rotated about its axis.* Further use of the graph to determine true gain with respect to a perfectly polarized antenna is described in Ref. 7.

Orthogonality

For any arbitrarily polarized antenna, there can be another antenna polarized so that it will not respond to the wave emanating from the first antenna. The polarizations of the two are said to be orthogonal. Using the convention of Fig. 23-6, two polarization ellipses will be orthogonal if they satisfy the relation

---

*Portions of this paragraph and Fig. 23-7 are reprinted with permission from publishers of *Microwave Journal*.

$$-E_{R1}E_{R2} = E_{L1}E_{L2} \qquad \textbf{(23-2)}$$

Figure 23-5 can be used to determine the orthogonal-polarization ellipse to a given ellipse by performing an inversion through the center of the chart and reading the phase-shift scale associated with the opposite predominant sense.

Reflection

When a vertically polarized wave is reflected from a smooth surface, there is no change in its character.* When a horizontally polarized wave is reflected from a smooth surface, because of the coordinate-system reversal in space when one looks in the reversed direction of propagation, there is a 180° phase change. When a circularly polarized wave is reflected from a smooth surface, its horizontal component is altered by 180°; hence the sense of the wave is reversed. For an elliptically polarized wave, reflection is equivalent to altering the differential phase shift (e.g., the phase difference between the horizontally and vertically polarized linear components) by 180°. The new polarization ellipse may be determined with the aid of Fig. 23-5 by inverting the original (before reflection) ellipticity across the horizontal diameter and reading the phase-shift scale corresponding to the opposite predominant sense.

Circularly polarized antennas are unique in being entirely unable to "see" their own images in any symmetrical reflecting surface, since the reflected wave has its sense reversed and is, therefore, orthogonal to the polarization of the antenna from which it originated.

Measurement Problems

The measurement of circularly polarized antennas is generally similar to that of other antenna types, except that it is often necessary to measure the state of polarization (degree of polarization ellipticity, or axial ratio) as well as the other more customary parameters. An experimentally convenient technique is to employ a linearly polarized antenna, arranged to rotate about its boresight axis, to probe the field. The resulting data will define the ellipticity but not its right-hand or left-hand sense, which must be separately determined.

Attention is called here to the unusual degree of sensitivity of polarization measurement to small distorting influences of secondary paths in a measuring setup. Figure 23-8 illustrates the conditions.

23-2 FREQUENCY REUSE BY POLARIZATION DECOUPLING

The expanded number of communications satellites in use has made it necessary to use the same frequency to communicate with closely spaced areas on the earth. Commonly called *frequency reuse,* this method requires a decoupling mechanism between

---

*Strictly speaking, this statement is correct only if the smooth surface is the interface between a normal propagation medium and one having an impedance of infinity. However, we are here interested only in the relative difference between the vertical and horizontal cases, so this detail will be overlooked.

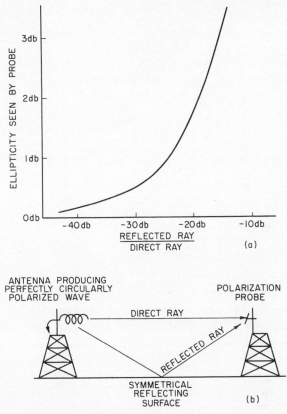

FIG. 23-8 Influence of the reflected wave on the measurement of a circularly polarized wave.

overlapping beams. Although careful shaping of the beams has been used in some cases, the use of orthogonal polarizations is more common. However, this puts stringent requirements on the amount of cross-polarization that can be present on any individual beam. Since the most common type of antenna used for satellite communications is a reflector antenna, the polarization properties of reflectors have been studied in great detail.[8-12]

Offset reflectors, which essentially eliminate aperture blockage, in particular have received a considerable amount of attention. Methods of computing the inter-beam isolation and thus the frequency-reuse performance have been developed.[13] For a symmetrical paraboloid, it has been shown that a physically circular feed with equal E- and H-plane amplitude and phase patterns will produce no cross-polarization;[14] however, in an offset configuration, cross polarization will be present because of the asymmetry. Chu and Turrin[11] have computed the maximum cross-polarization as a function of the offset angle θ_c and half of the reflector subtended angle θ_0 for linear polarization with a circular symmetric feed providing a 10-dB taper at the aperture edges. Their results, presented in Figs. 23-9 and 23-10, show that the amount of cross-

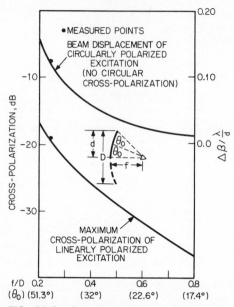

FIG. 23-9 Cross-polarization and beam displacement versus f/D ratio ($\theta_o = \theta_c$). (© 1973, IEEE.)

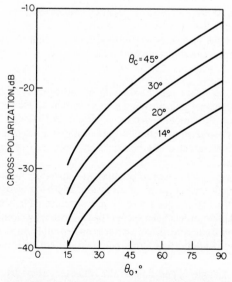

FIG. 23-10 Maximum cross-polarization of linearly polarized excitation. (© 1973, IEEE.)

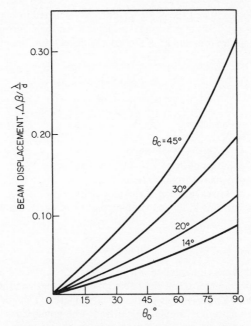

FIG. 23-11 Beam displacement of circularly polarized excitation (no circular cross-polarization). (© *1973, IEEE.*)

polarization decreases with increasing f/D and increases with increasing offset angle. In addition, they have shown that if the feed is perfectly circularly polarized, no cross-polarized energy will be found in the far-field pattern, although the position of the beam will be displaced from the axial direction because of a phase shift across the aperture. Figure 23-11 presents their results as a function of θ_0 with θ_c as a parameter.

The reduction of cross-polarized energy in offset configurations can be accomplished by careful feed design. The complex focal fields for an incoming plane wave can be calculated and a feed horn designed to match the focal plane by using multimode or hybrid modes (see Chap. 15). Alternatively, a polarizing grid can be used in front of the feed horn to reduce the cross-polarization.[15] In general, this grid will not consist of parallel wires but will be aligned to match the polarization of the primary field incident on the reflector surface.

Additional details on the subject of frequency reuse are found in Chaps. 17, 35, 36, and 45.

23-3 COMBINATIONS OF ANTENNAS

Circular polarization can be achieved by a combination of electric and magnetic antennas provided that the fields produced by these antennas are equal in magnitude and in time-phase quadrature. A simple case of this combination is a horizontal loop and a vertical dipole.[16] The time-phase-quadrature relationship is a fundamental relationship between the fields of a loop and a dipole when their currents are in phase. If the loop and dipole are oriented as in Fig. 23-12, the fields in the plane of the loop are given by

$$E_L = jCJ_1(ka)e^{j(\omega t - kr)} \qquad (23\text{-}3)$$

$$E_D = C_1 e^{j(\omega t - kr)} \qquad (23\text{-}4)$$

where C and C_1 = constants
$k = 2\pi/\lambda$
λ = free-space wavelength
a = radius of loop
r = distance from center of loop
J_1 = Bessel function of the first order
provided that the currents in the loop and dipole are in phase. Thus, if

$$CJ_1(ka) = C_1 \qquad (23\text{-}5)$$

the resulting field of the combination will be circularly polarized. Equation (23-5) will be true if the loop diameter is less than about 0.6 wavelength and the dipole length is less than a half wavelength. In this particular combination it should be noted that the resulting radiation pattern is circularly polarized at all points since the individual pattern of the loop and dipole are essentially the same. However, in practice this is difficult to obtain, except over narrow bandwidths, because of the different impedance characteristics of the loop and dipole.

A second combination is that shown in Fig. 23-13, consisting of two vertical one-half-wavelength-long cylinders in which vertical slots are cut.[16] Feeding the two vertical cylinders will give a vertically polarized omnidirectional pattern in the plane normal to the axis of the cylinders, while feeding the two slots will give a horizontally polarized pattern in the same plane. If the power to both feeding arrangements is adjusted to be equal and the phase adjusted by controlling the length of the feed lines

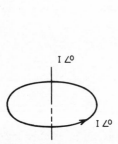

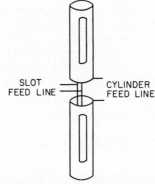

FIG. 23-12 Horizontal loop and vertical dipole.

FIG. 23-13 Slotted-cylinder circularly polarized antenna.

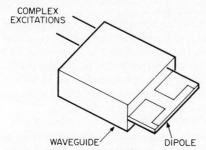

WAVEGUIDE   DIPOLE

FIG. 23-14 Slot-dipole element.

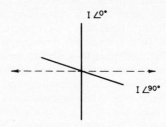

FIG. 23-15 Crossed dipoles—current in phase quadrature.

so that the two are in time-phase quadrature, the resulting pattern will be circularly polarized.

A combination of a slot and a dipole has been developed for a more directional antenna, e.g., as an element for a phased array.[17] The configuration, which is shown schematically in Fig. 23-14, uses a waveguide as the slot and a printed dipole as the complementary element. Since the waveguide and dipole radiation patterns track over a wide range of angles, this combined element can be used to generate any arbitrary polarization by controlling the complex excitations of the individual elements. Some overall degradation in the polarization performance will result from the difference in phase centers between the two individual elements.

The normal radiation (broadside) mode of a helix can also be considered as a combination of electric and magnetic antennas (dipoles and loops) producing circular polarization. This type of antenna has been discussed in Chap. 13.

A pair of crossed slots in the broad wall of a rectangular waveguide in which the field configurations in the waveguide are such that one slot is in time-phase quadrature with the other can also be considered as a combination of electric and magnetic antennas producing circular polarization. This combination will be discussed in greater detail below.

Two or more similar antennas when properly oriented in either time phase or space phase or a combination of both may be used to give a circularly polarized radia-

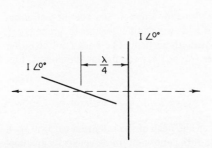

FIG. 23-16 Crossed dipoles—current in phase—$\lambda/4$ separation.

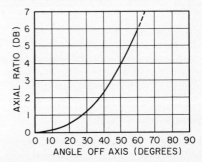

FIG. 23-17 Deviation of circularity as a function of off-axis angle for a pair of crossed dipoles.

tion field. A simple case is a pair of crossed half-wavelength dipoles. In Fig. 23-15, circular polarization is obtained by having the equal currents in the dipoles in phase quadrature. Radiation is right circularly polarized in one direction and left circularly polarized in the opposite direction. If the pair of crossed dipoles is fed in phase and separated in space by a quarter wavelength as shown in Fig. 23-16, circular polarization is again produced; however, the sense is the same in both directions. In both of these combinations the resulting field is circularly polarized only on axis. The deviation in circularity as a function of the off-axis angle is plotted in Fig. 23-17.

Another case of a simple combination of similar antennas to produce circular polarization is a pair of narrow slots at right angles and located at the proper point in the broad wall of a rectangular waveguide.[18] This may be explained by noting that the equations for the transverse and longitudinal magnetic fields of the dominant (TE_{10}) mode in rectangular waveguide (Fig. 23-18) are

$$H_x = H_0 \sqrt{1 - \left(\frac{\lambda}{2a}\right)^2} \sin\frac{\pi x}{a} \qquad (23\text{-}6)$$

$$H_z = -jH_0 \frac{\lambda}{2a} \cos\frac{\pi x}{a} \qquad (23\text{-}7)$$

From these two equations it may be seen that the fields are in phase quadrature and there are two values of x at which $|H_x| = |H_z|$. These values of x are given by

$$x = \frac{a}{\pi} \cot^{-1}\left[\pm \sqrt{\left(\frac{2a}{\lambda}\right)^2 - 1}\right] \qquad (23\text{-}8)$$

Two crossed slots at either of these points will then radiate circularly polarized energy. Figure 23-19 is a plot of x versus λ over the wavelength range between the cutoff of the TE_{10} and TE_{20} modes. The orientation of the slots is arbitrary, and they may be made resonant and thus radiate a large amount of power (Chap. 8). The theoretical axial ratio for $x = a/4$ is shown in Fig. 23-20, which gives circular polarization at a frequency for which $\lambda = 2a/\sqrt{2}$.

23-4 DUAL-MODE HORN RADIATORS

A conventional waveguide horn may be used for the radiation and beaming of circularly polarized waves provided that it is fed with waveguide capable of propagating vertically and horizontally polarized waves simultaneously. The horn may be either symmetrical or asymmetrical, that is, square (round) or rectangular (elliptical). Figure 23-21 illustrates two types.

Symmetrical Case

A circularly polarized field will be obtained on the peak of the radiation pattern when the horn is fed through the square waveguide with equal-amplitude vertically and horizontally polarized modes arranged to be in quadrature. The radiated field will not, in general, be circularly polarized at other points on the radiation pattern because the vertically and horizontally polarized radiation patterns will have different bandwidths.

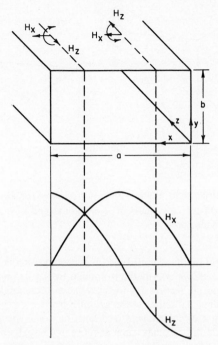

FIG. 23-18 Field configuration, TE_{10} mode.

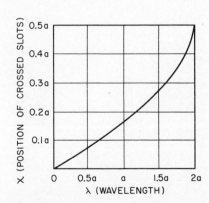

FIG. 23-19 Position of crossed slots for circular polarization versus wavelength.

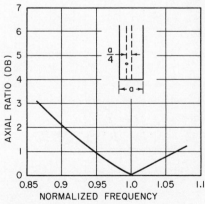

FIG. 23-20 Theoretical axial ratio for $x = a/4$.

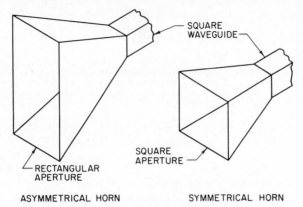

FIG. 23-21 Examples of symmetrical and asymmetrical dual-mode horns.

There are several methods useful in compensating for the two different beamwidths. Figure 23-22 shows two methods applied to the *azimuth plane* only of the horn. In actual practice, it is necessary to apply one of these methods (usually method *B*) to the elevation plane as well as to the azimuth plane.

When designing a circularly polarized horn for some specific radiation-pattern width, the standard design methods (such as those in Chap. 6 and Ref. 19) are appli-

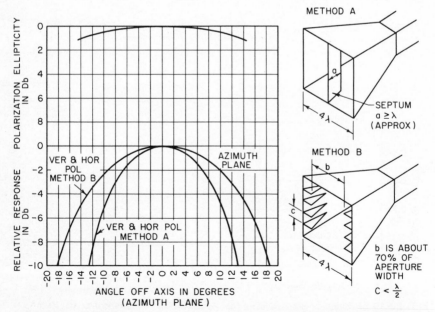

FIG. 23-22 Two methods of compensating dual-mode horns for improved circularity.

cable, bearing in mind, of course, that an *E*-plane dimension for one polarization is an *H*-plane dimension for the other and also bearing in mind the effects of any compensating scheme.

When the horn is designed along conventional pyramidal or sectoral lines, phase tracking for the vertically and horizontally polarized fields is usually not a problem; that is, the radiation centers of phase for the two polarizations are sufficiently close so that differential phase-shift variations over the radiation pattern usually do not exceed about 10 to 15° within the tenth-power points on the beam.

Horns are sometimes constructed with a wide flange around the aperture for mechanical reasons. On small horns especially, the flange width should be no larger than necessary to avoid degradation of phase tracking.

Simple conical horns have beamwidths more alike for the two principal polarizations than do pyramidal horns. Controlled excitation of TM_{11} modes can further refine beamwidth matching. However, such structures are not as conveniently connected to square or rectangular waveguide as pyramidal horns.[20]

Corrugated conical horns offer excellent radiation-pattern amplitude and phase tracking for the two principal polarizations over bandwidths of half an octave[3] or more.

Measurements reported in the literature[21] show that cross-polarization components suppressed 30 dB or more from 8 to 11 GHz over all angles within approximately ±45° of boresight. The conical-horn geometry is shown in Fig. 23-23. These structures tend to be larger for a given beamwidth than those of Fig. 23-22 and may limit the closeness of adjacent secondary-beam pointing when they are used as juxtaposed feeds.

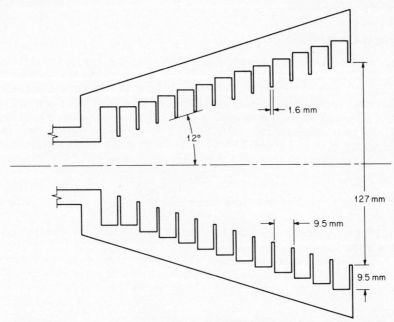

FIG. 23-23 Section through conical corrugated horn—9-GHz nominal design frequency.

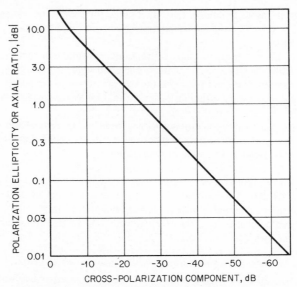

FIG. 23-24   Relationship between polarization ellipticity or axial ratio and cross-polarization component.

Figure 23-24 shows the relationship between polarization ellipticity and cross-polarization component.

Asymmetrical Case

The preceding discussion is also applicable to asymmetrical horns, except that circularly polarized fields on the peak of the radiation pattern will not be obtained unless allowance is made for the difference in phase velocity of the vertically and horizontally polarized waves within the asymmetrical-horn flare. This differential phase shift may attain quite large values. For example, a measured differential phase shift of about 220° has been observed at 2800 MHz in a horn having a flare length of about 14 in (356 mm), a width of 2.84 in (72.1 mm), and a height of 7.8 in (198 mm).

The magnitude of differential phase shift can be computed to an accuracy of perhaps 10 percent by evaluating the integral

$$\text{Differential phase shift} = \int_0^L \left[\beta(\ell) - \beta(\ell)\right] \, d\ell \atop \text{Vpol} \quad \text{Hpol} \qquad (23\text{-}9)$$

in which L is the flared length of the horn and $\beta(\ell)$ is based simply on the appropriate width of the flare and the operating wavelength. Figure 23-25 shows the information in detail. Note that the differential phase shift may vary fairly rapidly with frequency. In the example mentioned above, the differential phase shift increased by about 15° for every 100-MHz decrease in frequency between 2900 and 2700 MHz.

Method of Obtaining Quadrature

Quadrature phase relationship between two orthogonally polarized modes in round or square waveguide may be achieved by any of the following techniques:

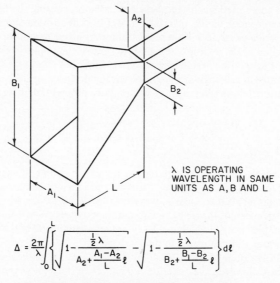

$$\Delta = \frac{2\pi}{\lambda} \left\{ \int_{0}^{L} \left[\sqrt{1 - \frac{\frac{1}{2}\lambda}{A_2 + \frac{A_1 - A_2}{L}\ell}} - \sqrt{1 - \frac{\frac{1}{2}\lambda}{B_2 + \frac{B_1 - B_2}{L}\ell}} \right] d\ell \right\}$$

Δ IS DIFFERENTIAL PHASE SHIFT IN RADIANS

FIG. 23-25 Method of determining differential phase shift in a sectoral horn.

1 Rectangular or elliptical cross section
2 Ridge guide
3 Dielectric slab
4 Multiple-lumped-element loading
5 Turnstile junction

The choice of method must be based on the application, keeping such things in mind as ease of design, bandwidth, power-handling capacity, ease of adjustment, and ease of fabrication. Table 23-3 shows the trend of these characteristics in order.

TABLE 23-3 Preference Trends for Methods for Obtaining Phase Quadrature

Method	Ease of design	Band-width	Power-handling capacity	Ease of adjustment	Ease of fabrication (small quantity)
Rectangular or elliptical cross-section guide	1	5	1	5	4
Ridge guide	4	3	2	2	3
Dielectric slab	5	2	5	3	1
Multiple-lumped-element loading	2	1	3	1	2
Turnstile junction	3	4	4	4	5

The first three methods listed above are based on causing the phase velocity of the vertically and horizontally polarized modes to differ within the same section of guide. This section of guide will then be of different phase length for the two polarizations. Selection of the appropriate length to furnish the requisite differential phase shift is the last step in the design process.

For example, a rectangular section of waveguide carrying TE_{01} and TE_{10} modes and having internal dimensions a and b will produce differential phase shift at the rate of

$$\frac{2\pi}{\lambda}\left[\sqrt{1-\left(\frac{\lambda}{2a}\right)^2} - \sqrt{1-\left(\frac{\lambda}{2b}\right)^2} \right] \quad \text{rad/unit length}$$

The values for a, b, and λ must be in the same units, of course. If such a section is to be connected to a square waveguide, it will be necessary to employ a suitable intermediate transformer section. Quarter-wavelength sections having an impedance equal to the geometric mean of the input and output impedances are satisfactory. Note that the differential phase shift (which can be computed with the aid of the above formula) of the transformer will contribute to the total.

Ridge guide can be handled in the same manner as the above rectangular guide. In the ridge case, however, the quantities $2a$ and $2b$ in the above expression should be replaced with λ_{CUTOFF} for TE_{01} and λ_{CUTOFF} for TE_{10} respectively.

Dielectric-slab loading in a square guide will also produce a difference in phase velocity for the two modes of interest. Figure 23-26 provides the pertinent information for a typical case. Just as above, it is necessary to provide means for obtaining an impedance match at each end of the phasing section. It is not sufficient as a rule to have the phasing section as a whole appear matched; it is important to have each end matched by itself. If this precaution is not observed, it will be very difficult to predict total phase shift as well as the change in phase shift with a change in geometry. In short, adjustment, or tailoring, is difficult. Power-handling capacity will be somewhat lower also.

Lumped-element loading in a square guide has also been used successfully. The use of probe pairs is particularly convenient. Pair doublets and pair triplets are the most convenient, with the latter providing greater bandwidth and greater power-handling capacity for a given differential phase shift. Figure 23-27 shows design data for

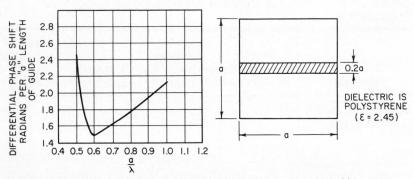

FIG. 23-26   Differential phase shift in a partially loaded square waveguide.

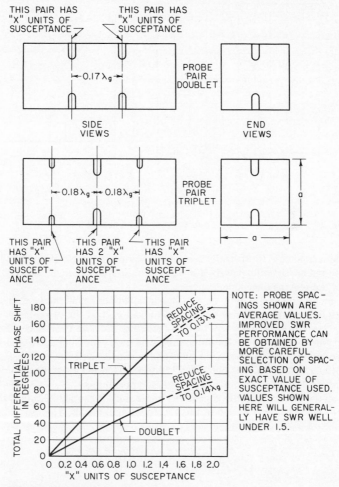

FIG. 23-27 Probe phase-shifter design.

the doublet and triplet cases. The length of the probe necessary for a given value of susceptance is shown in Fig. 23-28.

Circularly polarized waves can be generated directly in a waveguide junction known as a turnstile junction (not to be confused with a turnstile antenna). The general geometry is illustrated in Fig. 23-29.

Dual-Mode Generation

The generation of two orthogonally polarized modes in a section of square or round guide can be accomplished in several ways, of which the two most common are:

1 Coaxial cable probes at right angles

2 Conventional rectangular waveguide inclined at 45°

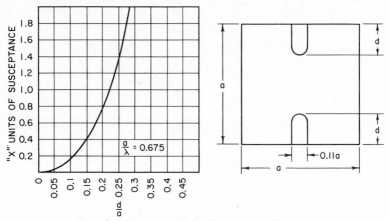

FIG. 23-28   Susceptance as a function of probe penetration.

The second case is especially convenient because the ratio of magnitudes of the vertically and horizontally polarized waves can be altered merely by altering the inclination of the rectangular guide.

In generating and guiding dual modes in square or round guide, care should be employed to keep the geometry symmetrical about each of the two principal longitudinal planes. Any asymmetry will tend to excite the TM_{11} or TE_{11} modes. For square or round guide, these two modes will usually not be very far from cutoff. For example, in square guide, with dimensions a high and a wide, cutoff for the TE_{01} and TE_{10} modes occurs at $\lambda = 2a$. The cutoff for TM_{11} and TE_{11} modes occurs at $\lambda = 2\sqrt{2}a$, or only about 41 percent above cutoff for the dominant modes.

In all the discussion above concerning phase shift, it has been assumed that there is an impedance match for each of the generated dual modes. The existence of mis-

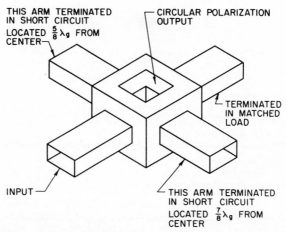

FIG. 23-29   Turnstile junction.

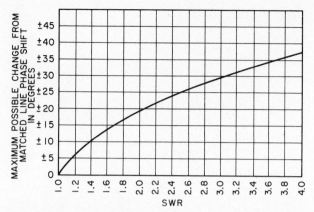

FIG. 23-30   Possible phase variation with standing-wave ratio.

match may produce an unexpected alteration in the phase shift. For example, if there is a perfect match for both the vertically and the horizontally polarized modes and the differential phase shift noted, the appearance of a mismatch in one of the modes may produce a change in differential phase shift as shown in Fig. 23-30. If the same magnitude of mismatch appears in both modes, the change may be as large as twice that of Fig. 23-29.

23-5   TRANSMISSION- AND REFLECTION-TYPE POLARIZERS

A transmission polarizer is an anisotropic propagation medium whose anisotropy is adjusted to achieve time (i.e., phase) quadrature for two waves whose (linear) polarization vectors are mutually orthogonal. Such polarizers operate independently of the nature of the source of the waves, are often relatively bulky and awkward to adjust, can be very inexpensive to manufacture, and can have very high power-handling capability.

The simplest type of polarizer is a parallel-metal-plate structure (Fig. 23-31). In this structure, the incident linearly polarized energy is inclined 45° to the metal-plate edges, so that there are two equal field components, one parallel to the plates and one perpendicular to the plates. The component perpendicular to the plates passes through the structure relatively undisturbed, while the component parallel to the plates experiences a phase shift relative to free-space propagation. If the spacing and length of the plates are adjusted so that the field parallel to the plates advances $\lambda/4$ with respect to the field perpendicular to the plates, the two fields at the exit of the plates result in a circularly polarized wave. This structure is commonly referred to as a *quarter-wave plate*.

Any anisotropic dielectric can be used as a transmission-type polarizer, provided only that positioning and properties result in two equal-amplitude, quadrature-phased linearly polarized waves at the exit port.

There are several ways to produce such anisotropy: (1) parallel metal plates,[22,23]

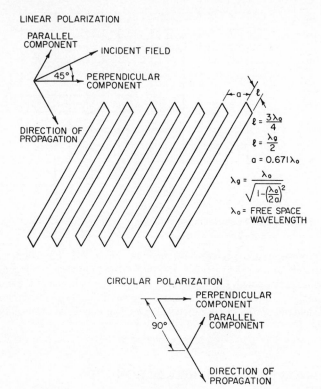

LINEAR POLARIZATION

PARALLEL COMPONENT

INCIDENT FIELD

45°

PERPENDICULAR COMPONENT

DIRECTION OF PROPAGATION

$\ell = \dfrac{3\lambda_0}{4}$

$\ell = \dfrac{\lambda_g}{2}$

$a = 0.671\lambda_0$

$\lambda_g = \dfrac{\lambda_0}{\sqrt{1-\left(\dfrac{\lambda_0}{2a}\right)^2}}$

$\lambda_0 = $ FREE SPACE WAVELENGTH

CIRCULAR POLARIZATION

PERPENDICULAR COMPONENT

PARALLEL COMPONENT

90°

DIRECTION OF PROPAGATION

FIG. 23-31 Parallel-plate polarizer.

(2) parallel dielectric plates[24] (Fig. 23-32), and (3) a lattice structure composed of strips or rods.[25,26] The first two methods are the heaviest, while the last two are the least frequency-sensitive. An unfortunate complication in design and adjustment of the simplest configurations is caused by imperfect impedance matching at the entrance and exit ports; in general, it is different for the two principal polarizations.

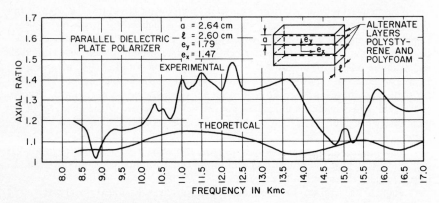

PARALLEL DIELECTRIC PLATE POLARIZER

$a = 2.64$ cm
$\ell = 2.60$ cm
$e_y = 1.79$
$e_x = 1.47$

EXPERIMENTAL

THEORETICAL

ALTERNATE LAYERS POLYSTY- RENE AND POLYFOAM

AXIAL RATIO

FREQUENCY IN Kmc

FIG. 23-32 Parallel-dielectric-plate polarizer.

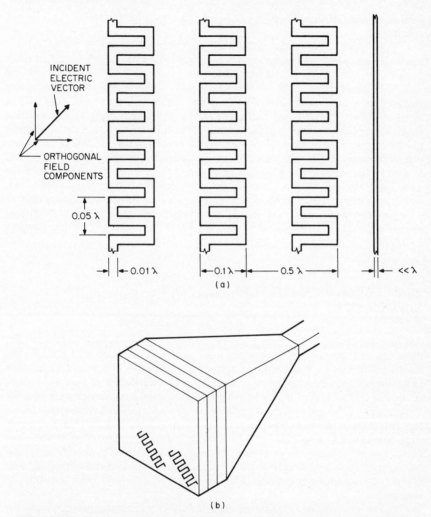

FIG. 23-33 Typical meander-line dimensions and radome applications. (*a*) Typical nominal dimensions of conductors on a layer (not to scale). (*b*) Four-layer polarizer-radome on a horn, illustrating orientation angle of typical conductors (not to scale).

Unwanted amplitude and phase effects result, including the result of a nonunity standing-wave ratio (SWR) inside the polarizer.

A variation of Method 3 above is the meander-line polarizer.[27,28,29,30] Figure 23-33a shows a typical meandering configuration of a thin metal conductor supported on a dielectric sheet. In general, a greater number of layers of such sheets yields greater bandwidths, just as in multiple discontinuities used for broadband transmission-line transformers.

Kotlarenko[27] reports bandwidths of 1.7 to 1 for axial ratios of less than 3 dB for incidence angles ranging over 30° when using a six-layer design. A five-layer design used at normal incidence yielded axial ratios under 3 dB for bandwidth exceeding an octave.

Epis[30] reports axial ratios under 3 dB over bandwidths of 1.75 to 1 by using a four-layer meander-line structure as a horn radome and a bandwidth exceeding 2.5 to 1 with axial ratios below 1.8 dB with six layers.

Reflection-type polarizers are essentially half-length transmission-type polarizers mounted on a conducting sheet. The simplest type of reflection-type polarizer is a series of closely spaced metal vanes, one-eighth wavelength high, on a conducting sheet. The incident energy is polarized at 45° to the vanes, so that the component parallel to the vanes is reflected by edges of the vanes and the component perpendicular to the vanes is reflected by the conducting sheet, thus delaying it 90° with respect to the parallel component and producing circular polarization. In general, any of the anisotropic propagating systems described above may be halved longitudinally and, with proper design, mounted above a conducting sheet to form a reflection polarizer.

Circular polarization may also be obtained via reflection from a conventional lossless dielectric when the dielectric constant e exceeds 5.8 and the incidence angle θ satisfies the condition $\sin^2 \theta = 2/(e + 1)$.[31]

23-6 RADAR PRECIPITATION-CLUTTER SUPPRESSION

Precipitation clutter is the name given to the radar echo from such targets as rain, snow, etc. It is not unusual for a radar to be rendered useless because precipitation clutter is sufficiently strong and extensive to hide the presence of a desired target (such as an aircraft). Because raindrops are substantially spherical (or at least much more so than an aircraft), they qualify as symmetrical reflectors. As mentioned in Sec. 23-1, a circularly polarized antenna is unable to see its own image in a symmetrical reflector. Therefore, if the radar antenna is circularly polarized, the echo from a symmetrical target such as a spherical raindrop will be circularly polarized with the wrong sense to be accepted by the antenna and will not be received. The echo from a composite target such as an aircraft will have scrambled polarization and will usually contain a polarization component to which the circularly polarized radar antenna can respond.

In other words, if the radar antenna is perfectly circularly polarized and the clutter is caused by spherical raindrops, the clutter cancellation will be perfect, enabling the presence of an otherwise-hidden composite target to be detected. When the radar antenna is elliptically polarized, the cancellation of an echo from symmetrical targets will not be complete; also, if the antenna is circularly polarized and the target is not symmetrical, cancellation will not be complete.

Cancellation Ratio

Cancellation ratio (CR) is defined at the ratio of radar power received from a symmetrical target when using a precipitation-clutter-suppression technique to the power received from the same target when not using the suppression technique. Cancellation ratio is not the same as cross-polarization ratio. Figure 23-34 shows cancellation ratio as a function of the polarization ellipticity of the antenna illuminating the symmetrical target.

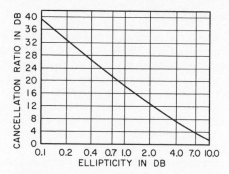

FIG. 23-34 Cancellation ratio as a function of polarization ellipticity.

Cancellation ratio, expressed in decibels, is related to the antenna polarization ellipticity by

$$\text{CR}_{\text{dB}} = 20 \log \left[\frac{1 - E^2}{1 + E^2} \right] \quad \textbf{(23-10)}$$

Note that the polarization ellipticity E is the ratio of minor to major axis (field quantities) and that the bracketed fraction is a voltage ratio. It is often convenient to measure ellipticity in relative power quantities when using a rotating linearly polarized polarization probe. The relative powers corresponding to the major and minor polarization axes are P_{max} and P_{min} respectively. In this case

$$\text{CR}_{\text{dB}} = 20 \log \left[\frac{P_{\text{max}} - P_{\text{min}}}{P_{\text{max}} + P_{\text{min}}} \right] \quad \textbf{(23-11)}$$

The fact that the bracketed fraction contains power quantities notwithstanding, the bracketed fraction is a voltage ratio.

Cancellation of Various Targets

Table 23-4 shows the extent of cancellation of various targets when the antenna is perfectly circularly polarized.

TABLE 23-4 Cancellation of Various Targets for Circular Polarization

Target	Return using circular polarization relative to return using linear polarization
Sphere	Complete cancellation
Disk facing radar	Complete cancellation
Large sheet facing radar	Complete cancellation
Wire grating facing radar (wires parallel to linear polarization for that case)	6-dB cancellation
Double-bounce corner reflector	Zero cancellation
Triple-bounce corner reflector	Complete cancellation

Integrated Cancellation Ratio

A suitable figure of merit for a circularly polarized monostatic radar antenna is the *integrated cancellation ratio* (ICR). The ICR is defined as the ratio of radar power received by a radar utilizing the nominally circularly polarized antenna (common for transmission and reception) to the radar power received by the same radar utilizing a linearly polarized (but otherwise identical to the above) antenna (again, common for transmission and reception) when the antenna, in both instances, is completely surrounded by a large number of randomly distributed small symmetrical targets.

The ICR places proper emphasis on the necessity for having the radar beam circularly polarized all over and not just on its peak. The rain cloud causing the clutter does not exist only on the beam peak; therefore, the complete beam must be circularly polarized if the most effective cancellation is to be obtained.

When calculating the ICR from antenna radiation-pattern measurements, recall that the backscatter from individual raindrops will have random phase and that the receiving antenna will sum their energy rather than their fields. If one-way measurements are made with constant signal-generator output and invariant receiver gain and the combined polarization and antenna gain data recorded in relative power units as P_{max} and P_{min}, the ICR may be calculated from

$$
\text{ICR}_{d\beta} = 10 \log_{10} \left\{ \frac{\displaystyle\sum_{\theta=0}^{360} \sum_{\phi=-90}^{90} (P_{max,\theta,\phi} - P_{min,\theta,\phi})^2 \, |\sin \phi| \, \Delta\phi\Delta\theta}{\displaystyle\sum_{\theta=0}^{360} \sum_{\phi=-90}^{90} (P_{max,\theta,\phi} + P_{min,\theta,\phi})^2 \, |\sin \phi| \, \Delta\phi\Delta\theta} \right\} \quad \textbf{(23-12)}
$$

$P_{max,\theta,\phi}$ denotes the relative power received in the rotating linearly polarized receiver as it lines up with the polarization-ellipse major axis, and $P_{min,\theta,\phi}$ similarly for the minor axis. The subscripts θ,ϕ denote that there is a separate pair of such measurements for each $\Delta\theta\Delta\phi$ solid-angle sample over the full range of θ and ϕ. This process will automatically include the effect of antenna polarization and gain parameters.

Measurements should cover the full spherical-coordinate system surrounding the antenna, with sample points spaced closely enough so that data are truly representative of antenna performance. If different angular increments are used on different parts of the pattern, the correspondingly correct entries for $\Delta\theta$ and $\Delta\phi$ must be made in Eq. (23-12).

With careful design, ICRs of 30 or 35 dB can be obtained with large reflector-type antennas. For operation over a 5 or 10 percent band, because of the nonideal shape of raindrops an ICR of 20 dB is usually considered adequate. It is fairly safe to assume that the ICR is the limit of performance that can be expected with a given antenna; usually the actual cancellation of the precipitation clutter will be somewhat poorer because of the nonideal shape of the water particles.

The use of circular polarization results in some diminution in the radar power received from an aircraft target also. The diminution can vary over wide limits, depending on aircraft type, aspect ratio, wavelength, etc., and by its nature it can only be described statistically. However, diminution values well under 10 dB are often observed. Additionally, dense precipitation usually does not completely surround a radar at all ranges but tends to occur in patches. The net improvement in apparent signal-to-noise ratio (that is, the desired aircraft-target signal to precipitation-clutter noise) obtained by using circular polarization for precipitation-clutter suppression is worthwhile, with net improvement in the 10- to 20-dB area being readily achievable.

REFERENCES

1 *IEEE Standard Test Procedures for Antennas,* IEEE Std. 149-1979, Institute of Electrical and Electronics Engineers, 345 East Forty-seventh Street, New York, N.Y. 10017.

2 W. B. Offutt, "A Review of Circular Polarization as a Means of Precipitation-Clutter Suppression and Examples," *Nat. Electron. Conf. Proc.,* vol. 11, 1955.

3 R. W. Kreutel, D. F. DiFonzo, W. J. English, and R. W. Gruner, "Antenna Techology for Frequency Re-Use Satellite Communications," *IEEE Proc.,* vol. 65, no. 3, March 1977. A comprehensive survey paper containing a good bibliography.

4 W. L. Stutzman and W. P. Overstreet, "Axial Ratio Measurements of Dual Circularly Polarized Antennas," *Microwave J.,* vol. 24, no. 10, October 1981, pp. 49–57. A method for polarization properties of antennas installed in operating position.

5 W. Sichak and S. Milazzo, "Antennas for Circular Polarization," *IRE Proc.,* vol. 36, no. 8, August 1948, pp. 997–1001. Derivation of expression for voltage induced in an arbitrarily polarized antenna by an arbitrarily polarized wave, plus typical applications.

6 L. Hatkin, "Elliptically Polarized Waves," *IRE Proc.,* vol. 38, no. 12, December 1950, p. 1455. Simplified derivation of expression for voltage induced in an arbitrarily polarized antenna by an arbitrarily polarized wave.

7 A. C. Ludwig, "A Simple Graph for Determining Polarization Loss," *Microwave J.,* vol. 19, no. 9, September 1976, p. 63.

8 I. Koffman, "Feed Polarization for Parallel Currents in Reflectors Generated by Conic Sections," *IEEE Trans. Antennas Propagat.,* vol. AP-13, no. 1, January 1966, pp. 37–40. Presentation of a method for determining the ideal feed polarization that should be incident on a conic reflector.

9 P. A. Watson and S. I. Ghobrial, "Off-Axis Polarization Characteristics of Cassegrainian and Front-Fed Paraboloidal Antennas," *IEEE Trans. Antennas Propagat.,* vol. AP-20, no. 6, November 1972, pp. 691–698. Comparison of the amount of cross-polarized energy for cassegrainian and front-fed paraboloids.

10 A. C. Ludwig, "The Definition of Cross Polarization," *IEEE Trans. Antennas Propagat.,* vol. AP-21, no. 1, January 1973, pp. 116–119. Presentation of a discussion of three definitions of cross-polarization with respect to several applications. It is noted that some confusion can exist depending on the definition.

11 T. S. Chu and R. H. Turrin, "Depolarization Properties of Offset Reflector Antennas," *IEEE Trans. Antennas Propagat.,* vol. AP-21, no. 3, May 1973, pp. 339–345. Presentation of results for both linearly and circularly polarized feeds. Measurements verify the analyses.

12 J. Jacobsen, "On the Cross Polarization of Asymmetric Reflector Antennas for Satellite Applications," *IEEE Trans. Antennas Propagat.,* vol. AP-25, no. 2, March 1977, pp. 276–283. Suggestions on how to design low cross-polarized feeds for offset-fed antennas.

13 Y. Rahmat-Samii and A. B. Salmasi, "Vectorial and Scalar Approaches for Determination of Interbeam Isolation of Multiple Beam Antennas—A Comparative Study," *Antennas Propagat. Int. Symp. Dig.,* vol. 1, 1981, pp. 135–139. Discussion of three approaches to calculating interbeam isolation and frequency-reuse characteristics of multiple-beam antennas.

14 A. C. Ludwig, "Antenna Feed Efficiency," Space Programs Summary 37-26, Jet Propulsion Laboratory, California Institute of Technology, 1965, pp. 200–208. Discussion of the factors relating to feed efficiency.

15 "30/20 GHz Spacecraft Multiple Beam Antenna System, Task II Report, Antenna System Concept for Demonstration Satellite," Cont. NAS 3-22499, TRW Defense and Space Systems Group, Apr. 22, 1981. Description of the design parameters of offset multiple-beam antennas.

16 C. E. Smith and R. A. Fouty, "Circular Polarization in F-M Broadcasting," *Electronics,* vol. 21, September 1948, pp. 103–107. Application of the slotted cylinder for a circularly polarized omnidirectional antenna.

17 R. M. Cox and W. E. Rupp, "Circularly Polarized Phased Array Antenna Element," *IEEE*

Trans. Antennas Propagat., vol. AP-18, no. 6, November 1970, pp. 804–807. Description and measurements of a combined waveguide and dipole element.

18   A. J. Simmons, "Circularly Polarized Slot Radiators," *IRE Trans. Antennas Propagat.*, vol. AP-5, no. 1, January 1957, pp. 31–36. Theoretical and experimental investigation of producing circular polarization by using two crossed slots in the broad wall of a rectangular waveguide.

19   D. R. Rhodes, "An Experimental Investigation of the Radiation Patterns of Electromagnetic Horn Antennas," *IRE Proc.*, vol. 36, no. 9, September 1948, pp. 1101–1105. Presentation containing design data for horn radiators.

20   P. D. Potter, "A New Horn Antenna with Suppressed Sidelobes and Equal Beamwidths," *Microwave J.*, vol. 6, no. 6, June 1963, pp. 71–78. Theory and design of a circularly symmetrical horn having steps to control the TE_{11} mode advantageously.

21   P. J. B. Clarricoats, A. D. Olver, and C. G. Parini, "Optimum Design of Corrugated Feeds for Low Cross-Polar Radiation," *Proc. Sixth European Microwave Conf.*, 1976, pp. 148–152.

22   J. Ruze, "Wide-Angle Metal Plate Optics," *IRE Proc.*, vol. 38, no. 1, January 1950, pp. 53–59. Derivation of the design equations for constrained metal-plate lenses.

23   J. F. Ramsay, "Circular Polarization for C. W. Radar," Marconi's Wireless Telegraph Co., Ltd., 1952. Proceedings of a conference on centimetric aerials for marine navigational radar held on June 15–16, 1950, in London.

24   H. S. Kirschbaum and L. Chen, "A Method of Producing Broadband Circular Polarization Employing an Anisotropic Dielectric," *IRE Trans. Microwave Theory Tech.*, vol. MTT-5, no. 3, July 1957, pp. 199–203. Description of a procedure whereby it is possible to design circular polarizers for both waveguide and window form to be used over a broad band of frequencies.

25   J. A. Brown, "The Design of Metallic Delay Dielectrics," *IEE Proc. (London)*, part III, vol. 97, no. 45, January 1950, p. 45. Development for the simplest case of a theory of metallic delay dielectrics based on an analogy with shunt-loaded transmission lines when the delay medium consists of an array of infinitely long conducting strips.

26   H. S. Bennett, "The Electromagnetic Transmission Characteristics of theTwo-Dimensional Lattice Medium," *J. App. Phys.*, vol. 24, no. 6, June 1953, p. 785.

27   I. Kotlarenko, "Analysis and Synthesis of Electromagnetic Wave Polarizers," *Proc. Eleventh IEEE Conf. (Israel)*, Oct. 23–25, 1979. Theoretical discussion and practical results of linear conductors and meander-line conductors in multilayer-array polarizers.

28   T. L. Blakney, J. R. Burnett, and S. B. Cohn, "A Design Method for Meander Line Circular Polarizers," *22d Ann. Symp. USAF Antenna R&D Prog.*, University of Illinois, Urbana, October 1972, pp. 1–15.

29   H. A. Burger, "A Dual Polarized Antenna System Using a Meander Line Polarizer," *Antennas Propagat. Int. Symp.*, May 1978, pp. 55–58.

30   J. J. Epis, "Theoretical Analysis of Meander Line Array Radome Polarizers," Western Div., GTE Sylvania Electronic Systems Group, December 1972. Theoretical establishment of the complex impedance discontinuities associated with meander-line arrays.

31   J. A. Stratton, *Electromagnetic Theory*, McGraw-Hill Book Company, New York, 1941, pp. 279–280, 499–500. Mathematical definition of an elliptically polarized wave and discussion of formation of a circularly polarized wave by total reflection from a dielectric interface.

BIBLIOGRAPHY

Ayres, W. P.: "Broadband Quarter Wave Plates," *IRE Trans. Microwave Theory Tech.*, vol. MTT-5, no. 4, October 1957, pp. 258–261. Design data on dielectric-slab quarter-wave plates for use in dual-mode waveguide.

Bahar, E.: "Full-Wave Solutions for the Depolarization of the Scattered Radiation Fields by Rough Surfaces of Arbitrary Slope," *IEEE Trans. Antennas Propagat.,* vol. AP-29, no. 3, May 1981, pp. 443–454. Article containing a particularly good bibliography of depolarization references.

———— and G. G. Rajan: "Depolarization and Scattering of Electromagnetic Waves by Irregular Boundaries for Arbitrary Incident and Scatter Angles-Full Wave Solutions," *IEEE Trans. Antennas Propagat.,* vol. AP-27, 1979, pp. 214–225.

Beckman, P.: *The Depolarization of Electromagnetic Waves,* Golem Press, Boulder, Colo., 1968.

————: *Radar Cross Section Handbook,* Plenum Press, New York, 1970.

———— and A. Spizzichino: *The Scattering of Electromagnetic Waves from Rough Surfaces,* The Macmillan Company, New York, 1963.

Brown, G. H., and O. M. Woodward, J.: "Circularly-Polarized Omnidirectional Antenna," *RCA Rev.,* vol. 8, June 1947, pp. 259–269. Theoretical and experimental investigation of a circularly polarized antenna system consisting of four inclined dipoles arrayed around a mast.

Chu, T. S., et al.: "Quasi-Optical Polarization Diplexing of Microwaves," *Bell Syst. Tech. J.,* vol. 54, December 1975, pp. 1665–1680.

Clarricoats, P. J. B., and P. K. Saha: "Propagation and Radiation Behavior of Corrugated Feeds," parts I and II, *IEE Proc.,* vol. 118, no. 9, September 1971, pp. 1167–1186.

Crandell, P. A.: "A Turnstile Polarizer for Rain Cancellation," *IRE Trans. Microwave Theory Tech.,* vol. MTT-3, no. 1, January 1955, p. 10. Description of turnstile polarizer in X-band radar antenna.

DiFonzo, D. F., and R. W. Kreutel: "Communication Satellite Antennas for Frequency Re-Use," *IEEE Antennas Propagat. Int. Symp.,* Sept. 22–24, 1971.

Fung, A. K., and H. J. Eom: "Multiple Scattering and Depolarization by a Randomly Rough Kirchhoff Surface," *IEEE Trans. Antennas Propagat.,* vol. 29, no. 3, May 1981, pp. 463–471. Theoretical exposition.

Heath, G. E.: "Bistatic Scattering Reflection Asymmetry, Polarization Reversal Asymmetry, and Polarization Reversal Reflection Symmetry," *IEEE Trans. Antennas Propagat.,* vol. AP-29, no. 3, May 1981, pp. 429–434. Discussion of bistatic theory and interpretation of target return.

King, A. P.: "The Radiation Characteristics of Conical Horn Antennas," *IRE Proc.,* vol. 38, March 1950, pp. 249–251.

Leupelt, U.: "Ursachen und Verringerung von Depolarisationeinflüssen bei dual polarisierten Mikrowellenantennen" (Causes and Dimunition of Depolarization Effects in Dual Polarized Microwave Antennas), *Frequenz,* vol. 35, no. 5, 1981, pp. 110–117.

Lorenz, R. W.: "Kreuzpolarisation fokusgespeister Parabolantennen" (Cross-Polarization of Front-Fed Paraboloid Antennas), *Technischer Bericht des Forschungsinstituts des Fernmeldetechnischen Zentralamt der Deutschen Bundespost,* FTZ 454 TBr 17, June 1973.

Ludwig, A. C.: "The Definition of Cross-Polarization, *IEEE Trans. Antennas Propagat.,* vol. AP-21, no. 1, January 1973, pp. 116–119.

Price, R.: "High Performance Corrugated Feed Horn for the Unattended Earth Terminal," *COMSAT Tech. Rev.,* vol. 4, no. 2, fall, 1974, pp. 283–302.

Ruze, J.: "Wide Angle Metal Plate Optics," Cambridge Field Sta. Rep. E5043, March 1949.

Schelkunoff, S. A., and H. T. Friis: *Antennas: Theory and Practice,* John Wiley & Sons, Inc., New York, 1952, pp. 390–393. Calculation of transmission between two elliptically polarized antennas; definition of a complex radiation vector to facilitate this calculation.

Scheugraf, E.: "New Broadband Circular Polarizer," *Int. URSI Symp. Electromagnetic Waves,* Munich, August 26–29, 1980.

Wait, J. R.: "The Impedance of a Wire Grid Parallel to a Dielectric Interface," *IEEE Trans. Microwave Theory Tech.,* vol. MTT-5, no. 4, April 1957, pp. 99–102.

Wolfson, R. I., and C. F. Cho: "A Wideband, Low-Sidelobe, Polarization-Agile Array Antenna," *IEEE Int. Radar Conf.,* 1980, pp. 284–287.

PART 3

Applications

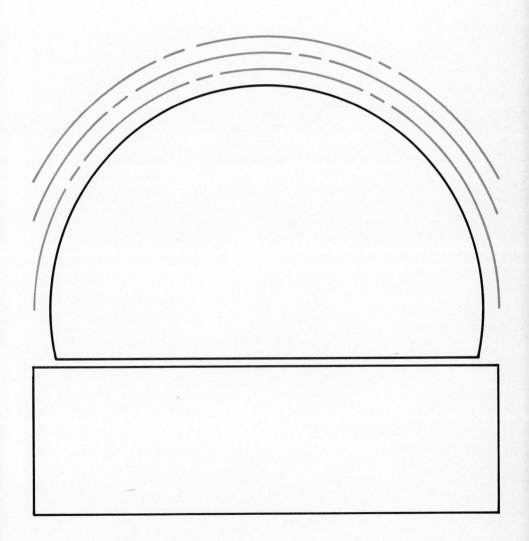

Chapter 24

Low-Frequency Antennas

Boynton G. Hagaman

Kershner & Wright Consulting Engineers

24-1 GENERAL APPLICATION*

The low-frequency (LF) portion of the radio-frequency (RF) spectrum is allocated to a number of special services to which the low attenuation rate and relatively stable propagation characteristics at low frequencies are of particular importance. Aeronautical and maritime navigational and communication services extend from approximately 10 to 500 kHz. Pulsed hyperbolic navigation systems operating near 100 kHz provide medium-range navigational aids. A phase-stable hyperbolic navigation system (Omega) transmitting within the 10.2- to 13.6-kHz region provides global navigation coverage. Other allocations include radio-location and maritime communications systems and LF fixed public broadcast. These services are of particular importance in the polar regions, which are subject to frequent and severe ionospheric disturbances.

Numerous strategic communication systems are currently operating within the region below 100 kHz, including a number of very-low-frequency (VLF) facilities capable of radiating hundreds of kilowatts within the 14.5- to 30-kHz band. These systems provide highly reliable service to airborne, surface, and subsurface transport.

VLF-LF transmitting systems involve large and expensive radiation systems and require extensive real estate and support facilities. Despite these drawbacks, activity within this portion of the spectrum is increasing, and many new systems have become operational within the past two decades.

24-2 BASIC CONFIGURATIONS

Transmitting Antennas

Early radiators for low-frequency transmitting application generally consisted of flat-top T or inverted-L arrangements using multiwire panels supported between two masts. As the operating frequency of interest moved further into the VLF region and available transmitter power increased, additional top loading was required. Additional masts and support catenaries were provided to suspend larger top-hat panels. These *triatic* configurations were employed on the majority of VLF radiators, although a variation termed the *umbrella type* was also successful. The top hat of the umbrella antennas consisted of several multiple-wire panels supported by a central mast and a number of peripheral masts. The top-hat panels were suspended by insulators, and the support masts were grounded. In some cases, the center support mast was insulated from ground at its base to reduce its effect on antenna performance.

Many examples of these configurations are still in use, although the T and inverted-L variations are used primarily for relatively low-power applications within the upper portion of the low-frequency band.

Base-insulated, freestanding, or guyed towers have come into general use as radiators for public broadcast service in the medium-frequency (MF) band (535 to 1605 kHz). At these frequencies, tower heights of one-sixth to five-eighths wavelength are practicable. In some cases, various top-loading arrangements were added to the guyed towers in an attempt to improve their performance. These experiments were quite successful.[1]

---

*10 to 300 kHz in this instance.

The most widely used variation of the base-insulated antennas for use at low frequencies is the top-loaded monopole. In this antenna, top loading is provided by additional "active" radial guys attached to the top of the tower and broken up by insulators at some point down from the top. In some cases, the top-loading radials also function as support guys.

Receiving Antennas

LF antennas designed for general receiving applications have very little requirement for efficiency. The receiving system is generally atmospheric-noise-limited, and air or ferrite loops, whips, or active probes are usually adequate. The design of low-frequency receiving antennas is not included within the scope of this chapter.

24-3   CHARACTERISTICS

Wavelengths within the low-frequency band range from 1 to 30 km. The physical size of low-frequency radiators is generally limited by structural or economic considerations to a small fraction of a wavelength. These antennas therefore fall into the *electrically small category* and are subject to certain fundamental limitations, clearly defined by Wheeler.[2] These limitations become increasingly apparent at very low frequencies and are a primary consideration in the design of practical VLF-LF radiation systems.

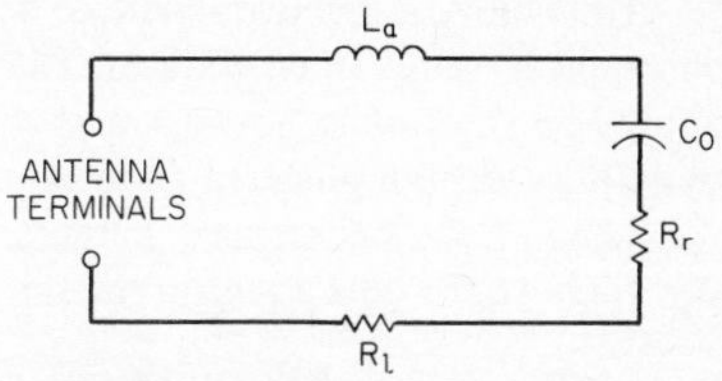

FIG. 24-1 Equivalent-circuit low-frequency antennas.

Since the low-frequency antenna is very small in terms of wavelength, its equivalent electrical circuit may be quite accurately represented by using lumped constants of inductance, capacitance, and resistance, as shown by Fig. 24-1,

where L_a = antenna inductance
C_0 = antenna capacitance
R_r = antenna radiation resistance
R_ℓ = radiation-system loss resistance

The antenna radiation resistance R_r accounts for the radiation of useful power. It is a function of the effective height H_e of the antenna:

$$R_r = 160\pi^2(H_e/\lambda)^2 \quad \Omega$$

The antenna effective height is equivalent to the average height of the electrical charge on the antenna. The effective height of an electrically short, thin monopole of uniform cross section is approximately equal to one-half of its physical height. For other, more practical antenna systems, the value of effective height is not simply related to the physical dimensions of the antenna structure. It is difficult to compute accurately the electrical height of complex multiple-panel antenna systems requiring numerous support masts and guy systems. Computer codes which will eventually make such analysis feasible are being developed. At present, the effective height and other basic design parameters of such an antenna are more effectively determined by scale-model measurement (see Sec. 24-8).

Radiation-system losses are often expressed in terms of equivalent resistance. The total loss resistance R_ℓ is the sum of all individual circuit losses:

$$R_\ell = R_g + R_t + R_c + R_d + R_{\text{misc}}$$

where R_g = ground loss
R_t = tuning loss
R_c = conductor loss
R_d = equivalent-series dielectric loss
R_{misc} = loss in structural-support system

The principal loss in most systems occurs in the ground system and in the tuning-network inductors. Other losses occur in antenna conductors, insulators, and miscellaneous items of the structural-support system. In some locations, antenna icing, snow cover, and frozen soil may also contribute to antenna losses.

Antenna efficiency η is determined by the relative values of radiation resistance and antenna circuit loss. It is expressed by

$$\eta = R_r/(R_r + R_\ell)$$

To achieve a given efficiency, there is a maximum value of radiation-system loss resistance R_ℓ that can be allowed. The individual-circuit losses, particularly those contributed by the tuning network and the ground system, can often be allocated during the detailed-design phase so as to minimize overall cost.

The lossless, or intrinsic, antenna bandwidth (to 3-dB points) is derived from the ratio f/Q, where Q is equal to the circuit reactance-to-resistance ratio. Therefore,

$$\text{BW (intrinsic)} = 1.10245 \times 10^{-7} C_0 H_e^2 f^4$$

where H_e = effective height, m
C_0 = static capacitance, μf
f = operating frequency, kHz

The intrinsic bandwidth is useful as a convenient means of antenna comparison.

The bandwidth of the radiation system (antenna, ground system, and tuning network) is inversely proportional to antenna efficiency:

$$\text{BW (radiation system)} = \text{BW (intrinsic)}/\eta$$

This expression defines the bandwidth of the radiation system. It neglects, however, the reflected resistance of the transmitter output amplifier R_t, which can increase the overall bandwidth substantially, depending upon the particular amplifier design and coefficient of antenna coupling. In practice, the actual operating bandwidth of a typical transmitting system driven from a vacuum-tube amplifier is 40 to 80 percent greater than that of the radiation system alone. This may not be the case when the power is generated by a solid-state amplifier of high efficiency.

VLF antennas are normally operated at a frequency below self-resonance, and the principal tuning component is a high-Q inductor, usually referred to as the *helix*. It is possible to increase the operating bandwidth of the radiation system appreciably by dynamically varying the effective inductance of the helix in a manner to accommodate the bandwidth requirements of data transmission. This method of synchronous tuning is accomplished with a magnetic-core *saturable reactor* shunted across a portion of the helix.

The inductance of the reactor is a function of the relative permeability of its magnetic core. The relative permeability is changed by regulating the amplitude of a bias current flowing through a separate winding on the magnetic core. A control signal derived from the data stream is processed and used to regulate the bias current and to synchronize the antenna tuning in accordance with the instantaneous requirements of the modulation system.

24-4 PRACTICAL DESIGN CONSIDERATIONS

The first task in the design of a low-frequency radiation system is the selection of the basic antenna configuration most suitable for the application. There are, at present, few choices capable of effective performance within the 10- to 300-kHz frequency range. The most common types are identified and illustrated by Fig. 24-2. Except for the conventional base-insulated monopole, they are all structural variations of a short, top-loaded current element. The principal radiated field is vertically polarized and essentially omnidirectional. The ultimate performance of each configuration is closely related to its effective height and volume. Therefore, antenna selection depends primarily on the frequency, power-radiating capability, and bandwidth requirement of the proposed system.

Vertical Radiators

The conventional base-insulated tower of Fig. 24-2a may be adequate for low-power application within the upper portion of the low-frequency spectrum. The theoretical radiation resistance for simple vertical antennas of this type, if linear and sinusoidal current distribution is assumed and metal-structure towers, guy wires, etc., are ignored, is shown by Figs. 24-3 and 24-4.

The input reactance is represented by the equation

$$X_a = -Z_0 \cot (2\pi\ell/\lambda) \qquad \Omega$$

which, for the case of a vertical cylindrical radiator of height h and radius ($a \ll h$), is usually evaluated by letting $\ell = h$ and

$$Z_0 = 60[\ln (2h/a) - 1] \qquad \Omega$$

If the tower is very short, greater accuracy is obtained by using[4]

$$Z_0 = 60[\ln (h/a) - 1] \qquad \Omega$$

Top-Loaded Monopoles

The base-insulated tower (Fig. 24-2b) in which top loading is provided by active radials or sections of the upper support guys is likely to be more cost-effective for low frequencies than the simple vertical radiator or the T or inverted-L types discussed later. A comprehensive parametric study of this antenna was completed by the Naval Electronics Laboratory in 1966. The study report contains numerous design curves that permit accurate evaluation of the performance of top-loaded monopoles of various configurations.[5] The data were acquired by scale-model measurement. The top-loading radials were held taut with little sag. A reduction of approximately 4 percent

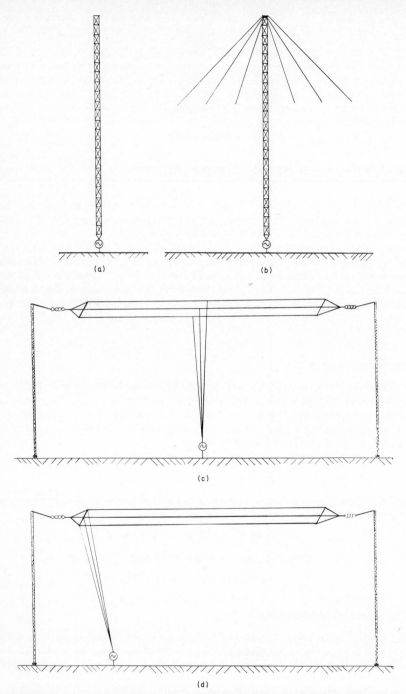

FIG. 24-2 Common types of low-frequency antennas. (*a*) Base-insulated monopole. (*b*) Top-loaded monopole. (*c*) T antenna. (*d*) Inverted-L antenna.

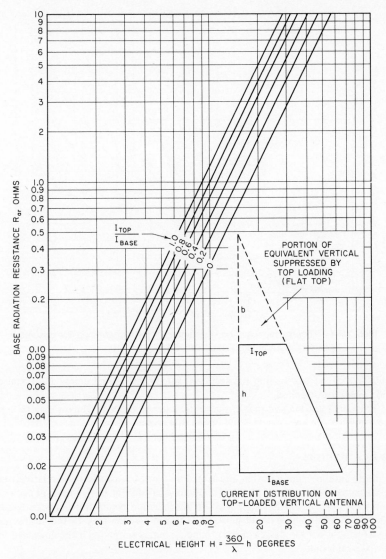

FIG. 24-3 Theoretical radiation resistance of vertical antenna for assumed linear current distribution. *(After Ref. 3.)*

should be applied to the effective height derived from the curves to account for the dead-load sag of a practical antenna.

T and Inverted-L Antennas

These antennas (Figs. 24-2*c* and *d*) require two masts from which to support the insulated top section, but they do not normally require base or guy insulators. The prac-

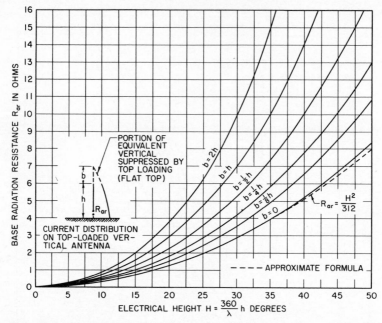

FIG. 24-4 Theoretical radiation resistance of vertical antenna for assumed sine-wave current distribution. *(After Ref. 1.)*

ticable area of the top-loading top section is somewhat limited unless additional end or side masts are supplied.

This configuration and that of the triatic antenna described below may be useful if the available site area is restricted in one dimension.

Triatic Antennas

The triatic-antenna configuration, illustrated by Fig. 24-5, consists of a relatively long antenna panel made up from parallel conductors. In addition to the masts at each end, the antenna panel is further supported at intermediate points from cross catenaries (triatics) and additional pairs of masts. The triatic antenna may be fed at any point along its length, but the design of the ground system is simplified if the downlead feed is near the midpoint.

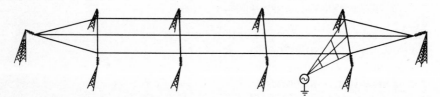

FIG. 24-5 Triatic-antenna configuration.

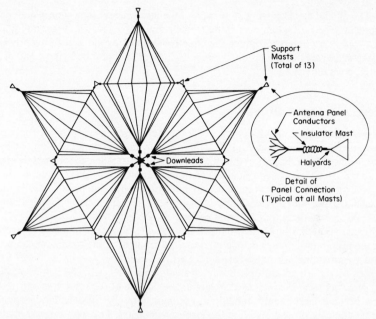

FIG. 24-6   Trideco-antenna configuration (plan view). Note: Individual down-leads are provided at each panel inner apex.

Trideco Antennas

The structural arrangement of a trideco antenna consists of one or more rhombic or triangular-shaped multiple-wire panels suspended from three or more masts. The individual downlead of each panel terminates at a common feed point (Fig. 24-6). This configuration has proved to be very effective for high-power applications, since it affords greater flexibility in the selection of both electrical-design and structural-design parameters. It is possible to design the top-hat panels so they can be individually lowered for maintenance without appreciably reducing antenna performance. This capability greatly simplifies antenna maintenance, thereby increasing system reliability and availability.

Valley-Span Antennas

These antennas have limited application. They are essentially T or inverted-L spans supported from natural geological formations such as those provided by a deep valley or mountainous ridges. The spans are interconnected and fed through a downlead extending to the valley floor. Typical examples of this type of antenna are in operation in Norway, Hawaii, and Washington State.

There are very few locations in the world where valley-spanning antennas can be effectively utilized. While the initial cost of such an antenna may be considerably less than that of a mast-supported system, the cost of ground-system maintenance and control of vegetation and erosion will generally be much greater.

The efficiency of existing valley-span antennas ranges from approximately 7 to 20 percent. Several factors contribute to their low efficiency. The topography of the typical valley site tends to reduce the effective antenna height and frequently limits the practicable area of the ground system. Soil cover may also be shallow and of poor effective conductivity.

Antenna-Design Voltage

The power-radiating capability of low-frequency antennas is proportional to the square of the antenna voltage ($P_{rad} \propto V^2$) and in most instances is limited by the allowable insulator voltage and the effective area of the antenna conductors. It appears that about 250 kV rms (rms values are commonly applied in VLF-LF antenna design) is the maximum operating voltage to which low-frequency antennas can presently be designed and put into service. This voltage level is not appreciably higher than that developed on VLF antennas 50 years earlier.

The voltage limitation is principally due to the physical limitation of available conductors and insulators. The potential surface gradient of the conductor can be controlled by providing adequate conductor surface. Insulator working voltages can be extended by the use of multiple-insulator assemblies. However, the structural design loads resulting from the increased weight and wind load of the insulators and antenna panels become increasingly difficult to accommodate. The problem is particularly severe in the case of insulated and top-loaded monopoles. The top-loading radials are generally insulated near the midpoint of their spans, at which point the insulator weight results in greatly increased sag or cable tension. The same is true of support-guy insulators, which must be capable of withstanding high tensile loads. The resulting insulator weight, distributed at several points along the guy, increases guy sag and reduces guy efficiency. The cumulative effect of insulator weight, increased cable diameters, wind load, and heavier towers is regenerative and eventually reaches a practical or economic limitation.

Triatic, trideco, and similar top-loaded antennas are somewhat less vulnerable to these problems because of their structural arrangement. The mast-support guys do not require insulators. The top-hat panel insulators can be attached relatively close to the supporting masts, where their weight is more easily supported.

Antenna Insulation

Virtually all insulators used on high-power LF-VLF antennas employ porcelain as the dielectric material. The dependability and long life of porcelain as an insulating material are well established. Reinforced-plastic insulators have been under development for many years and have recently shown promise in experimental service. They offer a considerable saving in weight.

The selection of the type of insulators to be used on the antenna depends to a large extent on structural considerations and the location of the insulator in the system. Three basic types are available: the strain, or "stick" type, in which the porcelain body is subjected to tensile loads; the fail-safe, or compression, type, in which the porcelain is supported in a frame or yoke arrangement which places the porcelain under compression; and a third type which does not subject the porcelain to any appreciable working loads. This type is a variation of the stick type in which tensile loads

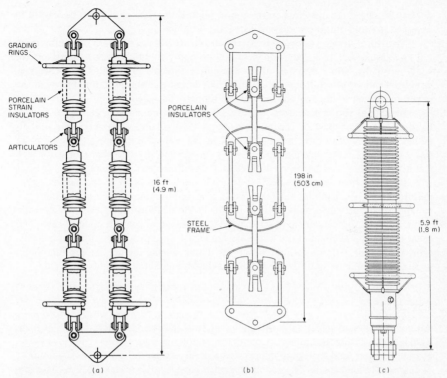

GRADING RINGS

PORCELAIN STRAIN INSULATORS

ARTICULATORS

16 ft (4.9 m)

PORCELAIN INSULATORS

198 in (503 cm)

STEEL FRAME

5.9 ft (1.8 m)

(a)   (b)   (c)

FIG. 24.7   Three types of insulators suitable for use in low-frequency antennas. (*a*) Strain type. (*b*) Fail-safe type. (*c*) Fiberglass-core type.

are accommodated by a high-strength material such as fiberglass, located within the interior of the porcelain body. These insulators are illustrated in Fig. 24-7.

The voltage-handling capability of the insulators must be adequate to ensure operating reliability at design power under normal environmental conditions. The published voltage rating of the insulators must be derated to allow for rain and condensation, surface contamination, overvoltage transients from atmospheric disturbances, and deterioration from service aging. A derating factor of 2:1 has been commonly applied; that is, a wet flashover rating twice the normal operating voltage is required. This factor has not always been adequate, primarily because of the difficulty of testing the insulator at its projected operating frequency rather than at 60 Hz and in an environment similar to that to be encountered in service. If the normal antenna working voltage exceeds about 150 kV rms, insulator assemblies should be tested at radio frequency to confirm their continuous wet-withstand capability at a voltage *at least* 50 percent above the working voltage.

Individual compression-type insulators are not available for operating voltages exceeding about 50 kV. Therefore, these insulators must often be connected in series to obtain the required voltage-withstand capability. Insulation "efficiency" decreases rapidly as insulators are added in series. Voltage-grading rings or cages are useful in this respect since they improve voltage distribution across the insulator string. These devices increase the insulator weight and wind-load area and must be carefully designed to avoid vibration and early structural failure.

Structural-design codes generally require structural guys to employ compression fail-safe insulators. This requirement may also apply to panel-support insulators, in which insulator failure may jeopardize the entire antenna structure. The active nonstructural radials of top-loaded monopoles may employ strain insulators provided the tower and primary antenna guy system are properly designed.

Individual strain insulators are capable of withstanding voltages up to several hundred kilovolts. However, the tensile strength of porcelain is relatively low, and individual insulators may have to be operated in parallel combinations of two or more series units in order to develop the necessary voltage and strength characteristics. The resulting insulator and hardware weight may be difficult to accommodate.

It may appear that the selection of insulators at this point in the design is premature. However, insulator selection has a considerable impact on the final antenna configuration and performance, since it affects the dead-load shape of the antenna spans, number and placement of guys, wind-load area, and connecting hardware. The usual practice is to conduct the initial design trades on typical insulator components, using the manufacturer's estimated weights. Later, during the detailed-design phase, as various factors are optimized, specific insulators and actual weights are used.

Conductors

An equally critical selection is the choice of conductors and structural cables to be used in constructing the antenna. Conductors are required for antenna panels or spans, current jumpers, and structural cables for halyards and guys. Low-frequency-antenna panels involve long spans. The conductors must have low RF resistance, adequate tensile strength, and minimum weight. The conductor diameter and total length must be sufficient to limit the surface potential to preclude corona.

The conductors are large and consist of multilayer cables stranded from aluminum or aluminum-covered steel wires. Figure 24-8 illustrates the construction of three types of conductors currently manufactured for electric power transmission and distribution systems. The ready availability of these conductors is an important consideration to the designer.

The conductor which has generally proved to be most suitable for antenna spans and panels is stranded from aluminum-clad, steel-core wires and is designated as AW (Fig. 24-8a). This conductor is listed by the manufacturer in diameters up to approximately 1⅛ in. It may be procured in diameters up to 2 in when arrangements are made for special stranding.

The size of the individual wires of the strand ranges from AWG No. 10 to AWG No. 4, depending upon the overall conductor diameter. The thickness of the outer layer of aluminum is 10 percent of the wire radius. This thickness may be less than the skin depth at the operating frequency, and RF losses are thereby increased. In applications such as downleads which are carrying a large current, it may be necessary to produce aluminum-clad wire cable with the outer strands of solid-aluminum wire (Fig. 24-8b). This type of conductor is designated AWAC, i.e., aluminum wire, aluminum-clad steel-wire core.

AWAC is not entirely satisfactory from a structural standpoint for applications in which the tensile load is high because of differences in the elasticity of the alumi-

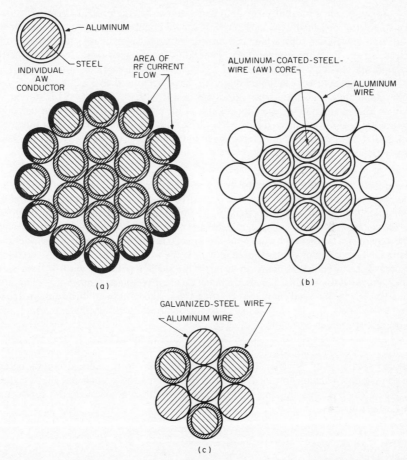

FIG. 24-8   Cross section of composite-cable conductors. (*a*) Concentric-lay stranded aluminum-clad steel conductor. (*b*) Aluminum-wire, aluminum-clad (AWAC) wire core. (*c*) Aluminum-conductor, steel-reinforced (ACSR).

num and aluminum-clad steel wires. This often results in the "basketing" of the outer wires when they are being handled or when pretension loads are relaxed.

A third type of composite conductor (Fig. 24-8*c*) is also in general use in power-transmission service. This conductor is made up in various combinations of aluminum and aluminized or galvanized steel wires and is designated ACSR, i.e., aluminum conductor, steel-reinforced. ACSR having steel wires exposed to the surface of the cable is not a suitable conductor for low-frequency conductors because of its relatively poor RF conductivity and the tendency of the aluminum wires to basket when tensile loads are released.

Conductors stranded of solid-aluminum wires are available and should be considered whenever high strength is not a factor. They are useful for current jumpers and for connections at feed-line terminal points.

When extreme flexibility and fatigue resistance are required in a jumper, stainless-steel wire rope should be considered. It should be fabricated from a nonmagnetic stainless alloy (Series 300), which will provide maximum skin depth and effective conductor area. Current jumpers are generally short, and their loss is usually insignificant.

Other specialized conductor materials have been used in high-power low-frequency antennas. Calsum bronze cables are in use at the U.S. Navy VLF station at Cutler, Maine, and hollow-core copper cable at the original VLF station at Annapolis. Future use of these specialized conductors is extremely unlikely, considering the present availability of aluminum-covered steel and the cost of procuring relatively short production runs of special cables.

Selection of Conductor Diameter The diameter of the conductors making up the antenna spans, panels, or downleads must be adequate to avoid corona as well as to develop the necessary tensile strength. The minimum conductor diameter required at the antenna design voltage should be established during the preliminary design since conductor weight and wind load are important factors in the structural design. Generally, a conductor size adequate to handle the working loads typical of very long spans will be of sufficient diameter to limit the surface potential gradient to a safe value, provided the antenna panels are made up of an adequate number of conductors.

Experience has confirmed that limiting the surface gradient of the antenna conductors to about 0.7 kV/mm will ensure corona-free operation under normal and extreme environmental conditions.

The usual expression for computing the surface gradient of an isolated conductor above ground is not applicable to a multiconductor transmitting-antenna panel. A method developed by Wheeler[6] (also discussed in Chap. 6) provides a solution based on the average current flowing from the wires toward the ground per unit of wire surface area. By using this method, the average voltage gradient on the conductor surface is found by

$$E_a \text{ [kV/mm]} = \frac{3\lambda^2}{2\pi H_e A_a} \sqrt{10P \text{ [watts]}} \times 10^{-6}$$

where E_a is the voltage gradient, kV/mm, and A_a is the total wire surface area, m^2.

When the antenna span or panels are made up of a number of conductors, the gradient at the outer wires is higher than that at the inner wires. The surface gradient of the wires of a multiwire panel can be partially equalized by varying their spacing with respect to each other, the more closely spaced wires being placed at the panel sides. The gradient on the outer wires can also be equalized by increasing their relative diameter.

Downleads

Base-insulated monopoles are generally fed at the base, and the antenna current is carried by the structural members of the tower. Virtually all low-frequency-antenna towers are fabricated of steel. The steel is usually galvanized except when a low-alloy steel is used. In any case, the thickness of the zinc is much less than a skin depth, and supplementary current buses must be provided up the tower to reduce losses.

It is recommended that triatics and similar wire-panel antennas be fed by one or more downleads terminating on a common feed point. A number of early low-frequency antennas used a system of multiple tuning, i.e., several downleads connected at various points to the antenna panel and separately tuned. Only one of the downleads was fed; the others terminated to the ground system. Multiple downleads substantially increase antenna cost and complicate tuning procedures. Their use has not proved to be effective in reducing antenna-system losses.

It is advantageous to use a multi-wire cage (Fig. 24-9) for the antenna download. The total antenna current flows in the download, and it is more practical to control the download loss and potential gradient by using a multiwire cage than to use a single conductor of large diameter.

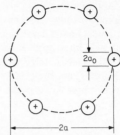

In addition, the effective diameter of a multiwire cage is much greater than that of its individual conductors. As a result, the download inductance is

FIG. 24-9 Multiwire cage for antenna download.

reduced, thereby increasing the self-resonant frequency of the antenna. The large effective diameter of the multiwire cage also reduces the potential gradient at the surface of the conductors. This is particularly important along the lower portion of the download where it approaches the earth and grounded structures.

The effective radius of a multiwire cage is given by[7]

$$a_{\text{eff}} = a(na_0/a)^{1/n}$$

Where a_{eff} = effective radius of cage, m
n = number of cage wires
a = cage radius, m
a_0 = individual wire radius, m

and, for practical purposes, the average potential gradient E_a on the individual cage wires is found by

$$E_a \text{ [V/m]} = V/na_0 \ln(2h/a_{\text{eff}})$$

where h = elevation of cage above ground, m
V = voltage on cage

Structural Cables

Non-current-carrying cables used for guys, halyards, and support cables are usually fabricated from bridge strand, which is available up to approximately 4 in in diameter. The strand is made up of a number of counterwound layers of galvanized high-strength steel wires. When flexibility is important, as in passing over winches and shives, wire rope fabricated from bundles of fine steel-wire strands over a hemp core is preferable to bridge strand.

Special Problems of Insulated Towers

Antennas requiring base-insulated towers can be advantageously employed in low- and medium-power applications, particularly within the upper portion of the low-frequency band and almost exclusively within the medium-frequency broadcast band.

They may not be cost-effective, however, in applications in which the antenna operating voltage is high. In addition to expensive guy and base insulators, such towers require a high-voltage obstruction-lighting isolation transformer with a voltage-withstand rating equal to that of the base insulator, a special provision for elevator power if an elevator is desired, and some provision for boarding the tower above the level of the base insulator.

In some locations, the heavy rainfall accompanying typhoons or hurricanes may flood the porcelains of base insulators and isolation transformers, causing flashover and operational interruptions. Isolated sections of tower guys may accumulate a high-potential static charge, triggering flashover of all insulators of the guy. These problems can be overcome, but the cost may not compare favorably with an alternative configuration using uninsulated masts and guys.

24-5 ANTENNA AND TUNING-NETWORK DESIGN PARAMETERS

The potential performance of a low-frequency antenna can be predicted from three basic parameters: effective height, antenna capacity, and self-resonant frequency. These values can be computed from the principal dimensions of the proposed antenna configuration or acquired by direct measurement on a scaled antenna model (Sec. 24-8). If the proposed antenna is similar to an existing system the properties of which are known, the design parameters may be appropriately scaled.

The principal performance characteristics of the antenna are related as follows. The power-radiating capability of the antenna for a chosen level of top-hat voltage V_t is equal to

$$P_{\text{rad}} = 6.95 \times 10^{-10} V_t [\text{kV rms}] \, C_0 [\mu f] \, H_e^2 f^4 [\text{kHz}] \qquad \text{kW}$$

The voltage at the antenna base or feed point V_b establishes the working-voltage requirement for the base insulator, download terminal, and entrance bushing:

$$V_b [\text{kV}] = V_t [\text{kV}] \{ 1 - (f [\text{kHz}]/f_r [\text{kHz}])^2 \}$$

where f = operating frequency

f_r = antenna self-resonant frequency

The antenna-base current required to radiate a specified power is given by $I_b = (P/R_r)^{1/2}$, where P is expressed in watts, the antenna radiation resistance R_r in ohms, and the antenna-base current I_b in amperes. The antenna-base current largely determines the current rating of the helix conductor.

The power input to the radiation system required to radiate a given power is

$$P [\text{watts}] = I_{\text{base}}^2 (R_r + R_\ell)$$

where R_r = antenna radiation resistance

R_l = antenna loss resistance, Ω

The antenna inductance L_a is expressed by

$$L_a [\mu H] = 1/4\pi^2 f_r^2 [\text{kHz}] \, C_0 [\mu f]$$

Since the total circuit inductance required to resonate the antenna at a given operating frequency f (below self-resonance) is $1/4\pi^2 f [\text{Hz}]^2 \, C_0 [\mu f]$, the effective tuning inductance required at the antenna feed point is

$$L_{tune} \ [\mu H] \ = \ 1/4\pi^2 f^2 \ [\text{Hz}] \ C_0 \ [\mu f] \ - \ L_a \ [\mu H]$$

If the antenna is to be designed for operation over a band of frequencies, the operating conditions at the lowest frequency will be the most severe and will establish the principal design requirements of the system. As the operating frequency is increased, the bandwidth will be improved and the current and voltage levels will be appreciably reduced.

Tuning Low-Frequency Antennas

Low-frequency antennas are generally designed to be operated well below their self-resonant frequency. The antenna impedance will then be capacitive, and the antenna may be tuned with a series inductor. A schematic drawing illustrating the arrangement of a typical tuning and matching network is given in Fig. 24-10. At frequencies above self-resonance the antenna impedance is inductive, and a series capacitor must be used. Capacitors of the size and rating required for series-tuning, high-power VLF antennas are not readily available. In addition, an inductor is inherently more reliable than a capacitor and less subject to failure from lightning or overvoltage transients. Therefore, in order to avoid the necessity of matching into an inductive reactance, the self-resonant frequency should be designed to be at least 10 percent and preferably 15 percent above the highest anticipated operating frequency. This objective may be difficult or impossible to achieve with very large antennas. In such cases, the downlead inductance should be minimized or alternative feed arrangements considered.

The helix is usually designed as a single-layer inductor. In this form, it is less difficult to install and to provide with means for taps. The voltage difference across such an inductor is quite evenly distributed. As a result, the turn-to-turn voltage is minimized and can be accommodated by means of normal turn-to-turn spacing (generally about two wire diameters). The helix is most conveniently fed at the bottom. Maximum voltage exists at the top, where it may be connected to a high-voltage exit bushing extending to the downlead.

Variometers

Low-frequency antenna-tuning circuits generally require one or more variable inductors termed *variometers*. Variometers consist of two concentric coaxial (partially

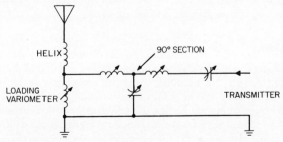

FIG. 24-10 Typical tuning circuit for use with a low-frequency antenna operating below self-resonant frequency.

spherical) coils connected in series. The smaller coil can be rotated about the common axis by means of a motor-driven shaft. The coefficient of coupling of the inner and outer coils is relatively high. Rotation of the inner coil through one-half revolution results in smooth variation of the effective inductance. A total inductance variation of about 8:1 can be achieved with large transmitting units.

Variometer performance can be estimated by computing the inductance, mutual inductance, and coefficient of coupling between the concentric inductors. Judgment is required in the selection of the effective radius and length of the coils, since the physical requirements of the variometer frame require the inductors to be incomplete spherical coils with few turns of varying pitch.

Inductance of helix (solenoid):

$$L = \pi\mu_0 a^2 n^2 / b + 0.9\ a$$

Mutual inductance of two concentric coaxial helices if inner coil is slightly smaller than outer coil:

$$L_{12} = L_2(a_1^2 n_1)/(a_2^2 n_2)$$

Coefficient of coupling of two concentric spherical coils:

$$K_{12} = (a_1/a_2)^{3/2}$$

where a = coil radius, m
b = $n \times$ pitch of winding, axial length, m
n = number of turns
μ_0 = 1.257×10^{-6} = magnetivity of free space, H/m
subscript 1 = inner coil
subscript 2 = outer coil
subscript 12 = mutual

24-6   RADIATION-SYSTEM LOSS BUDGET

The radiation resistance of a low-frequency radiator is relatively small; consequently, its efficiency is critically dependent upon antenna-system losses. It is useful to prepare a *loss budget* in which identifiable loss items are individually evaluated. The major losses are likely to occur in the ground system and tuning network. Among other losses are those in the antenna conductors and support guys and in the insulator dielectric. The sum total of all radiation-system losses must not exceed a *maximum allowable value* if the antenna efficiency goal is to be achieved.

Ground Losses

It is possible to some extent to allocate certain losses so as to minimize the overall cost of the radiation system. For example, if the site topography or soil geology makes installation of the ground system difficult and unusually expensive, it may be more economical to accept somewhat higher ground losses, provided the loss in the tuning network can be reduced accordingly. Calculation of ground loss is discussed in Sec. 24-7.

Tuning-Inductor (Helix) Loss

The principal loss in the tuning network will usually be in the main tuning inductor used to tune the antenna. It is possible to design a VLF inductor having a computed Q of several thousand, but the necessity of providing for taps and connectors makes it very difficult to achieve a practical value exceeding approximately 2500. Values of 1500 to 2000 are typical. The helix Q is primarily related to its overall size. Other factors include the conductor area and material, the inductor form factor, and the volume and material of the shield.

The most effective helix conductor is a composite cable stranded of Litz wire as illustrated in Fig. 24-11. The Litz-wire strands are made up of numerous insulated copper wires approximately a skin depth in diameter. The Litz strands are bundled together into larger groups and jacketed over a central jute core. The wire and wire bundles are transposed within the jacket so as to distribute the current density uniformly among the individual wires.

The number of individually insulated wires making up the cable is determined from the maximum value of the antenna current on the basis of 1000 circular mils per ampere. If the current requirement exceeds about 1500 A, the resulting cable diameter (typically 3½ in) becomes difficult to handle. In this case, two or three parallel cables of smaller diameter may be used. An excellent summary of the design of Litz inductors is given by Watt in Chap. 2.5 of *VLF Radio Engineering.*[8]

Copper tubing of appropriate size may also be used for low-frequency inductors in applications in which the additional losses may be tolerated and the power level permits the use of tubing of a reasonable diameter (6 in or less). Large-tubing inductors must be assembled from sections bolted or brazed together during installation.

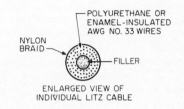

ENLARGED VIEW OF INDIVIDUAL LITZ CABLE

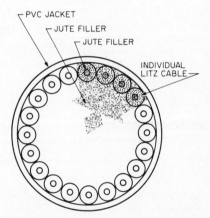

FIG. 24-11 Cross section of a large Litz conductor used in low-frequency inductor.

The helix-room shield may be fabricated from brazed or welded copper or aluminum sheets having a thickness of two to three skin depths at the lowest operating frequency. A square or rectangular helix room is usually more practical than one of cylindrical form.

The additional losses contributed by the shield will be small if the size of the helix building provides an inductor-to-shield separation of approximately one coil diameter.[9] The separation at the floor and ceiling may be somewhat less. It is advisable to limit the surface gradient of the Litz cable to a maximum of 0.5 kV/mm (rms).

Conductor Losses

Antenna current flows in downlead conductors, top-hat panels, or top-loading radials and, to a lesser extent, in guy cables and tower members. To compute the losses in these items, it is necessary to evaluate the RF resistance of each at the frequency of operation.

Concentric Lay-Stranded Aluminum-Clad Steel Conductors (AW)   This type of conductor is suitable for antenna panels and long spans. The individual wire size of the conductor depends upon the diameter of the finished cable and ranges from AWG No. 10 through AWG No. 4. The thickness of the aluminum cladding over the steel core is 10 percent of the finished wire radius, or approximately 0.25 mm for the largest-diameter wire (AWG No. 4). For frequencies below about 100 kHz, this thickness is less than a skin depth. The current penetration into the steel-wire core will be negligible owing to its relative permeability. Therefore, essentially all the current is forced to flow in the aluminum surface. This is illustrated by Fig. 24-8. Since the number and size of the individual wires are selected so as to provide a tightly grouped outer layer, the current will flow only in the outer half surface of each exposed wire.

For example, a nominal 1-in cable may consist of four layers of AWG No. 7 wire with 18 wires in the outer layer. The total effective cross-sectional area of the current-carrying aluminum is

$$A = (r_1 + r_2)(r_1 - r_2)\pi N \times 10^{-6}/2$$

where r_1 = radius of individual wire, mm
  r_2 = 0.9 r_1 (Thickness of aluminum on steel wire is 10 percent of wire radius.)
  N = number of wires in outer wrap (lay), or 18.04×10^{-6} m^2 in this example
The resistance per meter length is

$$R_{rf} = 1/\sigma A$$

where A = area, m^2
  σ = conductivity of aluminum (3.5×10^7 S/m)

$$R_{rf} = 1.6 \text{ m}\Omega/\text{m at 15 kHz}$$

Aluminum Wire, Aluminum-Clad Wire Core (AWAC)   In some applications, downlead cages for example, the loss in aluminum-clad steel wires may exceed the allowable loss budget. It is possible to procure cable similar to AW in which the outer lay consists of solid-aluminum wires. RF loss in this cable will be equivalent to that of all-aluminum strand.

Since the wire diameter will be very much greater than a skin depth, the resistance can be expressed as

$$R_{ac} = 1/\sigma \pi \, d\delta$$

Where σ = conductivity, S/m
  d = wire diameter, m
  δ = skin depth, m, at frequency of interest: $\delta = 1/2\pi(f\sigma\mu_r \times 10^{-7})^{1/2}$
  μ_r = relative permeability of conductor material

Bridge Strand

Guy cables are generally made up from bridge strand. The individual steel wires of bridge strand are protected from corrosion by a thin layer of zinc. The zinc is too thin to be a factor in lowering the resistance of the cable. If a value of 200 for relative permeability and 5×10^6 S/m for conductivity of the steel is assumed, the skin depth for bridge strand at 15 kHz is approximately equal to 0.13 mm. The RF resistance calculation is straightforward, taking into account skin depth and exposed-surface area.

Antenna Insulators

Low-frequency insulators employ porcelain dielectric, and the loss factor at operating frequency is relatively low. The insulator loss for the antenna system is estimated from the applied working voltage, dielectric power factor, and number of insulators in the system. The dielectric loss of the insulator can also be estimated by determining the porcelain heat rise during preproduction testing at low frequency.

24-7 DESIGN OF LOW-FREQUENCY GROUND SYSTEMS

The radiation resistance of electrically small antennas is very small, and if an antenna is to have a useful efficiency, system losses must be carefully controlled. The principal circuit losses occur in the tuning network and the ground system. Tuning losses can be minimized by using high-Q components. Ground losses within the site area can be reduced by providing a low-reactance, low-resistance ground system. This may consist of a radial network of wires buried a half meter or so below the surface and extending in all directions to some distance beyond the antenna.

Ground losses result from ground-return currents flowing through the lossy soil and from E-field displacement current in the soil or vegetation above the level of the ground wires. Generally, the E-field losses are not significant and may be neglected except in locations where the soil may freeze to a substantial depth or acquire a deep snow cover. E-field loss can be evaluated by methods outlined by Watt (Secs. 2.4.37 and 2.4.38).[8]

The radiation-system ground loss is derived by separately computing the E-field and H-field losses of small sectors within the radian circle (a circle having a radius of $\lambda/2\pi$) and summing them to get the total loss. The loss within the radian distance is attributed to ground-system loss; that beyond the radian circle is considered to be propagation loss. Since it is rarely practical to extend the ground system to the radian distance, ground losses can be further separated into *wired-area* and *unwired-area* losses.

Early VLF-LF ground systems varied widely in concept, depending upon the antenna configuration and whether single or multiple tuning was employed. In most cases, the ground wires were buried, although several designs returned earth currents from dispersed ground terminals on overhead conductor systems. Present practice is to design the ground system around a single-point feed by using an essentially radial configuration.

There are many ways in which ground-loss computations can be carried out. Since ground-conductivity values may vary considerably over the radian-circle area, a method that allows individual computation of relatively small areas is desirable. The appropriate conductivity values can then be applied. For example, the radian circle can be divided into 36 ten-degree sectors centered at the feed point, each sector being further divided into 50 or more segments. The loss is computed for each sector; when summed, these losses give the total loss.

Once the magnitude and orientation of the normalized* H field within the antenna region have been determined, the H-field losses are calculated in the following manner. The *longitudinal loss within the wired area* is found from the product of the square of the normalized longitudinal component of the H field, the resistance of a 1-m^2 area, and the area. The longitudinal loss is computed for each segment of the sector with the exception of the last few segments, which are treated as termination losses. The total loss for each sector is the sum of all its segments.

The unit resistance is the resistance component of the parallel impedance of the ground wires and the earth to a current flowing parallel to the ground wires. Losses due to the wire resistance (usually small) are covered by this method. The earth impedance, the wire impedance, and the parallel impedance are computed from Watt.[8]

The *transverse loss in the wired portion* is equal to the product of the square of the normalized transverse component of the H field, the earth resistance of 1 m^2, and the area (Watt,[8] Sec. 2.4.10). The transverse loss for a sector is the sum of the segment losses for all wired segments.

The *termination loss* occurs in the last few segments of the wired area, where the transition from the wired to the unwired area occurs. The number of segments is chosen to represent the end portion of the radial wires with a length equal to the skin depth of the earth. The resistance of a single wire of this length to ground is computed from Sunde[10] (Sec. 3.36). A computed or empirically derived mutual-resistance factor is employed to obtain the effective resistance of a single wire in proximity to adjacent wires. The resistance of the terminating ground rods, if used, is then computed from Sunde (Sec. 3.30). The termination resistance for the wire–ground-rod combination is then computed, using an estimated factor for the mutual resistance, and paralleling the two components. This figure is adjusted to a resistance per square meter and multiplied by the square of the longitudinal component of the normalized H field and by the area to provide the termination loss.

The *unwired losses* apply to the unwired area extending from the ground-system boundary to the radian circle. Losses are computed for each segment as the product of the H field squared, the earth resistance (Watt,[8] Sec. 2.4.10), and the area. The segment losses are summed to provide the total sector loss.

Calculations of ground-system resistance, if done manually, are laborious and time-consuming. As a result, it may be difficult to determine the most cost-effective configuration. The computations described above should be programmed for a computer, thereby enabling various ground-system configurations to be quickly evaluated.

---

*The H field is generally normalized for an antenna current of 1 A so as to make the loss resistance equal to the power loss in watts.

Practical ground systems are centered in the exit bushing or downlead terminal. The surrounding area is generally covered with a dense copper ground mat, lightly covered with earth and extending out to a distance of 50 to 100 m. The mesh area is enclosed by a rugged copper cable (secured by ground rods at corners and intermediate points), to which the ground radials are connected. All connections are brazed by hand or welded by using an exothermic process. The ground radials terminate to ground rods driven, if possible, to a depth of a few meters. The ground radials may be interconnected at various radii with circumferential buses to reduce transverse losses within the wired area. All support masts, foundation-reinforcing steel, and guy systems are connected to a number of ground-system radials. Guy-anchor foundations outside the ground-system area are grounded to multiple ground rods.

The ground system should be installed only after tower erection has been completed to avoid damage from heavy machinery. The ground system may be installed with a wire plow or placed manually, buried to a depth of approximately ½ m.

Soil-conductivity values generally available from published Federal Communications Commission (FCC) conductivity maps may not be suitable for use in VLF-LF ground-system design, since the skin depth in the earth at the frequency for which they are applicable is relatively shallow. The depth of penetration in average soil for frequencies within the VLF-LF region may extend to 50 m. Soils to this depth are usually composed of several layers having appreciably different values of conductivity. As a result, the effective conductivity may not be that of the surface layer but depend upon the conductivity and thickness of the individual layers at the frequency of interest.

A conductivity map of North America showing effective earth-conductivity values for a 10-kHz propagating wave has been prepared.[11] The work was extended at a later date to include the preparation of a worldwide map showing effective conductivity values of the earth at frequencies of 10 to 30 kHz.[12] The data presented in these reports are of a general nature. Specific conductivity values must be obtained at the proposed antenna location for use in the design of the ground system.

Ground conductivity can be determined within the antenna-site area from data acquired by the use of a four-terminal array.[13,14] Soil-conductivity values may vary greatly when the soil is frozen.[12] Sufficient measurement radials are required to establish the soil characteristics within the immediate antenna region and to establish an average value to apply to calculations beyond the wired area. The conductivity values can be entered onto the site map and tabulated into a data file for use with a computer.

24-8 SCALE-MODEL MEASUREMENTS

Scale-model measurements provide direct confirmation of the effective height, static capacity, and self-resonant frequency of a proposed antenna configuration from which the performance of the full-size antenna can be accurately predicted. Modeling is most useful as a supplement to the design of VLF-LF antennas having extensive top loading and numerous support masts.

The E-field and H-field distribution of a simple monopole can be accurately predicted from theory, and the corresponding ground-system design is thereby simplified.

This is not the case for a complex top-loaded antenna in which the field beneath the antenna is influenced by overhead-conductor currents and field distortion resulting from grounded masts and guys.

To evaluate ground losses within this high-current area, a knowledge of the *H*-field magnitude and orientation is necessary. These data can be acquired by probing the conductive ground plane upon which the model is erected with a small shielded loop.

Model Techniques

VLF-LF antennas operate over an image plane provided by the earth. To avoid excessive ground losses and to provide an effective image plane for the model, a large, flat, highly conductive surface is required. This can be simulated by using metal-wire mesh 1 cm or so per square, preferably galvanized after weaving. The area of ground plane required depends upon the model scale factor and antenna size. A scale factor of 100 is convenient for low-frequency antennas. The ground plane should extend beyond the antenna in all directions to accommodate *E*-field fringing and to approximate a semi-infinite surface.

It is advisable to elevate the ground plane to a height that will permit measurements to be taken from below. This will provide a degree of instrument shielding and avoid unnecessary disturbance of the antenna field.

The antenna model should be scaled as accurately as possible. Frame towers can be simulated with tubing of equivalent electrical diameter. Multiple-wire cages can be simulated with tubing or solid wire having an equivalent radius. Particular care should be taken to avoid stray shunt capacity at the antenna feed point, as this may reduce measurement accuracy.

A reference antenna with which to refer measurements of effective height should be provided. This can be a slender monopole located on its own ground plane some distance away.

A source antenna is necessary to provide an essentially plane radiation field of uniform intensity over the model ground plane for measurement of effective height. This may be provided by a nearby broadcast station if its frequency and field intensity are satisfactory or by a transmitting antenna set up several wavelengths distant. The measurement frequency should be scaled up approximately by the antenna model factor but need not correspond exactly to the design frequency. The general arrangement of an antenna range for scale-model measurements is shown in Fig. 24-12.

The effective height of the model antenna is derived from measurement of its open-circuit voltage. To reduce instrument loading, the voltage measurement should be taken directly at the antenna terminal by using a field-effect-transistor (FET) emitter follower or a similar very-high-impedance probe. The voltage at the reference antenna should be measured in a similar manner when each model reading is taken.

The effective height H_e of the antenna is determined by

$$H_e = K \frac{V_{oc} \text{ test model}}{V_{oc} \text{ reference model}} \times \text{scale factor}$$

where *K* is the range-calibration factor, which is determined by relating the open-circuit voltage of a reference monopole to the intensity of the incident field and its electrical height (the electrical height of a short, thin monopole is one-half of its physical height).

The antenna *static capacitance* C_0 is here defined as the apparent capacitance measured at the antenna terminal as the frequency approaches zero. In practice, the measurement can be taken at a frequency not greater than a few percent of the frequency of self-resonance without introducing a significant error. An accurate measurement can be taken by using a Q meter, work coil, and standard variable capacitor. The measurement value is multiplied by the scale factor to derive the full-size value.

The antenna self-resonant frequency is determined by noting the frequency at which the antenna terminal reactance is zero or at which the terminal current and voltage are in phase. The equivalent resonant frequency of the full-size antenna is equal to the model resonant frequency divided by the scale factor.

E- and *H*-Field Distribution

To obtain the *E*- and *H*-field distribution about the antenna, it is necessary to drive the model antenna with sufficient base current to permit measurements to be taken out to a distance well beyond the top-loaded panels.

The *H*-field probe can consist of a well-balanced shielded loop as small as possible ($<5 \times 10^{-4}\,\lambda$ diameter at the measurement frequency). The *H*-field magnitude, normalized to the antenna-base current, and *H*-field orientation with respect to a chosen reference point should be measured along the surface of the ground plane. Current direction is found by rotating the loop to the null position which corresponds to 90° from current maximum.

E-field distribution (normalized to base-current amperes) can be taken with a short *E*-field probe calibrated at the measurement frequency.

Potential-Gradient Evaluation

The scale model can also be used to identify areas of the antenna structure in which the potential gradient is higher than average. This condition may be due to panel-conductor configuration or conductor proximity to grounded structural members. The scale model is raised to a high alternating-current potential and observed or photographed after dark by using an appropriate time exposure. Successive photographs are taken as the voltage is increased in increments to the corona level. The test voltage may be provided by a variable high-voltage transformer or by a high-potential test set. Test voltages up to approximately 35 kV rms may be required.

These tests may reveal "hot spots" that will require additional grading. Scaling the corona on-set voltage to actual operating conditions, however, is not practical.

Electrolytic-Tank Model

Similitude modeling can also be conducted on a scale model immersed in an electrolytic tank. This technique has been found to be most useful in confirming the proper position of insulators in the support guys of insulated towers.

A scaled model (1 : 1000 is a convenient scale factor for large antennas) is erected

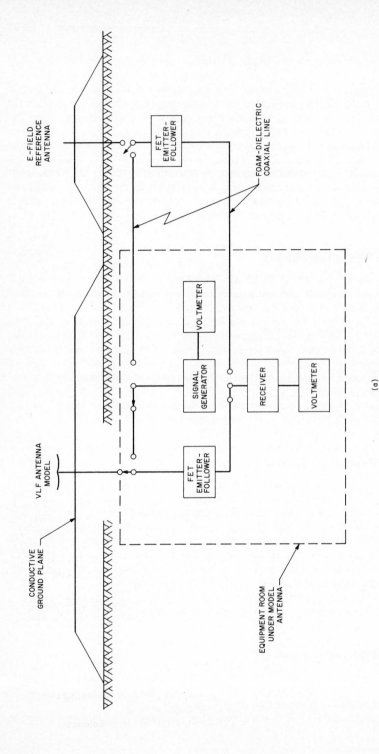

FIG. 24-12 Antenna range for scale-model measurements. (*a*) Sectional schematic. (*b*) Plain view.

(a)

24-28

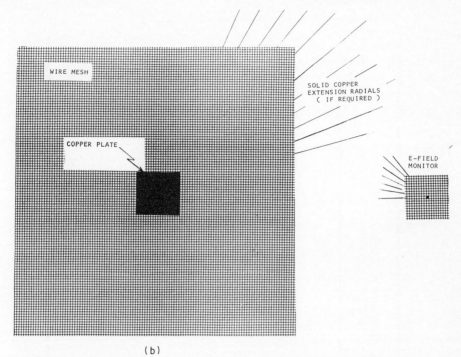

(b)

FIG. 24-12 (*Continued.*)

over a conductive sheet which forms the base for the antenna and guy system. Insulators can be simulated by small glass beads. The antenna is immersed to the bottom of the electrolytic tank and alternating-current-measurement voltages applied between the antenna-base insulator and the ground plane. The measurement voltage must be sufficient to enable accurate measurement. A convenient measurement frequency is 1 kHz. Ordinary tap water is usually a satisfactory electrolyte.

The voltage, relative to ground, on each side of each insulator is measured by using a probe that is insulated except at the tip. The number and location of the guy insulators can be varied as necessary to ensure that all units are working at their rated voltage.

Since the model simulates a static condition, this technique is applicable only to electrically small antennas.

24-9 TYPICAL VLF-LF ANTENNA SYSTEMS

The principal characteristics of a number of current VLF-LF antenna systems are summarized in Table 24-1. The table includes examples of monopoles, top-loaded monopoles, multiple-panel triatics, valley-spans, tridecos, and several hybrid varia-

TABLE 24-1 Characteristics of Typical VLF-LF Antenna Systems

Example	Location	Nominal operating frequency, kHz	Effective height, m	Static capacity, μF	Self-resonant frequency, kHz	Radiated power
Monopole, 800 ft (244 m)	Annapolis, Md.	51	104	0.0031	270.0	20 kW at 51 kHz
Top-loaded monopole, base-insulated tower, 366 m	Trelew, Argentina	10.2–13.6 (Omega)	193	0.0279	53.2	10 kW at 10.2 kHz
Hybrid top-loaded monopole, grounded tower, 427 m	St-Denis, Réunion	10.2–13.6 (Omega)	163	0.0348	27.17	10 kW at 10.2 kHz
Valley span, six 1920-m spans	Haiku, Hawaii	10.2–13.6 (Omega)	169	0.047	39.0	10 kW at 10.2 kHz
Valley span, 10 spans, average length 2200 m	Jim Creek, Wash.	15–27	114	0.078	34.0	15 kW at 24.8 kHz
Array of two top-loaded monopoles, base-insulated, 457 m	Lualualei, Hawaii	15.5–30	192	0.0399	37.5	460 kW at 23.4 kHz
Top-loaded monopole, base-insulated, 366 m	Hawes, Calif.	27–60	228	0.01448	72.5	50 kW at 27 kHz
Top-loaded monopole, base-insulated tower, 213 m	Aberdeen, Md.	179	152	0.00636	153.0	40 kW at 179 kHz
Hybrid trideco-triatic, 266-m base-insulated tower, eleven 183-m grounded towers	Annapolis, Md.	15.5–28	148	0.053	33.5	266 kW at 21.4 kHz
Array of two multiple-panel tridecos	Cutler, Me.	15.5–30	152	0.225	38.2	1 MW at 17.8 kHz
Single-multiple-panel trideco	North West Cape, Australia	15.5–30	192	0.1626	34.19	1 MW at 22.3 kHz

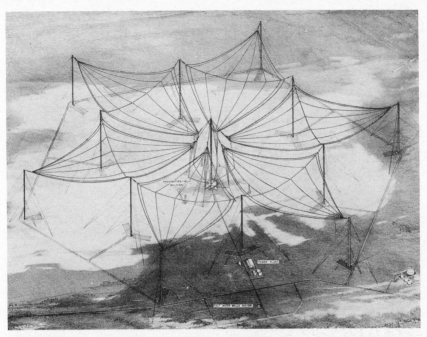

FIG. 24-13   U.S. Navy megawatt VLF antenna located at North West Cape, Australia.

tions. The frequency range represented extends from 10 to 179 kHz; input power, from 100 kW to 2 MW. A megawatt VLF antenna is illustrated in Fig. 24-13.

REFERENCES

1 C. E. Smith and E. M. Johnson, "Performance of Short Antennas," *IRE Proc.,* vol. 35, October 1947, pp. 1026–1038. Discusses characteristics of short vertical radiators. Presents low-frequency measurements of antenna impedance and field strength using an existing broadcast tower with various amounts of top-loading umbrella added.

2 H. A. Wheeler, "Small Antennas," chap. 6 in this book. Defines properties and limitations of small antennas.

3 E. A. Laport, *Radio Antenna Engineering,* McGraw-Hill Book Company, New York, 1952, chap. 1. Explains characteristics of small antennas. Describes low-frequency application of Beverage (wave) and Adcock antennas. Describes scale-model measurement of impedance characteristics. Presents measured impedance characteristics of several representative low-frequency configurations, some obtained from existing antennas and others from scale models.

4 E. C. Jordan, *Electromagnetic Waves and Radiating Systems,* Prentice-Hall, Inc., Englewood Cliffs, N.J., 1950, chap. 14. Discusses characteristics of short antennas.

5 T. E. Devaney, R. F. Hall, and W. E. Gustafson, "Low-Frequency Top-Loaded Antennas," R&D Rep. U.S. Navy Electronics Laboratory, San Diego, Calif., June 1966.

6 H. A. Wheeler, "Fundamental Relations in the Design of a VLF Transmitting Antenna," *IRE Trans. Antennas Propagat.,* vol. AP-6, January 1958, pp. 120–122. Defines concepts

and formulas for the fundamental relations that govern the design of a high-power VLF antenna.

7 S. A. Schelkunoff and H. T. Friis, *Antennas: Theory and Practice,* John Wiley & Sons, Inc., New York, 1952, app. I.

8 A. D. Watt, *VLF Radio Engineering,* Pergamon Press, New York, 1967. Textbook containing a detailed coverage of the fields involved in VLF radio engineering; a compendium of basic antenna, propagation, and system engineering.

9 A. G. Bogle, "Effective Resistance and Inductance of Screened Coils," *J. IEE,* vol. 87, 1940, p. 299.

10 E. D. Sunde, *Earth Conduction Effects in Transmission Systems,* D. Van Nostrand Company, Princeton, N.J., 1949.

11 R. R. Morgan and E. L. Maxwell, "Omega Navigational System Conductivity Map," Rep. 54-F-1, Office of Naval Research, December 1965.

12 R. R. Morgan, "Preparation of a Worldwide VLF Conductivity Map," Mar. 15, 1968, and "Worldwide VLF Effective Conductivity Map," Westinghouse Electric Corp., Environmental Science & Technology Department, Jan. 15, 1968. Gives VLF conductivity values for many major land areas of the world.

13 J. R. Wait and A. M. Conda, "On the Measurement of Ground Conductivity at VLF," *IRE Trans. Antennas Propagat.,* vol. AP-6, no. 3, July 1958.

14 G. V. Keller and F. C. Frischknecht, *Electrical Methods in Geophysical Prospecting,* Pergamon Press, New York, 1965.

BIBLIOGRAPHY

Abbott, F. R.: "Design of Optimum Buried-Conductor RF Ground System," *IRE Proc.,* vol. 40, July 1952, pp. 846–852. Derives a design procedure for a radial-wire ground system to obtain maximum power radiated per unit overall cost.

Alexanderson, E. F. W.: "Trans-Oceanic Radio Communications," *IRE Proc.,* vol. 8, August 1920, pp. 263–286. Explains the theory of multiple tuning and states the results obtained when applied to the New Brunswick, N.J., VLF antenna.

Ashbridge, N., H. Bishop, and B. N. MacLarty: "Droitwich Broadcasting Stations," *J. IEE (London),* vol. 77, October 1935, pp. 447–474. Contains a description in some detail and measured performance characteristics pertaining to a 1935, 150-kW, LF, BBC broadcast antenna of the T type. Explains the procedure used in the design of a broadbanding network and shows the performance of the network.

Bolljahn, J. R., and R. F. Reese: "Electrically Small Antennas and the Low-Frequency Aircraft Antenna Problem," *Trans. IRE Antennas Propagat.,* vol. AP-1, no. 2, October 1953, pp. 46–54. Describes methods of measuring patterns and effective height of small antennas on aircraft by means of models immersed in a uniform field.

Brown, G. H., and R. King: "High Frequency Models in Antenna Investigations," *IRE Proc.,* vol. 22, April 1934, pp. 457–480. Describes and justifies the use of small-scale models to investigate problems of vertical radiators.

Brown, W. W.: "Radio Frequency Tests on Antenna Insulators," *IRE Proc.,* vol. 11, October 1923, pp. 495–522. Discusses design and electrical tests of porcelain insulators for VLF antennas.

――― and J. E. Love: "Design and Efficiencies of Large Air Core Inductances," *IRE Proc.,* vol. 13, December 1925, pp. 755–766. Describes several designs and characteristics of VLF antenna-tuning coils.

Buel, A. W.: "The Development of the Standard Design for Self-Supporting Radio Towers for the United Fruit and Tropical Radio Telegraph Companies," *IRE Proc.,* vol. 12, February 1924, pp. 29–82.

Doherty, W. H.: "Operation of AM Broadcast Transmitter into Sharply Tuned Antenna Systems," *IRE Proc.*, vol. 37, July 1949, pp. 729–734. Shows the impairment of bandwidth of a broadcast transmitter caused by a high-Q antenna.

Feld, Jacob: "Radio Antennas Suspended from 1000 Foot Towers," *J. Franklin Inst.*, vol. 239, May 1945, pp. 363–390. Describes the mechanical design of a flattop antenna on guyed masts.

Grover, F. W.: "Methods, Formulas, and Tables for Calculation of Antenna Capacity," *Nat. Bur. Stds. Sci. Pap.*, no. 568. Discusses and employs Howe's average-potential method. Out of print but available in libraries.

Henney, K. (ed.): *Radio Engineering Handbook*, McGraw-Hill Book Company, New York, 1950. Section entitled "Low-Frequency Transmitting Antennas (below 300 Kc)," by E. A. Laport, pp. 609–623, explains characteristics of small antennas. Describes low-frequency application of Beverage (wave), Adcock, and whip antennas. Describes scale-model measurement of impedance characteristics.

Hobart, T. D.: "Navy VLF Transmitter Will Radiate 1,000 KW," *Electronics*, vol. 25, December 1952, pp. 98–101. Describes U.S. Navy VLF installation of J. R. Redman, "The Giant Station at Jim Creek."

Hollinghurst, F., and H. F. Mann: "Replacement of the Main Aerial System at Rugby Radio Station," *P.O. Elec. Eng. J.*, April 1940, pp. 22–27. Describes changes made to Rugby VLF antenna in 1927 and 1937.

Howe, G. W. O.: "The Capacity of Radio Telegraphic Antennae," *Electrician*, vol. 73, August and September 1914, pp. 829–832, 859–864, 906–909; "The Capacity of Aerials of the Umbrella Type," *Electrician*, vol. 75, September 1915, pp. 870–872; "The Calculation of Capacity of Radio Telegraph Antennae, Including the Effects of Masts and Buildings," *Wireless World*, vol. 4, October and November 1916, pp. 549–556, 633–638; "The Calculation of Aerial Capacitance," *Wireless Eng.*, vol. 20, April 1943, pp. 157–158. Discuss application of Howe's average-potential method for situations stated in the titles.

IRE Standards on Antennas, Modulation Systems, and Transmitters, Definitions of Terms, New York, 1948; *Standards on Antennas, Methods of Testing*, New York, 1948.

Jordan, E. C.: *Electromagnetic Waves and Radiating Systems*, Prentice-Hall, Inc., Englewood Cliffs, N.J., 1950, chap. 14. Discusses characteristics of short antennas.

Knowlton, A. E. (ed.): *Standard Handbook for Electrical Engineers*, 7th ed., McGraw-Hill Book Company, New York, 1941, sec. 13, p. 261. Gives design equation for sleet-melting current required on power transmission lines.

Lindenblad, N., and W. W. Brown: "Main Considerations in Antenna Design," *IRE Proc.*, vol. 14, June 1926, pp. 291–323. Discusses VLF antenna features and design problems and methods.

Miller, H. P., Jr.: "The Insulation of a Guyed Mast," *IRE Proc.*, vol. 15, March 1927, pp. 225–243. Discusses the value of mast and guy insulation in VLF antennas and describes procedures to determine their voltage duties and best placement.

Pender, H., and K. Knox (eds.): *Electrical Engineers' Handbook*, vol. 5: *Electric Communications and Electronics*, 3d ed., John Wiley & Sons, Inc., New York, 1936. Refers to sleet-melting facilities for antennas.

Pierce, J. A., A. A. McKenzie, and R. H. Woodward: *Loran*, MIT Rad Lab. ser., McGraw-Hill Book Company, New York, 1948, chap. 10. Contains a description and performance characteristics of temporary 1290-ft balloon-supported antenna used by the U.S. Air Force at 180 kHz.

Redman, J. R.: "The Giant Station at Jim Creek," *Signal*, January–February 1951, p. 15. Describes U.S. Navy 120-kW VLF installation.

Sanderman, E. K.: *Radio Engineering*, vol. I, John Wiley & Sons, Inc., New York, 1948, chap. 16. Treats antenna-coupling networks in general. Explains the procedure used in the design of the 1935 Droitwich broadbanding network. Describes mechanical design methods for flattops.

Shannon, J. H.: "Sleet Removal from Antennas," *IRE Proc.*, vol. 14, April 1926, pp. 181–195. Describes the sleet-melting and weak-link provisions for the Rocky Point VLF antenna.

Shaughnessy, E. H.: "The Rugby Radio Station of the British Post Office," *J. IEE (London),* vol. 64, June 1926, pp. 683–713. Contains detailed description and measured performance characteristics of a VLF antenna.

Stratton. J. A.: *Electromagnetic Theory,* McGraw-Hill Book Company, New York, 1941, sec. 9.3. Treats electrodynamic similitude and the theory of models.

Sturgis, S. D.: "World's Third Tallest Structure Erected in Greenland," *Civil Eng.,* June 1954, pp. 381–385. Describes structural details and installation procedures for a U.S. Air Force 1205-ft low-frequency vertical radiator.

Weagand, R. A.: "Design of Guy-Supported Towers for Radio Telegraphy," *IRE Proc.,* vol. 3, June 1915, pp. 135–159.

Wells, N.: "Aerial Characteristics," *J. IEE (London),* part III, vol. 89, June 1942, pp. 76–99. Contains a description and performance characteristics of a low-frequency antenna employing a variation of the multiple-tuning principle.

West, W., A. Cook, L. L. Hall, and H. E. Sturgess: "The Radio Transmitting Station at Criggion," *J. IEE (London),* part IIIA, vol. 94, 1947, pp. 269–282. Contains description and measured performance characteristics of a VLF antenna.

Chapter 25

Medium-Frequency Broadcast Antennas

Howard T. Head
John A. Lundin
A. D. Ring & Associates

25-1   INTRODUCTION

General

Medium-frequency broadcast transmitting antennas are generally vertical radiators ranging in height from one-sixth to five-eighths wavelength, depending upon the operating characteristics desired and economic considerations. The physical heights vary from about 150 ft (46 m) to 900 ft (274 m) above ground, making the use of towers as radiators practical. The towers may be guyed or self-supporting; they are usually insulated from ground at the base, although grounded shunt-excited radiators are occasionally employed.

Scope of Design Data

The design formulas and data in this chapter are applicable primarily to broadcast service (535- to 1605-kHz band). However, the basic design principles are valid for transmitting antennas for other services in the medium-frequency band (300 to 3000 kHz).

Characteristics of Radiators

Maximum radiation is produced in the horizontal plane, increasing with radiator height up to a height of about five-eighths wavelength. The radiated field from a single tower is uniform in the horizontal plane, generally decreases with angle above the horizon, and is zero toward the zenith. Radiators taller than one-half wavelength have a minor lobe of radiation at high vertical angles. For radiators with a height in excess of about 0.72 wavelength, the energy in this lobe is maximum, and there is a reduction in horizontal radiation with increasing height. For a height of one wavelength, negligible energy is radiated in the horizontal plane. Radiators taller than five-eighths wavelength may be utilized by sectionalizing the tower approximately each half wavelength (Franklin type) and supplying the current to each section with the same relative phase. Figure 25-1 shows vertical radiation patterns for several commonly employed antenna heights, both for constant power and for constant radiated field in the horizontal plane.

Ground Currents

Current return takes place through the ground plane surrounding the antenna. High earth-current densities are encountered and require metallic ground systems to minimize losses.

Choice of Plane of Polarization

Vertical polarization is almost universally employed because of superior ground-wave and sky-wave propagation characteristics. Ground-wave attenuation is much greater for horizontal than for vertical polarization, and ionospheric propagation of horizon-

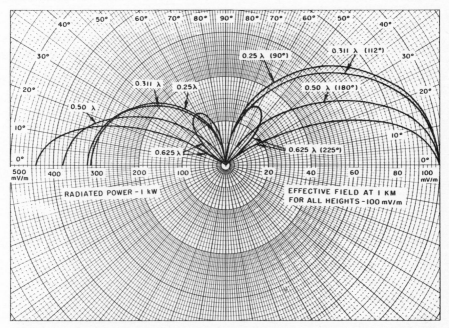

FIG. 25-1 Vertical radiation patterns for different heights of vertical antennas (FCC AM Technical Standards).

tally polarized signals is more seriously influenced by geomagnetic latitude and direction of transmission. Horizontal antennas immediately above ground produce negligible fields in the horizontal plane but radiate relatively large fields at high vertical angles for low antenna heights. This high-angle radiation is desirable only under special circumstances.

Performance Required of Medium-Frequency Broadcast Antennas

The performance required is determined by class of station and channel assignment. In North America three types of broadcast channels are established: clear, regional, and local. Class I stations, operating on clear channels, are assigned to provide ground-wave service during daytime and both ground-wave and sky-wave service at night. All other classes of station are assigned to render ground-wave service only. Class II stations are secondary stations operating on clear channels and employing directional antennas, when necessary, to protect the service areas of Class I stations. Class III stations are assigned to regional channels and employ directional antennas, when required, for mutual protection. Class IV stations are assigned to local channels; both directional and nondirectional antennas are authorized.

The station classes recognized by the Region 2 Hemispherical Agreement (North, Central, and South American countries) are Classes A, B, and C. A Class I station corresponds to Class A, Classes II and III correspond to Class B, and Class IV corresponds to Class C.

Minimum required antenna performance and power limitations for each class of broadcast station in the United States are established by the Federal Communications Commission (FCC) AM Technical Standards (Table 25-1).

Class I stations are assigned to provide both ground-wave and sky-wave service at night. The vertical radiation pattern for night operation of a Class I station should be designed to provide maximum sky-wave signal beyond the ground-wave service area and minimum sky-wave signal within the ground-wave service area (antifading antennas).[1]

TABLE 25-1   Antenna Performance and Power Limitations

Class of station	Required minimum effective field for 1-kW power (unattenuated field), mV/m		Minimum and maximum power, kW
	At 1 km*	At 1 mi	
I	362	225	10 –50
II	282	175	0.25–50
III	282	175	0.5 –5
IV	241	150	0.25– 1.0

*1 mV/m at 1 km = 1 V/m at 1 m.

25-2   CHARACTERISTICS OF VERTICAL RADIATORS

Assumptions Employed in Calculating Radiator Characteristics

The characteristics of tower antennas are ordinarily computed by assuming sinusoidal current distribution in a thin conductor over a perfectly conducting plane earth, with the wavelength along the radiator equal to the wavelength in free space. The effect of the earth plane is represented by an *image* of the antenna as shown in Fig. 25-2. These assumptions provide sufficiently accurate results for most purposes, but in the determination of base operating resistance and reactance the finite cross section of the tower must be taken into account. Also, the finite cross section modifies the vertical radiation patterns slightly. Generally, this effect is of significance only in antifading antennas for clear-channel stations and in certain directional-antenna systems (discussed in Sec. 25-4). Except as noted, all formulas and data in this chapter are based on the assumptions stated above.

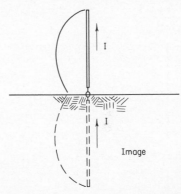

FIG. 25-2 Current distribution and image for a vertical antenna over a perfectly conducting plane.

Field Produced by Vertical Radiator

The field in the horizontal plane is a function of the current flowing and the electrical height. For uniform current in a vertical radiator over a perfectly conducting plane earth, the radiated field is 1.048 V/m (unattenuated field at 1 m) per degree-ampere. For other current distributions, the radiation is proportional to the maximum current and the *form factor K* of the antenna. For sinusoidal current distribution, the radiated field (unattenuated field, volts per meter, at 1 m) is

$$E_r = 60\, I_0\, (1 - \cos G) \qquad\qquad \textbf{(25-1)}$$

where I_0 = loop current, A
$(1 - \cos G)$ = K, form factor for sinusoidal current distribution in a vertical radiator
G = electrical height of radiator above ground

Radiation Resistance

The loop current I_0 is related to the radiated power P_r by the *loop radiation resistance* $R_r = P_r/I_0^2$. Figure 25-3 shows the radiated field for a power of 1 kW and the radiation resistance as a function of antenna height G. The radiation resistance may be calculated from

$$R_r = 15\,[4\cos^2 G\,\text{Cin}\,(2G) - \cos 2G\,\text{Cin}\,(4G)$$

$$- \sin 2G\,[2\,\text{Si}\,(2G) - \text{Si}\,(4G)]] \qquad \Omega \quad \textbf{(25-2)}$$

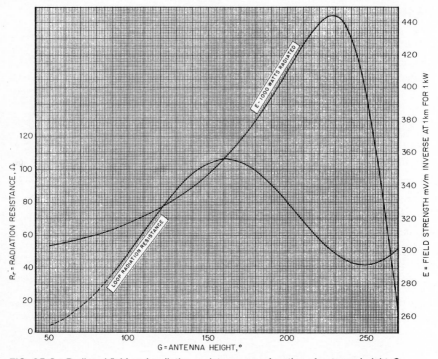

FIG. 25-3 Radiated field and radiation resistance as a function of antenna height *G*.

where Cin $(x) = \int_0^x \frac{1 - \cos x}{x}\, dx$ (cosine integral)

$$= \ln x + C - \text{Ci}\,(x)$$

Ci $(x) = -\int_x^\infty \frac{\cos x}{x}\, dx$ (cosine integral)

$C = 0.5772 \cdots$ (Euler's constant)

Si $(x) = \int_0^x \frac{\sin x}{x}\, dx$ (sine integral)

Operating Base Resistance

For sinusoidal current distribution, the *base radiation resistance* is related to the loop radiation resistance by $R_{r(\text{base})} = [R_{r(\text{loop})}]/\sin^2 G$. However, the actual base resistance of a practical tower radiator may vary widely from this value because of the finite cross section of the tower and other effects.[2,3]

Vertical-Radiation Characteristic

The relative field pattern in a vertical plane through the radiator is known as the *vertical-radiation characteristic*. It is defined as having unit value in the horizontal plane ($\theta = 0°$).* Based on the assumptions stated above, the vertical-radiation characteristic is

$$f(\theta) = \frac{\cos (G \sin \theta) - \cos G}{(1 - \cos G) \cos \theta} \tag{25-3}$$

General Formulas for Calculating Radiating Characteristics

The form factor and vertical-radiation characteristic establish the radiated field in the horizontal and vertical planes. For any current distribution, they may be determined from

$$Kf(\theta) = 1.048 \int_0^G I(z) \cos \theta \cos (z \sin \theta)\, dz \quad \text{V/m at 1 m} \tag{25-4}$$

where z = height of current element dz, electrical degrees
$I(z)$ = current at z, A
θ = elevation angle
G = antenna height
The radiation resistance is†

$$R_r = 60 \int_0^{\pi/2} [Kf(\theta)]^2 \cos \theta\, d\theta \quad \Omega‡ \tag{25-5}$$

Effects of Finite Cross Section

Current Distribution   The effect of finite tower cross section on the current distribution and vertical-radiation characteristic may be taken into account by assuming a current distribution of the following form suggested by Schelkunoff:[2]

$$I(z) = I_0 \sin [\gamma(G - z)] + jkI_0 (\cos \gamma z - \cos \gamma G) \tag{25-6}$$

where I_0 = maximum in-phase current, A
z = height of current element
$\gamma = \lambda_0/\lambda$, ratio of wavelength in free space to wavelength along tower
G = height of tower, electrical degrees
$k \approx 50/(Z_0 - 45)$, for antenna heights near $\lambda/2$
$Z_0 = 60 [\ln (2G/a_{eq}) - 1]$, characteristic impedance of tower   **(25-7)**
$a_{eq} = \sqrt{naa_0}$, equivalent tower radius
n = number of tower legs
a = cage radius (distance from tower center to tower leg)
a_0 = tower-leg radius

---

*The symbol θ is used in this chapter to denote angle above the horizontal plane.

†An alternative method of calculation is given by Carter.[4]

‡Note that the radiated power is equal to I^2R_r. For $E = Kf(\theta)$, mV/m at 1 km, P (watts) = $1/60 \int_0^{\pi/2} E^2 \cos \theta\, d\theta$.

When the cross section is vanishingly small, Z_0 is very large, $k \rightarrow 0$, and the current distribution approaches the simple sinusoidal.

For short towers, this current distribution is very nearly the simple sinusoidal; for a quarter-wave tower, the two are identical except for the reduced wavelength. For towers higher than a quarter wavelength, the departure becomes significant.

Vertical-Radiation Characteristic and Form Factor   Applying Eq. (25-4) to the Schelkunoff current distribution, the form factor K_2 and the vertical-radiation characteristic $f_2(\theta)$ become

$$K_2 f_2(\theta) = \frac{\gamma \cos \theta}{\gamma^2 - \sin^2 \theta} \left[\cos \left(G \sin \theta \right) - \cos \gamma G \right]$$

$$+ \, jk \cos \theta \left[\frac{\gamma \sin \gamma G \cos \left(G \sin \theta \right)}{\gamma^2 - \sin^2 \theta} \right.$$

$$\left. - \frac{\sin \theta \cos \gamma G \sin \left(G \sin \theta \right)}{\gamma^2 - \sin^2 \theta} \right.$$

$$\left. - \frac{\cos \gamma G \sin \left(G \sin \theta \right)}{\sin \theta} \right] \qquad \textbf{(25-8)}$$

The variations of the vertical-radiation characteristic, based on the sinusoidal and Schelkunoff assumptions, are relatively small and would have only minor practical effects in nondirectional radiators or in directional-antenna systems employing identical towers. However, for directional-antenna systems employing tall towers of different heights, the variations in amplitude and phase of the vertical-radiation characteristics must be taken into account [see Sec. 25-4, Eq. (25-45)] .

Shunt-Fed Radiators

Energy may be supplied to a grounded tower by shunt excitation, using a slant-wire feed as shown in Fig. 25-4. The dimensions h/λ and d/λ shown in the figure determine the impedance at the end of the slant wire.[5,6]

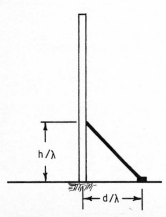

The current in the slant wires and between the tap point and ground modify the vertical radiation pattern of the shunt-fed tower and result in a slight nonuniformity in radiation in the horizontal plane. Notwithstanding this, the radiation field from a shunt-fed tower is essentially the same as that from a series-fed tower of the same height.

Top-Loaded Radiators

The radiating characteristics may be modified by altering the current distribution. Top loading is generally accomplished on guyed towers by connecting a portion of the top guy wires directly to the tower instead of having it connected through an insulator. Another method of top loading that is

FIG. 25-4   Shunt-fed element.

sometimes employed is a capacity "hat" mounted on the tower top. These methods of top loading have essentially the same effect (within limits) as increasing the tower height (Fig. 25-5). For guy-wire top loading, the length of the top portion of the guy wire connected directly to the tower can be used as a rough approximation for B.

A capacity disk of radius r at the tower top is equivalent (within practical limits) to an electrical length of

$$B = \frac{rZ_0}{15\pi} \qquad \textbf{(25-9)}$$

Z_0 is defined in Eq. (25-7).

The vertical-radiation characteristic for the top-loaded antenna is given by[7,8,9]

$$f(\theta) = \frac{\cos B \cos (A \sin \theta) - \sin \theta \sin B \sin (A \sin \theta) - \cos G}{\cos \theta (\cos B - \cos G)} \qquad \textbf{(25-10)}$$

where A = electrical height of vertical portion of antenna
B = equivalent electrical height of top loading
$G = A + B$

Sectionalized Radiators

Effects of Sectionalization   Antenna heights above five-eighths wavelength may be employed to obtain increased effective field by dividing the tower with insulators into sections of approximately one-half wavelength. The currents in each section are maintained in phase.

For heights less than one wavelength, the tower may be sectionalized in either of the arrangements shown in Fig. 25-6. The vertical-radiation characteristic for the two

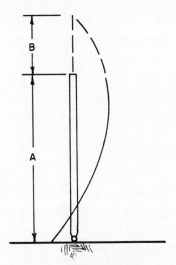

FIG. 25-5   Top-loaded element.

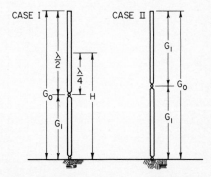

FIG. 25-6   Two methods of sectionalizing vertical radiators.

arrangements, based on simple sinusoidal current distribution, is

Case I: $f(\theta) = \dfrac{2 \cos (90 \sin \theta) \cos (H \sin \theta) + \cos (G_1 \sin \theta) - \cos G_1}{\cos \theta \, (3 - \cos G_1)}$ **(25-11)**

Case II: $f(\theta) = \dfrac{\cos (G_1 \sin \theta) \, [\cos (G_1 \sin \theta) - \cos G_1]}{\cos \theta \, (1 - \cos G_1)}$ **(25-12)**

If top loading is employed on a sectionalized antenna (Fig. 25-7), the vertical-radiation characteristic is expressed as[8,9]

$$f(\theta) = \frac{\sin \Delta \, [\cos B \cos (A \sin \theta) - \cos G] + \sin B \, [\cos D \cos (C \sin \theta) - \sin \theta \sin D \sin (C \sin \theta) - \cos \Delta \cos (A \sin \theta)]}{\cos \theta \, [\sin \Delta \, (\cos B - \cos C) + \sin B \, (\cos D - \cos \Delta)]}$$ **(25-13)**

where A = height of lower section
$\quad\quad B$ = equivalent top loading of lower section
$\quad\quad C$ = height of entire antenna
$\quad\quad D$ = equivalent top loading of top section
$\quad\quad G = A + B$
$\quad\quad H = C + D$
$\quad\quad \Delta = H - A$

Effects of Finite Cross Section of Sectionalized Radiators   The actual current distribution on sectionalized towers varies from simple sinusoidal in the manner discussed above. In this case, the upper section of the tower is a half-wavelength (or very nearly so), and the base termination is adjusted to provide the same current dis-

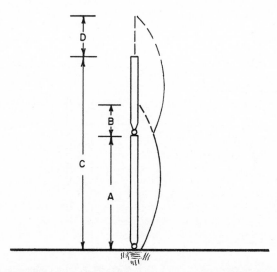

FIG. 25-7 Top-loaded sectionalized vertical radiator.

tribution on the lower section as on the upper section. For these conditions, the currents are (see Fig. 25-6, Case I)

$$I_1(z) \text{ (top section)} = I_0 \sin [\gamma(G_0 - z)]$$

$$+ jkI_0 \{1 + \cos [\gamma(G_1 - z)]\} \quad \textbf{(25-14a)}$$

$$I_2(z) \text{ (bottom section)} = I_0 \sin [\gamma(G_1 - z)]$$

$$+ jkI_0 \{1 + \cos [\gamma(G_1 - z)]\} \quad \textbf{(25-14b)}$$

where $\gamma = \lambda_0/\lambda$, ratio of wavelength in free space to wavelength along tower
   I_0 = maximum in-phase current
   G_0 = overall height of tower (both sections)
   z = height of current element
   $k \approx 50/(Z_0 - 45)$ for antenna heights near $\lambda/2$
See Eq. (25-7) for Z_0.

The form factor and the vertical-radiation characteristic are given by

$$
\begin{aligned}
K_0 f_0(\theta) = {} & \frac{\gamma \cos \theta}{\gamma^2 - \sin^2 \theta} \left[\cos (G_0 \sin \theta) - \cos \gamma G_1 \right. \\
& \left. - 2 \cos \gamma\pi \cos (G_1 \sin \theta) \right] \\
& + jk \cos \theta \left\{ \frac{\sin (G_0 \sin \theta)}{\sin \theta} \right. \\
& + \frac{\gamma}{\gamma^2 - \sin^2} \left[\sin \gamma G_1 - 2 \sin \gamma\pi \cos (G_1 \sin \theta) \right. \\
& + \left. \left. \frac{\sin \theta \sin (G_0 \sin \theta)}{\gamma} \right] \right\}
\end{aligned}
\qquad \textbf{(25-15)}
$$

25-3 GROUND SYSTEMS

General Requirements

The ground system for a medium-frequency antenna usually consists of 120 buried copper wires, equally spaced, extending radially outward from the tower base to a minimum distance of one-quarter wavelength. In addition, an exposed copper-mesh ground screen may be used around the base of the tower when high base voltages are encountered.

Wire size has a negligible effect on the effectiveness of the ground system and is chosen for mechanical strength; AWG No. 10 or larger is adequate. A depth of 4 to 6 in (102 to 152 mm) is generally adequate, although the wires may be buried to a depth of several feet if desired in order to permit cultivation of the soil. When such deep burial is required, the wires should descend to the required depth on a smooth, gentle incline from the tower base, reaching the ultimate depth some distance from the tower. If this precaution is observed, the deep burial will have relatively little effect on the effectiveness of the ground system for typical soil conditions.

Nature of Ground Currents

Ground currents are conduction currents returning directly to the base of the antenna. The total earth current flowing through a cylinder of radius x concentric with the antenna is known as the *zone current*. It is a function of tower height and is given by

$$I_{zone} = I_0 [\sin r_2 - \cos G \sin x + j (\cos r_2 - \cos G \cos x)] \quad \textbf{(25-16)}$$

where I_0 = loop antenna current
G = electrical height of antenna
$r_2 = \sqrt{x^2 + G^2}$

Effect of Ground-System Losses on Antenna Performance

Ground-system losses dissipate a portion of the input power and reduce the field radiated from the antenna. These losses are equivalent to the power dissipated in a resistor in series with the antenna impedance.[10] Computed values of radiated field based on an assumed series loss resistance of 2 Ω give results for typical installations in good agreement with actual measured effective field intensities. FCC rules require that the assumed series loss resistance for each element of a directional-antenna system shall be no less than 1 Ω. For directional-antenna systems, a series loss of 1 Ω is assumed for each radiator.

Effect of Local Soil Conductivity on Ground-System Requirements

A less elaborate ground system may be effective in soil of high local conductivity,[11,12] although adequate local conductivity data are rarely available.

For a seawater site (conductivity, approximately 4.6 S/m*), the salt water provides a adequate ground system; a submerged copper ground screen is employed to make contact with the salt water.

Ground Systems for Multielement Arrays

Individual ground systems are required for each tower of a multielement array. If the individual systems would overlap, the adjoining systems are usually terminated in a common bus. Complete systems may be installed around each tower when high ground currents are expected. Figure 25-8 shows a typical multielement ground system.

Recent theoretical studies have shown that the ground currents of directional antennas vary in both magnitude and direction from those which would be associated with single radiators considered individually. The magnitudes of the ground currents depend upon the directional-antenna parameters, and the directions of the ground currents vary over the individual radio-frequency cycle. This leads to the conclusion that the ground system for a directional antenna should in general be modified from the conventional nondirectional ground system in two ways: (1) the ground wires should be concentrated in the regions of higher current density, and (2) the paths of the

---

*Siemens/meter or mho/meter.

120 RADIALS SPACED EVENLY
AROUND EACH TOWER

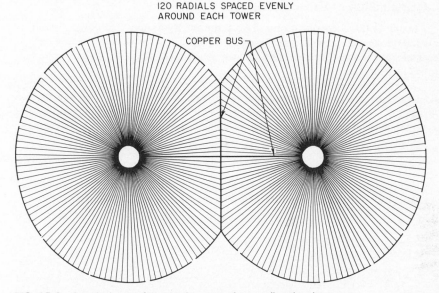

FIG. 25-8   A typical ground system for a two-element directional antenna.

ground wires should be chosen to favor the direction of maximum current flow at the
peak of the radio-frequency cycle.[13]

25-4   DIRECTIONAL ANTENNAS

Purpose of Directional-Antenna Systems

A directional antenna employs two or more radiators to produce radiation patterns in
the horizontal and vertical planes different from those produced by a single radiator.
Directional antennas are used principally to reduce the radiated signal toward other
stations on the same or adjacent channels in order to avoid interference, although the
resulting concentration of radiated signal in other directions may be utilized to
improve service to specific areas.

Permissible Values of Radiation

Methods of computing interference and establishing maximum permissible values of
radiation to avoid interference are described in detail in the FCC *Rules and Regulations,* Volume III, Part 73.[8] For interference from ground-wave signals, radiation in
the horizontal plane only is considered. For interference from sky-wave signals, radiation throughout a specified range of vertical angles must be considered. Suppression
of radiation is required over sufficient arc to subtend the protected station's service
area.

Computation of Pattern Shape

Fundamental Considerations: Two-Element Systems The principles underlying the computation of shape of radiation patterns from two-element arrays are fundamental and may be extended to the computations of shape of patterns from multielement arrays.

The field strength at any point from two radiators receiving radio-frequency
energy from a common source is the vector sum of the fields from each of the two
radiators. At large distances, the antenna system may be considered to be a point
source of radiation. Referring to Fig. 25-9 and considering tower 1 to be reference or
zero phase, the theoretical radiated field from the array at the angle ϕ, θ is

$$E_{th}(\theta) = E_1 f_1(\theta) \, \underline{/0} + E_2 f_2(\theta) \underline{/\alpha_2} \qquad (25\text{-}17)$$

where α_2, the difference in phase angle between the two fields, is the sum of the *time-
phase-angle* difference A_2 and the apparent *space-phase-angle* difference:

$$\alpha_2 = s \cos \phi \cos \theta + A_2 \qquad (25\text{-}18)$$

where E_1 = field radiated by element 1
E_2 = field radiated by element 2
$f_1(\theta)$ = vertical-radiation characteristic of element 1
$f_2(\theta)$ = vertical-radiation characteristic of element 2
s = spacing of element 2 from element 1
ϕ = angle between element line and azimuth of calculation
For towers of equal height with $F_2 = E_2/E_1$, Eq. (25-17) simplifies to

$$E_{th}(\theta) = E_1 f(\theta) \sqrt{2F_2} \left[\frac{1 + F_2^2}{2F_2} + \cos \left(s \cos \phi \cos \theta + A_2 \right) \right]^{1/2} \qquad (25\text{-}19)$$

The magnitude of E_{th} is a minimum when

$$\cos \left(s \cos \phi \cos \theta + A_2 \right) = -1 \qquad (25\text{-}20)$$

This occurs when $(s \cos \phi \cos \theta + A_2) = 180° \pm 360°$; if $F_2 = 1$, E_{th} will be zero.
For $F_2 = 1$, Eq. (25-19) simplifies to

$$E_{th}(\theta) = 2E_1 \cos \left(\frac{s}{2} \cos \phi \cos \theta + \frac{A_2}{2} \right) \qquad (25\text{-}21)$$

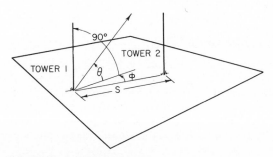

FIG. 25-9 Two-element array.

SPACING

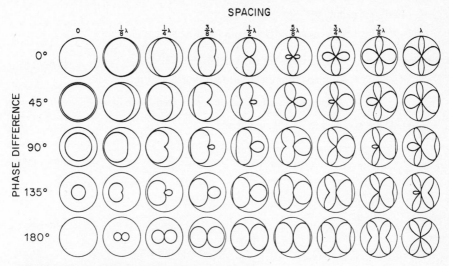

FIG. 25-10   Two-element horizontal plane patterns.

Combinations of equations of the two-element form are widely used in establishing and computing patterns for multielement arrays. Figure 25-10 shows horizontal plane patterns for two-element arrays for a variety of spacings and phasings.[14] The patterns shown are for equal fields in the two radiators ($F_2 = 1$); the effect of unequal fields is to "fill" the nulls of the pattern. For a given spacing, the angular position of the minimum is changed only by changing the phase angle. The value of $E_{th}(\theta)$ is unchanged by substituting the inverse for F_2, $F_2 = 1/F_2$.

Two or More Elements: General Equations   The pattern shape of directional arrays of two or more elements may be computed by an extension of Eq. (25-17):

$$E_{th}(\theta) = E_1 \sum_{i=1}^{n} F_i f_i(\theta) \underline{/\alpha_i} \qquad (25\text{-}22)$$

where E_1 = field radiated by element 1

F_i = field ratio of ith element, $F_i = E_i/E_1$

$f_i(\theta)$ = vertical-radiation characteristic of ith element

$\alpha_i = s_i \cos \phi_i \cos \theta + A_i$

s_i = spacing of ith element relative to a reference point (usually element 1)

ϕ_i = angle between orientation of ith element and azimuth of calculation

θ = vertical angle

A_i = phase angle of the ith element

Three or More Elements: Simplified Formulas for Special Cases   The following are formulas for many commonly used array configurations which, through the combination of element pairs, provide a convenient means for establishing the angular position and depth of the minima.

For these examples:

$$r_i = \text{assumed field ratio for two-element pair}$$

$$a_i = \text{phase angle for two-element pair}$$

$$F_i = \text{resulting field ratio for } i\text{th element in array}$$

$$A_i = \text{resulting phase angle for } i\text{th element in array}$$

Elements are assumed to be of equal height.

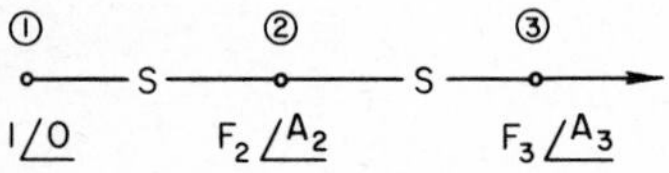

FIG. 25-11   Linear array of three elements.

$$E_{\text{th}}(\theta) = E_1 f(\theta) \sqrt{4r_2 r_3} \left\{ \left[\frac{1 + r_2^2}{2r_2} + \cos\left(s \cos \phi \cos \theta + a_2\right) \right] \right.$$

$$\left. \left[\frac{1 + r_3^2}{2r_3} + \cos\left(s \cos \phi \cos \theta + a_3\right) \right] \right\}^{1/2} \qquad \textbf{(25-23)}$$

$$F_2 \underline{/A_2} = r_2 \underline{/a_2} + r_3 \underline{/a_3}$$

$$F_3 \underline{/A_3} = r_2 r_3 \underline{/a_2 + a_3}$$

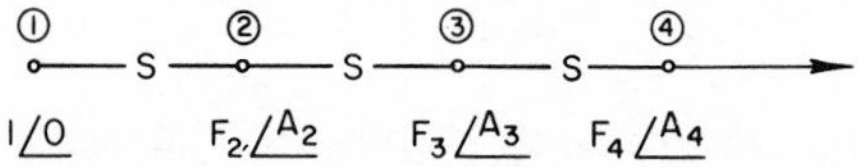

FIG. 25-12   Linear array of four elements.

$$E_{\text{th}}(\theta) = E_1 f(\theta) \sqrt{8r_2 r_3 r_4} \left\{ \left[\frac{1 + r_2^2}{2r_2} + \cos\left(s \cos \phi \cos \theta + a_2\right) \right] \right.$$

$$\left[\frac{1 + r_3^2}{2r_3} + \cos\left(s \cos \phi \cos \theta + a_3\right) \right]$$

$$\left. \left[\frac{1 + r_4^2}{2r_4} + \cos\left(s \cos \phi \cos \theta + a_4\right) \right] \right\}^{1/2} \qquad \textbf{(25-24)}$$

$$F_2 \underline{/A_2} = r_2 \underline{/a_2} + r_3 \underline{/a_3} + r_4 \underline{/a_4}$$

$$F_3 \underline{/A_3} = r_2 r_3 \underline{/a_2 + a_3} + r_2 r_4 \underline{/a_2 + a_4} + r_3 r_4 \underline{/a_3 + a_4}$$

$$F_4 \underline{/A_4} = r_2 r_3 r_4 \underline{/a_2 + a_3 + a_4}$$

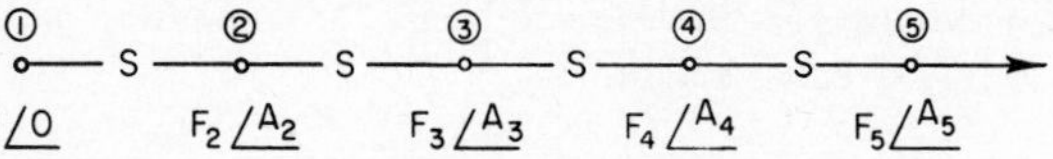

① —— S —— ② —— S —— ③ —— S —— ④ —— S —— ⑤
$1\underline{/0}$ $F_2\underline{/A_2}$ $F_3\underline{/A_3}$ $F_4\underline{/A_4}$ $F_5\underline{/A_5}$

FIG. 25-13 Linear array of five elements.

$$E_{th}(\theta) = E_1 f(\theta) \sqrt{16 r_2 r_3 r_4 r_5} \left\{ \left[\frac{1 + r_2^2}{2r_2} + \cos\left(s \cos\phi \cos\theta + a_2\right) \right] \right.$$

$$\left[\frac{1 + r_3^2}{2r_3} + \cos\left(s \cos\phi \cos\theta + a_3\right) \right]$$

$$\left[\frac{1 + r_4^2}{2r_4} + \cos\left(s \cos\phi \cos\theta + a_4\right) \right]$$

$$\left. \left[\frac{1 + r_5^2}{2r_5} + \cos\left(s \cos\phi \cos\theta + a_5\right) \right] \right\}^{1/2}$$

$$F_2\underline{/A_2} = r_2\underline{/a_2} + r_3\underline{/a_3} + r_4\underline{/a_4} + r_5\underline{/a_5} \qquad \textbf{(25-25)}$$

$$F_3\underline{A_3} = r_2 r_3\underline{/a_2 + a_3} + r_2 r_4\underline{/a_2 + a_4} + r_2 r_5\underline{/a_2 + a_5}$$
$$+ r_3 r_4\underline{/a_3 + a_4} + r_3 r_5\underline{/a_3 + a_5} + r_4 r_5\underline{/a_4 + a_5}$$

$$F_4\underline{/A_4} = r_2 r_3 r_4\underline{a_2 + a_3 + a_4} + r_2 r_3 r_5\underline{/a_2 + a_3 + a_5}$$
$$+ r_2 r_4 r_5\underline{/a_2 + a_4 + a_5} + r_3 r_4 r_5\underline{/a_3 + a_4 + a_5}$$

$$F_5\underline{/A_5} = r_2 r_3 r_4 r_5\underline{/a_2 + a_3 + a_4 + a_5}$$

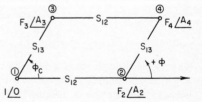

FIG. 25-14 Parallelogram array of four elements.

$$E_{th}(\theta) = E_1 f(\theta) \sqrt{4 F_2 F_3} \left\{ \left[\frac{1 + F_2^2}{2F_2} + \cos\left(s_{12} \cos\phi \cos\theta \right. \right.\right.$$

$$\left.\left. + A_2\right) \right] \left[\frac{1 + F_3^2}{2F_3} + \cos\left(s_{13} \cos\left[\phi - \phi_c\right] \cos\theta + A_3\right) \right] \right\}^{1/2} \qquad \textbf{(25-26)}$$

provided $F_4\underline{/A_4} = F_2 F_3\underline{/A_2 + A_3}$.

In these formulas, the field is the product of several two-element-array expressions and will be zero whenever any of the component expressions is zero. For $r_n \neq 1$, the position of the minima may be shifted slightly; however, this shift is usually small.

Formulas (25-23) through (25-26) are valid only for computing patterns in the vertical planes for towers of equal height.

Computation of Pattern Size

General   Formulas (25-17) through (25-26) establish the shape of radiation patterns. To determine the magnitude of the radiation, the reference field E_1 must be evaluated.

Total Resistance   The current I_1 in the reference element is related to the total radiated power P_r by the *total resistance* R_t of the array:

$$R_{t1} = \frac{P_r}{I_1^2} \quad \Omega \tag{25-27}$$

where the subscript 1 indicates that the total resistance is referred to the current in element 1. The radiated field is directly proportional to the current, and

$$E_1 = E_0 \left(\frac{R_{r1}}{R_{t1}} \right)^{1/2} \quad \text{mV/m at 1 km} \tag{25-28}$$

where E_0 is the effective field of the reference element, operating independently and without loss, and R_{r1} is the radiation resistance. The total resistance is calculated from the operating resistances of each element of the array, as discussed in the following subsection.

Mutual Impedance: Definition for Antenna Case   The mutual impedance between two antennas is defined in the usual manner:

$$Z_{12} = Z_{21} = V_1/I_2 = V_2/I_1$$

The mutual impedance is a complex quantity:

$$Z_{12} \underline{/\sigma_{12}} = R_{12} + jX_{12} = \sqrt{R_{12}^2 + X_{12}^2} \; \underline{/\tan^{-1} X_{12}/R_{12}}$$

The mutual impedance is a function of the spacing between the antennas and the height of the elements. The general equations for mutual resistance and reactance between two antennas of unequal height have been derived by Cox.[15]

Operating Resistance of Radiators in a Directional-Antenna System   The *operating resistance* of a radiator in a directional array is the sum of the radiation resistance R_r, an assumed loss resistance R_a (see Sec. 25-3), and the *coupled resistance* R_c.

$$R_0 = R_r + R_a + R_c \tag{25-29}$$

The coupled resistance of element 1 is

$$R_{c1} = M_{12}Z_{12} \cos (\mu_{12} + \sigma_{12}) + M_{13}Z_{13} \cos (\mu_{13}$$
$$+ \sigma_{13}) + \cdots M_{1n}Z_{1n} \cos (\mu_{1n} + \sigma_{1n}) \tag{25-30}$$

The coupled reactance is

$$X_{c1} = M_{12}Z_{12} \sin (\mu_{12} + \sigma_{12}) + M_{13}Z_{13} \sin (\mu_{13}$$
$$+ \sigma_{13}) + \cdots M_{1n}Z_{1n} \sin (\mu_{1n} + \sigma_{1n}) \quad \textbf{(25-31)}$$

where M_{1n} = current ratio between tower n and tower 1
   Z_{1n} = mutual impedance between tower n and tower 1
   μ_{1n} = phase angle of current between tower n and tower 1
   σ_{1n} = phase angle of Z_{1n}

Note that M_{12} is the *current* ratio of tower 2 referred to tower 1. The loop-current ratio and the *field* ratio are related by the ratio of the form factors. For loop-current reference,

$$\frac{M_{12}}{F_{12}} = (1 - \cos G_1)/(1 - \cos G_2) \quad \textbf{(25-32)}$$

The base resistance of a radiator in a directional array will usually be different from the operating resistance computed from Eq. (25-29). The determination of base operating resistance is discussed below.

Computation of Total Resistance: Formulas   The total resistance R_{t1} is

$$R_{t1} = R_{01} + M_{12}^2 R_{02} + M_{13}^2 R_{03} + \cdots M_{1n}^2 R_{0n} \quad \textbf{(25-33)}$$

Determination of Reference Field by Mechanical Integration (Hemispherical Integration)[*]   The value of E_1 may be established by integration of the power flow from the antenna system over the hemisphere. The resulting hemispherical root-mean-square (rms) value is then compared with an isotropic reference, in which the power flow is uniform in all directions over the hemisphere:

$$E_1 = 245 \sqrt{P} \qquad \text{mV/m at 1 km} \quad \textbf{(25-34)}$$

where P = power, kW. The hemispherical rms value of relative field strength e_h is determined by integration over the hemisphere. The integration may be approximated by the trapezoid method:[†]

$$e_h = \sqrt{\frac{\pi\Delta}{180} \frac{[e_{rms}(\theta = 0)]^2}{2} + \sum_{n=1}^{N} \{[e_{rms}(\theta)]^2 \cos \theta\}} \quad \textbf{(25-35)}$$

where Δ = interval in degrees of vertical angle θ for equally spaced calculations
   θ = vertical angle (0 = horizontal)
   $N = 90/\Delta - 1$

$$e_{rms}(\theta) = \sqrt{\sum_{i=1}^{k} \sum_{j=1}^{k} F_i f_i(\theta F_j f_j(\theta) \cos \mu_{ij} J_0(S_{ij} \cos \theta)} \quad \textbf{(25-36)}$$

---

[*]In the discussion which follows, lowercase e is used to denote relative fields. Actual electric field strengths are indicated by capital E.

[†]Other numerical approximations, such as Simpson's or Gauss's, may provide some improvement in accuracy.

where k = number of elements
$\quad\quad F$ = field ratio
$\quad f(\theta)$ = vertical-radiation characteristic
$\quad\quad \mu_{ij}$ = phase-angle difference between the ith and jth elements
$\quad\quad S_{ij}$ = spacing between ith and jth elements, rad
$\quad J_0(s)$ = zeroth-order Bessel function
The no-loss value of E_1 is then calculated from

$$E_{01} = \frac{E_i \sqrt{P}}{e_h} \quad\quad \text{mV/m at 1 km} \quad\quad\quad (25\text{-}37)$$

Power Losses   If the individual currents are known (i.e., the operating resistances are known), the total power lost in an array by dissipation in the antenna system (ground system, element-coupling system, etc.) can be calculated from

$$P_\ell = \frac{\displaystyle\sum_{i=1}^{k} R_{ai} I_{oi}^2}{1000} \quad\quad \text{kW} \quad\quad\quad (25\text{-}38)$$

where k = number of elements
$\quad\quad R_{ai}$ = assumed loss resistance of ith element
$\quad\quad I_{oi}$ = loop current of ith element (base current if ith-element height is less than $\lambda/4$)
The reference field E_1 is directly proportional to the current in the reference tower.

If, on the other hand, the individual operating resistances are not known, as would be the case when e_h is determined by hemispherical integration, the losses may be calculated and the reference field E_1 with loss determined by reference to the root-sum-square (rss) currents of the elements in the array. The rss currents are related to the rss fields by the form factor K [see Eq. (25-1)] of the individual elements. If all the elements in the array are of the same height, the two are directly proportional. To determine the losses for this condition, first determine the horizontal-plane no-loss rss of the fields from the elements of the array:

$$E_{\text{rss}} \text{ (no loss)} = \frac{245 \sqrt{P}}{e_h} \sqrt{\sum_{i=1}^{k} F_i^2} \quad\quad\quad (25\text{-}39)$$

Next, determine the no-loss *current* rss of the array by using the following:

$$I_{\text{rss}} \text{ (no loss)} = \frac{4.08 \sqrt{P}}{e_h} \sqrt{\sum_{i=1}^{k} \left(\frac{F_i}{K_i}\right)^2} \quad\quad\quad (25\text{-}40)$$

where K_i = form factor of element i.
For towers of different heights, the form factor K_i must be applied for each height of tower in determining I_{rss}. The reference field E_l with loss may be determined from

$$E_l \text{ (loss)} = E_l \text{ (no loss)} \left(\frac{P}{P + I_{\text{rss}}^2 \text{ (no loss)} \dfrac{R_a}{1000}} \right) \quad\quad\quad (25\text{-}41)$$

Base Operating Impedance of Radiators

General   The base operating resistance of a tower in a directional-antenna array is usually different from the computed operating resistance. The base operating resistance and reactance may be estimated as follows:

Estimate the base resistance (R_base) and reactance (X_base) of the tower. Then assume the base mutual impedance to be $Z_{m(\text{base})} = Z_m(R_\text{base}/R_r)$,[16] and substitute these values in Eqs. (25-30) and (25-31).

Parasitic (Zero-Resistance) Elements   A radiating element operating at zero resistance and not supplied with power by connections from the distribution circuits is said to be parasitic. This condition obtains when the negative coupled resistance is equal to the sum of the radiation resistance and loss resistance. A parasitic element properly tuned will operate in phase-and-field-ratio relationships approximating those computed. In critical multielement arrays, independent control of phase and amplitude is required and parasitic radiators should be avoided. However, they may be employed in antennas designed primarily for power gain.

Negative-Resistance Elements   A negative-resistance element receives more power by coupling to the other elements than is required to obtain the desired field from the element. The excess power is sometimes dissipated in a resistor but is usually returned to the positive-resistance elements through the power-distributing circuits.

Effective Field

Definition   The rms value of the radiation pattern in the horizontal plane ($\theta = 0$, in unattenuated field at 1 km) is referred to as the *effective field*. The effective field of a directional antenna is modified from that for a single radiator by directivity at vertical angles and higher ground and circuit losses.

Calculation of Directional-Antenna Effective Field   The rms field of a directional-antenna pattern at any vertical angle may be calculated from

$$E_\text{eff} = E_1 e_\text{rms}(\theta) \qquad \text{mV/m at 1 km} \qquad \textbf{(25-42)}$$

where E_1 is the reference field and $e_\text{rms}(\theta)$ is as defined in Eq. (25-36).

FCC Standard Radiation   In the United States the FCC *Rules and Regulations* require that the radiation from a medium-wave directional-antenna system be depicted by a standard pattern. The standard radiation pattern is an envelope around the theoretical radiation pattern. It is intended to provide a tolerance within which the actual operating pattern can be maintained. The standard radiation values are calculated from

$$E_\text{std}(\theta) = 1.05\sqrt{[E_\text{th}(\theta)]^2 + [Qg(\theta)]^2} \qquad \text{mV/m at 1 km} \qquad \textbf{(25-43)}$$

where $E_\text{th}(\theta)$ = theoretical radiation [Eqs. (25-17) through (25-26)]
$\quad Q$ = (10.0) or (10$\sqrt{P}$) or (0.025E_rss), whichever is largest
$\quad P$ = power, kW
$\quad f_s(\theta)$ = vertical-radiation characteristic of shortest element in array

$g(\theta) = f_s(\theta)$ if shortest element is shorter than $\lambda/2$

$g(\theta) = \dfrac{\sqrt{[f_s(\theta)]^2 + 0.0625}}{1.030776}$ if shortest element is taller than $\lambda/2$

$E_{rss} = \left| E_1 \right. \sqrt{\displaystyle\sum_{i=1}^{n} F_i^2}$, horizontal plane pattern root sum square

n = total number of elements
F_i = field ratio of ith element
E_l = reference field

FCC rules include a provision to modify or augment the standard pattern to accommodate actual operating patterns when radiation is in excess of the standard radiation pattern. Radiation is augmented over a specified azimuthal span and is calculated from

$$E_{aug}(\theta) = \sqrt{[E_{std}(\theta)]^2 + A\left[g(\theta)\cos\left(\frac{180 D_A}{S}\right)\right]^2} \qquad (25\text{-}44)$$

where $E_{std}(\theta)$ = standard radiation [See Eq. (25-43).]
$A = \{[E_{aug}(\theta)]^2 - [E_{std}(\theta)]^2\}$ at central azimuth of augmentation
S = azimuthal span for augmentation. The span is centered on central azimuth of augmentation.
D_A = absolute difference between azimuth of calculation and central azimuth of augmentation. D_A cannot exceed $S/2$ for augmentation within a particular span.

$g(\theta)$ is as defined above.

If there are overlapping spans of augmentation, the augmentations are applied in ascending order of central azimuth beginning with true north ($0°$ true). In this case, there is in essence an augmentation of an augmentation. If the central azimuth of an earlier augmentation overlaps the central azimuth of a later augmentation, the value of A is adjusted in the latter case to provide for the specified resulting radiation at the later central azimuth.

Choice of Orientation and Spacing in Directional-Antenna Design

The required placement of towers in a directional array is determined by the general shape of pattern desired. In-line arrays produce patterns having line symmetry; other configurations may or may not exhibit symmetry, depending upon the operating parameters. Closely spaced towers have mutual impedances which are relatively large compared with the self-resistances. This may result in low operating resistances and high circulating currents, which make for instability and high losses and should be avoided. In general, these effects may occur when spacings less than about one-quarter wavelength are introduced.

Choice of Tower Height

The choice of tower height is governed by the effective field required, the need for adequately high base resistances, the desired vertical patterns, aeronautical restrictions, and economic limitations. These requirements are usually met by heights on the order of one-quarter wavelength. Shorter towers are sometimes employed on the lower frequencies for practical reasons, as long as the required FCC minimum effective field is obtained and adequately high base resistances are provided. Taller towers produce higher effective field strengths and may reduce radiation at high vertical angles.

Effect of Finite Radiator Cross Section on Directional Antennas

The modified current distribution due to the finite cross section of practical tower radiators results in a complex vertical-radiation characteristic different from that computed by using the simplified assumptions. When towers of unequal height are employed, the difference in amplitude (and sometimes in phase) of vertical-radiation characteristics may be taken into account. The radiation from an array employing towers of unequal height may be computed by using Eqs. (25-17) and (25-22). The difference in phase angle requires the substitution for α_n with the angle α'_n, which includes an additional term δ_n:

$$\alpha'_n = s_n \cos \phi_n \cos \theta + A_n + \delta_n \qquad (25\text{-}45)$$

where δ_n = phase-angle difference between $f_1(\theta)$ and $f_n(\theta)$ at angle θ.

25-5 CIRCUITS FOR SUPPLYING POWER TO DIRECTIONAL AND NONDIRECTIONAL ANTENNAS

General

Radio-frequency power must be supplied to the individual radiators of the directional-antenna system in the proper proportions and phase-angle relationships to produce the desired radiation patterns. Means for controlling the current ratios and phase angles are required to permit adjustment and maintenance of the patterns. The circuits must provide a load into which the transmitter will operate properly.

The required functions are shown in block form in Fig. 25-15. The antenna-tuning units transform the operating base impedance of the radiators to the characteristic impedance of the transmission lines and provide a portion of the required phase shift. Additional phase shift is introduced by the transmission lines. The phase-control networks contain variable components for phase control. The power-dividing network supplies variable voltages to each line for power control. The networks should be

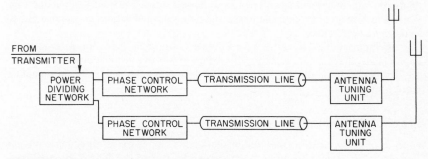

FIG. 25-15 Basic functions of array feed system.

designed for minimum power loss and for a broadband-frequency response. (See Chap. 43 of this handbook for more detailed information.)

Transmission-Line Requirements

Transmission lines may be either of the concentric or the open-wire, unbalanced type. Concentric lines have a lower characteristic impedance, usually requiring less transformation between tower and line. Their complete shielding eliminates any radiation from the line.

25-6 ADJUSTING DIRECTIONAL-ANTENNA ARRAYS

General Requirements

A directional antenna must be adjusted to produce a radiation pattern substantially in accordance with its design. In the United States, the construction permit issued by the FCC specifies maximum permissible values of radiation in pertinent directions.

Sampling System

An indication of the field-ratio and phase-angle relationships among the radiators is needed to adjust a directional-antenna system. Depending on the tower heights, this is provided by permanently installed *sampling transformers* in the antenna-tuning unit or *sampling loops* mounted on the towers. These devices are connected by *sampling lines* to an antenna monitor.

The individual sampling lines may be cut to the same length, making the phase delay on all lines equal. Excess line should be stored so as to be exposed to the same weather conditions as the longer portions of the other lines. Section 73.68 of the FCC *Rules and Regulations* specifies the requirements for an approved sampling system for medium-wave directional-antenna systems in the United States.

Initial Adjustment

The initial adjustment of a directional-antenna system is generally accomplished by setting the reactances of the antenna-tuning and power-dividing components to their computed values and supplying power to the antenna system. Usually, the phase and field-ratio indications will be different from those desired. The adjustable tuning components are then varied until the phase and field-ratio indications correspond closely to the computed values.

Preliminary field-intensity measurements are then made to determine the approximate shape of the pattern. These data are analyzed to determine if changes are required to bring the operating pattern shape into agreement with the computed pattern shape.

Proof of Performance

A *proof of performance* is required in the United States for all directional-antenna systems before regular operation is authorized. Part 73 of the FCC *Rules and Regu-*

lations describes the procedures and required measurements for an acceptable proof of performance.

After the pattern has been properly adjusted and the adjustment confirmed by the proof of performance, *monitoring points* are established in directions specified by the construction permit. The field intensities for directional operation are measured at each monitor point at regular intervals to provide an indication of pattern performance and stability.

25-7 MISCELLANEOUS PROBLEMS

Guy-Wire Insulation

Guy wires supporting tower radiators must be insulated from the tower and from ground and must be broken up into sections sufficiently short so that the induced currents do not distort the radiation pattern. Strain insulators are installed at the guy anchor, the point of attachment to the tower, and at intervals along the guy wire. The maximum length of any individual guy-wire section should not exceed $\lambda/8$ to $\lambda/10$. Occasionally, portions of the topmost guy wires are not insulated from the tower in order to provide top loading to the element.

Circuits across Base Insulators

It is usually necessary to cross the base insulator of a tower antenna with alternating-current power circuits. Power for aeronautical-obstruction lighting of the tower or other purposes may be supplied by means of chokes or transformers. Lighting chokes are wound of ordinary insulated copper wire on a suitable form, with a sufficient number of turns to provide a reactance at the operating frequency, which is high compared with the tower-base impedance. Radio-frequency bypass condensers are installed between individual windings.

Alternating current can be supplied by an Austin transformer, consisting of linked toroidal cores mounting the primary and secondary windings. There is an air gap of several inches between the two cores, and the only effect at the medium frequency is that due to the shunt capacity.

FM and television transmission lines (and other metallic circuits) may be isolated from the tower at the medium frequency by a quarter-wave isolation section of line. The outer conductor of the FM-TV coaxial cable is grounded immediately before the rise up the tower and is supported on insulated hangers to a point approximately one-quarter wavelength from ground, where it is connected to the tower. The outer conductor of the FM-TV line and the tower constitute a shorted quarter-wavelength transmission line at the medium frequency, resulting in a high impedance at the tower base. A method for towers shorter than $\lambda/4$ is to install the quarter-wavelength section along the ground rather than up the tower.

Simultaneous Use of Single Tower at Two or More Frequencies

It is occasionally desired to use a single tower radiator simultaneously at two or more different frequencies. This may be done by employing suitable filters to isolate the transmitters from each other.

Selection of Transmitter Sites

Transmitter sites should be selected in an area providing sufficient ground which is reasonably flat and level, of high local conductivity, free of obstructions which might interfere with the proper functioning of the radiating system, and so located as to provide maximum signal to the principal city and the service area. This last requirement, applied to operations with a directional-antenna system, usually dictates a choice of site which will place the main radiation lobe in the direction of the city.

Effect of Signal Scattering and Reradiation by Nearby Objects

General   Structures and terrain features near the transmitter site may reflect the signal from the antenna or may reradiate sufficient signal to affect the performance of the antenna. Large buildings near the transmitter site, mountains, or rugged terrain may distort the radiation pattern of directional and nondirectional antennas. The effects may be serious in the case of a directional-antenna system requiring a high degree of signal suppression, particularly if the buildings or hills are in the main radiation lobe of the antenna. Such objects are usually too irregular to permit application of analytic methods to a determination of reradiation. Their effect may often be estimated on the basis of experience with similar objects.

Tall Towers: General   It is frequently desired to erect tall towers to support FM and television transmitting antennas or for other purposes in the immediate vicinity of a medium-frequency antenna. These structures are usually of sufficient electrical height to be capable of substantial reradiation in high incident fields.

Tall Towers: Control of Reradiation   The tower location should be chosen to have minimum effect on the medium-frequency antenna. If the tower is to be installed in the immediate vicinity of a medium-frequency directional antenna, the field intensities should be computed at a number of locations, and a position chosen for the tower where the incident field is a minimum.

The reradiation from a tower of this height may be controlled by insulating the tower from ground and installing sectionalizing insulators at one or more levels. Guy wires must be insulated at suitable intervals, and transmission and alternating-current lines must be isolated.

Reradiation from tall towers may be controlled to some extent without sectionalizing insulators by insulating the tower from ground and installing a suitable reactor between the tower and ground.

Protection against Static Discharges and Lightning

In the absence of suitable precautions, static charges accumulate on towers and guy wires and may discharge to ground. This discharge ionizes the path, and a sustained radio-frequency arc may follow. Protection can be provided to minimize the accumulation or quench the arc, but it is difficult to provide protection against direct lightning hits. A lightning rod or rods extending above the beacon on the tower top will provide some protection to the beacon, and horn or ball gaps at the tower base will provide protection to the base insulator. However, depending on the magnitude of lightning current, some damage may result to the meters and tuning components.

A direct-current path from the tower to ground will minimize static accumulation. A separate radio-frequency choke, the tuning inductor, or the sampling-line inductor may be connected to maintain the tower at direct-current ground potential. Difficulty may occasionally be experienced with charges accumulating on the individual guy wires and arcing across the guy insulators. This may be eliminated by installing static drain resistors across each guy insulator. These resistors may have a value of 50,000 to 100,000 Ω and should have an insulation path somewhat longer than provided by the guy insulator.

REFERENCES

1 C. L. Jeffers, "An Antenna for Controlling the Nonfading Range of Broadcasting Stations," *IRE Proc.,* vol. 36, November 1948, pp. 1426–1431.
2 S. A. Schelkunoff, "Theory of Antennas of Arbitrary Size and Shape," *IRE Proc.,* vol. 29, September 1941, pp. 493–521.
3 G. H. Brown and O. M. Woodward, Jr., "Experimentally Determined Impedance Characteristics of Cylindrical Antennas," *IRE Proc.,* vol. 33, April 1945, pp. 257–262.
4 P. S. Carter, "Circuit Relations in Radiating Systems and Applications to Antenna Problems," *IRE Proc.,* vol. 20, June 1932, pp. 1004–1041.
5 J. F. Morrison and P. H. Smith, "The Shunt-Excited Antenna," *IRE Proc.,* vol. 25, June 1937, pp. 673–696.
6 P. Baudoux, "Current Distribution and Radiation Properties of a Shunt-Excited Antenna," *IRE Proc.,* vol. 28, June 1940, pp. 271–275.
7 G. H. Brown, "A Critical Study of the Characteristics of Broadcast Antennas as Affected by Antenna Current Distribution," *IRE Proc.,* vol. 24, January 1936, pp. 48–81.
8 *Rules and Regulations of the Radio Broadcast Series,* part 3, Federal Communications Commission, 1959.
9 *Final Acts of the Regional Administrative MF Broadcasting Conference (Region 2),* Rio de Janeiro, 1981.
10 C. E. Smith and E. M. Johnson, "Performance of Short Antennas," *IRE Proc.,* vol. 35, October 1947, pp. 1026–1038.
11 J. R. Wait and W. A. Pope, "The Characteristics of a Vertical Antenna with a Radial Conductor Ground System," *IRE Conv. Rec.,* part 1, *Antennas and Propagation,* 1954, p. 79.
12 F. R. Abbott, "Design of Optimum Buried-Conductor RF Ground System," *IRE Proc.,* vol. 40, July 1952, pp. 846–852.
13 O. Prestholdt, "The Design of Non-Radial Ground Systems for Medium-Wave Directional Antennas," NAB Eng. Conf., 1978, unpublished.
14 C. E. Smith, *Theory and Design of Directional Antennas,* Cleveland Institute of Radio Electronics, Cleveland, 1949.
15 C. R. Cox, "Mutual Impedance between Vertical Antennas of Unequal Heights," *IRE Proc.,* vol. 35, November 1947, pp. 1367–1370.
16 G. H. Brown: "Directional Antennas," *IRE Proc.,* vol. 25, January 1937, pp. 78–145.

BIBLIOGRAPHY

Brown, G. H.: "The Phase and Magnitude of Earth Currents near Radio Transmitting Antennas," *IRE Proc.,* vol. 23, February 1935, pp. 168–182.
———: "A Consideration of the Radio-Frequency Voltages Encountered by the Insulating Material of Broadcast Tower Antennas," *IRE Proc.,* vol. 27, September 1939, pp. 566–578.

————, R. F. Lewis, and J. Epstein: "Ground Systems as a Factor in Antenna Efficiency," *IRE Proc.*, vol. 25, June 1937, pp. 753–787.

Chamberlain, A. B., and W. B. Lodge: "The Broadcast Antenna," *IRE Proc.*, vol. 24, January 1936, pp. 11–35.

Effective Radio Ground-Conductivity Measurements in the United States, Nat. Bur. Stds. Circ. 546, February 1954.

Terman, F. E.: *Radio Engineers' Handbook,* McGraw-Hill Book Company, New York, 1943.

Fine, H.: "An Effective Ground Conductivity Map for Continental United States," *IRE Proc.*, vol. 42, September 1954, pp. 1405–1408.

Hansen, W. W., and J. G. Beckerley: "Concerning New Methods of Calculating Radiation Resistance, Either with or without Ground," *IRE Proc.*, vol. 24, December 1936, pp. 1594–1621.

King, R.: "Self- and Mutual Impedances of Parallel Identical Antennas," *IRE Proc.*, vol. 40, August 1952, pp. 981–988.

Laport, E. A.: *Radio Antenna Engineering,* McGraw-Hill Book Company, New York, 1952.

Moley, S. W., and R. F. King: "Impedance of a Monopole Antenna with a Radial-Wire Ground System," *Radio Sci. (MBS)*, vol. 68D, no. 2, February 1964.

Moulton, C. H.: "Signal Distortion by Directional Broadcast Antennas," *IRE Proc.*, vol. 40, May 1952, pp. 595–600.

Nickle, C. A., R. B. Dome, and W. W. Brown: "Control of Radiating Properties of Antennas," *IRE Proc.*, vol. 22, December 1934, pp. 1362–1373.

Norton, K. A.: "The Calculation of Ground-Wave Field Intensity over a Finitely Conducting Spherical Earth," *IRE Proc.*, vol. 29, December 1941, pp. 623–639.

Schelkunoff, S. A., and H. T. Friis: *Antennas: Theory and Practice,* John Wiley & Sons, Inc., New York, 1952.

Smeby, L. C.: "Short Antenna Characteristics: Theoretical," *IRE Proc.*, vol. 37, October 1949, pp. 1185–1194.

Spangenberg, K.: "Charts for the Determination of the Root-Mean-Square Value of the Horizontal Radiation Pattern of Two-Element Broadcast Antenna Arrays," *IRE Proc.*, vol. 30, May 1942, pp. 237–240.

Williams, H. P.: *Antenna Theory and Design,* vol. 2, Sir Isaac Pitman & Sons, Ltd., London, 1950.

Chapter 26

High-Frequency Antennas

Ronald Wilensky
William Wharton
Technology for Communications International

26-1 GENERAL DESIGN REQUIREMENTS

High-frequency (HF) antennas are used in the range of frequencies from 2 to 30 MHz for communications and broadcasting by means of ionospheric propagation. Ionospheric transmission over large distances usually involves high overall transmission losses, especially under unfavorable ionospheric propagation conditions. High initial and operating costs of transmitting equipment therefore make high-gain transmitting and receiving antennas desirable to provide reliable communications. High-gain antennas are also in common use at international broadcasting stations, where transmissions are beamed to specific geographical areas. In additition to providing an increased signal in the target area, the use of high-gain directional antennas decreases radiation in undesired directions, thereby reducing potential interference to other cochannel services.

Important Design Parameters

The important design parameters of an HF transmitting or receiving antenna include its frequency range, vertical and horizontal angles at which maximum radiation is desired, gain or directivity, voltage standing-wave ratio (VSWR), input power, and environmental and mechanical requirements.

The frequency range should be established by a propagation study to determine the optimum working frequency (OWF) for the path involved, which varies according to distance and location, time of day, season of the year, and sunspot activity. Details of such a study are beyond the scope of this book but may be found in the literature.[1,2] There are available several computer programs which determine path losses and optimum frequencies on the basis of stored ionospheric data.[3,4] Actual ionospheric characteristics may also be measured in real time by using commercially available ionospheric sounders and frequency-management systems.

The vertical angle of maximum radiation of the antenna is known as the takeoff angle (TOA). The range of vertical angles pertinent to ionospheric propagation depends on distance, effective layer height, and mode of propagation (i.e., one-hop, two-hop, etc.). Figure 26-1 is a graph showing estimated TOA for one-hop transmission as a function of distance and virtual layer heights. The height of the E layer may be considered constant at 100 km, but the height of the $F2$ layer varies widely according to time of day, season of the year, and location. A detailed propagation study is required to determine the required range of TOA. As a rough estimate, however, a range of $F1$ and $F2$ virtual layer height of approximately 250 to 400 km may be used in conjunction with Fig. 26-1 to determine approximate TOA range. Multiple-hop propagation can also be assessed approximately by using Fig. 26-1 for each hop. For transmission or reception over distances of more than 4000 km, maximum signal results from low-angle transmission in the range from 2 to 15°, with the lower angles generally providing better results. Propagation over very long paths (greater than 10,000 km) is more complicated than the simple multiple-hop theory, and sophisticated methods of propagation analysis must be used.[4]

The horizontal range of angles or beamwidth (defined as the azimuth angle between the points at which radiation is 3 dB below beam maximum) is determined by the geographical area to be covered. Wide areas require large beamwidths, and narrow areas or point-to-point circuits require small beamwidths. The minimum beamwidth for point-to-point coverage depends on irregularities in the ionosphere and effects of magnetic storms, which cause deviations from great-circle paths. Directions of arrival of HF signals may vary by as much as $\pm 5°$ in the horizontal plane because of these effects.[6] This makes the use of antennas with extremely narrow horizontal beamwidths undesirable, especially on circuits which skirt or traverse the auroral zone. International broadcasting requires that the beamwidth subtend the target area with due allowance for path-deviation effects.

Antenna directivity (the gain, neglecting losses) depends primarily on the vertical and azimuthal beamwidths of the radiation pattern. Antenna gain is the directivity multiplied by the antenna radiation efficiency. High gain is necessary when large effective radiated power is needed to overcome large ionospheric transmission losses.

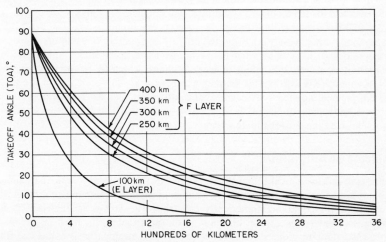

FIG. 26-1 Takeoff angles for one-hop ionospheric transmission as a function of virtual layer height. *(After Ref. 5.)*

High directivity is desirable for reception in the presence of interference. Low radiation efficiencies can generally be accepted for reception when there is a high level of ambient noise and the high antenna noise figure does not degrade performance. At very quiet receiving sites, high radiation efficiency (i.e., low noise figure) is necessary to prevent the antenna from degrading circuit performance.

For transmitting antennas, the acceptable level of reflected power from the antenna is usually determined by the characteristics of the transmitter. Reflected power is usually specified in terms of the VSWR. Modern solid-state transmitters operating in the 0.1- to 10-kW range tolerate maximum VSWR of 2.5:1. Many such transmitters will tolerate a higher VSWR of 3:1 or 4:1 by automatically reducing their output power. High-power broadcasting transmitters in the 100- to 500-kW range of carrier-power levels tolerate maximum VSWRs of 2:1. Older transmitters have much lower maximum VSWR limits of 1.5:1 or 1.4:1.

The power-handling capacity of a transmitting antenna is specified by both an average power and a peak envelope power (PEP). The average power determines the required current-carrying capacity of the antenna conductors. The peak power determines the voltages and electric fields on the antenna, which set requirements on insulator and conductor size to avoid arcing or corona discharge.

Mechanical requirements depend on environmental effects (wind, ice loading, temperature variation, corrosion-causing conditions), limitations on antenna size or tower height, special restrictions imposed by the site, or other considerations such as transportability.

Types of High-Frequency Antennas

Most HF antennas are very broadband and usually require little or no tuning. Several techniques are used to make HF antennas broadband: (1) lowering antenna Q, (2) using log-periodic arrays of monopole or dipole elements, and (3) automatic tuning. Antenna Q may be lowered by adding loss resistance or using "fat" radiators to decrease reactance. Examples of antennas with resistors are rhombics, terminated V's, and resistively loaded dipoles. Typical fattened antennas are fan dipoles, conical monopoles, and wide-bandwidth dipole arrays used in HF broadcasting. The bandwidth of a log-periodic antenna (LPA) is limited only by the number of radiators used, and it is easy to design such an antenna to operate over a 2- to 30-MHz range. Automatic tuning units are used with whips or loop antennas, which by themselves have unacceptably high VSWR over wide frequency ranges.

For special applications, narrow-bandwidth antennas, such as simple resonant dipoles or monopoles or Yagi-Uda arrays, are used because they are often smaller and less expensive than broadband antennas.

Analysis of Antennas

Analysis of most practical HF antennas is not possible by using simple mathematical techniques. Antennas are designed by using sophisticated computer programs which solve Maxwell's equations numerically for arbitrary antenna geometries.[7-10] Such programs calculate the current distribution on the antenna, the most important and hardest-to-calculate property of an antenna. The radiation patterns calculated by the most comprehensive computer programs are very accurate, and the results are acceptable substitutes for measured patterns. Driving-point impedances are also calculated well, but often to somewhat less accuracy than radiation patterns because impedance depends critically on the mathematical representation of the fields in the region of the driving point.

26-2   SIMPLE ANTENNAS MOUNTED ABOVE GROUND

Effect of the Ground Plane

HF antennas are usually operated within one or two wavelengths of the ground or, in the case of base-fed monopoles, directly on the ground. The proximity of the ground modifies the input impedance and the radiation pattern of an antenna. In most practical cases, an accurate evaluation of the effect of the ground is difficult. However, many fundamental properties of HF antennas can be understood in terms of the simple theory of idealized monopoles or dipoles situated over an infinite, flat ground plane. When the ground plane is perfectly conducting, it can be represented exactly by a single image (with the ground plane removed) as illustrated in Fig. 2-7. For ground of finite conductivity, simple image theory does not give an exact solution but will often provide useful approximate results.

Antenna Impedance

The currents in the conductors of an antenna induce currents in the ground plane, and these (or, alternatively, the currents in the image antenna) modify the antenna currents from the values that would occur in free space. The interaction between the antenna and the ground is given by the mutual impedance, which varies with the height of the antenna above ground. The input impedance of the antenna is given by

$$Z_{ant} = Z_{fs} + Z_m \qquad \textbf{(26-1)}$$

where Z_{fs} = self-impedance of antenna (impedance when isolated in free space)
   Z_m = mutual impedance between antenna and its image
The mutual impedance can be either positive or negative. Chapter 4 contains curves of mutual impedance for several simple but useful geometries. More complex arrangements are best analyzed by using antenna-analysis computer programs.

The mutual impedance of a thin center-fed half-wave vertical dipole whose lower point is at height $S/2$ above ground is given in Fig. 4-29. In this case, mutual impedance adds to self-impedance so that total antenna impedance increases as the height of the dipole decreases. The mutual impedance of a thin center-fed horizontal dipole at height $d/2$ above ground is given in Fig. 4-28. In this case, the currents in the image and the antenna are 180° out of phase so that the mutual impedance must be deducted from the self-impedance. For vanishingly small heights self-impedance and mutual impedance cancel, and radiation resistance approaches zero. This means that a horizontal dipole placed close to the ground has a low radiation resistance. Steps must be taken to modify its impedance so that the transmitter is presented with an acceptably low VSWR. The input resistance can be increased by adding series resistance, which improves the VSWR but reduces radiation efficiency (and hence gain). The radiation efficiency with added resistance is given by $R_{rad}/(R_{rad} + R_{loss})$, where R_{rad} is the radiation resistance and R_{loss} is the series loss resistance. Over ground of finite conductivity,

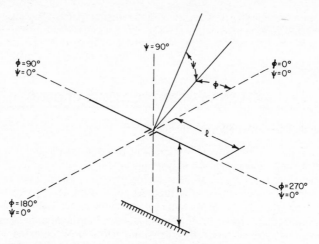

FIG. 26-2   Geometry of horizontal dipole above ground.

an extra resistance must be added in the denominator to account for losses in the ground.

Radiation Patterns of Horizontally Polarized Antennas

In terms of the geometry of Fig. 26-2, the free-space radiation pattern of a horizontal dipole of length 2ℓ is given by

$$d(\varphi') = \frac{\cos{(2\pi\ell \sin{\varphi'}/\lambda)} - \cos{(2\pi\ell/\lambda)}}{(1 - \cos{2\pi\ell/\lambda})\cos{\varphi'}} \qquad \textbf{(26-2)}$$

where φ' = angle between line of propagation and equatorial plane
$\sin{\varphi'} = \sin{\varphi} \cos{\psi}$
φ = azimuth angle
ψ = elevation angle
λ = wavelength

When the dipole is placed above ground as shown in Fig. 26-2, the radiation reflected from the ground adds to that from the dipole (i.e., the dipole and image fields add). For the practical situation $h/\lambda > 0.2$, the total radiation pattern,[11] with the small effect of mutual impedance ignored, is

$$D(\varphi,\psi) = d(\varphi')[1 + r_H^2 + 2r_H \cos{(\rho_H + 4\pi h \sin{\psi}/\lambda)}]^{1/2} \qquad \textbf{(26-3)}$$

where h is the height above ground and r_H and ρ_H are respectively the magnitude and the phase of the complex reflection coefficient[12]

$$r_H \underline{/\rho_H} = \frac{\sin{\psi} - (\epsilon' - \cos^2{\psi} - j60\sigma\lambda)^{1/2}}{\sin{\psi} + (\epsilon' + \cos^2{\psi} - j60\sigma\lambda)^{1/2}} \qquad \textbf{(26-4)}$$

where $\epsilon' = \epsilon/\epsilon_0$, the relative dielectric constant
σ = ground conductivity, S/m

At grazing incidence ($\psi = 0°$), $r_H = 1$ and $\rho_H = 180°$ for all ground constants, so that $D(\varphi,0) = 0$. In the HF band, for most practical values of ϵ' and σ, ρ_H is nearly equal to 180° for all elevation angles and r_H decreases from 1 at $\psi = 0°$ to about 0.5 to 0.6 directly overhead at $\psi = 90°$. In most HF applications the TOA is in the range $0 < \psi \leq 30°$, and the radiation pattern is given accurately by setting $r_H = 1$ and $\rho_H = 180°$ in Eq. (26-4), which is equivalent to assuming that the ground is perfectly conducting. Equation (26-3) therefore becomes

$$D(\varphi,\psi) = 2d(\varphi') \sin (2\pi h \sin \psi/\lambda) \qquad (26\text{-}5)$$

The maxima (TOA) and minima (nulls) of this expression occur in the equatorial plane ($\phi = 0$ and $180°$) at elevation angles given by

$$\psi_{max} = \sin^{-1} (n\lambda/4h) \qquad n = 1, 3, 5, \ldots \qquad (26\text{-}6)$$

$$\psi_{min} = \sin^{-1} (n\lambda/2h) \qquad n = 1, 2, 3, \ldots \qquad (26\text{-}7)$$

At the maxima, $D^2 = 4d^2$, so that the gain is 6 dB above that obtained in free space. This value is exact for perfectly conducting ground and is nearly exact for low elevation angles above ground of finite conductivity. Thus, low-angle radiation from horizontal antennas suffers negligible loss owing to the finite conductivity of the ground, and a radiation-pattern ground screen in front of the antenna is not required. For high values of TOA, the ground losses are not always negligible. As an example, consider an antenna mounted 0.25λ above ground, for which from Eq. (26-6) $\psi_{max} = 90°$. For a typical set of ground constants, $r_H = 0.5$ and $\rho_H = 170°$, so Eq. (26-3) becomes $D^2 = 2.24d^2$, which represents a reduction of gain of 2.24/4 or 2.5 dB compared with that for a perfect ground. However, because a high TOA is used for transmission over short paths, it is not generally necessary to use a ground screen to minimize the loss. An exception is a high-TOA, high-power HF broadcasting antenna which must deliver the maximum signal in the target service area, and in this situation, a ground screen is occasionally used.

Radiation Patterns of Vertically Polarized Antennas

The elevation pattern of a vertical dipole whose center point is at height h above ground has the same form as Eq. (26-3). For the vertical dipole the angles φ and ψ are interchanged in Eq. (26-2), and r_H and ρ_H in Eq. (26-3) are replaced respectively by r_V and ρ_V, the reflection coefficients for vertical polarization:[12]

$$r_V \underline{/\rho_V} = \frac{(\epsilon' - j60\sigma\lambda) \sin \psi - (\epsilon' - \cos^2 \psi - j60\sigma\lambda)^{1/2}}{(\epsilon' - j60\sigma\lambda) \sin \psi + (\epsilon' - \cos^2 \psi - j60\sigma\lambda)^{1/2}} \qquad (26\text{-}8)$$

This reflection coefficient behaves very differently from that for horizontal polarization [Eq. (26-4)]. The phase ρ_V varies from $-180°$ at grazing incidence to a value approaching $0°$ at high elevation angles. The magnitude r_V is 1 at grazing incidence, drops to a value of a few tenths, and then increases to the value equal to r_H at $\psi = 90°$. The elevation angle at which r_V reaches a minimum occurs when $\rho_V = 90°$ and is known as the pseudo-Brewster angle, which at HF is typically in the range $5° \leq \psi \leq 15°$.

The most important consequence of this behavior is that radiation at low elevation angles is drastically reduced over imperfectly conducting ground. Finite ground conductivity can cause a gain reduction of as much as 8 to 10 dB at low elevation angles, as shown later in Fig. 26-18 and in Ref. 12. For receiving antennas, losses of this magnitude are usually tolerable, but for transmitting antennas they are not. For transmission, the loss must be reduced by having the antenna radiate over seawater (which has a high conductivity) or by installing in front of the antenna a large metallic ground screen with a mesh small in relation to the skin depth of the ground underneath the mesh.

26-3 HORIZONTALLY POLARIZED LOG-PERIODIC ANTENNAS

Single Log-Periodic Curtain with Half-Wave Radiators

The basic curtain, illustrated in Fig. 26-3, comprises a series of half-wave dipoles spaced along a transposed transmission line. Successive values of the dipole length and spacing have a constant ratio τ. In a practical LPA, τ has a value between 0.8 and

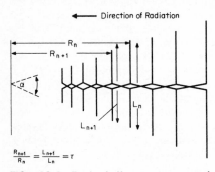

$$\frac{R_{n+1}}{R_n} = \frac{L_{n+1}}{L_n} = \tau$$

FIG. 26-3 Basic half-wave transposed dipole LPA.

0.95. The bandwidth of the LPA is limited only by the lengths of the longest and shortest dipoles. The number of radiators depends on the frequency range and the value of τ. The parameters τ and α, as discussed in detail in Chap. 14 and Refs. 13 and 14, determine the gain, impedance level, and maximum VSWR of the antenna. It is important to choose τ and α carefully because of the possibility of unwanted resonant effects. Sometimes, energy that normally radiates from the active region travels back to a second active region where the radiators are $3\lambda/2$ long. Excitation of this second active region causes undesirable impedance and pattern perturbations.

In a well-designed horizontally polarized LPA (HLPA), at any given frequency only dipoles whose lengths are approximately $\lambda/2$ long are excited, and these form a single active region. The active region moves along the curtain from the longest to the shortest dipole as the frequency is increased. Because of this movement, it is possible to mount an HLPA so that the electrical height of the active region is constant or variable in a controlled way, as shown in Fig. 26-4a and b. In other types of horizontal

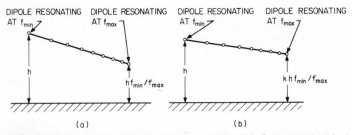

FIG. 26-4 Control of electrical height of LPA. (a) Constant electrical height. (b) Uniform variation of electrical height by a factor of K from f_{min} to f_{max}.

antennas, the physical height is constant, so the designer cannot control the variation of electrical height or TOA. A major advantage of the HLPA is, therefore, the ability to maximize the gain at the elevation angles required for reliable communications.

In the basic LPA shown in Fig. 26-3, the dipoles are conventional rods. It is also possible to form the dipoles into a sawtooth shape as shown in Fig. 26-5. The larger effective diameter of the sawtooth (compared with the rod) increases the bandwidth of each dipole so that a greater number are excited to form the active region at each frequency. This increases the gain and power-handling capacity and reduces the VSWR.

The basic HLPA using half-wave radiators has a gain of 10 to 12 dBi and an azimuthal beamwidth of 60 to 80°.

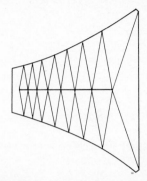

Clamped-Mode Log-Periodic Curtain with Full-Wave Radiators

The gain of a half-wave LPA can be increased by 2 to 3 dB by doubling its horizontal aperture to one wavelength, thereby decreasing its azimuthal beamwidth to 30 to 40°.

FIG. 26-5 LPA with sawtooth wire elements.

The obvious way to increase the horizontal aperture is to use two half-wave curtains in broadside array as shown in Fig. 26-6*a*, but the resulting structure is relatively complex. The same horizontal aperture can be achieved with a much simpler structure by use of the *clamped-mode technique*[15] illustrated in Fig. 26-6*b*.

The clamped-mode arrangement is equivalent to taking one of the LPA curtains in Fig. 26-6*a* and pulling it apart sideways, leaving its radiators connected to the same half of the transmission line. When properly designed, a clamped-mode antenna has a gain and radiation pattern nearly identical to those of the broadside array, but it requires only half the number of radiator and transmission-line wires.

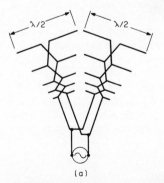

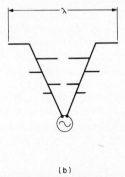

(a)   (b)

FIG. 26-6 Clamped-mode technique for increasing horizontal aperture of LPA. (*a*) Two LPA curtains in broadside array. (*b*) Clamped-mode LPA curtain with equivalent performance.

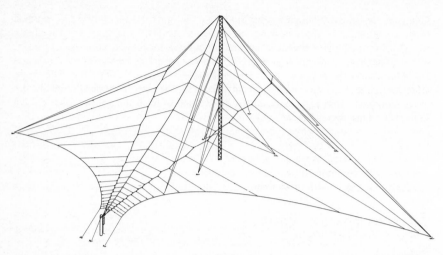

FIG. 26-7 Practical clamped-mode LPA using radiators made from single wires. *(Courtesy of TCI.)*

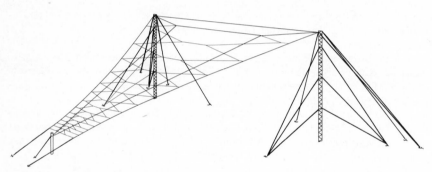

FIG. 26-8 Practical clamped-mode LPA using sawtooth radiators made from wires. *(Courtesy of TCI.)*

Two practical clamped-mode antennas are shown in Figs. 26-7 and 26-8. The antenna in Fig. 26-7 uses simple rod radiators made from single wires. It has only one tower in contrast to the two towers required for a conventional half-wave dipole HLPA, and consequently the radiators slope down toward the ground. This slope partially offsets the larger horizontal aperture and yields a gain of 12 dBi and an azimuthal beamwidth of 55 to 80°, the values for an equivalent conventional two-tower LPA. Figure 26-8 shows a clamped-mode antenna using sawtooth radiators and two towers. The horizontal curtain is flat, and the aperture is about 1.25λ, so the azimuthal beamwidth is 38° and the gain 14 to 15 dBi.

Multiple-Curtain Log-Periodic Arrays

LPAs can be stacked horizontally and/or vertically to decrease their beamwidths and increase their gain. Horizontal stacking can be done simply by using the clamped-mode technique described previously. Vertical stacking can be applied to simple LPAs with half-wave dipole elements, as shown in Fig. 26-9, or in conjunction with clamped-mode antennas, as in Fig. 26-10.

Vertical stacking raises the average radiating height of the antennas to a value given by the average electrical height of the curtains, thereby reducing the TOA and vertical beamwidth. Vertical stacking with spacings of about $\lambda/2$ also suppresses high-elevation secondary lobes, which are fully formed when the radiation height is large and can substantially reduce antenna gain. The number of elements in the vertical stack depends on the TOA and gain requirements and the maximum allowed tower height.

To achieve maximum gain in a vertically stacked LPA, it is necessary that the active regions of each curtain line up vertically at each frequency. This will not occur automatically because the lengths of the curtains are different, the upper curtains being longer than the lower curtains. This effect is most noticeable in antennas with more than two curtains in the stack. One technique for aligning the active regions is to control the velocity of propagation along the curtains, making the wave go more slowly on the shorter, lower curtains than on the longer, upper curtains. Wave velocity on a curtain can be controlled by adjusting the length, effective diameter, and spacing of the dipole elements.

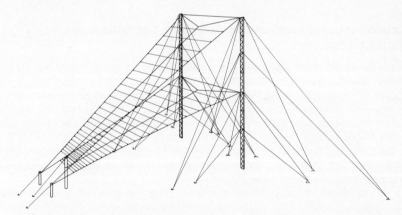

FIG. 26-9 Vertically stacked transposed dipole LPA curtains. *(Courtesy of TCI.)*

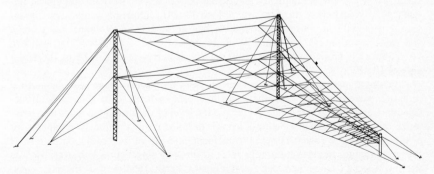

FIG. 26-10 Vertically stacked clamped-mode LPA curtains. *(Courtesy of TCI.)*

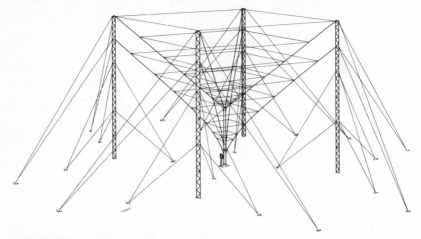

FIG. 26-11 Omnidirectional LPA with optimum TOA. *(Courtesy of TCI.)*

Omnidirectional Log-Periodic Array with Optimum Vertical Radiation Pattern

It has been pointed out that an HLPA can be arranged to give the optimum TOA at each frequency. A practical example of this technique is shown in Fig. 26-11, which illustrates an antenna designed for shore-to-ship and ground-to-air communication, for which an onmidirectional azimuth radiation pattern is normally required. Such an antenna must have a vertical radiation pattern in which the main lobe has a high TOA at low frequencies, dropping to a low TOA at high frequencies, in the HF band.[16] This is due to the fact that the low frequencies are used for short-distance communication, for which a high TOA is required, and the high frequencies are used for long-distance communication, for which a low TOA is required.

To meet these requirements, a number of horizontal loops which radiate omnidirectionally in azimuth are stacked vertically and connected in log-periodic form in the inverted-cone configuration shown in Fig. 26-11. The active region of the array, which comprises loops which are about two wavelengths in perimeter, is at the top of the array at the lowest frequency and at the bottom of the array at the highest frequency. A typical array operates over a 4- to 30-MHz-frequency range. The highest loop is placed 37 m above the ground, so that at 4 MHz the antenna has a broad vertical lobe with a TOA of about 30° and a gain of 7 dBi. The lowest loop is placed 10 m above the ground, so that the vertical lobe is narrow with a TOA of 10° and a gain of 10 dBi at 30 MHz.

A single inverted cone of loops suffers from the disadvantage that the vertical radiation pattern at high frequencies contains grating lobes. (See above, "Multiple-Curtain Log-Periodic Arrays.") To reduce the level of the unwanted lobes and maximize gain, a second inverted cone of loops is installed inside the first. The upper loops of the two inverted cones are nearly coincident, and the vertical radiation pattern at the lower frequencies is therefore the same as that of a single cone. However, because the lowest loops of the inner cone are at a greater height than those of the outer cone, the antenna behaves as two stacked loops at the higher frequencies. The stacking factor of the two loops modifies the vertical radiation pattern by suppressing the high-elevation grating lobes, thereby increasing the gain of the antenna.

Rotatable Log-Periodic Antenna

When a high-gain, steerable azimuthal beam is necessary, a rotatable LPA (RLPA) can be used. An example is illustrated in Fig. 26-12. This antenna is approximately 33 m high and has a boom length of about 33 m. Its operating frequency range is 4 to 30 MHz, its gain about 12 dBi, and its azimuthal beamwidth 60°. The antenna does not require a very large area for installation and provides complete azimuthal coverage. However, such antennas have major mechanical and electrical disadvantages. The major mechanical disadvantage is their complexity, resulting from the need to rotate a very large structure. These antennas require regular maintenance to ensure mechanical reliability and are prone to failure in harsh environments. The electrical disadvantage is that the antennas have constant physical height and, therefore, their electrical height increases with increasing frequency. As a result, grating lobes form at the higher frequencies, reducing gain in the principal lobe and introducing elevation nulls at TOAs required for communications.[16]

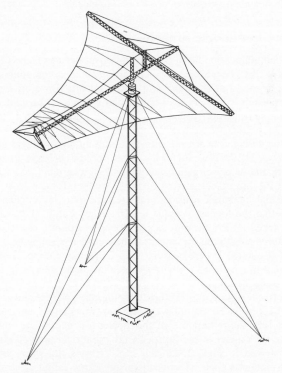

FIG. 26-12   Rotatable LPA.

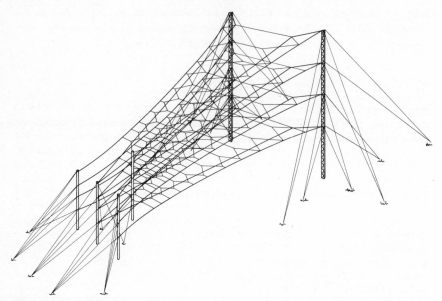

FIG. 26-13 Eight-curtain high-power LPA for HF broadcasting. *(Courtesy of TCI.)*

High-Power Log-Periodic Antenna

For most HF communication links, the transmitter power can be relatively low (usually 0.1 to 40 kW PEP) because high-gain receiving antennas and sophisticated receiving equipment are used at the other end of the circuit. However, for broadcasting much higher transmitter powers are necessary to provide an adequate signal for listeners using simple transistor receivers and low-gain antennas. As a result, transmitters with carrier powers between 100 and 500 kW are used, and the transmitting antenna must be designed to operate with the resulting high voltages and currents.

A high-power HLPA is useful for broadcast purposes because it has an extremely wide bandwidth and its power gain is not diminished by ground losses. In principle, a high-power HLPA can be designed to cover the whole HF band from 2 to 30 MHz. However, cost and land-area considerations require that the antenna size be minimized by selecting the highest possible low-frequency cutoff. A broadside array of horizontally stacked HLPA curtains can be slewed in azimuth electrically by setting up a phase difference between the stacks. If the electrical spacing between the radiation centers of the stacks is kept constant with frequency, the phase difference must also remain constant. Networks providing constant phase delay can be realized by using coupled transmission lines.

At the power levels used in HF broadcasting, the voltages and electric fields in an HLPA are very high. To minimize the radiator-tip voltages, which can be as high as 30 to 35 kV rms, the radiation Q of the dipoles (defined in Sec. 26-7) is reduced by constructing the dipoles in the form of a fat cylindrical cage of wires. A simpler technique, illustrated in Figs. 26-10 and 26-13, forms the wires into a sawtooth arrangement to increase their effective diameter. Large radial electric fields on the surface of the wires can initiate arcing and corona discharge, which can burn through the wire and cause structural failure of the antenna curtain. These fields can be reduced to safe levels by using wires with the largest diameter acceptable on structural grounds. All insulators used in the antenna, and particularly those at the radiator tips, must be able to withstand high voltages and fields with a large factor of safety.

Well-designed two- and four-curtain HLPAs can handle between 100- and 250-kW carrier power with 100 percent amplitude modulation. Eight- and sixteen-curtain HLPAs can handle up to 500-kW carrier power. The VSWR of these antennas is less than 2:1 over their entire frequency range and is generally less than 1.5:1 at most frequencies. As can be seen from Fig. 26-13, HLPAs with more than four curtains are very complicated and therefore are difficult to install and are expensive.

26-4 VERTICALLY POLARIZED ANTENNAS

Vertically polarized antennas are useful when a broad target area must be covered with a wide azimuthal beamwidth. An omnidirectional radiation pattern is obtainable with a single monopole. A directional pattern with wide beamwidth is obtainable by using vertical log-periodics. However, these antennas have a vertical null which militates against short-distance sky-wave communications. A very important disadvantage of vertical antennas is that, unlike horizontally polarized antennas, their power gain is diminished by losses in the surrounding ground. To overcome these losses partially, high ground conductivity or large ground screens are necessary, particularly if low TOAs are required for long-range communications.

Conical Monopole and Inverted Cone

Examples of the inverted cone and conical monopole are shown in Fig. 26-14a and b; their principal use is to provide omnidirectional coverage over a wide frequency band. The azimuthal pattern of these antennas is nearly circular at all frequencies, but the elevation pattern changes with frequency as shown in Fig. 4-14. The impedance bandwidth of the antennas is very large. The 22-m-high inverted cone illustrated in Fig. 26-14a has a VSWR of less than 2:1 over the entire 2- to 30-MHz range. The major disadvantage of this antenna is that the maximum radius of the cone is approximately equal to the height. The necessary support structure must consist of up to six nonconducting poles, which are usually expensive, and occupies a relatively large ground area.

The inverted-cone antenna can also be realized by using a single metallic support tower instead of the six nonconducting poles. The tower is placed along the cone axis on an insulated base. When designing the single-support antenna, great care must be taken to avoid internal resonances between the tower and the cone.

A more compact antenna, which also uses a single metallic mast and can be accommodated in a smaller ground area, is the conical monopole illustrated in Fig. 26-14b. The radiating structure comprises two cones, one inverted and one upright, connected at their bases. Here, the tower is grounded and the lower cone wires are insulated and fed against the ground. Midway up the tower the cone wires are connected to the tower to form an inductive loop. This acts as a built-in impedance-matching circuit, allowing the antenna to present a low VSWR when it is only 0.17λ tall. A

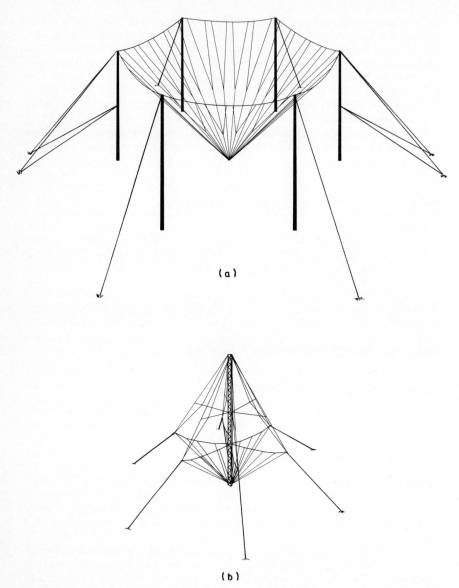

(a)

(b)

FIG. 26-14 Broadband monopole antennas. (*a*) Inverted cone. (*b*) Conical monopole. (*Courtesy of TCI.*)

26-16

conical monopole covering 2 to 14 MHz is 25 m high and has a maximum radius of 20 m.

All base-fed antennas usually require an impedance ground screen, which typically consists of 60 radials, each about $\lambda/4$ long at the lowest operating frequency.

Monopole Log-Periodic Antenna

At low frequencies in the HF band the height of a half-wave vertical dipole becomes undesirably large. Thus, for operation at the lower end of the HF band and in cases when directive gain is required, an LPA comprising quarter-wave monopoles operated over an impedance ground screen is a convenient antenna. A typical LPA covering 2 to 30 MHz is about 43 m tall and has a directive gain of 9 to 10 dBi and an azimuthal beamwidth of 140 to 180°.

Dipole Log-Periodic Antenna

When greater tower height can be accommodated (81 m for a 2- to 30-MHz antenna), an LPA comprising half-wave dipoles can be used, as shown in Fig. 26-15. The larger vertical aperture yields a gain of 12 to 13 dBi, which is about 3 dB greater than that of a monopole LPA. In contrast to the requirements of the monopole LPA, an impedance ground screen is not essential to achieve an acceptable VSWR. To achieve higher power gain and lower TOAs, a long radiation-pattern ground screen is necessary.

Monopole-Dipole Hybrid Log-Periodic Antenna

In an LPA covering the 2- to 30-MHz HF band, mast height can be minimized by using monopoles for the 2- to 4-MHz radiators. Above 4 MHz, dipole radiators can

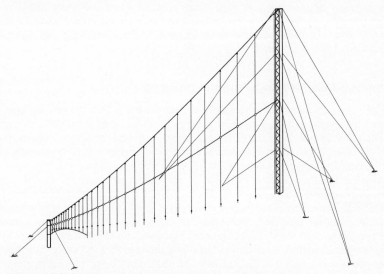

FIG. 26-15 Vertically polarized dipole LPA. *(Courtesy of TCI.)*

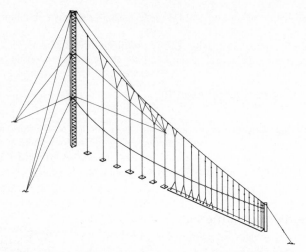

FIG. 26-16   Vertically polarized di-monopole hybrid LPA.
(Courtesy of TCI.)

be used because they are fairly short and have more gain than the monopoles. A *di-monopole*[17] hybrid LPA can be constructed (see Fig. 26-16) in which the low-frequency elements are monopoles, the high-frequency elements are dipoles, and the elements operating at intermediate frequencies are unbalanced dipoles in which the upper conductor is longer than the lower conductor. Lumped capacitors and inductors are used at the base of the lower conductor to equalize the resonant frequencies of the upper and lower radiators. The design of a di-monopole hybrid LPA must be done very carefully because unequal impedances between upper and lower radiators can cause unbalanced currents to flow along the balanced transposed feedline in the antenna. Such in-phase common-mode currents radiate overhead and can cause unwanted antenna resonances.

Extended-Aperture Log-Periodic Antenna

The principal advantage of a vertically polarized LPA is that it has very wide azimuthal beamwidth but high directivity. Its gain can be increased without decreasing its azimuthal beamwidth if the vertical aperture is extended by lengthening the radiators. However, if the radiator lengths are made longer than a half wavelength, there will be nulls in the current distribution along the active radiators which will produce undesirable radiation-pattern nulls and unacceptable variations in antenna impedance. Increased vertical aperture is effective only if steps are taken to maintain the current distribution characteristic of a half-wave dipole (or quarter-wave monopole) as the length of the radiator is increased. This can be achieved by use of the *extended-aperture technique*,[18] which is illustrated in Fig. 26-17 for the case of a single dipole. Capacitors are inserted in series with the dipole limbs so that the current distribution never goes to zero except at the radiator tips. An extended-aperture LPA using full-wave dipoles has about 2 to 3 dB more gain than a half-wave dipole LPA.

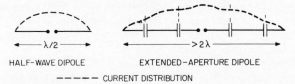

HALF-WAVE DIPOLE   EXTENDED–APERTURE DIPOLE

– – – – – CURRENT DISTRIBUTION

FIG. 26-17 Extended-aperture technique for eliminating current-distribution nulls on dipoles longer than $\lambda/2$. *(After Ref. 18.)*

Radiation-Pattern Ground Screens

The gain of a vertically polarized HF antenna is reduced and its TOA increased by the finite conductivity of the ground. To achieve maximum gain and the lowest possible TOA, a radiation-pattern ground screen extending several wavelengths in front of the antenna is necessary. The performance of a vertically polarized antenna depends on the linear dimensions of the ground screen and the conductivity of the ground. Practical ground screens are constructed by laying closely spaced wires on the earth. The wires of a ground screen must be dense enough so that the screen presents a very low resistive and reactive surface impedance. If a reactive impedance is presented, the radiation pattern is distorted because a quasi-trapped wave may be excited along the ground screen.[19]

A ground screen should not be buried more than a ̣w centimeters beneath the earth, nor should it be allowed to be covered by snow. Burial or snow cover can com-

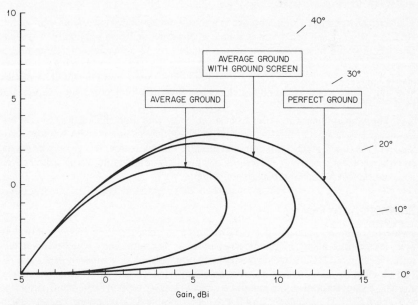

Gain, dBi

FIG. 26-18 Elevation pattern of a vertically polarized extended-aperture LPA at 10 MHz.

pletely mitigate the effectiveness of a ground screen and in some instances exacerbate the deleterious effects of the losses in the earth. There are no simple formulas for determining the radiation pattern of a vertical antenna with a ground screen. However, most of the available wire-antenna-analysis computer programs can calculate radiation patterns for antennas over lossy earth both with and without a polygonal ground screen of arbitrary size. Figure 26-18 shows the computed elevation pattern of an extended-aperture vertically polarized LPA (VLPA) over perfectly conducting ground and over ground of average conductivity, with and without a large ideal ground screen. Even with a large ideal ground screen there is a significant loss of gain owing to ground losses. Similar results are obtained for all vertically polarized antennas.

Impedance Ground Screens

For a monopole antenna, a ground screen is essential to obtain an input impedance giving an acceptable VSWR and to minimize conductive losses in the earth underneath the antenna. The input impedance of a vertical dipole is not affected by finite earth conductivity, so an impedance ground screen is not required. Impedance ground screens generally enclose an area extending no further than about $\lambda/4$ (at the lowest operating frequency) from the antenna.[20]

26-5   OTHER TYPES OF COMMUNICATIONS ANTENNAS

Vertical Whip

The vertically polarized whip antenna is a monopole 2 to 3 m long and of small diameter. It is widely used because it is simple to construct and easy to erect on a vehicle. Because the whip has narrow impedance bandwidth, an adjustable matching unit is required to tune the whip at each frequency. The less expensive matching units are preset to several frequencies. More expensive units tune automatically. The power-handling capacity of a whip and matching unit is usually limited to 1 kW average and PEP.

The whip antenna can suffer from severely low radiation efficiencies at lower HF frequencies because its radiation resistance is very low compared with the loss resistances of the tuner, whip conductor, and ground. The whip ground is usually a vehicle, and the resulting radiation pattern is not precisely predictable but is usually well behaved enough for adequate communication.

Rhombic Antenna

In the past, the horizontally polarized rhombic antenna (see Sec. 11-3) was widely used for point-to-point communication links. It has now been almost completely superseded by the HLPA, which is smaller in size with equal or superior performance. The rhombic is a large traveling-wave antenna which is terminated at its far end by a resistive load. Radiation efficiency of a typical rhombic is 60 to 80 percent, compared with more than 95 percent for an HLPA. The rhombic behaves approximately like a matched transmission line, and this results in a nearly constant input impedance over a very wide bandwidth. However, the gain and radiation pattern vary significantly with frequency, and this reduces the useful operational bandwidth to about 2:1 (one octave). The HF band covers a frequency range of up to 15:1, so three or four rhombics are required to give complete frequency coverage. In contrast, a single HLPA can be designed to cover the whole HF band.

Fan Dipole

A simple form of wideband transmitting antenna is a horizontal dipole in the form of a fan as illustrated in Fig. 26-19. This antenna is a broadband radiator based on the same principle as the inverted-cone monopole. With careful design it is possible to achieve a VSWR of less than 2.5:1 over the whole HF band from 2 to 30 MHz. Since the height above ground (25.9 m) is fixed, the vertical radiation pattern varies with frequency. At low frequencies there is a broad vertical lobe with a high TOA suitable for short-distance transmission. As the frequency increases, the TOA decreases and the antenna becomes suitable for transmission over longer distances. The azimuth radiation pattern at low frequencies and high elevation angles is nearly omnidirectional. This is advantageous when communication with a number of short-range targets spaced randomly in azimuthal bearing is required.

For transportable applications, it is possible to use a single-mast version (height, 12.2 m) of the fan dipole as illustrated in Fig. 26-20, which is closer to the ground than the conventional fan dipole. The lower height increases the negative mutual coupling to the ground, which results in decreased radiation resistance, particularly at low frequencies. To compensate for this, it is necessary to connect resistive loads between the ends of the fan and the ground. The resistive loading yields a low VSWR over the whole HF band but results in a loss of gain of 5 to 13 dB, the greatest loss occurring at the low frequencies.

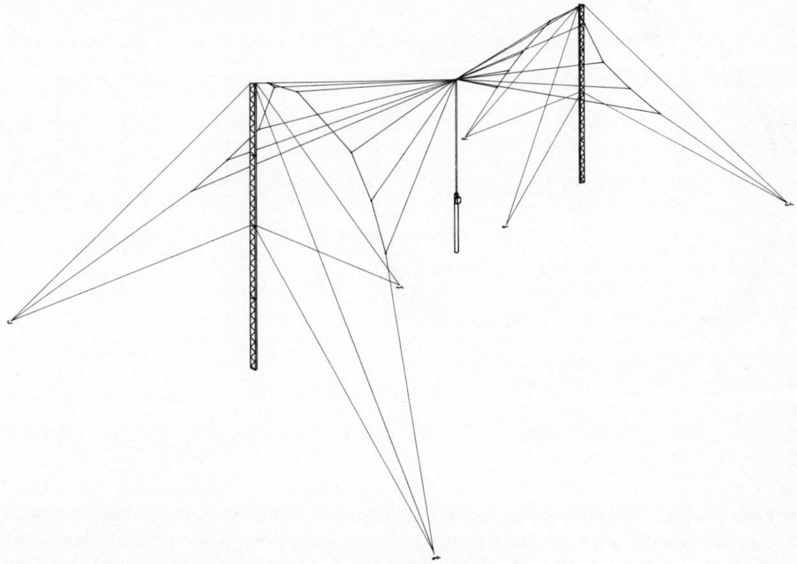

FIG. 26-19 Broadband fan dipole. *(Courtesy of TCI.)*

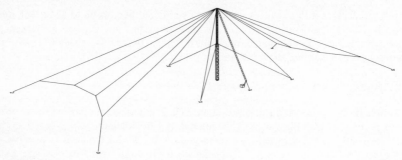

FIG. 26-20 Transportable fan dipole. *(Courtesy of TCI.)*

Transmitting Loop

When space is very restricted and an unobtrusive antenna is essential, it is possible to use a small-loop transmitting antenna. A small vertical loop in free space has radiation characteristics independent of frequency, and as is shown in Fig. 26-21 *a*, the azimuth radiation pattern becomes more circular with increasing elevation angle. Thus, unlike the whip and other monopole antennas, this antenna has no vertical null to militate against short-range communication. In practice, the loop must be mounted on the ground or possibly on the roof of a building. The performance of the free-space loop can be closely approximated by a half loop mounted on a conducting plane as shown in Fig. 26-21 *b*. The radiation resistance of the loop is proportional to $A^2 f^4$, where A

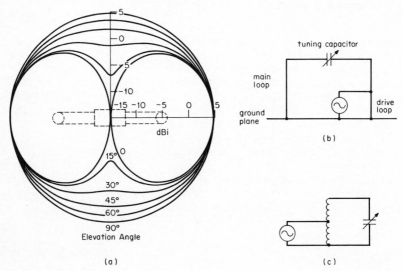

FIG. 26-21 Vertically polarized transmitting loop antenna. (*a*) Azimuth radiation patterns at different elevation angles. (*b*) Half loop mounted on ground plane. (*c*) Equivalent circuit of half loop on ground plane.

is the area of the loop and f is the frequency. For small loops, therefore, the radiation resistance is very low at low frequencies. The minimum acceptable size of the loop is determined by the need to achieve a bandwidth of around 2 kHz at the lowest operating frequency, which can be 3 MHz for many applications. This requirement can be met by a loop 1 m high and 2 m wide—a small and compact device. The loop has a narrow impedance bandwidth and must be tuned at each operating frequency, as illustrated in Fig. 26-21 c. A variable vacuum capacitor tunes the main radiating loop to resonance. An autotransformer, comprising a small drive loop coupled into the main radiating loop, transforms the resulting resistance to a value that maintains a VSWR of less than 2:1 over a frequency band of 3 to 24 MHz. It is possible to do this tuning automatically, using, for example, a microprocessor-based system which (1) measures the transmitter frequency, (2) monitors the current in the drive and main loops, (3) sets the capacitor to an appropriate value, using a stepping motor, and (4) determines the capacitor's position by means of a variable resistor mounted on its shaft.

The small radiation resistance of the loop at low frequencies makes it inevitable that resistive losses cause a reduction in radiation efficiency. Losses are minimized by using a high-Q vacuum capacitor and by ensuring that the radio-frequency (RF) paths in the loop and ground screen have very high conductivity. At 3 MHz the radiation efficiency of a well-designed and carefully constructed loop will be about 5 percent. Because the radiation resistance of the loop increases with the fourth power of frequency but the resistive losses increase with the square root of frequency, efficiency rises rapidly with increasing frequency, achieving a value of about 50 percent at 10 MHz and 90 percent above 18 MHz.

26-6 HIGH-FREQUENCY RECEIVING ANTENNAS

For a passive linear antenna (one without nonlinear elements or unidirectional amplifiers) the gain, radiation pattern, and impedance are the same for transmission and reception.[21] This is a consequence of the principle of reciprocity, which applies to all passive linear four-pole networks. However, the relative importance of these antenna parameters is different for transmission and reception. For transmission it is important to maximize field strength in the target area, so the antenna should be as efficient as possible. For reception the most important parameter is the signal-to-noise ratio (SNR) at the receiver output terminals, which obviously should be as large as possible. The SNR at the receiver output depends on the gain and mismatch losses of the antenna and feeder, the internal noise generated by the receiver, and the external atmospheric, human-made, and galactic noise levels. Because the external noise level at HF is often much larger than the internal noise levels in the receiver and antenna, it is possible in these instances to obtain acceptable SNR by using inefficient receiving antennas.

Noise Figure

For a simple analysis the receiver can be assumed to be linear, and the output SNR therefore equal to the input SNR. In calculating the output SNR, the signal and noise powers can therefore be referred to the receiver input and the available noise power at the output of an HF receiver taken to be

$$n_{\text{out}} = \text{onf} \cdot kT_0 b \qquad\qquad \textbf{(26-9)}$$

where n_{out} = available noise power, W
    onf = operating noise figure
    k = Boltzmann's constant, 1.38×10^{-23} J/K
    T_0 = reference temperature, 288 K
    b = effective noise bandwidth of receiver, Hz

The operating noise figure[22] (or factor) is a power ratio which accounts for the total internal and external noise of the receiving system and is given by

$$\text{onf} = f_e - 1 + f_a f_t f_r \qquad\qquad \textbf{(26-10)}$$

where f_e = total external noise power divided by $kT_0 b$
    f_a = antenna noise figure
    f_t = transmission-line noise figure
    f_r = receiver noise figure

The product $f_t f_r$ may be replaced by a single noise figure f_{rt} of a generalized receiving system containing additional devices such as filters, preamplifiers, and power dividers and combiners. In the rest of this section, as is common practice, the decibel equivalent of these noise figures will be denoted by capital letters.

The external noise figure F_e is the level of atmospheric, human-made, and galactic noise in a 1-Hz bandwidth in decibels above kT_0, where $10 \log_{10} kT_0 = -204$ dBW. Atmospheric noise is generated by lightning discharges which produce enormous amounts of HF energy that propagate great distances. Human-made noise is produced by various kinds of electrical equipment such as motors, vehicular ignition systems, electric welders, and fluorescent lamps. Galactic noise is generated by extraterrestrial radio sources. F_e can be determined from published noise maps[22] or by direct measurement by using a calibrated receiver and antenna. Figure 26-22 shows a typical example of the frequency variation of atmospheric noise, which usually drops from a high value at low frequencies to a low value at the upper end of the HF band. Atmospheric noise level also varies with geographical location, season, and time of day. Human-made noise level depends on whether the receiving site is in an urban, suburban, or rural location; in urban areas it can equal or exceed the atmospheric noise level.

The receiver noise figure F_r is usually obtainable from the receiver's specifications table. The transmission-line noise figure F_t is the reciprocal of the attenuation, expressed in decibels. For systems with filters, preamplifiers, etc., the noise figure F_{rt} can be determined by computing the noise figure for the equivalent cascaded network.

The antenna noise figure F_a is given by

$$F_a = D - G + L_M \qquad\qquad \textbf{(26-11)}$$

where D = directivity, dBi
    G = gain, dBi
    L_M = mismatch loss = $10 \log_{10} (1 - |\rho|^2)$, dB

The mismatch loss is given in terms of the reflection coefficient ρ at the input terminals of the antenna, where $|\rho| = (\text{VSWR} - 1)/(\text{VSWR} + 1)$. The quantity $D - G$ can be expressed in terms of the efficiency η by using the relation $D - G = -10 \log_{10} \eta$. The efficiency must include the resistive losses of both the antenna and the sur-

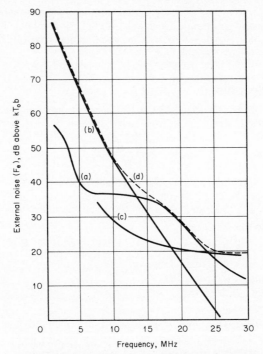

FIG. 26-22 Example of an external atmospheric and galactic noise figure. (*a*) Atmospheric daytime noise. (*b*) Atmospheric nighttime noise. (*c*) Galactic noise. (*d*) Worst noise condition.

rounding ground. For horizontally polarized antennas the ground losses are negligible. For vertical antennas ground losses may be significant, so they should be considered when there is no ground screen and earth conductivity is low.

Resistively loaded antennas, such as the small fan dipoles discussed in Sec. 26-5, usually have a VSWR of less than 2.5:1, so the mismatch loss is less than 1 dB. For these antennas the noise figure is determined primarily by the efficiency. For small nonresonant loop antennas, which will be discussed later, the mismatch loss is extremely large because the antenna has an extremely small input resistance and much larger reactance. For small loops both mismatch loss and low efficiency contribute to the noise figure. For small loops, monopoles, and dipoles the mismatch loss decreases by approximately 12-dB-per-octave increase in frequency.

Equation (26-10) has two important limiting cases. When $f_e \gg f_a f_t f_r$, the receiving system is said to be externally noise-limited because the external noise exceeds the internal noise. In this case the system performance cannot be improved by reducing antenna, feeder, or receiver noise figures. When $f_e \ll f_a f_t f_r$, the system is internally noise-limited, internal noise being greater than external noise. In this case system performance is improved by reducing the antenna or equipment noise figures.

In determining whether a receiving system is internally or externally noise-limited, it is important that actual external noise data be used. The values shown in Fig.

26-22 are for illustrative purposes only, actual variations being too extensive to display in one curve. In many geographical locations atmospheric and human-made noise levels are extremely low, and therefore receiving systems must have very low noise figures to avoid being internally noise-limited.

Signal-to-Noise Ratio

The SNR at the receiver output is the ratio s_{out}/n_{out} of the available signal and noise powers. The available signal power is given by

$$s_{out} = \frac{2.08e^2\lambda^2 d 10^{-15}}{\pi^2} \quad \text{W} \qquad \textbf{(26-12)}$$

where e = field strength of arriving signal, $\mu V/m$
d = antenna directivity as a power ratio
Dividing this by Eq. (26-9) and expressing the quantities in decibels gives the following equation for the SNR:

$$\text{SNR} = E + D - F - \text{ONF} - B + 97 \text{ dB} \qquad \textbf{(26-13)}$$

where $E = 20 \log_{10} e$, dBμ (dB relative to $1\mu V/m$)
D = antenna directivity, dBi
$F = 20 \log_{10} f$ (in MHz)
$\text{ONF} = 10 \log_{10} (\text{onf})$, dB
$B = 10 \log_{10} b$ (in Hz)

Practical Receiving Antennas

HF receiving antennas are divided into two classes. One class comprises antennas that are designed for transmission but are also used for reception. These antennas are very efficient, and their noise figures are often less than 3 dB, making them suitable for use at very quiet receiving sites. Full-sized vertically polarized antennas, when used for reception, do not require ground screens unless they are used in an emitter-locating system in which terrain variations must be neutralized by placing a metallic ground mesh under and in front of the antennas. The absence of a ground screen narrows the vertical beamwidth of the radiation pattern, thereby increasing directivity. However, ground losses reduce the gain by 5 to 10 dB and thus increase the antenna noise figure by the same amount. The resulting overall system noise level is nevertheless lower than most external noise levels, so a vertical antenna without a ground screen makes a highly suitable receiving antenna for many locations.

The second class of antennas consists of those specifically designed for reception. Again, because a high HF receiving-antenna noise figure is usually acceptable, a receive-only antenna can be inefficient and have high VSWR. Consequently, it can be small and, if necessary, be loaded resistively to improve its performance.

Typical acceptable values of antenna noise figure are given in the following illustrative analysis of a radioteletype receiving system. Assume that signal strength E is 34 dBμ (50 $\mu V/m$) and antenna directivity is 5 dBi. The required SNR for moderate character error rate is 56 dB in a 1-Hz bandwidth. The receiver noise figure F_r is 13 dB, and feeder noise figure F_t is 3 dB, both reasonable values for HF devices. The maximum allowable ONF is obtained by substituting E, D, and SNR in Eq. (26-13). The corresponding allowable antenna noise figure F_a is obtained by substituting the power ratios on f, f_e, f_r, and f_t into Eq. (26-10). In this example the external noise level F_e is assumed to be given by the dotted curve in Fig. 26-22. The accompanying table summarizes the results for several frequencies:

f, MHz	ONF, dB	F_e, dB	F_a, dB
2	74	80	· · ·
5	66	66	· · ·
10	60	46	44
15	56	37	40
20	54	29	38
30	50	19	34

Below 5 MHz the required SNR will not be obtained because the external noise level is too high. This can be overcome only by moving to a quieter site; improving equipment noise figures has no effect. Above 5 MHz the external noise level is smaller, and the required SNR can be obtained by using an antenna with a maximum noise figure at each frequency as given in the table. If the antenna VSWR is low so that the mismatch loss can be neglected, allowable antenna efficiency ranges from 0.004 percent at 10 MHz to 0.04 percent at 30 MHz.

Receiving Loops

A practical antenna having a noise figure near the upper limit allowable for most situations is the small vertical balanced loop. A typical loop is about 1.5 m in diameter and is mounted about 2 m above the ground on either fixed posts or transportable tripods. The antenna is usually made from a large-diameter metallic tube, which also serves to shield the feeder that runs through the tube to the feed point. The vertical radiation pattern of the loop has good response at both low and high TOAs, making the antenna suitable for both long- and short-distance reception of sky-wave signals. The loop also responds well to ground waves.

The noise figure of the loop depends strongly on its area and the operating frequency. A loop of 1.5-m diameter has a noise figure of about 50 dB at 2 MHz and 20 dB at 30 MHz. The noise figure can be reduced by making the area of the loop larger. However, if the loop perimeter is about a wavelength long, the loop will resonate and the radiation pattern will exhibit irregularities. These can be avoided in large loops by feeding the loop at several points.

A number of loops can be arrayed to provide a variety of directional radiation patterns. In most arrangements of small loops, mutual coupling between the loops is negligible and simple array theory can be used to predict radiation-pattern shape. Beams are formed by bringing the coaxial cable from each loop into a beam-forming unit located near the antenna. The most common arrangement places eight loops in an end-fire array with spacing between loops of 4 m, as shown in Fig. 26-23. The array can be made bidirectional by splitting the signal from each loop two ways and feeding each half into a separate beam former. Four such bidirectional arrays can be arranged in a circular rosette which provides eight independent beams covering 360° in azimuth.

FIG. 26-23   End-fire array of eight receiving loops. *(Courtesy of TCI.)*

Active Antennas

Many receiving antennas, including loop arrays, are made into *active antennas* by interposing a broadband low-noise amplifier between the output of the antenna or its beam former and the feeder line to the receiver. For some antennas the amplifier can improve the impedance match and thus reduce mismatch loss and antenna noise figure. However, for electrically small antennas little improvement in noise figure is possible with a practical amplifier.[23] The amplifier also sets the antenna-system noise figure at the point where the amplifier is placed. The amplifier has a beneficial effect in this case only if its noise figure (usually 1 to 5 dB) is less than the feeder noise figure and its gain (usually 10 to 30 dB) is much higher. The amplifier can have a deleterious effect on receiving-system performance if high ambient signal levels cause it to produce spurious intermodulation products. These are likely to occur if there are strong medium-frequency (MF), HF, or very-high-frequency (VHF) signals from nearby transmitting antennas. It is a simple matter to filter out the MF and VHF signals by using, respectively, high- and low-pass filters ahead of the amplifier. Unwanted HF signals must be rejected by using notch filters so as not to impair the usefulness of the antenna in the HF receiving system.

The availability of cheap microprocessors has made it possible to use active antenna elements as the basis of adaptive-antenna arrays (see Chap. 22). In an adaptive system the amplitudes and phases of the signals from a multielement array are adjusted continually to give a radiation pattern maximizing the wanted signal and minimizing interfering cochannel and adjacent-channel signals. Optimum weightings of component signals must be determined from a careful study of the actual signals with which the antenna must contend. These studies are not straightforward, and designing a suitable algorithm is difficult in many cases.

Circular Arrays

Arrays of circularly disposed antenna elements are used in monitoring or emitter-locating systems (see Chap. 39). In these applications there is quite often a need to receive signals from all azimuth angles and to be able to connect a number of receivers to any of a number of beams. Antennas of this type must operate over a large part of the HF band and must have highly predictable radiation patterns with low sidelobe levels. It is often advantageous to make each element of the array an LPA. Vertical LPAs should be used if the array must respond to ground waves or to low-TOA sky waves. Horizontal LPAs, placed close to ground, should be used if high-TOA sky-wave signals from short distances are to be received.

A typical array comprises 18 to 36 elements. The output of each element is fed into a beam former, containing suitable delay lines and power splitters, which forms a narrow azimuthal beam. It is possible to generate N beams simultaneously in an N-element array by splitting the power from each element N ways. The N beams are uniformly spaced around the circle. By using multicouplers, each beam can be accessed by several receivers.

In arrays of LPAs it is best to point the elements inward so that their direction of maximum sensitivity is toward the array center. This configuration keeps the electrical diameter of the array nearly constant, thereby making the azimuthal beamwidths nearly independent of frequency and eliminating unwanted azimuthal grating lobes. The disadvantage of the inward-looking array is that each element "fires through" those opposite, and this complicates the calculation of the radiation pattern.

26-7 BROADBAND DIPOLE CURTAIN ARRAYS

Broadband dipole curtain arrays are used for high-power (100- to 500-kW) HF ionospheric broadcasting. These antennas consist of square or rectangular arrays of dipoles, usually a half wavelength long, mounted in front of a reflecting screen, as shown in Fig. 26-24. Dipole arrays have many excellent performance characteristics, such as independent selectability of vertical and horizontal patterns, high power gain, slewability of beam in azimuth or elevation, wide impedance bandwidth, low VSWR, and high power-handling capacity, which have made them virtually the standard antenna at short-wave broadcasting stations.

Standard Nomenclature

Dipole curtain arrays are described by the internationally agreed nomenclature HRRS $m/n/h$,

where H denotes horizontal polarization

R denotes a reflector curtain

R (if not omitted) denotes that the direction of radiation is reversible

S (if not omitted) denotes that the beam is slewable

m is the width of the horizontal aperture in half wavelengths at the design frequency

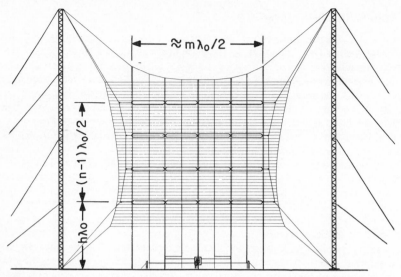

FIG. 26-24   Basic HF dipole curtain array. (Antenna shown is designated HRS 4/4/h.)

 n is the number of the dipoles in a vertical stack
 h is the height of the lowest dipole above ground, in wavelengths at the design
    frequency

The design frequency is $f_0 = \frac{1}{2}(f_1 + f_2)$, where f_1 and f_2 are the lower and upper frequency limits; λ_0 is the corresponding design wavelength. In dipole arrays which use half-wave dipoles, the width parameter m is the number of dipoles in a horizontal row.

Radiation Pattern and Gain

The TOA and first null at frequency f depend on the average height of the dipoles in the vertical stack and are given by

$$\text{TOA} = \sin^{-1}\left(f_0\lambda_0/4fH_{avg}\right) \qquad (26\text{-}14)$$

$$\text{Null} = \sin^{-1}\left(f_0\lambda_0/2fH_{avg}\right) \qquad (26\text{-}15)$$

where $H_{avg} = (H_1 + H_2 + \cdots + H_n)/n$ and $H_1, H_2, \ldots, H_n$ are the physical heights above ground of the n dipoles in the stack. The TOAs at f_0 are given for several configurations in Table 26-1. The level of the unwanted minor elevation lobes is determined by the number of dipoles in the stack and their spacing.

   The azimuthal half-power beamwidth (HPBW) depends primarily on the width of the array but also depends weakly on the TOA. The beamwidth at f_0 is given in Table 26-2. Beamwidth at frequency f is obtained approximately by multiplying the value in this table by f_0/f.

   The gain at f_0 of an array of half-wave dipoles is given in Table 26-3. For other configurations the gain is given approximately by $G = \log_{10}(27{,}000/A_1A_2)$, where A_1

TABLE 26-1 Takeoff Angle of Dipole Array with Reflecting Screen

Number of elements in vertical stack n, half-wavelength spacing	Height above ground of lowest element in wavelengths, h			
	0.25	0.5	0.75	1.0
1	45°*	29°	19°	15 and 48°†
2	22°	17°	14°	11°
3	15°	12°	10°	9°
4	11°	10°	8°	7°
5	9°	8°	7°	6°
6	7°	7°	6°	5°

*90° without reflector.

†Two lobes present.

and A_2 are the vertical and horizontal beamwidths in degrees. The approximate gain at frequencies different from f_0 can be obtained by adding 20 log (f/f_0) to the values in Table 26-3.[24]

Slewing

Early forms of dipole arrays used full-wave dipoles as shown in Fig. 26-25. These antennas contained a very-narrow-bandwidth feed system and thin dipoles, so they were capable of operating only in one or two broadcast bands. The azimuthal beam of the antenna was slewed by using a tapped feeding arrangement. This sets up a phase difference between the two halves of the array, which slews the beam up to $\pm 10°$ in azimuth. The slew angle is limited to a small value because the dipole centers are separated by one wavelength. If greater slew angles are attempted, the horizontal pattern contains large secondary lobes which reduce gain by up to 3 dB and may interfere with cochannel transmissions.

TABLE 26-2 Horizontal Beamwidth of Dipole Array*

Number of elements in vertical stack n, half-wavelength spacing	Number of half-wave elements wide m		
	1	2	4
1	76°	54°	26°
2	74°	50°	24°
3	74°	49°	24°
4	73°	49°	24°

*Between half-power points.

TABLE 26-3 Gain in dBi of Dipole Array with Perfect Reflecting Screen Spaced 0.25λ behind Dipoles

Number of elements in vertical stack n, half-wavelength spacing	Number of half-wave elements wide m															
	1				2				3				4			
	Height above ground of lowest element in wavelengths, h															
	0.25	0.50	0.75	1.0	0.25	0.50	0.75	1.0	0.25	0.50	0.75	1.0	0.25	0.50	0.75	1.0
1	12.5	13.2	14.0	13.6	13.5	14.4	15.3	14.8	15.4	16.4	17.1	16.8	16.1	17.2	17.9	17.5
2	14.0	15.0	15.6	15.8	15.4	16.5	17.0	17.2	17.5	18.4	19.0	19.2	18.2	19.3	19.8	20.0
3	15.6	16.4	16.9	17.2	17.1	17.9	18.4	18.7	19.0	19.8	20.3	19.6	19.8	20.7	21.2	21.5
4	16.7	17.3	17.9	18.2	18.2	18.9	19.4	19.7	20.1	20.8	21.3	21.6	21.0	21.7	22.2	22.5

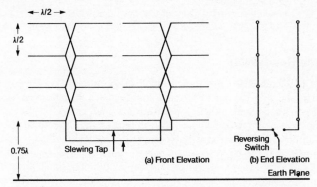

FIG. 26-25   Early form of narrowband HRS 4 / 4 / .75 dipole array using full-wave dipoles.

To achieve larger slew angles it is necessary to reduce the horizontal spacing between the dipole stacks. In these arrays the dipole length is slightly less than $\lambda_0/2$, and the spacing between the stacks is $\lambda_0/2$. For these antennas the slewing system shown in Fig. 26-26 enables slew angles to be as large as $\pm 30°$. It is possible, in principle, to slew any dipole array with two or more stacks. In practice, two-wide arrays do not slew effectively over wide frequency ranges because of unfavorable mutual impedances between dipoles. However, four-wide arrays slew very well, and the slew has no deleterious effect on either gain or impedance.

Dipole arrays can also be slewed vertically by introducing phase shifts between vertical elements in each stack. This is usually done by introducing a 180° phase reversal between pairs of elements. Vertical slewing systems are more complicated

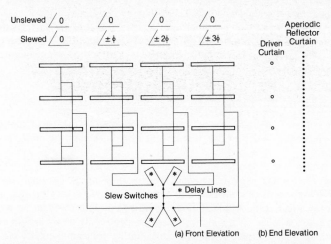

FIG. 26-26   Slewing system and corporate feed of a broadband HRS 4 / 4 / *h* dipole array. Dipole length = 0.46 λ_0, dipole spacing = 0.50 λ_0 center to center, and screen-to-dipole spacing = 0.25 λ_0.

than horizontal slewing systems. The impedance bandwidth of a vertically slewed antenna is generally smaller than that of a horizontally slewed array.

Impedance Bandwidth

The VSWR of a high-power broadband dipole array must be low for two important reasons. First, the voltages on the antenna and feeder lines become excessive if the VSWR is too high. Second, although most modern high-power HF transmitters will accept a VSWR of 2:1, the feeder line and switching systems introduce discontinuities so that the antenna itself must usually have a VSWR of less than 1.6:1 or 1.5:1. Low VSWR requires that the input impedance of the antenna must remain substantially constant over its entire frequency range. The bandwidth of a dipole array depends on the inherent bandwidth of the dipoles, the bandwidth of the interdipole feeder system, and the type of reflector placed behind the dipole array.

The inherent dipole bandwidth, by analogy with simple resistance-inductance-capacitance (RLC) resonant circuits, can be characterized by a radiation $Q = f_R/\Delta f$, where f_R is the resonant frequency of the dipole and Δf is the bandwidth between the frequencies at which dipole resistance and reactance are equal. In the broadband dipole array, a very low Q of about 2 is achieved by making the dipole fat. The thin dipoles used in early arrays have Q's of about 10.

The feed system which interconnects the dipoles is made broadband by constructing it in the form of a *corporate feed,* as shown in Fig. 26-26. Wide-bandwidth transmission-line transformers, often of the optimal Chebyshev type, are used whenever impedance transformations are necessary.[25]

Early dipole arrays used parasitic dipoles in the reflector curtain. Such antennas have very narrow bandwidth because the parasites act as reflectors over only a small range of frequencies, a phenomenon familiar to designers and users of Yagi-Uda arrays. Broadband dipole arrays use an aperiodic reflecting screen consisting of closely spaced horizontal wires placed about $0.25\lambda_0$ behind the dipoles.

A well-designed broadband dipole array has a VSWR of 1.5:1 or less over a one-octave bandwidth (i.e., $f_2/f_1 = 2$). This bandwidth permits a single antenna to operate over four or five adjacent HF broadcast bands, in contrast to the one or two bands of early dipole arrays.

Practical Considerations

The design of modern dipole arrays is greatly facilitated by the use of comprehensive computer programs which enable the current distribution on the entire array to be analyzed.

To achieve low dipole Q, the half-wave dipoles are constructed as fat multiple-wire cages which are either flat, rectangular boxes or cylinders. The dipoles are usually "folded" to step up their impedance to a more useful level and provide additional impedance compensation. A folded dipole operates simultaneously in two modes. The radiating antenna mode, or unbalanced mode, depends only on the length and equivalent diameter of the dipole cage. The nonradiating transmission-line, or balanced, mode describes the currents which flow in a loop around the dipole and through the familiar short circuits at the end of the folded dipole. The impedance variations of the two modes tend to cancel, so the transmission-line mode can be used to compensate and tune the antenna mode.

This compensating effect is maximized by moving the position of the folded-dipole short circuit away from the end of the dipole and toward the feed point. The optimum position can be determined in a straightforward manner by using an antenna-analysis computer program.

The driving-point impedance of each folded dipole is about 600 Ω when mutual impedances from other dipoles are included. This allows the impedance levels in the branch feeder to be in the range 300 to 600 Ω, which is easy to realize in practice.

HRS 4/4 arrays can be designed to have a power-handling capacity of 750 kW average and 4000 kW PEP, which is large enough to accommodate two fully modulated 250-kW transmitters fed into the antenna simultaneously by using a diplexer. Smaller dipole arrays, such as HR 2/2, can handle up to 750 kW average and 2000 kW PEP, corresponding to the power output of a fully modulated 500-kW transmitter. The wire diameters in the dipole array must be large enough to carry the high currents which flow and to prevent corona discharge. Insulators must be carefully chosen to withstand voltages of up to 45 kV rms, which can occur in the high-voltage points in the antenna.

The one-octave bandwidth of a dipole array allows three such antennas, 6/7/9/11, 9/11/13/15/17, and 13/15/17/21/26 MHz, to cover the whole international broadcast spectrum, with multiple-antenna coverage of many of the bands.

26-8   SITING CRITERIA FOR HIGH-FREQUENCY ANTENNAS

In the preceding sections it has been assumed that the antenna is mounted over a smooth, level ground plane of infinite extent. However, an actual antenna site is seldom smooth and level, and sites which approximate ideal conditions usually do so only over a limited area. In addition, the effects of co-sited antennas and of natural or artificial obstructions can result in degradation of antenna performance. This section assesses the effects of imperfect sites following closely Refs. 26 and 27, and presents practical criteria for obtaining acceptable performance.

Fresnel Zone and Formation of Antenna Beam

The main lobe of a transmitting or receiving antenna (the following discussion applies to both) is formed by the interaction of the direct radiation of the antenna and the reflected radiation from the ground plane. Reflections occurring near the antenna are of greater importance than those occurring far away. With a directional antenna, radiation at or near boresight angles is more important than that occurring at azimuth angles away from boresight. From these facts, it follows that there is an elliptically shaped area (with the major axis in the direction of the main lobe, as shown in Fig. 26-27) in which the ground must be level and smooth if the first lobe is to be formed without appreciable distortion.

From simple geometrical ray theory it is clear that as the TOA is decreased, the lengths of the major and minor axes of the elliptical area increase. The curvature of the earth also affects the size of the axes, but it can be ignored if the TOA is more than 3°

The ellipse dimensions are different for vertical and horizontal polarization and

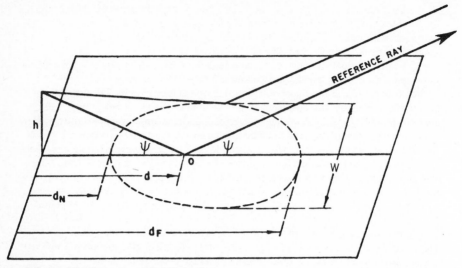

FIG. 26-27   Geometry of first Fresnel zone.

in the former case are difficult to evaluate. Fortunately, the dimensions for horizontal polarization, which are easy to evaluate, give a reasonable approximation for the vertically polarized case. The ellipse dimensions correspond to the first Fresnel zone, which is the region in front of the antenna in which direct and reflected radiation differ in phase by 180° or less. For horizontal polarization, if the antenna has a TOA of ψ_0, the radiation will appear to come from a height above ground of

$$h = \lambda/4 \sin \psi_0 \qquad \textbf{(26-16)}$$

The distance from this radiation point to the near and far edges of the first Fresnel zone, denoted by d_N and d_F respectively, are given by

$$d_N = \frac{h}{\tan \psi_0} \left(3 - \frac{2\sqrt{2}}{\cos \psi_0} \right) \qquad \textbf{(26-17)}$$

$$d_F = \frac{h}{\tan \psi_0} \left(3 + \frac{2\sqrt{2}}{\cos \psi_0} \right) \qquad \textbf{(26-18)}$$

The maximum width of the Fresnel ellipse is

$$w = 4\sqrt{2}h = 5.66h \qquad \textbf{(26-19)}$$

Equations (26-16) to (26-19) may be used to calculate, to a very good approximation, the area of flat, unobstructed land required in front of a directive antenna to ensure that the main lobe is fully formed. For very low TOA the Fresnel zone may extend for several kilometers, and it will not usually be possible to contain the zone within the boundaries of the antenna site. If the size of the controlled area is to be reduced, it is possible to limit it to the region in which the phase difference between direct and reflected radiation is 90° or less. In this case, there may be a loss of gain of up to 3 dB, but the dimensions of the ellipse will be reduced as follows:

d_F may be reduced to 0.6 times the full value.

d_N may be increased to 1.6 times the full value.

w may be reduced to 0.7 times the full value.

Roughness in First Fresnel Zone

In the zone with dimensions given in Eqs. (26-16) to (26-19), the main lobe is approximately fully formed provided the ground is flat and smooth. However, if the ground is rough or has natural or artificial obstructions or depressions, reflection will not be specular. The reflected wave will be scattered, and there will be a loss of gain and distortion of the beam. Rayleigh's criterion indicates that the transition between specular and scattered reflection occurs when the maximum height H of deviations above and below the average terrain profile does exceed

$$H = \lambda_0/16 \sin \psi_0 = h/4 \qquad \textbf{(26-20)}$$

This criterion may be relaxed if additional degradation in performance is acceptable. The relaxed criteria are based on the reasonable assumption that the permissible size of the obstruction may increase with distance from the antenna. It is usual to divide the first Fresnel zone into three regions, within which the heights of obstructions and depths of depressions should not exceed the values in the accompanying table.

Limit of departure from average smooth terrain	Region of first Fresnel zone
$H/4$	d_N to $0.2\ d_F$
$H/2$	$0.2\ d_F$ to $0.6\ d_F$
H	$0.6\ d_F$ to $1.0\ d_F$

It is difficult to predict with precision the degradation that occurs when the above criteria are applied, partly because the shapes of the obstructions have been ignored. However, it is reasonable to expect gain losses of several decibels. These criteria apply to cases in which the entire first Fresnel zone is rough. When a major portion of it is smooth, somewhat larger obstacles can be tolerated if they do not cover more than 5 to 10 percent of the zone. Caution must be exercised, however, when compromises in controlled-area size and roughness are made simultaneously because gain reductions may then become large.

Horizon Obstructions

Obstructions beyond the first Fresnel zone at distance $D_0 \gg d_F$ can reduce radiation at low angles even if the first Fresnel zone is perfectly smooth. If the height of the obstruction is H_0, antenna performance will not be noticeably degraded if the angle subtended by the obstruction as viewed from the antenna, $\psi_{obs} = \tan^{-1}(H_0/D_0)$, is less than 0.5 times the lower half-power point of the antenna elevation pattern. For TOAs of less than 30°, this criterion is given by $\psi_{obs} \leq \psi_0/4$.

Terrain Slope

If the ground at an antenna site slopes down in the direction of radiation by an angle b, the TOA is decreased from ψ_0 (the value for a flat site) to $\psi_0 - b$. Conversely, if the ground slopes up in the direction of radiation, the TOA becomes $\psi_0 + b$. Fresnel-zone theory can be applied to sloping sites by using the following modified antenna height factor:

$$h_{\text{slope}} = \lambda/4 \sin (\psi_0 \pm b) \qquad \textbf{(26-21)}$$

in Eqs. (26-16) to (26-19). The $+$ sign is used for upward slopes; the $-$ sign, for downward slopes.

Ground Conductivity

For vertically polarized antennas, high ground conductivity is important for the effective operation of transmitting antennas, as explained in Sec. 26-2. Vertical transmitting antennas mounted very near seawater, which has a high conductivity of $4\ S/m$, do not require ground screens for radiation-pattern enhancement. For most types of soil, conductivity is significantly smaller than this value, so metallic ground screens are necessary. For horizontally polarized antennas, good ground conductivity is not important unless the TOA is very high, as explained in Sec. 26-2.

Coupling between Co-Sited Antennas

When transmitting, an antenna transfers some of its radiated energy to any other antenna in its vicinity, and this transferred energy may affect the performance of the other antenna. *Antenna coupling* is defined as the ratio of the power delivered into a matched load terminating the other antenna (called for convenience the *receiving* antenna) and the input power to the *transmitting* antenna:

$$C = 10 \log_{10} (P_r/P_t) \qquad \text{dB} \qquad \textbf{(26-22)}$$

where P_r = power dissipated in matched load terminating the *receiving* antenna

P_t = input power to *transmitting* antenna

Coupling can be calculated accurately by solving the fundamental electromagnetic equations for the several antennas, using a comprehensive antenna-analysis computer program. However, good approximate coupling values can be obtained from two simple formulas, which assume that the antennas are separated by at least one wavelength λ. The magnitude of the coupling depends on the polarization of the antennas. For vertically polarized antennas the coupling factor is

$$C_v = 20 \log_{10} (\lambda/8\pi d) + G_t + G_r \qquad \text{dB} \qquad \textbf{(26-23)}$$

where d = spacing between antennas

G_t = gain of transmitting antenna, dBi

G_r = gain of receiving antenna, dBi

G_t and G_r are the gains of the two antennas in the directions toward each other. The formula neglects ground loss, which is generally negligible when d is less than a few kilometers. For horizontal polarization the coupling factor is

$$C_h = 20 \log_{10} \left[\frac{\lambda \sin (\pi L/\lambda)}{8\pi d} \right] + G_t + G_r \qquad \text{dB} \qquad \textbf{(26-24)}$$

where $L = [(H_t + H_r)^2 + d^2]^{1/2} - [(H_t - H_r)^2 + d^2]^{1/2}$
   $H_t = \lambda/4 \sin \psi_t$
   $H_r = \lambda/4 \sin \psi_r$
ψ_t and ψ_r are the TOAs of the transmitting and receiving antennas. λ, d, G_t, and G_r are defined in Eq. (26-23), except that G_t and G_r are the values at the TOAs ψ_t and ψ_r. For large values of d, C_h decreases by 12 dB each time that d is doubled.

Coupling between antennas can cause a number of effects, and one or more of the following may be relevant in a particular case.

Intermodulation   If both antennas are transmitting, the energy coupled by one into the other may give rise to intermodulation products in the transmitter connected to the *receiving* antenna. The amount of coupling that can be tolerated depends on the characteristics of the transmitter, but typical values are -20 to -25 dB.

VSWR   When energy from another antenna is coupled into a transmitting antenna, the VSWR presented to the transmitter will be changed because the apparent reflected power will be altered by the incoming energy. The reflection coefficient of the receiving antenna, ρ', has a maximum magnitude given by

$$\rho' = \left[\frac{\rho^2 P_r + C_r P_t}{P_r + C_r P_t} \right]^{1/2} \qquad \textbf{(26-25)}$$

where ρ = reflection coefficient of *receiving* antenna in absence of the other transmitter, as defined after Eq. (26-11)
   C_r = coupling factor between antennas expressed as a power ratio
P_t and P_r are the powers of the transmitters connected to the *transmitting* and *receiving* antennas. The apparent VSWR, in the worst case, is given by $V = (1 + \rho')/(1 - \rho')$.

Radiation-Pattern Distortion   Some of the power transferred from the *transmitting* antenna to the *receiving* antenna will be reradiated and may cause distortion of the *transmitting* antenna's radiation pattern. The amount of power reradiated depends on the terminating impedance of the *receiving* transmitter, which is the value presented by its output stage. In the worst case this power can be up to 4 times the coupled power, which is the power radiated by the *transmitting* antenna multiplied by the coupling factor expressed as a power ratio. However, in practice the reradiated power will seldom be this large, and it is reasonable to assume that it will be equal to the coupled power. The reradiated energy changes the gain of the *transmitting* antenna by the following approximate amount

$$\Delta G = \pm 20 \log_{10} [1 + (C_r g_r/g_t)^{1/2}] \qquad \text{dB} \qquad \textbf{(26-26)}$$

where g_r, g_t = maximum gains of *transmitting* and *receiving* antennas expressed as power ratios.
This formula gives only the minima and maxima of the pattern perturbation. The detailed shape of the perturbed pattern and the locations of the azimuth angles at

which the minima and maxima occur depend on the phase of the reflected energy and the relative positions of the antennas; they are difficult to calculate approximately.

Side-by-Side Antennas Horizontally polarized antennas with directive radiation patterns can be placed side by side, often using common towers, with a coupling factor of less than -20 dB being achieved. Equations (26-23) and (26-24) do not apply in this situation because the antennas are too close. Coupling must be calculated by using a comprehensive antenna-analysis computer program.

REFERENCES

1 K. Davies, *Ionospheric Radio Propagation,* Dover Publications, Inc., New York, 1966.
2 *C.C.I.R. Interim Method for Estimating Sky-Wave Field Strength and Transmission Loss at Frequencies between the Approximate Limits of 2 and 30 MHz,* CCIR Rep. 252-2, International Telecommunications Union, Geneva, 1970.
3 G. W. Haydon, M. Leftin, and R. K. Rosich, *Predicting the Performance of High Frequency Skywave Telecommunications Systems,* Rep. OT 76-102, Office of Telecommunications, U.S. Department of Commerce, September 1976.
4 J. L. Lloyd, G. W. Haydon, D. L. Lucas, and L. R. Teters, *Estimating the Performance of Telecommunication Systems Using the Ionospheric Transmission Channel,* draft report, Institute for Telecommunications Sciences, U.S. Department of Commerce, March 1978.
5 E. A. Laport, *Radio Antenna Engineering,* McGraw-Hill Book Company, New York, 1952.
6 C. B. Feldman, "Deviations of Short Radio Waves from the London–New York Great Circle Path," *IRE Proc.,* vol. 27, October 1939, p. 635.
7 R. L. Tanner and M. G. Andreasen, "Numerical Solution of Electromagnetic Problems," *IEEE Spectrum,* September 1967.
8 R. F. Harrington, *Field Computation by Moment Methods,* The Macmillan Company, New York, 1968.
9 G. A. Thiele, "Wire Antennas," in *Computer Techniques for Electromagnetics,* Pergamon Press, New York, 1973, chap. 2.
10 W. L. Stutzman and G. A. Thiele, *Antenna Theory and Design,* John A. Wiley & Sons, Inc., New York, 1982, chap. 7.
11 E. K. Miller et al., "Analysis of Wire Antennas in the Presence of a Conducting Half-Space. Part II: The Horizontal Antenna in Free Space," *Can. J. Phys.,* vol. 50, 1972, pp. 2614–2627.
12 E. C. Jordan and K. G. Balmain, *Electromagnetic Waves and Radiating Systems,* 2d ed., Prentice-Hall, Inc., Englewood Cliffs, N.J., 1968, pp. 630–644.
13 Ibid., Chap. 15.
14 Stutzman and Thiele, op. cit., pp. 287–303.
15 U.S. Patent 3,257,661.
16 H. D. Kennedy, "A New Approach to Omnidirectional High-Gain Wide-Band Antenna Design: The TCI Model 540," Tech. Note 7, Technology for Communications International, April 1979.
17 U.S. Patent 3,594,807.
18 U.S. Patent 3,618,110.
19 J. R. Wait, "On the Radiation from a Vertical Dipole with an Inductive Wire-Grid Ground System," *IEEE Trans. Antennas Propagat.,* vol. AP-18, July 1980, pp. 558–560.
20 W. L. Weeks, *Antenna Engineering,* McGraw-Hill Book Company, New York, 1968, pp. 44–52.
21 Stutzman and Thiele, op. cit., pp. 40–44.

22 *World Distribution and Characteristics of Atmospheric Radio Noise,* CCIR Rep. 322, International Telecommunications Union, Geneva, 1964.

23 R.A. Sinaiti, "Active Antenna Performance Limitation," *IEEE Trans. Antennas Propagat.,* vol. AP-30, November 1982, pp. 1265–1267.

24 Comité Consultatif International des Radiocommunications, *Antenna Diagrams,* International Telecommunications Union, Geneva, 1978.

25 G. L. Matthei, L. Young, and E. M. T. Jones, *Microwave Filters, Impedance-Matching Networks, and Coupling Structures,* McGraw-Hill Book Company, New York, 1964, chap. 6.

26 Comité Consultatif des Radiocommunications, *Handbook on High-Frequency Directional Antennae,* International Telecommunications Union, Geneva, 1966.

27 W. F. Utlaut, "Siting Criteria for HF Communications Centers," Tech. Note 139, National Bureau of Standards, April 1962.

Chapter 27

VHF and UHF Communication Antennas

Brian S. Collins

C & S Antennas, Ltd.

27-1 INTRODUCTION

The very-high-frequency (VHF) and ultrahigh-frequency (UHF) bands are used for private and public-access services carrying speech, data, and facsimile information. The ends of a link may be installed at fixed locations or in vehicles (including ships and aircraft) or may be carried in an operator's hand. The length of a link may vary from a few tens of meters up to the maximum over which reliable communication can be obtained. This wide variety of applications generates a need for many different types of antennas. In this chapter, we examine the selection of antennas to perform various tasks, together with aspects of reliability, siting, and economics.

27-2 SYSTEM-PLANNING OBJECTIVES

The design of antennas for a radio link must provide an adequate signal-to-noise ratio at the receiver. The necessary signal-to-noise ratio will depend on the nature of the information to be transmitted and the grade of service which is required. A power budget must be drawn up to determine the total antenna gain and input power needed in the system. The antenna engineer must decide how the necessary gain can be obtained and how it should be divided between the two ends of the link. In many point-to-point applications the most economical design is obtained by using transmitting and receiving antennas of equal gain. Mobile or portable stations do not generally allow the use of high-gain antennas, so as much gain as possible must be obtained from the base-station antenna.

The electromagnetic spectrum is a limited resource, and its use is controlled by restricting the field strength which may be laid down outside the area where communication is required. This often implies a limitation on the permitted effective radiated power (ERP) both inside and outside the main-beam direction of the transmitting antenna. The sensitivity of receiving antennas must be restricted in directions outside the main beam to prevent interference being caused by the reception of signals from stations other than that intended which use the same or an immediately adjacent frequency. The system designer and antenna engineer must acquaint themselves with the requirements of the regulatory authority (Federal Communications Commission or other government agency) to make sure that a new system will work without suffering or causing interference.

To allow the largest possible number of links to be established in a given geographical area on a particular frequency band, it is desirable that each station use a very low transmitter power together with a highly directive high-gain antenna. By this means the area over which any station lays down a field which may cause interference to others is limited. A specification template for a radiation pattern defines the minimum acceptable radiation-pattern performance for an antenna and leaves scope for the designer to decide how to achieve it.

Reliability

A communication link or system must provide an adequate level of reliability. A link may become unusable if the signal-to-noise ratio falls below the design level; it is important that the design objectives for a system specify the fraction of time for which this may occur. A downtime of 0.01 percent or even less may be necessary for a link to a lifesaving emergency service, but 1 percent downtime may be as little as can be economically justified for a radiotelephone in a boat used for leisure-time fishing.

Fading due to statistical fluctuations in the propagation path is usually guarded against by a fade margin in the power budget. In a severe case a diversity system may be used to reduce the impact of fading on system reliability. This takes advantage of low correlation between fading events over two physically separate paths, at two frequencies, or for two polarizations. Other important causes of system failure are given below.

Wind-Induced Mechanical Failures   The oscillating loads imposed by wind on antennas and their supporting structures cause countless failures. Aluminum and its alloys are very prone to fatigue failure, and the antenna engineer must be aware of this problem. To achieve real reliability:

1   Examine available wind-speed data for the location where the antenna is to be used.

2   Use derated permissible stress levels to allow for fatigue.

3   Check antenna designs for mechanical resonance.

4   Damp out, stiffen up, or guy parts of the antenna system which are prone to vibration or oscillation.

Reference 1 provides information on a wide range of the mechanical aspects of antenna design.

Corrosion   The effects of corrosion and wind-induced stresses are synergistic, each making the other worse. They are almost always responsible for the eventual failure of any antenna system. Every antenna engineer should also be a corrosion engineer; it is always rewarding to examine old antennas to see which causes of corrosion could have been avoided by better design. The essence of good corrosion engineering is:

1   Selection of suitable alloys for outdoor exposure and choice of compatible materials when different metals or alloys are in contact. A contact potential of 0.25 V is the maximum permissible for long life in exposed conditions.

2   Specification of suitable protective processes—electroplating, painting, galvanizing, etc.

There is an enormous variety in the severity of the corrosion environment at different locations, ranging from dry, unpolluted rural areas to hot, humid coastal industrial complexes.

Plastics do not corrode, but they degrade by oxidation and the action of ultraviolet light. These effects are reduced by additives to the bulk materials. References 2 and 3 provide detailed information on corrosion mechanisms and control.

Ice and Snow   The accumulation of ice and snow on an antenna causes an increase in the input voltage standing-wave ratio (VSWR) and a reduction in gain. The severity of these effects, caused by the capacitive loading of antenna elements and absorption of radio-frequency (RF) energy, increases as the frequency rises.

The fundamental design precaution is to ensure that the antenna and its mounting are strong enough to support the weight of snow and ice which will accumulate on them. This is very important, as even when the risk of a short loss of service due to the electrical effects of ice can be accepted, the collapse of even a part of the antenna is certainly unacceptable. Ice falling from the upper parts of a structure onto antennas below is a major cause of failure; safeguard against it by fitting lightweight antennas above more solidly constructed ones or by providing vulnerable antennas with shields to deflect falling ice.

In moderate conditions, antennas may be provided with radomes to cover the terminal regions of driven elements or even whole antennas. As conditions become more severe, heaters may be used to heat antenna elements or to prevent the buildup of ice and snow on the radome. A wide range of surface treatments has been tried to prevent the adhesion of ice; some of these show initial promise but become degraded and ineffective after a period of exposure to sunlight and surface pollution. Flexible radome membranes and nonrigid antenna elements have been used with some success.

Breakdown under Power   An inadequately designed antenna will fail by the over-heating of conductors, dielectric heating, or tracking across insulators. The power rating of coaxial components may be determined from published data, but any newly designed antenna should be tested by a physical power test. An antenna under test should be expected to survive continuous operation at 1.5 times rated power and at 2 times rated peak voltage; for critical applications even larger factors of safety should be specified.

Lightning Damage   Antennas mounted on the highest point of a tower are particularly prone to lightning damage. The provision of a solid, low-inductance path for lightning currents in an antenna system reduces the probability of severe damage to the antenna. Electronic equipment is best protected by good antenna design and system grounding, supplemented by gas tubes connected across the feeder cables. Figure 27-1 shows a typical system with good grounding to prevent side-flash damage and danger to personnel. For detailed guidance see Refs. 4, 5, and 6.

Precipitation and Discharge Noise   This is caused when charged raindrops fall onto an antenna or when an antenna is exposed to an intense electric field in thunderstorm conditions. Precipitation noise can be troublesome at the lower end of the VHF band and may be experienced frequently in some locations. When problems arise, antenna elements may be fitted with insulating covers. These prevent the transfer of the charges from individual raindrops into the antenna circuit and reduce the energy coupled to the antenna when a charge passes between drops.

Choice of Polarization

Base stations for mobile services use vertical polarization because it is then simple to provide an omnidirectional antenna at both the mobile terminals and the base station. There is sometimes an advantage in using horizontal polarization for obstructed point-to-point links in hilly terrain, but the choice of polarization is often determined by the need to control cochannel interference. Orthogonal polarizations are chosen for antennas mounted close together in order to increase the isolation between them.

It has been found that the use of circular polarization (CP) reduces the effects of destructive interference by reflected multipath signals, so CP should be considered for any path where this problem is expected. CP has been used with success on a

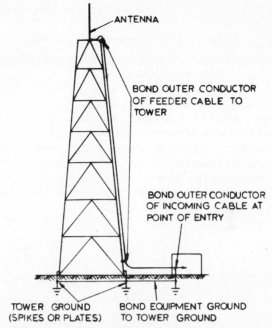

ANTENNA

BOND OUTER CONDUCTOR
OF FEEDER CABLE TO
TOWER

BOND OUTER CONDUCTOR
OF INCOMING CABLE AT
POINT OF ENTRY

TOWER GROUND   BOND EQUIPMENT GROUND
(SPIKES OR PLATES)   TO TOWER GROUND

FIG. 27-1 Typical example of good grounding practice.
(© C & S Antennas, Ltd.)

number of long grazing-incidence oversea paths where problems with variable sea-surface reflections had been expected to be troublesome. Each end of a CP link must use antennas with the same sense of polarization.

Meeting Cost Objectives

The designer of a communications system must strive to provide the necessary overall performance for the lowest cost. A 100 percent reliability is often very difficult and costly to achieve and is only necessary for a small number of services. By comparison, 99 percent availability will entirely satisfy many users and can be provided much more readily; the user cannot justify the high cost of that extra 1 percent.

Cost-effective design is only obtained by:

1 Identifying the availability needed
2 Determining the environment at both ends of the link
3 Estimating the propagation characteristics of the path and judging the reliability of the estimates
4 Selecting the right equipment and antennas for the link to meet the communications and reliability objectives

Trade-Offs The interdependence of various parameters deserves careful consideration. For any major scheme the following checklist should always be worked through.

1   Examine the interactions of structure height, transmitter power, feeder attenuation, and antenna gain.

2   Consider using split antennas and duplicate feeders to increase reliability.

3   Consider the use of diversity techniques to achieve target availability instead of a single system with higher powers and gains.

4   Review the propagation data, especially the probability of multipath or cochannel interference. Don't engineer a system with 99.9 percent hardware availability and find 3 percent outage due to cochannel problems. Check the cost of antennas designed to reduce cochannel problems by nulling out the troublesome signals.

5   Visit the chosen sites. General wind data are useless if the tower is near a cliff edge, and a careful estimate of actual conditions must be made. Similarly, a nearby industrial area may mean a corrosive environment, and nearness to main roads indicates a high electrical noise level. Look for local physical obstructions in the propagation path.

5   Don't overdesign to cover ignorance. Find out!

27-3   ANTENNAS FOR POINT-TO-POINT LINKS

Yagi-Uda Antennas   Yagi-Uda antennas are very widely used as general-purpose antennas at frequencies up to at least 900 MHz. They are cheap and simple to construct, have reasonable bandwidth, and will provide gains of up to about 17 dBi, or more if a multiple array is used. At low frequencies the gain which can be obtained is limited by the physical size of the antenna, but in the UHF band the limiting factor tends to be the accuracy with which the fed element and feeder system can be constructed.*

Yagi-Uda antennas provide unidirectional beams with moderately low side and rear lobes. The characteristics of the basic antenna can be modified in a variety of useful ways, some of which are shown in Fig. 27-2. The basic antenna (a) can be arrayed in linear or planar arrays (b). When the individual antennas are correctly spaced, an array of N antennas will have a power gain N times as large as that of a single antenna, less an allowance for distribution feeder losses. Table 27-1 indicates typical gains and arraying distances for Yagi-Uda antennas of various sizes. Different array spacings may be used when it is required to provide a deep null at a specified bearing, but the forward gain will be slightly reduced.

The bandwidth over which the front-to-back ratio is maintained may be increased by replacing a simple single reflector rod by two or three parallel rods (c). The back-to-front ratio of a simple Yagi-Uda antenna may be increased either by the addition of a screen (d) or by arraying two antennas with a quarter-wavelength axial displacement, providing a corresponding additional quarter wavelength of feeder cable to the forward antenna (e). A well-designed screen will provide a back-to-front ratio of as much as 40 dB, while that available from the quadrature-fed system is about 26 dB.

Circular polarization can be obtained by using crossed Yagi-Uda antennas: a

---

*For information on the design of Yagi-Uda antennas see Chap. 12. References 7 and 8 provide a useful understanding of the operation of Yagi-Uda and other surface-wave antennas.

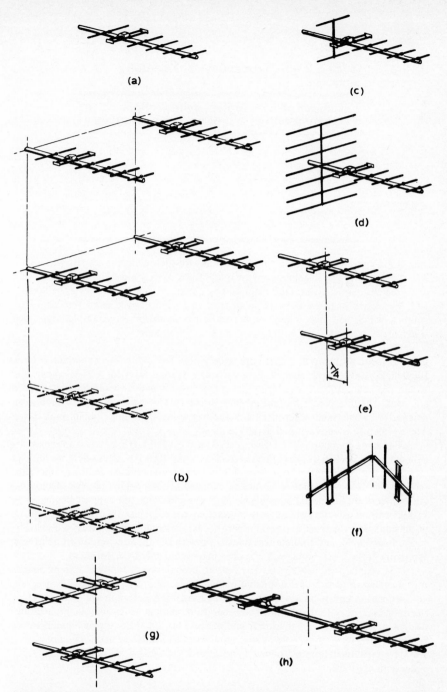

FIG. 27-2 Configurations of Yagi-Uda antennas. (*a*) Standard six-element antenna. (*b*) Stacked and bayed arrays. (*c*) Double reflector rods. (*d*) Reflector screen. (*e*) Increased *F*/*b* ratio by λ/4 offset. (*f*), (*g*), (*h*) Arrangements to produce aximuth radiation patterns for special applications. (© *C & S Antennas, Ltd.*)

TABLE 27-1 Typical Data for Yagi-Uda Antennas

Number of elements	Typical gain, dBi	Spacing for arraying, λ
3	7	0.7
4	9	1.0
6	10.5	1.25
8	12.5	1.63
12	14.5	1.8
15	15.5	1.9
18	16.5	2.0

pair of antennas mounted on a common boom with their elements set at right angles. The two antennas must radiate in phase quadrature, so they must be fed in quadrature or be fed in phase and mutually displaced by a quarter wavelength along the boom.

There has been an increase in interest in arrays derived from the Yagi-Uda antenna which use long, closed forms for their elements, for example, rings and squares.[9]

Log-Periodic Antennas These are widely used for applications in which a large frequency bandwidth is needed. The gain of a typical VHF or UHF log-periodic antenna is about 10 dBi, but larger gains can be obtained by arraying two or more antennas. The disadvantage of all log-periodic designs is the large physical size of an antenna with only modest gain. This is due to the fact that only a small part of the whole structure is active at any given frequency.

The most common type of design used on the VHF-UHF bands is the log-periodic dipole array (LPDA) described in Sec. 13-4 and Ref. 10. After selecting suitable values for the design ratio τ and apex angle α, the designer must decide on the compromises necessary to produce a practical antenna at reasonable cost. The theoretical ideal is for the cross-sectional dimensions of the elements and support booms to be scaled continuously along the array; in practice, the elements are made in groups by using standard tube sizes, and the support boom is often of uniform cross section. The stray capacitances and inductances associated with the feed region are troublesome, especially in the UHF band, and can be compensated only by experiment.

The coaxial feed cable is usually passed through one of the two support booms, thus avoiding the need for a wideband balun.

Printed-circuit techniques can be applied to LPDA design in the UHF band, as the antenna is easy to divide into two separate structures which may be etched onto two substrate surfaces. At the lower end of the VHF band the dipole elements may be constructed from flexible wires supported from an insulating catenary cord.

A typical well-designed octave-bandwidth LPDA has a VSWR less than 1.3:1 and a gain of 10 dBi.

Helices A long helical antenna has an easily predicted performance and is easy to construct and match. A VSWR as low as 1.2:1 can be obtained fairly easily over a frequency bandwidth of 20 percent. Conductive spacers may be used to support the helical element from the central support boom, so the antenna can be made very simple and robust.

Helices may be arrayed for increased gain; to obtain correct phasing the start position of each helix in the array must be the same. For further information on the design of helices see Chap. 13 or Ref. 11.

Panel Antennas   An antenna which comprises a reflecting screen with simple radiating elements mounted over it, in a broadside configuration, is generally termed a *panel antenna*. An array may comprise one or more panels connected together.

Typical panels use full-wavelength dipoles, half-wave dipoles, or slots as radiating elements (Fig. 27-3). They have several advantages over Yagi-Uda antennas:

1 More constant gain, radiation patterns, and VSWR over a wide bandwidth—up to an octave

2 More compact physical construction

3 Very low coupling to the mounting structure

4 Low side lobes and rear lobes

Panel antennas for frequencies in the UHF band lend themselves to printed-circuit design methods, as the radiating structures, feed lines, and matching system may all be produced by using stripline techniques. At lower frequencies the radiating elements are often mounted at voltage minimum points by using conducting supports, so a strong, rigid construction can be produced. A really solidly built but lightweight panel for a military application is shown in Fig. 27-4*a*. Here an all-welded aluminum frame and a skeleton-slot radiator are used so that the antenna will resist rough use in the field.

Multiple Arrays   Reference has been made to the use of a number of antennas arrayed together. Figure 27-5 shows a variety of simple cable harnesses which may be made from standard 50-Ω and 75-Ω coaxial cable. More complex harnesses may be constructed by combining several of the simple designs shown. For high-power applications rigid fabricated coaxial transformers are used. These may be designed to combine up to eight antenna feeds, and their impedance bandwidth may be increased by using two or three series quarter-wave matching sections designed to give Chebyshev or other chosen characteristics.[12]

Corner Reflectors   Well-designed corner-reflector antennas are capable of providing high gain and low sidelobe levels, but below 100 MHz they are mechanically cum-

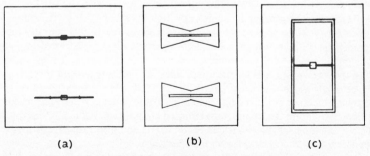

(a)   (b)   (c)

FIG. 27-3   Panel antennas. (*a*) Two full-wave dipole elements. (*b*) Two batwing slot elements. (*c*) Skeleton-slot elements. (© *C & S Antennas, Ltd.*)

FIG. 27-4 Robustly constructed antennas for military use. (*a*) Skeleton-slot-fed panel (225–400 MHz). (*b*) Grid paraboloid (610–1850 MHz).

bersome. Before using a corner reflector, make sure that the same amount of material could not be more effectively used to build a Yagi-Uda antenna, or perhaps a pair of them, to do the job even better.

In the UHF band, corner reflectors may be very simply constructed from solid or perforated sheet. The apex of the corner is sometimes modified to form a trough (Fig. 27-6). The provision of multiple dipoles extends the antenna aperture and increases the available gain.

Paraboloids The design of a high-gain antenna may be reduced to a problem of illuminating the aperture necessary to develop the specified radiation patterns and gain. The size of the aperture is determined only by the gain required, whatever type of elements is used to fill it. As the cost of the feed system and the radiating elements doubles for each extra 3-dB gain, a stage is reached at which it becomes attractive to use a single radiating element to illuminate a reflector which occupies the whole of the necessary antenna aperture. The design task is reduced to choosing the size and shape of the reflector and specifying the radiation pattern of the illuminating antenna. If the antenna aperture is incompletely filled or its illumination is nonuniform, the gain which is realized decreases. The ratio of the achieved gain to the gain obtainable from the same aperture when it is uniformly illuminated by lossless elements is termed the *aperture efficiency* of the antenna. In a receiving context, this quantity represents the proportion of the power incident on the aperture which is delivered to a matched load at the terminals of the antenna.

In the VHF and UHF bands, a reflector may be made of solid sheet, perforated sheet, wire netting, or a series of parallel curved rods. As the wavelength is large, the mechanical tolerance of the reflector surface is not very demanding, and various methods of approximating the true surface required are possible. Table 27-2 indicates some of the combinations of techniques currently in use and illustrates the diversity of the methods which are successful for various purposes.

Grid paraboloids are attractive to produce because the curvature of all the rods

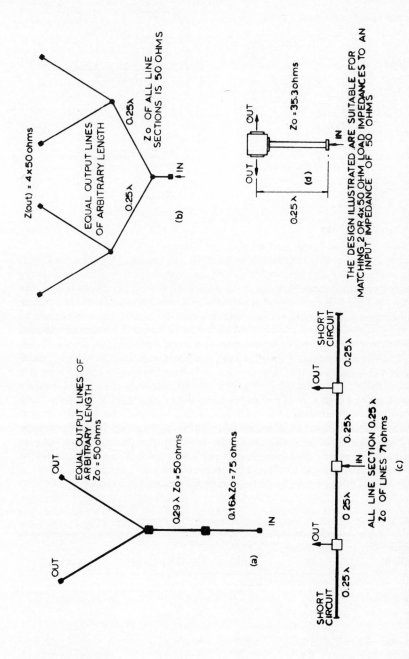

FIG. 27-5 Simple branching feeder systems. (*a*) Two-way. (*b*) Four-way. (*c*) Two-way, compensated. (*d*) Two-way, high-power. (© *C & S Antennas, Ltd.*)

THE DESIGN ILLUSTRATED ARE SUITABLE FOR MATCHING 2 OR 4×50 OHM LOAD IMPEDANCES TO AN INPUT IMPEDANCE OF 50 OHMS

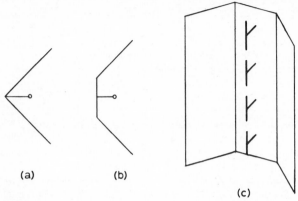

(a)   (b)

(c)

FIG. 27-6 Corner and trough reflectors. (© *C & S Antennas, Ltd.*)

is exactly the same; only their length varies across the antenna. A typical example is shown in Fig. 27-4*b*. The main deficiency of grid paraboloids is the leakage of energy through the surface, restricting the front-to-back ratio which can be achieved. For example, at 1500 MHz a front-to-back ratio of −30 dB is a typical limit. If a greater front-to-back ratio is needed, a continuous skin of solid or perforated sheet must be used and the consequent increase in wind-loaded area accepted as a necessary penalty. A mathematical treatment of grid reflectors appears in Ref. 13.

Radomes are frequently fitted to feeds or complete antennas in order to reduce the effects of wind and snow. They may be made from fiberglass or in the form of a tensioned membrane across the front of the antenna. In severe climates it is possible to heat a radome with a set of embedded wires, but this method can be applied only to a plane-polarized antenna.

Point-to-point links using tropospheric-scatter propagation require extremely high antenna gains and generally use a reflector which is an offset part of a full paraboloidal surface constructed from mesh or perforated sheet. Illumination is provided by a horn supported at the focal point by a separate tower.

For a full discussion of the design of reflector antennas refer to Chap. 17.

TABLE 27-2 Typical Paraboloid-Antenna Configurations

Frequency, MHz	Diameter, m	f/d ratio	Construction	Feed type
200	10.0	0.5	Mesh paraboloid	4-element Yagi-Uda
700	3.0	0.25	Solid skin	Dipole and reflector
900	7.0	0.4	Perforated steel sheet	Horn
610–960	1.2	0.25	Grid of rods	Slot and reflector
1500	2.4	0.25	Solid skin	Dipole and disk
1350–1850	1.2	0.25	Grid of rods	Slot and reflector

27-4   BASE-STATION ANTENNAS

Simple Low-Gain Antennas   The simplest types of base-station antennas will provide truly omnidirectional azimuth coverage only when mounted in a clear position on top of a tower. Figure 27-7 shows standard configurations for ground-plane and coaxial dipole antennas and demonstrates that these forms are closely related. They are cheap and simple to construct and may be made to handle high power. Exact dimensions must be determined by experiment, as the stray inductance and capacitance associated with the feed-point insulator cannot be neglected. The use of a folded feed system can provide useful mechanical support and gives better control over the antenna impedance. The satisfactory operational bandwidth of the coaxial dipole d depends critically on the characteristic impedance Z_0, of the lower coaxial section formed by the feed line (radius r) inside the skirt (radius R). If this section has too small a Z_0, radiating currents will flow on the outside of the feeder line unless the skirt length is exactly $\lambda/4$. The impedance, gain, and radiation pattern of the antenna then becomes critically dependent on the positioning of the feed line on the tower, severely limiting the useful bandwidth of the antenna.

Discone Antennas   The discone and its variants are the most commonly used low-gain wideband base-station antennas. The useful lower frequency limit occurs when the cone is a little less than $\lambda/4$ high, but the upper frequency limit is determined almost entirely by the accuracy with which the conical geometry is maintained near the feed point at the apex of the cone.

Discones may be made with either the disk or the cone uppermost. The support for the upper part of the antenna usually takes the form of low-loss dielectric pillars or a thin-walled dielectric cylinder, fitted well outside the critical feed region.

Variants of the basic discone use biconical forms in place of the conventional cone and replace the disk by a cone with a large apex angle. At the lower end of the VHF band the antennas may be mounted at ground level, so a minimal skeleton disk which couples to the ground may be used if some loss of efficiency can be accepted.

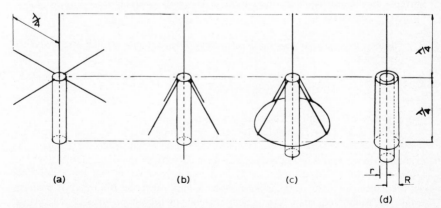

FIG. 27-7   Low-gain base-station antennas. (*a*) Standard ground plane with radials. (*b*) Ground plane with sloping radials. (*c*) Ground plane with closed ring. (*d*) Coaxial dipole. (© *C & S Antennas, Ltd.*)

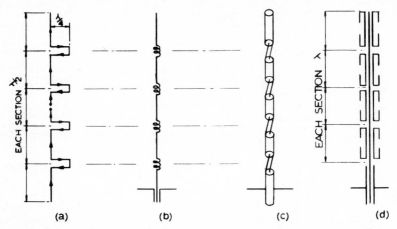

FIG. 27-8 Collinear dipole arrays. (*a*) Franklin array. (*b*) Array with meander-line phase reversal. (*c*) Array with transposed coaxial sections. (*d*) Alternative coaxial form. (© *C & S Antennas, Ltd.*)

Collinear Arrays The ground plane and coaxial dipole have about the same gain as a half-wavelength dipole. When more gain is needed, the most popular omnidirectional antennas are simple collinear arrays of half-wave dipoles. The original array of this type is the Franklin array shown in Fig. 27-8*a*. This design is not very convenient owing to the phase-reversing stubs which project from the ends of each half-wave radiating section, but various derivatives are now widely used. The arrangement at *b* uses noninductive meander lines to provide phase reversal, while those at *c* and *d* use coaxial-line sections. Arrangements such as these may be mounted in fiberglass tubes to provide mechanical support. An input-matching section transforms the input impedance of the lower section, which may be $\lambda/2$ or $\lambda/4$ long, to 50 Ω. A set of quarter-wavelength radial elements or a quarter-wavelength choke is used to suppress currents on the outside of the feeder cable. The gain available from these arrays is limited by two factors:

1 There is mechanical instability in a very long antenna with a small vertical beamwidth.

2 The available excitation current diminishes away from the feed as a result of the power lost by radiation from the array.

The practical upper limit of useful gain is about 10 dBi.

In the case of the coaxial-line designs, each section is shorter than a free-space half wavelength so that the correct phase shift is obtained inside the section. The examples shown would typically provide a gain of 9 dBi at the design frequency. The useful bandwidth of these antennas is inherently narrow because of the phase error between successive radiating sections which occurs when the frequency is changed from the design frequency. The typical behavior of the major lobes of the vertical radiation pattern of these arrays is shown in Fig. 27-9.

Dipoles on a Pole Much ingenuity has been applied to the design of simple wide-band high-gain omnidirectional antennas. A simple offset pole-mounted array is

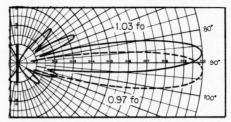

FIG. 27-9   Vertical radiation pattern of a typical end-fed collinear array.

shown in Fig. 27-10*a*. This will provide about 10-dBi gain in the forward direction but typically only 4 dBi rearward. An attempt to avoid this problem is shown in Fig. 27-10*b*, but this type of antenna has distorted vertical radiation patterns caused by the phase shifts which result from the displacement of the dipoles; gain is also reduced to about 6 dBi for the four-element array shown.

The solution in Fig. 27-10*c*, in which dipoles are placed in pairs and are cophased, is more satisfactory, as the phase center of each tier is concentric with the supporting pole. However, the antenna is relatively expensive, as eight dipoles provide only 6-dB gain over a single dipole.

One possibility is to use the in-line stacked array in Fig. 27-10*a* and place the base station toward the edge of the service area. The rearward illumination may be improved if the spacing between the dipoles and the pole is optimized for the pole size and frequency to be used.

Antennas on the Body of a Tower   Figure 27-11*a* shows a measured horizontal radiation pattern for a simple dipole mounted from one leg of a mast of 2-m face width. The distortion of the circular azimuthal pattern of the dipole is very typical

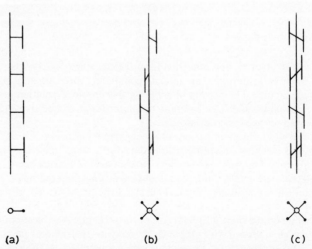

(a)   (b)   (c)

FIG. 27-10   Pole-mounted dipoles. (*a*) In line. (*b*) Four dipoles spaced around a pole. (*c*) Eight dipoles spaced around a pole. (© *C & S Antennas, Ltd.*)

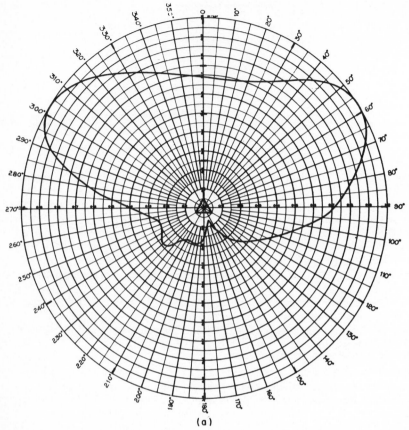

(a)

FIG. 27-11 Typical azimuth patterns of (*a*) VHF dipole mounted off one leg of a triangular mast and (*b*) three dipole panels mounted on the same structure.

and is caused by blocking and reflection from the structure. By contrast, Fig. 27-11*b* shows what can be achieved by an antenna comprising three dipole panels mounted on the same structure. The penalty of adopting this improved solution lies in the cost of the more complex antenna, so before an optimum design can be arrived at, the value of the improved service must be assessed.

The horizontal radiation pattern of a complete panel array is usually predicted from measured complex radiation-pattern data for a single panel, using a suitable computer program. For each azimuth bearing, the angle from each panel axis is found, and the relative field in that direction is obtained. The radiated phase is computed from the excitation phases and physical offsets of the phase centers of the individual panels.

Depending on the cross-sectional size of the structure, the most omnidirectional coverage may be produced with all panels driven with the same phase or by a phase rotation around the structure; for example, on a square tower the element current phases would then be 0, 90, 180, and 270°. When phase rotation is used, the individual

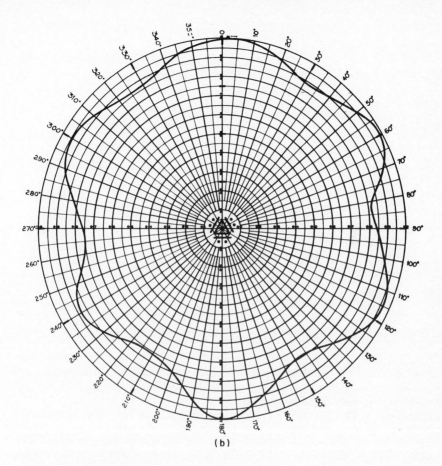

(b)

elements may be offset from the centerlines of the faces of the structure to give a more omnidirectional azimuth pattern, as in Fig. 27-12.

A well-designed panel array comprising four tiers, each of four panels, is an expensive installation, but if properly designed it can have a useful bandwidth of as much as 25 percent. This allows several user services to be combined into the same antenna, each user having access to a very omnidirectional high-gain antenna.

Special-Purpose Arrays   For applications in which the largest possible coverage must be obtained, the azimuth radiation pattern of the antenna must be shaped to concentrate the transmitted power in the area to be served, for example, an airway, harbor, or railroad track. Energy radiated in other directions is wasted and is a potential cause of interference to others.

Antennas with cardioidal azimuth radiation patterns are useful for a wide range of applications. Simple two-element arrays (dipole plus passive reflector) or dipoles mounted off the face of a tower may be adequate, but a wider range of patterns is available if two driven dipoles are mounted on a single supporting boom and excited with suitably chosen currents and phases. (See Chap. 3.)

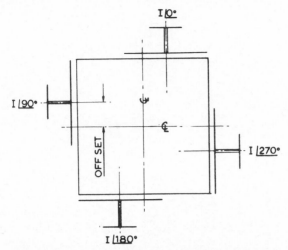

FIG. 27-12 Plan arrangement of an omnidirectional panel array on a large tower. (© *C & S Antennas, Ltd.*)

When a signal must be laid down over an arbitrarily shaped area of terrain, Yagi-Uda arrays may be arranged as at Fig. 27-2*f* and *g*. Due allowance must be made for the separation of the phase centers of the antennas when computing the radiation patterns. As an approximate guide, the phase center of a Yagi-Uda antenna lies one-third of the way along the director array, measured from the driven element.

Further tiers of antennas may be used to increase the gain of the system without modifying the azimuth radiation patterns.

When designing a complex array, estimation of gain can present a confusing problem. If an array contains sufficient elements to fill it, the gain of the array depends only on the size of the array aperture and not on the type of elements chosen. The gain of a filled array of identical tiers may be estimated by multiplying the vertical aperture power gain by the ratio of maximum power to mean power in the azimuth plane (the maximum-mean ratio). The vertical aperture gain may be assumed to be 1.15 times per wavelength of aperture relative to a half-wave dipole. The maximum-mean ratio may be obtained by integration of the azimuth radiation pattern, dividing the area of the pattern into the area of a circle which just encloses it (Fig. 27-13). Graphical integration is easily carried out by using a planimeter, but it is now quite simple to write integration routines for programmable pocket calculators. Once an array has been selected, its horizontal and vertical radiation patterns may be computed and used to predict the array gain more accurately.

27-5 MOBILE ANTENNAS

A road vehicle is not an ideal environment for an antenna. To make matters worse, the owner of a vehicle usually does not want antennas to be fixed in the most electrically favored positions like the center of the roof, but expects them to work when

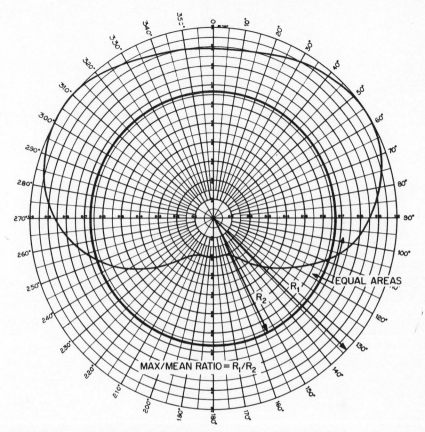

FIG. 27-13 The maximum-mean ratio of an azimuth radiation pattern.

mounted on gutters, fenders, or bumpers. The antenna is usually a severe compromise between what is ideally required and what is convenient. Figure 27-14 shows typical radiation patterns for a whip antenna measured with different mounting positions on a medium-sized automobile.

On both road vehicles and boats, antennas are subjected to severe mechanical shock, vibration, and exposure to all kinds of weather. Coaxial dipoles are widely used as VHF antennas on ships of all sizes. They can be encapsulated in a dielectric tube which protects the antenna from corrosion by seawater, and they have a vertical beamwidth which is large enough to avoid problems when the ship rolls. The most popular form of antenna for road vehicles is the simple base-fed whip. Quarter-wavelength whips require no loading, but they are inconveniently long at frequencies below about 100 MHz. Base or center loading can be used to shorten the physical length of a whip, but the efficiency of the antenna falls as the height becomes a smaller fraction of a wavelength. At higher frequencies it becomes possible to increase the gain of the antenna by using a five-eighths-wavelength whip, which has a small input coil to provide an input impedance suitably close to 50 Ω. Antennas with higher gain can be used

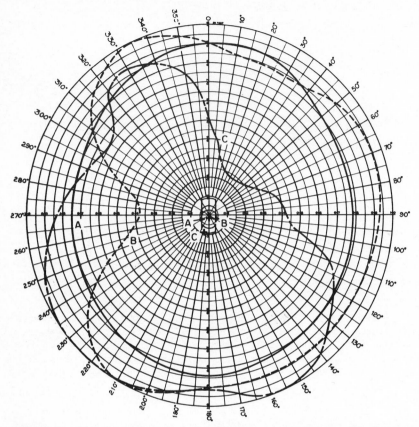

FIG. 27-14 Radiation patterns for a whip antenna on a typical automobile. Some of the data used in the preparation of this figure are from P. A. Ratliff, "VHF Mobile Radio Communications—A Study of Multipath Fading and Diversity Reception," Ph.D. thesis, University of Birmingham, 1974.

in the UHF band; they are typically short versions of the collinear arrays described in Sec. 27-4.

A variety of low-profile antennas are used on trains, buses, and security vehicles. These are usually derivatives of the inverted L or the annular slot (Fig. 27-15). Various discontinuities on a vehicle can be driven as slot radiators, although it is difficult to provide omnidirectional azimuth coverage. Antennas built into external mirrors or printed onto windows are used for applications when no conspicuous antennas must be carried.

During recent years much attention has been given to antennas which use active devices for matching or modification of radiating currents. As some of these antennas are physically small in terms of a wavelength, they are of particular interest for mobile use. Multielement antennas which provide steered beams or nulls offer promising lines of development, especially if control of the antenna is adaptively managed to optimize the received signal. Such techniques will become increasingly important with the growth of data links to vehicles.

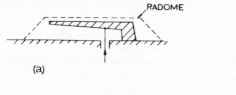
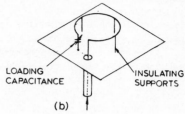

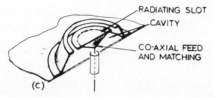

FIG. 27-15 Low-profile antennas. (*a*) Inverted L. (*b*) Hula hoop. (*c*) Annular slot. (© *C & S Antennas, Ltd.*)

Antennas for personal hand-held transceivers usually take the form of rigid telescopic whips, wires positioned in carrying straps, or short semiflexible whips. The helical whip is increasingly preferred, as rigid antennas are inconvenient and the performance of carrying-strap antennas varies greatly with their position. When designing these systems, the developing literature on biological hazards should be consulted.[14,15]

27-6 SYSTEM CONSIDERATIONS

Mounting Arrangements When mounting any antenna, it is important not to impair its performance by the influence of the supporting structure. The inevitable effect of the supporting structure on the radiation pattern of a dipole has been referred to in Sec. 27-4. This effect is accompanied by a modification of the input impedance, which may be unwelcome if a low VSWR is needed. In any critical application the change of the radiation patterns and gain must be taken into account when estimating system performance. Impedance matching of the antenna must be undertaken in the final mounting position or a simulation of it.

If Yagi-Uda arrays are mounted with their elements close to a conducting structure, they too will suffer changes of radiation patterns and impedance. The effects will be greatest if members pass through the antenna, as they do when an array is mounted on clamps fitted at the center of the cross boom. If at all possible, when an array is center-mounted, the member to which it is clamped should lie at right angles to the elements of the array.

Currents induced in diagonal members of the supporting structure will cause reradiation in polarization planes other than that intended. This will result in the cross-polar discrimination of the antenna system being reduced from that which would be measured on an isolated antenna at a test range. When polarization protection is important, the tower should be screened from the field radiated by the antenna with a cage of bars spaced not more than $\lambda/10$ apart, lying in the plane of polarization. (A square mesh is used for circular polarization.) Panel antennas are designed with an integral screen to reduce coupling to the mounting structure.

Long end-mounted antennas are subjected to large bending forces and turning moments at their support points. These forces can be reduced by staying the antenna, using nylon or polyester ropes for the purpose to avoid degrading its electrical characteristics.

In severe environments antennas may be provided with radomes or protective paints. It is very important that the antennas be tested and set up with these measures already applied, especially if the operating frequency is in the UHF band.

Coupling A further consideration when planning a new antenna installation on an existing structure is the coupling which will exist between different antennas. When a transmitting antenna is mounted close to a receiving antenna, problems which can arise include:

Radiation of spurious signals (including broadband noise) from the transmitter

Blocking or desensitization of the receiver

Generation of cross-modulation effects by the receiver

Radiation of spurious signals (intermodulation products and harmonics) due to non-linear connections in the transmitting antenna or generation of spurious signals and cross-modulation effects caused by the same mechanism in the receiving antenna

The last of these problems must be avoided by good antenna design—avoiding any rubbing, unbonded joints. The other effects depend critically on the isolation between the antennas and on parameters of the transmitters and receivers; these parameters should be specified by their manufacturers.

The isolation between two antennas may be predicted from Fig. 27-16 or from standard propagation formulas. Antenna isolations may be increased by using larger spacings between them or by using arrays of two or more antennas spaced to provide each with a radiation-pattern null in the direction of the other.

An alternative method of increasing the isolation between the antenna-system inputs is to insert filters. If a suitable filter can be constructed, the antenna isolation may be reduced until, in the limit, a single antenna is used with all equipment, transmitters and receivers, coupled to it through filters. When receivers are connected to a common antenna, the signal from the antenna is usually amplified before being divided by a hybrid network. For information on filter design, consult Refs. 17 and 18. The number of services which use a single antenna can be extended to six or more, provided adequate spacings are maintained between the frequencies allocated to different users. The whole system is expensive, but the cost may be justified if the antenna itself is large or if tower space is limited.

Coverage In free-space conditions the intensity of a radio wave diminishes as the distance from the transmitter increases in accordance with the inverse-square law. Terrestrial links are not usually in free-space conditions, and the field diminishes more rapidly with distance. Reference 16 provides a large variety of basic data and curves for planning point-to-point, aeromobile, and other services. References 19 and 20 pro-

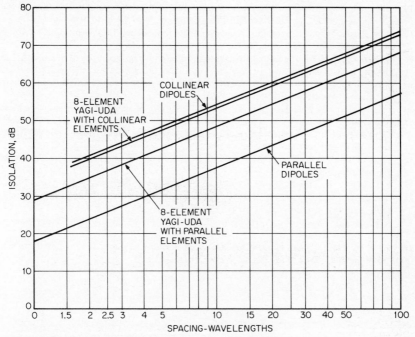

FIG. 27-16 Typical isolations between Yagi-Uda antennas.

vide further examples of the use of the data and also provide additional information on the methods to be adopted for dealing with obstructed paths.

REFERENCES

1 C. J. Richards (ed.), *Mechanical Engineering in Radar and Communications,* Van Nostrand Reinhold Company, New York, 1969.
2 V. R. Pludek, *Design and Corrosion Control,* The Macmillan Company, New York, 1977.
3 H. H. Uhlig, *Corrosion and Corrosion Control,* 2d ed., Interscience Publishers, a division of John Wiley & Sons, Inc., New York, 1971.
4 *Lightning Protection Code,* ANSI C5.1-1969, American National Standards Institute, New York, 1969.
5 *The Protection of Structures against Lightning,* British Standard Code of Practice CP 326:1965, British Standards Institution, London, 1965.
6 *Manual on Lightning Protection,* Australian Standard 1768–1975, Standards Association of Australia, Sydney, 1975.
7 H. W. Ehrenspeck and H. Poehler, "A New Method for Obtaining Maximum Gain from Yagi Antennas," *IRE Trans. Antennas Propagat.,* vol. AP-7, October 1959, pp. 379–385.
8 B. M. Thomas, "The Precise Mechanism of Radiation from Surface Wave Aerials," *J. Inst. Eng. Aust.,* September 1964, pp. 225–238.
9 M. Kosugi, N. Inasaki, and T. Sekiguchi, "Design of an Array of Circular-Loop Antennas with Optimum Directivity," *Electron. Commun. Japan,* vol. 54-B, no. 5, 1971, pp. 67–76.

10 H. V. Rumsey, *Frequency Independent Antennas,* Academic Press, Inc., New York, 1966.

11 J. D. Kraus, *Antennas,* McGraw-Hill Book Company, New York, 1950.

12 H. J. Riblet, "General Synthesis of Quarter-Wave Impedance Transformers," *IRE Trans. Microwave Theory Tech.,* vol. MTT-5, January 1957, pp. 36–43.

13 R. Neri and T. S. M. Maclean, "Receiving and Transmitting Properties of Small Grid Paraboloids by Moment Method," *IEE Proc.,* vol. 126, no. 12, December 1979, pp. 1209–1219.

14 *Safety Levels of Electromagnetic Radiation with Respect to Personnel,* C95, United States of America Standards Institute, New York, 1966.

15 J. R. Swanson, V. E. Rose, and C. H. Powell, "A Review of International Microwave Exposure Guides," in *Electronic Product Radiation and the Health Physicist,* Pub. BRH/DEP70-26, U.S. Department of Health, Education, and Welfare, Washington, 1970, pp. 95–110.

16 Texts of Thirteenth Plenary Assembly, Comité Consultatif International des Radiocommunications, Geneva, 1975.

17 A. I. Zverev, *Handbook of Filter Synthesis,* John Wiley & Sons, Inc., New York, 1967.

18 G. L. Matthaei, L. Young and E. Jones, *Microwave Filters, Impedance Matching Networks and Coupling Structures,* McGraw-Hill Book Company, New York, 1964.

19 A. Picquenard, *Radio Wave Propagation,* Philips Technical Library, John Wiley & Sons, Inc., New York, 1974.

20 M. P. M. Hall, *Effects of the Troposphere on Radio Communication,* Peter Peregrinus Ltd., London, 1979.

Chapter 28

TV and FM Transmitting Antennas

Raymond H. DuHamel

Antenna Consultant

Television broadcast services are located within four bands: the lower very-high-frequency (VHF) bands of 54 to 72 and 76 to 88 MHz, the upper VHF band of 174 to 216 MHz, and the ultrahigh-frequency (UHF) band of 470 to 890 MHz. The FM band is 88 to 108 MHz. The bandwidths of the TV and FM channels are 6 and 0.2 MHz respectively. A stringent requirement for broadcast antennas is that the voltage standing-wave ratio (VSWR) should be less than 1.1:1 over the band. The bandwidth is 0.2 percent for FM and varies from 10 to about 1 percent for TV. If several channels are multiplexed into a single antenna, achieving the required VSWR is even more difficult. In many cases, another stringent requirement is that the antenna be capable of handling power inputs of 50 kW or more.

The *Technical Standards* of the Federal Communications Commission (FCC) specify the maximum effective radiated power (ERP) which TV and FM stations can radiate. The maximum power[1,2] varies with regions or zones of the United States and with the height of the antenna above ground. The height is not specified, but large heights are desired to increase coverage. Heights of 1000 to 2000 ft (305 to 610 m) are commonly used. The antennas are usually supported on guyed towers or tall buildings.

The majority of station requirements call for omnidirectional azimuth patterns. The circularity of the pattern depends upon the type of antenna when top-mounted and also on the configuration of the support structure when side-mounted. Other requirements call for various types of azimuth patterns, such as cardioid, skull-shape, peanut-shape, etc., to protect other stations or reduce radiation in low-population areas. It is desirable to have a large antenna gain so as to reduce the required transmitter power. Since the gain of the azimuth pattern is low, it is necessary to use a large vertical aperture to increase the gain. For antennas mounted at large heights, the vertical beamwidth should not be less than about 1°. This implies a vertical aperture of about 50 wavelengths, which is practical only in the UHF band. Wind loading and cost considerations limit the vertical aperture to about 120 ft (36.6 m) for FM and the other TV bands. Thus, the minimum elevation beamwidths are about 7, 4, and 2° for the lower VHF, FM, and upper VHF bands respectively. For these narrow beamwidths and certain combinations of terrain and antenna heights, it may be desirable to tilt the beam and/or fill in several nulls below the beam. Null fill is desired when the antenna is located near a residential area.

The antenna structure usually consists of a vertical array of radiating elements, such as dipoles, loops, and slots, or of one or more bays of vertical traveling-wave antennas such as helices, zigzags, and rings. It is desirable to mount the antennas on top of the tower or support structure, but in many cases it is necessary to side-mount them. Side-mounted antennas present an interesting challenge to the antenna designer to control the azimuth pattern. The elements of a vertical array are commonly referred to as *bays*. The array-feed techniques are conventional, such as a corporate feed or an end feed in a matched or resonant manner.

Beam tilt is achieved by phasing the bays of the vertical array and/or by mechanically tilting the array. Combinations of electrical and mechanical beam tilt may be used to achieve a beam tilt which varies with azimuth direction for special applications.

One simple technique for null fill is to shift the phase of one or more elements in the center of the array. The currents in these elements may be considered as the super-position of in-phase and quadrature-phase currents with respect to the rest of the array. The quadrature-phase currents produce the null fill. Another technique is to increase and decrease the current in upper and lower elements by the same amount. By using superposition again, the radiation from the difference component is in quadrature with the sum component and produces null fill. A combination of these two techniques may be used to produce null fill only below the horizon. An exponential aperture distribution in an end-fed antenna also produces null fill. Null fills of 5 percent and 20 percent reduce the gain of the array by about 0.2 and 0.6 dB respectively.

Originally, the *Technical Standards* of the FCC specified that TV and FM stations radiate horizontal polarization. Then, in the 1960s, the FCC permitted FM stations to use circular polarization. This provided improved reception, especially for vehicles with whip antennas, which are predominantly vertically polarized. In 1977 the FCC allowed TV stations to radiate right-hand circular polarization and to use the maximum ERP for horizontal polarization so as to maintain the field strength existing before conversion to circular polarization. Thus, the same ERP can be used for vertical polarization, which means doubling transmitter power. This has provided greatly improved reception for receivers with indoor antennas such as monopoles and rabbit ears. In addition, circularly polarized receiving antennas can be used to reduce "ghosting," since reflections from buildings and other objects tend to have the opposite sense of circular polarization.

With the exception of bandwidth, the requirements for TV and FM broadcast are very similar. In multiple-station FM antennas, there is little difference. Thus, in the following discussion antennas will be classified as circularly and horizontally polarized antennas.

The gain of TV and FM antennas is specified with respect to a horizontally polarized half-wave dipole, which has a gain of 1.65 with respect to an isotropic antenna. The gain of omnidirectional horizontally polarized antennas varies from about 0.9 to 1.1 times the vertical aperture in wavelengths. The gain for circularly polarized antennas is one-half of this.

28-2 PANEL-TYPE ANTENNAS

In many cases, antenna elements must be placed near the sides of a triangular or square tower because of physical restrictions. Reflecting panels are placed on the faces of the tower to prevent reflections from tower members, which lead to erratic azimuth patterns. With proper design, each panel has a unidirectional pattern with the beam direction normal to the panel. The 6-dB beamwidth should be 90° for square towers when an omnidirectional pattern is desired, since the gain should be down 6 dB in the crossover direction. For a triangular tower, a 120° 6-dB beamwidth is desired. By proper design of the antenna element and/or the addition of parasitic elements, reflecting sheets, or cavities, it is possible to vary the 6-dB beamwidth over the range of about 80 to 130°. (This is discussed in later sections.)

With equal power fed to the sides of the tower and with the antenna elements placed in the center of the sides as shown in Fig. 28-1, an omnidirectional-type pattern is obtained with a maximum-minimum ratio that increases with the face width of the tower. The short lines represent radiating elements such as dipoles, rings, and zigzags.

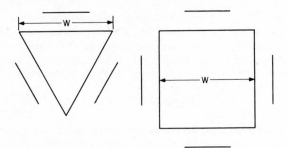

FIG. 28-1 Panel-type antennas for triangular and square towers.

Figure 28-2 shows this ratio for both square and triangular towers for optimum panel beamwidths and in-phase excitation of the panels. For good omnidirectional patterns, tower width should not be much greater than one wavelength. The null directions occur on each side of the crossover directions where the radiation from adjacent panels is not phased properly.

The elements may be displaced as shown in Fig. 28-3 to the point where quadrature phasing of adjacent panels is required to achieve in-phase radiation in the crossover direction. This requires different line length from the power splitter to the panels and provides reflection cancellation at the splitter, which in turn allows a much lower VSWR at the splitter than at the panels. An adverse effect is that the waves reflected back toward the elements distort the excitation of the elements and therefore the azimuth pattern. Azimuth patterns are about the same as for center-placed elements. The same technique may be used for triangular towers with a 120° differential phasing.

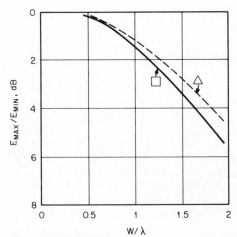

FIG. 28-2 Maximum-minimum ratio versus tower width in wavelengths for triangular and square towers.

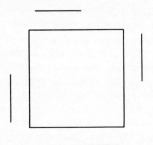   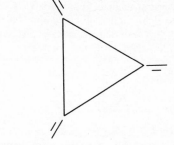

FIG. 28-3 Offset panel-type radiators on a square tower for reflection cancellation.   **FIG. 28-4** Skewed panel-type radiators for triangular towers.

Skewed-panel antennas[3] may be placed on the corners of large-width towers (such as those measuring five wavelengths), as shown in Fig. 28-4, in which the antenna element is placed on a narrow panel. The panels are skewed so that the crossover direction occurs in the line of the tower face. Thus, the relative phase of the radiation from adjacent antennas varies more slowly near the crossover direction than in a nonskew arrangement. Theoretically, this greatly improves the azimuth pattern. However, reflections from the tower members and backs of the panels degrade the patterns. For large vertical arrays, it is difficult to achieve better than 6-dB nulls.

Panel-type antennas are also very useful for directional applications. A wide variety of azimuth patterns may be achieved by controlling the magnitude and phase of excitation of the panels around the tower and the orientation of the panels.

28-3   CIRCULARLY POLARIZED ANTENNAS

The FCC allowance of circularly polarized broadcast transmission has led to the introduction of a wide variety of new transmitting antennas. The antenna types usually take the form of crossed dipoles, circular arrays of slanted dipoles, helical structures, and traveling-wave ring configurations.

Helix Antennas

The multiarm helix[4,5] is a versatile antenna for radiating circularly polarized waves. Figure 28-5 shows a three-arm helix with a pitch angle ψ wrapped around a conducting cylinder, which forms the support for the antenna and allows space for a transmission-line feed network for several bays of helices. For broadside radiation the turn length of an arm is equal to M wavelengths, where M is an integer and defined as the mode number. For an N-arm helix the arms are fed with equal powers and a phase progression of $360 \times N/M°$ such that the currents in the arms along a directrix of the helical cylinder are in phase. To obtain a low axial ratio and satisfactory radiation patterns, the number of arms N should be larger than the mode number M by a factor in the range of 1.5 to 2.0.

The left-hand helix of Fig. 28-5 radiates left-handed circular polarization toward the zenith, right-handed circular polarization to the nadir, and horizontal polarization at an elevation angle of approximately $\psi\,°$. Figure 28-6 shows the variation of the axial ratio for right-handed elliptical polarization in the broadside direction with the pitch angle ψ. It is seen that the average pitch angle should be about 40° to achieve an axial ratio of less than 3 dB. The calculation of this curve was based on a sheath model of the helical currents and neglects the effect of the cylinder. If the cylinder circumference in wavelengths is greater than about $M - 1$ for M greater than 1, then reflections from the cylinder produce a phase error between the vertical and horizontal polarizations which degrades the axial ratio by more than that shown in Fig. 28-6. For mode 1, the cylinder circumference should be less than one-half wavelength.

The uniform helix is a traveling-wave antenna with an exponential attenuation rate which is a complex function of M, N, ψ, cyl-

FIG. 28-5 Three-arm helical antenna.

inder diameter, and arm diameter. Attenuation increases with arm diameter and decreases with pitch angle. Attenuation rates of up to 6 dB per axial wavelength may be achieved. To approximate a uniformly illuminated aperture, the pitch angle may be varied along the aperture (keeping the turn length constant), which leads to a spiral-type structure. If $2\,M/N$ is not an integer, the reflected wave from the end of the helix will radiate a beam in an elevation direction other than broadside. If this is not the case, the reflected wave will radiate a broadside beam which produces scallops in the azimuth pattern with $2M$ lobes. The axial ratio is not degraded because the sense of circular polarization for the reflected wave is the same as that for the incident wave in the broadside direction. This effect may be reduced by terminating the helix with radiating loads or resistors. The helix is usually designed so that one-way attenuation is about 16 dB.

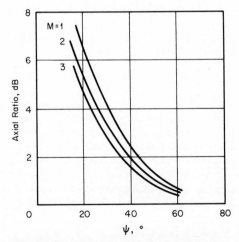

FIG. 28-6 Axial ratio of helical antennas in broadside direction versus pitch angle.

Because the helix is a traveling-wave antenna, the impedance bandwidth is large, especially if $2\,M/N$ is not an integer, since reflection cancellation occurs at the input to the feed network. However, the pattern bandwidth is limited by beam scan with frequency since it is equivalent to an end-fed array. For desirable pitch angles, the beam of a helix bay scans about 1° for a 1 percent frequency change. For Channels 2 to 6, this limits the bay length to about two to three wavelengths. Thus, two or more bays are generally used. A three-arm mode 1 helix may be used for these channels and has less wind loading than a horizontally polarized batwing antenna. For Channels 7 to 13, bay lengths of about six wavelengths may be used. Three- or four-arm mode 2 helices are used. In the UHF band, bay lengths may be in the range of 16 wavelengths, and mode numbers of five or more are used, with the number of arms being greater than the mode number.

Because of their symmetry, helical and spiral antennas have an excellent omnidirectional pattern, with a circularity of less than ± 1 dB. The axial ratio is about 2 dB for the low VHF channels and even less for the other channels. The arms of a helical antenna have a characteristic impedance similar to that of a rod over a ground plane, with the height equal to the spacing of the arm from the support cylinder. Thus, the impedance is in the order of several hundred ohms for practical arm diameters. Special techniques, such as inductance-capacitance tuners or transformers, are required to match this impedance to the outputs of the power splitter in the feed network for the multiarm helix.

A novel feature of the higher-order-mode multiple-arm helical antenna is that it may be placed around triangular or square towers and still produce an omnidirectional pattern. This occurs because the waves radiated toward the support structure with a $M \times 360°$ azimuth phase variation enter a cutoff region in a manner similar to that for radial waveguides. Thus, the waves are reflected, and the support structure has little effect on the radiation pattern if the mode number is about 5 times larger than the tower diameter in wavelengths.

Slanted-Dipole Antennas

Many circularly polarized FM and TV broadcast antennas are based on the concept of a circular array of slanted dipoles. The dipoles may be linear, V-shaped, curved, or of a similar configuration. Each dipole radiates linear polarization, but the slant angles and diameter of the circular array are adjusted so that an omnidirectional circularly polarized radiation pattern is obtained. The term *circular array* is used here to include one or more dipoles placed on a circle with rotational symmetry. Actually, the array of slanted dipoles may be considered as a short length of an N-arm helical antenna with a standing-wave rather than a traveling-wave current distribution. The cross section of the helical structure need not be circular to produce circular polarization. This approach is in contrast to that in which unidirectional circularly polarized radiators (such as crossed dipoles placed in front of a reflecting screen) are placed around a mast or tower to obtain omnidirectional or other azimuth patterns.

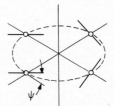

FIG. 28-7 Circular array of four slanted dipoles.

Figure 28-7 shows a circular array of four slanted dipoles[6] with a slant angle ψ (which is similar to the pitch angle of a helical antenna). If the dipoles are assumed to be fed in phase, the following is a simplified explanation of how circular polarization is achieved. In a direction in line with opposite dipoles, the phase of the radiation from the vertical and horizontal components of the current in the front dipole lead those of the rear. The vertically polarized components are vectorally added, whereas the horizontal components are subtracted, which produces a 90° phase difference between the two polarizations. The slant angle is adjusted to produce equal magnitudes of vertical and horizontal polarization, taking into account radiation from the other two dipoles, which results in circular polarization. To generalize, N slanted dipoles may be placed in a circular array and excited in mode M (an integer) to radiate omnidirectional circular polarization[7,8] in the plane of the array where the current in the nth dipole is given by

$$I_n = \exp(j2\pi nM/N) \qquad \textbf{(28-1)}$$

The circumferential spacing of the dipoles must be about one-half wavelength for mode 0 and less than that for other modes to obtain omnidirectional patterns and circular polarization. The pitch angle ψ, which depends on the shape of the dipole, is approximately

$$\psi = -\tan^{-1}\left(\frac{J_M'(\beta\rho)}{J_M(\beta\rho)}\right) \qquad \textbf{(28-2)}$$

for curved dipoles lying on the surface of a circular cylinder, where ρ is the radius of the circular array, $J_M(\beta\rho)$ is a Bessel function, and the prime represents the derivative. The effect of a vertical mast which introduces a phase error between the vertical and horizontal polarization components is not included Eq. (28.2). The slant angle for right-hand circular polarization is positive for mode 0 and negative for the other modes when the array circumference is equal to or less than M wavelengths.

Figure 28-8 illustrates a simple, compact version of this concept in which two V dipoles[9] are supported by a horizontal mast. With an included angle of about 90°, the V dipoles perform approximately as a four-dipole circular array. One-half of each dipole is shunt-excited from the center of the support mast. If the dipole length is about one-half wavelength, then the current on the parasitic arms will be about the same as the current on the shunt-driven arms of the dipoles. The antenna is matched by adjusting the positions of the shunt feeds and dipole lengths. Alternatively, one-half of each dipole may be series-fed as illustrated in Fig. 28-9, in which the two dipoles are supported in a T arrangement. The internal coaxial feeds are connected to gaps in the monopoles. Since the impedance bandwidth is on the order of 1 percent or less, these antennas are most useful for FM applications. A multiplicity of bays with wavelength spacing may be fed by and supported by a vertical transmission line.

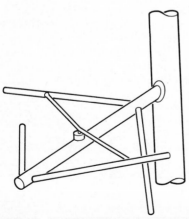

FIG. 28-8   Two shunt-fed slanted V-dipole antennas.

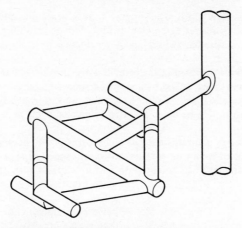

FIG. 28-9   Series-fed slanted dipoles.

The array is usually supported on the side of a mast or tower. Reflections from the support distort the azimuth patterns, especially for vertical polarization, and degrade the axial ratio. Parasitic dipoles may be added to reduce these effects.

A circular array of four curved dipoles[10] is shown in Fig. 28-10. The dipoles form a short section of a four-arm helix antenna and are approximately one-half wavelength long. The array circumference is approximately one wavelength. Thus, the overlap of the dipoles provides approximately the equivalent of a constant circular current distribution for both the horizontal and the vertical components. The four dipoles are shunt-fed asymmetrically by four rods emanating from the center of the array. The rods are connected to the center conductor of a coaxial feed enclosed in the horizontal support structure. The same approach may be used for circular arrays of two or three curved dipoles when the array circumference is approximately one-half and three-fourths of a wavelength respectively. The pattern circularity in free space is ±1 dB,

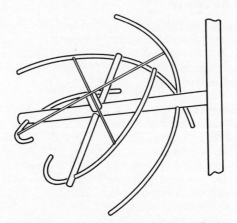

FIG. 28-10   Four shunt-fed helical-type dipoles.

and the axial ratio is about 3 dB. The support mast or tower degrades these values by several decibels. The power rating and bandwidth increase with the number of dipoles in the array. An 11 percent bandwidth has been achieved for a four-element FM array with 2-in-diameter arms.

A single dipole may be bent in the form of a one-turn helical antenna to produce circular polarization. It may be fed by a slotted coaxial-line balun[11] or inductive loop coupling. A disadvantage is that it radiates up and down so that bay spacings of less than one wavelength should be used. For two or more dipoles in each bay there are nulls up and down.

The degradation of azimuth patterns and axial ratio for side-mounted antennas may be eliminated by placing the circular array of slanted dipoles around the mast and adding shunt dipoles[8,12] to compensate for the vertical currents flowing on the

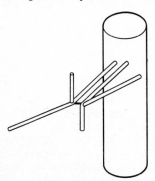

mast. Figure 28-11 shows one element of a circular array which consists of a slant half-wave dipole and a short vertical dipole that are fed and supported by a balun structure. For mode 0 and an array diameter of about one-half wavelength, only three elements are needed in the circular array to produce a circularity of $\pm$ 1.5 dB and an axial ratio of less than 3 dB. The elements are fed in phase with equal power. The slant angle is approximately that given by Eq. (28-2) (without mast reflections), and the length of the vertical shunt dipole is adjusted to achieve a low axial ratio. A bandwidth greater than 10 percent may be achieved

FIG. 28-11 Slant dipole with a parallel-connected short vertical dipole.

with a bay spacing of 0.8 wavelength. Thus, the antenna may be used for both TV and FM applications. Since the wind loading is equal to or less than that for the batwing antenna, it may be used to replace the batwing on existing towers for conversion to circular polarization.

Crossed-Dipole Antennas

A common technique for producing circular polarization has been to place two linear dipoles at right angles in front of a reflecting screen and to feed them with equal voltage magnitudes and a 90° phase difference. However, the azimuth beamwidth for horizontal and vertical polarization is about 60° and 120° respectively. Thus, the axial ratio is low only for directions near the normal to the screen. This deficiency may be corrected in several ways.

V dipoles as illustrated in Fig. 28-12 may be used to increase the azimuth beamwidth for horizontal polarization. Three crossed V-dipole panels may be placed around a triangular tower to obtain a circularity of ± 2 dB and a maximum axial ratio of 4 dB. The crossed dipoles may be identical and fed in phase quadrature or be unequal in length and fed in phase with the lengths adjusted to produce quadrature currents in the dipoles. Another version of this type of antenna has three reflecting panels placed in a Y configuration and supported by a central mast. Three crossed V dipoles are phased in the 120° sectors formed by the panels. This provides a more compact structure than the triangular tower.

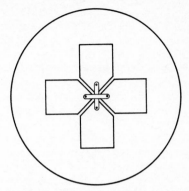

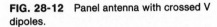

FIG. 28-12   Panel antenna with crossed V dipoles.

FIG. 28-13   Crossed broadband dipoles in a cylindrical cavity.

A better technique for equalizing the azimuth beamwidths for vertical and horizontal polarization is to enclose crossed planar dipoles in a cylindrical cavity[13] as shown in Fig. 28-13. The length-to-width ratio of the planar dipoles is about 3 and provides a bandwidth of 10 percent with a VSWR less than 1.1:1. The cavity depth is 0.25 wavelength. Cavity diameters of 0.65 and 0.8 wavelength produce azimuth 6-dB beamwidths of 120° and 90° respectively, which are desired for triangular and square towers respectively. The dipoles are fed in phase quadrature by two baluns forming a four-tube support structure. The circularity is ±2 dB, and the axial ratio is less than 2 dB.

Another approach is to place four half-wavelength dipoles in a square arrangement, with the side of the square being somewhat larger than a half wavelength. The vertical and horizontal dipoles are fed in phase quadrature. Since the 6-dB azimuth beamwidths are about 90°, four panels around a square tower may be used for omnidirectional applications.

Ring-Panel Antennas

The ring-panel antenna[14] consists of a multiplicity of ring radiators fed in series by a transmission line. Figure 28-14 shows two circular rings formed by strips over a panel and connected by rods over the panel which provide simple low-radiation transmission lines. The ring circumference is approximately one wavelength, as is the spacing between rings. The antenna is designed so that the characteristic impedance of the ring strip over ground is equal to the characteristic impedance of the transmission-line rod over ground. A practical value of the characteristic impedance is 140 Ω. By using a resistive termination on the last ring and/or special tuning techniques, it is possible to achieve a traveling-wave type of antenna. A traveling wave on a ring of one-wavelength circumference as shown in Fig. 28-14 radiates right-hand circular polarization. The ring is equivalent to four quarter-wave dipoles placed on a square with −90° progressive phasing. The azimuth beamwidth for horizontal polarization is usually about 10° less than that for vertical polarization. The beamwidths may be equalized and changed by means of parasitic elements such as monopoles on each side of the rings and dipoles in front of the rings. A cavity is not required.

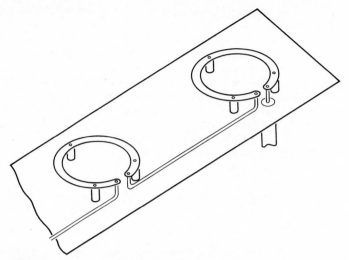

FIG. 28-14   Two elements of a series-fed traveling-wave ring-panel antenna.

The radiation from, or the attenuation through, a traveling-wave ring increases with the height of the ring above the panel and decreases with the characteristic impedance of the strip. The attenuation ranges from about 0.5 to 8 dB for ring heights of 0.05 to 0.2 wavelengths, respectively. Because of this attenuation, it is necessary to increase the height of the rings as one progresses from the feed point in order to approximate a uniform array.

Since the distance along the transmission line and ring between similar points on adjacent rings is two wavelengths for broadside radiation, the beam direction will scan 1.15° for a 1 percent change in frequency. This limits the number of end-fed rings to 3 for Channel 2 and to about 10 for the UHF channels.

The axial ratio may be reduced to a very low level by introducing reflections on the transmission-line rods which radiate left-hand circularly polarized waves. The magnitude and phase of the reflections may be controlled by the size and position of the reflecting device to cancel undesired left-hand circular polarization from other parts of the antenna.

Several panels may be stacked vertically to achieve the desired gain and beamwidth. Beam tilt and null fill may be achieved by control of the excitation of the panels and/or control of the amplitude and phase of the radiation from each ring by the height of the ring and the transmission line-rod length respectively.

Axial-Ratio Measurement

The axial ratio of elliptically polarized antennas may be measured by rotating a linearly polarized transmit antenna and receiving with the test antenna. If there are negligible multipath reflections, as in an anechoic chamber, then the axial ratio is simply the ratio of the maximum to the minimum field received by the test antenna. For TV and FM antennas it is not practical to eliminate ground reflections. However, it is possible to design some long ranges so that the reflected wave is at an angle much smaller than the Brewster angle. In this case the reflection coefficients for vertical and horizontal polarization are nearly identical, and the rotating linearly polarized antenna technique[15] can be used. In many other cases, such as a shorter range and/or variable moisture conditions, the reflection coefficients for the two polarizations will not be the same. However, the range can be calibrated and the axial ratio can be computed as follows. A linearly polarized antenna with a clockwise 45° slant angle, as viewed from the transmit site, is placed at the test site. The linearly polarized transmit antenna is then rotated, and the maximum–minimum ratio S_0 and the null angle of the transmit antenna ψ_0 are measured.

A convenient form of these antennas is a rotatable dipole in a cylindrical cavity. The process is then repeated by using the test antenna to measure S_1 and ψ_1, the maximum–minimum ratio and null angle for the transmit antenna. The following calculations are performed:

$$K = j\frac{1 + \Gamma}{1 - \Gamma} \qquad (28\text{-}3)$$

where

$$\Gamma = \frac{S_0 - 1}{S_0 + 1}\exp(j2\psi_0) \qquad (28\text{-}4)$$

and

$$\rho = \left|\frac{K(1 - P) - (1 + P)}{K(1 - P) + (1 + P)}\right| \qquad (28\text{-}5)$$

where

$$P = \frac{S_1 - 1}{S_1 + 1}\exp(j2\psi_1) \qquad (28\text{-}6)$$

and finally the axial ratio AR is

$$AR = \frac{1 + \rho}{1 - \rho} \qquad (28\text{-}7)$$

Another approach is to use a crossed-dipole transmit antenna with a switchable and adjustable feed network so as to produce right circular and left circular polarization at the test site. The feed circuit is adjusted so that the received signal of a rotating linearly polarized antenna is uniform for each of the transmit polarizations. The axial ratio is given by Eq. (28-7), where ρ is the ratio of the right circular to left circular polarization signals.

28-4   HORIZONTALLY POLARIZED ANTENNAS

Reference may be made to the first edition of the *Antenna Engineering Handbook* for descriptions of loop, cloverleaf, V, and other types of horizontally polarized antennas which were popular before the widespread use of circular polarization.

Dipole Antennas

A top view of a dipole panel-type antenna is illustrated in Fig. 28-15. The support arms are electrically connected to the dipole arms and form a part of the radiating structure and also aid in the matching of the dipole. The dipole is usually about a half wavelength long and spaced about a quarter wavelength from the panel. The dipole may be fed from a balanced transmission line connected to the center of the dipole or

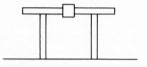

FIG. 28-15 Dipole panel antenna.

from a coaxial line entering one of the support arms and extending to the central feed gap, forming a balun. Bandwidths of 10 percent may be achieved with shunt stub-matching techniques. The panels may be placed on square or triangular towers to obtain omnidirectional or directional patterns.

Batwing Antennas

The batwing antenna[16,17,18] illustrated in Fig. 28-16 is the most popular horizontally polarized VHF TV antenna. It consists of several bays of turnstile configurations of broadband planar dipoles. The dipoles are formed by a grid of rods and have a length and width of about one-half wavelength. Each half of a dipole is supported by a spacer rod which is shorted to the supporting mast at the top and bottom of the dipole. Oppo-

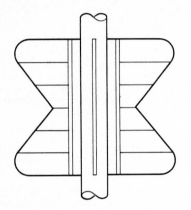

FIG. 28-16 Batwing, or superturnstile, antenna.

site halves of each dipole are fed out of phase from a power divider with equal-length coaxial transmission lines by feeding one-half from a coaxial line grounded to the mast and the center conductor connected to the center of the spacer and the other half from a coaxial line running along and connected to the spacer with the center conductor connected to the mast at the center of the spacer. The quadrature dipoles are fed 90° out of phase to provide an azimuth pattern with a circularity of ±2 dB. The 90° phasing may be obtained by different line lengths for the quadrature dipoles or by a quadrature hybrid. In the latter case, the visual and aural transmitters may be diplexed into the antenna system through

the hybrid. The isolation between the transmitters is about the same as the return loss from the dipoles, which is greater than 26 dB. Each half dipole has an impedance of 75 Ω, and bandwidths of 20 percent with a return loss greater than 26 dB have been achieved. The dipoles have nulls in the nadir and zenith directions because of their width or height when viewed from the horizontal plane. Thus the bays are spaced by one wavelength to achieve maximum gain. Two to six bays are normally used for Channels 2 to 6 and up to 18 bays for Channels 7 to 13. Peanut-shaped azimuth patterns may be achieved by an unequal power split between the orthogonal dipoles.

Slot Antennas

Both resonant and nonresonant (or standing-wave and traveling-wave respectively) end-fed arrays of slots are used for TV broadcasting. The resonant arrays are restricted to UHF applications because of their limited bandwidth.

The traveling-wave slot antenna[19] illustrated in Fig. 28-17 is a large end-fed coaxial transmission line with a slotted outer conductor. The slots are arranged in pairs at each layer, with the pairs separated by a quarter wavelength along the length of the antenna. Adjacent pairs occupy planes at right angles to each other. The slot pairs, which are approximately one-half wavelength long, are fed out of phase by the coaxial line by capacitive probes projecting radi-ally inward from one side of each slot so as to produce a figure-eight pattern. The probes are placed on opposite sides of adjacent in-line slots which are spaced one-half wavelength to provide in-phase excitation. The quarter-wavelength separation of lay-ers in conjunction with the space-quadra-ture arrangement of successive layers of slots effects a turnstile-type feed which pro-duces a horizontally polarized azimuth pat-tern with a circularity of ± 1 dB for VHF applications. An equal percentage of the power in the coaxial line is fed to each layer of slots, which results in an exponential aperture distribution that provides null fill.

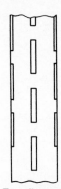

FIG. 28-17   Traveling-wave slot antenna.

Reflections from adjacent layers tend to cancel, which allows the traveling-wave oper-ation. The top slots are strongly coupled to the line to reduce reflections. For high-gain applications, one-half of the slots may be eliminated, resulting in a one-wave-length spacing of in-line slots. The pipe diameter is 10 to 20 in for Channel 7–13 applications.

The standing-wave antenna consists of layers or bays of one or more axial slots spaced by one wavelength and fed by a coaxial line with the slotted pipe forming the outer conductor. For UHF applications, the pipe diameter may vary from 6 to 18 in, depending on the gain and frequency. Azimuth patterns are controlled by the number of slots per bay. One slot per bay produces a skull-shaped pattern, two slots a peanut-shaped pattern, and three slots a trilobe pattern. Four or more slots per bay are usually required for an omnidirectional pattern with a circularity of ± 1 dB. The slots may be coupled to the coaxial line by several means such as capacitive probes or bars[20] con-nected to one side of the slot, rods connecting one side of the slot to the center con-ductor, or balanced loops connected to the slot with a variable orientation to control the magnetic coupling. The symmetry of the slots must be controlled to prevent exci-tation of propagating higher-order modes in the coaxial line. The length of the slot and the coupling mechanism are usually adjusted so that the shunt reactance of each bay of slots is zero and the shunt resistance is $NZ_0/2$, where N is the number of end-fed bays and Z_0 is the characteristic impedance of the coaxial line. This produces an overloaded resonant array[21] which provides larger bandwidth with the proper input-matching device. For antennas with a gain of more than 20, it is necessary to center-feed the slotted array. This may be accomplished by using a triaxial line in the lower half of the antenna.

Parasitic elements such as monopoles may be added to the cylinder to shape the azimuth pattern.

Zigzag Antennas

The panel type of zigzag antenna[22] shown in Fig. 28-18 is a simple type of traveling-wave antenna which may be placed around triangular or square towers to produce a wide variety of azimuth patterns. The antenna consists of a wire or rod that is bent at

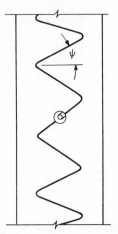

half-wavelength intervals to form the zigzag structure. This provides a broadside beam with horizontal polarization since the radiation from the vertical components of the currents in adjacent half-wavelength segments cancels, whereas that for the horizontal components adds. The azimuth 6-dB beamwidth is about 90°, which is desirable for a square tower. The beamwidth may be increased for triangular-tower configurations by adding arrays of parasitic monopoles along each side of the zigzag or bending the panel and zigzag in a V configuration. It is preferable to feed the zigzag in a balanced manner at the center of one rod as shown, which eliminates radiation from the feed structure. If an unbalanced feed is used, e.g., at the bend, the feed radiation will distort the azimuth pattern.

FIG. 28-18 Zigzag antenna with balanced feed.

The zigzag may be designed on the basis of a leaky-wave antenna, in which the attenuation per wavelength due to radiation increases with distance from the feed point. Figure 28-19 shows the attenuation per axial wavelength versus the height of the zigzag above the panel for several pitch angles ψ. The rod diameter and band radius are 0.01 and 0.03 wavelength respectively. To simulate a uniform aperture dis-

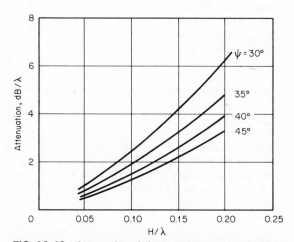

FIG. 28-19 Attenuation of zigzag antenna versus height.
(Courtesy of Cetec Antennas.)

tribution, the pitch angle decreases with distance from the feed point and is adjusted along with the height of the zigzag to give a one-way attenuation of about 15 dB for the current on the zigzag. It is preferable and simpler to use a constant-height zigzag. Since the beam direction of the incident and reflected waves on each half of the zigzag scan with frequency but in opposite directions, the length of a zigzag panel is limited by the bandwidth of the channel. The average pitch angle is usually about 35°, which produces a beam scan of 1° per 1 percent change in frequency. For UHF, panel lengths of 16 wavelengths may be used.

Since reflections occur at each bend, it is necessary to compensate for these in order to achieve a traveling-wave antenna. For a bend radius of 0.05 wavelength, the reflection coefficient varies over the range of 0.1 to 0.2 for practical pitch angles and heights of the zigzag. These reflections may be reduced considerably by placing the support insulators one-eighth wavelength before each bend as viewed from the feed point. It is usually necessary to add shunt capacitive tuners along the zigzag to achieve a traveling-wave condition, which is required for a wide impedance bandwidth.

Beam tilt may be achieved by displacing the feed from the center of the middle zigzag element or changing the lengths of the rods from the feed to the upper and lower zigzags.

Helix Antennas

Figure 28-20 shows a single bay of a single-arm right-hand and left-hand helix fed in phase at the center so that the vertically polarized components of the two helices cancel in the broadside direction. The pitch angle is about 12° so that the vertically polarized radiation from each helix is about 10 dB down from the horizontally polarized radiation, which produces about 0.5-dB loss in gain due to cross-polarization radiation. Since the beam of each helix scans about 2.7° per 1 percent change in frequency, the bay length is limited to about six wavelengths for Channels 7 to 13. For these channels the mechanical requirements of the supporting mast usually result in a mode 2 helix with a turn length of two wavelengths for broadside radiation. The mast diameter is chosen so that the one-way attenuation through each helix is 24 dB. Because of this, the circularity of the horizontal pattern is less than ±1.5 dB.

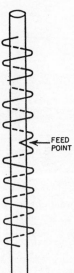

FIG. 28-20   Horizontally polarized helical antenna.

28-5   MULTIPLE-ANTENNA INSTALLATIONS

It is desirable to have a number of TV and FM stations share the same supporting structure for their antennas. This reduces the costs of the supporting structures and reduces or eliminates the need for rotatable receiving antennas.

FIG. 28-21   Candelabra antenna in San Francisco *(Courtesy of RCA Corp.)*

It is most economical to stack the antennas vertically. Unfortunately, each broadcaster wants to be "king of the hill," i.e., to have the top antenna, even though there is only a fraction of a dB difference in coverage for tall towers. This has resulted in candelabra installations in which antennas are mounted on the corners of a triangular support structure with a separation of 50 to 100 ft (15 to 30 m). Figure 28-21 shows the San Francisco Mount Sutro structure,[23] which supports five VHF, three UHF, and four FM antennas. A problem[24,25] with candelabra installations is that reflections from the other antennas produce ripples of several decibels in the azimuth pattern with the ripple directions being frequency-sensitive, which can lead to distortions in the received picture, especially for color-TV stations. The problem is tolerable for horizontally polarized antennas but more difficult for circularly polarized antennas, where reflections for vertical polarization from the antenna support structures are much stronger.

REFERENCES

1   Henry Jasik (ed.), *Antenna Engineering Handbook,* 1st ed., McGraw-Hill Book Company, New York, 1961, chap. 23.

2 *Reference Data for Radio Engineers,* 6th ed., Howard W. Sams & Co., Inc., Indianapolis, 1975, chap. 30.

3 J. Perini, "A Method of Obtaining a Smooth Pattern on Circular Arrays of Large Diameters with $\cos^n \phi$ Elements," *IEEE Trans. Broadcast.,* vol. BC-14, no. 3, September 1968, pp. 126–136.

4 R. H. DuHamel, "Circularly Polarized Helix and Spiral Antennas," U.S. Patent 3,906,509, Sept. 16, 1975.

5 O. Ben-Dov, "Circularly Polarized, Broadside Firing Tetrahelical Antenna," U.S. Patent 3,940,722, Feb. 24, 1976.

6 G. H. Brown and O. M. Woodward, Jr., "Circularly Polarized Omnidirectional Antenna," *RCA Rev.,* vol. 8, no. 2, June 1947, pp. 259–269.

7 O. M. Woodward, Jr., "Circularly Polarized Antenna Systems Using Tilted Dipoles," U.S. Patent 4,083,051, Apr. 4, 1978.

8 R. H. DuHamel, "Circularly Polarized Antenna with Circular Arrays of Slanted Dipoles Mounted around a Conductive Mast," U.S. Patent 4,315,264, Feb. 9, 1982.

9 Peter K. Onnigian, "Circularly Polarized Antenna," U.S. Patent 3,541,570, Nov. 17, 1970.

10 M. S. Suikola, "New Multi-Station Top Mounted FM Antenna," *IEEE Trans. Broadcast.,* vol. BC-23, no. 2, June 1977, p. 56.

11 R. D. Bogner, "Improve Design of CP FM Broadcast Antenna," *Communications News,* vol. 13, no. 6, June 1976.

12 M. S. Suikola, "A Circularly Polarized Antenna Replacement for Channel 2–6 Superturnstiles," *Broadcast Eng.,* February 1981.

13 R. E. Fisk and J. A. Donovan, "A New CP Antenna for Television Broadcast Service," *IEEE Trans. Broadcast.,* vol. BC-22, no. 3, September 1976, p. 91.

14 R. H. DuHamel, "Circularly Polarized Loop and Helix Panel Antennas," U.S. Patent 4,160,978, July 10, 1979.

15 J. A. Donovan, "Range Measurement Techniques for CP Television Antennas," *IEEE Trans. Broadcast.,* vol. BC-24, no. 1, March 1978, p. 4.

16 G. H. Brown, "A Turnstile Antenna for Use at Ultra-High Frequencies," *Electronics,* March 1936; "The Turnstile Antenna," *Electronics,* April 1936.

17 R. F. Holtz, "Super Turnstile Antenna," *Communications,* April 1946.

18 H. E. Gihring, "Practical Considerations in the Use of Television Super Turnstile and Super-Gain Antennas," *RCA Rev.,* June 1951.

19 M. S. Suikola, "The Traveling-Wave VHF Television Transmitting Antenna," *IRE Trans. Broadcast Transmission Syst.,* October 1957.

20 O. O. Fiet, "New UHF-TV Antenna," part 1: "Construction and Performance Details of TFU-24B UHF Antennas," *FM & Television,* July 1952; part 2: "TFU-24B Horizontal and Vertical Radiation Characteristics," *FM & Television,* August 1952.

21 S. Silver, *Microwave Antenna Theory and Design,* McGraw-Hill Book Company, New York, 1949, sec. 9.20.

22 O. M. Woodward, Jr., U.S. Patent 2,759,183, August 1956.

23 "Bay Area TV Viewers Turn to Sutro Tower," *RCA Broadcast News,* vol. 151, August 1973, pp. 19–31; H. H. Westcott, "A Closer Look at the Sutro Tower Antenna Systems," *RCA Broadcast News,* vol. 152, February 1974, pp. 35–41.

24 M. S. Suikola, "Predicting Characteristics of Multiple Antenna Arrays," *RCA Broadcast News,* vol. 97, October 1957, pp. 63–68.

25 M. S. Suikola, "Size and Performance Tradeoff Characteristics in Multiple Arrays of Horizontally and Circularly Polarized TV Antennas," *IEEE Trans. Broadcast.,* vol. BC-22, no. 1, March 1976, pp. 5–12.

Chapter 29

TV Receiving Antennas

Edward B. Joy

Georgia Institute of Technology

29-1 FCC TELEVISION FREQUENCY ALLOCATIONS

The Federal Communications Commission (FCC) frequency allocations for commercial television consist of 82 channels, each having a 6-MHz width.[1] Channels 2 through 6 are known as the low-band very high frequency (VHF) and span 54 to 88 MHz, Channels 7 through 13 are known as the high-band VHF and span 174 to 216 MHz, and Channels 14 through 83 are known as the ultrahigh-frequency (UHF) band and span 470 to 890 MHz. Channels 70 through 83 are designated for translator service. Channel designations and frequency limits are given in Table 29-1.

TABLE 29-1 Designations and Frequency Limits of Television Channels in the United States

Channel desig-nation	Frequency band, MHz	Channel desig-nation	Frequency band, MHz	Channel desig-nation	Frequency band, MHz
2	54–60	30	566–572	57	728–734
3	60–66	31	572–578	58	734–740
4	66–72	32	578–584	59	740–746
5	76–82	33	584–590	60	746–752
6	82–88	34	590–596	61	752–758
		35	596–602	62	758–764
7	174–180	36	602–608	63	764–770
8	180–186	37	608–614	64	770–776
9	186–192	38	614–620	65	776–782
10	192–198	39	620–626	66	782–788
11	198–204	40	626–632	67	788–794
12	204–210	41	632–638	68	794–800
13	210–216	42	638–644	69	800–806
		43	644–650		
14	470–476	44	650–656	70	806–812
15	476–482	45	656–662	71	812–818
16	482–488	46	662–668	72	818–824
17	488–494	47	668–674	73	824–830
18	494–500	48	674–680	74	830–836
19	500–506	49	680–686	75	836–842
20	506–512	50	686–692	76	842–848
21	512–518	51	692–698	77	848–854
22	518–524	52	698–704	78	854–860
23	524–530	53	704–710	79	860–866
24	530–536	54	710–716	80	866–872
25	536–542	55	716–722	81	872–878
26	542–548	56	722–728	82	878–884
27	548–554			83	884–890
28	554–560				
29	560–566				

29-2 TV SIGNAL-STRENGTH ESTIMATION

The FCC limits the maximum effective radiated power (antenna input power times antenna gain) of commercial television stations to 100 kW for low-band VHF, 316 kW for high-band VHF, and 5 MW for the UHF band. Maximum TV transmitting-antenna heights are limited to 2000 ft (609.6 m). Figure 29-1 presents predicted field-strength levels for Channels 7 to 13 versus transmitting-antenna height and distance from the transmitting antenna for 1 kW radiated from a half-wavelength dipole antenna in free space. The figure predicts field strength 30 ft (9.14 m) above ground that is exceeded 50 percent of the time at 50 percent of the receiving locations at the specified distance.

Table 29-2 shows correction factors to be used with Fig. 29-1 to determine predicted field-strength levels for Channels 2 through 6 and Channels 14 through 69. The table presents correction factors for a 1000-ft transmitting antenna, but it is reasonable to use these factors for all transmitting-antenna heights. The table shows that field-strength levels for the low-VHF band and the UHF-band channels are typically lower than those for the high-VHF-band channels. Ground roughness between the transmitting and receiving antennas is also important in predicting field-strength levels for receiving sites more than 6 mi (9.6 km) from the transmitting antenna. It is measured along a line connecting the transmitting antenna and the receiving antenna that begins 6 mi from the transmitting antenna and terminates either at the receiving-antenna location or 31 mi (49.9 km) from the transmitting antenna, whichever is least. The roughness ΔH, measured in meters, is the difference between the elevation level exceeded by 10 percent of the elevations along the line and the elevation level not reached by 10 percent of these elevations. The value of roughness assumed in the formulation of Fig. 29-1 is 50 m. The correction factor ΔF, given in decibels, to be applied to the field-strength value of Fig. 29-1 for a frequency f in megahertz is

$$\Delta F = C - 0.03\Delta H (1 + f/300) \qquad \textbf{(29-1)}$$

where $C = 1.9$ for Channels 2 to 6

TABLE 29-2 Correction Factors for Fig. 29-1 for an Antenna Height of 1000 Ft

Distance, mi	Channels 2–6, correction, dB	Channels 14–69, correction, dB
1	0	0
5	−1	−1.5
10	−3	−3
20	−3	−3.5
30	−3	−6
40	−2.5	−9
50	−1.5	−10
60	−1	−9
80	0	−6.5
100	0	−4
200	0	−2.5

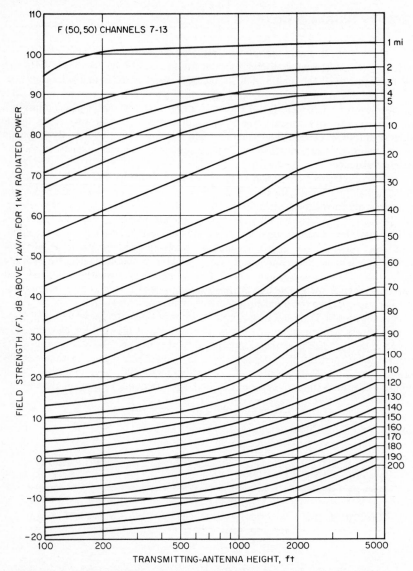

FIG. 29-1   Television Channels 7 to 13; estimated field strength exceeded at 50 percent of the potential receiver locations for at least 50 percent of the time at a receiving antenna height of 30 ft (9.14 m).

$C = 2.5$ for Channels 7 to 13
$C = 4.8$ for Channels 14 to 69

It can be seen that the correction factor lowers predicted field strength with increasing roughness and increasing frequency, reaching -39.2 dB for a roughness of 400 m and a frequency of 800 MHz.

Ground reflections cause the field strength to increase approximately linearly with height; i.e., the field strength is 6 dB less at a height of 15 ft (4.6 m) and 6 dB more at 60 ft (18.3 m). The rms signal voltage delivered to a 75-Ω load at the terminals of an antenna with gain G_A relative to a half-wavelength dipole immersed in a field with an rms field strength of E V/m at a frequency of f MHz is given by

$$V = \frac{48.5E\sqrt{1.64G_A}}{f} \qquad (29\text{-}2)$$

As an example, $f = 689$ MHz (Channel 50), receiving-antenna height = 10 ft (3 m), receiving-antenna gain = 12 dB ($G_A = 15.8$), transmitting-antenna height = 1000 ft (305 m), transmitting effective radiated power = 5 MW, distance = 40 mi (64.4 km), and surface roughness = 50 m. Let us find the rms received voltage. From Fig. 29-1, E at a 30-ft (9.14-m) height is 83 dB above 1 μV/m = 14,125 μV/m (it is noted that surface roughness is negligible); E at 10 ft is 4708 μV/m. The voltage received 50 percent of the time at 50 percent of such locations is greater than 1687 μV.

The received-signal voltage at a given receiving location can be increased by increasing the height and gain of the receiving antenna. In addition, the effect of building attenuation must be considered for TV receiving antennas installed indoors.[2] Measured attenuation values for a single wall between the receiving and the transmitting antennas range from 8 to 14 dB.

29-3   TV RECEIVING SYSTEMS

A TV receiving system is composed of an antenna, a transmission line, and a TV receiver. Baluns and splitters may be employed at the antenna or antennas and the TV receiver for impedance matching and for the separation or the combination of signals. Two TV receiving systems are typical: (1) an indoor antenna system using a small indoor antenna mounted on or near the TV receiver and (2) an outdoor antenna system using a larger antenna mounted high on a mast outdoors and connected to the TV receiver by an extensive transmission line. Indoor antenna systems typically are employed in high-signal-strength areas where multipath propagation is not severe. Outdoor antenna systems become necessary at a large distance from the transmitting antenna.

Overview of TV Antenna Types

A TV receiving antenna should have sufficient gain and a good impedance match to deliver a signal to a transmission line and subsequently to a TV receiver to produce a single clear TV picture and sound. Depending on its location, the antenna must reject reflected signals and other extraneous signals arriving from directions well off the direction to the transmitting antenna. Signal suppression should be particularly effective in the hemisphere directly opposite the transmitting antenna, as even small reflecting surfaces in this region can produce undesirable "ghost" images. TV receiving antennas should provide these properties over all TV channels of interest, which may include the complete 54- to 890-MHz range. They can be designed to receive all TV channels, or groups of channels such as all low-band VHF channels, all high-band

VHF channels, all VHF channels, and all UHF channels, or a single VHF or UHF channel.

TV receiving antennas fall into two major categories, indoor antennas and outdoor antennas. The most common configuration of an indoor antenna consists of two antennas, one for all VHF channels and one for all UHF channels. The most popular indoor VHF antennas are extendable monopole and dipole rods (rabbit-ear antennas). These antennas have measured average VHF gains of −4 dB with respect to a half-wavelength dipole and normally must be adjusted in length and orientation for best signal strength and minimum ghosting for each channel.[3] Rabbit-ear antennas are available with 75- or 300-Ω impedance. There are several popular indoor UHF antennas, including the circular loop, triangular dipole, and triangular dipole with reflecting screen. The loop and triangular-dipole antennas have low gain; the triangular dipole with reflecting screen has increased gain and a greatly improved ability to reject signals arriving from behind the reflecting screen. Indoor UHF antennas are most commonly designed with a balanced 300-Ω impedance. A popular indoor VHF-UHF combination antenna consists of a VHF rabbit-ear dipole antenna and a UHF loop antenna mounted on a fixture containing a switchable impedance-matching network. If a preamplifier is included as an integral part, the antenna is known as an *active antenna*.

Outdoor antenna systems can provide up to a 15-dB increase in antenna gain (including transmission-line loss), typically 15- to 20-dB greater rejection of ghost signals, and greater immunity to electrical interference over indoor antenna systems. This advantage, combined with a typical signal-strength increase due to antenna height of 14 dB [6 to 30 ft (1.8 to 9.1 m)] plus the removal of a typical 11-dB building attenuation, could result in an overall signal increase at the TV receiver of up to 40 dB. The most common outdoor antenna configuration is a combined VHF and UHF antenna. The combined antenna usually consists of two separate antennas mounted together to form a single structure. The most common VHF antenna is some variation of the log-periodic dipole array (LPDA). This antenna may be designed for a 300-Ω balanced or a 75-Ω unbalanced input impedance. There are several common types of UHF antennas including the LPDA, the broadband Yagi-Uda parasitic dipole array, the corner reflector, the parabolic reflector, and an array of triangular dipoles with a flat reflecting screen. Most UHF antenna types are designed with a balanced 300-Ω input impedance. The UHF LPDA may also be designed for an unbalanced 75-Ω impedance. Later sections treat these antenna types.

Transmission Lines for TV Receiving Systems

The standard impedances for the transmission lines used for TV and FM receiving systems are 75 Ω unbalanced and 300 Ω balanced. RG-59/U and RG-6/U types are the most commonly used 75-Ω lines, and twin-lead flat, foam, tubular, and shielded types are the most common 300-Ω lines. Table 29-3 shows average measured transmission-line losses for 100 ft (30.5 m) of the various types of lines.[4] All measurements were performed on new dry lines suspended in free space. The standard deviation of the measurements within the types was approximately 0.5 dB at VHF frequencies and 1.5 dB at UHF frequencies except for the flat 300-Ω twin-lead type, which was 0 dB at all frequencies, and for the 300-Ω shielded twin-lead type, which was 0.5 dB at VHF and increased to 6 dB at the high UHF frequencies.

The voltage standing-wave ratio (VSWR) measured on 100-ft sections (with

TABLE 29-3 Average Transmission-Line Insertion Loss, dB, for a 100-Ft Length of New Dry Line

Cable \ Channel	4	9	14	24	34	44	54	64	74	83
RG-59/U types	2.4	4.1	6.8	7.2	8.0	8.5	8.8	8.8	9.2	9.5
RG-6/U types	1.7	3.0	5.1	5.5	5.8	6.1	6.5	6.8	7.1	7.4
Flat twin-lead types	0.9	1.6	3.0	3.3	3.5	3.8	4.0	4.2	4.4	4.6
Foam twin-lead types	1.0	1.9	3.5	3.8	4.1	4.3	4.6	4.8	5.0	5.3
Tubular twin-lead types	0.9	1.7	3.2	3.5	3.7	4.0	4.2	4.5	4.7	5.0
Shielded twin-lead types	2.7	4.9	8.3	9.0	9.6	10.5	11.5	12.8	13.5	15.3

appropriate terminations) of the various types of lines is fairly independent of frequency on the VHF and UHF bands; it is shown in Table 29-4. Additional line losses and increased VSWR are reported for 50-ft (15.2-m) lengths of unshielded twin-lead transmission line because of proximity to metal and wetness. Table 29-5 presents these data for three cases: (1) 10 ft (3 m) of the 50-ft line parallel to and touching a metal mast, (2) 10 ft of the 50-ft line wrapped around a metal mast, and (3) a 50-ft section which is spray-wet.

Impedance-mismatch VSWR increases transmission loss. Such losses may occur at all junction points from the antenna to the TV receiver, including junctions at baluns, splitters, couplers, connectors, and preamplifiers. Impedance-mismatch loss is given by

$$\text{Mismatch loss (dB)} = -10 \log_{10} \left[1 - \left(\frac{\text{VSWR} - 1}{\text{VSWR} + 1} \right)^2 \right] \quad \textbf{(29-3)}$$

Mismatch loss is 0.2 dB for a VSWR of 1.5, 0.5 dB for a VSWR of 2.0, 1.9 dB for a VSWR of 4.0, and 4.0 dB for a VSWR of 8.0.

Baluns are used to match a 75-Ω unbalanced transmission line to a 300-Ω balanced line. Most commercially available baluns used for TV receiving systems employ bifilar transmission-line windings around a ferrite core to achieve broadband operation. Desirable features of baluns are minimum insertion loss and VSWR. Table 29-6 shows averaged measured balun insertion loss and VSWR for several TV channels. The standard deviation of these measurements is approximately 0.3 dB for insertion loss; it is 0.1 for VSWR at Channel 4, increasing to 0.5 at Channel 64.

TABLE 29-4 Average Measured Transmission-Line VSWR

Line	Average VSWR
RG-59/U types	1.2
RG-6/U types	1.2
Flat 300-Ω twin-lead types	2.0
Foam 300-Ω twin-lead types	1.8
Tubular 300-Ω twin-lead types	1.6
Shielded 300-Ω twin-lead types	2.4

TABLE 29-5 Average Additional Transmission-Line Loss and VSWR Because of Proximity to Metal and Wetness

Line	Additional loss, dB			VSWR increase		
	10 ft parallel to mast	10 ft wrapped on mast	50 ft spray-wet	10 ft parallel to mast	10 ft wrapped on mast	50 ft spray-wet
Flat 300-Ω twin-lead types	5.8	5.8	8.4	2.1	1.2	1.7
Foam 300-Ω twin-lead types	0.8	2.0	3.4	0.1	0.8	0.7
Tubular 300-Ω twin-lead types	0.8	3.0	0.8	0.1	0.4	0.3

Signal splitters are used to separate the VHF, FM, and UHF signals present on a transmission line at the receiver. They are also used as combiners to combine the signals from separate VHF, FM, and UHF antennas onto a single transmission line. Most commercially available splitters use three- to five-element high-pass and low-pass LC filters to accomplish the frequency separation. Some splitters contain input and/or output baluns to achieve 75- to 300-Ω impedance transformations. Desirable features of signal splitters, in addition to frequency selectivity, are low insertion loss and VSWR. Table 29-7 shows the average measured insertion loss and VSWR of a selection of commercially available signal splitters. The standard deviation of these measurements is approximately 1.0 dB for the insertion loss and 0.5 for the VSWR.

Transmission-line connectors can cause significant VSWR on a transmission line. Properly installed F-type connectors provide reasonably low VSWR for RG-59/U and RG-6/U coaxial lines for a limited number of matings. The connector problem is particularly severe for 300-Ω-twin-lead transmission lines. Standard twin-lead plugs and sockets and screw-connector terminal boards were found to have VSWRs near 1.5. Lower VSWR for twin-lead lines can be achieved by direct soldering of the lines in a smooth and geometrically continuous manner.

The noise figure for a lossy transmission-line system is equal to its total attenuation from input to output; 5 dB of attenuation means a noise figure of 5 dB. The total transmission-system loss has a significant impact on the noise figure of the TV receiving system.

TABLE 29-6 Average Measured Balun Insertion Loss and VSWR

Channel	4	9	24	44	64
Insertion loss, dB	0.8	0.6	1.1	1.2	1.7
VSWR	1.4	1.6	1.5	1.8	2.0

TABLE 29-7 Average Measured Splitter
Insertion Loss and VSWR

Channel	4	9	24	44	64
Insertion loss, dB	1.3	2.5	1.5	1.5	2.1
VSWR	1.8	2.1	1.7	1.7	2.5

TV Receivers

Typically, TV receivers have separate UHF and VHF inputs which are directly connected to the respective UHF and VHF tuners.[5] These inputs are either 75-Ω coaxial with a Type F connector or 300-Ω twin lead with a twin-lead screw-connector terminal board. The trend is toward the 75-Ω coaxial input, as it is more compatible with cable and other video systems which use the TV receiver as a display unit. The noise figure of a typical TV receiver is 6 dB for the low-VHF band and 8 dB for the high-VHF band; for the UHF band, the noise figure increases from 11 to 13 dB with increasing frequency. The input VSWR is approximately 2.5 for both the VHF and the UHF inputs.

29-4 TV RECEIVER-SYSTEM NOISE

The required signal strength for TV reception must be measured with respect to the TV receiver-system noise, including antenna noise. Figure 29-2 shows average observer picture-quality ratings versus the signal-to-random-noise ratio.[6] From this figure it is seen that a signal-to-noise ratio of approximately 42 dB is required to produce a rating of excellent by the average observer. The figure also shows that a signal-to-noise ratio of 10 dB produces an unusable picture.

The equivalent noise voltage at the terminals of a TV receiving antenna when

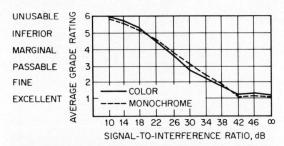

FIG. 29-2 Average observer rating versus signal-to-interference ratio. The solid line indicates color; the dashed line, monochrome. [*Source: "Engineering Aspects of Television Allocations," Television Allocation Study Organization (TASO), report to FCC, Mar. 16, 1959.*]

connected to a transmission-line–TV-receiving system is the rms sum of the antenna noise voltage and the equivalent noise voltage of the transmission-line–TV-receiving system.

Antenna noise voltage can be estimated by assuming that one-half of the gain of the antenna is receiving cosmic noise and the other half is receiving blackbody radiation from the earth, which is assumed to be at a temperature of 290 K.

The rms antenna noise voltage delivered to a matched impedance is given by

$$V_n = \sqrt{0.82 G_A kRB(T_g + 290°)} \qquad\qquad (29\text{-}4)$$

where G_A = gain of antenna with respect to a half-wavelength dipole
$\quad k$ = Boltzmann's constant = 1.38×10^{-23} W/Hz·K
$\quad T_g$ = equivalent noise temperature of sky, K
$\quad R$ = matched input resistance of antenna, Ω
$\quad B$ = bandwidth, Hz = 6×10^6 Hz

T_g is a function of frequency and varies primarily because of solar activity. Figure 29-3 gives measured maximum and minimum sky temperatures versus frequency.[7] It shows that sky temperature is larger than ambient temperature for VHF frequencies and smaller than ambient temperature for UHF frequencies. The rms noise voltage for Channel 2 (using a sky temperature of 10,000 K) impressed on a 75-Ω transmission line by a 6-dB-gain ($G_A = 4$) antenna is 14.5 μV.

The equivalent rms noise voltage of a transmission-line–TV-receiver system with an overall noise figure N is given by

$$V_n = \sqrt{NBRTk} \qquad\qquad (29\text{-}5)$$

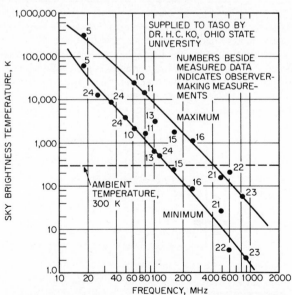

FIG. 29-3 Sky temperature as a function of frequency. [*Source: "Engineering Aspects of Television Allocations," Television Allocations Study Organization (TASO), report to FCC, Mar. 16, 1959.*]

where T is the temperature of the system in degrees Kelvin and is assumed to be 290 K. The parameters B, R, and k are as defined for Eq. (29-4). The overall noise figure of two systems connected in cascade can be calculated if the noise figures of both systems are known and the gain of the first system is known. The transmission-line system and the TV receiver form two systems in cascade as viewed from the terminals of the antenna. Let N_T represent the noise figure of the transmission line, G_T the gain of the transmission line (actually $G_T < 1$ for transmission lines), and N_{TV} be the noise figure of the TV receiver. The overall noise figure N_{T-TV} of this cascade system is given by

$$N_{T-TV} = N_T + \frac{N_{TV} - 1}{G_T} \qquad \textbf{(29-6)}$$

A Channel 2 outdoor antenna system might have a transmission-line system with an attenuation of 4 dB ($G_T = 0.4$) and thus a noise figure of 4 dB ($N_T = 2.5$). By using a noise figure for the TV receiver of 6 dB, the overall noise for the transmission-line–TV-receiver system is 9.95, or approximately 10 dB. From Eq. (29-5) the equivalent rms noise voltage is therefore 4.24 μV. The rms sum of this noise voltage and the antenna noise voltage for this Channel 2 example is 15.1 μV. The required signal voltage for an excellent picture as defined above is therefore 1901 μV at the terminals of the antenna. The Channel 2 example showed that the noise voltage is due primarily to the high sky temperature and the low transmission-line-system loss and low TV-receiver-noise figure. A Channel 69 example is quite different. Figure 29-3 shows a sky temperature of approximately 10 K, yielding an rms noise voltage for a 10-dB-gain, 75-Ω antenna of 3.96 μV. If transmission-line-system losses are 10 dB and the noise figure of the UHF tuner is 13 dB, the total-system noise figure is 23 dB, and the equivalent rms noise voltage for the voltage-line–TV-receiver system is 18.9 μV. The total equivalent system noise voltage at the terminals of the antenna is 19.3 μV. The Channel 69 example shows that antenna noise is small compared with TV receiving-system noise. An excellent-quality picture would therefore require a 2430-μV signal at the terminals of the antenna.

A preamplifier mounted at the terminals of the antenna with a noise figure of 5 dB ($NF = 3.16$) and a gain of 25 dB ($G = 316$) can significantly improve the noise figure of the UHF TV receiving system. Consider a cascade system composed of two sections. The first section is the preamplifier, and the second section is the above-mentioned UHF transmission-line–TV-receiver system with a noise figure of 23 dB. The overall noise figure of this cascade system, using Eq. (29-6) with the variables relabeled for the new cascade, is

$$N = 3.16 + \frac{199.5 - 1}{316} = 3.79 = 5.8 \text{ dB}$$

The equivalent noise voltage is 2.6 μV and combined with the antenna noise is 4.7 μV. The required signal at the terminals of the antenna for an excellent picture is now 592 μV, or 12.3 dB less than without the preamplifier. It is noted that greater preamplifier gain would not significantly reduce the required signal strength and might impair the large signal performance of the preamplifier.

29-5   TRIANGULAR-DIPOLE ANTENNAS

A dipole formed from two triangular sheets of metal is sufficiently broadband with respect to gain and VSWR (with respect to 300 Ω) for all-UHF-channel reception. The flare angle α and the length $2A$ of the triangular dipole (also known as the bowtie antenna) are shown in Figure 29-4. The triangular dipole has many of the same

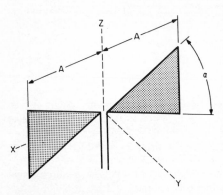

pattern and input-impedance characteristics of the biconical dipole discussed in Chap. 4, but it is lighter in weight and simpler to construct. Further simplification of the triangular dipole to a wire outline of the two triangles results in significant degradation of its broadband performance. However, the metal triangles can be approximated with wire mesh, provided the mesh spacing is less than one-tenth wavelength. The input impedance of a triangular monopole versus length A for a variety of flare angles α is shown in Ref. 8. The larger the values of A and α, the further the pattern in the XZ plane deviates from those in the

FIG. 29-4   Triangular dipole.

XY plane and the more predominant are the sidelobes. On the basis of pattern and impedance characteristics, compromise values of α between 60 and 80° may be used for values of A up to 210° (or 0.58λ). Figure 29-5 shows the measured-gain characteristics of a triangular dipole over the UHF band.[9] The triangular dipole can also be mounted approximately one-quarter wavelength in front of a flat reflecting screen to increase gain and decrease back radiation; typically, it is stacked vertically in two- or four-bay configurations for increased gain. The measured gain for several triangular dipoles with flat-screen-reflector configurations is also shown in Fig. 29-5. Measurements of several commercially available, vertically stacked multibay triangular dipoles with flat-screen-reflector arrays show that the VSWR is typically less than 2.0, the front-to-back ratio is greater than 15 dB, and sidelobe levels are less than 13 dB below the peak gain over 90 percent of the UHF band.[4]

29-6   LOOP ANTENNAS

The single-turn circular-loop antenna is a popular indoor UHF antenna primarily because of its low cost. The single-turn loop in free space is discussed in detail in Sec. 5-3 in Chap. 5. Figure 5-10a shows the typical configuration used for television reception when the impedance of the balanced feed line is 300 Ω. Although the loop is a resonant structure, entire-UHF-band operation is possible by using a 20.3-cm-diameter single loop where the circumference varies across the band from 1.0 wavelength at 470 MHz to 1.7 wavelengths at 806 MHz. Figure 5-16 shows the directivity of a single-turn loop versus circumference and thickness. This figure shows that the directivity is above 3.5 dB for a loop circumference between 1.0 and 1.7 wavelengths. Fig-

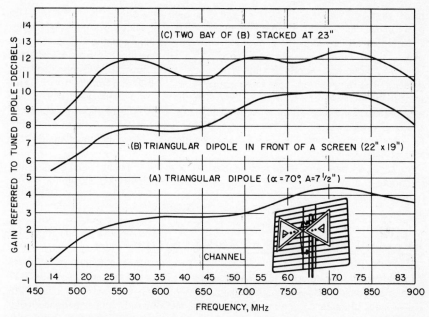

FIG. 29-5 Measured-gain characteristics. (*a*) Triangular dipole ($\alpha = 70°$, $A = 7\frac{1}{2}$ in). (*b*) Triangular dipole in front of screen (22 by 19 in). (*c*) Two bays of antenna *b* stacked 23 in apart.

ures 5-11 and 5-12 show the input resistance and reactance, respectively, of a single-turn circular loop. These figures show that the input resistance of a loop with a conductor thickness parameter* equal to 10 ranges from 100 to 520 Ω over the circumference range of 1.0 to 1.7 wavelengths, and the reactance is less than 100 Ω for a circumference between 1.0 and 1.45 wavelengths but increases to 210 Ω between 1.45 and 1.7 wavelengths. Thus, across the 1.0- to 1.7-wavelength band the input VSWR on a 300-Ω feed line is less than 3 and is near 1 for a circumference near 1.3 wavelengths. The far-zone patterns of a one-wavelength loop with the thickness parameter equal to 10 are shown in Figs. 5-14 and 5-15 *b*. These figures show that a one-wavelength loop has a bidirectional pattern with maximum directivity along the loop axis and that a vertical loop fed at the bottom is horizontally polarized. Measurements made on commercially available single-turn loops for UHF TV reception show that midband gain with respect to a half-wavelength dipole including mismatch losses is near 3 dB and falls off to −1 dB at both ends of the band.[4] The measured VSWR is close to 1 near the center of the band, increasing to 4 at both ends of the band. Measurements also show that concentric groupings of loops, sometimes configured so that the loops can be turned, have lower gain but better VSWR characteristics than the single loop.

A planar reflector mounted parallel to the loop can be used to increase directivity and greatly reduce the sensitivity of the loop to signals arriving from behind the reflec-

---

*The thickness parameter is 2 ln $(2\pi b/a)$, where *b* is the loop radius and *a* is the conductor radius.

tor. Figure 5-17 shows the geometry and directivity of a loop with a reflector. This figure shows that directivities greater than 8 dB are possible for a one-wavelength-circumference loop when the separation between the loop and the reflector is in a range from 0.1 to 0.2 wavelength. Figure 5-18 shows the input resistance and reactance for a one-wavelength-circumference loop versus separation from the planar reflector. The impedance is seen to be primarily real, increasing from 50 to 130 Ω for separations increasing from 0.1 to 0.2 wavelength. Figure 5-19 shows measured far-zone patterns for a loop with a reflector. This figure and Fig. 5-18 show the measured effects of variations in reflector size, suggesting a reflector size in the range of 0.6 to 1.2 wavelength on a side. Figure 5-23b shows a method of feeding the antenna with a coaxial line.

Yagi-Uda arrays of loops can be used for both single-channel and whole-band reception. Figure 5-21 shows the configuration and directivity achievable for single-frequency operation.

29-7   LOG-PERIODIC DIPOLE ARRAY ANTENNAS

The log-periodic dipole array (LPDA) antenna is the most commonly used all-VHF-channel antenna; it also is becoming a popular all-UHF-channel antenna. The LPDA is a broadband antenna capable of a 30:1 constant gain and input-impedance bandwidth. It has a gain of 6.5 to 10.5 dB with respect to a half-wavelength dipole, with most practical designs being limited to gains of 6.5 to 7.5 dB. The LPDA evolved from a broadband spiral antenna and is discussed more fully in Chap. 14. Figure 29-6 shows a schematic diagram of the LPDA and defines the angle α, dipole lengths ℓ_n, element diameters d_n, and element locations with respect to the apex of the triangle x_n. τ is shown as the ratio of a dipole's length or location to the length or location of the next larger dipole. The ratio τ is constant throughout the array. This figure also shows that all dipoles are connected to a central transmission line with a phase reversal between dipoles. In practice, the central transmission line takes two forms, a high-impedance

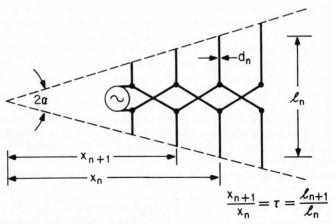

$$\frac{x_{n+1}}{x_n} = \tau = \frac{\ell_{n+1}}{\ell_n}$$

FIG. 29-6   Log-periodic dipole array.

form and a low-impedance form. The high-impedance form designed for a 300-Ω balanced input inpedance consists of a single boom insulated from all array dipoles. The transmission line in this case is a high-impedance open two-wire line extending the length of the antenna with crisscross connections from dipole to dipole to accomplish the phase reversal. The high-impedance form should be limited to lower frequencies, where the required spacing between the two wires of the line is a small fraction of a wavelength. The low-impedance form designed for an unbalanced 75-Ω input impedance is composed of two parallel conducting booms extending the length of the antenna, forming a low-impedance two-conductor transmission line. Phase reversal from dipole to dipole is accomplished by alternating the attachment of the dipoles along the two-boom transmission line. A coaxial feed line may be placed within one of the booms and connected to the short-dipole end of the array. In this configuration the array acts as a balun.

Figure 14-26 in Chap. 14 is a graph of equal-gain contours for the LPDA with a line showing optimal design conditions. This figure determines values of σ and τ on the basis of the desired gain. The angle α is calculated from σ and τ as

$$\alpha = \tan^{-1}\left[\frac{1-\tau}{4\sigma}\right] \tag{29-7}$$

The desired bandwidth B is given as the ratio of the highest frequency to the lowest frequency of operation of the array. A larger design bandwidth is determined to accommodate the active region of the array at the high-frequency end of the array. The design bandwidth B_S is given by

$$B_S = B[1.1 + 7.7(1-\tau)^2 \cot \alpha] \tag{29-8}$$

The required number of dipoles N in the array is given by

$$N = 1 + \frac{\ln (B_S)}{\ln (1/\tau)} \tag{29-9}$$

The length of the longest dipole ℓ_1 of the array is equal to the half wavelength at the lowest frequency in the desired bandwidth. The location of the dipole x_1 from the apex of the triangle is given by

$$x_1 = \frac{\ell_1}{2} \cot \alpha \tag{29-10}$$

The other dipole lengths ℓ_n and locations x_n are given by

$$\ell_n = \ell_1\tau^{(n-1)} \quad 2 \le n \le N$$
$$x_n = x_1\tau^{(n-1)} \quad 2 \le n \le N \tag{29-11}$$

Ideally, the length-to-diameter ratio of each dipole K_n should be identical. In practice, this is not usually the case. The primary effect of the variation of the length-to-diameter ratio is variation of the input impedance versus frequency. The impedance of the central transmission line Z_0 is next designed to transform the impedance of the active-region dipoles Z_A to the desired input resistance R_0. The impedance of the active-region dipoles is a function of the average K_n values of the elements K_{AVG} and is given by

$$Z_A = 120[\ln (K_{\text{AVG}}) - 2.25] \tag{29-12}$$

The characteristic impedance of the central transmission line Z_0 is given by[10]

$$Z_0 = R_0 \left[\frac{\sqrt{\tau}R_0}{8\sigma Z_A} + \sqrt{\left(\frac{\sqrt{\tau}R_0}{8\sigma Z_A} \right)^2 + 1} \right] \qquad (29\text{-}13)$$

The characteristic impedance Z_0 may be achieved with a two-conductor transmission line in which each conductor has a diameter D and the center-to-center spacing S of the conductors is given by

$$S = D \cosh (Z_0/120) \qquad (29\text{-}14)$$

Because of the wide bandwidth of the antenna, the accuracy of construction of an LPDA is not critical.

Several variations of the basic LPDA are also common. A large percentage of commercially available VHF LPDA receiving antennas are designed with reduced gain at the lower VHF because of the cost of the longer elements and the high signal strengths at the lower VHF frequencies. The low-frequency gain is normally reduced by elimination of the lower-frequency dipoles. A VHF LPDA can also be designed by using dipoles forming a V. This configuration allows the operation of the dipoles in their half-wavelength and ¾-wavelength modes and therefore eliminates the need for the higher-frequency dipoles. An LPDA array designed with V dipoles for the low-VHF band will operate in the ¾-wavelength mode for the high-VHF band, as the frequency ratio of these two bands is approximately 3.

The high-frequency gain of an LPDA can be enhanced with the addition of parasitic directors at the short-dipole end of the array. Most home-use LPDA antennas are designed for a balanced 300-Ω input impedance, while most master-antenna-television (MATV) and community-antenna-television (CATV) applications use the 75-Ω unbalanced input-impedance configuration.

Measured performance[4] of 19 home-use VHF LPDA antennas indicated an average low-VHF-band gain of 4.5 dB with a standard deviation of 1.5 dB with respect to a half-wavelength dipole at the same height above ground. The average high-VHF-band gain is 7.2 dB with a standard deviation of 1.4 dB. VSWR for the VHF band averages 1.9 with a standard deviation of 0.7. Three measured 300-Ω UHF LPDA antennas show an average gain of 7.1 dB with a standard deviation of 2.7 dB and an average VSWR of 2.7 with a standard deviation of 1.3. These averages are taken over frequency within the respective bands and over the antennas tested.

A circularly polarized LPDA can be formed by combining two linearly polarized LPDAs. The low-impedance configuration requires four booms, two for the horizontal LPDA and two for the vertical LPDA. All dipole lengths and locations of the second LPDA are obtained by multiplication of the respective dipole lengths and locations by $\tau^{1/2}$. The second LPDA should therefore have the same apex location, τ, α, and gain as the first. Multiplication by $\tau^{1/2}$ accomplishes the required frequency-independent 90° phase shift of the second LPDA with respect to the first.

29-8 SINGLE-CHANNEL YAGI-UDA DIPOLE ARRAY ANTENNAS

The Yagi-Uda dipole array antenna is a high-gain, low-cost, low-wind-resistance narrowband antenna suitable for single-TV-channel reception. Such antennas are popular in remote locations where high gain is required or where only a few channels are to be received. The Yagi-Uda dipole array is also discussed in Chaps. 12 and 27.

An empirically verified design procedure for the design of Yagi-Uda dipole arrays which includes compensation of dipole element lengths for element and metallic-boom diameters is presented.[11] Optimum Yagi-Uda arrays have one driven dipole, one reflecting parasitic dipole, and one or more directing parasitic dipoles as shown in Fig. 29-7. The design procedure is given for six different arrays having gains of 7.1, 9.2, 10.2, 12.25, 13.4, and 14.2 dB with respect to a half-wavelength dipole at the same height above ground. The driven dipole in all cases is a half-wavelength folded dipole which is empirically adjusted in length to achieve minimum VSWR at the design frequency. The length of the driven dipole has little impact on the gain of the array. Table 29-8 lists the optimum lengths and spacings for all the parasitic dipoles for each of the six arrays. The dipole lengths in this table are for elements which have a diameter of 0.0085 wavelength. Figures 29-8 and 29-9 are used to adjust these lengths for other dipole diameters and the diameter of a metallic boom, respectively. The frequency used in the design of single-channel Yagi-Uda arrays should be 1 per-

TABLE 29-8 Optimized Lengths of Parasitic Dipoles for Yagi-Uda Array Antennas of Six Different Lengths

$d/\lambda = 0.0085$	Length of Yagi-Uda array, λ					
$s_{12} = 0.2\lambda$	**0.4**	**0.8**	**1.20**	**2.2**	**3.2**	**4.2**
Length of reflector, ℓ_1/λ	0.482	0.482	0.482	0.482	0.482	0.475
Length of director, ℓ_i/λ: ℓ_3	0.424	0.428	0.428	0.432	0.428	0.424
ℓ_4		0.424	0.420	0.415	0.420	0.424
ℓ_5		0.428	0.420	0.407	0.407	0.420
ℓ_6			0.428	0.398	0.398	0.407
ℓ_7				0.390	0.394	0.403
ℓ_8				0.390	0.390	0.398
ℓ_9				0.390	0.386	0.394
ℓ_{10}				0.390	0.386	0.390
ℓ_{11}				0.398	0.386	0.390
ℓ_{12}				0.407	0.386	0.390
ℓ_{13}					0.386	0.390
ℓ_{14}					0.386	0.390
ℓ_{15}					0.386	0.390
ℓ_{16}					0.386	
ℓ_{17}					0.386	
Spacing between directors (s/λ)	0.20	0.20	0.25	0.20	0.20	0.308
Gain relative to half-wave dipole, dB	7.1	9.2	10.2	12.25	13.4	14.2
Design curve	(*A*)	(*B*)	(*B*)	(*C*)	(*B*)	(*D*)
Front-to-back ratio, dB	8	15	19	23	22	20

*SOURCE: P. P. Viezbicke, "Yagi Antenna Design," NBS Tech. Note 688, National Bureau of Standards, Washington, December 1968.

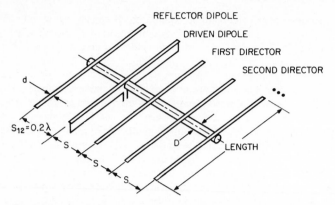

FIG. 29-7 Yagi-Uda array antenna.

cent below the upper frequency limit of the channel for VHF channels because the bandwidth of the Yagi-Uda is not symmetrical and gain falls rapidly on the high-frequency side of the band. The design frequency for UHF channels should be the center frequency of the channel. On the basis of desired gain, the lengths of the parasitic elements are found in one of the columns of Table 29-8. The reflector spacing from the driven element for all six configurations is 0.2 wavelength, and all directors are equally spaced by the spacing shown in the table. The table also specifies the design curve A, B, C, or D to be used in Fig. 29-8. The parasitic-dipole lengths are next adjusted for the desired dipole diameter other than the diameter of 0.0085 wavelength. The reflector length for the desired diameter can be read directly from one of the two upper curves in Fig. 29-8. Likewise, the length of the first director can be read directly from the designated director design curve for the desired diameter. This first director length is marked on the curve for subsequent use. The length of the first direc-

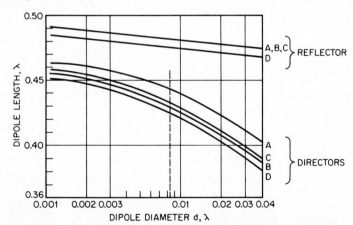

FIG. 29-8 Design curves to determine dipole lengths of Yagi-Uda arrays. *(Source: P. P. Viezbicke, "Yagi Antenna Design," NBS Tech. Note 688, U.S. Department of Commerce–National Bureau of Standards, October 1968.)*

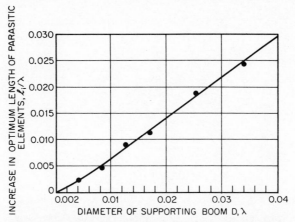

FIG. 29-9 Increase in optimum length of parasitic elements as a function of metal-boom diameter. *(Source: P. P. Viez-bicke, "Yagi Antenna Design," NBS Tech. Note 688, U.S. Department of Commerce–National Bureau of Standards, December 1968.)*

tor for a diameter of 0.0085 wavelength is also marked on the designated design curve. The arc length along the design curve between the two marks for the first director length is measured and used to adjust the remaining director lengths as follows. The other director lengths are plotted on the designated design curve without regard to their 0.0085-wavelength diameter. Each of the points is moved in the same direction and arc length as required for the first director. The new length is the diameter-compensated length for each of the remaining directors. If a metallic boom is used to support the reflector and directors, Fig. 29-9 is used to increase the length of each by the amount shown in the figure based on the diameter of the boom.

A circularly polarized Yagi-Uda dipole array can be formed by combining two linearly polarized Yagi-Uda dipole arrays on the same boom. The dipoles of one array might be horizontal and those of the other vertical. One array is positioned a quarter wavelength ahead of the other along the boom to accomplish the required 90° phase shift, and the feed position is midway between the two driven dipoles. Such a circularly polarized Yagi-Uda dipole array for Channel 13 is shown in Fig. 29-10.

29-9 BROADBAND YAGI-UDA DIPOLE ARRAY ANTENNAS

Several approaches to widening the inherently narrow bandwidth of the Yagi-Uda array are discussed. Yagi-Uda arrays with no more than five or six elements can be broadbanded by shortening the directors for high-frequency operation, lengthening the reflector for low-frequency operation, and selecting the driven dipole for midband operation. Figure 29-11 shows the measured gain versus frequency for a five-element single-channel and a five-element broadband Yagi-Uda array.[9]

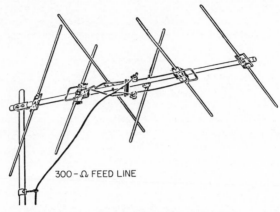

FIG. 29-10 Circularly polarized Yagi-Uda dipole array.
(*Channel Master Corp.*)

The driven dipole of a Yagi-Uda array can be replaced with a two-element ordinary end-fire driven array. The driven array is designed for midband operation, the directors for high-frequency operation, and the reflector for low-frequency operation. An extension of this concept is the replacement of the driven element with a log-periodic dipole array. The parasitic elements are then used primarily to increase gain at the upper and lower frequency limits of the log-periodic dipole array.

Two Yagi-Uda dipole arrays of different design frequencies may be designed on the same boom and use the same driven element. This is possible when the design frequency ratio of the two Yagi-Uda arrays is 3:1 as in the high-VHF and low-VHF bands. The interlaced parasitic elements have only a small impact on one another, and the driven element is used simultaneously in the first and third half-wavelength modes.

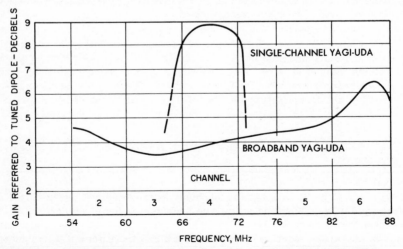

FIG. 29-11 Measured gain of five-element Yagi-Uda. (*a*) Single-element Yagi-Uda.
(*b*) Broadband Yagi-Uda.

The reflector dipole of a Yagi-Uda array may be replaced with a corner reflector or a paraboloidal reflector. The reflector and driven element are designed for low-frequency and midfrequency operation, and the directors are designed for high-frequency operation.

The broadband Yagi-Uda is not commonly used for all-VHF-channel operation, but it is commonly used, generally with a corner reflector, for all-UHF-channel reception. Measurements of commercially available all-UHF-channel Yagi-Uda antennas typically show large variation in gain and input impedance as the frequency is varied across the UHF band.[4] Typical of such an antenna is an average gain of 7 dB with respect to a half-wavelength dipole at the same height above ground with a variation of ±8 dB across the UHF band. The VSWR of such an antenna is found to vary within the range of 1.1 to 3.5 across the band.

29-10   CORNER-REFLECTOR ANTENNAS

The corner-reflector antenna is very useful for UHF reception because of its high gain, large bandwidth, low sidelobes, and high front-to-back ratio. (Corner-reflector antennas are discussed further in Sec. 17-1 in Chap. 17.) A 90° corner reflector, constructed in grid fashion, is generally used. One typical design is shown in Fig. 29-12. A wideband triangular dipole of flare angle 40°, which is bent 90° along its axis as shown in Fig. 29-13 so that the dipole is parallel to both sides of the reflector, is a good choice for the feed element.[9] Experimental results indicate that an overall length of 14¼ in (362 mm) for the dipole gives the optimum value of average gain over the band. The dipole has a spacing of about $\lambda/2$ at midband from the vertex of the corner reflector. A larger spacing would cause a split main lobe at the higher edge of the band, where the spacing from the vertex becomes nearly a wavelength.

Figures 29-13 and 29-14 are experimental results useful in designing grid-type corner reflectors. Figure 29-13 shows the relation between grid length L and gain at three different frequencies. It is seen that beyond 20 in (508 mm) of grid length very little is gained. Figure 29-14 shows the relation between gain and grid width W. At 700 and 900 MHz, the improvement in gain for a grid over 20 in wide

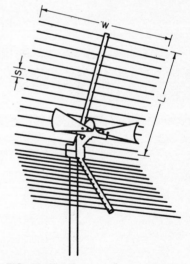

FIG. 29-12 Corner-reflector antenna.

is rather insignificant. To reduce the fabrication cost, a width of 25 in (635 mm) is considered a good compromise. In Fig. 29-15 the spacing for grid tubing of ¼-in (6.35-mm) diameter can be determined from the allowable level of the rear lobe in percentage of the forward lobe at the highest frequency, 900 MHz. If 10 percent is the allowable value, the spacing should be slightly under 1½ in (38 mm). Below 900 MHz, the rear pickup will be less than 10 percent provided that the grid screen is wide

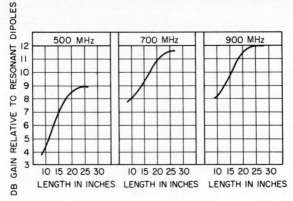

FIG. 29-13 Measured-gain characteristics versus grid length *L*.

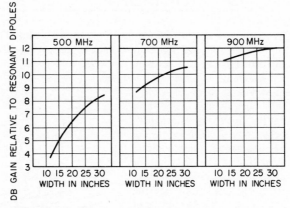

FIG. 29-14 Measured-gain characteristics versus grid width *W*.

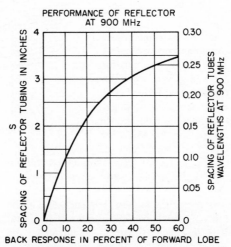

FIG. 29-15 Relation between the rear lobe in percentage of the forward lobe and the grid spacing *S* at 900 MHz.

$Z_0 = 280$

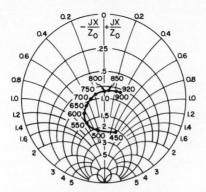

FIG. 29-16 Impedance characteristics of an ultrahigh-frequency corner reflector.

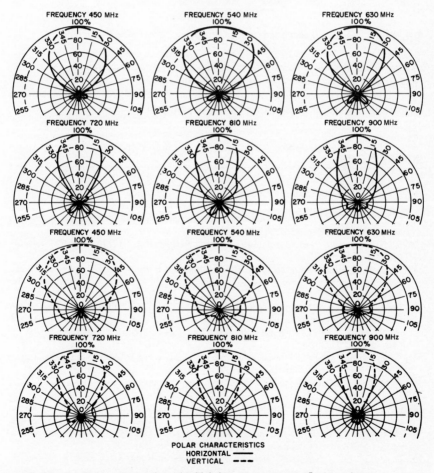

POLAR CHARACTERISTICS
HORIZONTAL ———
VERTICAL – – –

FIG. 29-17 Field patterns of an ultrahigh-frequency corner-reflector antenna.

29-23

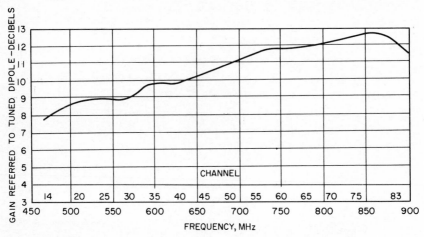

FIG. 29-18 Measured gain of an ultrahigh-frequency corner-reflector antenna.

enough. Thus, the width is determined by the lowest frequency, while the spacing of the grid tubing is determined by the highest frequency.

Figures 29-16 and 29-17 show impedance characteristics and field patterns respectively. Figure 29-18 shows the measured gain over the band.

REFERENCES

1 Federal Communications Commission, "Broadcast Television," *Rules and Regulations,* September 1972, sec. 73.6.

2 J. B. Snider, "A Statistical Approach to Measurement of RF Attenuation by Building Materials," NBS Rep. 8863, July 1965.

3 R. G. Fitzgerrell, "Indoor Television Antenna Performance," NTIA Rep. 79/28, NBS-9104386 Rep., 1979.

4 W. R. Free, J. A. Woody, and J. K. Daher, "Program to Improve UHF Television Reception," final rep. on FCC Proj. A-2475, Georgia Institute of Technology, Atlanta, September 1980.

5 D. G. Fink (ed.), *Electronics Engineers' Handbook,* McGraw-Hill Book Company, 1975. See sec. 21, "Television Broadcasting Practice," by J. L. Stern, and "Television Broadcasting Receivers," by N. W. Parker; sec. 18, "Antennas and Wave Propagation," by W. F. Croswell and R. C. Kirby.

6 "Engineering Aspects of Television Allocations," Television Allocations Study Organization (TASO) rep. to FCC, Mar. 16, 1959.

7 H. C. Ko, "The Distribution of Cosmic Radio Background Radiation," *IRE Proc.,* January 1959, p. 208.

8 G. H. Brown and D. M. Woodward, Jr., "Experimentally Determined Radiation Characteristics of Conical and Triangular Antennas," *RCA Rev.,* vol. 13, December 1952, p. 425.

9 H. T. Lo, "TV Receiving Antennas," in H. Jasik (ed.), *Antenna Engineering Handbook,* 1st ed., McGraw-Hill Book Company, New York, 1961, chap. 24.

10 G. L. Hall (ed.), *The ARRL Antenna Book,* American Radio Relay League, Newington, Conn., 1982.

11 P. P. Viezbicke, "Yagi Antenna Design," NBS Tech. Note 688, National Bureau of Standards, Washington, December 1976.

BIBLIOGRAPHY

Balinis, C. A.: *Antenna Theory Analysis and Design,* Harper & Row, Publishers, Incorporated, New York, 1982.

Elliott, R. S.: *Antenna Theory and Design,* Prentice-Hall, Inc., Englewood Cliffs, N.J., 1981.

Fink, D. G. (ed.): *Television Engineering Handbook,* McGraw-Hill Book Company, New York, 1957.

Fitzgerrell, R. G., R. D. Jennings, and J. R. Juroshek: "Television Receiving Antenna System Component Measurements," NTIA Rep. 79/22, NTIA-9101113 Rep., June 1979.

Free, W. R., and R. S. Smith: "Measurement of UHF Television Receiving Antennas," final rep. on Proj. A-2066 to Public Broadcasting Service, Engineering Experiment Station, Georgia Institute of Technology, Atlanta, February 1978.

Kiver, M. S., and M. Kaufman: *Television Simplified,* Van Nostrand Reinhold Company, New York, 1973.

Martin, A. V. J.: *Technical Television,* Prentice-Hall, Inc., Englewood Cliffs, N.J., 1962.

Stutzman, W. L., and G. A. Thiele: *Antenna Theory and Design,* John Wiley & Sons, Inc., New York, 1981.

Television Allocations Study Organization: *Proc. IRE* (special issue), vol. 48, no. 6, June 1960, pp. 989–1121.

UHF Television: *Proc. IEEE* (special issue), vol. 70, no. 11, November 1982, pp. 1251–1360.

Weeks, W. L.: *Antenna Engineering,* McGraw-Hill Book Company, New York, 1968.

Wells, P. I., and P. V. Tryon: "The Attenuation of UHF Radio Signals by Houses," OT Rep. 76-98, August 1976.

Chapter 30

Microwave-Relay Antennas

Charles M. Knop
Andrew Corporation

Typical Microwave-Relay Path

The typical microwave-relay path consists of transmitting and receiving antennas situated on towers spaced about 30 mi (48 km) or less apart on a line-of-sight path. If we denote P_T, G_T, P_R, and G_R as the transmitted power, transmitting-antenna gain, received power, and receiving-antenna gain respectively, then these are related by the well-known free-space Friis transmission law (neglecting multipath effects):

$$P_R = P_T G_T G_R / (4\pi R/\lambda)^2 \qquad (\textbf{30-1})$$

where R = distance between receiving and transmitting antennas
λ = operating wavelength in free space

A typical modern-day telecommunications link is designed to operate with $P_T = 5$ W ($+37$ dBm) and $P_R = 5$ μW (-23 dBm). (Note that -23 dBm is the operating level. Typical receivers still give adequate signal-to-noise ratios for fade margins 40 dB below this level, i.e., down to -63 dBm.) Thus, at a typical microwave frequency in the range of 2 to 12 GHz (say, 6 GHz) and $G_T = G_R$ (identical antennas at both sites), we see that antenna gains of about 10^4 (40 dBi) are required if no other losses exist. However, a typical installation (usually having several antennas on a given tower; see Fig. 30-1) has the antenna connected to a waveguide feeder running down the length of the tower (typically 200 ft, or 61 m, high) into the ground equipment. The total power received is then less than that given by Eq. (30-1) because of the losses in the waveguide feeder in the antenna (typically 0.25 dB) and in the tower feeder (typically 3.0 dB). This raises the above antenna gain requirement to about 43 to 44 dBi (all these values are at 6 GHz and change accordingly over the common-carrier bands, since the free-space loss and feeder loss increase with frequency).

FIG. 30-1 Typical tower with radio relay antennas. *(Courtesy Andrew Corp.)*

Specifications for Microwave-Relay Antennas

In addition to the above high gains, the radiation-pattern envelope (RPE; approximately the envelope obtained by connecting the peaks of the sidelobes) should be as narrow as possible, and the front-to-back ratio should be high to minimize interference with adjacent routes. Typical Federal Communications Commission (FCC) RPE specifications are shown in Fig. 30-2 (all patterns discussed here are horizontal-plane patterns; hence the E-plane pattern is horizontally polarized, and the H-plane is vertically polarized). Figure 30-2 also shows the RPE of a conical-cornucopia antenna, which will be discussed.

The requirement of simultaneous high gain and low sidelobes is, of course, inconsistent in antenna operation; hence some trade-off is necessary. Since the gain of an antenna is given by $G = \eta 4\pi A/\lambda^2$, where η is the total aperture efficiency (typically 0.50 to 0.70) and A is the physical area, we see that for circular apertures of diameter D, we require $D/\lambda \doteq 60$ for 43-dB gain and for $\eta = 0.5$; i.e., $D \doteq 10$ ft (3 m) at 6 GHz (with corresponding sizes at other bands). This means that the beamwidth between the first nulls, θ null, will be about θ null $= 120/(D/\lambda) \doteq 2°$. Thus, the supporting structure and tower must be very stable to ensure no significant movement of the beam (and consequent signal drop and cross-polarization degradation) even during strong winds (125 mi/h, or 201 km/h).

A further specification not found in radar or other communication systems is the extremely low voltage-standing-wave-ratio (VSWR; typically 1.06 maximum) requirement at the antenna input. This is to reduce the magnitude of the round-trip echo (due to the mismatch at both the antenna and the equipment ends) in the feeder

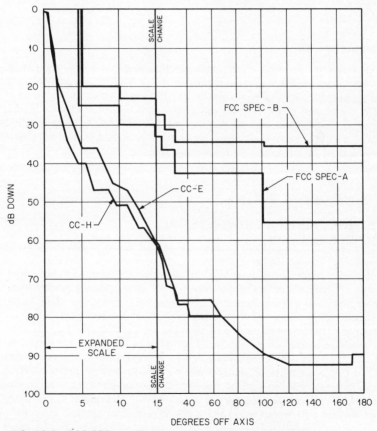

FIG. 30-2 FCC RPE specifications and conical-cornucopia RPE.

line. This echo produces undesirable effects: (1) cross talk in the baseband of a frequency-division-multiplexed-frequency modulation (FDMFM) analog system, especially in the upper telephone channels,[1] (2) ghosts on a TV link, and (3) an increase in the bit error probability in a digital system (which does, however, tolerate higher VSWR). Also, for dual-polarization operation at a given frequency, polarization purity expressed as cross-polar discrimination (XPD) between the peak of the main copolar beam and the maximum cross-polar signal is typically required to be -25 dB minimum across an angle twice the 3-dB beamwidth of the copolar pattern. Additionally, bandwidths of up to 500 MHz at any common-carrier band are required (these bands extend from 1.7 to 13.25 GHz, with FDMFM up to 2700 channels at 3.7 to 11.7 GHz and up to 960 channels of digital system above 10.7 GHz).

Added to the above are a low-cost requirement, a necessity to operate under virtually all weather conditions, and the low near-field coupling specifications between antennas mounted on the same tower (typically -120 dB back to back and -80 dB minimum side to side). Finally, no loose contacts or other nonlinear-contact phenomena (the so-called rusty-bolt effect) are allowed, because if signals of frequencies A and B arrive simultaneously, then a third signal of $2A$-B is generated. These $2A$-B products must be down to about -120 dBm or more.

Antennas in Use

The most common antennas in use today which meet all the above requirements are the symmetrical prime-fed circular paraboloidal dish (and its grid equivalent, though this is suitable for only single polarization and has much worse RPE), the offset paraboloid, the conical and pyramidal cornucopias, and the dual reflector (Cassegrain and gregorian). The principles of operation and designs of these antennas will now be reviewed, with major emphasis on the symmetrical prime-fed paraboloid, which is by far the most frequently used microwave-relay antenna.

30-2 THE PRIME-FED SYMMETRICAL PARABOLOIDAL DISH

Many excellent treatments describing the operation and design of this antenna are reviewed in Chap. 17 and in the open literature[2-13] (see especially Silver,[2] Ruze,[3] and Rusch-Potter[4]). Almost all these references, however, were written prior to the ease of accessibility (for most antenna engineers) to the minicomputer (this, in turn, required analytical approximations to be made; for example, the feed-horn patterns were described by $\cos^n \psi$-type functions) and prior to the full evolvement of Keller's geometric theory of diffraction (GTD) to the uniform geometric theory of diffraction (UGTD) by Kouyoumijian[14] and the concept of equivalent rim currents of Ryan and Peters.[15] Alternatively, Rusch's asymptotic physical optics (APO)[16] in its corrected form (CAPO)[17,18] can be used instead of UGTD.[19-20] Because of space limitations, we present here only results based on the above[21] and other analyses.

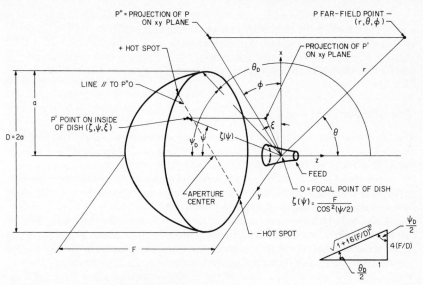

FIG. 30-3 Symmetrical-paraboloidal-dish geometry.

Basic Principle of Operation

Consider, then, the typical geometry of Fig. 30-3, depicting a prime feed illuminating a circular paraboloid. We start with the known horn E- and H-plane patterns in amplitude and phase in a computer file and assume that the horn produces a symmetrical pattern [see Eq. (1) of Ref. 17]. The surface current K produced on the dish surface is then found by using geometric optics (GO), i.e., by assuming that this surface is an infinite sheet and that the dish is in the far field of the horn, so that the E and H fields of the horn are related by the usual free-space impedance relationship. It is known that this GO approximation is questionable near the dish edge,[22] but this only has a consequence on the fields radiated in the 120° vicinity.[18]

The electric field E scattered by the dish is then obtained by integration of K over the *illuminated* surface of the dish up to and *including its edges*. The resulting expressions are considered in three regions: region 1, $0 \leq \theta \leq \theta_1 \doteq$ fifth sidelobe; region 2, $\theta_1 \leq \theta \leq \theta_2 \doteq 175°$; and region 3, $175° \leq \theta \leq 180°$. The region 1 expressions are double integrals [see Ref. 17, Eq. (2)], which can be reduced to a single integral via Bessel's orthogonality relation (Ref. 2, page 337). The region 2 UGTD (or CAPO) expressions are essentially asymptotic integrations (as given explicitly in Rusch[16] with the diffraction coefficients replaced by those of UGTD[19,20] or CAPO[18]) and come from the "hot spots" on the dish edge* (see Fig. 30-3). The region 3 fields come from the equivalent rim currents[15,19,20,22-24] [and are given by Eq. (6) of Ref. 23]. Also, in the range of $0 \leq \theta \leq \theta_D$ horn superposition is required.

---

*For the observation point in the horizontal plane, these hot spots are on the left and right edges of the dish.

The above expressions then give the patterns from the perfectly focused feed in a perfect paraboloid. To obtain the gain we compute the field on axis ($\theta = 0$), obtain the angular power density, and then divide by the total feed power over 4π sr. This also gives the aperture efficiency η. A program that computes both patterns and gain takes a few minutes to run on a minicomputer such as a DEC VAX 11/780.

The 100 Percent Feed Horn

Prior to considering actual feed horns, it is very instructive to create a hypothetical horn file having the pattern value* $1/\cos^2(\psi/2)$ for $0 \le \psi \le \psi_D$ and 0 for $\psi > \psi_D$ (as depicted in Fig. 30-4 for an $F/D = 0.30$ dish). This exactly negates the $(1/\rho)$ free-space loss [since $\rho = F/\cos^2(\psi/2)$ for a perfect paraboloid], so that the magnitude of the fields illuminating the dish surface is then constant. Hence, the aperture amplitude and phase are constant, giving $\eta = 100$ percent.

Using the expressions for the above three regions gives the patterns of Fig. 30-5a (near-in pattern) and 30-5b (wide-angle RPE†) for the E plane. The H-plane pattern is not shown because of lack of space, but it is similar except in the dish shadow $\theta > \theta_D$, where it is lower than the E-plane pattern since the H-plane diffraction coefficients here are smaller than those of the E plane. One notes that for small θ the patterns are close to $2J_1(U)/U$, where $U = C\sin\theta$, $C = \pi D/\lambda$.

Trade-Off between Gain and Radiation-Pattern Envelope

The above uniform case gives the highest possible gain, but its RPE is not low enough for most applications, since the dish edge illumination is 4.6 dB *above* that on the horn axis. Hence, a feed having lower edge illumination but still giving reasonable efficiency is required.

The opposite extreme of the 100 percent efficiency case is to illuminate the dish with a horn pattern having very low edge illumination. Now from Kelleher (see Chap. 17) we know that it is a universal characteristic of conventional single-mode horns (be they circular, rectangular, square, smooth-wall, corrugated, or whatever) that if their 10-dB beamwidths (i.e., let ψ_{10} be the angle off axis at which the power is 10 dB down from that on axis) are known, their power patterns from the 0- to about the 20-dB-down level can be described fairly well by Kelleher's formula:

$$\text{DBE}(\psi) = 10(\psi/\psi_{10E})^2 \tag{30-2}$$

$$\text{DBH}(\psi) = 10(\psi/\psi_{10H})^2$$

where the subscripts E and H denote E and H planes respectively.

The above characteristic allows us to investigate quickly the trade-off between high gain and RPE. Four examples (all assumed to have back radiation of -40 dB and $\psi_{10E} = \psi_{10H} = \psi_{10}$ and to have zero phase error) are shown in Fig. 30-4. These

---

*E and H planes are assumed to be equal and with zero phase error.

†Note that for $10° \lesssim \theta \le \theta_D = 100°$, the RPE of Fig. 30-5b is unrealistic, since this ideal feed has no spillover.

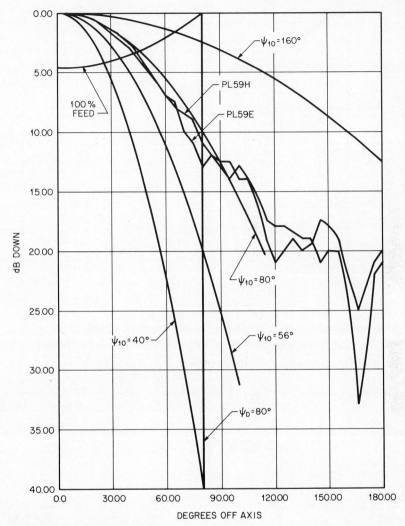

FIG. 30-4   Various feed patterns.

feeds produce the E-plane patterns also shown in Fig. 30-5a and b. A plot of η percent versus ψ_{10}/ψ_D (also in Fig. 30-5b) confirms the rule (e.g., see Silver,[2] page 427) of choosing about a 10-dB edge taper to achieve peak gain.* One notes from Fig. 30-5b that for $\theta > \theta_D$ (here $\theta_D \doteq 100°$) the 100 percent feed case has the highest RPE, whereas from Fig. 30-5a it has the narrowest main beam and the highest gain. Also,

---

*Note that actual gains, because of feed defocusing, dish rms, feed-support and strut scattering, etc., are lower than that of Fig. 30-5b by typically 5 to 10 percent, as discussed below.

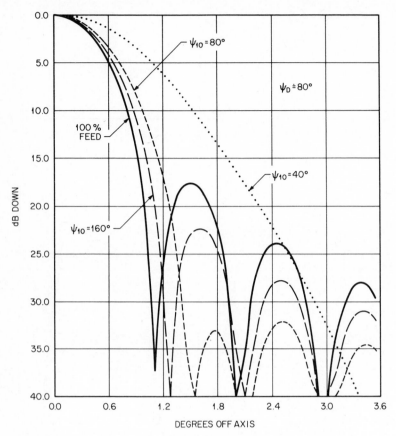

FIG. 30-5a   On- and near-axis *E*-plane dish patterns for various feeds.

it is seen from Fig. 30-5*b* that an excellent (wide-angle, $\theta \gtrsim 10°$) RPE but (from Fig. 30-5*a*) a very low gain are provided by the extremely underilluminated feed (ψ_{10} = 40°).

Practical Feeds

Actual feeds[25] have main-beam patterns close (with 1 to 2 dB at the 20-dB-down level) to the Kelleher description; some examples are given in Table 30-1. The most common feed used for terrestrial antennas consists of the circular-pipe TE_{11}-mode feed, sometimes with a recessed circular ground plate having quarter-wave-type chokes. These chokes control the external *E*-plane wall current that normally would flow on the horn's exterior surface, and they tend to make the *E*-plane pattern equal to that of the *H* plane, with both being flatter over the dish and having a faster falloff near the

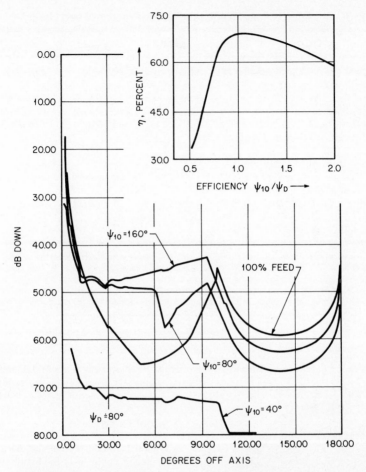

FIG. 30-5b *E*-plane RPE and aperture efficiency of dish with various feeds.

dish edge, causing improved efficiency. This horn has been patented by Yang-Hansen;[26] similar work by Wohlleben has been reported (see Love[25]). The rectangular horn is also commonly used since ℓ and w can be adjusted (at one band) to give almost equal *E*- and *H*-plane patterns.

Required *F/D*

For terrestrial radio work, the F/D ratios usually fall in the range of 0.25 (i.e., $\psi_D = 90°$, focal-plane dish) to about 0.38. For deeper dishes ($F/D < 0.25$), the requirement of building a feed having $\psi_{10} > 90°$ with low spillover is difficult. For shallower dishes

TABLE 30-1 Approximate 10-dB Angles for Practical Feed Horns*

Horn	Mode	ψ_{10E} °	ψ_{10H} °
Rectangular $\ell \times w$ (**E** perpendicular to ℓ)	TE_{10}	$42/(w/\lambda)$	$59/(\ell/\lambda)$
Circular,* $C \geq 1.84$	TE_{11}	$52/(DH/\lambda)$	$62/(DH/\lambda)$
Circular[25] (Potter), $C \geq 3.83$	$TE_{11} + TM_{11}$	$62/(DH/\lambda)$	$62/(DH/\lambda)$
Circular (corrugated, "$\lambda/4$ HE_{11} teeth") zero or small (less than 10°) phase error $C \geq 2.40$	HE_{11}	$62/(DH/\lambda)$	$62/(DH/\lambda)$
Circular (corrugated, "$\lambda/4$ HE_{11} teeth") large (180° or more) phase error, half angle α_0 ($0 \leq \alpha_0 \leq 85°$)	HE_{11}	$\alpha_0/\sqrt{2}$ (approximate)	$\alpha_0/\sqrt{2}$ (approximate)
Diagonal[25] (A. W. Love), $D/\lambda \geq 0.50$ (D = side length of aperture $D \times D$); (small phase error)	Two TE_{11}	$50.5/(D/\lambda)$	$50.5/(D/\lambda)$

*All circular horns have inner-diameter DH, $C = \pi DH/\lambda$. All zero- or small-phase-error horns have their phase center on or very close to the aperture; the large-phase-error horns have it close to the horn apex. Note that the E- and H-plane phase centers are usually not exactly coincident; their average position is made to coincide with the dish focal point. As seen, the zero- or small-phase-error horns are *diffraction-limited;* i.e., they have beamwidths which decrease with increasing frequency, whereas the large-phase-error corrugated horn has equal E- and H-plane patterns which are frequency-independent and hence is called a *scalar feed* (see Chap. 15).

($F/D > 0.38$, $\psi_D < 67°$), the feed is too far from the dish, which makes it impractical to use a radome; feed losses also are higher. Thus, F/D typically is about 0.3, and we therefore seek feeds having $\psi_{10} \doteq 80°$.

FIG. 30-6 A standard buttonhook feed in a paraboloidal dish. *(Courtesy Andrew Corp.)*

Typical Case

A typical feed is a rectangular horn (as described in Table 30-1) fed by a button-hook feed (similar to that shown for a circular-horn feed in Fig. 30-6). The feed patterns (the phase patterns of which are virtually flat over the dish and hence are not shown) of this rectangular horn are shown in Fig. 30-4 (averaged over both sides in the absence of the buttonhook) and are designated by PL59 (we note they have about a 10-dB edge taper, so the horn has $w/\lambda \doteq 0.53$, $\ell/\lambda \doteq 0.74$). The resulting measured and predicted E-plane dish patterns are shown in Fig. 30-7. We see that the prediction and measurements are close in the main beam, in the first few sidelobes, and in the rear, but in the approximately -10-dBi (here

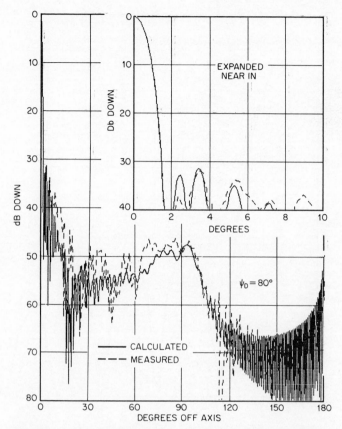

FIG. 30-7 Measured and predicted *E*-plane patterns of a parabo-loidal dish with a buttonhook feed.

the 0-dBi level $\doteq$ 43 dB down) level ($10° \lesssim \theta \lesssim 70°$) the measured fields are typically ± 5 dB in range about the predicted envelope. The major reasons for these discrepancies are twofold: dish imperfections and feed-support scattering, as discussed below.

Pattern and Gain Dependence on Distance

The above far-field formulations are (by definition) for an infinite-observation distance. We wish to determine the effects of a finite distance r in the range of 0.2 (D^2/λ) $\leq r \leq \infty$, since such short paths are necessary in many applications. To do this we use the double-integral (exact-phase) expressions (the same result, off axis, can also be obtained by GTD[27]). The *E*-plane results are shown in Fig. 30-8*a* for the case of the typical buttonhook feed; the results for the 100 percent feed are shown in Fig. 30-8*b*. Figure 30-9 shows the on-axis gain drop relative to that at infinity for these

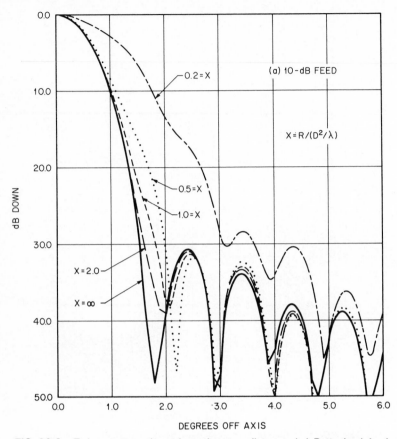

FIG. 30-8   *E*-plane-pattern-shape dependence on distance. (*a*) Buttonhook feed.

cases owing to this finite-observation distance. The gain drops because the main beam widens, the sidelobes rise, and the nulls fill in. In most practical installations, *r* is greater than D^2/λ; thus, the gain is degraded by at most 0.3 dB. These results are in agreement with those of Silver[2] (page 199, for the uniform case) and Yang[28] (for $\cos^n$ ψ-type feeds). If two identical antennas are used, the total gain drop is then double the amount given by Fig. 30-9.

Effect of Feed Displacement

In general, the feed is inadvertently defocused (that is, the feed's phase center is not coincident with the focal point of the dish). This comes about because in most cases the feed is attached to a waveguide feeder, which in turn is attached to the hub of the dish, which may be distorted slightly (especially axially) from paraboloidal. It has been shown (e.g., by Imbriale et al.; see Love[25]) that this defocusing effect can be

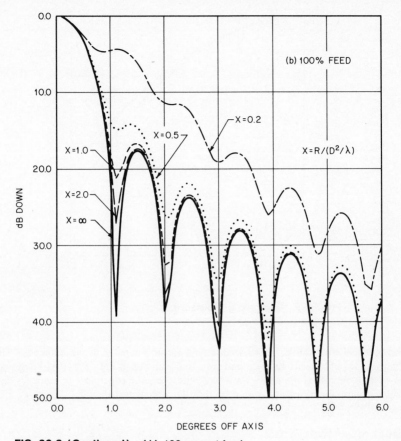

FIG. 30-8 (Continued)   (*b*) 100 percent feed.

accounted for by inserting a phase-error term, $\exp(j\Phi_D)$, multiplying the dish surface current, where (letting $k = 2\pi/\lambda$)

$$\Phi_D = k \cdot DT \sin \psi \cos (\xi - \xi_D) - k\, DZ \cos \psi \qquad \textbf{(30-3)}$$

where DT, ξ_D, and DZ define the feed displacement and are respectively the transverse ($DT > 0$), azimuthal angular, and axial displacement ($DZ > 0$ for feed movement away from the dish, and $DZ < 0$ for feed movement toward the dish).

The pattern degradation and gain drop resulting from axial shifting of the above typical buttonhook feed (in the above $F/D = 0.30$, $D/\lambda = 60$ dish) are shown in Figs. 30-10*a* and 30-11 respectively.[29] It is seen that the pattern distortion results in intolerable (≥ 0.2-dB) gain drops for $|DZ|/\lambda \gtrsim 0.15$. For transverse (lateral) displacements, the beam squints, but the pattern shape is not distorted as much for the same wavelength displacement as for the axial case (see Fig. 30-10*b*). The corresponding gain drops are also shown in Fig. 30-11, and it is seen that for $DT/\lambda < 0.15$ the gain

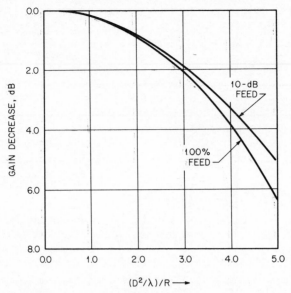

FIG. 30-9 Gain drop with distance.

drop is less than 0.04 dB. These results hold for any angular displacement ξ_D. The beam squint ($D\theta°$) for the subject antenna is also shown in Fig. 30-11. Although not shown here, repeating the above for a feed with the same edge taper in a deeper dish results in more distortion and gain drop than for a shallower dish.*

Effect of Surface Errors

By measuring at a sufficient number of points the normal difference Δn (ψ,ξ) between an actual dish and a fixture representing the paraboloid to be made, it is found that these surfaces are locally parallel. As such, a phase error of amount $360\Delta n$ $(\psi,\xi)/\lambda$ $\cos (\psi/2)$, representing the additional path length to the reflector surface, must be included in **K**. This phase error degrades the gain, RPE, and XPD. Performing the necessary *double* integration reveals that the resulting gain drop is very close to that given by Ruze's formula[30,10] (see also Zucker[31]):

$$\Delta G_{rms} = -10 \log_{10} [\exp (-4\pi\epsilon_{rms}/\lambda)^2] \quad \text{dB} \quad \textbf{(30-4)}$$

where ϵ_{rms} is the root-mean-square deviation (in the normal direction) of the measured data points from the best-fit paraboloidal surface of these points [actually, Eq. (30-4) is based on a large F/D dish and hence is an upper bound on the gain drop]. Since we

---

*Similarly, for a given F/D dish and a given feed displacement, a feed with higher edge illumination produces more pattern distortion and gain drop than one with lower edge illumination (not shown here).

strive for $\Delta G_{rms} \leq 0.15$ dB, we must make $\epsilon_{rms} \leq \lambda/70$ [e.g., $\epsilon_{rms} < 0.030$ in (0.76 mm) at 6 GHz] for gain purposes. However, use of the Δn (ψ,ξ) file and double integration of many antennas reveals a rapid degradation of RPE (only in the $\theta < \theta_D$ region) with ϵ_{rms} (since this file contains many spatial harmonics[32]). This work and some analytical work by Dragone and Hogg[33] using this approach, as shown in Fig. 30-12 (for a 10-dB feed and 15ψ harmonics representing the actual surface), discloses that deviations from the parabolic should be held within about $\pm\lambda/100$ [i.e., $\epsilon_{rms} \lesssim \lambda/300$ or $\epsilon_{rms} \lesssim 0.007$ in (0.1778 mm) at 6 GHz] if approximately 3-dB or less increases from the perfect-case RPE are desired at about the 60-dB-down level (see also Fig. 30-7 for this increase in RPE in the $10°$ to $70°$ region due to rms). The XPD for realistic dishes is typically more than $+50$ dB down from the main beam.*

Front-to-Back Ratio

Using the above analytical results to evaluate the front- and back-field expressions explicitly for the general case, one obtains the expression for f/b:[23]

$$f/b = G + T + K - G_{\text{HORN}} \quad \text{dB} \qquad \textbf{(30-5)}$$

where $G = $ gain of dish $= 10 \log_{10} (\eta C^2)$, $T = $ average taper of feed at dish edge, $K = 20 \log_{10}\{\sqrt{1 + 16(F/D)^2}/4(F/D)\}$, $G_{\text{HORN}} = $ on-axis gain of horn, all in decibels. Since typical values are $T = 10$ dB, $F/D = 0.3$, and $G_{\text{HORN}} \doteq 5$ dB, we arrive at the simple result

$$f/b \doteq G + 7 \qquad \textbf{(30-6)}$$

which has been empirically known for many years[34] and is good, in practice, within 2 to 3 dB.[23]

In many installations, f/b ratios of 65 dB or larger are required; we then see from Eq. (30-6) that (for typical 40- to 45-dB-gain dishes) it is impossible to achieve this requirement with standard circular paraboloids. To overcome this fact, *edge-geometry schemes* are employed (here the circular edge is replaced with a polygonal one so as to destroy, in the back of the dish, the in-phase addition from each increment of the edge). Alternatively, an absorber-lined cylindrical shroud is used (see Fig. 30-1); this decreases the RPE significantly in the shadow of the shroud.

Effect of Vertex Plate

A plane circular plate (vertex plate[2,35] of diameter DV and thickness T having $DV = 2\sqrt{\lambda F/3}$ and $T = 0.042\lambda$; these dimensions are usually fine-tuned experimentally) is invariably used at the center of the dish (see Fig. 30-6) to minimize the VSWR contribution from the dish. The vertex-plate reflections essentially cancel the remaining dish reflections. This plate also affects the pattern and gain of the dish antenna since it introduces a leading phase error. The result is to raise the close-in radiation levels

---

*This is due to rms effects only; other effects (especially mechanical-feed asymmetry) increase this to 25 to 30 dB down.

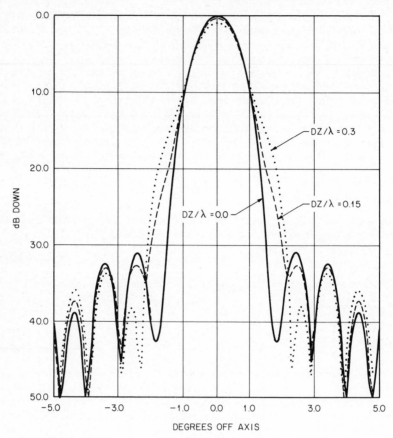

FIG. 30-10a Pattern change with axially displaced feed.

slightly and to cause a slight drop in gain. In some cases, edge diffraction from the plate degrades the RPE somewhat.

Effect of Feed-Support and "Strut" Scattering

The waveguide-bend feed support (see Fig. 30-6) is seen to interfere with the horn fields prior to their striking the dish; this results in amplitude, phase, and symmetry changes. Also, upon reflection, the dish fields pass by the same feed support and the vibration dampers (usually thin wires), where they are again diffracted or reflected. Analyses of these effects have been attempted[36] with some success, but in most cases they are usually best corrected by empirically determining the best positioning of absorber layers on the waveguide feed support and by making the vibration dampers from insulating material.

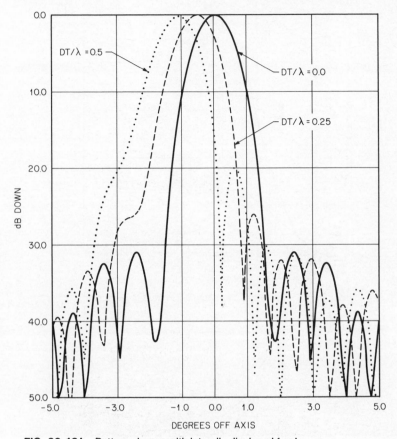

FIG. 30-10*b*   Pattern change with laterally displaced feed.

Effect of Radome

The radome is used to protect the antenna against the accumulation of ice, snow, and dirt. General radome effects are discussed in Chap. 44, but in relay applications the fundamental problems are (1) the VSWR produced by the radome, (2) its total losses, and (3) its pattern influence. In the case of a planar radome (as seen on the antennas in Fig. 30-1), the VSWR increase is cured by tilting the radome a few degrees (moving the top further from dish) to prevent specular return to the feed horn; the loss and RPE influence are held low by using thin, low-loss material (e.g., Teglar, a fiberglass material coated on both sides with Teflon which sheds water readily[37]). Fiberglass conical or paraboloidal radomes are also used; these reduce wind loading, although since they are thicker than planar radomes, they have higher loss and degrade the RPE somewhat.

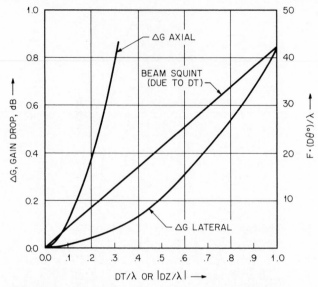

FIG. 30-11 Gain drop and beam squint due to defocusing.

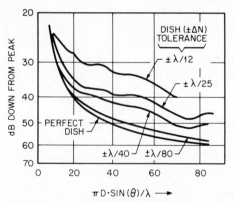

FIG. 30-12 Typical effect of surface errors on RPE. *(From Dragone and Hogg.[33])*

30-3 THE PARABOLOIDAL GRID REFLECTOR

Here metal (usually aluminum) tubes of outer radius r and center-to-center spacing s (where $s/\lambda \leq 0.3$ and $s/r \gtrsim 5$) and of paraboloidal contour replace the solid metal sheet (see Fig. 30-13) and act as a moderately good reflector (see Chap. 46). Typically, $s/\lambda = 0.3$ (maximum spacing possible so as to reduce cost) and $s/r = 5.5$ to 6

FIG. 30-13 Paraboloidal grid antenna. *(Courtesy Andrew Corp.)*

(to allow use of commercial aluminum tubes). Moullin's theory[38] based on thin wires (for which s and r are adjusted to have the self-inductance and mutual inductance cancel) has been shown[39] to give good results; however, it is now known (see Chap. 46) that it is not essential to meet his criterion. The efficiencies of these grid reflectors is typically 5 or so percent below that of a solid reflector (of the same F/D and D and fed with the same horn), and the front-to-back ratio is typically 3 dB lower.* The grid openings cause the wide-angle RPE (especially for $\theta > \theta_D$) to be quite inferior to that of a solid paraboloid (also the tubes have a small kr, where $k = 2\pi/\lambda$, and, as such, scatter readily to the rear). These antennas are used only in remote, uncongested areas not requiring narrow RPE and have the advantage of overall lower cost since their wind loading is much smaller than that of a corresponding solid reflector. This allows less expensive towers (the grid antenna is, however, actually more expensive than its solid counterpart). These antennas are not used above about 3 GHz because of achievable fabrication tolerances and are usable only for a single polarization.

---

*Fortunately, however, when the grid is mounted on a typical triangular cross-section tower, the back signal is "smeared," bringing the f/b ratio back up to about its solid-dish value.

30-4   THE OFFSET PARABOLOID

To eliminate feed blockage (and its consequent drop in efficiency of about 8 to 10 percent due primarily to the degradation of the prime-feed pattern by the waveguide feed support), one can use the prime-fed offset paraboloid, the geometry of which is shown in Chaps. 17 and 36. This antenna has been known for some time, and analyses by Chu-Turrin and later by Rudge (see Chap. 36 for these references) are available for predicting the front-region fields. Using Rudge's work and trying different feeds, one again arrives[40] at the conclusion that a good compromise between high gain and narrow RPE is about a 10- to 12-dB edge taper. The copolar patterns for a typical case are similar to those of the prime-fed symmetrical paraboloid, but they have lower near-in sidelobes owing to the absence of blockage. The penalty paid for the removal of blockage is that the asymmetrical plane (i.e., the horizontal plane for terrestrial antennas) now has a rabbit-ear cross-polar pattern with the peaks of the rabbit ears being about 20 dB down from the copolar peak and being located at about the 6-dB-down angle of the copolar pattern (see Chaps. 17 and 36). Dual-polarization operation is, however, still achievable with XPD $\doteq$ -30 dB at the boresight null (the depth of this null being quite sensitive to reflector-surface error). The offset also has the disadvantage of high cost and higher I^2R loss associated with the typically longer feed support.

30-5   CONICAL AND PYRAMIDAL CORNUCOPIAS

These antennas have been used for some time[10,41,42] and probably represent the ultimate in terms of the best wide-angle RPE attainable. In recent years, the conical cornucopia (CC) has found preference over the pyramidal (primarily for economic reasons; also, the pyramidal horn has undesirably high diffraction lobes at 90° which require edge blinders to be suppressed[43,44]). The CC geometry is depicted in Chap. 17, and a modern-day CC is shown in Fig. 30-14. The CC geometry has a feed offset of 90° (but also has a conical shield) and hence can be handled by the same analysis as in the previous section. However, the classic earlier analysis of Hines et al. (see Ref. 10) is generally used. Briefly, this work shows that the E fields in the projected circular aperture are the conformal transformation of the E fields in the circular aperture of the conical horn. That is, circles (and their perpendiculars) are mapped into circles (and their perpendiculars), except that concentric circles in the conical aperture are no longer concentric in the projected aperture. Double integration of the projected-aperture fields then shows that this produces cross-polarization rabbit ears in the horizontal (asymmetric) plane (again, like the general offset, of typically 20 dB down at the 6-dB angle of the copolar beam).

In most real cases, a shield with absorber lining is used to capture the horn spillover and to reduce (almost eliminate) multiple-scattering effects at wide angles. This absorber introduces a small gain drop (at most, ½ dB). Typically, total efficiencies are about 65 percent; a representative RPE for 6 GHz (also a specification for these antennas) is shown in Fig. 30-2, which is seen to be vastly superior to the A or B FCC specifications. The cornucopia has the advantage of narrow RPE, relatively high efficiency, and simultaneous multiband operation at typically 4-, 6-, and 11-GHz bands (and possibly 2-GHz as well[44]), though it is very costly.

30-6 SYMMETRICAL DUAL REFLECTORS

This antenna type is covered in Chaps. 17 and 36 and hence will only be briefly mentioned here. An old (but very useful) basic "back-of-the-envelope design" Cassegrain configuration is depicted in Fig. 30-15.* Typically, the horn is chosen (see Table 30-1) to produce an 8- to 10-dB taper at the hyperboloidal subreflector edge α_E, where $DE/\lambda \gtrsim 5$. Then a Potter flange is used to capture some of the subreflector spillover energy and effectively use it to scatter back to the dish.[45,46] In this way, directive efficiencies in the mid-60s to low 70s are achievable. A typical design based on Fig. 30-15 for a focal-plane dish of $D/\lambda = 60$ is as follows: choose $DE/\lambda = 6.84$, hence $DE/D = 0.114$; choose M (magnification factor) $= 2.81$, which gives $a/\lambda = 0.996$, $c/\lambda = 2.097$, and $\alpha_E = 39.2°$; choose a horn from Table 30-1 (either a corrugated or a Potter horn having zero phase error) of $DH/\lambda = 1.58$ so that $\psi_{10} = \alpha_E = 39.2°$; get $\theta_E = 19.1°$; choose $LE/\lambda = 2$ (not critical); therefore $DO/\lambda = 10.61$ and $DO/D = 0.176$, which is large but acceptable blockage. At 6 GHz this design gives over 67 percent measured directivity, and its patterns meet FCC specification A. To date, gregorians have found little terrestrial use; they are best suited for antennas with larger F/D ratios. All reasonably efficient dual reflectors suffer from large

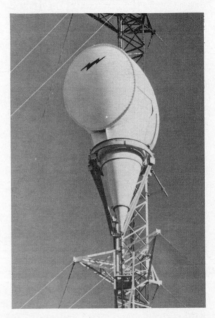

FIG. 30-14 Modern-day conical cornucopia. (*Courtesy Andrew Corp.;* for RPE see Fig. 30-2.)

subreflector blockage problems, since for sharp drop-off of the subreflector pattern near the dish edge one should make $DO/\lambda \gtrsim 5$ (preferably $DO/\lambda \gtrsim 8$, though one should limit blockage to 20 percent or less, $DO/D \lesssim 0.20$). This causes high first (typically 12 to 14 dB), third, etc., sidelobes. To achieve better results (higher efficiencies and better RPEs by minimizing spillover and VSWR) shaping of the subreflector and main reflector is mandatory[47] (see Chaps. 17 and 36).

---

*This figure does not show a small cone on the hyperboloid tip used to reduce the VSWR. The cone dimensions are usually determined empirically.

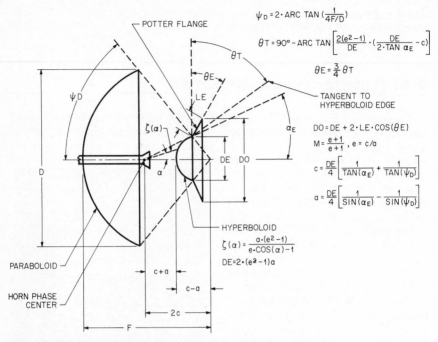

$$\psi_D = 2 \cdot \text{ARC TAN}\left(\frac{1}{4F/D}\right)$$

$$\theta_T = 90° - \text{ARC TAN}\left[\frac{2(e^2-1)}{DE} \cdot \left(\frac{DE}{2 \cdot \text{TAN } \alpha_E} - c\right)\right]$$

$$\theta_E = \frac{3}{4}\theta_T$$

$$DO = DE + 2 \cdot LE \cdot \cos(\theta_E)$$

$$M = \frac{e+1}{e+1}, e = c/a$$

$$c = \frac{DE}{4}\left[\frac{1}{\text{TAN}(\alpha_E)} + \frac{1}{\text{TAN}(\psi_D)}\right]$$

$$a = \frac{DE}{4}\left[\frac{1}{\text{SIN}(\alpha_E)} - \frac{1}{\text{SIN}(\psi_D)}\right]$$

$$\zeta(\alpha) = \frac{a \cdot (e^2-1)}{e \cdot \cos(\alpha)-1}$$

$$DE = 2 \cdot (e^2-1)a$$

FIG. 30-15   "Back-of-the-envelope-design" Cassegrain configuration.

30-7   MISCELLANEOUS TOPICS

Most antennas usually have pressurized feeders (including the waveguide run from the antenna to the equipment building at ground location) in order to prevent moisture accumulation on the waveguide's inner walls and its ultimate increase in attenuation. Alternatively, for 2 GHz and lower, foam-filled coaxial-cable feeders are used. On paths which do not allow frequent repeater installation (e.g., when mountains or lakes intervene between two sites), gain is of the utmost importance. Here, oversized circular waveguide feeders are used (they, typically, have one-third of the attenuation of dominant-mode-size feeders). Being oversized, they can propagate higher-order modes (typically TM_{01} and TE_{21}), which are removed by mode filters. These paths invariably use Cassegrain antennas (with their typically 10 percent higher efficiency); this combination can give up to 3-dB extra gain per path.

30-8   COMPARISON OF ANTENNA TYPES

A concise summary roughly categorizing the above microwave-relay antennas regarding their efficiency, relative RPE, and cost is given in Table 30-2. Most antennas are designed to operate at their peak efficiency for a specified RPE so as to reduce their

TABLE 30-2 Summary of Microwave-Relay Antennas

Antenna	Total efficiency[a]	RPE[b]	Near-in[c] sidelobes	Cost[d]
Solid paraboloid (symmetric prime-fed)	55–60	3	2	1
Grid paraboloid (symmetric prime-fed)	50–55	4	2	2
Offset paraboloid (prime-fed)	60–70	2	1	4
Conical cornucopia	65[e]	1[e]	3	5
Cassegrain	65–78[f]	3	4	3

[a]Including all I^2R loss, feed loss, rms loss, etc.

[b]Excluding near-in sidelobe levels, 1 = lowest energy level.

[c]1 = lowest energy level.

[d]1 = lowest cost.

[e]With absorber in cylindrical shield.

[f]Shaped.

required dish diameter, since the total cost is approximately proportional to the diameter squared or more.[48]

REFERENCES

1 W. R. Bennett, H. E. Curtiss, and S. O. Rice, "Interchannel Interference in FM and PM Systems under Noise Loading Conditions," *Bell Syst. Tech. J.,* vol. 34, May 1955, pp. 601–636.

2 S. Silver (ed.), *Microwave Antenna Theory and Design,* MIT Rad. Lab. ser., vol. 12, Boston Technical Publishers, Inc., Lexington, Mass., 1964, chap. 12.

3 J. Ruze, "Antennas for Radar Astronomy," in J. Evans and T. Hagford (eds.), *Radar Astronomy,* part 2, McGraw-Hill Book Company, New York, 1970, chap. 8.

4 W. V. T. Rusch and P. Potter, *Analysis of Reflector Antennas,* Academic Press, Inc., New York, 1970.

5 D. L. Sengupta and R. E. Hiatt, "Reflectors and Lenses," in I. Skolnik (ed.), *Radar Handbook,* McGraw-Hill Book Company, New York, 1970, chap. 10.

6 C. J. Sletten, "Reflector Antennas," in R. E. Collin and F. J. Zucker (eds.), *Antenna Theory,* part 2, McGraw-Hill Book Company, New York, 1969, chap. 17.

7 L. Thourel, *The Antenna,* John Wiley & Sons, Inc., New York, 1960, chap. 12.

8 A. Z. Fradin, *Microwave Antennas,* Pergamon Press, New York, 1961, chap. VII.

9 N. Bui-Hai, *Antennes micro-ondes,* Masson, Paris, 1978. (In French.)

10 A. W. Love (ed.), *Reflector Antennas,* IEEE Press, New York, 1978.

11 R. C. Hansen, *Microwave Scanning Antennas,* vol. I, Academic Press, Inc., New York, 1966.

12 E. A. Wolff, *Antenna Analysis,* John Wiley & Sons, Inc., New York, 1966, pp. 312–335.

13 P. Clarricoats and G. Poulton, "High Efficiency Microwave Reflector Antennas—A Review," *IEEE Proc.,* vol. 65, no. 10, October 1977, pp. 1470–1504.

14 R. G. Kouyoumjian and P. H. Pathak, "A Uniform Geometrical Theory of Diffraction for an Edge in a Perfectly Conducting Screen," *IEEE Proc.,* vol. 62, November 1974, pp. 1448–1461.

15 C. E. Ryan and L. Peters, "Evaluation of Edge Diffracted Fields Including Equivalent Currents for the Caustic Regions," *IEEE Trans. Antennas Propagat.,* vol. AP-17, no. 3, May 1969, pp. 292–299; see also vol. AP-18, no. 2, March 1970, p. 275.

16 W. V. T. Rusch, "Physical Optic Diffraction Coefficients for a Paraboloid," *Electron. Lett., IEE (England),* vol. 10, no. 17, Aug. 22, 1974, pp. 358–360.

17 C. M. Knop, "An Extension of Rusch's Asymptotic Physical Optics Diffraction Theory of a Paraboloid Antenna," *IEEE Trans. Antennas Propagat.,* vol. AP-23, no. 5, September 1975, pp. 741–743.

18 C. M. Knop and E. L. Ostertag, "A Note on the Asymptotic Physical Optic Solution to the Scattered Fields from a Paraboloidal Reflector," *IEEE Trans. Antennas Propagat.,* vol. AP-25, July 1977, pp. 531–534; see also correction, vol. AP-25, no. 6, November 1977, p. 912.

19 R. G. Kouyoumjian and P. A. J. Ratnasiri, "The Calculation of the Complete Pattern of a Reflector Antenna," Int. Electron. Conf., sess. 31, *Preconf. Dig.,* Pap. 69313, October 1969.

20 P. A. J. Ratnasiri, R. G. Kouyoumjian, and P. H. Pathak, "The Wide Angle Side Lobes of a Reflector Antenna," Tech. Rep. 2183-1, Ohio State University, Columbus, March 1970, pp. 17–25; also ASTIA Doc. AD707105, March 1970.

21 C. M. Knop and E. L. Ostertag, "The Complete Radiation Fields of a Parabolic Dish Antenna," Andrew Corp. Eng. Rep., July 1975.

22 G. L. James, *Geometrical Theory of Diffraction,* Peter Peregrinus Ltd., London, 1976, p. 148.

23 C. M. Knop, "On the Front to Back Ratio of a Parabolic Dish Antenna," *IEEE Trans. Antennas Propagat.,* vol. AP-24, no. 1, January 1976, pp. 109–111.

24 W. D. Burnside and L. Peters, Jr., "Edge Diffracted Caustic Fields," *IEEE Trans. Antennas Propagat.,* vol. AP-22, July 1974, pp. 620–623.

25 A. W. Love, *Electromagnetic Horn Antennas,* IEEE Press, New York, 1976.

26 "The Yang-Hansen Efficiency Plate," U.S. Patent 3,553,701, issued to Andrew Corporation, 1971.

27 M. S. Narisimhan and K. M. Prasad, "G.T.D. Analysis of Near-Field Patterns of a Prime-Focus Symmetric Paraboloidal Reflector Antenna," *IEEE Trans. Antennas Propagat.,* vol. AP-29, no. 6, November 1981, pp. 959–961.

28 R. F. Yang, "Quasi-Fraunhofer Gain of Parabolic Antennas," *IRE Proc.,* vol. 43, no. 4, April 1955, p. 486.

29 C. M. Knop and Y. B. Cheng, "The Radiation Fields Produced by a Defocused Prime-Fed Parabola," Andrew Corp. Eng. Rep., May 29, 1980.

30 J. Ruze, "The Effect of Aperture Errors on the Antenna Radiation Pattern," *Supplemento al Nuovo Cimento,* vol. 9, no. 3, 1952, pp. 364–380.

31 H. Zucker, "Gain of Antennas with Random Surface Deviations," *Bell Syst. Tech. J.,* October 1968, pp. 1637–1651.

32 N. I. Korman, E. B. Herman, and J. R. Ford, "Analysis of Microwave Antenna Sidelobes," *RCA Rev.,* September 1952, pp. 323–334.

33 C. Dragone and D. C. Hogg, "Wide-Angle Radiation Due to Rough Phase Fronts," *Bell Syst. Tech. J.,* September 1963, pp. 2285–2296.

34 A. Wojnowski and L. Hansen, Andrew Corporation, private communication.

35 H. Cory and Y. Leviatan, "Reflection Coefficient Optimization at Feed of Parabolic Antenna Fitted with Vertex Plate," *Electron. Lett.,* vol. 16, no. 25, Dec. 4, 1980, pp. 945–947.

36 P. Brachat and P. F. Combes, "Effects of Secondary Diffractions in the Radiation Pattern of the Paraboloid," *IEEE Trans. Antennas Propagat.,* vol. AP-28, no. 5, September 1980, pp. 718–721.

37 C. A. Sillar, Jr., "Preliminary Testing of Teflon as a Hydrophobic Coating for Microwave Radomes," *IEEE Trans. Antennas Propagat.,* vol. AP-27, no. 4, July 1979, pp. 555–557.
38 E. B. Moullin, *Radio Aerials,* Oxford University Press, New York, 1949, pp. 203–207.
39 E. F. Harris, "Designing Open Grid Parabolic Antennas," *Tele-Tech & Electronic Ind.,* November 1956.
40 C. M. Knop and Y. B. Cheng, "Analysis and Design of Corrugated Fog-Horn Feed Antennas," Final Rep. RP-883, Andrew Corp. Eng. Rep., Dec. 12, 1978.
41 H. T. Friis, "Microwave Repeater Research," *Bell Syst. Tech. J.,* April 1948.
42 R. W. Friis and A. S. May, "A New Broad-Band Microwave Antenna System," *AIEE Proc.,* March 1958, pp. 97–100.
43 D. T. Thomas, "Design of Multiple-Edge Blinders for Large Horn Reflector Antennas," *IEEE Trans. Antennas Propagat.,* vol. AP-21, March 1973, pp. 153–158.
44 J. E. Richards, "Horn Reflector Antenna Performance at 2 GHz with Simultaneous Operation in the 4, 6, 11 GHz," *IEEE Antennas Propagat. Symp. Dig.,* 1979.
45 P. D. Potter, "Application of Spherical Wave Theory to Cassegrain-Fed Paraboloids," *IEEE Trans. Antennas Propagat.,* vol. AP-15, no. 6, November 1967, pp. 727–736.
46 P. D. Potter, "Unique Feed System Improves Space Antennas," *Electronics,* vol. 35, June 22, 1962, pp. 36–40.
47 C. M. Knop, E. L. Ostertag, and H. J. Wiesenfarth, "The Analysis and Design of Advanced-Symmetrical-Dual-Shaped Reflector Antennas," Andrew Corp. Eng. Rep., October 1979.
48 T. Charlton, E. L. Brooker, and E. Book, Andrew Corporation, private communication.

BIBLIOGRAPHY

Beckmann, P., and A. Spizzichino: *The Scattering of Electromagnetic Waves from Rough Surfaces,* Pergamon Press, New York, 1963.
Croswell, W. F., and R. C. Kirby: "Antennas and Wave Propagation," in D. G. Fink (ed.), *Electronic Engineers' Handbook,* McGraw-Hill Book Company, New York, 1975, sec. 18, par. 51–53.
Lee. S. H., and R. C Ruddock: "Numerical Electromagnetic Code (NEC)—Reflector Antenna Code," *Code Manual,* part II, Electroscience Laboratory, Ohio State University, Tech. Rep. 784508-16, September 1979.
Semplak, R. S.: "A 30 GHz Scale Model Pyramidal Horn-Reflector Antenna," *Bell Syst. Tech. J.,* vol. 58, no. 6, July–August 1979, pp. 1551–1556.
Swift, C. T.: "A Note on Scattering from a Slightly Rough Surface," *IEEE Trans. Antennas Propagat.,* vol. AP-19, no.4 , July 1971, pp. 561–562.

Chapter 31

Radiometer Antennas

William F. Croswell

Harris Corporation

M.C. Bailey

NASA Langley Research Center

31-1 INTRODUCTION

Radiometer antennas are widely used as part of remote-sensing systems to infer physical properties of planetary atmospheres and the surface.[1] Such antennas are similar in type to those used in radio astronomy except for several important differences. In the case of radio astronomy, many of the sources to be measured are stable in time for minutes or longer. In addition, the so-called radio stars or other natural sources are point sources or are of limited angular extent and immersed in a cold sky of several degrees kelvin.[2,3] For most remote-sensing applications, the radiometric system observes a distributed target of large angular extent and warm in temperature. For example, the ocean brightness temperatures in the microwave bands are in the order of 120 K, while land surfaces can be 200 K or more.[1] Hence, antenna systems for downward-looking radiometric systems must have small, close-in sidelobes out to the angles where the surface brightness diminishes. Such techniques as interferometry used in radio astronomy to improve resolution at the expense of high sidelobes are of little value in the remote sensing of atmospheres and surfaces in a downward-looking observation.

Radiometer-antenna design requires much more precise consideration of different parameters than most antenna engineers are accustomed to. For example, beam efficiency, antenna losses, and antenna physical temperature are extremely important as well as directional gain, low sidelobes, mismatch, and polarization. Therefore, this chapter will include an extended section on basic principles and a description of basic system types in present or proposed use on spacecraft or aircraft, along with basic radiometer-antenna types commonly in use.

31-2 BASIC PRINCIPLES

For microwave remote-sensing applications, the radiometer antenna is used with a very sensitive receiver to detect and provide a measurement of the electromagnetic radiation emitted by downwelling radiation and the earth's surface. The downwelling radiation is from the cosmic-background and the sky-background radiation due to atmospheric properties including moisture.[4,5,6] Depending upon knowledge of the roughness and dielectric properties of the surface and the relationship of the dielectric properties to the physical properties of the surface, an indirect or remote measurement of the physical property is feasible.

The received power detected by the radiometric system is by Swift:[6]

$$P = KT_A f \qquad (31\text{-}1)$$

where K is Boltzmann's constant, f is the radio-frequency bandwidth of the radiometric receiver, and T_A is the antenna temperature.

Antenna Temperature

The antenna temperature, stated in a manner suppressing polarization, is given by an expression of the form

$$T_A = \frac{\int_0^{2\pi} \int_0^{\pi} f(\theta, \phi)\, T_B(\theta, \phi)\, \sin\theta\, d\theta\, d\phi}{\int_0^{2\pi} \int_0^{\pi} f(\theta, \phi)\, \sin\theta\, d\theta\, d\phi} \qquad \textbf{(31-2)}$$

where $f(\theta, \phi)$ is the normalized radiation pattern of a perfect plane-polarized antenna and $T_B(\theta, \phi)$ is the brightness temperature of the scene observed by the radiometer antenna with the same polarization. A typical scene in microwave remote sensing in which the antenna is pointed toward the surface is given in Fig. 31-1. T_{sky} is the downwelling sky radiation in the specular direction and includes thermal radiation from the atmosphere and the cosmic background. The brightness temperature observed by the radiometer antenna, therefore, is the sum of the brightness temperature T_{Bs} emitted by the surface and the amount of energy from the sky radiation scattered by the surface in the specular direction. This brightness temperature $T_B(\theta, \phi)$ is given by

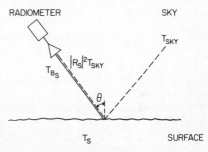

FIG. 31-1 Typical radiometer scene in microwave remote sensing.

$$T_B(\theta, \phi) = T_{Bs} + /R_s/^2 T_{\text{sky}} \qquad \textbf{(31-3)}$$

where $/R_s/^2$ is the plane-wave reflection coefficient for the air-surface interface.

The emissivity of the surface is given by

$$\varepsilon_s = 1 - /R_s/^2 \qquad \textbf{(31-4)}$$

and the brightness temperature T_{Bs} is related to the physical temperature T_s by the relation

$$T_{Bs} = \varepsilon_s T_s \qquad \textbf{(31-5)}$$

From Eqs. (31-3), (31-4), and (31-5),

$$T_B(\theta, \phi) = (1 - /R_s/^2) T_s + /R_s/^2 T_{\text{sky}} \qquad \textbf{(31-6)}$$

It is emphasized that these equations were written in a conceptual form and do not include the entire vector phasor property of the antenna and the observed surface. The polarization property of the antenna and its relationship to the surface-polarization property are discussed in the next subsection.

Polarization

To include polarization properly in the antenna temperature expressions the polarization properties of both the antenna and the surface must be included. The polarization is defined relative to the plane of the aperture in the case of the antenna and to the incident plane in the case of the surface. Consider the following coordinate

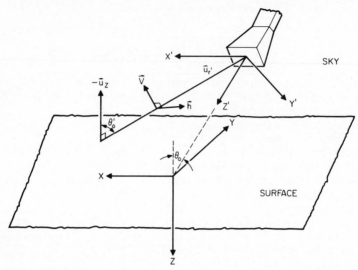

FIG. 31-2   Definition of polarization of a radiometer antenna with the surface and sky scene.

system given in Fig. 31-2, where the observed surface is in the xy plane and the antenna aperture is in the $x'y'$ plane. In this instance,

$$\mathbf{h} = \frac{\mathbf{u}_z \times \mathbf{u}_r'}{\sin \theta'} \qquad \mathbf{v} = \mathbf{h} \times \mathbf{u}_r$$

and ψ = the angle between $\mathbf{h}$ and $\mathbf{u}_\phi$. Also note that

$$\mathbf{h} \cdot \mathbf{u}_\phi' = \cos \psi = \frac{\sin \theta' \cos \theta_0 - \cos \theta' \sin \theta' \sin \theta_0}{\sin \theta_0'}$$

and

$$\mathbf{v} \cdot \mathbf{u}_v = \sin \psi = - \frac{\sin \theta_0 \cos \phi}{\sin \theta_0'}$$

so that

$$\begin{aligned}\mathbf{h} &= \cos \psi \, \mathbf{u}_\phi' - \sin \psi \, \mathbf{u}_\phi' \\ \mathbf{v} &= \sin \psi \, \mathbf{u}_\phi' - \cos \psi \, \mathbf{u}_\phi'\end{aligned} \qquad \textbf{(31-7)}$$

It should also be noted that $\mathbf{h}$ and $\mathbf{v}$ are the vector directions commonly referred to as horizontal and vertical polarization. This should allow the conversion of available reflection-coefficient data on natural surfaces to emissivity and produce brightness temperatures through the use of Eq. (31-6). If we assume that the far field from an aperture in the $x'y'$ plane can be broken into vertical and horizontal components, then

$$\mathbf{E} = \mathbf{h}[-\sin \psi \, E_\theta + \cos \psi \, E_\phi] + \mathbf{v}[\cos \psi \, E_\theta + \sin \psi \, E_\phi]$$

Using these results, the antenna temperature is defined as

T_A

$$= \frac{\int_0^{2\pi} \int_0^{\pi} \{[-\sin \psi \, E_\theta + \cos \psi \, E_\phi]^2 T_{BH} + [\cos \psi \, E_\theta + \sin \psi \, E_\phi]^2 T_{BV}\} \sin \theta d\theta d\phi}{\int_0^{2\pi} \int_0^{\pi} \{[-\sin \psi \, E_\theta + \cos \psi \, E_\phi]^2 + [\cos \psi \, E_\theta + \sin \psi \, E_\phi]^2\} \sin \theta d\theta d\phi}$$

(31-8)

If we assume, as given in Peake's derivation,[7] that $E_\theta = C\sqrt{f_\theta} \, e^{j\alpha}$ and that $E_\theta = C\sqrt{f_\phi} \, e^{j(\alpha + \delta)}$, where f_ϕ and f_θ are the normalized radiation patterns as a function of θ and ϕ and polarized in the θ and ϕ directions respectively, the **h** and **v** terms in Eq. (31-6) are given by

$$E_h = Ce^{j\alpha}[-\sin \psi \, \sqrt{f_\theta} = \cos \psi \, \sqrt{f_\phi} \, e^{j\gamma}]$$

$$E_v = Ce^{j\alpha}[\cos \psi \, \sqrt{f_\theta} + \sin \psi \, \sqrt{f_\phi} \, e^{j\gamma}]$$

(31-9)

Using these expressions, the antenna temperature is given by[8]

$$T_A = \frac{\int_0^{2\pi} \int_0^{\pi} \{[f_\theta \sin^2 \psi + f_\phi \cos^2 \psi] T_{BH} + [f_\theta \cos^2 \psi + f_\phi \sin^2 \psi] T_{BV} \\ + (T_{BV} - T_{BH}) f_\theta f_\phi \sin 2\psi \cos \gamma\} \sin \theta d\theta d\phi}{\int_0^{2\pi} \int_0^{\pi} (f_\theta + f_\phi) \sin \theta d\theta d\phi}$$

(31-10)

The form of Eq. (31-10) is a little different for other choices of coordinate systems, but the basic properties are the same. Notice in Eq. (31-10) that, in addition to the contributions from the clearly horizontal and vertical surface-polarization terms, there is a third term which contributes to antenna temperature that is related to the difference in horizontal and vertical brightness temperatures at given angles, the relative amplitudes of the radiation pattern, and the *phase difference between the polarization terms in the antenna pattern. If the cross-polarization level is small, this term can be neglected.* Depending upon the application, cross-polarization levels of -30 dB or more are usually satisfactory. Further discussion of this subject is given in Refs. 9 and 10 and in later sections of this chapter. It is interesting to note that for some surfaces[4,5,9] T_{BH} and T_{BV} have properties such that $T_{BH} - T_{BV} \approx 0$ out to 20 or 30°. Hence, if a narrow-beam low-sidelobe antenna is used for near-nadir observations, the antenna temperature as given in Eq. (31-10) will be independent of the polarization of the radiometer antenna. An additional reference useful in interpreting the importance of polarization is Classen and Fung.[11] For the antenna designer, the message at this point is rather clear. A radiometer antenna must be designed so that sidelobes, cross-polarization lobes, and backlobes do not pick up unwanted contributions. Common design parameters for comparing radiometer antennas are the beam efficiency and the cross-polarization index.

Beam Efficiency

The beam efficiency of an antenna is defined as

$$BE = \frac{\text{power radiated in cone angle } \theta_1}{\text{power radiated in } 4\pi \text{ sr}} \qquad \textbf{(31-11)}$$

or

$$BE = \frac{\displaystyle\int_0^{2\pi} \int_0^{\theta_1} f(\theta, \phi) \sin \theta\, d\theta\, d\phi}{\displaystyle\int_0^{2\pi} \int_0^{\pi} f(\theta, \phi) \sin \theta\, d\theta\, d\phi} \qquad \textbf{(31-12)}$$

It should be noted that polarization is not included in Eq. (31-12) for simplicity, but it will be included later. One can observe from Eqs. (31-11) and (31-12) that to relate one antenna to another in terms of beam efficiency a choice of the angle θ_1 must be made in a standard manner. A common choice for θ_1 is the angle between the beam axis and the first null. Another choice for comparing antennas, such as the wide-angle corrugated horn or the multimode horn in which the first null is poorly defined, is to use the criterion for θ_1 as 2½ times the half angle of the 3-dB beamwidth.

The beam efficiency of aperture antennas with variable amplitude distributions has received much attention in the past, curves to compare one aperture distribution to another being readily available.[12,13,14]

Examples of the effects that the shape of the amplitude distribution has upon beam efficiency are given in Fig. 31-3 for rectangular and circular apertures as a function of U ($U = Ka \sin \theta$). For uniform circular and rectangular amplitude distributions, the first several sidelobes cause pronounced ripples in the beam-efficiency curves, and values of $BE = 95$ percent are not achieved for large values of U. As more tapered distributions are assumed, the beam efficiency rises to large values rapidly and independently of the aperture shape.

This characteristic of ideal aperture distributions to approach large beam-efficiency values can be misleading in practical applications since the wide-angle sidelobes for many tapered aperture distributions are very small.[12,14] Indeed, some antennas such as the multimode, exponential, and corrugated horns have very low sidelobes and backlobes and hence excellent beam efficiencies. On the other hand, reflector antennas are sometimes used as radiometer antennas, and in this case the overall beam efficiency is influenced by spillover, cross-polarization, and blockage in addition to aperture illumination and reflector-surface roughness. These properties will be discussed later on.

The effects of polarization on beam efficiency can be stated as BE_{dp} (direct polarization):

$$BE_{dp} = \frac{\text{power at angle } \theta_1, \text{ direct polarization}}{\text{total received power, both polarizations}} \qquad \textbf{(31-13)}$$

Stated in equation form,

$$BE_{dp} = \frac{P_{\theta_1}, dp}{P_{dp} + P_{op}} \qquad \textbf{(31-14)}$$

where P_{dp} is the total power received in the direct polarization and P_{op} is the total power received in the orthogonal polarization. The effect of the orthogonal-polariza-

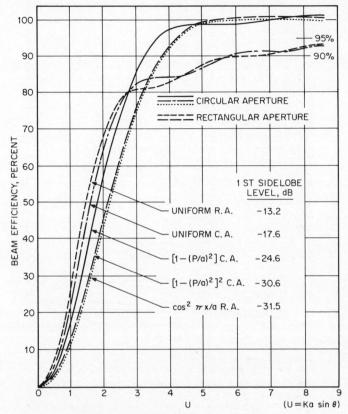

FIG. 31-3 Comparison of beam efficiencies of rectangular and circular apertures with various aperture distributions.

tion energy therefore is to decrease the direct-polarization beam efficiency of the antenna. A cross-polarization index (CPI) can be defined as[10]

$$\text{CPI} = BE_{dp}\frac{P_t}{P_{op}} \qquad\qquad \textbf{(31-15)}$$

For example, for $BE_{dp} = 85$ percent and on the assumption that the antenna has the property $P_{op}/P_t = 26$ dB, the cross-polarization isolation index is 25 dB.

The significance of cross-polarization levels in a radiometer antenna can be vividly demonstrated by reviewing the earlier discussion on polarization resulting in Eq. (31-10). The brightness temperature of a particular scene is polarization- and angle-dependent. In general, except at near-nadir angles the brightness temperature of a given surface will be different for, say, vertical and horizontal polarization at the same observation angle. To demonstrate the effect of cross-polarization properties of an antenna upon the radiometric measurement of a scene the curves in Fig. 31-4 are presented. The bias error ΔT_A in this figure is the error in brightness temperature produced by the integrated cross-polarization lobes observing the scene. This approx-

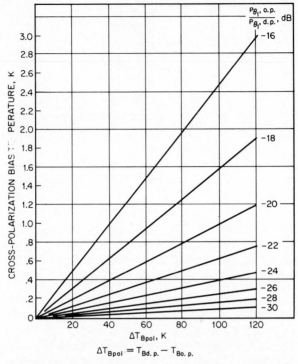

$$\Delta T_{Bpol} = T_{Bd.\,p.} - T_{Bo.\,p.}$$

FIG. 31-4   Effect of antenna cross-polarization power on radiometer bias as a function of scene polarization difference temperature.

imation is valid for only narrow-beam antennas in which emission from the surface at a given polarization is constant over 2½ times the 3-dB beamwidths of the antenna. For precision measurements of ocean temperature in which accuracies of 0.3 K are desired, values of cross-polarization energy less than 28 dB are required.

Stray Radiation

For narrow-beam antennas, $T_B\,(\theta,\,\phi)$ is relatively constant over the main beam, and Eq. (31-2) can be written as

$$T_A = T_B(\theta,\,\phi)BE_{\theta_1} + \frac{\displaystyle\int_0^{2\pi}\int_{\theta_1}^{\pi} f(\theta,\,\phi)\,T_B(\theta,\,\phi)\,\sin\theta\,d\theta\,d\phi}{\displaystyle\int_0^{2\pi}\int_0^{\pi} f(\theta,\,\phi)\,\sin\theta\,d\theta\,d\phi} \qquad \textbf{(31-16)}$$

If the nonphysical assumption that $T_B(\theta,\,\phi)$ is constant in the second term of Eq. (31-16) is made, this equation may be written as

$$T_A = T_B(\theta,\,\phi)BE_{\theta_1} + T_B(\theta,\,\phi)[1 - BE_{\theta_1}] \qquad \textbf{(31-17)}$$

The second term in Eq. (31-17) is the so-called stray-radiation contribution sometimes used in radio-astronomy applications. For such applications, the near-in and relatively far-out sidelobes observe cold sky surrounding the radio source, thus justifying the assumptions made in Eq. (31-17). For radio-astronomy applications, therefore, much design emphasis must be placed upon backlobes which point toward the hot earth. For remote-sensing applications, the close-in and wide-angle sidelobes observe brightness temperatures in the same range as the main beam (100–270 K), while in many configurations the backlobes point to the cold sky. Hence, for remote-sensing applications the stray-radiation term has added significance. In general, such concepts as stray radiation are useful in that they give some indication of the ultimate accuracy of the T_A measurement. Simplistic forms to compute stray radiation such as Eq. (31-17) should be avoided unless great care has been exercised to understand the amplitude and angular distribution of brightness temperatures in the scene.

Ohmic and Reflection Losses

The ohmic losses in the antenna will modify the apparent temperature observed by the radiometric system. These ohmic losses will modify the observed temperature by the relation[5]

$$T_a = (1 - \ell)T_A + \ell T_0 \qquad (31\text{-}18)$$

where T_a is the apparent antenna temperature of a source whose lossless observed antenna temperature of the scene is T_A, T_0 is the physical temperature of the antenna, and ℓ represents the fractional power loss in the antenna.

The significance of physical losses in the antenna upon the absolute accuracy of remotely sensed surface properties may be obtained from the following example. Assume that antenna physical temperature is that of room temperature, i.e., $T_0 \approx 300$ K. Typical ocean and land scenes exhibit lossless antenna temperatures between 100 and 300 K. Using this information, the effect of antenna losses upon measurement accuracy is given in Fig. 31-5. For some ocean-temperature applications ($T_A \approx 120$ K) absolute accuracies of 0.3 K are of importance. Hence, very small losses (≤ 0.005 dB) are of interest. For antennas in which losses are significant, antenna loss is generally treated as a fixed bias. This method is acceptable, but in such cases the physical temperature must be known and maintained to great precision. The Potter horn built for a precision S-band radiometer[5] exhibits a loss of ≤ 0.1 dB. Studies of corrugated and multimode horns[15] indicate small but significant losses when these are used as radiometric antennas. Hence, for calibration purposes both the loss and the antenna physical temperature must be known. To ensure stability in calibration, antennas may be enclosed in a thermally stable box with a very-low-loss radome.

The effect of mismatch in the input to the radiometer upon the observed antenna temperature can be expressed as[5]

$$T_a = (1 - \ell - \rho)T_A + \ell T_0 + \rho T_R \qquad (31\text{-}19)$$

where ρ is the reflection coefficient of the antenna and T_R is the microwave temperature seen looking into the receiver. The effects of small mismatches are very important for precision measurements. Stability of this mismatch will allow one to treat this error as a bias. Again, as in the case of physical loss, the thermal stability of the antenna impedance is important.

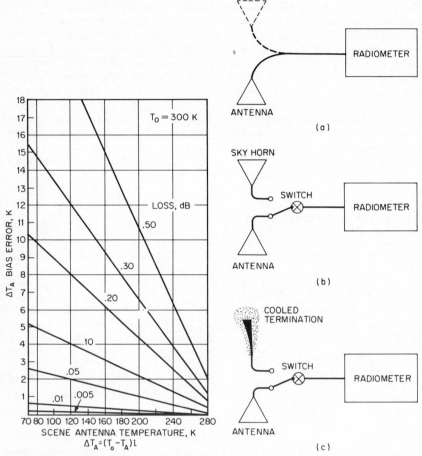

FIG. 31-5 Effects of losses upon scene antenna temperature.

FIG. 31-6 Radiometer calibration methods. (*a*) Steerable antenna. (*b*) Reference antenna and switch. (*c*) Cooled termination and switch.

Calibration

To provide a precision method of calibration, the normal procedure is to calibrate the radiometer antenna and radiometer as a single unit. This allows the user to obtain a relatively simple calibration factor for the entire system. Three simple methods of calibration, all of which have certain advantages and disadvantages, are given in Fig. 31-6. The steerable-antenna method (see Fig. 31-6*a*) requires that the radiometer antenna be pointed at the galactic pole periodically for a stable, cold reference temperature. This method has the additional advantage that the antenna can be pointed at a warm, stable surface feature such as a deep, thermally stable lake for an additional calibration point. The disadvantage of this method for spacecraft applications is that the spacecraft must be rotated. For ground applications the atmospheric attenuation is too large to use this method for upper-microwave or millimeter-wave frequencies. The switchable two-antenna method (see Fig. 31-6b) normally uses a so-called sky horn pointing to the galactic center and therefore eliminates the spacecraft-control problem. The primary disadvantage of this method is that the extra switch has losses that may change as a function of time, depending upon the switch technology and thermal control. A secondary disadvantage is that the two antennas are usually dissimilar with different losses. The third method (see Fig. 31-6c), the so-called cooled-termination method, has the advantage of not requiring a sky horn and may be very useful for millimeter- and higher-frequency ground or airborne systems in which atmospheric effects are very significant. Of course, the switch losses are common to the methods in Fig. 31-6b and c. A disadvantage of the system is the required provisions for a cooled transmission-line load.

A fourth calibration method, which is a modification of the first method, is to point the radiometer antenna toward a special free-space load[5] constructed from a porous microwave absorber located in a container filled with liquid nitrogen. This type of load has been successfully used in airborne radiometers when the radiometers have been designed to be very stable over long periods of time. Such a method also has advantages for millimeter and submillimeter applications.

31-3   SYSTEM PRINCIPLES

Radiometer Types

A variety of radiometer types have been used for radio astronomy and remote sensing. A very thorough tutorial discussion of radiometer designs is given by Hidy et al.[5] The three basic types commonly in use for remote sensing are described in simplistic block-diagram form in Fig. 31-7, which is adapted from Fig. 9.1 of Ref. 5.

The simplest radiometer, the absolute-power type, is shown in Fig. 31-7a. The output voltage of this radiometer, which is not electronically modulated, can be expressed as

$$V = G(T_A + T_N) \qquad (31\text{-}20)$$

where T_A is the antenna temperature, T_N is the system noise temperature referred to the antenna input terminals, and G is the gain of the receiver system. The field calibration of this system is dependent upon maintaining stability after the calibration procedure by using such sources as a hot and cold load. If the latest technology in radio-frequency components is employed, this system is good but usually is not adequate to achieve accuracies of 1 K over a period of time. For some spacecraft systems the calibration of this system is improved by spinning the spacecraft antenna so that the cold sky is observed during each revolution as in the calibration method indicated in Fig. 31-6a. This spinning-spacecraft-antenna system is equivalent to the signal-modulated Dicke radiometer discussed next, except that the switching circulator is eliminated and the modulation frequency is reduced to much lower frequencies.[16,17]

The next radiometer type commonly used is the Dicke radiometer,[18] as shown simplified in Fig. 31-7b. The basic improvement of the signal-modulated Dicke radiometer is that the stability of the noise from the receiver T_N is eliminated so that the output voltage is given by

$$V = G[T_A(1 - \ell_s) - T_{\text{ref}}(1 - \ell_R) + (\ell_s - \ell_R)T_0] \qquad (31\text{-}21)$$

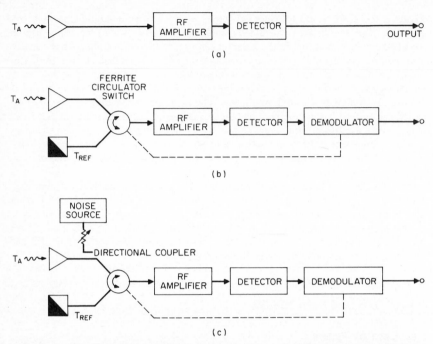

FIG. 31-7 Simplified diagram of basic radiometer types. (*a*) Absolute-power radiometer. (*b*) Dicke signal-modulated radiometer. (*c*) Noise-injection signal-modulated radiometer.

where ℓ_s and ℓ_r are the ohmic losses in the signal and reference arms and T_0 is the physical temperature of the radiometer components.

The next and latest improvement in radiometers is the noise-injection signal-modulated type shown in simplified form in Fig. 31-7*c*. The noise-injection method eliminates gain instability by allowing known amounts of noise to be put into the input of the radiometer. Various methods of performing this variable noise injection, including variable calibrated attenuators as shown in Fig. 31-7*c*, can be used. Variable-height pulse modulation of a noise diode feedback circuit can also be employed, as discussed in Ref. 5. A digital version of the noise-injection modulated radiometer has been devised and analyzed[19] and implemented.[20]

Radiometer-System Types

The most common radiometer system used in remote sensing is the scanning-beam-antenna type depicted in Fig. 31-8. This method of scanning on a spacecraft or an aircraft is usually achieved by bidirectional scanning of a reflector-antenna-feed-radiometer system relative to a stable platform. The advantage of this system is that wide-swath coverage of the surface scene can be obtained. The spatial resolution of this bidirectional scanning system is limited by an antenna size that can fit within an aircraft or launch-vehicle fairings. When finer spatial resolutions are required, deployable antennas which are limited in scan rate and can require the spacecraft to spin with the radiometer antenna as a single unit may be employed. Owing to the combination

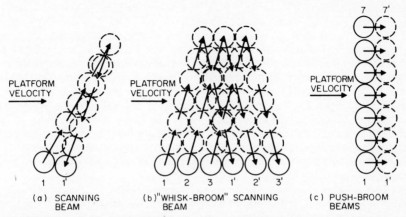

(a) SCANNING BEAM

(b) "WHISK-BROOM" SCANNING BEAM

(c) PUSH-BROOM BEAMS

FIG. 31-8 Ground spot patterns of radiometer-antenna systems.

of scan rate and spacecraft velocity, a single-beam radiometer antenna of the scanning type may not produce contiguous resolution cells on the surface at the beginning and end of each rotation. To produce this contiguous coverage, the so-called whisk-broom radiometer-antenna system can be employed, as shown in Fig. 31-8 b. Here each beam in the whisk broom requires an independent antenna port and radiometer.[21] The beam patterns can be produced by multiple feeds in a reflector system.

As even larger radiometer antennas are required to obtain better surface resolution, physical movement to obtain swath coverage may not be possible. To provide swath-width coverage for nonscanning systems, the so-called push-broom radiometer system may be used, as depicted in Fig. 31-8 c. In this design, an independent radiometer is connected to each antenna port. Movement of this push-broom beam along the surface can be time-gated to produce a surface-radiometric-brightness-temperature map.

Because of requirements to separate surface parameters such as salinity and temperature for the ocean, multiple frequencies are commonly required in radiometric systems that must be integrated into the scanning,[16] whisk-broom, or push-broom radiometric systems. Such multifrequency systems are usually designed as multiple feeds in a main reflector. Combining frequencies in a single broadband corrugated horn is feasible over nearly 2:1 bands with low-loss properties. A very broadband horn with over 5:1 bandwidth has been designed and used in space with a scanning system.[17] This antenna exhibits large losses, which must be compensated for as very large bias errors.

31-4 ANTENNA TYPES

Horns

A straightforward design for a moderate-beamwidth (10 to 30°) radiometer antenna is the electromagnetic horn; however, because of the requirement for high beam efficiency (and, therefore, for low sidelobes) specialized horn designs are necessary for radiometer applications. Radiometer horn antennas are designed so that the normally high E-plane sidelobes are reduced to an acceptable level. The H-plane sidelobes (typically −23 dB) are already low enough to achieve a sufficiently high beam efficiency for moderate beamwidths. The E-plane-sidelobe reduction can be accomplished by several techniques. The simplest technique is to utilize multimodes[22] in the radiating-horn aperture to provide sidelobe cancellation in the far-field pattern. The higher-order modes are excited by a step or sudden change in the cross section of the waveguide or horn taper. The bandwidth of a multimode or dual-mode horn is limited to a few percent because of the difference between the modal phase velocities in the tapered waveguide horn section. However, a shortened version[23] of the Potter horn can exhibit good pattern characteristics over a bandwidth approaching 10 percent. The performances of the square multimode horn and the conical dual-mode horn are quite similar.

For radiometer applications requiring a larger bandwidth, the class of corrugated horns[24] is attractive. As with the multimode-type horn, the purpose of the corrugated-horn design is to reduce the normally high E-plane sidelobes. This is accomplished in the corrugated horn by designing the corrugations so as to decrease the current along the corrugated wall, thus producing a tapered (approximately cosine) E-plane aperture field distribution. This tapered distribution can be maintained over a bandwidth approaching 2:1 but is usually limited to the operating bandwidth of the feed waveguide.

Another type of horn which shows promise as a wideband radiometer antenna is the exponential horn with a specially flared aperture.[25] A similar wideband horn is one which is flared like a trumpet.[26] The design approach of these wideband horns is to eliminate sharp discontinuities and provide a smooth transition between the horn modes and free space.

Beam-efficiency calculations are given in Fig. 31-9 for a circular aperture with a radial aperture distribution equivalent to that of the TE_{11}-mode H-plane distribution and with a quadratic phase taper, as being representative of the beam efficiencies obtained from radiometer horns. The figure also illustrates that, in order to achieve high beam efficiencies, the horn must be designed for small phase taper, either by decreasing the horn flare angle or by using a phase-correcting aperture lens.

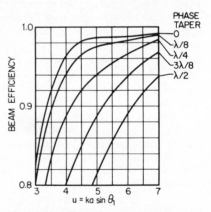

FIG. 31-9 Effect of aperture phase taper on beam efficiency for a conical corrugated or dual-mode horn.

Horn-Reflector Systems

For radiometer applications requiring narrow beamwidths, reflector antennas are more appropriate since highly tapered reflector illuminations can yield the much lower sidelobes necessary for high beam efficiency. The feed elements should be designed for spillover minimization, which represents additional beam-efficiency degradation.

The reflector surface must be constructed very accurately (usually a machined and possibly polished surface is required) to minimize loss in beam efficiency due to statistical roughness.

The beam efficiency for a reflector antenna can be expressed as

$$BE = BE_s BE_f BE_\delta \qquad (31\text{-}22)$$

where BE_s is the beam efficiency obtained by integration of the reflector secondary pattern, neglecting the back radiation, BE_f is the beam efficiency of the feed evaluated at the angle of the edge illumination, and BE_δ is the reduction factor due to reflector-surface roughness. Calculations of beam efficiencies BE_s for a circular aperture with a parabola-on-a-pedestal and with a parabola-on-a-pedestal-squared distribution are presented in Figs. 31-10 and 31-11 as representative of reflector illumination with an edge taper of -20, -10.5, -8, and -6 dB. The dashed curve for uniform illumination is included for reference. Figure 31-12 shows the effect of edge taper on the feed or spillover efficiency BE_f. The calculations in the figure are for a conical dual-mode horn with no phase error as being typical of feed patterns used in horn-reflector radiometer antennas. The reduction in beam efficiency BE_δ due to reflector-surface roughness can be obtained from the analysis of Ruze[27] as

$$BE_\delta = BE_s \exp{(-\delta^2)} + \Delta BE \qquad (31\text{-}23)$$

where $\delta = 4\pi\epsilon/\lambda$, ϵ is the rms surface roughness, and ΔBE is a correction term which accounts for the nonzero correlation length c of the surface error; i.e.,

$$\Delta BE = \exp{(-\delta^2)} \sum_{n=1}^{\infty} (\delta^{2n}/n!)[1 - \exp{(-(uc/D)^2/n)}] \qquad (31\text{-}24)$$

where D is the reflector diameter. Figure 31-13 shows a plot of the reduction factor versus rms surface roughness and correlation length. It should be noted from Eq. (31-

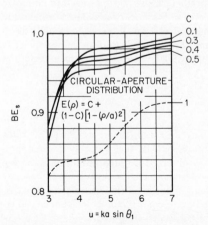

FIG. 31-10 Beam efficiency for circular aperture with parabola-on-a-pedestal distribution.

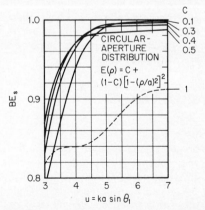

FIG. 31-11 Beam efficiency for circular aperture with a parabola-on-a-pedestal-squared distribution.

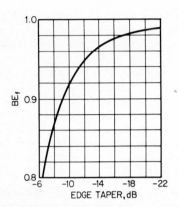

BE$_f$

1.0

0.9

0.8

-6   -10   -14   -18   -22
EDGE TAPER, dB

BEAM EFFICIENCY

1.0

0.9

0.8

0.7

0.6

0.5

0.4

0.3

0.2

0.1

0

BE$_s$

1.0
0.9
0.8
0.7

$$BE_\delta = \overline{BE_s \exp(-\overline{\delta^2})} + \Delta BE$$

$\frac{u_c}{D}$

0.5
0.4
0.3
0.2

0   0.02   0.04   0.06
RMS SURFACE ROUGHNESS, λ

FIG. 31-12 Feed-beam-efficiency factor for a reflector antenna.

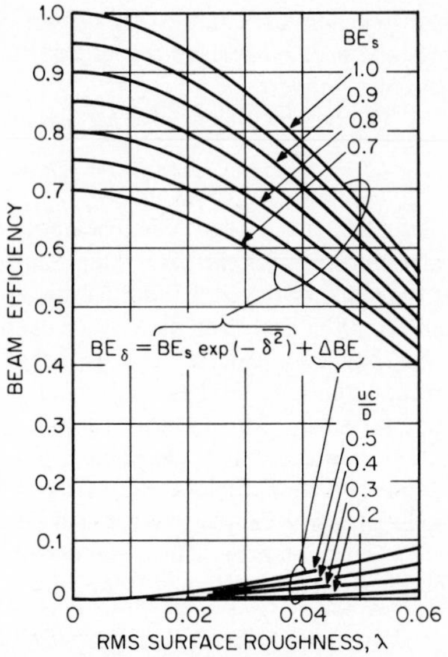

FIG. 31-13 Beam-efficiency factor of a random rough-surface reflector.

23) that for a beam efficiency greater than 90 percent the reflector surface should be $\lambda/50$ or smoother.

An additional concern for offset reflectors is the cross-polarization level, which could complicate the interpretation of antenna temperature data, as discussed earlier. The cross-polarization for reflectors decreases for larger focal-length-to-diameter ratios and smaller offset;[28] however, these parameters must be optimized to minimize feed- and spar-blockage effects. The BE_s curves in Figs. 31-10 and 31-11 neglect the effects of feed and spar blockage. Indeed, these effects cannot be treated simply as an aperture blockage as in gain calculations, and the secondary-pattern beam efficiency BE_s therefore should be recomputed with the wide-angle scattering from feed and spars included.

Phased Arrays

Phased arrays have been successfully employed as radiometer antennas[29,30] when beam scanning is required or volume constraints indicate that an array is the appropriate antenna type. High beam efficiency is obtained through amplitude tapering within the feed-distribution network. In the design of feed networks for radiometer array antennas, internal line losses should be minimized. If line losses are not excessive, stabilization of the losses through temperature control may be used in combination with calibration techniques to correct the radiometric temperature data. Resistors and terminations within the feed network should be avoided or used with discretion, since noise emitted by such components into the receiver could negate the radiometric measurement.

REFERENCES

1 D. H. Staelin and P. W. Rosenkranz, "High Resolution Passive Microwave Satellites," MIT Res. Lab. Electronics, Apr. 14, 1977.
2 P. W. Bounton, R. A. Stokes, and D. T. Wilkinson, "Primeval Fireball at λ = 3 mm," *Phys. Rev. Lett.,* vol. 21, 1968, p. 462.
3 J. R. Shakeshaft and A. S. Webster, "Microwave Background in Steady State Universe," *Nature,* vol. 217, 1968, p. 339.
4 W. H. Peak, "The Microwave Radiometer as a Remote Sensing Instrument," Electrosci. Lab. Rep. 1907-8, Ohio State University, Columbus, Jan. 17, 1969.
5 G. M. Hidy, W. F. Hall, W. N. Hardy, W. W. Ho, A. C. Jones, A. W. Love, J. Van Melle, H. H. Wang, and A. E. Wheeler, "Development of a Satellite Microwave Radiometer to Sense the Surface Temperature of the World Oceans," NASA CR-1060, National Aeronautics and Space Administration, Washington, February 1972.
6 C. T. Swift, "Passive Microwave Remote Sensing of the Ocean—A Review," *Boundary-Layer Meteorology,* vol. 18, 1980, pp. 25–54.
7 W. H. Peake, "Radar Return and Radiometric Emission from the Sea, Electrosci. Lab. Rep. 3266-1, Ohio State University, Columbus, October 1972.
8 C. T. Swift, private notes, 1981.
9 F. B. Beck, "Antenna Pattern Corrections to Microwave Radiometer Temperature Calculations," *Radio Sci.,* vol. 10, no. 10, October 1975, pp. 839–845.
10 W. H. Kummer, A. T. Villeneuve, and A. F. Seaton, "Advanced Microwave Radiometer Antenna System Study," NASA Cont. NAS 5-20738, Hughes Aircraft Company, Antenna Department, Culver City, Calif., August 1976.
11 J. P. Classen and A. K. Fung, "An Efficient Technique for Determining Apparent Temperature Distributions from Antenna Temperature Measurements," NASA CR-2310, National Aeronautics and Space Administration, Washington, September 1973.
12 A. F. Sciambi, "The Effect of the Aperture Illumination on the Circular Aperture Antenna Pattern Characteristics," *Microwave J.,* August 1965, pp. 79–31.
13 R. T. Nash, "Beam Efficiency Limitations of Large Antennas," *IEEE Trans. Antennas Propagat.,* vol. AP-12, November 1964, pp. 691–694.
14 J. Ruze, "Circular Aperture Synthesis," *IEEE Trans. Antennas Propagat.,* vol. AP-12, November 1964, pp. 691–694.
15 R. Caldecott, C. A. Mentzer, L. Peters, and J. Toth, "High Performance S-Band Horn Antennas for Radiometer Use," NASA CR-2133, National Aeronautics and Space Administration, Washington, January 1973.
16 T. Walton and T. Wilheit, "LAMMR, a New Generation Satellite Microwave Radiometer—Its Concepts and Capabilities," IEEE International Geoscience and Remote Sensing Society, Washington, June 8–10, 1981.
17 E. G. Njoku, J. M. Stacey, and F. T. Banath, "The Sea Sat Scanning Multichannel Microwave Radiometer (SMMR): Instrument Description and Performance," *IEEE J. Oceanic Eng.,* vol. OE-5, no. 2, April 1980, pp. 100–115.
18 R. H. Dicke, "The Measurement of Thermal Radiation at Microwave Frequencies," *Rev. Sci. Instrumen.,* vol. 17, 1946, pp. 268–275.
19 W. D. Stanley, "Digital Simulation of Dynamic Processes in Radiometer Systems," Final Rep., NASA Cont. NASI-14193 No. 46, Old Dominion University, Norfolk, Va., May 1980.
20 R. W. Lawrence, "An Investigation of Radiometer Design Using Digital Processing Techniques," master's thesis, Old Dominion University, Norfolk, Va., June 1981.
21 "A Mechanically Scanned Deployable Antenna Using a Whiskbroom Feed System," Cont. NAS 5-26494, Harris Corporation, Melbourne, Fla., February 1982.
22 P. D. Potter, "A New Horn Antenna with Suppressed Sidelobes and Equal Beamwidths," *Microwave J.,* June 1963, pp. 71–78.

23   M. C. Bailey, "The Development of an L-Band Radiometer Dual-Mode Horn," *IEEE Trans. Antennas Propagat.,* vol. AP-23, May 1975, pp. 439–441.

24   R. E. Lawrie and L. Peters, Jr., "Modification of Horn Antennas for Low Sidelobe Levels," *IEEE Trans. Antennas Propagat.,* vol. AP-14, September 1966, pp. 605–610.

25   W. D. Burnside and C. W. Chuang, "An Aperture-Matched Horn Design," *IEEE Antennas Propagat. Symp.,* Quebec, June 2–6, 1980, pp. 231–234.

26   J. C. Mather, "Broad-Band Flared Horn with Low Sidelobes," *IEEE Trans. Antennas Propagat.,* vol. AP-29, November 1981, pp. 967–969.

27   J. Ruze, "Antenna Tolerance Theory—A Review," *IEEE Proc.,* vol. 54, April 1966, pp. 633–640.

28   T.-S. Chu and R. H. Turrin, "Depolarization Properties of Offset Reflector Antennas," *IEEE Trans. Antennas Propagat.,* vol. AP-21, May 1973, pp. 339–345.

29   T. Wilheit, *The Electronically Scanning Microwave Radiometer (ESMR) Experiment: The Nimbus 5 User's Guide,* NASA Goddard Space Flight Center, Greenbelt, Md., November 1972.

30   B. M. Kendall, "Passive Microwave Sensing of Coastal Area Waters," AIAA Conf. Sensor Syst. for 80s, Colorado Springs, Colo., Dec. 2–4, 1980.

Chapter 32

Radar Antennas

Paul E. Rawlinson
Harold R. Ward

Raytheon Company

Radar (radio detection and ranging) is a technique for detecting and measuring the location of objects that reflect electromagnetic energy. This technique was first demonstrated in practice by Christian Hülsmeyer in 1903 by detecting radio reflections from ships.[1] With the development of the magnetron by the British in 1940 and the incentives of World War II, the radar principle was used effectively in many applications. These early radars generally operated at microwave frequencies, using a pulsed transmitter and a single antenna that was shared for transmission and reception.

The radar antenna is an important element of every radar system and is intimately related to two fundamental parameters: coverage and resolution. A radar's volume of coverage, in particular its maximum range, is one of its most basic parameters. The radar-range equation shows that the product of transmitted power and antenna gain squared is proportional to the fourth power of range.[2] Because of this dependency, economic considerations in the design of a radar usually conclude that the most cost-effective system will have 20 to 40 percent of the system cost budgeted for the antenna.

Resolution, the ability to recognize closely spaced targets, is another important radar property. The better the radar's resolution, the better it is able to separate desired returns from the returns of other objects. The size of the radar antenna measured in wavelengths is inversely proportional to its beamwidth and hence determines the radar's angular resolution. While radar applications vary, antenna beamwidths typically fall between 1 and 10°.

The radars developed to date span a wide range of size and importance. Sizes range from proximity fuzes used in artillery shells to phased-array radars housed in multistory buildings for detecting and tracking objects in space. In any one application the size of the radar and also its cost may be limited either by the physical space available or by the importance of the radar information in relation to other competing techniques or operational alternatives.

The radar applications listed in Table 32-1 illustrate the wide variety of problems that radar has been called upon to solve. Few of these applications have required large production quantities, so the radar industry is characterized by a great diversity of models and small production quantities of each model. A result is an industry that is development-intensive and requires considerably more engineering personnel than most other electronic industries. The many applications combined with a large community of development engineers has resulted in many antenna developments originating for radar applications. The phased arrays developed for satellite surveillance, weapon control, and precision approach control for aircraft landing are but a few examples.

Most of the radars designed for the applications listed in Table 32-1 operate in the microwave-frequency band. While certain radars, such as long-wave over-the-horizon radar and millimeter-wave radar for short-range applications, operate outside this region, the majority of the systems operate between 1 and 10 GHz. The considerations that bound the frequency choice for a given radar application are antenna size and the angular resolution at the low end and atmospheric attenuation and the availability of radio-frequency (RF) power at the high end. The particular frequency bands allocated for radar use by international agreement are listed in Table 32-2. The letter designations, originating during World War II for security reasons, have been

TABLE 32-1   Radar Applications

Acquisition	Navigation
Air defense	Over-the-horizon applications
Air search	Personnel detection
Air traffic control	Precision approach
Airborne early warning	Remote sensing
Airborne intercept	Satellite surveillance
Altimeter	Sea-state measurement
Astronomy	Surface search
Ballistic-missile defense	Surveillance
Civil marine applications	Surveying
Doppler navigation	Terrain avoidance
Ground-controlled interception	Terrain following
Ground mapping	Weather avoidance
Height finding	Weather mapping
Hostile-weapon location	Weapon control
Instrumentation	

used in the years since then as familiar designators of the particular radar-band segments.

Because of the great variety of radar applications, radar antennas are required to operate in many different environments. Each of these environments, listed in Table 32-3, has a special impact on a radar's antenna design and the parameters listed in Table 32-4. Land-based systems are classed as fixed-site, transportable, or mobile. At

TABLE 32-2   Standard Radar-Frequency Letter-Band Nomenclature*

Band designation	Nominal frequency range	Specific radio-location (radar) bands based on International Telecommunications Union assignments for Region 2
HF	3–30 MHz	
VHF	30– 300 MHz	138–144 MHz
		216–225 MHz
UHF	300–1000 MHz	420–450 MHz
		890–942 MHz
L	1000–2000 MHz	1215–1400 MHz
S	2000–4000 MHz	2300–2500 MHz
		2700–3700 MHz
C	4000–8000 MHz	5250–5925 MHz
X	8000–12,000 MHz	8500–10,680 MHz
Ku	12–18 GHz	13.4–14.0 GHz
		15.7–17.7 GHz
K	18–27 GHz	24.05–24.25 GHz
Ka	27–40 GHz	33.4–36.0 GHz
mm	40–300 GHz	

*From Ref. 1.

TABLE 32-3 Radar Environments

Location	Climate
Surface-based	Arctic
Airborne	Desert
Space-borne	Marine
Mobility	Electromagnetic environment
Fixed	Electromagnetic interference (EMI)
Transportable	Electromagnetic pulse (EMP)
Mobile	Electronic countermeasure (ECM)
Portable	

TABLE 32-4 Radar-Antenna Parameters

Peak power	Cross-polarization rejection
Average power	Scan volume
Gain	Scan time
Beamwidths	Pointing accuracy
Sidelobe levels	Size
Bandwidth	Weight
Loss	Environment
Mismatch	Cost
Polarization	

fixed sites the larger radar antennas are often protected by a radome, especially in arctic regions that experience heavy winds, ice, and snow. Transportable systems generally require that the antenna be disassembled for transport. Mobile systems are required to move rapidly from place to place and usually do not allow time for antenna disassembly.

Marine radar systems carried by surface craft also have special requirements. The lack of space above deck often forces the radar to a smaller antenna and larger transmitter than would be used in a comparable land-based application. The rolling platform leads to requirements for broad elevation beams or stabilization in the elevation coordinate in addition to the demands of a salt and smoke environment. Airborne and space-borne radar antennas also have special requirements peculiar to their operating environments.

In addition to the physical and climatological requirements, the electromagnetic environment has an important impact on radar-antenna design. There is a distinct contrast between radar systems designed for civil applications and those designed for military applications. While the electronic environment for civil systems consists of unintentional interference with radars operating at assigned frequencies, in military applications the threat of electronic countermeasures increases the need for an antenna to have wide bandwidth and low sidelobes.

Antennas designed for radar applications have many special requirements, some of which are not encountered in other applications. Most radar applications employ pulsed transmitters that allow the radar to resolve distant targets from nearby clutter. Peak powers from a few tens of kilowatts to a few megawatts and operating at duty ratios from 0.1 to 10 percent are typical. These high peak powers require special attention to the RF path through the antenna to the point at which power is distributed over the antenna aperture. Pressurized waveguide and feed horns are often used to prevent RF breakdown. In active array antennas the high peak power is not concentrated in a single microwave path, thus lessening the severity of this requirement.

Most radar applications require an antenna capable of scanning its beam, either mechanically or electronically or both, to search for targets in a volume of space. Scanning techniques for single or multiple beams may be classified as either mechanical or electronic. The early radar antennas developed during and after World War II relied primarily on mechanical scanning. More recently the development of high-power RF phase shifters and frequency-scanning techniques have allowed the beams to be scanned more rapidly by avoiding the inertia associated with moving mechanical components.

Accurate pointing is a requirement inherent in all radar applications that measure target location. Radar requirements are generally more demanding than those of communications antennas, especially when precision measurements of target positions are required.

Radar applications tend to be divided between two functions, search and track. The search function requires that the radar examine a volume of space at regular intervals to seek out targets of interest. In this case the volume must be probed at intervals ranging typically from 1 to 10 s and every possible target location examined.

The radar tracking function operates in a manner quite different from search. Here one or more targets are kept under continuous surveillance so that more accurate and higher-data-rate measurements may be made of the target's location. Often the interesting targets detected by a search radar will be assigned to and acquired by a tracking radar. Certain radar systems combine search and track functions by time-sharing the agile beam of a phased-array antenna.

The remainder of this chapter describes a few examples of the more common radar applications which are indicative of today's state of the art. Too few are included to give a proper perspective of the great variety of radar applications. However, those described do illustrate the differences between some of the basic types. For greater depth on the subject of radar antennas the reader is directed to Refs. 1 through 5.

32-2 SEARCH-RADAR ANTENNAS

Search radars scan one or more beams through a volume in space at regular intervals to locate targets of interest. The search function is sometimes referred to as *acquisition,* as in a missile-defense system in which detected targets are eventually tracked and intercepted. It is also called *surveillance* when targets are being monitored, as in the control of air traffic. But regardless of the term used to describe the search function, the common requirement is to scan a specified region of space at regular intervals.

The most important parameters determining the size and cost of the search radar are the coverage volume, scan time, and target size. For a given application with fixed target size and scan time, it has been shown that the requirement may be met by specifying the product of average transmitter power and antenna-aperture area.[2] It is a fundamental property of search radars that their coverage capability is proportional to their power-aperture product and does not depend on the radar's operating frequency. The benefits of operating at a lower frequency are that (1) higher-power RF transmitters are available, (2) antenna tolerances are less severe for a given antenna size, and (3) the peak power-handling capacity of waveguide and components in the high-power RF path is higher. Radars designed to perform only the search function therefore tend to operate at the lower microwave frequencies, and just how low is often determined by a trade-off between antenna size and desired angular resolution.

This section describes four examples of air search radars in operation today. These examples of current technology fall into two categories: two-dimensional (2D) and three-dimensional (3D) air search radars. The 2D systems use a fan beam that is broad in elevation and narrow in azimuth because only the target range and azimuth coordinates are required. In the 3D systems elevation is also measured to determine target altitude, and in this case the antenna's beams must be relatively narrow in both angular coordinates.

2D Systems

The two examples chosen to illustrate 2D search radars are both designed for air-traffic-control applications. In these applications, the radar must scan the air space every few seconds. Aircraft targets are detected and their range and azimuth measured. Modern systems use computers to track the aircraft from scan to scan to provide continuity in the radar data provided to the air traffic controllers.

Radar antennas used in air traffic control usually have an elevation beam with a pattern similar to that shown in Fig. 32-1. The particular pattern shape is matched to the desired radar coverage. Since aircraft have a maximum altitude, the pattern is shaped to provide a constant altitude cutoff; however, this is usually modified at shorter ranges to compensate for the use of sensitivity time control in the receiver.

The beamwidth in the azimuth coordinate is typically between 1 and 2° to provide sufficient resolution with today's air traffic densities. Since the beam has a much larger beamwidth in the elevation coordinate, it is clear that the width of the antenna aperture must be larger than its height.

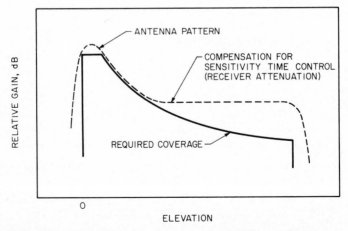

FIG. 32-1 Typical elevation pattern of a 2D surveillance radar antenna.

FIG. 32-2 ARSR-3 2D antenna. *(Courtesy of Westinghouse Electric Corp.)*

Other antenna requirements in air traffic control usually include selectable polarization and a second, high-elevation, receive-only beam. Circular polarization, if used during rainy weather, can provide between 10- and 20-dB rejection of rain return. However, circular polarization also has an associated target loss (about 3 dB), so a switch is often provided to allow the system to be operated with maximum sensitivity when rain rejection is not required.

A high-elevation receive-only beam is in common use in air-traffic-control radar antennas to give improved rejection of ground return at short range. An electronic switch connects the receiver to the high beam in the ground-clutter regions and then switches to the lower-elevation transmit beam pattern for the remainder of active range. Both polarization switching and beam switching add complexity to the feed of the feed-reflector antennas typically used in this application. Two examples which illustrate current designs follow.

ARSR-3

The ARSR-3 radar system was developed by Westinghouse for use in the en route air-traffic-control system of the Federal Aviation Administration (FAA).[6,7] The L-band radars (1.25 to 1.35 GHz) are used to maintain surveillance over aircraft in the high-altitude air routes between terminals. They provide 2D search data to a range of 366 km with a 12-s data interval. For this application, an antenna must have a shaped elevation coverage, a second high-elevation receive beam, and selectable linear or circular polarization.

Figure 32-2 shows the horn-reflector antenna used in the ARSR-3. The 6.9- by 12.8-m reflector forms a fan beam 1.25° wide in the azimuth coordinate. The antenna and pedestal are supported by a steel open-structure tower and are protected from the external environment by a space-frame radome.

The shaped elevation coverage, shown in Fig. 32-3, is formed by a doubly curved reflector. The upper portion of the reflector is nearly paraboloidal to direct energy to

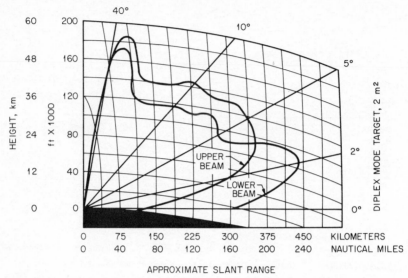

FIG. 32-3 ARSR-3 vertical coverage. *(Courtesy of Westinghouse Electric Corp.)*

the peak of the low-beam pattern, while the lower portion of the reflector departs from the parabola to direct more energy to elevation angles between 5 and 40°. The antenna features a rapid beam cutoff of the lower vertical pattern to minimize the lobing caused by surface reflections.

The dual-feed horns shown in Fig. 32-2 are used to provide two receive beams. One is a low-elevation beam used during transmission and reception, while the second is a higher-elevation beam which is switched in only during the time of arrival of the short-range returns in order to suppress land-clutter returns. For ranges beyond the clutter, the receiver is connected to the low-beam port to give better low-elevation coverage at far range. An important requirement of a two-beam design is the allowable gain difference in the direction of the horizon. If the gain difference is too large, the horizon range of the high-elevation beam will be shorter than the range of the farthest clutter, and hence the high beam cannot be fully utilized. This gain ratio is held to 16 dB in the ARSR-3 design by using the outputs of both horns shown in Fig. 32-2 to form the high-elevation beam.

Circular polarization (CP), needed to cancel rain return, complicates antenna and feed design. A switch is required to select either vertical or circular polarization; but, more important, symmetry is required between horizontal and vertical polarizations in order to achieve a high integrated cancellation ratio from the CP mode. An integrated cancellation ratio of better than 18 dB is realized by this design.

AN/TPN-24

The AN/TPN-24 is an airport surveillance radar developed by Raytheon for the U.S. Air Force as part of the AN/TPN-19 ground-controlled approach system.[8] This S-band radar (2.7 to 2.9 GHz) provides 2D surveillance of the air space in the vicinity of an airfield. Air traffic controllers seeing this 2D information displayed on a plan position indicator (PPI) vector landing aircraft onto the runway approach while communicating with the pilots via radio. For this application, a shorter-range and higher-data-rate radar is required than in the en route surveillance application described above, so the AN/TPN-24 gives coverage to 111 km with a 4-s data interval.

Figure 32-4 shows the AN/TPN-24 radar. The shelter contains all the radar electronics as well as a display, a radio, and microwave-relay equipment. The antenna mounted atop the shelter is disassembled and stowed inside the shelter for transport. When erected, the antenna reflector is 4.27 m wide and 2.4 m high, which at S band gives an azimuth beamwidth of 1.6°. The elevation beam pattern is shaped to match the desired elevation coverage, as shown in Fig. 32-5.

The requirement for transportability provides an incentive to use the smallest possible vertical aperture that will give the desired elevation pattern. In this antenna the entire reflector has a paraboloidal shape, and the elevation-pattern shaping is achieved by controlling the amplitude and phase of the power distributed to each of the 12 feed horns.

The particular feed illumination is determined by the microwave power divider mounted on the back of the reflector. Incorporated into this divider is a pin-diode switch that allows the amount of power coupled to the first feed horn (the horn contributing to the pattern near the horizon) to be varied during the receive mode. The

FIG. 32-4 AN/TPN-24 2D antenna. *(Courtesy of Raytheon Company.)*

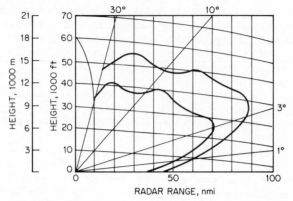

FIG. 32-5 AN/TPN-24 vertical coverage for single-channel and two-channel diversity. *(Courtesy of Raytheon Company.)*

switch allows the beam to be raised to reduce clutter at short range. The underside of the elevation beam pattern is tailored to produce a rapid cutoff and low sidelobes, thus minimizing lobing effects.

Linear or circular polarizations are selected by polarizers in series with each of the 12 feed horns. The polarizers are ganged and rotated mechanically to change polarization. To achieve the 20-dB integrated cancellation ratio provided by the antenna, mutual coupling between horns is minimized and compensated so as to equalize the feed pattern in both vertical and horizontal polarizations.

3D Systems

The need for 3D search radars originates in the requirements of military air defense systems. These systems protect an air space by detecting intruders and vectoring defending aircraft to intercept these intruders. This process requires that the interceptor know the altitude of an intruder, and therefore an elevation measurement by the radar is needed. Early systems used a 2D radar in combination with a separate height-finder radar that was designated to selected targets to measure altitude.[1] More recent systems combine the 2D search and height finding into a single 3D radar system with the capability of measuring a target's position in all three coordinates.

The antenna for a 3D radar must have a narrow receive beamwidth in the elevation as well as in the azimuth coordinate. While the required vertical coverage is about the same as that shown in Fig. 32-3 for the ARSR-3, the antenna must be capable of forming a narrow elevation beam within that coverage. Since the operational requirement is for a fixed-altitude accuracy, it is possible to allow the elevation error (and hence the elevation beamwidth) to increase at higher elevation angles.

Since these 3D radars are for military applications, the antennas must have low-azimuth sidelobes and wide bandwidth to operate effectively in an electronic-countermeasure (ECM) environment. The low sidelobes reduce jamming received from directions other than that of the target, and the wide bandwidth allows the use of frequency agility, which forces the jammer to dilute its power by spreading it over a wider band.

Various antenna techniques have been used to satisfy the 3D requirement. Two common types are stacked-beam and pencil-beam antennas. Stacked-beam systems transmit through a fan beam to illuminate the entire vertical coverage region and then receive returns through a stack of simultaneous receive beams. The pencil-beam system has a single beam, narrow in azimuth and elevation, which is used for both transmit and receive to scan the required elevations sequentially. The examples described below illustrate these 3D techniques.

AN/TPS-43

The AN/TPS-43 is a stacked-beam 3D surveillance radar developed by Westinghouse for the U.S. Air Force. It is one of the most successful 3D systems in use today, with over 120 radars manufactured and operating in 19 countries throughout the world.[9]

The antenna, shown in Fig. 32-6, is an important element of this S-band system. It has a 6.2-m-wide by 4.3-m-high paraboloidal reflector that forms a beam 1.1° wide in azimuth and 1.5° high in elevation. The reflector is illuminated by a stack of 15 vertically polarized horns combined to form a stack of six separate elevation receive beams covering from the horizon to 20° elevation. In the transmit mode, power is divided among the horns so as to illuminate the entire vertical coverage. On receive, the six beam outputs are received and processed in parallel to detect targets and measure their elevation, from which altitude is computed. The antenna completes a 360° azimuth scan every 10 s.[10,11]

This stacked-beam antenna requires considerably more microwave hardware above the rotary joint than is needed in a 2D antenna. The feed network, located at the base of the antenna structure, contains 13 duplexers and 6 RF receivers along with a transmit power divider and a receive power-combiner matrix. The received signals

FIG. 32-6 AN/TPS-43 3D antenna. *(Courtesy of Westinghouse Electric Corp.)*

are down-converted to intermediate frequency (IF) so that they may be brought by slip rings to the stationary portion of the antenna pedestal.

This AN/TPS-43 antenna also contains a separate secondary radar antenna mounted below the S-band feed.

AN/TPS-59

The AN/TPS-59 is a long-range 3D surveillance radar developed by General Electric for the U.S. Marine Corps. This L-band radar uses a single pencil beam formed by an active phased array to scan elevation sequentially as the antenna rotates mechanically in azimuth. The system's most distinctive feature is its solid-state transmitter, which is distributed throughout the active array antenna.[12,13]

The antenna shown in Fig. 32-7 is 9.1 m high and 4.6 m wide. It is made up of 54 rows of horizontally polarized elements and their associated feed networks and row electronics. Each row contains an azimuth power divider, a solid-state row transmitter, two RF receivers, and two low-power phase shifters. Column feed networks combine the row outputs to form the pencil-shaped transmit beam and a three-beam monopulse cluster for receive. A receiver and an RF exciter are also included on the rotating portion of the antenna so that only power, IF, and control signals need be transferred to the antenna pedestal.

FIG. 32-7   AN/TPS-59 3D antenna. *(Courtesy of General Electric Co.)*

32-3 TRACKING RADAR

In radar applications for which target data with greater accuracy or a higher data rate than can be provided by a search radar are required, a tracking radar is used. The tracking-radar antenna usually has a pencil beam that follows a target in track to provide continuous measurement of the target's position in range, azimuth, and elevation. In some applications target signature data in the form of amplitude and phase variations of the target return are also measured. Targets are usually first detected by a search radar and then designated to the tracker for measurement. To acquire a target, the tracker must search a small volume about the designated point. When the target is detected, range and angle tracking loops close, and the range gate and beam are held centered on the target until target data are no longer required and the track is broken off.

Tracking radars tend to operate at higher microwave frequencies than search radars. It has been shown that tracking-radar sensitivity is proportional to average transmitter power times the antenna aperture area squared times the operating frequencies squared.[14] The increased importance of aperture and frequency compared with the search-radar case provides a greater incentive to use larger apertures and to operate at higher frequencies. An upper limit on the choice in these parameters is often set by atmospheric attenuation or antenna cost in relation to transmitter cost. The exceptions are radar applications for which target signature or propagation data are needed at a specific frequency, independently of cost considerations.

Fundamental to all tracking-radar antennas is the ability to measure a target's position within the beam so that the beam may be centered on the target. The two techniques used for this purpose are conical scan and monopulse. Conical scan consists of rotating the pencil beam in a circular pattern around the target in track.[1] If the target lies off the center of the circular scan pattern, the target return is amplitude-modulated by the scanning beam. The phase and amplitude of this modulation are detected and used to measure the target's position within the beam. This process requires that at least one revolution of the beam be completed before the measurement can be made. Monopulse is another technique for measuring the target's position in the beam that is capable of measuring azimuth and elevation on a single pulse return, but it requires two additional receiver channels.[1] Monopulse uses a three-beam cluster consisting of a pencil beam plus two superimposed difference beams to measure the target off-axis error in both angular coordinates.

For examples of current tracking-radar antenna designs the reader is referred to Chap. 34, "Tracking Antennas." The tracking function as incorporated into multifunction phased arrays is discussed in the next section.

32-4 MULTIFUNCTION ARRAYS

Radar search and track functions are normally realized with separate antennas, optimized in frequency and aperture size. However, with the development of high-power RF phase shifters and advanced digital-processing techniques, phased arrays have been employed to furnish nonmechanical means of antenna beam scanning and, because of the highly agile characteristics of the beam positioning, to provide an

TABLE 32-5   Features of Multifunction Arrays

Versatility: search and track
Multiple independent beams
Inertialess beam agility
Computer control
Planar or conformal type
Blast resistance
Independent control of transmit and receive illumination
Potential use of distributed power amplifiers
Graceful degradation

opportunity to combine search and track functions in a single radar. The digital-processing and beam-steering aspects of the multifunction phased array are beyond the scope of this text but are significant contributors to both the successful operation of the radar and its cost.[15,16] For the most severe target environments, the use of the multifunction array can be justified, but only after careful consideration of alternatives such as mechanically scanning antennas and hybrid approaches. Table 32-5 lists some of the features available from the multifunction array. The primary disadvantages are high cost, complexity, and compromise in the choice of frequency between the optima for the search and track functions.

The high cost of these arrays has been significant in restricting the systems to which this technology has been applied. Nevertheless, there has been gradual, though sometimes unsteady, growth in the number of phased arrays in operation. Sperry Corp.'s AN/FPS-85 at ultrahigh frequency (UHF) used a separate transmit and thinned-receive aperture[17,18] to obtain a high target resolution at reduced costs. Aperture thinning was also used by Raytheon for COBRA DANE at L band and the all-solid-state PAVE PAWS at UHF. Raytheon's mobile Patriot radar at C band uses an optical RF power-distribution system (space feed) to reduce both cost and weight.

For applications in which beam scanning in one or both planes is less than 120°, several radiating elements can share one phase shifter, as in the AN/TPQ-37 by Hughes, or hybrid scanning techniques such as a combination of a reflector and an array can be utilized, as in the AN/GPN-22 by Raytheon. The Sperry dome antenna[17] incorporates a planar array space-feeding a dome and achieves hemispheric coverage with a significant reduction in radiating elements and phase shifters.

From an antenna engineer's viewpoint, costs are controlled by optimization of the aperture, using the minimum number of radiating elements and phase shifters to accomplish the radar mission. After the array has been optimized, the next task is to design a cost-effective radiating-element and feed-distribution network (including the phase shifters, power- and signal-distribution network, and beam-steering computer). For examples of current phased-array design practice the reader is referred to Chap. 20, "Phased Arrays." Examples of phased-array radar applications are discussed in the following subsections. For additional information the reader is directed to the Refs. 15 through 19.

COBRA DANE

The COBRA DANE system, designated the AN/FPS-108, is a large phased-array radar installed on Shemya Island, Alaska, near the western end of the Aleutian

Islands chain. With its single radiating antenna face directed northwest toward the Bering Sea, it has as its prime mission collecting data on missile systems launched toward the Kamchatka Peninsula and the north Pacific Ocean. It is also capable of providing early warning of ballistic-missile attack on the continental United States as well as satellite detection and tracking. It is capable of detecting and tracking targets as small as 1 m² at ranges exceeding 8000 km. The system replaces the AN/FPS-17 and AN/FPS-80, individual search and track radars. Development was undertaken in 1973 by Raytheon with system testing completed in 1976.[11,20]

The building housing the radar, located on the northwest corner of Shemya (as shown in Fig. 32-8), is 34 m high, 33 m wide, and 26 m deep. The array face is 29 m in diameter and contains 34,769 elements, of which 15,360 are active radiators. A near-field horn radiator and screening fence, which are used to monitor the phase and gain of the active elements, can be seen on the left side of Fig. 32-8.

The radar-system characteristics are shown in Table 32-6, and a simplified radar-system block diagram is presented in Fig. 32-9.[21] The active array elements are density-tapered (thinned) across the aperture to provide a 35-dB Taylor weighting on transmit. This thinning technique[18] provides lower near-in sidelobes and a narrower beamwidth than would have been possible from a full array with the same number of uniformly illuminated active radiators. The inactive (dummy) elements provide a constant mutual-coupling environment for the active elements. Work published prior and subsequent to the development of COBRA DANE indicates that a small reduction in gain and an increase in the sidelobe level results if these dummy elements are eliminated.[19,22]

The active elements are arranged into 96 subarrays, each containing 160 active elements. Because of density tapering, the number of dummy elements per subarray varies from a minimum near the center of the array to a maximum for the peripheral subarrays. Each of the 96 traveling-wave transmitter tubes is fed through a coaxial line to a 1:160 microwave equal-power splitter which drives the elements of a subarray.

The received signals are combined by the same 160:1 splitter-combiner. The

FIG. 32-8   COBRA DANE. *(Courtesy of Raytheon Company.)*

TABLE 32-6   COBRA DANE System Parameters[21]

Frequency	
Narrowband	1215–1250 MHz
Wideband	1175–1375 MHz
Antenna	
Total number of elements	34,769
Number of active elements	15,360
Type of feed	Corporate
Aperture diameter	29 m
Scan coverage	120° cone
Transmitter	
Peak power	15.4 MW
Average power	0.92 MW
Power amplifier	96 traveling-wave tubes

transmit-receive duplexing function is achieved by changing the phase by 180° between transmit and receive for half of the elements within any subarray. The received signal thus appears on the difference port of the final 2:1 waveguide magic T, where it is amplified by a low-noise amplifier. The receiver outputs are combined in quadrantally symmetric subarrays. Each group of four symmetric subarrays is combined in monopulse comparators to form a sum- and two difference-channel signals. The transmit-and-receive sum beam-illumination taper is achieved by the density taper of the active elements. The difference-channel monopulse beams are further tapered through the use of fixed attenuators at the outputs of the 24 monopulse comparators.

Phase steering of the array in both planes is accomplished through the use of 3-bit diode phase shifters at each active element and time-delay steering, for wideband operation, at each subarray. Since the COBRA DANE array diameter is considerably greater than the desired range resolution, some form of time delay is required to com-

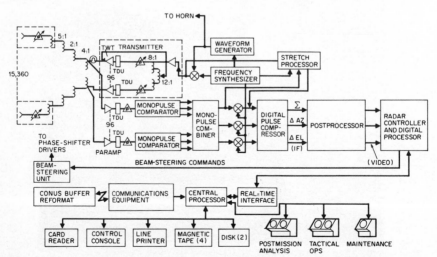

FIG. 32-9   COBRA DANE block diagram.

pensate for the time dispersion across the array for off-axis targets. Different time-delay units are used in the transmitting and receiving microwave feeds. Each time delay has two parts, 5 bits of time delay totaling one wavelength used as a phase shifter for narrowband subarray steering and 6 bits of time delay with a lowest significant bit of one wavelength.

Patriot

The Patriot radar, AN/MPQ-53, designed and built by Raytheon is a mobile C-band phased-array radar which performs target search and track, missile search and track, communications during midcourse guidance, and target-via-missile (TVM) terminal guidance. This tactical air defense system is designed to operate in a hostile ECM environment as a replacement for the improved Hawk and Nike-Hercules systems, which typically require up to nine radars.[11] Studies for this system were initiated in 1965, with an engineering development phase beginning in 1972. Initial production was funded in 1979, with a planned additional 12 units in 1982 and 18 per year through fiscal 1986.

The unique features of the radar antenna can be seen in Fig. 32-10. The antenna consists of a space-fed main array, a smaller TVM array, several ECM arrays, and an identification, friend or foe (IFF), antenna.[23,24] The main array, approximately 2.5 m in diameter, is filled with 5161 elements, which are contained in an array lens structure that is stowed in a horizontal position during transport and is erected hydraulically for system operation. Each element consists of a circular-waveguide dielectrically loaded front radiator, a flux-driven latching ferrite phase shifter, and a rectangular-waveguide dielectrically loaded rear radiator.

FIG. 32-10   Patriot-array lens. *(Courtesy of Raytheon Company.)*

FIG. 32-11 Patriot feed assembly. *(Courtesy of Raytheon Company.)*

The phase shifter utilizes a dielectrically loaded nonreciprocal garnet formed into a toroid and located in the center of the rectangular waveguide. Phase commands are in the form of row and column start-and-stop pulses whose time intervals are equivalent to 4-bit phase commands. Prior to any phase command, the garnet is driven into saturation to establish a reference phase from which all phase increments are set. The amount of phase shift is obtained by applying a voltage pulse of constant amplitude and variable width.

The feed assembly shown in Fig. 32-11, which is located approximately 2.5 m from the array, consists of transmitter horns and a multimode, multilayer receive feed. Through this arrangement, the duplexing function is implemented by the space separation of the feeds rather than through the use of conventional duplexers. The receive feed is located on the array axis. Phase-shift commands for transmit and receive differ to account for the different placements of the transmit-receive feeds. The receive feed is a five-layer (*E*-plane), multimode (*H*-plane) monopulse horn with independent control of the sum-and-difference patterns. Dielectric lenses in the feed aperture provide phase and amplitude correction of the *H*-plane excitation.

A TVM array (shown below and to the right of the main array in Fig. 32-10) produces a receive sum output from the TVM downlink signals and provides main-array sidelobe blanking and cancellation. There are 253 elements arranged in an aperture approximately half a meter in diameter. The array utilizes a stripline corporate feed. Uniform illumination is provided through a primary-distribution-network 12:1 power combiner, which in turn is connected to an aperture illumination network. This is followed by a transition to the waveguide phase-shift element, which is identical to that used in the main array.

Up to five ECM arrays are provided, three along the bottom of the antenna and two, one on each side, below the main array. These arrays each contain 51 elements identical to those used in the main and TVM arrays. The corporate feed is a single stripline layer which provides uniform amplitude and in-phase signals at each element.

The IFF antenna is located just below the main array.

Limited-Scan Arrays

Many radar applications for which antenna beam scanning in the order of $\pm 10°$ is required, such as ground-controlled approach and weapon locating, can use multi-function array antennas that are considerably less complex than those previously described. Limited-scan or limited-field-of-view (LFOV) antennas have been designed to reduce the number of phase shifters in the array.[25,26] The most common approach is either to provide one phase shifter for a subarray of elements or an oversized element or to use a small array in combination with either a reflector or a lens. It is interesting to note that this latter hybrid approach of using a lens to reduce the number of radiating elements and phase shifters also promises to reduce the cost of wide-angle scanning (hemispheric), as demonstrated in the Sperry dome antenna.[27]

Two examples of the implementation of multifunction arrays to the limited-scan application are given. In the first, the AN/GPN-22 utilizes a hybrid approach, and in the second, the AN/TPQ-37, one phase shifter per several elements is used in one plane of scan.

AN/GPN-22[28]

The antenna used by Raytheon for the ground-controlled approach AN/GPN-22 system is similar in its concept to the AN/TPN-25 antenna. The antenna and its radar shelter are shown in Fig. 32-12. After the development of the mobile version, AN/TPN-25, in 1969 and the subsequent production of 11 systems, the antenna system

FIG. 32-12 AN/GPN-22 precision approach radar. *(Courtesy of Raytheon Company.)*

TABLE 32-7   AN/GPN-22 Antenna
Parameters[28]

Type	Limited-scan phased array
Gain	42 dB
Beamwidth	
Azimuth	1.4°
Elevation	0.75°
Scan volume	
Azimuth	±10°
Elevation	8°
Polarization	Circular
Array elements	443
Phase-shifter bits	3
Reflector size	4 m by 4.7 m

was redesigned electrically and mechanically in 1975 to reduce costs further for fixed-site applications and was redesignated the AN/GPN-22. Approximately 50 of these systems have been produced. The radar searches a volume of 20° in azimuth and 8° in elevation out to a range of 36.5 km while simultaneously tracking up to six targets.

For this antenna, the parameters of which are listed in Table 32-7, a small space-fed array illuminates a large reflector. The antenna gain and beamwidths are determined by the reflector size, while the number of beam positions is determined by the number of array elements. The small phased array illuminates the reflector and, by scanning on the reflector surface with properly adjusted phase shifts, can cause the antenna's far-field beam to scan over a limited sector. Since the AN/GPN-22 system is to be used at fixed sites, the antenna reflector was made larger than in its mobile version. This, along with a reduced scan volume, reduced the number of array elements from 824 to 443. At the same time, the array was relocated above the reflector.

The antenna is on a pedestal mounted on a concrete base separate from the shelter. The pedestal base is capable of rotating the antenna through 280° of azimuth to permit its use for multiple-runway coverage. The feed array consists of 443 three-bit ferrite phase shifters space-fed from a monopulse multimode horn. The array RF monopulse receivers and phase-shifter power supplies are mounted in the base of the antenna.

AN/TPQ-37

The AN/TPQ-37 radar system is a tactical S-band phased-array system capable of being transported by surface vehicles or by helicopter lift. The combination of the AN/TPQ-37 and the AN/TPQ-36, which has a similar architecture but uses different antenna techniques at a higher frequency, is referred to as the Firefinder. The combined system provides automatic first-round location of hostile artillery positions and is designed to locate simultaneous fire from numerous weapons on the battlefield. The AN/TPQ-37 will normally be sited behind the battle area to locate opposing long-range artillery fire. Emplacement time is estimated at 30 min, with a displacement time of 15 min. Initial development was undertaken by the Hughes Aircraft Company in 1973, with limited production authorized in 1976 and an expected production of 72 systems.[11]

FIG. 32-13   AN/TPQ-37 radar set. *(Courtesy of Hughes Aircraft Company.)*

The antenna, shown in its erected position above its trailer in Fig. 32-13, remains stationary in normal operation but has phase scanning in both planes to provide a 90° azimuth sector scan and an elevation scan of a few degrees. Since the weapon-locating mission can be accomplished with a limited elevation scan, the number of phase shifters in the elevation plane is greatly reduced by having one phase shifter feed a vertical subarray of six elements.

Figure 32-14 is a block diagram of the antenna feed network.[29] The number of elements in the phased array is 2154, consisting of 359 vertical subarrays of 6 elements each. A vertical subarray module is shown in Fig. 32-15. Each module contains three separate microstrip diode phase shifters with the following phase states:

Phase shifter	Phase states
A	22.5°, 45°, 90°, 180°
B	25°, 50°
C	25°

The phase shifter is fabricated by using thick film techniques on an alumina substrate. The rest of the module is a 1:6 power divider using ring hybrids and trans-

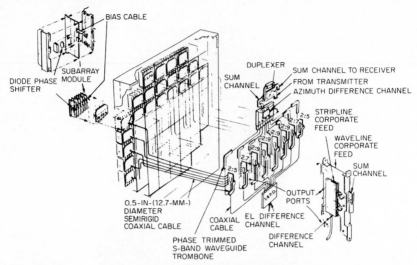

FIG. 32-14 AN/TPQ-37 antenna-feed block diagram. *(Courtesy of Hughes Aircraft Company.)*

mission lines to the six dipole feeds. These transmission-line lengths are designed to provide a progressive phase shift between dipoles, causing the radar beam to be tilted in the elevation plane.

Subarrays are arranged in groups of six fed by 1:6 air stripline power dividers. These groups are in turn stacked vertically into eight columns. The number of groups

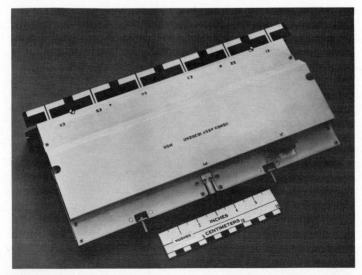

FIG. 32-15 AN/TPQ-37 antenna subarray module. *(Courtesy of Hughes Aircraft Company.)*

in a column increases from five at the edges to nine in the center. The 1:6 power dividers for subarrays in the last two columns on each side have unequal power split. Signals from the groups in any one column are connected through semirigid coaxial cable to a column-waveguide corporate feed with two outputs per column. One output, the sum arm, possesses even symmetry, and the other output, the difference, possesses odd symmetry in the elevation plane. The column difference signal is amplitude-weighted in a stripline feed assembly to control the elevation difference sidelobe level.

The difference signals for all eight columns are connected by a coaxial line to an 8:1 corporate feed with amplitude weighting for azimuth sidelobe control to form an elevation difference beam. The column sum signals which have been amplitude-weighted in their column networks to control the elevation sum sidelobe level are connected by waveguide to an 8:1 waveguide corporate feed to form a sum beam. The azimuth difference beam is formed by taking the difference of the eight column signals and combining them in the same manner as the elevation difference beam.

REFERENCES

1 M. I. Skolnik, *Introduction to Radar Systems,* 2d ed., McGraw-Hill Book Company, New York, 1980.

2 D. K. Barton, *Radar System Analysis,* Prentice-Hall, Inc., Englewood Cliffs, N.J., 1964.

3 M. I. Skolnik (ed.), *Radar Handbook,* McGraw-Hill Book Company, New York, 1970.

4 C. J. Richards (ed.), *Mechanical Engineering in Radar and Communications,* Van Nostrand Reinhold Company, London, 1969.

5 S. Silver, *Microwave Antenna Theory and Design,* MIT Rad. Lab. Ser., vol. 12, Boston Technical Publishers, Inc., Lexington, Mass., 1964.

6 P. C. Ratliff and L. F. Meren, "Advanced Enroute Air Traffic Control Radar System (ARSR-3)," *Eascon Conv. Rec.,* Institute of Electrical and Electronics Engineers, New York, 1973.

7 *The ARSR-3 Story,* Westinghouse Electric Corp., Baltimore, Md., n.d.

8 H. R. Ward, C. A. Fowler, and H. I. Lipson, "GCA Radars: Their History and State of Development," *IEEE Proc.,* vol. 62, no. 6, 1974.

9 P. L. Klass, "TPS-43 Radar Improvements Tested," *Aviation W.,* Aug. 18, 1980.

10 "Military Electronics," *Electronics,* Oct. 16, 1967.

11 R. T. Pretty (ed.), *Jane's Weapon Systems, 1980–81,* Jane's Publishing Company Ltd., London, 1980.

12 C. M. Lain and E. J. Gersten, "AN/TPS-59 Overview," *IEEE Int. Radar Conf. Rec.,* Institute of Electrical and Electronics Engineers, New York, 1975.

13 L. E. Bertz and L. J. Hayes, "AN/TPS-59—A Unique Tactical Radar," *IEEE Mechanical Eng. Conf. Radar,* Institute of Electrical and Electronics Engineers, New York, 1977.

14 D. K. Barton, "Radar Equations for Jamming and Clutter," *Eascon Conv. Rec.,* Institute of Electrical and Electronics Engineers, New York, 1967.

15 E. Brookner (ed.), "Practical Phased-Array Systems," *Microwave Journal Intensive Course,* Dedham, Mass., 1975.

16 P. J. Kahrilas, *Electronic Scanning Radar Systems (ESRS) Design Handbook,* Artech House, Inc., Dedham, Mass., 1976.

17 E. Brookner, *Radar Technology,* Artech House, Inc., Dedham, Mass., 1977.

18 R. E. Willey, "Space Tapering of Linear and Planar Arrays," *IRE Trans. Antennas Propagat.,* vol. AP-10, July 1962.

19 A. A. Oliner and G. H. Knittel (eds.), *Phased Array Antennas,* Artech House, Inc., Dedham, Mass., 1972.

20 R. W. Coraine, "COBRA DANE (AN/FPS-108) Radar System," *Signal,* May-June 1977.

21 E. Filer and J. Hartt, "COBRA DANE Wideband Pulse Compression System," *Eascon Conv. Rec.,* Institute of Electrical and Electronics Engineers, New York, 1976.

22 F. Beltran and F. King, "Elimination of the Dummy Elements in Thinned Phased Arrays," *IEEE Antennas Propagat. Int. Symp. Rec.,* New York, 1981.

23 E. J. Daly and F. Steudel, "Modern Electronically Scanned Array Antennas," *Electronic Prog. (Raytheon Co.),* winter, 1974.

24 D. R. Carey and W. Evans, "The PATRIOT Radar in Tactical Air Defense," *Eascon Conv. Rec.,* Institute of Electrical and Electronics Engineers, New York, 1981.

25 J. M. Howell, "Limited Scan Antennas," *IEEE Antennas Propagat. Int. Symp. Rec.,* Institute of Electrical and Electronics Engineers, New York, 1974.

26 R. J. Mailloux (chairman), "Antenna Techniques for Limited Sector Coverage," *IEEE Antennas Propagat. Int. Symp. Rec.,* Institute of Electrical and Electronics Engineers, New York, 1976.

27 P. M. Liebman, L. Schwartzman, and A. E. Hylas, "Dome Radar—A New Phased Array System," *IEEE Int. Radar Conf. Rec.,* Institute of Electrical and Electronics Engineers, New York, 1975.

28 H. R. Ward, "AN/TPN-25 & AN/GPN-22 Precision Approach Radars," *IEEE Int. Radar Conf. Rec.,* Institute of Electrical and Electrical Engineers, New York, 1980.

29 D. A. Ethington, "The AN/TPQ-36 and AN/TPQ-37 Firefinder Radar Systems," *Eascon Conv. Rec.,* Institute of Electrical and Electronics Engineers, New York, 1977.

Chapter 33

Microwave Beacon Antennas

Jean-Claude Sureau

Radant Systems, Inc.

33-1 INTRODUCTION

Microwave beacon systems are used whenever there is a need to enhance the target return signal with regard to strength and/or information content. As such, these systems are highly reliable and accurate surveillance systems and, in most cases, provide some data-link capability.

Beacon systems typically consist of transponders and interrogators. *Transponders* are the active devices associated with the targets and provide the enhanced echo. On the basis of certain criteria which are system-dependent, they will selectively reply to interrogations by using a recognizable encoded message format with different degrees of information content. Transponders are used either on moving targets to assist in their surveillance or at surveyed points for self-location of the interrogator. *Interrogators* are the devices which elicit and process the replies for surveillance and message decoding; they represent the users of the beacon system, and they tend to be functionally sophisticated.

By far the most widely deployed beacon-system complex is the military identification, friend or foe (IFF), Mark X and Mark XII system and its civilian surveillance-system derivative, the air traffic control radar beacon system (ATCRBS), also known as secondary surveillance radar (SSR). More recently, these have been evolving in the United States into the discrete-address beacon system (DABS) and, in the United Kingdom, into a similar system called address-selective (ADSEL) SSR. All these systems share a common frequency allocation, 1030 MHz for interrogation and 1090 MHz for reply, as well as other features in waveform format.[1,2,3] The antennas associated with transponders are typically omnidirectional on receive and transmit. From a design viewpoint, they offer no unique issues other than those related to their installation. On aircraft, for example, they are frame-mounted blades or annular slots which are located so as to minimize shadowing.[4,5] In contrast, interrogator antennas are generally more complex in their operation and, as a result, offer unique design features. These constitute the central focus of this chapter.

33-2 INTERROGATOR ANTENNAS: DESIGN PRINCIPLES

The principal system requirements imposed on interrogator antennas fall into three basic categories:

1 Support each one-way link from a power-budget as well as a time-on-target viewpoint.

2 Elicit replies only from main-beam interrogations, and process replies received only in the main beam.

3 Provide target-bearing estimates from the replies.

For ground-based interrogators, minimization of ground multipath as it relates to these three categories has been a central design consideration. Constraints resulting from colocation with a primary radar also are often factors. Such constraints become the principal issue in airborne interrogators because they impose limitations on possible performance.

Vertical Pattern Design

In ground-based (and shipborne) interrogators, link reliability can be radically affected by multipath-induced lobing in the vertical plane. An example of this lobing is shown in Fig. 33-1 for a linear array with a vertical aperture similar to that widely used for civilian air traffic control (ATC) and contrasted with that for a typical large vertical aperture providing a sharper horizon cutoff.

While the structure of the lobes is dependent on the height of the antenna above ground, the envelope of the amplitude at the minima is not but depends on the shape of the elevation pattern near the horizon. A simplistic but useful to first-order measure of this shape is the slope of the pattern, measured in decibels per degree at the horizon. The common practice has been to point antennas so that the horizon is at the -6-dB point whenever possible. Figure 33-2 shows the resulting envelope of the lobing minima as a function of this horizon cutoff rate. Since higher cutoff rates require larger vertical apertures, reduction of lobing is a major trade-off issue.

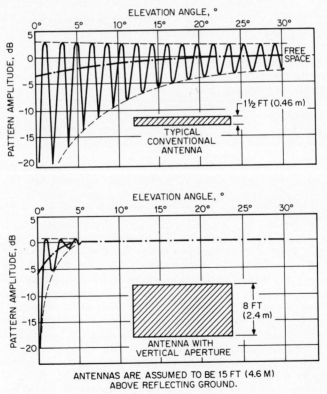

ANTENNAS ARE ASSUMED TO BE 15 FT (4.6 M) ABOVE REFLECTING GROUND.

FIG. 33-1 Influence of vertical aperture on elevation lobing patterns.

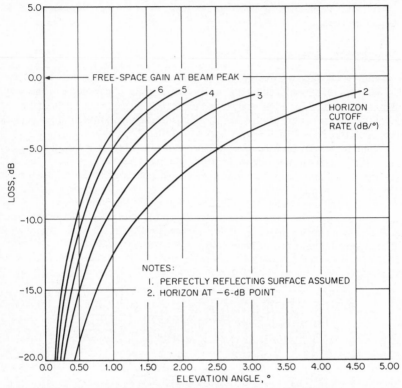

FIG. 33-2 Envelope of lobing minima for various horizon cutoff rates.

In some installations which require high cutoff rates, interrogator antennas with a large vertical aperture are directly implemented, while in others this feature is obtained by sharing the usually large primary radar reflector with the beacon system. These are called *integral beacon feed systems;* examples of both types will be discussed later in this chapter.

The most significant contribution of the development of interrogator antennas to the theory of antenna design has been the refinement of techniques for synthesizing sharp cutoff beam patterns, i.e., patterns which have constant gain over a prescribed angular sector (this tends to be the preferred shape for beacon operation). Specifically, these refinements have examined the relationship between the aperture size (or number of elements) and the rate of pattern roll-off at the beam edges as a function of sidelobe levels and gain ripple within the main beam. Evans[6] has adapted a procedure originally developed for the synthesis of digital filters. The procedure he describes allows independent specification of sidelobe level and gain ripple and results in a maximum roll-off slope for a given number of elements. For engineering estimates, Lopez[7] has provided a normalized curve (independent of aperture size), shown in Fig. 33-3, that exhibits the impact of sidelobe level and gain ripple on the roll-off rate. This rate is standardized as being measured at the -6-dB point and is given in units of decibels per degree per (D/λ), where D is the aperture size and λ is the wavelength. For exam-

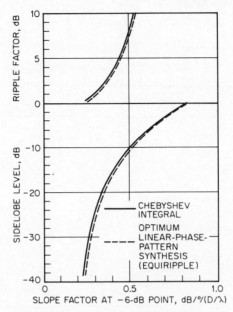

FIG. 33-3 Normalized horizon cutoff rate.
(From Ref. 7, © 1979 IEEE.)

ple, if a -24-dB sidelobe level and a 3-dB ripple-factor design are desired, the maximum slope factor is estimated by averaging the slope factors for the -24-dB sidelobe level (0.30) and the 3-dB ripple factor (0.38); the result is a slope factor of 0.34 dB/ $°/(D/\lambda)$.

Sidelobe Suppression

The procedure by which replies are elicited only from main-beam interrogations is called *interrogation sidelobe suppression* (SLS). It is generally accomplished through the sequential transmission of a pulse over the directional beam, followed by another pulse over a "control" pattern.[1] The transponder replies only when the relative amplitudes meet predetermined criteria which indicate that it is being interrogated in the main beam. A variation called *improved sidelobe suppression* (ISLS) attempts to go one step further and force suppression of transponders in sidelobes by transmitting the first pulse jointly over the sum (Σ) and control (Ω) beams. For DABS interrogations, suppression is obtained by masking the synchronization phase reversal through simultaneous transmission over the Σ and Ω beams. Some beacon systems also include receive sidelobe suppression (RSLS), in which the interrogator multichannel receiver compares, for each pulse, the amplitudes corresponding to the directional and control beams and determines whether or not the pulse should be accepted as being received through the main beam.

When the directional and control beams are derived from the same aperture *(integral suppression pattern),* they have common phase centers. Consequently, the multipath-induced vertical lobing will be the same in both beams and, to first order,

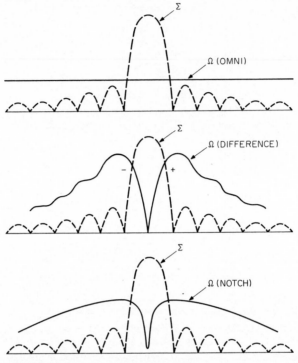

FIG. 33-4   Generic control-pattern types.

no false SLS operation results. However, when the phase centers are vertically displaced, the vertical lobing patterns do not track and faulty SLS operation results. This can take the form of false suppression and/or false interrogation when the directional beam is in a fade. This problem is mitigated when both beams have elevation patterns with sharp horizon cutoff even if they have displaced phase centers. The various types of control patterns which are typically used are illustrated in Fig. 33-4.

Omnidirectional Pattern   This pattern is implemented with a separate antenna (with a few exceptions). Its primary advantage is that it does not need to rotate and therefore does not require an additional channel in the rotary joint. The vertical pattern is often designed to match that of the directional antenna. Unfortunately, such installations tend to have vertically displaced phase centers and hence exhibit differential lobing. In addition, nearness of the directional antenna causes some blockage, which is modulated by scanning. Such systems often exhibit marginal SLS operation.[8,9]

Difference Pattern   This pattern is used primarily in airborne military interrogators because it is convenient, given installation constraints, and also because the interrogation window can be controlled, giving the effect of beam sharpening. It provides the desired narrow interrogation resolution that is somewhat better matched to the resolution of the associated radar than its inherent broad beam. This pattern is also

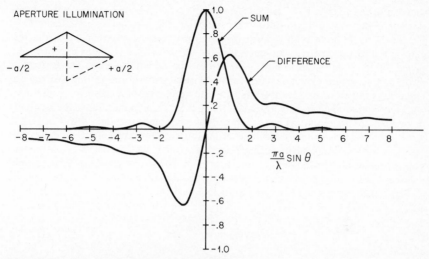

FIG. 33-5   Typical sum-difference SLS-pattern pair.

used in some integral-feed-type ground-based interrogators. Because of the central null and because such systems have no differential lobing (common phase centers), false main-beam suppression is virtually eliminated. Sidelobe breakthrough, resulting in false interrogations, can still occasionally occur. This type of suppression beam is usually implemented by taking the difference between the two halves of the aperture while the directional beam is taken as the sum. Whereas this arrangement generally results in suboptimal conventional monopulse performance, it is ideal for SLS purposes, as illustrated in Fig. 33-5.

Notch Pattern   This type offers the best available combination of common phase center, central null, and low probability of sidelobe punch-through. It can be implemented only by array-type interrogators because of the way in which such patterns are generated: the central element of the array provides the pattern's broad coverage and is combined, in antiphase, with an appropriately attenuated array directional pattern to create the central notch. The notch-and-difference-type control patterns often need to be complemented by a back-fill auxiliary radiator to cover the directional beam's backlobes.

Bearing Estimation

Estimation of the bearing to the transponder is determined by the interrogator antenna and the associated receiver system by one of two techniques, beam-split and monopulse. The beam-split technique requires that while the antenna is scanning through the transponder, several valid replies be received. Various standard algorithms for such angle estimation are used.

The monopulse technique used is inherently of the off-boresight type, and an estimate can be made on each reply pulse. Since the system typically operates at a high signal-to-noise ratio, estimation errors are due predominantly to inaccurate monopulse calibration. Experience indicates this error to be between one-fiftieth and one-hundredth of the beamwidth.[10] In particular, the monopulse slope tends to be dependent on elevation angle (approximately proportional to its cosine). This phenomenon is a direct geometric consequence of the definition of *bearing angle* as the projected direction into a horizontal plane of the true direction vector. The estimation errors are therefore elevation-dependent and also proportional to the off-boresight angle. It is therefore an advantage for the estimation algorithm to favor replies near boresight if more than one are available from a given transponder. From the viewpoint of antenna design, optimization of monopulse performance is standard with regard to slope maximization while low sidelobes are maintained for interference minimization.

For electronically phase-scanned fan-beam interrogators, the previously mentioned geometric effect takes on more serious proportions since the beam boresight exhibits a bearing angle ϕ which is elevation-angle-(α)-dependent (this phenomenon is then called *coning*) and is related to the nominal zero-elevation boresight angle Φ_0 by

$$\sin \Phi(\alpha) = \frac{\sin \Phi_0}{\cos \alpha} \qquad \textbf{(33-1)}$$

If an elevation correction is not included, bearing errors will be made for both split and monopulse systems.

33-3   INTERROGATOR ANTENNAS: PRACTICE

Ground-Based Systems

For many years the standard configuration for interrogators (ATC-IFF) consisted of a linear array referred to as a *hog trough* because of its shape and a stationary omnidirectional antenna. The array consists of a vertical-dipole-excited sectoral horn fed by a corporate power divider with an azimuth beamwidth of 2.3°; the omnidirectional antenna is typically a biconical horn. Figure 33-6 shows such an installation at a Federal Communications System (FAA) site. A different version, developed for the

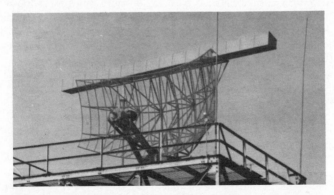

FIG. 33-6   Typical hog-trough antenna. *(Courtesy of Department of Transportation, Federal Aviation Administration.)*

FIG. 33-7 Typical ADSEL-compatible SSR antenna. *(Courtesy Cossor Electronics, Ltd.)*

ADSEL system[11] and shown in Fig. 33-7, features an integral notch-type SLS pattern and also a difference pattern for monopulse azimuth estimation and receive sidelobe suppression. The military versions of the hog trough, as exemplified by that shown in Fig. 33-8, are often much shorter but provide azimuth beam sharpening and sidelobe suppression by a difference pattern. All these linear arrays have little elevation beam shaping and a horizon cutoff rate near zero, and they are subject to different degrees, depending on the SLS pattern, to the problems of multipath.

To reduce multipath, recent designs have emphasized increased vertical aperture. The *open-array concept*,[12] an implementation of which is shown in Fig. 33-9, simultaneously satisfies the need for near-horizon shaping and physical compatibility with the existing primary radar (weight, wind loading, etc.). The use of a resonant design for the ground plane permits a wider-than-usual separation between ground-plane reflector elements and minimizes wind loading. This antenna features a notch-type SLS pattern with a back-fill radiator and a low sidelobe monopulse difference pattern. All three beams (Σ, Δ, and Ω) have the same vertical pattern, with a sharp horizon cutoff and a common phase center. Typical radiation patterns for this antenna are shown in Figs. 33-10 and 33-11. This open array is being substituted for the hog trough at many FAA terminal surveillance sites.[13]

Another approach to obtaining a sharp horizon cutoff has been to use the associated primary radar reflector (integral-feed implementation). The two principal design problems associated with this approach are the constraints on beacon patterns

FIG. 33-8 Typical military IFF antenna. *(Courtesy Hazeltine Corporation.)*

FIG. 33-9 Open-array SSR antenna. *(Courtesy Texas Instruments, Inc.)*

imposed by the radar reflector and the physical integration of beacon feed with radar feed. The latter problem is more acute with L-band radars than with S-band radars. At S band this technique has been implemented primarily for military installations (ATC-IFF), in which the 4° beacon beamwidth (inherent because of the standard S-band radar-antenna size) is effectively reduced through beam sharpening by using a difference-type SLS pattern. An example of such an implementation is shown in Fig.

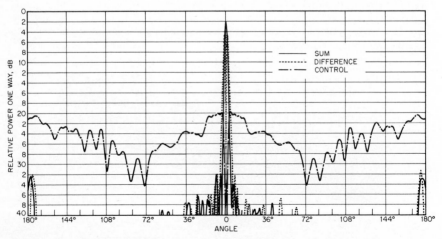

FIG. 33-10 Open-array azimuth patterns. *(Courtesy Texas Instruments, Inc.)*

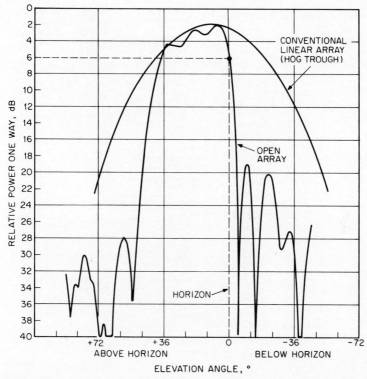

FIG. 33-11 Open-array elevation pattern. *(Courtesy Texas Instruments, Inc.)*

33-12. The horizon cutoff rate obtained is about 2 dB/°. The reflector, shaped for the radar, tends to produce an elevation pattern whose cosecant-squared upper-angle behavior is sometimes undesirable. By vertically stacking several beacon feeds (see Fig. 33-13), thereby obtaining some form of multiple-beam synthesis, the effective elevation pattern can be modified to resemble more closely a sector-type shape.

The integration of a beacon feed with an L-band radar is more difficult because of the closeness of the two frequency bands. The approach taken for the ARSR-3 radar (see Figs. 33-14 and 33-15) has been to offset the beacon feed sideways. This results in an azimuth beam shift between the radar and the beacon which must be compensated for in the information processing, since both skin and transponder returns are usually combined into one surveillance report. The feed elements are slots, several of which are vertically stacked for elevation-pattern control. The SLS function is provided by a separate omnidirectional antenna mounted on top of the reflector (it rotates with the rest of the system) and with an elevation pattern shaped to provide optimal sidelobe coverage at all elevation angles. Since the spoiled reflector has worse azimuth sidelobes at high elevation angles (see Fig. 33-16), the omnidirectional pattern cannot be allowed to drop off as much as the directional-beam principal-plane pattern.

FIG. 33-12 Tactical S-band radar antenna with integral IFF feed. *(Courtesy Thomson-CSF.)*

FIG. 33-13 Integral beacon feed for elevation-pattern shaping. *(Courtesy AEG-Telefunken.)*

For the TPS-43, a tactical air defense radar, a feed-mounted linear array (see Fig. 33-17) has been adopted. The use of a difference-type suppression pattern together with a back-fill radiator provides an effective 4° interrogation beamwidth as desired (shown in Fig. 33-18).

A limited number of beacon antennas featuring electronic scanning have taken the form of cylindrical arrays. The version shown in Fig. 33-19, implemented by Hazeltine Corporation, Wheeler Laboratory,[14] is designed to be compatible with an S-band radar tower installation. It features an integral omnidirectional SLS pattern obtained by a uniform excitation of all columns and a monopulse dif-

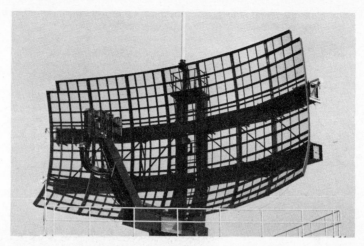

FIG. 33-14 ARSR-3 long-range surveillance radar antenna. (*Courtesy Westinghouse Electric Corp.*)

FIG. 33-15 Integral beacon feed for the ARSR-3. (*Courtesy Westinghouse Electric Corp.*)

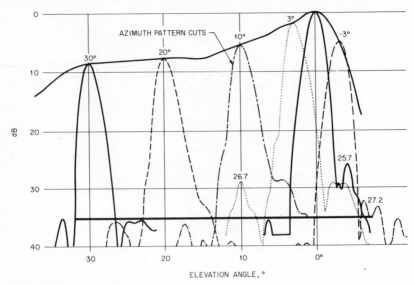

FIG. 33-16 Typical radiation patterns for the ARSR-3 beacon antenna. *(Courtesy Westinghouse Electric Corp.)*

FIG. 33-17 IFF array for the TPS-43 antenna. *(Courtesy Westinghouse Electric Corp.)*

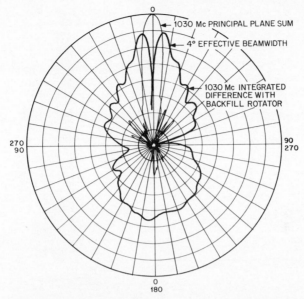

FIG. 33-18 Sum-difference suppression patterns for the TPS-43 IFF array. *(Courtesy Westinghouse Electric Corp.)*

ference pattern. The vertical pattern is of the sector beam type with a sharp horizon cutoff. In addition, there is an electronic hop-over feature in which the elevation beam can be lifted in specified azimuth directions to avoid multipath-producing obstacles. The azimuth beamwidth is maintained nearly constant as a function of elevation angle (see Fig. 33-20) by a multiple-angle collimation scheme. Scanning, performed by a

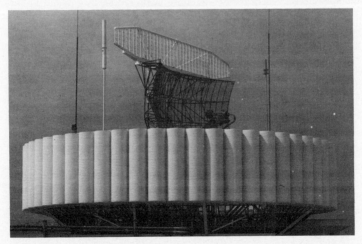

FIG. 33-19 E-SCAN cylindrical array. *(Courtesy Hazeltine Corporation.)*

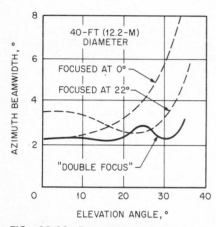

FIG. 33-20 Beamwidth variation of E-SCAN array versus elevation angle. *(Courtesy Hazeltine Corporation.)*

FIG. 33-21 Tactical IFF-ATM cylindrical array. *(Courtesy Bendix Communication Division.)*

combination of switches (sector steps) and phase shifters (fine incremental steps), can be either of the uniform continuous type to emulate a mechanically scanned system (for compatibility with existing processing) or of the agile type to support more advanced interrogation-management schemes.

The array shown in Fig. 33-21 has been developed for tactical applications. Scanning in increments of 5.6° of the 65° azimuth beam is done by a modal Butler-matrix type of feed in which all eight dipole columns are excited. The SLS and azimuth monopulse functions share the same difference-type pattern synthesized to provide backlobe coverage. All beams have a common elevation pattern with a high horizon cutoff rate.

The cylindrical array shown in Fig. 33-22 has been developed for shipborne ATC-IFF application. The directional beam, complemented by a stationary omnidirectional SLS beam, is positioned by a Lockheed-proprietary Trimode Scanner System.[16]

Airborne Systems

Traditionally, airborne beacon interrogation systems have been military IFF systems. These are typically used in conjunction with a radar which is often at X band. As mentioned earlier, the design problem is primarily one of real estate. Figure 33-23 shows a typical implementation in which L-band dipoles for the beacon interrogation antenna are mounted on the surface of the radar reflector. The dipoles use resonant techniques to make them invisible at the radar frequency. Beam sharpening and SLS are obtained by a difference pattern.

FIG. 33-22 Shipborne ATC-IFF cylindrical array AN/OE 120/UPX. *(Courtesy Lockheed Electronics Co.)*

Recently some airborne interrogators have been implemented for the civilian traffic-alert and collision-avoidance system (TCAS-II), which is based on air-derived surveillance of an aircraft's near airspace. A TCAS-II antenna is shown in Fig. 33-24. This antenna provides a capability to form either a directional beam or a notched beam in any one of eight azimuth positions. During the beacon interrogation the normal sidelobe-suppression pulse is transmitted over the notch pattern so that only transponders in a 45°-wide sector reply. The TCAS-II design can also transmit omnidirectional interrogations. Transponder replies are received omnidirectionally with a 360° instantaneous direction-finding capability to an 8° rms accuracy. Top-loaded

FIG. 33-23   IFF array for airborne fire-control radar. *(Courtesy Hazeltine Corporation.)*

FIG. 33-24 Fuselage-mounted interrogator array for collision-avoidance system. *(Courtesy Dalmo Victor Operations, Textron Inc.)*

monopoles are utilized in this design to provide a low-drag antenna. Four monopoles spaced approximately one-quarter wave apart are used. Antenna-mode control is provided by using stripline circuitry consisting of hybrid couplers, pin-diode switches, and phase shifters.

REFERENCES

1 M. I. Skolnik (ed.), "Beacons," *Radar Handbook,* McGraw-Hill Book Company, New York, 1970, chap. 38.
2 P. R. Drouilhet, "The Development of the ATC Radar Beacon System: Past, Present and Future," *IEEE Trans. Com.,* vol. Com-21, 1973, pp. 408–421.
3 D. Boyle, "DABS and ADSEL—The Next Generation of Secondary Radar," *Interavia,* March 1977, pp. 221–222.
4 K. J. Keeping and J.-C. Sureau, "Scale Model Pattern Measurements of Aircraft L-Band Beacon Antennas," MIT Lincoln Lab. Proj. Rep. ATC-47 (FAA-RD-75-23), Apr. 4, 1975.
5 G. J. Schlieckert, "An Analysis of Aircraft L-Band Beacon Antenna Patterns," MIT Lincoln Lab. Proj. Rep. ATC-37 (FAA-RD-74-144), Jan. 15, 1975.
6 J. E. Evans, "Synthesis of Equiripple Sector Antenna Patterns," *IEEE Trans. Antennas Propagat.,* vol. AP-24, May 1976, pp. 347–353.
7 A. R. Lopez, "Sharp Cut-Off Radiation Patterns," *IEEE Trans. Antennas Propagat.,* vol. AP-27, November 1979, pp. 820–824.
8 D. L. Sengupta and J. Zatkalik, "A Theoretical Study of the SLS and ISLS Mode Performance of Air Traffic Control Radar Beacon System," *Int. Radar Conf. Proc.,* April 1975, pp. 132–137.
9 N. Marchand, "Evaluation of Lateral Displacement of SLS Antennas," *IEEE Trans. Antennas Propagat.,* vol. AP-22, July 1974, pp. 546–550.
10 D. Karp and M. L. Wood, "DABS Monopulse Summary," MIT Lincoln Lab. Proj. Rep. ATC-72 (FAA-RD-76-219), Feb. 4, 1977.
11 M. C. Stevens, "Cossor Precision Secondary Radar," *Electron. Prog.,* vol. 14, 1972, pp. 8–13.
12 P. Richardson, "Open Array Antenna for Air Traffic Control," *Texas Instruments Equip. Group Eng. J.,* September–October 1979, pp. 31–39.

13 C. A. Miller and W. G. Collins, "Operational Evaluation of an Air-Traffic Control Radar Beacon System Open Array Antenna," *Eascon Conv. Rec.,* Institute of Electrical and Electronics Engineers, New York, 1978, pp. 176–185.

14 R. J. Giannini, J. Gutman, and P. Hannan, "A Cylindrical Phased Array Antenna for ATC Interrogation," *Microwave J.,* October 1973, pp. 46–49.

15 J. H. Acoraci, "Small Lightweight Electronically Steerable Antenna Successfully Utilized in an Air-Traffic Management System," *IEEE Nat. Aerosp. Electron. Conf.,* vol. 1, 1979, pp. 43–49.

16 U.S. Patent 3,728,648, Apr. 17, 1973.

Chapter 34

Tracking Antennas

Josh T. Nessmith
Willard T. Patton
RCA Corporation

34-1 INTRODUCTION

Tracking antennas are used in communications, direction finding, radio telescopes, and radar. The first three applications are covered in Chaps. 36, 39, and 41 respectively. This chapter is concerned with radar tracking antennas whose function is to provide accurate estimates of target location in two orthogonal angular coordinates. The reflector tracking antenna and the planar phased-array tracking antenna will be used for illustration.

Angle Estimation

Accuracy in estimating target direction is a fundamental requirement for the tracking antenna. It depends on systematic and random errors in the design of the antenna and mount, correction of these errors by calibration, and stability of the antenna and mount with respect to time. In addition, receiver linearity, the signal-to-noise ratio developed out of the signal processor, and the leads or lags in the difference-signal estimate, introduced by filters in the data processor, also affect the overall radar-system angular accuracy. Only the errors associated with the antenna and its mount are considered here.

Tracking a target requires a real-time feedback system that will direct the antenna beam to the target's angular direction when there is a difference* between the electrical reference direction, or boresight axis, and the actual target direction, as shown in Figs. 34-1 and 34-2.

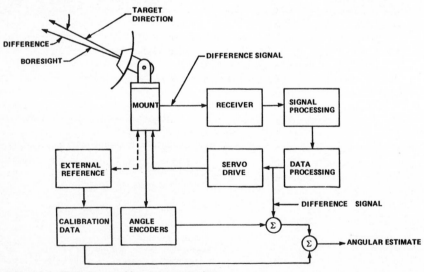

FIG. 34-1 Reflector-tracking-antenna system.

---

*Although the term *error* has been used in much of the literature to describe this quantity, the term *difference* will be used in this chapter to avoid confusion when discussing errors in tracking-antenna output data.

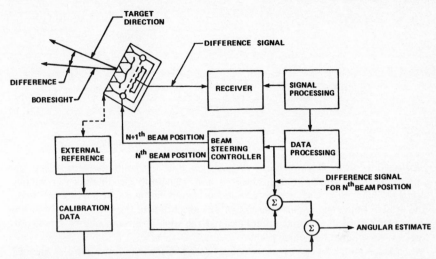

FIG. 34-2 Phased-array-tracking-antenna system.

The difference signal derived from the tracking antenna provides the information to drive the antenna boresight axis toward the target direction and also provides a correction factor[1] in the angular estimate when boresight direction leads or lags target direction. A reflector-tracking-antenna system has an output that is the sum of its mechanical orientation, as measured by angle encoders, and the electrical-angle difference, derived from the location of the target with respect to the boresight axis. Both the mechanical position and the electrical-angle-difference data are corrected as necessary from calibration data. The phased-array tracking system provides an angle estimate that is the calibration-corrected sum of the beam-steering order, the electrical-angle difference developed between the boresight axis and the target direction, and the mechanical orientation of the array face.

Because of inertia, the reflector tracking antenna usually is limited to tracking a single target, although angle estimates may be made on several targets that are in the same beam as the target being tracked. The phased-array antenna can provide accurate angle estimates on a large number of targets at widely different angular spacing. The limitation on the number of targets that may be tracked by the phased-array tracking antenna is due not to the antenna itself but to the radio-frequency (RF) energy available, the amount of energy required to provide the accuracy of estimation for each target, and the dwell time required for each target.

Tracking Techniques

Two basic classes of tracking techniques are used to obtain the angular estimate: sequential-lobe comparison and simultaneous-lobe comparison. Both provide the magnitude and direction of the difference between target and beam position necessary to perform the tracking function. Although tracking involves the entire radar system, the type of tracking is defined by the method of sensing angular-difference magnitude and direction by the antenna.

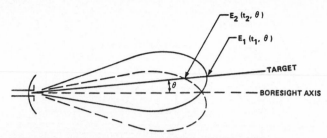

FIG. 34-3   Sequential-lobe comparison of squinted beams.

Sequential-Lobe Comparison   Sequential lobe operates by displacing the antenna beam on successive transmissions and measuring the difference in relative signal strength, expressed by $\Delta = E(t_1,\theta) - E(t_2,\theta)$. The signal strength E is a function of both time t and angular position θ. The amplitude of the difference Δ indicates the magnitude of the angular difference, and the sign, positive or negative, indicates angular direction in the lobing plane.

Sequential lobing can be performed, as shown in Fig. 34-3, by using a single feed with symmetrical mechanical or electrical displacements from the boresight axis to give the two beam positions or by electrically or mechanically switching between two offset feeds.

Conical scanning is another sequential technique that can be used. One method is to use a feed that is electrically or mechanically displaced from the boresight axis. The feed maintains this displacement as it is nutated about the boresight axis.[2] A constant return occurs for a target on the boresight or mechanical axis, and a sinusoidal return occurs for a target that is off axis. The amplitude of the sinusoid indicates the magnitude of the difference between boresight axis and target position, and the feed position at peak response provides the angular direction of the difference. Conical scanning also can be accomplished by fixing the feed and rotating a tilted reflector, as is sometimes done in small tracking systems, or by electronically combining the outputs from a monopulse feed.

Because the sequential technique is time-dependent, changes in target-signal amplitude with time can modulate the difference signal. The influence of such amplitude fluctuations can be minimized by speeding up the beam switching or the conical-scan rate. Both the switching and the scan rates, however, are limited by the pulse-repetition period of the radar.

Simultaneous-Lobe Comparison   Simultaneous-lobe comparison utilizes two different beams at the same time. Two identical beam shapes can be generated by using a reflector antenna and two feed horns that are symmetrically displaced from the boresight axis, as shown in Fig. 34-4. These "squinted" beams are identical to those received sequentially, but reception is simultaneous. Only a single signal channel following the antenna is required to make the difference estimate for sequential lobing, but two or more signal channels are required for simultaneous lobing.

If a microwave hybrid comparator is added to the feed network, a sum response, $\Sigma = E_1(t_1,\theta) + E_2(t_1,\theta)$, and a difference response, $\Delta = E_1(t_1,\theta) - E_2(t_1,\theta)$, are generated simultaneously as shown in Fig. 34-5a and b. Depending upon the comparator used, the Σ and Δ signals may occur in phase or in quadrature. If they occur in quadrature, a 90° phase shift is added to one of the signals to bring the signals in

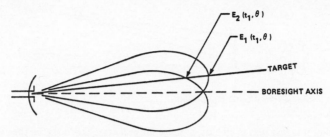

$E_2(t_1,\theta)$

$E_1(t_1,\theta)$

TARGET

— — BORESIGHT AXIS

FIG. 34-4 Simultaneous-lobe comparison of squinted beams.

phase before comparison in a product detector. In the tracking technique referred to as *amplitude monopulse*,[3] the product detector forms the vector dot product of the two normalized signals, $\Sigma/|\Sigma| \cdot \Delta/|\Sigma|$, as shown in Fig. 34-5c. Normalization often is obtained by using the sum signal for gain control of both the sum and the difference channels. Estimates out of the product detector are dependent only on the angle off axis and are independent of amplitude fluctuations of the target.

Certain objections exist to describing monopulse in terms of beams $E_1(t_1,\theta)$ and $E_2(t_1,\theta)$ because only the Σ and Δ beams are generated at the antenna ports. These expressions, however, serve as a convenient model for basic understanding. The term *difference pattern* resulted from taking the difference between the patterns of E_1 and E_2.

The separation of the two feed horns in order to obtain an amplitude comparison introduces a small phase difference when the target is off the boresight axis. If the two feed horns are separated far enough on the same aperture or if two separate antennas on a common mount are used, then an estimate of the angle between boresight direction and target direction may be derived by comparing the relative phase of the two signals received simultaneously.[3] In the dual-aperture implementation of this technique, known as phase-comparison monopulse, the two beams are pointed in the same direction rather than having an angular displacement, as in the amplitude-comparison

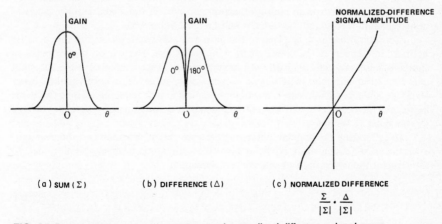

(a) SUM (Σ)   (b) DIFFERENCE (Δ)   (c) NORMALIZED DIFFERENCE

$$\frac{\Sigma}{|\Sigma|} \cdot \frac{\Delta}{|\Sigma|}$$

FIG. 34-5 Amplitude monopulse pattern and normalized difference-signal curve.

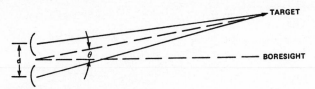

FIG. 34-6   Phase comparison.

method. For a target that is off axis, a phase difference occurs between the two signals, as shown in Fig. 34-6. This phase difference can be expressed as

$$\Delta\phi = \frac{2\pi}{\lambda}\, d \sin\theta \qquad (34\text{-}1)$$

where d is the distance between antennas, λ is the wavelength, and θ is the angle between boresight and target direction. For small angles, $\sin\theta$ is equivalent to θ, and the phase difference may be used as a linear estimate of the difference between the boresight and target direction.

Choice of Tracking Technique

The investment required to design and fabricate a phased-array antenna and the limited time available in most systems point to the choice of monopulse as the tracking technique but does not preclude any other choice of technique. The monopulse technique provides Σ and Δ outputs that may be used either as an amplitude or as a phase monopulse system or, with specific orders to the beam-steering controller, for sequential lobing or conical scan.

For reflector systems, the choice is most often a trade-off between costs and accuracy. Monopulse was developed as a means of eliminating the scintillation or amplitude noise effects on tracking accuracy. Although targets with scintillation may be tracked with sequential-lobing systems, accuracy is considerably degraded, and very high scintillation may cause loss of track. This phenomenon has been the basis for radar-jamming techniques because the modulation that is introduced enables an enemy intercept receiver to readily interpret the type of radar and employ the jamming techniques to which the radar would be most vulnerable.

Consideration must be given to the gain and sidelobe levels achievable because both are influenced by the choice of tracking technique. There is an improvement in signal-to-noise ratio of 2 to 3 dB with monopulse for targets on or near the boresight axis compared with lobe switching or conical scan. Monopulse also can achieve lower levels of sidelobes than can conical scanning or other sequential-lobing methods because of the designer's ability to optimize separately the sum and difference patterns of a monopulse system.

In estimating the angular difference between target direction and the boresight axis, the standard deviation of the error in the estimate is inversely proportional[4] to the slope of the normalized angular-difference signal:

$$\sigma_{A,E} = \frac{\theta_B}{K_m\sqrt{2\cdot S/N}} \cdot \sqrt{1 + \left(K_m\frac{\theta}{\theta_B}\right)^2} \qquad (34\text{-}2)$$

where $\sigma_{A,E}$ = standard deviation in estimate of azimuth or elevation angular difference

K_m = normalized monopulse difference slope

S/N = integrated signal-to-noise ratio

θ_B = half-power, one-way sum-pattern beamwidth (scan-angle-dependent for phased arrays)

θ = angle between boresight and target direction

Data in Skolnik[3] show that monopulse has a difference slope potential greater than conical scan, and a more precise estimate of the angular difference may be made for the same beamwidth and signal-to-noise ratio.

Although the tracking technique of choice appears to be monopulse, sequential techniques can be used when cost is the major factor and if the targets are reasonably well behaved. The same accuracy can be achieved with sequential techniques as for monopulse for targets with high signal-to-noise ratios, low scintillation, and low dynamic performance. If the targets have considerable angle noise or glint, all techniques suffer in accuracy. The attributes of monopulse in such cases may not override the economical advantages of sequential scan.

34-2 REQUIREMENTS

In specifying tracking-antenna requirements, such parameters as gain, sidelobes, and losses are often secondary to accuracy and are specified at values that support the accuracy requirements. Accuracy is expressed in terms of the allowable angular error in each axis between the true target direction, with respect to tracking-antenna location, and the direction estimate derived from antenna pattern and mount data. The specified error is generally subdivided into a budget that identifies major contributors to the error and the size for each contribution.

Errors

The term *precision,* defined as "the quality of being exactly or sharply defined or stated,"[5] often is incorrectly substituted for the term *accuracy,* which is "the quality of correctness or freedom from error." Thus, a tracking antenna may be precise but not accurate, whereas an accurate tracking antenna must be precise.

In specifying error in each axis, the root-sum-square (rss) value of both residual systematic errors and random errors is used, although in error budgets individual errors are expressed in terms of root-mean-square (rms) values. In much of the literature, the term *bias* is substituted for *systematic,* and *noise* is substituted for *random.* In this chapter, the terms *systematic* and *random* will be used, and electrical noises due to thermal effects will be discussed as a subset of random errors.[6]

Systematic Errors A *systematic error* is a constant or a variable error that can be predicted as a function of antenna beam position, frequency of operation, or environmental factors. Such errors may be measured by calibration procedures and can be expressed as algorithms or in tables to correct antenna output data. Because only a finite number of measurements can be made, algorithms or tables developed for correction are approximations. Consequently, systematic residual errors remaining after data correction include error due to approximation or interpolation, error inherent in instrumentation, and error on the part of the observer. The rms value of each residual error for each contributor in each angular coordinate is the value given in the tracking-antenna error budget under systematic residual errors. The rms value of each error contributor and the rss value of the total are usually expressed in milliradians (mrad). Sometimes these values are expressed in millisines for errors that are a function of the scan angles in phased arrays. Care must be taken in summing error budgets so that the dimensions are not intermixed. In this chapter, values will be expressed in milliradians.

Random Errors   The angular error due to thermal noise can be predicted on the basis of the known electromagnetic characteristics of the system. Maximum amplitude of backlash in data gearing or coupling to an encoder can be measured, but since it cannot be related to an angular position, it must be treated as a random error. Errors in a phased array due to phase-shifter quantization or rounding off in steering orders are also treated as random errors.

Random errors cannot be calibrated out of the system. They are grouped, and their rms values are squared and summed. The square root of this sum is the rss value for random error given in the antenna specification for each coordinate. The permitted error in each coordinate is the square root of the sum of the squares of total systematic residual and random errors. Though the process may be open to question from a strict mathematical interpretation, such procedures appear to be valid since measurements of targets of known trajectories and the use of regression analysis lead to similar estimates for the total error in each of the coordinates.

Thermal-Noise Errors   Although the tracking error due to thermal noise is a component of the random errors, the fact that it may be the dominant error in the tracking-antenna system makes it of special importance. In Eq. (34-2), the standard deviation or error in the single-pulse angular estimate due to target noise at the boresight is inversely proportional to the square root of the signal-to-noise ratio. The signal-to-noise ratio from the radar-range equation is

$$S/N = \frac{P_t \cdot G_t \cdot A_r \cdot \sigma}{(4\pi)^2 \cdot R^4 \cdot k \cdot T_0 \cdot \Delta f \cdot \overline{NF} \cdot L_r \cdot L_p} \qquad \textbf{(34-3)}$$

where P_t = peak power of transmitter

G_t = gain of transmitted pattern (scan-angle-dependent in phased arrays)

A_r = effective receiving-aperture area (scan-angle-dependent in phased arrays)

σ = target cross section

R = range of target

k = Boltzmann's constant ($1.38 \cdot 10^{-23}$ J/K)

T_0 = temperature in degrees Kelvin (290 K reference)

Δf = bandwidth of receiving channel

$\overline{NF}$ = noise figure of system

L_r = ohmic and nonohmic receiving losses

L_p = propagation losses (two-way)

The standard deviation of the target-angle estimate due to the signal-to-noise ratio is often used as a figure of merit for tracking radars and is also used to specify antenna performance. It is often used incorrectly in the literature to describe the accuracy of a radar system. This is correct only when thermal noise is the dominant error, for instance, when the radar tracks small targets at long ranges and available power and directivity limit the signal-to-noise ratio.

Error Budget for Reflector Tracking Antennas

Such parameters as gain, frequency, bandwidth, and losses enter into the error budget by means of Eqs. (34-2) and (34-3). Requirements on sidelobes are generally specified to minimize multipath for instrumentation tracking systems and also to minimize sidelobe-jammer impact for tactical tracking systems. Barton[7] points out that the difference-pattern sidelobes must be down by at least 27 dB when there is a ground-reflection coefficient of 0.3 so that the multipath return will be within a reasonable value to meet an overall accuracy requirement of 0.1 mrad.

Elements of the error budget that are electrical in nature are discussed in Sec. 34-3. Mechanical errors in the budget for an elevation-over-azimuth mount are discussed in Sec. 34-4.

Servodrive requirements for reflector-tracking-antenna systems usually are such that the torque available and the mechanical bandwidth of the antenna and mount are sufficient to keep a target of maximum dynamics well within the 3-dB sum-pattern beamwidth or within the linear region of the normalized difference signal. Because of the $\sqrt{1 + (K_m\theta/\theta_B)^2}$ factor of Eq. (34-2) and because of cross-polarization errors to be discussed later, it is desirable to track the target as near the boresight as possible.

A typical error budget for azimuth and elevation for a highly accurate reflector tracking antenna is given in Tables 34-1 and 34-2. The first column in each table gives the systematic residual errors after calibration. The second column indicates random errors that cannot be reduced by calibration. Both columns represent figures obtainable under favorable conditions, including a firm foundation to which the antenna mount is attached. The systematic residual errors are primarily a function of the calibration procedure, the capability of the observer, and the stability of the system. If the environment changes, the systematic residual errors will change, and recalibration should be initiated.

Error Budget for Phased-Array Tracking Antennas

Many of the mechanical errors associated with the determination of the tracking axis of a reflector antenna are also present in determining the direction of the normal to the array face. In addition, the phased array has errors associated with the angular displacement of the boresight axis from the direction of the array normal by means of electronic scanning. These latter errors are both dependent and independent of the amount of the displacement commanded by the beam-steering controller. However, the variation in the systematic dependent errors with scan can be accommodated in the calibration process. The measurement process by which these and the independent errors are determined is not dependent upon the scan angle, and therefore the systematic residual error remaining after the scan-dependent systematic error is corrected is independent of scan angle.

The reference system for a phased-array tracking antenna is shown in Fig. 34-7.

TABLE 34-1 Reflector-Tracking-Antenna Azimuth Error Budget, 0–85° Elevation

Contributor*	Systematic residuals, mrad rms	Random, mrad rms
Azimuth-axis tilt	0.02	
Azimuth encoder to true north	0.03	0.01
Optical-elevation-axis nonorthogonality	0.02	
Elevation-azimuth-axis nonorthogonality	0.02	
Depth of null, 10° rms postcomparator error		0.02
Optical-RF-axis collimation	0.02	
RF-elevation-axis nonorthogonality		
Mechanical	0.02	
Electrical (frequency)	0.02	
RF axis to target position	0.02	Eq. (34-2)
Cross-polarization		0.04
Wind	0.02	0.02
Thermal-mechanical	0.02	
rss	0.065	0.05†
rss total	0.082†	

*Described in Secs. 34-3 and 34-4.

†Exclusive of S/N, multipath, and refraction.

TABLE 34-2 Reflector-Tracking-Antenna Elevation Error Budget, 0–85° Elevation

Contributor*	Systematic residuals, mrad rms	Random, mrad rms
Azimuth-axis tilt	0.02	
Elevation encoder to true vertical	0.02	0.01
Depth of null, 10° postcomparator error		0.02
Optical-RF-axis collimation	0.02	
RF-axis mechanical droop	0.02	
RF axis to target position	0.02	Eq. (34-2)
Cross-polarization		0.04
Wind	0.02	0.02
Thermal-mechanical	0.02	
rss	0.051	0.050†
rss total	0.072†	

*Described in Secs. 34-3 and 34-4.

†Exclusive of S/N, multipath, and refraction.

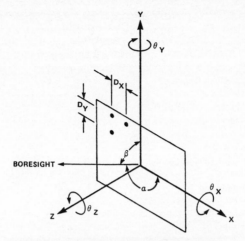

FIG. 34-7 Phased-array coordinate system.

The array face is in the X, Y plane with the Z axis as the normal to the array. The angle α defines the direction of the boresight position, or difference-pattern null, with respect to the X axis, while the angle β defines the direction of the boresight position with respect to the Y axis. On the Z axis, $\alpha = \beta = \pi/2$. The input to command the direction of the boresight position is given as α_i and β_i. The error in boresight direction is then $\alpha - \alpha_i$ and $\beta - \beta_i$.

The signals received in the monopulse difference channels are voltages V_α and V_β, representing the difference between the direction cosines ($\cos \alpha$, $\cos \beta$) of the boresight axis and the target direction relative to the X and Y axes in the plane of the array. These differences are added to the commanded boresight-direction cosines ($\cos \alpha_i$, $\cos \beta_i$) along with calibration data to form an estimate of the direction cosines of the target direction relative to the array-normal direction. The direction angles of the target are then combined with the direction of the array normal to estimate the direction of the target in earth-reference coordinates. By using these angles with target ranges, target position may be calculated in a rectilinear system. Predictions are then made as to the next target position. Target direction is transformed back from earth-reference to array-reference coordinates and into the α_i and β_i orders for the next beam position. The orders take into account the frequency of operation and temperature of the array.

Although thermal-noise error is not measured directly during alignment and calibration, the normalized slope of the difference patterns must be measured to estimate that error. Thermal-noise error can be estimated by Eq. (34-2) if it is interpreted to be the standard deviation in the direction cosine of the target direction and the parameters θ_B and K_m are those for the monopulse beam steered normal to the array ($\alpha_i = \beta_i = \pi/2$). In this direction the meaning of Eq. (34-2) is the same for a reflector antenna and for a phase-steered array. This estimate of error can be converted to angle measure by dividing by the square of the cosine of the angle off the array normal ($C^2 = 1 - \cos^2 \alpha - \cos^2 \beta$) for use in Table 34-4. By energy management of the phased array, that is, by making the power available proportional to $1/C^2$, the thermal-noise error is also made independent of scan angle. Other random errors dependent on scan angle are converted to angle measure by dividing by the cosine of the angle off the array normal (C).

TABLE 34-3 Phased-Array-Tracking-Antenna Array Face to Reference Error Budget

Contributor*	Systematic residuals, mrad rms	Random, mrad rms
Array face to true-north reference		
Alignment/calibration	$0.3/\cos\psi$	
Wind	0.1	0.1
Thermal	<0.1	
Tilt		
Alignment/calibration	0.1	
Wind	0.1	0.1
Thermal	<0.1	
Rotation		
Alignment/calibration	<0.1	
Thermal	<0.1	

*Described in Sec. 34-4.

TABLE 34-4 Phased-Array-Tracking-Antenna Array Face to Target-Position Error Budget, $[\pi/6 \leq (\alpha \text{ or } \beta) \leq 5\pi/6; \cos^2\alpha + \cos^2\beta \leq 0.75]$

Contributor*	Systematic residuals, mrad rms	Random, mrad rms
Beam command to boresight		
Position ($\alpha - \alpha_i, \beta - \beta_i$)		
Uncorrelated illumination (phase-shifter quantization, phase-shifter roundoff, phase-shifter insertion, rf path insertion, array element position, radome)		$0.2/C$
Precomparator and postcomparator phase (depth of null)	0.1	
Cross-polarization (cross coupling)		$<0.05/C$
Frequency	0.1	
Thermal	0.1	
Boresight position to target	0.1	Eq. (34-2)/C^2
rss	0.20	$0.21/C$†
rss total	$0.20(1 + 1.06/C^2)^{1/2}$†	

*Described in Secs. 34-3 and 34-4.

†Exclusive of S/N, multipath, and refraction.

Just as for the reflector tracking antenna, the gain, frequency, and beam-switching time are specified at levels that support total system requirements with the error budgets to be met under the environment specified. Receiving sidelobe levels in a tactical phased array are generally dictated by anticipated jamming levels rather than by multipath requirements. An illustrative budget for a ground-based tactical system is given in Tables 34-3 and 34-4. The budget is for an array of equal dimensions in X and Y. Otherwise Table 34-4 should be broken into separate α and β tables.

It will be noted that the major error contributor is in the α plane. As indicated in Sec. 34-4, this is due to the use of a north-seeking gyro to provide the reference direction and make the tactical phased array independent of an external reference. The use of an external reference such as employed for the instrumentation reflector tracking system (Sec. 34-5) can reduce this error.

34-3   ELECTRICAL-DESIGN CONSIDERATIONS

Aperture-Distribution Design

A monopulse pattern is a set of two or more patterns, formed simultaneously, that provides both the angular position and the signal-characteristic data from a source. Generally, the aperture distribution for one of these beams, designated the Σ, or sum, beam, will be designed to maximize the gain of the antenna in the principal direction while meeting sidelobe objectives in other directions. The aperture distributions for the other beams, designated Δ, or difference, beams, will be designed to produce a null response along one of two perpendicular planes* intersecting in the boresight-axis direction, to maximize the slope of the response with increasing angle away from the null plane in a direction normal to the plane, and to satisfy gain and sidelobe-level objectives in other directions. The design considerations governing the sum beam, such as gain, beamwidth, and sidelobe level, are similar to those of other antenna requirements. These have been covered adequately elsewhere in this handbook. In this section we will consider the special requirements of the difference beams in a monopulse cluster.

A measure of performance of the difference pattern is its slope at the central null. This slope may be normalized in at least two ways. The first, the difference slope K, refers only to the difference pattern. It is the slope of the pattern at the boresight null when the excitation is adjusted to radiate unit power relative to the response of a uniform aperture of the same size. In general terms,

$$K = \frac{\dfrac{\partial}{\partial u} \iint d(x,y)e^{-jk(ux+vy)} \, dxdy \,|\, u, v = 0}{\sqrt{A \iint d^2(x,y) \, dxdy}} \qquad \textbf{(34-4)}$$

or

$$K = \frac{-jk \iint x d(x,y) \, dA}{\sqrt{A \iint d^2(x,y) \, dA}} \qquad \textbf{(34-5)}$$

---

*For a phased array, these planes become conical surfaces about perpendicular axes in the array face.

where u (cos α) and v (cos β) are the direction cosines of the observation or target direction, $k = 2\pi/\lambda$, and $d(x,y)$ is the excitation aperture distribution.

The difference slope is maximized when the aperture has an odd linear distribution of illumination amplitude. The exact value of the slope depends upon the shape of the aperture. Two special values of K are of interest for reference purposes. For odd linear difference illumination, the difference slope for a rectangular aperture is

$$K_s = \pi/\sqrt{3} \quad \text{V/V/SBW}$$

and for a circular aperture is

$$K_s = \pi/2 \quad \text{V/V/SBW}$$

where K_s is K times the standard beamwidth of the aperture (SBW), in radians (equal to the wavelength divided by the width of the antenna). These maximum-slope functions are seldom used in practical tracking systems because of the high sidelobe levels, -8.3 dB for the rectangular aperture and -11.6 dB for the circular aperture. The aperture distributions (given by Taylor[8] for the sum pattern and by Bayliss[9] for the difference pattern), which maximize gain or slope for a given sidelobe level, are more practical models in designing aperture illumination for a monopulse tracking antenna. Approximate values for the difference slope K_s, derived from the Bayliss distribution for a square aperture, are given in Fig. 34-8.

The second way to normalize the difference slope, generally preferred in the literature of radar systems, is known as the normalized monopulse difference slope K_m, referred to in Sec. 34-2. This difference slope is derived from the difference slope K by normalizing the difference voltage to that of the sum beam and normalizing the angle to the half-power beamwidth (HPBW) of the sum beam. Thus,

$$K_m = K \frac{\theta_B}{\sqrt{\eta_\Sigma}} \tag{34-6}$$

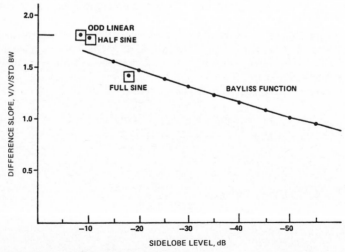

FIG. 34-8 Difference-pattern slope.

where θ_B = HPBW of sum beam

η_Σ = efficiency factor for sum beam, including only aperture illumination, spillover, and blockage

Feed Systems

The feed system distributes the RF excitation over the aperture and establishes the level of excitation at each point in the aperture. The special feature of the monopulse feed is its ability to establish three orthogonal distributions of excitation simultaneously.

A feed is constrained when the excitation proceeds from a beam port through a branching transmission line or waveguide network to the individual elements of the radiating aperture. Space feeds, or optical feeds, are those feed systems that employ a few elements, in a primary feed system, that radiate RF excitation to be intercepted and collimated by a larger secondary aperture. In this subsection, the term *optical feed* is used for this class of feed.

Constrained Feed Systems for Phased Arrays The simplest kind of monopulse feed system for a tracking array is formed by dividing the array into equal quadrants, as illustrated in Fig. 34-9. The sum of all four quadrants forms the sum beam, and the difference between the sums of the top pair and the bottom pair forms an elevation difference beam. The sum of the left pair is subtracted from the sum of the right pair to form the azimuth difference pattern. A fourth orthogonal beam, sometimes called the Q beam, can be formed by taking the difference of the differences between the quadrants. This beam is usually terminated in a matched load to prevent signals received off axis from being reflected into a difference channel.

This simple feed system does not allow for sidelobe-level optimization or error-slope optimization among the three beams of the monopulse cluster. Large tracking

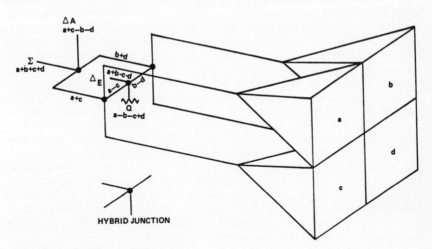

FIG. 34-9 Simple monopulse feed.

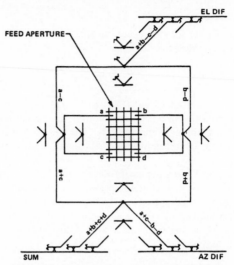

FIG. 34-10 A feed network providing complete independent control of the sum and both difference illumination distributions.

array antennas will usually employ a feed system that will allow independent optimization of the sum-and-difference-beam illumination.

Feed systems that provide sufficient degrees of freedom to allow independent optimization of the monopulse aperture distributions must take full advantage of the symmetry inherent in these distributions. Array elements or subarrays equally displaced from the monopulse axes in the array face all receive the same excitation amplitude for any given function, differing only in the pattern of phase (algebraic sign) for the different beams. A network similar to that shown in Fig. 34-9 can be used to connect symmetrically located elements or subarrays. This circuit is illustrated in Fig. 34-10. The sum ports of these may be combined to provide the array sum illumination distribution. The elevation difference ports may be independently combined to optimize the elevation difference pattern, and similarly the azimuth difference pattern may be independently optimized.

An alternative arrangement that simplifies the feed network with some loss of independence in design of the illumination function is shown in Fig. 34-11. This network allows complete independence for the elevation difference pattern, but the column sum distribution is shared between the sum and azimuth difference distributions. Because the column functions of these two distributions are similar, a compromise design based upon this sharing has little impact on the array pattern.

A row- or column-oriented feed network, subject to the compromise described for the alternative parallel feed, can be implemented with a serial monopulse ladder network as shown in Fig. 34-12. This network can be used in place of the column network or the row network, or both. There does not appear to be a generally applicable choice between these feed types, so that both types must be considered for the requirements of a specific tracking antenna.

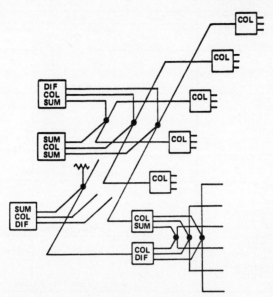

FIG. 34-11   Row-column-array monopulse feed.

Optical Feed Systems   The design of a monopulse feed system for an antenna with optical magnification, such as a reflector, a lens, or a transmission lens array, is complicated by the fact that there are two apertures to consider, the main aperture and the feed aperture. Feed apertures differ for the different beams, and there are fewer degrees of freedom available for aperture-distribution control. Optical monopulse feed systems are excellent examples of the art of design compromise.

The design of a monopulse optical feed, as also the design of a single-beam optical feed, involves a careful balance of aperture illumination efficiency, spillover efficiency, and, in case of a feed for a reflector antenna, blockage. The monopulse feed design is complicated by the need to balance the requirements of the difference beams with those of the sum beam. The optimization of the difference feed differs from that

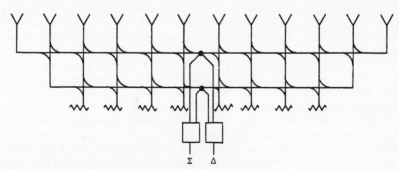

FIG. 34-12   Center-fed serial monopulse ladder network.

FEED EXCITATION WITH $(\frac{aA}{2\lambda F})$ AND $(\frac{bB}{2\lambda F})$ FOR MAXIMUM SUM GAIN	SUM		AZ DIF		EL DIF	
	GAIN	SPILLOVER EFFICIENCY	SLOPE	SPILLOVER EFFICIENCY	SLOPE	SPILLOVER EFFICIENCY
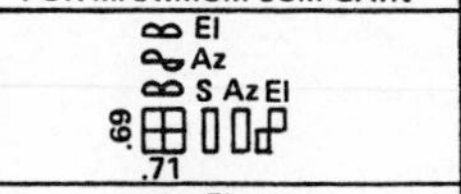 (a) FOUR-HORN	.58	.66	.943	.28	.87	.24
 (b) TWO-HORN DUAL-MODE	.75	.84	1.23	.50	1.00	.31
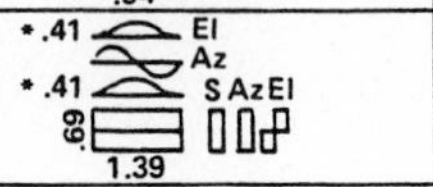 (c) TWO-HORN TRIPLE-MODE	.75	.83	1.47	.80	1.00	.31
 (d) TWELVE-HORN	.58	.66	1.29	.63	1.22	.62
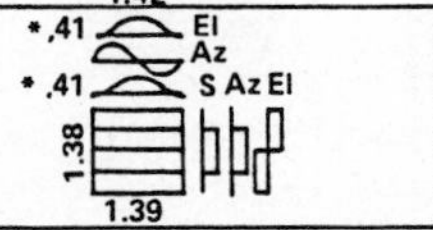 (e) FOUR-HORN TRIPLE-MODE	.75	.83	1.47	.80	1.36	.78

* RATIO OF MODE AMPLITUDES
El - ELEVATION
Az - AZIMUTH

FIG. 34-13 Monopulse feed characteristics. *(After Ref. 10.)*

for an array. Spillover efficiency for an optical feed is poor for an odd linear distribution, giving it a smaller angular sensitivity than a distribution that results from placing a greater part of the primary (feed-system) difference pattern on the focusing aperture. The optimum size of the difference feed is larger than the optimum size of the sum feed. Hannan[10] suggests that the linear dimension of the optimum difference aperture, in the plane of the difference pattern, is twice that of the sum pattern. Thus, an optimum monopulse feed may have 3 to 4 times the aperture area of an optimum single-beam feed for the same optics. As a result, a monopulse feed will have substantially larger aperture blockage than the optimum single-beam feed.

The performance of a conventional four-horn monopulse feed is poor because it requires a compromise between optimum size for sum-beam performance and for difference-beam performance. Several configurations that provide additional freedom to improve monopulse feed performance include both multibeam and multimode feeds (see Fig. 34-13). The multihorn feeds have network configurations similar to monopulse phased-array networks. For both multibeam and multimode feeds, the increased design complexity requires an increase in the minimum electrical size of the feed, making them more appropriate for large f/D optics.

Figure 34-13, which follows Hannan, represents comparative performance parameters of different feed configurations that are optimized on the basis of a collimating aperture of a rectangular cross section with dimensions A by B, located at a distance F from the feed aperture of dimensions a by b. The spillover efficiency given in this figure is the ratio of the energy radiating from the feed aperture intersecting

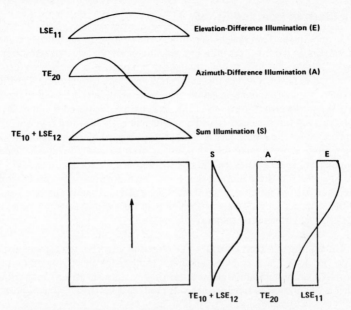

FIG. 34-14   Single-horn multimode feed-aperture distribution.

the collimating aperture to the total energy radiated by the feed in that mode. A major source of the improved performance of the more complex multihorn and multimode feeds is the reduction of spillover loss in the difference beams.

The multimode feed combines the network simplicity that is characteristic of the four-horn feed with the greater design freedom of the multihorn feed. A multimode feed uses several waveguide modes in combination to achieve a desired aperture illumination. A single mode with odd symmetry is used for the monopulse difference function. The sum-beam illumination is usually achieved by using a linear combination of two different modes, each with even symmetry. This has the desirable effect of substantially increasing the primary-pattern beamwidth and thereby improving aperture illumination efficiency at the reflector. Since two modes of different orders will have different velocities of propagation, they can be held in the proper phase relationship at the aperture over only a limited bandwidth. In practice, however, multimode feeds often compare favorably in bandwidth with more complex multihorn feeds.

An example of a dual-polarized multimode design is shown in Fig. 34-14. This feed makes use of two waveguide modes, the TE_{10} and the LSE_{12} modes, to form the sum-beam illumination. The azimuth difference-beam illumination uses the TE_{20}, and the elevation difference-beam illumination uses the LSE_{11} mode. These feed-aperture illumination distributions are illustrated in Fig. 34-14. The bandwidth of the multimode feed is limited by the difference in phase velocity between the TE_{10} and LSE_{12} modes forming the sum distribution. This feed design generates the LSE_{12} mode in phase opposition with the TE_{10}. The length of the tapered horn section is then set by the requirement to provide a 180° differential phase shift between these two modes.

Contributors to Pointing Errors

Defects in an antenna feed system as it affects the sum-beam performance, measured by such factors as gain. beamwidth, and sidelobe level, are treated elsewhere in this handbook. Defects that affect the tracking performance of a monopulse feed system are considered here. Included in this subsection are those errors in the distribution of illumination over the antenna aperture that have relatively little correlation in the aperture. When the error* locations in the aperture are separated by more than a few wavelengths, they often can be considered independently of other errors. Errors that are correlated over the entire part of the aperture, served by one port of the comparator, are treated as precomparator and postcomparator errors. Finally, the dependence of monopulse angle estimation on the polarization of the received signal will be considered. The effect of these illumination errors is an apparent shift in the position of the difference-pattern null, or a boresight shift, and a change in the relative slope of the difference pattern, or a gain shift, which may vary with frequency, time, temperature, or other environmental factors.

Uncorrelated Illumination Errors   The effects of uncorrelated errors must be considered in the design of a tracking antenna, whether for specifying either reflector-surface tolerance or allowable element errors in a phased array. A distribution of phase errors $\xi(x,y)$ across the aperture will produce a voltage V_e that is proportional to

$$\iint \xi(x,y) \ d(x,y) \ dA \approx V_e$$

which to the first order of approximation is in phase with the difference voltage which would exist in the absence of uncorrelated errors. This voltage may be divided by the difference slope in Eq. (34-5) to determine the shift in the apparent position of the difference-pattern null:

$$\Delta u = \frac{\iint \xi(x,y) \ d(x,y) \ dA}{k \iint x d(x,y) \ dA} \tag{34-7}$$

This expression is valid for an arbitrary distribution of any correlation interval across the array, including such periodic phase errors as those resulting from phase quantization errors in some beam-steering equipment.

The specific distribution of phase errors in the difference-pattern illumination for an antenna often is unknown, but we can infer by means of specification or measurement the statistics of such distributions that are averaged over an ensemble of similar antennas. The null shift given in Eq. (34-7) can be treated as a random variable, and the expected value of the null shift is dependent upon the distribution of the expected value of the phase error at each point in the aperture:

$$\overline{\Delta u} = \frac{\iint \overline{\xi}(x,y) \ d(x,y) \ dA}{k \iint x d(x,y) \ dA} \tag{34-8}$$

Usually the phase error, and therefore the null shift, will have a zero mean (expected) value. If the error distribution is such that the rms phase error σ (radians) is the same for every independent uncorrelated region of the antenna, the rms of the null shift (boresight) error is

---

*Error as used here is that phase or amplitude error in the final aperture illumination function which contributes to the beam null shift or pointing error in Tables 34-1, 34-2, and 34-4.

$$\sigma_u = \frac{\sigma}{K\sqrt{N_e}} \qquad \textbf{(34-9)}$$

where N_e is the effective number of uncorrelated independent regions in the aperture with that phase-error statistic. If the antenna is a phased array and the statistic refers to phase errors that are independent from element to element, then N_e is just the total number of elements in the array. If the phase errors are attributable to some higher organizational level in the array, N_e is the number of components in that level.

Nester[11] has shown that, for practical ranges of error values, the phase error has the predominant effect on null shift. He has also shown the changes in the error slope due to phase and amplitude errors. These are second-order because both the position of the target off boresight and the error in the normalized difference slope are small, independent random variables.

The illumination error distribution associated with the feed system of a phased-array antenna is invariant with scan. This error contribution can be significantly reduced by alignment. The illumination error distribution may also have a component due to a setability error in the phase shifters that is dependent upon scan. This error is difficult to remove by monopulse calibration. Although it might be removed by calibration at the element level, the computational effort makes this approach unattractive for arrays of more than a few elements.

Precomparator and Postcomparator Errors   The accuracy of the monopulse estimation of target location is influenced by both precomparator and postcomparator errors. Precomparator errors affect the symmetry of the patterns, producing a modified sum-pattern error in the difference beam and a modified difference-pattern error in the sum beam. If the modified sum-pattern error is in phase with the difference pattern (resulting from a precomparator phase error), a shift in the angle of the difference null will be observed. If the modified sum-pattern error is in quadrature with the difference pattern, no shift will occur in difference-pattern-null location, but the depth of the null will be reduced. Techniques employed to improve effective null depth make use of the relative phase between the signals from the sum and the difference channels. These techniques make the monopulse performance near the null sensitive to postcomparator phase error in the presence of quadrature precomparator error. The relation between these error sources depends both upon the microwave network employed and upon the technique used in the signal processor to extract angle information.

An example is given in Fig. 34-15 of the phase errors in a monopulse feed system when the difference-pattern error d_ϵ is in quadrature with the difference pattern d. Figure 34-15a illustrates the situation in which no postcomparator phase error exists; d_ϵ is a phasor slowly varying with the angle to the source, while d varies linearly (approximately) with the source angle. In this instance, the ratio

$$\frac{d' \cos \theta}{s} = \frac{d}{s} \qquad \textbf{(34-10)}$$

is the best estimator of the source angle. In Fig. 34-15b, the effect of the postcomparator phase error ϕ is illustrated. Here, the estimator of source angle

$$\frac{d' \cos (\theta + \phi)}{s} = \frac{d}{s} \frac{\cos \theta \cos \phi - \sin \theta \sin \phi}{\cos \theta} \qquad \textbf{(34-11)}$$

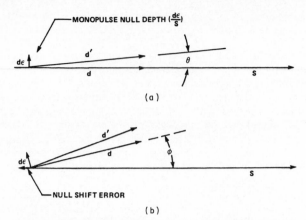

FIG. 34-15 Monopulse comparator phase errors. (*a*) Precomparator phase error. (*b*) Precomparator and postcomparator phase error.

will produce a shift in the apparent null position (boresight) such that

$$e_{dn} = \frac{\theta_B}{K_m} \frac{d}{s} = \frac{-\theta_B}{K_m} \frac{d_\epsilon}{s} \tan \phi \qquad \textbf{(34-12)}$$

Precomparator errors that are in phase with the difference pattern produce a null shift error in source-angle prediction. Postcomparator errors produce a null shift error if precomparator errors are in quadrature with the difference pattern. Design or alignment action to reduce these errors depends upon identifying the source of the phase error. Determining the behavior with frequency will often help with this identification. When the alignment has reduced this error to an acceptable level, it may be further reduced by calibrating and correcting the observed angle data. The boresight error due to this source will be a function of frequency.

Cross-Polarization Angle Crosstalk The normalized output voltage from the two difference channels of a monopulse feed can be related to the displacement of the target from the RF axis of the antenna by the matrix equation

$$\begin{bmatrix} V_u \\ V_v \end{bmatrix} = \begin{bmatrix} K_{uu} & K_{uv} \\ K_{vu} & K_{vv} \end{bmatrix} \begin{bmatrix} \Delta u \\ \Delta v \end{bmatrix} \qquad \textbf{(34-13)}$$

where the main diagonal terms are the difference slope coefficients from Eq. (34-2), and the cross-diagonal terms are cross-coupling terms due to nonorthogonality between the monopulse null planes (or cones). These coefficients are generally a function of frequency and of the relative polarization of the received signal. If the receiving system does not sense the polarization of the incoming signal, this cross-polarization angle crosstalk represents an uncorrected error in the angle estimate, which can result in loss of track.[12,13]

An example of the difference-voltage equation applied to the elevation-over-azimuth pedestal can be obtained by means of coordinate conversions:

$$\Delta v = \Delta E$$

$$\Delta u = \cos E \, \Delta A$$

giving

$$\begin{bmatrix} V_u \\ V_v \end{bmatrix} = \begin{bmatrix} K_{uu} \cos E & 0 \\ 0 & K_{vv} \end{bmatrix} \begin{bmatrix} \Delta A \\ \Delta E \end{bmatrix} \qquad \textbf{(34-14)}$$

when no angle crosstalk occurs and perfect alignment exists between the RF null planes and the pedestal coordinate planes. The angle estimates in pedestal coordinates are obtained by inverting the matrix, giving the relation

$$\begin{bmatrix} \Delta A \\ \Delta E \end{bmatrix} = \begin{bmatrix} 1/(K_{uu} \cos E) & 0 \\ 0 & 1/K_{vv} \end{bmatrix} \begin{bmatrix} V_u \\ V_v \end{bmatrix} \qquad \textbf{(34-15)}$$

The cross-polarization angle crosstalk is a consequence of the polarization characteristics of the feed system. It tends to be most pronounced in optical feed systems but is also present in small phased-array antennas. Crosstalk is not observed in large arrays.

A classic explanation of cross-polarization crosstalk is derived by considering the feed for a reflector antenna with polarization characteristics identical to that of a linear electric dipole. The electric vector at the surface of the reflector will be directed along the intersection of the parabolic surface and the plane containing the feed dipole axes and a point on the reflector surface. The copolarization of the antenna is determined by the plane through the feed dipole and the vertex of the reflector. Most of the energy collimated by the reflector is polarized parallel to the principal polarization at the vertex, but each quadrant of the reflector aperture has a cross-polarized component of excitation.[14] During sum-beam excitation, the phase of this cross-polarized energy alternates from quadrant to quadrant, as shown in Fig. 34-16, producing a cross-polarized Q beam. In the principal polarization elevation difference beam, the

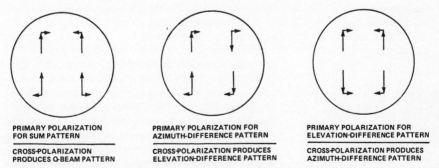

PRIMARY POLARIZATION
FOR SUM PATTERN

CROSS-POLARIZATION
PRODUCES Q-BEAM PATTERN

PRIMARY POLARIZATION FOR
AZIMUTH-DIFFERENCE PATTERN

CROSS-POLARIZATION PRODUCES
ELEVATION-DIFFERENCE PATTERN

PRIMARY POLARIZATION FOR
ELEVATION-DIFFERENCE PATTERN

CROSS-POLARIZATION PRODUCES
AZIMUTH-DIFFERENCE PATTERN

FIG. 34-16 Cross-polarization crosstalk electric dipole feed on parabolic reflector.

cross-polarization response is that of an azimuth difference pattern. Thus, a cross-polarized source displaced in azimuth would drive the beam in the elevation direction. The same situation occurs in the principal polarization azimuth difference beam, in which the cross-polarized response is that of an elevation difference pattern.

If this feed were to be used as an optical feed for a planar transmission lens array, no such cross-polarization effect would be observed. This source of crosstalk depends upon both the feed polarization and the surface of the focusing element. Koffman[15] has determined the "ideal" feed polarization for each type of surface derived from a conic section by showing that, in general, the strength of the electric dipole polarization should equal the strength of the magnetic dipole polarization, multiplied by the ellipticity of the focusing surface. Thus, a magnetic dipole is optimum for a sphere, an electric dipole is optimum for a plane, and a Huygens source (having equal electric and magnetic dipole strength) is optimum for the paraboloid.

Optical polarization conversion, however, is not the only source of crosstalk; it can arise because of cross coupling in the feed itself. An example of this effect in a five-horn monopulse antenna has been studied by Bridge.[16] The particular feed he investigated was designed for dual polarization and had a direct circuit for measuring the coupling or isolation between the polarizations in the monopulse network.

Multimode feeds are particularly susceptible to cross-polarization crosstalk because the relatively large size of the single aperture can support higher-order cross-polarized modes, and the feed system usually will excite them unless particular care is exercised. Multiple-aperture, single-polarization feeds may exhibit some relatively low-level cross-polarization because of edge diffraction (if we assume that the individual apertures will support only one polarization). This source of angle crosstalk becomes negligible as the size or number of apertures in the feed array becomes large.

34-4 MECHANICAL-DESIGN CONSIDERATIONS

Antenna Mounts

The mount not only must support the antenna and direct it in angle but must accurately provide that direction under dynamic conditions that are introduced by target and/or antenna-mount-platform motion or stress that is introduced by wind. The requirements on the mount are interdependent with the requirements on the antenna. The mechanical bandwidth of the tracking antenna as well as its mount depends upon the resonance characteristics of the antenna structure and its mount as well as upon the coupled resonance. A discussion of bandwidth and design of servo systems to drive tracking-antenna systems is given by Humphrey.[17]

Many reflector tracking antennas use the mount of Fig. 34-21 below, though tracking coverage in this mount is limited near zenith. For a target (see Fig. 34-17) traveling normal to the line of sight and perpendicular to the axis of rotation of a tracking antenna at a ground range R_g and velocity v, the angular velocity of the tracking antenna reaches v/R_g. The acceleration about the azimuth axis of rotation reaches a value of $0.65v^2/R_g^2$ at $\pm 30°$ from the normal. The transfer function that describes the angular velocity and acceleration of the tracking antenna about the azimuth axis has a "pole" when the ground range approaches zero

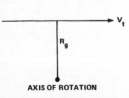

FIG. 34-17 Target geometry.

and the target is near overhead. A loss of track can occur. The three-axis mount described in Skolnik[3] removes the pole at the expense of increasing weight, complexity, and cost.

Even though coverage is sacrificed, a two-axis mount is often chosen for reflector-antenna tracking systems for fixed earth stations or for installation on aircraft or ship platforms. The mount is oriented so that the loss in mission coverage due to the pole is minimized and has little impact on overall operation.

Phased-array-antenna tracking systems that must provide hemispherical coverage for different missions but limited coverage for a single mission may also use the elevation-over-azimuth mount or may use an azimuth-only mount. Dynamic performance requirements are usually limited, but accuracy of positioning must still be maintained.

Mechanical Errors in Reflector Tracking Antennas

This subsection discusses mechanical errors in using, for example, an elevation-over-azimuth mount. The errors are treated as being independent. Though not strictly true, the assumption appears to be valid for the size of errors expected in an accurate system. These formulas may be used as corrective algorithms to the elevation and azimuth angular estimates once the constants have been determined by calibration. More exact expressions for some of the errors will be found in Rozansky.[18] In a new design or in one whose error models are not well known, these algorithms may be used initially, but the final algorithm should be adjusted to match the particular system.

Encoder Errors   The angular position of the antenna is translated into digital or analog data by using angle encoders. Most designs today incorporate digital encoders. The output of these encoders is subject to both systematic and noise errors within the encoder itself; in the data gear train, if used; and in the coupling of the encoder to the mechanical takeoff from the antenna. The state of the art is such that the latter two are usually the greater source of error. If the coupling is rigid, the relative motion between the driving axis and the encoder shaft introduces bending movements in the shaft of the encoder and cyclic errors in the output. Multicyclic errors can be produced by gear trains.

When a flexible coupling between the axis takeoff and the encoder is used, mechanical eccentricities that produce systematic cyclic variations in output data may occur. Backlash produces random errors. Noise in the encoder also produces a random error. With the encoder shaft held fixed, a variation of ± 2 bits in a high-resolution (19-bit) system is not unexpected.

Torquing of setscrews or like fasteners can produce bending moments that may distort the encoder shaft. Rather than adjusting the encoders and creating such an error, provision can be made in the calibration process to introduce the values by which the encoders are offset from the reference positions.

The equation for a combined cyclic and encoder offset error with respect to the reference position is

$$E_{Ac0} = M \cos (A_m - A) + A_0 \qquad \textbf{(34-16)}$$

where M = maximum amplitude of cyclic error
A = azimuth reading of encoder

A_m = azimuth reading of cyclic-error maximum amplitude

A_0 = azimuth reading for external azimuth reference

This equation is also valid for the elevation encoder when the elevation values E, E_m, E_0 are substituted for A, A_m, and A_0. The use of this equation, when M and A_m (or E_m) are measured during the calibration process as a correction algorithm, leads to the level of systematic residual error given in the error budgets in Tables 34-1 and 34-2.

Azimuth-Axis Tilt Leveling a reflector tracking antenna can be accomplished with three adjustable legs, using two orthogonal gravitational sensors that are parallel to the azimuth plane. An accurate tracking system is usually leveled to within 0.1 to 0.2 mil of the vertical, preferably under cloudy conditions or at night to eliminate temperature gradients. Leveling to a tighter tolerance is not productive, since day-to-day variations of the pedestal vertical, as shown in Fig. 34-22 below, can be greater. In addition, the local vertical may not be the true vertical because of gravitational anomalies, and calibration with respect to the true vertical may be required.

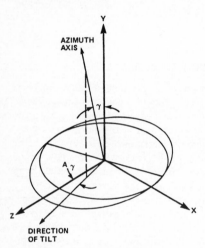

FIG. 34-18 Azimuth-axis tilt.

When the azimuth axis is tilted at an angle γ from the vertical reference as shown in Fig. 34-18, errors in both azimuth and elevation that are generated may be approximated by

$$e_{A_\gamma} = \gamma \sin (A_\gamma - A) \tan E$$

$$e_{E_\gamma} = \gamma \cos (A_\gamma - A)$$

(34-17)

where γ = tilt angle

A_γ = azimuth direction of tilt

A = azimuth encoder output

E = elevation encoder output

Boresight-Telescope Nonorthogonality The boresight telescope used for system alignment and calibration usually will not be exactly orthogonal to the elevation axis. An error in azimuth, as shown in Fig. 34-19, is produced for the optical axis. Correction can be made through physical adjustment to the body of the telescope or through correction of the data by measuring the angle ζ_0 of nonorthogonality. The error-correction algorithm is approximated by

$$e_{A\zeta0} = \zeta_0 \sec E$$

(34-18)

where E is the elevation angle. This equation is needed to derive the RF-axis nonorthogonality to the elevation axis and is not used directly in correcting tracking-antenna azimuth data.

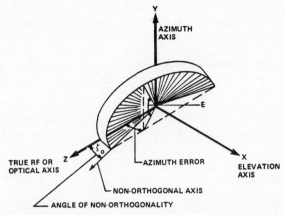

FIG. 34-19 RF- or optical-axis to elevation-axis nonorthogonality error.

RF-Axis Nonorthogonality   The same type of error is produced in azimuth data if the RF axis is not orthogonal to the elevation axis. However, RF-axis-nonorthogonality errors may be a combination of mechanical and electrical boresight shifts, with the latter being a function of frequency of operation.

$$e_{A\zeta r} = \zeta_r \sec E \qquad\qquad (34\text{-}19)$$

Steps in the calibration process require that the optical nonorthogonality be determined first and that the RF nonorthogonality as a function of frequency then be determined with the use of a target that is visible both optically and at RF.

Elevation-Axis Nonorthogonality (Skew)   If the elevation axis is nonorthogonal or skewed to the azimuth axis by a small angle ϵ_0, as shown in Fig. 34-20, the resultant azimuth error and correction algorithm is approximated by

$$e_{A_\epsilon} = \epsilon_0 \tan E \qquad\qquad (34\text{-}20)$$

for both the optical and the RF axis.

Droop   Droop is an elevation error produced by gravitational force on the boresight telescope and on the RF feed. The errors are at a maximum at 0 and 180° elevation and at a minimum at 90°. They are normally constant with azimuth position. The droop in the telescope can be approximated as a linear equation:

$$e_{Ed0} = d_0 (E - \pi/2) \qquad\qquad (34\text{-}21)$$

where d_0 is the telescope droop constant. The RF boresight-axis droop is approximated by

$$e_{Edr} = d_r \cos E \qquad\qquad (34\text{-}22)$$

where d_r is the RF feed constant. The telescope constant is developed first during the

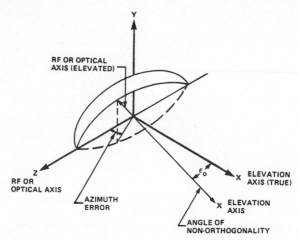

FIG. 34-20   Elevation-axis to azimuth-axis nonorthogonality error.

calibration process and then, with a common target, used to determine the RF-axis droop constant.

Collimation   *Collimation* is defined as "making parallel" or "adjusting accurately the line of sight of [a telescope]." In this chapter, collimation refers to the relationship between the RF boresight axis and the boresight telescope that is used as a reference.

The collimation error between the RF boresight axis and the optical axis is determined by the resolution of the boresight telescope and the optical recording, the depth of the null for the RF difference pattern, multipath error, and the rms phase error of the receiving system. If a common target has a high RF signal-to-noise ratio, optical visibility, reflectivity characteristics that do not depolarize the signal return, high elevation angle, and low velocity and acceleration so that the target can be tracked at or near the RF null, then the collimation error can be reduced to 0.02 mrad rms in each axis. In elevation, the residual error is the error remaining after accounting for mechanical droop, electrical boresight shift, multipath, and the differential offset constant between the optical axis and the RF axis. In azimuth the residual error is the error remaining after accounting for the electrical boresight shift of the RF axis and the nonorthogonality of both the optical axis and the RF axis with respect to the elevation axis.

Thermal Effects   Thermal effects on accuracy primarily are those due to solar radiation producing thermal gradients in the reflector tracking antenna and mount. Differential heating of the mount base will produce an azimuth tilt error. Unequal expansion in the trunion arms supporting the elevation axis can produce an elevation-axis nonorthogonality error. Droop, collimation, and RF-axis-nonorthogonality errors can be accentuated or decreased by gradients in the feed supports. For highly accurate systems, material with low thermal coefficients of expansion should be used. This is particularly true for feed supports. The use of white paint for the entire antenna and mount for nontactical systems and sunshields for the mount will considerably reduce thermal gradients.

Thermal effects are long-term when compared with the time for most instrumentation radar missions. A system designed for rapid premission calibration can reduce the impact of such errors.

Wind   Wind produces both systematic and random errors. A steady-state wind having a velocity V produces a small systematic deflection error:[7]

$$e_{wd} = K_w K_s V_w^2 \qquad\qquad (34\text{-}23)$$

where K_w is an aerodynamic constant that relates the torque produced about one of the angular axes to the average wind speed and $V_w K_w$ is a function of the aspect angle of the antenna with respect to wind direction. The mechanical spring constant of the antenna K_s is a measure of the deflection of the antenna about the axis per unit of torque.

Gusty winds produce a small random error with a standard deviation of

$$\sigma_{wd} = 2 K_s K_w V_w \sigma_v \qquad\qquad (34\text{-}24)$$

where σ_v is a measure of the variation in average wind velocities.

Consideration can be given to the use of a radome to reduce wind-load errors, but a trade-off must be made between radome and wind-induced pointing errors.

Mechanical Errors in Phased-Array Tracking Antennas

Rotation   Angular displacement ($\Delta\theta_z$) about the Z axis or the array normal produces errors in both α and β proportional to $(\beta - \pi/2) \cdot \Delta\theta_z$ and $(\alpha - \pi/2) \cdot \Delta\theta_z$, respectively, where $\Delta\theta_z$ is small. A highly damped tiltmeter, located parallel to the X axis, can be used to level the array initially and automatically monitor for subsequent rotation displacement. High damping is necessary to prevent response of the tiltmeter to vibration and to short-term wind loads but will still permit measurement of permanent shifts.

Tilt   The array is usually tilted backward about the X axis to obtain elevation coverage consistent with tracking demands at the horizon. Angular displacement about the X axis can be measured with a highly damped tiltmeter mounted perpendicularly to that axis with the center of the tiltmeter range at the desired tilt angle. Angular displacement errors about the X axis directly contribute to errors in β, with a minimum amount of coupling into α. Both the rotational and the tilt sensors provide the reference to the local gravitational vertical rather than the true vertical.

Array Face to True North   Alignment of the array-face normal so that its relationship is known with respect to the plane established by the local vertical and true north can be accomplished through a north-seeking gyro designed as an integral subsystem of the array. Gyros are available with an error in pointing of less than 0.3 mrad rms divided by the cosine of the latitude. They require settling times of 5 to 15 min to reach such values.

Thermal Effects   In a planar phased array, a shift in boresight can occur when the overall array-face temperature changes and when there is a uniform differential expansion or contradiction between element distances D_x and D_y. At an α of 45°, a 25-ppm thermal-expansion coefficient and a 40°C differential temperature produce a 1-mrad shift. The formula for the error before correction is

$$e_{\alpha t, \beta t} = L_t \, (T_c - T_0) \cot (\alpha_t, \beta_t) \qquad (34\text{-}25)$$

where L_t is the coefficient of expansion, T_c is the calibration temperature, and T_0 is the operational temperature.

A temperature gradient across the face of the array will produce errors dependent on the thermal-gradient pattern. Such errors will be difficult to determine and correct by calibration. Consequently, the thermal design of the array face and structure should minimize such gradients.

An additional thermal error is that due to both absolute temperature and thermal gradients of the feed system. It also will be difficult to determine and correct by calibration. Thermal design must consider the feed architecture and techniques to maintain small thermal gradients between parallel elements in order to minimize differential phase errors.

Wind   The potential for error due to wind and wind gusts is generally greater in a tactical than in an instrumentation tracking-antenna system. The instrumentation system usually has a solid base to which the antenna mount can be rigidly attached, whereas the tactical radar tracking-antenna support structure rests on the ground. The weight of the phased array or wind loads can produce pressure, which in turn compresses or deforms the surface supporting the mount, giving errors about one or more of the array-face axes. Such errors caused by displacement can be read out by tiltmeters and gyros that have been designed as an integral part of the array. In addition to the angular displacement caused by wind loads in θ_X and θ_Z, unless there is means of lateral bracing, angular displacements that are primarily in θ_Y can take place.

In addition to the small permanent shifts, which can be measured by the gyros and tiltmeters, shorter-term shifts due to wind deflection of the mount structure can occur. High and gusty winds can cause large errors, particularly for a system with a low resonant frequency.

34-5   SYSTEM CALIBRATION

Calibration serves two functions: (1) to provide direction for realignment or correction of faults after assembly or after transport in which the antenna may be subject to mechanical stresses; and (2) to provide constants for the algorithms to correct the systematic errors, permitting the system to operate within the accuracy required. To provide maximum accuracy, the system should be calibrated under conditions as near as possible to the conditions under which it will operate. Knowledge of absolute temperature is important for phased-array calibration but is of limited value for reflector-antenna tracking systems. For thermal gradients that distort the antenna and pedestal for reflector tracking antennas used in highly accurate instrumentation systems, it is desirable to calibrate as close to mission time as possible.

The first method of calibration involves direct measurement to determine the errors and then correction of the errors as a function of angular direction and frequency. The second method is the tracking of targets whose trajectories are accurately known, such as satellites, and the use of regression analysis to determine the errors as a function of direction and frequency. This latter approach is generally limited to large power-aperture product systems.

The use of sensors having automatic readout under control of a computer appears to give the fastest, most consistent results and removes the possibility of observer error. When an observer is involved, the design of the readout should reduce the requirements on the observer to a minimum. Physical stress due to temperature extremes, muscular strain due to position, and lighting all contribute to potential observer error.

Frequency of calibration and number of sample measurements depend on user requirements and the stability of the system. It should be noted that in general the larger the number of independent sample points chosen for a particular calibration procedure, the more precisely one may model any particular error. Initially, a tracking-antenna system may require a large number of samples to establish a model or algorithm for the error and a lesser number on subsequent calibrations. A tracking antenna with a specified accuracy of 0.1 mrad (exclusive of thermal noise, glint, multipath, refraction, and clutter) may require calibration for each mission. A tracking antenna with a specified accuracy of 0.5 mrad (exclusive of thermal noise, glint, multipath, refraction, and clutter) may require only leveling and the use of automatic sensors to measure changes.

Calibration for Reflector Tracking Antennas

The order of calibration is chosen to minimize coupling of errors. The order, or technique, used may not be appropriate for all systems, but the principles involved may be applied to develop those required for a specific tracking antenna and user.

Optical-Axis Calibration   Accurate angle encoder readouts with respect to the antenna azimuth and elevation axes of the antenna mount of Fig. 34-21 are fundamental in establishing angular errors in a tracking antenna. A multifaceted optical flat (with 17 facets, for example), temporarily mounted perpendicular to and centered on the axis of rotation and used with an optical collimator, can establish the cyclic errors in the encoder within 0.02 mrad rms. This readout should be performed after installation of the encoder in the mount. Each measurement should be made by rotating about the axis in the same direction to eliminate backlash errors. An encoder readout is made when the reflection of the collimator cross hairs from the optical flat coincides with the collimator cross hairs. Backlash can be determined by reversing the direction of antenna rotation about the axis and reading the encoder at any one of the optical flat positions to establish the backlash random error. The random error due to granularity of the encoder can be read by leaving the antenna stationary and reading the encoder output variation.

The next step in calibration for a reflector antenna after encoder cyclic errors have been established is the use of on-mount levels to bring the azimuth axis parallel to the local vertical. After leveling in one azimuth position, readings are taken at a number of azimuth locations, giving results similar to that of Fig. 34-22.

The error models or algorithms of Sec. 34-4 are used to correct the data output.

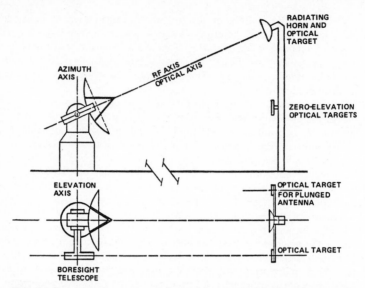

FIG. 34-21 Elevation- over azimuth-axis tracking antenna and boresight tower.

The difference estimates that are furnished by use of these algorithms and true position results in residual errors that are small. Figure 34-22, for example, gives two sets of measurements made on different days. Peak-to-peak error approached 0.4 mil* on March 31 and 0.2 mil on April 3. Correction of the data left a residual error of less than 0.02 mrad for each day. If a second calibration had not been made on April 3, peak errors in excess of 0.1 mil would have been introduced, indicating the need to recalibrate highly accurate systems on a mission basis.

With an antenna capable of plunging, i.e., with an elevation motion greater than 180°, the optical-axis nonorthogonality to the elevation axis may be established by reading the azimuth position of the zero-elevation optical targets of Fig. 34-21 with the antenna in both normal and plunged positions. The difference in azimuth encoder readout should be exactly π rad. One-half of any difference is the optical- to elevation-axis nonorthogonality.

To determine the nonorthogonality or skew constant of the elevation-to-azimuth axis, azimuth encoder readings are then made in the normal and plunged portions of the upper optical targets, taking into account the calibration data from the encoder and the optical-elevation-axis nonorthogonality measurements. The readings should be π rad apart. If not, the skew constant ϵ for the correction algorithm is the difference in reading divided by 2 and then divided by the tan E of the elevation of the upper optical targets. The use of the zero-elevation optical targets precludes the coupling of the optical-elevation-axis-nonorthogonality error into the measurement.

Once the nonorthogonality error of the elevation to azimuth axis has been estab-

---

*This error was taken on a radar using the artillery mil., i.e., ⅟₆₄₀₀ of a circle. This value should not be confused with the milliradian or the artillery-spotting mil (1 yd subtended at 1000 yd). The abbreviation of mil for milliradian is sometimes seen in the literature. Many older systems still utilize the artillery mil.

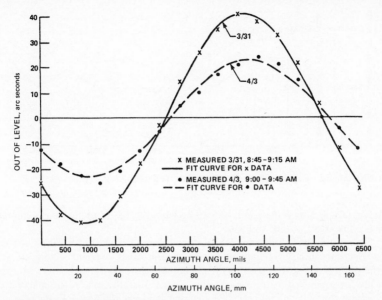

FIG. 34-22   Antenna tilt. *(After Ref. 13.)*

lished, the azimuth encoder offset may also be established by using the upper bore-sight-tower targets as a reference, which usually is surveyed with respect to true north. If the positions of both sets of optical targets and the position of the centerline of the telescope axis and antenna elevation axis are accurately known by an external survey, the droop constant d_0 for the telescope may be determined by the difference in elevation encoder readings between the lower and upper optical targets. Once the droop constant has been established, the elevation encoder offset with respect to the optical axis may be entered into the calibration data.

RF-Axis Calibration   Once the errors have been established for the optical axis and correction made, the relation of the RF axis to the optical axis must be determined. In azimuth, this is relatively simple. Usually there is no azimuth multipath. A bore-sight-tower RF horn radiating a polarized signal and an optical target separated by exactly the same distance as the RF boresight axis of the antenna and the optical axis of the telescope are used, as shown in Fig. 34-22. The RF axis is found at minimum reading from the difference channel in azimuth and elevation. If the depth of the null in either axis does not meet the error-budget requirements, it should be corrected since further measurements will be contaminated by postcomparator phase errors.

The elevation- to azimuth-axis nonorthogonality or skew error is the same for the RF axis as for the optical axis. The azimuth error introduced by RF boresight-axis-to elevation-axis nonorthogonality will be a function of both the mechanical position of the feed and the variation of the boresight axis in azimuth as a function of frequency. Since the skew error is known, the boresight tower can then be used to determine the RF boresight-axis to elevation-axis nonorthogonality as a function of frequency. The test results in Fig. 34-23 show this nonorthogonality to be approximately 0.025 mil and the peak-to-peak variation in azimuth of the boresight axis to be approx-

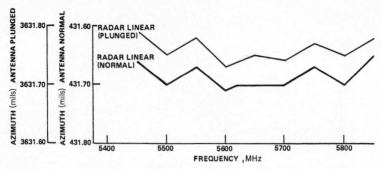

FIG. 34-23   Azimuth RF boresight shift versus frequency. *(After Ref. 13.)*

imately 0.06 mil across the band. As can be seen, the RF to optical-axis collimation error is quite small, <0.01 mil rms.

The multipath, droop in the RF feed, and frequency shift of boresight are combined when measured at a single elevation. The droop is constant at that elevation, but the multipath and frequency shift of the boresight axis are variable. Multipath error may be separated by tracking a balloon-borne metallic sphere; the sphere is centered within the balloon and launched near the tracking antenna. RF tracking is initiated. Corrected data from the radar and optical recording of the boresight-telescope field of view are compared when the elevation angle is above that at which multipath is a factor. A set of 10 frequencies across the band is used for 10 different elevation angles. From this measurement, the boresight shift in elevation with frequency may be determined separately from multipath. Once the droop constant and the boresight shift with frequency have been determined, the RF- to optical-axis offsets may be determined. The remaining error is that due to the collimation error, that is, the limits of telescope resolution, depth of null, and observer.

Off-Axis Calibration   Once the electrical boresight axis has been established, off-boresight positions are calibrated. A plot of error versus angle offset is given in Fig. 34-24 in both elevation and azimuth for a single frequency. These measurements were made by using a boresight-tower RF horn radiating a vertically polarized signal. There may be some cross-polarization, multipath, and droop contamination, but the amount should be negligible, since the angular change during measurement was much smaller than the beamwidth.

Star Calibration of the Optical Axis   The most accurate method for optical-axis calibration of the antenna mount is that of star calibration.[19] The steps are similar to those used by Meeks,[20] except that the sensitivity of most tracking antennas does not permit direct calibration of the RF axis. A set of stars which is visible (brightness magnitude greater than 3.8) at the time of calibration[21] and distributed uniformly over the celestial sphere between 15 and 70° with respect to the local vertical is chosen as the reference base. Under computer control the azimuth and elevation positions of each star selected are measured by using the optical telescope. A timing error of ±0.2 s contributes about 0.005 mrad rms to the instrumentation error. A telescope with a resolution of 4 s contributes about the same error.

After the measurements have been taken and corrections for refraction made, a

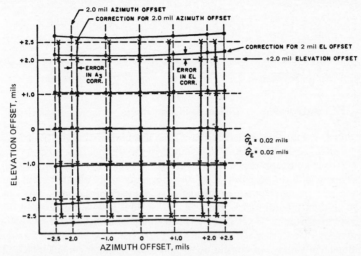

FIG. 34-24   RF off-axis error. *(After Ref. 13.)*

regression analysis will permit determination of the error coefficients for each of the mechanical-error-correction algorithms. The entire process, which may be limited by cloud cover, takes only a few minutes and offers an opportunity to maintain the mechanical system at its highest level of accuracy.

The use of stars as a reference has the advantage of determining true vertical as compared with local vertical—a difference that could be as much as 0.3 mrad in certain locations. It further provides a means of assessing errors as a function of azimuth rather than just the normal and plunged positions. A boresight tower and balloon tracks are still required to establish the relationships of the RF to the optical axis.

Calibration for Phased-Array Tracking Antennas

Calibration techniques for phased arrays depend on the size, configuration, and use of the arrays. If the design is in a limited-scan phased array utilizing a two-axis mount, the mount can be calibrated by using techniques similar to those used for reflector antennas. Once the mount and the face of the array have been established with respect to the azimuth and elevation axis, the array itself may be calibrated by using the mount encoder output and a boresight tower. The same limitation in accuracy by multipath applies here as for the reflector antenna.

If the final phased-array tracking system does not use a mount but the antenna is of a size so that it can be placed on a two-axis mount, then the angular position of the array face to the boresight tower can be established with respect to a theodolite attached to the antenna. A scan-angle order can be given, and the angular position to which the antenna face must be rotated to obtain a difference-pattern null can be established by the theodolite. This angle is then compared with that ordered by the beam-steering controller. An example of the error expected is given in Fig. 34-25. If corrected as a function of frequency, the residual systematic error is less than 0.1 mrad rms.

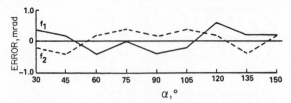

FIG. 34-25 $\alpha - \alpha_i$ error with scan angle.

Another technique that has recently been demonstrated is the use of near-field measurements. The antenna is first mounted in an anechoic chamber. A radiating probe antenna is raster-scanned across the face of the array at a constant distance from the array with the polarization of the probe antenna parallel to that of the array. Measurements are made of amplitude and phase for each of the probe positions at the sum-and-difference output ports. Once the near field is known, the far-field sum-and-difference patterns may be calculated. This technique has been validated by far-field measurements that were made with the same antenna. The near-field technique has the additional advantage of low multipath and radio-frequency interference effects.

The use of the near-field techniques provides data for the sum-and-difference pattern for any one scan angle, frequency, or temperature value. Calculations of the field pattern provide the calibration data for correcting both boresight and off-boresight errors.

A sufficient number of beam positions, both on and off the cardinal planes, must be measured to develop the error-correction algorithms. The same beam positions should be repeated with different frequencies to assure that the beam-steering algorithms in the controller are correct and that errors present in beam position as a function of frequency can be established and corrected. In addition, a set of patterns should also be measured at a different temperature if possible, to develop error-correction algorithms as a function of temperature.

When the antenna is large and cannot be assembled at the factory but must be installed and aligned on site, the position of the array face in the local coordinate system is determined by a survey. By using optical flats and prisms on the antenna face, the orientation of the face with respect to true north, tilt with respect to true vertical, and the rotation about the normal to the array face are determined.

Such an antenna is generally large enough to track such objects as satellites having well-established trajectories. A track of a sufficient number of objects, over the band of operating frequencies, permits a regression analysis that provides for calibration as a function of scan angle and frequency. Scanning about the target establishes the difference-pattern slope. Thermal and wind conditions affecting the array must be recorded concurrently so that the errors produced can be correlated with these conditions and data-correction equations can be developed.

REFERENCES

1 G. M. Kirkpatrick, "Final Engineering Report on Angular Accuracy Improvement," August 1952; reprinted in D. K. Barton, *Monopulse Radar,* Artech House, Inc., Dedham, Mass., 1974.

2 H. Jasik (ed.), *Antenna Engineering Handbook,* 1st ed., McGraw-Hill Book Company, New York, 1961, sec. 25.5.

3 M. I. Skolnik, *Introduction to Radar Systems,* 2d ed., McGraw-Hill Book Company, New York, 1980.

4 D. K. Barton and H. R. Ward, *Handbook of Radar Measurement,* Prentice-Hall, Inc., Englewood Cliffs, N.J., 1969.

5 *IEEE Standard Dictionary of Electrical and Electronics Terms,* 2d ed., Institute of Electrical and Electronics Engineers, New York, 1977.

6 E. B. Wilson, *An Introduction to Scientific Research,* McGraw-Hill Book Company, New York, 1952.

7 D. K. Barton, *Radar System Analysis,* Prentice-Hall, Inc., Englewood Cliffs, N.J., 1964.

8 T. T. Taylor, "Design of Circular Apertures for Narrow Beamwidth and Low Sidelobes," *IRE Trans. Antennas Propagat.,* vol. AP-8, January 1960, pp. 17–22.

9 E. T. Bayliss, "Design of Monopulse Antenna Difference Pattern with Low Sidelobes," *Bell Syst. Tech. J.,* vol. 47, May 1968, pp. 623–650.

10 P. Hannan, "Optimum Feeds for All Three Modes of a Monopulse Antenna in Theory and Practice," *IRE Trans. Antennas Propagat.,* vol. AP-9, no. 5, September 1961, pp. 444–461.

11 W. H. Nester, "A Study of Tracking Accuracy in Monopulse Phased Arrays," *IRE Trans. Antennas Propagat.,* vol. AP-10, no. 3, May 1962, pp. 237–246.

12 M. I. Skolnik (ed.), *Radar Handbook,* McGraw-Hill Book Company, New York, 1970, pp. 21-50, 21-53.

13 R. Mitchell et al., "Measurements and Analysis of Performance of MIPIR (Missile Precision Instrumentation Radar Set AN/FPQ-6)," final rep., Navy Cont. NOW 61-0428d, RCA, Missile and Surface Radar Division, Moorestown, N.J., December 1964.

14 S. Silver (ed.), *Microwave Antenna Theory and Design,* McGraw-Hill Book Company, 1949, p. 419.

15 I. Koffman, "Feed Polarization for Parallel Currents in Reflectors Generated by Conic Sections," *IEEE Trans. Antennas Propagat.,* vol. AP-14, no. 1, January 1966, pp. 37–40.

16 W. M. Bridge, "Cross Coupling in a Five Horn Monopulse Tracking System," *IEEE Trans. Antennas Propagat.,* vol. AP-20, no. 4, July 1972, pp. 436–442.

17 W. M. Humphrey, *Introduction to Servomechanism System Design,* Prentice-Hall, Inc., Englewood Cliffs, N.J., 1973, pp. 234–289.

18 M. Rozansky, "Exact Target Angular Coordinates from Radar Measurements, Corrupted by Certain Bias Errors," *IEEE Trans. Aerosp. Electron. Syst.,* vol. AES-12, no. 2, March 1976, pp. 203–209.

19 J. T. Nessmith, "Range Instrumentation Radars," *IEEE Trans. Aerosp. Electron. Syst.,* vol. AES-12, no. 6, November 1976, pp. 756–766.

20 M. L. Meeks, J. A. Ball, and A. B. Hull, "The Pointing Calibration of Haystack Antenna," *IEEE Trans. Antennas Propagat.,* vol. AP-16, no. 6, November 1968, pp. 746–751.

21 A. Becvar, *Atlas Coeli II: Katalog 1950.0,* Nakladatelství Československé Akademie věd, Prague, 1960.

Chapter 35

Satellite Antennas * +

Leon J. Ricardi

Lincoln Laboratory
Massachusetts Institute of Technology

*This work is sponsored by the Department of the Air Force.

†The United States government assumes no responsibility for the information presented.

The use of satellites in communication and remote-sensing systems has increased tremendously in the past decade or two, resulting in designers who specialize in satellite antennas. The uniqueness of the platform, the environment, and the applications requires antenna designers to interact with several disciplines including mechanics, structural analysis, thermal analysis, surface charging and radiation effects, and communications theory. This chapter will address fundamental spacecraft-stabilization characteristics and the related considerations of antenna design and polarization. Earth-coverage antennas will be described, and the use of higher-gain multiple-beam antennas will be discussed. Finally, a comparison of two types of cross-link antennas is presented. Although these discussions may apply to antennas used in other types of systems, the communications satellite (COMSAT) will be in mind unless otherwise specified.

35-1   SPACECRAFT-STABILIZATION AND FIELD-OF-VIEW CONSIDERATIONS

Not unlike an airplane, an automobile, a ship, and a tower, spacecraft have their unique "platform" characteristics, which determine some of the antenna's more important characteristics. In particular, a spacecraft derives its "stability" from the gyroscopic action produced by rotating all or some of its mass. Consequently, all satellites are designed so that all or a significant part of their mass rotates so as to establish a reference axis, plane, or sphere. In the early years of the satellite age, all satellites were intentionally rotated about the axis with the largest inertial moment; these satellites are commonly referred to as *spinners*. The orientation of this axis was often maintained by applying external forces to the satellite by various means, such as gas propulsion, emission of ionized particles of a solid, and interaction with the earth's magnetic[1] or gravitational fields.

With the development of lubricated bearings that can be operated satisfactorily in the vacuum of space, three-axis stabilization became possible, resulting in a *despun platform* similar to a tower on which the antennas of a microwave land link are mounted. Recently it has become popular to use an *inertia wheel* (i.e., a three-axis gyro) to establish the needed reference system for a three-axis stabilized satellite. Of course, in all cases (except perhaps one-axis stabilization) controlling external forces are applied, in cooperation with pointing and attitude error-sensing devices, to maintain the satellite in the desired orbit or at a particular point in the orbit. The latter function is often referred to as *station keeping*.

Since the prime purpose of a communication satellite is to provide a relay between two or more communication terminals, the earth-bound satellite must have an antenna capable of transmitting signals to and receiving signals from the earth. The degree of stabilization and whether or not the satellite is a spinner determine the fundamental conditions under which it must perform this function. In particular, an antenna on a spinning satellite must have a radiation pattern that is uniform in the plane perpendicular to the spin axis, or it must be designed so that its radiation pattern is "despun" mechanically or electrically. Even an antenna on a three-axis stabilized vehicle must be designed to compensate for satellite attitude and station-keeping errors, again either by modification of its radiation pattern or by physical reorientation of its structure. Clearly, using the tumbling satellite as a platform presents the great-

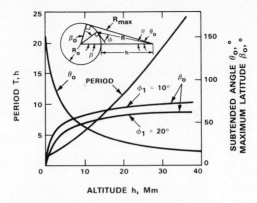

FIG. 35-1 Satellite-earth characteristics.

est challenge to the antenna designer, and the three-axis stabilized station-kept satellite allows for the most sophisticated antenna design.

Communication satellites are often placed in a nearly circular orbit and hence the angle θ_0, subtended by the earth and measured at the satellite, is relatively constant. Since the communications requirement usually involves a relay between widely separated points on the earth, the satellite's field of view (FOV) is often defined by θ_0 and is given by

$$\theta_0 = \sin^{-1}(R_0/R_0 + h) \qquad \textbf{(35-1)}$$

Referring to Fig. 35-1, we see that θ_0 varies from about 145° for a low-attitude satellite to 17.3° for a satellite in a synchronous (i.e., 24-h-period) orbit.

Although most communication satellites make use of the geostationary properties of the equatorial synchronous orbit, other considerations may dictate an elliptical orbit and the concomitant change in θ_0. For an elliptical orbit, the earth is located at one of the foci of the ellipse traced by the orbit and is defined as the barycenter. The eccentricity ϵ of the ellipse is given by

$$\epsilon = [1 - (b/a)^2]^{0.5} \qquad \textbf{(35-2}a\textbf{)}$$

where a and b are the semimajor and semiminor axes of the elliptical orbit, for a circular orbit $a = \text{b}$ and $\epsilon = 0$. The orbital period T is independent of ϵ (i.e., as long as the orbit does not intersect the earth), and it is given by

$$T = 2\pi[(R_0 + h)^3/\alpha]^{0.5} \qquad \text{h} \qquad \textbf{(35-2}b\textbf{)}$$

where $R_0 = 6378$ km is the radius of the earth, h is the maximum altitude of the satellite above the earth's surface (measured in kilometers), and $\alpha = 5164 \times 10^{12}$.

An earth terminal located at the edge of the satellite's FOV views the satellite at an elevation angle $\phi = 0°$; that is, the satellite is on the local earth horizon. System performance usually requires $\phi > 10°$; hence the satellite's FOV ϕ_0' is somewhat less than that described by θ_0. The angle θ between the earth-satellite axis and the terminal-satellite direction determines the satellite elevation angle ϕ, as shown in Fig. 35-2. The relation between ϕ and θ is plotted for a satellite at synchronous altitude and at binary multiples of synchronous altitude. The corresponding satellite period T is also indicated for ease of reference. Clearly, $\phi = 0$ when $\theta = \theta_0$ and $\phi = 90°$ when

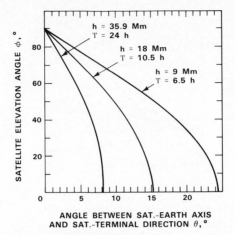

FIG. 35-2 Elevation angle of the satellite measured at the earth terminal.

$\theta = 0$. Using Fig. 35-2, β_0, the maximum latitude of a terminal, was computed by using

$$\beta_0 = 90 - \phi_1 - \theta_1 \qquad \qquad \textbf{(35-3}a\textbf{)}$$

and

$$\theta_1 = \sin^{-1}[R_0 \sin (90 - \phi_1)/(R_0 + h)] \qquad \qquad \textbf{(35-3}b\textbf{)}$$

where ϕ_1 is the minimum value of ϕ and θ_1 is the corresponding value of θ for the $\phi = \phi_1$ (see Fig. 35-2). With reference to Fig. 35-1, β_0 is plotted for $\phi_1 = 10°$ and $20°$ and indicates that satellites in an equatorial orbit cannot operate with terminals located at north or south latitudes in excess of about $70°$ unless their altitude above the earth exceeds 40 Mm.

Inclined 12-h elliptical and 24-h polar orbits and the equatorial synchronous orbits form what might be considered a primary set of orbits that have most desirable characteristics. The geostationary orbit is most often used by systems with terminals located in the tropical and temperate zones. Satellites in the polar and, most often, the inclined elliptical orbit are used by terminals in the temperate and polar regions. Systems with satellites in both orbits provide worldwide coverage; worldwide service is obtained through the use of satellite-to-satellite links (i.e., cross links) or earth terminals as gateways. Gateway terminals must be in the FOV of at least two satellites. Cross-link-antenna design[2] will be discussed in Sec. 35-7; however, it is of importance to note that θ_0 and h of a critically inclined 12-h elliptical orbit[3] vary as shown in Fig. 35-3. This orbit is often referred to as the Molniya orbit since it was first used by the Russians to provide "commercial" communications capacity to all populated parts of the Soviet Union. For the critically inclined orbit, solar pressure will not cause it to precess. The times indicated in Fig. 35-3 are measured either prior to or after apogee. A pair of satellites spaced about 6 h along the same orbit provide continuous coverage in the temperate regions and one polar region. Note that θ_0 and h vary by approximately 30 percent over the 6-h useful portion of the orbit, that is, during the period from 3 h before to 3 h after the satellite is at its highest altitude above the earth. It is

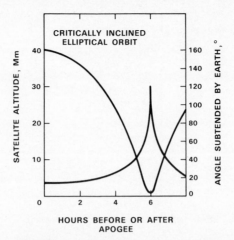

FIG. 35-3 Characteristics of a 12-h ellipti-cal orbit.

the indicated change in θ_0 and the relative direction to the center of the earth that affect antenna-design considerations.

Before considering antenna designs as they interact with various stabilization methods, it is important to point out that satellite-antenna-pattern coverage, as viewed from the angular space conventionally used to describe antenna performance, becomes distorted when radiation-pattern contours are plotted on conventional (e.g., Mercator, gnomonic, etc.) projections of the earth. Presentation of antenna-pattern contours of constant gain on earth maps are referred to as *beam footprints*. The beam-contour coordinates ϕ and θ can be transferred to earth latitude and longitude to determine the beam footprint by first assuming that the satellite's nadir is located at a longitude $= a°$ and at $0°$ latitude. Then use

$$\phi = \cos^{-1} (1 + h/6378) \sin \theta \qquad (35\text{-}4a)$$

$$\beta = 90 - \phi - \theta \qquad (35\text{-}4b)$$

to compute β. The coordinates of a point P on the earth at longitude ζ and latitude ξ are related to θ and ψ through

$$\xi = \sin^{-1} (\sin \beta \cos \psi) \qquad (35\text{-}5a)$$

$$\zeta = a + \tan^{-1} (\tan \beta \sin \psi) \qquad (35\text{-}5b)$$

where ψ is the angle measured from the plane containing the satellite and the earth's polar axis to the plane determined by the earth-satellite axis and the line from the satellite to the point P (Fig. 35-4). When the satellite's nadir is not on the equator, the transformation is given by

$$\xi = \sin^{-1} [\sin (b + \beta) + \cos b \sin \beta (\cos \psi - 1)] \qquad (35\text{-}6)$$

$$\zeta = a + \sin^{-1} (\sin \psi \sin \beta / \cos \xi) \qquad (35\text{-}7)$$

where b is the latitude of the satellite's nadir and a is its longitude.

The foregoing is presented to point out that the satellite's stabilization method

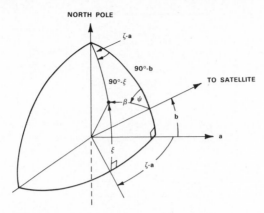

FIG. 35-4   Transformation coordinate system.

and its orbit have a serious impact on its antennas and their FOV. It is beyond the scope of this chapter to discuss these interactive design relationships in detail; however, some designs will be described to indicate methods of accommodating these characteristics.

35-2   DESPUN-ANTENNA SYSTEMS

Many earth satellites are cylindrically or spherically shaped bodies rotating about the axis for which their inertial moment is maximum. If the orientation of this axis is uncontrolled, the antenna has to produce a radiation pattern that is uniform in the plane perpendicular to the spin axis. Alternatively, one could use a directional antenna whose radiation pattern is either despun by varying the excitation of the antenna (i.e., as in the case of an array) or by mechanically rotating the antenna at the same angular rate but in a direction opposite to the satellite's rotation. Clearly if the antenna radiation pattern is not despun and the spin-axis orientation is not controlled (one-axis stabilization), the antenna pattern must be independent of direction. This leads to an *isotropic* antenna and the concomitant directivity and directive gain of 0 dB.

Telstar, the first active communication satellite, employed antennas[5] having a nearly isotropic radiation pattern. The satellite rotated on its axis of maximum moment of inertia and had some attitude control; it was, therefore, a two-axis stabilized spacecraft. It was spherically shaped with an approximate radius of 35 in (889 mm). It had antennas operating at very high frequency (VHF), at 4 and at 6 GHz. All radiation patterns were uniform in a plane perpendicular to the spin axis, providing the necessary despinning function. Except for the surface occupied by the antennas, solar cells covered the satellite's outer surface.

The first Lincoln experimental satellite (LES-1) had two-axis stabilization; the orientation of the spin axis was maintained by force derived from an interaction with the earth's magnetic field. One of the major purposes of LES-1 was to demonstrate the feasibility of electronically despinning the radiation pattern by switching among eight low-directivity (about 7 dB) antennas distributed uniformly over the satellite duodecahedron shape. This experimental satellite also demonstrated the feasibility of a simple magnetic attitude-control system that was subsequently used[6] on LES-4 and enabled the use of antennas with higher directivity (about 14 dB).

With the exception of a VHF antenna on Telstar, communication antennas on these earlier satellites operated in the superhigh-frequency (SHF) range (i.e., 4, 6, and 8 GHz). This enabled the antenna designer to realize moderate antenna gain without significantly compromising the other satellite functions and systems. However, the SHF ground terminals were necessarily large and hence few in number, resulting in communication satellites which served only a few earth terminals or direct users. In the late 1960s, LES-5 and LES-6 were placed[7] in synchronous orbit and operated in the ultra high-frequency (UHF) communications band to service a wide variety of military communication needs because only modest-size earth terminals were required. The antenna systems on LES-5 and LES-6 had approximately the same radiation pattern as Telstar and LES-4 respectively. It is of interest to compare the methods of achieving these radiation patterns.

Telstar employed a large number of open-ended waveguide radiators spaced one-half wavelength on center along the satellite's equator. Two such antenna arrays, one operating at 4.17 GHz and one at 6.39 GHz, were used; each was fed so as to excite each element in phase and with the same signal amplitude. This symmetry guarantees a symmetrical radiation pattern in the plane perpendicular to the satellite's spin axis. Uniformity of the radiation pattern is controlled by the number of individual radiators per wavelength along the satellite's equator and the phase and amplitude equality of the signals exciting each radiator in the array. These antennas had an equatorial-plane radiation pattern that was uniform within ± 1 dB. The feed network[5] provided the desired excitation amplitude and phase within 0.5 dB and 5° respectively.

LES-5 was a cylindrically shaped satellite[7] about one wavelength (λ) long and one wavelength in diameter; that is, it was much smaller than Telstar in terms of their respective operating wavelengths. The communication antenna consisted of two circular arrays. One array was made up of eight full-wave dipoles spaced about $\lambda/2$ circumferentially on the satellite's cylindrical surface and excited in phase with equal-amplitude signals. The other array consisted of two rings of eight half-wave cavity-backed slots colocated with the dipole array. The slots were also excited in phase with equal-amplitude signals but in-phase quadrature with the dipole array so that the combined dipole-slot pair would radiate and receive circularly polarized signals. The antenna radiation pattern was uniform, within ± 1 dB, in planes perpendicular to the satellite's spin axis. Unlike Telstar, LES-5's radiation pattern in a plane containing the spin axis approximated that of a linear array of two $\lambda/2$ dipoles spaced about $\lambda/2$ on center along the satellite's spin axis.

It is of interest to note that the configuration of each satellite (Telstar and LES-5) prohibited placing the antennas at the center of the satellite, where the small single radiator could have been placed to obtain the desired circularly symmetric radiation patterns in the satellite's equatorial plane (i.e., such as that obtained in a plane perpendicular to the axis of a dipole). In fact, most of the volume within the satellite was reserved for other subsystems of the satellite. This required that the antenna be distributed around the outside surface of the satellite and appropriately excited through circuits, made up of transmission lines and n-way power dividers, integrated with the other subsystems located within the satellite's surface. Virtually any one- or two-axis stabilized satellite without a despun antenna must use this general type of antenna configuration to meet the usual requirements dictated by earth-bound stations using a communication satellite. The antenna configuration is often similar to that of public television broadcast antennas mounted on a high tower.

Occasionally a dipole, stub, biconical, or slot antenna that will produce the desired despun radiation pattern can be mounted at one or both poles of the satellite. Alternatively, a somewhat simple antenna can be used when the small volume described by a thick disk coincident with the satellite's equator is available exclusively as the antenna site. In this case, a radial transmission line (i.e., parallel disks excited at their center) exciting a ring of open-ended waveguide radiators[8] represents that case when antenna performance is essentially uncompromised by its necessary integration into the satellite.

Electronically despun antennas are similar in that they usually consist of an array of low- to medium-gain antennas (i.e., 5- to 15-dB directivity) dispersed uniformly around the spin axis of the satellite and excited through a switch network that connects the transmitter or receiver to only those antennas that point most closely toward the center of the earth. An alternative, less common, electronically despun antenna consists of a circular array of low-gain elements simultaneously excited with equal power and phased to produce a unidirectional beam. The phase is varied to despin the beam as the satellite rotates. However, this type of antenna must occupy a complete cylindrical section of the satellite, usually on one end of the spacecraft. Consequently, electronically despun antennas usually consist of an array of several antennas which are sequentially switched to despin the radiation pattern. The details of these antennas are very well described in the literature.[6,7,9]

Before leaving this topic, it is important to discuss mechanically despun antennas used on Intelsat-III[10] and ATS-III[11] (Fig. 35-5a and b). Both of these employ reflector antennas and despin the radiation pattern by counterrotating the reflector (with respect to the satellite sense of rotation) so that its focal axis always points toward the center of the earth. The ATS-III antenna system (Fig. 35-4) consists of two antennas, one operating at 6.26 GHz and one operating at 4.85 GHz. Each antenna employs a parabolic cylinder illuminated by a collinear array of dipoles located on the focal line of the reflector. The latter reflects and converges the incident energy into a beam coincident with the focal axis of the parabola that generates the reflector's surface. Since the radiation pattern of a collinear array of dipoles is uniform in a plane perpendicular to the axis of the array, rotating the reflector about the dipole array rotates the antenna's beam in a plane perpendicular to the axis of the dipole array. The latter is installed coincident with the satellite spin axis so that counterrotation of the reflector results in a despun antenna. A unique feature of this antenna permits the reflector to be ejected if the rotary mechanism fails. Without the reflector the antenna has an omnidirectional radiation pattern permitting continued operation but with about 13-dB reduction in directivity.

Intelsat-III also uses a rotating reflector to despin its radiated beam; however, the reflector surface is flat and oriented at a 45° angle with respect to the spin axis of the satellite. A horn antenna produces a high-directivity (∼21-dB) beam pointed along the spin axis of the satellite toward the reflector. The latter redirects the beam perpendicular to, and causes it to rotate about, the satellite spin axis. This antenna, the ATS-III despun, and similar mechanically scanned devices continuously despin the radiation pattern with essentially no variation in the antenna's directive gain. Electronically despun antennas usually scan the radiated beam in a stepwise fashion; this usually produces a significant (from 1- to 3-dB) variation in the antenna's directive gain as it would be measured by a fixed terminal on the earth.

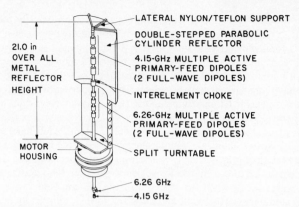

FIG. 35-5*a*   ATS-III mechanically despun antenna.

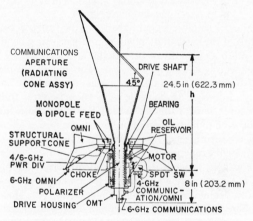

FIG. 35-5*b*   ITS-III mechanically despun antenna.

35-3   ANTENNA POLARIZATION

Before considering three-axis stabilized satellites, it is important to discuss the antenna's polarization and why it plays an important role in satellite-antenna design. Recall that an antenna is defined by the polarization of the electromagnetic (EM) energy that it radiates. It is important to measure this polarization in the far zone of the antenna, that is, at distances sufficiently far from the antenna so that a further increase in this distance will not change the measured polarization. A distance $R = 2D^2/\lambda$ is customarily chosen as adequate for measuring the antenna's polarization and directive gain, where D is the antenna-aperture size and λ is the operating wavelength. The electric field direction defines the polarization of the EM energy.

Although essentially all polarization properties of EM waves play a role in satellite-antenna design, let us review those which are most important. For example, a linearly polarized (LP) antenna such as a dipole, oriented with its axis vertical (with respect to the earth's surface), will radiate and receive vertically polarized signals. Conversely, it will neither radiate nor receive horizontally polarized signals. This phenomenon is commonly referred to by stating that an antenna will not radiate or receive cross-polarized signals, or that orthogonally polarized signals are rejected. This statement is not limited to LP antennas; circular and elliptically polarized EM waves and antennas have copolarized and cross-polarized properties identical to those of LP waves and antennas.[12] Circularly polarized (CP) waves have a right-hand sense (i.e., RHCP) if the electric field vector rotates in a clockwise sense as the wave is propagating away from the observer. The electric field vector of a left-hand circularly polarized (LHCP) wave rotates in a counterclockwise sense for receding waves. Changing both the direction of propagation (i.e., receding to approaching) and the sense of rotation (i.e., clockwise to counterclockwise) does not alter the polarization. Elliptically polarized waves result when the strength of the electric field varies as its direction rotates. In summary, virtually all antennas are, in fact, elliptically polarized, but the ellipticity is such that referring to them as CP or LP is an adequate descriptor. The important point is that copolarized antennas couple well to one another and cross-polarized antennas tend to reject one another's signals.

Now consider earth-satellite signal links when the frame of reference (i.e., vertical and horizontal) of the earth station will not, in general, coincide with the frame of reference (i.e., north and south) of the satellite. Since the satellite usually serves many users simultaneously and its antenna can assume only one polarization at any instant of time, it follows that when LP antennas are used, the earth station must adjust its frame of reference to coincide with the satellite's frame of reference. Although this is possible, it is far simpler to use CP satellite and earth-terminal antennas and remove the need to align them in order to maximize coupling between them. Consequently, it is not surprising that most satellite antennas are circularly polarized.

When an LP satellite antenna is used, orientation of the associated EM waves is altered as they propagate through the earth's ionosphere,[13,14] a phenomenon often referred to as the Faraday rotation effect. This rotation of LP waves is usually negligible (less than a few degrees) at frequencies above a few GHz. However, at frequencies below 1 GHz Faraday rotation effects can rotate the wave polarization more than 360°. Fortunately, the polarization of a CP wave is not altered by the Faraday rotation effect. Change in polarization due to transverse "static" magnetic fields along the propagation path is much smaller; therefore, circular polarization is preferred because CP waves propagate through the ionosphere with no essential change in polarization.

Most spacecraft and earth-terminal antennas are shared by the associated transmitter and receiver. Consequently, diplexer (or duplexer for a radar system) filtering is required. The use of antennas that are orthogonally polarized for transmitted and received radiation enhances the isolation between the transmitter and the receiver. For this and the foregoing reasons it is customary for satellite antennas to be opposite-sense circularly polarized for simultaneous transmit and receive functions.

35-4   THREE-AXIS STABILIZED SATELLITES

It is quite common to station-keep satellites whose attitude is completely controlled (i.e., orientation of the satellite's three principal axes is controlled). The satellite then takes on many of the characteristics of a relay station installed on a tower between two earth terminals. In particular, high-gain directional antennas can be used to increase the link's data rate, or reliability, and decrease its vulnerability to external noise sources. The antenna can become much more sophisticated than antennas mounted on a satellite whose total mass is spinning. Suffice it to say, the antennas can be as common as a narrow-beam paraboloidal reflector mounted on a two-axis pedestal or as complicated and unique as the multiple-frequency, multiple-beam antenna[15,16] that makes the National Aeronautics and Space Administration's ATS-6 such a unique satellite.

Three-axis stabilization is achieved by despinning a "platform," or part of the spacecraft, or by including an inertia wheel as an integral part of the satellite. The gyroscopic action of the spinning spacecraft or inertia wheel can maintain the satellite's attitude within about 0.1° of a given frame of reference. The addition of a propulsion device keeps the spacecraft at a prescribed location in space.

35-5 EARTH-COVERAGE ANTENNA

Probably the most common spacecraft antennas are those that have a broad pattern with approximately the same directive gain over the earth as it is viewed from the satellite. These are called *earth-coverage antennas* and vary from simple dipole arrays to horn antennas, shaped-lens antennas, or shaped-reflector antennas. For low-altitude satellites, the angle θ_0 subtending the earth is relatively large (see Fig. 35-1); consequently the associated antenna may be a dipole, a helix, a log-periodic dipole, a backfire antenna, or a spiral antenna. Higher-altitude satellites usually use horn-type earth-coverage antennas. In all cases, however, the desired antenna pattern $P_0(\theta)$ should provide the same signal strength at its output port (terminal) as a constant-power signal source is moved on the surface of the earth within the satellite's FOV. At frequencies below about 5 GHz atmospheric attenuation is negligible, and $P(\theta)$ depends only on the variation in path length R (Fig. 35-1) from the satellite to the points on the earth's surface that are within the satellite's FOV. At frequencies above 10 GHz atmospheric attenuation becomes significant, and $P(\theta)$ should be adjusted accordingly. Atmospheric attenuation A depends on absorption by the water-vapor and oxygen molecules and particulate scattering principally due to rain. Total attenuation due to atmospheric effects is, to a first order, proportional to the path length L_a through the atmosphere. Since the height h_a of the atmosphere in the vicinity of an earth terminal is nearly constant, atmospheric attenuation is approximately inversely proportional to the cosine of the satellite elevation angle ϕ (Fig. 35-1).

The desired earth-coverage-antenna pattern $P(\theta)$ must equalize the change in R and A over the FOV; hence

$$P(\theta) \propto (R(\theta)/h)^2 A(\theta) \qquad \textbf{(35-8)}$$

To determine $A(\theta)$ refer to Fig. 35-6 and note that the path length L_a through the earth's atmosphere is given by

$$L_a = [(R_0 + h_a)^2 + R_0^2 - 2R_0(R_0 + h_a)\cos\alpha]^{1/2} \qquad \textbf{(35-9)}$$

$$L_a \approx h_a[1 + 4(R_0/h_a)^2 \sin^2(\alpha/2)]^{1/2} \qquad \textbf{(35-10}a\textbf{)}$$

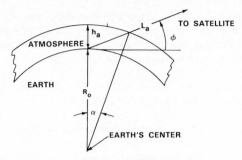

FIG. 35-6 Atmospheric path L_a.

where R_0/h_a is assumed $\gg 1$ and

$$\alpha \approx 90° - \phi - \sin^{-1}(\cos \phi) \qquad \textbf{(35-10b)}$$

Hence,

$$A(\theta) = e^{A_0(1 - L_a/h_a)} \qquad \textbf{(35-11)}$$

where A_0 is the attenuation when $\phi = 90°$. It remains to relate θ and ϕ and determine $R(\theta)$.

Referring to Fig. 35-1, note that

$$R = [R_0^2 + (R_0 + h)^2 - 2R_0(R_0 + h) \cos \beta]^{1/2} \qquad \textbf{(35-12)}$$

which reduces to

$$R = h\{1 + 4[(R_0/h)^2 + R_0/h] \sin^2 (\beta/2)\}^{1/2} \qquad \textbf{(35-13a)}$$

where

$$\beta = 90 - \theta - \phi \qquad \textbf{(35-13b)}$$

and

$$\phi = \cos^{-1}[(1 + h/R_0) \sin \theta] \qquad \textbf{(35-15c)}$$

The directivity of an antenna with a radiation pattern given by Eq. (35-8) can be calculated from

$$D = \frac{4\pi}{2\pi \int_0^{\theta_1} P(\theta) \sin \theta d\theta} \qquad \textbf{(35-14)}$$

where θ_1 is the edge of the coverage zone and $P(\theta) = 0$ for θ greater than θ_1. Note that $P(\theta)$ is identical in all planes containing the $\theta = 0$ axis and that θ_1 is given by

$$\theta_1 = \sin^{-1}[R_0 \cos \phi_0/(R_0 + h)] \qquad \textbf{(35-15)}$$

where ϕ_0 is the minimum satellite elevation angle within the desired FOV.

Having determined the optimum earth-coverage-antenna pattern $P(\theta)$, it is of interest to calculate the corresponding antenna directivity and directive gain and define a figure of merit F_e for assessing the performance of a specific earth-coverage antenna.

Toward this end, consider an earth-coverage antenna on a synchronous satellite (i.e., $h = 3.9$ Mm). Operational experience indicates that $\phi_0 > 20°$ prevents intolerable atmospheric-related diffraction effects and does not include intolerable atmospheric attenuation (i.e., atmospheric attenuation less than 1.5 dB at 10 GHz and less than 3 dB at 45 GHz) at the higher frequencies. For $\phi_0 = 19°$, Eq. (35-15) gives θ_0 = 8.2. By using these values for θ_0 and ϕ_0, the directivity of an optimum earth-coverage pattern is shown in Fig. 35-7 for $A_0 = 0$ and 1.0 dB. Note that increasing the atmospheric attenuation, in the zenith direction, from 0 to 1 dB changes the directivity of the ideal earth-coverage pattern by about 0.4 dB. However, the directivity D (8.2°) toward the edge of the coverage area increases by 1.4 dB in order to overcome the increase in atmospheric attenuation.

A shaped earth-coverage pattern is shown to indicate a feasible approximation to the ideal pattern. Finite antenna-aperture size will permit only an approximation to the cusp (at $\theta = \theta_0$) in the ideal pattern. Nevertheless, pattern shaping that increases $D(-\theta_0)$ will enhance the overall system performance even if $D(0)$ is reduced by 2 dB, as indicated by the shaped pattern (Fig. 35-7).

A typical conical-horn pattern is shown (Fig. 35-7) to indicate a primitive yet commonly used earth-coverage antenna. Note that for the shaped pattern $D(\theta_0)$ is about 4 dB greater than for the horn pattern. On the other hand, $D(0)$ is about 1 dB greater for the horn than for the shaped pattern. This conflict in performance points out the need for a figure of merit F_e that will aid in assessing the performance of an earth-coverage antenna. For this reason, it is suggested that F_e be set equal to the maximum difference in the antenna directivity $D_a(\theta)$ and the directivity $D(\theta)$ of the ideal antenna pattern. That is,

$$F_e = \max[D_a(\theta) - D(\theta)] /\big|_{\theta=0}^{\theta=\theta_0} \quad \textbf{(35-16)}$$

where $D(\theta)$ and $D_a(\theta)$ are expressed in decibels. Therefore, F_e is negative with a maximum possible value of zero.

To estimate the maximum possible F_e, assume that the antenna aperture is

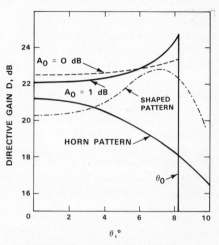

FIG. 35-7 Directive gain of an earth-coverage antenna on a synchronous satellite.

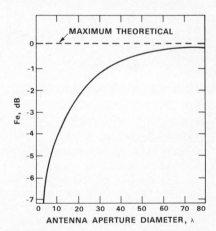

FIG. 35-8 Earth-coverage-antenna figure of merit.

inscribed in a sphere of radius a and that a set of spherical waves, with wave number $N < 2\pi a/\lambda$, is used to approximate $P(\theta)$. The directivity $D_0(\theta)$ of the pattern generated by the finite set of spherical waves is then used to compute $D_a(\theta)$ and F_e. Since the synthesis procedure guarantees a least-mean-square fit to $P(\theta)$ and the antenna-aperture excitation is prevented from exciting supergain waves[17] (i.e., $N < 2\pi a/\lambda$), the computed F_e is maximum for the given-size antenna aperture $D = 2a$. The maximum value of F_e shown in Fig. 35-8 was calculated by using this procedure.

35-6 SPOT-BEAM COVERAGE

Satellite antennas designed to provide high gain to a point within view of the satellite are required to have directivity greater than that of a simple fixed-beam earth-coverage antenna. In the case of a satellite in synchronous orbit, the theoretical maximum directivity of an earth-coverage antenna is about 24 dB; practical considerations limit its directivity to about 20 dB. Hence, for a directivity greater than 20 dB and an earth FOV, the synchronous-satellite antenna must produce a beam that scans or steps over the earth or the FOV. If a sequential raster scan is desired, the antenna might be steered mechanically or electronically over the FOV. However, for pseudo-random coverage of the FOV, step scan at microsecond rates is usually preferred. When step scan is preferred, the FOV is covered by several overlapping beam footprints. The minimum gain provided by N beams, designed to cover the FOV, is required to estimate system performance and is the subject of this section.

Since the cross section of a high-gain antenna beam is usually circular, the following analysis assumes that N beams with circular cross section are arranged in a triangular grid with beam-axis-to-beam-axis angular separation θ_b (see Fig. 35-9). A circular FOV is assumed; however, this analytic method can be readily extended to

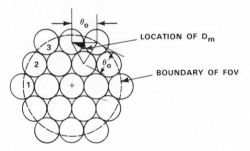

FIG. 35-9 Spot-beam coverage of the earth.

any FOV with a noncircular boundary. The analysis considers the black area in Fig. 35-9 and assumes that the directivity variation over this shaded area is replicated over the entire FOV. An edge-correction effect will be discussed following detailed consideration of the black area.

The analysis assumes a gaussian-shaped beam and determines θ_b and beam directivity D_0 that maximize the minimum directivity D_m over the black area. It follows that D_m is also the maximum value of the minimum directivity over the entire FOV with the possible exception of some small areas along the boundary of the FOV.

Since the useful directivity D of a beam will be within 5 dB of the beam directivity D_0, the directivity of all beams can be adequately represented by the function

$$D = D_0 e^{-\alpha(2\theta/\theta_1)^2} \qquad (35\text{-}17)$$

where θ_1 = the half-power beamwidth (HPBW) of the beam. It is further assumed that when θ_1 is expressed in radians,

$$D_0 = \eta \frac{4\pi}{\theta_1^2} \qquad (35\text{-}18)$$

where η ranges from about $\frac{1}{2}$ to $\frac{3}{4}$. (Experimental data indicate that the antenna efficiency of a center-fed parabola is about 93η percent.) By noting that $D = D_0/2$ when $\theta = \theta_1/2$ and using Eq. (35-17),

$$\alpha = \ln 2 = 0.693 \tag{35-19}$$

and

$$D = D_0 e^{-2.77(\theta/\theta_1)^2} \tag{35-20}$$

Using Eqs. (35-18) and (35-20) gives an alternative expression for D, namely,

$$D = D_0 e^{-0.22\theta^2(D_0/\eta)} \tag{35-21}$$

Over the FOV (not including the boundary), D_m occurs at the center of the equilateral triangle formed by the center of three adjacent beam footprints. The angle θ_m between the axis of the three adjacent beams and the direction of minimum direction gain is given by

$$\theta_m = \frac{\theta_b}{\sqrt{3}} \tag{35-22}$$

Hence, from Eqs. (35-22) and (35-21)

$$D_m = D_0 e^{-0.074\theta_b^2(D_0/\eta)} = D_0 e^{-BD_0} \tag{35-23}$$

Differentiating D_m with respect to D_0 gives

$$\frac{\partial D_m}{\partial D_0} = (1 - BD_0)e^{-BD_0} \tag{35-24}$$

where $B = 0.074\theta_0^2/\eta$. From Eq. (35-24) D_m is maximum when

$$D_0 = \frac{\eta}{0.074\theta_0^2} \tag{35-25}$$

$D_{\max}$, the maximum value of D_m, is given by

$$D_{\max} = D_0 e^{-1} \tag{35-26}$$

Expressing Eq. (35-26) in decibels gives

$$D_{\max} = D_0 \text{ (dB)} - 4.34 \text{ dB} \tag{35-27}$$

Hence, for maximum gain over the FOV, the crossover level between three adjacent beams is -4.34 dB with respect to D_0, the directivity of a beam.

By returning to Eq. (35-21), the crossover level D_2 between two adjacent beams is given by

$$D_2 = D_0 e^{-0.22(D_0/\eta)(\theta_b/2)^2} \tag{35-28}$$

Substituting Eq. (35-25) into Eq. (35-28) gives

$$D_2 = D_0 e^{-3/4}$$

or

$$D_2 \text{ (dB)} = D_0 \text{ (dB)} - 3.25 \text{ dB} \tag{35-29}$$

Hence the crossover level between two adjacent beams is 1.1 dB higher than the crossover level between three adjacent beams.

Statistical Distribution of D

Having knowledge of D_m is not always satisfying to the system designer because it represents a worst case. Whenever the estimated performance indicates that D_m may not be large enough, it is important to address the probability that the worst case will occur. With this probability placed in perspective with the foregoing analysis, it is important to determine the probability that $D > D' \geq D_m$. Toward this end, consider an enlargement (Fig. 35-10) of the black area in Fig. 35-9. The probability that $D > D_2$ is given by

$$P(D \geq D_2) = \frac{2(\pi/12)(\theta_0/2)^2}{(\theta_b/2)(\theta_b/2\sqrt{3})} = 0.907 \qquad \textbf{(35-30)}$$

Hence less than 10 percent of the FOV has a directivity less than $D_2 = D_0 - 3.25$ dB. The probability that $D > D'$ can be found for $D' > D_2$ [i.e., $\theta_1 \leq \theta_b/2$)] by first finding the probability that $\theta < \theta_1$. That is,

$$P(\theta < \theta_1) = P(D > D') = 3.628 \left(\frac{\theta'}{\theta_b}\right)^2 \qquad \textbf{(35-31)}$$

From Eq. (35-21)

$$D' = D_0 e^{-0.22(D_0/\eta)\theta'^2} \qquad \textbf{(35-32)}$$

Solving Eq. (35-32) for θ'^2, substituting in Eq. (35-31), and using Eq. (35-25) give

$$P(D \geq D') = 1.22 \ln\left(\frac{D_0}{D'}\right) \qquad \textbf{(35-33)}$$

By using Eq. (35-33) and a linear interpolation for $\theta_m > \theta > \theta_0$, the probability that $D \geq D'$ is given in Fig. 35-11.

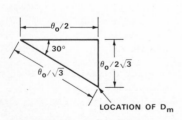

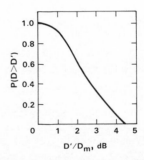

FIG. 35-10 Characteristic area of spot-beam coverage.

FIG. 35-11 Distribution of directivity of a multiple-beam antenna.

The foregoing derives D_{max} in terms of D_0. It remains to determine D_0 in terms of N, the number of beams required to cover the FOV. By assuming a hexagonal FOV, N is given by

$$N = 1 + \sum_{m=1}^{(M-1)/2} 6m \qquad \textbf{(35-34)}$$

where M, the maximum number of beams in a row (Fig. 35-9), is given by

$$M = \frac{\psi}{\theta_b} + 1 \qquad\qquad (35\text{-}35)$$

and ψ is the angle subtended by the major diagonal of the FOV and is measured at the antenna. From Eqs. (35-25) and (35-35)

$$M = \text{int}(1 + \psi\sqrt{0.07D_0/\eta}) \qquad\qquad (35\text{-}36)$$

hence,

$$D_0 = \left(\frac{M-1}{\psi}\right)^2 \frac{\eta}{0.074} \qquad\qquad (35\text{-}37)$$

For a synchronous satellite, the earth subtends an angle $\psi = 17.2°$ (0.3 rad); therefore, from Eq. (35-37) the maximum directivity of a multiple-beam antenna designed to cover the earth FOV with spot beams is given by

$$D_0 = 150(M-1)^2\eta \qquad\qquad (35\text{-}38)$$

In accordance with Eq. (35-27) the corresponding minimum gain $D_m = 55(M - 1)^2\eta$.

The foregoing derivation considered only those minimum-directivity points within the FOV. Usually the desired FOV does not conform to that described by the outer boundary of the footprints of the beams that cover the edge of the FOV. For example, for a circular FOV, arranging the beam footprints on a triangular grid results in the minimum number of beams to cover the circular area, but the resultant array is not hexagonal. To determine the number of beams required and their configuration, refer to Fig. 35-9 and note the location of D_m. Since the hexagonal array has sixfold symmetry, a point between beams 2 and 3 determines the edge of the FOV where $D = D_m$. As more beams are added to the array (i.e., as the FOV gets larger or as larger values of D_m are desired), the hexagonal grid gives poorer coverage along the edge of the FOV. Improved coverage is obtained by adding beams just at those points along the edge of the FOV where $D < D_m$ rather than by completing another ring of the hexagonal array of beams. The maximum directivity D_0 and the minimum directivity D_m, for a set of N beams arranged on a triangular grid and designed to cover the earth's disk from a synchronous satellite, are given in Table 35-1. The number of beams M in a diagonal of the hexagon pattern from which the set of N beams is derived is also given. For example, the 31-beam array was obtained by adding 2 beams to each side of a 19-beam array; hence $M = 5$ (for 19 beams) and $N = 19 + 2 \times 6 = 31$. The beam spacing θ_b is obtained from K in Table 35-1 and the relationship

$$\theta_b = 17.2°/K \qquad\qquad (35\text{-}39)$$

In summary, coverage of an area or FOV by exciting any one of N beams can be achieved by arranging the beams in a triangular grid with angular spacing θ_b between adjacent beams. The minimum directivity D_m over the FOV is maximized when $\theta_b = 3.67\sqrt{\eta/D_0}$; D_m is 4.3 dB less than D_0, the directivity of a single beam. If adjacent beams are excited simultaneously,[18,19] D_m can increase by more than 2 dB.

TABLE 35-1 Estimated
Directivity, decibels

| M | N | K | $\eta = 0.5$ | |
			D_0	D_{min}
3	7	2.30	26.0	21.7
3	13	3.06	28.4	24.1
5	19	4.16	31.1	26.8
5	31	5.04	32.8	28.5
9	37	5.76	34.0	29.7
9	43	6.10	34.5	30.2
9	55	7.02	35.7	31.4
11	61	7.36	36.1	31.8
11	73	8.08	36.9	32.6
11	85	9.02	37.8	33.5
11	91	9.22	38.0	33.7
11	97	9.44	38.3	34.0
13	109	10.26	39.0	34.7
13	121,127	11.0	39.6	35.3
15	139	11.36	39.9	35.6

35-7 CROSS-LINK ANTENNAS

Communication links between satellites are becoming more common for a number of reasons. Where initially satellites were used to relay communications between two earth terminals worldwide or even continent to continent, communications require more than a single satellite relay station. If communication between satellites is carried out through an earth station (i.e., a gateway), the round-trip delay, the vulnerability of the earth station, and the inherent insecurity of the earth-satellite link versus that of a satellite cross link motivate and often justify the use of cross links. Consequently, this section is addressed to a description of cross-link-antenna characteristics and some considerations essential to their design.

Although satellite cross links can operate at other frequencies (e.g., UHF), a 1-GHz band at 60 GHz (designated by international agreement) set aside for satellite-to-satellite communications is most attractive. Incidentally, operating a radiating system at frequencies different from those designated by international agreement and national law is acceptable to all provided there is no interference with those who have been allocated a frequency band. Consequently, it is wise to choose an appropriate frequency band within the designated frequency bands and obtain a frequency allocation.

The 60-GHz band has much more bandwidth than lower-frequency bands and can accommodate very-high-data-rate communications. In addition, the earth's atmosphere serves to reduce interference of human origin if not to eliminate it. For example, the attenuation of EM waves at frequencies between 55 and 65 GHz is on the order of 10 dB/km (see Fig. 35-12) at the earth's surface and decreases with increasing altitude above mean sea level. Wave attenuation becomes negligible at altitudes

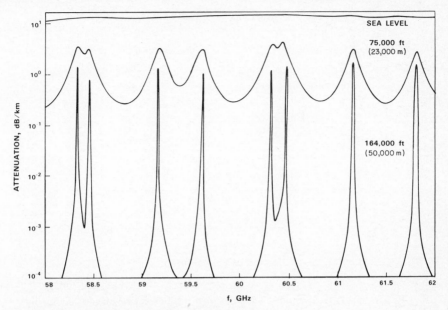

FIG. 35-12 Atmospheric attenuation of EM waves.

higher than 500 km (310 mi). Consequently, the cross links between satellites at altitudes greater than 500 km are shielded from signals or noise generated on or near the earth's surface. Since airplanes fly at altitudes of less than 25 km (82,000 ft), even high-flying platforms can experience substantial atmospheric attenuation unless the spacecraft is near the aircraft's zenith direction.

The natural shield provided by the earth's atmosphere is enhanced by the directivity of cross-link antennas. The short wavelength permits directive ($\sim$1°-HPBW) beams from a relatively small- ($\sim$0.5-m-) diameter antenna aperture. Thus, bandwidth, wave attenuation, and antenna directivity considered together with the difficulty with which high-power RF signals can be generated lead to 60 GHz as a preferred cross-link frequency band. Similar reasoning leads to the conclusion[20] that extremely high frequency (EHF) is the preferred operating frequency of military (and perhaps nonmilitary) satellite communication systems.

Having chosen a frequency of operation, the designer must determine the type of antenna and its gain. If it is assumed that the antenna will be steered in the direction of the intended satellite station, it is necessary to develop a trade-off algorithm capable of evaluating an array, a paraboloid, or some other design. For the purposes of this chapter, a conventional paraboloid will be compared with an array of elements, each of which has a beam pattern that defines the desired FOV. The paraboloid is steered by pointing it in the desired direction; the array is steered by appropriate phasing. Furthermore, the array aperture is square with side d, and the paraboloid has a circular aperture with diameter d. The number of elements in the array depends on the FOV. If the elements are assumed to be horn antennas arranged on a square grid, the number of elements versus antenna size is given in Fig. 35-13 for various angular (α) fields of view.

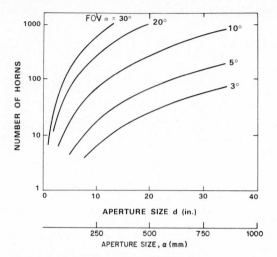

FIG. 35-13 Number of array elements.

By assuming a receiver noise temperature of 1500 K, a 64-Mm (40,000-mi) link between satellites, an energy-per-bit-to-noise power-density ratio $E_b/N_0 = 13$ dB, and an antenna system efficiency of 10 percent for the array and 20 percent for the paraboloid, the antenna and transmitter weight and power were estimated as a function of the aperture size and the data rate. The results shown in Figs. 35-14 and 35-15 indicate that the paraboloid requires about 40 percent more bus power but can be substantially lighter than the array. However, it is definitely clear (Fig. 35-14) that the paraboloid both is lighter and requires much less transmitter power than the array when the data rate exceeds about 10,000 bits per second (b/s). It is further clear that

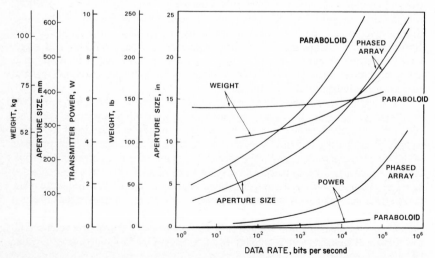

FIG. 35-14 Phased-array and paraboloid characteristics.

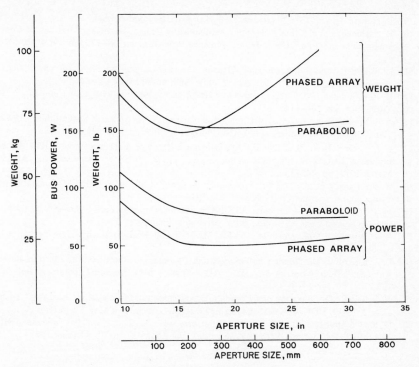

FIG. 35-15 Weight and aperture size of 10^5-b/s system.

if an aperture size of about 0.5 m is acceptable, the paraboloid is probably the best choice for a cross-link antenna.

REFERENCES

1   K. T. Alfriend, "Magnetic Attitude Control Systems for Dual Spin Satellites," *AIAA J.,* vol. 13, no. 6, June 1975, pp. 817–822.
2   W. C. Cummings, "Satellite Crosslinks," Tech. Note 1978-25, MIT Lincoln Laboratory, Aug. 4, 1978.
3   Y. L. Lo, "Inclined Elliptical Orbit and Associated Satellite Field of View," Tech. Note 1978-62, MIT Lincoln Laboratory, Sept. 26, 1979.
4   G. A. Korn and T. M. Korn, *Mathematical Handbook for Scientists and Engineers,* McGraw-Hill Book Company, New York, 1961, par. 3.1-12.
5   J. T. Bangert et al., "The Spacecraft Antenna," *Bell Syst. Tech. J.,* July 1963.
6   J. B. Rankin et al., "Multifunction Single Package Antenna System for Spin-Stabilized Near-Synchronous Satellite," *IEEE Trans. Antennas Propagat.,* vol. AP-17, July 1969, pp. 435–442.
7   M. L. Rosenthal et al., "VHF Antenna Systems for a Spin-Stabilized Satellite," *IEEE Trans. Antennas Propagat.,* vol. AP-17, July 1969, pp. 443–451.
8   W. F. Croswell et al., "An Omnidirectional Microwave Antenna for Use on Spacecraft," *IEEE Trans. Antennas Propagat.,* vol. AP-17, July 1969, pp. 459–466.

9   E. Norsell, "SYNCOM," *Astronaut. Aerosp. Eng.,* September 1963.

10   E. E. Donnelly et al., "The Design of a Mechanically Despun Antenna for Intelsat-III Communications Satellite," *IEEE Trans. Antennas Propagat.,* vol. AP-17, July 1969, pp. 407–414.

11   L. Blaisdell, "ATS Mechanically Despun Communications Satellite Antenna," *IEEE Trans. Antennas Propagat.,* vol. AP-17, July 1969, pp. 415–427.

12   R. E. Collin and F. J. Zucker, *Antenna Theory,* part I, McGraw-Hill Book Company, New York, 1969, pp. 106–114.

13   J. A. Wick, "Sense Reversal of Circularly Polarized Waves on Earth-Space Links," *IEEE Trans. Antennas Propagat.,* vol. AP-15, November 1967, pp. 828–829.

14   I. J. Kantor, D. B. Rai, and F. DeMendonca, "Behavior of Downcoming Radio Waves Including Transverse Magnetic Field," *IEEE Trans. Antennas Propagat.,* vol. AP-19, March 1971, pp. 246–254.

15   A. Kampinsky et al., "ATS-F Spacecraft: A EMC Challenge," 16th EM Compatibility Symp., San Francisco, Calif., July 16–18, 1974.

16   V. J. Jakstys et al., "Composite ATS-F&G Satellite Antenna Feed," *Seventh Inst. Electron. Eng. Ann. Conf. Commun. Proc.,* Montreal, June 1971.

17   R. Harrington, *Time Harmonic Fields,* McGraw-Hill Book Company, New York, 1961, pp. 307–311.

18   L. J. Ricardi et al., "Some Characteristics of a Communications Satellite Multiple-Beam Antenna," Tech. Note 1975-3, DDC AD-A006405, MIT Lincoln Laboratory, Jan. 28, 1975.

19   A. R. Dion, "Optimization of a Communication Satellite Multiple-Beam Antenna," Tech. Note 1975-39, DDC AD-A013104/5, MIT Lincoln Laboratory, May 27, 1975.

20   W. C. Cummings et al., "Fundamental Performance Characteristics That Influence EHF MILSATCOM Systems," *IEEE Trans. Com.,* vol. COM-27, no. 10, October 1979.

Chapter 36

Earth Station Antennas

James H. Cook, Jr.

Scientific-Atlanta, Inc.

36-1 INTRODUCTION AND GENERAL CHARACTERISTICS

An earth-station-antenna system consists of many component parts such as the receiver, low-noise amplifier, and antenna. All the components have an individual role to play, and their importance in the system should not be minimized. The antenna, of course, is one of the more important component parts since it provides the means of transmitting signals to the satellite and/or collecting the signal transmitted by the satellite. Not only must the antenna provide the gain necessary to allow proper transmission and reception, but it must also have radiation characteristics which discriminate against unwanted signals and minimize interference into other satellite or terrestrial systems. The antenna also provides the means of polarization discrimination of unwanted signals. The operational parameters of the individual communication system dictate to the antenna designer the necessary electromagnetic, structural, and environmental specifications for the antenna.

Antenna requirements can be grouped into several major categories, namely, electrical or radio-frequency (RF), control-system, structural, pointing- and tracking-accuracy, and environmental requirements, as well as miscellaneous requirements such as those concerning radiation hazards, primary-power distribution for deicing, etc. Only the electrical or RF requirements will be dealt with in this chapter.

The primary electrical specifications of an earth station antenna are gain, noise temperature, voltage standing-wave ratio (VSWR), power rating, receive-transmit group delay, radiation pattern, polarization, axial ratio, isolation, and G/T (antenna gain divided by the system noise temperature). All the parameters except the radiation pattern are determined by the system requirements. The radiation pattern should meet the minimum requirements set by the International Radio Consultative Committee (CCIR) of the International Telecommunications Union (ITU) and/or national regulatory agencies such as the U.S. Federal Communications Commission (FCC).

Earth station antennas operating in international satellite communications must have sidelobe performance as specified by INTELSAT standards or by CCIR Recommendation 483 and Report 391-2 (see Fig. 36-1).

The CCIR standard specifies the pattern envelope in terms of allowing 10 percent of the sidelobes to exceed the reference envelope and also permits the envelope to be adjusted for antennas whose aperture is less than 100 wavelengths (100λ). The reference envelope is given by

$$G = [52 - 10 \log (D/\lambda) - 25 \log \theta] \quad \text{dBi} \qquad D \leq 100\lambda$$

$$= (32 - 25 \log \theta) \quad \text{dBi} \qquad\qquad D > 100\lambda$$

This envelope takes into consideration the limitations of small-antenna design and is representative of measured patterns of well-designed dual-reflector antennas.

Earth station antennas can be grouped into two broad categories: single-beam antennas and multiple-beam antennas. A single-beam earth station antenna is defined as an antenna which generates a single beam that is pointed toward a satellite by means of a positioning system. A multiple-beam earth station antenna is defined as an antenna which generates multiple beams by employing a common reflector aperture with multiple feeds illuminating that aperture. The axes of the beams are determined by the location of the feeds. The individual beam identified with a feed is pointed toward a satellite by positioning the feed without moving the reflector.

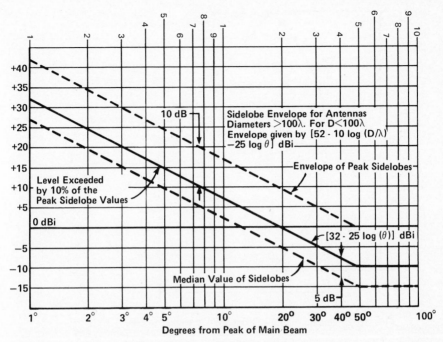

FIG. 36-1   Sidelobe envelope as defined by CCIR Recommendation 433 and Report 391-2.

36-2   SINGLE-BEAM EARTH STATION ANTENNAS

Single-beam antenna types used as earth stations are paraboloidal reflectors with focal-point feeds (prime-focus antenna), dual-reflector antennas such as the Cassegrain and gregorian configurations, horn reflector antennas, offset-fed paraboloidal antennas, and offset-fed multiple-reflector antennas. Each of these antenna types has its own unique characteristics, and the advantages and disadvantages have to be considered when choosing one for a particular application.

Axisymmetric Dual-Reflector Antennas

The predominant choice of designers of earth station antennas has been the dual-reflector Cassegrain antenna. Cassegrain antennas can be subdivided into three primary types:

   1   The classical Cassegrain geometry[1,2] employs a paraboloidal contour for the main reflector and a hyperboloidal contour for the subreflector (Fig. 36-2). The paraboloidal reflector is a point-focus device with a diameter D_p and a focal length f_p. The hyperboloidal subreflector has two foci. For proper operation, one of the two foci is the real focal point of the system and is located coincident with the phase center of the feed; the other focus, the virtual focal point, is located coincident with the focal

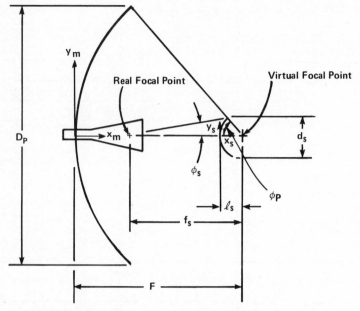

FIG. 36-2   Geometry of the Cassegrain antenna system.

point of the main reflector. The parameters of the Cassegrain system are related as follows:

$$\phi_p = 2 \tan^{-1} (0.25 D_p/F_p) \tag{36-1}$$

$$f_s/d_s = 0.5(\cot \phi_p + \cot \phi_s) \tag{36-2}$$

$$\ell_s/f_s = 0.5(1 - \{\sin [0.5(\phi_p - \phi_s)]/\sin [0.5(\phi_p + \phi_s)]\}) \tag{36-3}$$

In a typical design, the parameters f_p, D_p, f_s, and ϕ_s are chosen, and the remaining three parameters are then calculated.

The contours of the main reflector and subreflector are given by

Main reflector: $y_m^2 = 4F_p x_m$ $\tag{36-4}$

Subreflector: $(y_s/b)^2 + 1 = (x_s/a + 1)^2$ $\tag{36-5}$

where
$$a = (f_s/2e) \qquad b = a\sqrt{e^2 - 1}$$
$$e = \sin [0.5(\phi_p + \phi_s)]/\sin [0.5(\phi_p - \phi_s)]$$

The quantities a, b, and e are half of the transverse axis, half of the conjugate axis, and the eccentricity parameters of the hyperboloidal subreflector respectively.

2   The geometry of this type consists of a paraboloidal main reflector and a special-shaped quasi-hyperboloidal subreflector.[3,4,5,6,7] The geometry in Fig. 36-2 is appropriate for describing this antenna. The main difference between this design and the classical Cassegrain above is that the subreflector is shaped so that the overall efficiency of the antenna has been enhanced. This technique is especially useful with antenna diameters of approximately 60 to 300 wavelengths. The subreflector shape may be solved for by geometrical optics (GO) or by diffraction optimization, and then, by comparing the required main-reflector surface to a paraboloidal surface, the best-fit paraboloidal surface is found. Aperture efficiencies of 75 to 80 percent can be realized by a GO design and efficiencies of 80 to 95 percent by diffraction optimization of the subreflector (Fig. 36-3).

   3   This type is a generalization of the Cassegrain geometry consisting of a special-shaped quasi-paraboloidal main reflector and a shaped quasi-hyperboloidal subreflector.[9,10,11,12] Green[8] observed that in dual-reflector systems with high magnification—essentially a large ratio of main-reflector diameter to subreflector diameter—the distribution of energy (as a function of angle) is largely controlled by the subreflector curvature. The path length or phase front is dominated by the main reflector (see Fig. 36-4). Kinber[9] and Galindo[10,11] found a method for simultaneously solving for the main-reflector and subreflector shapes to obtain an exact solution for both the phase and the amplitude distributions in the aperture of the main reflector of an axisymmetric dual-reflector antenna. Their technique, based on geometrical optics, is highly mathematical and involves solving two simultaneous, nonlinear, first-order, ordinary differential equations. Figure 36-5 gives the geometry showing the path of a single ray. The feed phase center is located as shown, and the feed is assumed to have a power radiation pattern $I(\theta_1)$. The parameters α and β represent respectively the distance of the feed phase center from the aperture plane and the distance between

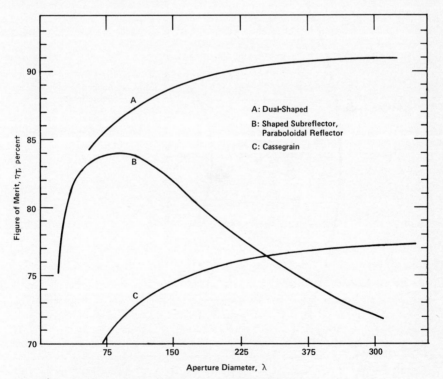

FIG. 36-3   Antenna of figure of merit versus aperture diameter.

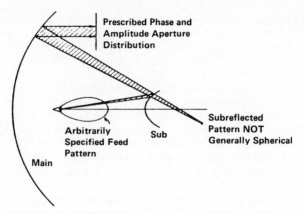

FIG. 36-4 Circularly symmetric dual-shaped reflectors.

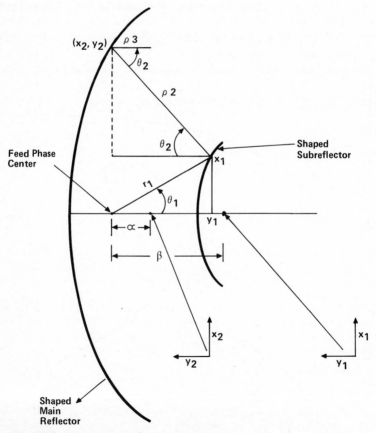

FIG. 36-5 Dual-shaped reflector geometry.

the feed phase center and the back surface of the subreflector. The constraints to the dual-reflector system are as follows:

a The phase distribution across the main-reflector aperture plane will be uniform, or

$$r_1 + p_2 + p_3 + C_p(\theta_1) = \text{constant} \qquad \textbf{(36-6)}$$

for $0 \leq \theta_1 \leq \theta_{1\text{max}}$. $C_p(\theta_1)$ represents the phase distribution across the primary-feed radiation pattern in units of length.

b The feed energy, or ray bundles intercepted and reflected by the subreflector, is conserved and redistributed according to a specified aperture distribution, or

$$I(\theta_1) \sin (\theta_1) \, d\theta_1 = C \cdot I(X_2)X_2 \, dX_2 \qquad \textbf{(36-7)}$$

where $I(X_2)$ represents the power radiation distribution across the main-reflector aperture and C represents a constant which is determined by applying the conservation-of-power principle.

$$\int_0^{\theta_{1\text{max}}} I(\theta_1) \sin \theta_1 \, d\theta_1 = C \int_{X_{2\text{min}}}^{X_{2\text{max}}} I(X_2)X_2 \, dX_2 \qquad \textbf{(36-8)}$$

The lower limit of integration over the main reflector can be arbitrarily chosen so that only an annular region of the main reflector is illuminated.

c Snell's law must be satisfied at the two reflecting surfaces, which yields

$$\frac{dY_1}{dX_1} = \tan \left(\frac{\theta_1 - \theta_2}{2} \right) \qquad \textbf{(36-9)}$$

$$\frac{dY_2}{dX_2} = -\tan \left(\frac{\theta_2}{2} \right) \qquad \textbf{(36-10)}$$

Solving Eqs. (36-7) through (36-10) simultaneously results in a nonlinear, first-order differential equation of the form

$$\frac{dY_1}{dX_2} = f(\theta_1, \theta_2, \alpha, \beta, \text{etc.}) \qquad \textbf{(36-11)}$$

which leads to the cross sections of each reflector when subject to the boundary condition $Y_1 (X_2 = X_{2\text{max}}) = 0$, where X_2 is the independent variable. Equation (36-11) can be solved numerically by using an algorithm such as a Runge-Kutta, order 4.

The above procedure is based on GO, but it is evident that the assumptions of GO are far from adequate when reflectors are small in terms of wavelengths. An improvement in the design approach is to include the effects of diffraction. Clarricoats and Poulton[7] reported a gain increase of 0.5 dB for a diffraction-optimized design over the GO design with a 400λ-diameter main reflector and 40λ-diameter subreflector.

Prime-Focus-Fed Paraboloidal Antenna

The prime-focus-fed paraboloidal reflector antenna is also often employed as an earth station antenna. For moderate to large aperture sizes, this type of antenna has excellent sidelobe performance in all angular regions except the spillover region around the edge of the reflector. The CCIR sidelobe specification can be met with this type of antenna. The reader is referred to Chap. 17 for more information on paraboloidal antennas.

Offset-Fed Reflector Antennas

The geometry of offset-fed reflector antennas has been known for many years, but its use has been limited to the last decade or so because of its difficulty in analysis. Since the advent of large computers has allowed the antenna engineer to investigate theoretically the offset-fed reflector's performance, this type of antenna will become more common as an earth station antenna.

The offset-fed reflector antenna can employ a single reflector or multiple reflectors, with two-reflector types the more prevalent of the multiple-reflector designs. The offset front-fed reflector, consisting of a section of a paraboloidal surface (Fig. 36-6), minimizes diffraction scattering by eliminating the aperture blockage of the feed and feed-support structure. Sidelobe levels of $(29 - 25 \log \theta)$ dBi can be expected from this type of antenna (where θ is the far-field angle in degrees) with aperture efficiencies of 65 to 80 percent. The increase in aperture efficiency as compared with an axisymmetric prime-focus-fed antenna is due to the elimination of direct blockage. For a detailed discussion of this antenna, see C. A. Mentzer.[13]

Offset-fed dual-reflector antennas exhibit sidelobe performance similar to that of the front-fed offset reflector. Two offset-fed dual-reflector geometries are used for

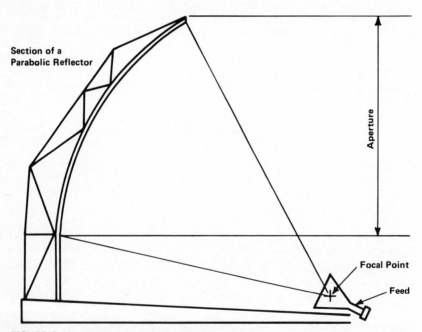

FIG. 36-6   Basic offset-fed paraboloidal antenna.

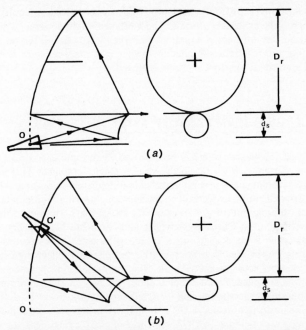

FIG. 36-7 Offset dual-reflector geometries. (*a*) Double-offset geometry (feed phase center and paraboloidal vertex at 0). (*b*) Open cassegrainian geometry (feed phase center located at 0; paraboloidal vertex, at 0).

earth station antennas: the double-offset geometry shown in Fig. 36-7*a* and the open Cassegrain geometry introduced by Cook et al.[14] of Bell Laboratories and shown in Fig. 36-7*b*. In the double-offset geometry, the feed is located below the main reflector, and no blocking of the optical path occurs. In contrast, the open Cassegrain geometry is such that the primary feed protrudes through the main reflector; thus it is not completely blockage-free. Nevertheless, both of these geometries have the capability of excellent sidelobe and efficiency performance.

The disadvantage of offset-geometry antennas is that they are asymmetric. This leads to increased manufacturing cost and also has some effect on electrical performance. The offset-geometry antenna, when used for linear polarization, has a depolarizing effect on the primary-feed radiation and produces two cross-polarized lobes within the main beam in the plane of symmetry. When it is used for circular polarization, a small amount of beam squint whose direction is dependent upon the sense of polarization is introduced. The beam squint is approximately given by[15] $\psi_s = $ arc sin $[\lambda \sin (\theta_0)/4\pi F]$, where θ_0 is the offset angle, λ is the free-space wavelength, and F is the focal length.

During the past few years, considerable analysis has been performed by Galindo-Israel, Mittra, and Cha[16] for this geometry for high-aperture-efficiency applications. Their analytical techniques are reported to result in efficiencies in the 80 to 90 percent range. Mathematically, the offset geometry, when formulated by the method used in designing axisymmetric dual reflectors, results in a set of partial differential equations for which there is no exact GO solution. The method of Galindo-Israel et al. is to solve the resultant partial differential equations as if they were total differential equations. Then, by using the resultant subreflector surface, the main-reflector surface is perturbed until a constant aperture phase is achieved.

36-3   MULTIPLE-BEAM EARTH STATION ANTENNAS

During the past few years there has been an increasing interest in receiving signals simultaneously from several satellites with a single antenna. This interest has prompted the development of several multibeam-antenna configurations which employ fixed reflectors and multiple feeds. The antenna engineering community, of course, has been investigating multibeam antennas for many years. In fact, in the middle of the seventeenth century, Christian Huygens and Sir Isaac Newton first studied the spherical mirror, and the first use of a spherical reflector as a microwave antenna occurred during World War II. More recently, the spherical-reflector, the torus-reflector, and the dual-reflector geometries, all using multiple feeds, have been offered as antennas with simultaneous multibeam capability. Chu[17] in 1969 addressed the multiple-beam spherical-reflector antenna for satellite communication, Hyde[18] introduced the multiple-beam torus antenna in 1970, and Ohm[19] presented a novel multiple-beam Cassegrain-geometry antenna in 1974. All three of these approaches, as well as variations of scan techniques for the spherical reflector, are discussed below.

Spherical Reflector

The properties, practical applications, and aberrations of the spherical reflector are not new to microwave-antenna designers. The popularity of this reflector is primarily due to the large angle through which the radiated beam can be scanned by translation and orientation of the primary feed. This wide-angle property results from the symmetry of the surface. Multiple-beam operation is realized by placing multiple feeds along the focal surface. In the conventional use of the reflector surface, the minimum angular separation between adjacent beams is determined by the feed-aperture size. The maximum number of beams is determined by the percentage of the total sphere covered by the reflector. In the alternative configuration described below, these are basically determined by the f/D ratio and by the allowable degradation in the radiation pattern.

In the conventional use of the spherical reflector, the individual feed illuminates a portion of the reflector surface so that a beam is formed coincident to the axis of the feed. The conventional multibeam geometry is shown in Fig. 36-8. All the beams have similar radiation patterns and gains, although there is degradation in performance in comparison with the performance of a paraboloid. The advantage of this antenna is that the reflector area illuminated by the individual feeds overlaps, reducing the surface area for a given number of beams in comparison with individual single-beam antennas.

The alternative multibeam-spherical-reflector geometry is shown in Fig. 36-9. For this geometry, each of the feed elements points toward the center of the reflector,

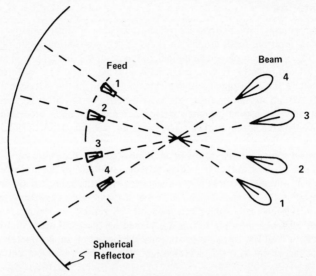

FIG. 36-8 Conventional spherical multibeam antenna using extended reflector and multiple feeds.

with the beam steering accomplished by the feed position. This method of beam generation leads to considerable increase in aberration, including coma; therefore, the radiation patterns of the off-axis beams are degraded with respect to the on-axis beam. This approach does not take advantage of the spherical-reflector properties that exist in the conventional approach. In fact, somewhat similar results could be achieved with a paraboloidal reflector with a large f/D.

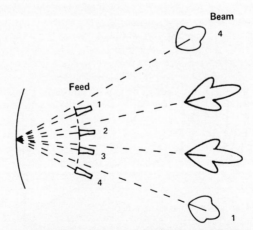

FIG. 36-9 Alternative spherical multibeam antenna using minimum reflector aperture with scanned beam feeds.

Torus Antenna

The torus antenna is a dual-curvature reflector, capable of multibeam operation when it is fed with multiple feeds similar to those of the conventional spherical-reflector geometry. The feed-scan plane can be inclined to be in the orbital-arc plane, allowing the use of a fixed reflector to view geosynchronous satellites. The reflector has a circular contour in the scan plane and a parabolic contour in the orthogonal plane (see Fig. 36-10). It can be fed in either an axisymmetric or an offset-fed configuration. Offset geometry for use as an earth station antenna has been successfully demonstrated by COMSAT Laboratories.[20] The radiation patterns meet a (29-25 log θ)-dBi envelope.

The offset-fed geometry results in an unblocked aperture, which gives rise to low wide-angle sidelobes as well as providing convenient access to the multiple feeds.

The torus antenna has less phase aberration than the spherical antenna because of the focusing in the parabolic plane. Because of the circular symmetry, feeds placed anywhere on the feed arc form identical beams. Therefore, no performance degradation is incurred when multiple beams are placed on the focal arc. Point-focus feeds may be used to feed the torus up to aperture diameters of approximately 200 wavelengths. For larger apertures, it is recommended that aberration-correcting feeds be used.

The scanning or multibeam operation of a torus requires an oversized aperture to accommodate the scanning. For example, a reflector surface area of approximately 214 m² will allow a field of view (i.e., orbital arc) of 30° with a gain of approximately 50.5 dB at 4 GHz (equivalent to the gain of a 9.65-m reflector antenna). This surface area is equivalent to approximately three 9.65-m antennas.

Offset-Fed Multibeam Cassegrain Antenna

The offset-fed multibeam Cassegrain antenna is composed of a paraboloidal main reflector, a hyperbolic oversized subreflector, and multiple feeds located along the scan

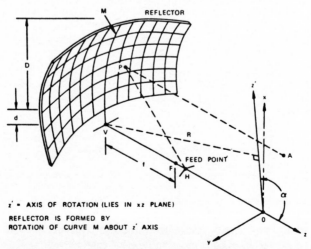

FIG. 36-10 Torus-antenna geometry. (*Copyright 1974,* COM-SAT Technical Review. *Reprinted by permission.*)

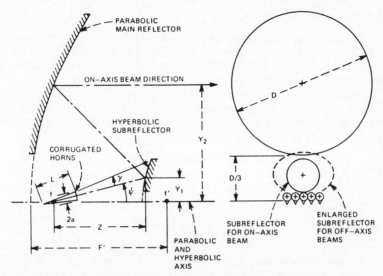

FIG. 36-11 Geometry of the offset-fed multibeam Cassegrain antenna. *(Copyright 1974, American Telephone and Telegraph Company. Reprinted by permission.)*

axis, as shown in Fig. 36-11. The offset geometry essentially eliminates beam blockage, thus allowing a significant reduction in sidelobes and antenna noise temperature. The Cassegrain feed system is compact and has a large focal-length-to-diameter ratio (f/D), which reduces aberrations to an acceptable level even when the beam is moderately far off axis. The low-sidelobe performance is achieved by using a corrugated feed horn which produces a gaussian beam.

A typical antenna design consisting of a 10-m projected aperture would yield half-power beamwidths (HPBWs) and gain commensurate with an axisymmetric 10-m antenna, 0.5° HPBW, and 51-dB gain at 4 GHz. The subreflector would need to be approximately a 3- by 4.5-m elliptical aperture. The feed apertures would be approximately 0.5 m in diameter. The minimum beam separation would be less than 2°, which is more than sufficient to allow use with synchronous satellites with orbit spacings of 2° or greater. For the ±15° scan, the gain degradation would be approximately 1 dB, and the first sidelobe would be approximately 20 to 25 dB below the main-beam peak.

36-4 ANGLE TRACKING

Automatic tracking of satellite position is required for many earth station antennas. Monopulse, conical-scan, and sequential-lobing techniques described in Chap. 34 are applicable for this purpose. Two other types of tracking techniques are also commonly employed: single-channel monopulse (pseudo monopulse) and steptrack. The single-channel monopulse is a continuous angle-tracking scheme, and steptrack is a time-sequencing, signal-peaking technique.

Single-Channel Monopulse

This technique utilizes multiple elements or modes to generate a reference signal, an elevation error signal, and an azimuth error signal. The two error signals are then combined in a time-shared manner by a switching network which selects one of two phase conditions (0° and 180°) for the error signal. The error signal is then combined with the reference signal, with the resultant signal then containing angle information. This allows the use of a single receiver for the tracking channel. This technique is equivalent to sequential lobing in which the lobing is done electrically and can be adjusted to any desired fixed or variable scan rate.

Steptrack

Steptracking is a technique employed primarily for maintaining the pointing of an earth station's antenna beam toward a geosynchronous satellite. Since most geosynchronous orbiting satellites have some small angular box of station keeping, a simple peaking technique can be used to keep the earth station beam correctly pointed. The signal-peaking, or steptrack, routine is a software technique that maneuvers the antenna toward the signal peak by following a path along the steepest signal-strength gradient or by using some other algorithm that accomplishes the same result.[21,22] Any method of direct search for maximization of signal is applicable; one such method calculates the local gradient by determining the signal strength at three points (A, B, C; see Fig. 36-12). From the signal strength at points A, B, and C, an angle α_2, representing the unit's gradient vector angle relative to the azimuth axis, is computed. Once C_1 has been determined, A_2 is set equal to C_1 and another angle θ_2 is defined by $\theta_2 = \theta_1 + \alpha_2$ ($\theta_1 = 0$). The angle is used to relate pedestal movements to a fixed-pedestal coordinate system. The procedure is repeated a number of times until the antenna boresight crosses throughout the peak. At this time the step size is reduced

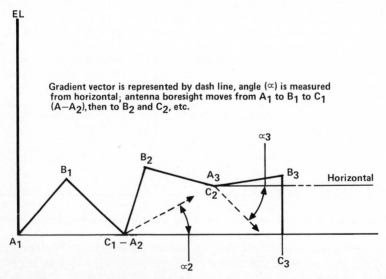

Gradient vector is represented by dash line, angle (α) is measured from horizontal; antenna boresight moves from A_1 to B_1 to C_1 ($A-A_2$), then to B_2 and C_2, etc.

FIG. 36-12 Signal-strength gradients.

by one-half, and for each time that the peak is crossed it is again reduced by one-half until the peaking resolution is accomplished.

36-5 POLARIZATION

Many satellite communication systems are dual-polarized (frequency reuse) and are therefore susceptible to interference from the intended cross-polarized signals when the medium of propagation is such that the ellipticity of the signals is changed. This condition can occur during atmospheric conditions such as rain and Faraday rotation in the ionosphere. Therefore, in order to maintain sufficient polarization discrimination, special precautions must be taken in regard to the polarization purity of dual-polarization antennas; indeed, adaptive polarization-correcting circuits may be necessary.[23,24]

Several types of polarization-discrimination-enhancement schemes may be used. They include:

1 Simple rotation of the polarization major-ellipse axis to correct for rotation due to the ionosphere (applicable for linear polarizations). Transmit and receive rotations are in opposite directions.

2 Suboptimal correction of depolarizations; ellipticity correction with respect to one of the signals but not orthogonalizing the two signals (differential phase).

3 Complete adaptive correction including orthogonalization (differential phase and amplitude).

The polarization-enhancement schemes may be implemented at the RF frequencies, or they may follow a down-conversion stage and be implemented at intermediate frequencies (IF). In either case the circuitry must operate over the full bandwidth of the communication channel. Kreutel et al.[24] treat the implementation of RF frequencies in detail. The reader is also referred to Gianatasio[25,26] for both RF and IF circuitry and experiments to verify performance of the circuitry.

REFERENCES

1 P. W. Hannah, "Microwave Antennas Derived from the Cassegrain Telescope," *IRE Trans. Antennas Propagat.,* vol. AP-9, March 1961, pp. 140–153.
2 P. A. Jensen, "Designing Cassegrain Antennas," *Microwave J.,* December 1962, pp. 10–16.
3 G. W. Collins, "Shaping of Subreflectors in Cassegrainian Antennas for Maximum Aperture Efficiency," *IEEE Trans. Antennas Propagat.,* vol. AP-21, no. 3, May 1973, pp. 309–313.
4 S. Parekh, "On the Solution of Best Fit Paraboloid as Applied to Shaped Dual Reflector Antennas," *IEEE Trans. Antennas Propagat.,* vol. AP-28, no. 4, July 1980, pp. 560–562.
5 J. W. Bruning, "A 'Best Fit Paraboloid' Solution to the Shaped Dual Reflector Antenna," Symp. USAF Antenna R&D, University of Illinois, Urbana, Nov. 15, 1967.
6 P. J. Wood, "Reflector Profiles for the Pencil-Beam Cassegrain Antenna," *Marconi Rev.,* vol. 35, no. 185, 1972, pp. 121–138.

7 P. J. B. Clarricoats and G. T. Poulton, "High-Efficiency Microwave Reflector Antennas—A Review," *IEEE Proc.*, vol. 65, no. 10, October 1977, pp. 1470–1504.

8 K. A. Green, "Modified Cassegrain Antenna for Arbitrary Aperture Illumination," *IEEE Trans. Antennas Propagat.*, vol. AP-11, no. 5, September 1963.

9 B. Y. Kinber, "On Two Reflector Antennas," *Radio Eng. Electron. Phys.*, vol. 6, June 1962.

10 V. Galindo, "Design of Dual Reflector Antenna with Arbitrary Phase and Amplitude Distributions," PTGAP Int. Symp., Boulder, Colo., July 1963.

11 V. Galindo, "Synthesis of Dual Reflector Antennas," Elec. Res. Lab. Rep. 64-22, University of California, Berkeley, July 30, 1964.

12 W. F. Williams, "High Efficiency Antenna Reflector," *Microwave J.*, July 1965, pp. 79–82.

13 C. A. Mentzer, "Analysis and Design of High Beam Efficiency Aperture Antennas," doctoral thesis 74-24, 370, Ohio State University, Columbus, 1974.

14 J. S. Cook, E. M. Elam, and H. Zucker, "The Open Cassegrain Antenna: Part I. Electromagnetic Design and Analysis," *Bell Syst. Tech. J.*, September 1965, pp. 1255–1300.

15 A. W. Rudge and N. A. Adatia, "Offset-Parabolic-Reflector Antennas: A Review," *IEEE Proc.*, vol. 66, no. 12, December 1978, pp. 1592–1611.

16 V. Galindo-Israel, R. Mittra, and A. Cha, "Aperture Amplitude and Phase Control of Offset Dual Reflectors," USNC/URSI & IEEE Antennas Propagat. Int. Symp., May 1978.

17 T. S. Chu, "A Multibeam Spherical Reflector Antenna," *IEEE Antennas Propagat. Int. Symp.: Program & Dig.*, Dec. 9, 1969, pp. 94–101.

18 G. Hyde, "A Novel Multiple-Beam Earth Terminal Antenna for Satellite Communication," *Int. Conf. Comm. Rec.*, Conf. Proc. Pap. 70-CP-386-COM, June 1970, pp. 38-24–38-33.

19 E. A. Ohm, "A Proposed Multiple-Beam Microwave Antenna for Earth Stations and Satellites," *Bell Syst. Tech. J.* vol. 53, October 1974, pp. 1657–1665.

20 G. Hyde, R. W. Kreutei, and L. V. Smith, "The Unattended Earth Terminal Multiple-Beam Torus Antenna," *COMSAT Tech. Rev.*, vol. 4, no. 2, fall, 1974, pp. 231–264.

21 M. J. D. Powell, "An Efficient Method for Finding the Minimum of a Function of Several Variables without Calculating Derivatives," *Comput. J.*, vol. 7, 1964, p. 155.

22 J. Kowalik and M. R. Osborne, *Methods for Unconstrained Optimization Problems*, American Elsevier Publishing Company, Inc., New York, 1968.

23 T. S. Chu, "Restoring the Orthogonality of Two Polarizations in Radio Communication Systems, I and II," *Bell Syst. Tech. J.*, vol. 50, no. 9, November 1971, pp. 3063–3069; vol. 52, no. 3, March 1973, pp. 319–327.

24 R. W. Kreutel, D. F. DiFonzo, W. J. English, and R. W. Gruner, "Antenna Technology for Frequency Reuse Satellite Communications," *IEEE Proc.*, vol. 65, no. 3, March 1977, pp. 370–378.

25 A. J. Gianatasio, "Broadband Adaptively-Controlled Polarization-Separation Network," *Seventh Europ. Microwave Conf. Proc.*, Copenhagen, September 1977, pp. 317–321.

26 A. J. Gianatasio, "Adaptive Polarization Separation Experiments," final rep., NASA CR-145076, November 1976.

Chapter 37

Aircraft Antennas

William P. Allen, Jr.

Lockheed-Georgia Company

Charles E. Ryan, Jr.

Georgia Institute of Technology

37-1 INTRODUCTION

The design of antennas for aircraft differs from other applications. Aircraft antennas must be designed to withstand severe static and dynamic stresses, and the size and shape of the airframe play a major role in determining the electrical characteristics of an antenna. For this latter reason, the type of antenna used in a given application will often depend on the size of the airframe relative to the wavelength. In the case of propeller-driven aircraft and helicopters, the motion of the blades may give rise to modulation of the radiated signal. Triboelectric charging of the airframe surfaces by dust or precipitation particles, known as p static, gives rise to corona discharges, which may produce extreme electrical noise, especially when the location of the antenna is such that there is strong electromagnetic coupling to the discharge point. In transmitting applications, corona discharge from the antenna element may limit the power-handling capacity of the antenna. Also, special consideration must frequently be given to the protection of aircraft antennas from damage due to lightning strikes.

37-2 LOW-FREQUENCY ANTENNAS

The wavelengths of frequencies below about 2 MHz are considerably larger than the maximum dimensions of most aircraft. Because of the inherently low radiation efficiency of electrically small antennas and the correspondingly high radio-frequency (RF) voltages required to radiate significant amounts of power, nearly all aircraft radio systems operating at these lower frequencies are designed so that only receiving equipment is required in the aircraft.

Radiation patterns of aircraft antennas in this frequency range are simple electric or magnetic dipole patterns, depending upon whether the antenna element is a monopole or a loop. In considering first electric dipole antennas, it can be shown with reference to Fig. 37-1 that while the pattern produced by a small monopole antenna will always be that of a simple dipole regardless of location, the orientation of the equivalent-dipole axis with respect to the vertical will depend upon location. The figure shows the electric field fringing produced by an airframe for incident fields polarized in the three principal directions: vertical, longitudinal, and transverse. A small antenna element placed on the airframe would respond to all three of these field components, indicating that the dipole moment of the antenna-airframe combination has projections in all three directions.

The sensitivity of low-frequency (LF) antennas is customarily expressed in either of two ways, depending upon whether the antenna is located on a relatively flat portion of the airframe such as the top or bottom of the fuselage or at a sharp extremity such as the tip of the vertical stabilizer. In the first case it can be assumed (at least when the antenna element is small relative to the surface radii of curvature) that the antenna performs as it would on a flat ground plane, except that the incident-field intensity which excites it is greater than the free-space incident-field intensity because of the field fringing produced by the airframe (Fig. 37-1). The effect of the airframe on antenna sensitivity is hence expressed by the ratio of the local-field intensity on the airframe surface to the free-space incident-field intensity. For a vertically polarized incident field (which is the case of primary importance in LF receiving-antenna

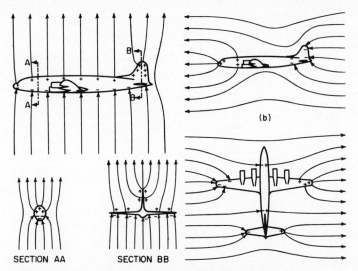

SECTION AA   SECTION BB

FIG. 37-1 Low-frequency field fringing due to the airframe.

design) this ratio is designated as F_v and is called the *curvature factor for vertical polarization*. Survey data giving the equivalent-dipole tilt angle and the factor F_v for top and bottom centerline locations on a typical airframe are shown in Fig. 37-2.

Effective height and capacitance data for antennas installed in locations for which this method is applicable are usually obtained by measurements or calculations for the antenna on a flat ground plane. The effective height (for vertically polarized signals) of the antenna installed on the airframe is then estimated by multiplying the flat-ground-plane effective height by the factor F_v appropriate to the installation, while the capacitance may be assumed to be the same as that determined with the antenna on a flat ground plane. The presence of a fixed-wire antenna on the aircraft may have a significant effect on F_v. Because of the shielding effect due to a wire, LF antennas are seldom located on the top of the fuselage in aircraft which carry fixed-wire antennas.

Flat-ground-plane data for a T antenna are shown in Fig. 37-3. The capacitance curves shown apply to an antenna made with standard polyethylene-coated wire [0.052-in- (1.32-mm-) diameter conductor and 0.178-in- (4.52-mm-) diameter polyethylene sheath]. Two antennas of this type are frequently located in close proximity on aircraft having dual automatic-direction-finder (ADF) installations. The effects of a grounded T antenna on h_e and C_a of a similar nearby antenna are shown in Fig. 37-4. To determine the significance of capacitance interaction it is necessary to consider the Q's of the input circuits to which the antennas are connected and the proximity of the receiver frequencies.

Sensitivity data for two flush antennas and a low-silhouette antenna consisting of a relatively large top-loading element and a short downlead are shown in Figs. 37-5, 37-6, and 37-7 respectively. The antenna dimensions shown in these figures are not indicative of antenna sizes actually in use but rather are the sizes of the models used in measuring the data. Both h_e and C_a scale linearly with the antenna dimensions.

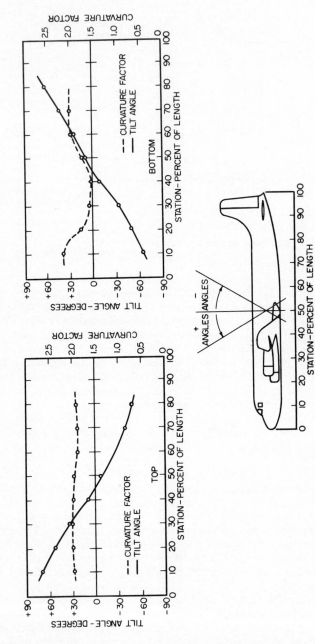

FIG. 37-2 Low-frequency-antenna survey data for a DC-6 aircraft.

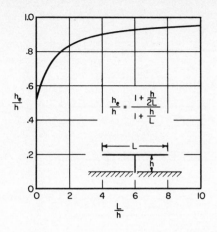

$$\frac{h_e}{h} = \frac{1 + \frac{h}{2L}}{1 + \frac{h}{L}}$$

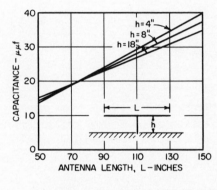

FIG. 37-3 Effective height and capacitance of a T antenna.

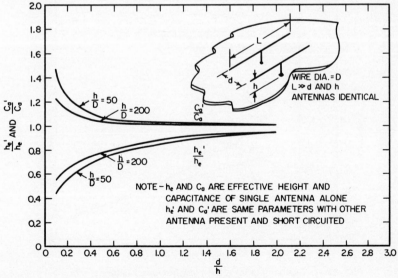

NOTE – h_e AND C_a ARE EFFECTIVE HEIGHT AND
CAPACITANCE OF SINGLE ANTENNA ALONE
h_e' AND C_a' ARE SAME PARAMETERS WITH OTHER
ANTENNA PRESENT AND SHORT CIRCUITED

WIRE DIA. = D
L ≫ d AND h
ANTENNAS IDENTICAL

FIG. 37-4 Shielding effect of one T antenna on another.

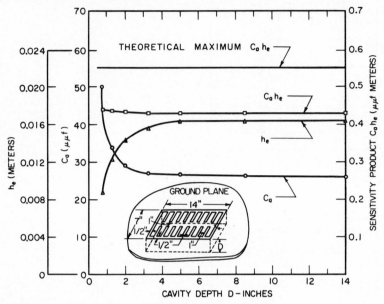

FIG. 37-5 Design data for a flush low-frequency antenna.

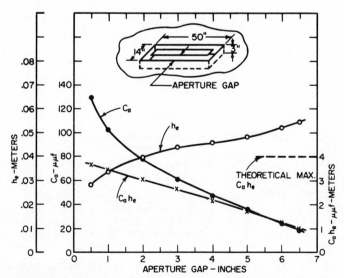

FIG. 37-6 Design data for a flush low-frequency antenna.

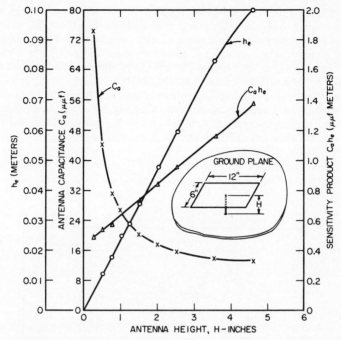

FIG. 37-7 Design data for a heavily top-loaded low-frequency antenna with flat-plate top loading.

A rule-of-thumb design limit for flush antennas may be derived on the basis of a quasi-static analysis of LF antenna performance. If any electric dipole antenna is short-circuited at its feed terminals and placed in a uniform electrostatic field with its equivalent-dipole axis aligned parallel to the field, charges of $+q$ and $-q$ will be induced on the two elements of the antenna. It may be shown that the product of the LF parameters h_e and C_a is related to the induced charge by the equation

$$h_e C_a = q/E \qquad (37\text{-}1)$$

where h_e and C_a are expressed in meters and picofarads respectively, q is expressed in picocoulombs, and the incident-field intensity E is expressed in volts per meter. The quantity q/E is readily calculated for a flush cavity-backed antenna of the type shown in Fig. 37-6, at least for the case in which the antenna element virtually fills the cutout in the ground plane. In this case, the shorted antenna element will cause practically no distortion of the normally incident field, and the number of field lines terminating on the element will be the same as the number which would terminate on this area if the ground plane were continuous. The value of q/E is hence equal to $\varepsilon_0 a$, where ε_0 = 8.85×10^{-12} F/m and a is the area of the antenna aperture in square meters. For a flush antenna with an element which nearly fills the antenna aperture we have

$$h_e C_a = \varepsilon_0 a \qquad (37\text{-}2)$$

The line labeled "theoretical maximum h_eC_a" in the curves of Figs. 37-5 and 37-6 was calculated from Eq. (37-2). With practical antenna designs, there will be regions of the aperture not covered by the antenna element, so that some of the incident-field lines will penetrate the aperture and terminate inside the cavity. As a result, the induced charge q will be smaller than that calculated above, and the product h_eC_a will be smaller than the value estimated from Eq. (37-2).

The helicopter presents a special antenna-design problem in the LF range as well as in other frequency ranges because of the shielding and modulation caused by the rotor blades. Measured curves of F_v obtained with an idealized helicopter model to demonstrate the effects of different antenna locations along the top of the simulated fuselage are shown in Fig. 37-8.

It is characteristic of rotor modulation of LF signals that relatively few modulation components of significant amplitude are produced above the fundamental rotor

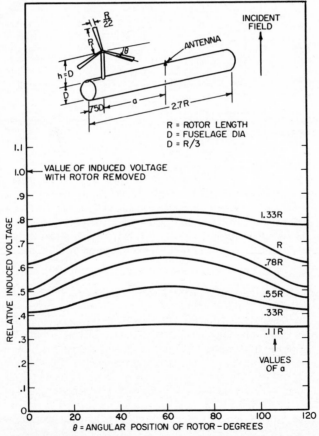

FIG. 37-8 Rotor modulation for a low-frequency receiving antenna on an idealized helicopter model.

modulation frequency, which is equal to the number of blade passages over the antenna per second. This fundamental frequency is of the order of 15 Hz for typical three-blade single-rotor helicopters. It should be noted that new sidebands will be generated about each signal sideband and about the carrier. It is hence not a simple matter to predict the effects of rotor modulation on a particular system unless tests have been made to determine the performance degradation of the airborne receiver due to these extra modulation components.

In the case of loop antennas, it is necessary to consider the distortion caused by the airframe in the magnetic field component of the incident wave. Unlike the electric field lines, the magnetic field lines distort in such a way that they avoid entering the conducting airframe. The field-line sketches in Fig. 37-9 illustrate the airframe effect for two cases in which the ground station is respectively to the side of the aircraft and

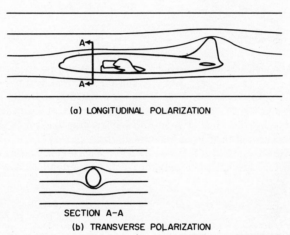

(a) LONGITUDINAL POLARIZATION

SECTION A–A

(b) TRANSVERSE POLARIZATION

FIG. 37-9 Magnetic field distortion caused by a conducting airframe for (*a*) longitudinal polarization and (*b*) transverse polarization.

ahead of the aircraft. For most locations near the top and bottom centerline, the local-field intensity is greater than the incident-field intensity in both cases. This field enhancement is important because it serves to increase the signal induced in a loop antenna and also affects the bearing accuracy of the direction-finder system. The ratio of local magnetic field intensity on the airframe surface to the incident magnetic field intensity is designated as a_{xx} for the case in which the incident field is transverse to the line of flight (the signal arrives from the front or rear of the aircraft) and as a_{yy} when the incident field is along the line of flight (the signal arrives from the side of the aircraft). These two coefficients give all the essential data for estimating the performance of a direction-finder loop antenna designed to take bearings on LF ground-wave signals, provided the loop is placed on the top or bottom centerline of the airframe. The amplitude and direction of the local field on the airframe surface may be calculated for any incident-field amplitude and direction (provided the latter is horizontal), once the coefficients a_{xx} and a_{yy} are known, by simply resolving the incident field into x and y components, multiplying these components by a_{xx} and a_{yy} respectively, and recombining the components.*

The ratio of local-field intensity to incident-field intensity, which in this case is a function of the angle of arrival of the wave, may be used as a curvature factor for estimating loop-antenna sensitivity in the same way that the factor F_v is used for monopole-antenna calculations.

The ADF system, which determines the direction of arrival of the signal by rotating its loop antenna about a vertical axis until a null is observed in the loop response (Fig. 37-10), is subject to bearing errors because of the difference in direction of the local incident magnetic fields. The relationship between the true and the apparent directions of the signal source is given by the equation

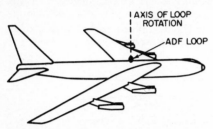

I AXIS OF LOOP
I ROTATION

ADF LOOP

$$\tan \phi_t = \frac{a_{xx}}{a_{yy}} \tan \phi_a \qquad (37\text{-}3)$$

FIG. 37-10   ADF loop antenna on aircraft.

where ϕ_t and ϕ_a are respectively the true and the apparent bearings of the signal source. A curve of the bearing error ($\phi_t - \phi_a$) as a function of ϕ_t with the ratio $a_{xx}/a_{yy} = 2$ is shown in Fig. 37-11a. Figure 37-11b shows graphs of the maximum bearing error $|\phi_t - \phi_a|$ and the value of ϕ_t (in the first quadrant) at which the maximum bearing error occurs as functions of the ratio a_{xx}/a_{yy}.

Survey data showing the coefficient a_{xx} and the maximum bearing error for loop antennas along the top and bottom centerline of a DC-4 aircraft are shown in Fig. 37-12.

37-3   ADF ANTENNA DESIGN

The automatic-direction-finder loop antenna is normally supplied as a component of the ADF system, and the problem is to find a suitable airframe location. The two restrictions which govern the selection of a location are (1) that the cable between the loop antenna and receiver be of fixed length, since the loop and cable inductances form a part of the resonant circuit in the loop amplifier input, and (2) that the maximum bearing error which the system can compensate be limited to about 20°. As can be seen from the bearing-error data in Fig. 37-12, positions on the bottom of the fuselage forward of the wing meet the minimum bearing-error requirements on the aircraft type for which the data are applicable (and on other comparable aircraft). Such positions are consistent with cable-length restrictions in most cases since the radio-equipment racks are usually just aft of the flight deck.

---

*For locations off the centerline, an incident field polarized in one of the principal directions may cause local-field components in both principal directions (as well as in the z direction), and hence coefficients a_{xy}, a_{yx}, a_{zy}, and a_{zx} are needed in the general case to describe the local-field amplitude and direction in terms of the incident-field amplitude and direction.

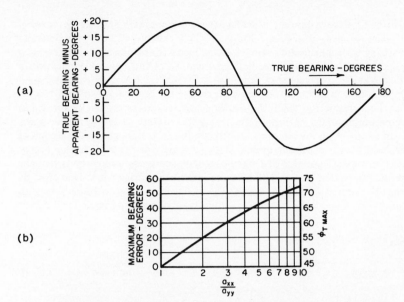

(a)

(b)

FIG. 37-11 (*a*) Bearing-error curve for $a_{zz}/a_{yy} = 2$. (*b*) The maximum bearing error and the true bearing at which the maximum bearing error occurs as functions of a_{zz}/a_{yy}.

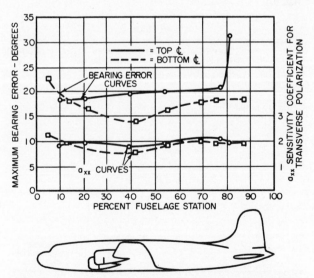

FIG. 37-12 Low-frequency loop-antenna survey data for a DC-4 aircraft showing maximum bearing error and sensitivity coefficient for transverse polarization for top and bottom centerline locations.

Bearing-error compensation is accomplished in most ADF systems by means of a mechanical compensating cam.[1,2] Standard flight-test procedures[1,3] are generally used in setting the compensating cams. In some loop designs[4] electrical compensation is achieved by modifying the immediate environment of the actual loop element in order to equalize the coefficients a_{xx} and a_{yy}.

Unlike the loop antenna, the sense antenna must be located within a limited region of the airframe for which the equivalent-dipole axis is essentially vertical in order to ensure accurate ADF performance as the aircraft passes over or near the ground station to which the receiver is tuned.[5] In Fig. 37-13 are shown the regions of confusion which exist above the ground station for an airborne ADF having a non-vertical sense-antenna pattern. The significance of such a confusion zone is that the ADF needle attempts to reverse and hence to indicate a bearing which is in error by 180° when the aircraft is within the zone. The intersection of the confusion zone with a surface of constant altitude is a circle; various types of needle behavior for different

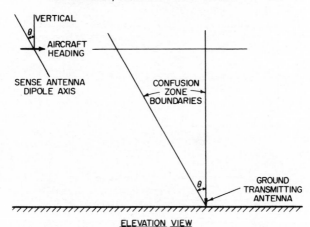

NOTE: SMALL ARROWS SHOW DIRECTIONS WHICH ADF NEEDLE
TRIES TO INDICATE AT VARIOUS POINTS ALONG A GROUP
OF PARALLEL, COPLANAR FLIGHT TRACKS

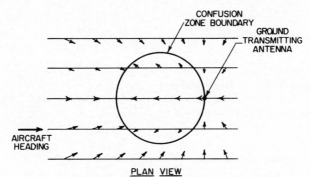

NOTE: ANGLE BETWEEN CONFUSION ZONE BOUNDARIES
IS EQUAL TO SENSE ANTENNA TILT ANGLE

FIG. 37-13   ADF overstation confusion zone.

amounts of course offset at the same altitude are illustrated in Fig. 37-13. Actually, the needle has a finite reversal time, so that the responses to the various reversal signals which it receives as the aircraft traverses a flight path near the station become superimposed to an extent which depends upon the aircraft altitude and speed as well as on the sense-antenna tilt angle. As a result, it is virtually impossible for the pilot to make an accurate determination of the time of station passage unless the confusion zone is made quite small.

The size of the zone is also dependent on the phase difference between the loop and sense signals at the point where they are mixed in the receiver. It can be shown[5] that, by introducing a controlled phase error of the proper sense at this point, the size of the confusion zone may be reduced very substantially, thereby relaxing the requirement for proper sense-antenna placement. Existing ADF receivers do not incorporate this feature, however, and hence a small sense-antenna tilt angle is usually a design requirement.

A typical flush ADF sense-antenna installation is illustrated by the C-141 aircraft.[6] The antenna is a cavity built into the upper fuselage fairing just aft of the wing crossover. It was decided during the design phase to use the Bendix ADF-73 automatic-direction-finder system, which to some extent determined the sense-antenna requirements. The basic antenna requirements were (1) antenna capacitance of 300 pF, (2) a minimum quality factor of 1.4, and (3) a radiation-pattern tilt angle between 0 and 10° downward and forward. Figure 37-14 shows the patterns of the dual sense antenna.

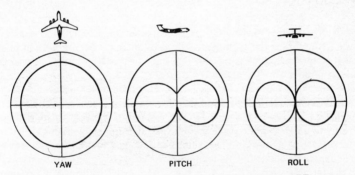

YAW   PITCH   ROLL

FIG. 37-14 Voltage radiation patterns for the C-141 ADF sense antenna.

The sensitivity of the ADF system varies directly with the sense-antenna quality factor, which is defined as $h_e C_a$, where h_e is the antenna effective height in meters and C_a is the antenna capacitance in picofarads. In actual practice the effective height is determined by a measurement of the antenna open-circuit voltage in microvolts divided by the received-signal field intensity in microvolts per meter.

The antenna effective height, for a top-loaded design, is directly proportional to the distance separating the loading element from the ground plane. Antenna capacitance is directly proportional to the element area and inversely proportional to the separation from the ground plane. Thus, as the separation is increased, h_e increases while C_a decreases. It can be shown that the quality factor will also increase for an external design. In the case of a flush-type cavity, the quality factor increases asymptotically with separation. With a limited separation, it is apparent that the desired quality factor may be obtained only by providing sufficient element area. Owing to aerodynamic and structural considerations, there is a limit to the area or separation which may be economically utilized, and a compromise is made.

The sense-antenna input of the ADF receiver is designed to receive an antenna circuit with a nominal capacitance of 3000 pF. The transmission line is composed of 60 ft (18.3 m) of RG-11/U. This represents 1230 pF, since RG-11/U has a capacitance of 20.5 pF per foot. The sense-antenna coupler (susceptiformer) transforms the remaining 1770 pF to the sense-antenna capacitance of 300 pF.

For the C-141, the electrical center of the aircraft (pattern tilt angle of zero) was initially located through radiation-pattern measurements by using a small model configuration; a tilt-angle change was noted when the feed point for the cavity ADF antenna was moved. A zero tilt angle was obtained, and it was found that an approximate 10-in (254-mm) movement of the feed point aft produced a $+10°$ tilt angle. Since a tilt angle of 0 to $10°$ is required, the measurements permitted selection of an optimum feed point ($+5°$ tilt). Figure 37-14 shows the yaw, pitch, and roll radiation patterns obtained with the optimum antenna feed point.

37-4   MOMENT-METHOD ANALYSIS OF LOW-FREQUENCY AND HIGH-FREQUENCY ANTENNA PATTERNS

The method of moments (MOM)[7] is a computer analysis method which can be employed to determine the radiation patterns of antennas on aircraft and missiles. A discussion of this technique is beyond the scope of this handbook, but user-oriented computer programs are available.[8,9]

The method of moments is based upon a solution of the electromagnetic-wave-scattering integral equations by enforcing boundary conditions at a number N of discrete points on the scattering surface. The resulting set of equations is then solved numerically to determine the scattered fields. This is typically accomplished by modeling the surface as a wire grid for which the self- and mutual impedances between the wires can be calculated. This technique, for a set of N wire elements which represents the aircraft such as shown in Fig. 37-15, results in an $N \times N$ impedance matrix. For a given antenna excitation, inversion of the $N \times N$ matrix yields the solution for the wire currents and hence the radiated fields of the antenna-aircraft system.

Computer programs which implement this technique for wire scatterers[7,10,11] and for surface patches[12] have been written. Some codes which employ special computer techniques are presently capable of handling matrices on the order of (200×200) to (1200×1200) elements. The number of wire or patch elements which must be employed to ensure a convergent numerical solution varies as a function of the particular code implementation. Typically, a wire element less than $\lambda/8$ in length must be used so that a $1\lambda^2$ surface patch requires a "square" wire-grid model containing 144 wires. Thus it is seen that as the surface area increases, the required matrix increases roughly as the square of the area which, in practice, restricts the MOM to electrically small- to moderate-sized geometries. Codes based on body-of-revolution approximations are also available[13] and are particularly applicable to helicopters and missiles.

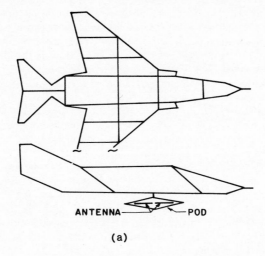

(a)

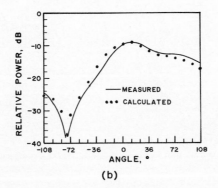

(b)

FIG. 37-15 MOM wire-grid analysis of low-frequency aircraft antenna patterns. (*a*) MOM wire-grid model of an RF-4C aircraft. (*b*) MOM calculated roll-plane pattern for a pod-mounted dipole array on the RF-4C aircraft. *(From Ref. 11, © 1977 IEEE.)*

A wire-grid model which was used for the pattern analysis of a pod-mounted high-frequency (HF) folded-dipole array antenna on an RF-4C aircraft is shown in Fig. 37-15*a*. An MOM code employing piecewise sinusoidal basis functions[10,11] was used to calculate the roll-plane patterns for the array, and these results are compared with measured data in Fig. 37-15*b*.

A surface-patch method has also been employed to analyze the patterns of antennas on helicopters.[12] In this analysis, the helicopter surface was subdivided into a collection of curvilinear cells. A result of the analysis, which employed 94 surface-patch cells, is shown in Fig. 37-16 for the pitch and yaw planes. Generally, good results were obtained for this complex-shaped airframe. Also, we note that, by representing the helicopter rotor blades as wires, the rotor effects on the patterns can be computed.

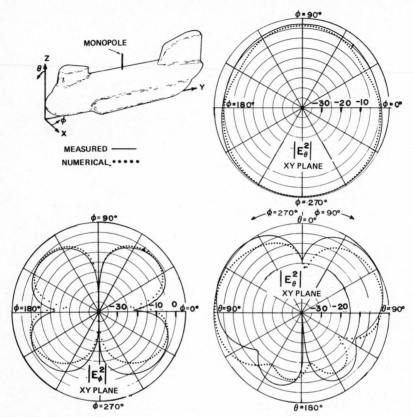

FIG. 37-16 MOM surface-patch-calculated patterns for the CH-47 helicopter. *(From Ref. 12, © 1972 IEEE.)*

For some LF applications, a simplified "stick model" of an aircraft may be sufficient,[14] and a model of an aircraft is shown in Fig. 37-17. The fuselage, wings, and empennage can be represented by either single or multiple thin wires if their maximum electrical width is approximately less than $\lambda/10$. Figure 37-17 shows a calculated result for a monopole mounted on a stick model, and the effects of the fuselage and wings can be seen by comparison with the unperturbed-monopole pattern.[15]

37-5 HIGH-FREQUENCY COMMUNICATIONS ANTENNAS

Aircraft antennas for use with communications systems in the 2- to 30-MHz range are required to yield radiation patterns which provide useful gain in directions significant to communications. Also, impedance and efficiency characteristics suitable for acceptable power-transfer efficiencies between the airborne equipment and the

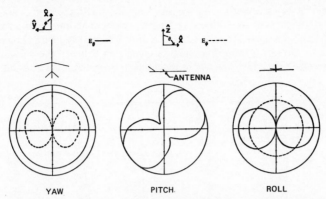

FIG. 37-17 MOM calculated voltage patterns for a blade antenna (tilted monopole) on a "stick model" aircraft.

radiated field are needed. A receiving antenna which meets these requirements will deliver to the input of a matched receiver atmospheric noise power under noise field conditions which is many times greater than the input-circuit noise in the communications receiver. When this is the case, no improvement in signal-to-noise ratio can be achieved by further refinement of the antenna design. Since the transmitting mode of operation poses the more stringent requirements, the remainder of the discussion of HF antennas is confined to the transmitting case. Sky-wave propagation is always an important factor at these frequencies, and because of the rotation of polarization which is characteristic of reflection from the ionosphere, polarization characteristics are usually unimportant; the effective antenna gain can be considered in terms of the total power density. At frequencies below 6 MHz, ionosphere (and ground) reflections make almost all the radiated power useful for communication at least some of the time, so that differences between radiation patterns are relatively unimportant in comparing alternative aircraft antennas for communication applications in the 2- to 6-MHz range. In this range, impedance matching and efficiency considerations dominate. For frequencies above 6 MHz, pattern comparisons are frequently made in terms of the average power gain in an angular sector bounded by cones 30° above and below the horizon.

In the 2- to 30-MHz range, most aircraft have major dimensions of the order of a wavelength, and currents flowing on the skin usually dominate the impedance and pattern behavior. Since the airframe is a good radiator in this range, HF antenna design is aimed at maximizing the electromagnetic coupling to the airframe. The airframe currents exhibit strong resonance phenomena that are important to the impedance of antennas which couple tightly to the airframe.

Wire Antennas

Wire antennas, supported between the vertical fin and an insulated mast or trailed out into the airstream from an insulated reel, are reasonably effective HF antennas on lower-speed aircraft. Aerodynamic considerations limit the angle between a fixed wire and the airstream to about 15°, so that fixed-wire antennas yield impedance characteristics similar to moderately lossy transmission lines, with resonances and antiresonances at frequencies at which the wire length is close to an integral multiple of $\lambda/4$. Figure 37-18 shows the input impedance of a 56-ft (17-m) fixed-wire antenna on a C-130 Hercules aircraft. Lumped reactances connected between the wire and the fin produce an effect exactly analogous to reactance-terminated lossy transmission lines.

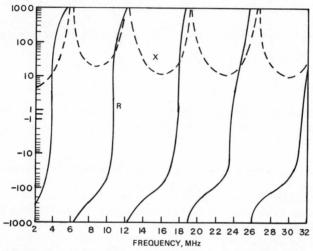

FIG. 37-18 Impedance of the C-130E right-hand long-wire antenna.

The average directive gain of these antennas in the sector $\pm 30°$ relative to the horizon (i.e., the fraction of the total radiated energy which goes into the sector bounded by the cones 30° above and below the horizontal plane) remains near 60 percent from 6 to 24 MHz. The efficiency of wire antennas is not high because of resistance loss in the wire itself and dielectric loss in the supporting insulators and masts. The resistance of commonly used wires is of the order of 0.05 Ω/ft at 4 MHz. The RF corona breakdown threshold of fixed-wire antennas is a function of the wire diameter and the design of the supporting fittings. To minimize precipitation static a wire coated with a relatively large-diameter sheath of polyethylene is frequently used, with special fittings designed to maximize the corona threshold. Even with such precautions, voltage breakdown poses a serious problem with fixed-wire antennas at high altitudes. Measurements indicate that standard antistatic strain insulators have an RF corona threshold of about 11 kV peak at an altitude of 50,000 ft (15,240 m)[16] and a frequency of 2 MHz. Such an insulator, placed between the vertical stabilizer and the aft end of the open-circuited antenna would go into corona at this altitude if the antenna were energized with a fully modulated AM carrier of about 150 W at 2 MHz.

Isolated-Cap Antennas

The airframe can be effectively excited as an HF antenna by electrically isolating a portion of the fin tip or a wing tip to provide antenna terminals. Figure 37-19a shows

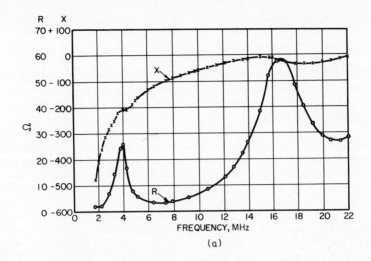

(a)

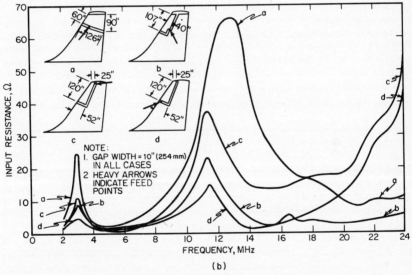

(b)

FIG. 37-19 Impedance characteristics of tail-cap antennas. (*a*) Input impedance of a 7-ft (2.1-m) tail-cap antenna on a DC-4 aircraft. (*b*) Effect of changing feed configurations on the input resistance of tail-cap antennas.

the impedance curves for a 7-ft (2.1-m) tail-cap antenna on an aircraft. The isolating gap was cut straight across the aircraft extremity, and the gap width was 6 in (152 mm). The sharp peak of input resistance at 4 MHz is due to coupling to the $\lambda/2$ resonance of the currents on the airframe, which at this frequency flow predominantly along a path extending from the fin tip to the fuselage, thence to the wing root, and along the wings to the tips. This resonance dominates the lower-frequency impedance characteristics of all cap-type antennas. The resistance peaks at higher frequencies are associated with current resonances on the wings and fuselage and, in the tail-cap cases, the empennage.

Probe Antennas

A variation of the isolated-cap antenna is the probe or bullet-probe antenna. The radiation-pattern and impedance-resonance effects of the empennage, fuselage, and wings also affect probe antennas. The marked advantage of the probe antenna over the tail-cap antenna lies in the relative ease of incorporation into the structure of the aircraft empennage. The C-141 bullet-probe antenna shown in Fig. 37-20[18] makes use of a portion of the structural intersection of the horizontal and vertical stabilizers in the T-tail configuration. The antenna is composed of the forward portion of the aerodynamic bullet with a probe extension. The plastic isolation gap in the bullet extends forward 11 in (279 mm), and the overall length of the antenna including the isolation band is 112 in (2845 mm). The maximum structural gap width was 13 in (330 mm), and the minimum gap width to withstand RF potentials was 3 to 4 in (76 to 102 mm). To obtain maximum airframe coupling, a maximum resistive component is required at the lowest expected resonant frequency. This resonance occurred at 2.4 MHz, which corresponds to a half-wavelength resonant condition caused by the wing-tip-to-horizontal-stabilizer-tip distance. Another parameter considered was the capacitive reactance of the antenna at 2.0 MHz. This reactance had to remain below 1000 Ω if existing couplers were to be used. This resulted in a gap width of 11 in for satisfactory performance. Figure 37-20 shows the impedance curves for the C-141 forward bullet-probe antenna.

A solid bulkhead was installed just aft of the isolation band for two reasons: (1) the metallic bulkhead acts as an electrostatic shield between the antenna and other

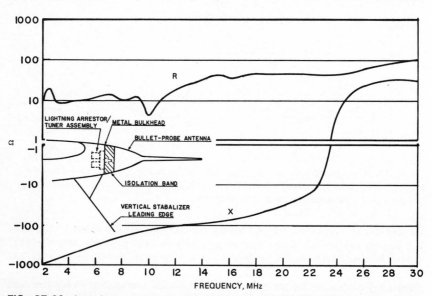

FIG. 37-20 Impedance curves for the C-141 aircraft forward bullet-probe antenna. *(From Ref 18.)*

metallic objects, and (2) the antenna tuner will mount on the bulkhead. Connection to the antenna is made through a lightning arrester which protrudes through the bulkhead into the gap area. The lightning arrester has a "birdcage" terminal which plugs into a flared tube tied electrically to the probe and metal portions of the forward bullet.

Shunt or Notch High-Frequency Antennas

Shunt- and notch-type HF antennas are becoming increasingly popular on modern high-speed aircraft because of drag reduction and higher reliability. The basic difference in the shunt and notch configurations is the ratio of length to width of the cutout (dielectric) portion of the aircraft. Figure 37-21 shows the basic configuration.[19]

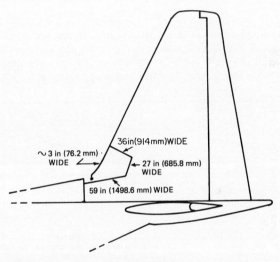

FIG. 37-21 Basic configuration of the C-130 notch high-frequency antenna. *(From Ref. 19.)*

The shunt or notch antenna is a kind of transmission line that uses the vertical stabilizer or wing to feed the remainder of the aircraft. It can be located at various points on the airframe, but since it functions as an inductive coupler, it should be located at a point of low impedance or high current. The shunt-fed loop current causes the airframe itself to act as the antenna. The particular portion of the airframe supporting the current path is a function of the frequency and physical lengths of parts of the airframe involved. For example, the wing-tip-to-horizontal-tip half-wavelength resonance on the C-5 aircraft occurs at approximately 2.5 MHz. The coupler feed point should always be located at the end of the shunt feed closest to the center of the aircraft.

The design of the antenna involves first finding a suitable location for the dielectric area and then arranging the dimensions of the shunt feed and dielectric area to provide an antenna that can be efficiently matched to 50 Ω with a relatively simple antenna coupler. Guidelines to observe are as follows: (1) The first choice for location on the airframe from a radiation-pattern and impedance standpoint is the root of the

vertical stabilizer. An alternative is the wing root, but in this case the radiated vertically polarized energy is reduced. (2) The antenna coupler should be connected to the
shunt feed near the stabilizer or wing root to provide maximum efficiency. (3) The
dielectric material used to fill the opening should have low-loss characteristics. (4)
The exact size and shape of the dielectric area is not a critical design parameter, but
the dielectric area should be as deep as possible for maximum coupling to the airframe. Radiation efficiency is generally proportional to the dielectric area.

To avoid the need for a complicated antenna-tuning network, the antenna reactance at 2.0 MHz should be no less than $+j18$ Ω, and ideally parallel resonance should
occur between 20 and 30 MHz. Series resonance should be completely avoided, and
antenna reactance should be no lower than $-j100$ Ω at 30 MHz. To ensure good
coupling to the airframe the parallel component of the antenna impedance should be
less than 20,000 Ω.

Figure 37-21 shows the shunt-fed notch antenna for the C-130 aircraft. A portion of the dorsal and a portion of the leading-edge fairing are combined to yield a
notch in the root of the vertical stabilizer. The total distance around the notch is
approximately 160 in (4064 mm), or slightly smaller than the ideal periphery of
approximately 200 in (5080 mm). This was necessary because of interference with the

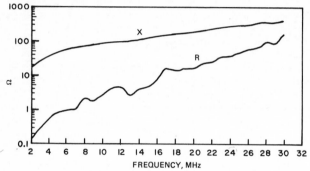

FIG. 37-22 Impedance of the shunt-fed HF notch antenna on the
C-130 aircraft. *(From Ref. 19.)*

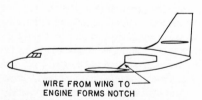

WIRE FROM WING TO
ENGINE FORMS NOTCH

FIG. 37-23 Shunt-fed HF notch-antenna
configuration on the Lockheed JetStar.
(From Ref. 20.)

leading-edge deicing system. The impedance curve for this antenna is shown in
Fig. 37-22.

A unique application of the shunt-
fed HF antenna is found on the Lockheed
JetStar aircraft.[20] Figure 37-23 is a
sketch of the antenna. The antenna is fed
from a coaxial feed on the upper surface
of the wing, where a 32-in (813-mm)
piece of phosphor-bronze stranded cable
connects the center conductor to the lead-
ing lower edge of an aft-mounted engine nacelle. Excellent omnidirectional patterns
are obtained, and the impedance is tuned by using an off-the-shelf coupler inside the
pressurized area of the fuselage.

37-6   UNIDIRECTIONAL VERY-HIGH-FREQUENCY ANTENNAS

The marker-beacon, glide-slope, and radio-altimeter equipments require relatively narrowband antennas with simple patterns directed down or forward from the aircraft. This combination of circumstances makes the design of these antennas relatively simple. Both flush-mounted and external-mounted designs are available in several forms.

Marker-Beacon Antennas

The marker-beacon receiver operates on a fixed frequency of 75 MHz and requires a downward-looking pattern polarized parallel to the axis of the fuselage. A standard external installation employs a balanced-wire dipole supported by masts. The masts may be either insulated or conducting. Low-drag and flush-mounted designs are sketched in Fig. 37-24. The low-drag design is a simple vertical loop oriented in the

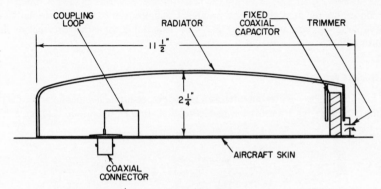

(a) SCHEMATIC OF THE COLLINS 37X-I MARKER BEACON ANTENNA

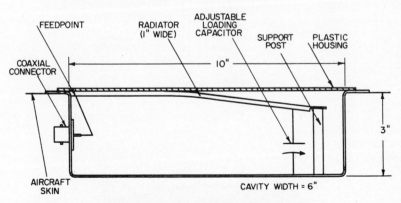

(b) SCHEMATIC OF THE ELECTRONIC RESEARCH INC. AT-134/ARN FLUSH MARKER BEACON ANTENNA

FIG. 37-24   Typical marker-beacon antennas.

longitudinal plane of the aircraft. The feed line is inductively coupled to the loop, which is resonated by a series capacitor, and the antenna elements are contained in a streamlined plastic housing. The flush design is electrically similar, but in this case the structure takes the form of a conductor set along the longitudinal axis of the open face of a cavity. To achieve the desired impedance level, the antenna conductor is series-resonated by a capacitor and the feed point tapped partway along the antenna element.

Glide-Slope Antennas

The glide-slope receiver covers the frequency range from 329 to 335 MHz and requires antenna coverage only in an angular sector 60° on either side of the nose and 20° above and below the horizon. This requirement can be met by horizontal loops or by vertical slots. Because of the narrow bandwidth, the antenna element need not be electrically large. Two variations on the loop arrangement are sketched in Fig. 37-25. Configuration a is a simple series-resonant half loop which can be externally mounted on the nose of the aircraft or within the nose radome. A variation of this antenna has two connectors for dual glide-slope systems. Configuration b, which is similar to the cavity-marker-beacon antenna of Fig. 37-24, is suitable for either external or flush mounting.

Altimeter Antennas

The radio altimeter, which operates at 4300 MHz, requires independent downward-looking antennas for transmission and reception. Proper operation requires a high

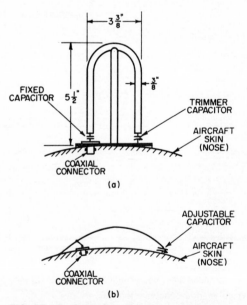

FIG. 37-25 Protruding glide-slope antennas.

degree of isolation between the transmitting and receiving elements, and horns are typically used, although microstrip antennas are becoming common.

37-7   OMNIDIRECTIONAL VERY-HIGH-FREQUENCY AND ULTRAHIGH-FREQUENCY ANTENNAS

In the frequency range in which the airframe is electrically large, the achievement of omnidirectional patterns, such as those used for short-range communications, is complicated by airframe effects. Since very-high-frequency (VHF) and ultrahigh-frequency (UHF) antennas of resonant size are structurally small, the required impedance characteristics can be achieved with fixed matching networks. Shadowing and reflection by the airframe result in major distortions of the primary pattern of the radiating element.

Airframe Effects on VHF and UHF Patterns

The tip of the vertical fin is a preferred location for omnidirectional antennas in the VHF and UHF range because antennas located there have a relatively unobstructed "view" of the surrounding sphere. Figure 37-26 shows the principal-plane patterns of a λ/4 stub on the fin tip of a B-50 aircraft at 1000 MHz. At this frequency, at which the mean chord of the fin is 10 wavelengths, the principal effect of the airframe on the radiation patterns is a sharply defined "optical" shadow region, indicated by the dashed lines in the figure. At the lower end of this frequency range and on smaller aircraft, the effect of the airframe on the patterns is more complicated. Deep nulls in the forward quadrants can occur because of the destructive interference between direct radiation and radiation reflected from the fuselage and wings. The latter contribution is more important for small aircraft since the ground plane formed by the surface of the fin tip is now sufficiently small to permit strong spillover of the primary pattern, which also can create lobing in the transverse-plane pattern. In this frequency range the null structure is strongly influenced by the position of the radiating element along the chord of the fin, and careful location of the antenna along the chord may result in improvements in the forward-horizon signal strength.

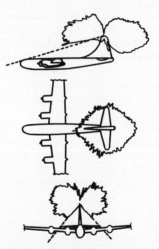

FIG. 37-26 Principal-plane patterns of a 1000-MHz stub antenna on the tip of the vertical stabilizer of a B-50 aircraft.

Most external antennas are located on the top or bottom centerline of the fuselage in order to maintain symmetry of the radiation patterns. Pattern coverage in such locations is limited by the airframe shadows. Figure 37-27 shows the patterns of a

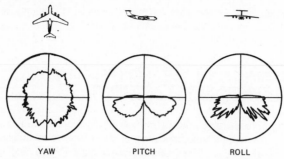

YAW   PITCH   ROLL

FIG. 37-27 Voltage radiation patterns of a 1000-MHz monopole on the bottom of a C-141 aircraft.

1000-MHz monopole on the bottom centerline of a C-141 aircraft. It is apparent that
coverage is limited to the hemisphere below the aircraft. UHF antennas on the top of
the fuselage yield patterns confined to the upper hemisphere, with a null aft owing to
the shadow cast by the vertical fin. In many applications, as, for example, in sched-
uled-airline operations, these pattern limitations are acceptable, and fuselage locations
are frequently used.

The deep lobing in the roll-plane pattern of Fig. 37-27 is due to reflections from
the strongly illuminated engine nacelles. In some locations similar difficulties are
encountered because of reflection from the wing flaps when they are extended. Shadows and lobing due to the landing gear, when extended, are frequently troublesome
for bottom-mounted antennas.

GTD Antenna Analysis

The MOM technique is limited to surfaces which are relatively small in terms of
square wavelengths. However, as the frequency increases and the surface becomes
large in terms of square wavelengths, the propagation of electromagnetic (EM) energy
can be analyzed by using the techniques of geometrical optics. For example, if an
incident field is reflected from a curved surface, ray optics and the Fresnel reflection
coefficients yield the reflected field.

The geometrical theory of diffraction (GTD) uses the ray-optics representation
of EM propagation and incorporates both *diffracted rays* and *surface rays* to account
for the effects of edge and surface discontinuities and surface-wave propagation. In
the case of edge diffraction, the diffracted rays lie on a cone whose half angle is equal
to the half angle of the edge tangent and incidence vector. For a general curved surface, the incident field is launched as a surface wave at the tangent point, propagates
along a geodesic path on the surface, and is diffracted at a tangent point toward the
direction of the receiver. Each of these diffraction, launching, and surface propagation
processes is described by appropriate complex diffraction and attenuation coefficients
which are discussed in the literature.[21] In practice, these coefficients are implemented
as computer algorithms which can be used as building blocks for a specific analysis.

In the GTD analysis, the received far- or near-zone field is composed of rays
which are directly incident from the antenna, surface rays, reflected rays, and edge-
diffracted rays. Also *higher-order effects* such as multiple reflections and diffractions

can occur. Several computer models have been implemented to perform the required differential-geometry and diffraction calculations.[22,23,24] It is anticipated that these methods will become more user-oriented and powerful with the advent of increasingly sophisticated computer-graphics systems.

Figure 37-28 shows a model of a missile composed of a conically capped circular cylinder with a circumferential-slot antenna.[25] In this case, the radiated field is due to direct radiation from the slot, surface rays which encircle the cylinder, and diffracted rays from the cone-cylinder and cylinder-end junctions and from the cone tip. The computed results shown in Fig. 37-28 were obtained by employing the solution for a slot on an infinite circular cylinder to represent the direct- and surface-ray fields and by computing the diffracted fields from the junctions and tip as corrections to the infinite-cylinder result. The principal-plane pattern shown is in agreement with measured data, and comparable results have been obtained for the off-principal planes.[26]

An analysis due to Burnside et al.[22] approximates the fuselage as two joined spheroids, and the wings and empennage are modeled by arbitrarily shaped flat plates. In addition, the engines can be modeled by finite circular cylinders. Figure 37-29 shows a result obtained from this analysis of a $\lambda/4$ monopole on a KC-135 aircraft for the roll plane. Reference 23 presents a complete 4π-sr plot for a monopole on a Boeing 737 aircraft. The results, which compare well with measured data, illustrate the accuracy of the GTD analysis for VHF and UHF antennas.

It should be noted that a complete GTD model may not always be required for an engineering assessment of antenna performance. For example, if wing reflection and blockage effects are dominant, one can use a simpler model consisting of two

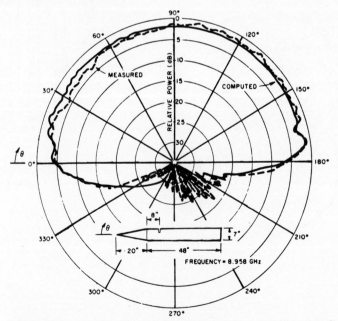

FIG. 37-28 Pattern of a circumferential slot on a conically capped circular cylinder. *(From Ref. 25, © 1972 IEEE.)*

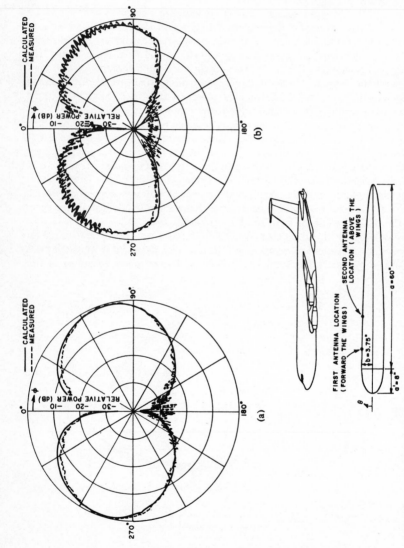

FIG. 37-29 (a) Roll-plane pattern (E_ϕ) for a 1:25-scale model of a KC-135 aircraft with a $\lambda/4$ monopole on the fuselage forward of the wings at a frequency of 34.92 GHz (model frequency). (b) Roll-plane pattern (E_ϕ) for a $\lambda/4$ monopole over the wings. (*From Ref. 22, © 1975 IEEE.*)

submodels: a submodel for an antenna on a finite (or perhaps an infinite) cylindrical fuselage and a flat-plate submodel for the wings.[24] In this case, the antenna-on-cylinder pattern results are used as the antenna illumination for calculation of the wing blockage, reflection, and diffraction.

Antennas for Vertical Polarization

The monopole and its variants are the most commonly used vertically polarized VHF and UHF aircraft antennas. The antenna typically has a tapered airfoil cross section to minimize drag. Simple shunt-stub matching networks are used to obtain a voltage standing-wave ratio (VSWR) below 2:1 from 116 to 156 MHz.

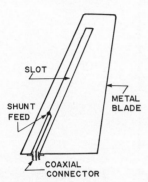

FIG. 37-30 Basic configuration of an all-metal VHF blade antenna.

The basic configuration of an all-metal VHF blade antenna is shown in Fig. 37-30. This antenna employs a shunt-fed slot to excite the blade. The advantage of this configuration is its all-metal construction, which provides lightning protection and eliminates p static. Figure 37-31 shows a low-aerodynamic-drag sleeve monopole for the 225- to 400-MHz UHF communications system. The cross section is diamond-shaped, with a thickness ratio of 5:1. A feature of this antenna[27] is the introduction into the impedance-compensating network of a shunt stub such that its inner conductor serves as a tension member to draw the two halves of the antenna together. In addition to providing mechanical strength, this inner conductor forms a direct-current path from the upper portion of the antenna to the aircraft skin, thereby providing lightning-strike protection.

A number of monopole designs have been developed for installation on a fin in which the top portion of the metal structure has been removed and replaced by a suitable dielectric housing. Various other forms of vertically polarized UHF and VHF radiators have been designed for tail-cap installation. The pickax antenna, consisting of a heavily top-loaded vertical element, has been designed to provide a VSWR of less than 3:1 from 110 to 115 MHz with an overall height of 15½ in (393.7 mm) and a length of 14 in (355.6 mm).

The basic flush-mounted vertically polarized element for fuselage mounting is the annular slot,[28] which can be visualized as the open end of a large-diameter, low-characteristic-impedance coaxial line. As seen from the impedance curve of Fig. 37-32, such a structure becomes an effective radiator only when the circumference of the slot approaches a wavelength. The radiation patterns have their maximum gain in the plane of the slot only for very small slot diameters and yield a horizon gain of zero for a slot diameter of 1.22.[29] This pattern variation is illustrated in Fig. 37-33. For these reasons and to minimize the structural difficulty of installing the antenna in an aircraft, the smallest possible diameter yielding the required bandwidth is desirable. For the 225- to 400-MHz band, the minimum practical diameter, considering construction tolerances and the effect of the airframe on the impedance, is about 24 in (610 mm).

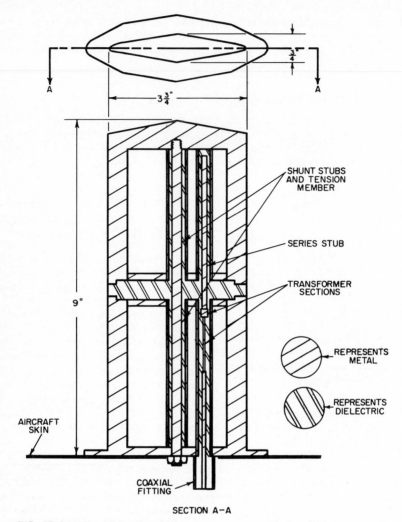

FIG. 37-31 The AT-256A antenna.

A VSWR under 2:1 can be obtained with this diameter and a cavity depth of 4.5 in (114.3 mm). Figure 37-34 shows a design by A. Dorne together with its approximate equivalent circuit. In the equivalent circuit the net aperture impedance of the driven annular slot and the inner parasitic annular slot is shown as a series resistance-capacitance (RC) circuit. The annular region 1, which is coupled to the radiating aperture through the mutual impedance between the two slots, and the annular region 2, which is part of the feed system, are so positioned and proportioned that they store primarily magnetic energy. The inductances associated with the energy storage in these regions are designated as L_1 and L_2 in the equivalent circuit. The parallel-tuned circuit in the equivalent circuit is formed by the shunt capacitance between vane 3 and horizontal

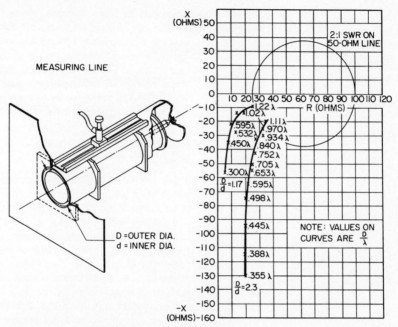

FIG. 37-32 Impedance of an annular-slot antenna.

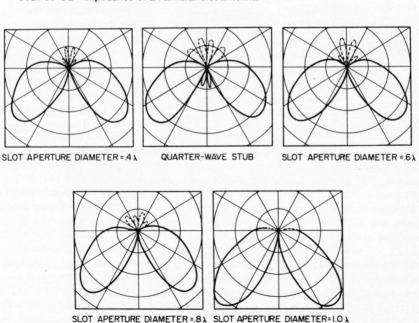

SLOT APERTURE DIAMETER =.4 λ   QUARTER-WAVE STUB   SLOT APERTURE DIAMETER =.6λ

SLOT APERTURE DIAMETER =.8λ SLOT APERTURE DIAMETER=1.0 λ

FIG. 37-33 Radiation patterns of an annular-slot antenna and a quarter-wave stub on a 2½-wavelength-diameter circular ground plane.

disk 4, together with the shunt inductance provided by four conducting posts (5) equally spaced about the periphery of 4, which also serve to support 4 above 3. From this element inward to the coaxial line, the base plate is cambered upward to form a conical transmission-line region of low characteristic impedance. A short additional section of low-impedance line is added external to the cavity to complete the required impedance transformation.

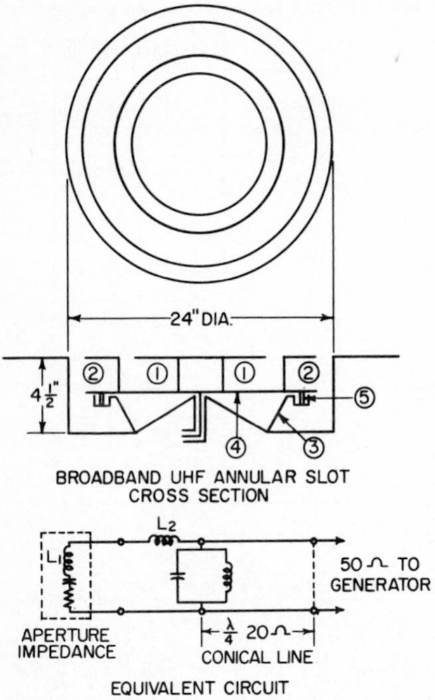

BROADBAND UHF ANNULAR SLOT CROSS SECTION

EQUIVALENT CIRCUIT

FIG. 37-34 Annular-slot antenna for the 225- to 400-MHz band and its equivalent circuit.

Antennas for Horizontal Polarization

There are three basic antenna elements which yield omnidirectional horizontally polarized patterns: the loop, the turnstile, and the longitudinal slot in a vertical cylinder of small diameter. All three are used on aircraft, and all suffer from a basic defect. Because they must be mounted near a horizontal conducting surface of rather large extent (i.e., the top or bottom surface of the aircraft), their gain at angles near the horizon is low. The greater the spacing from the conducting surface, the higher the horizontal gain. For this reason, locations at or near the top of the vertical fin are popular for horizontally polarized applications, particularly for the VHF navigation system (VOR) which covers the 108- to 122-MHz range.

Figure 37-35 shows an E-fed cavity VOR antenna designed into the empennage tip of the L-1011 aircraft.[30] The antenna system consists of a flush-mounted dual E-slot antenna and a stripline feed network mounted on the forward bulkhead of the cavity between the two antenna halves. The antenna halves are mirror-image assemblies consisting of 0.020-in- (0.508-mm-) thick aluminum elements bonded to the inside surface of honeycomb-fiberglass windows. The feed network consists of a ring hybrid–power-divider device with four RF ports fabricated in stripline. The two input ports for connecting to the antenna elements are out-of-phase ports. The two output ports connect to separate VOR preamplifiers. The operating band of the antenna is 108 to 118 MHz with an input VSWR of less than 2:1 referred to 50 Ω. Radiation coverage is essentially omnidirectional in the horizontal plane. The principal polarization is horizontal with a cross-polarization component of more than 18 dB below horizontal polarization. The antenna gain is 3.6 dB below isotropic at band center. Figure 37-36 shows the principal-plane voltage radiation patterns.

The VOR navigation system is particularly vulnerable to the modulation effects of a helicopter rotor since, with this system, angular-position information is contained

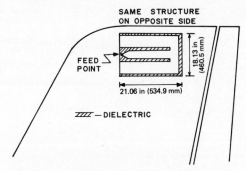

FIG. 37-35 *E*-slot VOR antenna for the L-1011 aircraft. *(From Ref. 30.)*

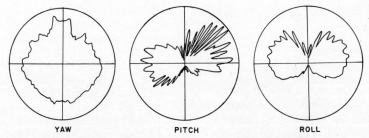

YAW   PITCH   ROLL

FIG. 37-36 Voltage radiation patterns for the *E*-slot VOR antenna on the L-1011 aircraft. *(From Ref. 30.)*

in a 30-Hz modulation tone which corresponds closely to the third harmonic of the fundamental blade-passage frequency on a typical helicopter. In Fig. 37-37 are shown two VOR antenna installations on an H-19 helicopter and the horizontal-plane radiation patterns of each. The fine structure on these patterns shows the peak-to-peak variation in signal amplitude due to passage of the rotor blades. The percentage of modulation of the signal received on the horizontal loop antenna is seen to be lower than that received on the ramshorn antenna. This is due partly to the shielding afforded the loop by the tail boom and partly to the fact that the loop antenna has inherently less response to scattered signals from the blades because its pattern has a null along its axis while the ramshorn antenna receives signals from directly above it very effectively.

Multiple-Antenna Systems

The limitations on omnidirectional coverage due to shadowing by the airframe can be overcome by the use of two or more antennas. A variety of diversity schemes are possible. If the pattern coverage of the two antennas is complementary, or at least approximately so, and if the separation between the two is a large number of wavelengths, the antennas can be connected directly together without the use of diversity techniques.[31] The resultant pattern is characterized by a large number of narrow lobes, with deep nulls only at those angles where the patterns of the two antennas have nearly equal amplitudes, as shown in Fig. 37-38. In view of the dynamic nature of the air-

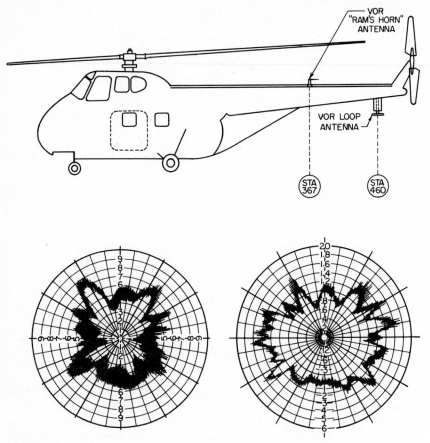

FIG. 37-37 Two VOR-antenna locations on the H-19 helicopter and horizontal-plane radiation patterns.

to-ground communications problem, it is easy to see that the time interval in which a given ground station is within one of the nulls is small, especially if the two patterns overlap in the directions broadside to the aircraft.

37-8 HOMING ANTENNAS

Airborne homing systems permit a pilot to fly directly toward a signal source. Although the navigational data supplied by a homing system are rudimentary, in the sense that only the direction and not the amount of course correction required are provided, the system does not require special cooperative ground equipment. Because of the symmetry of airframes, satisfactory homing-antenna performance can be obtained in frequency ranges in which direction-finder antennas are unusable because of airframe effects.

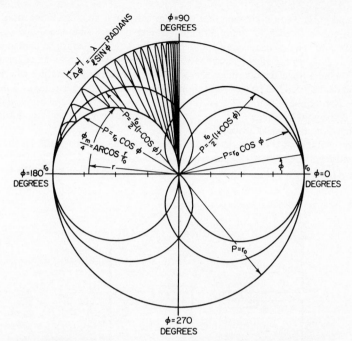

FIG. 37-38 Resultant of two cardioid patterns connected in parallel; r_0 = maximum range, ℓ = spacing between antennas, ϕ_m = total angle for which the range $\geqq r$.

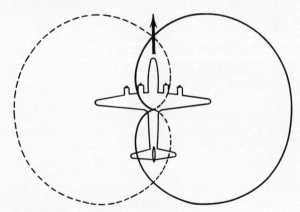

FIG. 37-39 Cardioid homing patterns.

The principle of operation used in airborne homing systems is illustrated in Fig. 37-39. Two patterns, which are symmetrical with respect to the line of flight and which ideally are cardioids, are generated alternately in time either by switching between separate antennas or by alternately feeding a symmetrical antenna array in two modes. The homing system compares the signals received under these two conditions and presents an indication that the pilot is flying a homing course or, if not, in which direction to turn to come onto the homing course. The equisignal condition which leads to the on-course indication can also arise for a reciprocal course leading directly away from the signal source. The pilot can resolve this ambiguity by making an intentional turn after obtaining the on-course indication and noting whether the direction of the required course correction shown by his or her instrument is opposite the direction of the intentional turn (in which case the pilot is operating about the correct equisignal heading) or in the same direction (in which case the pilot is operating about the reciprocal heading).

One of the basic problems in the design of a homing-antenna system is the proper compromise between directivity patterns and system sensitivity. Consider the idealized homing array shown in Fig. 37-40, which is assumed to be in free space. With mutual

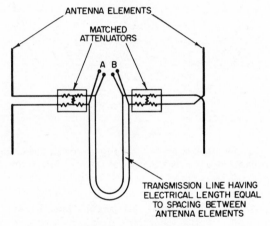

NOTE: FEEDING TERMINAL A GIVES CARDIOID
PATTERN POINTING TO LEFT
FEEDING TERMINAL B GIVES CARDIOID
PATTERN POINTING TO RIGHT

FIG. 37-40   A homing array.

impedance between the two elements neglected, it is readily shown that, by arranging the feed system as indicated, the array will have a null in one direction along the line joining the midpoints of the two elements and that the null direction reverses as the feed is switched between the two sets of feed terminals. Figure 37-41 shows the horizontal-plane patterns for such an array with various element spacings. For small spacings, the pattern becomes a cardioid. The pattern quality, as determined by the pattern slope at the on-course heading (a steep slope is desirable since this will lead to a clear indication of course error due to a small deviation from the homing course) and by the difference in response between the left-hand and right-hand patterns for courses other than the homing course and the reciprocal course, becomes poorer as the spacing is increased. The relative system sensitivity for the homing-course direction, on the other hand, increases as the element spacing is increased, being proportional to the sin $(\pi s/\lambda)$, where s/λ is the spacing between elements in wavelengths. Since most homing systems are designed for relatively wide frequency bands, a compromise element spac-

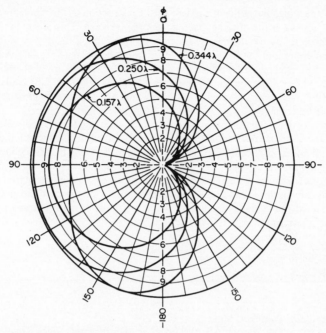

FIG. 37-41 The effect of dipole spacing on the pattern of a homing array.

ing which will avoid severe pattern deterioration at the high-frequency end while retaining sufficient system sensitivity at the low-frequency end is chosen. The sensitivity of the system is increased by increasing the directive gain of the individual elements, and hence broadband elements of resonant size are used wherever possible. At the lower frequencies, when resonant-size elements are structurally too large, the largest elements practicable are used.

Since it is difficult to predict accurately the airframe effects on homing-antenna patterns, the design is usually a step-by-step process in which experimental-model measurement data are used to supplement the free-space antenna-design concepts.

At sufficiently low frequencies, it is possible to use a symmetrical pair of balanced vertical elements as illustrated in Fig. 37-42 to achieve essentially free-space antenna characteristics. The elements are decoupled from currents and charges

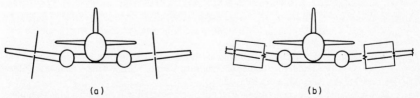

(a)   (b)

FIG. 37-42 Aircraft homing array. (*a*) Vertical elements. (*b*) Resistance-loaded loops.

induced in the wings by virtue of the balanced construction of the elements and from fuselage effects by virtue of their spacing from the fuselage. This technique is limited at low frequencies by the increasing difficulty of maintaining the degree of symmetry between the upper and lower halves of each element necessary to isolate the antenna from airframe resonance effects[32] and at high frequencies because of scattering from the fuselage and empennage. For frequencies as high as 10 MHz, balanced dipoles of this type have been found to provide patterns which are remarkably close to the free-space dipole pattern.

The use of separate antennas to generate the right- and left-hand cardioids is a sound design procedure for frequencies that are above the range for which the homing-array technique is applicable. A homing system using two resistance-loaded loop antennas is illustrated in Fig. 37-42. The balanced configuration is used to isolate the antennas from airframe resonance effects, and the elements may be placed well outboard from the fuselage. Since the low-signal portions of the cardioid patterns are oriented toward the scattering sources on the airframe, this technique can provide clean homing patterns.

REFERENCES

1 *Instruction Book for Type NA-1 Aircraft Navigation System*, Bendix Radio Div., Bendix Aviation Corp., Baltimore, Md.

2 L. R. Mullen, "The Marconi AD 7092 Series of ADF Receivers," *IRE Trans. Aeronaut. Navig. Electron.*, vol. AN-2, no. 4, December 1955.

3 J. T. Bolljahn and R. F. Reese, "Electrically Small Antennas and the Low-Frequency Aircraft Antenna Problem," *IRE Trans. Antennas Propagat.*, vol. AP-1, no. 2, October 1953, pp. 46–54.

4 A. A. Hemphill, "A Magnetic Radio Compass Antenna Having Zero Drag," *IRE Trans. Aeronaut. Navig. Electron.*, vol. AN-2, no. 4, December 1955.

5 H. H. Ward, 3d, "Analysis of the Over-Station Behavior of Aircraft Low-Frequency ADF Systems," *IRE Trans. Aeronaut. Navig. Electron.*, vol. AN-2, no. 4, December 1955, pp. 31–41.

6 P. M. Burdell, "C-141A ADF Sense Antenna Development," Eng. Rep. 5832, Lockheed-Georgia Company, May 28, 1963.

7 R. F. Harrington, *Field Computation by Moment Methods,* The Macmillan Company, New York, 1968.

8 G. L. Burke and A. J. Poggio, "Numerical Electromagnetic Code (NEC)—Method of Moments, Part 1: Program Description—Theory," Tech. Doc. 116, Naval Electronic Systems Command (ELEX 3041), July 18, 1977.

9 Computer Program Librarian, Ohio State University ElectroScience Laboratory, 1320 Kinnear Road, Columbus, Ohio 43212.

10 J. H. Richmond, "A Wire-Grid Model for Scattering by Conducting Bodies," *IEEE Trans. Antennas Propagat.*, vol. AP-14, November 1976, pp. 782–786.

11 J. J. Wang and C. E. Ryan, Jr., "Application of Wire-Grid Modelling to the Design of a Low-Profile Aircraft Antenna," *IEEE Antennas Propagat. Int. Symp. Dig.*, Stanford University, June 20–22, 1977, pp. 222–225.

12 D. L. Knepp and J. Goldhirsh, "Numerical Analysis of Electromagnetic Radiation Properties of Smooth Conducting Bodies of Arbitrary Shape," *IEEE Trans. Antennas Propagat.*, vol. AP-20, May 1972, pp. 383–388.

13 J. R. Mantz and R. F. Harrington, "Radiation and Scattering from Bodies of Revolution," *J. App. Sci. Res.*, vol. 20, 1969, p. 405.

14 Edmund K. Miller and Jeremy A. Landt, "Direct Time Domain Techniques for Transient Radiation and Scattering from Wires," *IEEE Proc.,* vol. 68, no. 11, November 1980, pp. 1396–1423.

15 V. K. Tripp, Engineering Experiment Station, Georgia Institute of Technology, private communication, March 1982.

16 R. L. Tanner, "High Voltage Problems in Flush and External Aircraft HF Antennas," *IRE Trans. Aeronaut. Navig. Electron.,* vol. AN-1, no. 4, December 1954, pp. 16–19.

17 O. C. Boileau, Jr., "An Evaluation of High-Frequency Antennas for a Large Jet Airplane," *IRE Trans. Aeronaut. Navig. Electron.,* vol. AN-3, no. 1, March 1956, pp. 28–32.

18 B. S. Zieg and W. P. Allen, "C-141A HF Antenna Development," Eng. Rep. ER 5101, Lockheed-Georgia Company, July 1963.

19 P. M. Burdell, "C-130 HF Notch Antenna Design and Development," Eng. Rep. LG 82 ER 0036, Lockheed-Georgia Company, March 1982.

20 B. S. Zieg, "JetStar HF Antenna," *Tenth Ann. USAF Antenna R&D Symp.,* University of Illinois, Monticello, Oct. 3–7, 1960.

21 R. C. Hansen (ed.), *Geometric Theory of Diffraction,* selected reprint ser., IEEE Press, Institute of Electrical and Electronics Engineers, New York, 1981.

22 W. D. Burnside, M. C. Gilreath, R. J. Marhefka, and C. L. Yu, "A Study of KC-135 Aircraft Antenna Patterns," *IEEE Trans. Antennas Propagat.,* vol. AP-23, May 1975, pp. 309–316.

23 C. L. Yu, W. D. Burnside, and M. C. Gilreath, "Volumetric Pattern Analysis of Airborne Antennas," *IEEE Trans. Antennas Propagat.,* vol. AP-26, September 1978, pp. 636–641.

24 W. P. Cooke and C. E. Ryan, Jr., "A GTD Computer Algorithm for Computing the Radiation Patterns of Aircraft-Mounted Antennas," *IEEE Antennas Propagat. Int. Symp. Prog. Dig.,* Laval University, Quebec, June 2–6, 1980, pp. 631–634.

25 C. E. Ryan, Jr., "Analysis of Antennas on Finite Circular Cylinders with Conical or Disk End Caps," *IEEE Trans. Antennas Propagat.,* vol. AP-20, July 1972, pp. 474–476.

26 C. E. Ryan, Jr., and R. Luebbers, "Volumetric Patterns of a Circumferential Slot Antenna on a Conically-Capped Finite Circular Cylinder," ElectroScience Lab., Ohio State Univ. Res. Found. Rep. 2805-3, Cont. DAAA21-69-C-0535, 1970.

27 H. Jasik, U.S. Patent 2,700,112.

28 A. Doring, U.S. Patent 2,644,090.

29 A. A. Pistolkors, "Theory of the Circular Diffraction Antenna," *IRE Proc.,* vol. 36, no. 1, January 1948, p. 56.

30 N. R. Ray, "L-1011 VOR Antenna System Design and Development," Eng. Rep. ER-10820, Lockheed-Georgia Company, Aug. 13, 1970.

31 A. G. Kandoian, "The Aircraft Omnidirectional Antenna Problem for UHF Navigational Systems," *Aeronaut. Eng. Rev.,* vol. 12, May 1953, pp. 75–80.

32 P. S. Carter, Jr., "Study of the Feasibility of Airborne HF Direction Finding Antenna Systems," *IRE Trans. Aeronaut. Navig. Electron.,* vol. AN-4, no. 1, March 1957, pp. 19–23.

BIBLIOGRAPHY

Bennett, F. D., P. D. Coleman, and A. S. Meier: "The Design of Broadband Aircraft Antenna Systems," *IRE Proc.,* vol. 33, October 1945, pp. 671–700.

Bolljahn, J. T., and J. V. N. Granger: "The Use of Complementary Slots in Aircraft Antenna Impedance Measurements," *IRE Proc.,* vol. 39, no. 11, November 1951, pp. 1445–1448.

Granger, J. V. N.: "Shunt Excited Flat Plate Antennas with Applications to Aircraft Structures," *IRE Proc.,* vol. 38, no. 3, March 1950, pp. 280–287.

————: "Design Limitations on Aircraft Antenna Systems," *Aeronaut. Eng. Rev.,* vol. 11, no. 5, May 1952, pp. 82–87.

————: "Designing Flush Antennas for High-Speed Aircraft," *Electronics,* vol. 11, March 1954.

———— and T. Morita, "Radio-Frequency Current Distributions on Aircraft Structures," *IRE Proc.,* vol. 39, no. 8, August 1951, pp. 932–938.

Haller, G. L.: "Aircraft Antennas," *IRE Proc.,* vol. 30, no. 8, August 1942, pp. 357–362.

Hurley, H. C., S. R. Anderson, and H. F. Keary: "The Civil Aeronautics Administration VHF Omnirange, *IRE Proc.,* vol. 39, no. 12, December 1951, pp. 1506–1520.

Kees, H., and F. Gehres: "Cavity Aircraft Antennas," *Electronics,* vol. 20, January 1947, pp. 78–79.

Lee, K. S. H., T. K. Liu, and L. Marin: "EMP Response of Aircraft Antennas," *IEEE Trans. Antennas Propagat.,* vol. AP-26, no. 1, January 1978, pp. 94–99.

Moore, E. J.: "Factor of Merit for Aircraft Antenna Systems in the Frequency Range 3–30 Mc," *IRE Trans. Antennas Propagat.,* vol. AP-3, August 1952, pp. 67–73.

Raburn, L. E.: "A VHF-UHF Tail-Cap Antenna," *IRE Proc.,* vol. 39, no. 6, June 1951, pp. 656–659.

————: "Faired-In ADF Antennas," *IRE Conv. Rec.,* vol. 1, part 1, March 1953, pp. 31–38.

Sinclair, G., E. C. Jordan, and E. W. Vaughan: "Measurement of Aircraft Antenna Patterns Using Models," *IRE Proc.,* vol. 35, no. 12, December 1947, pp. 1451–1462.

Tanner, R. L.: "Shunt-Notch-Fed HF Aircraft Antennas," *IRE Trans. Antennas Propagat.,* vol. AP-6, no. 1, January 1958, pp. 35–43.

Chapter 38

Seeker Antennas

James M. Schuchardt

The Bendix Corporation

Dennis J. Kozakoff

Millimeter Wave Technology, Inc.

Maurice M. Hallum III

U.S. Army Missile Command
Redstone Arsenal

38-1 INTRODUCTION

This chapter will focus on antennas forward-mounted in a missile functioning in the role of seeker of target emissions. Such a seeker antenna is a critical part of the entire airborne guidance system, which includes the missile radome, seeker antenna, radio-frequency (RF) receiver, antenna gimbal, autopilot, and airframe. The seeker has several functions: to receive and track target emissions so as to measure line of sight and/or line-of-sight angular rate, to measure closing velocity, and to provide steering commands to the missile autopilot and subsequently to the control surfaces.[1-6] The signals received by the missile-mounted seeker antenna or antennas are thus utilized in a closed-loop servocontrol system to guide the missile to the target.

The seeker RF elements, radome and antenna, initially discriminate in angle through the seeker antenna's pencil beam. This beam can be steered mechanically (the whole antenna moves), electromechanically (an antenna element such as a subreflector moves), or electronically (there is a phased-array movement). In some situations, the antenna is stationary and only forward-looking; thus beam motion occurs only if the entire missile airframe rotates. Fixed seeker-antenna beams are used when a pursuit navigation (or a variant) guidance algorithm is used. The use of movable seeker-antenna beams occurs when a proportional navigation (or a variant) guidance algorithm is used.

38-2 SEEKER-ANTENNA RADIO-FREQUENCY CONSIDERATIONS

Seeker antennas are defined as forward-mounted antennas in a missile or a projectile which function in the role of reception of target emissions. The target emissions result from either target-reflected radar signals, antiradiation homing (ARH) in the case of an emitting target such as ground or airborne radar, or passive (radiometric) emissions which occur because of natural background radiation in accordance with the Planck equation.

Table 38-1 presents a summary of commonly employed seeker-antenna types; detailed performance and design criteria for many of the basic antenna radiator types are found in the appropriate chapters in this handbook. General information on many of these antenna elements relative to various seeker categories is presented below.

Many system factors must also be considered in the antenna-design process. Antenna beamwidth determines whether spatial resolution of multiple targets occurs.[7] The earlier that resolution takes place in a flight path, the more time the missile will have to correct errors induced by tracking the multiple-target centroid. Low sidelobes are important because they decrease electronic-jamming effects. This is particularly important if the missile gets significantly close to the jammer during the engagement. Noise injected in this manner can increase the final miss distance.

The body-fixed antenna configuration confronts designers with unique problems. The beam may be very broad for a large field of view (FOV), or it may be rapidly steered to form a tracking beam.[4] The broad-FOV approach requires that the airframe and the autopilot be more restricted in their responsiveness. The steered-beam approach requires rapid beam forming and signal processing to isolate the missile-body rotational motion properly from the actual target motions.

TABLE 38-1 Missile-Borne Seeker-Antenna Summary

Basic radiator	Body-fixed				Gimballed				Comments
	Monopulse	Sequential lobing	Conscan*	Multimode	Monopulse	Sequential lobing	Conscan	Multimode	
Stripline	X	X	X		X	X			Narrowband, intermediate power, dual-polarization capability
Microstrip	X	X	X	X	X	X		X	Narrowband, intermediate power, dual-polarization capability
Spiral				X	X	X		X	Broadband, circular polarization, low power
Ring array					X	X			Narrowband, high power, linear polarization
Paraboloid			X		X	X	X		High power, dual-polarization capability
Lens			X		X	X	X		High power, dual-polarization capability
Slots	X	X	X	X	X	X			Narrowband, linear polarization
Monopoles	X	X			X	X			Narrowband, linear polarization

*Achieved with a rolling airframe and squinted beams.

It is often necessary to be able to integrate and test key receiver front-end elements and the RF elements of the transmitter as part of the seeker-antenna assembly. For example, it is generally possible to integrate in one package the feed antenna, monopulse comparator, mixer or mixers, local oscillator, and transmitter oscillator.[8]

38-3   IMPACT OF AIRFRAME ON ANTENNA DESIGN

The missile-borne guidance antenna enters the picture as depicted in Fig. 38-1. The figure shows the major functional features present in missiles and projectiles with on-board guidance. For a body-fixed on-board sensor antenna, the body isolation and beam steering are not always present. The implementations of these functions vary from system to system. Factors such as operational altitude and range, targets to be engaged, missile speed, and terrain over which missions are to be performed also influence the design.[9,10]

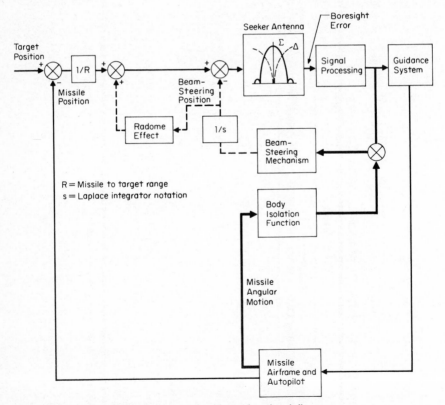

FIG. 38-1   Seeker antenna and overall guidance functional diagram.

The airframe and the autopilot steer the missile according to the guidance-computer commands to minimize the final miss distance at impact. The dynamic behavior of the missile in response to commands consists of attitude changes that cause the antenna to receive through an ever-changing portion of the radome. This, coupled with the natural geometry changes that occur as an engagement takes place, makes the antenna appear to have time-varying directivity and gain properties. The response characteristics of the airframe and the autopilot generate a pseudo-noise influence of the radome-antenna combination.

38-4 SEEKER-ANTENNA MECHANICAL CONSIDERATIONS

Generally, the seeker antenna is of the maximum size possible within the limits of the missile-body diameter in order to maximize antenna gain and minimize the antenna beamwidth. Alternately conformal- and flush-mounted antennas (most often forward-mounted) utilize the entire conical forward area to achieve maximum gain.[11]

Gimballed antenna systems often must mechanically steer not only the antenna structure but also the attendant beam-forming network and critical transmitter and receiver elements. As the mechanical-steering rates become excessive, the conformal-mounted antenna with electronically steered beams must be used.[10-16] When all the seeker-antenna and associated hardware are gimballed together, the use of lightweight materials may achieve the desired results. Lightweight techniques include:[8,17]

1 The use of lightweight honeycomb materials (including low-density dielectric foams) for both filler and structural-load-bearing surfaces. Foam reflectors and waveguide elements including feed horns can be machined and then metallized by vacuum deposition, plating, or similar techniques.

2 The use of stripline techniques. Using multilayer techniques, one can have a layer with microstrip radiating elements, a layer with beam-forming networks, and a layer with beam-switching and signal-control elements such as a diode attenuator and phase shifters. The use of plated-through holes and/or pins to make RF connections can eliminate cables and connectors. Excess substrate can be removed and printed-circuit-board edges plated in lieu of the use of mode-suppression screws.

3 The use of solid dielectric lenses, above Ku band, with no significant weight penalty. Lenses can also be zoned to remove excess material.[18]

38-5 APERTURE TECHNIQUES FOR SEEKER ANTENNAS

Precise control of both amplitude and phase to permit aperture illumination tailoring is necessary for achieving a well-behaved antenna pattern.[19-25] Aperture tapering of rotationally symmetric illuminations starts with the feed itself and continues by varying the energy across the aperture by several methods: corporate-power splitting, aperture-element proportional size or spacing, surface control of the reflector or lens surface, and element thinning. Several of these aspects are discussed below.

Reflector or Lens Antennas

A consideration in the design of monopulse reflector antennas is the four-horn-feed design and the effects of aperture blockage. One criterion posed for an optimum monopulse feed configuration is maximizing the product of the sum times the derivative of the difference error signal.[26] The monopulse horn size for optimum monopulse sensitivity is shown in Fig. 38-2. The impact of optimizing monopulse feed design on antenna performance is illustrated in Fig. 38-3, in which a 5.5-in- (139.7-mm-) diameter aperture is assumed in the calculations. These data illustrate that at millimeter wavelengths (above 30 GHz) the effects of aperture blockage on antenna performance are small for this size of antenna even when a reflector is used.

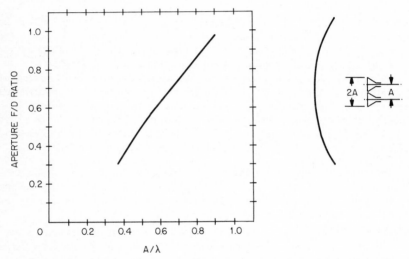

FIG. 38-2 Monopulse antenna horn size or spacing for optimum monopulse sensitivity.

Other criteria, such as the requirement for equal E- and H-plane antenna beamwidths, may enter into the details of monopulse feed design. An example of a monopulse-feed antenna operating near 95 GHz that meets equal-beamwidth requirements is shown in Fig. 38-4.[27] Here, the subaperture dimensions are very small [0.080 by 0.100 in (2.032 by 2.54 mm)], and an electroformed process was required in the feed fabrication.

A wide-angle-scan capability can be achieved by using a twist-reflector concept with a rotatable planar mirror.[28,29] In this configuration, the forward nose of the missile can be approximately paraboloidal-shaped, or the paraboloid can be located inside a higher-fineness-ratio radome. This paraboloid is composed of horizontal metal strips. The movable planar twist reflector uses 45°-oriented strips ¼λ above the planar metallic reflector. (Grids may also be used to combine widely separated frequencies to permit dual-band operation.[30])

Beam steering is obtained by moving the planar mirror. The steering is enhanced because for every 1° that the mirror moves the beam moves 2°. By using a parallel-

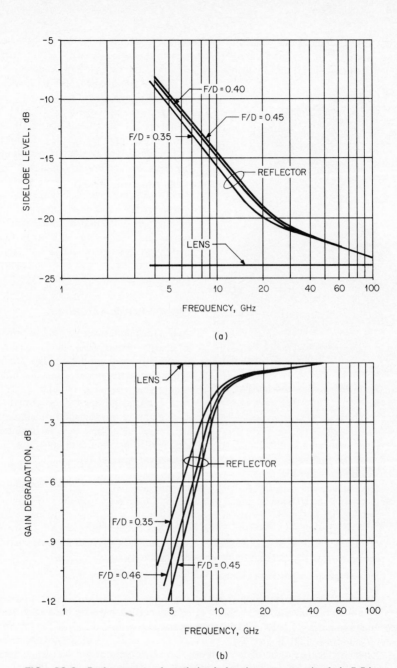

FIG. 38-3 Performance of optimized four-horn monopulse-fed 5.5-in-(139.7-mm-) diameter apertures. (*a*) Sidelobe level. (*b*) Gain degradation.

ogram gimbal to move the planar mirror, beam accelerations of $20,000°/s^2$ have been demonstrated.[29] In this situation a 42° beam scan was achieved with 21° of mirror motion, and less than 2 dB of sidelobe degradation of the sum pattern occurred.

By using a corrugated horn with a rotationally symmetric feed pattern providing a −17-dB edge taper (capable of achieving a −30-dB sidelobe level), one can further taper by varying the inner- and outer-surface contour of a collimating lens to achieve a circular Taylor amplitude distribution ($\bar{n}$ = 7) with a −40-dB sidelobe level.[18] Practically, one can make the lens out of a material having an ε_r = 6.45 (titania-loaded polystyrene) and follow up with a surface-matching layer made out of a material having an ε_r = 2.54 (Rexolite-polystyrene).

FIG. 38-4 Monopulse-feed antenna (95-GHz operation). (*After Ref. 27.*)

Planar-Array Antennas

Array techniques can be used to provide a nearly planar (flat-plate) aperture. Commonly used are arrays that provide symmetric patterns and offer a maximum use of the available circular aperture with reduced grating-lobe potential. In such an array, higher aperture efficiency with low sidelobes can be achieved because of reduced aperture blockage and spillover. In monopulse applications, it is noted that the ring array contains no central element and exhibits quadrantal symmetry, as indicated in Fig. 38-5.

The choices in a ring-array design include the number of rings and the number of elements in each ring and the feed network.[31,32] Constraints include:

1 Minimum distance of the outer elements to the antenna edge

2 Minimum spacing of the elements as impacted by excess mutual coupling (between elements and quadrants and feed geometry)

3 Desired radiation pattern (as impacted by the density tapering)

Optimization of small ring arrays (having diameters $<10\lambda$) is often carried out heuristically or nonlinearly. Examples of array geometries that have been analytically determined to have −17- to −20-dB sidelobes are given in Table 38-2 for the array geometry shown in Fig. 38-5a.

Array elements can be of many types.[33-40] A two-dimensional array of waveguide slots (Fig. 38-5b) excited by a network of parallel waveguides forming the antenna structural supports has been used.[39] Waveguide slot arrays have demonstrated low cross-polarized response as well. The bandwidth of a waveguide slot array is inversely proportional to the size of the array. A variety of outer contours can be utilized to conform to the space available.

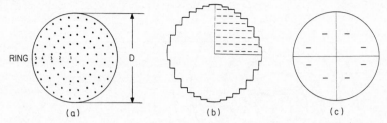

FIG. 38-5   Monopulse planar-array geometries. (*a*) Five-ring monopulse array.
(*b*) Waveguide slot array. (*c*) Thinned-slot (image) ring array.

TABLE 38-2   Data on Five-Ring Monopulse Uniformly Excited Array*

Ring number	Case A† Directivity = 27.6 dB Maximum sidelobe level = −20 dB Radius, λ	Case B† Directivity = 28.1 dB Maximum sidelobe level = −17 db Radius, λ
r_1	0.45	0.50
r_2	1.25	1.33
r_3	2.08	2.16
r_4	3.06	2.98
r_5	3.72	3.72

*After Ref. 31.

†The maximum outer diameter = 8.16λ for both cases.

Techniques for thinning these types of arrays reduce complexity.[41] By using an image-element approach, a reduction of the number of elements by over 90 percent is possible (Fig. 38-5 *c*). This method also readily allows for integration of the monopulse comparator.

Multimode and Single-Mode Spirals

The multimode spiral antenna is a very broadband, broad-beamed, circularly polarized antenna which is amenable to ARH applications.[42] Such an antenna can also be synthesized by using a circular array.[43] The circuit to resolve monopulse sum-and-difference patterns is shown in Fig. 38-6; typical angular coverage is from ±30 to 40°. The rather wide instantaneous FOV makes these antennas attractive for body-fixed applications.

The printed-circuit construction of four-arm spirals limits their use to low-power applications. Thus, this type of antenna is almost always used for passive (nontransmitting) applications. Loading techniques can be employed to reduce size, permitting operation at low frequencies.[44] Current fabrication technology permits high-frequency operation to above 40 GHz. Single-mode spirals are also very useful as elements in small arrays.

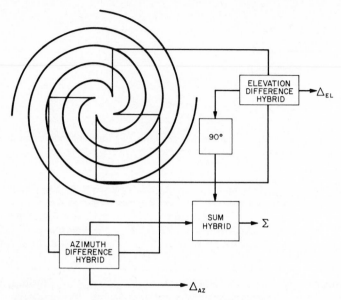

FIG. 38-6   Monopulse circuit for a four-arm spiral antenna.

38-6   SMALL ARRAYS FOR SEEKER ANTENNAS

Arrays of two, three, four, and five elements can be used for seeker antennas in small interferometers. These types of arrays are usually fixed-mounted (nongimballed) and used when pursuit navigation algorithms are suitable. With the addition of a gyroscope to sense missile angular motion, line-of-sight data can be derived.[45]

Good sum-and-difference or phase monopulse angle-tracking performance can be achieved when the array elements, the array beamwidth, and the array element spacing are properly chosen. Broadband performance is often achievable by using antenna elements such as spirals or log-periodic antennas.

The receiving arrays as shown in Fig. 38-7 are either two-element (rolling missile) or four-element (stabilized missile). The third or fifth element can be used for transmission. The equations for the receiving voltage pattern are

$$\Sigma_4 = \left[1 + \cos\left(\frac{\pi d}{\lambda} \sin\theta \right) \right] f(\theta)$$

$$\Sigma_2 = \left[\cos\left(\frac{\pi d}{\lambda} \sin\theta \right) \right] f(\theta)$$

$$\Delta_2 = \left[\sin\left(\frac{\pi d}{\lambda} \sin\theta \right) \right] f(\theta)$$

where $f(\theta)$ = element pattern [$\cos^m(\theta)$ is often used.]
      d = center-to-center element spacing
      λ = operating wavelength

These equations can be used to determine either the principal-plane (0°) or the intercardinal-plane (45°) pattern. Figure 38-8 indicates the appropriate array dimen-

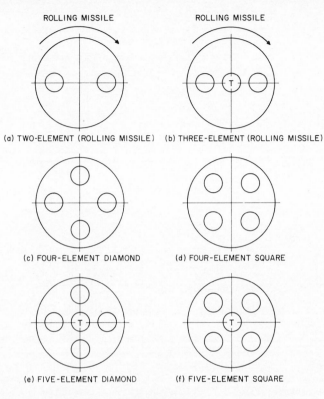

T = TRANSMIT ONLY.

FIG. 38-7 Small-array geometries.

sions and equations to use for each case. Also shown are the two comparator networks used for either diamond or square arrays. Small arrays can be used by themselves, or they can illuminate apertures such as reflectors or lenses. Control of the array element pattern is important in reducing energy spillover beyond the difference-pattern lobes. Often one can achieve this control by using end-fire techniques with the elements. For example, dielectric rods can be used for waveguide, printed-circuit, and spiral elements. An example of this approach is shown in Fig. 38-9.

A four-element array can serve as an RF seeker by utilizing a phase processor. The effects of system errors for such an array on antenna patterns, S curves, and angular sensitivity as ascertained analytically can be used to form a catalog of results that can be employed to diagnose fabrication errors or tolerances based on measured data.[46]

38-7 RADOME EFFECTS

The introduction of a protective radome over a seeker antenna results in an apparent boresight error (BSE) because of the wavefront phase modification over the seeker-

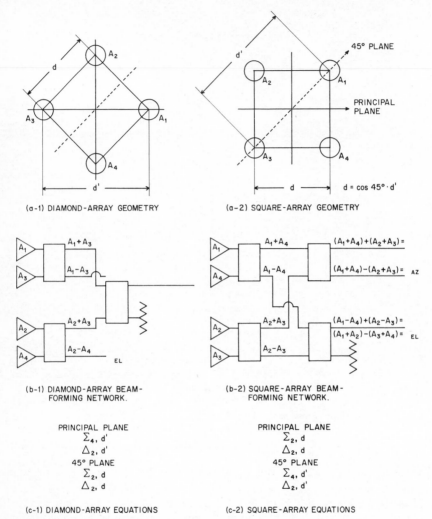

(a-1) DIAMOND-ARRAY GEOMETRY

(a-2) SQUARE-ARRAY GEOMETRY

$d = \cos 45° \cdot d'$

(b-1) DIAMOND-ARRAY BEAM-
FORMING NETWORK.

(b-2) SQUARE-ARRAY BEAM-
FORMING NETWORK.

PRINCIPAL PLANE
Σ_4, d'
Δ_2, d'
45° PLANE
Σ_2, d
Δ_2, d

PRINCIPAL PLANE
Σ_2, d
Δ_2, d
45° PLANE
Σ_4, d'
Δ_2, d'

(c-1) DIAMOND-ARRAY EQUATIONS

(c-2) SQUARE-ARRAY EQUATIONS

FIG. 38-8 Square or diamond interferometer array parameters.

antenna aperture. (Radome-analysis methods are discussed in greater detail in Chap. 44.) These effects need to be carefully considered in the selection of a radome design since they impact seeker-system performance.[9,47-49]

In the case of pursuit guidance, the miss distance is proportional to the BSE. Conversely, in the case of proportional guidance, it is the boresight-error slope (BSES) that more directly impacts miss distance. The relationship between BSE or BSES and miss distance is not straightforward but generally is quantified only by a hardware-in-the-loop (HWIL)[50,51] or computer simulation that takes into account a variety of scenarios and flight conditions. A BSE requirement of better than 10 mrad (or a BSES

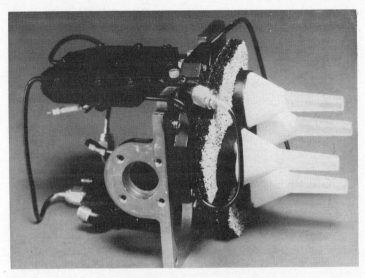

FIG. 38-9 Four-element seeker array using dielectric rods and spiral antennas. *(Photo courtesy Eaton Corporation, AIL Division.)*

requirement better than 0.05 deg/deg and maximum transmission loss of 1.0 dB are typical for many radome types.

The method most often employed for the evaluation of radome effects on seeker-antenna performance utilizes open-loop testing in a suitable anechoic-chamber facility.[52] This requires a precalibration of the monopulse error-channel sensitivity in volts per degree without the radome and subsequently measuring the error-channel outputs with the radome over the seeker antenna.

Broadband ARH antennas generally require the use of multilayer radome walls to obtain a required radome bandwidth commensurate with antenna performance. Figure 38-10 illustrates hemispherical-radome performance for half-wave, full-wave, A and B sandwich walls. Half-wave radomes generally have a useful bandwidth on the order of 5 to 20 percent, depending on dielectric constant and radome-fineness ratio. Higher-order wall radomes have considerably narrower useful bandwidths and thus are limited to narrowband-antenna applications. Figure 38-10 also demonstrates transmission loss versus frequency for a half-wave wall radome when radome-fineness ratio is taken as a parameter.

38-8 EVALUATION OF SEEKER ANTENNAS

Seeker-antenna testing proceeds from conventional methods utilizing anechoic-chamber techniques with the antenna alone.[53-57] Next, open-loop testing including critical elements of the seeker electronics (S curves) is most often performed. Ultimately a variety of simulations are used to ascertain the seeker-antenna performance in both benign and complex environments.

The many nonlinear elements in the guided missile make complete closed-form

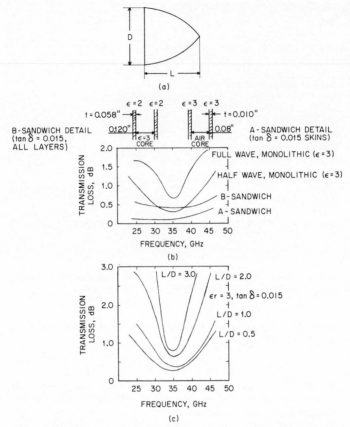

FIG. 38-10 Missile-radome transmission data. (*a*) Radome geometry.
(*b*) Hemispherical-radome (*L / D* = 0.5) performance for a low-velocity
(<Mach 1) missile. (*c*) Monolithic half-wave radome performance for
various fineness (*L / D*) ratios.

analytical simulations unsuitable for final analyses. The precise simulation of these
elements is also difficult. The final analyses are best done with an HWIL simulation.[49-51] As many hardware items as possible should be inserted, leaving only the
propulsion, aerodynamics, and instrument feedback to be simulated analytically.

There are several concepts for implementing a simulation with seeker hardware.
Placing the seeker in an anechoic chamber with the target of appropriate characteristics moved about according to the engagement geometry is one method. Flying a
target with the seeker on an orientation table is another.

The HWIL simulation serves as a tool not only for statistical performance analysis but for provision of validity information for an improved all-analytical model of
the system. The HWIL method provides the only valid present means of evaluating
the interaction of the radome-antenna combination in a missile-guidance-seeker
environment.

REFERENCES

1   J. F. Gulick, "Overview of Missile Guidance," *IEEE Eascon Rec.,* September 1978, pp. 194–198.

2   W. Kelley, "Homing Missile Guidance—A Survey of Classical and Modern Techniques," *Southcon Tech. Prog.,* January 1981.

3   C. F. Price et al., "Performance Evaluation of Homing Guidance Laws for Tactical Missiles," final rep. on N00014-69-C-0391 (AD761626), January 1973.

4   P. G. Savage, "A Strapdown Phased Array Radar Tracker Loop Concept for a Radar Homing Missile," *AIAA Guid. Cont. Flight Conf.,* August 1969, pp. 1–8.

5   G. M. Siouris, "Noise Processor in a Homing Radar Seeker," *NTZ J.,* July 1973, pp. 321–323.

6   G. L. Slater and W. R. Wells, "Optimal Evasive Tactics against a Proportional Navigation Missile with Time Delay," *J. Spacecr. Rockets,* May 1973, pp. 309–313.

7   H. G. Oltman and M. E. Beebe, "Millimeter Wave Seeker Technology," *AIAA Guid. Cont. Conf.,* August 1978, pp. 148–158.

8   C. R. Seashore et al., "MM-Wave Radar and Radiometer Sensors for Guidance Systems," *Microwave J.,* August 1979, pp. 41–59.

9   P. Garnell and D. J. East, *Guided Weapons Control Systems,* Pergamon Press, London, 1977.

10   R. A. Chervolk, "Coherent Active Seeker Guidance Concepts for Tactical Missiles," *IEEE Eascon Rec.,* September 1978, pp. 199–202.

11   R. C. Hansen (ed.), *Conformal Antenna Array Design Handbook,* AIR310E, U.S. Navy, Washington, September 1981.

12   P. C. Bargeliotes and A. F. Seaton, "Conformal Phased Array Breadboard," final rep. on N0019-76-C-0495, ADA038-350.

13   T. W. Bazire et al., "A Printed Antenna/Radome Assembly (RADANT) for Airborne Doppler Navigation Radar," *Fourth Europ. Microwave Conf. Proc.,* Montreux, September 1974, pp. 494–498.

14   H. S. Jones, Jr., "Some Novel Design Techniques for Conformal Antennas," *IEE Conf. Antennas Propagat.,* London, Nov. 28–30, 1978, pp. 448–452.

15   Naval Air Systems Command, "Conformal Antennas Research Program Review and Workshop," AD-A015-630, April 1975.

16   A. T. Villeneuve et al., "Wide-Angle Scanning of Linear Arrays Located on Cones," *IEEE Trans. Antennas Propagat.,* vol. AP-22, January 1974, pp. 97–103.

17   J. S. Yee and W. J. Furlong, "An Extremely Lightweight Electronically Steerable Microstrip Phased Array Antenna," *IEEE Antennas Propagat. Int. Symp. Dig.,* May 1978, pp. 170–173.

18   D. K. Waineo, "Lens Design for Arbitrary Aperture Illumination," *IEEE Antennas Propagat. Int. Symp. Dig.,* October 1976, pp. 476–479.

19   E. T. Bayliss, "Design of Monopulse Antenna Difference Patterns with Low Sidelobes," *Bell Syst. Tech. J.,* May–June 1968, pp. 623–650.

20   L. J. Du and D. J. Scheer, "Microwave Lens Design for a Conical Horn Antenna," *Microwave J.,* September 1976, pp. 49–52.

21   P. W. Hannan, "Optimum Feeds for All Three Modes of a Monopulse Antenna (Parts I and II)," *IRE Trans. Antennas Propagat.,* vol. AP-9, September 1961, pp. 444–464.

22   R. W. Kreutel, "Off-Axis Characteristics of the Hyperboloidal Lens Antenna," *IEEE Antennas Propagat. Int. Symp. Dig.,* May 1978, pp. 231–234.

23   S. Pizette and J. Toth, "Monopulse Networks for a Multielement Feed with Independent Control of the Three Monopulse Modes," *IEEE Microwave Theory Tech. Int. Symp. Dig.,* April–May 1979, pp. 456–458.

24   R. L. Sak and A. Sarremejean, "The Performance of the Square and Conical Horns as Monopulse Feeds in the Millimetric Band," *IEE Ninth Europ. Microwave Conf. Proc.,* September 1979, pp. 191–195.

25 P. A. Watson and S. I. Ghobrial, "Off-Axis Polarization Characteristics of Cassegrainian and Front-Fed Antennas," *IEEE Trans. Antennas Propagat.,* vol. AP-20, November 1972, pp. 691–698.

26 D. R. Rhodes, *Introduction to Monopulse,* McGraw-Hill Book Company, New York, 1959.

27 D. J. Kozakoff and P. P. Britt, "A 94.5 GHz Variable Beamwidth Zoned Lens Monopulse Antenna," *IEEE Southeastcon Proc.,* April 1980, pp. 65–68.

28 E. O. Houseman, Jr., "A Millimeter Wave Polarization Twist Antenna," *IEEE Antennas Propagat. Int. Symp. Dig.,* May 1978, pp. 51–53.

29 D. K. Waineo and J. F. Koneczny, "Millimeter Wave Monopulse Antenna with Rapid Scan Capability," *IEEE Antennas Propagat. Int. Symp. Dig.,* June 1979, pp. 477–480.

30 L. Goldstone, "Dual Frequency Antenna Locates MM Wave Transmitters," *Microwave Syst. N.,* October 1981, pp. 69–74.

31 D. A. Huebner, "Design and Optimization of Small Concentric Ring Arrays," *IEEE Antennas Propagat. Int. Symp. Dig.,* May 1978, pp. 455–458.

32 A. R. Lopez, "Monopulse Networks for Series-Feeding an Array Antenna," *IEEE Trans. Antennas Propagat.,* vol. AP-16, July 1968, pp. 68–71.

33 S. W. Bartley and D. A. Huebner, "A Dual Beam Low Sidelobe Microstrip Array," *IEEE Antennas Propagat. Int. Symp. Dig.,* June 1979, pp. 130–133.

34 C. W. Garven et al., "Missile Base Mounted Microstrip Antennas," *IEEE Trans. Antennas Propagat.,* vol. AP-25, September 1977, pp. 604, 616.

35 C. S. Malagisi, "Microstrip Disc Element Reflect Array," *IEEE Eascon Rec.,* September 1978, pp. 186–192.

36 S. Nishimura et al., "Franklin-Type Microstrip Line Antenna," *IEEE Antennas Propagat. Int. Symp. Dig.,* June 1979, pp. 134–137.

37 P. K. Park and R. S. Elliott, "Design of Collinear Longitudinal Slot Arrays Fed by Boxed Stripline," *IEEE Trans. Antennas Propagat.,* vol. AP-29, January 1981, pp. 135–140.

38 *Flat Plate Antennas,* Rantec Div., Emerson Electric Co., June 1969.

39 G. G. Sanford and L. Klein, "Increasing the Beamwidth of a Microstrip Radiating Element," *IEEE Antennas Propagat. Int. Symp. Dig.,* June 1979, pp. 126–129.

40 D. H. Schaubert et al., "Microstrip Antennas with Frequency Agility and Polarization Diversity," *IEEE Trans. Antennas Propagat.,* vol. AP-29, January 1981, pp. 118–123.

41 W. H. Sasser, "A Highly Thinned Array Using the Image Element," *IEEE Antennas Propagat. Int. Symp. Dig.,* June 1980, pp. 150–153.

42 J. D. Dyson, "Multimode Logarithmic Spiral Antennas," *Nat. Electron. Conf.,* vol. 17, October 1961, pp. 206–213.

43 J. M. Schuchardt and W. O. Purcell, "A Broadband Direction Finding Receiving System," *Martin Marietta Interdiv. Antenna Symp.,* August 1967, pp. 1–14.

44 T. E. Morgan, "Reduced Size Spiral Antenna," *IEE Ninth Europ. Microwave Conf. Proc.,* September 1979, pp. 181–185.

45 E. R. Feagler, "The Interferometer as a Sensor for Missile Guidance," *IEEE Eascon Rec.,* September 1978, pp. 203–210.

46 L. L. Webb, "Analysis of Field-of-View versus Accuracy for a Microwave Monopulse," *IEEE Southeascon Proc.,* April 1973, pp. 63–66.

47 G. Marales, "Simulation of Electrical Design of Streamlined Radomes," *AIAA Summer Computer Simulation Conf.,* Toronto, July 1979, pp. 353–354.

48 A. Ossin et al., "Millimeter Wavelength Radomes," Rep. AFML-TR-79-4076, Wright-Patterson Air Force Base, Dayton, Ohio, July 1979.

49 K. Siwak et al., "Boresight Errors Induced by Missile Radomes," *IEEE Trans. Antennas Propagat.,* vol. AP-27, November 1979, pp. 832–841.

50 R. F. Russell and S. Massey, "Radio Frequency System Simulator," *AIAA Guidance Cont. Conf.,* August 1972, pp. 72–861.

51 D. W. Sutherlin and C. L. Phillips, "Hardware-in-the-Loop Simulation of Antiradiation Missiles," *IEEE Southeastcon Proc.,* Clemson, S.C., April 1976, pp. 43–45.

52 J. M. Schuchardt et al., "Automated Radome Performance Evaluation in the RFSS Facility at MICOM," *Proc. 15th EM Windows Symp.,* Atlanta, June 1980.

53 R. C. Hansen, "Effect of Field Amplitude Taper on Measured Antenna Gain and Side-lobes," *Electron. Lett.,* April 1981, pp. 12–13.

54 R. C. Hansen et al., "Sidewall Induced Boresight Error in an Anechoic Chamber," *IEEE Trans. Aerosp. Electron. Syst.,* vol. AES-7, November 1971, pp. 1211–1213.

55 D. J. Kaplan et al., "Rapid Planar Near Field Measurements," *Microwave J.,* January 1979, pp. 75–77.

56 A. S. Thompson, "Boresight Shift in Phase Sensing Monopulse Antennas Due to Reflected Signals," *Microwave J.,* May 1966, pp. 47–48.

57 R. D. Monroe and P. C. Gregory, "Missile Radar Guidance Laboratory," *Range Instrumentation–Weapons System Testing and Related Techniques,* AGARD-AG-219, vol. 219, 1976.

Chapter 39

Direction-Finding Antennas and Systems

Hugh D. Kennedy
William Wharton

Technology for Communications International

39-1 INTRODUCTION

Radio Direction-Finding Systems

A radio direction-finding (DF) system is basically an antenna-receiver combination arranged to determine the azimuth of a distant emitter. In practice, however, the objective of most DF systems is to determine the location of the emitter.[1] Virtually all DF systems derive emitter location from an initial determination of the arrival angle of the received signal. Figure 39-1*a* shows how the location of the emitter is found if azimuth angles are measured at two DF stations connected by a communication link and separated by a distance that is comparable with the distance to the emitter. The determination of emitter location by using azimuth angles measured at two or more DF stations is known as *horizontal or azimuth triangulation.*

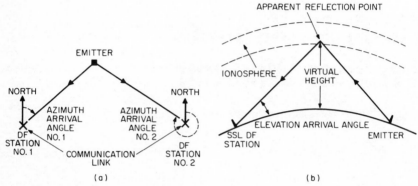

FIG. 39-1 Emitter location by triangulation. (*a*) Horizontal (azimuth) triangulation. (*b*) Vertical triangulation (single-station location).

In the high-frequency (HF) band (approximately 2 to 30 MHz), emitter location by using only a single DF station is possible. Such an arrangement is known as a *single-station-location (SSL) system.*[2] A DF-SSL system can only be used when the signal arrives at the DF station after refraction through the ionosphere (i.e., via sky-wave propagation), as illustrated in Fig. 39-1*b*. The DF-SSL system must be able to measure both the azimuth and the elevation angles[3] of the signal arriving at the DF station. Furthermore, the height of the ionosphere must be either known or determined. The measured elevation angle, in conjunction with knowledge of the height of the ionosphere, enables the distance to the emitter to be established. This process is known as *vertical triangulation.* Emitter location is then calculated by using azimuth and distance.

Applications of Direction-Finding Systems

There are three principal applications for DF systems which influence the details of their design:

1 DF systems that are designed to determine the unknown location of an emitter. Such systems may be fixed or movable.

2 Navigation systems designed to determine the location of the DF system itself with respect to emitters of known location. The DF system is on a moving vehicle (ship or aircraft), and azimuth angles are measured to two or more emitters at the same time or at successive times on the same emitter. The determination of the position of the vehicle is the complement of the horizontal triangulation process illustrated in Fig. 39-1 *a*.

3 Homing systems that are designed to guide a vehicle carrying a DF system toward an emitter which may be either a beacon of known location or an emitter of unknown location.

System Approach to Direction Finding

The essential components of any DF system are shown in Fig. 39-2. They comprise:

1 An antenna system to collect energy from the arriving signal

2 A receiving system to measure the response of the antenna system to the arriving signal

3 A processor to derive the required DF information (for example, azimuth and elevation angles and emitter location) from the output of the receiver

4 An output device to present the required DF information in a form convenient to the user

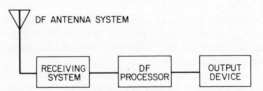

FIG. 39-2   Essential components of a DF system.

In the past, each component of a DF system was often regarded as a separate design problem. Attention was focused mainly on the antenna because in general it imposed the major constraints on performance. Some form of symmetry was usually required, and the choice of antenna type was usually quite limited. However, the advent of small high-speed digital computers has meant that DF-system designers now have more freedom in their choice of DF antennas and that full use can now be made of antenna-system responses. That is, most older systems made use of either amplitude response or phase comparison but not both. Furthermore, most older systems were restricted to the use of only one linear polarization. The employment of real-time signal-processing techniques now makes it possible to use both phase and amplitude responses, to use antennas of any polarization, and to respond to signals of any polarization. It is therefore desirable that a DF system for all but the simplest applications be designed as a complete integrated system to take full advantage of these possibilities. This chapter will discuss antenna-design principles in the context of complete DF systems.

39-2 RADIO-WAVE PROPAGATION

Propagation Characteristics

The performance of a DF system depends on the nature of the signals arriving at the DF antenna. In turn, the nature of these signals depends on the mode of propagation, which will fall into one of two broad categories:

1 Signals that arrive directly at the DF antenna from the emitter with near-zero elevation angle. Included in this category are low-frequency (LF), medium-frequency (MF), and HF surface waves which will be vertically polarized and very-high-frequency (VHF), ultrahigh-frequency (UHF), and microwave direct-wave signals which may be of any polarization.

2 HF signals that are refracted from the ionosphere and arrive at the DF antenna with an elevation angle that may vary from a few degrees to nearly 90°. Irrespective of their transmitted polarization, these signals will be elliptically polarized upon arrival at the DF antenna system. That is, the two polarization components will vary independently with time.[4,5]

Whether the signals arriving at the DF antenna fall into category 1 or category 2 or include both categories depends primarily on the frequency band and secondarily on the daily and seasonal variations of the ionosphere.

Frequencies in the Low- and Medium-Frequency Bands

In these bands, during the day only a vertically polarized surface wave is propagated since sky waves are absorbed in the D layer of the ionosphere. However, at night the D layer disappears, and signals are refracted from the ionosphere so that sky-wave signals can arrive at the DF antenna.

The surface wave is attenuated with distance so that, depending on time of day, transmitter power, ground conductivity, and distance of the DF site from the emitter, three types of received signal may exist:

1 Near the emitter the vertically polarized surface wave will predominate because the sky wave will always be small with respect to it.

2 With increasing distance, the surface wave becomes attenuated. At night there will be a zone (known as the *fading zone*) in which sky wave and surface wave are comparable in magnitude. Fading occurs owing to interaction between surface and sky waves.

3 At distances beyond the fading zone the surface wave becomes highly attenuated so that the only receivable signal at such distances will be the night-time sky wave.

Magnitude of the surface wave (which primarily determines the distance of the fading zone from the emitter) increases with increasing ground conductivity and decreasing frequency. Thus, in the LF and lower MF bands the signal available at the DF site is, in most cases, predominantly a surface wave arriving at zero elevation angle. But in the upper MF band the signal may be a combination of surface and sky waves or, alternatively, a sky wave alone, especially at night.

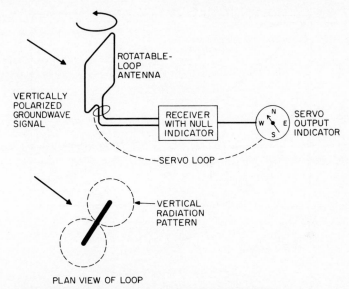

PLAN VIEW OF LOOP

FIG. 39-3 Sample of a system using a rotatable loop.

A simple rotating-loop arrangement, as shown in Fıg. 39-3, has an amplitude response to vertically polarized signals shaped like a figure of eight. During daylight hours, when only a vertically polarized field is present, the two nulls will give an accurate determination of bearing if means are provided to remove the 180° ambiguity. However, the amplitude response of the loop to horizontally polarized signals is also a figure of eight, but rotated 90° in space in relation to the vertically polarized response. At night, in the presence of a sky wave, which will have both vertically and horizontally polarized field components, there will be no nulls or maxima with a fixed relationship to the loop orientation. Consequently, an accurate measurement of bearing is impossible. This phenomenon is referred to as *night effect* or *polarization error.*[6] It illustrates a fundamental difficulty that arises when sky waves are received by DF systems that depend only on the amplitude of the signal and when the antenna system is responsive to horizontal polarization.

Frequencies in the High-Frequency Band

In this band the attenuation of the surface wave over land is high, as is illustrated in Fig. 39-4. In many cases only a sky wave arrives at the DF site. However, over seawater the surface and sky waves will be comparable in magnitude at much greater distances.

At distances up to about 2000 km the sky-wave signal may arrive via one refraction from the ionosphere (a *one-hop mode*). But at any except the very shortest distances multihop signals, in addition to the one-hop signal, can occur. The arriving signal will then consist of a number of components of different amplitudes, of random phase, and with different elevation angles. As these components vary with time in both amplitude and phase, the arriving wavefront will be distorted and the azimuth and

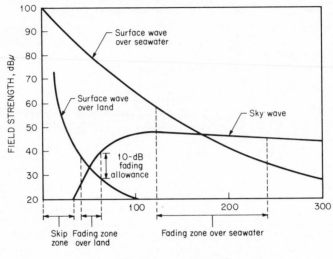

DISTANCE FROM EMITTER, km

FIG. 39-4   Approximate variation of HF field strength with distance for 1-kW emitter power.

elevation angles of the signal will vary.[7] When DF measurements on such complex signals are necessary, simple amplitude-only DF systems are inadequate. However, systems employing digital computers and using suitable analytical programs can provide accurate measurements of bearing, elevation angle, and emitter location under these conditions (see Sec. 39-4).

Frequencies in the Very-High-Frequency and Ultrahigh-Frequency Bands

At frequencies above about 50 MHz the ionosphere becomes transparent to radio transmission. Propagation is by direct wave, and polarization is determined by the transmitting antenna. It may be vertical or horizontal or may contain both vertical and horizontal components. Furthermore, range is limited because the signal is rapidly attenuated beyond the line of sight.

39-3   DIRECTION-FINDING-SYSTEM PLANNING

Planning a DF system (which may comprise a single station or a network of stations) must take into account a number of factors of major importance:

- Geographic area in which the emitters are located in relation to the DF stations.
- Frequencies of the emitters.
- Number of DF stations necessary to ensure adequate accuracy of measurement.

- Response time (the minimum time in which a measurement must or can be made).
- Physical limitations of the system. These may include the space available for antennas, the need (or not) for a movable system, and limitations on equipment size, weight, and complexity.

Geographic Area and Frequency Coverage

The geographic area in which emitters are located in relation to the available locations of the DF stations will determine the maximum and minimum distances over which signals must be received. These maximum and minimum distances, together with the frequency range in which the emitters operate, enable the most likely propagation modes to be identified so that the optimum DF system can be chosen.

Emitter-Location Systems   If the area in which the emitters are situated is known and will remain fixed and if all the emitters are within propagation range of the proposed DF sites, a fixed-site emitter-location DF system will be chosen. Fixed-site DF systems have the advantage that it is possible to use electrically large antennas and antenna arrays, which are faster and more accurate than small-aperture systems. Furthermore, electrically large antenna elements have low noise figures, enabling DF measurements to be made on relatively low-field-strength signals unless the received signal-to-noise ratio is limited by external noise.

On the other hand, the area of the emitters may vary or be located so that it is not possible to receive adequate signals from the emitters at fixed DF sites. In this case it may be necessary to use a movable DF system with electrically small antenna elements. (But movable DF systems can employ large-aperture arrays. See Sec. 39-5.)

Navigation and Homing Systems   If the objective is aid to navigation or homing, the DF system will be mounted on a moving vehicle. This will impose a serious limitation on the antenna system. As an offset to the antenna disadvantage, such systems normally have the advantage of operating over short ranges in a fixed and narrow frequency band using direct or surface waves. The need for electrically large antennas is not so great as would be the case if sky waves were involved.

Number and Location of Stations in a Direction-Finding Network

The number and location of stations in a DF network have a direct effect on the accuracy with which an emitter can be located. The limitations of radio propagation as well as the usual geometric limitations on the accuracy of triangulation must be considered.

Horizontal Triangulation (Azimuth)   Under optimum conditions, measurements of bearing from two separated DF sites are sufficient to determine location. In practice, location of the DF sites may not be optimum so that three sites are the practical minimum. By using just two sites, minimum location error will occur when the two lines of bearing intersect at 90°. With reference to Fig. 39-5, if radius R is the standard deviation of error for two lines of bearing intersecting at right angles, then R' will be the major-axis radius of an ellipse of error when the two lines intersect at less than 90°.

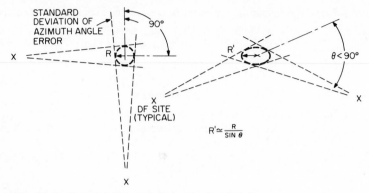

FIG. 39-5 Horizontal triangulation, using two DF sites.

If the two sites are not optimally sited or if one site is unable to determine a line of bearing for any reason, a third site will prove valuable. A third site is likely to increase location accuracy in any case, but only a small further increase in accuracy will likely result from more than three sites. (See Fig. 39-6.)

The minimum spacing between DF sites should depend on the expected distance to the emitter. For a given site spacing, there is a maximum range at which an emitter can be located with any reasonable accuracy. This means that, quite apart from the limitation of DF range imposed by propagation conditions, a limitation is also imposed by the spacing between the DF sites.

Ideally, the sites of a DF system would be placed uniformly round the edge of the area containing the emitters, as illustrated in Fig. 39-7a, but geographic features, national boundaries, and communication problems between the DF sites sometimes make this impossible. In such cases the best arrangement is to locate the sites of a DF network as close as possible to the area containing the emitters, keeping the sites as far apart as possible.

The following general rules, which are illustrated in Fig. 39-7b, c, and d, give a useful guide for DF siting:

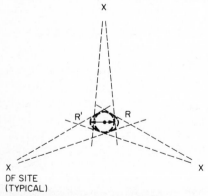

FIG. 39-6 Horizontal triangulation, using three DF sites.

- As shown in Fig. 39-7b, DF sites should be evenly spaced, and the distance between adjacent sites should be at least equal to the distance across the area containing the emitters.

- If none of the DF sites can be located close to the area containing the emitters, the distance between sites should be approximately equal to the distance between the centroid of the area and the centroid of the DF sites, as shown in Fig. 39-7c.

If one of the DF sites is so close as to be in the sky-wave skip zone of one of the emitters (so that no signal is available), the foregoing spacing criteria will maximize the probability that one or more of the other sites will receive a signal. This assumes that the emitter is transmitting a signal to a location in the area containing the other emitters.

In addition to the foregoing DF-station-siting criteria, the various DF sites should be located with a nonparallel baseline, as illustrated in Fig. 39-7d. Arranging the sites in a triangle or a square will minimize azimuth triangulation error.

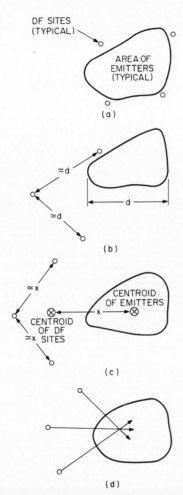

Vertical Triangulation (Single-Station Location) Single-station location (SSL) against sky-wave signals by definition requires only one site. Azimuth measurement is used to determine direction to the emitter, and elevation measurement, in conjunction with the height of the ionosphere, is used to determine distance to the emitter.[8] Location error of the emitter will depend on the accuracies with which the DF system can determine direction and distance. These two parameters are subject to different causes of error:

● The accuracy with which direction can be determined will be identical to that of a single conventional DF system.

● The accuracy with which distance can be determined will depend on the errors inherent in measurement of elevation angle and on the accuracy with which the height of the ionosphere can be calculated or measured. (See Fig. 39-8.)

The accuracy of a DF-SSL system will be higher for azimuth than for distance. Although single-station location is generally less accurate than multistation triangulation, SSL can offer significant advantages of reliability, simplicity, and speed of operation since interstation communications and correlation of separate measurements are not necessary.

A number of DF-SSL sites can be used in a network for improved emitter-

FIG. 39-7 Distribution of DF sites for minimum horizontal triangulation error. (*a*) Ideal distribution of DF sites. (*b*) Approach to ideal distribution of DF sites. (*c*) Optimum distribution if no DF site can be near the target area. (*d*) DF site location with respect to other fixed sites. Sites should be located on nonparallel baselines and as nearly in a triangle or a square as possible.

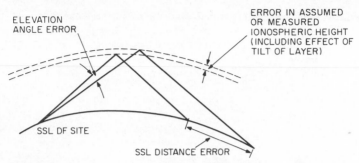

FIG. 39-8 Effect of angular and ionospheric errors on SSL distance measurement.

location accuracy. The form of the location process will be different from the conventional azimuth triangulation process. (See Sec. 39-6.)

Measurement Accuracy and Speed

The measurement accuracy of all DF and emitter-location systems is affected to a first order by six main parameters:

1 Aperture of the antenna system

2 Instrumental accuracy of the measuring system,[9] which includes the effects of errors in antenna performance, site error due to local scattering and ground irregularity, equipment and processing error, and operator error in the case of manually operated DF systems

3 Ionospheric behavior in the case of systems operating in the upper part of the MF band and in the HF band

4 Received signal-to-noise ratio

5 Integration time, that is, the time over which the signal is averaged in the measurement process

6 The distance of the emitter from the DF site or sites

Because the effects of noise and of some forms of ionospheric error are decreased with integration time, measurement speed (inverse of the minimum time that is required for a measurement) and accuracy (the accuracy with which an emitter can be located) are linked together. There is a speed-times-accuracy product that is relatively stable for a given DF system (see Sec. 39-7). Large-aperture systems tend to be fast and accurate, whereas small-aperture systems tend to be slow and less accurate.

For this reason, it is desirable to use the largest practicable aperture for skywave HF DF antenna systems. The advantages of large antenna aperture are:

1 The effect of noise and interference can be reduced (i.e., the signal-to-noise ratio available at the antenna output for a given field strength can be increased) if the aperture of the array is used to form beams. Such arrays are able to discriminate against noise and cochannel interference arriving from directions other than that of the emitter.

2 The effect of sky-wave error is reduced. Sky-wave error occurs when more than one sky-wave mode is present and results in a standing-wave pattern which changes shape slowly with time (in terms of seconds or minutes). This is often referred to as wave interference.[10] See Fig. 39-9.

3 As a result of improved signal-to-noise ratio and reduction of sky-wave error, the integration time required for a given measurement accuracy is reduced.

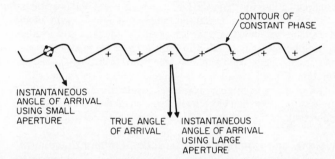

FIG. 39-9 Instantaneous angle-of-arrival errors of small and large aperture under wave-interference conditions. • = sampling ports of a small-aperture system (such as an Adcock). + = sampling ports of a large-aperture system.

4 Errors due to reradiation from obstructions in the vicinity of the DF site are reduced by the directivity of formed beams in the same way that the effect of external noise is reduced. This is true even if beams are not formed since errors contributed by scattered signals at individual elements of the array tend to be random and tend to cancel. Thus, large well-filled apertures are less sensitive to reradiation than are small or sparsely filled apertures.

The use of a large-antenna aperture therefore increases speed, accuracy, and resolution[11] (the ability of the system to distinguish between emitters that are in relatively close proximity).

Physical Limitations

Available Antenna Space   The size of the antenna array of a DF system will be determined by its application (whether fixed or movable), by the frequency band of the emitters, and by the required accuracy and speed of measurement. The largest antenna arrays are required in the HF and upper-MF bands where sky-wave signals are received and variations of phase and amplitude of the arriving wavefront are at a maximum (see Fig. 39-9). In the LF and lower-MF bands, surface-wave signals which have stable phase and amplitude characteristics are received, so that relatively small-aperture antenna arrays provide adequate speed and accuracy. In the VHF and UHF bands, antennas having a large electrical aperture require relatively little physical space.

Thus, antenna arrays may be small when high accuracy and high speed are not primary operating requirements and when sky-wave signals are not likely to be received.

System Mobility   If a DF system is to be fully mobile (meaning that the whole operational system including the antennas is vehicle-mounted), the size of the antenna system will be limited by the size of the vehicle. This restriction is most demanding for land-based systems, less so for ships and aircraft. The restriction is acceptable in the LF and MF bands, permits a reasonable electrical aperture in the VHF and UHF bands, but is often unacceptable in the HF band. Land-based mobile HF DF systems will have relatively low accuracy, low speed, and low resolution. Their use will be limited to applications in which the DF vehicles can be sited near the target emitters.

If the DF system is to be transportable, as opposed to fully mobile, the restrictions on antenna-system size are far less severe. If the land area available at each DF location is sufficient, it is possible to form a large antenna array by spacing out a number of small loop or whip antennas (see below, Fig. 39-17). The accuracy, speed, and resolution of such a system will be greater than is possible with a fully mobile system. However, transportable antenna elements will be electrically small and consequently less sensitive than electrically large elements, so that low-level signals may not be detectable.

Size, Weight, and Complexity of Direction-Finding Equipment   The advent of solid-state electronics permits the design of sophisticated measuring, processing, and output-display equipment that is light and portable. Thus, the size and weight of such equipment is not usually a limiting system-design consideration. It is customary to use modular design so that units can be added to the basic DF system to satisfy more complex operational requirements such as networking, signal acquisition, signal monitoring, and signal processing.

39-4   DIRECTION-FINDING-SYSTEM DESIGN

Types of Direction-Finding Systems

DF systems can be grouped into two broad classifications in accordance with the way in which their antenna systems obtain information from the arriving signals of interest and in which this information is subsequently processed. Scalar DF systems obtain and use only scalar numbers about the signal of interest. Vector or phasor systems obtain vector numbers about these signals. Scalar systems work with either amplitude or phase, while vector systems work with both amplitude and phase.

Scalar Systems   The simplest scalar system is the rotary loop illustrated in Fig. 39-3, which depends on the figure-of-eight symmetry of its vertically polarized amplitude response. Most scalar systems employing amplitude response depend upon some form of symmetry. Examples are the Adcock and Watson-Watt systems (see Sec. 39-5) and any circularly disposed system using a rotating goniometer.

Scalar systems employing phase response are typified by interferometers and Doppler systems (see Sec. 39-5). These systems employ multiport antenna systems that provide phase differences.

Scalar systems are capable of measuring either azimuth or elevation angles of arrival, or both. However, scalar systems of reasonable size and complexity are not well suited for the resolution of the individual angles of arrival of multimode signals (such as signals received in the fading zone).

Vector or Phasor Systems Vector systems have the ability to obtain and use vector or phasor information from the arriving signals of interest. That is, they make use of both amplitude and phase. Vector systems require the use of multiport antennas and at least two amplitude- and phase-coherent receivers. Wavefront-analysis systems, described in the following paragraphs, are a special class of vector system especially intended for the resolution of multicomponent wave fields.[12,13,14]

Wavefront-Analysis Direction-Finding Systems Wavefront-analysis (WFA) systems employ multiport antenna arrays and a suitable receiver-processor-output system so that the defining parameters of the incident wave or waves can be determined. That is, a single incident wave (or a single arriving ray; either wave or ray theory may be assumed) is defined by four parameters: arriving azimuth and elevation angles and relative amplitude and phase (i.e., polarization) of the electric vector.[15] Similarly, a two-component wave (i.e., of two modes) is defined by 10 parameters. Thus, a WFA system must have a sufficient number of antenna ports and sufficient measuring and processing capability so that the desired number of unknown parameters can be resolved.

In practical systems, the received signals are usually processed in digital form, fully preserving relative amplitude and phase, so that the unknown wave parameters can be extracted from the set of antenna-port responses. Given the contemporary state of minicomputer capacity, two wave (or two ray) fields can be resolved.

WFA systems require that the phase and amplitude response to signals of any arriving angle and any polarization be known for every antenna port. Such responses can be determined with considerable precision for any arbitrary antenna element or array of elements.[16] Thus, WFA systems have the important advantage that they may employ virtually any arbitrary array of antenna elements provided only that:

1 The aperture of the array is consistent with the desired speed and accuracy of the system.

2 The number of elements is large enough so that the unknown wave parameters can be determined.

3 The antenna angular patterns, polarization responses, and sensitivity are consistent with the desired application.

Consequently WFA systems can employ antenna arrays of virtually any shape or form, symmetrical or not, including circular or randomly placed land-based arrays or arrays mounted on irregular vehicles (such as ships or aircraft).

Receivers

Receivers comprise the measuring equipment of a DF system. They are used as RF voltmeters to measure antenna responses and to provide responses to the DF processor. In systems in which the azimuth angle is determined by observation of a null or a beam maximum, a single receiver can be used, since the processor will require only information concerning the amplitude of the response pattern of the antenna system.

However, in any system in which measurement is based on an amplitude and/or phase comparison, a number of alternative receiver arrangements are possible (see Fig. 39-10):

1 Single-channel receiver with switch to compare the outputs of two or more antenna ports sequentially (Fig. 39-10a)
2 Dual-channel receiver without switch to compare the outputs of two antenna ports simultaneously (Fig. 39-10b)

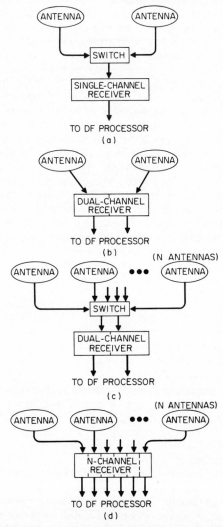

FIG. 39-10 Alternative receiver arrangements.

3   Dual-channel receiver with switch to compare the outputs of three or more antenna ports sequentially and allow any two to be compared simultaneously (Fig. 39-10*c*)

4   *N*-channel receiver without switch to compare the outputs of *N* antenna ports simultaneously (Fig. 39-10*d*)

Dual-channel and multichannel receivers are usually arranged to operate from a common frequency-synthesizer source, so that the phases of the output signals will be in the same relationship as the phases of the incoming signals. Dual-channel and multichannel receivers may also be matched in gain and/or phase. Alternatively, they may be made gain- and phase-stable over their measurement bandwidth and their relative gain and phase normalized for each measurement.

DF receivers are usually provided with selectable-measurement bandwidths. Narrowing the bandwidth can reduce the effect of adjacent channel interference but usually requires an increase in measurement time.

Direction-Finding Processor

The function of the processor is to calculate the required DF and emitter location on the basis of the signal voltages at the output of the receiver system. The complexity of the processor varies with the nature of the calculations required to deduce emitter location. At one extreme, the processor may be as simple as an angle scale on a rotating loop. At the other extreme, a digital processor such as in a multiport WFA system may be required.[17]

Processors for WFA systems usually consist of small digital computers having sufficient speed and memory for their applications. The software for such systems is usually modular, so that a variety of antenna systems can be accommodated and optional capabilities such as networking, signal acquisition, and signal monitoring can be provided.

Figure 39-10*c* and *d* shows two antenna-receiver arrangements suitable for use with a WFA system. The output ports of the antenna elements are connected to the switch for routing to the receivers. Alternatively, the antenna-element responses may be combined into groups to form directive beams.

The wavefront-analysis technique[18] is essentially the inverse of the antenna-port-response computation as previously described.[16] That is, the WFA system, under the control of its processor, measures the responses of the antenna ports to the incoming wave field. Since the current distribution on the antenna array and the resulting antenna-port responses are uniquely determined by the incoming wave field and since the antenna-port responses have previously been computed with precision, the processor is able to solve for the wave-field parameters by inverting the matrix consisting of the measured and the computed antenna-port responses.

The computed wave-field parameters consist of the arrival angles (azimuth and elevation) and the instantaneous sense of polarization of up to two waves (or rays). Each computation also results in a complex correlation coefficient which indicates the quality of that particular computation. Thus, a means for eliminating results seriously contaminated with noise or interference is provided.

The DF processor may also be provided with software so that the measured elevation angle, in conjunction with either stored or measured ionospheric data, can be used to compute a distance to the target emitter, resulting in SSL capability. The software and hardware associated with real-time ionospheric-data systems can become complex and extensive if the ultimate in SSL accuracy is desired.

The WFA technique is fully automatic (human intervention is not required), and it can be very fast. Depending on the configuration of the particular WFA system, individual measurements (or *cuts*) can be made in as little as 0.1 s.

Output-Display Arrangements

The purpose of output-display equipment is to provide the required DF information in a form suitable to the user. The simplest form of display is the scale on the control of a rotating-loop antenna. Scalar DF systems usually employ some form of analog display such as a cathode-ray tube (CRT) or graphic plotter. Some scalar systems are equipped with automatic bearing processors so that numerical values of bearing angle can be used remotely.

Vector (such as WFA) systems often provide a variety of local and remote displays. Visual displays in graphic and tabular form can be furnished, as can hard copies of both. Interactive displays can also be provided so that operators can edit and enhance the accuracy and usefulness of results. Data can also be easily transmitted to remote locations.

39-5 DIRECTION-FINDING ANTENNA ELEMENTS AND ARRAYS

In general, a DF antenna system is an array of individual antenna elements arranged to provide the responses required by the particular system. The antenna elements and arrays are of standard types that are used in other radio-communications applications, and their basic theory of operation is explained elsewhere in this handbook (principally in Chaps. 2 through 6). In this section, therefore, only those properties of the elements and arrays that are of importance in DF applications are discussed in detail.

Antenna Elements

An antenna element for a DF system is essentially a single-port (two-terminal) arrangement of conductors for the interception and collection of radio-frequency energy. A wide variety of element types are used in DF systems. These include monopoles, dipoles, loops, log-periodic antennas, and current sensors. The choice of a particular type of element is dictated by the DF application and the frequency band in which the target emitters operate. Those scalar HF DF systems which measure only the amplitude of the signal incident on the antenna must use antenna elements that are sensitive only to vertically polarized signals (monopoles, vertically polarized dipoles, and vertically polarized log-periodic antennas). Otherwise, polarization error will occur in the presence of sky waves.

Monopoles Monopoles are vertically polarized elements operated over a ground plane. Since they economize in height (they are shorter than a vertical dipole resonating at the same frequency), they are suitable for applications in the LF, MF, and HF bands.

DF systems using arrays of monopole elements are sensitive only to the vertically polarized component of the signal and are thus suitable for scalar (amplitude-only) systems. However, all monopoles have a pattern null at zenith and are not well suited for short-range, high-angle signals. Sleeve monopoles (sometimes referred to as *elevated-feed monopoles*)[19] provide a high sensitivity to low-elevation-angle signals over a wide bandwidth because there is no reverse loop of current when the monopole length exceeds a half wavelength.

Dipoles   Dipoles are used at frequencies in the HF, VHF, and UHF bands at which their length can be accommodated. Vertically polarized dipoles have most of the same advantages and disadvantages of vertical monopoles. Horizontally polarized dipoles can be arranged to be sensitive to signals arriving at high elevation angles. Horizontal dipoles typically have low sensitivity to signals arriving at low elevation angles so that they minimize the effect of local human-made interference, which is usually propagated by surface wave.

Loops   Broadband (untuned) single-turn loop antennas are particularly suitable for wideband DF applications in the HF band. Loop sizes from about 2 m^2 to 4 m^2 represent a good compromise between sensitivity (a larger size is better) and freedom from reradiation when used in arrays and ease of handling when used in transportable systems (a smaller size is better).

Loops are responsive to both vertical and horizontal polarization. Whereas this is a serious disadvantage with amplitude-only systems when sky waves are present, it is a distinct advantage when used with WFA systems. A pair of crossed loops, suitably phased, is responsive to radiation of at least one sense of polarization from virtually all angles of arrival in half space. That is, such loops are sensitive to ground waves and to sky waves from all but very low elevation angles.

Loops, being electrically small and broadband, have a high noise factor (a substantial amount of internal noise is generated by conductor, mismatch, and earth losses[20]). Thus, when the external noise level is low (at the higher frequencies in the HF band and at quiet locations), a DF system with loop antennas may become internally noise-limited. With weak signals longer integration time may be required to provide acceptable measurement accuracy.

Current Sensors   In shipboard and airborne DF-SSL systems, the incoming wave induces RF currents over the entire hull or aircraft body. If the vehicle has dimensions of the order of a wavelength or more, the vehicle itself becomes part of the antenna system, and its effect cannot be ignored. However, for any incoming wave (or set of waves) the current distribution on the vehicle will be unique and will therefore define the incoming wave. Thus, the whole hull and superstructure of the ship or the aircraft body can be used as the DF antenna if the current distribution is sampled by a wavefront-analysis DF system. A convenient form of current sensor can be constructed from a small ferrite core with a coil winding mounted on a backing plate that can be secured to the hull. An essential step in designing such a system is to determine the response of every sensor, wherever located on the vehicle, to both amplitude and phase of waves arriving from any direction.

Arrays of Direction-Finding Antenna Elements

Factors to be considered in the selection of antenna arrays are:

- Coverage (the range and azimuth sector over which the target emitters are located). This will determine the form of the array.
- Expected propagation modes of the arriving signals (surface-wave, direct-wave, sky-wave, or multimode). These will determine the required elevation response of the antenna array and the type of antenna elements to be used.
- Combination of measurement speed and measurement accuracy. These will determine the required aperture of the antenna array.
- Physical requirements of the DF system (fixed or movable and the related limitation on space available for the array). These will affect the physical form of the elements and the size of the array.

DF antenna arrays can be conveniently divided into three categories:

1 Antenna arrays with a single or multiport output that are rotated either mechanically or electrically so that azimuth can be determined by a scalar system based on knowledge of the amplitude of the directional responses

2 Antenna arrays with a multiport output in which the azimuth and elevation angle are determined by measuring the phase difference of the signals at the output ports

3 Antenna arrays with a multiport output from which azimuth angles, elevation angles, and polarizations of the components of a multimode signal can be determined by a vector system

Single-Port or Multiport Antennas with Rotation of a Directional Pattern

This type of antenna system requires a scalar system to measure the amplitude of the signal. The antenna radiation pattern will be symmetrical about some axis, forming in general either a directive beam or a sharp null. In the DF operation the antenna array may be rotated mechanically (if it is sufficiently small), in which case it will have a single port. Alternatively, the antenna array may be rotated electrically by means of a goniometer (an electromechanical device that takes a weighted sum of the outputs of a number of fixed antenna elements and produces a response pattern that can be rotated). This type of antenna will inevitably have a multiport output, and the azimuth of the emitter will be determined by the maximum response (if the beam maximum is used) or the minimum response (if a null is used). It is often convenient to use the beam maximum when searching for the emitter and the null to establish the precise azimuth.

An example of a DF system using an antenna array with a symmetrical beam of known response is the Adcock, which is shown in two-port fixed form in Fig. 39-11. A basic form of the Adcock antenna array[21,22] comprises four vertical monopoles connected so that a figure-of-eight response pattern is produced. The pattern is rotated by a goniometer until the null is directed toward the emitter. The Adcock array differs from the rotating loop of Fig. 39-3 in that there are no unshielded horizontal conductors, so that polarization error due to the horizontally polarized component of sky waves is eliminated. Since there are two nulls in the figure-of-eight response pattern, there will be an ambiguity of 180° in the determination of azimuth. It is necessary to switch the monopoles to provide a cardioid radiation pattern to determine the correct azimuth. Thus, the measurement is made in two stages, the first to determine two precise azimuths by means of a null output and the second to determine which of these azimuths is the correct one.

The Adcock antenna array is typical of a narrow-aperture DF system and suffers from the disadvantages of low speed and/or accuracy, as discussed in Sec. 39-3. There are many variations of the basic Adcock design, most of which employ more than four elements with the objective of increasing aperture without increasing spacing error.[23] But all these must be considered narrow-aperture systems.

An alternative to the conventional Adcock system is the Adcock Watson-Watt arrangement,[24] one form of which is illustrated in Fig. 39-12. In this arrangement, the two antenna ports are connected (by means of two receivers) to the X and Y plates of a CRT so that the display is a line indicating the azimuth of the emitter on a surrounding circular scale calibrated in degrees. The advantage of this arrangement over the Adcock

HORIZONTAL CONDUCTORS AND SHIELDED CABLES INSTALLED UNDER GROUND SCREEN TO MINIMIZE PICKUP OF HORIZONTALLY POLARIZED FIELDS

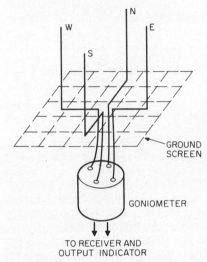

FIG. 39-11 Two-port, four-element narrow-aperture Adcock DF antenna.

with a goniometer is that an instantaneous reading of azimuth rather than the intermittent reading provided by the goniometer is obtained. The Watson-Watt display is subject to jitter and variation caused by modulation, fading, and noise so that considerable integration is required for accurate measurement.

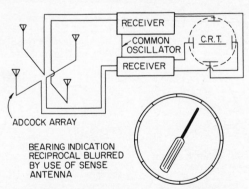

FIG. 39-12 Adcock antenna with a Watson-Watt instantaneous cathode-ray-tube display.

Some DF systems with a rotating directional pattern are capable of high speed and high sensitivity. These are wide-aperture multiport antennas of many elements disposed in a circle. Rotation of the response pattern is obtained by a rotating capacity-coupled goniometer. The goniometer usually combines responses from adjacent antenna elements so as to form highly directive sum-and-difference beams. An example of a wide-aperture DF system using a circular array of elements, together with goniometer pattern rotation, is the Wullenweber,[25] which is illustrated in Fig. 39-13. As in the case of the Adcock, the Wullenweber system with a goniometer measures only the amplitude (and hence only the azimuth angle) of the signal. To minimize polarization error the elements are monopoles.

The monopoles are arranged symmetrically around a cylindrical screen, and each is connected to a stator segment of the goniometer. The rotor segments span about 100° of arc and are connected to the switch by delay lines D_1, D_2, and D_3, whose lengths are equal to the free-space delays of the signal, as shown in Fig. 39-13. The signals from the antenna elements in operation at any moment combine in phase at the receiver and produce a sharp beam, since the arrangement functions as a broadside array. As shown in Fig. 39-13, the delay lines can be optionally split into two groups which are connected in antiphase, thus producing a rotating null as opposed to a rotating beam. Either the sum or the difference output can be connected to the receiver input. The output of the receiver is connected to a CRT display with a synchronized rotating time base so that the response pattern of the antenna appears as a polar display centered on the direction of the emitter. When searching for an emitter, the sum mode is normally used, but when the emitter has been identified, the difference mode is used so that the sharp null in the response pattern can display azimuth angle with maximum accuracy.

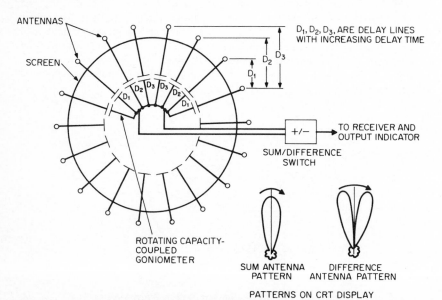

FIG. 39-13   Wullenweber wide-aperture antenna.

Multiport Antennas Using Phase Difference   Measurements of phase differences between the ports of a multielement antenna enable both azimuth and elevation angle of the arriving signal to be determined. One system of this type uses the Doppler technique.[26] In principle, an antenna element could be moved in a circular path so that the instantaneous frequency of the received signal would be modified. In practice, it is usually inconvenient to rotate an antenna element physically so the quasi-Doppler antenna arrangement shown in Fig. 39-14a is used instead. A rotating commutator is used to couple a receiver in rapid sequence to the elements of the array, thereby introducing a frequency shift on the received signal which is extracted by a frequency discriminator. As illustrated in Fig. 39-14b, the frequency shift is proportional to sin (θ − θ_0) cos ϕ_0, where θ is the angular position of the rotor, θ_0 is the azimuth angle of the received signal, and ϕ_0 is the elevation angle of the received signal. By using this expression, the azimuth angle is given by the angular position of the rotor at which zero instantaneous frequency shift occurs. Ambiguity can be removed by taking account of the angles at which maximum positive and negative frequency shifts occur. Also by using this expression, the peak-to-peak amplitude of the instantaneous fre-

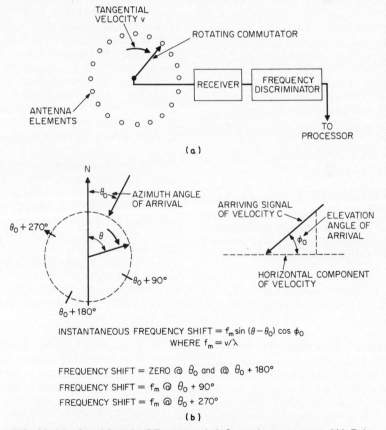

(a)

$$\text{INSTANTANEOUS FREQUENCY SHIFT} = f_m \sin (\theta - \theta_0) \cos \phi_0$$
$$\text{WHERE } f_m = v/\lambda$$

FREQUENCY SHIFT = ZERO @ θ_0 and @ θ_0 + 180°

FREQUENCY SHIFT = f_m @ θ_0 + 90°

FREQUENCY SHIFT = f_m @ θ_0 + 270°

(b)

FIG. 39-14   Quasi-Doppler DF system. (*a*) General arrangement. (*b*) Relationship between frequency shift and angles of arrival.

quency shift is proportional to the cosine of the elevation angle so that the latter can be determined.

In practice, it is found that the quasi-Doppler arrangement will give accurate measurements of azimuth angle with short integration times. However, accurate measurement of elevation angle with multimode HF signals requires that one of the modes be dominant. The required integration time will depend on the relative levels of the minor modes. Furthermore, azimuth-angle accuracy drops off at high elevation angles as the amplitude of the instantaneous frequency shift approaches zero.

Another DF system making use of phase differences between the signals at the ports of a multiport antenna is the interferometer,[27] one form of which is shown in Fig. 39-15. This arrangement comprises five identical antenna elements, which can be loops, crossed loops, or monopoles. Since the distance between elements A and B is large in terms of a wavelength, the phase difference at their output ports will be large and will be very sensitive to angle of arrival relative to baseline AB. But just because AB is greater than $\lambda/2$, different arrival angles will produce the same phase difference (see Fig. 39-15). This ambiguity can be resolved by the less accurate measurement of phase difference between elements A and E, provided their spacing is less than $\lambda/2$ at the highest frequency of operation. The same principles apply to the line of elements A, D, C.

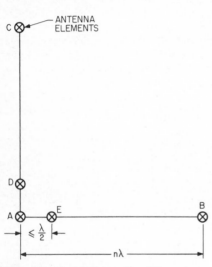

FIG. 39-15 Interferometer DF system. Example: If $n = 3$, arrival angles of 43°, 66.5°, and 86° relative to line AB will all result in a phase difference of 70° between A and B. The phase difference between A and E will resolve this ambiguity.

The loci of the angles of arrival with respect to the two baselines will be cones in half space, with the baselines as their axes. The line of intersection of the half cones about the two baselines will be the direction of arrival of the signal. The processor will compute the line of intersection, which will indicate both azimuth and elevation angles of arrival.

Scalar systems such as interferometers are not generally able to resolve the individual components of a multimode signal such as will occur under conditions of wave interference.

Multiport Antennas for Wavefront-Analysis Direction-Finding Systems

When wide-aperture systems are required, the Wullenweber type of antenna shown in Fig. 39-13 can be used, but the land area required is large. This is due to the fact that a ground screen extending well beyond the antenna elements is required for control of the near-field environment so as to minimize site errors. A more efficient wide-aperture system which achieves equal or better performance over a greater bandwidth and has a much smaller radius is shown in Fig. 39-16. This system comprises a symmetrical ring of quasi-log-periodic antennas (LPAs) arrayed so that their main lobes point inward (toward the center of the ring), thus conserving space. The array can

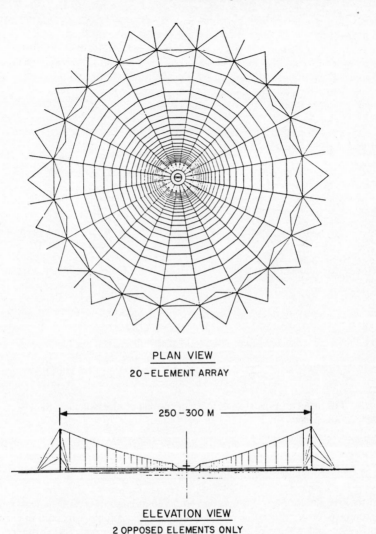

PLAN VIEW
20-ELEMENT ARRAY

250-300 M

ELEVATION VIEW
2 OPPOSED ELEMENTS ONLY

FIG. 39-16 Circular array of quasi-log-periodic antennas, suitable for HF
DF-SSL and monitoring, horizontally and/or vertically polarized.

comprise vertically and/or horizontally polarized LPAs, as illustrated in Fig. 39-16. The ground screen need not extend beyond the outer antenna elements because each LPA faces inward toward a controlled and symmetrical environment in its near field.

Loops are also used for medium-aperture WFA arrays suitable for fixed or movable use. Loops are preferred over monopoles because they do not have a vertical pattern null and are therefore sensitive to high-angle sky waves. If space for the array is limited and if the number of individual elements is to be minimized, loops or crossed loops, as illustrated in Fig. 39-17, can be used. With this type of array, beams usually are not formed. The crossed-loop responses can be combined, or they may be used individually.

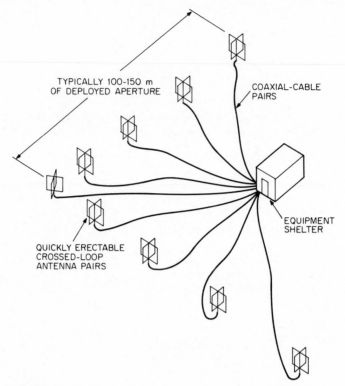

TYPICALLY 100-150 m
OF DEPLOYED APERTURE

COAXIAL-CABLE
PAIRS

EQUIPMENT
SHELTER

QUICKLY ERECTABLE
CROSSED-LOOP
ANTENNA PAIRS

FIG. 39-17   Typical deployment of one unit of a transportable HF DF-SSL system.

39-6   EMITTER-LOCATION ACCURACY

Emitter-location accuracy depends on two different kinds of error. The first, angular-measurement error, affects the accuracy of emitter location when azimuth triangulation is used and when SSL techniques are used. The second, SSL distance error, affects the accuracy of emitter location only when SSL techniques are used.

Angular-measurement error is the fundamental performance parameter of all DF systems.[28] SSL distance error is applicable only to sky-wave propagation.

Emitter-location error (or accuracy) can be affected by both angular-measurment error and (if used) SSL distance error. Figure 39-18 shows the hierarchy of emitter-location errors.

Angular-Measurement Error

Angular-measurement accuracy is usually taken to mean the ability of an ideal DF system to measure *direction to the target emitter*. In actuality, DF systems measure *direction of arrival* of the signal radiated by the emitter. These two directions may be sensibly identical for ground-wave and direct-wave signals, but for sky-wave signals

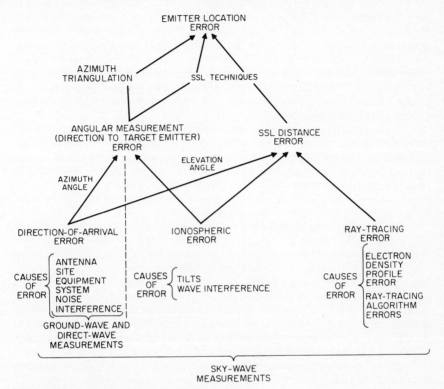

FIG. 39-18 Hierarchy of emitter-location error.

they will generally be different, often by a significant amount because of ionospheric error.

Thus, angular-measurement error must account for (1) direction-of-arrival errors such as those contributed by antenna, equipment, site, noise, and interference and (2) ionospheric errors.

Direction-of-Arrival Error   This type of error can be caused by antenna siting or construction errors, unequal cable lengths, unequal gain or phase match, antenna-pattern errors, computational or interpolation errors, reradiation and scattering from nearby objects, noise errors, and cochannel interference errors. For well-designed large aperture systems installed at good sites, the sum of all the errors caused by system design, construction, and siting should not exceed 0.1 to 0.5°.

Noise errors can be reduced to insignificance with sufficient integration time. Cochannel interference errors can be reduced by employing filtering in the domains of frequency, time, or space or some combination thereof. In view of the effect of operational choice upon the reduction of noise and interference errors, it is not practical to give a numerical allowance for them.

Ionospheric Error   This error results from wave interference, from large-scale ionization gradients (tilts) caused by hourly and seasonal solar variation, and from traveling ionospheric disturbances. The magnitude of ionospheric error varies with geographic location, with azimuth direction of signal propagation, and with distance to the emitter. Some of these errors are systematic, while others are random. The component of angular-measurement error due to ionospheric error can be reduced to insignificance with very long integration times (hours or days or seasons), usually much longer than is practical. Short-term statistical allowances for ionospheric errors have been suggested,[29],[30] but these are of little operational use. It is possible to apply a tilt correction either by measuring tilt in near real time or by forecasting tilt on the basis of large-scale stored ionospheric data.

Sky-wave DF systems present a special problem in angular terminology because they do not measure azimuth angles directly (in the conventional meaning of terrestrial geometry) except for signals arriving at 0° elevation angle. In fact, signals arriving from zenith (90° elevation angle) convey no information at all about azimuth. For these reasons, it is convenient to use the concept of *great-circle degrees* when discussing direction-of-arrival accuracy (but not when discussing azimuth angular-measurement accuracy). Figure 39-19 defines this concept.

The ionospheric-signal path from the true position of the emitter to the DF site can be idealized as two straight-line segments joined at their apparent reflection point in the ionosphere. If the ionosphere is spherical and geocentric, these two line segments would define a great-circle plane. But if the ionosphere is tilted, the two line segments from *true emitter position* to DF site will not define a great-circle plane, although a great-circle plane would be defined by the two line segments from *apparent emitter position* to DF site. The direction-of-arrival error can be defined as the difference between the actual and the ideal straight-line segments from their reflection points to the DF site. For small error angles, great-circle error angles can be converted to azimuth error angles by dividing by the cosine of the elevation angle.

The great-circle-degree concept of angular error is necessary so that azimuth error results of signals arriving from different elevation angles can be compared. For example, if the horizontal great-circle angle-of-arrival error of three signals is, say, 1.0° (great circle) and if their elevation angles happen to be 0°, 30°, and 60°, their errors expressed in azimuth degrees will be 1.0°, 1.15°, and 2.0°, respectively.

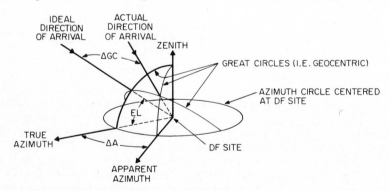

ΔGC = ERROR EXPRESSED IN GREAT-CIRCLE DEGREES

ΔA = ERROR EXPRESSED IN AZIMUTH DEGREES

$\Delta A = \dfrac{\Delta GC}{\cos EL}$ (FOR SMALL-ERROR ANGLES)

FIG. 39-19 Definition of direction-of-arrival error, expressed in great-circle degrees.

Lateral Error   *Lateral error* is defined as the location error, perpendicular to the measured azimuth, at the true distance (see Fig. 39-20). This is the component of location error caused by angular-measurement error.

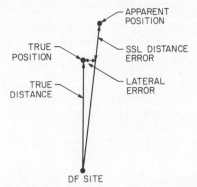

For surface waves and direct waves, in which the signal arrives at zero elevation angle, lateral error is directly proportional to angular-measurement error. If, for example, total angular-measurement error were 1°, lateral error would be as in Table 39-1.

The result of angular-measurement error as a function of true distance is very different for sky waves, in which the signal arrives at an elevation angle determined by the true distance and the height of the ionosphere. Figure 39-21 shows percent lateral and SSL distance errors as a function of true distance, assuming 1.6° (great-circle) direction-of-arrival error allowance and an ideal spherical-mirror ionosphere of 300-km height. The lateral error shown in Fig. 39-21 is presented in Table 39-2.

FIG. 39-20   Geometry of emitter-location error.

TABLE 39-1

True distance, km	Lateral error, km	Lateral error, % of true distance
50	0.9	1.8
100	1.8	1.8
500	9	1.8

SSL Distance Error

SSL distance error will be a function of elevation angle error and of the ability to trace the direction of the arriving ray backward through its zone of ionospheric refraction to the location of the emitter. Of these two error components, elevation angle error will be the lesser. Well-designed DF systems are able to measure elevation angle as accurately as they are able to measure azimuth angle except at very low elevation angles, at which the projected vertical aperture of most DF antenna systems approaches zero.

The ability of a DF-SSL system to trace the direction of the arriving ray backward to the location of the emitter will depend on the accuracy of the available ionospheric data, the quality of the ray-tracing algorithm, and the degree of uncorrected ionospheric tilt.

In their simplest form, ionospheric data may consist of stored hourly median

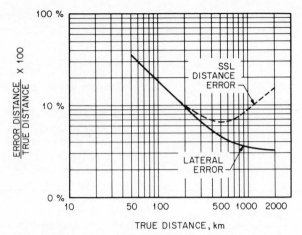

TRUE DISTANCE, km

FIG. 39-21 Percent lateral and SSL distance errors as a function of true distance for sky-wave signals reflected from an ideal spherical-mirror ionosphere. Assumptions: (1) Perfect-mirror ionosphere at 300-km height. (2) 1.6° (great-circle) direction-of-arrival error.

TABLE 39-2

True distance, km	Lateral error, km	Lateral error, % of true distance
50	17.5	35
100	18	18
500	22	4.4
1000	34	3.4
2000	65	3.25

values published by numerous government agencies.[31] Such data can be used with or without tilt correction. In their most complex form ionospheric data may consist of near-real-time electron-density profiles obtained by a network of ionospheric sounders throughout the geographic area of interest and combined into a map of electron-density contours. Such a map will include tilt information.

Ray-tracing algorithms of varying complexity are available to convert elevation angles into distances. Most of these algorithms provide a one-hop distance based on a single measured value of elevation angle. More complex ray-tracing algorithms combined with WFA systems capable of resolving multiple angles of arrival show promise of an ability to provide multiple-hop SSL distances.

By referring again to Fig. 39-21, it is clear that SSL distance error is equal to or greater than lateral error for reasons of geometry. That is, for distances greater than about 200 km (for a 300-km-ionosphere) distance error is more sensitive to a given arrival-angle error than is lateral error. Furthermore, Fig. 39-21 does not account for ray-tracing error, which will further degrade SSL distance error. The SSL distance error shown in Fig. 39-21 is presented in Table 39-3.

Figure 39-22 shows the data from Fig. 39-21 and Tables 39-2 and 39-3 plotted out to 1000 km.

TABLE 39-3

True distance, km	SSL distance error, km	SSL distance error, % of true distance
50	17.5	35
100	18	18
500	33.5	6.7
1000	91	9.1
2000	310	15.5

Emitter-Location Error

As can be seen from Fig. 39-18, emitter-location error depends upon how emitter location is to be accomplished. Azimuth triangulation using a network of DF sites and SSL are the two techniques to be considered.

Azimuth Triangulation with a Network   With azimuth triangulation, location error will depend on the lateral error of the individual lines of bearing, on the number of sites reporting, and on locations of the sites with respect both to the emitter and to each other.[7] (See Sec. 39-3.)

If two DF sites are optimally situated with respect to the emitter, that is, if their lines of bearing intersect at 90°, then emitter-location error will approximately equal lateral error. If additional well-situated DF sites are used and if the individual lateral errors are random, then an increase in the number of DF sites will reduce emitter-location error.

Use of SSL Techniques   SSL techniques are clearly of greatest utility if network operation is not possible for some reason. But if the individual DF sites of a working network do provide SSL results, overall emitter-location accuracy can be improved. This can be accomplished at the network control center where the lines of bearing and SSL

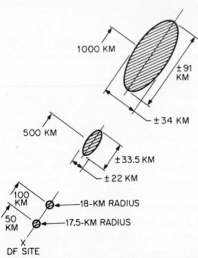

FIG. 39-22 Uncertainty of emitter location when SSL technique is used, employing data from Fig. 39-21.

results from all the sites are processed to give a best-point estimate (BPE). The most rational procedure is to require each DF-SSL site to furnish a quality rating for each line of bearing and for each SSL result. The BPE algorithm should be designed to test the available results for convergence, giving weight to the various quality ratings. This procedure, making best use of whatever results are available, will, on the average, produce the most reliable and accurate emitter-location results.

39-7   PERFORMANCE OF DIRECTION-FINDING SYSTEMS

Specifying and testing the performance of DF systems are more complex than might be supposed, especially for systems operating against sky waves. This section suggests how performance can be specified and what quality of results may be expected with several kinds of systems.

Factors Affecting Performance

Emitter-location accuracy of DF-SSL systems operating against sky waves is affected by at least the six kinds of parameters first given in Sec. 39-3. That is, the quality of results will be influenced to a first order by all of the following:

1 Antenna aperture
2 Instrument or system accuracy
3 Ionospheric behavior
4 Received S/N ratio
5 Integration time
6 Distance to emitter

The first two of these parameters are under the control of the system designer. For any particular system and for any frequency or band of frequencies these parameters may be regarded as fixed. Instrument or system accuracy is the parameter most likely to be specified by most system suppliers, even though it will seldom be the determining parameter with respect to emitter-location accuracy. Furthermore, instrument or system accuracy is difficult to measure accurately, especially with sky-wave DF-SSL systems.

The third parameter, ionospheric behavior, varies continuously. Corrections can be applied by the use of statistical ionospheric data or by the use of near-real-time ionospheric data. The last three parameters, S/N ratio, integration time, and distance, can obviously vary widely and are primarily under the control of the operator of the emitter.

It follows that a performance specification should take account of all six parameters. Furthermore, a testing program should permit separation of the different contributions to emitter-location error so the system can be rationally evaluated.

Performance Comparison of Direction-Finding Systems

It is useful to have a method for comparing the accuracy of different DF systems or of the same system under different conditions. But in view of the six parameters having a first-order effect on DF accuracy, fully rigorous comparisons are likely to be too complex to be useful.

Figure 39-23 suggests a practical method of azimuth accuracy comparison for different kinds of systems operating under widely different conditions. It is not meant to be rigorous, but it is useful for comparative purposes. It shows the nature of the relations between accuracy and integration time and between accuracy and aperture.

Figure 39-23 accounts for the first two of the six parameters, namely, aperture and instrument or system accuracy, by means of the four kinds of systems, A to D. Regarding the third parameter, ionospheric behavior, Fig. 39-23 implies that there is some amount of ionospheric error which is unlikely to be eliminated within the integration times of interest, but the figure does not account for large-scale solar-induced tilts.

The fourth parameter, the received S/N ratio, is assumed to be at least high enough to yield plausible results at the shortest integration times shown for each kind of system. The slope of measurement error versus integration time reflects the fact that each doubling of integration time is equivalent to increasing the S/N ratio by 3 dB and that a 6-dB S/N improvement reduces the effect of random errors by a factor of approximately 2. The fifth parameter, time, is the independent variable. The figure is approximately independent of the sixth parameter, distance, because the independent variable (measurement-error angle) is expressed in great-circle degrees.

Figure 39-23 is useful in visualizing the following ideas:

1 Large-aperture systems are faster and more accurate than small-aperture systems.

2 Manual systems (wherein a human operator makes the bearing estimate) probably require 5- to 10-s minimum integration to obtain any plausible result.

3 Fully automatic systems, combined with wide-aperture antennas, are between one and two orders of magnitude faster than manual systems.

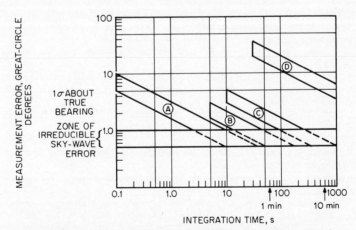

FIG. 39-23 Comparison of sky-wave DF-system performance. A = wide-aperture fully automatic WFA systems. B = medium-aperture goniometric systems such as Wullenwebers and wide-aperture interferometers. C = Adcock and Watson-Watt systems (narrow aperture). D = vehicular-mounted systems (zero aperture).

REFERENCES

1 F. E. Terman, *Radio Engineering,* McGraw-Hill Book Company, New York, 1947, p. 817.
2 G. S. Sundaram, "Ground-Based Radio Direction-Finding System," *Int. Def. Rev.,* January 1981.
3 P. J. D. Gething, J. G. Morris, E. G. Shepherd, and D. V. Tibble, "Measurement of Elevation Angles of H.F. Waves," *IEE Proc.,* vol. 116, 1969, pp. 185–193.
4 J. Ames, "Spatial Properties of the Amplitude Fading of Continuous HF Radio Waves," *Radio Sci.,* vol. 68D, 1964, pp. 1309–1318.
5 B. H. Briggs and G. J. Phillips, "A Study of the Horizontal Irregularities of the Ionosphere," *Phys. Soc. Proc.,* vol. B63, 1950, pp. 907–923.
6 R. L. Smith-Rose and R. H. Barfield, "The Cause and Elimination of Night Errors in Radio Direction Finding," *J. IEE,* vol. 64, 1926, pp. 831–838.
7 P. J. D. Gething, *Radio Direction-Finding and the Resolution of Multicomponent Wave-Fields,* Peter Peregrinus Ltd., Stevenage, England, 1978, chap. 14.
8 R. F. Treharne, "Vertical Triangulation Using Skywaves," *Proc. Inst. Radio Electron. Eng.,* vol. 28, 1967, pp. 419–423.
9 P. J. D. Gething, "High-Frequency Direction Finding," *IEE. Proc.,* vol. 113, no. 1, January 1966, pp. 55–56.
10 Gething, 1978, p. 35.
11 Gething, 1978, chaps. 8 and 9.
12 D. Cawsey, "Numerical Methods for Wavefront Analysis," *IEE. Proc.,* vol. 119, 1972, pp. 1237–1242.
13 P. J. D. Gething, "Analysis of Multicomponent Wave-Fields," *IEE. Proc.,* vol. 118, 1971, pp. 1333–1338.
14 J. M. Kelso, "Measuring the Vertical Angles of Arrival of HF Skywave Signals with Multiple Modes," *Radio Sci.,* vol. 7, 1972, pp. 245–250.
15 Gething, 1978, p. 175.
16 R. L. Tanner and M. G. Andreasen, "Numerical Solution of Electromagnetic Problems," *IEEE Spectrum,* vol. 4, no. 9, September 1967, pp. 53–61.
17 H. V. Cottony, "Processing of Information Available at the Terminals of a Multiport Antenna," *Antennas Propagat. Int. Symp.,* Sendai, Japan, 1971, pp. 9–10.
18 R. L. Tanner, "A New Computer-Controlled High Frequency Direction-Finding and Transmitter Locating System," paper presented to NATO-AGARD Symposium of the Electromagnetic Wave Propagation Panel, Lisbon, May–June 1979.
19 R. W. P. King, *Theory of Linear Antennas,* Harvard University Press, Cambridge, Mass., 1956, p. 407.
20 H. A. Wheeler, "Fundamental Limitation of Small Antennas," *IRE Proc.,* September 1947.
21 F. Adcock, "Improvements in Means for Determining the Direction of a Distant Source of Electro-Magnetic Radiation," British Patent 130,490, 1919.
22 F. Adcock, "Radio Direction Finding in Three Dimensions," *Proc. Inst. Radio Eng. (Australia),* vol. 20, 1959, pp. 7–11.
23 Gething, 1966, p. 51.
24 R. A. Watson-Watt and J. F. Herd, "An Instantaneous Direct-Reading Goniometer," *J. IEE (London),* vol. 64, 1926, p. 11.
25 H. Rindfleisch, "The Wullenweber Wide Aperture Direction Finder," *Nachrichtenech. Z.,* vol. 9, 1956, pp. 119–123.
26 C. W. Earp and R. M. Godfrey, "Radio Direction Finding by Measurement of the Cyclical Difference of Phase," *J. IEE (London),* part IIIA, vol. 94, March 1947, p. 705.
27 W. Ross, E. N. Bradley, and G. E. Ashwell, "A Phase Comparison Method of Measuring the Direction of Arrival of Ionospheric Radio Waves," *IEE Proc.,* vol. 98, 1951, pp. 294–302.
28 Gething, 1966, pp. 58, 59.

29 T. B. Jones and J. S. B. Reynolds, "Ionospheric Perturbations and Their Effect on the Accuracy of HF Direction Finders," *Radio Electron. Eng.*, vol. 45, no. 1–2, January–February 1975.

30 A. D. Morgan, "A Qualitative Examination of the Effect of Systematic Tilts in the Ionosphere on HF Bearing Measurements," *J. Atm. Ter. P.*, vol. 36, 1974, pp. 1675–1681.

31 *Ionospheric Communications Analysis and Prediction Program (IONCAP),* version 78.03, Institute for Telecommunication Sciences, Boulder, Colo., March 1978.

Chapter 40

ECM and ESM Antennas

Vernon C. Sundberg

GTE Systems

Daniel F. Yaw

Westinghouse Defense and Electronic Systems

40-1   INTRODUCTION

Electronic-countermeasures (ECM) and electronic-support-measures (ESM) systems place a great variety of requirements on the antenna or antennas used, depending on the specific function being performed by the system, the characteristics of the target, and the environment within which the system is functioning. The breadth of these requirements is such that virtually any type of antenna, from a simple monopole to the most sophisticated phased array, may eventually be used in an ECM or ESM system. This chapter describes a few "typical" antenna subsystems used in ECM or ESM systems.

The functions to be performed by an antenna in an ESM system can generally be divided into two broad categories: (1) to monitor the environment for signals and (2) to determine the direction of arrival of an incoming signal (direction finding or tracking). The primary antennas on an ECM system are used to direct a significant amount of jammer power toward the threat emitter. ECM systems also use passive antennas for direction finding.

40-2   SURVEILLANCE ANTENNAS

The first and primary function of an ESM system is that of monitoring the environment for signals or, more particularly, for certain specific signals or types of signals. The knowledge that certain signals are being radiated can be of great importance to the battlefield commander or the combatant himself. A few typical examples of surveillance antennas are described in the following paragraphs.

Biconicals

One of the most useful omnidirectional antennas is the biconical horn,[1,2] particularly since the advent of meander-line-array[3] aperture polarizers. The most popular method of feeding the biconical for broadband operation is with coaxial line. This method produces a vertically polarized antenna. Cylindrical-aperture polarizers can be used to convert the polarization to either circular or slant-45° linear polarization.

By using meander-line-array polarizers to produce circular polarization, axial ratios can typically be held to less than 3 dB. For broadband operation, a flat disk of resistance card in the biconical is needed to absorb the cross-polarized reflections off the polarizer, however small these may be. If these low-level reflections were not absorbed, they would be re-reflected by the biconical and would radiate as the undesired sense of circular polarization and rapidly degrade the axial-ratio performance. At the present state of development, meander-line-array polarizers are limited to two octaves or less, which becomes the controlling factor for a circularly polarized biconical.

Slant-45° linear polarizers can provide broader bandwidths of operation. Bandwidths of 12:1 have been achieved on biconicals by using this type of polarizer. The slant-45° linear polarizer consists of a series of cylindrical printed-circuit-board (PC-board) layers, each with a grid of linear parallel lines. The PC boards are spaced in the radial dimension by approximately a quarter wavelength, with the innermost layer having its lines oriented orthogonally to the polarization of the antenna. On each succeeding sheet, the grid is tilted at an increased angle relative to the innermost board until the last sheet has its lines at 45° to the innermost sheet and orthogonally to the desired slant-linear-polarization orientation. Usually this is accomplished with four or five layers.

Another useful circularly polarized, omnidirectional antenna is the Lindenblad,[4] which covers octave bandwidths with broader elevation patterns.

Radial-Mode Horns

Radial-mode horns[5] can provide increased gain relative to an omnidirectional antenna when the azimuthal angular sector to be monitored is less than 360°. Radial-mode horns (Fig. 40-1) consist of a wave-launching section and a radial section in which a cylindrical wave moves out and radiates from the cylindrical aperture. The azimuthal (or E-plane) pattern is relatively flat-topped with a beamwidth of 0.95 θ, independent of frequency, provided that the wave-launcher slot width w is less than 0.5λ. Useful angles of θ have been found to vary from approximately 90° up to at least 200°. The radius R, measured from the point at which the back walls would intersect, must be greater than some minimum length (which is a function of θ) to maintain the flat-topped-pattern shape. The minimum radius R required increases rapidly as the angle θ decreases to 90°.

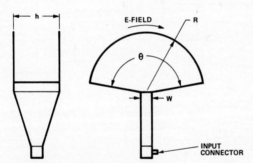

FIG. 40-1 *E*-plane radial-mode horn.

The elevation (or H-plane) pattern of this antenna varies with frequency in the normal manner of an aperture with a cosine amplitude distribution. Bandwidths of up to 2.25:1 have been achieved with this antenna.

The radial-mode horn itself generates a polarization that is perpendicular to the cylindrical axis. This linear polarization can be converted to either sense of circular polarization by using a cylindrical section of meander-line-array polarizer or to slant-45° linear polarization by the polarizer previously described. The polarization has also been converted to the orthogonal linear polarization very successfully by using essentially a double thickness of meander-line-array polarizer.

Figure 40-2 is an azimuthal (E-plane) pattern of an E-plane radial-mode horn with a meander-line polarizer taken with a linearly polarized transmit antenna rotating in polarization.

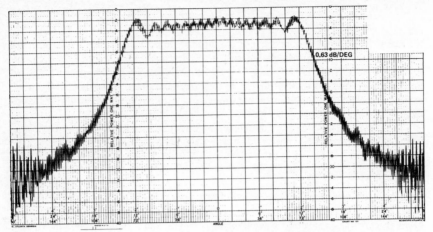

FIG. 40-2 Azimuthal pattern of an *E*-plane radial mode horn ($R = 8.9\lambda$, $\theta = 179°$).

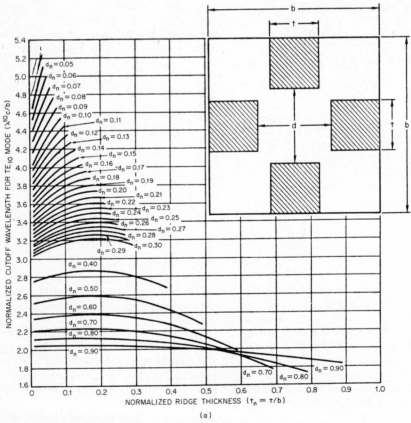

(a)

FIG. 40-3 Cutoff wavelength versus ridge thickness for various gap widths. (*a*) TE$_{10}$ mode. (*b*) TE$_{30}$ mode.

are used in the horns for phase correction in the azimuth plane. Meander-line polarizers convert the vertical linear polarization in the horns to right-hand circular polarization in the radiated field. A horizontal absorber fence reduces the effects of the metal surfaces of the horns on the conical spiral patterns. DF accuracy of this system against emitters with high signal strength and amplitude stability varied from 5° at the low frequency to 0.4° at the high frequency of each horn.

This system used a five-channel rotary joint to bring the five RF bands off the rotating platform.

Shaped-Beam Antennas

Another form of rotating-beam antenna uses a specially shaped reflector to control the elevation pattern coverage for ESM purposes. The doubly curved reflector is used to shape the main-beam radiation pattern in the elevation plane for specific system requirements while focusing the beam in the azimuth plane for maximum directivity. For instance, for an air or surface search system, a constant-power return from the target as a function of elevation or depression angle is optimum. The amplitude pattern for such an optimization is nominally proportional to $csc^2 \theta$. Another application of the shaped-reflector system for shipborne ESM systems is provided when the system is designed to maintain a fixed elevation-plane beamwidth across the frequency range. This type of system is often installed on unstabilized shipborne platforms. In this case the beam must be sufficiently broad to compensate for the pitch and roll of the platform and maintain a constant illumination of the potential threats.

The doubly curved reflector has no exact solution, and any solution obtained is not unique. The design methods outlined in the literature[13,14,15] are based on an iterative geometrical-optics process that tends to converge rapidly. The geometrical-optics method transforms the primary radiation pattern of the feed into the desired secondary radiation pattern in the elevation plane. Although this method of surface determination was originally developed for narrowband systems, it is useful for broadband ESM applications provided that the illuminating primary pattern is relatively constant over the frequency range of interest. Therefore, the shaped-beam technique works well with primary feeds such as log-periodics, spirals, and, to some extent, scalar electromagnetic horns, whose beamwidths are a slowly varying function over at least a waveguide bandwidth.

Measured radiation patterns of doubly curved reflectors closely approximate the desired theoretical radiation-pattern shape and hence validate the geometrical-optics design process. As one might expect, the larger the reflector in terms of wavelengths, the better the conformance to the theoretical pattern shape. However, even for small reflectors (0.6- to 0.75-m diameter) typical of some ESM applications, pattern-shaping techniques are extremely useful. Frequency ranges of 18:1 are entirely feasible by using this technique. Such an antenna (0.75-m diameter) is shown in Fig. 40-5. The antenna is linearly polarized at 45° and provides a minimum elevation beamwidth of 18° to 18 GHz. As the frequency of this antenna decreases, a crossover point is reached where the physical aperture size dominates and the elevation-plane beamwidth becomes inversely proportional to frequency. At the low end of the frequency range, the beamwidth increases to 24°. Some aperture gain is sacrificed even at the low end of the frequency range because of the phase distortion associated with shaping the radiation pattern for the high end of the frequency range.

FIG. 40-5 Shaped-beam antenna (1 to 18 GHz). *(Courtesy of Condor Systems, Inc.)*

Amplitude Comparison

Amplitude-comparison DF requires two or more overlapping antenna patterns whose peaks are separated by some angular extent in the azimuth angle. By measuring the relative amplitude on the two patterns and comparing the measured data with a priori data of the pattern shapes, the angle between the antenna boresight and the emitter's bearing can be determined. Narrower beamwidths provide better DF accuracy, but only over narrower azimuthal sectors. Hence, a trade-off must be made between gain and DF accuracy against sector coverage or number and volume of antennas required.

DF accuracy is reduced by the polarization sensitivity of the pattern. Because the polarization of the incident signal is normally unknown, any variation of the pattern shape as a function of polarization will introduce error in the angle determination. Pattern changes that are a function of frequency can be accommodated by storing data in the computer as a function of frequency, but polarization effects cannot be similarly accommodated unless the incoming-signal polarization is measured.

Pairs of phase- and amplitude-matched conical spirals, arrayed in a frequency-independent manner, can be used in amplitude-comparison DF systems. In a circular array of spiral pairs, the summed outputs of each pair can be compared to obtain a DF bearing. DF is often accomplished by using a circular array of single spirals (either conical or flat cavity-backed), but the pairs provide a narrower azimuthal beamwidth and hence better accuracy. Corrugated horns are also useful in amplitude-comparison systems. A circular array of almost any type of directive antenna elements can be used for amplitude comparison; however, to obtain similar DF accuracy over a broad bandwidth, the pattern performance of the elements must be relatively independent of frequency.

40-4 TRACKING ANTENNAS

A tracking antenna is designed to provide the information required by the electronic system to change periodically the pointing of the antenna in both azimuth and elevation so that the antenna boresight is repositioned to the target emitter's location. When this is accomplished sufficiently often, the antenna will track the emitter in a smooth motion. Three main techniques are used: conical scan, sequential lobing, and monopulse.[16] These techniques have been described in Chap. 34. Examples of how these techniques are applied in broadband systems will be described in the following paragraphs.

Conical Scan

Conical scan is achieved by mechanically moving the feed in a circle about the focal point in the focal plane of a paraboloidal reflector. Moving the phase center of the feed antenna off the focal axis causes the secondary beam to squint off the focal axis in the opposite direction; moving the feed in a circle then causes the peak of the secondary beam to move on a conical surface. If the feed maintains a fixed plane of polarization as the feed rotates, the feed is said to *nutate* as compared with a *rotating* feed, in which the plane of the polarization rotates with the feed. For ESM applications, for which linear polarization is required, the antenna must have a nutating feed because rotating the feed would modulate the signal at twice the scan rate, destroying the tracking information.

A conical-scan tracking antenna,[17] which operates over the 1- to 11-GHz band, is shown in Fig. 40-6a. The feed is a linearly polarized, ridge-loaded horn. A dielectric lens is used in the aperture to correct phase error. The decrease in secondary beamwidth with frequency is limited by the fact that above approximately 4 GHz the feed horn illuminates only a portion of the reflector. As the frequency increases above 4 GHz, the area illuminated is reduced sufficiently as a function of frequency to maintain essentially a constant secondary beamwidth (Fig. 40-6b). This allows the feed to be nutated at a constant radius about the focal point, with the crossover level being maintained within 4 dB of the beam peaks. Gain and boresight shift as measured on this antenna are also shown in Fig. 40-6b. While the constant-beamwidth technique results in a system whose gain does not increase with frequency, it does result in a system with a constant acquisition angle, which minimizes acquisition time.

FIG. 40-6a   1- to 11-GHz conical-scan tracking antenna.

Sequential-Lobing Antennas

Sequential lobing is very similar to conical scan, the difference being that whereas in conical scan the beam is moved continuously in a conical fashion, in sequential lobing the beam is switched between four positions. The beam is stepped from beam right to beam up, to beam left, to beam down. Signal strength is sampled at each position and compared with the opposite position to obtain tracking information.

Figure 40-7a shows a four-horn sequential-lobing tracking antenna. The horns are doubly ridged and hence provide linear polarization which can be rotated by remote control. Dielectric lenses correct any phase error in the aperture.

The beam-forming network used with this antenna system is shown in Fig. 40-7b. In the normal tracking mode, all four horns are combined with the proper phase to form the beam in one of the desired beam positions, the phase being introduced by the comparator quadrature hybrids. The comparator hybrids inherently provide a crossover level of approximately 3 dB on boresight independently of frequency. Diode switches select the proper path for each horn to achieve or change the beam position. Because the horns are separated by a fixed distance, grating lobes which can provide a multiple-tracking axis develop at the higher frequencies. To ensure that the true boresight is achieved, an acquisition mode is provided. Each horn is physically pointed 5° off the boresight direction. By combining only the horns on the right side, a beam squinted to the right is obtained. The width of the beam is the same as the individual horn beamwidth and hence is sufficiently wide to cover all grating lobes. Coaxial switches bypass the comparator hybrids for this mode. This mode provides a sufficient

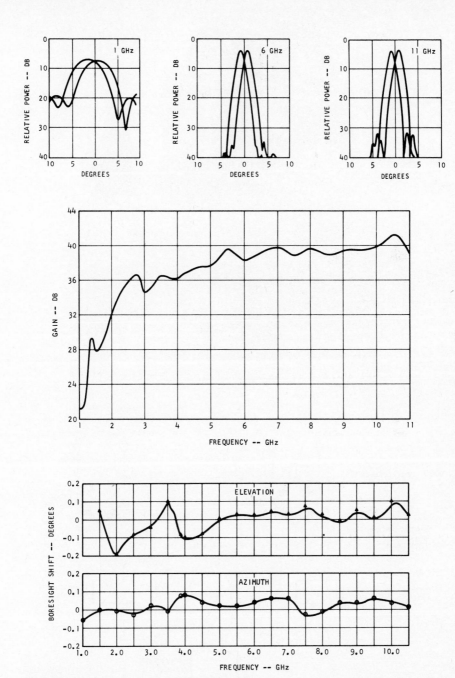

FIG. 40-6*b* Patterns, gains, and boresight shifts of a 1- to 11-GHz conical-scan tracking antenna.

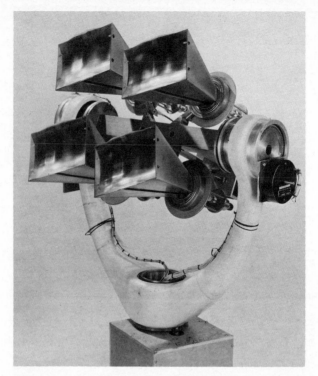

FIG. 40-7a Four-horn sequential-lobing tracking antenna (1 to 5 GHz).

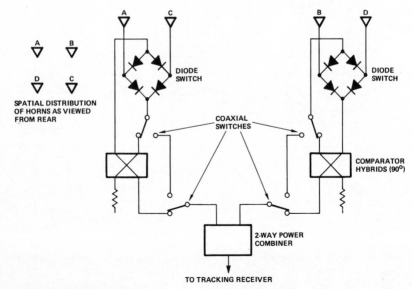

FIG. 40-7b Beam-forming network of the four-horn sequential-lobing tracking antenna (1 to 5 GHz).

error gradient to ensure that the system will lock on at the true boresight when it is switched to the normal tracking mode for better tracking accuracy.

Two-Channel Monopulse

Figure 40-8 shows a two-channel monopulse feed consisting of eight log-periodic monopole arrays on an octagonal cone. The outputs from the eight arrays are combined in a beam-forming network to form circularly polarized sum and difference patterns. For two-channel monopulse, both patterns are figures of revolution, with the difference pattern having a point null. The phase of the sum pattern changes from 0 to 360° around the boresight axis; for the difference pattern, the phase changes from 0 to 720°. Thus, the phase difference between the patterns defines the angle ϕ about the boresight axis (relative to an established reference plane) from which the target signal is arriving (Fig. 40-9), while the sum-to-difference-amplitude ratio defines the angle θ off the axis.

FIG. 40-8   Two-channel monopulse feed.

Two-channel monopulse systems lend themselves to broadband (36:1 bandwidth) tracking systems because beam-forming-network errors result in boresight shift rather than null fill-in, as is the case for three-channel systems. Null fill-in degrades tracking accuracy owing to degradation of error slope, while boresight shift in two-channel systems produces bias errors which in principle can be eliminated by calibration techniques.

Two-channel monopulse patterns are also inherent in multimode spiral antennas[18] in addition to the multiple log-periodic monopole array configuration. However, since spiral patterns turn about the boresight axis as a function of frequency, the reference plane at which the sum and difference channels are in phase also turns

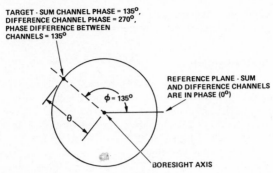

FIG. 40-9 Target location using two-channel monopulse techniques.

about the axis as a function of frequency, whereas in the log-periodic array configuration this plane is independent of frequency. Phase-compensation networks can be employed to stabilize the reference plane when spirals are used.

It should also be noted that, given a priori knowledge of the two-channel monopulse system's patterns and the orientation of the reference plane, this system can also be used for direction finding by measuring the amplitude ratio and phase difference between the sum and difference channels.

40-5 ECM TRANSMIT ANTENNAS

The antenna has been considered to be a transformer used to couple efficiently signals to or from free space. In that context, the problem for ECM-antenna designers has been to impedance-match the coupling device so that maximum power is transferred. Controlled distribution of the signal in space is a requirement of equal importance imposed upon the ECM antenna. Octave bandwidths or greater, medium to high radio-frequency (RF) power-carrying capacity, multiple-polarization response, and a low voltage standing-wave ratio (VSWR) have of course been additional requirements placed upon the ECM-antenna designer.

Antenna Elements

Many different types of antennas have been used in ECM systems. Some of the more common are broadband dipoles, broadband monopoles, spirals, horns and slots, logperiodics, and reflector types. During the past several decades, the RF power-carrying capacity and bandwidth of these basic types have been increased. As an example, the transmitting horns (shown in Fig. 40-10) used in an existing system exhibit a broad frequency bandwidth, high-power capability, high-aspect-ratio beam coverage, and circular-polarization response. Three-to-one-bandwidth, high-aspect-ratio circularly polarized horns have also been developed with satisfactory results.

The basic design problem associated with such horn configurations is the maintenance of a 90° phase differential between two (traveling-wave) quadrature fields in a waveguide structure. Frequency-compensated insertion-phase design techniques

FIG. 40-10 Multiband ECM transmitting horns. *(Courtesy of Transco, Inc.)*

applicable to the problem have been defined. A simplified analysis of the design of an octave-band circularly polarized horn will illustrate the approach applicable to a horn that uses a dielectric slab to obtain the 90° relative phase shift for the quadrature fields. The phase shift in the plane perpendicular to the wide horn aperture (*Y* axis, with the phasing plate) and the phase shift in the orthogonal plane (*X* axis) as illustrated in Fig. 40-11 were calculated; the results are shown in the figure. The calculations are approximate because average horn dimensions were assumed and the lens

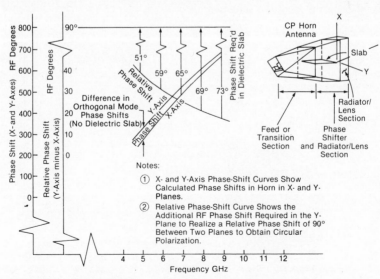

FIG. 40-11 Calculated phase shifts in an asymmetric-beam horn.

portion of the radiator section was assumed to be a waveguide section with solid-dielectric loading. The phase shift in a waveguide section loaded with dielectric is given by

$$\Delta \theta = 30.5X \left(F^2 K_e - \frac{34.9}{a^2} \right)^{1/2} \quad {}^\circ$$

where X is the length of the waveguide section in inches, F is the frequency in gigahertz, K_e is the relative permittivity of the material within the guide if dielectric material is used, and a is the guide width in inches. ($K_e = 1$ for air-filled guide. The formula is derived from the standard waveguide-wavelength relationship.) An additional phase delay that increases with frequency in a complementary fashion is required in the Y plane to achieve circularly polarized operation (90° relative phase shift) over the complete RF band. The increase in differential-phase-shift requirements at the higher frequencies is compatible with the dielectric-slab phase-shift characteristics.

Extremely broadband monopoles are now available. Using thick (bladelike) configurations and special impedance matching, a 2:1 bandwidth design is relatively straightforward. One company is engaged in research and development to achieve up to a 6:1 bandwidth. Many broadband blades or monopoles have been used in ECM systems.

Fixed-Beam Arrays

An array can be used to create simultaneous fixed beams or to form beams with a high aspect ratio and of a desired shape. One example of a fixed-beam array is presented to illustrate current designs. Twelve horn elements are arranged in the arc of a circle, as illustrated in Fig. 40-12, to form a beam fixed in space that has broad azimuth coverage and a narrower elevation extent. This configuration is used to combine multiple traveling-wave-tube (TWT) powers in space. The azimuth beamwidth, in the plane of the four-horn subgroups, is broad as shown; the elevation beamwidth

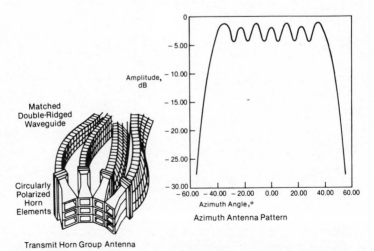

FIG. 40-12 Fixed-beam array for multiple power-amplifier application.

is dependent on the vertical (three-element) array configuration. The individual horn aperture size, both in azimuth and in elevation, must be selected with care. The horn beamwidths are chosen so that approximately one-fourth of the total desired coverage sector is "illuminated" by each horn; the flared section of the horn must be long to reduce aperture phase error, which minimizes phase center spacings. This reduces out-of-phase interference at angles on either side of the in-phase-beam-crossover points. The calculated pattern in Fig. 40-12 shows an approximate 3-dB variation in amplitude; in view of the wide variations in several "factors" affecting jammer-to-radar-signal ratios in a typical ECM application, this pattern variation is quite acceptable. The RF power of 12 TWTs feeding the horn group can be added in space if phase-tracking requirements are satisfied.

Arrays with Beam Control

Combinations of broad- and narrow-beam antenna response are often required. Narrow, or high-gain, beams can be directed to difficult-to-jam high-effective-radiated-power (ERP) pulse radars on a very-short-burst sequential basis, whereas the broad-coverage beam, used most of the time, can be employed against the high-duty-cycle, lower-ERP radars that may be spatially separated. High-quality narrow- and broad-beam patterns have been achieved over an octave bandwidth.

A patented[19] matrix-hybrid array configuration is shown in Fig. 40-13. The feed-system loss depends on the number of radiating elements; amplifiers are inserted near the radiators to negate this feed-loss characteristic. Simultaneous high-ERP beams with fixed positions in space can be created with the design. Beam switching and broad-beam operation with all inputs fed in phase are also possible by using this approach.

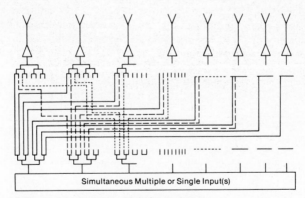

Simultaneous Multiple or Single Input(s)

FIG. 40-13 Matrix multiple-beam array.

Another multibeam, multifeed array configuration now receiving considerable attention for both airborne and shipborne applications is the Rotman lens.[20] RF amplifiers are normally inserted in the assembly between the lens feed and each radiating element. The Rotman lens antenna has several advantages: frequency-independent beam pointing is achieved, printed-microwave-circuit technology can be used to fabricate the lens, and very rapid control of high-ERP beams is possible because beam switching is exercised at the low-RF drive power. By using external-polarization-

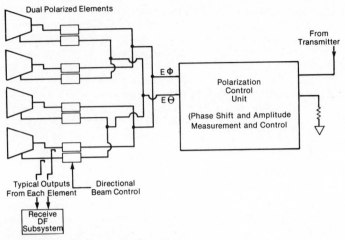

FIG. 40-14 Corporate-feed array.

adjustment techniques, a circularly polarized version of this antenna has been designed for a 3:1 bandwidth. (See Chap. 16.)

A corporate-feed array is illustrated in Fig. 40-14. This array offers some significant advantages: it is compatible with a single-output, high-power jammer; a zoom, or broad-beam pattern response, is available by adjusting the phase shifters in the array to yield a circular phase front; frequency-independent beam pointing is available if nondispersive phase shifters are used; and continuously pointed beams in any direction are available with fine phase-shifter adjustments. For applications in which single-aperture transmit-receive operation is needed, directional couplers can be added near the radiation elements for direction finding and beam control. A polarization-control unit is shown on the right side of the diagram. By using hybrids and phase shifters in the polarization section, any polarization response can be attained.

A concept for very-high-power amplifier paralleling and antenna switching is shown in Fig. 40-15. With the input switch in the position shown, the signals passing through the amplifiers will exhibit a phase front as illustrated; all the power is thus directed to the bottom horn antenna. Low-drive-power switching at the input can direct very large amounts of power to high-gain horns pointed at different angles at the outputs. Without switching, the back-to-back lens configuration can be used as a high-power broadband switch and/or power combiner.

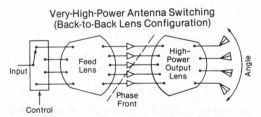

FIG. 40-15 Parallel-amplifier configurations.

40-6 ANTENNA ISOLATION IN JAMMING SYSTEMS

Isolation between transmitting and receiving antennas is an important consideration in an ECM system. Analyses and expressions for the calculation of field-strength propagation beyond obstacles have been published. Propagation around a curved surface (cylinders) and beyond knife edges (diffraction) is documented in Ref. 21.

Another antenna-coupling-analysis and measurements program[22] addressed the problem of defining the isolation between antennas separated by two conducting planes forming a corner with a variable included angle (see Fig. 40-16). The expressions for antenna coupling, determined by physical-optics analysis and by experiment, are

$$\text{Isolation} = \frac{K_1 G_T G_R \lambda^3}{x^2 Y} \text{ for } x > Y$$

$$= \frac{K_1 G_T G_R \lambda^3}{x Y^2} \text{ for } Y > x$$

K_1 is -36.5 dB for circularly polarized horns and spiral antennas and -39.5 dB for monopole antennas. x and Y are defined in Fig. 40-16. G_T and G_R are the transmit and receive antenna gains, each antenna being pointed at the common corner edge. The gain values are in power units with the gain values associated with the polarization perpendicular to the edge surface. These formulas yield reasonably accurate answers for antennas

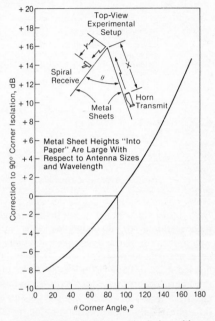

FIG. 40-16 Variation of isolation with corner angle for antenna coupling beyond a corner obstacle.

moved 10 to 20° (measured to the corner) away from the ground planes. Measurements beyond 20° were not performed. The investigation was carried out in the 2- to 4-GHz range. Measured isolations for several different corner angles and antenna types correspond to the calculated isolations to within 2 dB for large variations in x and Y.

40-7 ADVANCED ARRAY SYSTEMS

Solid-state arrays offer hope for significant improvements in ECM systems.[23] By using a building-block approach, significant ERPs can be obtained because the gain of the array increases with the number of elements; also the total power increases with the number of elements (one amplifier per element). Moreover, transmission-line losses will be reduced because the power amplifiers are located near the radiators. An all-solid-state array configuration, using distributed field-effect-transistor (FET) amplifiers, with potential for full-polarization-diversity capability is shown in Fig. 40-17.

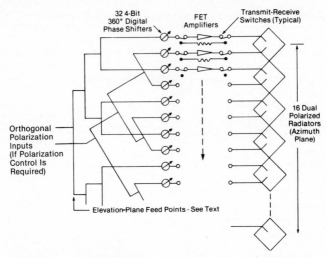

FIG. 40-17   30-kW-EIRP, 8 × 16 solid-state array with full-polarization-diversity capability.

An 8- by 16-element array is represented; only one layer of 16 elements in the azimuth plane is indicated. The orthogonal-polarization inputs near the left side of the diagram would, in turn, be fed by an elevation-plane feed network (not shown). Eight layers of antennas (one of which is shown), each layer including its own drive amplifier, are needed to achieve the 30-kW effective isotropic radiated power (EIRP) indicated in the figure. Appropriate feed-subsystem losses have been assumed. The eight layers in elevation would be fed by using a constant circular phase front so that beam control in azimuth only and direction finding in azimuth only would be required. When cost, element power output, and FET efficiencies are improved, this concept will be used in many applications.

REFERENCES

1   J. D. Kraus, *Antennas,* McGraw-Hill Book Company, New York, 1950, pp. 217–229.
2   H. Jasik, *Antenna Engineering Handbook,* 1st ed., McGraw-Hill Book Company, New York, 1961, pp. 3-10–3-13.
3   L. Young, L. A. Robinson, and C. A. Hacking, "Meander-Line Polarizer," *IEEE Trans. Antennas Propagat.,* vol. AP-21, no. 3, May 1973, pp. 376–378.
4   N. E. Lindenblad, "Antennas and Transmission Lines in the Empire State Television Station," *Communications,* vol. 21, no. 4, April 1941, pp. 13–14.
5   J. J. Epis and F. E. Robles, "Partial Radial-Line Antennas," U.S. Patent 3,831,176, 1974.
6   K. L. Walton and V. C. Sundberg, "Broadband Ridged Horn Design," *Microwave J.,* March 1964, pp. 96–101.
7   T. Sexon, "Quadruply Ridged Horn," Rep. EDL-M1160, Cont. DAAB07-67-C-0181, par. 2.1, 1968.
8   D. K. Barton, *Radar Systems Analysis,* Prentice-Hall, Inc., Englewood Cliffs, N.J., 1964, pp. 51–57, 275–310, 317–347.

9   M. I. Skolnik, *Introduction to Radar Systems,* McGraw-Hill Book Company, New York, 1962, pp. 476–482.

10 L. G. Bullock, G. R. Oeh, and J. J. Sparagna, "Precision Broadband Direction Finding Techniques," *Antennas Propagat. Int. Symp. Proc.,* University of Michigan, Ann Arbor, Oct. 17–19, 1967, pp. 224–232.

11 L. G. Bullock et al., "An Analysis of Wide-Band Microwave Monopulse Direction Finding Techniques," *IEEE Trans. Aerosp. Electron. Syst.,* vol. AES-7, January 1971, pp. 188–203.

12 N. M. Blachman, "The Effect of Noise on Bearings Obtained by Amplitude Comparison," *IEEE Trans. Aerosp. Electron. Syst.,* September 1971, pp. 1007–1009.

13 S. Silver, *Microwave Theory and Design,* McGraw-Hill Book Company, New York, 1949, sec. 13.8.

14 A. S. Dunbar, "Calculations of Doubly Curved Reflectors for Shaped Beams," *IRE Proc.: Waves and Electron. Secs.,* October 1948, pp. 1289–1296.

15 A. Brunner, "Possibilities of Dimensioning Doubly Curved Reflectors for Azimuth Search Radar Antenna," *IEEE Trans. Antennas Propagat.,* vol. AP-19, no. 1, January 1971, pp. 52–57.

16 Skolnik, op. cit, pp. 164–189.

17 K. L. Walton and V. C. Sundberg, "Constant Beamwidth Antenna Development," *IEEE Trans. Antennas Propagat.,* vol. AP-16, September 1968, pp. 510–513.

18 G. A. Deschamps and J. D. Dyson, "The Logarithmic Spiral in a Single-Aperture Multimode Antenna System," *IEEE Trans. Antennas Propagat.,* vol. AP-19, no. 1, January 1971, pp. 90–96.

19 E. Kadak, "Conformal Array Beamforming Network," U.S. Patent 3,968,695.

20 W. Rotman and R. F. Turner, "Wide-Angle Microwave Lens for Line-Source Applications," *IEEE Trans. Antennas Propagat.,* vol. AP-11, November 1963.

21 J. L. Bogdner, M. D. Siegel, and G. L. Weinstock, "Air Force ATACAP Program: Intra-Vehicle EM Capability Analysis," AFAL-TR-71-155, July 1971.

22 D. F. Yaw and J. G. McKinley, "Broadband Antenna Isolation in the Presence of Obstructions," Westinghouse Tech. Mem. EVTM-75-124, December 1975.

23 D. F. Yaw, "Electronic Warfare Antenna Systems—Past and Present," *Microwave J.,* September 1981, pp. 22–29.

Chapter 41

Radio-Telescope Antennas

John D. Kraus

The Ohio State University

41-1 RADIO TELESCOPES: DEFINITION

A radio telescope consists of an antenna for collecting celestial radio signals and a receiver for detecting and recording them. The antenna is analogous to the objective lens or mirror of an optical telescope, while the receiver-recorder is analogous to the eye-brain combination or a photographic plate. By analogy the entire antenna-receiver-recorder system may be referred to as a *radio telescope,* although it may bear little resemblance to its optical counterpart. Radio telescopes are used in much the same manner as optical telescopes for the observation and study of celestial objects. However, the appearance of the sky at radio wavelengths is very different from the way in which it looks optically. Thus, the sun is much less bright. On the other hand, the Milky Way radiates with tremendous strength, and the rest of the sky is dotted with radio sources almost entirely unrelated to any objects visible to the unaided eye (see Fig. 41-1).

The earth's atmosphere and ionosphere are opaque to electromagnetic waves with two principal exceptions, a band or window in the optical region and a wider window in the radio part of the spectrum. This radio window extends from a few millimeters to tens of meters in wavelength, being limited on the short-wavelength side by molecular absorption and on the long-wavelength side by ionospheric reflection. Because of this wide range of wavelengths many forms of antennas are used.

41-2 POSITION AND COORDINATES

The accurate position of a radio source is necessary to distinguish the source from others and to assist in its identification with optical objects when possible. The position is conveniently expressed in *celestial equatorial coordinates: right ascension* α and *declination* δ. The poles of this coordinate system occur at the two points where the earth's axis, extended, intersects the celestial sphere. Midway between these poles is the *celestial equator,* coinciding with the earth's equator, expanded.

The declination of an object is expressed in degrees and is the angle included between the object and the celestial equator. It is designated as a positive angle if the object is north of the equator and negative if south. For example, at the earth's equator a point directly overhead (the zenith) has a declination of 0°, while at a north latitude of 40° the declination of the zenith is +40°.

The *meridian* is a great circle passing through the poles and a point directly overhead (the zenith). The *hour circle* of an object is the great circle passing through the object and the poles. The *hour angle* of the object then is the arc of the celestial equator included between the meridian and the object's hour circle. This angle is usually measured in hours.

A reference point has been chosen on the celestial equator. It is called the *vernal equinox.* The arc of the celestial equator included between the vernal equinox and the object's hour circle is termed the *right ascension* of the object. It is measured eastward from the vernal equinox and is usually expressed in hours, minutes, and seconds of time.

The right ascension and declination of an object define its position in the sky, independent of the earth's diurnal rotation. However, because of the earth's preces-

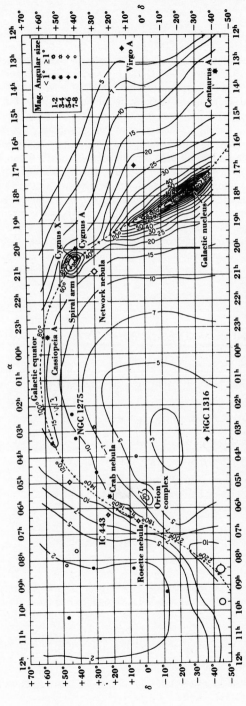

FIG. 41-1 Radio sky at 250 MHz as observed with the Ohio State University 96-helix radio telescope. (*After Ko and Kraus.*[1])

41-3

sion, there is a gradual change in these coordinates for a fixed object in the sky, the change completing one cycle in 26,000 years. Thus, the right ascension and declination of an object will again be the same as they are now in 26,000 years. To be explicit, it is necessary to specify the date to which the right ascension and declination refer. This date is called the *epoch*. At present the epoch 1950.0 is commonly used (that is, the right ascension and declination are those of January 1, 1950), but epoch 2000.0 is being used increasingly.

Sometimes the positions of celestial objects are given in *galactic coordinates,* which are based on the geometry of our galaxy. These coordinates are independent of the earth and hence require no date or epoch. However, there are two systems, *old* (used before 1960) and *new* (used after 1960).[2] Their poles differ by about 1.5°.

By placing an antenna on an *equatorial* or *polar mounting,* that is, on axes one of which is parallel to the earth's axis and the other perpendicular to it, a source can be tracked as it moves across the sky by motion in only one coordinate (right ascension). If the antenna is mounted on vertical and horizontal axes, tracking requires motion in two coordinates. The coordinates in this case are *altitude* (or elevation) and *azimuth* (or horizontal angle around the horizon), and hence this type of mounting is called an *altazimuth* mounting.

41-3 BRIGHTNESS AND FLUX DENSITY

Radiation over an extended area of the sky is conveniently specified in terms of its brightness, that is, the power per unit area per unit bandwidth per unit solid angle of sky. Thus,

$$B = \text{brightness (Wm}^{-2}\text{Hz}^{-1}\text{sr}^{-1})$$

Solid angle may be expressed in steradians (sr) or in square degrees (deg²):

$$1 \text{ sr} = 57.3^2 \text{ deg}^2 = 3283 \text{ deg}^2$$

The integral of the brightness B over a given solid angle of sky yields the power per unit area per unit bandwidth received from that solid angle. This quantity is called the power *flux density, S.* Thus

$$S = \int \int B \, d\Omega = \text{power flux density (Wm}^{-2} \text{ Hz}^{-1})$$

where B = brightness, $\text{Wm}^{-2} \text{ Hz}^{-1} \text{ sr}^{-1}$
$d\Omega$ = element of solid angle, sr

Integrating the power flux density from a radio source with respect to frequency yields the *total power flux density* S_T in the frequency band over which the integration is made. Thus,

$$S_T = \int S \, df = \text{total power flux density (Wm}^{-2})$$

where S = power flux density, $\text{Wm}^{-2} \text{ Hz}^{-1}$
df = element of bandwidth, Hz

41-4   TEMPERATURE AND NOISE

A resistor of resistance R and temperature T matched to a receiver by means of a lossless transmission line as in Fig 41-2a delivers a power to the receiver given by

$$P = kT \,\Delta f \qquad \qquad \textbf{(41-1)}$$

where P = power received, W
 $\quad k$ = Boltzmann's constant = 1.38×10^{-23} J·K^{-1}
 $\quad T$ = absolute temperature of resistor, K
 $\quad \Delta f$ = bandwidth, Hz
 If the resistor is replaced by an antenna of radiation resistance R, this equation also applies. However, the radiation resistance is not at the temperature of the antenna structure but at the effective temperature T of that part of the sky toward which the antenna is directed, as in Fig. 41-2b. In effect, the radiation resistance is distributed over that part of the sky included within the antenna acceptance pattern. From this point of view the antenna and receiver of a radio telescope may be regarded as a *bolometer* (or heat-measuring device) for determining the effective temperature of distant regions of space coupled to the system through the radiation resistance of the antenna.
 The power received by a radio telescope from a celestial source is

$$P = SA_e \,\Delta f \qquad \qquad \textbf{(41-2)}$$

where P = power received, W
 $\quad S$ = source flux density, W m^{-2} Hz^{-1}
 $\quad A_e$ = effective aperture of telescope antenna, m^2
 $\quad \Delta f$ = bandwidth, Hz

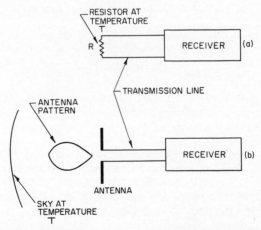

FIG. 41-2 Receiver connected to matched resistor and to antenna.

If this power equals that of a matched (calibration) resistor at temperature T connected to the receiver in place of the antenna, we have by equating Eqs. (41-1) and (41-2)

$$S = \frac{kT}{A_e} \tag{41-3}$$

However, if the source is unpolarized (as is true of most celestial sources), only half of its power will be received by any antenna, so the flux density of the source is twice the above value, or

$$S = \frac{2kT}{A_e} \tag{41-4}$$

where T = temperature of celestial object, K.

It should be noted that the celestial object is not necessarily at the thermal temperature T. Rather, T is the temperature that a blackbody radiator would need to have to emit radiation equal to that actually emitted by the object at the wavelength of observation. Hence, T is called the *equivalent temperature* of the object. This temperature is the same as the thermal temperature that the reference or calibration resistor must have to give the same response.

Sensitivity is a function of aperture size and also a system temperature and other factors. Thus, the sensitivity, or *minimum detectable (power) flux density,* is given by

$$S_{\min} = \frac{2k}{A_e} T_{\min} \tag{41-5}$$

where k = Boltzmann's constant = 1.38×10^{-23} J·K^{-1}
$T_{\min}$ = minimum detectable temperature, K
A_e = effective aperture = $\epsilon_{ap}A_p$, m^2
where ϵ_{ap} = aperture efficiency, dimensionless ($0 \leq \epsilon_{ap} \leq 1$)
A_p = physical aperture = $\pi(D^2/4)$ for circular aperture, m^2
where D = diameter of aperture.
The *minimum detectable temperature*

$$T_{\min} = \frac{K_s}{\sqrt{\Delta f t}} T_{\text{sys}} \tag{41-6}$$

where K_s = receiver sensitivity constant, dimensionless ($1 \leq K_s \leq 2.8$)
Δf = intermediate-frequency bandwidth, Hz
t = output time constant, s
T_{sys} = system temperature, K

The *system temperature* depends on the noise temperature of the sky, the temperature of the ground and environs, the antenna pattern, the antenna thermal efficiency, the receiver noise temperature, and the efficiency of the transmission line or waveguide between the antenna and the receiver. The system temperature at the antenna terminals is given by

$$T_{\text{sys}} = T_A + T_{AP}\left(\frac{1}{\epsilon_1} - 1\right) + T_{LP}\left(\frac{1}{\epsilon_2} - 1\right) + \frac{1}{\epsilon_2} T_R \tag{41-7}$$

where T_A = antenna noise temperature, K
  T_{AP} = physical temperature of antenna, K
   ϵ_1 = antenna (thermal) efficiency, dimensionless ($0 \leq \epsilon_1 \leq 1$)
  T_{LP} = physical temperature of transmission line or waveguide between antenna and receiver, K
   ϵ_2 = line or guide efficiency, dimensionless ($0 \leq \epsilon_2 \leq 1$)
   T_R = receiver noise temperature, K

The *antenna temperature* is given by

$$T_A = \frac{1}{\Omega_A} \int \int T(\theta,\phi) P_n(\theta,\phi) \, d\Omega \qquad \textbf{(41-8)}$$

where Ω_A = antenna-pattern solid angle or antenna beam area, deg^2 or sr
  $T(\theta,\phi)$ = noise temperature of sky and environs as a function of position angle at wavelength of operation, K
  $P_n(\theta,\phi)$ = normalized antenna power pattern, dimensionless
   $d\Omega$ = elemental solid angle, deg^2 or sr

The *elemental solid angle* is given by

$$d\Omega = \sin \theta \, d\theta \, d\phi \qquad \textbf{(41-9)}$$

where θ = angle from zenith (assuming antenna is pointed at zenith), deg or rad
   ϕ = azimuthal angle, deg or rad

The *antenna beam area* is given by

$$\Omega_A = \oiint P_n(\theta,\phi) \, d\Omega \qquad \textbf{(41-10)}$$

The *receiver noise temperature* is given by

$$T_R = T_1 + \frac{T_2}{G_1} + \frac{T_3}{G_1 G_2} + \cdots \qquad \textbf{(41-11)}$$

where T_1 = noise temperature of first stage of receiver (usually a low-noise preamplifier), K
   T_2 = noise temperature of second stage, K
   T_3 = noise temperature of third stage, K
   G_1 = gain of first stage
   G_2 = gain of second stage

Additional terms may be required if the temperature is sufficiently high and the gain sufficiently low on additional stages.

Radio telescopes are often used at one end of a communication circuit, and for this application the *signal-to-noise ratio* is of prime importance. When a radio telescope is the receiving station, we have from the Friis transmission formula that

$$P_r = \frac{P_t A_{et} A_e}{r^2 \lambda^2} \qquad \textbf{(41-12)}$$

where P_r = power received, W
  A_{et} = effective aperture of transmitting antenna, m^2
  A_{er} = effective aperture of receiving or radio-telescope antenna, m^2
   r = distance between transmitting and receiving station, m
   λ = wavelength of operation, m

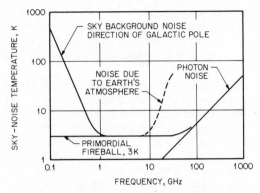

FIG. 41-3 Antenna sky noise temperature as a function of frequency. It is assumed that the antenna is pointed at the zenith, that the beam angle (HPBW) is less than a few degrees, and that the beam efficiency is 100 percent. *(After Kraus and Ko[3] below 1 GHz; after Penzias and Wilson[4] above 1 GHz.)*

TABLE 41-1   Radio Astronomy Units

Dimension or quantity	Symbol	Description	Unit
Brightness	B	$\dfrac{\text{Power flux density}}{\text{Solid angle}}$	$\mathrm{Wm^{-2}Hz^{-1}sr^{-1}}$
Power flux density	S	$= \displaystyle\int\int B\,d\Omega$	$\mathrm{Wm^{-2}Hz^{-1}}$
Total power flux density	S_T	$= \displaystyle\int S\,df$	$\mathrm{Wm^{-2}}$
Power (in terms of temperature)	P	$= kT\,\Delta f$	W
Brightness temperature	T	$= \dfrac{B\lambda^2}{2k}$	K
Power flux density (in terms of temperature)	S	$= \dfrac{2kT}{A_e}$	$\mathrm{Wm^{-2}Hz^{-1}}$
Minimum detectable flux density	S_{mim}	$= \dfrac{2kT_{\min}}{A_e}$	$\mathrm{Wm^{-2}Hz^{-1}}$
Minimum detectable temperature	$T_{\min}$	$= \dfrac{K_s}{\sqrt{\Delta ft}}$	K

NOTE: W = watts; m = meters; Hz = hertz; sr = steradians; K = kelvins = degrees celsius (°C) above absolute zero; df = infinitesimal bandwidth; Δf = finite bandwidth = $f_2 - f_1$; λ = wavelength; J = joules; K_s = receiver constant.

By dividing this received power by the minimum detectable power of the radio telescope we obtain the signal-to-noise ratio as

$$\frac{S}{N} = \frac{P_r}{S_{min}A_{er}\Delta_f} = \frac{P_t A_{et} A_{er} \sqrt{t}}{r^2 \lambda^2 k T_{sys} \sqrt{\Delta f}}$$ **(41-13)**

where P_t = transmitter power, W
A_{et} = effective aperture of transmitting antenna, m^2
A_{er} = effective aperture of receiving antenna, m^2
t = output time constant, s
r = distance between transmitting and receiving stations, m
λ = wavelength of operation, m
k = Boltzmann's constant = 1.38×10^{-23} J·K^{-1}
T_{sys} = system temperature of radio telescope, K
Δf = intermediate-frequency bandwidth, Hz
It is assumed that the receiver is a total-power type ($K_s = 1$) and that the radio telescope is matched to both the polarization and the bandwidth of the transmitted signal.

The antenna temperature T_A (which contributes to the system temperature T_{sys}, as discussed above) includes the contribution of the sky background and the temperature of the ground to the entire antenna pattern (main lobe and backlobes). In no case can T_A be less than 3 K, the limiting temperature of the sky resulting from the primordial fireball. The sky background temperature versus frequency is presented in Fig. 41-3.

Radio-astronomy units are summarized in Table 41-1.

41-5 RESOLUTION

The limiting resolution of an optical device is usually given by *Rayleigh's criterion*. According to this criterion, two identical point sources can just be resolved if the maximum of the diffraction pattern of source 1 coincides with the first minimum of the pattern of source 2.

Assuming a symmetrical antenna pattern as in Fig. 41-2a, Rayleigh's criterion applied to antennas states that the resolution of the antenna is equal to one-half of the beamwidth between first nulls; that is,

$$R = \frac{\text{BWFN}}{2}$$ **(41-14)**

where R = Rayleigh resolution or Rayleigh angle
BWFN = beamwidth between first nulls
An antenna pattern for a single-point source is shown in Fig. 41-4a. The pattern for two identical point sources separated by the Rayleigh angle is given by the solid curve in Fig. 41-4b, with the pattern for each source when observed individually shown by the dashed curves. It is to be noted that the two sources will be resolved provided that the half-power beamwidth (HPBW) is less than one-half of the beamwidth between first nulls, as is usually the case.

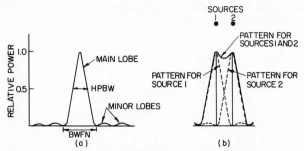

FIG. 41-4   Antenna power pattern (*a*) and power patterns for two identical point sources (*b*) separated by the Rayleigh angle as observed individually (dashed) and together (solid).

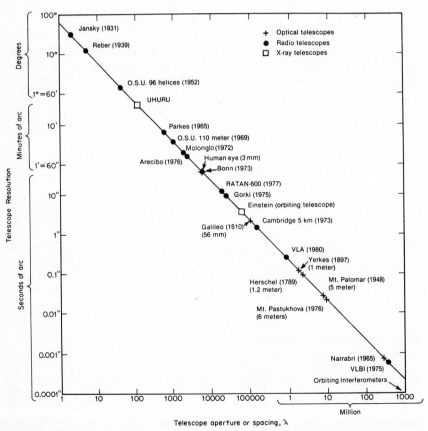

FIG. 41-5   Resolution of radio, optical, and x-ray telescopes as a function of the telescope aperture or interferometer spacing. *(From J. D. Kraus.[6])*

TABLE 41-2

Uniform Apertures	Rayleigh resolution (BWFN/2)	HPBW
Rectangular (length L_λ wavelengths)	$57.3°/L_\lambda$	$50.3°/L_\lambda$
Circular (diameter D_λ wavelengths)	$70°/D_\lambda$	$59°/D_\lambda$

It may be shown[5] that the beamwidth between first nulls for a broadside antenna with a uniform aperture many wavelengths long is given by

$$\text{BWFN} = \frac{114.6}{L_\lambda} \quad °$$ **(41-15)**

These results are summarized in Table 41-2, which also gives the beamwidths for uniform circular apertures. The HPBWs are some 15 percent less than the Rayleigh resolution angles for both rectangular and circular apertures.

For two-element interferometers the resolution is given by $57.3°/S_\lambda$, where S_λ is the spacing in wavelengths between the elements. For large spacings, extremely fine resolution can be obtained as indicated in Fig. 41-5, which presents the resolution of radio, optical, and x-ray telescopes.

41-6 PATTERN SMOOTHING

A unique pattern (or far-field pattern) of an antenna is obtained when the radiator is a point source situated at a sufficient distance from the antenna. The distance is sufficient if an increase produces no detectable change in the pattern. (See Sec. 41-8 for a more detailed discussion.) Let this pattern of a receiving antenna be as shown in Fig. 41-6a. If the point source is replaced by an extended source at the same distance, the observed pattern is modified as suggested in Fig. 41-6b. The extended source results in a broadened pattern with reduced minor lobes.

The patterns in Fig. 41-6 are in polar coordinates. In rectangular coordinates, the antenna pattern is as shown in Fig. 41-7a. The extended source distribution is shown in Fig. 41-7b. The observed pattern is then as shown in Fig. 41-7c. These three patterns are superposed in Fig. 41-7d to facilitate intercomparison.

It is to be noted that the observed pattern is only an approximation of the actual source pattern or distribution, the sharp shoulders of the source distribution

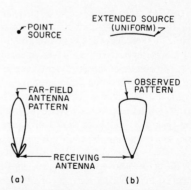

FIG. 41-6 Antenna patterns in polar coordinates for a point source and for an extended source.

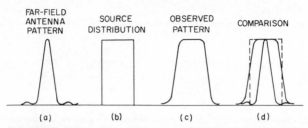

FIG. 41-7   Pattern in rectangular coordinates for antenna (*a*) and for uniform source distribution (*b*). The resultant observed pattern is at *c*. The three patterns are superposed for comparison in *d*.

being rounded off. It is said that the source distribution has been smoothed or blurred by the observing technique. The broader the antenna beamwidth compared with the source pattern, the greater the smoothing effect. On the other hand, the narrower the antenna beamwidth compared with the source pattern, the more nearly is the observed pattern an exact reproduction of the source distribution. However, there is always some smoothing, and one of the important problems of radio astronomy is to reconstruct, insofar as possible, the true source distribution from the observed pattern. It turns out that it is not possible to reconstruct the true source distribution since certain of the finer source details have no effect on the observed pattern and are irretrievably lost. However, partial reconstruction is usually possible, so that a source distribution that is more nearly like the exact source distribution than the observed pattern can be obtained. Pattern smoothing and related topics are discussed in detail in Kraus.[2]

41-7   CELESTIAL RADIO SOURCES FOR PATTERN, SQUINT, AND EFFICIENCY MEASUREMENTS OF ANTENNAS

Celestial radio sources are often useful for the measurement of far-field patterns, squint, and aperture efficiency of antennas, especially large-aperture antennas operating at short wavelengths when the distance to the far field is very large (10s or 100s of kilometers). (See Sec. 41-8 concerning the magnitude of this distance.)

For pattern measurements, the radio source should have a small angular extent (considerably less than the antenna HPBW), be relatively strong, and be well isolated from nearby sources. For aperture-efficiency measurements, accurate flux densities are required over a wide frequency range, and generally the source should be essentially unpolarized (less than 1 or 2 percent). For squint measurements, the position of the sources should be accurately known.

The upper part of Table 41-3 lists a few selected radio sources which meet most or all of the above requirements, that is, they are relatively strong, ½ to 1° from the nearest neighboring sources, of small angular extent, and essentially unpolarized and have accurate positions and also accurate flux densities over a wide frequency range. The lower part of the table lists three sources which do not meet all the above requirements but nevertheless may be useful for some purposes.

TABLE 41-3 Radio Sources for Pattern, Squint, and Efficiency Measurements

Source	Position (epoch 1950.0) Right ascension h	min	s	s error	Declination °	Arc min	Arc s	Arc s error	Size, arc min	Isolation, °	Flux density, (Jy)† 38 MHz	178 MHz	750 MHz	1400 MHz	2695 MHz	5000 MHz	Polarization, %	Distance, light-years
3C20	00	40	19.6	0.4	51	47	09	2	1	1	112	43	17	11.3	6.4	4.2	1.8	8×10^9 ($z = 0.9$)
3C196	08	09	59.4	0.2	48	22	07	2	0.25	1	166	68	23	13.9	7.7	4.4	1.2	3×10^9 ($z = 0.16$)
3C273	12	26	32.9	0.1	02	19	39	2	0.35		140	63	45	40	42	45	2	7.5×10^9 ($z = 0.85$)
3C295	14	09	33.6	0.3	52	26	14	2	0.3	0.5	94	83.5	35	22.4	12	6.5	0.4	3×10^9 ($z = 0.15$)
3C348	16	48	40.1	0.6	05	04	28	5	2	1	1,690	351	84	44.5	22.4	11.9	1.5	7×10^9 ($z = 0.7$)
3C380	18	28	13.4	0.2	48	42	40	2	0.25	0.5	211	59	22	14.4	10	7.5	1.2	10^9
Cygnus A	19	57	44.5	0.5	40	35	02		1.6	...*	22,000	8,700	2,980	1,590	785	371	0.2	
3C123	04	33	55.2	0.2	29	34	14	2	0.1	...	57⁻	189	72	45.8	27	16	1.2	
3C286	13	28	49.7	0.2	30	46	02	2	0.04	0.5	32	24	18	14.6	10	7.5	9	
Cas A	23	21	07		58	33	48		5	...*	37,200	11,600	3,880	2,410	1,470	910	...	11,000

*In complex galactic plane region.
†1 Jy = 1 jansky = 10^{-26} W m^{-2} Hz^{-1}.

41-13

The source designation refers to the third Cambridge (3C) catalog. Distances are given in light-years. The red shift z is also given. The red shift is the physical quantity measured from which a distance is inferred. Cygnus A is an exploded galaxy at a distance of 1 billion (10^9) light-years. Cassiopeia A (Cas A) is a supernova remnant (plasma cloud from an exploded star) at a distance of 11,000 light-years. Cas A is the strongest source in the sky except for the sun. Positions are from Bridle, Davis, Fomalont, and Lequeux,[7] and flux densities are from Kellermann, Pauliny-Toth, and Williams.[8] Flux densities at frequencies between those given in the table can be interpolated from the values given.

The source isolation (proximity of neighboring sources) was determined by inspection of the radio maps of the Ohio Sky Survey.[9] This survey is in seven parts with two supplements.

From the measurement of the antenna temperature difference (ΔT in °C) produced by the observed radio source, the effective aperture (A_e) of the antenna is given by

$$A_e = \frac{2k\Delta T}{S} \qquad \textbf{(41-16)}$$

where S = flux density of source at frequency of measurement, $Wm^{-2} Hz^{-1}$
$\qquad k = 1.38 \times 10^{-23} \, J \cdot K^{-1}$
The aperture efficiency is then

$$\epsilon_{ap} = \frac{A_e}{A_p} \qquad \textbf{(41-17)}$$

where A_e = effective aperture, m^2
$\qquad A_p$ = physical aperture m^2
For a detailed discussion of the measurement of celestial radio sources see Kraus.[2]

41-8   NEAR-FIELD–FAR-FIELD CONSIDERATIONS; RADEP

As has been mentioned, the far-field pattern of an antenna is obtained when measurements are made at a sufficiently great distance that an increase in distance produces no detectable change in the pattern. At shorter distances the pattern is different, its shape becoming a function of the distance.

The transition to this near-field pattern or patterns from the far-field pattern occurs when the source is sufficiently close that the phase of the incident wavefront across the antenna aperture departs significantly from a constant phase. Thus, if the distance δ in Fig. 41-8 is one-half wavelength, the incident field will be in opposite phase at the edges of the aper-

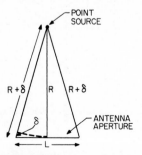

FIG. 41-8 Geometric relations for distance requirement.

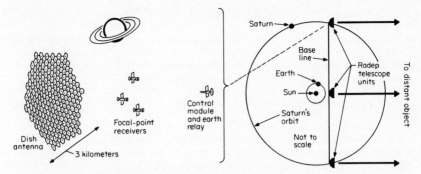

FIG. 41-9 The radep interferometer telescope proposed by a group of Soviet scientists could measure the distance of objects out to the full radius of the universe (15 billion light-years). Three units, each like the one shown at the left, would be deployed in space, two near the orbit of Saturn and one near the Earth.

ture as compared with the phase at the center. This could cause a very significant change in the observed antenna pattern. Thus, for accurate far-field patterns δ should be considerably less, and a value of one-sixteenth wavelength is commonly regarded as desirable. However, if we take δ = one-eighth wavelength, the relation for the required or critical distance becomes very simple; that is,

$$r \geq \frac{L^2}{\lambda}$$
(41-18)

where r = required or critical distance, m
   L = aperture dimension, m
   λ = wavelength, m
For a more detailed discussion see Kraus.[5]

It has been suggested[10] that the transition between the near-field and far-field conditions be utilized to measure the distance of remote astronomical objects. This passive radio technique for measuring distance, called *radep* for *ra*dio *dep*th, requires radio telescopes in space to avoid atmospheric effects and also to permit the construction of telescopes of sufficient size.

It has been proposed that an interferometer radep telescope of three units be constructed, two units being deployed near the orbit of Saturn with the third near the Earth.[10] Operating at centimeter wavelengths, this radep array (see Fig. 41-9) has the potential, in principle, of measuring the distance of celestial objects out to the full radius of the universe (15 billion light-years).

REFERENCES

1 H. C. Ko and J. D. Kraus, "A Radio Map of the Sky at 1.2 Meters," *Sky Telesc.,* vol. 16, February 1957, p. 160.

2 J. D. Kraus, *Radio Astronomy,* McGraw-Hill Book Company, New York, 1966; Cygnus-Quasar Books, Powell, Ohio, 1982.

3 J. D. Kraus and H. C. Ko, "Celestial Radio Radiation," Ohio State Univ. Radio Observ. Rep. 7, May 1957.

4 A. A. Penzias and R. W. Wilson, "A Measurement of Excess Antenna Temperature at 4080 MHz," *Astrophys. J.,* vol. 142, 1965, p. 419.

5 J. D. Kraus, *Antennas,* McGraw-Hill Book Company, New York, 1950.

6 J. D. Kraus, *Our Cosmic Universe,* Cygnus-Quasar Books, Powell, Ohio, 1980.

7 A. H. Bridle, M. M. Davis, E. B. Fomalont, and J. Lequeux, "Flux Densities, Positions and Structures for a Complete Sample of Intense Radio Sources at 1400 MHz," *Astronom J.,* vol. 77, August 1972, p. 405.

8 K. I. Kellerman, I. I. K. Pauliny-Toth, and P. J. S. Williams, "The Spectra of Radio Sources in the Revised 3C Catalogue," *Astrophys. J.,* vol. 157, July 1969, p. 1.

9 J. D. Kraus, "The Ohio Sky Survey and Other Radio Surveys," *Vistas Astron.,* vol. 20, 1977, p. 445. This article gives a summary of the Ohio Sky Survey and includes a master map and complete references to all installments and supplements.

10 N. Kardashev, J. Shklovsky, and others, *Academy of Sciences USSR Report PR-373,* Space Research Institute, Moscow, 1977.

PART 4

Topics Associated with Antennas

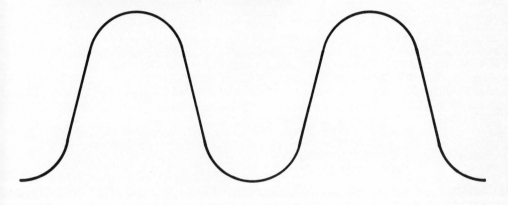

Chapter 42

Transmission Lines and Waveguides

Roderic V. Lowman

AIL Division
Eaton Corporation

42-1 GENERAL*

In the applications of antennas, it is necessary to use some form of transmission line to connect the antenna to a transmitter or receiver. It is essential to know the propagation characteristics of the more common forms of line.

The transmission lines in Secs. 42-2 to 42-6 are almost always used in the transverse electromagnetic (TEM) mode of propagation and therefore possess a unique characteristic impedance. In air, the guide wavelength λ_g is equal to the free-space wavelength λ. Propagation in this mode exists at any frequency, although above a certain frequency higher modes may also exist. It is assumed that the line loss is small and may be neglected in calculating the characteristic impedance.[1]

The waveguides considered in Secs. 42-7 and 42-8 use modes of propagation having a longitudinal component and do not possess a unique characteristic impedance.[2] Propagation may only take place above a certain unique cutoff frequency defined by the mechanical dimensions. The guide wavelength λ_g is related to the free-space wavelength λ by $1/\lambda_g^2 = 1/\lambda^2 - 1/\lambda_c^2$, where λ_c is the cutoff wavelength.

In all cases, it will be assumed that the skin depth is small compared with the dimensions of the conductors and that these conductors are nonmagnetic. The medium surrounding the conductors is often air, but for other dielectrics a loss will be introduced, and this is independent of the line dimensions. In the case of TEM modes this loss is

$$\alpha_D = \frac{\pi}{\lambda_D} \tan \delta \quad \text{Np/m}$$

where λ_D ($= \lambda/\sqrt{\epsilon}$) is the intrinsic wavelength in the medium and $\tan \delta$ is the loss tangent of the dielectric. ϵ is the relative dielectric constant. For waveguide modes, the wavelength in the guide must be considered, and the dielectric loss becomes

$$\alpha_D = \frac{\pi\lambda_g}{(\lambda_D)^2} \tan \delta \quad \text{Np/m}$$

In circuits having discontinuities, such as bends, the open types of line will have additional loss because of radiation at these points, and therefore shielded types should generally be used for complex circuits.

The power-handling capacity of transmission lines in the following sections is given in terms of the maximum allowable field intensity E_a in the dielectric, whereas for waveguides the power is given directly for air dielectric. For such calculations, $E_a = 3 \times 10^6$ V/m is the theoretical maximum for air dielectric, at normal temperature and atmospheric pressure, but for proper derating a value of 2×10^6 V/m is more practical. Other dielectrics having a higher dielectric strength may be used to increase the power limit, but in the case of solid dielectrics, the increased field strength in air pockets, which are very likely to exist, imposes a serious limitation. A common method of increasing the power limit is to use pressurized air, the maximum power being proportional to the square of the absolute pressure.

---

*A list of symbols is presented in Sec. 42-9.

The conditions discussed above apply to a matched line. The maximum power is inversely proportional to the standing-wave ratio.

Important transmission-line equations are shown in Table 42-1 for both TEM and non-TEM lines and waveguides.

42-2 OPEN-WIRE TRANSMISSION LINES

Open-wire transmission lines are an arrangement of wires whose diameters generally are small compared with the spacings involved. This arrangement is sometimes used in conjunction with a ground plane to which the wires are parallel. Such transmission lines have the advantage of simplicity and economy. The spacing between the wires and between the wires and the ground plane is very much less than a wavelength.

The most common of the open-wire lines is the two-wire line. For two wires of diameter d, spaced at a center-to-center distance D, the characteristic impedance is

$$Z_0 = \frac{120}{\sqrt{\epsilon}} \cosh^{-1} \frac{D}{d} \quad \Omega$$

This relation is plotted in Fig. 42-1. In the case of unequal wire diameters d_1 and d_2, d is replaced by $\sqrt{d_1 d_2}$.

The conductor loss is

$$\alpha_c = 2.29 \times 10^{-6} \sqrt{\frac{f\epsilon}{\sigma}} \frac{1}{d \log_{10}(2D/d)} \quad \text{Np/m}$$

The power-handling capacity is

$$P = \frac{E_a^2 d^2 \sqrt{\epsilon}}{240} \cosh^{-1} \frac{D}{d} \quad \text{W}$$

where E_a is the maximum allowable field intensity in the dielectric (Sec. 42-1).

Other configurations of open-wire lines are shown in Fig. 42-2, together with formulas for the characteristic impedance. Lines near a ground plane are also included. Other types of open-wire lines can be found in the literature.

42-3 WIRES IN VARIOUS ENCLOSURES

It is often advantageous to tailor a transmission line to fit a specific application when one of the more common forms is not convenient. Furthermore, with many antenna configurations a nonstandard form of line has definite advantages.

Various forms of transmission line comprising wires in different-shaped enclosures are shown in Fig. 42-3, together with formulas for characteristic impedance. It should be noted that these lines are shielded, with the exception of the corner enclosure, provided that those that are physically open have an opening less than one-half wavelength across, extend beyond the wire at least one-half wavelength, and have opposing surfaces that are maintained at the same potential.

TABLE 42-1 Transmission-Line Equations

Quantity	General line		Ideal line		Approximate results for low-loss lines	
	TEM line	Other	TEM line	Other	TEM line	Other
Propagation constant γ $= \alpha + j\beta$	$\sqrt{(R + j\omega L)(G + j\omega C)}$		$j\omega\sqrt{LC}$	$j\dfrac{2\pi}{\lambda_g}$	(See α and β below)	(See α and β)
Phase constant β	$\mathrm{Im}(\gamma)$	$\mathrm{Im}(\gamma)$	$\omega\sqrt{LC} = \dfrac{\omega}{v} = \dfrac{2\pi}{\lambda}$	$\dfrac{2\pi}{\lambda_g}$	$\omega\sqrt{LC}\left[1 - \dfrac{RG}{4\omega^2 LC} + \dfrac{G^2}{8\omega^2 C^2} + \dfrac{R^2}{8\omega^2 L^2}\right]$	$\dfrac{2\pi}{\lambda_g}$
Attenuation constant α	$\mathrm{Re}(\gamma)$	$\mathrm{Re}(\gamma)$	0	0	$\dfrac{R}{2Z_0} + \dfrac{GZ_0}{2} \approx \dfrac{\pi}{\lambda}\tan\delta^*$	$\dfrac{\pi}{\lambda}\lambda_g \tan\delta^*$
Characteristic impedance Z_0	$\sqrt{\dfrac{R + j\omega L}{G + j\omega C}}$	Geometry-dependent	$\sqrt{\dfrac{L}{C}}$	Geometry-dependent	$\sqrt{\dfrac{L}{C}}\left[1 + j\left(\dfrac{G}{2\omega C} - \dfrac{R}{2\omega L}\right)\right]$	Geometry-dependent

Input impedance Z_i	$Z_0\left[\dfrac{Z_L\cosh\gamma l + Z_0\sinh\gamma l}{Z_0\cosh\gamma l + Z_L\sinh\gamma l}\right]$	$Z_0\left[\dfrac{Z_L\cos\beta l + jZ_0\sin\beta l}{Z_0\cos\beta l + jZ_L\sin\beta l}\right]$	$Z_0\left[\dfrac{\alpha l\cos\beta l + j\sin\beta l}{\cos\beta l + j\alpha l\sin\beta l}\right]$
Impedance of shorted line	$Z_0\tanh\gamma l$	$jZ_0\tan\beta l$	$Z_0\left[\dfrac{\cos\beta l + j\alpha l\sin\beta l}{\alpha l\cos\beta l + j\sin\beta l}\right]$
Impedance of open line	$Z_0\coth\gamma l$	$-jZ_0\cot\beta l$	
Impedance of quarter-wave line	$Z_0\left[\dfrac{Z_L\sinh\alpha l + Z_0\cosh\alpha l}{Z_0\sinh\alpha l + Z_L\cosh\alpha l}\right]$	$\dfrac{Z_0^2}{Z_L}$	$Z_0\left[\dfrac{Z_0+Z_L\alpha l}{Z_L+Z_0\alpha l}\right]$
Impedance of half-wave line	$Z_0\left[\dfrac{Z_L\cosh\alpha l + Z_0\sinh\alpha l}{Z_0\cosh\alpha l + Z_L\sinh\alpha l}\right]$	Z_L	$Z_0\left[\dfrac{Z_L+Z_0\alpha l}{Z_0+Z_L\alpha l}\right]$
Voltage along line $V(z)$	$V_i\cosh\gamma z - I_0 Z_0\sinh\gamma z$	$V_i\cos\beta z - jI_iZ_0\sin\beta z$	
Current along line $I(z)$	$I_i\cosh\gamma z - \dfrac{V_i}{Z_0}\sinh\gamma z$	$I_i\cos\beta z - j\dfrac{V_i}{Z_0}\sin\beta z$	
Reflection coefficient ρ	$\dfrac{Z_L-Z_0}{Z_L+Z_0}$	$\dfrac{Z_L-Z_0}{Z_L+Z_0}$	
Standing-wave ratio	$\dfrac{1+|\rho|}{1-|\rho|}$	$\dfrac{1+|\rho|}{1-|\rho|}$	

Symbols:

R, L, G, C = distributed resistance, inductance, conductance, and capacitance per unit length.

l = length of line.

Subscript i denotes input-end quantities.

Subscript L denotes load-end quantities.

z = distance along line from input end.

λ = wavelength measured in dielectric.

v = phase velocity of line equals velocity of light in dielectric of line for an ideal line.

λ_g = guide wavelength.

$\tan\delta$ = loss tangent of dielectric.

*Case of low conductor loss.

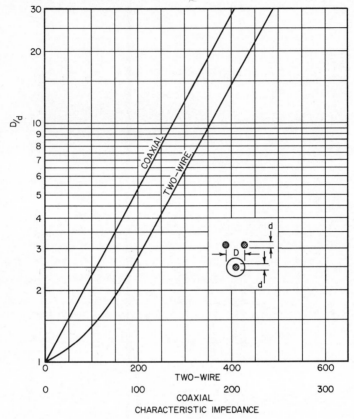

FIG. 42-1   Characteristic impedance of common lines.

42-4   STRIP TRANSMISSION LINES

In strip transmission lines the conductors are flat strips that most frequently are photoetched from a dielectric sheet which is copper-clad on one or both sides. Although
not often used as uniform transmission lines, they are very useful for antenna feeds,
networks, printed arrays, and slots. They are also useful for a number of microwave
integrated circuits such as amplifiers, filters, and high-speed digital logic circuits.

   A number of different types of these transmission lines have been suggested, but
a good variety of useful characteristics are supplied by four basic types that complement each other well: microstrip, slotline, coplanar waveguide, and coplanar strip.[4]
Cross sections of these types are shown in Fig. 42-4, along with methods for computing
Z_0 and λ_g or effective dielectric constant. Microstrip is made of a dielectric sheet with
a narrow conducting pattern of copper on one side and a copper sheet serving as a
ground plane on the other. Transmission is mostly of the TEM mode with a good part
of the electric field between the broad face of the conductor and the large ground
plane.

WIRES IN SPACE	CHARACTERISTIC IMPEDANCE	WIRES NEAR GROUND	CHARACTERISTIC IMPEDANCE
1. 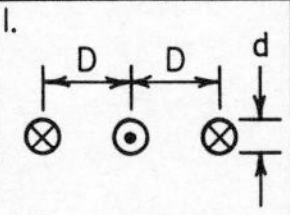 (⊗ ⊙ ⊗, spacing D, D, diameter d)	$Z_0 = \dfrac{207}{\sqrt{\epsilon}} \log_{10} 1.59 \dfrac{D}{d}$	4. 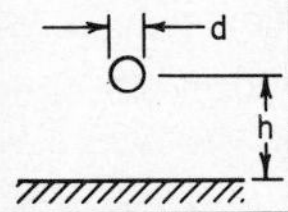 (single wire, diameter d, height h above ground)	$Z_0 = \dfrac{138}{\sqrt{\epsilon}} \log_{10} \dfrac{4h}{d}$
2. 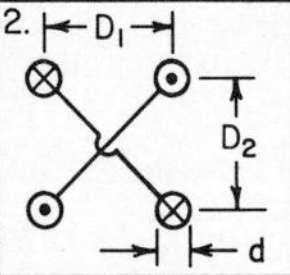 (crossed configuration, D_1, D_2, d)	$Z_0 = \dfrac{138}{\sqrt{\epsilon}} \log_{10} \dfrac{2D_2}{d\sqrt{1+(D_2/D_1)^2}}$	5. BALANCED PAIR 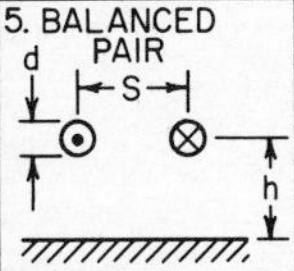 (d, spacing S, height h)	$Z_0 = \dfrac{276}{\sqrt{\epsilon}} \log_{10} \dfrac{4h}{d\sqrt{1+(2h/S)^2}}$
3. 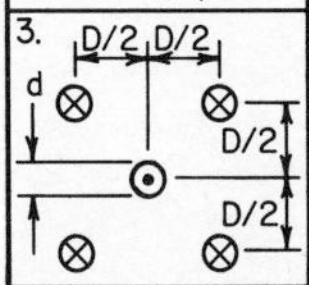 (D/2, D/2 configuration, d)	$Z_0 = \dfrac{173}{\sqrt{\epsilon}} \log_{10} 1.14 \dfrac{D}{d}$	6. PARALLEL PAIR 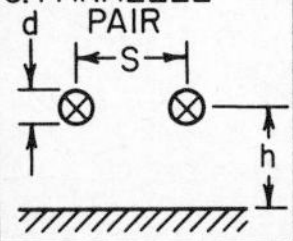 (d, spacing S, height h)	$Z_0 = \dfrac{69}{\sqrt{\epsilon}} \log_{10} \dfrac{4h}{d}\sqrt{1+(2h/S)^2}$
TWIN LEAD 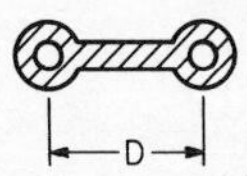	Z_0 DEPENDS ON WIRE DIAMETER, SPACING, AND THE SHAPE OF THE DIELECTRIC, USUALLY POLYETHYLENE. IT IS MOSTLY MADE IN 300 Ω.	7. 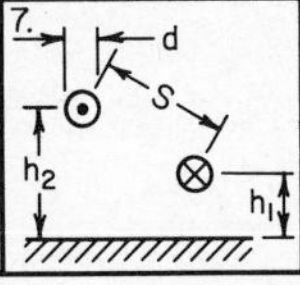 (d, S, h_2, h_1)	$Z_0 = \dfrac{276}{\sqrt{\epsilon}} \log_{10} \dfrac{2S}{d\sqrt{1+S^2/4h_1 h_2}}$

FIG. 42-2 Open-wife lines. Note that d is small compared with other dimensions.

SINGLE WIRE	CHARACTERISTIC IMPEDANCE	BALANCED WIRE PAIR	CHARACTERISTIC IMPEDANCE
1. 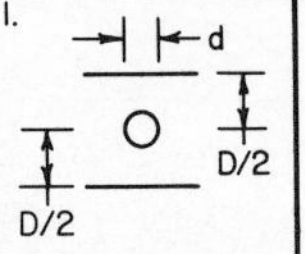 (d, D/2, D/2)	$Z_0 = \dfrac{138}{\sqrt{\epsilon}} \left[\log_{10} \dfrac{4D}{\pi d} - \dfrac{0.0367(\frac{d}{D})^4}{1 - 0.355(\frac{d}{D})^4} \right]$	5. 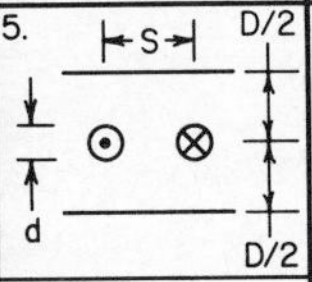 (S, D/2, D/2, d)	$Z_0 = \dfrac{276}{\sqrt{\epsilon}} \log_{10} \dfrac{4D}{\pi d} \tanh \dfrac{\pi S}{2D}$
2. 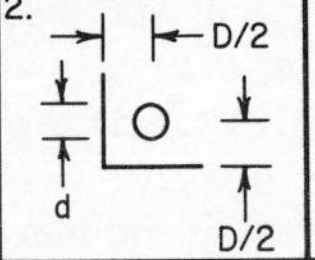 (D/2, D/2, d)	$Z_0 = \dfrac{138}{\sqrt{\epsilon}} \log_{10} \dfrac{\sqrt{2}D}{d}$	6. 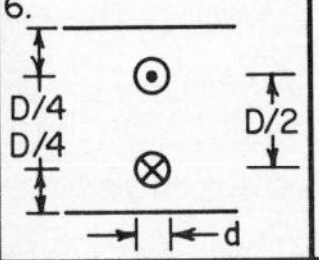 (D/4, D/4, D/2, d)	$Z_0 = \dfrac{276}{\sqrt{\epsilon}} \log_{10} \dfrac{2D}{\pi d}$
3. 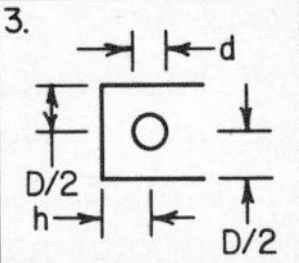 (d, D/2, h, D/2)	$Z_0 = \dfrac{138}{\sqrt{\epsilon}} \log_{10} \dfrac{4D}{\pi d} \tanh \dfrac{\pi h}{D}$ FOR EXACT FORMULA SEE REF. 3	7. 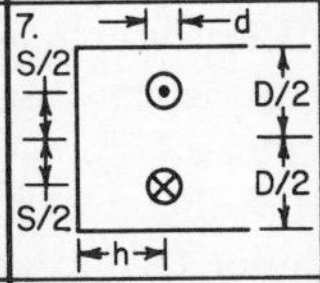 (d, S/2, D/2, S/2, D/2, h)	$Z_0 = \dfrac{276}{\sqrt{\epsilon}} \log_{10} \dfrac{2D}{\pi d\sqrt{A}}$ $A = \csc^2\left(\dfrac{\pi S}{d}\right) + \operatorname{csch}^2\left(\dfrac{2\pi h}{d}\right)$
4. 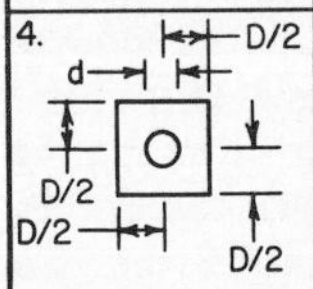 (d, D/2, D/2, D/2)	$Z_0 = \dfrac{138}{\sqrt{\epsilon}} \log_{10} 1.08 \dfrac{D}{d}$	8. 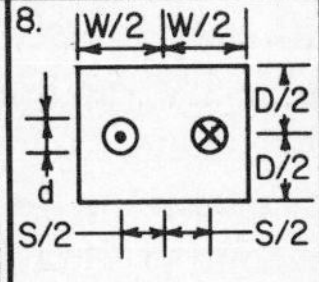 (W/2, W/2, D/2, D/2, d, S/2, S/2)	$Z_0 = \dfrac{276}{\sqrt{\epsilon}} \left(\log_{10} \dfrac{4D}{\pi d} \tanh \dfrac{\pi S}{2D} - \sum_{n=1}^{\infty} \log_{10} \dfrac{1+u_n^2}{1-v_n^2} \right)$ $u_n = \dfrac{\sinh\frac{\pi S}{2D}}{\cosh\frac{n\pi W}{2D}}, \quad v_n = \dfrac{\sinh\frac{\pi S}{2D}}{\sinh\frac{n\pi W}{2D}}$

FIG. 42-3 Wires in various enclosures. Note that d is small compared with other dimensions except in case 1.

TRANSMISSION LINE	CHARACTERISTIC IMPEDANCE	EFFECTIVE DIELECTRIC CONSTANT
MICROSTRIP 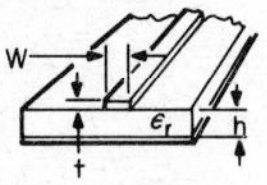	WIDE STRIP, W/h > 1 $Z_0 = \dfrac{377}{\sqrt{\varepsilon_{re}}}\left\{\dfrac{W_e}{h}+1.393+0.667\ell n\left(\dfrac{W_e}{h}+1.444\right)\right\}$ NARROW STRIP, W/h < 1 $Z_0 = \dfrac{377}{2\pi\sqrt{\varepsilon_{re}}}\ell n\left\{\dfrac{8h}{W_e}+0.25\dfrac{W_e}{h}\right\}$ WHERE $\dfrac{W_e}{h}=\dfrac{W}{h}+\dfrac{1.25}{\pi}\dfrac{t}{h}\left(1+\ell n\dfrac{4\pi W}{t}\right)$ FOR $\dfrac{W}{h}<\dfrac{1}{2\pi}$ and $\dfrac{W_e}{h}=\dfrac{W}{h}+\dfrac{1.25}{\pi}\dfrac{t}{h}\left(1+\ell n\dfrac{2h}{t}\right)$ FOR $\dfrac{W}{h}\geqslant\dfrac{1}{2\pi}$	$\varepsilon_{re}=\dfrac{\varepsilon_r+1}{2}+\left(\dfrac{\varepsilon_r-1}{2}\right)F\left(\dfrac{W}{h}\right)-C$ $F\left(\dfrac{W}{h}\right)=\sqrt{1+12\dfrac{h}{W}}+0.04\left(1-\dfrac{W}{h}\right)^2$ FOR $\dfrac{W}{h}\leqslant 1$ $=\sqrt{1+12\dfrac{h}{W}}$ FOR $\dfrac{W}{h}\geqslant1$ AND $C=\dfrac{\varepsilon_r-1}{4.6}\dfrac{t/h}{\sqrt{W/h}}$
COPLANAR STRIPS 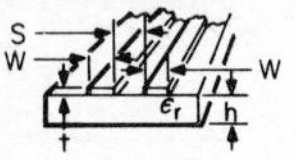	$Z_0 = \dfrac{377}{\sqrt{\varepsilon_{re}^t}}\dfrac{K(ke)}{K'(ke)}$ WHERE $ke=\dfrac{Se}{Se+2We}\approx k-\dfrac{(1-k^2)\Delta}{2W}$ AND $\Delta=\left(\dfrac{1.25t}{\pi}\right)\left[1+\ell n\left(4\pi\dfrac{W}{t}\right)\right]$	$\varepsilon_{re}^t=\varepsilon_{re}-\dfrac{1.4(\varepsilon_{re}-1)\,t/s}{\frac{K'(k)}{K(k)}+1.4\frac{t}{s}}$ WHERE $\varepsilon_{re}=\dfrac{\varepsilon_r+1}{2}\left[\tanh\left\{1.785\log\left(\dfrac{h}{W}\right)+1.75\right\}+\right.$ $\left.\dfrac{kW}{h}\left\{0.04-0.7k+0.01(1-0.1\varepsilon_r)(0.25+k)\right\}\right]$ AND $k=S/(S+2W)$
COPLANAR WAVEGUIDE 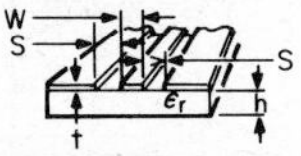	$Z_0 = \dfrac{30\pi}{\sqrt{\varepsilon_{re}^t}}\dfrac{K'(ke)}{K(ke)}$ and ke, & ε_{re} are as above. K IS THE COMPLETE ELLIPTIC OF THE FIRST KIND, K'(k)=K(k), AND k'=(1−k²)^{1/2}	$\varepsilon_{er}^t=\varepsilon_{er}-\dfrac{0.7(\varepsilon_{er}-1)\,t/W}{\frac{K(k)}{K'(k)}+1.4\frac{t}{W}}$
SLOTLINE 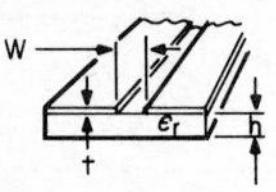	WIDE SLOTS 0.2 ⩽ W/h ⩽ 1.0 $Z_0=113.19-53.55\log\varepsilon_r+1.25\,W/h$ $(114.59-51.88\log\varepsilon_r)+20(W/h-0.2)$ $(1-W/h)$ $-[0.15+0.23\log\varepsilon_r+W/h(0.79$ $+2.07\log\varepsilon_r)]$ $\cdot[10.25-5\log\varepsilon+W/h(2.1-1.42\log\varepsilon_r)$ $-h\lambda_0\times10]^2$ NARROW SLOTS 0.02 ⩽ W/h ⩽ 0.2 $Z_0=72.62-35.19\log\varepsilon$ $+50\dfrac{(W/h-0.02)(W/h-0.1)}{W/h}$ $+\log(W/h\times10^2)[44.28-19.58\log\varepsilon_r$ $-[0.32\log\varepsilon_r-0.11+W/h(1.07\log\varepsilon_r+1.44]$ $\cdot(11.4-6.07\log\varepsilon_r-h/\lambda_0\times10^2)^2$	$\varepsilon_{er}=[0.987-0.483\log W/h(0.111-0.0022\varepsilon_r)$ $-(0.121+0.094\,W/h-0.0032\varepsilon_r$ $\log(h/\lambda_0\times10^2)]^{-2}$ $\varepsilon_{er}=[0.923-0.448\log\varepsilon_r+0.2\,W/h$ $-(0.29\,W/h+0.047)\log(h/\lambda_0\times10^2)]^{-2}$

FIG. 42-4 Stripline characteristics.

Microstrip is similar to a wire above ground that has been flattened. The field pattern is more complex, as only a part of it is in the dielectric sheet and the rest is in the air above the dielectric sheet. The other three are also mostly TEM, but their electric field is between the edges of the conductor pattern.

Coplanar strip is similar to two-wire line with flattened wires. It, too, is complex, with the fields partly in the dielectric and partly in the air above and below the sheet. The impedance is high because the capacity per unit length is low owing to the edge-to-edge position of the conductors.

Coplanar waveguide is really two coplanar strips in parallel. As might be expected, the impedance is lower for the same dimensions and dielectric. Slotline is similar to coplanar strips, with very wide conductor patterns.

The ranges of Z_0 available and the losses with the different types of strip transmission lines are compared in Table 42-2. These show that the microstrip and the coplanar waveguide are most useful for low-impedance circuits and that the slotline and coplanar strips are more appropriate for high-impedance circuits.[5]

The characteristics of microstrip are shown in Fig. 42-5. The dotted curve is the plot of Z_{0m}^a for airline. The solid curves are for finding the square root of the effective

TABLE 42-2 Comparison of Z_0 Limits and Loss for the Various Lines
($\epsilon = 10.0$, $h = 25$ mil, or 0.635 mm, and frequency $= 10$ GHz)

	Z_0 range		Loss, dB/cm	
Transmission line	**Minimum**	**Maximum**	**50**	**100**
Microstrip	20 (m)	110 (d)	0.04	0.14
Coplanar waveguide	25 (m, d)	155 (m, d)	0.08*	0.28*
Slotline	55 (d)	300 (m)	0.15†	
Coplanar strips	45 (m, d)	280 (m, d)	0.83*	0.13*

NOTE: (m) $= Z_0$ limited by mode; (d) $= Z_0$ limited by small dimensions.

*$h/W = 2$.

†$\epsilon = 16$, $Z_0 = 75$ Ω.

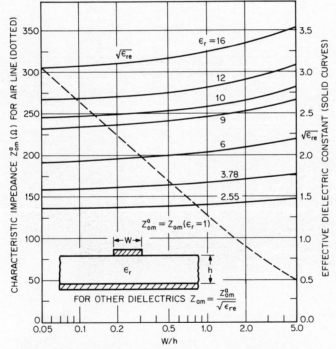

FIG. 42-5 Characteristic impedance and effective dielectric constant of microstrip lines.

dielectric constant. The Z_{0m} for other dielectrics is $Z_{0m}^a / \sqrt{\epsilon_{re}}$. Figure 42-6 shows the losses in microstrip on different substrates. Line wavelengths and characteristic impedance of coplanar strips and of coplanar waveguide are shown in Figs. 42-7 and 42-8. These same characteristics for slotline are shown in Fig. 42-9.

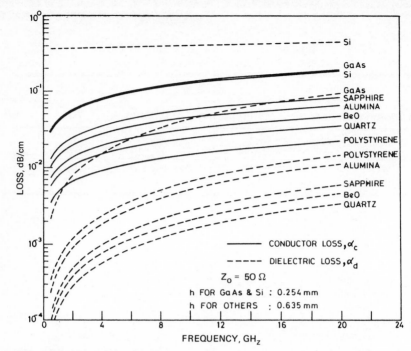

FIG. 42-6 Conductor and dielectric losses when microstrip is constructed on different substrates.

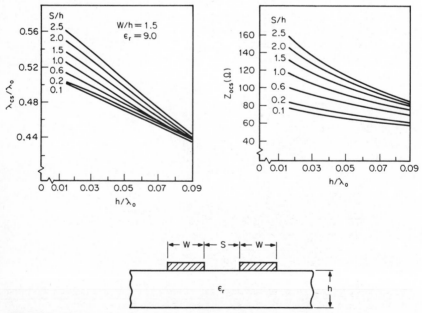

FIG. 42-7 Line wavelength and characteristic impedance of coplanar strips.

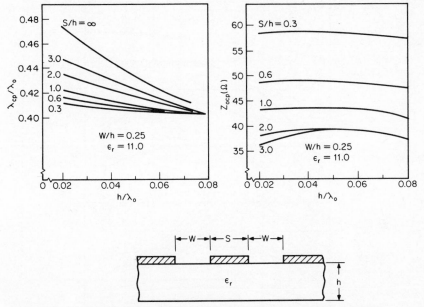

FIG. 42-8 Line wavelength and characteristic impedance of coplanar waveguide.

42-5   COAXIAL LINES: SOLID CONDUCTOR

Coaxial transmission lines using a cylindrical center conductor within a cylindrical tubular outer conductor are widely employed for the propagation of microwave power. Although more costly than open-wire transmission lines, they completely enclose the electromagnetic fields, preventing radiation losses and providing shielding from nearby circuits.

Coaxial-Line Parameters

For a coaxial line with inner-conductor diameter $2a$ and outer conductor diameter $2b$, the characteristic impedance is

$$Z_0 = \frac{138}{\sqrt{\epsilon}} \log_{10} \frac{b}{a} \quad \Omega$$

which is plotted in Fig. 42-1.

The cutoff wavelength of the first higher mode is[2]

$$\lambda_c = F\pi \sqrt{\epsilon}\,(a + b) \qquad F \cong 1$$

The conductor loss for the dominant mode is

$$\alpha_c = 1.14 \times 10^{-6} \sqrt{\frac{f\epsilon}{\sigma}} \left(\frac{1}{a} + \frac{1}{b}\right) \frac{1}{\log_{10} \dfrac{b}{a}} \quad \text{Np/m}$$

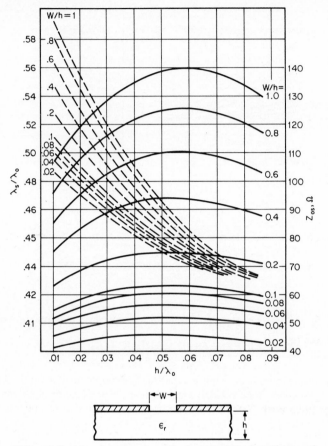

FIG. 42-9 Line wavelength and characteristic impedance of slotline.

The minimum conductor loss occurs at the limiting size for the first higher mode, in which case $\sqrt{\epsilon}\, Z_0 = 92.6\ \Omega$ and

$$\alpha_c = 0.637 \frac{\epsilon}{\lambda_c^{3/2}\, \sqrt{\sigma}} \qquad \text{Np/m}$$

For a fixed outer-conductor size, the minimum conductor loss occurs for $\sqrt{\epsilon}\, Z_0 = 77\ \Omega$ and is

$$\alpha_c = 0.164 \frac{1}{b} \sqrt{\frac{\epsilon}{\sigma\lambda}} \qquad \text{Np/m}$$

The power transmitted by the line is

$$P = \frac{E_a^2 a^2 \sqrt{\epsilon}}{52.2} \log_{10} \frac{b}{a} \qquad \text{W}$$

where E_a is the electric field intensity at the center conductor (Sec. 42-1). The maximum power-handling capacity occurs for a 44.4-Ω line operating at the limit of the first higher mode and is

$$P_{max} = 6.53 \times 10^{-5} \frac{E_a^2 \lambda_c^2}{\sqrt{\epsilon}} \quad W$$

For a fixed outer-conductor size, the maximum power-handling capacity occurs for $\sqrt{\epsilon}\, Z_0 = 30\ \Omega$ and is

$$P'_{max} = 1.53 \times 10^{-3} E_a^2 b^2 \sqrt{\epsilon} \quad W$$

The effect of dimensional tolerances on the characteristic impedance may be found from

$$\frac{\Delta Z_0}{Z_0} = \frac{60}{Z_0 \sqrt{\epsilon}} \left(\frac{\Delta b}{b} - \frac{\Delta a}{a} \right)$$

while the effect of eccentricity is to change the characteristic impedance to

$$Z_0 = \frac{138}{\sqrt{\epsilon}} \log_{10} \left[\frac{b}{a} \left(1 - \frac{e^2}{b^2} \right) \right] \quad \Omega \qquad \frac{e}{b} < \log_{10} \frac{b}{a} < 0.7$$

where e is the off-center distance.

For a balanced coaxial line (sheathed two-wire line) having two center conductors spaced at a distance s within a single outer conductor, the characteristic impedance is given by

$$Z_0 = \frac{276}{\sqrt{\epsilon}} \log_{10} \left(\frac{s}{a} \frac{4b^2 - s^2}{4b^2 + s^2} \right) \quad \Omega$$

Coaxial Line with Helical Center Conductor

Coaxial lines with a helical inner conductor are sometimes useful to obtain a high characteristic impedance or slow propagation velocity. Design relations are given in Ref. 6.

Coaxial Line, Bead-Supported

In order to support the center conductor of an air-dielectric coaxial line, insulating beads are often used. These beads will introduce discontinuities and, if not properly designed, produce large reflection losses. The design procedure for these support beads depends upon the frequency range of application, a broad range requiring a more complex design. The more common procedures are summarized in Fig. 42-10, and others can be found in the references.[7]

In the bead-supported line, the cutoff wavelength in the bead section should be kept below the operating wavelength. The power-handling capacity is about 0.033 times that of the theoretical maximum for the unsupported line.

Coaxial Line, Stub-Supported

In many applications, particularly when high power is to be transmitted, it is desirable to support the center conductor of an air-dielectric coaxial line by means of stubs. The

TYPE	CONFIGURATION	DESIGN FORMULAS	REMARKS
1. HALF WAVE	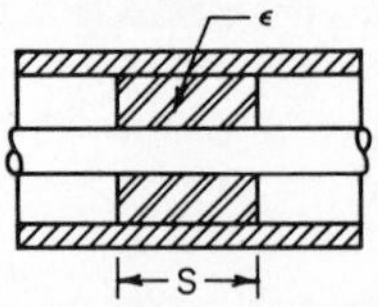	$S = \dfrac{\lambda}{2\sqrt{\epsilon}}$	FREQUENCY-SENSITIVE
2. SPACED PAIR	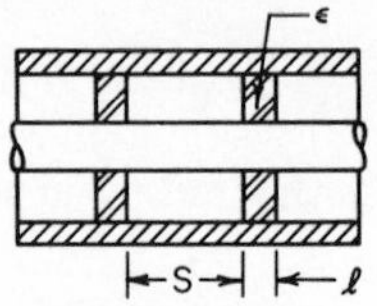	$S = \dfrac{1}{\beta}\arctan\left[\dfrac{2\sqrt{\epsilon}}{1+\epsilon}\cot\beta\ell\sqrt{\epsilon}\right]$ $\beta = \dfrac{2\pi}{\lambda}$	FREQUENCY-SENSITIVE
3. SIMPLE UNDERCUT	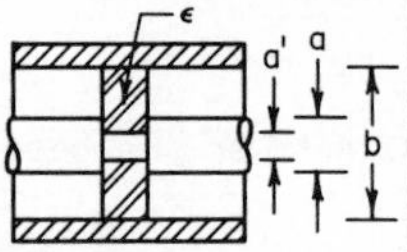	$a' = b\left(\dfrac{b}{a}\right)^{-\sqrt{\epsilon}}$	GOOD AT LOW FREQUENCY. NEGLECTS FRINGING CAPACITANCE AT UNDERCUT.
4. COMPENSATED UNDERCUT	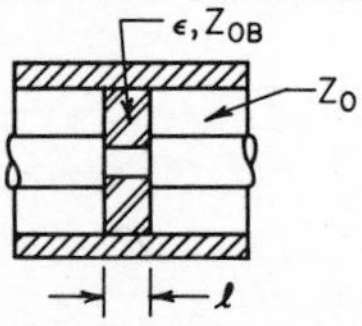	$\ell = \dfrac{1}{\beta}\arctan\left[\dfrac{2\omega C Z_{OB}}{(\omega C Z_{OB})^2+\left(\dfrac{Z_{OB}}{Z_0}\right)^2-1}\right]$ $\beta = \dfrac{2\pi\sqrt{\epsilon}}{\lambda}$	COMPENSATES FRINGING CAPACITANCE AT UNDERCUT. MATCH AT $f = \omega/2\pi$, LOW REFLECTION AT LOWER FREQUENCIES.
5. COMPOUND UNDERCUT (REF. 4)	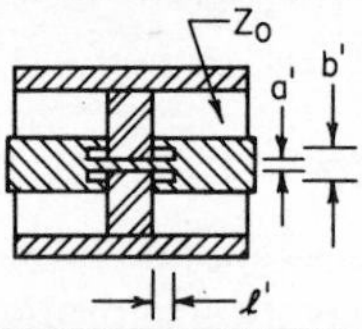	$\sqrt{L/C} = Z_0$ $L = \dfrac{138\ell'}{v}\log_{10}\dfrac{b'}{a'}$ FIND a' AS IN 3) ABOVE $v = 3 \times 10^8$ m/s	VERY LOW REFLECTION IF DISCONTINUITIES ARE OF INFINITESIMAL EXTENT ($\ell' < \lambda/20$). GOOD WHERE MANY BEADS MUST BE USED.

FIG. 42-10 Coaxial-line beads. Capacitive discontinuities at an abrupt change in diameter (as in items 4 and 5) can be found from this curve if the discontinuities are separated by at least the space between the inner and outer conductors in the intervening line. If the ratio of the inner and outer conductors is less than 5 before and after the discontinuity, the curves are accurate to better than 20 percent.

power-handling capacity of a stub-supported line is about 0.15 times that of the theoretical maximum for the unsupported line. Such stubs take the form of a short length of short-circuited transmission line in shunt with the main transmission line. For frequencies at which the stub is an odd number of quarter wavelengths long, it presents zero admittance and therefore does not affect the main line. Obviously, it is often necessary to maintain a low stub admittance over a broad range of frequencies, and this may be effected by using an impedance transformer on the main line at the junction with the stub. The design of stubs and transformers can be found in the literature.[7]

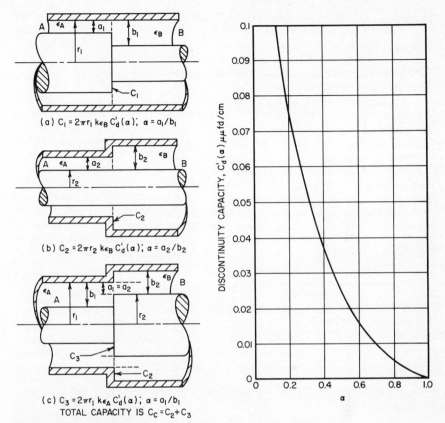

(a) $C_1 = 2\pi r_1 k\epsilon_B C_d'(\alpha)$; $\alpha = a_1/b_1$

(b) $C_2 = 2\pi r_2 k\epsilon_B C_d'(\alpha)$; $\alpha = a_2/b_2$

(c) $C_3 = 2\pi r_1 k\epsilon_A C_d'(\alpha)$; $\alpha = a_1/b_1$
TOTAL CAPACITY IS $C_C = C_2 + C_3$

FIG. 42-10   (Continued)

Semiflexible Coaxial Lines

There are several commercially available coaxial lines having a limited permissible bending radius but exhibiting a loss only slightly greater than that of rigid coaxial lines. They are characterized by a helical-, ribbed-, or foamed-dielectric support for a copper center conductor and a solid aluminum tube for the outer conductor. An analysis of the propagation characteristics of this type of line may be found in the literature.[8]

42-6   FLEXIBLE COAXIAL LINES

Flexible coaxial line using a solid or stranded inner conductor, a plastic-dielectric support, and a braided outer conductor is probably the most common means of connecting together many separated components of a radio-frequency system. Although the transmission loss is relatively high because of the dielectric loss, the convenience often outweighs this factor in applications for which some loss is tolerable. Such lines are commercially available in a wide variety of size and impedance.

Table 42-3 summarizes the characteristics of the more common lines. The attenuation is shown in Fig. 42-11, and the power-handling capacity in Fig. 42-12.

Leakage of electrical energy through the braided outer conductor is sometimes a problem. The amount of leakage has been evaluated in two ways. One is to measure the equivalent coupling impedance to the outside environment, often another cable.[9,10] Typical coupling impedance for an RG-8/U (new equivalent cable is RG-213/U) cable has been reported as 150 $\mu\mu$h/ft from 100 to 5500 MHz. The other is to measure the shielding factor of the cable as the decibel loss through the cable-shield braid.[11,12] The coupling loss increases linearly with length (and the shielding decreases with length) and increases slightly at higher frequencies, where the openings between wires in the shield braid become a bigger part of a wavelength. Single-braid cable typically provides 30- to 40-dB shielding from 1 ft of cable. For longer lengths shielding in decibels is reduced by 20 log (length in feet). Double-braid cable provides 60 to 80 dB. Special braid of flat metal strip provides 80 to 90 dB. Semirigid lines which have

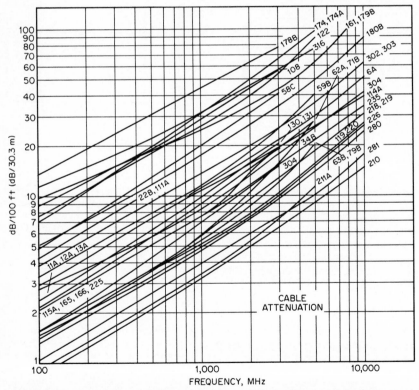

FIG. 42-11 Attenuation of flexible coaxial lines. (RG numbers are indicated on each curve.)

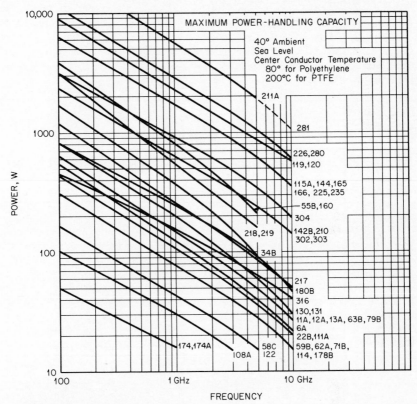

FIG. 42-12 Power-handling capacity of flexible coaxial lines. (RG numbers are indicated on each curve.)

a solid sheath provide from 300 to several thousand decibels at the higher frequencies, where the skin depth becomes very small and the field inside penetrates to only a small part of the solid metal sheath.

Frequently used cable connectors are listed in Table 42-4.

42-7   HOLLOW-TUBE WAVEGUIDES

Propagation

Electromagnetic energy can be propagated down hollow metal tubes if the tubes are of sufficient size and are properly excited. The size of the tubes required usually limits their use to the very-high frequency (VHF) region or above.

The energy can be propagated in a number of different types of waves as described below. In the usual case, waveguide devices are designed for transmission of a single wave type (most often the dominant wave or that having the lowest cutoff frequency) because the design problems are very greatly simplified.

TABLE 42-3 Flexible Coaxial Lines

RG/U type	Inner conductor, in (cm)	Dielectric	Diameter, dielectric, in (cm)	Number and type of shield braid	Jacket material	Outer diameter, in (cm)	Weight, lb/ft (kg/m)	Nominal impedance, Ω	Nominal capacitance, pF/ft (pF/m)	Maximum temperature range, °C	Maximum operating volts, rms	Comments
6A	0.0285 CCS (0.0724)	PE	0.185 (0.469)	2: inner SC, outer BC	PVC-IIA	0.332 (0.843)	0.082 (0.122)	75.0	20.6 (67.6)	−40 +80	2,700	Good attenuation; stability
11A	0.0477 (0.121) 7/0.0159 TC	PE	0.285 (0.724)	1: BC	PVC-IIA	0.405 (1.029)	0.096 (0.143)	75.0	20.6 (67.6)	−40 +80	5,000	Use up to 1000 MHz
12A	0.0477 (0.121) 7/0.0159 TC	PE	0.285 (0.724)	1: BC	PVC-IIA with armor	0.475 (1.207)	0.141 (0.210)	75.0	20.6 (67.6)	−40 +80	5,000	Use up to 1000 MHz
13A	0.0477 (0.121) 7/0.0159 TC	PE	0.280 (0.711)	2: BC	PVC-IIA	0.420 (1.067)	0.126 (0.187)	74.0	20.8 (68.2)	−40 +80	5,000	May prefer RG 216
22B	2 × 0.0456 (0.116) 7/0.0152 BC	PE	0.285 (0.724)	2: TC	PVC-IIA	0.420 (1.067)	0.151 (0.225)	95.0	16.0 (52.5)	−40 +80	1,000	Balanced; twisted conductor
23A	2 × 0.0855 (0.217) 7/0.0285 BC	2 PE	0.380 (0.965)	2: separate inner com. outer BC	PVC-IIA	0.945 (2.400)	0.490 (0.729)	125.0	12.0 (39.4)	−40 +80	3,000	Balanced; dual coaxial
24A	2 × 0.0855 (0.217) 7/0.0285 BC	2 PE	0.380 (0.965)	2: separate inner com. outer BC	PVC-IIA with armor	1.034 (2.626)	0.670 (0.997)	125.0	12.0 (39.4)	−40 +80	3,000	Armored RG 23A
25A	0.0585 (0.149) 19/0.0117 TC	Rubber E	0.288 (0.732)	2: TC	Rubber-IV	0.505 (1.283)	0.205 (0.305)	48.0	50.0 (164.1)	−40 +80	10,000	

Type	Inner conductor dia. in. (mm)	Insulation	Dia. over insulation in. (mm)	Shield (braids)	Jacket	OD in. (mm)	Weight lb/ft (kg/m)	Imped. (ohms)	Capacitance pF/ft (pF/m)	Temp. °C	Max. operating voltage (V)	Remarks
26A	0.0585 (0.149) 19/0.0117 TC	Rubber E	0.288 (0.732)	1:TC	Rubber-IV with armor	0.505 (1.283)	0.189 (0.281)	48.0	50.0 (164.1)	−40 +80	10,000	Armored RG 25A
27A	0.0925 (0.235) 19/0.0185 TC	Rubber D	0.455 (1.156)	1:TC	Rubber-IV with armor	0.670 (1.702)	0.304 (0.452)	48.0	50.0 (164.1)	−40 +80	15,000	
28A	0.0925 (0.235) 19/0.0185 TC	Rubber D	0.455 (1.156)	2:TC,GS	Rubber-IV	0.750 (1.905)	0.370 (0.551)	48.0	50.0 (164.1)	−40 +80	15,000	
34B	0.0747 (0.190)	PE	0.460 (1.168)	1:BC	PVC-IIA	0.630 (1.600)	0.224 (0.333)	75.0	20.6 (67.6)	−40 +80	6,500	Use up to 1000 MHz
35B	0.1045 (0.265) BC 7/0.0249 BC	PE	0.680 (1.727)	1:BC	PVC-IIA with armor	0.945 (2.400)	0.525 (0.781)	75.0	20.6 (67.6)	−40 +80	10,000	Armored RG 164
58C	0.0355 (0.090) 19/0.0071 TC	PE	0.116 (0.295)	1:TC	PVC-IIA	0.195 (0.495)	0.029 (0.043)	50.0	30.8 (101.1)	−40 +80	1,900	Flexible RG 58B
59B	0.0230 CCS (0.058)	PE	0.146 (0.371)	1:BC	PVC-IIA	0.242 (0.615)	0.032 (0.048)	75.0	20.6 (67.6)	−40 +80	2,300	Use up to 1000 MHz
62A	0.0253 CCS (0.064)	Air and PE	0.146 (0.371)	1:BC	PVC-IIA	0.242 (0.615)	0.038 (0.057)	93.0	13.5 (44.3)	−40 +80	750	Low capacitance
63B	0.0253 CCS (0.064	Air and PE	0.285 (0.724)	1:BC	PVC-IIA	0.405 (1.029)	0.083 (0.124)	125.0	10.0 (32.8)	−40 +80	1,000	Low capacitance
64	0.0585 (0.149) 19/0.0117 TC	Rubber D	0.308 (0.782)	2:TC	Rubber-IV	0.495 (1.257)	0.225 (0.335)	48.0	60.0 (196.9)	−40 +80	10,000	
64A	0.0585 (0.149) 19/0.0117 TC	Rubber E	0.288 (0.732)	2:TC	Rubber-IV	0.475 (1.207)	0.205 (0.305)	48.0	50.0 (164.1)	−40 +80	10,000	
65A	0.0080 (0.020) 0.128-diameter helix	PE	0.285 (0.724)	1:BC	PVC-IIA	0.405 (1.029)	0.096 (0.143)	950	44.0 (144.4)	−40 +80	1,000	High impedance; video delay
71B	0.0253 CCS (0.064)	Air and PE	0.146 (0.371)	2:TC	PE-IIIA	0.250 (0.635)	0.046 (0.068)	93.0	13.5 (44.3)	−40 +80	750	Low capacitance
79B	0.0253 CCS (0.064)	Air and PE	0.285 (0.724)	1:BC	PVC-IIA with armor	0.475 (1.207)	0.136 (0.202)	125.0	10.0 (32.8)	−40 +80	1,000	Low capacitance

TABLE 42-3 Flexible Coaxial Lines (*Continued*)

RG/U type	Inner conductor, in (cm)	Dielectric	Diameter, dielectric, in (cm)	Number and type of shield braid	Jacket material	Outer diameter, in (cm)	Weight, lb/ft (kg/m)	Nominal impedance, Ω	Nominal capacitance, pF/ft (pF/m)	Maximum temperature range, °C	Maximum operating volts, rms	Comments
81	0.0625 BC (0.159)	Mag. oxide G	0.321 (0.815)	None	Copper tube	0.325 (0.826)	0.172 (0.256)	50.0	37.0 (121.4)	250	3,000	
82	0.1250 BC (0.318)	Mag. oxide G	0.650 (1.651)	None	Copper tube	0.750 (1.905)	0.698 (1.039)	50.0	36.0 (118.1)	250	5,000	
86	2 × 0.0855 (0.217)	PE	0.300 × 0.650 (0.762 × 1.651)	None	0.300 × (0.762 ×	0.650 (1.651)	0.100 (0.149)	200.0	7.8 (25.6)	−55 +80	10,000	Twin lead
108A	2 × 0.0378 (0.096) 7/0.0126 TC	PE	0.079 (0.201)	1: TC	PVC-IIA	0.235 (0.597)	0.032 (0.048)	78.0	24.5 (80.4)	−40 +80	1,000	Balanced line
111A	2 × 0.0456 (0.116) 7/0.0152 BC	PE	0.285 (0.724)	2: TC	PVC-IIA with armor	0.490 (1.245)	0.146 (0.217)	95.0	16.0 (52.5)	−40 +80	1,000	Armored RG 22B
114A	0.0070 CCS (0.018)	Air and PE	0.285 (0.724)	1: BC	PVC-IIA	0.405 (1.029)	0.087 (0.129)	185.0	6.5 (21.3)	−40 +80	1,000	Low capacitance
115A	0.0840 (0.213)	Taped PTFE	0.255 (0.648)	2: SC	FG braid-V	0.415 (1.054)	0.180 (0.268)	50.0	29.4 (96.5)	−55 +250	5,000	Flexible RG 225
119	0.1020 BC (0.259)	PTFE	0.332 (0.843)	2: BC	FG braid-V	0.465 (1.181)	0.225 (0.335)	50.0	29.4 (96.5)	−55 +250	6,000	Use up to 1000 MHz
120	0.1020 BC (0.259)	PTFE	0.332 (0.843)	2: BC	FG braid-V with armor	0.525 (1.334)	0.282 (0.420)	50.0	29.4 (96.5)	−55 +250	6,000	Armored RG 119
122	0.0300 (0.076) 27/0.0050 TC	PE	0.096 (0.244)	1: TC	PVC-IIA	0.160 (0.406)	0.116 (0.173)	50.0	29.4 (96.5)	−40 +80	1,900	Use up to 1000 MHz
130	2 × 0.0855 (0.217) 7/0.0285 BC	PE	0.472 (1.199)	1: TC	PVC-I	0.625 (1.588)	0.220 (0.327)	95.0	17.0 (55.8)	−40 +80	3,000	RG 57 with twisted pair

No.	Conductor	Dielectric	Dielectric diam.	Shield	Jacket	Diam. 1	Diam. 2	Z	Value	Temp.	Voltage	Description
131	2 × 0.0855 (0.217)	PE	0.472 (1.199)	1: TC	PVC-I with armor	0.710 (1.803)	0.290 (0.432)	95.0	17.0 (55.8)	−40 +80	3,000	Armored RG 130
133A	7/0.0285 BC (0.064)	PE	0.285 (0.724)	1: BC	PVC-IIA	0.405 (1.029)	0.094 (0.140)	95.0	16.2 (53.2)	−40 +80	4,000	95-Ω RG 8
142B	0.0390 SCCS (0.099)	PTFE	0.116 (0.295)	2: SC	FEP	0.195 (0.495)	0.050 (0.074)	50.0	29.4 (96.5)	−55 +200	1,900	High-temperature
144	0.0537 (0.136) 7/0.0179 SCCS	PTFE	0.285 (0.724)	1: SC	FG braid-V	0.410 (1.041)	0.137 (0.204)	75.0	19.5 (64.0)	−55 +250	5,000	High-temperature RG 11A
156	0.0855 (0.217) 7/0.0285 TC	PE	0.285 (0.724)	3: TC,GS,TC	PVC-IIA	0.540 (1.372)	0.211 (0.314)	50.0	32.0 (105.0)	−40 +80	10,000	Triaxial-pulse cable
157	0.1005 (0.255) 19/0.0201 TC	PE	0.455 (1.156)	3: TC,GS,TC	PVC-IIA	0.725 (1.842)	0.317 (0.472)	50.0	38.0 (124.7)	−40 +80	15,000	Triaxial-pulse cable
158	0.1988 (0.505) 37/0.0284 TC	PE	0.455 (1.156)	3: TC,GS,TC	PVC-IIA	0.725 (1.842)	0.380 (0.565)	25.0	78.0 (255.9)	−40 +80	15,000	Triaxial-pulse cable
161	0.012 (0.030) 7/0.004 SCCadBR	PTFE	0.057 (0.145)	1: SC	Nylon	0.082 (0.208)	0.015 (0.022)	70.0	20.0 (65.6)	−60 +120	1,000	Miniature
164	0.1045 BC (0.265)	PE	0.680 (1.727)	1: BC	PVC-IIA	0.870 (2.210)	0.490 (0.729)	75.0	20.6 (67.6)	−40 +80	10,000	RG 35B without armor
165	0.0960 (0.244) 7/0.0320 SC	PTFE	0.285 (0.724)	1: SC	FG braid-V	0.410 (1.041)	0.121 (0.180)	50.0	29.4 (96.5)	−55 +250	5,000	One braid RG 225
166	0.0960 (0.244) 7/0.0320 SC	PTFE	0.285 (0.724)	1: SC	FG braid-V with armor	0.460 (1.168)	0.144 (0.214)	50.0	29.4 (96.5)	−55 +250	5,000	Armored RG 165
174	0.0189 (0.048) 7/0.0063 CCS	PE	0.060 (0.152)	1: TC	PVC-I	0.100 (0.254)	0.008 (0.012)	50.0	30.8 (101.1)	−40 +80	1,500	Miniature data transmission
177	0.195 BC (0.495)	PE	0.680 (1.727)	2: SC	PVC-IIA	0.895 (2.273)	0.470 (0.699)	50.0	30.8 (101.1)	−40 +80	11,000	High frequency RG 218

TABLE 42-3 Flexible Coaxial Lines (*Continued*)

RG/U type	Inner conductor, in (cm)	Dielectric	Diameter, dielectric, in (cm)	Number and type of shield braid	Jacket material	Outer diameter, in (cm)	Weight, lb/ft (kg/m)	Nominal impedance, Ω	Nominal capacitance, pF/ft (pF/m)	Maximum temperature range, °C	Maximum operating volts, rms	Comments
178B	0.0120 (0.030) 7/0.0040 SCCS	PTFE	0.034 (0.086)	1: SC	FEP-IX	0.075 (0.191)	0.0054 (0.0080)	50.0	29.4 (96.5)	−55 +200	1,000	
179B	0.0120 (0.0305) 7/0.0040 SCCS	PTFE	0.063 (0.160)	1: SC	FEP-IX	0.100 (0.254)	0.010 (0.015)	75.0	19.5 (64.0)	−55 +200	1,200	
210	0.0253 SCCS (0.0643)	Air and PTFE	0.146 (0.371)	1: SC	FG braid-V	0.242 (0.615)	0.040 (0.060)	93.0	13.5 (44.3)	−55 +250	750	High temperature; low capacitance*
211A	0.1900 BC (0.4826)	PTFE	0.620 (1.575)	1: BC	FG braid-V	0.730 (1.854)	0.641 (0.954)	50.0	29.4 (96.5)	−55 +250	7,000	High temperature; high power
212	0.0556 SC (0.1412)	PE	0.185 (0.470)	2: SC	PVC-IIA	0.332 (0.843)	0.083 (0.124)	50.0	29.4 (96.5)	−40 +80	3,000	Use up to 10,000 MHz
213	0.0888 (0.2256) 7/0.0296 BC	PE	0.285 (0.724)	1: BC	PVC-IIA	0.405 (1.029)	0.099 (0.147)	50.0	30.8 (101.1)	−40 +80	5,000	Use up to 1000 MHz
214	0.0888 (0.2256) 7/0.0296 SC	PE	0.285 (0.724)	2: SC	PVC-IIA	0.425 (1.080)	0.126 (0.187)	50.0	30.8 (101.1)	−40 +80	5,000	Use up to 10,000 MHz
215	0.0888 (0.2256) 7/0.0296 BC	PE	0.285 (0.724)	1: BC	PVC-IIA with armor	0.475 (1.207)	0.121 (0.180)	50.0	30.8 (101.1)	−40 +80	5,000	Armored RG 213
216	0.0477 (0.1212) 7/0.0159 BC	PE	0.285 (0.724)	2: BC	PVC-IIA	0.425 (1.080)	0.114 (0.170)	75.0	20.6 (67.6)	−40 +80	5,000	Use up to 1000 MHz
217	0.106 BC (0.269)	PE	0.370 (0.940)	2: BC	PVC-IIA	0.545 (1.384)	0.201 (0.299)	50.0	30.8 (101.1)	−40 +80	7,000	Use up to 1000 MHz

Type RG	Inner conductor	Dielectric	Dielectric dia.	Shielding	Jacket	O.D.		Z₀	Capacitance	Temp. °C	Volts	Remarks
218	0.195 BC (0.495)	PE	0.680 (1.727)	1: BC	PVC-IIA	0.870 (2.210)	0.460 (0.684)	50.0	30.8 (101.1)	−40 +80	11,000	Use up to 1000 MHz
219	0.195 BC (0.495)	PE	0.680 (1.727)	1: BC	PVC-IIA with armor	0.945 (2.400)	0.585 (0.870)	50.0	30.8 (101.1)	−40 +80	11,000	Armored RG 218
223	0.035 SC (0.089)	PE	0.116 (0.295)	2: SC	PVC-IIA	0.216 (0.549)	0.034 (0.051)	50.0	30.8 (101.1)	−40 +80	1,900	Usable to 10,000 MHz
225	0.0936 (0.238)	PTFE	0.285 (0.724)	2: SC	FG braid-V	0.430 (1.092)	0.180 (0.268)	50.0	29.4 (96.5)	−55 +250	5,000	See RG 393 for FEP jacket
226	7/0.0312 SC 0.1270 (0.323)	PTFE	0.370 (0.940)	2: BC	FG braid-V	0.500 (1.270)	0.445 (0.662)	50.0	29.4 (96.5)	−55 +250	7,000	
230	19/0.0254 SC 0.1988 (0.505)	Rubber D	0.455 (1.156)	3: TC,GS,GS	Rubber-IV	0.740 (1.880)		25.0	100.0 (328.1)	−40 +80	15,000	
235	37/0.0284 TC 0.0852 (0.216)	Taped PTFE	0.255 (0.648)	2: SC	SIL/DAC-VI	0.470 (1.194)	0.160 (0.238)	50.0	29.5 (96.8)	−55 +250	5,000	RG 115 with VI jacket
266	7/0.0284 SC 0.0113 Spiral on 0.144 mag. core (0.029)	PE	0.285 (0.724)	75 spiral cond. 68 BC and 7 insulated	PVC-I	0.400 (1.016)	0.120 (0.179)	1530.0	53.0 (173.9)	−40 +80	5,000	Delay cable 50 ns/ft
280	0.1144 BC (0.291)	PTFE taped	0.327 (0.831)	2: SC	FEP-IX	0.468 (1.189)	0.200 (0.298)	50.0	25.4 (83.3)	−55 +200	3,000	Low loss; high frequency
281	0.1890 (0.480)	Taped PTFE	0.500 (1.270)	2: SC	SIL/DAC-VI	0.750 (1.905)	0.400 (0.595)	50.0	25.4 (83.3)	−55 +150	4,000	Low loss; high power
296	19/0.0378 SC 0.2352 (0.597)	SIL	0.906 (2.301)	1: SC	Neoprene	1.190 (3.023)		50.0	36.4 (119.4)	−55 +80	13,800	
302	37/0.0336 SC 0.0250 SCCS (0.064)	PTFE	0.146 (0.371)	1: SC	FEP-IX	0.206 (0.523)	0.031 (0.046)	75.0	19.5 (64.0)	−55 +200	2,300	FEP jacket RG 140
303	0.0390 SCCS (0.099)	PTFE	0.116 (0.295)	1: SC	FEP-IX	0.170 (0.432)	0.030 (0.045)	50.0	29.4 (96.5)	−55 +200	1,900	FEP jacket RG 141A
304	0.0590 SCCS (0.150)	PTFE	0.185 (0.470)	2: SC	FEP-IX	0.280 (0.711)	0.088 (0.131)	50.0	29.4 (96.5)	−55 +200	3,000	FEP jacket RG 143A
307A	0.0290 (0.074) 19/0.0058 SC	Foam PE	0.146 (0.371)	2: SC PUR interlayer	PE-IIIA	0.270 (0.686)	0.070 (0.104)	75.0	16.7 (54.8)	−55 +80	1,000	Triaxial; use to 100 MHz

TABLE 42-3 Flexible Coaxial Lines (*Continued*)

RG/U type	Inner conductor, in (cm)	Dielectric	Diameter, dielectric, in (cm)	Number and type of shield braid	Jacket material	Outer diameter, in (cm)	Weight, lb/ft (kg/m)	Nominal impedance, Ω	Nominal capacitance, pF/ft (pF/m)	Maximum temperature range, °C	Maximum operating volts, rms	Comments
316	0.0201 (0.051) 7/0.0067 SCCS	PTFE	0.060 (0.152)	1: SC	FEP-IX	0.102 (0.259)	0.012 (0.018)	50.0	29.4 (96.5)	−55 +200	1,200	FEP jacket RG 188A
328	0.4850 TC braid (1.232)	Rubber H,J,H	1.065 (2.705)	3: TC,GS,TC	Neoprene	1.460 (3.708)	1.469 (2.186)	25.0	85.0 (278.9)	−55 +80	20,000	
174A	0.0189 (0.0480) 7/0.0063 CCS	PE	0.060 (0.152)	1: TC	PVC-IIA	0.100 (0.254)	0.008 (0.012)	50	30.8 (101.1)	−40 +80	1,500	Miniature data transmission

Abbreviations:

Dielectric

    PE = solid polyethylene
    PTFE = solid polytetrafluoroethylene
    PIB = polyisobutylene, Type B per MIL-C-17
    Rubber per MIL-C-170
    SIL = silicone rubber
    PS = polystyrene

Conductors and braid materials

    BC = bare copper
    SC = silver-covered copper
    CCS = copper-covered steel
    TC = tinned copper
    SCCS = silver, copper-covered steel
    SCCadBr = silver-covered cadmium bronze
    GS = galvanized steel
    TCCS = tinned copper-covered steel
    SSC = silver-covered strip

Jacket material

    PVC-I = black polyvinylchloride, contaminating, Type I, per MIL-C-17E
    PVC-II = gray polyvinylchloride, noncontaminating, Type II, per MIL-C-17E
    PVC-IIA = black polyvinylchloride, noncontaminating, Type IIA, per MIL-C-17E
    PE-III = clear polyethylene
    PE-IIIA = high-molecular-weight black polyethylene, Type IIIA, per MIL-C-17E
    FG braid-V = fiberglass-impregnated, Type V, per MIL-C-17E
    FEP-IX = fluorinated ethylene propylene, Type IX, per MIL-C-17E
    PUR = polyurethane, black specific compounds
    SIL/DAC-VI = dacron braid over silicon rubber, Type VI, per MIL-C-17E
    Rubber per MIL-C-17E

TABLE 42-4 Characteristics of Frequently Used Cable Connectors

	BNC	TNC	C	SC	N
Cable size, outer diameters	in 0.150–0.250 cm 0.381–0.635	0.150–0.250 0.381–0.635	0.300– 0.550 0.762– 1.40	0.300– 0.550 0.762– 1.40	0.330–0.550 0.762–1.40
Coupling type					
Bayonet (quick disconnect)	B		B		
Screw		7/16–28		11/16–24	5/8–24
Maximum operating volts	500	500	1000	1500	1000
Frequency range, GHz, DC to ——	4	11	11	11	11
RF leakage, dB	−55	−60	−55	−90	−90
Insertion loss, dB	0.2	0.2	0.5	0.15	0.15
At GHz	3	3	10	10	10

Cable types	Use	UG-	M39012/	UG-	M39012/	UG-
RG-58,	Plug (m)	88/U	26-0001			
141,	Plug, right-angle	913/U	30-0001			
142,	Cable jack (f)	89/U	27-0001			
223/U	Panel jack (f)	262/U	29-0001		Flange	
		909/U	28-0001	704C/U	nut	
RG-59,	Plug (m)	260/U	26-0002			
62,	Plug, right-angle		30-0002			
71,	Cable jack (f)	261/U	27-0002			
140,	Panel jack (f)	291/U	29-0002		Flange	
210/U		910/U	28-0002	631B/U	nut	
	Panel jack (f)	290/U	32-0001	Flange		
		625/U	31-0001	nut		
RG-6,*	Plug (m)			626C/U	35-0001	18/U
212,	Plug, right-angle			710C/U	39-0001	
143/U†	Cable jack (f)			633B/U		20/U
	Panel jack (f)		Flange	629B/U		159B/U
			nut	630B/U	40-0001	
RG-213,	Plug (m)			573C/U	35-0002	21/U
214,	Cable jack (f)			572B/U		23/U
216*	Panel jack (f)		Flange	571B/U		160C/U
393/U			nut	570B/U	40-0002	594/U
	Panel jack (f)		Flange	568A/U	41-0001	58/U
			nut	706B/U	42-0001	680/U
			nut	569B/U	43-0001	

NOTE: Underscored cables are not matched to connectors. m = male; f = female.

*Fits Types C and SC only.

†Fits Type N only.

TE$_{mn}$ waves: In the transverse electric waves, sometimes called H_{mn} waves, the electric vector is always perpendicular to the direction of propagation.

TM$_{mn}$ waves: In the transverse magnetic waves, sometimes called the E_{mn} waves, the magnetic vector is always perpendicular to the direction of propagation.

The propagation constant γ_{mn} determines the amplitude and phase of each component of a wave as it is propagated along the waveguide. Each component may be represented by $A \exp(jwt - \gamma_{mn}z)$, where A is a constant, z is the distance along the direction of propagation, and $\omega = 2\pi f$. When γ_{mn} is real, there is no phase shift along the waveguide, but there is high attenuation. In fact, no propagation takes place, and the waveguide is considered below cutoff. The reactive attenuation L along the waveguide under these conditions is given by

$$L = \frac{54.58}{\lambda_c} \left[1 - \left(\frac{\lambda_c}{\lambda} \right)^2 \right]^{1/2} \quad \text{dB/unit length}$$

where λ is the wavelength in the unbounded medium and λ_c is the cutoff wavelength of that wave (a function of waveguide dimensions only). Waveguides are often used at frequencies far below cutoff as calibrated attenuators, since the rate of attenuation is determined by cross-section dimensions of the waveguide and the total attenuation in decibels is a linear function of the displacement of the output from the input.[13]

When γ_{mn} is imaginary, the amplitude of the wave remains constant but the phase changes with z and propagation takes place. γ_{mn} is a pure imaginary quantity only for lossless waveguide. In a practical case, γ_{mn} has both a real part α_{mn} which is the attenuation constant, and an imaginary part β_{mn}, which is the phase constant; that is, $\gamma_{mn} = \alpha_{mn} + j\beta_{mn}$.

The wavelength in a uniform waveguide is always greater than the wavelength in the unbounded medium and is given by

$$\lambda_g = \frac{\lambda}{[1 - (\lambda/\lambda_c)^2]^{1/2}}$$

The phase velocity is the apparent velocity, judging by the phase shift along the guide. Phase velocity, $v = c\,(\lambda_g/\lambda)$, is always greater than that in an unbounded medium.

The group velocity is the velocity of energy propagation down the guide. Group velocity, $u = c\,(\lambda/\lambda_g)$, is always less than that in an unbounded medium.

For air-filled guide and guides filled with dielectric having very low loss, the attenuation is mainly due to conductor losses in the walls. For any particular type of wave, the loss is high near cutoff and decreases as the frequency is increased. For all excepting the TE$_{0n}$ waves in circular waveguide, the attenuation reaches a minimum value for that wave and that waveguide, then increases with frequency. For most waves, this minimum is slightly above $2f_c$. To avoid high loss near cutoff and the complexity of multiwave transmission, the useful band is usually considered to lie between $1.3f_c$ of the desired mode and $0.9f_c$ of the next higher mode. For rectangular waveguides having a width equal to twice the height, the useful range is about $1.5{:}1$.

Rectangular Waveguides

For TE$_{mn}$ waves in rectangular waveguides, m and n may take any integer value from 0 to infinity, except for the case $m = n = 0$. For the TM$_{mn}$ waves, m and n may take any value from 1 to infinity. The m and n denote the number of half-period variations of the electric field for TE waves or magnetic field for TM waves in the direction of the small and large dimensions of the waveguide respectively. Field patterns for some of the simpler waves are shown in Fig. 42-13.

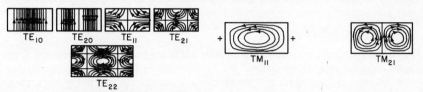

TE$_{10}$ TE$_{20}$ TE$_{11}$ TE$_{21}$ TM$_{11}$ TM$_{21}$

TE$_{22}$

E LINES FOR TE MODES   H LINES FOR TM MODES

FIG. 42-13   Field configurations for rectangular waveguide.

The propagation constant for rectangular guides is given by

$$\gamma_{mn} = \sqrt{\left(\frac{m\pi}{a}\right)^2 + \left(\frac{n\pi}{b}\right)^2 - \omega^2\mu\epsilon} = \frac{2\pi}{\lambda_g}$$

where a is the wide dimension, b is the narrow dimension, ϵ is the dielectric constant, and μ is the permeability of the dielectric in the waveguide. Since propagation takes place only when the propagation constant is imaginary, the cutoff frequency for rectangular waveguide is

$$f_c = \frac{c}{2\sqrt{\mu\epsilon}} \sqrt{\left(\frac{m}{a}\right)^2 + \left(\frac{n}{b}\right)^2}$$

and

$$\lambda_c = \frac{2\sqrt{\mu\epsilon}}{\sqrt{(m/a)^2 + (n/b)^2}}$$

Most frequently, operation is limited to the TE$_{10}$ or dominant wave in rectangular waveguide. For this simplified case, the important formulas reduce to

$$\lambda_c = 2a\sqrt{\mu\epsilon}$$

$$\gamma = 2\pi\sqrt{(1/4a) - f^2\mu\epsilon}$$

In order to relate the waveguide properties to similar properties of low-frequency circuits, the impedance concept has been developed. Three characteristic impedances can be defined, differing from each other by a constant:[14]

$$Z_{VI} = \frac{V}{I} = 377\,\frac{\pi}{2}\frac{b}{a}\frac{\lambda_g}{\lambda} = 592\,\frac{b}{a}\frac{\lambda_g}{\lambda}$$

$$Z_{PV} = \frac{V^2}{2P} = 754\,\frac{b}{a}\frac{\lambda_g}{\lambda} = 754\,\frac{b}{a}\frac{\lambda_g}{\lambda}$$

$$Z_{PI} = \frac{2P}{I^2} = 377\,\frac{\pi^2}{8}\frac{b}{a}\frac{\lambda_g}{\lambda} = 465\,\frac{b}{a}\frac{\lambda_g}{\lambda}$$

Any one of the three is reasonably satisfactory if used consistently throughout, since the most frequent use is in determining mismatch at waveguide junctions and it is the ratio of impedance that matters. Ratios involving only different values of b give accurate indication of impedance mismatch. Differences in a give ratios nearly correct for small changes in a from the usual cross-section dimensions of rectangular waveguide, but errors are appreciable for large differences in a. Z_{PV} is most widely used, but Z_{VI} is found to be more nearly correct in matching coaxial line to waveguide.

Circular Waveguides

The usual coordinate system is ρ, θ, z, where ρ is in the radial direction, θ is the angle, and z is the longitudinal direction.

For TE_{mn} waves in circular waveguides m denotes the number of axial planes along which the normal component of electric field vanishes and n the number of cylinders including the boundary of the guide along which the tangential component of electric field vanishes. The number m may take any integral value from 0 to infinity, and n may take any integral value from 1 to infinity. The dominant wave in circular waveguide is the TE_{11}. For TM_{mn} waves, m denotes the number of axial planes along which the magnetic field vanishes and n the number of cylinders to which the electric field is normal. The number m may take any integral value from 0 to ∞, and n may take any integral value from 1 to ∞. Of the circularly symmetrical waves, the TM_{01} has the lowest cutoff frequency.

Field patterns for some of the simpler waves in circular guides are shown in Fig. 42-14.

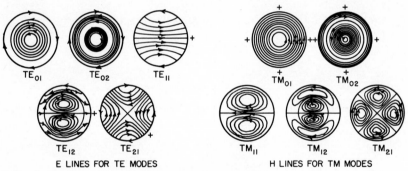

TE$_{01}$ TE$_{02}$ TE$_{11}$ TM$_{01}$ TM$_{02}$

TE$_{12}$ TE$_{21}$ TM$_{11}$ TM$_{12}$ TM$_{21}$

E LINES FOR TE MODES   H LINES FOR TM MODES

FIG. 42-14   Field configurations for circular waveguide.

The cutoff wavelength in lossless circular guide is given by

$$\lambda_c = \sqrt{\mu\epsilon}\, D_{mn}a$$

where a is the radius, ϵ the relative dielectric constant, and μ the relative permeability of the dielectric, and the constant D_{mn} is as shown in Table 42-5.

A summary of the cutoff and attenuation constant formulas for circular and rectangular waveguides is given in Table 42-6,

where $A = \dfrac{\sqrt{c/\lambda}}{\sqrt{1 - (\lambda/\lambda_c)^2}}$

c = velocity of light in free space $\approx 3 \times 10^8$ m/s

TABLE 42-5 Cutoff Constants for Circular Waveguide

D_{mn} for TE_{mn} waves

n \ m	0	1	2	3
1	1.640	3.412	2.057	1.496
2	0.896	1.178	0.937	0.764
3	0.618	0.736	0.631	0.554
4	0.475	0.54	0.48	0.44

D_{mn} for TM_{mn} waves

n \ m	0	1	2	3
1	2.613	1.640	1.224	0.966
2	1.139	0.896	0.747	0.644
3	0.726	0.618	0.541	0.482
4	0.534	0.475	0.425	0.388

TABLE 42-6 Summary of Cutoff and Attenuation Constant Formulas

TYPE OF GUIDE	RECTANGULAR TE_{MO}	TM_{OI}	TE_{II}	TE_{OI}
CUTOFF WAVELENGTH	$\dfrac{2a}{m}$	$2.613a$	$3.412a$	$1.640a$
ATTENUATION CONSTANT (NEPERS/METER) DUE TO CONDUCTION LOSSES ONLY	$\dfrac{4a_0 A}{a}\left(\dfrac{a}{2b}+\dfrac{\lambda^2}{\lambda_c^2}\right)$	$2\dfrac{a_0 A}{a}$	$2\dfrac{a_0}{a}A\left(0.415+\dfrac{\lambda^2}{\lambda_c^2}\right)$	$2\dfrac{a_0}{a}A\dfrac{\lambda^2}{\lambda_c^2}$
ATTENUATION FAR BELOW CUTOFF	27.28 DB PER GUIDE WIDTH	41.78 DB PER DIAMETER	31.98 DB PER DIAMETER	66.56 DB PER DIAMETER

For copper and air, $\alpha_0 = 3.5 \times 10^{-8}$ Np/m. To convert nepers per meter to decibels per 100 ft, multiply by 264.

Standard Waveguide Sizes

The waveguide sizes which have become standardized are listed in Table 42-7, together with the flanges used in connecting them together.

TABLE 42-7 Standard Rectangular Waveguides and Flanges

EIA designation WR-	DOD part no. M85/1-	AN designation	Material	Inside dimensions, in (mm)	Tolerance, in (mm)	Wall Thickness, in (mm)	Frequency range, GHz for TE_{10}	f_c, GHz for TE_{10}	λ_c, mm	Attenuation dB/100 ft (dB/30.5 m) — Lowest and highest	Theoretical peak power, MW (highest)	Theoretical maximum continuous wave, kW	Cover	Choke
650	017-	69/U	B	6.500 × 3.250 (165.1 × 82.55)	±0.008 (0.20)	0.080 (2.03)	1.12–1.70	0.908	330.2	0.316–.0209	41.3–59.7	80.5–122	323/U	322/U
	018-	103/U	A							0.273–0.180		88.5–136	1720/U*	
510	023-	337/U	B	5.100 × 2.550 (129.5 × 64.77)	±0.008 (0.20)	0.080 (2.03)	1.45–2.20	1.154	259.1	0.440–0.299	26.2–37.0	47.9–70.4	1715/U*	
	025-	338/U	A							0.380–0.258		53.2–78.3	1717/U*	
430	029-	105/U	B	4.300 × 2.150 (109.2 × 54.61)	±0.008 (0.20)	0.080 (2.03)	1.70–2.60	1.375	218.4	0.502–0.334	18.2–26.3	35.3–53.1	437/U*	
	031-	104/U	A							0.583–0.387		31.7–47.7	435A/U*	
340	035-	113/U	B	3.400 × 1.700 (86.36 × 43.18)	±0.008 (0.20)	0.080 (2.03)	2.20–3.30	1.737	172.7	0.682–0.474	11.9–16.4	21.7–31.3	554/U*	
	037-	112/U	A							0.791–0.550		19.5–28.1	553A/U*	
284	041-	75/U	B	2.840 × 1.340 (72.14 × 34.04)	±0.006 (0.15)	0.080 (2.03)	2.60–3.95	2.080	144.3	0.950–0.651	7.65–10.9	13.4–19.6	584/U	585A/U
	043-	48/U	A							1.10–0.754		12.1–17.6	53/U	54B/U
229	047-	341/U	B	2.290 × 1.145 (58.16 × 29.08)	±0.006 (0.15)	0.064 (1.63)	3.30–4.90	2.577	116.3	1.21–0.858	5.48–7.55	8.99–12.7	1727/U*	
	049-	340/U	A							1.40–0.996		8.08–11.4	1726/U*	
187	053-	95/U	B	1.872 × 0.872 (47.55 × 22.15)	±0.005 (0.13)	0.064 (1.63)	3.95–5.85	3.155	95.10	1.79–1.24	3.30–4.70	5.17–7.45	407/U	406B/U
	055-	49/U	A							2.07–1.44		4.64–6.69	149A/U	148C/U
159	059-	344/U	B	1.590 × 0.795 (40.39 × 20.19)	±0.005 (0.13)	0.064 (1.63)	4.90–7.05	3.705	80.77	1.99–1.49	2.79–3.72	4.20–5.62	1731/U./U*	
	061-	343/U	A							2.31–1.72		3.77–5.05	247/U	248/U
137	065-	106/U	B	1.372 × 0.622 (34.85 × 15.80)	±0.004 (0.10)	0.064 (1.63)	5.85–8.20	4.285	69.70	2.53–2.00	1.98–2.53	2.90–3.67	441/U	440B/U
	067-	50/U	A							2.94–2.32		2.60–3.29	344B/U	343B/U
112	071-	68/U	B	1.122 × 0.497 (28.50 × 12.62)	±0.004 (0.10)	0.064 (1.63)	7.05–10.0	5.260	57.00	3.55–2.76	1.28–1.70	1.79–2.30	138/U	137B/U
	073-	51/U	A							4.11–3.20		1.61–2.07	51/U	52B/U
90	077-	67/U	B	0.900 × 0.400 (22.86 × 10.16)	±0.004 (0.10)	0.050 (1.27)	8.20–12.4	6.560	45.72	5.54–3.83	0.758–1.12	0.959–1.39	135/U	136B/U
	079-	52/U	A							6.42–4.45		0.862–1.25	39/U	40B/U
75	083-	347/U	A	0.750 × 0.375 (19.05 × 9.53)	±0.003 (0.08)	0.050 (1.27)	10.0–15.0	7.847	38.10	6.55–4.58	0.622–0.903	0.737–1.06		
	085-	346/U	B							7.60–5.31		0.662–0.948		
62	089-	91/U	B	0.622 × 0.311 (15.80 × 7.90)	±0.0025 (0.06)	0.040 (1.02)	12.4–18.0	9.490	31.60	9.58–7.04	0.457–0.633	0.451–0.614	419/U	541A/U
	090-	349/U	A							8.26–6.07		0.502–0.683	1665/U	1666/U
51	093-	107/U	SC	0.510 × 0.255 (12.95 × 6.48)	±0.0025 (0.06)	0.040 (1.02)	15.0–22.0	11.54	25.91	6.91–5.08	0.312–0.433	0.602–0.818		
	096-	353/U	B							13.1–9.48		0.290–0.400		
	097-	351/U	A							11.3–8.17		0.323–0.445	419/U	541/U

42	102-	53/U	B	0.420 × 0.170 (10.67 × 4.32)	±0.002 (0.05)	0.040 (1.02)	18.0–26.5	14.08	21.34	20.5–15.0	0.171–0.246	0.157–0.213	595/U	596A/U
	103-	121/U	A							17.7–13.0		0.174–0.237	597/U	598A/U
	106-	66/U	SC							14.8–10.9		0.209–0.284	595/U	596A/U
34	109-	354/U	B	0.340 × 0.170 (8.64 × 4.32)	±0.002 (0.05)	0.040 (1.02)	22.0–33.0	17.28	17.27	25.0–17.4	0.139–0.209	0.118–0.169	1530/U*	
	110-	355/U	A							21.6–15.0		0.131–0.188		
	113-	357/U	SC							16.2–11.3		0.175–0.252		
28	114-	96/U	S	0.280 × 0.140 (7.11 × 3.56)	±0.0015 (0.04)	0.040 (1.02)	26.5–40.0	21.10	14.22	24.6–16.8	0.096–0.146	0.212–0.310	599/U	600A/U
	117-	271/U	SC							22.0–15.1		0.242–0.347		
22	118-	97/U	S	0.224 × 0.112 (5.69 × 2.84)	±0.0010 (0.03)	0.040 (1.02)	33.0–50.0	26.35	11.38	34.5–23.5	0.064–0.097	0.134–0.197	383/U*	
	121-	272/U	SC							31.0–21.1		0.149–0.219		
19	124-	358/U	SC	0.188 × 0.094 (4.78 × 2.39)	±0.0010 (0.03)	0.040 (1.02)	40.0–60.0	30.69	9.550	38.0–27.3	0.048–0.070	0.111–0.155	1529/U*	
15	125-	98/U	S	0.148 × 0.074 (3.76 × 1.88)	±0.0010 (0.03)	0.040 (1.02)	50.0–75.0	39.90	7.518	64.2–43.9	0.030–0.044	0.052–0.077	385/U*	
	128-	273/U	SC							57.6–39.3		0.058–0.085		
12	129-	99/U	S	0.122 × 0.061 (3.10 × 1.55)	±0.0005 (0.01)	0.040 (1.02)	60.0–90.0	48.40	6.198	87.8–58.9	0.020–0.030	0.047–0.067	387/U*	
	132-	274/U	SC							78.7–52.7		0.049–0.074		
10	135-	359/U	SC	0.100 × 0.050 (2.54 × 1.27)	±0.0005 (0.01)	0.040 (1.02)	75.0–110	58.85	5.080	101–71.0	0.014–0.020	0.032–0.045	1528/U*	
8	138-	278/U	SC	0.080 × 0.040 (2.03 × 1.02)	±0.0003 (0.01)	0.020 (0.51)	90.0–140	73.84	4.064	154–98.7	0.009–0.013	0.015–0.024	1527/U*	
7	141-	276/U	SC	0.065 × 0.0325 (1.65 × 0.83)	±0.00025 (0.01)	0.020 (0.51)	110–170	90.85	3.302	214–135	0.006–0.009	0.010–0.016	1525/U*	
5	144-	275/U	SC	0.051 × 0.0255 (1.30 × 0.65)	±0.00025 (0.01)	0.020 (0.51)	140–220	115.8	2.591	308–194	0.004–0.006	0.006–0.01	1524/U*	
4	147-	277/U	SC	0.043 × 0.0215 (1.09 × 0.546)	±0.0002 (0.01)	0.020 (0.51)	170–260	137.5	2.184	377–251	0.003–0.005	0.005–0.007	1526/U*	
3	152-	139/U	S	0.034 × 0.0170 (0.864 × 0.432)	±0.0002 (0.01)	Round	220–325	173.3	1.727	512–341	0.0004–0.0005	0.005–0.008		

Materials	Resistivity, µΩ · cm
A = aluminum alloy 1100	2.90
B = brass	3.90
SC = silver-clad copper	
S = silver	1.63

*These flanges mate with themselves.

Flexible Waveguides

Flexible rectangular waveguides are made to match most of the standard waveguides. These differ primarily in mechanical construction. Waveguides using seamless corrugations, spiral-wound strip with adjacent edges crimped and soft-soldered, and spiral-wound strip with heavier crimping to provide sliding contact, as well as vertebra type (consisting of cover-choke wafers held in place by a rubber jacket), are available. The first two will bend in either plane, stretch, or compress but will not twist. The other two twist as well as bend and stretch. When the guide is flexed during operation or pressurized, it is nearly always covered with a molded-rubber jacket. When unjacketed, all are subject to a minimum bending radius (of the guide centerline) of 2 to 3 times the outer guide dimension in the plane of the bend, and when jacketed, to about 4 to 6 times the outer dimensions. The mismatch between rigid and flexible guide is small when straight and designed for lower frequencies. Mismatch increases as the waveguide size decreases, since the depth of convolutions cannot be decreased as fast as the waveguide dimensions (mismatch also increases as the bending radius is decreased). Similarly, attenuation which is only slightly greater than that in rigid guide for low frequencies becomes about twice as great at 40 GHz. Power capacity is nearly equal to that of rigid guide.

Hollow-Tube Waveguides with Other Cross Sections

One of the most useful of the many cross sections that might be used is the ridged waveguide as shown in Fig. 42-15, which is useful in wideband transmission. For ridged waveguide[15-17] (and waveguide of arbitrary cross section), the best method of obtaining cutoff wavelength is by resonance in the cross section. A convenient longitudinal plane is chosen. At cutoff, the susceptance looking into the shorted parallel-plate guide to the right of this plane is equal in amplitude and opposite in phase to that looking to the left. If the guide is symmetrical, only half of the guide need be used since Y at the center is zero. For the ridged guide this gives

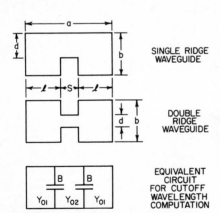

FIG. 42-15   Ridged-waveguide configurations.

$$\cot \frac{2\pi l}{\lambda_c} - \frac{B}{Y_{01}} = \frac{b}{a} \tan \frac{2\pi S}{\lambda_c 2}$$

where B is the capacitive discontinuity at the height change.

The loading in the center of the guide lowers the cutoff frequency of the dominant mode so that a useful bandwidth of over 4:1 may be obtained with single-mode transmission. The impedance is reduced by the loading and can be adjusted by proportioning the ridges for impedance matching of waveguides to coaxial line, for example.

The ridged guide for a given frequency band is smaller than the regular guide, but the losses are higher.

Designs exist, and there are government standards for both single- and double-ridge guide with bandwidth ratios of 2:4 and 3:6. The loss of the 3:6-bandwidth guide approaches that of some of the low-loss coaxial lines and has not had wide acceptance. The 2:4-bandwidth double ridge is being used above 3.5 GHz, and the characteristics of the more popular sizes are listed in Table 42-8.

Fin-Line Guide   Shown in Fig. 42-16, fin-line guide uses central loading as in ridge guide, but the central-loading fin is an etched pattern on a suitable plastic or ceramic and may be insulated from the rest of the structure. It has the same wide bandwidth and concentration of the electric field at the guide center as the ridge guide, providing excellent excitation of the etched circuit patterns and any lumped circuit elements mounted on the center fin. Although not widely used as a uniform transmission line, fin-line guide is very useful when it is desirable to build filters, mixers, and other circuit elements right into the antenna feed.[18,19]

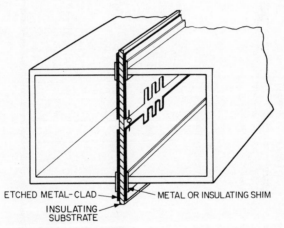

ETCHED METAL-CLAD
INSULATING SUBSTRATE
METAL OR INSULATING SHIM

FIG. 42-16   Fin-line guide.

42-8   MISCELLANEOUS TYPES OF WAVEGUIDES

Trough Waveguide[20]

Trough waveguide having the cross section shown in Fig. 42-17 is derived from the lowest TE mode on symmetrical stripline by inserting a longitudinal electric wall at the center of the strip. Energy is transmitted in the TE modes in which the electric fields are symmetrical about the center vane. The cross section is such that probes, tuning, or attenuating devices may easily be inserted and moved from the open side with a minimum of electrical disturbance. Trough waveguide is a broadband transmission device since the cutoff of the second propagating mode is 3 times that of the dominant mode instead of twice as in rectangular waveguide. It is easily fed from an end-on transition from coaxial or symmetrical stripline. The outer conductor of the TEM line is connected to the outer walls of the trough guide. The inner conductor of the line is connected at a point on the center vane of the trough guide at a distance S

TABLE 42-8 Double-Ridge Waveguide: Bandwidth Ratio, 2.4:1

Type designation WRD no.	DOD part no. M23351/4-	Material	Inside dimensions, in (mm)	Tolerance, in (mm)	Outside dimensions, in (mm)	Tolerance, in (mm)	Gap, in (mm)	Frequency range, GHz	Cutoff frequency, GHz TE₁₀	TE₂₀	Theoretical attenuation, dB/100 ft (dB/30.5m)	Theoretical power, kW	Flange, UG-
350	029	A	1.480 × 0.688 (37.59 × 017.48)	±0.003 (0.08)	1.608 × 0.816 (40.84 × 20.73)	±0.004 (0.10)	0.2920 ±0.002 7.417	3.50–8.20	2.915	8.620	3.07	151.3	1574/U
	030	B									3.03		1575/U
	031	C									2.04		1575/U
	032	S									2.18		1575/U
475	033	A	1.090 × 0.506 (27.69 × 12.85)	±0.003 (0.08)	1.190 × 0.606 (30.23 × 15.39)	±0.003 (0.08)	0.215 ±0.002 (5.46) (±0.05)	4.75–11.00	3.961	11.705	4.87	83.72	1577/U
	034	B									4.81		1578/U
	035	C									3.24		1578/U
	036	S									3.47		1578/U
750	037	A	0.691 × 0.321 (17.55 × 8.15)	±0.003 (0.08)	0.791 × 0.421 (20.09 × 10.69)	±0.003 (0.08)	0.136 ±0.002 (3.45) (±0.05)	7.50–18.00	6.239	18.464	9.64	33.58	1580/U
	038	B									9.51		1581/U
	039	C									6.41		1581/U
	040	S									6.86		1581/U
110	041	A	0.471 × 0.219 (11.96 × 5.56)	±0.003 (0.08)	0.551 × 0.299 (14.00 × 7.59)	±0.003 (0.08)	0.0930 ±0.002 (2.362) (±0.05)	11.00–26.50	9.363	27.080	17.1	15.63	1583/U
	042	B									16.9		1584/U
	043	C									11.4		1584/U
	044	S									12.2		1584/U
180	045	A	0.288 × 0.134 (7.32 × 3.40)	±0.003 (0.08)	0.368 × 0.214 (9.35 × 5.44)	±0.003 (0.08)	0.0570 ±0.002 (1.448) (±0.05)	18.00–40.00	14.495	44.285	35.8	5.834	1586/U
	046	B									35.3		1587/U
	047	C									23.8		1587/U
	048	S									25.5		1587/U

Materials   Resistivity, $\mu\Omega \cdot$ cm

A = aluminum (6061)   4.0
B = brass   3.9
C = copper   1.77
S = coin silver   2.03

NOTE: Attenuation is computed at $f = 3f_{c10}$. Actual attenuation may be considerably higher, depending on operating frequency and temperature. Actual power handling may be considerably less, depending on temperature, altitude, operating frequency, etc. For further information on both attenuation and power-handling capacity and on other design characteristics, see Ref. 18.

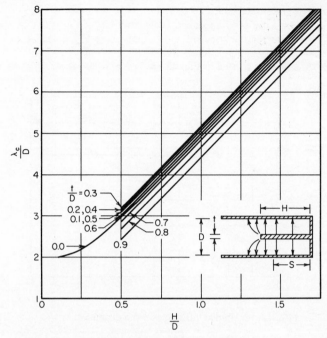

FIG. 42-17   Cutoff wavelength of trough waveguide.

up from the bottom of the center vane, as determined from the relation for the power-voltage characteristic impedance at that point,

$$Z_{PV}(S) = 754 \frac{D}{\lambda_c} \frac{\lambda_g}{\lambda} \sin^2 \left(\frac{2\pi S}{\lambda_c} \right)$$

Similarly, a crystal may be attached at an appropriate point to provide a crystal mount. The cutoff wavelength of the dominant mode in trough waveguide may be determined from the graph in Fig. 42-17.

Radial Line and Biconical Guide

Two circular parallel conducting plates, separated by a dielectric and fed at the center or outer edge, form a line in which the transmission is radial. This type of line is frequently used in choke junctions and resonant cavities such as microwave oscillator tubes. The simplest wave transmitted by this type of line is a TEM wave. The phase front of this wave is a circle of ever increasing or decreasing radius. The radial current in one plate returns radially through the other plate. With radial lines it is very useful to know the input impedance with (1) known termination, (2) output shorted, and (3) output open. Input impedance is

$$Z_i = Z_{0i} \left[\frac{Z_L \cos (\theta_i - \psi_L) + j Z_{0L} \sin (\theta_i - \theta_L)}{Z_{0L} \cos (\psi_i - \theta_L) + j Z_L \sin (\psi_i - \psi_L)} \right]$$

where Z_i = input impedance, Ω
  Z_{0i} = characteristic impedance, Ω, at input (Z_0 of Fig. 42-18 at $r = r_i$)
  Z_{0L} = characteristic impedance at output (Z_0 of Fig. 42-18 at $r = r_L$)
  Z_L = terminating impedance, Ω
$\theta_i, \theta_L, \psi_i, \psi_L$ = angles as plotted in Fig. 42-18

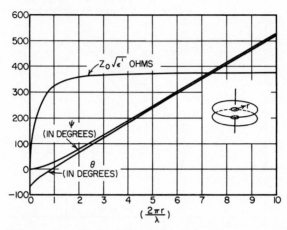

FIG. 42-18 Radial transmission-line quantities.

In one special case of this, $Z_L = 0$; then

$$Z_i = jZ_{0i}\frac{\sin(\theta_i - \theta_L)}{\cos(\psi_i - \theta_L)}$$

In another, $Z_L = \infty$; then

$$Z_i = jZ_{0i}\frac{\cos(\theta_i - \psi_L)}{\sin(\psi_i - \psi_L)}$$

Many higher-order modes are possible. Those with variations in ϕ only will propagate with any spacing of plates. Those having variations in z propagate only if the plate separation is greater than a half wavelength. More complete descriptions can be found in Refs. 1, 2, and 21.

Two cones with their apices facing and fed by a balanced input at the center as shown in Fig. 42-19 form a biconical waveguide. This structure also simulates a dipole antenna and certain classes of cavity resonators.

One important wave transmitted by this type of guide has no radial components and propagates with the velocity of light along the cones. It is analogous to the TEM wave in cylindrical systems. The ratio of voltage to current or characteristic impedance is

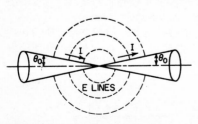

FIG. 42-19 Biconical waveguide.

$$Z_0 = 120 \log_e \cot\frac{\theta}{2} \quad \Omega$$

where θ is the conical angle. From this it is seen that the characteristic impedance is independent of radius and not variable as in the radial line.

Many higher modes can propagate on this system. These will all propagate at a velocity different from that of light. For transmission of higher modes, see Refs. 1 and 2.

Dielectric Waveguide

Electromagnetic waves will propagate along a dielectric rod[22] if the rod is of sufficient size. At low frequencies, there is little advantage to dielectric waveguides since low loss in metal waveguide makes them a convenient shielded carrier of microwave energy. At frequencies above 20 GHz, the lower loss possible with dielectric wave-guides makes their use attractive. For lossless dielectric waveguides, the propagation constant can never be real, so there is no cutoff frequency as with metal waveguides. For a given waveguide, VHF energy is confined entirely within the dielectric. The velocity of propagation and the loss correspond to that in the waveguide dielectric. As the frequency is reduced, more of the field is outside the waveguide, and the velocity and loss approach that of the surrounding air, but the dielectric ceases to guide the wave. The only TE and TM modes possible in a circular rod are those having axial symmetry. One hybrid mode, the HE_{11} (hybrid because it has both E_z and H_z components), is particularly well adapted for microwave transmission. It can be small (it is the only mode which can be propagated when the ratio of diameter to wavelength is low; less than 0.6 for polystyrene). It can be launched from the dominant TE metallic waveguide mode, and it has low loss. The approximate field configuration of this mode is shown in Fig. 42-20.

Figures 42-21 to 42-23 show variation of guide wavelength with diameter, loss as a function of diameter, and a waveguide launcher for the HE_{11} hybrid mode. The polarization of the hybrid mode in circular rod is subject to rotation because of internal stresses, dimensional

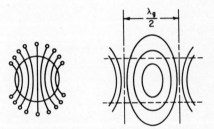

FIG. 42-20 Approximate E-field configuration of HE_{11} mode on a dielectric rod waveguide.

nonuniformity, and bends. A rectangular or oval cross section prevents this depolarization. Measurements of loss and radius of field extent (radius at which field decreases to $1/e$ times that at surface) for cross sections of the oval type are shown in Table 42-9 for 24 and 48 GHz.

Transmission with dielectric tubes as well as rods is possible. Tubes can have lower loss than rods, and theoretical calculations indicate[23] that with a polystyrene tube attenuation at 30 GHz compares with TE_{01} guide.

Other calculations shown in Fig. 42-23 indicate that losses of a few decibels per kilometer can be attained up to the infrared region by choice of the proper diameter.[24,25]

The image line[26-28] is a variation of the dielectric guide which simplifies the support problem. The HE_{11} mode, being symmetrical, can be split longitudinally and one-half replaced by a metal plate as shown in Fig. 42-24. Losses at bends are lower for concave bends than for convex. Losses in the image plane are lower than those in the dielectric for all commonly used materials.

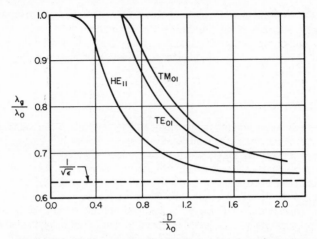

FIG. 42-21 Guide wavelength versus diameter for polystyrene waveguide.

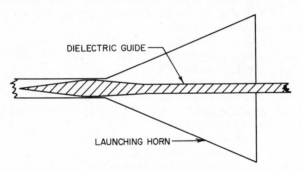

FIG. 42-22 Waveguide launcher for HE_{11} mode.

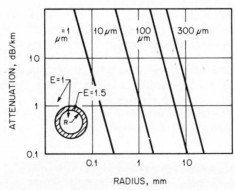

FIG. 42-23 Attenuation of dielectric tube wavelength with a dielectric of 1.5 for the HE_{11} mode.

TABLE 42-9 Loss for HE$_{11}$ Mode in Dielectric Waveguide

MATERIAL POLYSTYRENE	DIMENSIONS in (cm)	E DIRECTION	RADIUS OF FIELD EXTENT, in (cm)	LOSS dB/km	FREQ. GHz
SHEET STOCK	.095 x .156 (.241 x .396)		1.5 (3.8)	164	
EXTRUDED	.086 x .155 (.218 x .394)		4 (10.2)	164	24
EXTRUDED	.086 x .155 (.218 x .394)		7 (17.8)	82	
EXTRUDED	.056 x .142 (.142 x .361)		.4 (1.0)	3281	
EXTRUDED	.056 x .142 (.142 x .361)		.8 (2.0)	886	48
EXTRUDED	.038 x .114 (.097 x .290)		1.5 (3.8)	148	
EXTRUDED	.032 x .096 .081 x .244)		3 (7.6)	23	

A variation of the image line has been suggested for use in integrated circuits. In this version[29] a rectangular dielectric strip is laid on top of an integrated circuit board which is copper-clad on the lower surface.

DIELECTRIC

COPPER PLATE

FIG. 42-24 Image line using HE$_{11}$ mode.

Multimode Waveguide

Normally waveguides are used in the lowest mode, and the higher modes are avoided. At millimeter-wave frequencies losses in the lowest-mode-only waveguide become very high because of conductor losses. By using low-loss modes in circular waveguide far above cutoff, most of the energy is carried in the fields in the air inside and losses are reduced. Operating far from cutoff also reduces the dispersion and increases the signal bandwidth. Special efforts must be taken to prevent excitation of unwanted modes. Multimode circular guide has demonstrated a signal bandwidth of 40 to 117 GHz and transmission losses of less than 1 dB/km.[30]

Beam Waveguide

Another method of avoiding conductor loss is to use lenses to focus the energy into a converging beam or at least a nondiverging beam, refocusing periodically along the transmission path. For more information on this type see Ref. 31.

Optical-Fiber Waveguide

Optical fiber is usually a single strand of glass designed to hold an electromagnetic wave closely bound to the glass signal path. Since it is not a carrier of the fundamental signal frequency (a microwave signal would ordinarily be modulated on a light beam), it is not covered in detail here. It has many of the advantages of the multimode waveguide, the beam waveguide, and the dielectric waveguide, principally low loss and large bandwidth. A small amount of dispersion limits the distance that a wideband signal can go before some frequencies overtake others and cause information degradation. However, operational use has demonstrated losses well under 1 dB/km and bandwidths of over 1 GHz divided by the distance in kilometers.[32] It will undoubtedly become a major transmission medium for wideband signals.

Elliptical Waveguide

Waveguide with an elliptical cross section is very useful for antenna feeds. It has many of the same characteristics as rectangular waveguide, including a similar waveguide mode (so it mates easily with rectangular guide), and about the same loss and guide wavelength. It does have a narrower bandwidth. It is formed with corrugations along its length, which gives it high transverse stability and crush strength yet allows bending in both planes and a small amount of twist. This makes installation easier than with rectangular guide. It is usually covered with tough black polyethylene to protect it during the hanging, installation, and use. Elliptical waveguide is made in long lengths or is assembled in desired sections with an adapting flange to rectangular guide. Although not yet standardized, it is made by several manufacturers. Most give it a prefix indicating elliptical waveguide such as EW or WE, followed by the lowest frequency or a midfrequency in tenths of a gigahertz. Characteristics of typical sizes are shown in Table 42-10.

42-9   LIST OF SYMBOLS

f = frequency, Hz

ω = $2\pi f$

c = velocity of light $\approx 3 \times 10^8$ m/s

λ = wavelength

ϵ = relative dielectric constant

μ = relative permeability

σ = conductivity, S/m

α = attenuation constant, Np/m (Np = 8.686 dB)

E_a = breakdown of air, V/m

P = power, W

tan δ = loss tangent or dissipation factor

The properties of some commonly used metals will be found in Chap. 46. The conductivity σ can be found from the resistivity values by the relationship

$$\sigma(\text{S/m}) = \frac{10^8}{\text{resistivity } (\mu\Omega \cdot \text{cm})}$$

Values for ϵ and tan δ of some commonly used dielectrics will be found in Chap. 46.

TABLE 42-10 Typical Characteristics of Elliptical Waveguide

Type	Maximum potential operating range, GHz	TE$_{11}$-mode cutoff frequency, GHz	Major and minor dimensions over jacket, in (mm)	Bending-radii minimum, in (mm) E plane	H plane	Recommended twist,° /ft (m)
EW20	1.9–2.7	1.60	5.02 × 2.83 (127.5 × 71.9)	26 (660)	71 (1800)	0.25 (0.75)
EW28	2.6–3.5	2.20	3.65 × 2.33 (92.5 × 59.2)	22 (560)	52 (1320)	0.25(0.75)
EW37	3.3–4.3	2.81	2.90 × 1.86 (73.7 × 47.2)	17 (430)	41 (1040)	0.5 (1.5)
WE37	3.4–4.2	2.87	3.04 × 1.93 (71.0 × 49.0)	19.7 (500)	39.4 (1000)	0.5 (1.5)
EW44	4.2–5.1	3.58	2.31 × 1.59 (58.7 × 40.4)	15 (380)	32 (810)	0.5 (1.5)
WE44	4.2–5.0	3.58	2.24 × 1.59 (520 × 43.0)	15.7 (400)	47.2 (1200)	0.5 (1.5)
WE52	4.6–6.425	3.63	2.25 × 1.31 (57.2 × 33.3)	12 (305)	32 (810)	1 (3)
WE56	5.4–6.425	4.08	1.97 × 1.30 (50.0 × 33.0)	19.7 (500)	39.4 (1000)	1 (3)
WE59	5.9–7.15	4.34	1.85 × 1.22 (47.0 × 31.0)	11.8 (300)	23.5 (600)	1 (3)
EW63	5.85–7.125	3.96	2.01 × 1.16 (51.1 × 29.5)	10 (260)	29 (740)	1 (3)
EW64	5.3–7.75	4.36	1.91 × 1.12 (48.5 × 28.4)	10 (260)	27 (685)	1 (3)
WE64	6.4–7.75	4.16	1.84 × 1.14 (49.0 × 29.0)	11.8 (300)	23.6 (600)	1 (3)
WE71	7.1–8.6	5.23	1.54 × 0.98 (390 × 25.0)	11.8 (300)	23.6 (600)	1 (3)
EW77	6.1–8.5	4.72	1.72 × 1.00 (43.6 × 25.4)	9 (230)	25 (635)	1 (3)
EW85	7.7–10.0	6.55	1.32 × 0.90 (33.5 × 22.9)	8 (200)	19 (480)	1 (3)
WE85	8.5–10.2	5.83	1.38 × 0.94 (35.0 × 24.0)	88 (200)	16 (400)	1 (3)
EW90	8.3–11.7	6.50	1.32 × 0.80 (33.5 × 20.3)	7 (180)	19 (480)	2 (6)
WE107	8.5–11.7	6.80	1.18 × 0.79 (30.0 × 20.0)	6 (150)	16 (400)	2 (6)
WE122	12.2–13.25	7.29	1.09 × 0.71 (28.0 × 18.0)	6 (150)	16 (400)	2 (6)
EW122	10.0–13.25	8.46	1.07 × 0.72(27.2 × 18.3)	6 (150)	15 (380)	2 (6)
EW132	11.0–15.35	9.33	0.96 × 0.61 (24.4 × 15.5)	5 (130)	14 (360)	2 (6)

REFERENCES

1 S. Ramo, J. R. Whinnery, and T. Van Duzer, *Fields and Waves in Communication Electronics,* John Wiley & Sons, Inc., New York, 1965.
2 C. G. Montgomery, R. H. Dicke, and E. M. Purcell (eds.), *Principles of Microwave Circuits,* McGraw-Hill Book Company, New York, 1948. A good general reference work for theoretical background.
3 R. M. Chisholm, "The Characteristic Impedance of Trough and Slab Lines," *IRE Trans. Microwave Theory Tech.,* vol. MTT-4, July 1956, pp. 166–172.
4 K. C. Gupta, R. Garg, and I. J. Bahl, *Microstrip Lines and Slot Lines,* Artech House, Inc., Dedham, Mass., 1979. Includes a very comprehensive bibliography on striplines.
5 A. F. Hinte, G. V. Kopcsay, and J. J. Taub, "Choosing a Transmission Line," *Microwaves,* December 1971, pp. 46–50.
6 W. Sichak, "Coaxial Line with Helical Conductor," *IRE Proc.,* vol. 42, August 1954, pp. 1315–1319; also correction, *IRE Proc.,* vol. 43, February 1955, p. 148.
7 G. L. Ragan (ed.), *Microwave Transmission Circuits,* McGraw-Hill Book Company, New York, 1948. A good general reference for applications.
8 J. W. E. Griemsmann, "An Approximate Analysis of Coaxial Line with Helical Dielectric Support," *IRE Trans. Microwave Theory Tech.,* vol. MTT-4, January 1956, pp. 13–23.
9 S. Greenblatt, J. W. E. Griemsmann, and L. Birenbaum, "Measurement of Energy Leakage from Cables at VHF and Microwave Frequencies," AIEE Conf. pap., winter general meeting, 1956.
10 A. A. Smith, Jr., *Coupling of External Electromagnetic Fields to Transmission Lines,* John Wiley & Sons, Inc., New York, 1977.
11 H. W. Ott, *Noise Reduction Techniques in Electronic Systems,* John Wiley & Sons, Inc., New York, 1976.
12 *R.F. Transmission Line Catalog & Handbook,* Times Wire and Cable Company, Wallingford, Conn., 1972.
13 H. A. Wheeler, "The Piston Attenuator in a Waveguide below Cutoff," Wheeler Monog. 8, Wheeler Laboratories, Inc., Great Neck, N.Y., 1949.
14 S. A. Schelkunoff, *Electromagnetic Waves,* D. Van Nostrand Company, Inc., Princeton, N.J., 1943, p. 319.
15 Radio Research Laboratory staff, *Very High Frequency Techniques,* McGraw-Hill Book Company, New York, 1947.
16 S. Hopfer, "The Design of Ridged Waveguides," *IRE Trans. Microwave Theory Tech.,* vol. MTT-3, October 1955, pp. 20–29.
17 Tsung-Shan Chen, "Calculation of the Parameters of Ridge Waveguides," *IRE Trans. Microwave Theory Tech.,* vol. MTT-5, January 1957, pp. 12–17.
18 P. J. Meier, "Integrated Fin-Line Millimeter Components, *IEEE Trans. Microwave Theory Tech.,* vol. MTT-22, December 1974, part II, pp. 1209–1216.
19 P. J. Meier, "Millimeter Integrated Circuits Suspended in the E-Plane of Rectangular Waveguide," *IEEE Trans. Microwave Theory Tech.,* vol. MTT-26, October 1978, pp. 726–732.
20 K. S. Packard, "The Cutoff Wavelength of Trough Waveguide," *IRE Trans. Microwave Theory Tech.,* vol. MTT-6, no. 4, October 1958, pp. 455–456.
21 N. Marcuvitz (ed.), *Waveguide Handbook,* McGraw-Hill Book Company, New York, 1951.
22 M. T. Weiss and E. M. Gyorgy, "Low Loss Dielectric Waveguides," *IRE Trans. Microwave Theory Tech.,* vol. MTT-2, September 1954, pp. 38–44.
23 *Arch. Elek. Übertragang,* vol. 8, no. 6, June 1954.
24 M. Miyagi and S. Nishida, "A Proposal of Low-Loss Leaky Waveguide for Submillimeter Wave Transmission," *IEEE Trans. Microwave Theory Tech.,* vol. MTT-28, April 1980.
25 M. Miyagi and S. Nishida, "Transmission Characteristics of Dielectric Tube Leaky Waveguide," *IEEE Trans. Microwave Theory Tech.,* vol. MTT-28, June 1980.

26 D. D. King, "Circuit Components in Dielectric Image Lines," *IRE Trans. Microwave Theory Tech.,* vol. MTT-3, December 1955, pp. 35–39.

27 D. D. King, "Properties of Dielectric Image Lines," *IRE Trans. Microwave Theory Tech.,* vol. MTT-3, March 1955, pp. 75–81.

28 S. Shindo and T. Itanami, "Low Loss Rectangular Dielectric Image Line for Millimeter-Wave Integrated Circuits," *IEEE Trans. Microwave Theory Tech.,* vol. MTT-26, October 1978, pp. 747–751.

29 W. V. McLevige, T. Itoh, and R. Mittra, "New Waveguide Structures for Millimeter-Wave and Optical Integrated Circuits," *IEEE Trans. Microwave Theory Tech.,* vol. MTT-23, October 1975, pp. 788–794.

30 "Millimeter Waveguide Systems," report on international conference in London, *Microwave J.,* March 1977, pp. 24–26.

31 J. A. Arnaud and J. T. Ruscio, "Guidance of 100 GHz Beams by Cylindrical Mirrors," *IEEE Trans. Microwave Theory Tech.,* vol. MTT-23, April 1975, pp. 377–379.

32 K. Shirahata, W. Susaki, and H. Mamizaki, "Recent Developments in Fiber Optic Devices," *IEEE Trans. Microwave Theory Tech.,* vol. MTT-30, February 1982, pp. 121–130.

Chapter 43

Impedance Matching and Broadbanding

David F. Bowman

RCA Corporation

Impedance matching is the control of impedance for the purpose of obtaining maximum power transfer or minimum reflection. This chapter describes circuits and techniques used for impedance matching with emphasis on those most suitable for broadband operation. The impedance-matching methods of this chapter are limited to the use of linear, passive, and reciprocal elements.

Impedance Matching for Maximum Power Transfer

A variable load impedance connected to a source will receive the maximum possible power from the source when it is adjusted to equal the complex conjugate of the impedance of the source (Fig. 43-1 a). The load impedance and source impedance are then matched on a conjugate-impedance basis.*

A fixed source impedance and a fixed load impedance may be coupled for maximum power transfer by a properly proportioned network (Fig. 43-1 b) interposed between them. This "matching" network transforms the source impedance to the conjugate of the load impedance, and vice versa. Thus a conjugate-impedance match occurs at the input to the network and at the output of the network. It is true of any lossless transmission circuit that if a conjugate-impedance match is obtained at any point along the transmission path, then a conjugate-impedance match is obtained at all other points along the path.

Impedance Matching for Minimum Reflection

A length of lossless transmission line may form one link in a lossless transmission circuit. Lines commonly used in antenna systems have characteristic impedances that are purely or nearly real. Characteristic impedances designated by Z_0 in this chapter are real unless otherwise noted. Electrical length is denoted by $\beta\ell$. Maximum power transfer will occur when the circuit is adjusted for a conjugate-impedance match (Fig. 43-1 c). However, in general, the line will be subject to two waves, a direct and a reflected wave, propagating in opposite directions. The interference between these waves creates standing-wave patterns of voltage and current. These effects may be tolerable, but usually they are undesirable.

If the line has finite attenuation, maximum power transfer from the source to the load is obtained only when the following conditions are met:

1 The generator is loaded by the conjugate of its internal impedance.

2 The line is terminated in its characteristic impedance.

The first condition provides maximum power delivery from the source. The second condition provides minimum power dissipation in the line by eliminating the reflected wave on the line.

---

*This is distinct from impedance matching on an image-impedance basis in which the source and load impedances are equal. Image-impedance matching is sometimes used for convenience, often when the impedances are nearly resistive. However, it does not result in the maximum-power-transfer condition unless the impedances are purely resistive.

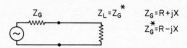

A. A SOURCE AND A LOAD MATCHED ON A CONJUGATE-
    IMPEDANCE BASIS

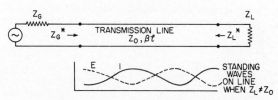

B. A SOURCE AND A LOAD MATCHED ON A CONJUGATE-
    IMPEDANCE BASIS BY A MATCHING NETWORK

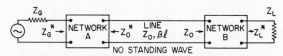

C. A SOURCE, LOSSLESS LINE AND A LOAD MATCHED ON
    A CONJUGATE-IMPEDANCE BASIS

D: A SOURCE, A LINE AND A LOAD MATCHED ON A CONJU-
    GATE IMPEDANCE AND CHARACTERISTIC IMPEDANCE
    BASIS BY NETWORKS

FIG. 43-1 Impedance matching of transmission circuits.

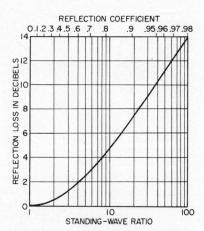

FIG. 43-2 Reflection loss.

Figure 43-1d shows a "matched" system, that is, a system matched both on a conjugate-impedance basis and on a characteristic-impedance basis. This represents an ideal condition.

The importance of matching can be seen by an examination of the detrimental effects of a mismatch. A measure of mismatch at the load junction is the voltage reflection coefficient.

$$\rho = \rho \exp{(2\psi)} = \frac{Z_L - Z_0}{Z_L + Z_0}$$

This defines vectorially the reflected voltage wave for a unit wave incident on the junction. The power in the load is thus reduced from the maximum available power by the ratio of $1 - \rho^2$.

The reduction in transmission is called a *reflection loss* or *transition loss*. Reflection loss expressed in decibels is shown as a function of the standing-wave ratio (SWR)

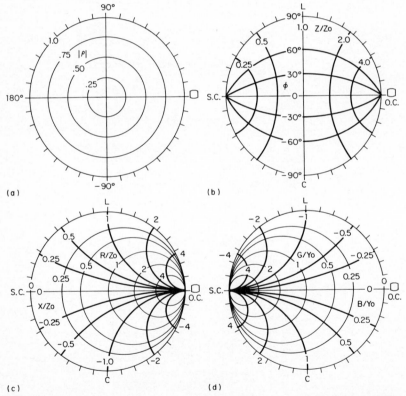

FIG. 43-3 Reflection coefficient charts in hemisphere form; positive real axis to the right. (*a*) Reflection coefficient in polar coordinates of $|\rho|$ and 2ψ. (*b*) Impedance coordinates of $|z| = |Z/Z_0|$ and ϕ, the Carter chart. (*c*) Impedance coordinates of $|r| = |R/Z_0|$ and $|x| = |X/Z_0|$, the Smith chart. (*d*) Admittance coordinates of $|g| = |G/Y_0|$ and $|b| = |B/Y_0|$, a variant of the Smith chart.

in Fig. 43-2. The reflected wave combines with the incident wave on the line to form a standing wave having an SWR (maximum to minimum) of

$$S = \frac{1 + \rho}{1 - \rho}$$

The presence of the standing wave increases the maximum voltage and current limits on the line for the same delivered power by the ratio $\sqrt{S}$. The efficiency η of the line may be expressed by

$$\eta = \frac{S_2 - 1/S_2}{S_1 - 1/S_1}$$

$$= \frac{4S_1}{10^{0.1A}(S_1 + 1)^2 - 10^{-0.1A}(S_1 - 1)^2}$$

where subscripts 1 and 2 refer to the load and source ends of line, respectively, and A is the normal line attenuation in decibels.

Reflection-Coefficient Charts

The reflection coefficient is a vector quantity related directly to impedance ratio and mismatch. Its magnitude does not exceed unity, so a plot of all possible reflection ratios for passive impedances may be charted within a circle of unit radius (see Fig. 43-3a). Furthermore, the representations of a mismatch referred to different distances along the length of a line lie at a constant radius from the center of the chart at angles proportional to the distances. Coordinate systems of immittance normalized to the characteristic or reference immittance are superimposed to give the useful Carter and Smith charts,[2,3] shown in skeleton form in Fig. 43-3b, c, and d. Usage varies on the orientation of the axes. Positive interpretation is aided by labels for open- and short-circuit points and inductive and capacitive sides.

43-2 IMPEDANCE MATCHING WITH LUMPED ELEMENTS

Throughout the lower range of radio frequencies, it is convenient to use lumped reactance elements such as coils and capacitors in impedance-matching networks. In higher frequency ranges, pure inductance or capacitance may not always be obtained from practical elements, but it is convenient to analyze circuits in terms of their lumped reactance (and resistance) elements.

Any two complex impedances may be matched by a simple L section of two reactance elements. If, in addition to matching, it is necessary to maintain given phase relationships between the source and load voltages and currents, a third element must be used to form a T or π section. A lattice section using four elements is more convenient for some applications. The primary use of the simple sections is for matching at a single frequency, although it is possible to obtain matching at two or more separate frequencies by replacing each reactance element of the basic network with a more complex combination giving the required reactance at each specified frequency. Harmonic reduction can be effected in a similar manner by introducing high series reactances or low shunt reactances at the harmonic frequencies.

L Section

The expressions for the required reactance values for an L section providing a match between pure resistances is shown in Fig. 43-4a. (Here and in the remainder of this chapter, X and B will designate reactance and susceptance values respectively.) If X_s is positive, the network will delay the phase of the wave by angle θ. If X_s is negative, it will advance the phase by that angle. The reflection coefficient of R_1 with respect to R_2 is reduced by a factor of at least 10 over a frequency ratio of at least 1.1 by introducing the proper L section of one capacitor and one inductor for R_2/R_1, up to 2.0.

If two complex terminations are to be matched, the series reactance X'_s is made to include a compensation for the series reactance of the right-hand termination and the shunt susceptance B'_p is made to include the susceptance of the left-hand termination as shown in Fig. 43-4b. For this case θ represents the phase between the current in the left- to the voltage on the right-hand termination.

T and π Sections

Design expressions are given in Fig. 43-5 for T and π sections matching between resistive terminations. If the terminations contain reactive or susceptive components, it is

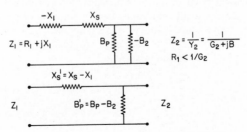

A.   L SECTION FOR MATCHING BETWEEN RESISTANCES

B.   L SECTION FOR MATCHING BETWEEN COMPLEX IMPEDANCES

FIG. 43-4 Impedance matching with reactive L sections.

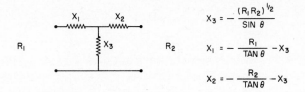

$$X_3 = -\frac{(R_1 R_2)^{1/2}}{SIN\ \theta}$$

$$X_1 = -\frac{R_1}{TAN\ \theta} - X_3$$

$$X_2 = -\frac{R_2}{TAN\ \theta} - X_3$$

A. T SECTION FOR MATCHING BETWEEN RESISTANCES

$$B_3 = -\frac{(G_1 G_2)^{1/2}}{SIN\ \theta}$$

$$B_1 = -\frac{G_1}{TAN\ \theta} - B_3$$

$$B_2 = -\frac{G_2}{TAN\ \theta} - B_3$$

B. π SECTION FOR MATCHING BETWEEN RESISTANCES

FIG. 43-5 Impedance matching with reactive T and π sections.

necessary to include an appropriate compensating component in the end elements as in the case of the L section.

L, T, and π sections may be used in unbalanced circuits as shown or in balanced circuits by moving one-half of each series element to the opposite conductor of the line. The lattice section is inherently balanced.

Lattice Section

The elements of a lattice section as well as T and π sections are given in Fig. 43-6 in terms of an equivalent transmission line having the same characteristic impedance and phase shift. The required equivalent line may be determined as described in latter parts of this section.

Inductive Coupling

A pair of inductively coupled coils is useful in a wide variety of impedance-matching circuits. Figure 43-7 shows two possible equivalent circuits of lossless coupled coils. The first is expressed in terms of reactance elements, including the mutual reactance X_m. The second is in terms of susceptance elements including the transfer susceptance B_T. (The transfer susceptance is the susceptance component of the transfer admittance, which is the ratio of the current induced in the short-circuited secondary to the voltage applied to the primary.) If capacitive tuning reactances are added, the equivalent circuits can be proportioned in accordance with the T and π matching sections described above. The capacitive reactances required for (1) series tuning and (2) parallel tuning are given in Fig. 43-8 for matching from R_1 to R_2.

The series and parallel capacitors may be used for tuning out the series reactance or shunt susceptance, respectively, of a complex termination.

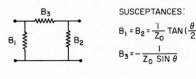

A.   T   SECTION

REACTANCES:

$X_1 = X_2 = Z_0 \, \mathrm{TAN} \left(\frac{\theta}{2} \right)$

$X_3 = \dfrac{-Z_0}{\mathrm{SIN} \, \theta}$

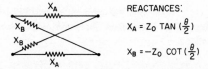

B. π SECTION

SUSCEPTANCES:

$B_1 = B_2 = \dfrac{1}{Z_0} \, \mathrm{TAN} \left(\frac{\theta}{2} \right)$

$B_3 = -\dfrac{1}{Z_0 \, \mathrm{SIN} \, \theta}$

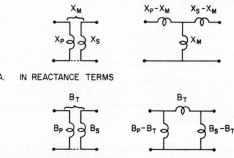

C.   LATTICE   SECTION

REACTANCES:

$X_A = Z_0 \, \mathrm{TAN} \left(\frac{\theta}{2} \right)$

$X_B = -Z_0 \, \mathrm{COT} \left(\frac{\theta}{2} \right)$

FIG. 43-6   T, π, and lattice equivalents of transmission-line section of characteristic impedance Z_0 and electrical length θ.

Lumped Matching Reactance

The standing wave on a transmission line may be eliminated on the source side of a matching reactance that is properly proportioned and positioned, as shown in Fig. 43-9. The choice of series or parallel connections, inductive or capacitive elements, may be determined by the position of the standing-wave pattern and other practical considerations.

A.   IN REACTANCE TERMS

B.   IN SUSCEPTANCE TERMS

FIG. 43-7   Inductively coupled circuits and equivalent T and π sections.

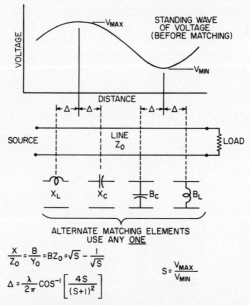

A. SERIES TUNING

$$X_M \gtreqless (R_1 R_2)^{1/2}$$

$$X_{CI} = -X_P - R_1 \left(\frac{X_M}{R_1 R_2} - 1 \right)^{1/2}$$

$$X_{C2} = -X_P - R_2 \left(\frac{X_M}{R_1 R_2} - 1 \right)^{1/2}$$

$$\theta = \sin^{-1} \left(-\frac{\sqrt{R_1 R_2}}{X_M} \right)$$

B. PARALLEL TUNING

$$X_{CI} = -\frac{1}{B_{CI}}$$

$$X_{C2} = -\frac{1}{B_{C2}}$$

$$G_1 = \frac{1}{R_1}$$

$$G_2 = \frac{1}{R_2}$$

$$\theta = \sin^{-1} \left(-\frac{\sqrt{G_1 G_2}}{B_M} \right)$$

$$B_{CI} = -B_P - G_1 \left(\frac{B_M}{G_1 G_2} - 1 \right)^{1/2}$$

$$B_{C2} = -B_S - G_2 \left(\frac{B_M}{G_1 G_2} - 1 \right)^{1/2}$$

$$B_P = \frac{1}{X_P}$$

$$B_S = \frac{1}{X_S}$$

$$B_M = \frac{X_M}{X_M^2 - X_P X_S}$$

FIG. 43-8 Tuned inductively coupled circuits for matching between resistances.

43-3 IMPEDANCE MATCHING WITH DISTRIBUTED ELEMENTS

In many frequency ranges, it is desirable to use sections of transmission line having distributed reactances rather than lumped reactances in the form of coils and capac-

$$\frac{X}{Z_0} = \frac{B}{Y_0} = BZ_0 = \sqrt{S} - \frac{1}{\sqrt{S}}$$

$$\Delta = \frac{\lambda}{2\pi} \cos^{-1} \left[\frac{4S}{(S+1)^2} \right]$$

$$S = \frac{V_{MAX}}{V_{MIN}}$$

FIG. 43-9 Determination of lumped reactance to match a transmission line.

itors. The lines are usually proportioned to yield negligible loss so that the following expressions for impedance relationships in lossless lines apply:

$$Z_{in} = Z_0 \frac{Z_L \cos \beta\ell + jZ_0 \sin \beta\ell}{Z_0 \cos \beta\ell + jZ_L \sin \beta\ell}$$

(43-1)

= input impedance of line Z_0, $\beta\ell$ terminated in Z_L

For a short-circuited line, $Z_L = 0$, so that

$$Z_{in} = Z_{oc} = jZ_0 \tan \beta\ell$$

(43-2)

For an open-circuited line, $Z_L = \infty$, so that

$$Z_{in} = Z_{oc} = -jZ_0 \cot \beta\ell$$

(43-3)

Transmission-Line Stubs

Lengths of transmission line short-circuited or open-circuited at one end are often used as reactors in impedance-matching circuits. Inspection of Eqs. (43-2) and (43-3) will show that the designer by choice of characteristic impedance and line length has control of reactance value and slope (with respect to frequency) at any given frequency. Alternatively, the designer has control of the value of reactance at any two frequencies. The available slope of reactance is always greater than unity, the value obtained from a single lumped reactance element. Unfortunately, the available slope of reactance is always positive, although a negative slope would be ideal in many applications. However, an effect similar to that of a series-connected or parallel-connected reactance having the unattainable negative slope may often be obtained over a limited frequency range by the use respectively of a parallel-connected or series-connected reactance.

General Line Transformer

The expression for Z_{in} [Eq. 43-1] may be recast to relate generator and load impedances ($Z_G = R_G + jX_G$, $Z_L = R_L + jX_L$) to the characteristic impedance and electrical length of a line section, providing a perfect match between them. The new expressions are

$$Z_0 = \sqrt{\frac{R_L|Z_G|^2 - R_G|Z_L|^2}{R_G - R_L}}$$

$$= \sqrt{|Z_G Z_L| \frac{\frac{R_L}{R_G}\left|\frac{Z_G}{Z_L}\right| - \left|\frac{Z_L}{Z_G}\right|}{1 - \frac{R_L}{R_G}}}$$

$$\tan \beta\ell = \frac{Z_0(R_L - R_G)}{R_L X_G - R_G X_L}$$

$$= \frac{Z_0(X_G - X_L)}{R_L R_G + X_L X_G - Z_0^2}$$

For Z_0 to be a positive finite real number, it is necessary that

$$\frac{1}{2} < \frac{\log \left| \dfrac{Z_L}{Z_G} \right|}{\log \left(\dfrac{R_L}{R_G} \right)} < \infty$$

Instead of using the above expression for tan $\beta\ell$, one may determine $\beta\ell$ from a plot of Z_L/Z_0 and Z_G^*/Z_0 on a Carter chart or a Smith chart once Z_0 has been determined. The two points will be at the same radius. The electrical angle measured clockwise from Z_L/Z_0 to Z_G^*/Z_0 is $\beta\ell$. (Note that $Z_G^* = R_G - jX_G$.)

  The Z_0 and $\beta\ell$ quantities for the required line section may be used to compute T, π, and lattice sections of lumped reactances to perform the same function by using expressions given in Fig. 43-6.

Line Transformer for Matching to Resistance

For the frequently encountered case in which either the load impedance or the source impedance is a pure resistance, a solution for the required matching line may be

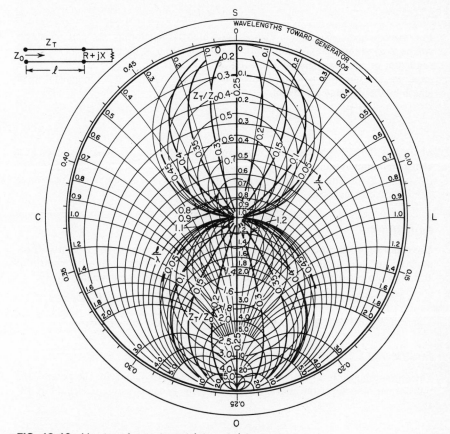

FIG. 43-10 Line transformer to match to a resistance.

obtained from Fig. 43-10. The complex impedance $R + jX$ is located on the Smith chart in terms of its normalized components R/Z_0 and X/Z_0, where Z_0 is the resistance to which a match is desired. The point so found is located within an area having a second set of coordinates if a solution is possible. These coordinates Z_T/Z_0 and ℓ/λ give the characteristic impedance and length of the required line.

Quarter-Wave Transformer

The useful quarter-wave transformer results from the general line transformer when $\beta\ell = \pi/2$. It has an impedance-inverting property, as seen from an inspection of

$$Z_{in} = \frac{Z_0^2}{Z_L}$$

The input impedance is thus proportional to the reciprocal of the load impedance. The phase angle of the input impedance is the negative of that of the load impedance. The quarter-wave line can be used, for example, to transform an inductive low impedance to a capacitive high impedance.

The quarter-wave transformer is often used to match between different resistance levels. In this case

$$Z_0 = \sqrt{R_1 R_2}$$

In a mismatched coaxial line a quarter-wave transformer formed by a simple sleeve can be used to tune out the reflection. The sleeve forms an enlargement of the inner conductor or a constriction of the outer conductor which may be stationed where required. The characteristic impedance of the line is reduced over the quarter-wave length by the sleeve to a value of

$$\frac{Z_0}{\sqrt{S}}$$

where S is the initial SWR of the line. The load end of the sleeve is positioned at a voltage minimum on the standing-wave pattern. The impedance, here looking toward the load, is Z_0/S. The transformer transforms this to Z_0 so that the line on the source side is perfectly matched.

Z_0 Trimmer

The transformation ratio of a quarter-wave section can be adjusted by a reactive trimming immittance, as shown in Fig. 43-11, to obtain a match when other control of circuit values is not convenient. The immittance required is approximated by

$$X/Z_0 = B/Y_0 = 2.29 \log{(R_1 R_2/Z_0)}/\sin A \quad \text{for } 2/3 \le R_1 R_2/Z_0 \le 3/2$$

The resulting network is nearly equivalent to a line for which $Z_0 = R_1 R_2$.

Frequency Sensitivity of Line Transformer

The mismatching effect of a departure from the design frequency for the general line transformer (and hence the quarter-wave transformer) can be estimated from Fig. 43-12.

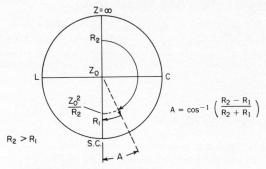

DIAGRAM FOR B_C TRIMMER

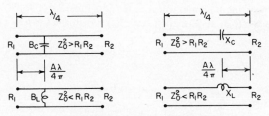

ALTERNATE FORMS OF TRIMMER

FIG. 43-11 Reactive transformation trimmer for a quarter-wave transformer.

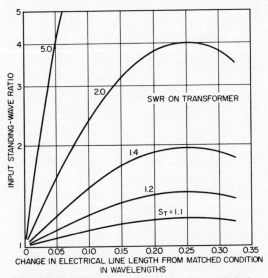

FIG. 43-12 Frequency sensitivity of a transmission-line transformer.

Cascaded Quarter-Wave Transformers[4,5]

A number of quarter-wave transmission-line sections may be arrayed in cascade to realize great improvement in wideband performance over a single-section transformer. The characteristic impedances of the successive sections are proportioned to divide the overall transformation systematically. Figure 43-13 defines the terms that will be used in the discussion below.

Binomial Transformer

The binomial (or binomial-coefficient) distribution gives almost maximally flat performance. In this distribution the logarithms of the impedance ratios of the steps between sections are made to be in the ratio of the binomial coefficients.

Table 43-1 may be used to determine the characteristic impedance Z_n of the nth section in an N-section binomial transformer as a function of R_2/R_1.

The input SWR of an N-section binomial transformer can be expressed by

$$S = 1 + (\cos\theta)^N \ln\frac{R_2}{R_1}$$

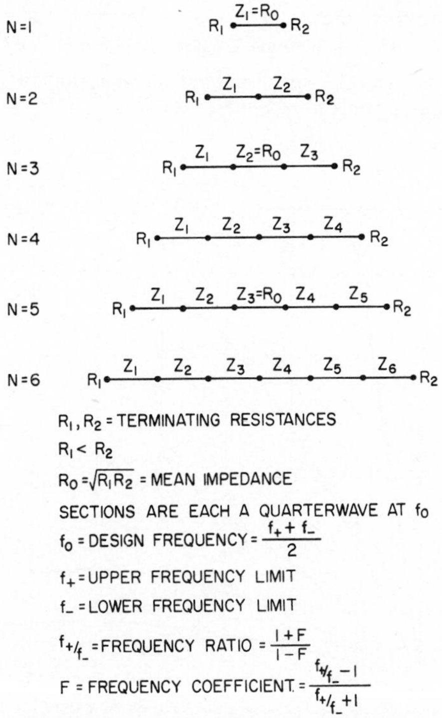
N=1
R1 Z1=R0 R2
N=2
R1 Z1 Z2 R2
N=3
R1 Z1 Z2=R0 Z3 R2
N=4
R1 Z1 Z2 Z3 Z4 R2
N=5
R1 Z1 Z2 Z3=R0 Z4 Z5 R2
N=6
R1 Z1 Z2 Z3 Z4 Z5 Z6 R2
R1, R2 = TERMINATING RESISTANCES
R1 < R2
R0 = √(R1R2) = MEAN IMPEDANCE
SECTIONS ARE EACH A QUARTERWAVE AT f0
f0 = DESIGN FREQUENCY = (f+ + f−)/2
f+ = UPPER FREQUENCY LIMIT
f− = LOWER FREQUENCY LIMIT
f+/f− = FREQUENCY RATIO = (1+F)/(1−F)
F = FREQUENCY COEFFICIENT = (f+/f− − 1)/(f+/f− + 1)

FIG. 43-13 Cascaded quarter-wave transformers.

TABLE 43-1 Design Ratios for Binomial Transformer

$N =$ \ $n =$	1	2	3	4	5	6	7
1	1/2						
2	1/4	3/4					
3	1/8	4/8	7/8				
4	1/16	5/16	11/16	15/16			
5	1/32	6/32	16/32	26/32	31/32		
6	1/64	7/64	22/64	42/64	57/64	63/64	
7	1/128	8/128	29/128	64/128	99/128	120/128	127/128

$$\text{Ratios} = \frac{\log (Z_n/R_1)}{\log (R_2/R_1)} = 1 - \frac{\log (R_2/Z_n)}{\log (R_2/R_1)}$$

where θ is the electrical length of each section. This expression is subject to the assumptions of small steps, zero-discontinuity capacitance,[6,7] and equal lengths of the sections.

The two-section transformer has maximally flat bandwidth curve for all transformation ratios. For other values of N the performance approximates the maximally flat curve for transformation ratios near unity.

Chebyshev Transformer

If a certain maximum reflection coefficient ρ_m may be tolerated within the operating band, an optimum design which allows the reflection coefficient to cycle between 0 and ρ_m within the band and to increase sharply outside the band is possible. This is called the Chebyshev transformer, since Chebyshev polynomials are used in its design.

The input SWR of the Chebyshev transformer is given by

$$S = 1 + \ln \left(\frac{R_2}{R_1}\right) \frac{T_N \left(\dfrac{\cos \theta}{\cos \theta_-}\right)}{T_N \left(\dfrac{1}{\cos \theta_-}\right)}$$

where $T_N(x)$ is the Chebyshev polynomial of mth degree defined by

$$T_0(x) = 1$$

$$T_1(x) = x$$

$$T_2(x) = 2x^2 - 1$$

$$T_{m+1}(x) = 2x T_m(x) - T_{m-1}(x)$$

and where θ is the electrical length of each section and θ_- is the length at f_-. The expression is subject to the same assumptions as were made in the bionomial case.

Another expression of Chebyshev transformer performance, which does not require the assumption of small steps, is

$$P_L = 1 + \frac{\left(\dfrac{R_2}{R_1} - 1\right)^2 T_N^2\left(\dfrac{\cos \theta}{\cos \theta_-}\right)}{\dfrac{4R_2}{R_1}} T_N^2\left(\dfrac{1}{\cos \theta_-}\right)$$

where P_L is the power-loss ratio defined by $1/(1 - |\rho|^2)$.

The above equation may be converted to an implicit relationship among four principal quantities, namely,

R_2/R_1, the transformation ratio

ρ_m, the maximum tolerable reflection coefficient within the band

N, the number of sections

F, the frequency coefficient

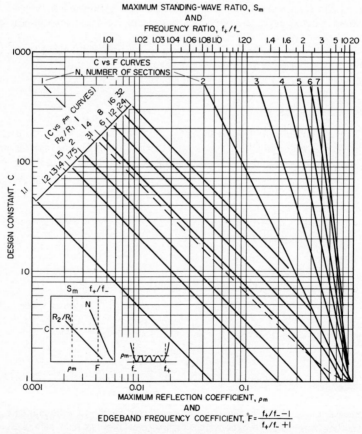

FIG. 43-14 Performance of Chebyshev transformer.

TABLE 43-2   Design Ratios for Chebyshev Transformer

N	$\dfrac{\log\,(R_2/R_1)}{\log\,(Z_1/R_1)}$	$\dfrac{\log\,(Z_2/R_1)}{\log\,(Z_1/R_1)}$	$\dfrac{\log\,(Z_3/R_1)}{\log\,(Z_2/R_1)}$
2	$4 - 2C^2$		
3	$8 - 6C^2$		
4	$16 - 16C^2 + 2C^4$	$5 - 4C^3$	
5	$32 - 40C^2 + 10C^4$	$6 - 5C^2$	
6	$64 - 96C^2 + 36C^4 - 2C^6$	$7 - 6C^2$	$22 - 30C^2 + 9C^4$

NOTE: $C = \cos\theta = \cos\,[(1 - F)\,\pi/2]$

Any one of these may be determined by reference to Fig. 43-14 if the other three are known. Design constant C of Fig. 43-14 can be used as a measure of difficulty of transformation. It may be expressed as a function of R_2/R_1 and ρ_m or as a function of N and F:

$$C(R_2/R_1, \rho_m) = (R_2/R_1 - 1)\ \sqrt{(1/\rho_m^2 - 1)R_1/4R_2}$$

$$C(N,F) = T_N\,(1/\cos\,(\pi/2(1 - F))) \geq C\,(R_2/R_1, \rho_m)$$

The characteristic impedances may be calculated from functions given in Table 43-2. The impedance of the center section when N is odd is simply $\sqrt{R_1 R_2}$ and is independent of F.

Dual-Band Transformers[8]

Multisection transformers can be synthesized for dual or multiple passbands by systematic distribution of the available reflection-coefficient zeros over the passbands. The series of impedance levels of the sections becomes nonmonotonic for a large frequency ratio between the two passbands. A nonmonotonic design for one operating band can be made with an arbitrarily short length by placing a second dummy passband at an arbitrarily high frequency.

Short-Step Transformer

A two-section transformer, as shown for $N = 2$ in Fig. 43-13 but with a much shorter overall length at midband, is useful for narrow bandwidth or for transformation ratios near unity. $Z_1 Z_2$ is set equal to $R_1 R_2$. The electrical length of each section at the exact-match or design frequency f_0 is set to

$$\beta\ell = \tan^{-1}\,\sqrt{\dfrac{R_2/R_1 - 1}{(Z_1/R_1)^2 - Z_2/Z_1}}$$

The length can be made as short as desired by selecting Z_1/Z_2 sufficiently high within limits imposed by increasing effects of abrupt discontinuities. Within the ranges $R_2/R_1 < 2$ and $Z_1/Z_2 < 2$, the total bandwidth $2\,\Delta f$ is equal to $kf_0\rho m/\rho_0$, where ρ_m/ρ_0 is the ratio of maximum inband reflection coefficient to the reflection coefficient without the transformer, and k is a coefficient greater than unity. The coefficient becomes much greater than unity for $\beta\ell = \pi/8$ or more. The special case with $Z_1 = R_2$ and $Z_2 = R_1$ is a handy design for matching two dissimilar transmission lines simply by the use of sections of those lines. For this special case, $\beta\ell \leqq \pi/12$ and $k \geqq 1.1$.

Much advanced theory has been developed for lumped-constant circuits. Useful adaptation to distributed-constant circuits is often possible by transformation of the frequency variable.[9]

43-4   TAPERED LINES

If the characteristic impedance of a long section of transmission line varies gradually with distance along the line, a nearly perfect match between resistive terminations may be obtained. Since a line of such length and such small taper can seldom be afforded, much attention has been directed toward the design of tapered lines giving acceptable transforming performance over short lengths and broad bandwidths.[10-21]

Exponential Line

The exponential line is a simple form of tapered line that may be analyzed exactly with simple relations. Sections of exponential line are useful both as simple impedance transformers and as cascaded elements in a more complexly tapered line. The name comes from the function describing the magnitude of the characteristic impedance:

$$K_0(x) = \sqrt{L_x/C_x} = K_0(0) \exp \delta x$$

where L_x/C_x is the ratio of the per-unit-length inductance and capacitance of the line, δ is the taper constant, and x is the length coordinate that is positive in the direction of increasing impedance. Usually, the product $L_x C_x$ is constant.

Above a critical cutoff frequency, the exponential line supports the propagation of independent characteristic waves traveling in opposite directions. The wave voltages are determined by the boundary conditions at the ends of the line. The impedance at any point or the reflection coefficient at that point may be found by transforming the known reflection coefficient at some other point. The analysis is like that for a uniform line except for several significant differences:

1   The characteristic impedance, which is the impedance seen by a traveling wave, is complex. The magnitude varies exponentially with x. The impedance phase varies with frequency and reverses sign with direction of travel.

$$Z_0(x) = K_0(x) \exp \pm j\phi_0$$

$$|\phi_0| = \sin^{-1}(\delta/2\beta)$$

where $\beta = 2\pi/\lambda$ is the phase constant for $L_x C_x$. The sign of ϕ_0 is positive for a wave traveling in the negative-x, or decreasing-impedance, direction and negative for the other direction. The characteristic impedances for oppositely traveling waves constitute a conjugate pair.

2   The characteristic waves propagate with a higher phase velocity and lower phase constant than on a uniform line with the same $L_x C_x$. The phase constant is $\beta_0 = \beta \cos \phi_0$.

3   The cutoff condition occurs for $\delta/2\beta = 1$, $|\phi_0| = \pi/2$, and $\beta_0 = 0$. Traveling-wave analysis is not useful below the cutoff frequency.

4   A more general expression for the reflection coefficient of load Z_L must be used.

$$\rho = \frac{Z_L - Z_{0L}}{Z_L + Z_{0G}} = \frac{Z_L - Z_{0G}^*}{Z_L + Z_{0G}}$$

where Z_{0L} and Z_{0G} are the complex characteristic impedances seen looking toward the load and toward the generator, respectively. The (voltage) reflection coefficient at x with respect to $Z_0(x_1)$ for load Z_L at $x = x_1$ is

$$\rho_1 = \frac{Z_L - K_0(x_1) \exp \mp j\phi_0}{Z_L + K_0(x_1) \exp \pm j\phi_0}$$

where the upper or lower signs apply when the direction of looking into Z_L from the line is in the positive- or negative-x direction, respectively.

The corresponding reflection coefficient with respect to $Z_0(x_2)$ at $x = x_2$ a distance ℓ toward the generator is

$$\rho_2 = \rho_1 \exp - j2\beta_0\ell$$

5   The input or driving point impedance at x_2 is

$$Z_{\text{in}}(x_2) = K_0(x_2) \exp(\mp j\phi_0) \frac{1 + \rho_2 \exp(\pm j2\phi_0)}{1 - \rho_2}$$

where the signs follow the same rule.

6   The traveling waves experience changing impedance, voltage, and current with x, but the reflection coefficient changes only in phase angle, not in magnitude. Thus, if a length of exponential line is well matched at the load end to a first-impedance level, the source end presents a well-matched impedance at another impedance level, related by the ratio exp $\delta\ell$ to the first.

The exponential line is commonly employed to transform between two resistance levels equal to the magnitudes of characteristic impedance at the ends. Equal mismatches of opposite sense occur at the ends so that reflection zeros exist for section lengths of $n\pi$ rad where n is an integer. The reflection coefficient in the passband is

$$\rho = 0.5 \ln (R_2/R_1) \sin (\beta_0\ell)/\beta_0\ell$$

The mismatches at the ends may be improved by the use of reactive elements.[10,11]

Other Tapered Lines

Several tapered lines[21] are attractive for special simplicity in analysis. Optimum highpass tapers have been described in Refs. 13 and 14. The taper of Klopfenstein[13] is optimum in the sense that for a given taper length the input reflection coefficient has the minimum peak magnitude or that for a specified peak magnitude the taper has the minimum length. The design is based on the multisection Chebyshev transformer, in which the number of sections is increased without limit to raise the upper cutoff frequency without limit. The resulting reflection coefficient in the passband is

$$\rho \exp(j\beta\ell) = 0.5 \ln (R_2/R_1) \cos [(\beta\ell)^2 - A^2]/\cosh A$$

where $\beta\ell$ is the electrical length. At the low-frequency cutoff, $\beta\ell = A$, and the length is approximately 0.019 (20 log cosh A) + 0.1 wavelength. Maximum inband reflection coefficient is

$$\rho_m = 0.5 \ln (R_2/R_1)/\cosh A$$

The characteristic impedance taper over the section length ℓ against coordinate x, measured from the midpoint of ℓ toward the high-impedance (or R_2) end, is

$$Z_0(x,A) = (R_1/R_2)^{1/2} (R_2/R_1)^{1/L}$$

Values of the function L for a wide range of cosh A are given in Table 43-3. The table is based on the following expressions evaluated by a simple routine of Grossberg:[19]

$$L(x) = -L(-x) = A^2 \phi(2x/\ell,A)/\cosh A$$
$$\phi(2x/\ell,A) = -\phi(-2x/\ell,A) = \int_0^{2x/\ell} \frac{I_1 (A\sqrt{1 - y^2})}{A\sqrt{1 - y^2}} \, dy$$
$$2x/\ell \leq 1.0$$

where I_1 is the first kind of modified Bessel function of the first order.

TABLE 43-3 Values of the Function L for Klopfenstein Taper

$2x/\ell$	20 log (cosh A)					
	20	30	40	50	60	70
0.0	0.00000	0.00000	0.00000	0.00000	0.00000	0.00000
	0.058765	0.073036	0.084713	0.094889	0.104038	0.112422
0.10	0.117280	0.145568	0.168607	0.188594	0.206487	0.222811
	0.175301	0.217099	0.250881	0.279970	0.305820	0.329230
0.20	0.232583	0.287148	0.330773	0.367944	0.400634	0.429934
	0.288894	0.355261	0.407580	0.451551	0.489702	0.523442
0.30	0.344006	0.421013	0.480668	0.529960	0.572017	0.608598
	0.397706	0.484017	0.549490	0.602495	0.646819	0.684608
0.40	0.449793	0.543931	0.613597	0.668653	0.713613	0.751053
	0.500080	0.600460	0.672641	0.728109	0.772175	0.807880
0.50	0.548399	0.653361	0.726384	0.780718	0.822535	0.855369
	0.594597	0.702447	0.774696	0.826505	0.864957	0.894082
0.60	0.638543	0.747585	0.817555	0.865656	0.899908	0.924807
	0.680125	0.788700	0.855039	0.898497	0.928012	0.948487
0.70	0.719251	0.825770	0.887318	0.925470	0.950013	0.966150
	0.755851	0.858829	0.914647	0.947108	0.966722	0.978843
0.80	0.789876	0.887959	0.937349	0.964007	0.978975	0.987577
	0.821300	0.913291	0.955802	0.976799	0.987594	0.993283
0.90	0.850116	0.934994	0.970429	0.986124	0.993355	0.996772
	0.876338	0.953277	0.981675	0.992607	0.996955	0.998725
1.00	0.900000	0.968377	0.990000	0.996838	0.999000	0.999683

43-5   COMBINATIONS OF TRANSFORMERS AND STUBS

Transformer with Two Compensating Stubs

The simple quarter-wave transformer has rather poor performance over a wide band. Its wideband performance may be improved greatly by the addition of a compensating stub at each end, as shown in Fig. 43-15. The reactance introduced by each stub counteracts the variation with frequency of one-half of the transformer length at and near the design frequency. For the impedance proportions shown, the performance is maximally flat and comparable with that of a three-section binomial transformer.

Equiripple performance within the band, for a maximum reflection coefficient of ρ_m, is obtained with modified stub impedances. Table 43-4 lists some examples of equiripple performance, showing an improvement in bandwidth by a factor of about 1.2 over the maximally flat case for $\rho_m = 0.1\,\rho_0$.

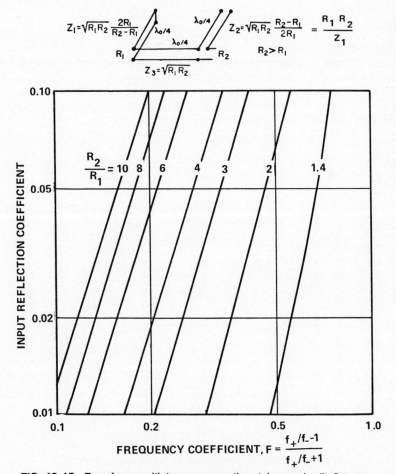

$$Z_1 = \sqrt{R_1 R_2}\ \frac{2R_1}{R_2 - R_1} \qquad Z_2 = \sqrt{R_1 R_2}\ \frac{R_2 - R_1}{2R_1} = \frac{R_1 R_2}{Z_1}$$

$$Z_3 = \sqrt{R_1 R_2} \qquad R_2 > R_1$$

FIG. 43-15   Transformer with two compensating stubs; maximally flat case.

TABLE 43-4 Transformer with Two Compensating Stubs, Equiripple Case, $\rho_m = 0.1\rho_0$

R_2/R_1	$Z_2/Z_3 = Z_3/Z_1$	Design F
2	0.6572	0.515
4	2.1205	0.351
6	3.6308	0.286
8	5.1574	0.248

Transformer with Single Compensating Stub

If it is convenient to use only one compensating stub with the transformer, one of the circuits shown in Fig. 43-16 may be used. In this equiripple design the allowable maximum standing-wave ratio S_m affects the choice of Z_1 and Z_2.

Tapped-Stub Transformer

The tapped-stub transformer, shown in Fig. 43-17, is a useful circuit for matching between two widely different resistive terminations. The design chart shows the rela-

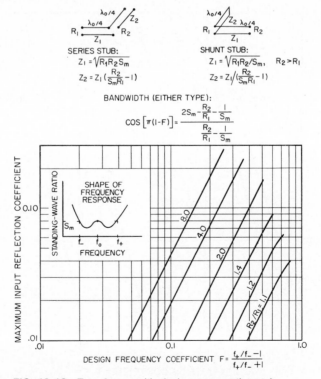

FIG. 43-16 Transformer with single compensating stub.

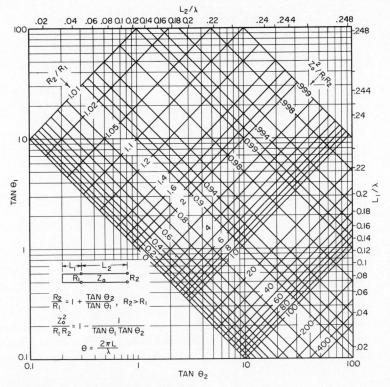

FIG. 43-17   Tapped-stub transformer.

tionship existing between R_2/R_1, Z_0^2/R_1R_2, and the two line lengths L_1 and L_2 for the matched condition. The total stub length is a minimum of one-quarter wavelength and increases with increasing Z_0. The chart is based on a lossless transmission line. For very high transformation ratios the effect of even a small loss is appreciable and must be accounted for separately. For moderate transformation ratios, a 5 or 10 percent bandwidth over which the maximum reflection coefficient is no more than $0.1(R_2 - R_1)/(R_2 + R_1)$ usually can be obtained.

43-6   BALUNS

A balun is an impedance transformer designed to couple a balanced transmission circuit and an unbalanced transmission circuit. The impedance transformation may be accomplished usually with established techniques.[22,23] The conversion between a balanced mode and an unbalanced mode, however, requires special techniques. Several basic techniques are exemplified by the balun types shown in Fig. 43-18.

A balanced circuit is obtained in type 1 by introducing a high impedance of a

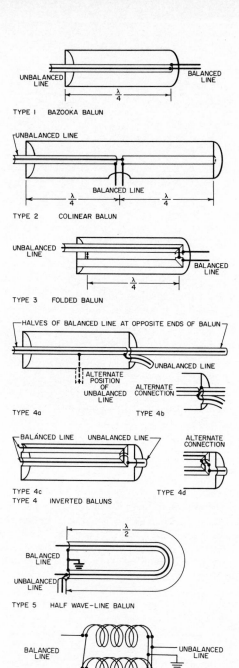

TYPE 1   BAZOOKA BALUN

TYPE 2   COLINEAR BALUN

TYPE 3   FOLDED BALUN

TYPE 4a   TYPE 4b

TYPE 4c   TYPE 4d
TYPE 4   INVERTED BALUNS

TYPE 5   HALF WAVE-LINE BALUN

TYPE 6   BIFILAR-COIL OR ELEVATOR BALUN

FIG. 43-18 Basic types of baluns.

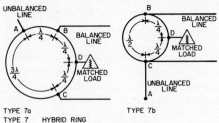

TYPE 7a   TYPE 7b
TYPE 7   HYBRID RING

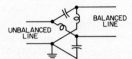

TYPE 8   LATTICE

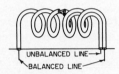

TYPE 9   COILED-CABLE BALUN

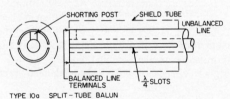

TYPE 10a   SPLIT-TUBE BALUN

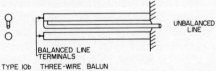

TYPE 10b   THREE-WIRE BALUN

TYPE 11

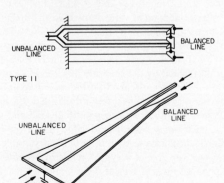

TYPE 12 TAPERED BALUN

resonant coaxial-choke structure between the outer conductor of the unbalanced coaxial circuit and ground. There is no counterpart of the choke impedance in the other side of the balanced circuit. Hence the balun does not present a well-balanced impedance at any frequency for which the choke impedance is not high.

The shortcoming of type 1 is overcome in types 2 and 3. Symmetry of the balanced circuit is maintained independently of frequency by a two-conductor choke (type 3) or by two identical opposed coaxial-choke cavities (type 2). The first three types have been widely used because of their simplicity and mechanical adaptability. The impedance bandwidth is limited by the shunting effect of the choke. Thus it is advantageous to keep the characteristic impedance of the choke lines as high as possible for wideband operation.

Type 4 represents an improvement for many applications, particularly those involving impedance-level transformation. Unlike types 2 and 3, type 4 has the balanced circuit in the innermost line. Coupling is attained through a gap between line sections, as before. The necessary choke cavity, however, is around this gap in the unbalanced side of the structure. An unbalanced or single-ended cavity can be used without disturbing the perfect balance. The connections at the gap may be either of the series type, as in types 4a and 4c, or of a parallel type, as in types 4b and 4d. The parallel connection results in a balanced-to-unbalanced impedance ratio of 4.

An additional impedance control is available in any of the type 4 baluns at some expense of bandwidth by application of the principle of the tapped-stub transformer. This is illustrated by the alternative position of the unbalanced line in the drawing of type 4a. The desired impedance transformation ratio determines the length of coaxial-choke line on each side of the unbalanced-line tap position in accordance with Fig. 43-17. Broadband compensation techniques[22] may be applied to types 2, 3, and 4.

The type 5 balun employs a half-wave delay line. It is connected between the two balanced-line terminals as shown, forcing the potentials to ground to be equal and opposite at the design frequency. A balanced-to-balanced impedance ratio of 4 is obtained.

Type 6 employs helically wound transmission lines. At the unbalanced end, they are connected in parallel. Enough line length is used to develop high impedances to ground at the opposite ends. There the lines are connected in series to form the balanced terminals. A balanced-to-unbalanced impedance ratio of 4 is obtained.

Type 7a is a conventional $6\lambda/4$ hybrid ring which operates similarly to type 5. The path length from A to C is one-half wave longer than the path from A to B. Each path is an odd number of quarter waves long so that impedance transformation may be incorporated by proper selection of wave impedances. Furthermore, a load resistor at D is connected to terminals B and C by quarter-wave lines. This arrangement tends to dissipate the energy in an unbalanced wave without affecting a balanced wave.

Type 7b is a $4\lambda/4$ hybrid ring that operates similarly, except that the impedance-matching function is performed by the separate quarter-wave line AC.

Type 8 is the familiar lattice circuit. In it the incoming unbalanced wave energy is divided equally between two channels, one providing a 90° lead, the other a 90° lag. The output voltage is balanced with respect to ground and is in quadrature with the input voltage. The lattice may be proportioned to match any two resistance values.

Type 9 is similar in principle to type 6. It is most suitable for operation at the longer wavelengths where the size of a type 4 tends to become excessive. It may be connected for an impedance ratio of 1 or of 4.

Type 10a is a convenient balun type for feeding balanced dipoles. The slotted portion of transmission line supports two modes of transmission simultaneously. Energy in a coaxial mode from the unbalanced input is substantially unchanged by the presence of the slots. The field of this mode is almost completely confined within the outer conductor. There is also a balanced mode in which the halves of the slotted cylinder are at opposite potentials and the center conductor is at zero potential. The field is not confined to the space within the slotted cylinder. (If leakage and radiation are to be minimized, another cylinder may be added overall for shielding.) The connection strap between the inner conductor and one-half of the slotted cylinder requires that the coaxial-mode voltage be equal and opposite to that half of the balanced-mode voltage. Thus a balanced-to-unbalanced impedance ratio of 4 is obtained.

Type 10*a* will give almost perfect balance over a wide frequency range if the slot width is kept small and symmetry is maintained at the strap end. A variation of this basic type of balun is the three-wire balun shown in type 10*b*. It is sufficiently well balanced for use in noncritical applications.

Type 11 illustrates another method by which an impedance transformation may be obtained by a connection independently of frequency. Two coaxial lines are connected in parallel at the unbalanced terminal and in series at the balanced terminal. A third cylinder is added to preserve symmetry. The impedance transformation ratio for this structure having two lines is 2^2, or 4. More coaxial lines could be added to give impedance transformation ratios of 9, 16, etc., but practical limits are soon reached in this direction. Bifilar coils can also be used in a balun operating on this principle.

Baluns, for example, those of types 1 through 4, 6, 9, and 11, can be improved for very wideband operation by the inclusion of such high-permeability materials as ferrite or powdered iron. The material in the form of toroidal cores, cylindrical cores, pot cores, or beads, placed in the magnetic field of a low-frequency-limiting shunt element, increases the length of that element, raises its impedance, and improves the low-frequency match. The influence of the material is much less at the high-frequency end, which can be as high as 1 GHz, because of a reduced effective permeability. Compact baluns and transmission-line transformers of the bifilar-coil and coiled-cable types, having ferrite cores, typically operate over bandwidths of two to three decades.[24,25] Core loss is low because the core does not interact strongly with the transmission-line mode.

The wideband balun of type 12 is a half wave or more of tapered transmission line that converts gradually, in cross-sectional characteristics, from an unbalanced line (coaxial line or stripline) at a first impedance at one end to a balanced two-conductor line at a second impedance at the other end, using a Chebyshev taper.[26,27]

43-7   BROADBANDING

The most general problem of broadbanding is that of synthesizing a circuit to match one arbitrary impedance to another arbitrary impedance over a prescribed frequency range to within a prescribed tolerance. Fortunately, the broadbanding problems found in practice involve impedances that are not completely arbitrary, inasmuch as they are composed of a combination of physically realizable inductances, capacitances, and resistances. A further simplification of the broadbanding problem often occurs because at least one of the impedances to be matched usually is the characteristic impedance of a transmission line and thus is either constant and resistive or nearly so.

Nevertheless, the impedances to be matched are functions of frequency that usually are too complex for simple analysis. An iterative approach is often taken. The subject impedance, prescribed by theoretical or experimental data, is examined to perceive a suitable step toward optimum broadbanding. Graphical representations of the function are usually helpful for this. The matching network or a portion of it is added to the circuit. The resulting impedance is found by calculation or experiment. The steps are repeated until the required broadband match is obtained. Modern computational tools and measuring instruments minimize the tedious detail and impart the needed accuracy and so make the iterative approach rapid and efficient.

The problem of matching two resistive impedances is treated in earlier parts of this section. The problem of matching a load of frequency-varying impedance to a constant resistance is so common that the following discussion will be limited to this case.

Several broad rules that apply to broadbanding practice are given, as follows:

1 The difficulty of obtaining a prescribed tolerance of match increases with the required bandwidth. Refer to performance charts of a Chebyshev transformer for an example of this and of the following rule.

2 The difficulty increases with transformation ratio (expressed as a quantity greater than 1).

3 The difficulty increases with the electrical length of the transmission circuit between the load and the first point of control.

4 Improvement of match throughout one sector of the frequency range will generally be accompanied by an increase in mismatch in other sectors of the frequency range.

5 Any physically realizable passive impedance plotted on the reflection-coefficient plane, using the conventions of Fig. 43-3, displays a circular or spiral motion having a clockwise sense of rotation with frequency.

The effect of a length of mismatched transmission line between the load terminals and the first impedance-matching control is to introduce an additional variation with frequency that is seldom favorable over an appreciable bandwidth (Rule 3). For this reason it is advantageous to conduct impedance-matching control at a position close to the load terminals. In fact, if control of the load impedance is available from within the load itself, advantage should be taken of this circumstance to select the most suitable shape and position of the impedance locus on the reflection-coefficient plane.

Transformation of the given impedance locus on the chart to one that is compactly situated about the desired impedance point requires, first, a method of moving the impedance locus and, second, a method of compensating for the variation with frequency inherent in the original impedance locus and that introduced by the moving process. Means for moving the impedance locus may include lumped or distributed constant elements forming shunt or series reactances, cascade transformers, or a combination of these. Selection is made to introduce as little adverse variation of impedance with frequency as possible.

Compensation of variation of impedance with frequency is usually limited to a band including less than one complete convolution of the original impedance locus if a high degree of compensation is required. Typically, a sector of the original impedance locus which may be made to appear similar to those shown in Fig. 43-19 is

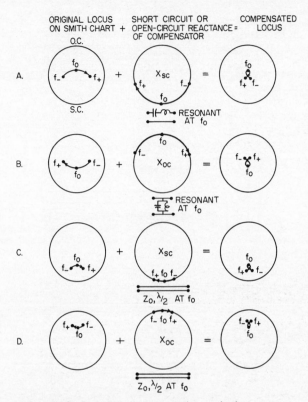

FIG. 43-19 Broadband compensation methods.

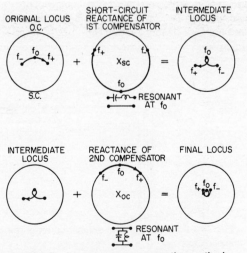

FIG. 43-20 A two-stage compensation method.

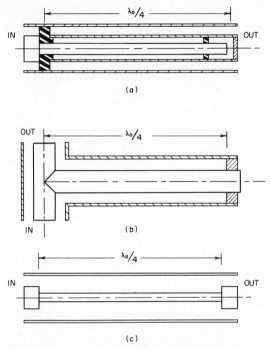

FIG. 43-21 Possible physical embodiments of Fig. 43-19.

selected. Then a matching circuit having a variation of reactance of opposite sense (and of the proper magnitude) is added to yield the tightly knotted transformed locus as shown. Essentially all the reactance variation may be eliminated; however, a small reactance variation and a larger resistance variation remain. In certain instances it may be advantageous to utilize two stages of compensation, as illustrated in Fig. 43-20, in order to accomplish a doubly knotted transformed locus.

There are many ways of forming these matching circuits into physical structures.

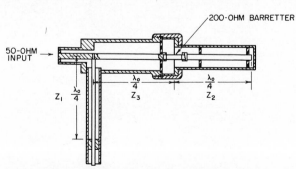

FIG. 43-22 Structure to match barretter to 50-Ω impedance.

A few simple examples are shown in Fig. 43-21. A more complex structure utilizing a transformer with two compensating stubs in a circuit described earlier in this chapter is shown in Fig. 43-22.

Resonant lengths of the transmission line may be used in place of the lumped-constant resonant circuits. The LC product or the length of the line is selected for resonance at f_0, the midfrequency of the sector. The L/C ratio or the characteristic impedance of the line section is selected to give the proper amount of compensating reactance at f_+ and f_-, the edge-band frequencies.

REFERENCES

1   W. L. Everitt and G. E. Anner, *Communication Engineering,* 3d ed., McGraw-Hill Book Company, New York, 1956.
2   P. S. Carter, "Charts for Transmission-Line Measurements and Computations," *RCA Rev.,* vol. 3, no. 3, January 1939, pp. 355–368.
3   P. H. Smith, "An Improved Transmission Line Calculator," *Electronics,* vol. 17, January 1944, pp. 130–133, 318–325.
4   S. B. Cohn, "Optimum Design of Stepped Transmission-Line Transformers," *IRE Trans. Microwave Theory Tech.,* vol. MTT-3, April 1955, pp. 16–21.
5   H. J. Riblet, "General Synthesis of Quarter-Wave Impedance Transformers," *IRE Trans. Microwave Theory Tech.,* vol. MTT-5, January 1957, pp. 36–43.
6   J. R. Whinnery and H. W. Jamieson, "Equivalent Circuits for Discontinuities in Transmission Lines," *IRE Proc.,* vol. 32, February 1944, pp. 98–116.
7   J. R. Whinnery, H. W. Jamieson, and T. E. Robbins, "Coaxial-Line Discontinuities," *IRE Proc.,* vol. 32, November 1944, pp. 695–709.
8   A. G. Kurashov, "Multiband Stepped Taper," *Radio Eng. Electron. Phys.,* vol. 10, February 1979, pp. 10–15.
9   A. I. Grayzel, "A Synthesis Procedure for Transmission Line Networks," *IRE Trans. Circ. Theory,* vol. CT-5, September 1958, pp. 172–181.
10 C. R. Burrows, "The Exponential Transmission Line," *Bell Syst. Tech. J.,* vol. 17, October 1938, pp. 555–573.
11 H. A. Wheeler, "Transmission Lines with Exponential Taper," *IRE Proc.,* January 1939, pp. 65–71.
12 H. Kaufman, "Bibliography of Nonuniform Transmission Lines," *IRE Trans. Antennas Propagat.,* vol. AP-3, October 1955, pp. 218–220.
13 R. W. Klopfenstein, "A Transmission Line Taper of Improved Design," *IRE Proc.,* vol. 44, January 1956, pp. 31–35. (See Refs. 16, 19, and 20.)
14 R. E. Collin, "The Optimum Tapered Transmission Line Matching Section," *IRE Proc.,* vol. 44, April 1956, pp. 539–547. (See Ref. 16.)
15 E. F. Bolinder, "Fourier Transforms and Tapered Transmission Lines," *IRE Proc.,* vol. 44, April 1956, p. 557.
16 Correspondence re Refs. 13 and 14, *IRE Proc.,* vol. 44, August 1956, pp. 1055–1056.
17 C. P. Womack, "The Use of Exponential Transmission Lines in Microwave Components," *IRE Trans. Microwave Theory Tech.,* vol. MTT-10, March 1962, pp. 124–132.
18 A. H. Hall, "Impedance Matching by Tapered or Stepped Transmission Lines," *Microwave J.,* vol. 9, March 1966, pp. 109–114.
19 M. A. Grossberg, "Extremely Rapid Computation of the Klopfenstein Impedance Taper," *IEEE Proc.,* vol. 56, September 1968, pp. 1629–1630.
20 D. Kajfez and J. O. Prewitt, "Correction to a Transmission Line Taper of Improved Design," *IEEE Trans. Microwave Theory Tech.,* vol. MTT-21, May 1973, p. 364.
21 M. J. Ahmed, "Impedance and Transformation Equations for Exponential, Cosine-Squared,

and Parabolic Tapered Transmission Lines," *IEEE Trans. Microwave Theory Tech.*, vol. MTT-29, January 1981, pp. 67–68.

22 G. Oltman, "The Compensated Balun," *IEEE Trans. Microwave Theory Tech.*, vol. MTT-14, March 1966, pp. 112–119.

23 H. R. Phelan, "A Wide-Band Parallel-Connected Balun," *IEEE Trans. Microwave Theory Tech.*, vol. MTT-18, May 1970, pp. 259–263.

24 C. L. Ruthroff, "Some Broad-Band Transformers," *IRE Proc.*, vol. 47, August 1959, pp. 1337–1342.

25 J. Sevick, "Broadband Matching Transformers Can Handle Many Kilowatts," *Electronics*, vol. 49, no. 24, Nov. 25, 1976, pp. 123–128.

26 J. W. Duncan and V. P. Minerva, "100:1 Bandwidth Balun Transformer," *IRE Proc.*, vol. 48, February 1960, pp. 156–164.

27 M. Gans et al., "Frequency Independent Balun," *IEEE Proc.*, vol. 53, June 1965, pp. 647–648.

Chapter 44

Radomes

Gene K. Huddleston
Harold L. Bassett
Georgia Institute of Technology

Definition and Function

A radar dome, or radome, is a protective dielectric housing for a microwave or milli-meter-wave antenna.[1] The function of the radome is to protect the antenna from adverse environments in ground-based, shipboard, airborne, and aerospace applica-tions while having insignificant effect on the electrical performance of the enclosed antenna or antennas. The frequency band of application for radomes is approximately from 1 to 1000 GHz.

Radomes are generally composed of low-loss dielectrics of thickness comparable to a wavelength which are shaped to cover the antenna and, if necessary, to conform

FIG. 44-1 Two airborne radome applications. (*a*) Relatively blunt aircraft-nose radome. (*b*) Streamlined missile radome.

to aerodynamic streamlining (Fig. 44-1). Radomes are used with virtually all aperture-type airborne antennas and with many ground-based and shipboard aperture antennas which must withstand severe weather conditions, blast effects, or water pressures (submarine applications). In most applications, the radome must be large enough to allow scanning of the antenna inside the radome.

Electromagnetic windows and radants are variations of the radome. Originally, an electromagnetic window was a planar or slightly curved dielectric covering for a stationary antenna. Original applications included protective covers for communications and radar altimeter antennas on reentry spacecraft where temperatures in excess of 2000°F were encountered and ceramic materials were required.[2] A radant is a radome which includes the antenna as an integral part; e.g., antenna elements imbedded in a streamlined dielectric nose structure for an aircraft.[3] Currently, the term *electromagnetic window* applies to all types of radomes, windows, and radants.

Electrical Considerations

A radome always changes the electrical performance of the antenna because of wave reflections and refractions at interfaces between material media and because of losses in the radome materials. These changes manifest themselves as pattern distortion (Fig. 44-2), including changes in gain, sidelobe levels, beamwidth, null depth, and polarization characteristics.[4,5] For tracking antennas, boresight errors are invariably introduced by a streamlined radome; for proportional navigation systems, the rate of change of boresight error with antenna scan (boresight-error slope or rate) can be a major problem.[5,6] Excessive reflections from the radome may cause magnetron pulling. For high-power applications, excessive losses in the radome material may raise its

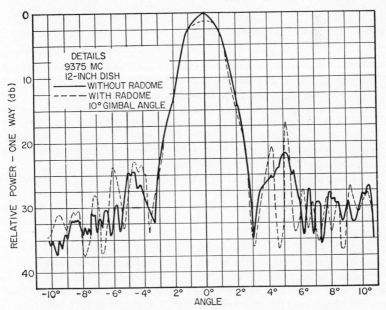

FIG. 44-2 Typical radiation patterns showing the effects of the radome.

temperature to a point at which its structural properties and electrical performance are degraded.[7] Radome losses also will raise the system noise temperature.

Radome effects can be qualitatively explained and understood in terms of TEM (plane) wave propagation through and reflection from planar dielectric panels. The curved radome wall is considered to be locally plane at each point. Waves emanating from the enclosed transmitting antenna are also considered to be locally plane at each point of incidence on the radome wall. The reflected and transmitted waves can then be approximated from plane-sheet theory,[8] and the resultant effects on overall antenna performance can be calculated. By reciprocity, the effect of the radome on the antenna as a receiver can be determined in the same manner.

The transmission properties of a plane dielectric sheet vary with frequency, incidence angle, and polarization of the incident plane wave. In Fig. 44-3, the plane of incidence is defined by the unit normal $\hat{n}_{0s}$ and the direction of wave propagation $\hat{k}$. The incidence angle is given by $\sin^{-1}(\hat{k} \times \hat{n}_{0s})$. Arbitrary wave polarizations are resolved into an electric field component perpendicular to the plane of incidence ($E_{i\perp}$) and a component parallel to the plane of incidence ($E_{i\parallel}$). The power-transmission coefficient (transmittance) and insertion phase delay (IPD) are generally different for the two polarizations, so that the polarization, power density, and phase of the transmitted wave are different from those of the incident wave. (The insertion phase is the phase of the transmitted wave relative to the phase of the incident wave at the same point if the panel were removed. IPD is the negative of the insertion phase.)

The electrical characteristics of flat panels are important because radome-wall design is based on them. Refinements in the wall design are often determined from subsequent analyses which account for antenna and radome geometry and for radome-wall curvature. A number of computer-aided analyses have been developed for this purpose and will be described below.

Environmental Considerations

The operating environment of a radome is the primary factor in determination of material type, wall design, and radome shape. Temperature, structural loads, vibration, wind, sand, hail, and rain are to be considered in an initial design.

For ground-based radomes, wind loading is the most important factor in the selection of a radome design. Other factors to be considered are humidity, blowing dust or sand, rain, ice, snow, and moisture buildup on the outer wall.[9] The radome configuration and mounting scheme evolve from these environmental considerations and the desired electrical performance. Typical ground-based structures include the metal space-frame,[10] sandwich, monolithic-wall, and inflatable radomes.[11]

Missile and aircraft radome designs are dictated primarily by aerodynamic loading and the thermal environment. The radome design becomes a trade-off of materials and shapes based on vehicle speed, trajectory, and desired electrical performance. Other design considerations include rain erosion, water absorption, and static electrification.

Mechanical stresses are produced in the radome by aerodynamic loading due to airflow, acceleration forces, and sudden thermal expansion due to aerodynamic heating (thermal shock).[12] In high-speed radomes, thermal shock is perhaps more important than other mechanical loading factors. The attachment of the radome to the airframe can be a critical design problem in high-temperature applications.

Airborne-radome electrical performance is affected by the elevated temperatures

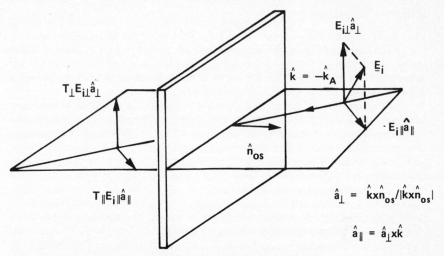

FIG. 44-3 Plane-wave propagation through a flat dielectric panel.

(up to 2300°F) caused by aerodynamic heating during flight. Dielectric constant and loss tangent normally increase or decrease with temperature, and the generally non-uniform heating of the radome produces changes in the boresight-error performance. Aerodynamic heating has been suspected of increasing boresight errors to a level to cause in-flight oscillations of a missile. Measurements made on Pyroceram radomes using concentrated solar energy for nonuniform radome heating have shown that measured changes in boresight error are significant.[13]

Rain erosion can be a severe problem in missile-radome operation and is a continuing problem in aircraft radomes for speeds over 250 mi/h (400 km/h). The radome shape, material type, and velocity influence rain erosion. Essentially all leading-edge surfaces are susceptible to raindrop impact, and a sharp-nose radome is less susceptible to rain damage than a blunted-nose radome. Rain erosion manifests itself as pitting of the radome surface and, in severe cases, catastrophic failure of the radome itself. Degradation of performance can result from rain erosion. Special rain-erosion coatings, metal tips, and aerodynamic spikes are used to provide resistance to rain erosion.

Water absorption by the radome increases the dielectric constant and loss tangent of the radome wall; hence, materials which do not readily absorb water or ones which have been treated with protective coatings are used. Protective coatings for high-speed radomes must be selected carefully because charred residues may hamper electrical performance during the terminal phase of flight.[14]

Static electrification of radomes by moving air can present a serious shock hazard. Thin antistatic coatings are used to provide a means of conducting static charge to nearby structures for its safe neutralization. Lightning strikes to aircraft radomes are not uncommon, and metallic lightning-diverter strips[15] are often employed on aircraft radomes. Antistatic coatings are usually thin at the wavelengths of interest, but they can produce some effects on electrical performance. Diverter strips also affect electrical performance.[16]

Radome Materials

Electrical and structural performance and application are the prime factors in selection of the radome material. A number of thermal plastic materials, some of which are listed in Table 44-1 along with other organic radome materials, are suitable for radome use.[17] A more comprehensive list of materials is presented in Chap. 46 and in Refs. 18 and 19.

The thermal plastic materials are limited to applications involving modest temperature increases, whereas the common laminates using the polyimide or polybenzimidazole resins can withstand temperatures near 750°F. For higher temperatures, either ceramic materials or ablative coatings are usually required. Fiber-loaded Teflon (Duroid) materials, depending on the specific application, can be used at very high temperatures for a short period of time.

The ceramic materials of prime interest are aluminum oxide, beryllium oxide, boron nitride (pyrolytic and hot-pressed), magnesium aluminate (spinel), magnesium oxide, silicon dioxide, sintered silica, and silicon nitride. The electrical properties of commonly used ceramic radome materials are tabulated in Table 44-2. The electrical parameters are a function of material density for the ceramics. The electrical properties of the organics (Table 44-1) are functions of both density and resin content.

A radome for multimode missile seekers requires materials that have low transmission losses at microwave, millimeter-wave, optical, and intermediate-range (IR) wavelengths. A number of materials offer the appropriate physical parameters for use at these wavelengths:[20] silica mullite, germanium mullite, boron aluminate, zinc germanate, thorium germanate, sapphire, $Mg F_2$, spinel, pollucite, hafnium titanate, aluminum nitride, calcium lanthanate sulfide, zirconia-toughened zirconia, and zirconia-toughened (Al, Cr) oxide.

For high-speed homing missiles, the most popular radome material has been Pyroceram 9606. Other materials in use include slip-cast fused silica and epoxy glass, of which the latter is used on lower-speed missiles. Duroid 5650 has been chosen as the

TABLE 44-1   Organic Radome Materials (8.5×10^9 Hz)

Material	Dielectric constant	Loss tangent
Thermal plastics		
Lexan	2.86	0.006
Teflon	2.10	0.0005
Noryl	2.58	0.005
Kydox	3.44	0.008
Laminates		
Epoxy-E glass cloth	4.40	0.016
Polyester-E glass cloth	4.10	0.015
Polyester-quartz cloth	3.70	0.007
Polybutadiene	3.83	0.015
Fiberglass laminate polybenzimidazole resin	4.9	0.008
Quartz-reinforced polyimide	3.2	0.008
Duroid 5650 (loaded Teflon)	2.65	0.003

TABLE 44-2 Ceramic Radome Materials (8.5×10^9 Hz)

Material	Density, g/cm³	Dielectric constant	Loss tangent
Aluminum oxide	3.32	7.85	0.0005
Alumina, hot-pressed	3.84	10.0	0.0005
Beryllium oxide	2.875	6.62	0.001
Boron nitride, hot-pressed	2.13	4.87	0.0005
Boron nitride, pyrolytic	2.14	5.12	0.0005
Magnesium aluminate (spinel)	3.57	8.26	0.0005
Magnesium aluminum silicate (cordierite ceramic)	2.44	4.75	0.002
Magnesium oxide	3.30	9.72	0.0005
Pyroceram 9606		5.58	0.0008
Rayceram 8		4.72	0.003
Silicon dioxide	2.20	3.82	0.0005
Silica-fiber composite (AS-3DX)	1.63	2.90	0.004
Slip-cast fused silica	1.93	3.33	0.001
Silicon nitride	2.45	5.50	0.003

best performer in a high-supersonic-rain environment. Silicon nitride also holds promise for such environments.

Polyurethane and fluoroelastomer coatings are generally used for rain-erosion protection and as antistatic coatings.[21,22,23] These are typically from 0.010 to 0.020 in (0.254 to 0.508 mm) thick. For higher-speed missiles, rain-erosion coatings of either polyurethane or neoprene are not used because of the higher temperatures involved. Duroid-material coatings appear to offer the radome designer a method to reduce rain erosion.[24,25] Duroid does ablate, and the resulting changes in wall thickness during flight should be considered in the radome electrical design. Ceramic coatings over plastic radomes have had limited success as rain-erosion coatings. Avcoat materials, which are ablative, can be used to provide thermal protection and rain-erosion resistance.

44-2 RADOME CONFIGURATIONS

Shape

Radome shape and wall structure comprise the radome configuration. Unstreamlined radome shapes include the cylinder, the sphere, and combinations thereof (Fig. 44-4). Since the incidence angles encountered in these shapes are usually less than 30°, such radomes are called normal-incidence radomes. The transmission properties of suitable wall structures are relatively independent of polarization and incidence angles between 0 and 30°, provided that small tolerances on thickness variations are maintained and that a narrow frequency band of operation is specified. Reflections are a problem in normal-incidence radomes since they may be returned directly to the antenna and increase the voltage standing-wave ratio (VSWR) in the feed system.[26]

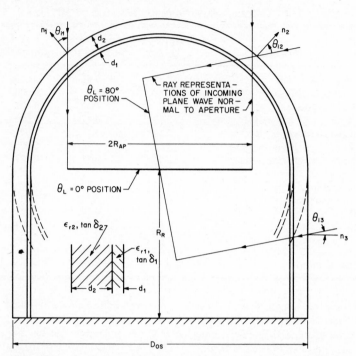

FIG. 44-4 Hemisphere-on-cylinder radome design for ground-based application.

Streamlined radome shapes are used primarily in airborne applications for aircraft and missiles. Common streamlined shapes are the ogive (Fig. 44-5), von Kármán, Sears-Haack, conical, ellipsoidal, power-series, and log-spiral.[27] Incidence angles range from 0 to 80° and higher, depending on the degree of streamlining as quantified by the ratio of radome length to diameter (fineness ratio). Fineness ratios typically range from 0.5 (hemisphere) to 3.26. Antenna-radome interaction (VSWR) is not significant in streamlined radomes, but focusing effects, Lloyd's mirror effect, and effects of trapped waves may cause significant resultant effects on electrical performance.[28]

Radome shape affects both aerodynamic and electrical performance. Generally, the more streamlined shapes have less drag (hence, greater range) and more resistance to rain damage; however, electrical performance usually improves for less streamlined shapes. Sound radome-design practice dictates that radome-shape selection be coordinated to satisfy both aerodynamic and electrical requirements.

Wall Structure

Radome wall structures combine materials technology and plane-sheet electrical principles; whereas the former is continually changing to meet ever-increasing requirements, the latter remains as unchanged as Maxwell's equations. Common wall struc-

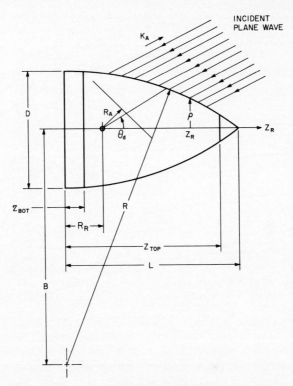

FIG. 44-5 Antenna-radome geometry for a tangent ogive radome.

tures include homogeneous single-layer (monolithic), multilayer, metallic, space-frame, and grooved structures.

A *monolithic*-wall structure consists of a single slab of homogeneous dielectric material whose thickness is less than approximately one-tenth wavelength (thin-wall) or is an integer multiple of one-half wavelength in the dielectric (*n*-order half-wave design). Thin-wall designs are applicable at lower frequencies, at which the permissible electrical thickness (Fig. 44-6) also provides adequate strength and rigidity.

For adequate strength at higher frequencies, the monolithic-wall thickness is chosen according to

$$d = \frac{n\lambda}{2(\epsilon_r - \sin^2\theta)^{1/2}} \qquad (44\text{-}1)$$

where n = integer
  λ = free-space wavelength
  ϵ_r = dielectric constant = ϵ/ϵ_0
  θ = incidence angle

The integer n is called the *order of the radome wall*. The case $n = 1$ is the *half-wave wall*. The angle θ at which Eq. (44-1) holds is called the *design angle*. For both per-

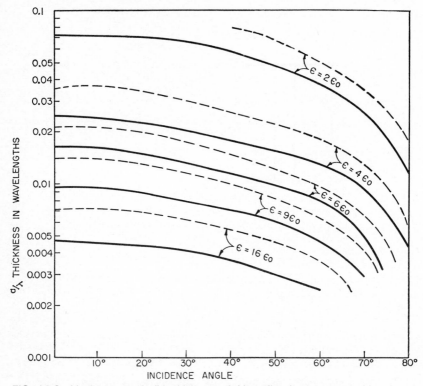

FIG. 44-6 Maximum permissible thickness of thin-wall radome versus incidence angle and dielectric constant for 95 percent power transmission (solid curves) and for 90 percent transmission (dashed).

pendicular and parallel polarization, reflection will be zero at this angle of incidence, maximum and equal transmittance will be obtained, and equal insertion phase delays will be introduced by the plane dielectric sheet. These features are illustrated in Fig. 44-7 for a low-loss alumina half-wave wall with design angle $\theta = 55°$. Equation (44-1) strictly holds for lossless materials but is a good approximation for the low-loss materials normally used for radomes. Complete transmission curves for monolithic wall designs can be found in Refs. 29 and 30.

The transmittance versus incidence angle for parallel polarization is always equal to or better than that for perpendicular polarization. In addition, at Brewster's angle defined by

$$\theta_B = \tan^{-1} \sqrt{\epsilon_r}$$

(44-2)

a transmittance of unity can be attained in the lossless case for parallel polarization and is independent of panel thickness. No similar condition exists for perpendicular polarization.

Monolithic radome walls are used in many ground-based and shipboard applications for which weight is not a problem and simplicity of design and construction is

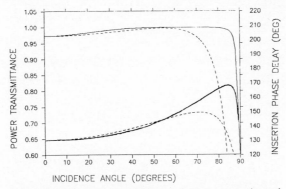

FIG. 44-7 Power transmittance (upper curves) and insertion phase delay versus incidence angle for alumina half-wave panel having a design angle of 55° for parallel (solid curves) and perpendicular (dash) polarizations (ϵ_r = 9.3, tan δ = 0.0003, d = 0.17λ).

desired. Half-wave and full-wave walls are prevalent in ceramic missile radomes. Lighter sandwich structures are used in many aircraft and lower-speed missile applications.

Multilayer structures include the A sandwich, B sandwich, C sandwich, and others. The A sandwich consists of three layers: two thin, dense high-strength skins separated by a low-density ($\sim$10-lb/ft^3) core material of foam or honeycomb. The B sandwich also consists of three layers: a dense core material with two lower-density skin materials which serve as quarter-wave matching layers. The C sandwich consists of two A sandwiches joined together for greater strength and rigidity.[31] Other multilayer structures utilize an arbitrary number of dielectric layers to achieve high transmittance over a broad frequency band.[32]

Design formulas and graphical data for A-sandwich structures are presented in Refs. 29 and 30. In the symmetrical sandwich, the inner and outer skins are identical and are thin with respect to wavelength. The spacing d_c of the skins (core thickness) is chosen so that the reflected wave from the second skin cancels the reflected wave from the first skin at a desired incidence angle. The angle θ at which this cancellation occurs is called the *core design angle*. The smallest value d_c at which cancellation occurs is called the *first-order spacing*. Salient transmission properties of an A-sandwich panel are shown in Fig. 44-8.

A-sandwich designs are used in many aircraft and in some missile radomes for both narrowband and broadband applications. Skins as thin as 0.005 in (0.127 mm) have been developed for use in radomes which achieve good circular polarization performances over an octave or greater bandwidth in the X/Ku frequency bands.[33] B-sandwich designs can be used in applications where the lower-density skins can withstand the environment; in some cases, the outer skin can be omitted to obtain a two-layer structure with good transmission properties (Fig. 44-4).

A *metal-inclusions* radome structure consists of dielectric layers containing metal inclusions which together form a high-frequency circuit element.[30] A *metallic radome*[34,35] uses a perforated metal outer skin. Metallic radomes offer advantages of

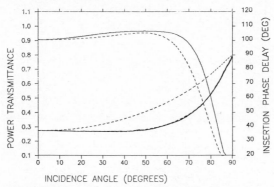

FIG. 44-8 Power transmittance (upper curves) and insertion phase delay versus incidence angle for an A-sandwich panel having a core design angle of 50° for parallel (solid curve) and perpendicular (dash) polarizations (ϵ_{rs} = 4.0, tan δ_s = 0.020, d_s = 0.025λ, ϵ_{rc} = 1.2, tan δ_c = 0.005, d_c = 0.293λ).

better rain-erosion resistance, high strength-to-weight ratio, decreased static-charge buildup, and ease of handling.

A *space-frame* radome consists of a network of structural beams covered with a electrically thin dielectric material.[10,36] It is often in the shape of a truncated sphere and is used in ground-based applications. The dielectric or metallic structural beams form triangular sections networked together to support the loads applied to the dielectric material. Although some antenna gain reduction is caused by the beams, their size and random orientation prevent unwanted beam deflection and boresight errors.

The *grooved* radome wall structure is a nonresonant panel consisting of a thick dielectric into which V-shaped grooves have been cut on both sides of the panel.[37] The grooves cut on one side can be oriented 90° with respect to those on the other side to help cancel the anisotropic effects of the structure and to equalize the transmission properties for orthogonal polarizations. The grooves provide a gradual transistion from air to dielectric and back again, thus furnishing a continuous match to the incident wave in a manner analogous to that used in pyramidal absorber materials. The depth of the grooves determines the lowest frequency of operation, and the spacing of the grooves determines the upper frequency limit. Panels have been built and tested to operate from 5 to 40 GHz for incidence angles up to 50°. Similar anisotropic dielectric structures have been used to improve the performance of a streamlined radome.[38]

A *broadband* radome wall structure is designed to provide high transmittance over a broad frequency band of operation. Loyet[39] has compared the broadband frequency properties of solid, sandwich, Oleesky, and gradient wall structures. The multilayer (Oleesky) and gradient walls are electrically superior, but their weight, thickness, and cost inhibit their use. The boresight error and error-slope performance of broadband radomes varies considerably over the bandwidth of high transmittance and is usually inferior to that obtainable with a solid half-wave wall in its optimum (6 percent) band of operation.

44-3   ELECTRICAL-DESIGN DATA

General

Theoretical electrical-design data are presented in this section for a tangent ogive, alumina (ϵ_r = 9.3), half-wave wall radome containing a gimballed, uniformly illuminated, vertically polarized monopulse X-band antenna. The data illustrate both the computer-aided radome-design procedure and the predicted electrical performance of the tracking antenna as functions of radome wall thickness, fineness ratio, wall-thickness variations, and operating frequency. Electrical performance is quantified in terms of boresight error and gain in two orthogonal (pitch and yaw) planes. Some comparisons with measured boresight-error data are also presented for corroboration.

The antenna-radome geometry of Fig. 44-5 is applicable with radome diameter $D = 6.6\lambda$, length $L = 14.4\lambda$, constant wall thickness $d = 0.17\lambda$, position of the gimbal point along the radome axis $R_R = 2.14\lambda$, antenna offset $R_A = 1.35\lambda$, and antenna diameter $D_A = 5.0\lambda$, where λ is the free-space wavelength. The pitch plane of radome scan is the plane of the figure, in which the positive gimbal angle corresponds to the upward scan of the radome tip with respect to R_A or, equivalently, the downward scan of the antenna. The yaw plane of radome scan is perpendicular to the plane of the figure and contains the gimbal point and the antenna axis R_A. The positive gimbal angle in the yaw plane corresponds to the scan of the radome tip out of the page toward the reader.

Boresight errors are expressed in milliradians and can have positive and negative values. The boresight error is defined to be the true direction to the target with respect to the antenna axis R_A when electrical boresight is indicated by the outputs of the elevation-difference and azimuth-difference ports of the monopulse antenna. For an observer positioned behind the antenna and looking in the direction of R_A as indicated in Fig. 44-5, a positive pitch error indicates that the actual target location is above the R_A axis; a positive yaw error indicates that the actual target location is to the left of R_A in the yaw plane. A complete description of the coordinate systems is contained in Ref. 40.

Design Procedure

The design procedure for the missile radome starts with the specified tangent ogive shape ($L/D = 2.18$). For high-speed applications, a ceramic radome material is dictated by operating temperatures. Alumina is chosen for its strength, resistance to rain erosion, and fabricability. A half-wave wall structure is chosen for simplicity in fabrication and superior boresight-error performance. Electrical design consists of choosing the uniform or tapered wall thickness which minimizes boresight error and gain loss as functions of gimbal angle and frequency in a narrow (6.5 percent) band centered at f_M.

Computer-aided radome analysis is used to determine the effects of the radome on boresight error and gain. A receiving formulation based on geometrical optics and Lorentz reciprocity is used.[40] Briefly, the antenna near fields are represented by a sampled data array of 16 × 16 points spaced equally in the x and y directions approximately $\lambda/4$ apart. A vertically polarized plane wave (target return) is assumed to be incident on the outside of the radome from the direction $\hat{k}_A$. From each sample point in the antenna aperture, a ray is traced in the direction $\hat{\mathbf{k}}_A$ to determine the point of intersection and the unit inward normal $\hat{\mathbf{n}}_R$ to the inner radome-wall surface. The unit vectors $\hat{\mathbf{k}}_A$, $\hat{\mathbf{n}}_R$ are used to transform the incident plane wave through the radome wall by using the insertion-voltage transmission coefficients for a plane dielectric sheet having the same cross section as that of the radome wall (Fig. 44-3). The voltage V_R received in each channel of the monopulse antenna is computed according to

$$V_R = \int_{S_1} (\mathbf{E}_T \times \mathbf{H}_R - \mathbf{E}_R \times \mathbf{H}_T) \cdot \hat{\mathbf{n}} \, da \qquad \textbf{(44-3)}$$

where $\mathbf{E}_T, \mathbf{H}_T$ = antenna near fields when transmitting (sum, elevation difference, or azimuth difference)

$\mathbf{E}_R, \mathbf{H}_R$ = fields incident on the antenna after traversing the radome

$\hat{\mathbf{n}}$ = unit normal to antenna aperture

The integration is carried out over the planar surface S_1 coinciding with the antenna aperture. It is the choice of the surface S_1 and the approximations used to find the fields on it which distinguish the various methods of radome analysis.

Gain loss caused by the radome is determined by comparing the maximum voltage received on the sum channel with the radome in place to that received for the case of a "free-space radome." Boresight error is determined in the pitch and yaw planes by finding the direction of arrival $\hat{\mathbf{k}}_0$ which produces a null indication in each tracking function defined by the ratio of difference-channel voltage to sum-channel voltage.

Electrical Effects

The effects of design angle (wall thickness) on boresight error and gain are illustrated in Figs. 44-9 and 44-10. It is clear from the figures that the smallest boresight errors and least gain loss are realized for the larger design angles of 55 and 75°. This result is expected from consideration of the incidence angles of rays drawn normally from the antenna aperture to the radome wall and the transmission characteristics of the half-wave wall. The difference in thickness of the walls with $\theta = 0°$ and $\theta = 75°$ is 0.009λ at band center f_M.

To demonstrate the accuracy of the predictions, Fig. 44-11 shows comparisons of measured and predicted boresight errors at three frequencies for the alumina radome having a uniform wall thickness of 0.17λ (design angle $\theta = 55°$ at midband). In each graph, the measured error is shown by the solid curve. The small dashed curve is the error predicted by using the ray-tracing radome analysis described above. The large dashed curve is the error predicted by using a surface-integration method of radome analysis based on Lorentz reciprocity and the Huygens-Fresnel principle.[41] Whereas the latter method required approximately 30 s per pattern point, the former required approximately 1 s on the same type of computing system. Boresight-error computations require at least two pattern points.

The predicted effects on boresight error of tapering the wall thickness from tip to base are illustrated in Fig. 44-12. For the positive taper, the radome wall is thicker at the tip than at the base: at the base, the wall thickness corresponds to a design angle of $\theta = 0°$; at the tip, the thickness corresponds to a design angle of $\theta = 75°$. For the negative taper, the thickness at the base corresponds to $\theta = 75°$, and the thickness at the tip corresponds to $\theta = 0°$. Examination of Fig. 44-12 shows an improvement in pitch error by using the positive taper; however, yaw error is made worse. Pitch and yaw gains were little affected by the positive taper, whereas the neg-

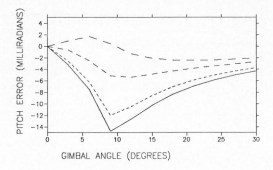

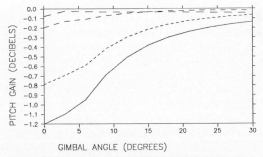

FIG. 44-9 Predicted boresight error and gain versus pitch-plane gimbal angle of alumina (L/D = 2.18) tangent ogive radome at frequency f_M for four design angles: θ = 0° (solid curve), 30°, 55°, and 75° (largest dash).

ative taper caused gain reductions of approximately 0.7 dB in both planes. Other wall-tailoring prescriptions may yield better boresight-error and error-slope performance than those considered here.

Boresight errors are also affected by the fineness ratio (L/D) of the radome, as shown in Fig. 44-13. Generally, the higher fineness ratios produce larger boresight errors over a given frequency bandwidth; however, at a single frequency the wall thickness can be optimized to give comparable boresight errors over a range of fineness ratios. In Fig. 44-13, the wall thickness is d = 0.17λ for each case, and no attempt has been made to optimize performance for each fineness ratio.

Variations in the wall thickness can have strong effects on the boresight errors, as illustrated in Fig. 44-11 for uniform thickness changes and as illustrated in Fig. 44-14 for a cosinusoidal thickness variation of 0.006λ from radome tip to base. The wide departure, especially in the yaw plane, of the boresight error due to such a thickness variation emphasizes the importance of mechanical tolerances in the manufacturing process.

At frequencies higher than X band, a solid-wall ceramic radome must often have a thickness greater than one-half wavelength to meet structural-strength requirements. Higher-order half-wave walls are inherently more frequency-sensitive than the

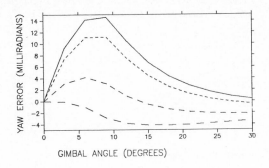

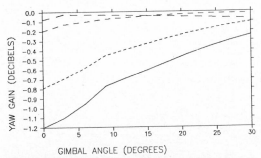

FIG. 44-10 Predicted boresight error and gain versus yaw-plane gimbal angle of alumina ($L/D = 2.18$) tangent ogive radome at frequency f_M for four design angles: $\theta = 0°$ (solid curve), 30°, 55°, and 75° (largest dash).

first-order ($n = 1$) wall, as illustrated by the data in Fig. 44-15. Although not illustrated, an n-order half-wave wall of material with high dielectric constant is more frequency-sensitive than the same-order wall of lower dielectric constant.

The boresight error and boresight-error slope produced by the radome can affect missile performance in different ways, depending on the type of guidance system employed.[42,43] Consequently, radome specifications may vary from one system to another. Some investigators have seriously questioned the necessity of tight specification on error and error slope.[44] Electronic compensation to correct radome boresight errors has been implemented and shown to be feasible at ambient temperatures;[45] however, questions have been raised concerning the accuracy of the technique at elevated operating temperatures and for wave polarizations other than those for which the boresight-error map of the radome was made.

From the foregoing discussion, the importance and usefulness of accurate, practical computer-aided radome-analysis methods should be apparent. Development of such methods has paralleled the development of airborne radar systems, whose complexities have increased over the years. Silver[46] illustrates the geometrical-optics approaches taken up to 1949 and developed during the previous war years. Kilcoyne[47]

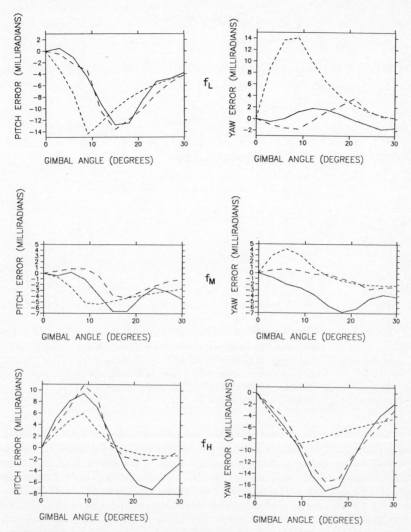

FIG. 44-11 Comparisons of measured boresight errors (solid curve) with those computed by using ray tracing (small dash) and surface integration (large dash) for alumina half-wave wall ($d = 0.17\lambda$), tangent ogive radome 6.6 wavelengths in diameter and 14.4 wavelengths long. X-band frequencies f_L and f_H are the edges of a 6.5 percent band; f_M is band center. (*After Ref. 41.*)

presents a two-dimensional ray-tracing method for analyzing radomes which utilizes the digital computer and is an extension of work done earlier by Snow[48] and by Pressel.[49] A more rigorous method of analysis, using an integral equation to compute fields inside the radome caused by an incident plane wave, was introduced in the same year by Van Doeren.[50] Tricoles[51] formulated a three-dimensional method of radome anal-

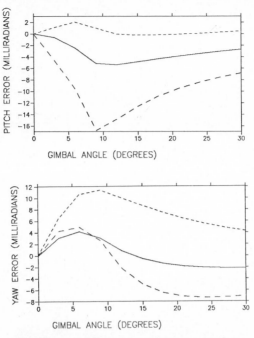

FIG. 44-12 Predicted boresight errors versus gimbal angle of alumina (L/D = 2.18) radome at frequency f_M for uniform wall thickness (solid) and for positive (small dash) and negative (large dash) taper of wall thickness.

ysis based on Shelkunoff's induction and equivalence theorems. Tavis[52] describes a three-dimensional ray-tracing technique to find the fields on an equivalent aperture external to an axially symmetric radome. Paris[53] describes a three-dimensional radome analysis wherein the tangential fields on the outside surface of the radome due to the horn antenna radiating inside the radome are found. Wu and Rudduck[54] describe a three-dimensional method which uses the plane-wave-spectrum (PWS) representation to characterize the antenna. Joy and Huddleston[55] describe a computationally fast, three-dimensional radome analysis which utilizes the plane-wave-spectrum representation and exploits the fast Fourier transform (FFT) to speed up the computer calculations. Chesnut[56] combined the program of Wu and Rudduck with the work of Paris to form a three-dimensional radome-analysis method. Huddleston[57] developed a three-dimensional radome-analysis method which uses a general formulation based on the Lorentz reciprocity theorem. Siwiak et al.[58] applied the reaction theorem to the analysis of a tangent ogive radome at X-band frequencies to determine boresight error. Hayward et al.[59] have compared the accuracies of two methods of analysis in the cases of large and small radomes to show that ray tracing does not accurately predict wavefront distortion in small radomes. Burks[60] developed a ray-tracing analysis which includes first-order reflections from the radome wall. Kvam and Mei[61] have applied the unimoment method to the solution of the radome problem.

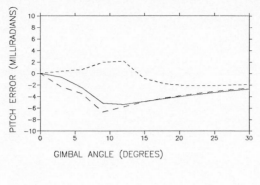

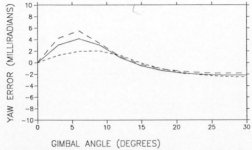

FIG. 44-13   Predicted boresight errors versus gimbal angle of alumina radome at frequency f_M for three fineness ratios: $L/D = 2.18$ (solid), $L/D = 1.5$ (small dash), $L/D = 2.5$ (large dash).

Aerothermal Analysis

Computer-aided analyses have also been developed to determine temperatures and mechanical stresses in the radome due to aerodynamic loading and heating as functions of trajectory, altitude profile, velocity, and maneuvers of the missile or aircraft. URLIM is one such computer program.[62] It calculates thermal stress, melting, dielectric changes, and maneuver loads versus trajectory time for hypersonic radomes. In addition, it estimates bending stresses at the base of the radome due to aerodynamic pressure and maneuver.

44-4   FABRICATION AND TESTING

Fabrication

Fabrication methods depend on the radome material, and it is often the availability of a reliable manufacturing method which makes the final selection of material. Typical construction methods for organic radomes include molding (vacuum bag, autoclave, pressure bag, matched die), filament winding, resin infiltration, and foam in place.[63] Injection molding is also used.[64] Inorganic (ceramic) radomes have been con-

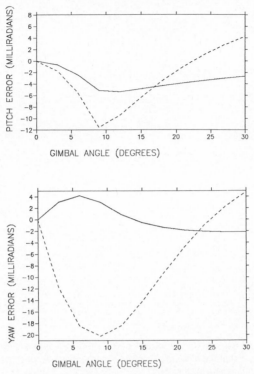

FIG. 44-14 Predicted boresight errors versus gimbal angle of alumina (L/D = 2.18) radome at frequency f_M for uniform wall thickness of d = 0.17λ (solid) and for cosinusoidal longitudinal thickness variation of 0.006λ (dash).

structed by using slip casting, cold pressing, injection molding, hot pressing, glass forming, and flame spraying.[65] Machining is often required for both organic and inorganic radomes to provide the necessary tolerances on wall thickness.

Mechanical and electrical tests are performed throughout production. Before fabrication, tests are performed on panels of candidate radome wall structures to ensure that the required mechanical and electrical properties can be attained reproducibly and within the margins required. During fabrication, quality-control tests are carried out to ensure correct shape, size, dielectric constant, loss tangent, and wall thickness. After fabrication, laboratory and operational tests are performed to help ensure successful radome performance in the field.

Mechanical Testing

Mechanical testing includes measurements of physical dimensions and testing of the structural, thermal, and rain-erosion properties of the complete radome. Ultrasonic- and microwave-interferometer techniques are used to measure wall thickness. Struc-

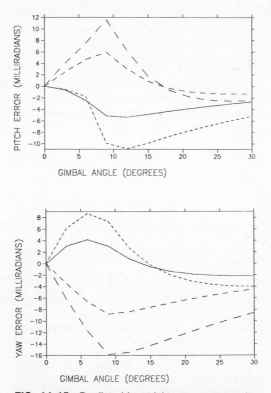

FIG. 44-15 Predicted boresight errors versus gimbal angle of alumina (L/D = 2.18) radome for two wall thicknesses at two frequencies: half-wave wall d = 0.17λ at f_M (solid); full-wave wall d = 0.34λ at f_M (smallest dash); half-wave wall at f_H = 1.032 f_M; full-wave wall at f_H (largest dash).

tural testing involves static loading of the radome, using mechanical, pressure, and vacuum methods in a manner simulating operational conditions, and is often done at design temperatures. High-temperature and thermal-shock testing has been carried out by using oxyacetylene torches, quartz lamps, rocket motors, ramjet engines,[18] solar furnaces,[12,66] and supersonic wind tunnels.[67] Rain-erosion testing has been carried out by using rocket sleds, rotating arms, and shotgun lead pellets.[68] Electrical tests are sometimes combined with thermal testing.[14]

Electrical Testing

Electrical testing of radomes includes measurements of complex permittivity of radome materials and electrical characteristics of the complete antenna-radome combination. The dielectric constant and loss tangent of material samples can be determined at microwave and millimeter waves by using free-space-bridge,[69] Fabry-Perot-resonator,[70] and short-circuit-waveguide[71] techniques. Below millimeter wavelengths, cavity and capacitance bridge techniques can be used.[72,73]

Radome electrical testing includes measurement of transmission efficiency, boresight error, boresight-error slope, beam deflection, radome reflection, antenna-pattern distortion, and depolarization as functions of antenna-radome orientation (scan) and frequency. Standard antenna range equipment,[74] with the addition of a radome-test fixture,[75] can be used to measure pattern distortion, transmission efficiency, reflection (VSWR), and depolarization. Boresight-error and beam-deflection measurements are more demanding and are made by using automated closed-loop[74,76] and electronically calibrated[77] systems.

REFERENCES

1 T. E. Tice (ed.), "Techniques for Airborne Radome Design," AFAL-TR-66-391, vol. 1, December 1966, p. 6.

2 J. D. Walton (ed.), "Techniques for Airborne Radome Design," AFAL-TR-66-391, vol. 2, December 1966, p. 450.

3 R. Timms and J. Kmetzo, "Traveling Wave Radants—Applications, Analysis, and Models," *Proc. Ninth USAF–Georgia Tech. EM Windows Symp.*, June 1968.

4 P. H. Dowsett, "Cross Polarization in Radomes: A Program for Its Computation," *IEEE Trans. Aerosp. Electron. Syst.*, vol. AES-9, no. 3, May 1973, pp. 421–433.

5 D. J. Yost, L. B. Weckesser, and R. C. Mallalieu, "Technology Survey of Radomes for Anti-Air Homing Missiles," JHU/APL Rep. FS-80-022, Cont. N00024-78-C-5384, March 1980.

6 F. D. Groutage, "Radome Development for a Broadband RF Missile Sensor," Rep. NELC/TR-2023, AD-A038 262/2ST, January 1977.

7 K. D. Hill and J. D. Kelley, "Radome Design for High Power Microwave Systems," *Proc. 21st USAF Antenna R&D Symp.*, Air Force Avionics Laboratory and University of Illinois, October 1971, p. 1–49.

8 J. H. Richmond, "Calculation of Transmission and Surface Wave Data for Plane Multilayers and Inhomogeneous Plane Layers," OSU Antenna Lab. Rep. 1751-2, Cont. AF33 (615)-1081, October 1963.

9 I. Anderson, "Measurements of 20-GHz Transmission through a Radome in Rain," *IEEE Trans. Antennas Propagat.*, vol. AP-23, no. 5, September 1975, pp. 619–622.

10 A. L. Kay, "Electrical Design of Metal Space Frame Radomes," *IEEE Trans. Antennas Propagat.*, vol. AP-13, March 1965, pp. 182–202.

11 M. B. Punnett, "C.N. Tower Microwave Radome," *Proc. 14th EM Windows Symp.*, Georgia Institute of Technology, Atlanta, June 1978, pp. 39–44.

12 L. K. Eliason et al., "A Survey of High Temperature Ceramic Materials for Radomes," Melpar Rep. ML-TDR-64-296, Cont. AF33 (657)-10519, AD607 619, September 1964.

13 L. B. Weckesser et al., "Aerodynamic Heating Effects on the SM-2 Block II Radome," JHU/APL Rep. FS-80-146, Cont. N00024-78-C—5384, October 1980.

14 J. D. Walton, Jr., and H. L. Bassett, "A Program to Measure Radome/Antenna Patterns at High Temperature," Georgia Tech. Rep., Cont. DAAH01-73-C-0617, U.S. Army Missile Command, March 1974.

15 G. W. Scott, "F-5F Shark Nose Radome Lightning Test," *Lightning Tech. Symp.*, NASA Conf. Pub. 2128, FAA-RD-80-30, April 1980, pp. 421–429.

16 S. W. Waterman, "Calculation of Diffraction Effects of Radome Lightning Protection Strips," *Proc. Fourth Int. EM Windows Conf.*, Direction des Constructions et Armes Navales de Toulon, Toulon, France, June 1981.

17 K. W. Foulke, "Thermal Plastic Radomes," *Proc. 12th EM Windows Symp.*, Georgia Institute of Technology, Atlanta, June 1974, pp. 151–155.
18 J. D. Walton, Jr. (ed.), *Radome Engineering Handbook: Design and Principles*, Marcel Dekker, Inc., New York, 1970.
19 W. B. Westphal and A. Sils, "Dielectric Constant and Loss Data," AFML-TR-72-39, April 1972.
20 H. E. Bennett et al., "Optical Properties of Advanced Infrared Missile Dome Materials," Tech. Rep. NWC TP 6284, Naval Weapons Center, July 1981.
21 C. L. Price, "Camouflage Colored Rain Erosion Resistant Antistatic Coatings for Radome," *Proc. 14th EM Windows Symp.*, Georgia Institute of Technology, Atlanta, June 1978, pp. 153–157.
22 K. W. Foulke, "Polyurethane Tape Erosion Boots," *Proc. 14th EM Windows Symp.*, Georgia Institute of Technology, Atlanta, June 1978, pp. 159–165.
23 P. W. Moraveck and P. W. Sherwood, "New Polyurethane Coatings for Radome Applications," *Proc. 14th EM Windows Symp.*, Georgia Institute of Technology, Atlanta, June 1978, pp. 147–152.
24 G. F. Schmitt, Jr., "Supersonic Rain Erosion Behavior of Ablative Fluorcarbon Plastic Radome Materials," *Proc. 14th EM Windows Symp.*, Georgia Institute of Technology, Atlanta, June 1978, pp. 87–96.
25 K. N. Letson et al., "Rain Erosion and Aerothermal Sled Test Results on Radome Materials," *Proc. 14th EM Windows Symp.*, Georgia Institute of Technology, Atlanta, June 1978, pp. 109–116.
26 R. M. Redheffer, "The Interaction of Microwave Antennas with Dielectric Sheets," Rad. Lab. Rep. 483-18, Mar. 1, 1946.
27 J. D. Walton, Jr. (ed.), *Radome Engineering Handbook*, pp. 47–50.
28 T. E. Tice (ed.), "Techniques for Airborne Radome Design," pp. 66–77.
29 Ibid., chap. 13.
30 M. I. Skolnik (ed.), *Radar Handbook*, McGraw-Hill Book Company, New York, 1970, chap. 14.
31 T. L. Norin, "AWACS Radar Radome Development," *Proc. 11th EM Windows Symp.*, Georgia Institute of Technology, Atlanta, August 1972, pp. 41–46.
32 D. Cope and J. Kotik, "Digital Optimization of Radome Walls," *Proc. OSU-WADC EM Windows Symp.*, vol. 1, Ohio State University, Columbus, June 1960.
33 L. Rader, "Thin-Skinned Broadband Radome Technology," *Proc. 13th EM Windows Symp.*, Georgia Institute of Technology, Atlanta, September 1976, p. 161.
34 E. L. Pelton and B. A. Munk, "A Streamlined Metallic Radome," *IEEE Trans. Antennas Propagat.*, vol. AP-22, November 1974, pp. 799–803.
35 W. R. Bushelle et al., "Development of a Resonant Metal Radome," *Proc. 14th EM Windows Symp.*, Georgia Institute of Technology, June 1978, pp. 179–185.
36 M. I. Skolnik (ed.), *Radar Handbook*, pp. 14-25–14-27.
37 D. G. Bodnar and H. L. Bassett, "Analysis of an Anisotropic Dielectric Radome," *IEEE Trans. Antennas Propagat.*, vol. AP-23, November 1975, pp. 841–846.
38 E. L. Rope and G. P. Tricoles, "Anisotropic Dielectrics: Tilted Grooves on Flat Sheets and on Axially Symmetric Radomes," *IEEE Antennas Propagat. Int. Symp. Dig.*, June 1979, pp. 610–611.
39 D. L. Loyet, "Broadband Radome Design Techniques," *Proc. 13th EM Windows Symp.*, Georgia Institute of Technology, Atlanta, September 1976, pp. 169–173.
40 G. K. Huddleston, H. L. Bassett, and J. M. Newton, "Parametric Investigation of Radome Analysis Methods: Computer-Aided Radome Analysis Using Geometrical Optics and Lorentz Reciprocity," final tech. rep., Grant AFOSR-77-3469, vol. 2 of 4, Georgia Institute of Technology, Atlanta, February 1981.
41 K. Siwiak, T. B. Dowling, and L. R. Lewis, "Boresight Error Induced by Missile Radomes," *IEEE Trans. Antennas Propagat.*, vol. AP-27, no. 6, November 1979, pp. 832–841.
42 T. E. Tice (ed.), "Techniques for Airborne Radome Design," chap. 4.

43 D. J. Yost et al., "Technology Survey of Radomes for Anti-Air Homing Missiles," app. A.

44 Ibid., app. B.

45 T. B. Dowling, L. R. Lewis, and A. R. Chinchillo, "Radome Computer Compensation," *IEEE Antennas Propagat. Int. Symp. Dig.,* vol. 2, June 1979, pp. 602–605.

46 S. Silver, *Microwave Antenna Theory and Design,* McGraw-Hill Book Company, New York, 1949, pp. 522–542.

47 N. R. Kilcoyne, "An Approximate Calculation of Radome Boresight Error," *Proc. USAF–Georgia Tech. EM Windows Symp.,* June 1968, pp. 91–111.

48 O. Snow, "Discussion of Ellipticity Produced by Radomes and Its Effects on Crossover Point Position for Conically Scanning Antennas," Rep. E15108, U.S. Naval Air Development Center, 1951.

49 P. I. Pressel, "Boresight Prediction Techniques," *Proc. OSU-WADC Radome Symp.,* 1956.

50 R. E. Van Doeren, "Application of an Integral Equation Method to Scattering from Dielectric Rings," *Proc. USAF–Georgia Tech. EM Windows Symp.,* June 1968, pp. 113–127.

51 G. P. Tricoles, "Radiation Patterns and Boresight Error of a Microwave Antenna Enclosed in an Axially Symmetric Dielectric Shell," *J. Opt. Soc. Am.,* vol. 54, no. 9, September 1964, pp. 1094–1101.

52 M. Tavis, "A Three-Dimensional Ray Tracing Method for the Calculation of Radome Boresight Error and Antenna Pattern Distortion," Rep. TOR-0059 (56860)-2, Air Force Systems Command, May 1971.

53 D. T. Paris, "Computer-Aided Radome Analysis," *IEEE Trans. Antennas Propagat.,* vol. AP-18, no. 1, January 1970, pp. 7–15.

54 D. C. F. Wu and R. C. Rudduck, "Plane Wave Spectrum–Surface Integration Technique for Radome Analysis," *IEEE Trans. Antennas Propagat.,* vol. AP-22, no. 3, May 1974, pp. 497–500.

55 E. B. Joy and G. K. Huddleston, "Radome Effects on Ground Mapping Radar," Cont. DAAH01-72-C-0598, AD-778 203/0, U.S. Army Missile Command, March 1973.

56 R. Chesnut, "LAMPS Radome Design," *Proc. 13th EM Windows Symp.,* Georgia Institute of Technology, Atlanta, September 1976, pp. 73–78.

57 G. K. Huddleston, H. L. Bassett, and J. M. Newton, "Parametric Investigation of Radome Analysis Methods," *IEEE Antennas Propagat. Int. Symp. Dig.,* May 1978, pp. 199–201.

58 K. Siwiak, T. Dowling, and L. R. Lewis, "The Reaction Approach to Radome Induced Boresight Error Analysis," *IEEE Antennas Propagat. Int. Symp. Dig.,* May 1978, pp. 203–205.

59 R. A. Hayward, E. L. Rope, and G. P. Tricoles, "Accuracy of Two Methods for Numerical Analysis of Radome Electromagnetic Effects," *Proc. 14th EM Windows Symp.,* Georgia Institute of Technology, Atlanta, June 1978, pp. 53–55.

60 D. G. Burks, E. R. Graf, and M. D. Fahey, "Effects on Incident Polarization on Radome-Induced Boresight Errors," *Proc. 15th EM Windows Symp.,* Georgia Institute of Technology, Atlanta, June 1980, pp. 1–5.

61 T. M. Kvam and K. K. Mei, "The Internal Fields of a Layered Radome Excited by a Plane Wave," *IEEE Antennas Propagat. Int. Symp.,* June 1981, pp. 608–611.

62 R. K. Frazer, "Use of the URLIM Computer Program for Radome Analysis," *Proc. 14th EM Windows Symp.,* Georgia Institute of Technology, Atlanta, June 1978, pp. 65–70.

63 J. D. Walton (ed.), *Radome Engineering Handbook,* pp. 135–160.

64 E. A. Welsh, A. Ossin, and R. A. Mayer, "Injection Molded Plastics for Millimeter Wave Radome Applications," *Proc. 14th EM Windows Symp.,* Georgia Institute of Technology, Atlanta, June 1980, pp. 192–197.

65 J. D. Walton (ed.), *Radome Engineering Handbook,* pp. 298–321.

66 R. K. Frazer, "Duplication of Radome Aerodynamic Heating Using the Central Receiver Test Facility Solar Furnace," *Proc. 15th EM Windows Symp.,* Georgia Institute of Technology, Atlanta, June 1980, pp. 112–116.

67 D. J. Kozakoff, "Aerodynamically Heated Radome Measurement Techniques," Georgia Tech. Rep. A-2945, Cont. CFSI 4-81 (F40600-81-0003), September 1981.

68 J. D. Walton (ed.), *Radome Engineering Handbook,* chap. 8.

69 H. L. Bassett, "A Free-Space Focused Microwave System to Determine the Complex Permittivity of Materials to Temperatures Exceeding 2000°C," *Rev. Sci. Instrum.,* vol. 42, no. 2, February 1971, pp. 200–204.

70 K. H. Breeden and J. B. Langley, "Fabry-Perot Cavity for Dielectric Measurements," *Rev. Sci. Instrum.,* vol. 40, no. 9, September 1969.

71 G. H. Sage, "Measurement of Electrical Properties of High Temperature Radome Materials by Means of the XB 3000A Dielectrometer," Ph.D. thesis, GE/EE65-21, Air Force Institute of Technology, Dayton, Ohio, November 1965.

72 W. B. Westphal, Tech. Rep. 182, MIT Laboratory for Insulation Research, October 1963.

73 R. E. Charles et al., Tech. Rep. 201, MIT Laboratory for Insulation Research, October 1966.

74 J. S. Hollis, T. J. Lyon, and L. Clayton, Jr. (eds.), *Microwave Antenna Measurements,* Scientific-Atlanta, Inc., Atlanta, July 1970.

75 D. O. Gallentine et al., "Radome Positioner for the RFSS," Georgia Tech. Rep., Cont. DAAK40-78-D-008, February 1978.

76 T. B. Dowling et al., "Automated Radome Test Facility," *Proc. 14th EM Windows Symp.,* Georgia Institute of Technology, Atlanta, June 1978, pp. 173–178.

77 J. M. Schuchardt et al., "Automated Radome Performance Evaluation in the Radio Frequency Simulation System (RFSS) Facility at MICOM," *Proc. 15th EM Windows Symp.,* Georgia Institute of Technology, Atlanta, June 1980, pp. 134–141.

Chapter 45

Microwave Propagation

Geoffrey Hyde

COMSAT Laboratories

45-1 INTRODUCTION

Microwave propagation as considered here deals with the propagation of radio waves through the atmosphere. Propagation through the earth, in tunnels in the earth, or in bound modes (indeed, other than in radio line of sight) will not be considered. In general, the range of frequencies given most detailed examination is 3 to 30 GHz, although propagation effects from 1 to 100 GHz will be treated. The effects on radio-wave propagation through the clear troposphere will be considered, and then the effects of rain will be examined. Finally, effects arising in the ionosphere will be touched upon.

Microwave antennas can be considered transducers coupling energy from a guided-wave system into the unbound-wave-propagation medium. As such, microwave antennas are necessarily affected by both regimes, but since the properties of the one can be controlled by the antenna designer and the other cannot, until recently propagation effects were given short shrift by antenna designers. More usually, these effects appear as entries in system link budgets or error budgets. The need for more careful control of where energy is directed to or received from (interference-coordination and security), for polarization control (dual-polarized systems), and, as frequencies in common use rise above 10 GHz, for limiting radio-frequency (RF) power requirements to practical values, has given rise to the need for information on propagation effects.

45-2 CLEAR-SKY-INDUCED PROPAGATION EFFECTS

At most antenna sites, for most of the time the path between two antennas within radio line of sight will be occupied by atmosphere and/or free space which is clear to the eye, that is, free of obstructions, clouds, rain or other hydrometeors, and dust or other particulates. This gaseous atmosphere[1,2] has two main constituents, nitrogen and oxygen. The proportions of these constituents are fixed, and their effects for a particular link are constant. But it is a third constituent, water vapor, whose proportions vary, that causes most of the clear-sky impairments. Table 45-1 lists the constituents

TABLE 45-1 Gases in the Atmosphere

Gases	Percent volume (dry air)*
Nitrogen	78.1
Oxygen	20.9
Argon	0.9
Other†	0.1

*Water vapor ≤ 4 percent near surface (very variable)
≤ 0.01 percent above 15 km.

†Carbon dioxide ~ 0.03 percent (variable).

of the normal atmosphere. Figure 45-1 shows mean temperature as a function of altitude. The pressure falls off approximately exponentially. This is modeled in the *U.S. Standard Atmosphere.*[3]

Only water vapor and oxygen cause appreciable absorption of radio waves in the range of frequencies of concern. Water-vapor distribution and gravity-induced variation of the density and pressure of the atmosphere as a function of altitude contribute to systematic ray bending through variation of the radio refractive index. However, because at times the atmosphere develops local inhomogeneities, which result in local changes in the radio refractive index, other impairments to radio waves at microwave frequencies may arise, among them ducting, fluctuations in angle of arrival, fluctuations of amplitude and phase, phase-delay variation, and defocusing. Absorption brings along with it the corollary effect of noise-temperature increase. The longer the path through the atmosphere, the more severe these impairments tend to be. Consequently, for nearly horizontal paths at low elevation angles, these impairments tend to be more severe. The situation is further complicated by the fact that the transmission properties of the atmosphere vary diurnally (throughout the day) and seasonally as well as geographically.

Antenna design is affected by absorption in the clear atmosphere A_a, as increased gain may be required to overcome this absorption, which is given by

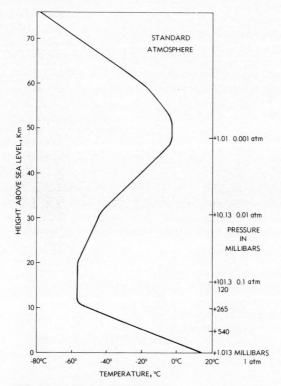

FIG. 45-1   Standard atmosphere.

$$A_a = \int_0^d \gamma_a(r) \, dr \qquad (45\text{-}1)$$

Over a horizontal path of length d, if it is assumed that the absorption coefficient γ_a dB/km is uniform along the path,

$$A_a = \gamma_a d = d[\gamma_o(f) + \gamma_w(f,\xi)] \qquad (45\text{-}2)$$

where γ_o = absorption coefficient due to O_2
$\quad\ \gamma_w$ = absorption coefficient due to water vapor
$\quad\ f$ = frequency
$\quad\ \xi$ = water-vapor concentration, g/m^3

To a first approximation, the effects of water vapor can be estimated as linear (except for the higher concentrations). Thus

$$\gamma_a(f,\xi) = \gamma_o(f) + (\xi/7.5)\gamma_w(f, 7.5) \qquad (45\text{-}3)$$

Curves for $\gamma_o(f)$ and $\gamma_w(f, 7.5)$ are given in Fig. 45-2. For the slant path γ_o, γ_w and hence γ_a are affected by the distribution of oxygen and water vapor as a function of altitude and pressure.[4,5] The effects of oxygen on slant-path absorption are reasonably straightforward to access, as oxygen distribution is not a strong function of climate and is estimated to be equivalent at zenith to 7 to 10 km of attenuation on a horizontal path, complicated by the effects of starting altitude.[5] But water-vapor absorption is quite a bit more complex.[4] Obviously, the slant path contains little water vapor above the 0° isotherm, which changes its altitude diurnally, seasonally, and geographically. Additionally, to a first approximation, water vapor at altitude is a function of water-vapor content at the surface. More approximately and simplistically, Fig. 45-3 gives the approximate dry gaseous absorption ($1 \le f \le 100$ GHz) and the approximate contribution due to water vapor for 7.5 g/m^3 at zenith. Down to 5° elevation angle,

$$\gamma_a(\theta) \cong \gamma_a(90°)/\sin \theta \qquad (45\text{-}4)$$

$$\theta \equiv \text{elevation angle}$$

$$\gamma_a(90°) \cong \gamma_o(90°) + \frac{\xi}{7.5}\,\gamma_w(90°) \qquad (45\text{-}5)$$

Ray bending arises because of the variation in the index of refraction of the atmosphere for radio waves. For lower-elevation-angle paths, the systematic ray bending caused by this has been calculated. The average radio refractive index for an exponential atmosphere is given in CCIR, Volume V, Recommendation 369-3[6] as

$$n(h) = 1 + N \times 10^{-6} = 1 + 315 \times 10^{-6} \times \exp(-0.136\,h) \qquad (45\text{-}6)$$

where h = altitude above sea level
$\quad\ N$ = radio refractivity = $77.6\,p/T + 3.73 \times 10^5\,p_\omega/T^2 \qquad (45\text{-}7)$
$\quad\ p$ = atmospheric pressure, mb
$\quad\ T$ = temperature, K
$\quad\ p_\omega$ = water-vapor pressure, mb

At any point on the ray path, ray curvature[7] $(1/\rho)$ is given by

$$\frac{1}{\rho} = \frac{\cos \theta}{n} \times \frac{dn}{dh} \qquad (45\text{-}8a)$$

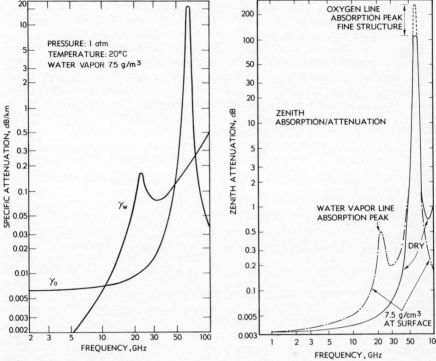

FIG. 45-2 Specific attenuation for water vapor and oxygen.

FIG. 45-3 Zenith absorption.

$$\cong \frac{dn}{dh} \text{ for low elevation angles} \qquad \textbf{(45-8b)}$$

For earth space (i.e., slant paths), this ray bending, in general downward, requires an apparent increase in elevation-angle pointing of the antenna. Detailed calculations of the required increment in elevation angle have been made by Crane,[8] who has calculated the average increments for different climate zones and also the day-to-day variations in the increment, all as a function of elevation angle.

Signal fluctuations caused by the clear troposphere impact antenna design because telecommunications systems are designed to have gain margin to overcome such fluctuations, part of which may be contributed by antenna gain, and because the antenna aperture size enters into fluctuation phenomenology directly.[8,9]

For line-of-sight paths, the atmosphere can behave in such a way that for some percentage of time rays are bent so that the earth appears to block transmission for certain path geometries, and transmission is then dominated by diffraction considerations[10] with an apparent loss of gain. Variations in atmospheric properties can cause the propagation-path bending owing to refraction due to the gradient in refractive index to change, thus changing the apparent angle of arrival of signals, i.e., the angle to which the antenna must be pointed. (Otherwise, the signals will appear to fluctuate slowly.) As might be expected, this occurs principally in the vertical plane;

variations up to 0.75° have been reported,[11,12] and the antenna pattern obviously enters into the fluctuation effects.[13] Another serious source of signal fluctuation is interference of signals traversing different paths from transmitter to receiver (i.e., multipath transmission). This is thought to arise because of local variations in water-vapor content and temperature, which imply local fluctuations in refractive index, which means that "rays" that start off in directions different from line of sight to the receive antenna can have trajectories that terminate at that antenna. As many as six such trajectories have been shown to exist simultaneously.[11] The probability of a deep multipath fade of depth F (dB) is given by the relationships[14,15]

$$F = 10 \log_{10} \frac{P}{P_0} \qquad (45\text{-}9)$$

($P_0 \equiv$ reference power level for no fade) and, obtained by regression from empirical data,

$$p(F) = k_c k_t \, 10^{-F/10} \, f \, d^3 \qquad (45\text{-}10)$$

where k_c, k_t are climate- and terrain-dependent parameters whose product has been evaluated empirically for the United States.[14,15] f is in GHz,

$$10^{-5} \le (k_c k_t) \, S_1^{1.3} \le 4 \times 10^{-5} \qquad (45\text{-}11)$$

where S_1 is the terrain-roughness factor (the standard deviation of terrain-elevation samples taken at 1-km intervals). Internationally, the frequency and path-length dependencies and terrain-roughness factors have not been fully reconciled. But it has been shown[16] that antenna beamwidth and path-clearance factors enter into those effects and must be taken into account in reconciling effects of antenna beamwidth of the different sets of measurements.

Multipath fade durations in the United States have been shown empirically to follow a log-normal distribution,[17,18] with the median fade duration given by

$$\bar{t} = 57 \, 10^{-F/20} \, (d/f)^{1/2} \qquad (45\text{-}12)$$

The use of two antennas, spatially separated, to overcome fading is called *diversity*. A vertical separation of several meters has been found to improve the fading characteristics of line-of-sight links with small ground reflections. Vigants[18] has proposed a formula for improvement:

$$\text{Improvement} \equiv I \cong (\Delta h)^2 \, f/d \, 10^{-F/10} \, 10^{\Delta G} \times k \qquad (45\text{-}13)$$

where Δh = vertical separation, m
ΔG = difference in antenna gains
k = constant $\cong 1.2 \times 10^{-3} \eta$
where η is the efficiency of the diversity-combining arrangement. It is thought that this equation is useful only for $I \ge 10$, $2 \le f \le 11$ GHz, $24 \le d \le 70$ km, and $5 \le h \le 15$ m.

Signal fluctuations also occur on the slant path. One mechanism is thought to be scattering of radio waves by small-scale spatial variations of refractive index arising under appropriate conditions. Measurements[19] seem to indicate that the presence of high humidity and some wind makes amplitude scintillation more probable. Analytic considerations[20] also show that amplitude scintillations of radio-wave signals may be produced by turbulent fluctuations of the troposphere. Tatarski[20] has shown that, in addition to the signal fluctuations, there is an apparent loss in gain arising from the averaging of signals over the antenna aperture. Crane and Blood[4] have developed a model for the variation of the rms signal-fluctuation level.

$$\left(\frac{\delta(f,\theta,D)}{\delta_{\text{Reference}}}\right) \propto f^{7/12}(\csc \theta)^{0.85}\left(\frac{G(R)}{G(R)_{\text{Reference}}}\right)^{1/2} \quad \text{dB} \quad \textbf{(45-14)}$$

where $R = \sqrt{\eta}D/2$ = effective antenna radius
η = antenna efficiency factor
$G(R)$ = Tatarski's aperture-averaging factor

Thus, it is seen that antenna aperture size and efficiency enter into signal-fluctuation modeling. Measurements made in the INTELSAT system[19] indicate that for elevation angles of 6.5° signal fluctuations of 4.5 dB or greater peak to peak occurred for about 0.01 percent of the time, and of 3 dB or greater for 0.3 percent of the time, for a single path (up at 6 GHz, down at 4 GHz). For the combined paths (up at 6 GHz, down at 4 GHz), signal fluctuations in the worst month (July 1974) exceeded 3 dB peak to peak for about 16 percent of the month.

Antenna gain loss due to defocusing has been demonstrated for low-elevation-angle slant paths.[21] But it appears that this loss is negligible (less than 0.2 dB) for paths with elevation angles above 4°.

45-3 RAIN-INDUCED PROPAGATION EFFECTS

Rain is a major cause of impairments to radio-wave transmission in the frequency ranges of concern. Rain absorbs radio waves and scatters them. Both actions extract energy from the incident wave, causing signal attenuation. Both cause a change of polarization state from that in the incident wave, causing signal depolarization. Both can cause a net increase in noise temperature to receiving systems. The scattering also diverts radio-wave energy into directions which might otherwise not receive energy, causing interference between telecommunications systems operating at the same frequency from common volume scattering. Attenuation affects antenna design, as increased antenna gain and/or antenna site diversity are means of ameliorating radio-wave attenuation and noise-temperature increase (although they are not the only means). Depolarization also can be mitigated at the antenna, in the microwave system behind the feed. Common-volume-scattering-induced interference is dependent on antenna gain and sidelobe characteristics.

To assess these impairments it is necessary to know the rainfall climate (rainfall, rain rate, drop-size distributions, etc.) in and around the sites of concern. Ideally one would like to know the rainfall characteristics for many years for each radio link of concern, but this information is not generally available. References 22, 23, 24, and 25 are good starting points for worldwide information (including the United States) on climate parameters in general and precipitation in particular. More detailed and fine-grained data are obtainable through weather services in the various countries. In the United States, this would be the National Weather Service Library, Silver Spring, Maryland, or the National Weather Service National Climatic Center, Asheville, North Carolina, from which various detailed records can be obtained. One very useful type of record is climatalogical data by state, by recording site, by month, and by year.

With enough back issues of such data, one can readily construct a data file on precipitation measurements near and around almost any desired site in the United States. In Canada, data can be obtained from the Atmospheric Environment Office, Downsview, Ontario.

Noise energy enters the antenna from the atmosphere, as is well known. The effects of the normal environment on "antenna" noise-temperature contributions are well understood and documented elsewhere. Portions of these contributions arise in the troposphere. The dependence of the contributions on atmospheric attenuation A (dB) is given by

$$\Delta T_s = T_m [1 - 10^{-A(\text{dB})/10}] \qquad \textbf{(45-15)}$$

where ΔT_s is the sky noise-temperature increase, K, and T_m is the effective temperature of the clear atmosphere, which has been determined empirically.[26]

$$T_m = [1.12 (T_{\text{surface}}) - 50] \qquad K \qquad \textbf{(45-16)}$$

for frequencies above 10 GHz.

Relationship (45-15) applies reasonably well for the noise-temperature contributions arising from hydrometeors in the troposphere, including and especially rain. It means that the noise-temperature increase due to rain approaches the temperature of the rain. But the rain itself is cool, generally between 273 K (freezing) and 283 K. Radiometric measurements tend to verify this. T_m(rain) of 278 to 280 K are commonly used.

Rain-induced attenuation and depolarization are functions of the impairment per unit path length and the path length through the medium. For example, for rain

$$A = L_e(R)\gamma(R) \qquad \textbf{(45-17)}$$

where A = attenuation, dB

$L_e(R)$ = equivalent path length in rain at rain rate R

$\gamma(R)$ = attenuation per unit path length at rain rate R

The impairment per unit path length is dependent on the cumulative rain-rate distribution,[27] the distribution of rain into various drop sizes,[28] and several other parameters. For attenuation, one can write[29]

$$\gamma(R) = aR^b \qquad \text{dB/km} \qquad \textbf{(45-18)}$$

where R = surface-point rain rate

a, b = constants

Figure 45-4 gives $\gamma(R)$ for $1 \leq R \leq 150$ mm/h for spherical raindrops. Unfortunately, not all raindrops are spherical.[30] For sizes larger than 0.5-mm radius, their shape evolves through spherical to oblate spheroidal, to oblate spheroids flattened on the down side (distorted by the air pressure when falling) to almost cardioids of revolution (equal-volume-sphere radius > 3 mm), as shown in Fig. 45-5. As a result, attenuation due to rain is polarization-sensitive,[31] and attenuation perpendicular to the inclination of the raindrop major axes is less than that parallel to them. In general, most nonspherical raindrops fall with their major axis only slightly inclined to the horizontal. Thus, a radio wave which is vertically polarized is less attenuated than one horizontally polarized, and this has been demonstrated repeatedly by measurements (for example, Ref. 32). The difference (in decibels) is termed *differential attenuation*.

There remains the matter of the path length through the medium. The problem is that rain does not fall everywhere on the path, whether the path links terrestrial

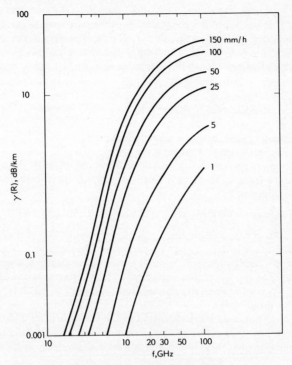

FIG. 45-4 Rain-induced specific attenuation.

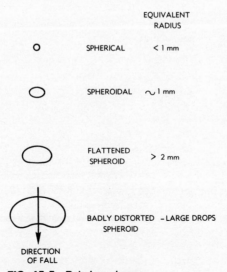

FIG. 45-5 Raindrop shapes.

sites or is a slant path into space. And where rain falls, the rain rate on the path is not uniform. This has led to the concept of effective path length, $L_e(R)$, such that if the rain were falling uniformly at a rain rate R on this portion of the total path, then Eq. (45-17) would give values of A consistent with measured data.

The cumulative probability in percentage of a surface-point rainfall rate R is given by[27,33]

$$p(R) = \frac{M}{87.66} [0.03\beta e^{-0.03R} + 0.2(1 - \beta) \cdot (e^{-0.258R} + 1.86e^{-1.63R})] \quad \textbf{(45-19)}$$

where M = mean annual rainfall accumulation, mm
$\beta \equiv$ Rice-Holmberg[27] thunderstorm ratio*

The effective path length has a number of representations. Based on data collected in western Europe and eastern North America,[34] the effective path length L_{e1} has been approximated by

$$L_{e1}(R,\theta) = [0.00741R^{0.766} + (0.232 - 0.00018R) \sin \theta]^{-1} \quad \textbf{(45-20)}$$

On the basis of data collected in Japan, curves given in Fig. 45-6 provide effective path length for 1.0, 0.1, 0.01, and 0.001 percent of the mean year, $L_{e2}(R, \theta)$. L_{e1} is most useful for a benign climate (lower thunderstorm ratios); L_{e2}, where there are tropical storms and the like. It has been found[34] that for attenuation calculations for $f < 10$ GHz, L_{e2} is more generally accurate, whereas for $f > 10$ GHz, $L_e = 0.75 L_{e2} + 0.25 L_{e1}$ has proved effective.

This subject has been much discussed in the literature.[34,35,36] More recently,[4,37] procedures based on climatological considerations have evolved. It is hoped, but not yet fully established, that these procedures will be both generally applicable and accu-

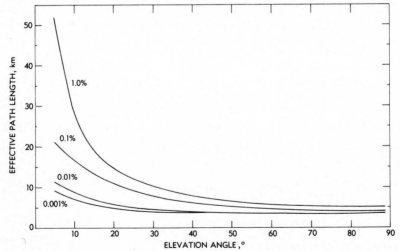

FIG. 45-6   Effective-path-length curves derived from measurements made in Japan.

---

*β is also defined as the fraction of M arising from rainfall for $R > 25$ mm/h.

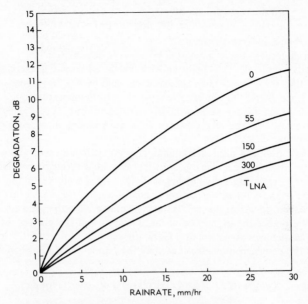

FIG. 45-7 Degradation of C/T versus rain rate.

rate. A plethora of measured data exists,[33,37,38] and new data are continually being presented.[39,40,41] Truly, one hopes for verified modeling.

In the end, the result is effectively a loss of signal power in both transmission and reception and a degradation of system noise temperature. These effects can be offset by increased antenna gain, or by increased transmitter power and decreased receiver-system noise temperature, or by site diversity. Figure 45-7[42] shows a typical trade-off between antenna gain and low-noise receiver noise temperature on a slant-path link. The trade-off between transmitter power and antenna gain is a 1-dB decrease in transmitter power required for each decibel increase in antenna gain on a terrestrial or slant-path link. The trade-off between receiver noise temperature and antenna gain shown in Fig. 45-7 arises out of the equations for carrier-to-noise ratio for the downlink.

$$\left(\frac{C}{T}\right)_{\text{downlink}} = \text{eirp} - L + \left(\frac{G}{T}\right)_{\text{sys}} \qquad \text{all in dB} \qquad \textbf{(45-21)}$$

where eirp = effective isotropic radiated power
$\quad\quad L$ = loss = $L_{fs} + L_a + L_w$, all in dB
$\quad\quad G$ = antenna gain
$\quad\quad T_{\text{sys}}$ = system noise temperature,* K

---

*T_{sys} for earth-space paths may be approximated by

$$T_{\text{sys}} \cong (0.95/L_a L_\omega) \, [T_c + T_m(L_a - 1)] + (T_0/L_\omega) \, [0.05 + (L_\omega - 1)] + T_R$$

where $T_c \equiv$ clear-sky noise-temperature contribution
$\quad\quad T_0 \equiv$ ambient temperature of equipment
$\quad\quad T_R$ = noise temperature of receiver

all in K.

L_{fs} = free-space-path loss, dB
L_a = loss in the atmosphere, dB, due to rain
L_ω = loss in feed system, dB

Under degraded conditions caused by rain, the degradation of C/T can be shown to appear to be equivalent to operating with a reduced G and an increased T. It can be shown[42] that the effect of increasing antenna gain by ΔG dB is independent of rain effects, whereas decreasing the receiver-system noise temperature by ΔT dB is most effective when there is no rain, and the heavier the rain rate (i.e., the higher the rain-induced attenuation), the less effective the improvement in receiver-system noise temperature.

Depolarization is a particularly damaging impairment, especially when encountered in a polarization-sensitive system such as a dual orthogonally polarized communications system using a dual orthogonally polarized feed system. As noted above, rain can change the polarization state of a radio wave, i.e., depolarize it. The departure of raindrops from spherical shapes[30] is the mechanism.[31] Further, the more intense the rain rate, the higher the proportion of large drops[28] and the greater the depolarization.[31] For frequencies above about 8 GHz, the relationship between attenuation due to rain and the depolarization also caused by the rain is well established[43,44] for earth-space paths:

$$XPD = U - V \log_{10} (\text{copolarized attenuation, dB}) \qquad \textbf{(45-22)}$$

where
$$U = \kappa^2 - 10 \log_{10} [1/2(1 - \cos(4\tau)e^{-\kappa_m^2})] \qquad \textbf{(45-23)}$$
$$+ 30 \log_{10}(f) - 40 \log_{10}(\cos\theta)$$

for $8 < f \leq 35$ GHz, and f in GHz
$10° \leq \theta \leq 60°$

$$V = 20 \text{ for } 8 < f \leq 15 \text{ GHz} \qquad \textbf{(45-24a)}$$

$$V = 23 \text{ for } 15 < f \leq 35 \text{ GHz} \qquad \textbf{(45-24b)}$$

where θ = elevation angle
τ = polarization tilt angle to horizontal (for circular polarization, $\tau = 45°$)
$\kappa \cong 0$ for conservative estimations of XPD
$\kappa_m \cong 0.25$ for conservative estimations of XPD

Thus, for frequencies above 8 GHz, calculations of XPD (cross-polar discrimination) can be made either from measured values of copolarized attenuation or from estimations made with Eq. (45-17) and its companion equations. Above 8 GHz, XPD is increasingly affected by differential attenuation, but below 8 GHz it is heavily dominated by differential phase shift, the difference in phase shift of a radio wave orthogonal to the principal axis of the spheroidal drops and parallel to this axis, i.e., almost vertical and almost horizontal,[31,45,46,47] and attenuation effects are much less important. A different approximation[36,43] has proved useful for earth-space paths for circular polarization over the range $1 \leq f < 8$ GHz, although it has some validity up to 35 GHz:

$$XPD = U - V \log_{10} R - 20 \log_{10} L_e \qquad \textbf{(45-25)}$$

where
$$U = 90 - 20 \log_{10} f - 40 \log_{10}(\cos\theta) \qquad \textbf{(45-26)}$$

$$V = 25 \qquad\qquad 1 \leq f \leq 15 \text{ GHz}$$

$$= 27 - 0.13f \quad 15 < f \leq 35 \text{ GHz}$$

$$\text{and } f \text{ in GHz, } R \text{ in mm/h} \quad \textbf{(45-27)}$$

$$L_e \cong 0.7 \, L_{e2} \text{ for } f < 10 \text{ GHz}$$

$$L_e \cong L_{e1} \quad \text{for } f \geq 10 \text{ GHz}$$

Statistics for depolarization measurements are available both above and below 10 GHz.[48,49]

For terrestrial paths, horizontal or vertical polarization has been shown to be better than circular polarization,[50] and the corresponding relationships are

$$\text{XPD} = U - V(f) \log_{10} (\text{copolar attenuation, dB}) \quad \textbf{(45-28)}$$

where
$$U = U_0 + 30 \log_{10} (f) \quad \textbf{(45-29)}$$

$$V = 20 \quad \textbf{(45-30)}$$

where for f in GHz $8 \leq f \leq 20$ GHz and $U_0 \cong 9$ for attenuation ≥ 15 dB.

Means of ameliorating the effects of rain are available. One approach is to use site diversity. Whereas for multipath fading [Sec. 45-2, Eqs. (45-9) to (45-13)] a diversity spacing of several meters is often effective, now spacings of the order of 5 to 50 km must be considered.[51,52,53] If this is done, it is very effective for ameliorating attenuation and depolarization effects for frequencies above 10 GHz, but at the cost of the extra site or sites. Below 10 GHz, attenuation is not so strong an effect, and it is possible to ameliorate XPD by depolarization compensation.[54,55]

45-4   OTHER EFFECTS

There are several other mechanisms, such as clouds and dust or sand, which can affect the propagation path and thus antenna design, but perhaps the most important remaining mechanism is the ionosphere for transionospheric links. Ionospheric effects[56,57] that may have to be considered above 1 GHz include Faraday rotation (the rotation of linearly polarized signals), propagation time delay and dispersion, and gigahertz ionospheric scintillation. Faraday rotation and propagation delay vary as $1/f^2$, and dispersion as $1/f^3$. The numbers in Table 45-2 were derived from JPL Publication 77-71.[58]

TABLE 45-2

	Frequency			
	1 GHz	**3 GHz**	**10 GHz**	**30 GHz**
Faraday rotation	108°	12°	1.1°	0.12°
Propagation delay	0.25 μs	0.028 μs	2.5 ns	0.28 ns
Dispersion	4×10^{-4} ps/Hz	1.5×10^{-5} ps/Hz	4×10^{-7} ps/Hz	1.5×10^{-8} ps/Hz

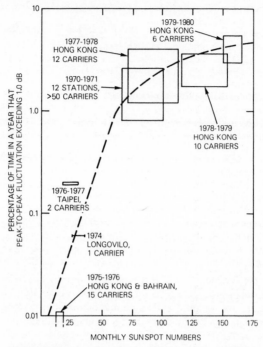

FIG. 45-8 Dependence of 4-GHz equatorial iono-spheric scintillation on monthly averaged Zürich sun-spot numbers. *(After Ref. 57.)*

The numbers given in Table 45-2 are based on an ionospheric total electron content of 10^{18} electrons/m^2, most likely to be encountered at low latitudes during the daytime in periods of high solar activity, and as such should be considered near the maximum values likely to be encountered for any length of time.

Gigahertz ionospheric scintillation[57] is encountered on earth-space paths typically near the spring and fall equinoxes and can vary from 1 dB peak to peak a few times an equinox during periods of low solar activity to 14 dB peak to peak at the peak of solar activity (both measured on 30-m-diameter antennas). It is inversely dependent on antenna diameter (the smaller the antenna, the worse the effects). Solar activity peaks every 11 years with a major peak every 22 years, the last major peak having occurred in 1981. A peak can be expected in 1992–1993, and a major peak in 2003–2004. Figure 45-8 gives data on peak-to-peak ionospheric scintillation over the half cycle 1970–1981.

REFERENCES

1 H. Riehl, *Introduction to the Atmosphere*, 3d ed., McGraw-Hill Book Company, New York, 1978.

2   R. G. Barry and R. J. Chorley, *Atmosphere, Weather and Climate,* Holt, Rinehart and Winston, Inc., New York, 1970.

3   *U.S. Standard Atmosphere Supplements,* U.S. Department of Commerce, Environmental Sciences Services Administration, 1966.

4   R. K. Crane and D. W. Blood, *Handbook for the Estimation of Microwave Propagation Effects—Link Calculations for Earth-Space Paths,* ERT Tech. Rep. 1, Doc. P-7376-TR1, NASA Cont. NASS-25341, June 1979.

5   H. J. Liebe, "Molecular Transfer Characteristics of Air between 40 and 140 GHz," *IEEE Trans. Microwave Theory Tech.,* vol. MTT-23, no. 4, April 1975, pp. 380–386.

6   International Telecommunications Union, Comité International des Radiocommunications, vol. V: *Propagation in Non-Ionized Media,* Geneva, 1978, pp. 68–69.

7   B. R. Bean and E. J. Dutton, *Radio Meterology,* Dover Publications, Inc., New York, 1966.

8   R. K. Crane, "Refraction Effects in the Neutral Atmosphere," in M. C. Weeks (ed.), *Methods of Experimental Physics,* vol. 12: *Astrophysics,* part B, "Radio Telescopes," Academic Press, Inc., New York, 1976.

9   R. K. Crane, "Low Elevation Angle Measurement Limitations Imposed by the Troposphere: An Analysis of Scintillation Observations Made at Haystack and Millstone," Tech. Rep. 518, MIT Lincoln Laboratory, Lexington, Mass., 1976.

10   A. Vigants, "Microwave Obstruction Fading," *Bell Syst. Tech. J.,* vol. 60, no. 6, 1981, pp. 785–801.

11   A. G. Crawford and W. C. Jakes, "Selective Fading of Microwaves," *Bell Syst. Tech. J.,* vol. 41, no. 1, 1952, pp. 68–90.

12   A. R. Webster and T. Ueno, "Tropospheric Microwave Propagation—An X-Band Diagnostic System," *IEEE Trans. Antennas Propagat.,* vol. AP-28, 1980, pp. 693–699.

13   H. T. Dougherty, "A Survey of Microwave Fading Mechanisms, Remedies and Applications," ESSA Tech. Rep. ERL-69-WPL-4, U.S. Department of Commerce, 1968 (available through National Technical Information Service, NTIS Access No. COM-1-50288).

14   W. T. Barnett, "Multipath Propagation at 4, 6, and 11 GHz," *Bell Syst. Tech. J.,* vol. 51, no. 2, February 1972, pp. 321–361.

15   A. Vigants, "Space Diversity Engineering," *Bell Syst. Tech. J.,* vol. 54, no. 1, January 1975, pp. 103–142.

16   D. D. Crombie, U.S. Department of Commerce, National Telecommunications and Information Administration, Institute of Telecommunication Sciences, Boulder, Colo., private communications.

17   S. H. Lin, "Statistical Behavior of a Fading Signal," *Bell Syst. Tech. J.,* vol. 50, no. 10, December 1971, pp. 3211–3270.

18   A. Vigants, "Number and Duration of Fades at 4 and 6 GHz," *Bell Syst. Tech. J.,* vol. 50, no. 3, March 1971, pp. 815–841.

19   J. M. Harris, "Tropospheric Scintillation of Microwaves Propagated on Low Elevation Angle Earth-Space Paths," COMSAT Lab. Tech. Mem. CL-54-77, 1977.

20   V. I. Tatarski, *The Effects of the Turbulent Atmosphere on Wave Propagation,* Nanka, Moscow; available in English through National Technical Information Service, No. N72-18163, 1971 (1967 for Russian original).

21   H. Yokoi, M. Yamada, and T. Satoh, "Atmospheric Attenuation and Scintillation of Microwaves from Outer Space," *Astronom. Soc. Japan,* vol. 22, no. 4, 1970, pp. 511–524.

22   *Climatological Normals (Clino) for Climate and Climate Ship Stations for the Period 1931–1960,* World Meterological Organization, 1971. These documents provide clino monthly precipitation (millimeters), clino annual precipitation, and number of days with precipitation greater than or equal to 1.0 mm.

23   *World Weather Records 1951–60,* vols. 1, 2, 3, 4, 5, and 6, U.S. Department of Commerce, Environmental Science Services Administration, 1965–1968. These books provide monthly and annual precipitation in millimeters by month and year for the decade 1951–1960 and mean (and sometimes clino) annual precipitation in millimeters.

24   H. E. Landsberg, (ed. in chief), *World Survey of Climatology,* vols. 1–15, Elsevier Pub-

lishing Company, Amsterdam, 1970. These books provide climate tables for many cities and other locations around the world (covering at least 10 years of data) and give mean precipitation by month and annually, maximum 24-h precipitation, average number of days with precipitation ≥ 0.1 mm, average number of days with thunderstorms, and much more. Typically, contours of equal mean annual precipitation (isohyetes) are given by region and lengthy discussions of climate by season and month.

25 J. A. Ruffner and F. E. Bair (eds.), *The Weather Almanac,* 2d ed., Gale Research Company, Detroit, 1977. This book provides weather information in detail on 108 United States cities, by month and annually, including 30 years of precipitation data (1936–1975) when available, and "normal" precipitation in inches; also weather information on 550 cities around the world covering 30 years (1931–1960) when available, including 50 cities in the continental United States, providing monthly and annual mean precipitation in inches.

26 K. H. Wulfsberg, "Apparent Sky Temperatures at Millimeter Wave Frequencies," Air Force Cambridge Res. Lab. Rep. AFCRL 64-590, 1964.

27 P. L. Rice and N. R. Holmberg, "Cumulative Time Statistics of Surface Point Rainfall Rates," *IEEE Trans. Com.,* vol. COM-21, no. 10, 1973, pp. 1131–1136.

28 J. O. Laws and D. A. Parsons, "The Relation of Raindrop Size to Intensity," *AGU Trans.,* vol. 24, 1943, pp. 452–460.

29 R. L. Olsen, D. V. Rogers, and D. B. Hodge, "The aR^b Relationship in the Calculation of Rain Attenuation," *IEEE Trans. Antennas Propagat.,* vol. AP-26, no. 2, 1978, pp. 318–319.

30 H. R. Pruppacher and P. L. Pitter, "A Semi-Empirical Determination of the Shape of Cloud and Raindrops," *J. Atmos. Sci.,* vol. 28, no. 1, 1971, pp. 86–94.

31 T. Oguchi and Y. Hosoya, "Scattering Properties of Oblate Raindrops and Cross-Polarization of Radio Waves Due to Rain (Part II): Calculation at Microwave and Millimeter Wave Regions," *J. Radio Res. Labs. (Japan),* vol. 21, 1974, pp. 191–259.

32 J. R. Harris and G. Hyde, "Preliminary Results of COMSTAR 19/29 GHz Beacon Measurements at Clarksburg, Maryland," *COMSAT Tech. Rev.,* vol. 7, no. 2, fall, 1977, pp. 599–623.

33 E. J. Dutton and H. T. Dougherty, "Year to Year Variability of Rainfall for Microwave Applications in the U.S.A.," *IEEE Trans. Com.,* vol. COM-27, no. 5, May 1979, pp. 829–832.

34 D. V. Rogers, "Propagation Analysis Package (PAP-2) Users Guide," COMSAT Labs. Tech. Mem. CL-9-80, January 1980, sec. 1.2.2.

35 L. J. Ippolito, R. D. Kaul, and R. G. Wallace, *Propagation Effects Handbook for Satellite System Design,* 2d ed., NASA Ref. Publ. 1082, December 1981, sec. 3.7, pp. 91–98.

36 International Telecommunications Union, Comité Consultatif International des Radiocommunications, vol. V: *Propagation in Non-Ionized Media,* Geneva, 1978, pp. 223–225.

37 International Telecommunications Union, Comité Consultatif International des Radiocommunications, vol. V: *Propagation in Non-Ionized Media,* Geneva, 1982. Report 564, especially pp. 354–371, and Report 338, especially pp. 309–313.

38 D. C. Hogg and T. S. Chu, "The Role of Rain in Satellite Communication," IEEE Proc., vol. 63, no. 9, September 1975, pp. 1308–1331.

39 D. Davidson and D. D. Tang, "Diversity Reception of COMSTAR SHF Beacons with the Tampa Triad 1978–1981," final rep. on JPL Cont. 956078, Jan. 1, 1982.

40 P. N. Kumar, "Precipitation Fade Statistics for 19/29 GHz COMSTAR Beacon Signals and 12 GHz Radiometric Measurements at Clarksburg, Maryland," *COMSAT Tech. Rev.,* vol. 12, no. 1, 1982, pp. 1–27.

41 D. C. Cox and H. W. Arnold, "Results from the 19 and 28 GHz COMSTAR Satellite Propagation Experiments at Crawford Hill," *IEEE Proc.* (invited paper, to be published)

42 J. Steinhorn, J. Harris, and G. Hyde, "Effects of Rain on Communications Satellite Systems Tradeoffs at 14/11 GHz and 29/19 GHz," COMSAT Labs. Tech. Mem. CL-32-76, May 1976.

43 R. L. Olsen and W. L. Nowland, "Semi-Empirical Relations for the Prediction of Rain Depolarization Statistics: Their Theoretical and Experimental Basis," *Antennas Propagat. Int. Symp. Proc.,* Sendai, Japan, 1978.

44 T. S. Chu, "Analysis and Prediction of Cross-Polarization on Earth-Space Links," *URSI (Comm. F) Int. Symp. Proc.,* Lennoxville, Quebec, 1980.

45 R. Taur, "Rain Depolarization Measurements on a Satellite-Earth Path at 4 GHz," *IEEE Trans. Antennas Propagat.,* vol. AP-23, 1975, pp. 854–858.

46 R. Taur, "Rain Depolarization: Theory and Experiment," *COMSAT Tech. Rev.,* vol. 4, no. 1, spring, 1974, pp. 187–190.

47 D. J. Fang, "Attenuation and Phase Shift of Microwaves Due to Canted Raindrops," *COMSAT Tech. Rev.,* vol. 5, no. 1, spring, 1975, pp. 135–156.

48 D. C. Cox, "Depolarization of Radiowaves by Atmospheric Hydrometeors in Earth-Space Paths: A Review," *Radio Sci.,* vol. 16, no. 5, September–October 1981, pp. 781–812.

49 D. J. Kennedy, "Rain Depolarization Measurements at 4 GHz," *COMSAT Tech. Rev.,* vol. 9, no. 2, pp. 629–668.

50 Rep. 338-3 (Mod I), CCIR Doc. 5/206-E, July 1980, p. 273.

51 D. V. Rogers and G. Hyde, "Diversity Measurements of 11.6 GHz Rain Attenuation at Etam and Lenox, West Virginia," *COMSAT Tech. Rev.,* vol. 9, no. 1, spring, 1979, pp. 243–254.

52 J. E. Allnutt, "Nature of Space Diversity in Microwave Communications Via Geostationary Satellites: A Review," *IEE Proc.,* vol. 125, no. 5, 1978, pp. 369–376.

53 K. Yasukawa and M. Yamada, "11 GHz Rain Attenuation and Site Diversity Effect at Low Elevation Angles," *URSI 20th General Assembly Abs.,* Washington, Aug. 10–19, 1981.

54 R. R. Persinger, R. W. Gruner, J. E. Effland, and D. F. DiFonzo, "Operational Measurements of a 4/6 GHz Adaptive Polarization Compensation Network Employing Up/Down-Link Correction Algorithms," *IEE Conf. Publ. 195,* part 2, pp. 181–187.

55 M. Yamada, H. Yuki, K. Inagaki, S. Endo, and N. Matsunaka, "Compensation Techniques of Rain Depolarization in Satellite Communications," *Radio Sci.,* vol. 17, no. 4, 1982.

56 Comité Consultatif International des Radiocommunications, Rep. 263-4, vol. VI, Geneva, 1978, table VI.

57 D. J. Fang and M. S. Pontes, "4/6 GHz Ionospheric Scintillation Measurements during the Peak of Sunspot Cycle 21," *COMSAT Tech. Rev.,* vol. 11, no. 2, fall, 1981, pp. 293–320.

58 E. K. Smith and E. E. Reinhart, "Sharing the 620–790 MHz Band Allocated to Terrestrial Television with an Audio Bandwith Social Service Satellite System," JPL Publication 77-71, Oct. 31, 1977. See Table C1.

Chapter 46

Materials and Design Data

Donald G. Bodnar

Georgia Institute of Technology

46-1 PROPERTIES OF MATERIALS

The design of many types of antennas, radomes, and impedance-matching devices requires the selection of appropriate dielectric materials. Strength, weight, dielectric constant, loss tangent, and environmental resistance are primary parameters of interest. Handbooks of mechanical properties are readily available[1,2] and will not be reproduced here. Instead, emphasis will be on electrical properties. The permittivity ϵ, permeability μ, and conductivity σ must be known at the operating frequency for a material to be completely specified electrically. For the dielectrics presented in this section, conductivity is zero, permeability equals that of free space ($4\pi \times 10^{-7}$ H/m), and permittivity is a complex number given by

$$\epsilon = \epsilon' - j\epsilon'' = \epsilon'\,(1 - j\tan\delta)$$

where ϵ'/ϵ_0 and $\tan\delta$ are the dielectric constant and loss tangent, respectively, of the material and ϵ_0 is the permittivity of free space (8.854×10^{-12} F/m). For the metals discussed in this section permeability equals that of free space, permittivity is unimportant, and skin depth equals $(\pi f\mu\sigma)^{-1/2}$, where f is the operating frequency.

Dielectrics

Foam dielectrics may be obtained in stock shapes or be foamed in place (Table 46-1). They are widely used as sealers for microwave components, spacers in twist reflectors or transreflectors, and low-cross-section target supports. Specially loaded foams can be used to create a variable dielectric constant which is useful in a beam-collimation device such as a Luneburg lens. Foams are typically used in low-power applications.

High-average-power applications (in the kilowatt range) rule out all known dielectrics except ceramics. In addition, high-speed missiles with their high surface and internal temperatures typically require ceramic radomes. Commonly used ceramics are listed in Table 46-1.

Reinforced plastics (Table 46-1) are widely used as structural members for antennas, feeds, and mounting structures. Conducting surfaces can be added by flame spraying, painting, etc., to form reflectors. Sandwich construction techniques are used in radomes and polarizers. Here, two or more thin skins are stabilized against buckling with a lightweight core. The core material can be rigid foam (Styrofoam), foam-in-place foam (Eccofoam S), or a honeycomb material (Hexcell HRP). The skins can be epoxy-fiberglass, Teflon, or Mylar. The thicknesses of the skins and core are chosen to make the sandwich transparent electrically. Careful control of tolerances, glue-line thickness, and materials can produce repeatably accurate devices. Photoetched printed-circuit boards are used in polarizers, spatial filters, and frequency-selective reflecting surfaces. Additional information on dielectric porperties is given in Von Hippel.[3]

Metals

Copper, brass, and aluminum are the most important metals in antenna construction today. If weight is not of primary importance, brass and copper are used extensively. These two metals machine, solder, and plate so easily, with little or no special equipment, that they are favorites of the model shop and the production shop alike. Complicated assemblies can be built with practically no radio-frequency (RF) discontinuities due to contact between parts. Aluminum equals or surpasses both copper and brass on all counts except plating. Although special equipment is required to weld aluminum, the resulting structure is lighter in weight than brass or copper. Very complicated shapes such as flat-plate arrays can be fabricated by dip-brazing aluminum. Table 46-2 lists the conductivity of commonly used metals.

46-2 ABSORBING MATERIALS

Microwave absorbing material is commercially available in sheet, pyramidal, block, and rod form. Pyramidal absorber is commonly used to create an RF anechoic chamber for indoor-antenna range measurements since it offers the best performance (reflected energy 50 dB down at normal incidence). Pyramidal absorber is usually constructed by loading a spongelike rubber with graphite, and it is available in 2- by 2-ft (0.6- by 0.6-m) pieces ranging in thickness from 3 in (76 mm) to 12 ft (3.7 m). It is rated in decibels of incident to reflected energy at stated frequency, polarization, and incidence angle (see Table 46-3). Performance degrades if the pyramidal tips are broken or if the absorber is exposed to weather for extended periods.

Flat sheets of absorber are available when space is limited or when the absorber must be exposed to weather, wear, or high power. The reflected energy for sheet absorber is 15 to 25 dB below the incident level. Blocks and special shapes of absorber are available for waveguide-load applications, or they can be machined to the desired shape.

The above types of absorber are currently available commercially from Advanced Absorber Products, Inc., Amesbury, Massachusetts, and Emerson and Cuming, Inc., Canton, Massachusetts.

46-3 NONSOLID SURFACES

Antenna structures are sometimes made from nonsolid metal in order to make the structure lighter, to reduce wind loading, or for special RF purposes (e.g., a frequency-selective reflector). Typical construction techniques employ metal wires, rods, strips, and patches, perforated metal, and expanded mesh. How large the "holes" in the structure can be depends on how much transmitted or reflected loss can be tolerated.

Gratings

An array of identical parallel metal wires or rods can be used to reflect energy polarized parallel to the wires and, consequently, to shield the region on the other side of the wires from parallel polarization. The perpendicularly polarized energy is negligibly affected by the wires for most geometries. The parallel and perpendicularly polarized power-transmission coefficients are given respectively by[4]

TABLE 46-1 Dielectric Data

		Frequency f, Hz; values of tan δ multiplied by 10^4				
		$f = 10^2$	$f = 10^4$	$f = 10^6$	$f = 10^8$	$f = 10^{10}$
Ceramics						
Alsimag 243	ϵ'/ϵ_0	6.30	6.28	6.22	6.10	5.76
	tan δ	12.5	4.0	3.7	3	8.5
Alsimag 393	ϵ'/ϵ_0	4.95	4.95	4.95	4.95	4.95
	tan δ	38	16	10	10	9.7
Aluminum oxide; Coors AI-200	ϵ'/ϵ_0	8.83	8.82	8.80	8.80	8.79
	tan δ	14	4.8	3.3	3.0	18
Beryllium oxide	ϵ'/ϵ_0	4.61	4.41	4.28	...	4.20
	tan δ	170	74	38	...	5
Magnesium oxide	ϵ'/ϵ_0	9.65	9.65	9.65	9.65	...
	tan δ	3	3	3	3	...
Porcelain, dry-process	ϵ'/ϵ_0	5.50	5.23	5.08	5.02	4.74
	tan δ	220	105	75	98	156
Silicon nitride	ϵ'/ϵ_0	...	...	...	...	5.50
	tan δ	...	...	...	...	30
Steatite 410	ϵ'/ϵ_0	5.77	5.77	5.77	5.77	5.7
	tan δ	55	16	7	6	22
Titania	ϵ'/ϵ_0	100	100	100	100	90
	tan δ	23	6.2	3	2.5	20
Glasses						
Fused quartz	ϵ'/ϵ_0	3.78	3.78	3.78	3.78	3.78
	tan δ	8.5	6	2	1	1
Fused silica	ϵ'/ϵ_0	3.78	3.78	3.78	3.78	3.78
	tan δ	6.6	1.1	0.1	0.3	1.7
E glass	ϵ'/ϵ_0	6.43	6.39	6.32	6.22	6.11
	tan δ	42	27	15	23	60
Low-density materials						
Eccofoam S	ϵ'/ϵ_0	1.05	...	1.18	...	1.47
	tan δ	20	...	40	...	70

Material		1	2	3	4	5
Hexcel HRP, 1/4–4.5; honeycomb	ϵ'/ϵ_0	1.09	...	...	...	...
	tan δ	30	...	...	...	...
Hexcel HRH-10, 3/16–4.0; honeycomb	ϵ'/ϵ_0	1.12	...	...	...	...
	tan δ	30	...	...	...	...
Styrofoam 103.7	ϵ'/ϵ_0	1.03	...	1.03	1.03	1.03
	tan δ	1.5	...	2	1	2
Plastics						
Bakelite	ϵ'/ϵ_0	3.52	4.4	5.4	6.5	8.2
	tan δ	366	770	600	630	1350
Duroid 5650	ϵ'/ϵ_0	2.65	...	...	...	...
	tan δ	30	...	...	...	...
Epoxy resin RN-48	ϵ'/ϵ_0	2.91	3.32	3.52	3.61	3.64
	tan δ	184	264	142	68	31
Fiberglass; laminated BK-174	ϵ'/ϵ_0	4.37	4.8	5.3	7.2	14.2
	tan δ	360	260	460	1600	2500
Lexan	ϵ'/ϵ_0	2.86	...	...	...	...
	tan δ	60	...	...	...	...
Plexiglas	ϵ'/ϵ_0	2.59	...	2.76	2.95	3.40
	tan δ	67	...	140	300	605
Polystyrene	ϵ'/ϵ_0	2.54	2.55	2.56	2.56	2.56
	tan δ	4.3	1	0.7	0.5	0.5
Rexolite 1422	ϵ'/ϵ_0	2.54	2.55	2.55	2.55	2.55
	tan δ	4.7	3.8	1.3	1	2.1
Polyethylene	ϵ'/ϵ_0	2.24	2.25	2.25	2.25	2
	tan δ	6.6	2	2	2	2
Polyvinyl chloride, W-176	ϵ'/ϵ_0	...	3.00	3.53	4.70	6.21
	tan δ	...	500	720	1070	730
Teflon	ϵ'/ϵ_0	2.08	2.1	2.1	2.1	2.1
	tan δ	3.7	2	2	3	5

NOTE: Properties are typical for a class of materials. Consult manufacturers for exact properties of their material.

SOURCE: Ref. 3.

TABLE 46-2 Conductivity of Various Metals

Material	Conductivity, s/m at 20°C
Aluminum, commercial hard-drawn	3.54×10^7
Brass, yellow	1.56×10^7
Copper, annealed	5.80×10^7
Copper, beryllium	1.72×10^7
Gold, pure drawn	4.10×10^7
Iron, 99.98 percent pure	1.0×10^7
Iron, gray cast	$0.05 - 0.20 \times 10^7$
Steel	$0.5 - 1.0 \times 10^7$
Lead	0.48×10^7
Nickel	1.28×10^7
Silver, 99.98 percent pure	6.14×10^7
Tin	0.869×10^7
Tungsten, cold-worked	1.81×10^7
Zinc	1.74×10^7
Titanium	0.182×10^7

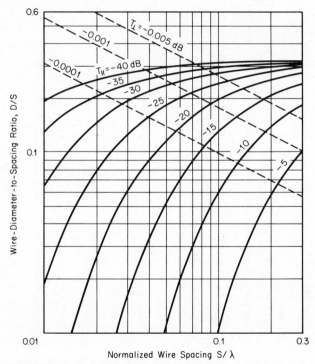

FIG. 46-1 Transmission coefficient for parallel polarization and for perpendicular polarization for a grid of round wires of diameter D and center-to-center spacing S.

TABLE 46-3 Reflection Coefficient of Typical Commercial Pyramidal Absorbing Material

Nominal thickness, in	Reflection Coefficient, dB									
	120 MHz	200 MHz	300 MHz	500 MHz	1 GHz	S band	C band	X band	Ku band	K band
3	..	..	..	..	..	..	30	40	45	50
5	..	..	..	..	..	30	40	45	50	50
8	..	..	..	..	30	40	45	50	50	50
12	..	..	..	..	35	40	45	50	50	50
18	..	..	..	30	40	45	50	50	50	50
24	..	..	30	35	40	50	50	50	50	50
45	..	30	35	40	45	50	50	50	50	50
70	30	35	40	45	50	50	50	50	50	50
106	35	40	45	50	50	50	50	50	50	50
144	40	45	50	50	50	50	50	50	50	50

$$T_{\parallel} = 1 - \cfrac{1}{1 + \left(\cfrac{2S}{\lambda}\ln\cfrac{S}{\pi D}\right)^2}$$

$$T_{\perp} = 1 - \cfrac{\left(\cfrac{\pi^2 D^2}{2\lambda S}\right)^2}{1 + \left(\cfrac{\pi^2 D^2}{2\lambda S}\right)^2}$$

Figure 46-1 is a plot of these equations, where the grating has wires of diameter D, center-to-center spacing of S, and wavelength λ. These curves are derived from Lamb's work[5] and agree well with Mumford's nomograph.[6] The grating can be envisioned as waveguides below cutoff. Thus to be effective, (1) the electric field of the incident wave must lie in the plane determined by the axis of the grating element and the direction of propagation, and (2) S must be less than $\lambda/(1 - \sin\theta)$, where θ is the angle between the direction of propagation and the normal to the axis of the grating element. Larger spacings result in undesirable secondary reflection lobes.[5]

The round wires may be replaced by thin metal strips of width W perpendicular

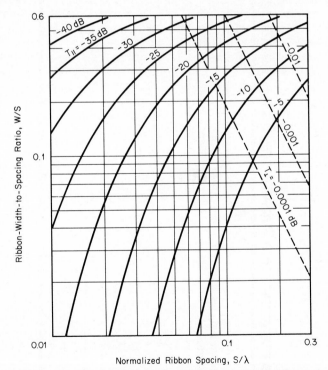

FIG. 46-2 Transmission coefficient for parallel polarization and for perpendicular polarization for a grid of broadside ribbons of width W and center-to-center spacing S.

to the direction of propagation by using metal bars or printed-circuit-board construction techniques. The transmission coefficients for the thin broadside strips are given in Fig. 46-2, which are plots of the equations[4]

$$T_{\parallel} = 1 - \cfrac{1}{1 + \left(\cfrac{2S}{\lambda} \ln \sin \cfrac{\pi W}{2S}\right)^2}$$

$$T_{\perp} = 1 - \cfrac{\left(\cfrac{2S}{\lambda} \ln \cos \cfrac{\pi W}{2S}\right)^2}{1 + \left(\cfrac{2S}{\lambda} \ln \cos \cfrac{\pi W}{2S}\right)^2}$$

Increased attenuation of the transmitted signal is obtained by orienting the strips edgewise to the direction of propagation. The length of parallel-plate waveguide below cutoff is greater, resulting in greater attenuation than for the round wires or broadside strips. Figure 46-3 gives the percentage of transmitted power for infinitely long edgewise strips of width W, center-to-center spacing S, and depth t.[7]

The polarization sensitivity of the preceding wire and strip gratings can be eliminated by using (1) a grid of wires composed of two crossed gratings, preferably contacting each other or at least closely spaced, (2) a metal sheet periodically perforated by round or square holes,[8] or (3) expanded metal mesh[9] (see Fig. 46-4).

The above structures are relatively broadband devices. Narrowband structures can be constructed by using a periodic array of strip-metal dipoles.[10] Operation below the resonant frequency of the dipoles results in little transmission loss versus essen-

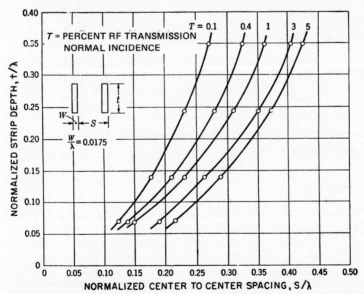

FIG. 46-3 Grating of edgewise strips: relation between strip depth and spacing for constant transmission. (*Ref. 7, page 450.*)

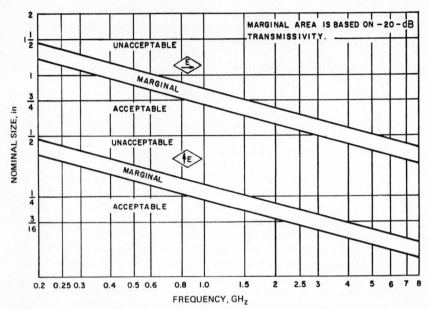

FIG. 46-4   Radio-frequency properties of expanded-metal-mesh screens. *(After Ref. 9.)*

tially total reflection at resonance. Dipole and crossed-dipole arrays are used in dichroic subreflectors for dual-frequency operation of a reflector antenna.

Transmission-Type Circular Polarizers[11]

Transmission-type polarizers are anisotropic media which, when placed over a radiating aperture, convert linearly polarized signals to circularly polarized signals. They may take several forms, including the following:

1   Parallel-plate structures using metal slats[12]
2   Alternating dielectric plates of differing materials[13]
3   Lattice structures using strips or rods[14]
4   Meander-line photoetched structures[15]
5   Etched-dipole arrays

These types of polarizers have been applied to linearly polarized antennas when other means of achieving circular polarization cannot be economically or easily achieved. For example, simple horns, biconical horns, and slot, horn, and dipole arrays have been successfully equipped with these devices. Recent advances in precision-photoetching technology permit close-tolerance photoetched structures which exhibit excellent uniformity of insertion phase and loss characteristics and which have been designed for use over greater-than-3 : 1-frequency bandwidths and with power-handling densities of up to 100 W/in^2 (average). Structures 1, 2, and 3 are older approaches; the reader is referred to the References for background design information.

Structures 4 and 5 above consist of bonded sandwiches of alternating etched sheets of Mylar or Teflon-fiberglass copper-clad dielectric material (e.g., MIL-P-631 Type G Mylar or MIL-P-13949 Type GT Teflon-fiberglass). (See Fig. 46-5.) Bonding is done by using a low-loss bonding film (e.g., 3M Type HX-1000) and spacer material such as Hexcell Type HRH-10 honeycomb or Eccofoam SH polyurethane foam. A dielectric layer is generally added to these sandwiches for outer-surface environmental protection; this may consist of epoxy-fiberglass Type G-10 approximately 0.010 to 0.015 in (0.254 to 0.381 mm) thick. Structures of this type have been used on both flat and cylindrical apertures with essentially identical performance.

In use, these structures are oriented so that the incident electric field is 45° to the slow axis of the polarizer. That component of the field which lies in the slow or perpendicular direction travels through an equivalent circuit which consists of a ladder network of shunt capacitive elements. The parallel-field component travels through a network of shunt inductive elements. In meander-line polarizers, which are in predominant use today, the shunt capacitance elements arise between the parallel meandering strips[16] and, for some very broadband designs, also from added dipole elements which can extend the frequency bandwidth to higher frequencies for a given meander line.

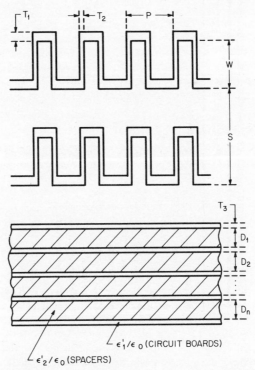

FIG. 46-5 Geometry of a meander-line circular polarizer showing meander-line and sandwich dimensions.

TABLE 46-4 Parameters of Meander-Line Polarizers Shown in Fig. 46-5

	Polarizer 1[18]	Polarizer 2[20]	Polarizer 3[21]
Frequency range	4.0–8.0 GHz	7.0–14.0 GHz	2.6–7.8 GHz
Axial ratio (maximum)	3.0 dB	3.0 dB	3.0 dB
Return loss increase (polarizer)	5.0 dB	3.0 dB	3.0 dB
Power handling (Average W/in²)	100	⋯	50
No. of circuits	4	5	6
Circuit board			
Material	⋯	⋯	MIL-P-631 Mylar
Thickness (T3)	0.010	⋯	0.003
Dielectric constant	2.52	⋯	3.00
Spacers			
Material	Foam	Styrofoam	HRH-10-OX Honeycomb
Thicknesses (D_n)	0.200, 0.300, 0.200	All 0.194	All 0.312
Dielectric constant	1.08	⋯	1.08

Outer circuit			
S	0.707	0.570	0.750
W	0.280	0.154	0.215
P	0.3535	0.118	0.058
T1	0.018	0.031	0.007
T2	0.018	0.017	0.007
Next circuit			
S	....		Same as outer circuit with dipoles* 0.063 by 0.500 and spaced 1.500 added on alternate meander lines
W	....	Same as outer circuit	
P	....		
T1	....		
T2	....		
Inner circuit			
S	0.707		Same as outer circuit with dipoles 0.063 by 0.500 and spaced 1.500 added on each meander line
W	0.370	Same as outer circuit	
P	0.3535		
T1	0.027		
T2	0.027		
Other features	....	....	0.015 G-10 fiberglass window attached

*Dipoles oriented in the w direction.

NOTE: All dimensions are in inches.

46-13

Shunt inductive elements are realized by meandering copper lines etched from the copper-clad material. Because of the broadband nature of the multilayer meander-line polarizer, tolerances on the etching, alignment, and bonding of the sandwich are somewhat looser than for earlier devices such as the metal-slat or dielectric polarizers.

Etched multilayer polarizers are usually designed to be optimum in plane-wave field conditions; they can achieve better than 2-dB axial ratios, insertion losses of less than 0.2 dB, and a return loss of better than 17 dB over more than octave bandwidths. When they are applied to apertures which are less than approximately one wavelength in size or when they are forced into the flares of small horns, performance can be expected to deteriorate seriously. In addition, as with all circular polarizers, the best axial-ratio performance will be achieved when provisions are made for dissipating the cross-polarized wave which scatters from the polarizer and reenters the antenna. The latter can be accomplished by the use of a terminated orthomode coupler or by the use of a resistive sheet inside the feeding horn and aligned with the *H* plane.

Reflection-type polarizers can be derived from the full-transmission-type polarizers described above by using half the number of layers, which are then mounted onto a metal sheet.[12] For narrower-band applications, simple etched-construction transmission-type polarizers can be designed by using narrow-line-width strips as in Lerner[17] or two-layer arrays of etched dipoles. In the latter, phase-shift quadrature is achieved by the shunt capacitance of a dipole medium only; the differential phase shift due to the longitudinal component of the incident field is negligible. For two layers, axial ratios of less than 0.5 dB can be achieved for bandwidths of a few percent.

Multilayer polarizers are best designed by formulating a model of the susceptance of the metallic circuits on each layer[18,19] and then iterating the parameters of a ladder network by using an ABCD-matrix (or equivalent) analysis on a computer. Analysis is performed for both parallel and perpendicular field components. The network's insertion phase and reflection properties are then optimized for a minimum reflection coefficient for both field components and for an insertion phase-shift differential as close as possible to the ideal 90°. Such an analysis can readily model the thickness of the sandwich materials, their electrical properties, the incidence angle, and the frequency-varying nature of the individual susceptances. Reference 18 is useful in this respect. Examples of three documented broadband polarizers are tabulated in Table 46-4 and Fig. 46-5.

Wind Loading

The primary forces to which an antenna is subjected are wind, ice, vibration, and acceleration. The wind-loading force is given in Fig. 46-6 for solid and open surfaces.[22] Ice loading increases the stress on a structure because of the increase in areas presented to wind and because of the additional weight due to the ice loading [approximately 57 lb/ft^3 (913 kg/m^3)].

46-4   RADIO-FREQUENCY-BAND DESIGNATIONS

Historically, letter band designations originated during World War II for secrecy reasons. These designations were continued after they had been made public, but different authors used different definitions for the band edges. One set of common-usage band

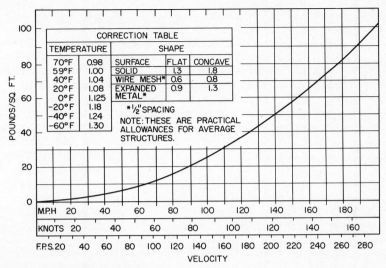

FIG. 46-6 Wind-loading data.

designations is listed in Table 46-5. Standardized bands have been defined by two official international organizations and are also listed in Table 46-5. The Institute of Electrical and Electronics Engineers (IEEE) standard is the preferred radar band designation. The International Telecommunications Union (ITU) assigns frequency allocations by international treaty. Specific frequency ranges within the ITU bands shown in Table 46-5 are assigned for radar use, and these ranges differ slightly depending on which region of the world is involved. ITU band designations are used as a general indication of the frequency of operation.

In addition, the U.S. Department of Defense has issued a directive concerning the designation of frequencies used in electronic-countermeasures (ECM) operations. Each of the 13 letter bands listed in Table 46-5 is further subdivided into 10 equal-bandwidth channels and given a phonetic and number designation such as Delta-4 (1.3 to 1.4 GHz). It is improper usage to designate a radar by an ECM band designation since the ECM designations only apply to ECM equipment.

46-5   ANTENNA-PATTERN CHARACTERISTICS

Both the initial design of an antenna and the validation of measured antenna data are greatly facilitated through the use of theoretically derived pattern characteristics. The patterns produced by a variety of one- and two-dimensional aperture distributions have appeared in the literature over the years. The task of engineers is to determine which theoretical aperture distribution best fits their physical situations so that they can assess antenna performance from the theoretical data. This can be done by measuring or calculating the actual aperture distribution and then comparing it with a variety of theoretical distributions.[23] The three primary factors of concern to antenna

TABLE 46-5 Frequency-Band Designations

IEEE radar bands*		ITU frequency bands†		Common-usage bands‡		Electronic-countermeasures bands§	
Band	Frequency range, GHz	Band	Frequency range, GHz	Band	Frequency range, GHz	Band	Frequency range, GHz
HF	0.003–0.03	HF	0.003–0.03	HF	0.003–0.03	A	0–0.25
VHF	0.03–0.3	VHF	0.03–0.3	VHF	0.03–0.3	B	0.25–0.5
UHF	0.3–1	UHF	0.3–3	UHF	0.3–1	C	0.5–1
L	1–2	SHF	3–30	L	1–2	D	1–2
S	2–4	EHF	30–300	S	2–4	E	2–3
C	4–8			C	4–8	F	3–4
X	8–12			X	8–12.4	G	4–6
Ku	12–18			Ku	12.4–18	H	6–8
K	18–27			K	18–26.5	I	8–10
Ka	27–40			Ka	26.5–40	J	10–20
mm	40–300			Q	33–50	K	20–40
				V	50–75	L	40–60
				W	75–110	M	60–100

*From Institute of Electrical and Electronics Engineers Standard 521–1976, Nov. 30, 1976.

†From International Telecommunications Union, Art. 2, Sec. 11, Geneva, 1959.

‡No official international standing.

§From AFR 55-44(AR 105–86, OPNAVINST 3430.9B, MCO 3430.1), Oct. 27, 1964.

designers are gain, beamwidth, and sidelobe level. This section presents data on these three parameters for both line-source and circularly symmetric distributions.

The gain of an antenna is usually compared with that of an antenna with the same aperture dimensions but having a constant phase and amplitude distribution (i.e., uniformly illuminated). The power gain G of the antenna then is

$$G = G_0\, \eta_a$$

where $G_0 = 4\,\pi A/\lambda^2$ = gain of uniformly illuminated aperture
   A = antenna aperture area
   λ = wavelength
   $\eta_a = \eta\eta_L$ = antenna efficiency
   η = aperture illumination efficiency
   η_L = all other efficiency factors

The aperture illumination efficiency represents the loss in gain resulting from tapering the aperture distribution in order to produce sidelobes lower than those achievable from a uniform illumination. For a circular aperture η is given by a single number. For separable rectangular aperture distributions it equals the product (sum in decibels) of the efficiencies in each of the two aperture directions.

The half-power beamwidth BW of an antenna is related to the beamwidth constant β by (Ref. 7, Chap. 6)

$$\text{BW} = 2\,\sin^{-1}\left(\frac{\beta\lambda}{2L}\right) \simeq \beta\,\frac{\lambda}{L} \qquad \text{(linear aperture)}$$

$$\text{BW} = 2\,\sin^{-1}\left(\frac{\beta\lambda}{2D}\right) \simeq \beta\,\frac{\lambda}{D} \qquad \text{(circular aperture)}$$

where L is the length of the linear aperture and D is the diameter of the circular aperture. The small-argument approximation for the arcsine is typically used for calculating BW. Values of β and η will be given as a function of sidelobe level for a number of distributions in the following subsections.

Continuous Line-Source Distributions

The problem of determining the optimum pattern from a line source has received considerable attention. The optimum pattern is defined as one that produces the narrowest beamwidth measured between the first null on each side of the main beam with no sidelobes higher than the stipulated level. Dolph[24] solved this problem for a linear array of discrete elements by using Chebyshev polynomials. If the number of elements becomes infinite while element spacing approaches zero, the Dolph pattern becomes the optimum continuous line-source pattern[25] given by

$$E(u) = \cos\sqrt{u^2 - A^2}$$

where $u = \dfrac{\pi L}{\lambda}\sin\theta$

   λ = wavelength
   θ = angle from the normal to aperture

The beamwidth constant β in degrees and the parameter A are given by

$$\beta = \frac{360}{\pi^2}\sqrt{[\text{arccosh}\,(R)]^2 - \left[\text{arccosh}\left(\frac{R}{\sqrt{2}}\right)\right]^2}$$

where $A = \text{arccosh}\,(R)$

$R = \text{main-lobe-to-sidelobe voltage ratio}$

This Chebyshev pattern provides a useful basis for comparison even though it is physically unrealizable since the remote sidelobes do not decay in amplitude. In fact, all sidelobes have the same amplitude in the Chebyshev pattern. The aperture distribution which produces the Chebyshev pattern has an impulse at both ends of the aperture[26] and produces very low aperture efficiency. In addition, the pattern is very sensitive to errors in the levels of these impulses.

Taylor[25] developed a method for avoiding the above problems by approximating arbitrarily closely the Chebyshev pattern with a physically realizable pattern. Taylor approximated the Chebyshev uniform sidelobe pattern close to the main beam but let the wide-angle sidelobe decay in amplitude. Taylor used a closeness-parameter $\bar{n}$ in his analysis. As $\bar{n}$ becomes infinite, the Taylor distribution approaches the Chebyshev distribution. By using the largest $\bar{n}$ that still produces a monotonic aperture distribution, one obtains the beamwidth constant and aperture efficiency shown in Figs. 46-7 and 46-8. Notice that the beamwidth from this Taylor distribution is almost as narrow as that from the Chebyshev distribution while still producing excellent aperture efficiency.

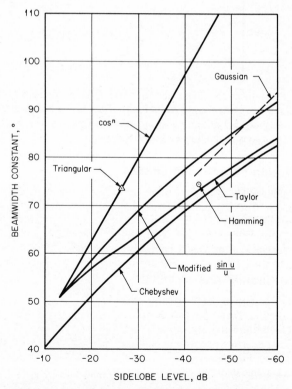

FIG. 46-7 Beamwidth constant versus sidelobe level for several line-source aperture distributions.

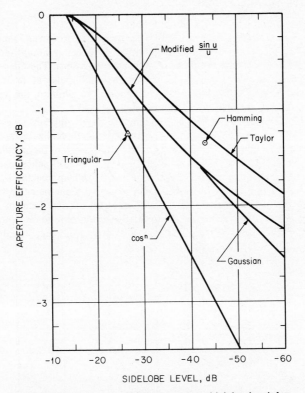

FIG. 46-8 Aperture efficiency versus sidelobe level for several line-source aperture distributions.

Several other common distributions[27,7] are also listed in Figs. 46-7 and 46-8. The advantage of the $\cos^n (\pi x/L)$ distribution and the $\sin (\sqrt{u^2 - B^2})/\sqrt{u^2 - B^2}$ pattern is that both the distribution and the pattern for them may be obtained in closed form. This mathematical convenience is obtained at the expense of poorer beamwidth and efficiency performance as compared with the Taylor distribution.

Continuous Circular-Aperture Distributions

The Chebyshev pattern of the preceding section can also be shown to be optimum for the circular aperture. Taylor has generalized his line-source distribution to the circular case,[28,29] and his pattern approaches the Chebyshev pattern as his closeness-parameter $\bar{n}$ for the circular aperture approaches infinity. The beamwidth constant and aperture efficiency shown in Fig. 46-9 and 46-10 are obtained by using the largest $\bar{n}$ that still produces a monotonic distribution.[30]

The Bickmore-Spellmire distribution[31] is a two-parameter distribution which can be considered a generalization of the parabola to a power distribution.[26] The Bickmore-Spellmire distribution $f(r)$ and pattern $E(u)$ are given by

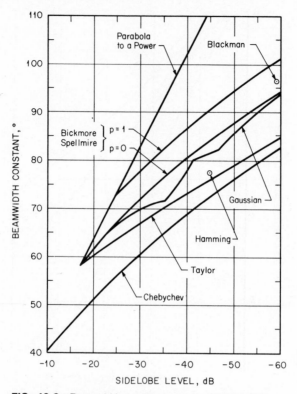

FIG. 46-9 Beamwidth constant versus sidelobe level for several circular-aperture distributions. *(After Ref. 30.)*

$$f(r) = p\left[1 - \left(\frac{2r}{D}\right)^2\right]^{p-1} \Lambda_{p-1}\left(jA\sqrt{1 - \left(\frac{2r}{D}\right)^2}\right)$$

$$E(u) = \Lambda_p(\sqrt{u^2 - A^2})$$

where p and A are constants that determine the distribution, Λ is the lambda function, and

$$u = \frac{\pi D}{\lambda}\sin\theta$$

The Bickmore-Spellmire distribution reduces to the parabola to a power distribution when $A = 0$ and to the Chebyshev pattern when $p = -\frac{1}{2}$.

A gaussian distribution[30] produces a no-sidelobe gaussian pattern only as the edge illumination approaches zero. In general, the aperture distribution must be numerically integrated to obtain the far-field pattern. The second sidelobe of this pattern is sometimes higher than the first, which accounts for the erratic behavior of β and η in Figs. 46-9 and 46-10.

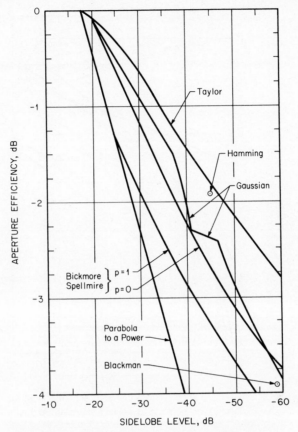

FIG. 46-10 Aperture efficiency versus sidelobe level for several circular-aperture distributions. *(After Ref. 30.)*

Blockage

The placement of a feed in front of a reflector results in blockage of part of the aperture energy. In the geometric-optics approximation, no energy exists where the aperture is blocked, and the undisturbed aperture distribution persists outside the blocked region.[32] Using a line source of length L with a $\cos^n$ aperture distribution and a centrally located blockage of length L_b produces the gain loss and the resulting sidelobe level shown in Fig. 46-11. The corresponding changes are shown in Fig. 46-12 for a circular aperture of diameter D and a parabola to a power distribution blocked by a centrally located disk of diameter D_b. Notice that the line-source blockage affects the pattern much more rapidly than in the circular case since it affects a larger portion of the aperture. Calculation of strut-blockage effects is given by Gray.[33]

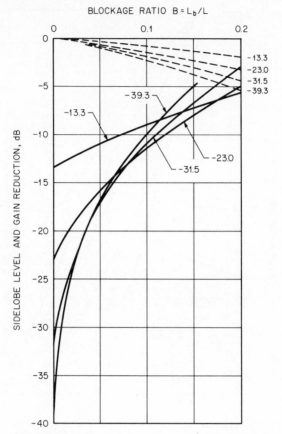

FIG. 46-11 Gain loss ($- - - -$) and resulting sidelobe level (————) for a centrally blocked line-source distribution having a specified unblocked sidelobe level.

46-6   ELECTRICALLY SMALL HORNS

Aperture theory does not accurately predict the pattern characteristics of electrically small horns (on the order of a wavelength or less in size). Diffraction effects from the flange or rim around the horn (also considered to be edge currents flowing on the outside surface of the horn) markedly influence the pattern. Experimental beamwidth data[7] for small horns without a flange are given in Fig. 46-13. These curves were obtained from measurements of a larger number of 10-dB beamwidths. Empirical formulas for these E- and H-plane curves are given respectively by

$$\text{BW}_E(10\ \text{dB}) = 88°\ \frac{\lambda}{B} \qquad \text{for}\ \frac{B}{\lambda} < 2.5$$

BLOCKAGE RATIO B = D_b/D

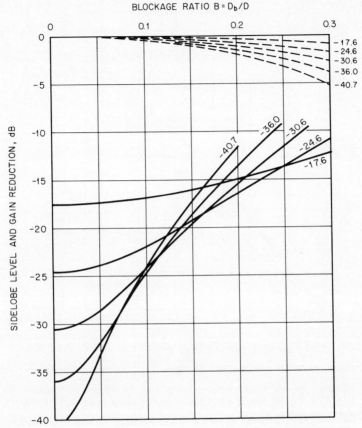

FIG. 46-12 Gain loss (— — —) and resulting sidelobe level
(————) for a centrally blocked circular-aperture distribution having a
specified unblocked sidelobe level.

$$\mathrm{BW}_H(10\ \mathrm{dB}) = 31° + 79° \frac{\lambda}{A} \qquad \text{for } \frac{A}{\lambda} < 3$$

where B and A are the E- and H-plane aperture dimensions respectively. The actual
beamwidth of any particular horn may vary from the above values since different flare
angles produce different phase variations over the aperture.

REFERENCES

1 J. D. Walton (ed.), *Radome Engineering Handbook,* Marcel Dekker, Inc., New York, 1970.
2 J. Agranoff (ed.), *Modern Plastics Encyclopedia,* McGraw-Hill Book Company, New
   York, 1981.

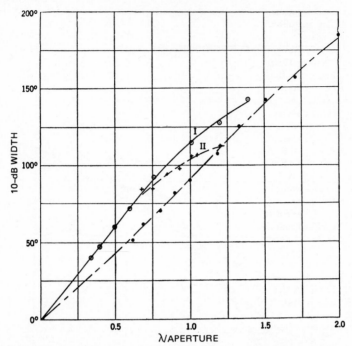

FIG. 46-13 Experimental 10-dB beamwidths of horns having small phase variations over the aperture $\left(\dfrac{\delta}{\lambda g} < \dfrac{1}{8}\right)$: ————— E plane; —————— H-plane sectoral horns; ———— H plane of compound horns with E-plane aperture equal to or greater than a wavelength. *(After Ref. 7, page 364.)*

3 A. R. Von Hippel (ed.), *Dielectric Materials and Applications,* The M.I.T. Press, Cambridge, Mass., 1954.

4 D. G. Bodnar, "SEASAT Antenna Study," Final Tech. Rep. EES/GIT, Proj. A-1617-000, Georgia Institute of Technology, Atlanta, August 1974, pp. 27–32.

5 T. Larsen, "A Survey of the Theory of Wire Grids," *IRE Trans. Microwave Theory Tech.,* May 1962, pp. 191–201.

6 W. W. Mumford, "Some Technical Aspects of Microwave Hazards," *IRE Proc.,* February 1961, pp. 427–445.

7 S. Silver (ed.), *Microwave Antenna Theory and Design,* M.I.T. Rad. Lab. ser., vol. 12, McGraw-Hill Book Company, New York, 1949.

8 P. W. Hannan and P. L. Burgmyer, "Metal-Grid Spatial Filter," Interim Tech. Rep. RADC-TR-79-295, AD No. A089756, Hazeltine Corporation, July 1980.

9 *ITE Antenna Handbook,* 2d ed., ITE Circuit Breaker Company, Philadelphia, p. 26, circa 1965.

10 Chao-Chun Chen, "Scattering by a Two-Dimensional Periodic Array of Conducting Plates," *IEEE Trans. Antennas Propagat.,* vol. AP-18, September 1970, pp. 660–665.

11 This subsection was written by John M. Seavey, Adams-Russell Company, Amesbury, Mass.

12 S. Cornbleet, *Microwave Optics,* Academic Press, Inc., London, 1976, pp. 309–323.

13 H. S. Kirschbaum and L. Chen, "A Method of Producing Broad-Band Circular Polarization Employing an Anisotropic Dielectric," *IRE Trans. Microwave Theory Tech.*, July 1957, pp. 199–203.

14 J. A. Brown, "Design of Metallic Delay Dielectrics," *IEE Proc. (London)*, part III, vol. 97, no. 45, January 1950, p. 45.

15 L. Young, L. A. Robinson, and C. A. Hacking, "Meanderline Polarizer," *IEEE Trans. Antennas Propagat.*, vol. AP-23, May 1973, pp. 376–378.

16 N. Marcuvitz, *Waveguide Handbook*, M.I.T. Rad. Lab. ser., vol. 10, McGraw-Hill Book Company, New York, 1951.

17 D. S. Lerner, "Wave Polarization Converter for Circular Polarization," *IEEE Trans. Antennas Propagat.*, vol. AP-13, January 1965, pp. 3–7.

18 T. L. Blakney, J. R. Burnett, and S. B. Cohn, "A Design Method for Meanderline Circular Polarizers," USAF Antenna Conf., Allerton Park, Ill., 1972.

19 A. C. Ludwig, M. D. Miller, and G. A. Wideman, "Design of Meanderline Polarizer Integrated with a Radome," *IEEE Antennas Propagat. Int. Symp. Dig.*, 1977, pp. 17–20.

20 D. A. McNamara, "An Octave Bandwidth Meanderline Polarizer Consisting of Five Identical Sheets," *IEEE Antennas Propagat. Int. Symp. Dig.*, 1981, pp. 237–240.

21 Data furnished courtesy of H. Schlegel, Adams-Russell Company Amesbury, Mass.

22 H. Jasik (ed.), *Antenna Engineering Handbook*, 1st ed., McGraw-Hill Book Company, New York, 1961, chap. 35.

23 R. C. Hansen, *Microwave Scanning Antennas*, vol. I, Academic Press, Inc., New York, 1964, chap. 1.

24 C. L. Dolph, "A Current Distribution for Broadside Arrays Which Optimizes the Relationship between Beam Width and Side-Lobe Level," *IRE Proc.*, vol. 34, June 1946, pp. 335–348.

25 T. T. Taylor, "Design of Line-Source Antennas for Narrow Beamwidth and Low Side Lobes," *IRE Trans. Antennas Propagat.*, vol. AP-3, January 1955, pp. 16–28.

26 J. W. Sherman III, "Aperture-Antenna Analysis," in M. I. Skolnik (ed.), *Radar Handbook*, McGraw-Hill Book Company, New York, 1970, chap. 9.

27 F. J. Harris, "On the Use of Windows for Harmonic Analysis with the Discrete Fourier Transform," *IEEE Proc.*, vol. 66, January 1978, pp. 51–83.

28 T. T. Taylor, "Design of Circular Apertures for Narrow Beamwidth and Low Sidelobes," *IRE Trans. Antennas Propagat.*, vol. AP-8, January 1960, pp. 17–22.

29 R. C. Hansen, "Tables of Taylor Distributions for Circular Aperture Antennas," *IRE Trans. Antenna Propagat.*, vol. AP-8, January 1960, pp 23–26.

30 A. C. Ludwig, "Low Sidelobe Aperture Distributions for Blocked and Unblocked Circular Apertures," RM 2367, General Research Corporation, April 1981.

31 W. D. White, "Circular Aperture Distribution Functions," *IEEE Trans. Antennas Propagat.*, vol. AP-25, September 1977, pp. 714–716.

32 P. W. Hannan, "Microwave Antennas Derived from the Cassegrain Telescope," *IRE Trans. Antennas Propagat.*, vol AP-9, March 1961, pp. 140–153.

33 C. L. Gray, "Estimating the Effect of Feed Support Member Blockage on Antenna Gain and Side-Lobe Level," *Microwave J.*, March 1964, pp. 88–91.

Index

About the Editors

Richard C. Johnson, Ph.D., has been actively engaged in applied research and development at the Engineering Experiment Station, Georgia Institute of Technology, since 1956. A fellow of the Institute of Electrical and Electronics Engineers, Dr. Johnson is a former president of the IEEE Antennas and Propagation Society and former editor of the society's newsletter. He is the author of more than 60 technical papers and reports. He received his Ph.D. degree from the Georgia Institute of Technology.

Henry Jasik (deceased) was formerly vice president of the AIL Division of Eaton Corporation and director of its Antenna Systems Division. A well-known authority on antenna engineering and a fellow of the Institute of Electrical and Electronics Engineers, Dr. Jasik was cited for his contributions to the theory and design of VHF and microwave elements. He organized and edited the first edition of *Antenna Engineering Handbook*.